Defects in Microelectronic Materials and Devices

Defects in Microelectronic Materials and Devices

Edited by

Daniel M. Fleetwood • Sokrates T. Pantelides
Ronald D. Schrimpf

CRC Press is an imprint of the
Taylor & Francis Group, an **informa** business

Figure on cover: Z-contrast image of a silicon-silicon dioxide-hafnium dioxide structure showing an isolated Hf atom in the SiO_2 interlayer (image courtesy of K. Van Benthem and S. J. Pennycook). The expanded image is an electron density plot for this structure (courtesy of A. G. Marinopoulos and S. T. Pantelides).

CRC Press
Taylor & Francis Group
6000 Broken Sound Parkway NW, Suite 300
Boca Raton, FL 33487-2742

First issued in paperback 2019

ISBN-13: 978-1-4200-4376-1 (hbk)
ISBN-13: 978-0-367-38639-9 (pbk)

Library of Congress Cataloging-in-Publication Data

Fleetwood, Daniel.
Defects in microelectronic materials and devices / Daniel Fleetwood, Sokrates T. Pantelides, and Ronald D. Schrimpf.
p. cm.
Includes bibliographical references and index.
ISBN 978-1-4200-4376-1 (alk. paper)
1. Microelectronics--Materials--Testing. 2. Metal oxide semiconductor field-effect transistors--Testing. 3. Integrated circuits--Defects. I. Pantelides, Sokrates T. II. Schrimpf, Ronald Donald. III. Title.

TK7871.F5485 2008
621.381--dc22 2008018722

Visit the Taylor & Francis Web site at
http://www.taylorandfrancis.com

and the CRC Press Web site at
http://www.crcpress.com

Contents

Preface

Defects in microelectronic materials can profoundly affect the yield, performance, long-term reliability, and radiation response of microelectronic devices and integrated circuits (ICs). This book provides a comprehensive survey of defects in silicon-based metal oxide semiconductor (MOS) field-effect transistor technologies, which dominate the worldwide microelectronics marketplace. This book also discusses the defects in linear bipolar technologies, silicon carbide based devices, and gallium arsenide materials and devices. An appendix is provided with supplemental material on highly cited papers on defects in these and other materials systems (e.g., GaN, ZnO, C) that are being investigated for present and future microelectronics technologies. The chapters described below summarize decades of experience in characterizing defect properties and their impact on microelectronic devices, and also look forward to the challenges that will have to be overcome as new materials (e.g., high-*K* gate dielectrics and high-mobility substrate materials) are incorporated into increasingly more highly scaled devices and ICs.

Chapters 1 through 4 focus on yield- and performance-limiting defects and impurities in the device silicon layer, and/or at the critical Si/SiO_2 interface. In Chapter 1, Law et al. discuss yield-limiting defects that are important to control in highly scaled technologies with ultra-shallow junctions. The impact of defects on diffusion and activation processes is emphasized. Defects in Si, Ge, and SiGe alloys associated with hydrogen are discussed by Peaker et al. in Chapter 2. The incorporation of hydrogen and its complexes in these materials are reviewed in detail, with examples provided from a broad range of experimental and theoretical work. In Chapter 3, Sun and Thompson describe dislocations in transistors that have been designed with highly strained layers to enhance carrier mobilities. In Chapter 4, Fischetti and Jin discuss the effects of ionized dopant atoms and interface roughness on electron transport in very highly scaled devices and ICs. An extensive review of mobility-limiting scattering processes is presented.

Chapters 5 through 8 describe electrical, analytical, spectroscopic, and state-of-the-art microscopic methods to characterize defects in MOS gate oxides. In Chapter 5, Schroder provides a comprehensive review of electrical and analytical techniques to estimate densities and energy distributions of defects in MOS gate dielectrics and/or at the Si/SiO_2 interface. A wide variety of examples are provided. These enable a comparison of the relative advantages and limitations of different characterization methods. In Chapter 6, Lenahan provides an extensive review of electron spin resonance studies that have provided significant insight into the microstructure of the dominant SiO_2 hole trap. This is the E' defect, a trivalent Si center in SiO_2 associated with an O vacancy. The dominant defect at the Si/SiO_2 interface is also characterized extensively; this is the P_b defect, a Si dangling bond that frequently is passivated by hydrogen during device processing, but when depassivated after electrical stress or ionizing radiation exposure can function as an interface trap. In Chapter 7, the properties of oxide and interface traps are reviewed by Fleetwood et al., with particular emphasis on separating the effects of "true" oxide and interface traps from the effects of near-interfacial oxide traps that exchange charge with the Si on the timescale of typical device measurements (border traps). Thermally stimulated current and low frequency noise techniques are described to estimate oxide and border trap densities and energy distributions. In Chapter 8, Pennycook et al. present the results of aberration-corrected scanning transmission electron microscopy on MOS structures; these techniques provide information on defects and impurities with sub-angstrom resolution.

Chapters 9 through 11 focus on defects in high-dielectric constant (high-K) materials that are under intense development to replace SiO_2 (more precisely, nitrided SiO_2) as the preferred gate dielectric material for Si-based ICs at or beyond the 45 nm technology node. In Chapter 9, Robertson et al. describe theoretical calculations of defect energy levels in high-K dielectrics, where it is found that the O vacancy is the main electrically active defect in HfO_2-based high-K gate dielectrics. In Chapter 10, Lucovsky surveys an extensive amount of spectroscopic data on transition metal oxides. These materials are compared and contrasted with SiO_2, and a critical review of the suitability of these materials is provided for incorporation into device manufacturing. In Chapter 11, Garfunkel et al. review experimental results and computational calculations of defects in HfO_2-based gate stacks. The effects of these defects on device electrical properties are emphasized.

In Chapter 12, Houssa et al. survey negative bias temperature instabilities (NBTIs) for high-K materials, and they separate the effects of interface and bulk traps on device electrical response. In Chapter 13, Tsetseris et al. discuss the role of hydrogen in NBTI in SiO_2, and summarize first-principles calculations of defect formation, defect dynamics, and defect annihilation in the Si/SiO_2 system. A comprehensive engineering model of NBTI is developed by Grasser et al. in Chapter 14. The effects of poststress relaxation are shown to be important to a complete understanding of the underlying mechanisms, as well as in developing techniques to predict device lifetimes from accelerated measurements.

Chapters 15 through 18 discuss defects in ultrathin oxides (SiO_2 and silicon oxynitride). In Chapter 15, Suehle describes the role of defects and hydrogen in SiO_2 on the time-dependent dielectric breakdown of MOS devices and ICs. The present understanding of defect generation processes is discussed, as are statistical models for the resulting device failure distributions. Defects associated with dielectric breakdown and conduction in oxides degraded by long-term, high-field stress are discussed in detail by Suñé and Wu in Chapter 16. The percolation model of breakdown is described, and the relative roles of holes and hydrogen in the breakdown process are evaluated experimentally and analytically. The effects of radiation and stress-induced defects in thin oxides are described by Cellere et al. in Chapter 17 and Touboul et al. in Chapter 18. Cellere et al. emphasize processes that lead to radiation and stress-induced leakage currents, which are especially important for nonvolatile memory cells that can be discharged via extremely low currents, and Touboul et al. discuss the physical damage caused by a high energy ion when it passes through a dielectric layer. The latter is of particular concern for the reliability of space systems.

The effects of radiation-induced defects on linear bipolar devices are discussed by Schrimpf et al. in Chapter 19. Even though similar defects are formed in the base oxides of linear bipolar transistors as in MOS structures (primarily oxide and interface-trap charge), the effects of these charges on device operation can be quite different for linear bipolar transistors than for MOS devices. One example of this is enhanced low-dose-rate sensitivity, which is the excess buildup of (primarily) interface traps during low-electric-field irradiations of typical base oxides that can be affected by trap densities and hydrogen concentrations of the structures.

In Chapter 20, Dhar et al. present an overview of defects in oxides on SiC wafers. The higher growth temperatures and the differing interfacial layers for SiC, relative to Si, lead to significant differences in defect properties. Most notably, dangling bond defects cannot be passivated as easily at the SiC/SiO_2 interface with hydrogen treatment as for the Si/SiO_2 interface. A comprehensive survey of defects in SiC is provided by Janzén et al. in Chapter 21. An extensive array of experimental data and theoretical calculations are tabulated and discussed in detail. Defects in GaAs are reviewed by Bourgoin and von Bardeleben in

Chapter 22. Particular emphasis is placed on As antisite defects associated with the EL2. Finally, the appendix lists a large number of highly cited journal articles related to defects in the Si/SiO_2 system, high-K dielectrics, GaAs, GaN, and ZnO. Brief synopses are provided of more than 450 highly cited articles; the interested reader can use these reference lists as a starting point to obtain more information on the nature and effects of defects and impurities in these or other semiconductor-based material systems.

This book was encouraged by the program managers of two multidisciplinary university research initiatives (MURIs) sponsored by the Air Force Office of Scientific Research, Gerald Witt and Kitt Reinhardt, on the effects of ionizing radiation on microelectronic materials and devices. We appreciate their support during these programs, as well as the efforts of all of our MURI collaborators—you will see many of their contributions in this book, as well as contributions from many other experts in the field who graciously agreed to provide chapters in their fields of specialty. We also wish to thank all our valued professional colleagues, research collaborators, and sponsors who have contributed so much to these efforts, as well as Jill Jurgensen, Allison Shatkin, and Shelley Kronzek at Taylor & Francis for their interest in and assistance with this book. We finally wish to thank Arun Kumar, Jennifer Smith, and the entire production team for their strong and timely support.

Daniel M. Fleetwood
Sokrates T. Pantelides
Ronald D. Schrimpf
Nashville, Tennessee

Editors

Daniel M. Fleetwood received BS, MS, and PhD degrees from Purdue University in 1980, 1981, and 1984, respectively. He joined Sandia National Laboratories as a member of the technical staff in 1984. In 1990, he was named a distinguished member of the technical staff in the Radiation Technology and Assurance Department at Sandia. Dan accepted a position as professor of electrical engineering at Vanderbilt University in 1999. Between 2001 and 2003, he served as an associate dean for research in the School of Engineering. In 2003, he was named the chairman of the Electrical Engineering and Computer Science Department at Vanderbilt University. He has authored or coauthored more than 300 papers on radiation effects and low frequency noise. He served as a guest editor of the December 1990 and April 1996 issues of the *IEEE Transactions on Nuclear Science,* and presently is the executive vice-chairman of the IEEE Nuclear and Plasma Sciences Society (NPSS) Radiation Effects Steering Group. Dan has received seven outstanding paper awards for the IEEE NSREC and three for the HEART Conference, as well as several meritorious conference paper awards. Dan is a fellow of the IEEE and the American Physical Society.

Sokrates T. Pantelides is the William A. and Nancy F. McMinn professor of physics at Vanderbilt University, Nashville, Tennessee. He holds a secondary appointment as a distinguished visiting scientist at Oak Ridge National Laboratory, Oak Ridge, Tennessee. He received his PhD in physics from the University of Illinois at Urbana-Champaign in 1973. After a postdoctoral appointment at Stanford University, he joined the IBM T. J. Watson Research Center in Yorktown Heights, New York, in 1975 where he carried out theoretical research in semiconductors and served as manager, senior manager, and program director. He joined Vanderbilt University in 1994. He has authored or coauthored more than 300 research articles and edited eight books. He is a fellow of the American Physical Society and the American Association for the Advancement of Science. His research focuses on the structure, defect dynamics, and electronic properties of electronic materials, radiation effects, transport in molecules and thin films, and catalysis.

Ronald D. Schrimpf received BEE, MSEE, and PhD degrees from the University of Minnesota in 1981, 1984, and 1986, respectively. He joined the University of Arizona in 1986, where he served as an assistant professor (1986), an associate professor (1991), and a professor (1996) of electrical and computer engineering. He joined Vanderbilt University in 1996 and was an invited professor at the University of Montpellier II, Montpellier, France, in 2000. Ron is currently a professor of electrical engineering at Vanderbilt University, where his research activities focus on microelectronics and semiconductor devices. In particular, he has a very active research program dealing with the effects of radiation on semiconductor devices and integrated circuits. Ron is also the director of the Institute for Space and Defense Electronics (ISDE) at Vanderbilt University. The engineering staff of ISDE performs design, analysis, and modeling work for a variety of space and defense-oriented organizations, as well as commercial semiconductor companies. Ron is a fellow of the IEEE and past chairman of the IEEE NPSS Radiation Effects Steering Group.

Contributors

M. Aoulaiche
IMEC
Leuven, Belgium

and

Department of Electrical Engineering
University of Leuven
Leuven, Belgium

H.J. von Bardeleben
Institut de NanoSciences de Paris
Université Paris 6
Paris, France

K. van Benthem
Materials Science and Technology Division
Oak Ridge National Laboratory
Oak Ridge, Tennessee

and

Center for Nanophase Materials Sciences
Oak Ridge National Laboratory
Oak Ridge, Tennessee

Gennadi Bersuker
Sematech International
Austin, Texas

Jacques Bonnet
Institut d'Electronique du Sud
UMR-CNRS 5214
University of Montpellier II
Montpellier, France

J.C. Bourgoin
Genie des Semiconducteurs (GESEC)
Research, Inc.
Paris, France

Renata Camillo-Castillo
IBM
Burlington, Vermont

Aminata Carvalho
Institut d'Electronique du Sud
UMR-CNRS 5214
University of Montpellier II
Montpellier, France

Giorgio Cellere
Department of Information
Engineering
University of Padova
Padova, Italy

M.F. Chisholm
Materials Science and Technology Division
Oak Ridge National Laboratory
Oak Ridge, Tennessee

S. Dhar
Department of Physics and Astronomy
Vanderbilt University
Nashville, Tennessee

and

Cree Incorporated
Durham, North Columbia

L. Dobaczewski
Institute of Physics
Polish Academy of Sciences
Warsaw, Poland

L.C. Feldman
Department of Physics and Astronomy
Vanderbilt University
Nashville, Tennessee

and

Institute for Advanced Materials, Devices
and Nanotechnology
Rutgers University
New Brunswick, New Jersey

M.V. Fischetti
Department of Electrical and Computer Engineering
University of Massachusetts
Amherst, Massachusetts

Daniel M. Fleetwood
Department of Electrical Engineering and Computer Science
Vanderbilt University
Nashville, Tennessee

A. Gali
Department of Atomic Physics
Budapest University of Technology and Economics
Budapest, Hungary

Eric Garfunkel
Departments of Chemistry
Rutgers University
Piscataway, New Jersey

Jacob Gavartin
Accelrys
Cambridge, United Kingdom

S. De Gendt
IMEC
Leuven, Belgium

and

Department of Chemistry
University of Leuven
Belgium

Jean Gasiot
Institut d'Electronique du Sud UMR-CNRS 5214
University of Montpellier II
Montpellier, France

Simone Gerardin
Department of Information Engineering
University of Padova
Padova, Italy

Wolfgang Goes
Christian Doppler Laboratory for TCAD in Microelectronics
Institute for Microelectronics
Wien, Austria

Tibor Grasser
Christian Doppler Laboratory for TCAD in Microelectronics
Institute for Microelectronics
Wien, Austria

G. Groeseneken
IMEC
Leuven, Belgium

and

Department of Electrical Engineering
University of Leuven
Leuven, Belgium

A. Henry
Department of Physics, Chemistry and Biology
Linköping University
Linköping, Sweden

M.M. Heyns
IMEC
Leuven, Belgium

and

Department of Electrical Engineering
University of Leuven
Leuven, Belgium

M. Houssa
IMEC
Leuven, Belgium

and

Department of Electrical Engineering and Physics
University of Leuven
Leuven, Belgium

I.G. Ivanov
Department of Physics, Chemistry and Biology
Linköping University
Linköping, Sweden

E. Janzén
Department of Physics, Chemistry and Biology
Linköping University
Linköping, Sweden

S. Jin
Department of Electrical and Computer Engineering
University of Massachusetts
Amherst, Massachusetts

Kevin S. Jones
Department of Materials Science and Engineering
University of Florida
Gainesville, Florida

Ben Kaczer
IMEC
Leuven, Belgium

Mark E. Law
Department of Electrical and Computer Engineering
University of Florida
Gainesville, Florida

Patrick M. Lenahan
Department of Engineering Science and Mechanics
The Pennsylvania State University
University Park, Pennsylvania

Gerald Lucovsky
Department of Physics
North Carolina State University
Raleigh, North Carolina

B. Magnusson
Department of Physics, Chemistry and Biology
Linköping University
Linköping, Sweden

and

Norstel AB
Norrköping, Sweden

Mathias Marinoni
Institut d'Electronique du Sud UMR-CNRS 5214
University of Montpellier II
Montpellier, France

A.G. Marinopoulos
Department of Physics and Astronomy
Vanderbilt University
Nashville, Tennessee

V.P. Markevich
School of Electrical and Electronic Engineering
University of Manchester
Manchester, United Kingdom

Alessandro Paccagnella
Department of Information Engineering
University of Padova
Padova, Italy

Sokrates T. Pantelides
Department of Physics and Astronomy
Vanderbilt University
Nashville, Tennessee

A.R. Peaker
School of Electrical and Electronic Engineering
University of Manchester
Manchester, United Kingdom

Ronald L. Pease
RLP Research
Los Lunas, New Mexico

S.J. Pennycook
Materials Science and Technology Division
Oak Ridge National Laboratory
Oak Ridge, Tennessee

and

Department of Physics and Astronomy
Vanderbilt University
Nashville, Tennessee

J. Robertson
Department of Engineering
Cambridge University
Cambridge, United Kingdom

the projected range region in nonamorphizing implants and the EOR damage region of amorphizing implants.

During annealing, the dopants diffuse. This diffusion for most dopants is accomplished on the microscopic scale by interaction with point defects. As a simple picture, the diffusivity is proportional to the excess point defect population. This leads to the damage from the implant, increasing the diffusivity until the damage is fully annealed, known in the literature as TED. This process is reasonably well understood. The implant-induced damage condenses into various extended defect structures, which are discussed in more detail subsequently. As these extended defects evolve, they release interstitials that drive the diffusion enhancement. These interstitials are thought to recombine at surfaces. The extended defect dissolution and evolution controls the diffusion enhancement duration and magnitude. This has driven the study of extended defect behavior in silicon. As mentioned earlier, interaction with defects can also control dopant activation processes. Complete understanding of the dopant, point defect, and extended defect system is still being sought.

Because of transient effects brought on by damage, the trend in annealing has been to reduce the time of the anneal. Furnace annealing in silicon is now mostly a thing of the past. Current annealing technologies are done in rapid thermal annealing (RTA) for times in the duration of seconds and temperatures between 900°C and 1200°C. In current operation, anneals are called spike anneals, which have no dwell time at temperature whatsoever. The entire anneal is done as a ramp up to peak temperature and an immediate cool down. These short time anneals present difficulty in understanding the dopant and defect system, as many of the "pseudoequilibrium" approximations break down. For example, regrowth processes of the amorphous layer become more important as time at temperature becomes shorter and no longer can be neglected.

In recent years millisecond annealing has come to the forefront in the semiconductor industry in its attempt to continue to drive Moore's law by reducing the total thermal budget imparted to the wafer. Commercially known as flash-assist rapid thermal process (fRTP), this process is designed to operate within the time gap between spike, rapid, and laser thermal processing techniques. The process offers three main advantages over conventional RTA systems that stem from the differences in the heating technology. Tungsten filament lamps are utilized in conventional RTAs compared to the water-walled arc lamps in flash systems. The arc lamps, which are very high-quality optical sources, deliver greater power; have a faster response time; and deliver short wavelength radiation that is more effective in heating the silicon substrate. A high-pressure argon plasma in an arc lamp, when heated to 12,000 K, produces radiation power of 1×10^6 W, enabling very high ramp rates that are four orders of magnitude higher than a conventional RTA, which has a resultant power of 1×10^3 W. The smaller thermal mass of the argon in the arc lamps also enables the arc lamps to respond approximately 10 times faster than tungsten lamps [7]. Since the transition from heating to cooling is also a function of the response time of the heat source, a wafer heated by flash techniques will transition much faster from heating to cooling. Finally, over 95% of the arc radiation is below the 1.2 μm band gap absorption of silicon compared to 40% for radiation generated by the tungsten lamps [7], hence it is more effective in heating the wafer.

The process uses a continuous arc lamp to heat the bulk of the wafer to an intermediate RTP (iRTP) temperature, where the dwell time is essentially zero. This heating is slower than the thermal conduction rate through the wafer, thus the entire wafer remains at approximately the same temperature [8]. The iRTP serves as the initial temperature of the flash anneal. Subsequently, a capacitor bank is discharged through an arc lamp, which adds additional power to the device side of the wafer at a rate much faster than the thermal conduction rate. Short time pulses allow for heating of the surface of the wafer to the peak flash temperatures (fRTP), while the substrate never attains these high temperatures. This is possible since the time constant of the flash, which is on the order 1 ms, is much shorter than the thermal time constant of the wafer (~10–20 ms). Therefore, a thin slice of the device side of the wafer is heated and cooled rapidly at rates on the order of 1×10^6 °C s^{-1}. The fast cooling is achieved since the bulk of the wafer acts as a heat sink, removing heat from the top layer via conduction much more efficiently and faster than can be accomplished in bulk cooling. The high absorbance of the reactor chamber also complements the cooling rates. However, as the flash time pulse approaches the time constant of silicon, the flash elevates the bulk temperature considerably, therefore only allowing for cooling by radiative and convective methods, resulting in much slower cooling rates. Heating rates up to the intermediate temperature are similar to conventional spike annealing (50°C s^{-1}–300°C s^{-1}), as are cooling rates during the bulk radiative cooling (up to ~150°C s^{-1}).

1.5 Defect Evolutionary Processes

In high concentrations found after implantation, free interstitials cluster to reduce free energy. The interstitial cluster configurations believed to occur in ion-implanted silicon include the di-interstitial, interstitial chain, {311} rod defect, and dislocation loops. Transmission electron micrograph (TEM) images of {311} defects and dislocation loops are shown in Figures 1.1 and 1.2, respectively. The free interstitial has strain energy associated with it because it is larger than any interstitial site. It also has a free energy of 1 eV from each unbonded orbital. A di-interstitial represents a more stable configuration compared to the free interstitial since it reduces the number of unbonded orbitals. Theoretically, by

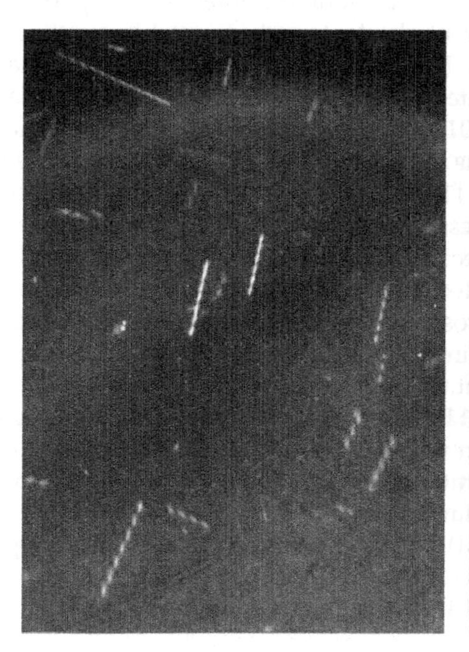
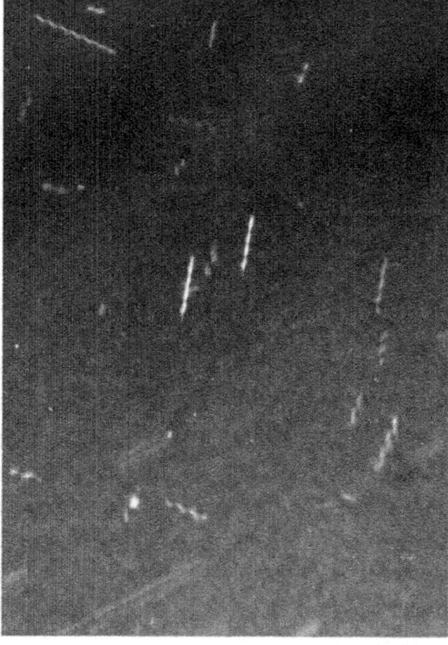

FIGURE 1.1
{311} defects imaged in plan-view, weak-beam dark-field TEM. The left picture is after a preanneal of 2 h at 750°C. The right picture is of an in situ anneal of the sample at left after 15 additional minutes at 770°C. (From Law, M.E. and Jones, K.S., *International Electron Device Meeting*, 2000, p. 511. With permission.)

Kim [12], Takeda [11], and Tan [10] have modeled the formation of the {311} defect. These models and empirical evidence [32,33] result in a consensus of $a/21$ <116> for the Burger's vector of the {311} defect. Few have observed or modeled the transformation of a {311} defect into a dislocation loop. Eaglesham [28] has observed that {311} defects can unfault to form Frank loops. The defect reaction proposed by Eaglesham

$$a/21 \langle 116\rangle + a/21 \langle 661\rangle = a/3 \langle 111\rangle$$

gives a plausible unfaulting reaction. Eaglesham then postulated that Frank loops may unfault to form Shockley dislocation loops through the common unfaulting reaction,

$$a/3 \langle 111\rangle + a/6 \langle 112\rangle = a/2 \langle 110\rangle,$$

which has been observed in many cubic lattices.

Li and Jones [35] have observed similar transformations of the {311} defect into dislocation loops via in situ TEM measurements. This transformation may occur while conserving the number of interstitials bound in the defect. If such conservation is maintained, then the defect transformation itself should not affect the supersaturation of the interstitial concentration in the surrounding crystal.

Dissolution of interstitials from {311} defects is another possible evolutionary path for the interstitials, as the defect becomes thermodynamically unstable. Eaglesham [30] proposed that the dissolution of interstitials from {311} defects is the source of interstitials that induce TED, since a correlation exists between both time constant and energy barrier for both TED and {311} dissolution. Such dissolution is expected since recombination, surface or bulk, represents a lower free energy of the system. At higher temperatures, {311} defects are relatively unstable and dissolve upon annealing. This occurs after only 3 min at 815°C [30].

In situ annealing in the TEM allows individual defect behavior to be observed and monitored so that the dissolution process can be observed [36]. A 100 keV Si 10^{14} cm^{-2} implant was used to damage a silicon wafer. These samples were then preannealed at 750°C for 2 h in a conventional furnace. The samples were then annealed in situ in the TEM at a variety of temperatures. Figure 1.1 shows the evolution at 0 and 15 min. In this time period, note that the defect in the lower left has completely dissolved. The longer defects are not more stable energetically than smaller defects. This work clearly shows that longer defects can dissolve much faster than shorter defects. Figure 1.6 shows the dissolution of nine different defects from the {311} ensemble. These dissolution curves are fit better by linear decay rates than exponential decay rates. Linear decay fits are shown in Figure 1.6, and the decay rates as a function of initial defect size are extracted. Capture and release of interstitials on the {311} defects occurs only at the end of the defects, and therefore is proportional to the number of defects, D_{311}. This provides two distinct results. First, individual defects dissolve at a nearly constant rate, since the dissolution is proportional only to the end size. The length of the defect does not determine the dissolution rate.

{311} defects may also unfault to form dislocation loops, while simultaneously releasing interstitials into the surrounding crystal. Figure 1.7 shows as an example a {311} defect in a 100 keV, 2×10^{14} cm^{-2} Si^{+} ion-implanted silicon wafer which had been preannealed in a furnace at 800°C for 5 min. After the furnace preanneal, this defect contains 6400 interstitials. The sample is then annealed in situ in a TEM. Over the 40 min in situ anneal, the {311} defect unfaults to form a dislocation loop. Over this same interval the number of interstitials bound by the defect decreases from 6400 to 1500.

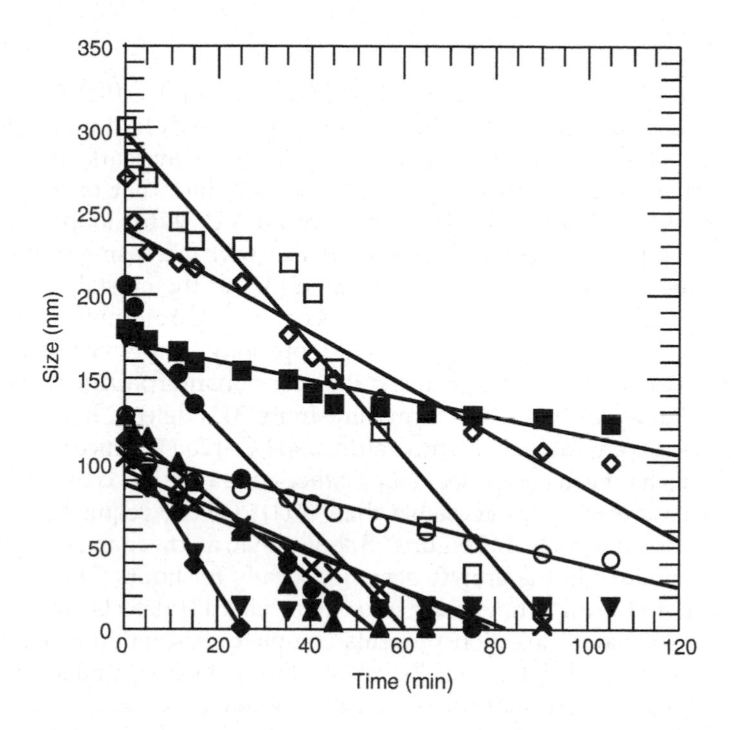

FIGURE 1.6
Size of individual defects as a function of in situ anneal time. Best fits are linear decays. The slowest decaying defect is about average size at time zero. (From Law, M.E. and Jones, K.S., *International Electron Device Meeting*, 2000, p. 511. With permission.)

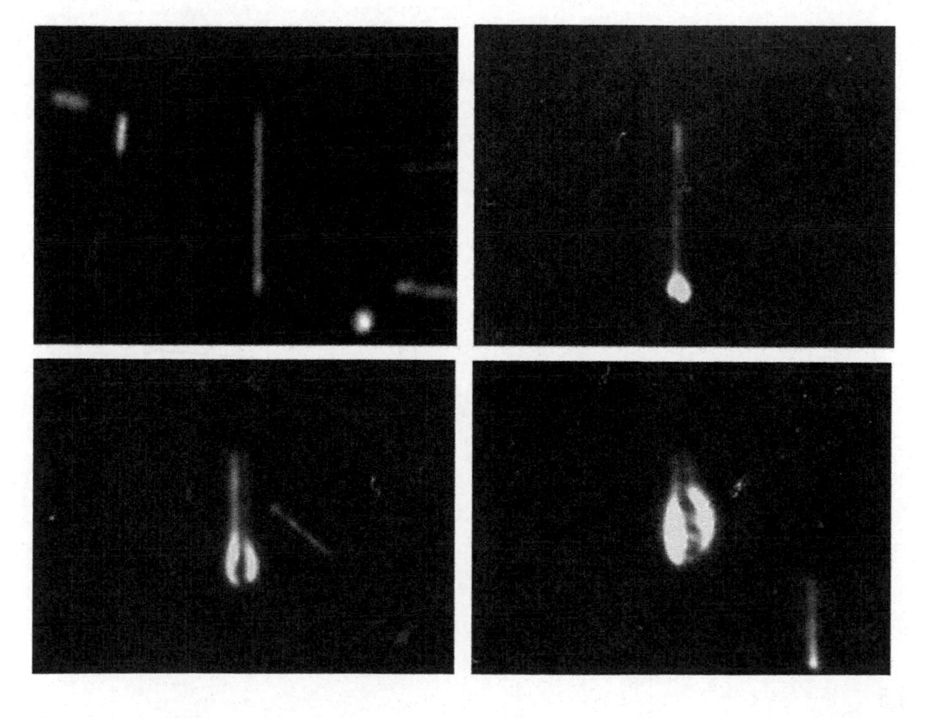

FIGURE 1.7
In situ annealing showing a {311} defect unfaulting into a dislocation loop.

1.5.4 Loops

The formation threshold for dislocation loops [34] (around 1×10^{14} cm^{-2} for implants below 100 keV) is considerably higher than the reported threshold dose for {311} defects of only 7×10^{12} cm^{-2} for 40 keV Si^+ implants [28]. Thus, it appears that the nucleation barrier for theformation of a dislocation loop is higher than that of a {311} defect. For higher-energy implants (380 keV–1 MeV), the threshold dose for loops can drop as low as 4×10^{13} cm^{-2} [37]. The decrease in the threshold dose with increasing energy is thought to be due either to the increase in damage deposition [37] in the crystal or to the increased separation of the Frenkel pairs [38–40] that reduces the *I–V* recombination efficiency. As mentioned previously, it has been proposed that loops may simply evolve from unfaulting of {311} defects [28]. Li and Jones [35] showed that, for nonamorphizing implants, all of the dislocation loops that were observed to form came from {311} defects. It has not been proven that the same process occurs at higher temperatures, 900°C–1200°C. Once the nucleation stage for dislocation loops has been completed, loops either remain stable, coarsen, or dissolve.

Dislocation loops are much more stable than {311} defects, requiring temperatures of 1000°C–1100°C to dissolve [34,41,42]. Figure 1.8 shows that, at lower annealing temperatures, dislocation loops continue in the growth stage kinetically for hours. This time scale is well beyond the {311} defect growth time scale, and occurs as {311} defects are dissolving. Often oxidation-induced interstitial injection prevents complete dissolution of dislocation loops, instead resulting in the growth of the loops into large stacking faults. Dislocation loop dissolution can provide interstitials for some diffusion enhancement [43], but because the temperature is so high, the relative enhancement C_I/C_I^* (C_I^* is the intrinsic concentration of interstitials) is not as large as the effect from {311} dissolution at lower temperatures.

The presence of dislocation loops in the silicon crystal represents an increase in the free energy of the lattice when compared to the equilibrium lattice. During annealing, the minimum size for a stable dislocation loop depends both on the temperature

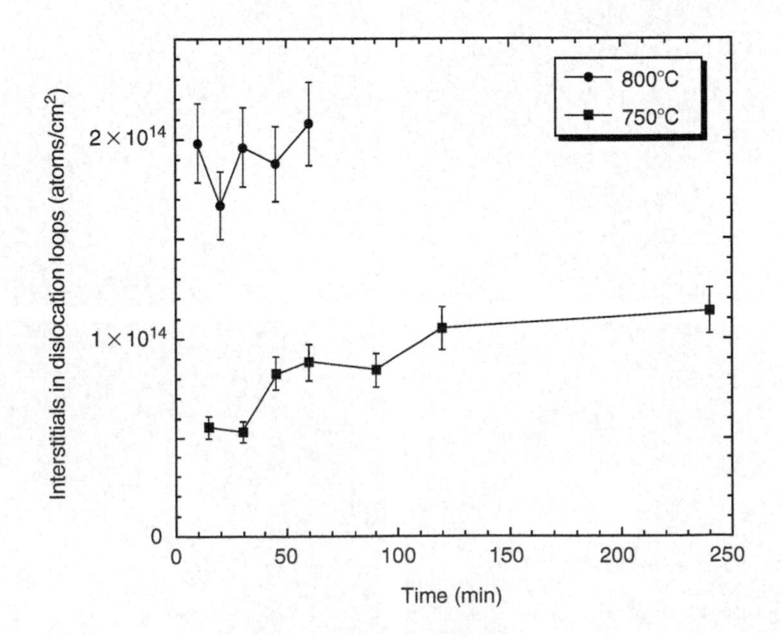

FIGURE 1.8
Density of silicon interstitials in dislocation loops as a function of annealing time at 750°C and 800°C for 120 keV, 1×10^{15} cm^{-2} Si^+ implant.

and the time-dependent supersaturation of interstitials. Due to the finite time required for the nucleation stage, there is a distribution of dislocation loop sizes at a given time. As thermal annealing continues, the dislocation loops move from a nucleation stage to a growth stage. During the growth stage all of the dislocation loops appear to increase in size, so the total number of interstitials bound by dislocation loops increases. After the growth stage, the loops then enter a coarsening stage where the number of dislocation loops decreases and the average loop size increases, while the concentration of bound interstitials in the dislocation loops remains relatively unchanged [44]. The coarsening phase has been modeled both by Liu [45] and Laanab [46]. In the coarsening phase the supersaturation of interstitials around the loops drops significantly [47]. Some authors [47–49] have termed this coarsening stage of dislocation loops "Ostwald ripening," since its behavior mimics that of second phase precipitates in many metals systems, where larger precipitates grow at the expense of smaller precipitates to reduce the surface free energy of the precipitates.

Typically loops enter a coarsening phase within 30–60 min at 800°C [50]. For room temperature implants, if the implant conditions are such that the defects form at depths greater than 1000 Å and the implant species is below its solid solubility, then it has been shown that high annealing temperatures (>1000°C for 24 h) are required to dissolve EOR defects completely [34,41]. These temperatures are, in general, well beyond the thermal budgets of modern integrated circuit (IC) processing. However, as the implant energies decrease, the implant temperature control increases and the dose rate increases the EOR defects become less stable [51–53]. This is attributed to a decrease in the net excess interstitials coming to rest in the EOR region, as opposed to a true surface proximity effect [54]. A review of the effect of implant and anneal conditions on the stability of these defects has been presented elsewhere [40,55].

Theoretically, no dislocation loop is thermodynamically stable in a single-crystal silicon, since it represents an increase in the free energy of the crystal. Kinetic limitations limit this theoretical thermodynamic outlook on the stability of dislocation loops in ion-implanted silicon. As a result of these limitations, many dislocation loops in ion-implanted silicon are, for all practical purposes, stable during postimplantation annealing. Dislocation loops that remain in submicron silicon transistors after processing can be detrimental to the device. Dislocation loops in silicon tend to get metal impurities, which disrupt the electron transport in the vicinity of the loop. Buck [56] and Landi [57] have shown that, if dislocation loops exist in the space charge region of a junction, they can cause high leakage currents. For these reasons, the electrical junction of a silicon transistor must form deep enough so that dislocation loops are never in the space charge region of the device.

1.6 Defects in Ultra-Shallow Processes

1.6.1 Surface Effects

In an effort to investigate the role of the surface on interstitial recombination, a series of experiments was conducted using preamorphized Si. Czochralski (CZ) grown (100) Si wafers were implanted with 1×10^{15} cm^{-2} germanium ions at 5, 10, and 30 keV. This produced amorphous layers of 120, 220, and 480 Å below the surface, respectively. After annealing at 750°C for varying times, TEM images were taken and the defects quantified. Figure 1.9 shows that quantification of the defects showed that, as the energy decreased to 5 keV, the dissolution kinetics rapidly accelerated. The question arises: Is this a result of the proximity to the surface, or the decrease in interstitial population in the end of range region? To study this further, lapping experiments were performed. A sample was implanted

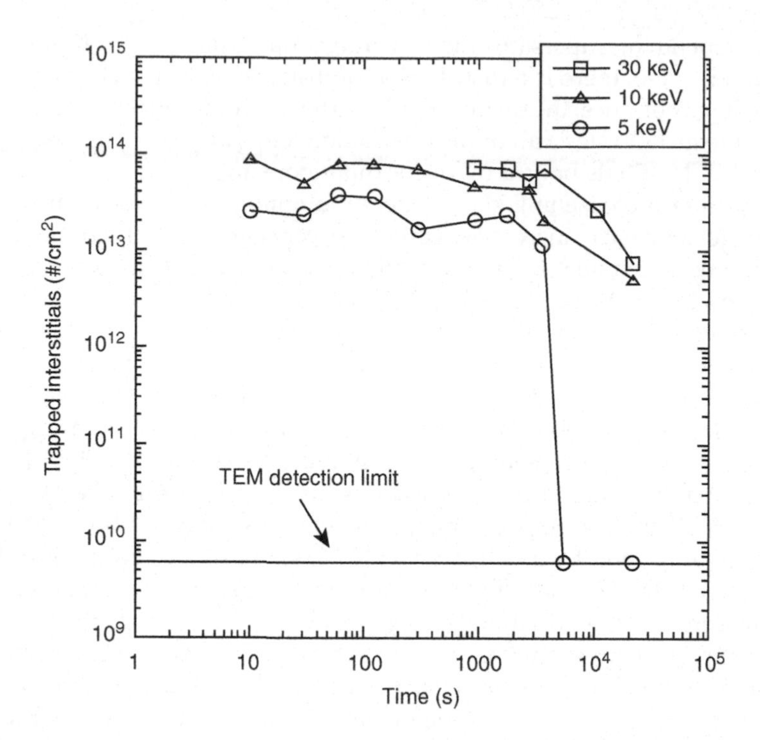

FIGURE 1.9
Dissolution kinetics of defects as the energy of the implant is reduced. At a threshold of 5 keV, the defects rapidly dissolve even though earlier kinetics is not significantly different.

with a 10 keV Ge 1×10^{15} cm^{-3} implant. Next, pieces of the wafer were chemically, mechanically polished to reduce the thickness for the amorphous layer from 180 Å down to 20 Å. The aforementioned 5 keV sample had an amorphous thickness of ~100 Å. Thus, if the surface is the reason for the rapid dissolution, then thinning the sample should have a dramatic effect on defect dissolution rates. After thinning, the samples were annealed at 750°C for two different times, and plan-view transmission electron microscopy (PTEM) was used to quantify the trapped interstitials in the defects. Figure 1.10 shows that, as the amorphous layer was thinned, no surface effect was observed until the amorphous layer was less than 60 Å. This means that the surface proximity did not have a measurable effect on the defect evolution between 60 and 180 Å. This is counter to most of the models, which consider the surface as an infinite sink for interstitials. Hence, there should exist a significant gradient of interstitials toward the surface. However, these results imply that the surface is not as strong a sink as previously thought.

1.6.2 Millisecond Annealing

Damage annealing in the millisecond time regime presents a unique opportunity to investigate the early stages of the damage evolution process, which were not previously possible with former available annealing processes such as RTA. The high temperatures attainable by the flash annealing technique, coupled with the extremely short anneal times, has enabled the defect evolution to be mapped from the early stages in the evolution to mature defect structures. These have been extensively characterized in the past, so this process allows for a more complete picture of the defect evolutionary processes. Such knowledge is crucial to an understanding of the interstitial concentrations, and hence the

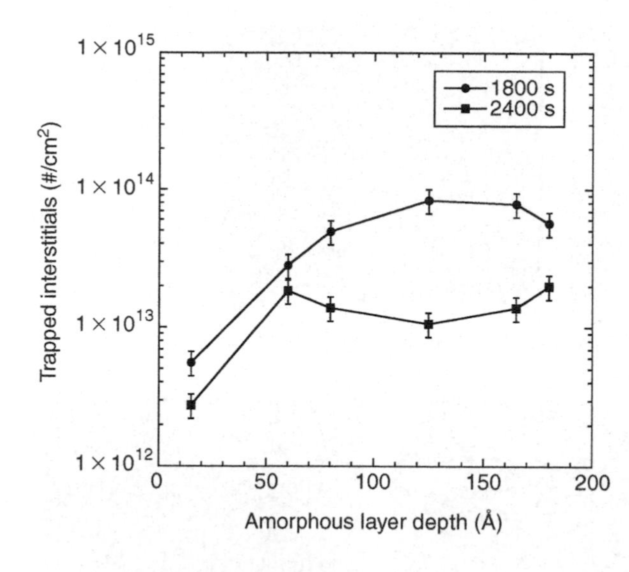

FIGURE 1.10
Number of trapped interstitials as a function of amorphous layer depth. Instead of varying energy, the surface was thinned with polishing.

mechanisms that are largely responsible for dopant diffusion in the silicon lattice during thermal processing cycles.

We now consider defect evolution for amorphizing germanium implants in silicon, during flash annealing. Traditionally defect evolution studies have been conducted by time-dependent studies using isothermal anneals [30,58–60] in which specific defect structures could be isolated. The nature of the flash annealing process does not permit for such investigations, since the anneal time is set by the full width at half maximum (FWHM) of the radiation pulse, which is on the order of milliseconds. Thus, to investigate the evolution, one must resort to isochronal anneals at different temperatures. Studies conducted on the effect of flash annealing on the evolution of the damage for amorphizing germanium implants into silicon reveal the presence of defects at different stages in their evolution, for different flash anneal temperatures. Such studies have been conducted on 30 keV, 1×10^{15} cm^{-3} germanium implants performed on (100) n-type CZ grown silicon wafers [61]. Peak flash temperatures of 1000°C, 1100°C, 1200°C, and 1300°C are investigated, for a constant iRTP anneal temperature of 700°C. Figure 1.11 illustrates the PTEM images of the damage observed as the flash temperature is varied. The 700°C iRTP anneal results in the formation of dot-like defect structures of very high density. Similar dot-like defects are present in the microstructure after the 1000°C and 1100°C flash anneals, but appear to be larger and of a lower density. As the flash temperature is increased to 1200°C and 1300°C, respectively, {311}-type defects and dislocation loops are evident.

The damage present in the structure is consistent with type II defects [34], commonly known as EOR damage, which occurs beyond the amorphous–crystalline interface. That different defect structures are identified at different annealing temperatures in this work is not surprising. Isochronal anneals, although a valid experimental approach to temperature-dependent studies, do not yield EOR defects in the same phase of their evolution. At a given isochronal annealing time, lower temperature anneals generate EOR defects in their earlier nucleation, growth and coarsening stages, while high-temperature anneals result in defects further along in their evolution, possibly in the dissolution regime [62]. The evolution of the type II damage observed in this study at the higher anneal temperatures concur with previous findings, as the EOR defects are observed to evolve from {311}-type defects into dislocation loops with increasing flash anneal temperature.

The existence of {311}-type defects and dislocation loops at such high temperatures can be explained by considering that the anneal times at these temperatures were extremely

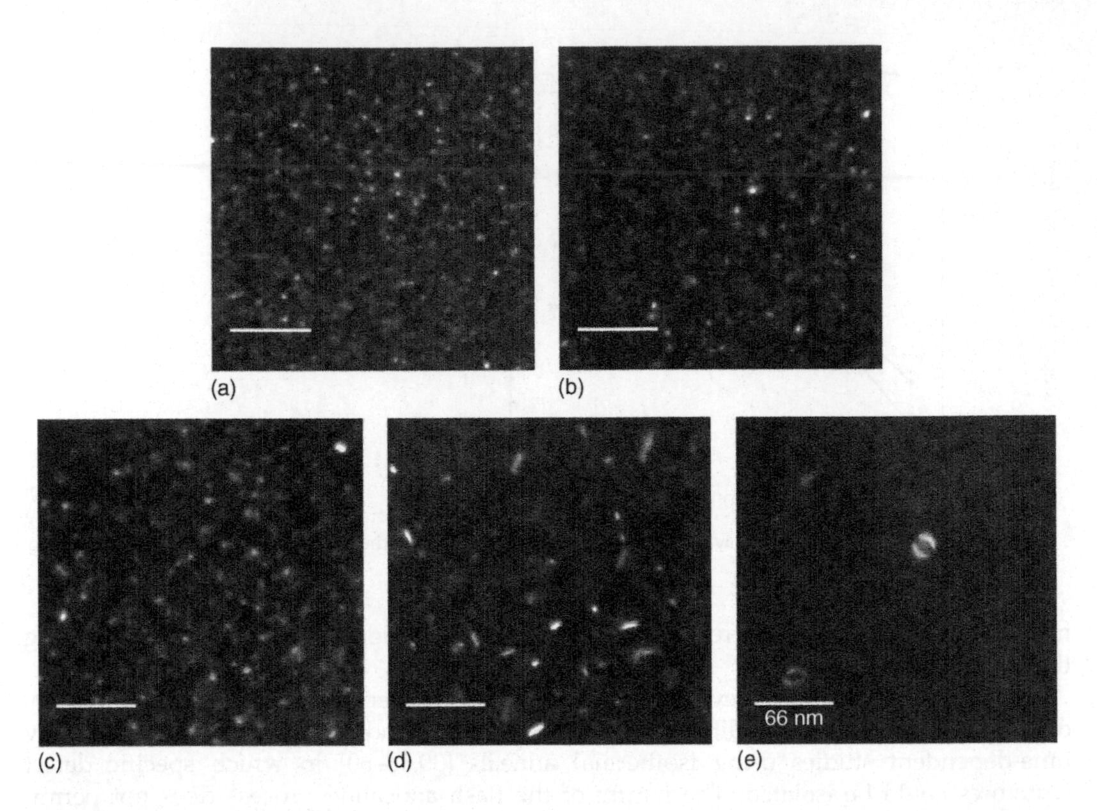

FIGURE 1.11
WBDF PTEM images of the EOR defects imaged under g_{220} two beam conditions of a 30 keV, 1×10^{15} cm^{-2} Ge amorphizing implant into Si (100): (a) 700°C iRTP; (b) 700°C iRTP, 1000°C fRTP; (c) 700°C iRTP, 1100°C fRTP; (d) 700°C iRTP, 1200°C fRTP; and (e) 700°C iRTP, 1300°C fRTP. (From Camillo-Castillo, R.A. et al., *Appl. Phys. Lett.*, 88, 232104, 2006. With permission.)

short. Keys [62] demonstrates that it is possible to affect similar defect structures and diffusion states at different anneal temperatures by determining an equivalent anneal time. Knowledge of the time, t_1, required for defect dissolution at a given anneal temperature, T_1, allows one to calculate the anneal time, t_2, required to affect similar defect structures at another temperature, T_2, from the ratio of the time constants for decay, τ, in accordance with the relation:

$$t_2 = \frac{\tau_1 t_1}{\tau_2} \tag{1.1}$$

in which τ is related to the anneal temperature, T, by an Arrhenius relation given by Equation 1.2 that includes an activation energy, E_a, a preexponential factor, τ_0, and the Boltzmann constant, k,

$$\tau = \tau_0 \exp\left(\frac{E_a}{kT}\right) \tag{1.2}$$

The value of the E_a used in the calculation is based on the dominant defect structure present in the microstructure. Thus, in the case of the {311}-type defect, approximately 5 ms at 1200°C is required to affect total {311}-type defect dissolution based on an activation energy of 3.7 eV [30,58]. This time exceeds the size of the radiation pulse produced by the flash lamps by an order of magnitude. Hence the presence of {311}-type defect in the microstructure after the 1200°C fRTP is plausible, since the time of the flash anneal was insufficient to affect dissolution based on the equivalent anneal time calculation. A similar argument holds for the occurrence of dislocation loops upon annealing at 1300°C. An anneal time of 0.3 ms is required to account for the differences in the interstitial densities between 1200°C and 1300°C flash anneals, which falls within the range of the FWHM of the flash pulse at 1300°C, confirming that the anneal time is sufficient to affect {311}-type defect dissolution, and explaining the presence of only dislocation loops in the microstructure.

There has been no previous experimental evidence of the EOR defects evolving from the dot-like structures to {311}-type defects. To confirm that these defects are in the early stages of their evolution, postflash thermal processing, consisting of a spike rapid thermal anneal at 950°C, is performed on the material containing these dot-like structures. This is based on the notion that the additional thermal budget should evolve the damage if it is in its infancy. Examination of the microstructure subsequent to the 950°C spike RTA reveals the presence of dislocation loops in the structure, which corroborates the theory that the dot-like defects evolve into more stable defect structures with subsequent annealing [61].

The exact configuration of small interstitial clusters has been the center of a number of investigations, yet very little is still known about them. Recent experimental and theoretical data [9,63,64] demonstrate that precise cluster sizes exhibit enhanced stability, indicated by the existence of minima and maxima in the cluster binding energy curve. However, considerable debate remains over the exact sizes of the stable clusters. Cowern [9] found that interstitial clusters that consist of more than 20 atoms have a similar differential formation energy to the {311}-type defect, suggesting that the interstitial clusters undergo a transition to {311}-type defects at a smaller cluster size. Other investigations [65,66] support this idea and suggest that the transition from small interstitial clusters to {311}-type defects occurs for interstitial clusters containing eight atoms. Hence, the small interstitial clusters observed at 700°C iRTP, and the 1000°C and 1100°C flash may in fact be {311}-type defects, since the total number of atoms in these structures exceeds eight atoms, and the smallest defect that can be imaged by a conventional TEM is approximately 100 atoms [67]. Other studies [60] of lower-energy germanium amorphizing implants propose that small interstitial clusters may exhibit defect morphology very similar to plate-like dislocation loops. These dislocation loops were shown to be very unstable, dissolving with an activation energy of 1.13 ± 0.14 eV [60]. If the dot-like defects in these experiments are analogous to small {311}-type defects alluded to by Cowern [9] or the loops observed by King [60], then any additional thermal budget applied to them should result in a defect dissolution behavior that adheres to the respective dissolution kinetics of these defects.

Figures 1.12 and 1.13, respectively, depict the defect densities and trapped interstitial populations as a function of the peak anneal temperature [7,61]. It is evident that both of these quantities decrease as the flash temperature is increased above the 700°C iRTP anneal, indicating interstitial loss from the EOR damage as the defects evolve over the course of the flash anneal. This loss of interstitials during the flash anneal process may be attributed to interstitial recombination at the amorphous–crystalline interface during regrowth of the amorphous layer, recombination at the surface on completion of the regrowth, and interstitial loss to the bulk of the material. These simultaneous factors suggest that the defects are in a coarsening regime in which a fraction of the interstitials is not recaptured by evolving defect structures. Consequently, during the flash anneal, the system is viewed as a "leaky box" from which interstitials are lost as the EOR defects undergo coarsening.

Investigations of the kinetics of the defect decay are traditionally performed by fitting the interstitial density with time over various temperatures to an exponential function of the

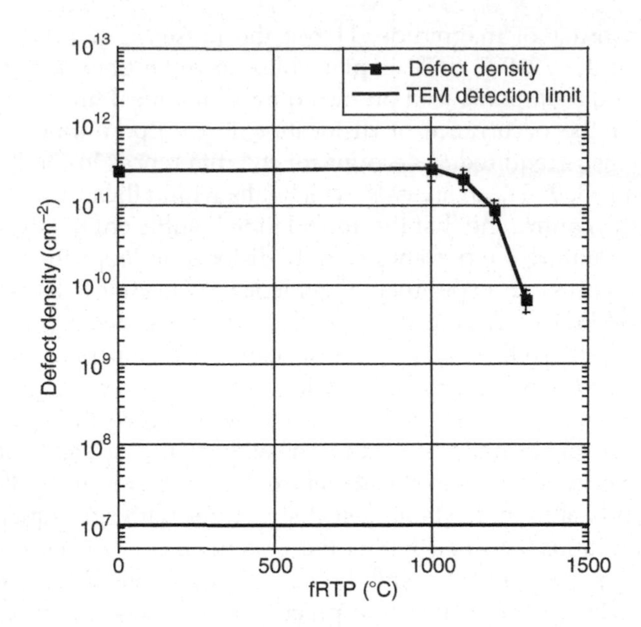

FIGURE 1.12
Defect density as a function of fRTP anneal temperature for a 1×10^{15} cm^{-2}, 30 keV Ge amorphizing implant into Si (100).

form of Equation 1.1. In these cases isolated defect structures such as {311}-type defects [30,58] and dislocation loops [59,60] were examined, enabling the kinetics for each defect type to be extracted. Since the nature of the flash anneal process limits the investigations

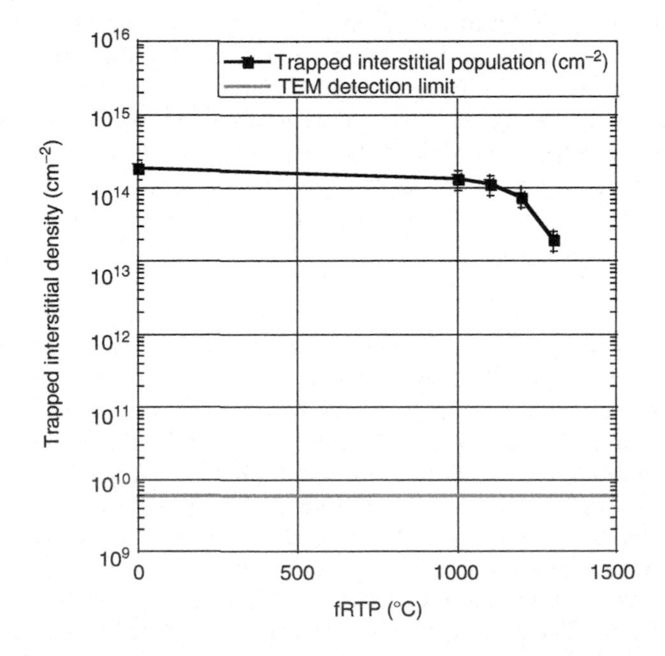

FIGURE 1.13
Trapped interstitial density as a function of fRTP anneal temperature for a 1×10^{15} cm^{-2}, 30 keV Ge amorphizing implant.

to isochronal anneals, which affects EOR defects at different stages in their evolution, conventional kinetic analyses methods in which specific defect structures are isolated cannot be employed to determine the rate of interstitial loss for single isolated defect structures. Rather the interstitial loss data for the duration of the flash anneal over differing defect structures is obtained. This necessitates the development of another approach to analyzing the attainable experimental data, so that meaningful comparisons can be made to past studies [61]. The adopted approach entails the selection of a reference anneal, from which the decay in the trapped interstitial density can be tracked with the flash anneal temperature, allowing for the kinetics of the interstitial decay to be extracted. The interstitial value corresponding to the 700°C iRTP anneal temperature can serve as such a reference for the initial trapped interstitial value. This is validated by the observation that all of the wafers are heated to this temperature prior to application of the flash, and hence allows for the decay in the trapped interstitials during the flash portion of the thermal profile, i.e., flash anneal, to be extracted such that the effect of the flash on the defects could be isolated.

Figure 1.13 noticeably demonstrates that the trapped interstitial decay follows an exponential relation as the flash temperature is increased. Hence, in accordance with kinetic rate theory, the interstitial decay can be approximated by the relation:

$$\frac{\partial C_{\text{interstitials}}}{\partial t} = -\frac{C_{\text{interstitials}}}{\tau} \tag{1.3}$$

where

$C_{\text{interstitials}}$ is the concentration of trapped interstitials (cm^{-2})

t(s) is the anneal time

τ(s) is the captured interstitial lifetime, which is related to the anneal temperature, T(K), by an Arrhenius expression that includes an activation energy, E_a

$$\tau = \tau_0 \exp\left(\frac{E_a}{kT}\right) \tag{1.4}$$

The process simulator FLOOPS [68] is utilized to calculate the trapped interstitial density for each flash anneal temperature from an initial trapped interstitial density (value for the 700°C iRTP anneal), by fitting the parameters τ_0 and E_a. The temperature–time variations for each anneal need to be incorporated into the simulation to allow for an accurate determination of τ as time and temperature are altered. This approach facilitates the precise integration of the trapped interstitials with time. The decay rates derived from the fits of the experimental trapped interstitial populations are illustrated in Figure 1.14. The interstitial decay rate is found to vary linearly with the inverse flash temperature, yielding an activation energy, E_a of 2.1 ± 0.05 eV and preexponential factor, K_0, of 3.3×10^{10} s^{-1}. We note that the interstitial decay rates varied over two orders of magnitude for the flash temperatures investigated, from approximately 100 s^{-1} at 1000°C, compared to 2000 s^{-1} at 1200°C. Such high decay rates had never been observed.

Comparison of the interstitial decay rates for {311}-type defects [30] and dislocation loops [59,60] demonstrates the much higher interstitial decay rates during the flash anneal compared to former studies. The interstitial decay rates for the flash anneals range two orders of magnitude between 1×10^2 and 1×10^3 s^{-1} for anneal temperatures from 1000°C to 1300°C, which are three orders of magnitude larger than the maximum rate previously reported, obtained by Seidel et al. [59] for the interstitial decay from stable dislocation loops. The interstitial decay rates for {311}-type defects and small dislocation loops were much lower. The temperature range of Seidel's experiments coincides with flash temperatures lower than 1200°C. Yet, the interstitial decay rates are vastly dissimilar for this

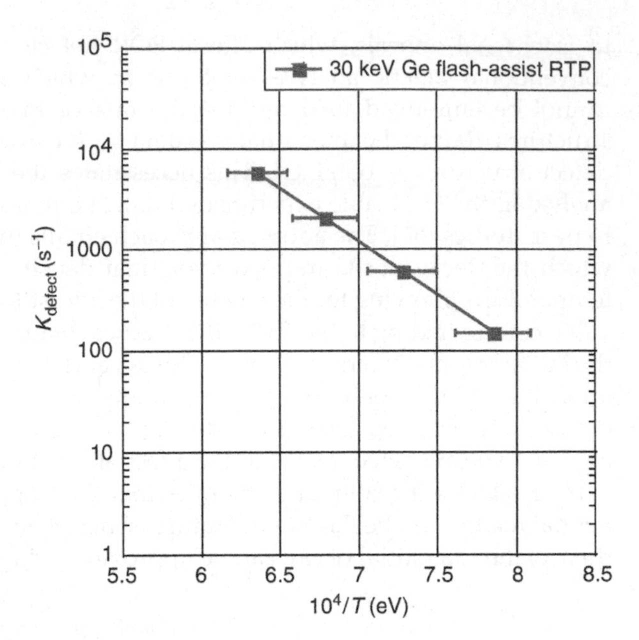

FIGURE 1.14
Arrhenius plot of the time constant derived for defect decay extracted from the simulated experimental data, indicating an activation energy, E_a, of 2.1 ± 0.05 eV for dissolution.

temperature regime, suggesting that the interstitial loss is most likely from a defect not similar to the dislocation loop. This suggests the existence of a highly unstable defect structure at these temperatures, and correlates with the dot defects observed at this temperature during the flash anneal.

Examination of the differences in the activation energies for defect dissolution in the literature provides further insight into the characteristics of the proposed highly unstable defect. The extracted activation energies for interstitial loss during the flash anneal are 2.7 and 1.6 eV smaller than the values obtained for dislocation loop and {311}-type defect dissolution, respectively. That the activation energy determined for the interstitial decay during the flash anneal is not similar to those previously extracted values supports the theory that the interstitial loss is not from comparable defect structures. It also clearly indicates that this defect is less stable than the {311} defect and dislocation loops.

That dot-like defects exist in the structure for flash anneal temperatures of 1100°C and lower show that these defects only exist in the early stages of annealing, and either evolve into {311}-type defects or dissolve. The decrease in the trapped interstitial concentration between 1100°C and 1200°C flash anneal temperatures suggests that some of these defects dissolved, losing interstitials to either the surface to the bulk of the material. The data imply that those dot-like defects that did not dissolve must have therefore evolved into the {311}-type defect detected after the 1200°C anneal, since they are no longer observed in the microstructure. This validates the supposition that the dot-like defect is a precursor for the {311}-type defect. Consequently, the extracted kinetics for the decrease in trapped interstitial density as a function of the temperature applies to the dissolution of this dot-like defect structure.

1.6.3 Regrowth-Related Defects

Regrowth-related defects arise upon recrystallization of an implantation-induced amorphous layer. These defects may arise if the amorphous–crystalline interface is rough. This can occur for lighter ions, at higher implant energies, or if the temperature of the wafer rises

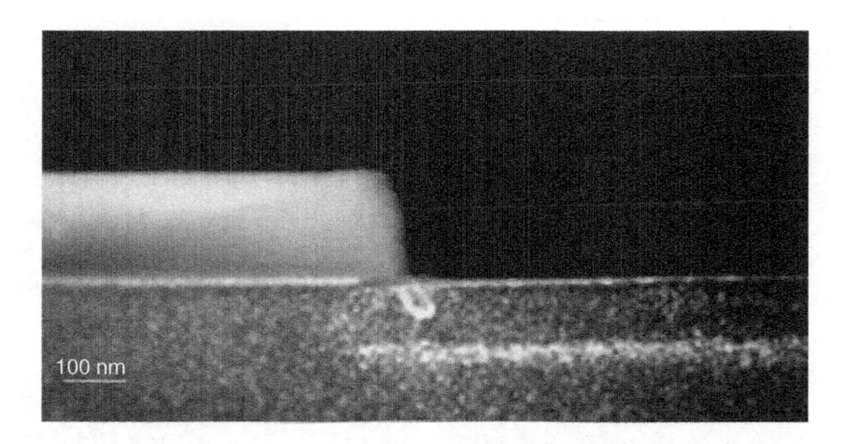

FIGURE 1.15
Half loop dislocation.

during the implant. These defects may be in the form of hairpin dislocations [69] or stacking faults and microtwins. Because the amorphous–crystalline interface is not planar, extended defects can form upon solid-phase recrystallization as the interface begins to propagate toward the surface. In addition, regrowth-related defects may arise in patterned structures such as at a gate edge. These defects arise when the lateral and vertical regrowth velocities are such that a pinch point forms where the two interfaces meet. This also requires the depth of the implant-induced amorphous layer to typically be over 700–800 Å thick [70].

Figure 1.15 shows an image of the extended defects that can form near the mask edge during regrowth. This defect is a half loop dislocation that propagates up to the surface. Research has shown that this defect occurs because of the way the amorphous layer is shaped during regrowth. Figure 1.16a shows a regrowth condition that leads to these mask edge defects. Figure 1.16b clearly shows the bottom corners beginning to pinch

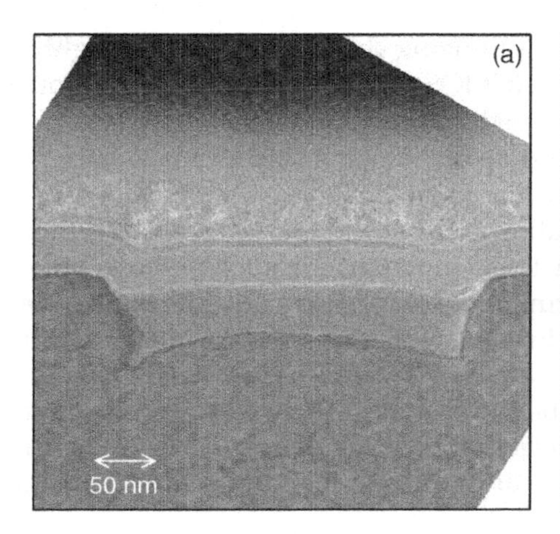

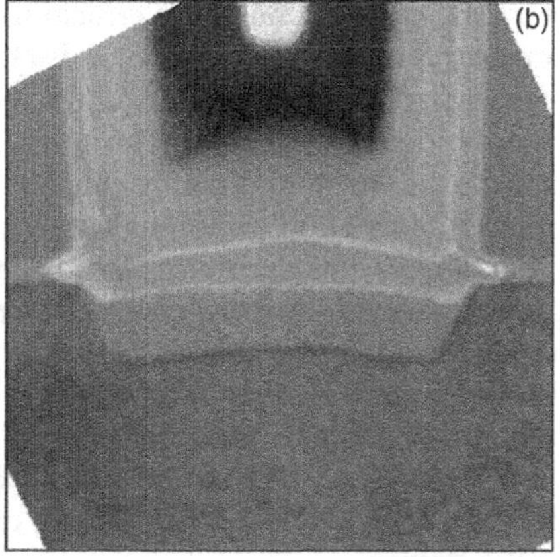

FIGURE 1.16
The amorphous layer in cross section during regrowth. Both had 40 keV 10^{15} Si^{+} implants, and were annealed at 750°C for 13 min. On the left there is no nitride pad and the pinching of the regrowth in the corner is evident. The right has a nitride pad and avoids pinching.

27. Davidson, S.M., Study of radiation damage in silicon using scanning electron microscopy, *Nature*, 227, 487, 1970.
28. Eaglesham, D.J. et al., Implant damage and transient enhanced diffusion in Si, *Nucl. Instrum. Meth. B*, 106, 191, 1995.
29. Liu, J. et al., The effect of boron implant energy on transient enhanced diffusion in silicon, *J. Appl. Phys.*, 81, 1656, 1997.
30. Eaglesham, D.J. et al., Implantation and transient B diffusion in Si: The source of the interstitials, *Appl. Phys. Lett.*, 65, 2305, 1994.
31. Pan, G.Z. et al., Microstructural evolution of {113} rod-like defects and {111} dislocation loops in silicon-implanted silicon, *Appl. Phys. Lett.*, 71, 659, 1997.
32. Ferreira Lima, C.A. and Howie, A., Defects in electron-irradiated germanium, *Phil. Mag. A*, 34, 1057, 1976.
33. Salisbury, I.G. and Loretto, M.H., {113} Loops in electron-irradiated silicon, *Phil. Mag. A*, 39, 317, 1979.
34. Jones, K.S. et al., A systematic analysis of defects in ion implanted silicon, *Appl. Phys. A*, 45, 1, 1988.
35. Li, J.-H. and Jones, K.S., {311} defects in silicon: The source of the loops, *Appl. Phys. Lett.*, 73, 3748, 1998.
36. Law, M.E. and Jones, K.S., A new model for {311} defects based on in-situ measurements, *International Electron Device Meeting*, San Francisco, CA, 2000, p. 511.
37. Schreutelkamp, R.J. et al., Pre-amorphization damage in ion-implanted silicon, *Mater. Sci. Rep.*, 6, 275, 1991.
38. Laanab, L. et al., Variation of end of range density with ion beam energy and the predictions of the excess interstitials model, *Nucl. Instrum. Meth. B*, 96, 236, 1995.
39. Cho, N.-H. et al., MeV ion induced damages and their annealing behavior in silicon, *Mater. Res. Soc. Symp.*, San Francisco, CA, 1996, p. 430.
40. Jones, K.S. and Gyulai, J., Annealing of implantation damage in silicon, *Ion Implantation Science and Technology*, Ziegler, J.F. (ed.), Ion Implantation Technology Co., Yorktown, 1996.
41. Jones, K.S. et al., Enhanced elimination of implantation damage upon exceeding the solid solubility, *J. Appl. Phys.*, 62, 4114, 1987.
42. Fair, R.B., Damage removal/dopant diffusion tradeoffs in ultra-shallow implanted p^+–n junctions, *IEEE Trans. Electron Dev.*, 37, 2237, 1990.
43. Michel, A.E., Anomalous transient diffusion of ion implanted dopants: A phenomenological model, *Nucl. Instum. Meth. B*, 37/38, 379, 1989.
44. Listebarger, J.K. et al., Use of type II (end of range) damage as detectors for quantifying interstitial fluxes in ion-implanted silicon, *J. Appl. Phys.*, 73, 4815, 1993.
45. Liu, J. et al., Evolution of dislocation loops in silicon in an inert ambient (I), *Solid State Electron.*, 38, 1305, 1995.
46. Laanab, L. et al., A model to explain the variations of end-of-range defect densities with ion implantation parameters, *Mater. Res. Soc. Symp.*, 1993, p. 302.
47. Bonafos, C.B. et al., TED of boron in presence of EOR defects: The role of the evolution of Si self-interstitial supersaturation between the loops, *Intl. Conf. Ion Imp. Tech.*, Austin, TX, 1997, p. 1.
48. Claverie, A. et al., Interactions between dopants and end of range defects in silicon, *Solid State Phen.*, 47–48, 195, 1996.
49. Bonafos, C. et al., The effect of the boron doping level on the thermal behavior of end-of-range defects in silicon, *Appl. Phys. Lett.*, 71, 365, 1997.
50. Listebarger, J.K. et al., Study of end of range loop interactions with B+ implant damage using a boron doped diffusion layer, *J. Appl. Phys.*, 78, 2298, 1995.
51. Ajmera, A.C. and Rozgonyi, G.A., Elimination of end-of-range and mask edge lateral damage in Ge^+ preamorphized, B^+ implanted Si, *Appl. Phys. Lett.*, 49, 1269, 1986.
52. Ajmera, A.C. et al., Point defect/dopant diffusion considerations following preamorphization of silicon via Si^+ and Ge^+ implantation, *Appl. Phys. Lett.*, 52, 813, 1988.
53. Prussin, S. and Jones, K.S., Role of ion mass, implant dose, and wafer temperature on end-of-range defects, *J. Electrochem. Soc.*, 137, 1912, 1990.

54. Ganin, E. and Marwick, A., Is the end-of-range loops kinetics affected by surface proximity or ion beam recoils distribution? *Mater. Res. Soc. Symp.*, San Diego, CA, 1989, p. 147.
55. Jones, K.S. and Rozgonyi, G.A., Extended defects from ion implantation and annealing, *Rapid Thermal Processing: Science and Technology*, Fair, R.B. (ed.), Academic Press, Orlando, FL, 1993, p. 123.
56. Buck, T.M. et al., Gettering rates of various fast-diffusing metal impurities at ion-damaged layers on silicon, *Appl. Phys. Lett.*, 21, 485, 1972.
57. Landi, E. and Solmi, S., Electrical characterization of p+/n shallow junctions obtained by boron implantation into preamorphized silicon, *Solid. State Electron.*, 29, 1181, 1986.
58. Stolk, P.A. et al., Physical mechanisms of transient enhanced dopant diffusion in ion-implanted silicon, *J. Appl. Phys.*, 81, 6031, 1997.
59. Seidel, T.E. et al., A review of rapid thermal annealing (RTA) of B, BF_2 and As ions implanted into silicon, *Nucl. Instrum. Meth. B*, 7/8, 251, 1985.
60. King, A.C. et al., Defect evolution of low energy, amorphizing germanium implants in silicon, *J. Appl. Phys.*, 93, 2449, 2003.
61. Camillo-Castillo, R.A. et al., Kinetics of the end-of-range damage dissolution in flash-assist rapid thermal processing, *Appl. Phys. Lett.*, 88, 232104, 2006.
62. Keys, P., Phosphorus-defect interactions during thermal annealing of ion implanted silicon, PhD dissertation, University of Florida, Gainesville, FL, 2001.
63. Chichkine, M.P., De Souza, M.M., and Sankara Narayanan, E.M., Growth of precursors in silicon using pseudopotential calculations, *Phys. Rev. Lett.*, 88, 085501, 2002.
64. Gilmer, G.H. et al., Diffusion and interactions of point defects in silicon: Molecular dynamics simulations, *Nucl. Instrum. Meth. B*, 102, 247, 1995.
65. Chichkine, M.P., De Souza, M.M., and Sankara Narayanan, E.M., Self-interstitial clusters in silicon, *Mater. Res. Soc. Symp.*, San Francisco, CA, 2000, p. 610.
66. Claverie, A. et al., Modeling of the Ostwald ripening of extrinsic defects and transient enhanced diffusion in silicon, *Nucl. Instrum. Meth. B*, 186, 281, 2002.
67. Robertson, L.S. et al., Annealing kinetics of {311} defects and dislocation loops in the end-of-range damage region of ion implanted silicon, *J. Appl. Phys.*, 87, 2910, 2000.
68. Law, M.E. and Cea, S.M., Continuum based modeling of silicon integrated circuit processing: An object oriented approach, *Comput. Mater. Sci.*, 12, 289, 1998.
69. Sands, T. et al., Influence of the amorphous-crystalline interface morphology on dislocation nucleation in pre-amorphized silicon, *XIII Intl. Conf. on Defects in Semiconductors*, The Metallurgical Society, Los Angeles, CA, 1984, p. 531.
70. Olson, C.R. et al., Effect of stress on the evolution of mask-edge defects in ion-implanted silicon, *J. Vac. Sci. Technol. B*, 24, 446, 2006.
71. Ross, C. and Jones, K.S., The role of stress on the shape of the amorphous–crystalline interface and mask-edge defect formation in ion-implanted silicon, *Mater. Res. Soc. Symp.*, San Francisco, CA, 2004, p. 810.
72. Rudawski, N.G. et al., Effect of uniaxial stress on solid phase epitaxial regrowth and mask edge defect formation in two-dimensional amorphized Si, *Mater. Res. Soc. Symp.*, San Francisco CA, 2006, p. 912.

2

Hydrogen-Related Defects in Silicon, Germanium, and Silicon–Germanium Alloys

A.R. Peaker, V.P. Markevich, and L. Dobaczewski

CONTENTS

2.1 Introduction

Hydrogen is a very common impurity in semiconductors and has been investigated in great detail in silicon. There is also a small amount of work on hydrogen in silicon–germanium and germanium. Czochralski-grown ingots of silicon and germanium contain almost no hydrogen. However, subsequent processing can introduce considerable concentrations in the near surface regions and in some cases deep in the bulk of the materials. How hydrogen is incorporated into silicon and germanium is discussed in detail in Section 2.2, so here it is sufficient to say that processes such as chemical etching, polishing, and plasma processing release considerable quantities of atomic hydrogen, which is free to diffuse and react with the semiconductor. In some technological processes hydrogen is deliberately incorporated, a classic case is where silicon metal-oxide semiconductor (MOS) devices are heated in forming gas at about 450°C to passivate interface traps or surface states, a process that was patented by Fowler in 1974 [1]. This passivation process is discussed in Chapter 7 of this book, but quite obviously there is the potential for hydrogen to diffuse into the semiconductor itself.

The incorporation of hydrogen in silicon and germanium is a rather complex process that depends on the form of the hydrogen and the doping and impurity content of the semiconductor. Hydrogen molecules diffuse in silicon, but more importantly, so do atomic defects of hydrogen. These can exist in positively or negatively charged states or, exceptionally, in neutral states. The charged single atomic species are extremely reactive and are commonly observed to passivate shallow donors and acceptors as well as deep-level defects. Much work has been devoted to studying the reaction products of various defect species with hydrogen. This reactivity makes the diffusion behavior of atomic hydrogen at lower temperatures somewhat complex; a key issue is its trapping at various impurity sites. The diffusion depends on the detail of the silicon doping, the saturation of capture sites, and the subsequent thermal release of the hydrogen from these trapping sites. In consequence, the solubility and diffusion of hydrogen in highly doped silicon is dramatically different than that of its behavior in undoped material.

Isolated hydrogen atoms are known to exist in silicon, silicon–germanium, and germanium at low temperatures; various techniques have been used to determine their precise position in the semiconductor lattice. Similarly detailed work has been conducted on hydrogen passivation and on electrically active hydrogen complexes. Although this chapter is about hydrogen, it is important to say that deuterium is often used independently or in combination in studies of hydrogen-related defects. The reasons for this are that (1) secondary ion mass spectrometry (SIMS) has a much higher sensitivity to deuterium than hydrogen and (2) that some techniques such as local mode spectroscopy can distinguish between hydrogen and deuterium, thereby making the identities of complexes (especially those containing more than one hydrogen atom) easier to resolve.

In general, the total concentration of hydrogen in silicon determined by SIMS is much higher than the concentration of hydrogen and hydrogen complexes determined their electrical or optical measurements. Combinations of theoretical studies and optical work have revealed that this is due to the presence of hydrogen molecules. These are neutral and, in their usual siting within the silicon lattice, are largely electrically and optically inactive. However, much work has now been done to resolve the weak optical activity of molecular species using Raman studies and optical absorption techniques. In recent years, this work has been given a particular stimulus by the development of the SmartCut process for producing silicon on insulator wafers [2]. In this technique, light ions (predominantly hydrogen) are implanted at high concentrations to produce gas bubbles just below the surface. Subsequent heat treating can then be used to defoliate a thin uniform layer. This behavior is now used extensively to produce silicon on insulator wafers for high performance, extremely scaled complementary MOS (CMOS), and indeed for many other applications. The technique can be applied to silicon, silicon–germanium, and germanium, and is now of considerable commercial importance. All these issues are discussed in detail in the subsequent sections of this chapter, but attention should be drawn to a number of previous reviews of hydrogen in silicon and related materials and also to conference proceedings focused specifically to hydrogen in silicon [3–8].

2.2 Incorporation, Diffusion, and Solubility of Hydrogen

2.2.1 Incorporation

Hydrogen can be incorporated into silicon in molecular or atomic form. In the latter case, it can exist in the positive, negative, or metastable neutral charge states. Hydrogen in its atomic form is very reactive and can combine with impurities in the silicon. As a consequence, the behavior of atomic hydrogen is strongly influenced, at least in the low-temperature regime (less than ~200°C), by the presence of impurities. All the above factors affect the incorporation, solubility, and diffusivity at different temperatures. Hence, the actual values for incorporation rate, diffusion, and solubility depend on the experimental conditions used. So, a very wide scatter exists in the experimental values in the literature. However, it is evident that hydrogen has a very high intrinsic diffusivity and a low equilibrium solubility.

The most fundamental mode of incorporation is diffusion at high temperatures (~1000°C) in an atmosphere of hydrogen gas. Several publications base solubility and diffusion studies on incorporation in this way. However, the detectivity of hydrogen in SIMS is rather low due to the high background levels of hydrogen in most SIMS machines. Thus, deuterium has been used as a hydrogen analog in several studies. The improvement in detectivity is machine specific, but more than two orders of magnitude improvement is stated in some publications resulting in densities $[D] \geq 5 \times 10^{13}\ cm^{-3}$ being quantifiable.

An important issue in relation to incorporation into silicon from hydrogen or deuterium gas is that the process does not depend linearly on the partial pressure p of the gas. The experiments in 1956 of Van Wieringen and Warmoltz [9] measured the transport of hydrogen through a thin-walled silicon cylinder and found the steady-state permeation at 1050°C to be a function of $p^{0.5}$, and at 1200°C, $p^{0.54}$. If diffusion is the rate-limiting step in these experiments, the most likely explanation is that the hydrogen diffuses predominantly as atoms rather than molecules. The dissociation energy of a hydrogen molecule is ~4 eV, but catalytic dissociation of hydrogen at temperatures ≤1050°C is well known and probably can occur at the silicon surface. Subsequent measurements of rapid-quenched samples by Newman et al. found the apparent solubility of both hydrogen and deuterium varied as $p^{0.54}$ at 1300°C [10]. The conclusion was also drawn that incorporation and diffusion occur in atomic form, although molecules were shown to be present after quenching and assumed to be due to recombination of the atomic species.

Another very widely used method of hydrogen and deuterium incorporation is from a discharge produced in the gas at a pressure of ~2 Torr by either a DC current or radio frequency plasma. Often the sample is remote from the discharge to minimize damage due to energetic ions. Atomic species reach the sample by drift in the gas flow. The atomic species often recombine before reaching the sample, a process enhanced by the proximity of the surface of the discharge tube. As a result, the concentration of atomic hydrogen at the sample surface is difficult to control and quantify. However, it is a very effective method for low-temperature incorporation, as shown in Figure 2.1, where the SIMS-determined deuterium profile is shown after exposure to a remote 70 W plasma for 30 min with the sample held at 150°C [11]. After deuteration, the samples were rapidly cooled to room temperature (RT). The figure shows high concentration (~$10^{20}\ cm^{-3}$) incorporation at the surface, while deeper into the silicon the concentration of deuterium is similar to the dopant concentration for the case of boron but not in the case of phosphorus. This rather complex behavior will be discussed later in this chapter.

Ion-beam sources are also used to incorporate hydrogen as H^+, although of course considerable damage is introduced into the silicon lattice by these protons. These techniques have been reviewed in detail by Seager [12], and more recently have been the basis

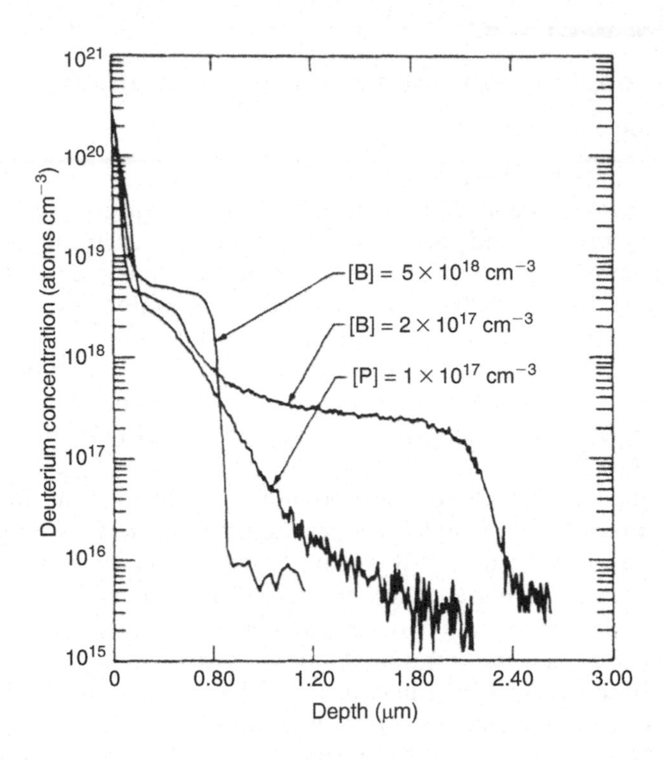

FIGURE 2.1
Deuterium depth profile in Czochralski-grown Si after plasma treatment at 150°C for 30 min using three different slices of material doped with (a) 5×10^{18} boron, (b) 2×10^{17} boron, and (c) 10^{17} phosphorus. (From Johnson, N.M. and Moyer, M.D., *Appl. Phys. Lett.*, 46, 787, 1985. With permission.)

of the commercially important SmartCut process [2]. These are processes where hydrogen is introduced intentionally; hydrogen incorporation into silicon also occurs as an unintentional consequence of plasma etching and related processes.

It is also possible to introduce hydrogen into silicon electrochemically at low temperature. This has been reported for hydrogen fluoride and oxidizing etches [13], and even water [14]. Essentially any process that produces atomic hydrogen at the surface of the silicon provides species that can diffuse rapidly into the semiconductor and react with dopant or other impurity species. Among the many technologically important processes in which hydrogen incorporation occurs are chemical mechanical polishing (CMP) of silicon and the fabrication of microelectromechanical systems (MEMS) structures by chemical or plasma etching.

2.2.2 Diffusion

Diffusion falls into two clearly distinct regimes. The first is high-temperature behavior in which the hydrogen migrates predominantly as atomic species via an interstitial mechanism. The second is a complex low-temperature behavior that again reflects the motion of atomic species, but where hydrogen binds to impurities in the silicon, forming stable or metastable defects. As a result, the hydrogen transports through the lattice via a hopping mechanism. Under these circumstances the effective diffusivity depends not only on the temperature, but also on the concentration of the impurities present, the concentration of the hydrogen, and in some cases the position of the Fermi level. The key issues are presented below; for more detail, the reader is referred to a comprehensive review of the early work on diffusion and solubility of hydrogen in crystalline silicon [15], published by Stavola in 1999.

Van Wieringen and Warmoltz's measurement [9] of the permeation of hydrogen through a thin-walled silicon cylinder provides a very reliable direct measurement of diffusivity over the temperature range 1090°C–1200°C. The relationship they derived is

$$D(T) = 9.4 \times 10^{-3} \exp(-0.48/kT)\ [\text{cm}^2\ \text{s}^{-1}]$$

where kT is expressed in eV.

These values and their extrapolation represent the upper limit of all published experimental data. This is understandable in the context of the experimental conditions in which the temperature is sufficiently high to render trapping processes in the transport insignificant.

The minimum energy position for H^+ is a bond-center (BC) site located midway between two first neighbor Si atoms. To diffuse, the hydrogen must jump between (BC) sites. However, the energies of other interstitial sites are sufficiently close so as to make it unlikely that there will be a unique path for hydrogen diffusion. In consequence, static total energy calculations could be unreliable. To overcome this limitation, Buda et al. [16] have undertaken ab initio molecular dynamics simulation of high-temperature diffusion of H in Si (1200–1800 K) and obtained diffusivity values that are very similar to the experimental values of Van Wieringen and Warmoltz [9], as do Blochl et al. using first-principles calculations based on rate-theory formalisms [17]. More recently, Panzarini and Colombo have been able to extend the molecular dynamics method to lower temperatures. They calculated the diffusivity over the range 800–1800 K [18], and obtain excellent agreement with Wieringen and Warmoltz's experiments over the range covered by the experiments, but predict lower diffusivities than the extrapolation would indicate because of changes in the jump length. It is important to point out that this is predicted to occur at temperatures above that at which trapping/detrapping would be expected to be important. No experimental data are available at the lower temperatures that these molecular dynamic calculations cover, but there is a trend toward the many experimental results that fall below the extrapolation of the Wieringen and Warmoltz data.

The above trends can be seen in Figure 2.2, which presents a comparison of diffusivity results from the literature. The line is an extrapolation of Wieringen and Warmoltz's work [9]. The Panzarini and Colombo results [18] can be seen as tending toward the main grouping of experimental results at slightly lower temperatures. However, it is interesting to note that two sets of low-temperature measurements [19,20] fit very well to the Wieringen and Warmoltz extrapolation. The data of Gorelkinskii and Nevinnyi [19] were taken over the range 126–143 K and measures the reorientation of a single jump of H^+, so that the retrapping considerations do not enter into the process. Because of this, the mechanisms of the high-temperature measurement are mimicked in this low-temperature electron paramagnetic resonance (EPR) work. A similar argument can be applied to the data of Kamiura et al. [20], who released hydrogen from CH pairs and observed its recapture at nearby P. They undertook this experiment for D as well as H and, as expected, observed a slightly lower diffusivity for D. Other work mostly reports rather lower diffusivities than the extrapolation at lower temperatures. Mogro-Campero et al. [21] calculated the diffusivity from the penetration depth of hydrogen derived from passivation studies under conditions that were likely to include both the effects of trapping and detrapping. Newman [22] used the interaction of H with O to determine the diffusivity by observing the relaxation of stress-induced dichroism of the 9 μm oxygen infrared (IR) absorption band. The measurements enabled the hydrogen diffusion coefficient to be estimated as $1.7 \times 10^2 \exp(-1.2\ \text{eV}/kT)\text{cm}^2\ \text{s}^{-1}$ for temperatures in the range 225°C–350°C.

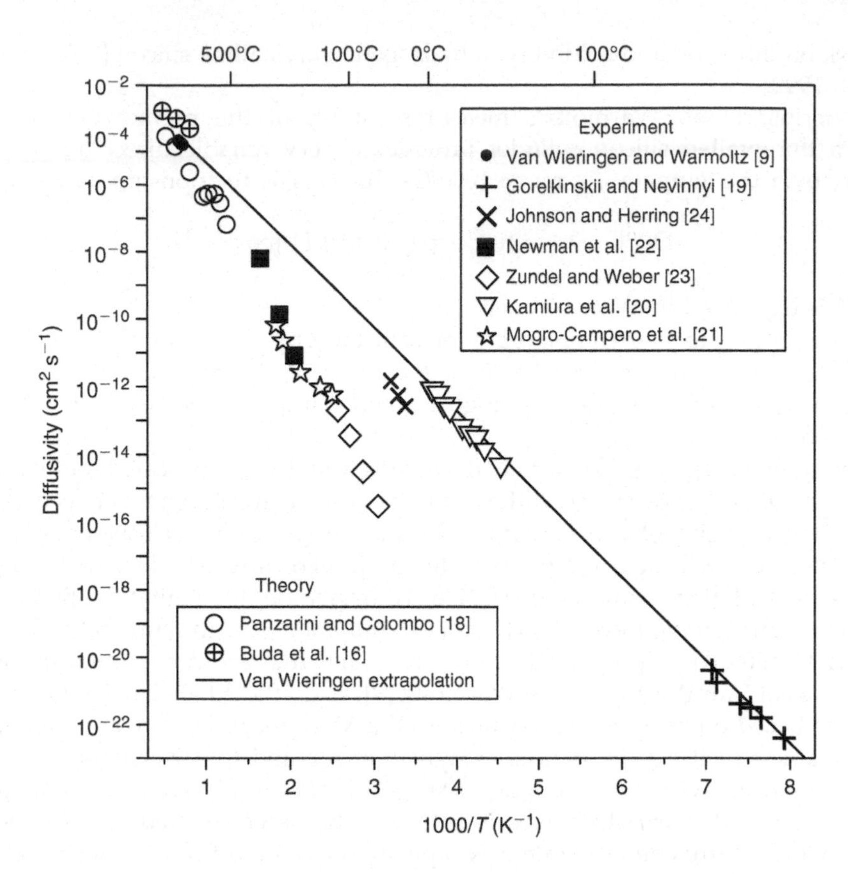

FIGURE 2.2
Compilation of diffusivity as a function of temperature from various publications. The solid line is an extrapolation of the data of Van Wieringen and Warmoltz [9]. Theoretical calculations and experimental data are as referenced in the figure.

At low temperatures where trapping is important, much experimental and modeling work has been done. Zundel and Weber [23] have studied the influence of boron concentration on the diffusivity over the range 60°C–140°C on the hydrogen profile by observing the passivation of the boron. They find that the effective hydrogen diffusion coefficient shows no dependence on the diffusivity of free hydrogen and is entirely trap-limited. Figure 2.3 shows an Arrhenius plot of the product of the effective diffusivity and acceptor concentration as a function of temperature, indicating that the effective diffusivity is reduced as the boron concentration increases. The measurements discussed above probably relate to H^+, or possibly in the case of Wieringen and Warmoltz, H^0 (although a widely held view is that H^0 is metastable due to the negative-*U* character of H in Si). Johnson and Herring [24] have reported measurements on D^- at low temperatures by dissociating P–D complexes; these are shown in Figure 2.2. These diffusivities are considerably lower than the Wieringen and Warmoltz extrapolation (only a reduction of $\sqrt{2}$ for D compared to H would be expected from simple diffusion theory), but are higher than most H^+ results at temperatures where trapping and hoping diffusion would be expected to dominate. In selecting data for Figure 2.2, we have omitted H^+ results that combine the effects of drift with diffusion, i.e., results in which the motion of H^+ has been observed under the action of an electric field. At RT the effects of drift can be very substantial compared to diffusion alone.

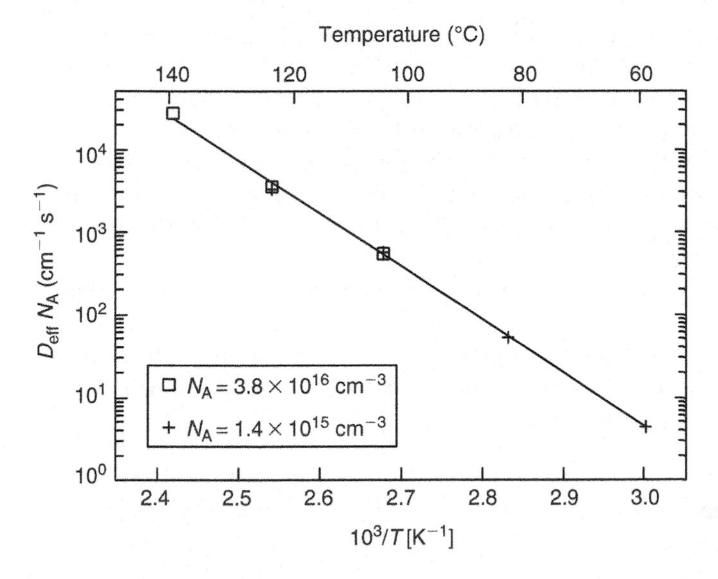

FIGURE 2.3
Low-temperature diffusivity showing dependence on acceptor concentration. (From Zundel, T. and Weber, J., *Phys. Rev. B*, 46, 2071, 1992. With permission.)

2.2.3 Solubility

The equilibrium solubility of hydrogen in silicon has been determined over the range 900°C–1300°C. Four sets of experimental results determined by different methods are shown in Figure 2.3. The early work of Van Wieringen and Warmoltz [9] described earlier measured concentration of molecules and so is scaled by a factor of 2 in Figure 2.4 to enable

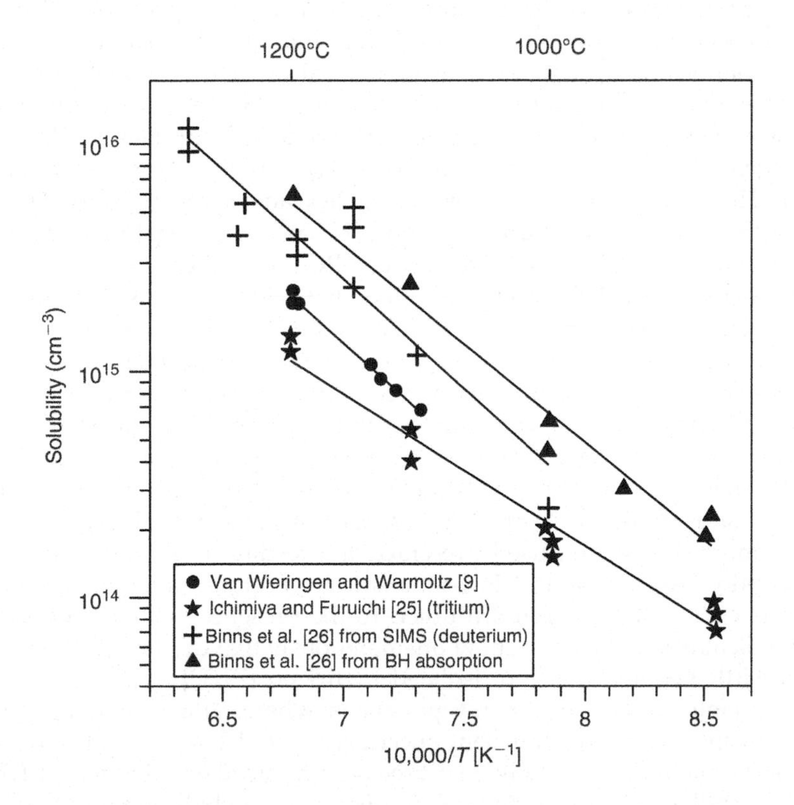

FIGURE 2.4
High-temperature solubility of hydrogen, deuterium, and tritium in silicon.

direct comparison with the other data that relate to atomic concentrations. Ichimiya and Furuichi [25] measured the concentration of tritium atoms using the emitted beta radiation as a measure of concentration. Binns et al. [26] heated high-resistivity silicon in D_2 and quenched to RT, measuring the deuterium concentration by SIMS. Also shown in Figure 2.3 are data derived by the same authors from optical absorption measurements associated with the BH or BD complex. In this case, the deuterium SIMS data were used as a calibration and extended to hydrogen. Experimental conditions were established such that all the hydrogen would be expected to form BH pairs.

The hydrogen and deuterium concentrations shown in Figure 2.1 after 150°C plasma incorporation are many orders greater than the equilibrium values at high temperature reported above. As will be discussed later the high surface concentration is attributed to hydrogen molecules, while the saturated values at the plateaux equal to the boron concentration are due to metastable boron complexes.

2.3 Isolated Hydrogen

The identification and lattice location of isolated hydrogen in silicon has been a much bigger challenge for experimentalists than for theorists. Initially, a point of particular interest has been the diffusion of hydrogen at the atomic level. As discussed above, hydrogen migrates, according to the present theoretical consensus [6], as isolated ionic species through interstitial sites of the silicon crystal. The potential-energy surfaces governing this migration have minima at the interstitial BC site, and at the interstitial tetrahedral site (T), giving rise to a donor level and an acceptor level, respectively. The pathway of migration is expected to depend on the position of these levels relative to the Fermi level. Ab initio calculations [27] predicted that the neutral BC hydrogen, H^0(BC), in an n-type material represents a local minimum in free energy. However, this minimum energy configuration is metastable inasmuch as either positive BC hydrogen, H^+(BC), or negative T-site hydrogen, $H^-(T)$, represents the global energy minimum for any position of the Fermi level. This implies that the donor level lies above the acceptor level, with the consequence that monatomic hydrogen should behave as a negative-U defect in silicon when the H^0 impurity can jump swiftly between BC and T sites.

The high reactivity and RT mobility of interstitial hydrogen makes it very difficult, or even impossible, to observe the H^+(BC), H^0(BC), and $H^-(T)$ configurations of hydrogen directly. There is much experimental work where this fact has been ignored and, as a result, the hydrogen–oxygen and hydrogen–carbon complexes have been erroneously interpreted as the isolated hydrogen. One way to avoid this problem is to introduce hydrogen by proton implantation at cryogenic temperatures where hydrogen and intrinsic defects are immobile. In such an approach, the structure of isolated hydrogen in silicon may be studied by in situ applications of standard spectroscopic techniques. The first report where this approach has been used presented the results of the local vibrational mode (LVM) absorption studies of the low-temperature proton-implanted silicon [28]. The observed 1990 cm^{-1} line has been attributed to the presence of hydrogen in the crystal; however, it took almost 20 years to show unambiguously that this line (1998 cm^{-1} precisely) refers to the H(BC) configuration of isolated hydrogen [29,30]. The same procedure has been used by Holm et al. [31] for n-type silicon where, after low-temperature proton implantation using deep-level transient spectroscopy (DLTS), a $E_c - E_t = 0.16$ eV donor level ($E3'$) has been observed. Finally, in p-type or intrinsic material where H^0(BC) converts into H^+(BC), Gorelkinskii and Nevinnyi [32] used band gap light to produce a steady-state population of H^0(BC) and observed the EPR signal, labeled $AA9$. The metastability of the

*E*3′ center was established by Bech Nielsen et al. [33], and its annealing behavior was found to match that of the *AA*9 signal. On this basis, the DLTS signal *E*3′ was finally assigned to electron emission from H^0(BC).

Further detailed studies performed by Bonde Nielsen et al. [34] showed that at least two kinds of BC hydrogen donors exist. One, H^0(BC), is associated with hydrogen at the BC site; the other, H^0(BC)–O_i, is associated with BC hydrogen weakly bound to interstitial oxygen. A key point in this interpretation is that H^+(BC)–O_i converts to H^+(BC) in oxygen-lean material, whereas in oxygen-rich material the reverse process occurs. Both centers are abundant after low-temperature proton implantation giving rise to two signals, denoted *E*′ and *E*3″, which are observed when DLTS is applied in situ to as-implanted samples. These signals are always present after implantation into the depletion layer of a reverse-biased diode (no free electrons available in the implantation region). They are indistinguishable as far as their electronic DLTS signals are concerned, but can be discerned on the basis of characteristic differences in their formation and annealing properties. The tracking of charge accumulated in the depletion layer shows that both centers can convert temporarily into singly negatively charged centers. These, so far hidden, centers are thermally stable up to ~250 K, where they both anneal.

As a consequence of the perturbation, the H(BC)–O_i center should have a lower symmetry than the H(BC) center. This has been confirmed by the application of the uniaxial-stress technique to the DLTS method [35]. Figure 2.5 depicts the splitting pattern of the *E*3′ emission signal which reveals the expected trigonal symmetry of H(BC), whereas the splitting under <100> stress of the *E*3″ signal into two components indicates a lower (monoclinic-I) symmetry as expected for H(BC)–O_i.

The position of the *T*-site acceptor level has been much more difficult to identify. The acceptor level has been defined as the energy difference between $H^0(BC) + e_c$ and $H^-(T)$ and placed slightly below midgap on the basis of a kinetic study [36,37]. As the first step in this work an initial amount of H^+ was released in a diode space-charge layer from hydrogen-passivated shallow donors by reverse-bias illumination and converted to H^- by the application of a filling pulse. Then the subsequent thermally induced removal of H^- was recorded as an exponential capacitance transient. This is ascribed to the first step of the process $H^-(T) \rightarrow H^0(BC) + e_c \rightarrow H^+(BC) + 2e_c$ and the activation enthalpy of this step obtained [36,37]. It was found, consistent with the anticipated negative-*U* property,

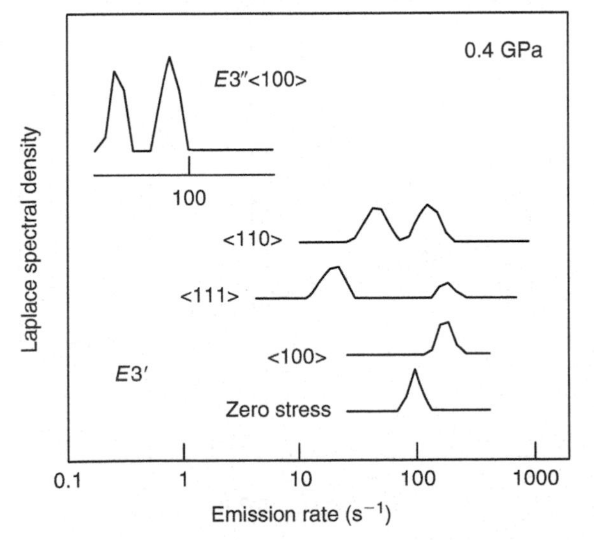

FIGURE 2.5
Laplace DLTS spectra revealing the trigonal symmetry of the center associated with the *E*3′ signal, and a lower symmetry of the center associated with the *E*3″ signal (inset). (Adapted from Bonde Nielsen, K. et al., *Physica B*, 308, 134, 2001. With permission.)

that this enthalpy was larger than that described above [31] for the second step of the process. The authors obtained the position of the acceptor level from detailed-balance considerations. In view of the evidence that at least two kinds of BC hydrogen donors exist [H(BC) and H^0(BC)–O_i] the obvious question arose whether a similar situation applied to the *T*-site acceptor. If so, two different configurations of H^-(*T*) with and without an oxygen atom nearby, should exist.

The main idea behind the H^-(*T*) (or H^-(*T*)–O_i) configuration detection was that this configuration annealing is triggered by a thermal electron emission process. The emission was then observed as a capacitance transient at temperatures slightly below 250 K, the temperature at which it is known that the H^-(*T*) centers anneal. As depicted in Figure 2.6, the single-shot capacitance-transient measurements for long decay times reveal the emission [38]. The transient shown has been recorded with an oxygen-rich hydrogen-implanted diode with about 55% of the implants initially present as H(BC) and the rest as H(BC)–O_i. Both centers were first converted into the negative hidden centers by bias removal at 110 K. The capacitance transient was then recorded after reapplication of the bias at 242 K. As can be seen, the transient consists of two exponential components. The fast decay is ascribed to electron emission from H^-(*T*) and labeled as AT'. The slow decay is analogously ascribed to emission from H^-(*T*)–O_i and labeled AT''. These assignments were based on detailed studies of the stabilities of both signals and on the analogy to the stabilities of the H(BC) and H^0(BC)–O_i configurations. From the transients AT' and AT'' measured at different temperatures, the Arrhenius plots are depicted in the inset to Figure 2.6. The remarkable agreement over many decades between the AT'' data and those obtained in Ref. [36] can be noted. The linear-regression analysis yielded the activation $\Delta H'' = 0.79 \pm 0.03$ eV for the emission from H^-(*T*)–O_i, and $\Delta H' = 0.65 \pm 0.10$ eV for the emission from H^-(*T*).

The electron emission enthalpies found for H^-(*T*)–O_i and H^-(*T*) exceed those of the corresponding BC donor states in accordance with the expected negative-*U* property. However, to demonstrate this property more directly, the correlation between the amount of charge bound as H^0(BC)–O_i and H^-(*T*)–O_i has been examined. For this purpose, a reverse-biased sample was prepared containing, predominantly, H^+(BC)–O_i. The H^+(BC)–O_i was then partially converted to H^-(*T*)–O_i by the application of a filling pulse with the number of the converted centers controlled by the pulse duration. The remaining

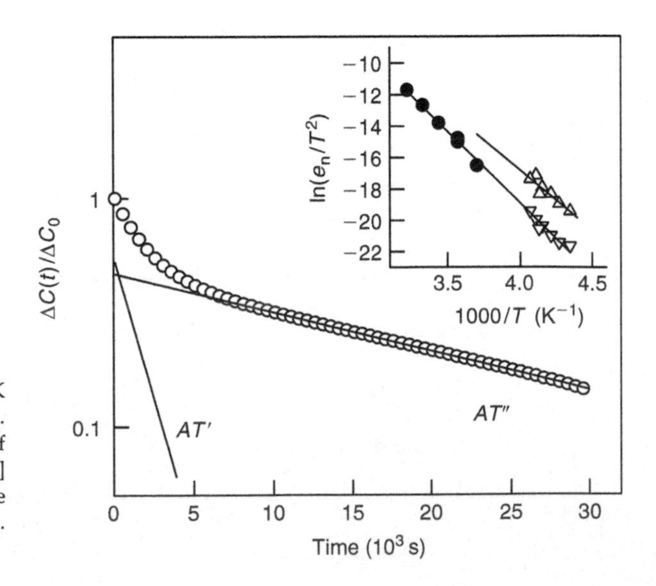

FIGURE 2.6
Single-shot capacitance transient at 242 K revealing two components AT' and AT'' [38]. The inset shows the Arrhenius analysis of $AT''(\nabla)$ and $AT'(\Delta)$. The data (●) of Ref. [36] have been included. (Adapted from Bonde Nielsen, K. et al., *Physica B*, 308, 134, 2001. With permission.)

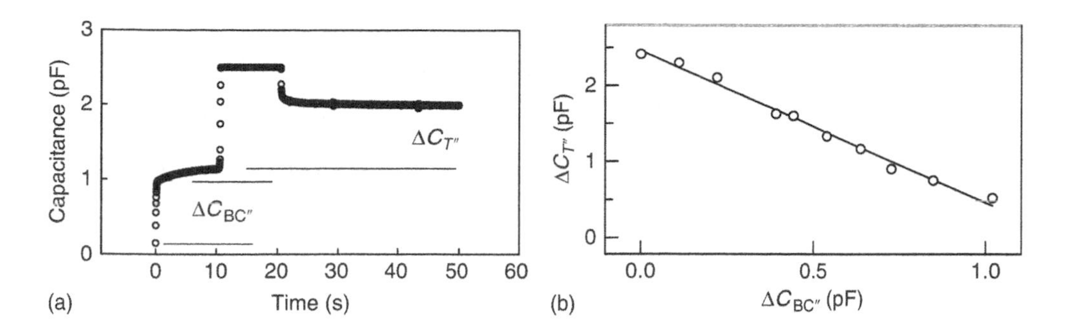

FIGURE 2.7
Illustration of the correlation between $E3''$ signal and the reservoir for the AT'' signal. (a) The pulsing sequence to obtain $\Delta C_{T''}$ and $\Delta C_{BC''}$ as explained in the text. (b) The $\Delta C_{T''}$ versus $\Delta C_{BC''}$ dependence showing that the $H^-(T)$–O_i configuration binds twice more electrons than H^0(BC)–O_i. (From Bonde Nielsen, K. et al., *Phys. Rev. B*, 65, 075205-1, 2002. With permission.)

H^+(BC)–O_i gives rise to a capacitance transient (the $E3''$ signal) following the filling pulse. The amplitude of this transient (see Figure 2.7a) is a measure of the amount of charge emitted in the process H^0(BC)–O_i → H^+(BC)–O_i and is denoted as $\Delta C_{BC''}$. An illumination pulse applied immediately after the $E3''$ transient has leveled off causes a shift in steady-state capacitance, which is denoted as $\Delta C_{T''}$. This capacitance shift measures the charge change as the $H^-(T)$–O_i configuration converts back to the H^+(BC)–O_i configuration. Then, choosing different filling pulse lengths to vary the relative amounts of H^0(BC)–O_i and $H^-(T)$–O_i, the linear correlation between $\Delta C_{T''}$ and $\Delta C_{BC''}$ depicted in Figure 2.7b is obtained. The slope of this correlation line reveals that two electrons are emitted when $H^-(T)$–O_i is converted optically to H^+(BC)–O_i. This result establishes the one-to-one correspondence $E3'' \leftrightarrow AT''$, and hence supports the assignment of AT'' to H(T)–O_i and evidences the negative-U behavior of the H(T)–O_i defect.

The electrical measurements performed for the $H^-(T)$ and $H^-(T)$–O_i configurations of hydrogen seem to be the only spectroscopic data where the properties of these species can be observed. The negatively charged hydrogen is diamagnetic and does not produce a signal suitable for the magnetic resonance technique. Despite the fact that the H^+(BC) configuration can be observed in the far-IR LVM absorption measurements, the $H^-(T)$ configuration is not observed by this method. The detailed studies on this problem performed recently by Pereira et al. [39] showed that negatively charged hydrogen in the antibonding position attached to one silicon atom can swiftly reconfigure between four equivalent interstitial positions as the barrier separating these positions in silicon is low. Therefore, the antibonding configuration of negatively charged hydrogen in silicon may be short-lived due to rapid jumps of hydrogen among these equivalent antibonding sites, which would broaden the LVM lines and hamper their detection.

2.4 Passivation

The term passivation is used generically to describe the removal of electrical activity. Hydrogen is a remarkably effective passivator in its atomic form due to its high reactivity. It migrates through the lattice and reacts with unsatisfied bonds of defects and impurities rendering them electrically inactive. This is the basis of the widely used technique of forming gas anneal to reduce the interface trap density in MOS devices. Passivation is an

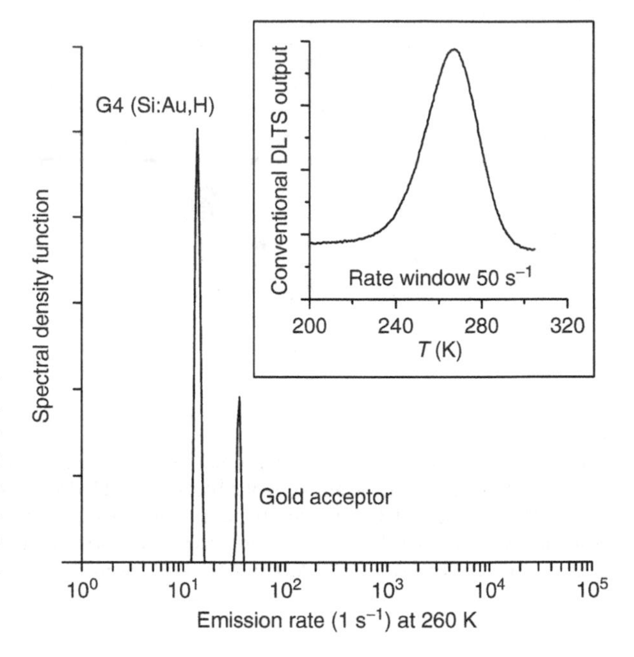

FIGURE 2.10
DLTS and Laplace DLTS spectra of hydrogenated silicon containing gold. The conventional DLTS spectrum is shown as an inset at the top right of the figure. The broad peak centered at 260 K is attributed to electron emission from the gold acceptor and G4. The main spectrum uses the Laplace technique and clearly separates the gold-acceptor level and the gold-hydrogen level G4. (Adapted from Deixler, P. et al., *Appl. Phys. Lett.*, 73, 3126, 1998. With permission.)

has been concluded that hydrogen in the complex is directly bonded to silicon, presumably in the antibonding position.

There is a great deal of data on the electrical studies of hydrogen passivation of gold in silicon. Following the work of Pearton et al. [51], Sveinbjörnsson and Engström [73] studied the complexes formed during hydrogenation using DLTS and reported that there are four electrically active deep levels (referred to as G1, G2, G3, and G4) resulting from the formation of AuH complexes. It is believed that G1, G2, and G4 are different charge states of the same AuH. Recent ab initio calculations embracing this defect system support this interpretation [87]. It is generally accepted that gold (without hydrogen) forms an acceptor which acts as a majority-carrier trap in n-type silicon. The G4 level appears to be very close in energy to the gold acceptor and has almost identical electron emission characteristics not resolved by conventional DLTS [73]. Figure 2.10 shows a comparison of the Laplace and conventional DLTS spectra obtained from the same Si:Au:H sample [74]. The Laplace DLTS spectrum shown in the main part of the figure reveals that there are two separate and quite distinct bound- to free-electron emission rates at the measurement temperature of 260 K. This is consistent with the previous interpretation of deconvolved DLTS spectra, and confirms unambiguously that the conventional DLTS peak at the Au-acceptor position in hydrogenated silicon consists of two contributions.

2.5.6 Pt–H_2

Platinum-related deep centers in silicon are widely used to control the minority-carrier lifetime in fast-switching semiconductor devices. However, unintentionally introduced hydrogen interacts with platinum, leading to the formation of new electrically active complexes. The existence of PtH_n (where $n = 1$–3) complexes have been detected by EPR [90–93], LVM [84,94], and DLTS [84–86,95]. The platinum–dihydrogen complex (PtH_2) is of particular interest from the point of view of application of platinum for minority-carrier lifetime reduction in power devices (increase the switching speed) because, among all platinum–hydrogen complexes, it has the shallowest electronic level. This makes the platinum–dihydrogen center an effective carrier recombination center. But as the electronic level is well away from the midgap position, the carrier generation via this defect is less effective, so the device leakage is reduced. In spite of a number of different studies by both theoretical [71,88] and experimental groups, there are some uncertainties relating to the PtH_2 structure. It is well known that it possesses C_{2v} symmetry, which is the result of a static Jahn–Teller distortion combining tetragonal and trigonal components analogous with the isolated negatively charged vacancy. This symmetry has been demonstrated with use of EPR [90–93], LVM [94], and DLTS [85,86] methods. However, the position of hydrogen atoms in platinum–hydrogen complexes is disputed. Some models assume that hydrogen atoms are attached directly to the platinum; however, it is more generally believed that two hydrogen atoms bond to two Si atoms neighboring the substitutional Pt and pointing away from the platinum.

2.6 Hydrogen Molecules

In addition to interacting with impurities and defects in semiconductors, H can also form H_2 molecules in the silicon lattice. Early theoretical calculations [96,97] predicted that the H_2 molecule in Si would sit at a tetrahedral interstitial (*T*) site where it should be electrically and optically inactive. It has also been suggested that H_2 is one of the most stable forms of hydrogen in silicon and plays an important role during the in-diffusion of hydrogen and in a variety of H-related reactions [3,4]. However, in spite of its proposed importance, the H_2 molecule was not observed in semiconductors until recently.

H_2 in crystalline silicon was observed first by Raman spectroscopy in hydrogen-plasma-treated Si samples [98]. A Raman line at 4158 cm^{-1}, which was assigned in Ref. [98] to H_2, was proved later to be caused by molecules associated with H-saturated platelets [99]. Subsequently a vibrational line due to the isolated interstitial H_2 at the *T*-site in the Si lattice was discovered in Raman [100] and IR absorption spectra [101]. The line associated with the isolated H_2 was found to be at 3601 cm^{-1} (Raman, RT) or 3618 cm^{-1} (IR, 10 K). Vibrational lines due to D_2 and HD were also observed. Furthermore, theoretical calculations found vibrational frequencies for isolated H_2 in Si that are in agreement with experiment [102–105]. It appears that the interaction between H_2 and the host lattice weakens the H–H bond, which results in a drop in its stretch frequency by about 550 cm^{-1} relative to that of the free molecule ($\sim$4160 cm^{-1}).

Properties of the isolated H_2 molecules in silicon have been studied in detail by two groups with the use of IR absorption [106–109] and Raman spectroscopies [110–112]. It is found that all of the experimental observations can be explained in terms of a nearly free rotator, which interacts with host atoms surrounding a tetrahedral cage and exchanges energy with the host crystal. In oxygen-rich Si crystals H_2 molecules can be trapped at *T*-sites near interstitial oxygen (O_i) atoms [113,114]. Careful annealing studies have shown that the binding energy of H_2 to O_i is 0.26 ± 0.02 eV, and the activation energy for diffusion of H_2 among interstitial sites is 0.78 ± 0.02 eV [114]. The diffusion of the molecule is expected to occur along the trigonal axis from the *T* to hexagonal and then to another *T*-sites. Multivacancy voids in Si crystals have also been found to be effective trapping sites for H_2 molecules [115,116].

The H_2^* defect, another hydrogen dimer with one hydrogen atom close to a BC site and the other one along the same trigonal axis at an antibonding site, was also predicted to be stable in semiconductors [117]. The H_2^* defect in Si was identified later in an IR absorption spectroscopy study combined with ab initio modeling [118]. Theory predicts that the relative stability of interstitial H_2 as compared to the H_2^* increases with the host lattice constant for the group IV semiconductors [119]. In Si, interstitial H_2 is expected to be the ground state of hydrogen. However, molecular dynamics simulations have shown that the interaction of the interstitial H_2 molecule with a vacancy and a self-interstitial can result in its dissociation and transformation to the H_2^* defect [120].

2.7 Hydrogen Clusters and Platelets

Under certain circumstances hydrogen can be incorporated into silicon in such high concentrations that it exists in a precipitated form. The formation of such defects is well-documented after exposure to a hydrogen plasma or after ion implantation. The latter process is of considerable commercial significance with the advent of the SmartCut process [2].

Weldon et al. have investigated the precipitation of hydrogen and the formation of microvoids and platelets after the implantation of hydrogen [121]. They have studied the evolution of the internal defect structure as a function of implanted hydrogen concentration and anneal temperature and found that the chemical action of hydrogen leaves the formation of (100) and (111) internal surfaces above 400°C via an agglomeration of the initial defect structure. In addition, between 200°C and 400°C molecular hydrogen is released. This is subsequently trapped in the microvoids bounded by the internal surfaces resulting in the buildup of internal pressure. This is the mechanism that leads to the blistering of silicon, and is the basis of the process used for layer transfer in the SmartCut process.

First-principles calculations of such platelets are difficult as they involve studying large defect structures using relatively small clusters. However, modeling work by Reboredo et al. [122], based on calculations of configuration energies using density functional theory, suggest that mechanisms for the nucleation and growth of aggregates of second nearest neighbor hydrogenated vacancies are energetically favorable. One of the complexes proposed is a second nearest neighbor pair of two VH_4 molecules. The bimolecular complex is predicted to accumulate progressively, forming the platelets. Such complexes are expected to have a preference for formation along {111} and {100} planes as is observed in experiments.

The formation and evolution of hydrogen platelets in silicon after exposure to remote plasma has been investigated in detail by Lavrov and Weber [123]. The structure of such platelets is rather complex, but electron microscopy studies suggest that they consist of {111} platelets with a hydrogen-saturated internal surface in which each silicon–silicon bond in a (111) plane is replaced by two Si–H bonds [124]. However, it does seem that other forms of platelets can exist [125]. In this case H_2^* aggregates are involved in the formation of the {111} platelets. The existence of two different structures has been confirmed by Lavrov and Weber using polarized Raman scattering spectroscopy [123]; the relative concentrations of the two species of platelet depend on the plasma conditions. They observe that the H_2^* structure predominates under low-temperature formation conditions and transforms into the $[2Si–H]_n + H_2$ structure as the temperature rises above 100°C.

2.8 Hydrogen in Germanium

Whereas the properties of hydrogen in silicon have been studied extensively in the past and are now reasonably well understood, understanding of hydrogen in germanium requires further efforts of researchers. Taking into account the similar properties of crystalline Si and Ge and very similar chemistries of {Si, H} and {Ge, H} compounds, one would not expect qualitative differences in the behavior of H in these materials. Indeed, some hydrogen-related phenomena like the passivation of deep centers and reactivation of neutral centers have been found to be very similar in Ge and Si. However, some experiments involving H, which were done on comparable Si and Ge samples, showed different results. For example, in contrast to Si, in Ge only one successful attempt of passivation of shallow acceptors has been reported [126], and passivation of shallow donors has never been realized. A brief review on the properties of hydrogen in Ge is presented in the following sections.

2.8.1 Hydrogen Incorporation, Diffusivity, and Solubility in Ge

The solubility and diffusivity of hydrogen near the melting point has been studied in the 1960s by permeation experiments [9,127]. According to Frank and Thomas [127], the diffusion coefficient D and solubility S of atomic hydrogen in the temperature range 800°C–910°C can be described by the following equations:

$$D_H = 2.7 \times 10^{-3} \exp(-0.38/kT)\,[\text{cm s}^{-1}] \quad (2.1)$$

$$S_H = 1.6 \times 10^{24} \exp(-2.3/kT)\,[\text{cm}^{-3}] \quad (2.2)$$

Hydrogen diffusivity at elevated temperatures is high and typical for interstitial diffusion. It indicates that trapping does not affect the movement of hydrogen in this temperature range. Near the melting point the S_H value is about 4×10^{14} cm^{-3}, but the extrapolation of Equation 2.2 yields a negligible equilibrium concentration of H near RT; e.g., a value of 10^5 cm^{-3} is obtained at 350°C. However, in practice, much higher values have been observed in high-purity Ge crystals [128]. This indicates that at lower temperatures there is a trapping enhanced increase of the hydrogen concentration in germanium. Further, experimental values of hydrogen diffusivity at lower temperatures have been found to be much lower than the values extracted from Equation 2.1, thus indicating that at lower temperatures, as in the case of silicon, hydrogen diffusion is trap-limited [4,129].

Hydrogen incorporation during growth and following cooling to RT seems to be very efficient in the case of Ge [128]. In contrast, hydrogen (deuterium) incorporation into germanium from H(D) plasma is not very effective, compared to this process for the case of Si. Thermal effusion (TE) measurements of D-plasma-treated Ge samples showed that out-diffusion of hydrogen occurs at 200°C, a significantly lower temperature than in Si (350°C–400°C) [130]. Further, the amount of D that penetrates into a Ge sample exposed to a D plasma decreases with the ambient temperature, while the opposite holds for Si. At 250°C all hydrogen was found to be located on or near the sample surface, even for long exposure times. It appears that the Ge surface constitutes an effective diffusion barrier or binding center for atomic hydrogen, thus preventing significant concentrations of H inside the bulk material. Ge–D and GeD_2 species (mono- and dideuterides) formed at the Ge surface were suggested as sources of the high surface concentrations [130]. GeD and GeD_2 species desorb at 300°C and 200°C, respectively.

2.8.2 Isolated Hydrogen Species and Complexes with Intrinsic Defects in Ge

Theory predicts negative-U properties for isolated hydrogen atoms in all semiconductor materials [131]. In Ge, the E(−/+) level of the isolated H atom is calculated [131] to be close to or within the valence band, leading to nonamphoteric behavior. For all Fermi-level positions, H^- should be the stable equilibrium configuration. Although one might expect

of different H_{BC}^{+} sites will correlate with the local strain conditions of the BC site that accommodates the hydrogen atom. In this context, the H_{BC}^{+} in dilute SiGe alloys is especially interesting, since the local strain imposed by the minority species may be probed by measuring the strain-induced LVM frequency shifts on H_{BC}^{+} sites neighbored by the minority species, with respect to the frequencies observed for isolated H_{BC}^{+} centers. Knowing the local strain properties is essential to clarify the origins of the unique properties of SiGe alloys and to develop the device potential of these materials.

Configurations and vibrational properties of elemental hydrogen centers in Si, Ge, and dilute SiGe alloys have been modeled recently by an ab initio supercell method [156]. Possible sites for an isolated hydrogen atom in dilute SiGe alloys are shown in Figure 2.9. According to results of the modeling [156], all of the bond-centered configurations for an isolated hydrogen atom are nearly degenerate in both the Si-rich and Ge-rich dilute SiGe alloys; i.e., the binding energies of the hydrogen atom to the minority species are supposed to be negligible. However, the calculated LVM frequencies have been found to be different for a hydrogen atom at the different BC sites [156]. It has been found that H^{-} ions may be trapped at low temperatures in the form of H_{AB}^{-}–Si–Ge units in dilute SiGe alloys. In a Si-rich host, the presence of soft Si–Ge bonds, when compared to the abundant and stiff Si–Si bonds, encourages the formation of Si–H bonds at the expense of the Si–Ge ones. On the other hand, in Ge-rich alloys, H_{AB}^{-}–Si–Ge defects are about 0.27 eV more stable than the isolated H_{AB}^{-}–Ge defect.

Different configurations of isolated hydrogen atoms have been observed in dilute Si-rich and Ge-rich SiGe alloys by means of LVM spectroscopy after low-temperature (20 K) proton (deuteron) implantation [39,157,158]. LVMs of $X_sH_{BCn}^{+}$ ($n = 1, 2, 3$) defects have been detected. For details of the LVM frequencies for different $X_sH_{BCn}^{+}$ configurations, readers are referred to Ref. [158]. In agreement with the results of ab initio modeling [156], the analysis of intensities of LVMs bands associated with different H_{BC}^{+}-related configurations has shown that there are no preferential sites for the H_{BC}^{+} species in the vicinity of Ge atoms in Si-rich SiGe alloys [158]. Furthermore, isochronal annealing studies have shown that thermal stabilities of different $X_sH_{BCn}^{+}$ defects are similar.

In Ge-rich SiGe alloys implanted with protons at low temperature, in addition to LVM lines due to bond-centered hydrogen atoms, three LVM lines at 815.8, 1430.2, and 1630.7 cm^{-1} have been observed and assigned to the negatively charged hydrogen atom located at an antibonding site and attached to a substitutional silicon atom [39]. It has been found that a conversion process takes place at temperatures in the range 20–125 K between negatively charged hydrogen at a germanium antibonding site to a silicon antibonding site. This provides direct evidence that silicon acts as an effective trap for negatively charged hydrogen in SiGe alloys. Further, it has been reported in Ref. [159] that Si atoms in Ge-rich SiGe are effective nucleation sites for the formation of a dominant H_2*-like defect at and above RT. This complex consists of a Ge–$H_{BC}\cdots$Si–H_{AB} structure along a <111> axis, where the H atoms have effectively broken a Si–Ge bond in the alloy. Several absorption lines have been assigned to LVMs of the complex: a stretch mode at 1850.7 cm^{-1}, along with a bend mode and its first overtone at 813.3 and 1592.6 cm^{-1}. The formation of Ge–$H_{BC}\cdots$Si–H_{AB} complex strongly reduces the trapping of H by vacancies and self-interstitials, which are major H-trapping centers in pure Ge [159].

In an in situ Laplace DLTS study of electrically active defects induced in Si-rich dilute SiGe samples by proton implantation at 60 K, in addition to an $E3'$ signal due to H^0(BC) $\rightarrow$ H^+(BC) transition, which is similar to that observed in pure silicon, another signal {$E3'$(Ge)} has been observed and assigned to a defect with a hydrogen atom in the Ge_sH_{BC2} configuration [160]. The activation energy of the electron emission from the $E3'$(Ge) donor has been found to be ~158 meV when extrapolated to zero electric field. This is slightly lower than the value of ~175 meV for the $E3'$ defect. It was deduced from the constructed configuration diagram of the Ge-strained site that alloying with ~1% Ge does not significantly influence the low-temperature migration of hydrogen as compared to elemental Si [160].

In summary, the recent results of the properties of hydrogen-related defects in dilute SiGe alloys [39,156–160] have shown that in Si-rich SiGe alloys the minority species (Ge atoms) are not effective sinks for hydrogen atoms and do not influence the diffusivity of hydrogen significantly. In Ge-rich SiGe alloys the binding energy of the H_{AB}^-–Si–Ge center has been found to be significantly higher than that for the isolated H_{AB}^-–Ge defect. Further, the formation of Ge–$H_{BC}\cdots$Si–H_{AB} complex is very effective in Ge-rich SiGe. This causes the suppression of the trapping of H atoms by other lattice defects, particularly by vacancies and self-interstitials.

2.10 Summary

This chapter has summarized the behavior of hydrogen in silicon, germanium, and silicon–germanium alloys in terms of its incorporation, diffusion, and defect reactions. It is evident that the behavior is very complex as a consequence of the high diffusivity and reactivity of the atomic species. Hydrogen can passivate electrically active defects or react with impurities to form new complexes that have electronic states within the band gap. Much work has been done in this field with the result that the literature contains detailed studies of most of the commonly observed defects, especially those of technological importance in silicon material and devices. In many cases a comprehensive understanding has been established of the electronic properties of the defect and its stability. In an increasing number of cases a combination of experimental and theoretical studies has resulted in an appreciation of the detailed structure of the defect and its energetics. Far less work has been done on germanium and silicon–germanium, but considerable effort is now being devoted to hydrogen in these materials. It is anticipated that the next few years will see a maturity of vision in relation to hydrogen-related defects in germanium and silicon–germanium comparable with our present understanding of hydrogen in silicon.

References

1. Fowler, A.B., Process for the elimination of interface states in MIOS structures, US Patent 3,849,204, 1974.
2. Aspar, B. et al., Basic mechanisms involved in the SmartCut® process, *Microelectron. Eng.*, 36, 233, 1997.
3. Pankove, J.I. and Johnson, N.M. (Eds.), *Hydrogen in Semiconductors* (Semiconductors and Semimetals, Vol. 34), Academic Press, Boston, 1991.
4. Pearton, S.J., Corbett, J.W., and Stavola, M., *Hydrogen in Crystalline Semiconductors* (Springer Series in Materials Science, Vol. 16), Springer-Verlag, Heidelberg, 1992.
5. Newman, R.C., Hydrogen atoms and complexes in monocrystalline semiconductors, *Philos. Trans. R. Soc. A*, 350, 215, 1995.
6. Estreicher, S.K., Hydrogen-related defects in crystalline semiconductors: A theorist's perspective, *Mater. Sci. Eng.*, R14, 314, 1995.
7. Willardson, R.K. and Beer, A.C. (Eds.), *Hydrogen in Semiconductors* (Semiconductors and Semimetals, Vol. 61), Academic Press, San Diego, 1999.
8. Van de Walle, C.G. and Neugebauer, J., Hydrogen in semiconductors, *Annu. Rev. Mater. Res.*, 36, 179, 2006.

63. Feklisova, O.V. et al., On the nature of hydrogen-related centers in p-type irradiated silicon, *Physica B*, 308–310, 210, 2001.
64. Johannesen, P., Bech Nielsen, B., and Byberg, J.R., Identification of the oxygen-vacancy defect containing a single hydrogen atom in crystalline silicon, *Phys. Rev. B*, 61, 4659, 2000.
65. Coutinho, J. et al., Effect of stress on the energy levels of the vacancy-oxygen-hydrogen complex in Si, *Phys. Rev. B*, 68, 184106, 2003.
66. Dobaczewski, L. et al., Piezospectroscopic studies of the re-orientation process of defect complexes, *Physica B*, 340–342, 499, 2003.
67. Andersen, O. et al., Electrical activity of carbon-hydrogen centers in Si, *Phys. Rev. B*, 66, 235205, 2002.
68. Endrös, A., Charge-state-dependent hydrogen-carbon-related deep donor in crystalline silicon, *Phys. Rev. Lett.*, 63, 70, 1989.
69. Kamiura, Y. et al., Electronic state, atomic configuration and local motion of hydrogen around carbon in silicon, *Defect Diffus. Forum*, 183, 25, 2000.
70. Hoffmann, L. et al., Weakly bound carbon-hydrogen complex in silicon, *Phys. Rev. B*, 61, 16659, 2000.
71. Jones, R. et al., The interaction of hydrogen with deep level defects in silicon, *Solid State Phen.*, 71, 173, 2000.
72. Yarykin, N. et al., Silver-hydrogen interactions in crystalline silicon, *Phys. Rev. B*, 59, 5551, 1999.
73. Sveinbjörnsson, E.Ö. and Engström O., Reaction kinetics of hydrogen-gold complexes in silicon, *Phys. Rev. B*, 52, 4884, 1995.
74. Deixler, P. et al., Laplace-transform deep-level transient spectroscopy studies of the G4 gold-hydrogen complex in silicon, *Appl. Phys. Lett.*, 73, 3126, 1998.
75. Rubaldo, L. et al., Gold-hydrogen complexes in silicon, *Mater. Sci. Eng. B*, 58, 126, 1999.
76. Knack, S. et al., Evolution of copper-hydrogen-related defects in silicon, *Physica B*, 308–310, 404, 2001.
77. Sadoh, T. et al., Deep levels of vanadium- and chromium-hydrogen complexes in silicon, *Mater. Sci. Forum*, 143–147, 939, 1994.
78. Jost, W., Weber, J., and Lemke, H., Hydrogen-induced defects in cobalt-doped n-type silicon, *Semicond. Sci. Technol.*, 11, 22, 1996.
79. Feklisova, O.V. et al., Dissociation of iron-related centers in Si stimulated by hydrogen, *Mater. Sci. Eng. B*, 71, 268, 2000.
80. Shiraishi, M. et al., DLTS analysis of nickel–hydrogen complex defects in silicon, *Mater. Sci. Eng. B*, 58, 130, 1999.
81. Sachse, J.-U., Weber, J., and Lemke, H., Deep-level transient spectroscopy of Pd-H complexes in silicon, *Phys. Rev. B*, 61, 1924, 2000.
82. Jost, W. and Weber, J., Titanium-hydrogen defects in silicon, *Phys. Rev. B*, 54, R11038, 1996.
83. Sadoh, T., Nakashima, H., and Tsurushima, T., Deep levels of vanadium and vanadium-hydrogen complex in silicon, *J. Appl. Phys.*, 72, 520, 1992.
84. Weinstein, M.G. et al., Pt-H complexes in Si: complementary studies by vibrational and capacitance spectroscopies, *Phys. Rev. B*, 65, 035206, 2001.
85. Kamiura, Y. et al., Effect of uniaxial stress on the electronic state of a platinum-dihydrogen complex in silicon and charge-state-dependent motion of hydrogen during stress-induced reorientation, *Phys. Rev. B*, 69, 045206, 2004.
86. Kolkovsky, V. et al., Electrical activity of the PtH_2 complex in silicon: High-resolution Laplace deep-level transient spectroscopy and uniaxial-stress technique, *Phys. Rev. B*, 73, 195209, 2006.
87. Resende, A. et al., Theory of gold-hydrogen complexes in silicon, *Mater. Sci. Forum*, 258–263, 295, 1997.
88. Hourahine, B. et al., Platinum and gold dihydrides in silicon, *Physica B*, 340–342, 668, 2003.
89. Evans M.L. et al., Vibrational spectroscopy of defect complexes containing Au and H in Si, *Mater. Sci. Eng. B*, 58, 118, 1999.
90. Huy, P.T. and Ammerlaan, C.A.J., Complexes of gold and platinum with hydrogen in silicon, *Physica B*, 302–303, 233, 2001.
91. Huy, P.T. and Ammerlaan, C.A.J., Electronic and atomic structure of transition-metal-hydrogen complexes in silicon, *Physica B*, 308–310, 408, 2001.

92. Höhne, M. et al., EPR spectroscopy of platinum-hydrogen complexes in silicon, *Phys. Rev. B*, 49, 13423, 1994.
93. Williams, P.M. et al., Structure-sensitive spectroscopy of transition-metal-hydrogen complexes in silicon, *Phys. Rev. Lett.*, 70, 3816, 1993.
94. Uftring, S.J. et al., Microscopic structure and multiple charge states of a PtH_2 complex in Si, *Phys. Rev. B*, 51, 9612, 1995.
95. Sachse, J.-U. et al., Electrical properties of platinum-hydrogen complexes in silicon, *Phys. Rev. B*, 55, 16176, 1997.
96. Mainwood, A. and Stoneham, A.M., Interstitial muons and hydrogen in crystalline silicon, *Physica*, 116B, 101, 1983.
97. Corbett, J.W. et al., Atomic and molecular hydrogen in the Si lattice, *Phys. Lett.*, 93A, 303, 1983.
98. Murakami, K. et al., Hydrogen molecules in crystalline silicon treated with atomic hydrogen, *Phys. Rev. Lett.*, 77, 3161, 1996.
99. Leitch, A.W.R., Alex, V., and Weber, J., H_2 molecules in c-Si after hydrogen plasma treatment, *Solid State Commun.*, 105, 215, 1998.
100. Leitch, A.W.R., Alex, V., and Weber, J., Raman spectroscopy of hydrogen molecules in crystalline silicon, *Phys. Rev. Lett.*, 81, 421, 1998.
101. Pritchard, R.E. et al., Isolated interstitial hydrogen molecules in hydrogenated crystalline silicon, *Phys. Rev. B*, 57, R15048, 1998.
102. Okamoto, Y., Sato, M., and Oshiyama, A., Comparative study of vibrational frequencies of H_2 molecules in Si and GaAs, *Phys. Rev. B*, 56, R10016, 1997.
103. Van de Walle, C.G., Energetics and vibrational frequencies of interstitial H_2 molecules in semiconductors, *Phys. Rev. Lett.*, 80, 2177, 1998.
104. Hourahine, B. et al., Hydrogen molecules in silicon located at interstitial sites and trapped in voids, *Phys. Rev. B*, 57, R12666, 1998.
105. Estreicher, S.K. et al., Dynamics of interstitial hydrogen molecules in crystalline silicon, *J. Phys.: Condens. Matter*, 13, 6271, 2001.
106. Chen, E.E. et al., Key to understanding interstitial H_2 in Si, *Phys. Rev. Lett.*, 88, 105507, 2002.
107. Chen, E.E. et al., Rotation of molecular hydrogen in Si: unambiguous identification of ortho-H_2 and para-D_2, *Phys. Rev. Lett.*, 88, 245503, 2002.
108. Fowler, W.B., Walters, P., and Stavola, M., Dynamics of interstitial H_2 in crystalline silicon, *Phys. Rev. B*, 66, 075216, 2002.
109. Shi, G.A. et al., Rotational-vibrational transitions of interstitial HD in Si, *Phys. Rev. B*, 72, 085207, 2005.
110. Lavrov, E.V. and Weber, J., Ortho and para interstitial H_2 in silicon, *Phys. Rev. Lett.*, 89, 215501, 2002.
111. Hiller, M., Lavrov, E.V., and Weber, J., Raman scattering study of H_2 in Si, *Phys. Rev. B*, 74, 235214, 2006.
112. Hiller, M., Lavrov, E.V., and Weber, J., Ortho–para conversion of interstitial H_2 in Si, *Phys. Rev. Lett.*, 98, 055504, 2007.
113. Pritchard, R.E. et al., Interactions of hydrogen molecules with bond-centered interstitial oxygen and another defect center in silicon, *Phys. Rev. B*, 56, 13118, 1997.
114. Markevich, V.P. and Suezawa, M., Hydrogen-oxygen interaction in silicon at around 50°C, *J. Appl. Phys.*, 83, 2988, 1998.
115. Ishioka, K. et al., Hydrogen molecules trapped by multivacancies in silicon, *Phys. Rev. B*, 60, 10852, 1999.
116. Weber, J. et al., Properties of hydrogen-induced voids in silicon, *J. Phys.: Condens. Matter*, 17, S2303, 2005.
117. Chang, K.J. and Chadi, D.J., Diatomic-hydrogen-complex diffusion and self-trapping in crystalline silicon, *Phys. Rev. Lett.*, 62, 937, 1989.
118. Holbech, J.D. et al., H_2^* defect in crystalline silicon, *Phys. Rev. Lett.*, 71, 875, 1993.
119. Estreicher, S.K., Roberson, M.A., and Maric, Dj.M., Hydrogen and hydrogen dimers in c-C, Si, Ge and α-Sn, *Phys. Rev. B*, 50, 17018, 1994.
120. Estreicher, S.K., Hastings, J.L., and Fedders, P.A., Radiation-induced formation of H_2^* in silicon, *Phys. Rev. Lett.*, 82, 815, 1999.

3

Defects in Strained-Si MOSFETs

Yongke Sun and Scott E. Thompson

CONTENTS

3.1 Introduction

Strained-Si channel metal-oxide-semiconductor field-effect transistors (MOSFETs) are used in nearly all 90 nm and smaller commercial logic technologies. Strained channels improve the electron and hole motilities by altering the semiconductor band structure. Strained semiconductor layers are known to produce dislocation defects and have been extensively studied on blanket wafers on which several materials science review papers have recently been published [1–4]. However, much less has been published on strained-Si defects relating to manufacturing commercial production technologies which we address in this chapter. The chapter is outlined as follows: Section 3.2 briefly reviews strained-Si MOSFETs in state-of-the-art production technologies, Section 3.3 covers the yield of integrated circuits, Section 3.4 focuses on defects in strained-Si layers, Section 3.5 discusses strain relaxation, Section 3.6 describes process flow thermal cycles and manufacturing tradeoffs, and Section 3.7 briefly introduces the potential benefits and problems involving incorporating strain in alternative wafer orientations.

3.2 State of the Art: Strained-Si MOSFETs

In this section, we briefly describe techniques used in commercial 90 and 45 nm logic technologies to introduce uniaxial stress into the Si channel; more process details are published elsewhere [5,6]. The techniques in production include high stress tensile and

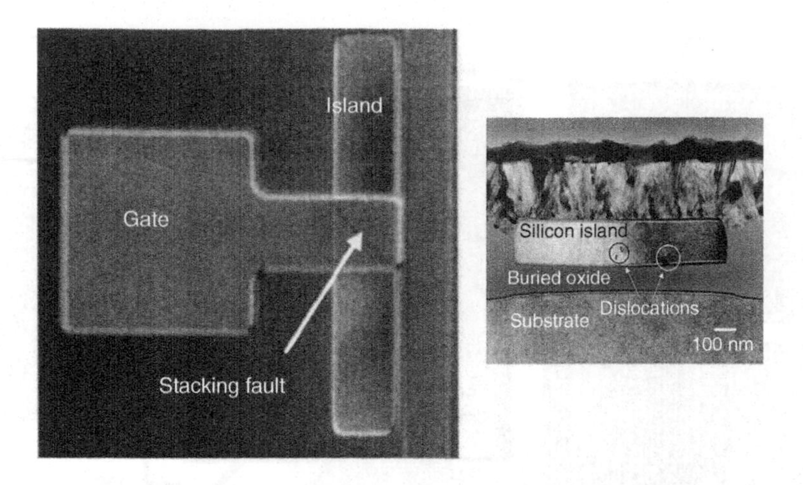

FIGURE 3.3
Examples of stacking fault and dislocation defects between source and drain of a MOSFET. (The left panel of this figure is reprinted with permission from Yang, J., Neudeck, G.W., and Denton, J.P., *Appl. Phys. Lett.*, 77, 20, 2000. The right panel of this figure is reprinted with permission from Sleight, J.W., Chuan Lin, and Gula, G.J., *IEEE Electron Devices Lett.*, 20, 5, 1999.)

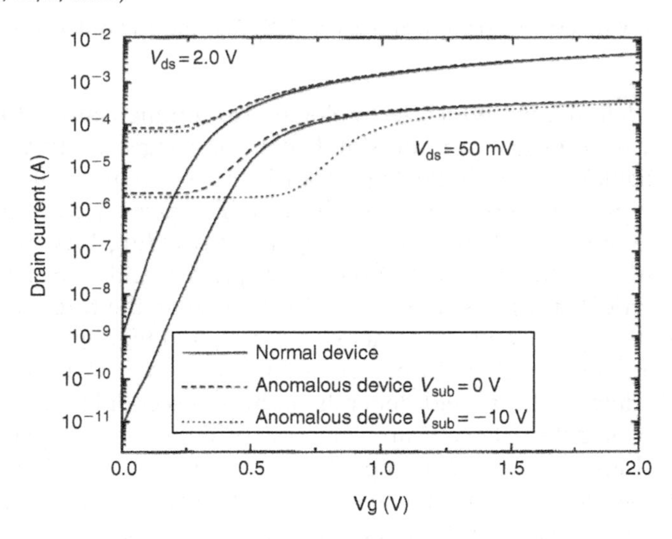

FIGURE 3.4
Current versus voltage curve for a normal and MOSFET with a source to drain defect. (Reprinted with permission from Sleight, J.W., Chuan Lin, and Gula, G.J., *IEEE Electron Device Lett.*, 20, 5, 1999.)

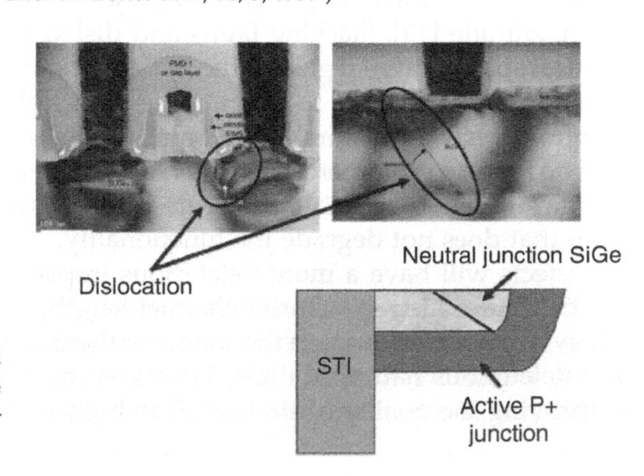

FIGURE 3.5
Examples of stacking faults in embedded SiGe source and drain. (From Pentium 4 prescott Microprocessor report, www.chipworks.com.)

3.4 Defects in Strained Layers

Misfit dislocations and stacking faults in the S/D SiGe epitaxy can be generated during epitaxial growth and thermal processes due to strain relaxation. The misfit dislocations are one-dimensional threading defects, while the stacking faults are two-dimensional planar defects. They are different yet closely related to each other. The formation of threading dislocations and stacking faults is strongly influenced by several factors, including strain level, lattice temperature, and implantation damage. Threading dislocations are further classified into two types, edge and screw dislocations, as shown in Figure 3.6. A dislocation can be visualized by imagining cutting a crystal along a plane and slipping one half across the other by a lattice vector. The halves will fit back together without leaving a defect. But if the cut only goes part way through the crystal, the boundary of the cut will leave a dislocation. The boundary of the cut is the dislocation line; the vector of the slip is the Burgers vector. When the Burgers vector is perpendicular (parallel) to the dislocation line, the dislocation is called edge (screw) dislocation. As we will see in the following discussion, the dominant dislocation type in strained-Si is neither pure edge nor screw, but is mixed.

A mixed-type dislocation is best visualized by the concept of an edge dislocation. An edge dislocation is formed by slipping one half of the crystal in the direction perpendicular to the cutting plane. Then the defect is created as if one plane of atoms is inserted to a regular crystal and ends at the defect line as shown in Figure 3.7. Here, even though there exists one extra plane of atoms in half of the crystal, except for the nearby area of the dislocation, the atoms still maintain the stacking order of crystal. Thus, the extra half plane is not a planar defect, in contrast to the stacking fault-induced edge dislocation that will be introduced next. It is easy to see that a dislocation line cannot end inside the crystal. Dislocations either end at the surface or form a ring inside the crystal. Most dislocations in strained-Si devices terminate at both ends of the shallow trench isolation (STI) that surrounds all devices. Likewise, in the process of crystal growth, the dislocations in the substrate tend to propagate into the newly grown crystal unless techniques are used to terminate the defects on the sides [4,20].

The stacking fault, just as its name implies, is caused by wrong stacking order of atoms in the process of crystal or film growth. A face-centered cubic (FCC) crystal lattice is formed by close packing of the atoms, which are shown in Figure 3.8a and b, where the close-packed plane is the (111) plane. If we take the base plane as the A-plane, the second

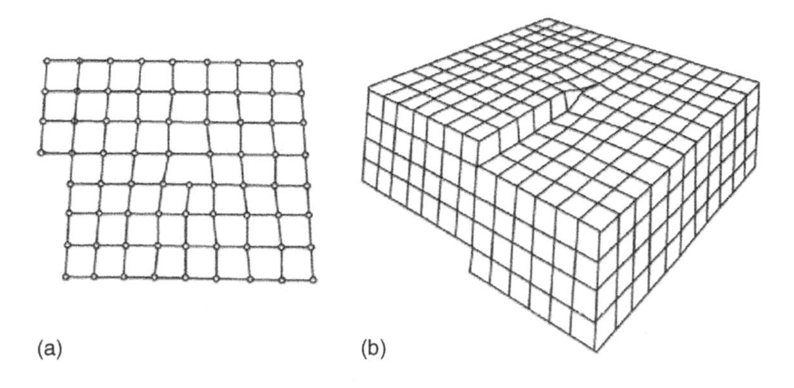

FIGURE 3.6
Two types of dislocations: (a) edge and (b) screw dislocation.

TABLE 3.1

Process Modification of Strained-Si Technology to a Standard CMOS Process Flow

Typical CMOS Process Flow	Strained-Si
Gate stack	
Gate patterning	
Offset spacers (optional)	
n-Type LDD+halo/pocket implants	
p-Type LDD+halo/pocket implants	
Spacer deposition and etch (STI)	
←	Recess and SiGe deposition (pMOSFETs)
n-Type HDD implants	
p-Type HDD implants	
RTA	
Silicide protection deposition and etch	
Silicidation	
←	High stress capping layer (nMOSFETs)

around 1000°C. Only slight modifications to a standard CMOS logic technology process flow are needed to insert the longitudinal compressive and tensile stress into the p- and nMOSFETs, respectively. For strained-Si pMOSFETs, a Si recess etch is inserted after spacer formation and followed by selective epitaxial chemical vapor deposition (CVD) of SiGe in the S/D region. The specific process flow steps are shown in Figure 3.11. Germanium concentration of 17%–35% is used, which creates a lattice spacing larger than Si. The mismatch in the SiGe to Si lattice causes the smaller lattice constant Si channel to be under compressive stress, the magnitude of which also depends on the spacing of the SiGe epitaxy to the channel in addition to the Ge concentration. A 30 nm recessed SiGe layer at the S/D location results in a 250 MPa stress in the middle of the channel; while the same SiGe layer, when present at the S/D extension location, stresses the channel to ~900 MPa. As an example, the three-dimensional finite element analysis in Figure 3.12 shows the stress distribution for a 45-nm gate length transistor after the SiGe deposition. Strain can also be engineered at other process steps like STI and silicide [32]. Integrating the SiGe layer in the S/D extension location also has a key advantage to reduce the effect of the SiGe deposition thermal budget on S/D. Furthermore, by confining the SiGe to the S/D region and introducing it late in the process flow, integration challenges, such as misfit dislocations, yield, and increased self-heating due to the low thermal conductivity of SiGe, are reduced.

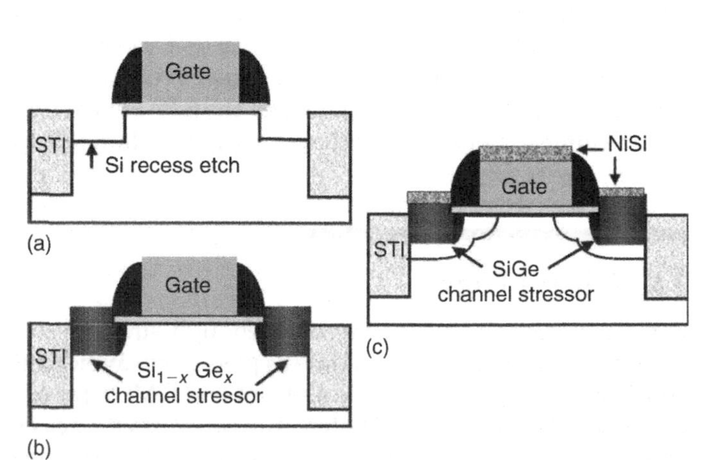

FIGURE 3.11
Strained-Si process flow.

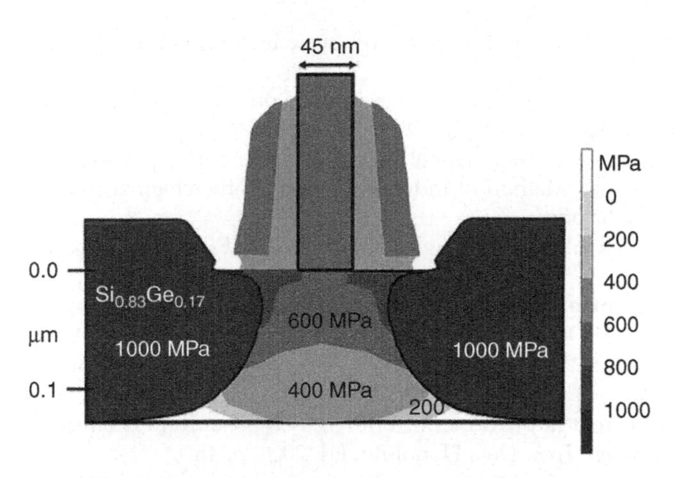

FIGURE 3.12
Stress distribution after SiGe deposition obtained by 3D finite element simulation for a 45 nm gate length transistor.

Longitudinal uniaxial tensile stress is introduced into the nMOSFETs by engineering the stress and thickness of the Si nitride-capping layer [32–35]. There are several techniques to nearly completely neutralize the capping layer strain on p-type MOSFETs; one is the use of a Ge implant and masking layer [33]. Another technique to relax the strain is to selectively remove the capping from the p-type transistors. Wafer-based strained-Si, e.g., Si channels grown on virtual SiGe substrates, also attracted a lot of attention in the last decade. However, compared to process-induced strain, the strain induced in the channel is biaxial tensile strain. This has been proven not to be as efficient as the longitudinal tensile strain for nMOSFETs [36], and even degrades the performance of the pMOSFETs [37,38]. Second, thermal annealing can easily cause strain relaxation in the strained-Si layer [39–41]. In contrast, a significant advantage of the uniaxial stress Si process flow is that, on the same wafer, compressive stress is introduced into the pMOSFETs and tensile stress in the nMOSFETs to improve both the electron and hole mobility. Since the nitride capping layer is already present to support unlanded contacts, only a few new process steps are introduced at less than a 2% wafer cost increase.

3.7 Strain and Alternative Wafer Orientation

Critical thickness and strain relaxation also relate to the wafer orientation. Although there are attempts to use wafers with different orientations, combined with epitaxial strain, to potentially enhance carrier mobility, the critical thickness for dislocation introduction is very low in non-(001) wafers, and kinetically suppressing dislocation nucleation is very difficult [42]. Diamond cubic and zinc blende structures create undesirable partial dislocation stacking in (111), (110), and (112), leading to relaxed layers with very high threading dislocation densities [43]. Thus, for CMOS technology incorporated with strain, at present, the best recommendation continues to be to use (001) wafers.

References

1. Mooney, P.M. and Chu, J.O., SiGe technology: Heteroepitaxy and high-speed microelectronics, *Annu. Rev. Mater. Sci.*, 30, 33, 2000.
2. Mooney, P.M., Strain relaxation and dislocations in SiGe/Si structures, *Mater. Sci. Eng. Rep.*, R17, 105, 1996.

3. Tsuya, H., Present status and prospect of Si wafers for ultra large scale integration, *Jpn. J. Appl. Phys.*, 43, 4055, 2004.
4. Taraschi, G., Pitera, A.J., and Fitzgerald, E.A., Strained-Si, SiGe, and Ge on-insulator: Review of wafer bonding fabrication techniques, *Solid-State Electron.*, 48, 1297, 2004.
5. Arghavani, R., Method of inducing stresses in the channel region of a transistor, patent (US). Available at: http://www.freshpatents.com/Method-of-inducing-stresses-in-the-channel-region-of-a-transistor-dt20051117ptan20050255667.php.
6. Thompson, S.E. et al., A 90-nm logic technology featuring strained-silicon, *IEEE Trans. Electron Devices*, 51, 1790, 2004.
7. Chipworks Report. Available at: http://www.chipworks.com/seamark.aspx?sm = s4%3BDatedfl14%3BDeviceCategory13%3BDigital + Logic&ss = no&tv = strained + Si&tn = 1ReportCode%2CTitle%2CDescription%2CWhatsInside%2CWhoShouldBuy%2CWhyToBuy%2CManufacturer%2CDeviceCategory&ns = 1.
8. Chidambaram, P.R. et al., 35% drive current improvement from recessed-SiGe drain extensions on 37 nm gate length PMOS, in *VLSI Symp. Tech. Dig.*, Honolulu, HI, 2004, p. 48.
9. Bai, P. et al., A 65 nm logic technology featuring 35 nm gate lengths, enhanced channel strain, 8 Cu interconnect layers, low-κ ILD and 0.57 μm^2 SRAM cell, in *IEDM Tech. Dig.*, 2004, p. 657.
10. Ohta, H. et al., High performance 30 nm gate bulk CMOS for 45 nm node with σ-shaped SiGe-SD, in *IEDM Tech. Dig.*, 2005, p. 4.
11. Horstmann, M. et al., Integration and optimization of embedded-SiGe, compressive and tensile stressed liner films, and stress memorization in advanced SOI CMOS technologies, in *IEDM Tech. Dig.*, 2005, p. 233.
12. Ang, K. et al., Enhanced performance in 50 nm N-MOSFETs with silicon-carbon source/drain regions, in *IEDM Tech. Dig.*, 2004, p. 1069.
13. Yang, M., Chan, V., and Ku, S.H., On the integration of CMOS with hybrid crystal orientations, in *VLSI Symp. Tech. Dig.*, 2004, p. 160.
14. Welser, J., Hoyt, J.L., and Gibbons, J.F., Electron mobility enhancement in strained Si N-type metal-oxide-semiconductor field-effect-transistors, *IEDM Tech. Dig.*, 1992, p. 1000.
15. Thompson, S.E. et al., A logic nanotechnology featuring strained-silicon, *IEEE Electron Device Lett.*, 25, 191, 2004.
16. Lee, W.H. et al., High performance 65 nm SOI technology with enhanced transistor strain and advanced-low-κ BEOL, in *IEDM Tech. Dig.*, 2005, p. 61.
17. Chen, C. et al., Stress memorization technique (SMT) by selectively strained-nitride capping for sub-65 nm high-performance strained-Si device application, in *VLSI Symp. Tech. Dig.*, 2004, p. 56.
18. Singh, D.V. et al., Stress memorization in high-performance FDSOI devices with ultra-thin silicon channels and 25 nm gate lengths, in *IEDM Tech. Dig.*, 2005, p. 505.
19. Braga, N. et al., Formation of cylindrical n/p junction diodes by arsenic enhanced diffusion along interfacial misfit dislocations in p-type epitaxial Si/Si(Ge), *Appl. Phys. Lett.*, 65, 1410, 1994.
20. Cohen, G.M. et al., Dislocation-free strained-silicon-on-silicon by in-place bonding, *Appl. Phys. Lett.*, 86, 251902, 2005.
21. Ball, C.A. and van der Merwe, J.H., The growth of dislocation-free layers, in *Dislocations in Solids*, Nabarro, F.R.N., Eds., North-Holland Press, Amsterdam, 1983, Chapter 27.
22. Matthews, J.W. and Blakeslee, A.E., Defects in epitaxial layers: I. Misfit dislocations, *J. Cryst. Growth*, 27, 118, 1974.
23. Houghton, D.C. et al., Equilibrium critical thickness for $Si_{1-x}Ge_x$ strained layers on (100) Si, *Appl. Phys. Lett.*, 56, 460, 1990.
24. Timbrell, P.Y. et al., An annealing study of strain relaxation and dislocation generation in SiGe heteroepitaxy, *J. Appl. Phys.*, 67, 6292, 1990.
25. Tsao, J.Y. et al., Critical thicknesses for Si_xGe_{1-x} strained-layer plasticity, *Phys. Rev. Lett.*, 59, 2455, 1987.
26. Ikeda, S. et al., The impact of mechanical stress control on VLSI fabrication process, in *IEDM Tech. Dig.*, 1997, p. 77.
27. Ha, D. et al., Anomalous junction leakage current induced by STI dislocations and its impact on dynamic random access memory devices, *IEEE Trans. Electron Devices*, 46, 940, 1999.

28. Damiano, J. et al., Characterization and elimination of trench dislocations, in *VLSI Symp. Tech., Dig.*, 1998, p. 212.
29. Ishimaru, K. et al., Mechanical stress induced MOSFET punch-through and process optimization for deep submicron TEOS-O_3 filled STI devices, in *VLSI Tech. Dig.*, 997, p. 123.
30. Park, M.H. et al., Stress minimization in deep sub-micron full CMOS devices by using an optimized combination of the trench filling CVD oxides, in *IEDM Tech. Dig.*, 1997, p. 669.
31. Jeon, C. et al., Generation of trench dislocation in 0.25 μm logic technology and its elimination, *Proc. 6th Int. Conf. on VLSI and CAD* (IVCC 1999), October, 26–27, 1999, p. 463.
32. Ito, S. et al., Mechanical stress effect of etch-stop nitride and its impact on deep submicrometer transistor design, in *IEDM Tech. Dig.*, 2000, p. 247.
33. Shimizu, A. et al., Local mechanical-stress control (LMC): A new technique for CMOS-performance enhancement, in *IEDM Tech. Dig.*, 2001, p. 433.
34. Ghani, T. et al., A 90-nm high volume manufacturing logic technology featuring novel 45-nm gate length strained-silicon CMOS transistors, in *IEDM Tech. Dig.*, 2003, p. 978.
35. Ge, C.H., Process-strained-Si (PSS) CMOS technology featuring 3-D strain engineering, in *IEDM Tech. Dig.*, 2003, p. 73.
36. Uchida, K. et al., Physical mechanisms of electron mobility enhancement in uniaxial stressed MOSFETs and impact of uniaxial stress engineering in ballistic regime, in *IEDM Tech. Dig.*, 2005, p. 129.
37. Rim, K. et al., Fabrication and mobility characteristics of ultra-thin strained Si directly on insulator (SSDOI) MOSFETs, in *IEDM Tech. Dig.*, 2003, p. 49.
38. Sun, Y., Thompson, S.E., and Nishida, T., Physics of strain effects in semiconductors and metal-oxide-semiconductor field-effect transistors, *J. Appl. Phys.*, 101, 104503, 2007.
39. Fiorenza, J.G. et al., Film thickness constraints for manufacturable strained-silicon CMOS, *Semicond. Sci. Technol.*, 19, L4, 2004.
40. Hirashita, N., Moriyama, Y., and Sugiyama, N., Misfit strain relaxation in strained-Si layers on silicon-germanium-on insulator substrates, *Appl. Phys. Lett.*, 89, 091916, 2006.
41. Samavedam, S.B. et al., Relaxation of strained-Si layers grown on SiGe buffers, *J. Vac. Sci. Technol., B*, 17, 4, 1999.
42. Kuan, T.S. and lyer, S.S., Strain relaxation and ordering in SiGe layers grown on (100), (111), and (110) Si surfaces by molecular-beam epitaxy, *Appl. Phys. Lett.*, 59, 2242, 1991.
43. Lee, M.L., Antoniadis, D.A., and Fitzgerald, E.A., Challenges in epitaxial growth of SiGe buffers on Si (111), (110), and (112), in *Proc. ICSI4 Conference*, Awaji, Japan, 2005.

recently witnessed regarding crystal growth, gettering, and annealing after implantation and other radiation-rich processing steps (such as reactive-ion etching [RIE]). Ignoring neutral defects such as those associated with dislocations and stacking faults—possibly returning to the fore as strained channel materials are becoming more and more commonly employed—more realistically, defects to be considered at present are intentionally induced defects. If we define, from an electron transport perspective, a defect as any entity which breaks the three-dimensional (3D) symmetry of the semiconductor lattice, the ionized impurities introduced by doping and interfaces are the major defects to be considered. Here, first we consider ionized impurities, discussing the models employed to describe their associated potential, various approximations used to account for dielectric screening, and two major competing approximations, the Conwell and Weisskopf (CW) [1] and the Brooks–Herring (BH) [2] models, employed to obtain the associated scattering and momentum relaxation rates. Then, we consider genuine defects, namely those interfacial defects commonly labeled as interface roughness. As charge carriers are squeezed harder against the interfaces by the increasingly large perpendicular electric fields present in thin insulator devices, scattering with the roughness is a growing concern. We limit our analysis to extensions to thin bodies and quantum wells of the continuum empirical model originally proposed by Prange and Nee [3], Saitoh [4], Sakaki [5], and Ando et al. [6].

4.2 Semiclassical Electron Transport

Before moving to the core of this chapter, we recall some basic tools employed in dealing with electronic transport in semiconductors.

4.2.1 Transport in Bulk Semiconductors

At the semiclassical level charge transport in semiconductors is described by the Boltzmann transport equation (BTE). Its validity at the 10 nm scale is a question that goes beyond the scope of this chapter, so we naively assume the BTE as valid.

The BTE describes the irreversible temporal evolution of the distribution function $f(\mathbf{r}, \mathbf{k}, \mu, t)$ (i.e., the density, f of the particles in band μ at position $\mathbf{r}$ with wave vector $\mathbf{k}$ at time t):

$$\frac{\partial f(\mathbf{k},\mathbf{r},\mu,t)}{\partial t} = -\frac{\mathrm{d}\mathbf{k}}{\mathrm{d}t}\cdot\nabla_{\mathbf{k}} f(\mathbf{k},\mathbf{r},\mu,t) - \frac{\mathrm{d}\mathbf{r}}{\mathrm{d}t}\cdot\nabla_{\mathbf{r}} f(\mathbf{k},\mathbf{r},\mu,t) + \left(\frac{\partial f(\mathbf{k},\mathbf{r},\mu,t)}{\partial t}\right)_{\mathrm{coll}}. \tag{4.1}$$

In words, this equation expresses the total time variation of the population of an infinitesimal volume in phase space as due to particles that leave the volume drifting away in $\mathbf{k}$-space under external forces (the term $(\mathrm{d}\mathbf{k}/\mathrm{d}t)\cdot\nabla_{\mathbf{k}}\mathbf{f}$), by diffusing away in real space (the term $(\mathrm{d}\mathbf{r}/\mathrm{d}t)\cdot\nabla_{\mathbf{r}}\mathbf{f}$), and due to particles leaving and entering the volume via collisions (the term $(\partial f/\partial t)_{\mathrm{coll}}$). The last term is the crucial and hard-to-treat collision term (often called collision integral or collision operator):

$$\begin{aligned}\left(\frac{\partial f(\mathbf{k},\mathbf{r},\mu,t)}{\partial t}\right)_{\mathrm{coll}} &= \sum_{\nu}\int\frac{\mathrm{d}\mathbf{k}'}{(2\pi)^3}S(\mathbf{k},\mu;\,\mathbf{k}',\nu;\,\mathbf{r})f(\mathbf{k}',\mathbf{r},\nu,t)[1-f(\mathbf{k},\mathbf{r},\mu,t)]\\ &\quad-\sum_{\nu}\int\frac{\mathrm{d}\mathbf{k}'}{(2\pi)^3}S(\mathbf{k}',\nu;\mathbf{k},\mu;\mathbf{r})f(\mathbf{k},\mathbf{r},\mu,t)[1-f(\mathbf{k}',\mathbf{r},\nu,t)],\end{aligned} \tag{4.2}$$

where $S(\mathbf{k}', \nu; \mathbf{k}, \mu; \mathbf{r})$ represents the probability per unit time that a particle will make a transition from a state $(\hbar\mathbf{k}', \mu)$ in reciprocal space into the infinitesimal volume centered around $(\hbar\mathbf{k}', \nu)$ because of a collision. These terms, aside from the less complicated band structure effects, contain the entire physics of the problem. This is where the knowledge of the defect enters when analyzing its effect on electronic transport.

Note that in Equation 4.1 the drift term $(d\mathbf{k}/dt) \cdot \nabla_{\mathbf{k}} f$ is determined by the electric field $\mathbf{F} = -\nabla_{\mathbf{r}}\varphi$, the electrostatic potential φ being determined by the Poisson equation

$$\nabla \cdot [\varepsilon(\mathbf{r})\nabla\varphi(\mathbf{r}, t)] = -\rho(\mathbf{r}, t) - \rho_d(\mathbf{r}), \tag{4.3}$$

where

$\varepsilon(\mathbf{r})$ is the position-dependent static permittivity of the material(s) constituting the device
$\rho_d(\mathbf{r})$ is the charge density of the ionized donors and acceptors, $eN_D^+(\mathbf{r}) - eN_A^-(\mathbf{r})$
$\rho(\mathbf{r}, t)$ is the charge density of the free carriers related to the distribution function $f(\mathbf{k}, \mathbf{r}, t)$ via its separate hole (valence bands) and electron (conduction bands) contribution:

$$\rho(\mathbf{r},t) = 2e \sum_{\substack{\mu \\ \text{valence}}} \int \frac{d\mathbf{k}}{(2\pi)^3} f(\mathbf{k}, \mathbf{r}, \mu, t) - 2e \sum_{\substack{\mu \\ \text{conduction}}} \int \frac{d\mathbf{k}}{(2\pi)^3} f(\mathbf{k}, \mathbf{r}, \mu, t), \tag{4.4}$$

having assumed, here and in the following, that the distribution function is normalized to the total local particle density:

$$2\sum_{\mu} \int \frac{d\mathbf{k}}{(2\pi)^3} f(\mathbf{k}, \mathbf{r}, \mu, t) = n(\mathbf{r},t).$$

Thus, Equations 4.1 and 4.3 are coupled since the equations of motion for a particle in phase space depend on the electrostatic potential $\varphi(\mathbf{r}, t)$ (assuming a homogeneous material),

$$\frac{d\mathbf{k}}{dt} = \mp \frac{e}{\hbar} \nabla_{\mathbf{r}}\varphi(\mathbf{r}, t), \tag{4.5}$$

$$\frac{d\mathbf{r}}{dt} = \frac{1}{\hbar} \nabla_{\mathbf{k}} E_\mu(\mathbf{k}), \tag{4.6}$$

the "minus" sign in Equation 4.5 holding for holes, the "plus" for electrons, e is the magnitude of the electron charge, $\hbar$ is the reduced Planck constant, and $E_\mu(\mathbf{k})$ is the dispersion relation for carriers in the semiconductor.

Solving this integro-differential equation (possibly nonlinear since the collision kernel $S(\mathbf{k}', \nu; \mathbf{k}, \mu; \mathbf{r})$ depends on f itself when accounting for dielectric screening and/or interparticle collisions) is best done with Monte Carlo techniques [7]. Below we present results using self-consistent Monte Carlo simulations employing 3D transport accounting for the full band structure of the semiconductor, coupled to a 2D solution of the Poisson equation in the electronic device [8].

4.2.2 Low-Field Mobility

Of particular interest—for both its simplicity and relatively straightforward comparison with experimental information—is to consider the homogeneous, steady-state, weak-field limit of the BTE. As stated above, thanks to the acceleration theorem [9], the driving term $d\mathbf{k}/dt$ in

Then, the electron charge density in Equation 4.4 is replaced by

$$\rho(\mathbf{r},t) = \rho^{(3D)}(\mathbf{r},t) + 2e\sum_{\mu,\alpha}\eta_\mu \int d\mathbf{K} f^{(2D)}(\mathbf{K},\mathbf{R},\mu,\alpha,t)|\zeta_{\mu\alpha}(\mathbf{K},z)|^2, \tag{4.16}$$

where η_μ is +1 for the valence bands (holes), −1 for the conduction bands (electrons), $\rho^{(3D)}$ is the charge density given by Equation 4.4 due to particles which are not confined (bulk or 3D particles), whose transport is described by the BTE (Equation 4.1), while the second component is due to the contribution of confined (2D or quantized particles) whose in-plane distribution function $f^{(2D)}$ obeys the semiclassical BTE:

$$\begin{aligned}\frac{\partial f^{2D}(\mathbf{K},\mathbf{R},\mu,\alpha,t)}{\partial t} &= -\frac{d\mathbf{K}}{dt}\cdot\nabla_{\mathbf{K}} f^{(2D)}(\mathbf{K},\mathbf{R},\mu,\alpha,t) - \frac{d\mathbf{R}}{dt}\cdot\nabla_{\mathbf{R}} f^{2D}(\mathbf{K},\mathbf{R},\mu,\alpha,t)\\ &\quad + \left(\frac{\partial f^{(2D)}(\mathbf{K},\mathbf{R},\mu,\alpha,t)}{\partial t}\right)_{\text{coll}},\end{aligned} \tag{4.17}$$

where the collision terms takes a form similar to Equation 4.2, with the appropriate scattering kernels reflecting the confined nature of the carriers via the low-dimensionality matrix elements involving the subband wave functions ζ (see Equation 4.20).

4.2.4 Collision Kernels

The collision kernels $S(\mathbf{k}'\nu';\, \mathbf{k}\nu;\, \mathbf{r})$ in Equation 4.2 are obtained typically by employing the lowest order coupling between the carriers and the scattering potentials (which are assumed to conserve spin and not to depend on this variable). In order to evaluate this lowest order coupling and evaluate the scattering rate from Fermi golden rule, we need the matrix element of the scattering potential between the initial and the final states, $M_s(\mathbf{k}'\nu', \mathbf{k}\nu)$. Let us denote by $V_s(\mathbf{r}) = \Sigma_{s,\mathbf{q}} V_{s,\mathbf{q}} e^{i\mathbf{q}\cdot\mathbf{r}}$, the scattering potential of type s, where s runs over phonon branches, Coulomb potential for scattering with ionized impurities or other carriers, alloy potential, etc. Then, proceeding in a standard way:

$$M_s(\mathbf{k}'\nu',\mathbf{k}\nu) \equiv \langle \mathbf{k}'\nu'|V_s|\mathbf{k}\nu\rangle = \frac{1}{N_c\Omega_c}\int d\mathbf{r}\, e^{-i\mathbf{k}'\cdot\mathbf{r}} u^*_{\nu'\mathbf{k}'}(\mathbf{r}) V_s(\mathbf{r}) u_{\nu k}(\mathbf{r}) e^{i\mathbf{k}\cdot\mathbf{r}},$$

where

N_c is the number of cells in the normalization volume

Ω_c the volume of a cell

$u^*_{\nu\mathbf{k}}(\mathbf{r})$ is the Bloch wave

We can now express the integral over the variable $\mathbf{r}$ as a sum over cells labeled by the index α and equal integrals over a cell, so that $\mathbf{r} \to \mathbf{R}_\alpha + \mathbf{r}$ and

$$M_s(\mathbf{k}'\nu',\mathbf{k}\nu) = \frac{1}{N_c}\sum_{\mathbf{q}}\sum_{\alpha} e^{i(\mathbf{k}-\mathbf{k}'+\mathbf{q})\cdot\mathbf{R}_\alpha} V_{s,\mathbf{q}} \frac{1}{\Omega_c}\int_{\text{cell}} d\mathbf{r}\, e^{i(\mathbf{k}-\mathbf{k}'+\mathbf{q})\cdot\mathbf{r}} u^*_{\nu'\mathbf{k}'}(\mathbf{r}) u_{\nu\mathbf{k}}(\mathbf{r}).$$

The sum over α yield a nonzero contribution (equal to N_c) only when $\mathbf{k}-\mathbf{k}'+\mathbf{q}=\mathbf{G}$, for some $\mathbf{G}$-vector of the reciprocal lattice. Therefore,

$$M_s(\mathbf{k}'\nu',\mathbf{k}\nu) = \sum_{\mathbf{G}} V_{s,\mathbf{k}-\mathbf{k}'+\mathbf{G}}\,\mathcal{I}(\mathbf{k}\nu,\mathbf{k}'\nu';\,\mathbf{G}), \tag{4.18}$$

where $\mathcal{I}$ ($\mathbf{k}\nu, \mathbf{k}'\nu'; \mathbf{G}$) is the overlap factor between Bloch waves,

$$\mathcal{I}(\mathbf{k}\nu, \mathbf{k}'\nu'; \mathbf{G}) = \frac{1}{\Omega_c} \int_{\text{cell}} d\mathbf{r}\, e^{i\mathbf{G}\cdot\mathbf{r}} u^*_{\nu'\mathbf{k}'}(\mathbf{r}) u_{\nu\mathbf{k}}(\mathbf{r}).$$

For phonon scattering, the only **G**-vector contributing to the sum is the particular vector, $\mathbf{G}_u$, which maps $\mathbf{k}' - \mathbf{k}$ to the first Brillouin zone. If $\mathbf{G}_u = 0$, we talk of Normal processes, of Umklapp processes otherwise. For the other interactions we are particularly interested here, such as Coulomb scattering (impurity, interparticle), in principle we should retain a sum over all possible **G**-vectors. We indeed retain these terms as a matter of computational convenience. In practice, the squared Coulomb matrix elements will decrease very fast (as G^{-4}) and only the first term will contribute. As matter of simplicity of notation, we drop the sum over **G**-vectors for now. Now, the scattering kernel associated to this process can be calculated from the Fermi golden rule (or first Born approximation):

$$S_s(\mathbf{k}'\nu'; \mathbf{k}\nu) = \frac{2\pi}{\hbar} |M_s(\mathbf{k}'\nu', \mathbf{k}\nu)|^2 \delta[E_\nu(\mathbf{k}) - E_{\nu'}(\mathbf{k}') \pm \hbar\omega_s(\mathbf{k}' - \mathbf{k})], \tag{4.19}$$

where $\hbar\omega_s(\mathbf{q})$ is the energy of the elementary excitation associated to the scattering potential, and the DOS of the material is implicitly accounted for by the delta function expressing energy conservation. The overall kernel S appearing in the collision integral is obtained by summing over all scattering processes s.

Ignoring the degeneracy correction $(1-f)$, which is accounted for in the approximation described in the following, the rate at which a carrier of crystal momentum **k** in band ν scatters with the scattering potentials of type s is obtained from:

$$\frac{1}{\tau_s(k,\nu)} = \sum_{k',\nu'} S_s(\mathbf{k}'\nu'; \mathbf{k}\nu) = \frac{2\pi}{\hbar} \frac{1}{(2\pi)^3} \sum_{\nu'} \int d\mathbf{k}' |M_s(\mathbf{k}'\nu', \mathbf{k}\nu)|^2 \delta[E_\nu(\mathbf{k}) - E_{\nu'}(\mathbf{k}') \pm \hbar\omega_s(\mathbf{k}' - \mathbf{k})]. \tag{4.20}$$

Scattering involving carriers in a low-dimensionality region is treated in a similar way, but the matrix element (Equation 4.18) is replaced by the matrix element between envelope wave functions in bands ν', ν (which in all practical situations are both identical and equal to the lowest conduction band, simplification which we assume in the following by dropping the band index) and subbands α', α:

$$M_s^{(2D)}(\mathbf{K}'\alpha', \mathbf{K}\alpha) \approx \frac{L_z}{N_c\Omega_c} \int d\mathbf{R} dz\, e^{-i\mathbf{K}'\cdot\mathbf{R}} \zeta^*_{\alpha'}(\mathbf{K}', z) V_s(\mathbf{R}, z) \zeta_\alpha(\mathbf{K}, z) e^{i\mathbf{K}\cdot\mathbf{R}}, \tag{4.21}$$

where L_z is the normalization length in the quantization direction z, the envelope wave functions have been normalized so that

$$\int_{L_z} dz\, \zeta^*_{\alpha'}(\mathbf{K}', z) \zeta_\alpha(\mathbf{K}, z) = \delta_{\mathbf{K}',\mathbf{K}},$$

and having approximated the overlap factor with unity, since only low energy states all lying in a small region of the Brillouin zone are involved. In the common case of a

impurities, as in Equation 4.25. This double-counting prompted Mahan to perform more accurate quantum-mechanical calculations. Using a variational approach, he found that the wave functions of the free carriers do indeed pile up around the impurity potential at low N_{imp} ($\sim 10^{17}$ cm^{-3} in Si at 300 K), but appear to be relatively insensitive to the presence of the impurities in the opposite limit of large N_{imp}. Thus, it appears that the BH model should be employed at low impurity concentrations, the CW model in the opposite limit of large N_{imp}. Ridley [22], with his statistical screening model, has attempted to merge the two approaches by considering only the potential due not to the impurity exhibiting the smallest impact parameter (as in the CW model), but considering the Yukawa-like screened BH potential due to the closest impurity. This leads to a rescaling of the BH scattering rate:

$$\frac{1}{\tau^{(\text{Ridley})}(\mathbf{k},\mathbf{r},\nu)} = \frac{v_{\mathbf{k}}}{d}\left\{1 - \exp\left[\frac{d}{\tau^{(\text{BH})}(\mathbf{k},\mathbf{r},\nu)v_{\mathbf{k}}}\right]\right\}, \tag{4.31}$$

where

$v_{\mathbf{k}}$ is the group velocity
$d = (2\pi N_{imp})^{-1/3}$ is the average interimpurity distance.

4.3.2 Dielectric Screening

We have made use before of some elementary concepts related to dielectric screening by free carriers in semiconductors. Since these concepts will be required below, both in treating screening in low-dimensionality situations as well as to deal with surface roughness (SR), we review here briefly the simple theory of linear screening.

Let us consider a free semiconductor, i.e., a semiconductor described by the free Schrödinger equation in a crystal $\mathbf{H}_0\psi_{\mathbf{k},\mu} = E_\mu(\mathbf{k})\psi_{\mathbf{k},\mu}$, where $\mathbf{H}_0$ is the free Hamiltonian and $\psi_{\mathbf{k},\mu}(\mathbf{r}) = \langle \mathbf{r}|\mathbf{k}, \mu\rangle$ the (Bloch) plane waves which are the eigenfunctions of the lattice Hamiltonian $\mathbf{H}_0$ with eigenvalues $E_\mu(\mathbf{k})$ for each band μ. If we apply an external perturbation described by the potential $\varphi^{(\text{ex})}(\mathbf{r}, t)$, the charges in the semiconductor will rearrange themselves, causing an additional polarization (or screening) potential $\varphi^{(\text{pol})}(\mathbf{r}, t)$ which modifies the original external potential. Thus, the actual potential in the material will be the sum of the external perturbation and of the polarization potential,

$$\varphi(\mathbf{r}, t) = \varphi^{(\text{ex})}(\mathbf{r}, t) + \varphi^{(\text{pol})}(\mathbf{r}, t). \tag{4.32}$$

We are interested in describing the total potential φ, since this is the effective perturbation acting on the material. We may think of $\varphi^{(\text{ex})}$ as the bare potential of a positively charged donor impurity in n-type Si, described by the $1/r$ Coulomb potential, while φ will be the screened potential felt by the conduction electrons, resulting from single positive charge of the impurity ion and from the negative cloud of free conduction electrons attracted by the impurity. Two approximations are very commonly made to reduce the complexity of the problem: first, the Fourier components of the various potentials are considered independently. Cross-terms mixing different wavelengths are ignored, on the grounds that their phases will vary wildly, resulting in their cancellation. This is the so-called random phase approximation (RPA). An equivalent way of stating this approximation is to say that each electron will respond to the average potential caused by the other electrons. We thus transform this many-body problem to a single-particle problem, thanks to this mean-field approximation. Second, we assume that both the external and the polarization potentials are weak perturbations of the free Hamiltonian $\mathbf{H}_0$. This is the so-called linear screening approximation which allows us to express the polarization charge in simple first-order perturbation theory. Thanks to the RPA, we can state the problem more simply. Let us expand the external perturbation in plane waves:

$$\varphi^{(\mathrm{ex})}(\mathbf{r},t) = \sum_{\mathbf{q}} \varphi_{\mathbf{q}}^{(\mathrm{ex})} \mathrm{e}^{-\mathrm{i}\mathbf{q}\cdot\mathbf{r}-\mathrm{i}\omega t}, \tag{4.33}$$

having assumed a simple harmonic time dependence. Then, we are interested in finding the dielectric function $\varepsilon^{(r)}(\mathbf{q}, \omega) = \varepsilon(\mathbf{q}, \omega)/\varepsilon_0$ where ε_0 is the vacuum permittivity, such that

$$\varphi_{\mathbf{q}} = \varphi_{\mathbf{q}}^{(\mathrm{ex})} + \varphi_{\mathbf{q}}^{(\mathrm{pol})} = \frac{\varphi_{\mathbf{q}}^{(\mathrm{ex})}}{\varepsilon^{(r)}(\mathbf{q},\omega)}. \tag{4.34}$$

The usefulness of this expression lies in the fact that, once we know the dielectric function $\varepsilon(\mathbf{q}, \omega)$, the knowledge of the simple external potential is sufficient to determine the net response of the system, without having to reevaluate the self-consistent redistribution of the charges internally to the system. We proceed as follows: first, we employ the approximation of linear response to determine the new wave functions of the system under the perturbation of the net potential φ, assuming that it is known. From the perturbed wave functions, we derive the polarization charge,

$$\rho^{(\mathrm{pol})}(\mathbf{r},t) = \sum_{\mathbf{q}} \delta\rho_{\mathbf{q}} \mathrm{e}^{-\mathrm{i}\mathbf{q}\cdot\mathbf{r}-\mathrm{i}\omega t}, \tag{4.35}$$

and from this the polarization potential. Summing the external and polarization potentials, we solve for the net potential self-consistently.

Let us now express the wave functions, $\phi_{\mathbf{k},\mu}(\mathbf{r}, t)$, of the perturbed system (lattice Hamiltonian plus perturbing potential) in terms of the unperturbed wave functions $\psi_{\mathbf{k},\mu}(\mathbf{r}, t) = \langle \mathbf{r}, t|\mathbf{k}, \mu\rangle$, using first-order perturbation theory:

$$\phi_{\mathbf{k},\mu}(\mathbf{r},t) = \psi_{\mathbf{k},\mu}(\mathbf{r},t) + \delta\psi_{\mathbf{k},\mu}(\mathbf{r},t) = \psi_{\mathbf{k},\mu}(\mathbf{r},t) + \sum_{\mathbf{k}'\mu'} \frac{e\langle \mathbf{k}',\mu'|\varphi|\mathbf{k},\mu\rangle}{E_\mu(\mathbf{k}) - E_{\mu'}(\mathbf{k}') + \hbar\omega + \mathrm{i}\hbar s} \psi_{\mathbf{k}',\mu'}(\mathbf{r},t), \tag{4.36}$$

and

$$\phi^*_{\mathbf{k},\mu}(\mathbf{r},t) = \psi^*_{\mathbf{k},\mu}(\mathbf{r},t) + \delta\psi^*_{\mathbf{k},\mu}(\mathbf{r},t) = \psi^*_{\mathbf{k},\mu}(\mathbf{r},t) + \sum_{\mathbf{k}',\mu'} \frac{e\langle \mathbf{k},\mu|\varphi|\mathbf{k}',\mu'\rangle}{E_\mu(\mathbf{k}) - E_{\mu'}(\mathbf{k}') - \hbar\omega - \mathrm{i}\hbar s} \psi^*_{\mathbf{k}',\mu'}(\mathbf{r},t). \tag{4.37}$$

From Equation 4.33 and some manipulations with Bloch waves, we have:

$$\langle \mathbf{k}',\mu'|\varphi|\mathbf{k},\mu\rangle = \varphi_{\mathbf{k}-\mathbf{k}'} \sum_{\mathbf{G}} \left\langle \mathbf{k}' + \mathbf{G},\mu' \middle| \mathrm{e}^{\mathrm{i}(\mathbf{k}-\mathbf{k}')\cdot\mathbf{r}} \middle| \mathbf{k},\mu \right\rangle. \tag{4.38}$$

The new charge density is obtained by summing the individual contributions $e|\phi_{\mathbf{k},\mu}|^2$ over all occupied states $\phi_{\mathbf{k},\mu}$. Denoting by $p(\mathbf{k}, \mu)$ the occupation of each state—not necessarily the equilibrium occupation number—we have:

$$\begin{aligned}\rho(\mathbf{r},t) &= e\sum_{\mathbf{k},\mu} p(\mathbf{k},\mu)|\phi_{\mathbf{k},\mu}(\mathbf{r},t)|^2 = e\sum_{\mathbf{k},\mu} p(\mathbf{k},\mu)|\psi_{\mathbf{k},\mu}(\mathbf{r},t) + \delta\psi_{\mathbf{k},\mu}(\mathbf{r},t)|^2 \\ &\simeq \rho_0(\mathbf{r},t) + e^2\sum_{\mathbf{k},\mu} p(\mathbf{k},\mu)\psi^*_{\mathbf{k},\mu}(\mathbf{r},t) \sum_{\mathbf{G},\mathbf{k}',\mu'} \frac{\varphi_{\mathbf{k}-\mathbf{k}'}\left\langle \mathbf{k}' + \mathbf{G},\mu' \middle| \mathrm{e}^{\mathrm{i}(\mathbf{k}-\mathbf{k}')\cdot\mathbf{r}} \middle| \mathbf{k},\mu \right\rangle}{E_\mu(\mathbf{k}) - E_{\mu'}(\mathbf{k}') + \hbar\omega + \mathrm{i}\hbar s} \psi_{k',\mu'}(r,t) + \mathrm{cc},\end{aligned}$$

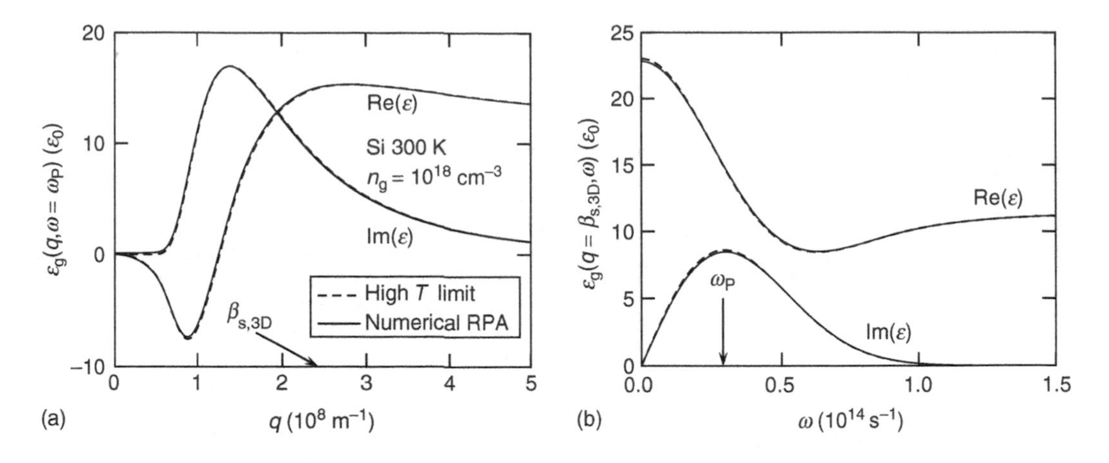

FIGURE 4.2
Real and imaginary part of the dielectric constant in Si evaluated as a function of q at $\omega = 0$ (top frame) and as a function of ω at $q = 0$ (bottom frame). The solid line has been obtained from the full RPA, the dashed line from the high-T, nondegenerate approximation. The labels $\beta_{s,3D}$ and ω P indicate the Thomas–Fermi/Debye–Hückel wavevector and the plasma frequency, respectively. (Reprinted from Fischetti, M.V., *J. Appl. Phys.*, 85, 7984, 1999. With permission.)

where

$\mathcal{I}(\mathbf{k}\nu, \mathbf{k}'\nu')$ is the overlap integral between the Bloch states

$\beta(q, 0)$ is the wave vector-dependent static screening parameter evaluated at the crystal momentum transfer $q = |\mathbf{k} - \mathbf{k}' + \mathbf{G}|$

The function h represents the phase-shift corrections to the first Born approximation, as tabulated by Mayer and Bartoli [25], at the values $y_{\pm} = \pm(1/2)kr_B$ for the dimensionless wave vector, and $s = 2/(r_B\beta_s)$ for the screening parameter rescaled so to satisfy Friedel's sum rule [26]. Here r_B is the effective Bohr radius, $\mathbf{k}$ is the magnitude of the wave vector, β_{DH} is the Debye–Hückel (or, where appropriate, Thomas–Fermi) screening parameter. The "+" sign is taken in the case of an attractive interaction (such as for electron-donor or

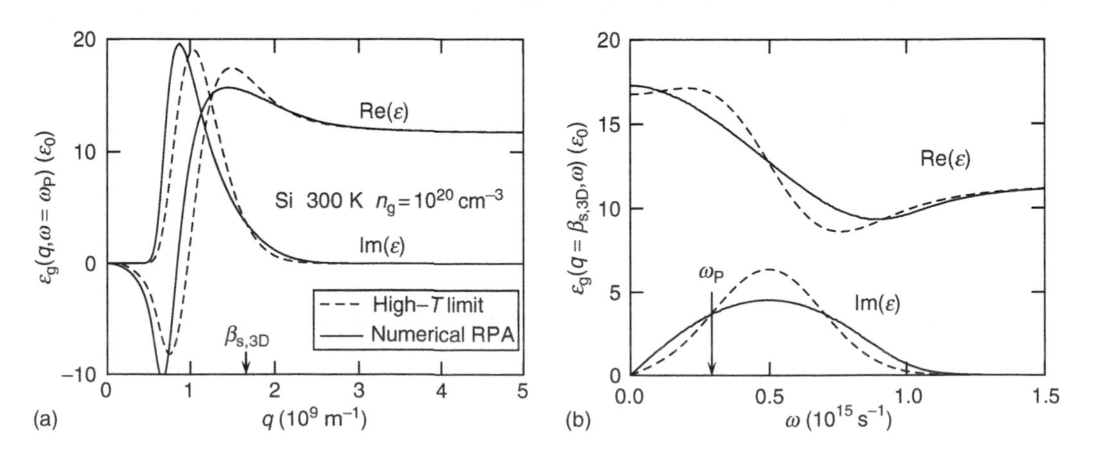

FIGURE 4.3
As in Figure 4.2, but at an electron density of 10^{20} cm^{-3}. Note that only at densities larger than ~10^{19} cm^{-3} the nondegenerate limit begins to differ significantly from the exact RPA result, as it is clearly evident here. Yet, even in this case, the nondegenerate approximation may be considered acceptable. On the other hand, in low-dimensionality situations the disagreement is more evident already at small sheet densities.

hole-acceptor scattering), the "−" sign for a repulsive interaction (such as for electron-acceptor and hole-donor scattering). The effect of degeneracy, represented by the factor $(1-f)$ in Equation 4.59 is included by replacing the distribution functions f with their equilibrium values,

$$f(\mathbf{k},\mu,\mathbf{r},t) \simeq f_0[\mathbf{k},\mu,T_{\mathrm{p}}(\mathbf{r},t),E_{\mathrm{F}}(\mathbf{r},t)],$$

where

$f_0(\mathbf{k},\mu,T,E_{\mathrm{F}}) = \{1+\exp[(E_\mu(\mathbf{k})-E_{\mathrm{F}})/(k_{\mathrm{B}}T)]\}^{-1}$ is the Fermi–Dirac function at temperature T for band μ and Fermi energy E_{F}

$T_{\mathrm{p}}(\mathbf{r},\mathbf{t})$ is the equivalent particle temperature yielding the average local particle density and energy estimated locally during the Monte Carlo simulation

The overlap integral $\mathcal{I}(\mathbf{k}\nu,\mathbf{k}'\nu')$ is computed from $\mathbf{k}\cdot\mathbf{p}$ perturbation theory [27], while screening is treated in the high-temperature, nondegenerate limit valid for a single parabolic and spherical valley, and in the long-wavelength limit (because of the dominance of small k-vectors in Equation 4.60, so that $\beta(\mathbf{q},\omega=0)\simeq\beta_{\mathrm{s}}$, where β_{s} is the Debye–Hückel screening parameter given by Equation 4.51 using the local particle density and temperature. A sum over the separate electron and hole contributions to our current use of Equation 4.51 must be understood.

In the actual calculation, a self-scattering algorithm is chosen. First, an appropriate multiple of Equation 4.63 is employed as a self-scattering rate. For particles undergoing this self-scattering event, a second self-scattering rate is computed by ignoring overlap factors and degeneracy:

$$\frac{1}{\tau_{\mathrm{imp,self}}(\mathbf{k},\mathbf{r},\nu)} = \frac{2\pi Z^2 e^3 N_{\mathrm{imp}}}{\hbar a^3 \varepsilon_{\mathrm{s}}^2} h(y\pm,S) \sum_{\mathbf{G}} \sum_{\text{energy conserving cubes } j,\nu'} \frac{S_{j,\nu'}^{\mathrm{GR}}(w_j)}{\left[|\mathbf{k}-\mathbf{k}_j+\mathbf{G}|^2+\beta_{\mathrm{s}}^2\right]^2}, \tag{4.61}$$

where $S_{j,\nu'}^{\mathrm{GR}}(w_j)$ is the DOS in cube j' and band ν' resulting from having discritized the entire B2 into cubes of side $\Delta\mathbf{k}=0.05\ (2\pi/a)$, where a is the lattice constant at the dimensionless final energy w_j, according to the Gilat–Raubenheimer scheme [28], when energies are measured in eV and crystal momenta in $(2\pi/a)$-units. The sum over **G** vectors is performed over the first 80–100 vectors, up to a maximum magnitude of about $\sqrt{11}(2\pi/a)$. The final state after collision is determined by ignoring phase-shift corrections (and, possibly, corrections due to implementation of the Ridley model), by selecting a random $\mathbf{k}_j$ with probability density given by the matrix element in Equation 4.61, and by performing a final rejection/acceptance step on degeneracy and overlap factors.

Whenever transport is simulated using the approximated analytical representation of the band structure, a self-scattering rate is obtained analytically by ignoring once more degeneracy and overlap-factor corrections, considering only intravalley processes, and including only the **G**-vector $\mathbf{G}_u$ required to bring $\mathbf{k}-\mathbf{k}'+\mathbf{G}_u$ inside the first BZ:

$$\frac{1}{\tau_{\mathrm{imp,self}}(\mathbf{k},\mathbf{r},\nu)} = \frac{e^4 Z^2 N_{\mathrm{imp}}}{2^{9/2}\pi m_{d,\nu}^{1/2}\varepsilon_{\mathrm{s}}^2} h(y\pm,s)\gamma(E)^{-3/2}(1-2\alpha E)\int_{-1}^{+1} \mathrm{d}\mu \left[(1-\mu)+\frac{E_\beta}{2\gamma(E)}\right]^{-2} \tag{4.62}$$

so that

$$\frac{1}{\tau_{\mathrm{imp,\,self}}(\mathbf{k},\mathbf{r},\nu)} = \frac{e^4 Z^2 N_{\mathrm{imp}}\gamma(E)^{1/2}}{2^{3/2}\pi m_{d,\nu}^{1/2}\varepsilon_{\mathrm{s}}^2 E_\beta^2} h(y\pm,s)\frac{1+2\alpha E}{1+4\gamma(E)/E_\beta}, \tag{4.63}$$

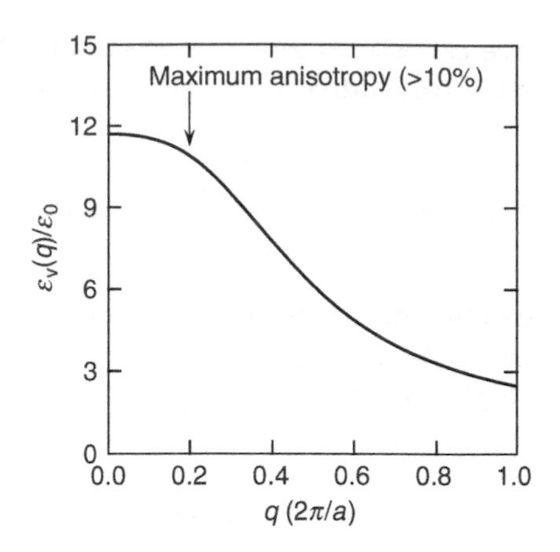

FIGURE 4.5
Isotropic static dielectric function, ε_v, for Si from Ref. [32] plotted as a function of the magnitude of the scattering wavevector q. Maximum anisotropy occurs at wavelengths of the order of the bond length, while an isotropic response is seen around the core ($q \to \infty$, where $\varepsilon \to \varepsilon_0$) and at large distances away from the impurity ($q \to 0$, where $\varepsilon \to \varepsilon_{sc}$). (Reprinted from Fischetti, M.V. and Laux, S.E., *J. Appl. Phys.*, 89, 1205, 2001. With permission.)

isotropy is occurs only about 10%. Thus, the overall correction to the BH model due to the anisotropy of the dielectric response of the valence electron is comparable to the corrections due to the differences in doping element, and so negligible in practice.

Additional corrections to the BH model abound in the literature. As we noted above, the BH model fails at large N_{imp} not only for the reasons considered by Mahan [21], but also because of the failure of the Born approximations. This is seen in an overestimation of the mobility. For example, in Si and at room temperature, this happens for $N_{imp} \approx n > 10^{17}$ cm^{-3}. More important, the Born approximation fails to distinguish between scattering with attractive (situation common in charge neutral samples) or repulsive (in the electron transport in the p-type base of bipolar transistors) scattering centers. Scattering of an electron from a positively charged center is enhanced with respect to the result of the Born approximation, because the electron wave function is more localized near the attractive potential. Vice versa, a negative center (such as an acceptor in p-type regions) pushes away the electron wave function and the scattering cross section is depressed. These corrections are sufficiently important in electronic devices to warrant some form of extension beyond the Born approximation (such as models based on partial-wave analysis [25,36]) in our transport models. Finally, a major assumption embraced almost universally (with notable exceptions) is that different scattering potentials due to different ions are added incoherently. This assumption is made in the spirit of the approximations on which the BTE is based. It is also justifiable whenever phase-destroying inelastic collisions, such as scattering with phonons, occur very frequently, so that the probability that a particle could remember its phase between successive impurity collisions is very small. Yet, it is not hard to envision situations where the coherent effect of multiple scattering events may yield noticeable effects in depressing the conductivity via destructive interference. Claims have been made that even in Si at 300 K, at large impurity concentrations multiple scattering should be accounted for in order to explain the observed ohmic mobility. The review paper by Chattopadhyay and Queisser [10], and the many publications cited there on heavy doping effects which have appeared in the literature should be consulted for details about these effects.

4.3.5 2D Model

As an example of impurity scattering in low-dimensional situations caused by quantum confinement, we consider the usual case of a Si channel at the Si–SiO_2 interface and consider scattering with ionized impurities located in the space charge region, or with charged defects located in the oxide, or at the interface, corresponding to ionized dopants, fixed oxide charges, and interface traps, respectively. We sketch here only the final equations, their derivation in terms of the Poisson Green's function of the system [37] having been given extensively in Refs. [38,39]. Using again the first Born approximation, considering again only first-order nonparabolic corrections, and ignoring phase-shift corrections, the Coulomb scattering rate for an electron of crystal momentum **K** in the (x, y)-plane of the interface located at $z=0$ and in subband ν can be written as [40]:

$$\frac{1}{\tau^{(r)}_{\mathrm{imp}}(\mathbf{K},\nu)} = \frac{e^4 N_r}{8\pi\hbar^3\varepsilon_r^2}\sum_{\mu} m_{d,\mu}\kappa_{\mu}[\Delta E_{\nu\mu}(\mathbf{K})]|\theta|[\Delta E_{\nu\mu}(\mathbf{K})]\int_0^{2\pi} \mathrm{d}\beta' \frac{\left\langle H^{(r)}_{\nu\mu}[Q(\beta')]\right\rangle_{\mathrm{screened}}}{G_{\mu}(\beta')Q(\beta')^2}, \tag{4.70}$$

where the index r runs over impurity, oxide, and interface charges, and where

$$G_{\mu}(\beta') = \frac{m_{d,\mu}}{m_{1,\mu}}\cos^2\beta' + \frac{m_{d,\mu}}{m_{2,\mu}}\sin^2\beta',$$

$$Q(\beta')^2 = K^2 - 2KK'(\beta')\cos(\beta-\beta') + K'(\beta')^2,$$

$$K'(\beta')^2 = \frac{2m_{d,\mu}\Delta E_{\nu\mu}(\mathbf{K})[1-\alpha\Delta E_{\nu\mu}(\mathbf{K}) - 2\alpha B_{\mu}]}{\hbar^2 G_{\mu}(\beta')},$$

$$\Delta E_{\nu\mu}(\mathbf{K}) = E_{\nu}(\mathbf{K}) - E_{\mu},$$

$$\beta = \arccos\left(\frac{K_1}{K}\right),$$

$$\kappa_{\mu}(E) = 1 - 2\alpha E - 2\alpha B_{\mu}$$

is a nonparabolic correction in terms of the parabolic subband bottom $E^{(0)}_{\mu}$ and of the expectation value $\langle V\rangle_{\mu}$ of the confining potential energy in the subband μ:

$$B_{\mu} = \left\langle E^{(0)}_{\mu} - V\right\rangle_{\mu} = E^{(0)}_{\mu} - \int_{-\infty}^{+\infty} \mathrm{d}z\zeta^*_{\mu}(z)V(z)\zeta_{\mu}(z),$$

which represents the average effective kinetic energy of an electron confined to the bottom of the subband ν. Finally, $m_{d,\mu}$, $m_{1,\mu}$, and $m_{2,\mu}$ are the DOS effective mass in subband μ and the masses along the 1- and 2-axis, respectively.

The unscreened form factor $H^{(r)}_{\mu\nu}$ is defined by

$$N_r\frac{e^2 H^{(r)}_{\nu\mu}(Q)}{\varepsilon_r^2 Q^2} = \int_{-\infty}^{+\infty} \mathrm{d}z_e N_r(z_e)\left|\varphi_{\nu\mu}(Q;\, z_e)\right|^2, \tag{4.71}$$

where the matrix elements of the Fourier–Bessel components of the potential due to charges at $z=z_e$ are defined below by Equations 4.84 through 4.87, so that, assuming a uniform distribution of charges,

$$H^{(r)}_{\nu\mu}(Q) = \int_0^{\infty} \mathrm{d}z \int_0^{\infty} \mathrm{d}z'\zeta_{\mu}(z)\zeta_{\mu}(z')I^{(r)}_{Q}(z,z')\zeta_{\nu}(z)\zeta_{\nu}(z'). \tag{4.72}$$

is the static screening parameter we have derived above, while the function $g_1(x)$ is given by Equation 4.58, in terms of the plasma dispersion function $\Phi(y)$, and

$$\beta_{s,\nu}^{(DH)} = \frac{e^2}{2\varepsilon_s} \frac{\partial n_s}{\partial E_F} \tag{4.93}$$

is the Debye–Hückel parameter for subband ν, a simple extension of Equation 4.53 in terms of the thermal population in the νth subband, n_ν, and of the thermal wavelength of electrons in subband ν, $L_\nu = [2\pi\hbar^2/(m_{c,\nu} k_B T)]^{1/2}$, where $m_{c,\nu}$ is the conductivity mass in the νth subband, and n_s is the electron sheet density.

4.3.7 Ballistic Limit and the DOS Bottleneck: The Role of Impurity Scattering

A reduction of the low-field mobility is the first obvious and well-studied effect of carrier-impurity scattering. We do not wish to discuss here the relevance of the low-field mobility to the performance of devices at the 10 nm scale. Some of these issues can be found in Refs. [42,43] for opposite views of the problem. It will suffice to say that the on-performance of short-channel device is largely insensitive to Coulomb scattering on any type, because of the weakening of the scattering rate as some inverse power of the kinetic energy of the carriers ($\sim E^{-1/2}$ in the parabolic-band approximation shown in Equation 4.27 or 4.62). Here, we would like to mention the recent controversy surrounding the possible role of scattering between electrons in the channel of a thin-insulator MOSFET and ionized impurities in the depleted region of the polycrystalline Si (poly-Si) gate [44]. The controversy is based on our fundamental inability to treat correctly dielectric screening in the highly inhomogeneous electron gas present in the depleted poly-Si gate [45]. The state of the art being represented by the approximate staircase (or piecewise uniform) approach recently followed by Ishihara [46].

A more important aspect of impurity scattering, this time an unexpected and somewhat surprising beneficial effect is related to the ability of momentum randomizing collisions—such as electron–impurity collisions—in the source of a device operating close to the ballistic-transport regime to supply carriers which can be injected in the channel. The discussion of this issue brings us back to the problem known as DOS bottleneck.

The search for faster devices able to sustain Moore's law is clearly focused on semiconductors with a small effective mass, since this translates both into a larger ballistic velocity v for a given kinetic energy E, $v \sim (2E/m^*)^{1/2}$, and into a reduced scattering rate. Indeed assuming at zeroth order that the scattering rate scales with the DOS [47], we would have $1/\tau \propto m^{*3/2}$ for parabolic bands. Therefore, III–V compound semiconductors with effective masses much smaller than that of Si have been always looked at with great interest. The problem with this picture is that what matters in determining the speed of a device in a circuit is the maximum current the device can sustain. In the ballistic limit, the current density, given by the product of the carrier density n (scaling as the DOS, $\propto m^{*3/2}$) and their velocity v, scales as $nv \propto m^*$. In words, in the ballistic limit a larger effective mass is desirable [48]. The problem actually dates back to 1991 [49]. In investigating the transconductance of nominally identical devices (MOSFETs), but built on different substrates (n-Si, p-Si, Ge, GaAs, InP, $In_{0.53}$ $Ga_{0.47}$ As) we found that—with the exception of In-containing alloys we discuss momentarily—as the channel length approached 50 nm the performance seemed unrelated to the low-field mobility (and so m^*) of the semiconductor employed. The explanation was two-fold. First, the DOS of face-centered cubic (fcc) semiconductors at an energy of roughly 0.5 eV above the band extremum (bottom of the conduction band or top of the valence band) is very similar, as the many satellite valleys contribute in roughly the same way (see Figure 4.6). Thus, GaAs does not appear to behave

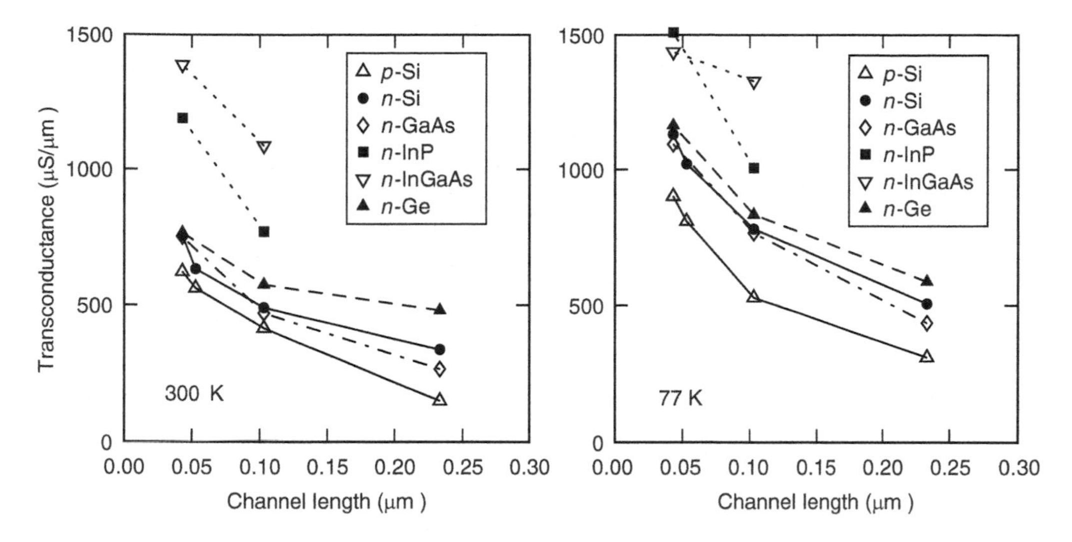

FIGURE 4.6
Large-signal transconductance g_m as a function of (metallurgical) channel length calculated in 1991 using the Monte Carlo program DAMOCLES for similarly designed bulk MOSFETs built on different semiconductors. Both at 300 K (*left*) and 77 K (*right*) only In-based devices exhibit improved performance when compared to the Si benchmark. (Reprinted from Fischetti, M.V. and Laux, S.E., *IEEE Trans. Electron. Dev.*, 38, 650, 1991. With permission.)

differently from Si as electron leave the low-mass bottom of the Γ valley and transfer to the heavier-mass satellite *L* (or even *X*) satellite valleys. Only semiconductors exhibiting high-energy satellite valleys (the In-containing materials in Figure 4.6) retain a noticeable advantage when compared to Si. Second, the low DOS of small-mass materials implies that a large swing of the Fermi level is required to change substantially the carrier density in the channel. This means that a large swing of gate voltage is required to change substantially the drain current of such a device and the resulting transconductance is depressed.

What was not considered at the time (when 50 nm-long devices were on the threshold of being considered science-fictional devices) was the behavior which devices would exhibit at even shorter dimensions approaching the ballistic limit. In this case, a second flavor of the DOS bottleneck is uncovered (as shown in more recent simulations performed all way to the 15 nm channel length [42]). As transport approaches the ballistic regime, calculations of the gate capacitance assuming an equilibrium distribution of carriers even close to the source/channel junction will necessarily result in optimistic overestimations. While streaming ballistically, carriers populate mainly states of wavevector **k** (or velocity) aligned along the transport direction (we call these states longitudinal **k** states). Momentum randomizing collisions are required to replenish transverse **k** states. In absence of collisions, the density drops significantly when compared to the equilibrium value. Fast, light-mass materials approach the ballistic regime at lengths longer than heavier-mass semiconductors. This results in the disappointing behavior also of In-based channels, as seen in Figure 4.7 for a variety of structures (bulk, silicon-on-insulator [SOI], and double gate [DG] FETs). This is indeed consistent with the idea that a large mass is actually beneficial as we approach the ballistic regime.

Coming finally to the role played by impurities, a third flavor of the DOS bottleneck, which may be labeled as the source starvation effect, emerges when we consider in more detail III–V-based devices at the 22 nm node [50]. A prototypical device is schematically shown in Figure 4.8. The channel is a 2.5 nm-thick $In_{0.53}$ $Ga_{0.47}$ As quantum well under a thin HfO_2 gate insulator layer and over a 2.5 nm-thick InP barrier layer on an $Al_{0.48}$ $In_{0.52}As$

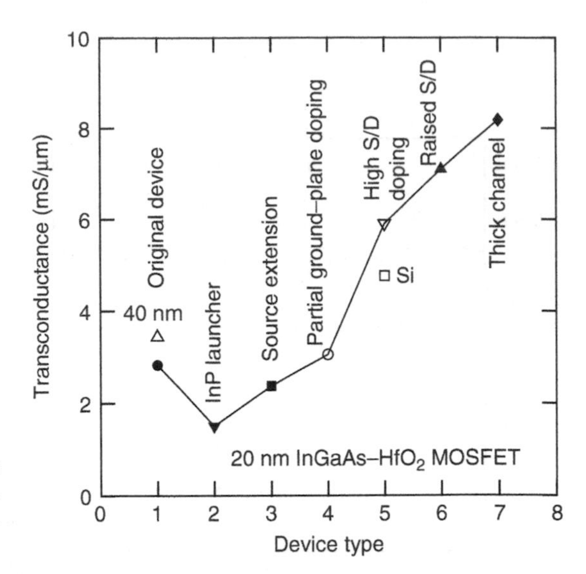

FIGURE 4.10
Calculated maximum large-signal transconductance extracted from the data shown in Figure 4.9. (Reprinted from Fischetti, M.V. et al., *IEDM Tech. Dig.*, 109, 2007. With permission.)

In conclusion, impurities seem to be highly desirable as we approach the ballistic regime. Models based on the assumption that carriers are injected from the source in equilibrium conditions [43] fail to capture this behavior.

4.4 Interface Roughness

Scattering with the roughness (at the atomic scale) of semiconductor–insulator is being blamed for the disappointing performance of devices at the 10 nm scale [51]. As devices shrink, so does the thickness of the gate insulator, t_{ox}. Since historically t_{ox} has been scaled at a rate much faster than other dimensions and supply bias (since it is the easiest knob to turn in order to increase the device performance), carriers are confined ever so harder against the semiconductor–insulator interface and scattering with roughness at this interface becomes a dominant effect.

A full atomistic (or ab initio) treatment of the atomic roughness requires complex calculations. Recently progress has been made in this area [52]. Evans and coworkers have employed density functional theory (DFT) at various levels of sophistication to investigate the structure of specific microscopic defects (suboxide bonds and oxygen protrusions)—which may be identified as roughness steps at the Si–SiO_2 interface—and Green's functions techniques to quantify their effect on the electron mobility, obtaining results in flattering agreement with experimental observations not only for the relaxed interface, but also for the strained-Si–SiO_2 interface which had emerged as a puzzle for the theory [53]. Here, wishing to avoid these complex techniques, we follow, instead and historically, a phenomenological model proposed originally by Ando [6] which—perhaps thank to the presence of adjustable parameters—has proven successful in explaining many features of this scattering process. Mainly, its dependence on the confining field (the scattering rate increasing roughly with the square of the surface field F_s). In its original form this model considered two major broad classes of effects. The first is related to the modification of the wave function $\zeta\nu(z)$ of carriers caused directly by the shift of the location of the interface at a position $\mathbf{r}$ on the plane of the interface. We refer to this shift as to a step at the interface. Microscopically it can be viewed as the presence of an atomic step between adjacent terraces on an imperfect interface. The matrix element for the scattering process is thus simply the energy-shift caused by a weaker or stronger confinement when the interface moves away or closer to the semiconductor. This effect is, sadly, the only effect usually considered in the literature, with minor exceptions. The second class of effects—which we show can play a role of the same magnitude—is related to the perturbation potential due to Coulomb effects. A shift of the charge density of the 2D electron gas which modifies the potential in the inversion layer at the step; the emergence of dipoles caused by both the shift of the polarization (or image) charge at the interface between the two dielectric with different dielectric constants and by the polarization of the vertical wall of the step induced by the inversion charges and their images. Clearly, the last effects can be safely ignored in the case of rough interfaces at heterojunctions between III and V compound semiconductors with very similar dielectric constants. This might explain the reasons why they have been ignored. Yet, in the interesting cases of the Si–SiO_2 (or even Si/high-k dielectrics) the potential due to these potentials can be very important.

In its original formulation, Ando's model was applicable only to a single interface, case appropriate to bulk MOS devices of the past. Scaled devices of the present technology fight short-channel effects mainly using thin semiconductor layers (as in SOI or DG devices). The presence of the bottom interface presents significant challenges from a theoretical perspective. Obviously, roughness at this interface yields an additional scattering mechanisms. But, even assuming a perfect bottom interface, the polarization (or image) charges at this interface modify the magnitude of the Coulomb terms proposed by Ando. In addition, the boundary conditions at the bottom interface affect also the magnitude of the modifications to the wave function caused by the step (the first class of processes in Ando's formulation). Thus, revisiting the problem and extending the model to the case of thin-body devices is of great interest. This is the major goal of this section which is based heavily on our recent work [20].

4.4.1 Effective Potential in Thin-Body Si

Our prototypical structure is an SOI MOSFET with two Si–SiO_2 interfaces at $z=0$ and t_s. We assume that the confinement direction is along the z-axis and a separable dispersion, so that for ideally smooth interfaces the wave function $\zeta\nu(z)$ and associated eigenenergy E_{ϖ} of the ϖth subband are obtained from the Schrödinger equation:

$$H_z \varsigma_\nu(z) = E_\nu \varsigma_\nu(z), \quad H_z = \frac{\hbar^2}{2} \frac{\partial}{\partial z} \frac{1}{m_z} \frac{\partial}{\partial z} + V(z) + \Phi(z), \tag{4.94}$$

where the mass along the quantization direction, m_z, is given by the Si effective mass m_z^{si} for $0<z<t_s$, and by the SiO_2 mass m_z^{ox} otherwise. The presence of the interfacial barriers is described by the barrier potential $\Phi(z) = V_{ox}[\Theta(-z) + \Theta(z-t_s)]$, where V_{ox} is the barrier height, and $\Theta(z)$ is the Heaviside step function. The effective potential energy $V(z)$ accounts for three terms: the Hartree potential, the image-potential (which requires some discussion in the case of thin layers as it differs from the usual case), and an exchange-correlation term:

$$V(z) = -e\phi(z) + V_{im}(z) + V_{xc}(z). \tag{4.95}$$

The Hartree term $-e\phi$ is obtained self-consistently from the Poisson equation. The image-potential arises from the discontinuity of the dielectric constants, ε_{si} and ε_{ox} in the Si layer and in the SiO_2 barriers, respectively, and it is given by [54]:

$$V_{im}(z) = \frac{e^2}{16\pi} \sum_{n=0}^{\infty} \frac{\tilde{\varepsilon}^{2n}}{\varepsilon_{si}} \left[\frac{\tilde{\varepsilon}}{z + nt_s} + \frac{\tilde{\varepsilon}}{t_s - z + nt_s} + \frac{2\tilde{\varepsilon}^2}{nt_s + t_s} \right], \tag{4.96}$$

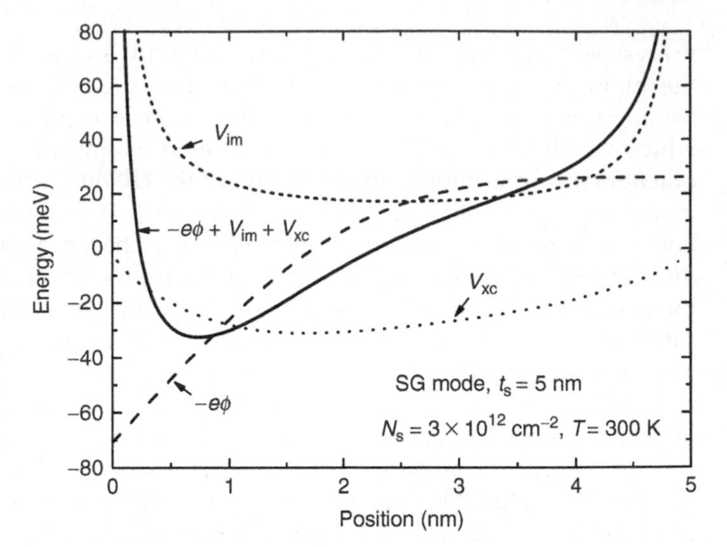

FIGURE 4.11
Effective potential energy V^0 inside the silicon body and its three components ($-e\phi$, V_{im}, and V_{xc}) when $t_s = 5$ nm, $N_s = 3 \times 10^{12}$ cm^{-2}, and $T = 300$ K. The device is operated in the SG mode. The reference of the electrostatic energy is the Fermi energy ($E_F = 0$ eV). (Reprinted from Jin, S., Fischetti, M.V., and Tang, T.-W., *IEEE Trans. Electron Dev.*, 54, 2191, 2007. With permission.)

where, in the usual notation, $\tilde{\varepsilon} = (\varepsilon_{si} - \varepsilon_{ox})/(\varepsilon_{si} + \varepsilon_{ox})$. Finally, for the exchange-correlation term, V_{xc}, here we employ the local-density approximation within the density functional formalism [55,56] in the form given by [57–60]:

$$V_{xc}(z) = -\frac{e^2}{8\pi\varepsilon_{si}a}\frac{2}{\pi\alpha r_s}\left[1 + 0.7734x\ln\left(1 + \frac{1}{x}\right)\right], \tag{4.97}$$

where
- $\alpha = (4/9\ \pi)^{1/3}$
- $r_s = a^{-1}[4\pi n(z)/3]^{-1/3}$
- $x = r_s/21$
- $n(z)$ is the electron volume density
- $a = 4\pi\varepsilon_{si}\hbar^2/(m^*e^2)$ is the effective Bohr radius

The anisotropy of the Si band structure close to the conduction band minima creates some concerns about the proper choice for the effective mass entering the expression for the effective Bohr radius. Lacking better justifications, here we choose the (average) conductivity effective mass $m^* = 3/(2m_t^{-1} + m_l^{-1})$, as suggested in Ref. [61].

These components of the effective potential energy and the resulting total potential are shown in Figure 4.11. Note that exchange-correlation effects lower the potential energy near the center of the well, while image-potential effects raise it close to the interfaces. The net result is a stronger confinement when image and exchange-correlation effects are accounted for compared to results obtained ignoring these terms.

4.4.2 Roughness as a Perturbation

Let us now move away from the ideal case of smooth interfaces and consider atomic roughness at the top interface. We describe a single atomic step at position $\mathbf{r}$ on the (x, y)-plane of the interface as a motion of the semiconductor–insulator boundary moves from $z = 0$ to $z = \Delta(\mathbf{r})$. Assuming that roughness is determined by random processes, it is safe to assume that the Fourier components $\Delta_{\mathbf{q}}$ of $\Delta_{\mathbf{r}}$ exhibit a power spectrum with zero mean characterized by the Gaussian model [6]:

$$\langle|\Delta_{\mathbf{q}}|\rangle^2 = \pi\Delta^2\Lambda^2 e^{-q^2\Lambda^2/4}, \tag{4.98}$$

or by an exponential model [62]:

$$\langle|\Delta_{\mathbf{q}}|\rangle^2 = \pi\Delta^2\Lambda^2(1 + q^2\Lambda^2/2)^{-3/2}, \tag{4.99}$$

depending on its autocorrelation function. The latter seems to better agree with the experimentally observed power spectrum [62] and will be employed here.

This model of the roughness changes directly the barrier potential to:

$$\Phi_{\mathbf{r}}(z) = V_{\mathrm{ox}}[\Theta(\Delta_{\mathbf{r}} - z) + \Theta(z - t_{\mathrm{s}})]. \tag{4.100}$$

This modification corresponds to the first class of processes discussed above. The second class is related to first-order modifications of the effective potential energy due to Coulomb effects associated with the distortion of the interface. In order to extract the leading-order modifications to the wave function (first class) and the leading-order Coulomb terms (second class), we consider the perturbed Schrödinger equation along the z direction:

$$H_{z\mathbf{r}}\zeta_{\nu\mathbf{r}}(z) = E_{\nu\mathbf{r}}\zeta_{\nu\mathbf{r}}(z). \tag{4.101}$$

Note that now the Hamiltonian $H_{z\mathbf{r}} = -\frac{\hbar^2}{2}\frac{\partial}{\partial z}\frac{1}{m_{z\mathbf{r}}}\frac{\partial}{\partial z} + V_{\mathbf{r}}(z) + \Phi_{\mathbf{r}}(z)$, the wave function $\zeta_{\nu\mathbf{r}}(z)$, and the energy level $E_{\nu\mathbf{r}}$ become $\mathbf{r}$-dependent. Here $m_{z\mathbf{r}} = m_z^{\mathrm{si}}$ if $\Delta_{\mathbf{r}} < z < t_s$, and $= m_z^{ox}$ otherwise, expressing the fluctuation of the effective mass. Retaining only terms of order $O(\Delta_{\mathbf{q}})$, we can express the $\mathbf{r}$-dependent quantities $V_{\mathbf{r}}(z)$, $\zeta\nu_{\mathbf{r}}(z)$, and $E_{\nu\mathbf{r}}$ in terms of their Fourier components:

$$V_{\mathbf{r}}(z) = V(z) + \sum_{\mathbf{q}} V_{\mathbf{q}}^1(z)\Delta_{\mathbf{q}}e^{i\mathbf{q}\cdot\mathbf{r}}, \tag{4.102}$$

$$\zeta_{\nu\mathbf{r}}(z) = \zeta_\nu(z) + \sum_{\mathbf{q}} \zeta_{\nu\mathbf{q}}^1(z)\Delta_{\mathbf{q}}e^{i\mathbf{q}\cdot\mathbf{r}}, \tag{4.103}$$

$$E_{\nu\mathbf{r}} = E_\nu + \sum_{\mathbf{q}} E_{\nu\mathbf{q}}^1\Delta_{\mathbf{q}}e^{i\mathbf{q}\cdot\mathbf{r}}. \tag{4.104}$$

The major task consists now in determining the unknown coefficients $V_{\mathbf{q}}^1(z)$, $\zeta_{\nu\mathbf{q}}^1(z)$, and $E_{\nu\mathbf{q}}^1$. Indeed the matrix element of the scattering potential is obtained from the difference

$$V_{\nu\mathbf{k}\nu'\mathbf{k}'}^{\mathrm{SR}} = H_{\nu\mathbf{k}\nu'\mathbf{k}'} - H_{\nu\mathbf{k}\nu'\mathbf{k}'}^0, \tag{4.105}$$

where

$$H_{\nu\mathbf{k}\nu'\mathbf{k}'}^0 = \int d\mathbf{r}\int dz\frac{e^{-i\mathbf{k}\cdot\mathbf{r}}}{\sqrt{A}}\zeta_\nu(z)H_{\mathbf{R}}^0\zeta_{\nu'}(z)\frac{e^{i\mathbf{k}'\cdot\mathbf{r}}}{\sqrt{A}}, \tag{4.106}$$

with

$$H_{\mathbf{R}}^0 = -\frac{\hbar}{2m_x}\frac{\partial^2}{\partial x^2} - \frac{\hbar^2}{2m_y}\frac{\partial^2}{\partial y^2} + H_z, \tag{4.107}$$

is the unperturbed Hamiltonian, while

$$H_{\nu\mathbf{k}\nu'\mathbf{k}'} = \int d\mathbf{r}\int dz\frac{e^{-i\mathbf{k}\cdot\mathbf{r}}}{\sqrt{A}}\zeta_{\nu\mathbf{r}}(z)H_{\mathbf{R}}\zeta_{\nu'\mathbf{r}}(z)\frac{e^{i\mathbf{k}'\cdot\mathbf{r}}}{\sqrt{A}}, \tag{4.108}$$

expression originally derived by Prange and Nee. In the limit of an infinitely high-potential barrier ($V_{ox} \to \infty$), one can show that (we omit the proof here, see one of the Appendices of Ref. [20]):

$$\lim_{V_{ox}\to\infty} \Gamma^{GPN}_{\nu\nu'} = \Gamma^{PN}_{\nu\nu'}, \tag{4.127}$$

where $\Gamma^{PN}_{\nu\nu'}$ is the well-known Prange–Nee term [3]:

$$\Gamma^{PN}_{\nu\nu'} = \frac{\hbar^2}{2m_z}\frac{\partial\zeta_\nu(0)}{\partial z}\frac{\partial\zeta_{\nu'}(0)}{\partial z}. \tag{4.128}$$

We see that the integral expression given by Equation 4.126 reduces to the well-known Prange–Nee expression, Equation 4.128, in the limit $V_{ox} \to \infty$. But in the general case of finite barrier heights, the two expressions $\Gamma^{GPN}_{\nu\nu'}$ and $\Gamma^{PN}_{\nu\nu'}$ differ and only the former provides the correct result. This difference is highlighted in Figure 4.12. Here, we show the diagonal matrix element Γ^{GPN}_{00} and Γ^{PN}_{00} in the case of finite barrier heights. Note that in this case the correct expression should follow dE_0/dt_s [65]. This is indeed the case for the new general expression Γ^{GPN}_{00}, but the conventional Prange–Nee term Γ^{PN}_{00} deviates from the correct behavior as $t_s \to 0$. This is due to the fact that the original Prange–Nee expression assumes that the energy of the ground state subband can be neglected compared to the barrier height, so that one can write the following approximation for the wave function inside the barrier [6,66]:

$$\zeta_\nu(z) \approx \zeta_\nu(0)\exp(z/L_\nu), \tag{4.129}$$

where $L_\nu = \hbar(2m_zV_{ox})^{-1/2}$. This eventually leads to the usual expression

$$V_{ox}\zeta_\nu(0)\zeta_{\nu'}(0) \approx \frac{\hbar^2}{2m_z}\frac{\partial\zeta_\nu(0)}{\partial z}\frac{\partial\zeta_{\nu'}(0)}{\partial z}. \tag{4.130}$$

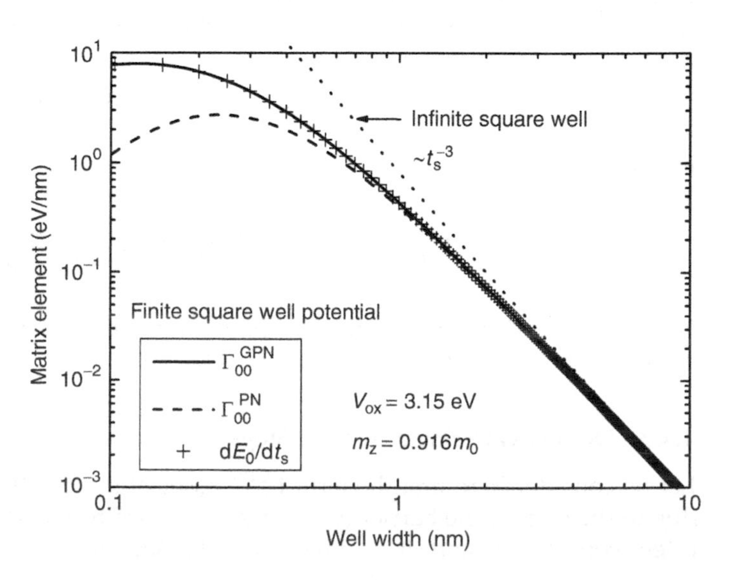

FIGURE 4.12
Comparison of the diagonal matrix element Γ^{GPN}_{00} and Γ^{PN}_{00} for the ground state of the finite square well potential. The exact solution dE_0/dt_s and the matrix element of the infinite square well potential are also shown. The potential barrier height is 3.15 eV and the effective mass is $0.916m_0$. (Reprinted from Jin, S., Fischetti, M.V., and Tang, T.-W., *IEEE Trans. Electron Dev.*, 54, 2191, 2007. With permission.)

More correctly, however, one should assume for L_ν:

$$L_\nu = \hbar[2m_z(V_{ox} - E_\nu)]^{-1/2}. \tag{4.131}$$

In so doing we see that the factor $V_{ox}\, L_\nu\, L_{\nu'}$ becomes larger than $\hbar^2/(2m_z)$, explaining why the usual Prange–Nee term gives a smaller matrix element.

Another point of view usually embraced to understand SR scattering in a quantum well is to consider the fluctuations of the subband energy as the body thickness fluctuates and look at these fluctuations as scattering agents. This point of view is already included in the generalized expression for Γ_{00}^{GPN}. Indeed the matrix element usually associated with thickness fluctuations is generally expressed as [5,67,68]:

$$V_{\nu\mathbf{k}\nu'\mathbf{k}'}^{FL} \approx \delta_{\nu\nu'} \frac{\partial E_\nu}{\partial t_s} \Delta_{\mathbf{q}}. \tag{4.132}$$

In the limit of a square well with infinitely high-potential barriers $\Gamma_{\nu\nu'}^{GPN}$ becomes:

$$\Gamma_{\nu\nu'}^{GPN} = \frac{\hbar^2\pi^2}{m_z} \frac{\nu\nu'}{t_s^3} = \left|\frac{\partial E_\nu}{\partial t_s}\right|^{1/2} \left|\frac{\partial E_{\nu'}}{\partial t_s}\right|^{1/2}, \tag{4.133}$$

also containing off-diagonal elements, which yields the well-known t_s^6 dependence of the mobility [5,68,69]. Our expression for Γ_{00}^{GPN} contain additional corrections to this term, corrections arising from the fact that the potential energy $V(z)$ is not constant (flat), but contains a hump especially noticeable at large carrier densities.

4.4.4 Scattering via Roughness-Induced Coulomb Terms

Let us now consider the second class of processes, those arising from the changes of the electrostatic potential (in turn due to modifications of the charge) due to the presence of an atomic step. These are accounted for by the term $\Gamma_{\nu\nu'}^{\mathbf{q}}$ in Equation 4.125. This scattering potential $V_{\mathbf{q}}^1(z)\Delta_{\mathbf{q}}$ results from three different terms: a term, $V_{\mathbf{q}}^n(z)$, due to the shift of the electron (free carrier) sheet density along the z direction as the interface goes through a step; a term, $V_{\mathbf{q}}^\sigma(z)$, due to the shift of the polarization charge; and a term, $V_{\mathbf{q}}^{im}(z)$, due to a change of the image-potential. Thus:

$$V_{\mathbf{q}}^1(z)\Delta_{\mathbf{q}} = \left[V_{\mathbf{q}}^n(z) + V_{\mathbf{q}}^\sigma(z) + V_{\mathbf{q}}^{im}(z)\right]\Delta_{\mathbf{q}}. \tag{4.134}$$

The first term is easily evaluated in terms of the Poisson Green's function appropriate to the geometry at hand. When the interface shifts by an amount $\Delta_{\mathbf{q}}$ the change of the sheet charge, $\delta n_{\mathbf{r}}(z) = n_{\mathbf{r}}(z) - n(z)$, can be approximated to first order as

$$\delta n_{\mathbf{r}}(z) \approx \frac{t_s}{t_s - \Delta_{\mathbf{r}}} n(\tilde{z}) - n(z) \approx \left[\left(\frac{z}{t_s} - 1\right)\frac{\partial n(z)}{\partial z} + \frac{n(z)}{t_s}\right]\Delta_{\mathbf{r}}. \tag{4.135}$$

For the time being we assume that the number of electrons per unit area remains fixed. This effect will be considered later when discussing screening effects. Then, the potential $V_{\mathbf{q}}^n$ due to the shift $\delta n_{\mathbf{r}}(z)$ is immediately given by

$$V_{\mathbf{q}}^n(z) = -\frac{e^2}{\varepsilon_{si}} \int_0^{t_s} dz' G_{\mathbf{q}}(z,z') \left[\left(\frac{z'}{t_s} - 1\right)\frac{\partial n(z')}{\partial z'} + \frac{n(z')}{t_s}\right], \tag{4.136}$$

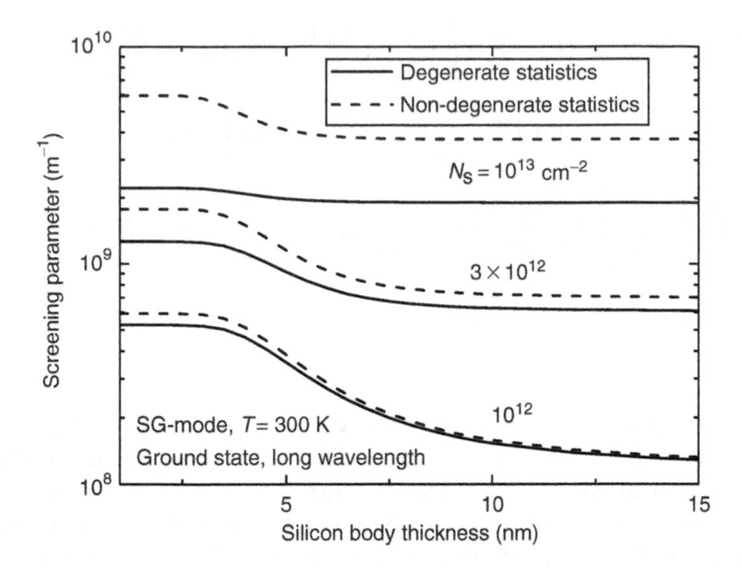

FIGURE 4.13
Screening parameter of the ground state in the long wavelength limit ($q \to 0$) as a function of the silicon body thickness for three different electron density ($N_s = 10^{12}, 3 \times 10^{12}, 10^{13}$ cm^{-2}). The device is assumed to be operated in the SG mode at 300 K. (Reprinted from Jin, S., Fischetti, M.V., and Tang, T.-W., *IEEE Trans. Electron. Dev.*, 54, 2191, 2007. With permission.)

where
$\rho_\mu = g_\mu m\ \mu/(\pi\hbar^2)$ is the DOS of μth subband
g_μ is the valley degeneracy factor
$m_\mu = \sqrt{m_x m_y}$ is the DOs effective mass
k_F is the Fermi wavevector
$C = 1$ if $q > 2k_F$ and $C = 0$ otherwise

Inserting Equation 4.149 into Equation 4.148, we obtain [73]:

$$\beta^T_{q;\mu\mu} = \frac{e^2\rho_\mu}{2\varepsilon_{si}}\left[1 + \tanh(y) - \int_0^z \mathrm{d}x \frac{\sqrt{1 - x/z}}{\cosh^2(x - y)}\right], \tag{4.150}$$

where
$y = \tilde{E}_F/(2k_B T)$
$z = \hbar^2 q^2/(16 m_{op} k_B T)$
$m_{op} = 2m_x m_y/(m_x + m_y)$ is the conductivity effective mass

The integration in this equation is performed numerically. Figure 4.13 shows $\beta^{300}_{0;00}$, the screening parameter of the ground state subband at 300 K in the long wavelength limit ($q \to 0$), as a function of the silicon body thickness for several electron sheet-densities, assuming that the device is operated in the single gate (SG) mode. We can see that the nondegenerate statistics significantly overestimates the screening parameter as the silicon thickness decreases even in the case of relatively moderate electron density ($N_s = 3 \times 10^{12}$ cm^{-2}).

4.4.6 Mobility Calculation

The low-field mobility for a 2D electron gas can be obtained from the linearization of the Boltzmann transport equation, as we have outlined above for the 3D case. The general expression for the xx component of the mobility tensor for νth subband can be derived from Equation 4.9 and has a form similar to the Kubo–Greenwood expression, Equation 4.12 [13,14,16]:

$$\mu_\nu^{xx} = \frac{e g_\nu}{2\hbar^2 \pi^2 k_B T N_\nu} \int_0^{2\pi} d\phi \int_{E_\nu}^{\infty} \frac{dEk}{|\partial E/\partial k_\mathbf{k}|} \left(\frac{\partial E}{\partial k_x}\right)_\mathbf{k}^2 \tau_{\nu\mathbf{k}}^x f_E(1 - f_E), \tag{4.151}$$

where $\mathbf{k}(\mathbf{E}, \phi)$ is the wave vector, $k = |\mathbf{k}|$, N_ν is the number of electrons in νth subband including the valley and spin degeneracy, $\tau_{\nu\mathbf{k}}^x$ is the anisotropic momentum relaxation time for the x component of the momentum in subband n, and $f_E = \{\exp[(E - E_F)/(k_B T) + 1]\}^{-1}$ is the Fermi–Dirac distribution function. In the parabolic, ellipsoidal band structure, this equation becomes:

$$\mu_\nu^{xx} = \frac{e\rho_\nu}{m_x k_B T N_\nu} \int_{E_\nu}^{\infty} dE(E - E_\nu) f_E(1 - f_E) \int_0^{2\pi} d\phi \frac{(m_\phi \cos\phi)^2}{\pi m_x m_y} \tau_{\nu\mathbf{k}}^x, \tag{4.152}$$

where $m_\phi = (\cos^2 \phi/m_x + \sin^2 \phi/m_y)^{-1}$ is the conductivity effective mass along the ϕ direction. The momentum relaxation time $\tau_{\nu\mathbf{k}}^x$ due to SR scattering can be written as [15,16]:

$$\frac{1}{\tau_{\nu\mathbf{k}}^x} = \frac{1}{2\pi\hbar^3} \sum_{\nu'} \Theta(E - E_{\nu'}) \int_0^{2\pi} d\phi' m_{\phi'} \left\langle |\tilde{V}_{\nu'\mathbf{k}'\nu\mathbf{k}}^{SR}|^2 \right\rangle \left[1 - \frac{\tilde{\tau}_{\nu'\mathbf{k}'}^x v_{\nu'\mathbf{k}'}^x}{\tilde{\tau}_{\nu\mathbf{k}}^x v_{\nu\mathbf{k}}^x}\right], \tag{4.153}$$

where

$\Theta(x)$ is, as before, the Heaviside step function
$v_{\nu\mathbf{k}}^x = \hbar k_\nu(E,\phi) \cos\phi/m_x$ is the velocity along the x-direction
$k_\nu(E,\phi) = \sqrt{2m_\phi(E - E_\nu)}/\hbar$ is the magnitude of the wave vector
$\tilde{\tau}_{\nu\mathbf{k}}^x$ is the momentum relaxation time due to all scattering processes

The dependence of $\tilde{\tau}_{\nu\mathbf{k}}^x$ inside the square bracket of Equation 4.153 complicates the calculation, and this dependence has usually been neglected so far. Although SR scattering is anisotropic, it is elastic and we simplify $\tilde{\tau}_{\nu'\mathbf{k}'}^x v_{\nu'\mathbf{k}'}^x / \tilde{\tau}_{\nu\mathbf{k}}^x v_{\nu\mathbf{k}}^x$ as $\approx v_{\nu'\mathbf{k}'}^x / v_{\nu\mathbf{k}}^x$ inside the square bracket. Finally, the total effective electron mobility is obtained from the weighted average of the electron mobility in each:

$$\mu_{\text{eff}}^{xx} = \frac{1}{N_s} \sum_\nu \mu_\nu^{xx} N_\nu, \tag{4.154}$$

where N_ν is the electron density of the νth subband.

4.4.7 Simulation Results and Discussion

We now present results regarding the SR-limited electron mobility of UTBSOI MOSFETs at room temperature ($T = 300$ K). We have assumed a substrate doping density of 2×10^{16} cm^{-3} and SR parameters $\Lambda = 1.3$ nm and $\Delta = 0.47$ nm in Equation 4.99 for both of the top and bottom interfaces. In most cases, we have considered SG SOI MOSFETs with a silicon dioxide insulator ($\varepsilon_{ox} = 3.9\varepsilon_0$), and a bottom oxide thick enough to render the

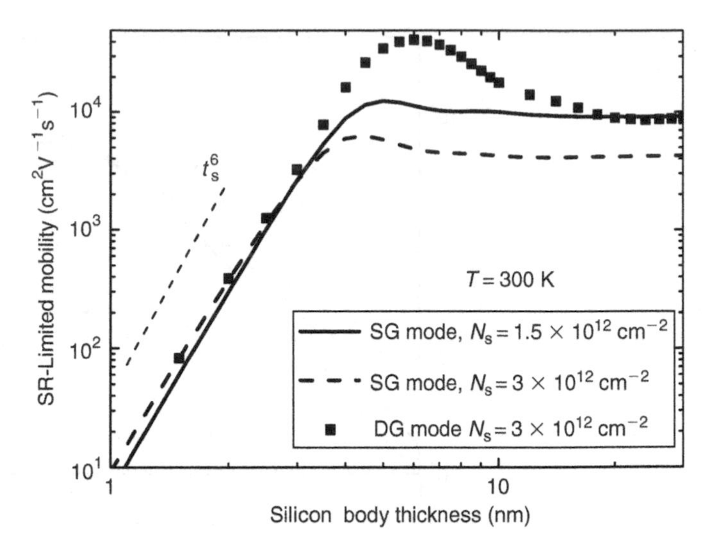

FIGURE 4.16
Comparison of the SR scattering-limited mobility of DG SOI MOSFETs and SG SOI MOSFETs as a function of silicon body thickness. (Reprinted from Jin, S., Fischetti, M.V., and Tang, T.-W., *IEEE Trans. Electron Dev.*, 54, 2191, 2007. With permission.)

We also show results regarding the influence of the effective field (electron sheet-density) on the SR-limited mobility, which shows well-known universal characteristics in bulk MOSFETs [75]. Here, we have defined the effective field as

$$E_{\text{eff}} = \frac{e}{N_s} \int_0^{t_s} \mathrm{d}z n(z) \left| \frac{\partial \phi(z)}{\partial z} \right|. \tag{4.159}$$

Figure 4.17 shows the SR-limited mobility as a function of E_{eff} for different silicon body thickness. As expected, the geometric confinement weakens the dependence of the SR-limited mobility on E_{eff} and the effective field ceases to be a good metric for

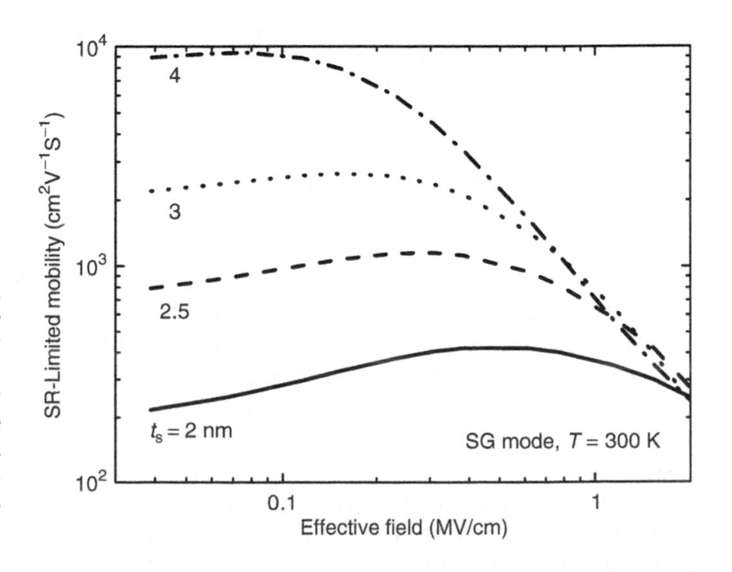

FIGURE 4.17
Calculated SR scattering-limited mobility as a function of effective field for four different silicon body thickness (t_s = 2, 2.5, 3, and 4 nm). The slight increase in the mobility in the low effective field region is due to the screening effect. (Reprinted from Jin, S., Fischetti, M.V., and Tang, T.-W., *IEEE Trans. Electron Dev.*, 54, 2191, 2007. With permission.)

4.4.6 Mobility Calculation

The low-field mobility for a 2D electron gas can be obtained from the linearization of the Boltzmann transport equation, as we have outlined above for the 3D case. The general expression for the xx component of the mobility tensor for νth subband can be derived from Equation 4.9 and has a form similar to the Kubo–Greenwood expression, Equation 4.12 [13,14,16]:

$$\mu_\nu^{xx} = \frac{eg_\nu}{2\hbar^2\pi^2 k_B T N_\nu} \int_0^{2\pi} d\phi \int_{E_\nu}^{\infty} \frac{dEk}{|\partial E/\partial k_{\mathbf{k}}|} \left(\frac{\partial E}{\partial k_x}\right)_{\mathbf{k}}^2 \tau_{\nu\mathbf{k}}^x f_E(1 - f_E), \tag{4.151}$$

where $\mathbf{k}(\mathbf{E}, \phi)$ is the wave vector, $k = |\mathbf{k}|$, N_ν is the number of electrons in νth subband including the valley and spin degeneracy, $\tau_{\nu\mathbf{k}}^x$ is the anisotropic momentum relaxation time for the x component of the momentum in subband n, and $f_E = \{\exp[(E - E_F)/(k_B T) + 1]\}^{-1}$ is the Fermi–Dirac distribution function. In the parabolic, ellipsoidal band structure, this equation becomes:

$$\mu_\nu^{xx} = \frac{e\rho_\nu}{m_x k_B T N_\nu} \int_{E_\nu}^{\infty} dE(E - E_\nu) f_E(1 - f_E) \int_0^{2\pi} d\phi \frac{(m_\phi \cos\phi)^2}{\pi m_x m_\nu} \tau_{\nu\mathbf{k}}^x, \tag{4.152}$$

where $m_\phi = (\cos^2 \phi/m_x + \sin^2 \phi/m_y)^{-1}$ is the conductivity effective mass along the ϕ direction. The momentum relaxation time $\tau_{\nu\mathbf{k}}^x$ due to SR scattering can be written as [15,16]:

$$\frac{1}{\tau_{\nu\mathbf{k}}^x} = \frac{1}{2\pi\hbar^3} \sum_{\nu'} \Theta(E - E_{\nu'}) \int_0^{2\pi} d\phi' m_{\phi'} \left\langle \left|\tilde{V}_{\nu'\mathbf{k}'\nu\mathbf{k}}^{SR}\right|^2 \right\rangle \left[1 - \frac{\tilde{\tau}_{\nu'\mathbf{k}'}^x v_{\nu'\mathbf{k}'}^x}{\tilde{\tau}_{\nu\mathbf{k}}^x v_{\nu\mathbf{k}}^x}\right], \tag{4.153}$$

where

- $\Theta(x)$ is, as before, the Heaviside step function
- $v_{\nu\mathbf{k}}^x = \hbar k_\nu(E,\phi) \cos\phi/m_x$ is the velocity along the x-direction
- $k_\nu(E,\phi) = \sqrt{2m_\phi(E - E_\nu)}/\hbar$ is the magnitude of the wave vector
- $\tilde{\tau}_{\nu\mathbf{k}}^x$ is the momentum relaxation time due to all scattering processes

The dependence of $\tilde{\tau}_{\nu\mathbf{k}}^x$ inside the square bracket of Equation 4.153 complicates the calculation, and this dependence has usually been neglected so far. Although SR scattering is anisotropic, it is elastic and we simplify $\tilde{\tau}_{\nu'\mathbf{k}'}^x v_{\nu'\mathbf{k}'}^x / \tilde{\tau}_{\nu\mathbf{k}}^x v_{\nu\mathbf{k}}^x$ as $\approx v_{\nu'\mathbf{k}'}^x / v_{\nu\mathbf{k}}^x$ inside the square bracket. Finally, the total effective electron mobility is obtained from the weighted average of the electron mobility in each:

$$\mu_{\text{eff}}^{xx} = \frac{1}{N_S} \sum_\nu \mu_\nu^{xx} N_\nu, \tag{4.154}$$

where N_ν is the electron density of the νth subband.

4.4.7 Simulation Results and Discussion

We now present results regarding the SR-limited electron mobility of UTBSOI MOSFETs at room temperature ($T = 300$ K). We have assumed a substrate doping density of 2×10^{16} cm^{-3} and SR parameters $\Lambda = 1.3$ nm and $\Delta = 0.47$ nm in Equation 4.99 for both of the top and bottom interfaces. In most cases, we have considered SG SOI MOSFETs with a silicon dioxide insulator ($\varepsilon_{ox} = 3.9\varepsilon_0$), and a bottom oxide thick enough to render the

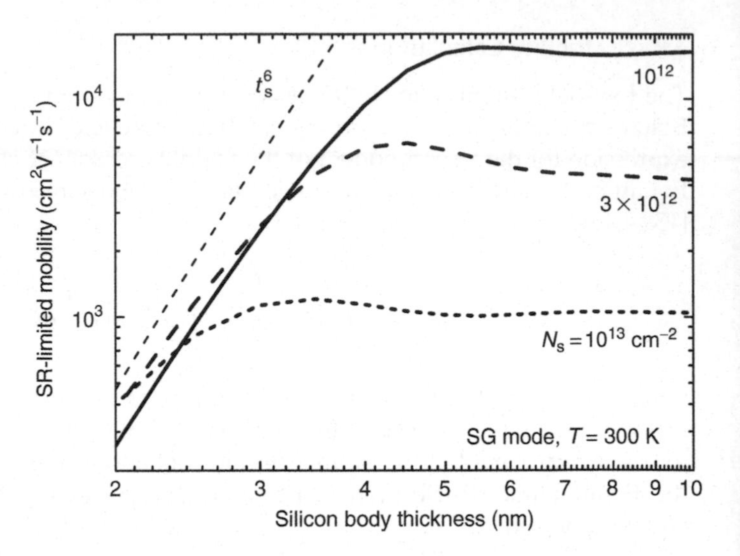

FIGURE 4.14
Calculated SR scattering-limited mobility as a function of silicon body thickness for three different electron densities ($N_s = 10^{12}$, 3×10^{12}, and 10^{13} cm^{-2}). The values of the effective field corresponding to these three electron densities are approximately 0.077–0.079, 0.23–0.24, and 0.77 MV/cm, respectively. (Reprinted from Jin, S., Fischetti, M.V., and Tang, T.-W., *IEEE Trans. Electron Dev.*, 54, 2191, 2007. With permission.)

electric field within it negligible. Here, we also assume an infinite potential barrier height ($V_{ox} \to \infty$). Although this assumption will underestimate the mobility when $t_s \to 0$, the results still illustrate the qualitatively correct behavior of the SR-limited mobility.

We present first results regarding the influence of the silicon body thickness on the SR-limited mobility. Figure 4.14 shows the calculated SR-limited mobility as a function of silicon body thickness for three different values of the electron density ($N_s = 10^{12}$, 3×10^{12}, and 10^{13} cm^{-2}). At all electron densities the SR-limited mobility exhibits a strong thickness dependence as $t_s \to 0$, but this dependence weakens, converging to the bulk mobility, when $t_s \to \infty$. In the region between these two limits the mobility exhibits a small maximum. Note also that in the first limit of $t_s \to 0$, the SR-limited electron mobility shows the expected t_s^6-like dependence due to thickness fluctuations, more evident at the small electron density of $N_s = 10^{12}$ cm^{-2}. For increasing electron densities the thickness dependence on the SR-limited mobility weakens, as the matrix element is now dominated by the fluctuations of the potential energy. The small peak of the mobility arises from two competing effects: the already mentioned mobility degradation due to the thickness fluctuation as $t_s \to 0$, and the mobility degradation due to the increasing occupation of the higher-mobility ground state unprimed valley at the expense of the low-mobility primed valleys. In Figure 4.15, we also plot the individual contribution of the top and bottom interfaces on the SR-limited mobility for the case of $N_s = 3 \times 10^{12}$ cm^{-2}. We have studied the cases of correlated, anticorrelated, and uncorrelated roughness at both interfaces. This is achieved by considering the expressions for $\Gamma_{\nu\nu'}^{(GPN)}$ (given by Equation 4.136), and the matrix elements corresponding to the potentials $V_{\mathbf{q}}^{n}$, $V_{\mathbf{q}}^{\sigma}$, and $V_{\mathbf{q}}^{\rm im}$, given by Equations 4.136, 4.141, and 4.142, respectively, for scattering at the bottom interface. These will be given by expressions above quite similar to those given above, but replacing the factors $(1 - z/t_s)$ with z/t_s and accounting for a few sign-changes.

To be explicit, the terms reflecting scattering with roughness at the bottom interface will be given by

$$\Gamma_{\nu\nu'}^{\rm GPN} = \frac{\hbar^2}{t_s}\int {\rm d}z \frac{\partial}{\partial z}\left[\frac{1}{m_2}\frac{\partial \zeta_{\nu'}(z)}{\partial z}\right] + \int {\rm d}z \zeta_\nu(z)\frac{\partial V(z)}{\partial z}\frac{z}{t_s}\zeta_{\nu'}(z)$$
$$+ (E_\nu - E_{\nu'})\int {\rm d}z \zeta_\nu(z)\frac{z}{t_s}\frac{\partial \zeta_{\nu'}(z)}{\partial z}, \tag{4.155}$$

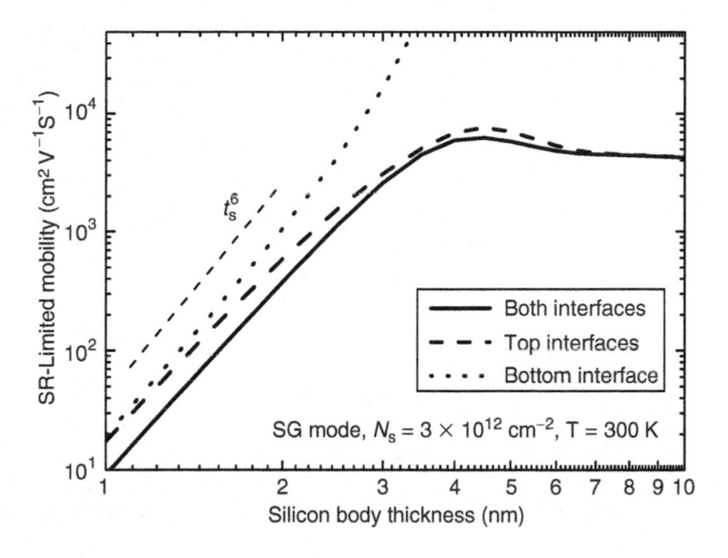

FIGURE 4.15
Calculated SR-limited mobility due to top, bottom, and both interfaces as a function of silicon body thickness for $N_s = 3 \times 10^{12}$ cm^{-2}. The contribution of the bottom interface on the SR-limited mobility is appreciable only for t_s smaller than 6 nm. We also show the mobility when the roughness at the top and bottom interfaces are fully correlated (the positions of the two interfaces fluctuate in the same direction, crosses), or anticorrelated (the positions fluctuate in opposite directions, triangles. Note that in the latter case thickness fluctuations vanish. (Reprinted from Jin, S., Fischetti, M.V., and Tang, T.-W., *IEEE Trans. Electron Dev.*, 54, 2191, 2007. With permission.)

$$V_{\mathbf{q}}^{n}(z) = \frac{e^2}{\varepsilon_{\mathrm{si}}} \int \mathrm{d}z' G_{\mathbf{q}}(z,z') \left[\frac{z'}{t_s} \frac{\partial n(z')}{\partial z'} + \frac{n(z')}{t_s} \right], \tag{4.156}$$

$$V_{\mathbf{q}}^{\sigma} = -e\tilde{\varepsilon} \frac{\partial \phi(t_s^-)}{\partial z} \frac{\mathrm{e}^{qz-qt_s} + \tilde{\varepsilon}\mathrm{e}^{-qz-qt_s}}{-\tilde{\varepsilon}^2 \mathrm{e}^{-2qt_s}}, \tag{4.157}$$

where $-\partial\phi(t_s^-)/\partial z$ is the electric field at the bottom interface, and, finally:

$$V_{\mathbf{q}}^{\mathrm{im}}(z) = \frac{e^2\tilde{\varepsilon}}{8\pi^3} \int \frac{\mathbf{dk}}{\varepsilon_{\mathrm{si}} + \varepsilon_{\mathrm{ox}}} \left[\frac{\varepsilon_{\mathrm{ox}}}{\varepsilon_{\mathrm{si}}} + \frac{\mathbf{k} \cdot \mathbf{k}}{kk'} \right] \left[\frac{\mathrm{e}^{kz-kt_s} + \tilde{\varepsilon}\mathrm{e}^{-kz-kt_s}}{1 - \tilde{\varepsilon}^2 \mathrm{e}^{-2kt_s}} \right] \left[\frac{\mathrm{e}^{k'z-k't_s} + \tilde{\varepsilon}\mathrm{e}^{-k'z-k't_s}}{1 - \tilde{\varepsilon}^2 \mathrm{e}^{-2k't_s}} \right]. \tag{4.158}$$

Denoting by $V_{\mathbf{q}}^{(\mathrm{top})}$ and $V_{\mathbf{q}}^{(\mathrm{bottom})}$ the Fourier components of the scattering potential for the top and bottom interfaces, respectively, in the evaluation of the momentum relaxation rate, we then consider either $|V_{\mathbf{q}}^{(\mathrm{top})} + V_{\mathbf{q}}^{(\mathrm{bottom})}|^2$ for correlated roughness, $|V_{\mathbf{q}}^{(\mathrm{top})} + V_{\mathbf{q}}^{(\mathrm{bottom})}|^2$ for anticorrelated roughness, or, finally, $|V_{\mathbf{q}}^{(\mathrm{top})} + V_{\mathbf{q}}^{(\mathrm{bottom})}|^2$ for uncorrelated roughness. As expected, the contribution of the bottom interface on the SR-limited mobility is appreciable only when t_s is smaller than 6 nm and correlated roughness results in a minimal mobility degradation since thickness fluctuations are absent in this case.

In Figure 4.16, we compare the SR-limited electron mobility for SG and DG SOI MOSFETs. In the $t_s \to 0$ limit both devices exhibit very similar mobilities, while the SR-limited mobility in DG SOI MOSFETs approaches the mobility of SG SOI MOSFETs at one-half the electron density in the $t_s \to \infty$ limit. This is expected from the consideration that in DG SOI MOSFETs charge transport occurs in two independent channels if t_s is sufficiently large. Note also that, compared to the case of SG SOI MOSFETs, the mobility enhancement of DG SOI MOSFETs in the intermediate range of t_s is quite noticeable, thanks to the volume inversion, characteristic which has been observed experimentally [74].

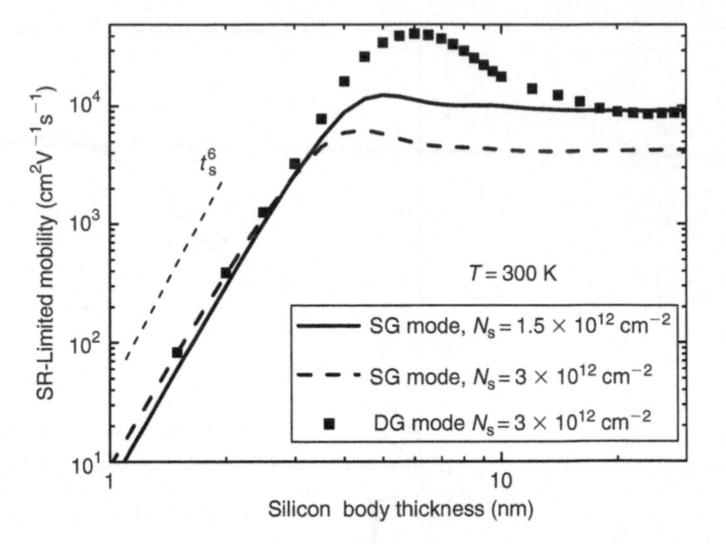

FIGURE 4.16
Comparison of the SR scattering-limited mobility of DG SOI MOSFETs and SG SOI MOSFETs as a function of silicon body thickness. (Reprinted from Jin, S., Fischetti, M.V., and Tang, T.-W., *IEEE Trans. Electron Dev.*, 54, 2191, 2007. With permission.)

We also show results regarding the influence of the effective field (electron sheet-density) on the SR-limited mobility, which shows well-known universal characteristics in bulk MOSFETs [75]. Here, we have defined the effective field as

$$E_{\text{eff}} = \frac{e}{N_s} \int_0^{t_s} \mathrm{d}z n(z) \left| \frac{\partial \phi(z)}{\partial z} \right|. \tag{4.159}$$

Figure 4.17 shows the SR-limited mobility as a function of E_{eff} for different silicon body thickness. As expected, the geometric confinement weakens the dependence of the SR-limited mobility on E_{eff} and the effective field ceases to be a good metric for

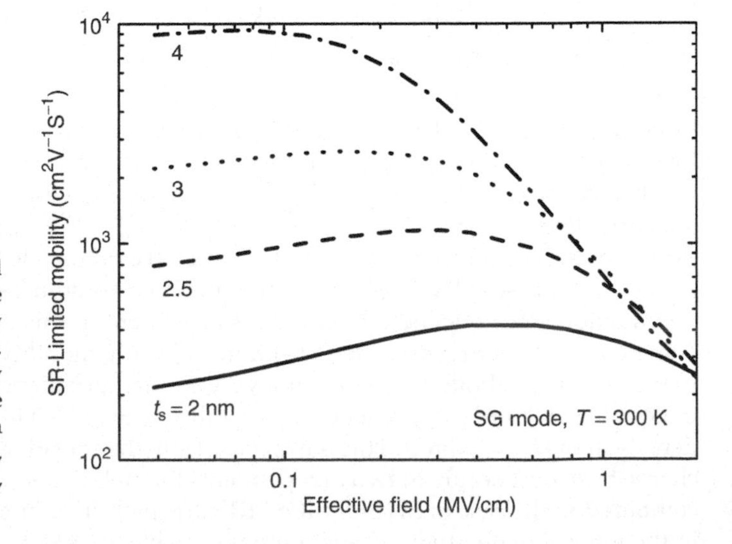

FIGURE 4.17
Calculated SR scattering-limited mobility as a function of effective field for four different silicon body thickness ($t_s = 2$, 2.5, 3, and 4 nm). The slight increase in the mobility in the low effective field region is due to the screening effect. (Reprinted from Jin, S., Fischetti, M.V., and Tang, T.-W., *IEEE Trans. Electron Dev.*, 54, 2191, 2007. With permission.)

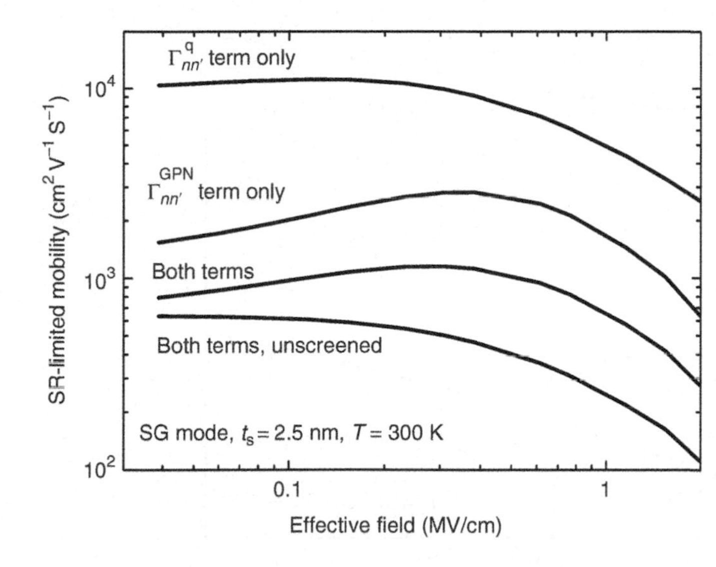

FIGURE 4.18
Calculated SR scattering-limited mobility due to the potential fluctuation term $\Gamma^{q}_{\nu\nu'}$, generalized Prange–Nee term $\Gamma^{GPN}_{\nu\nu'}$, and both terms as a function of effective field for $t_s = 2.5$ nm. Also shown is the SR scattering limited mobility when the screening is not taken into account. (Reprinted from Jin, S., Fischetti, M.V., and Tang, T.-W., *IEEE Trans. Electron Dev.*, 54, 2191, 2007. With permission.)

the characterization of the mobility in UTBSOI MOSFETs [66]. We also find a slight increase of the mobility in the low E_{eff} region for extremely thin devices. In Figure 4.18, we plot separately the SR-limited mobility due to the (generalized) Prange–Nee term $\Gamma^{GPN}_{\nu\nu'}$, to the potential (Coulomb) fluctuation term $\Gamma^{q}_{\nu\nu'}$, and to both terms as a function of effective field for $t_s = 2.5$ nm. The influence of the potential-fluctuation term on the mobility is smaller than the effect of the Prange–Nee term, but the potential (Coulomb) fluctuation term is never negligible, since it may reduce the mobility by about 50% with respect to the mobility limited by the Prange–Nee alone. Also shown is the unscreened SR-limited mobility, which, as a consequence of dielectric screening, decreases monotonically with increasing effective field.

Finally, interesting results have been obtained by considering the influence of the dielectric constant of the insulating material on the SR-limited mobility. This issue has not been addressed before, perhaps as a result of having neglected Coulomb terms (thee image-potential and the potential-fluctuation terms). Intuitively, however, we expect that the SR-limited electron mobility will be affected by the dielectric constant of the insulating material since the difference between the semiconductor and insulator dielectric constants appear in all Coulomb terms. In Equations 4.96, 4.141, and 4.142, we see that the image-potential V^{im} and the scattering potentials V^{σ}_{q} and V^{im}_{q} depend on $\tilde{\varepsilon} = (\varepsilon_{si} - \varepsilon_{ox})/(\varepsilon_{si} + \varepsilon_{ox})$. This parameter can span the range 1 to -1 as ε_{ox} is varied from 0 to ∞, with an obvious sign-change occurring when ε_{ox} equals ε_{si}. As ε_{ox} is increased from 0 to ε_{si}, the repulsive image potential decreases and finally disappear when $\varepsilon_{ox} = \varepsilon_{si}$. When $\varepsilon_{ox} > \varepsilon_{si}$, on the contrary, the image potential becomes attractive, and the magnitude increases with increasing ε_{ox}. Similarly, the scattering potential V^{σ}_{q} and V^{im}_{q} decrease with ε_{ox} increasing from 0 to ε_{si}, and it vanishes when $\varepsilon_{ox} = \varepsilon_{si}$. For even higher values of ε_{ox}, the sign of this term changes, thus canceling (i.e., dielectric screening due to polarization charges) all other scattering potentials terms. As a result of these considerations we expect that the SR-limited mobility will increase with increasing ε_{ox}. This expectation is confirmed by the results shown in Figure 4.19. We see that indeed the SR-limited mobility increases with increasing dielectric constant ε_{ox}. In Figure 4.20, we show the thickness dependence of the SR-limited mobility for two different dielectric materials (SiO_2 and HfO_2). It should be

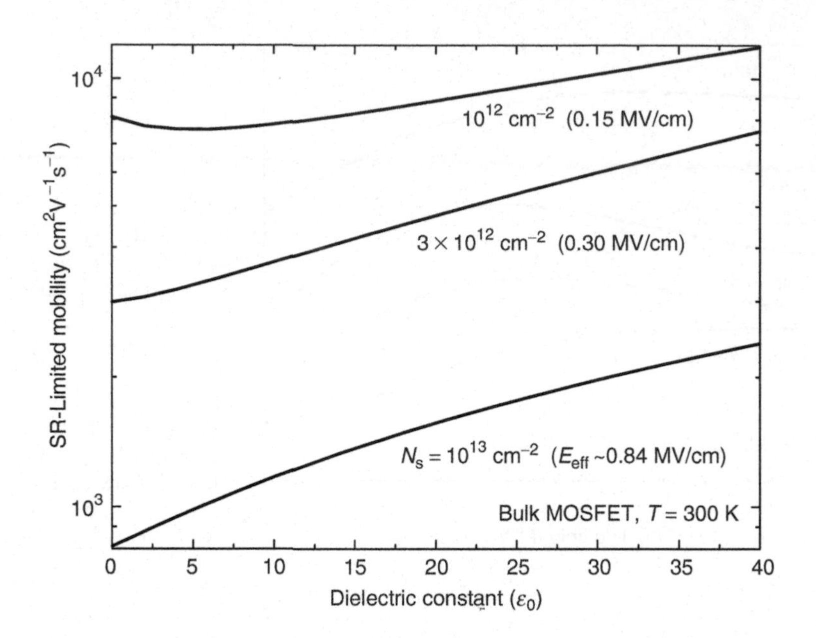

FIGURE 4.19
Calculated SR scattering-limited mobility of bulk MOSFETs as a function of the dielectric constant of the insulator (ε_{ox}) for three different electron densities ($N_s = 10^{12}$, 3×10^{12}, and 10^{13} cm^{-2}). (Reprinted from Jin, S., Fischetti, M.V., and Tang, T.-W., *IEEE Trans. Electron Dev.*, 54, 2191, 2007. With permission.)

noted in this discussion that this dependence of the mobility on the dielectric constant of the insulator should be expected only when the SR parameters remain fixed. Thus, if we can maintain the same quality of the interface, high-κ materials may exhibit higher SR-limited mobility then SiO_2. Experimental verification of this trend is not easily obtained, since the presence of additional scattering mechanisms present in high-κ-based MOSFETs [15] and the difficulty of obtaining smooth interfaces may mask the effect.

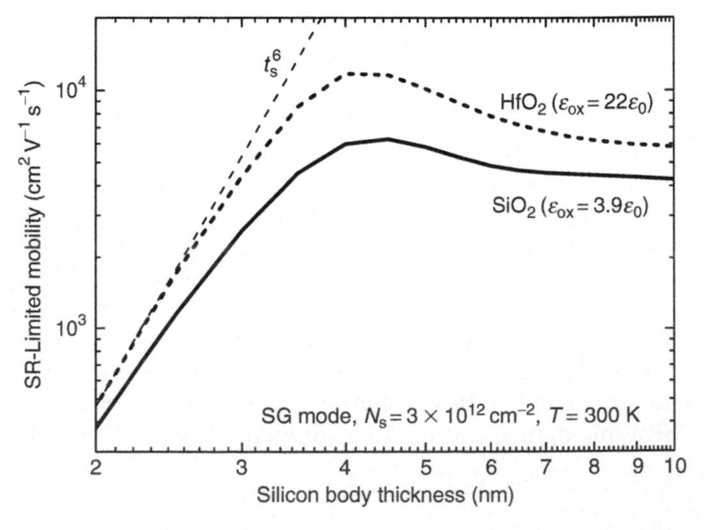

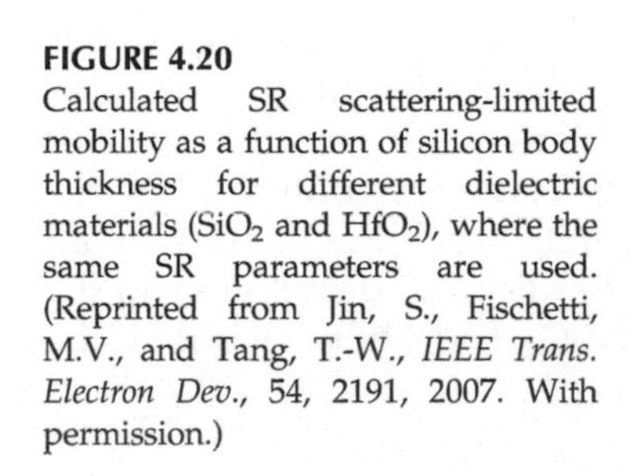

FIGURE 4.20
Calculated SR scattering-limited mobility as a function of silicon body thickness for different dielectric materials (SiO_2 and HfO_2), where the same SR parameters are used. (Reprinted from Jin, S., Fischetti, M.V., and Tang, T.-W., *IEEE Trans. Electron Dev.*, 54, 2191, 2007. With permission.)

4.5 Conclusions

We have considered two major types of defects in semiconductor devices (dopants and interface roughness) and have looked at some of the effects they have on electronic transport. Regarding the effect of dopants, bypassing the obvious effect that ionized impurities have on depressing the electron mobility, we have shown that impurity scattering in the source of devices at the 10 nm scale may be beneficial by allowing more electrons to be injected into the quasiballistic channel. This prevents the source from being starved, that is, from being depleted of carriers occupying those velocity states (longitudinal **k** states) properly oriented to allow injection into the channel. This is the aspect of the DOS bottleneck most evident in devices operating close to the ballistic transport regime. Regarding the effect of roughness at semiconductor/insulator interfaces, we have considered the case of thin Si bodies (of interest in today's UTBSOI and DG FETs) and extended the well-known Ando's model to this case. In particular, we have taken into account all scattering terms, including the often ignored Coulomb terms due to distortion of the inversion, polarization and image charges. The new model reduces to the usual (Ando's) model in the limit of infinite barrier heights (for the Prange–Nee term) and body thickness (for the Coulomb terms). We have studied the effect of body thickness, of scattering at both interfaces, and of different insulator dielectric constants. We have seen that the mobility drops significantly for small body thickness, that the Coulomb terms are significant and cannot be neglected, that the degree of correlation of the roughness at both interfaces matters significantly, and finally, that insulators with higher dielectric constants contribute to improve the SR-limited mobility (for a fixed roughness).

Acknowledgments

We would like to thank P. Asbeck, Y. Taur, L. Wang, B. Yu, and C. Sachs for discussion on the source starvation issue outlined in Section 4.3.7. This work has been partially supported by the Semiconductor Research Corporation. An SUR Grant from IBM Corporation is also acknowledged.

References

1. Conwell, E.M. and Weisskopf, V.F., Theory of impurity scattering in semiconductors, *Phys. Rev.*, 77, 388, 1950.
2. Brooks, H., Scattering by ionized impurities in semiconductors, *Phys. Rev.*, 83, 879, 1951.
3. Prange, R.E. and Nee, T.W., Quantum spectroscopy of the low-field oscillations in surface impedance, *Phys. Rev.*, 168, 779, 1968.
4. Saitoh, M., Warm electrons on the liquid ^{4}He surface, *J. Phys. Soc. Japan*, 42, 201, 1977.
5. Sakaki, H. et al., Interface roughness scattering in GaAs/AlAs quantum wells, *Appl. Phys. Lett.*, 51, 1934, 1987.
6. Ando, T., Fowler, A.B., and Stern, F., Electronic properties of two-dimensional systems, *Rev. Mod. Phys.*, 54, 437, 1982.
7. Jacoboni, C. and Reggiani, L., The Monte Carlo method for the solution of charge transport in semiconductors with application to covalent materials, *Rev. Mod. Phys.*, 55, 345, 1983.
8. Fischetti, M.V. and Laux, S.E., Monte Carlo analysis of electron transport in small semiconductor devices including band-structure and space-charge effects, *Phys. Rev. B*, 38, 9721, 1988.

67. Esseni, D. et al., Physically based modeling of low field electron mobility in ultrathin single-and double-gate SOI n-MOSFETs, *IEEE Trans. Electron Dev.*, 50, 2445, 2003.
68. Uchida, K. and Takagi, S., Carrier scattering induced by thickness fluctuation of silicon-on-insulator film in ultrathin-body metal-oxide-semiconductor field-effect transistors, *Appl. Phys. Lett.*, 82, 2916, 2003.
69. Gold, A., Electronic transport properties of a two-dimensional electron gas in silicon quantum-well structures at low temperature, *Phys. Rev. B*, 35, 723, 1987.
70. Jackson, J.D., *Classical Electrodynamics*, Wiley, New York, 1999.
71. Maldague, P.F., Many-body corrections to the polarizability of the two-dimensional electron gas, *Surf. Sci.*, 73, 296, 1978.
72. Stern, F., Polarizability of a two-dimensional electron gas, *Phys. Rev. Lett.*, 18, 546, 1967.
73. Gold, A. and Dolgopolov, V.T., Temperature dependence of the conductivity for the two-dimensional electron gas: Analytical results for low temperatures, *Phys. Rev. B*, 33, 1076, 1986.
74. Prunnila, M. et al., Gate bias symmetry dependence of electron mobility and prospect of velocity modulation in double-gate silicon-on-insulator transistors, *Appl. Phys. Lett.*, 85, 5442, 2004.
75. Takagi, S. et al., On the universality of inversion layer mobility in Si MOSFETs: Part I-effects of substrate impurity concentration, *IEEE Trans. Electron Dev.*, 41, 2357, 1994.

5

Electrical Characterization of Defects in Gate Dielectrics

Dieter K. Schroder

CONTENTS

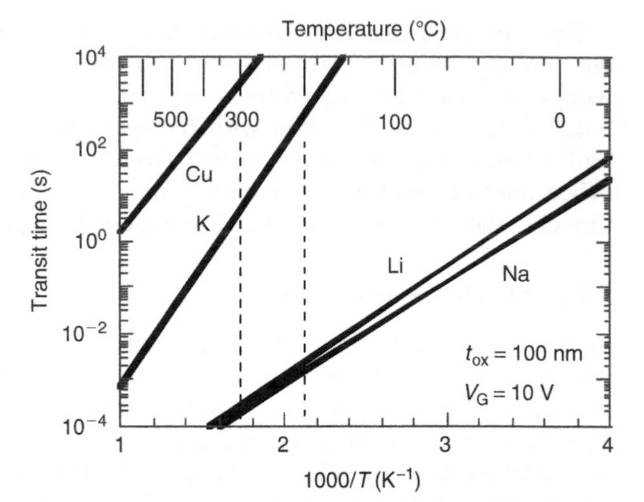

FIGURE 5.1
Transit times for Na, Li, K, and Cu for an oxide electric field of 10^6 V/cm. (Reprinted from Schroder, D.K., *Semiconductor Material and Device Characterization*, 3rd edn., Wiley-Interscience, New York, 2006. With permission.)

5.2.4 E′ Center

The E′ center, usually observed in irradiated MOS devices, consists of two Si atoms joined by a weak, strained Si–Si bond with a missing oxygen atom, sometimes referred to as an oxide vacancy, shown in Figure 5.2. It is one of the most dominant radiation-induced defects (Chapter 6). E′ centers also preexist in oxide films due to the amorphous nature of SiO_2 and thermodynamic considerations. Each Si atom is back bonded to three oxygen atoms. It is believed that when a positive charge is captured, the Si–Si bond breaks. Feigl et al. argued that the lattice relaxation is asymmetrical with the positively charged Si relaxing into a planar configuration, away from the vacancy and the neutral Si relaxing toward the vacancy [9]. The annealing characteristics of E′ centers have been correlated with positive oxide charge [10].

5.2.5 Neutral Electron Traps

Several models have been proposed to explain oxide breakdown (Chapters 15 through 17). One of these is the electron trap generation model, based on the principles of percolation theory [11]. This model, originally suggested by Massoud and Deaton [12] and later verified by other groups [13–15], assumes that neutral electron traps are randomly generated in the oxide during oxide stressing. It is assumed that traps are continuously generated during oxide stress until there are sufficient numbers of traps somewhere in the device that a continuous, conducting path is formed across the oxide and breakdown occurs. The percolation model can explain the reduced trap density required for breakdown and the reduced Weibull slope as the oxide becomes thinner. The latter has an important

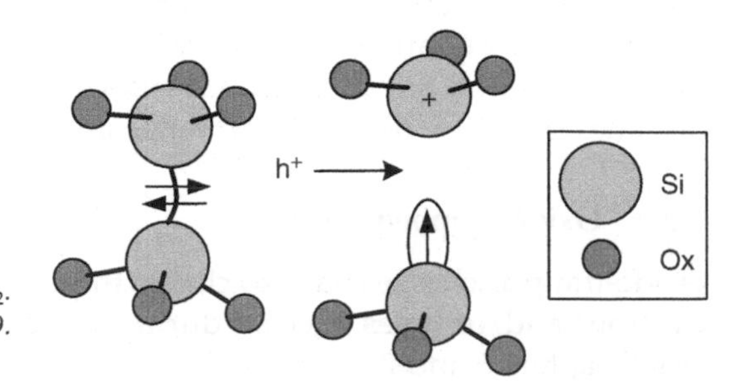

FIGURE 5.2
Model for hole trapping and E′ center formation in SiO_2. (After Caplan, P.J. et al., *J. Appl. Phys.*, 50, 5847, 1979. With permission.)

influence on the area dependence of breakdown. If the neutral electron traps capture electrons, they lead to flatband and threshold voltage shifts. That is, in fact, how they are detected, by filling them with electrons.

Oxide breakdown exhibits a surge in current or a sudden drop voltage during stress measurements. It has a "weakest link" character or extreme value statistics [16]. The statistical description is based on the Weibull distribution model in which the cumulative failure probability F is given by [17]:

$$F(t_{BD}) = 1 - \exp\left(-\left(\frac{t_{BD}}{\alpha}\right)^{\beta}\right) \tag{5.3}$$

where
- t_{BD} is the time-to-breakdown
- α is the t_{BD} at the 63rd percentile
- β is the Weibull shape factor or the Weibull slope

Equation 5.3 is usually written in the form:

$$\ln(-\ln(1 - F(t_{BD}))) = \beta \ln\left(\frac{t_{BD}}{\alpha}\right) \tag{5.4}$$

with a plot of $\ln(-\ln(1 - F))$ versus $\ln(t_{BD})$ yielding a straight line. Equations 5.3 and 5.4 also apply when the time-to-breakdown is replaced by the charge-to-breakdown Q_{BD}. The area dependence is

$$t_{BD}(A_1) = t_{BD}(A_2)(A_2/A_1)^{1/\beta} \tag{5.5}$$

where A_1 and A_2 correspond to two different areas. Equation 5.5 shows that the area dependence is not simply linear because it depends on the shape factor β, which in turn depends on the oxide thickness, i.e., as β decreases, the area dependence becomes stronger. This makes it very important to specify the area during breakdown measurements. The reduction of β with decreased oxide thickness is attributed to a reduced number of defects required to trigger a breakdown.

Two techniques have been used to measure the neutral electron trap density. An indirect measure is stress-induced leakage current (SILC) and a more direct measure is substrate hot electron injection (SHE) followed by a measurement of the threshold voltage or flatband voltage shift. Both are described in Section 5.3.

The nature of neutral electron traps is still under debate. A possible defect structure is the following. It is well established that the E′ center is formed by breaking the Si–Si bond in an oxygen vacancy defect, illustrated in Figure 5.3a. The bond breaking is facilitated by capture of a hole (Figure 5.3b), leaving a positively charged trap and one Si atom with a dangling orbital containing one unpaired electron. The resonant flipping of the spin of this unpaired electron gives rise to the E′ signal in electron spin resonance. Upon electron capture, the center can return to the E′ center or the electron from one of the Si atoms decays to a ground state by joining the unpaired electron of the other Si atom forming a neutral amphoteric trap (Figure 5.3c) [18]. Capturing a second electron leaves it negatively charged (Figure 5.3d). It is this electron trapping event that gives rise to the threshold voltage shifts associated with filled neutral electron traps following electron injection measurements [19]. Attempts to anneal neutral traps have been only partially successful. The usual 400°C–450°C/30 min forming gas anneal only anneals a portion of the traps.

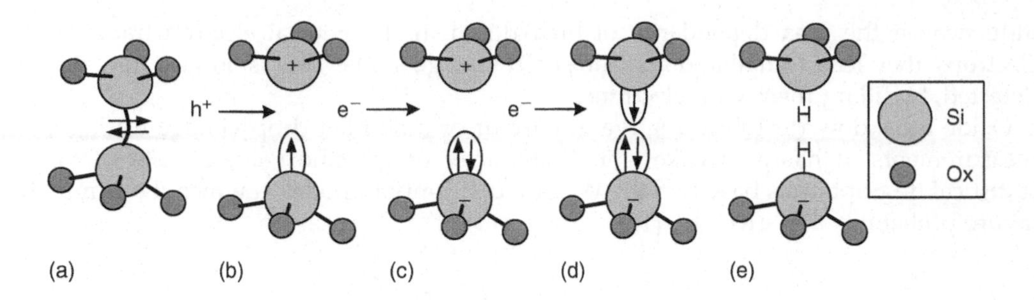

FIGURE 5.3
(a) Neutral hole trap (E′ center, weak Si–Si bond, oxygen vacancy), (b) positive charge ($E_\gamma{}'$ center), (c) neutral electron–hole trap, (d) negatively charged trap, and (e) hydrogen-annealed trap.

High-pressure forming gas is more successful in annealing most of the traps. The hydrogen-anneal model is illustrated in Figure 5.3e. Once such a trap is annealed by hydrogen capture, the Si–H bonds may break leading to trap creation, which may be a precursor to oxide breakdown.

5.2.6 Interface-Trapped Charge

Interface-trapped charges, also known as interface states, interface traps, and fast surface states (Chapter 7), exist at the SiO_2/Si interface. They are the result of a structural imperfection. Silicon is tetrahedrally bonded with each Si atom bonded to four Si atoms in the wafer bulk. When the Si is oxidized, the bonding configuration at the surface is as shown in Figure 5.4a and b with most Si atoms bonded to oxygen at the surface. Some Si atoms bond to hydrogen, but some remain unbonded. An interface trap is an interface trivalent Si atom with an unsaturated (unpaired) valence electron usually denoted by $Si_3 \equiv Si\bullet$, where the "≡" represents three complete bonds to other Si atoms (the Si_3) and the "•" represents the fourth, unpaired electron in a dangling orbital (dangling bond). Interface traps, also known as P_b centers [20], are designated as D_{it} (1/cm^2 eV), Q_{it} (C/cm^2), and N_{it} (1/cm^2). The P_b ESR spectrum was first observed by Nishi [21] and later identified by Poindexter et al. as a paramagnetic dangling bond [22,23].

On (111)-oriented wafers, the P_b center is situated at the Si/SiO_2 interface with its unbonded central atom orbital perpendicular to the interface and aimed into a vacancy in the oxide immediately above it, as shown in Figure 5.4a. On (100)-oriented Si, the four tetrahedral Si–Si directions intersect the interface plane at the same angle. Two defects, named P_{b1} and P_{b0} and shown in Figure 5.4b, have been detected by electron spin resonance. A recent calculation suggests the P_{b1} center to be an asymmetrically oxidized dimer, with no first neighbor oxygen atoms [24]. By 1999, it was unambiguously established that both P_{b0} and P_{b1} are chemically identical to the P_b center [25]. However, there is

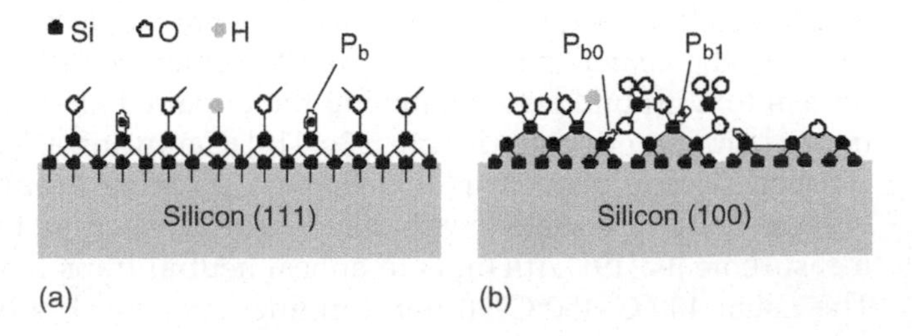

FIGURE 5.4
Structural model of the (a) (111) Si surface and (b) (100) Si surface. (Reprinted from Schroder, D.K., *Semiconductor Material and Device Characterization*, 3rd edn., Wiley-Interscience, New York, 2006. With permission.)

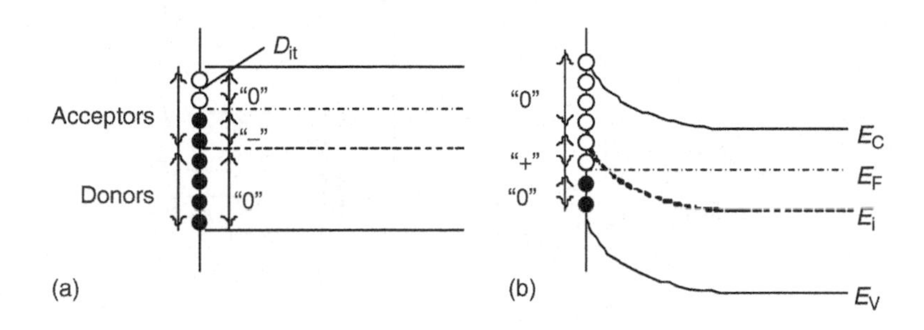

FIGURE 5.5
Band diagrams of the Si substrate of a p-channel MOS device showing the occupancy of interface traps and the various charge polarities for a p-substrate with (a) negative interface trap charge at flatband and (b) positive interface trap charge at inversion. Interface traps are either occupied by electrons (solid circle) or holes (open circle).

a charge state difference between these two centers indicating P_{b0} is electrically active, while some authors believe the P_{b1} to be electrically inactive [26].

Interface traps are electrically active defects with an energy distribution throughout the Si band gap. They act as generation and recombination centers and can contribute to leakage current, low-frequency noise, reduced mobility, drain current, and transconductance. Since electrons or holes occupy interface traps, they become charged and contribute to threshold voltage shifts. The surface potential dependence of the occupancy of interface traps is illustrated in Figure 5.5.

Interface traps at the SiO_2/Si interface are acceptor-like in the upper half and donor-like in the lower half of the band gap [27]. Hence, as shown in Figure 5.5a, at flatband, with electrons occupying states below the Fermi energy, the states in the lower half of the band gap are neutral (occupied donors designated by "0"). Those between midgap and the Fermi energy are negatively charged (occupied acceptors designated by "−"), and those above E_F are neutral (unoccupied acceptors). For an inverted p-MOSFET, shown in Figure 5.5b, the fraction of interface traps between midgap and the Fermi level is now unoccupied donors, leading to positively charged interface traps (designated by "+"). Hence interface traps in p-channel devices in inversion are positively charged, leading to negative threshold voltage shifts.

5.2.7 Border Traps

In 1980, a committee headed by Bruce Deal established the nomenclature for charges associated with the SiO_2/Si system, i.e., interface trapped, fixed oxide, mobile ionic and oxide-trapped charge [28]. In 1992, Fleetwood suggested that this list be augmented by including border traps, which also have been designated as slow states, near-interfacial oxide traps, E′ centers, switching oxide traps, and others (Chapter 7) [29,30]. He proposed border traps to be those near-interfacial oxide traps located within approximately 3 nm of the oxide/semiconductor interface. There is no distinct depth limit, however, border traps are considered to be those traps that can communicate with the semiconductor through capture and emission of electrons and/or holes on the time scale of interest of measurement.

Oxide, border, and interface traps are schematically illustrated in Figure 5.6a. Defects at or near the SiO_2/Si interface are distributed in space and energy and communicate with the Si over a wide range of time scales. While for interface traps, the communication of substrate electrons/holes with interface traps is predominantly by capture/emission,

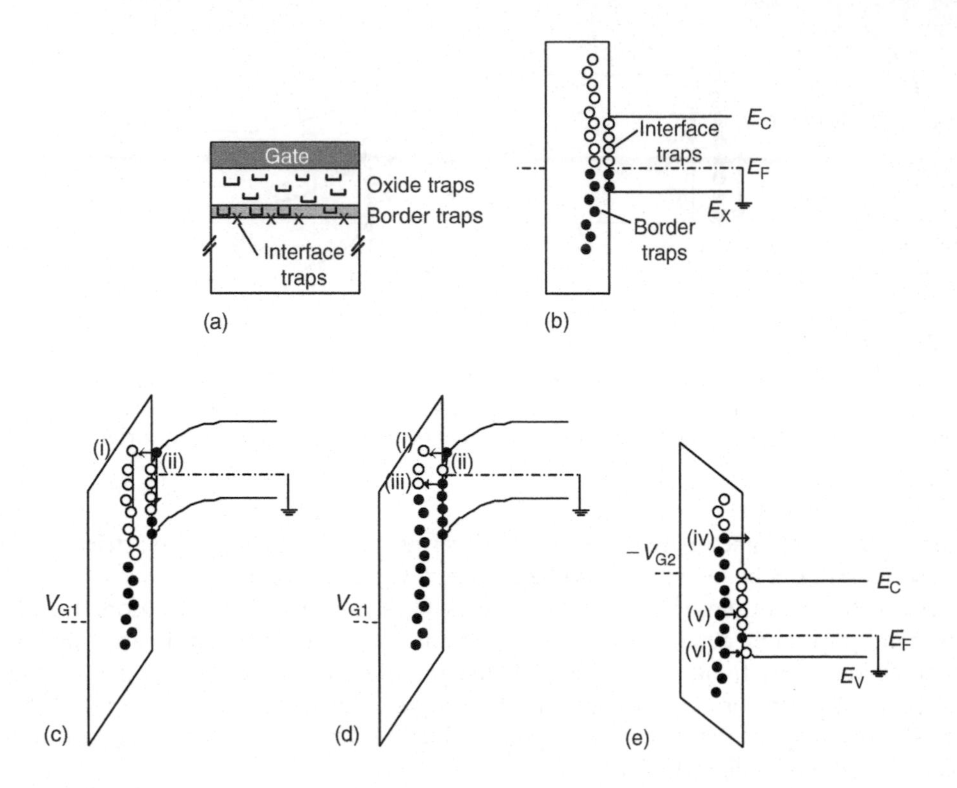

FIGURE 5.6
(a) Schematic illustration of oxide, border, and interface traps, (b) occupancy at flatband, (c) capture of electrons by interface traps and tunneling of electrons to border traps from conduction band, (d) border and interface trap occupied by electrons, and (e) electron tunneling from border traps. The solid circles represent occupied and the open circles unoccupied traps.

for border traps it is mainly by tunneling from the semiconductor to the traps and back. Figure 5.6b shows the flatband diagram with interface and border traps occupied by electrons to the Fermi level E_F. The border traps are shown over a wide energy range for illustrative purposes only. The band diagram in Figure 5.6c applies immediately after V_{G1} is applied, before unoccupied border and interface traps have captured electrons. Interface traps now capture electrons from the conduction band, indicated by (ii), and inversion electrons tunnel to border traps, indicated by (i). Tunnel process (i) is followed by electron capture of lower energy border traps. In Figure 5.6d, interface and border traps up to E_F are occupied by electrons through (ii) electron capture and (iii) tunneling. For $-V_{G2}$ in Figure 5.6e, electrons tunnel from border traps to the conduction band (iv), interface traps (v), and the valence band (vi). The insulator electric field, shown as constant in these figures, will, of course, distort as the border trap occupancy changes. This change is disregarded here to bring out the main points.

Inversion electron tunneling is a direct tunnel process with time constant [31]

$$\tau_t \approx \tau_0 \exp(x/\lambda), \quad \lambda = \frac{\hbar}{\sqrt{8m_t^* \phi_B}} \tag{5.6}$$

where
τ_0 is a characteristic time ($\approx 10^{-10}$ s)
λ is the attenuation length ($\approx 10^{-8}$ cm)
m_t^* is the tunneling effective mass
φ_B is the barrier height at the semiconductor/insulator interface

The value of τ_t varies from 0.01 to 1 s (100–1 Hz) for x varying from 1.8 to 2.3 nm. Hence, border traps can be determined to a depth of approximately 2.5 nm from the SiO_2/Si interface by measurements for frequencies as low as 1 Hz. Lower frequencies, of course, allow deeper traps to be characterized, showing that the trap depth that can be characterized depends on the measurement frequency. Such measurements include low-frequency noise, conductance, frequency-dependent charge pumping, and others. The valence band hole tunnel times are longer than for electrons due to the higher effective mass and barrier height. Tewksbury and Lee give a more detailed discussion of tunneling [32].

Tunneling from the conduction band into border traps was questioned, as measurements did not support energy dissipation in the oxide [33]. Tunneling from interface traps is a two-step process: the electron must be captured from the conduction band before it can tunnel [34]. The capture time is

$$\tau_c = \frac{1}{\sigma_n \nu_{th} n} \tag{5.7}$$

where
- σ_n is the capture cross section
- ν_{th} is the thermal velocity
- n is the inversion electron density

For strong inversion $n = 10^{18}$–$10^{19}/cm^3$, and using $\sigma_n = 10^{-16}/cm^2$ and $\nu_{th} = 10^7$ cm/s, $\tau_c = 10^{-9}$–10^{-10} s. Since the capture and tunnel processes proceed in series, to first order the time constant is

$$\tau_{it} = \tau_c + \tau_t \tag{5.8}$$

and the tunnel time constant dominates for all but the shallowest border traps.

5.2.8 Interface between Two Different Insulators

For a device consisting of two insulators of thicknesses t_1 and t_2 and dielectric constants K_1 and K_2 on a semiconductor, charge accumulates at the interface between the two insulators as a result of differing conductivities. When a gate voltage is applied to such a two-layer structure, the two dielectrics will begin to conduct with current densities J_1 and J_2. Since the conductivities of the two layers differ from each other, $J_1 \neq J_2$ initially, leading to interfacial charge density Q accumulation at the interface. Eventually, the system reaches steady state with the same current density flowing through the entire gate stack. This phenomenon is known as the Maxwell–Wagner instability [35]. With a Maxwell–Wagner instability, it is not the usual static dielectric constants that determine the electric fields of the two layers in steady state, but rather the requirement that the same current density flow through each. Thus, the field in dielectric 1 may differ from $VK_2(t_1K_2 + t_2K_1)$ by an amount depending on the magnitude of Q. The interfacial charge density Q, and from it the electric fields, could be calculated if the current densities J_1 and J_2 were known. Without this knowledge, $VK_2(t_1K_2 + t_2K_1)$ is only an approximation. However, when the dielectric constants of high-K dielectrics are determined experimentally, the interfacial charge is usually ignored, with the two layers of the gate stack being treated as capacitors in series and the "effective dielectric constants" include the effects of the interfacial charge.

In this brief defect discussion, radiation-induced defects have been mentioned several times. Such defects have been fertile ground for studying defects and developing characterization techniques because they can be introduced and annealed at will and their densities can be very high, facilitating measurements. In contrast, oxide defects in conventional ICs are generally of low density and frequently more difficult to characterize. Although radiation-induced defects often differ from those induced during normal IC processing, such processing can introduce its own form of radiation defects. After all, plasma and reactive ion etching and ion implantation are radiation sources that may introduce such defects.

5.3 Measurements

5.3.1 Capacitance–Voltage

5.3.1.1 Theory

Capacitance–voltage measurements have played an important role in MOS characterization. They can be found in some of the earliest MOS-related papers. Frankl, in 1961, proposed the use of MOS-C *C–V* curves to analyze such devices [36]. In 1962, Terman used *C–V* measurements to determine surface state densities [37]. One of the first comprehensive papers was by the Fairchild group in 1964, the same group that played a large role in understanding and developing MOS technology [38]. Figure 5.7 shows one of those early *C–V* curves. Barrier heights for various gate metals were characterized with *C–V* measurements by the same group [39]. A good early overview of the variety of material and device parameters that can be determined from *C–V* and *C–t* measurements is given by Zaininger and Heiman [40]. And, of course, the entire MOS-C field is very well covered in the well-known book by Nicollian and Brews [41].

Insulator-related parameters typically determined with *C–V* measurements are (i) oxide charge density, $\rho_{ox}(x)$, (ii) interface trap density, D_{it}, and (iii) gate–semiconductor work function difference, φ_{MS}. The various charges are illustrated in the device cross section in Figure 5.8a. They are determined from the flatband voltage

$$V_{FB} = \phi_{MS} - \frac{1}{K_{ox}\varepsilon_0}\int_0^{t_{ox}} x\rho(x)\mathrm{d}x \tag{5.9}$$

The fixed charge density Q_f and interface trap charge density Q_{it} are assumed to be located at the SiO_2/Si interface and the charge density $\rho_{ox}(x)$ in the SiO_2, leading to the flatband voltage expression

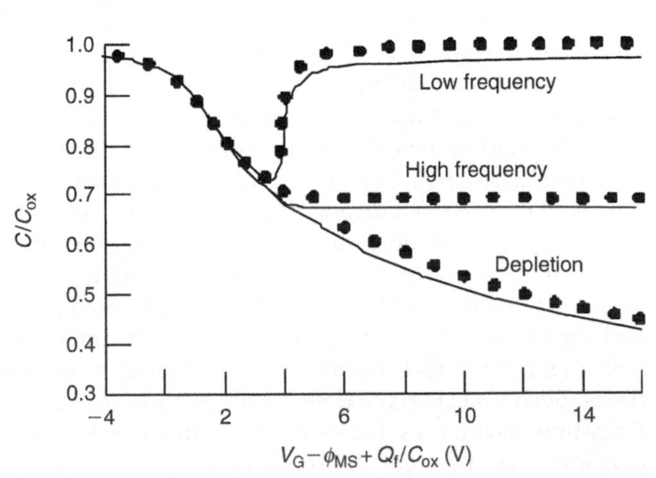

FIGURE 5.7
Capacitance–voltage characteristics of an MOS-C: lines, experiment; dots, theory. $N_A = 1.45 \times 10^{16}/\text{cm}^3$, $t_{ox} = 200$ nm. (After Grove, A.S. et al., *Solid State Electron.* 8, 145, 1965. With permission.)

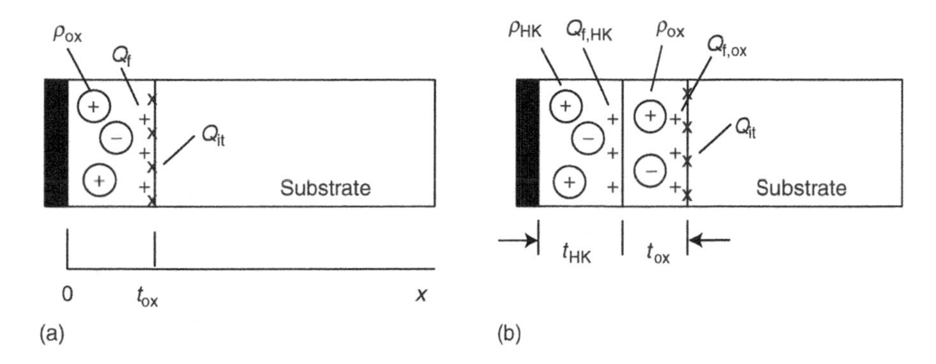

FIGURE 5.8
MOS-C cross section: (a) oxide only and (b) high-*K* and oxide. The various charges are indicated.

$$V_{FB} = \phi_{MS} - \frac{1}{K_{ox}\varepsilon_0}\int_0^{t_{ox}} xQ_f\delta(t_{ox})\mathrm{d}x - \frac{1}{K_{ox}\varepsilon_0}\int_0^{t_{ox}} x\rho_{ox}(x)\mathrm{d}x - \frac{1}{K_{ox}\varepsilon_0}\int_0^{t_{ox}} xQ_{it}(\phi_s)\delta(t_{ox})\mathrm{d}x \quad (5.10)$$

where
t_{ox} is the oxide thickness
δ is the delta function
Q_{it} is a function of surface potential φ_s, as shown in Figure 5.5

V_{FB} can be written as

$$V_{FB} = \phi_{MS} - \frac{Q_f}{K_{ox}\varepsilon_0}t_{ox} - \frac{1}{K_{ox}\varepsilon_0}\int_0^{t_{ox}} x\rho_{ox}(x)\mathrm{d}x - \frac{Q_{it}(\phi_s)}{K_{ox}\varepsilon_0}t_{ox} \quad (5.11)$$

The oxide charge consists of mobile oxide charge (Na, K, etc.), oxide-trapped charge (electrons and holes), and any other charge that may reside within the oxide.

For uniform oxide charge density, Equation 5.11 becomes

$$V_{FB} = \phi_{MS} - \frac{Q_f + Q_{it}(\phi_s)}{K_{ox}\varepsilon_0}t_{ox} - \frac{\rho_{ox}t_{ox}^2}{2K_{ox}\varepsilon_0} \Rightarrow \text{Intercept} = \phi_{MS} \quad (5.12)$$

Equation 5.12 is plotted in Figure 5.9a assuming constant Q_{it}. This plot is clearly nonlinear, making it difficult to extract the various charges. However, φ_{MS} is given by the V_{FB} intercept. Differentiating Equation 5.12 with respect to oxide thickness gives

$$\frac{\mathrm{d}V_{FB}}{\mathrm{d}t_{ox}} = -\frac{Q_f + Q_{it}}{K_{ox}\varepsilon_0} - \frac{\rho_{ox}t_{ox}}{K_{ox}\varepsilon_0} \Rightarrow \text{Intercept} = -\frac{Q_f + Q_{it}}{K_{ox}\varepsilon_0}, \quad \text{Slope} = -\frac{\rho_{ox}}{K_{ox}\varepsilon_0} \quad (5.13)$$

and plotted in Figure 5.9b, yielding $Q_f + Q_{it}$ and ρ_{ox}. The fixed oxide charge and interface trap densities cannot be determined independently, only their sum. However, Q_{it} can be measured independently by other techniques.

Frequently the oxide charge density in thermally grown SiO_2 in Equation 5.12 is low and can be neglected. In that case, the V_{FB} expression in Equation 5.12 becomes linear with t_{ox}.

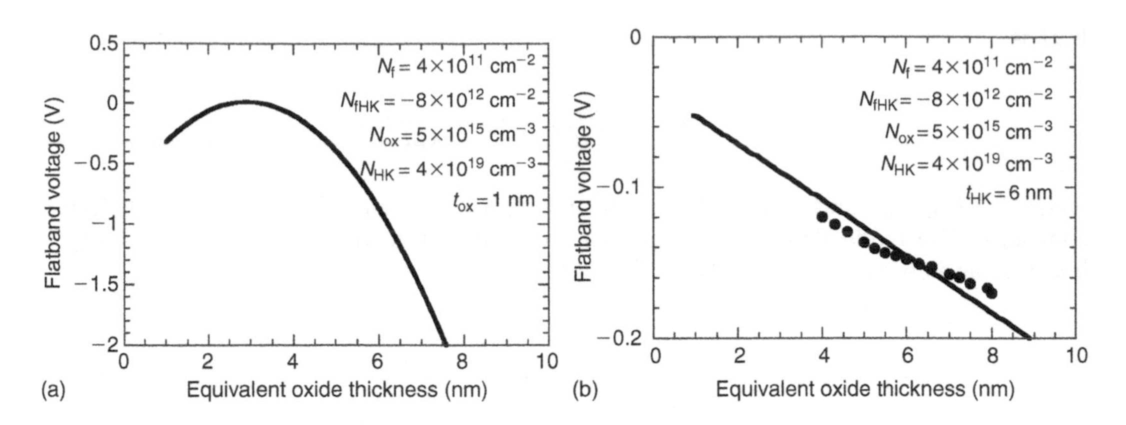

FIGURE 5.11
Flatband voltage versus equivalent oxide thickness with (a) t_{ox} fixed, vary t_{HK}, (b) t_{HK} fixed, vary t_{ox} for a TaN/TiN/HfSiO$_x$/SiO$_2$ structure. (Data in the inset after Kaushik, V.S. et al., *IEEE Trans. Electron Dev.*, 53, 2627, 2006. With permission.)

$$V_{FB}(HK) = -\frac{8\times 10^{-7}\times 0.78\times 10^{-7}}{3.45\times 10^{-13}} - \frac{8\times 6.1\times 10^{-15}}{6.9\times 10^{-13}} = -251\ \text{mV}$$

Clearly the HK charges contribute a much higher flatband voltage shift. Hence, the charges associated with the thin SiO_2 film are frequently ignored during flatband voltage analyses of high-*K* samples.

Equation 5.20 is plotted in Figure 5.11, where in Figure 5.11a t_{ox} is held constant and t_{HK} is allowed to vary, while in Figure 5.11b t_{HK} is constant and t_{ox} varies. The points in Figure 5.11b are experimental data with the extracted charges shown in the inset. Note the very different V_{FB}–EOT behavior for the two cases. This is mainly due to the different thicknesses for these two cases. For EOT = 8 nm, in Figure 5.11a t_{ox} = 1 nm and t_{HK} = 45 nm, while in Figure 5.11b t_{ox} = 7 nm and t_{HK} = 6 nm. The much thicker t_{HK} in Figure 5.11a has a more severe effect on V_{FB} than Figure 5.11b.

To measure these devices the oxide is sometimes grown to a certain thickness and then etched to varying thicknesses across the wafer. In one method, the oxidized wafer is immersed and slowly withdrawn from a 0.34% HF/H_2O solution at a constant withdrawal rate yielding a beveled oxide across the wafer [45]. In another method, the oxide is step etched [46–48].

5.3.1.2 *Interface Traps*

MOS capacitance measurements are made at high and low frequencies. Low-frequency *C–V* measurements are referred to as quasistatic measurements, first demonstrated in 1968–1970 [49] to measure interface traps. Here, a calculation of the effects of interface traps on C–V_G curves is shown to illustrate what one may expect. Figure 5.12a shows the assumed interface trap density as a function of surface potential for these calculations. This distribution approximates the interface trap density distribution at the SiO_2/Si interface with $D_{it,min}$ at midgap. Figure 5.12b shows the surface potential versus gate voltage behavior without and with interface traps. The discontinuity at $\varphi_s \approx 0.4$ V is the result of the assumption of D_{it} being donors in the upper half and acceptors in the lower half of the band gap. In real devices there is a more gradual transition. The interface trap density

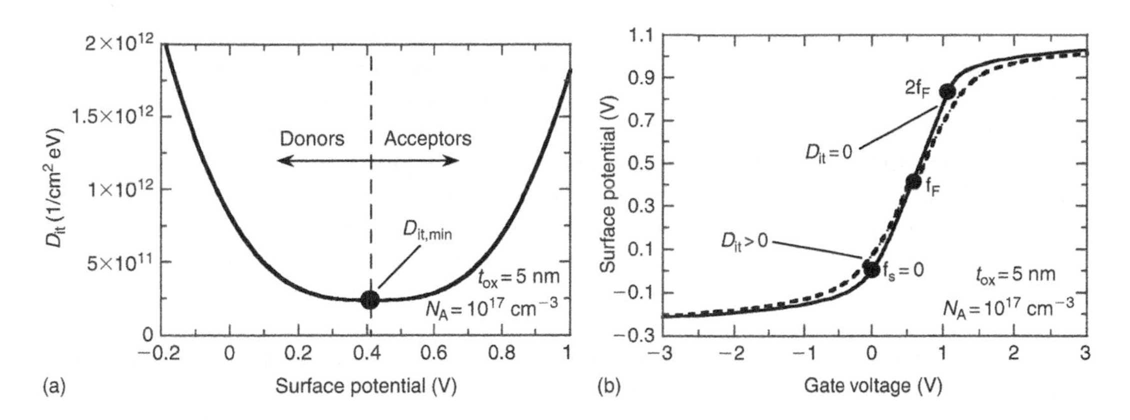

FIGURE 5.12
(a) Interface trap distribution and (b) surface potential versus gate voltage. $N_A = 10^{17}/\text{cm}^3$, $t_{ox} = 5$ nm, $D_{it,min} = 2.4 \times 10^{-11}/\text{cm}^2$ eV.

results in a "stretch-out" of the φ_s–V_G characteristic because charged interface traps lead to gate voltage shifts.

Figure 5.13 shows the effects of D_{it} on (a) low- and (b) high-frequency C–V_G curves. Here it is assumed that $Q_{ox} = 0$ in these calculations in order to bring out the effects of interface traps. Q_{ox} would lead to a parallel shift of the curves along the V_G axis. It is also assumed

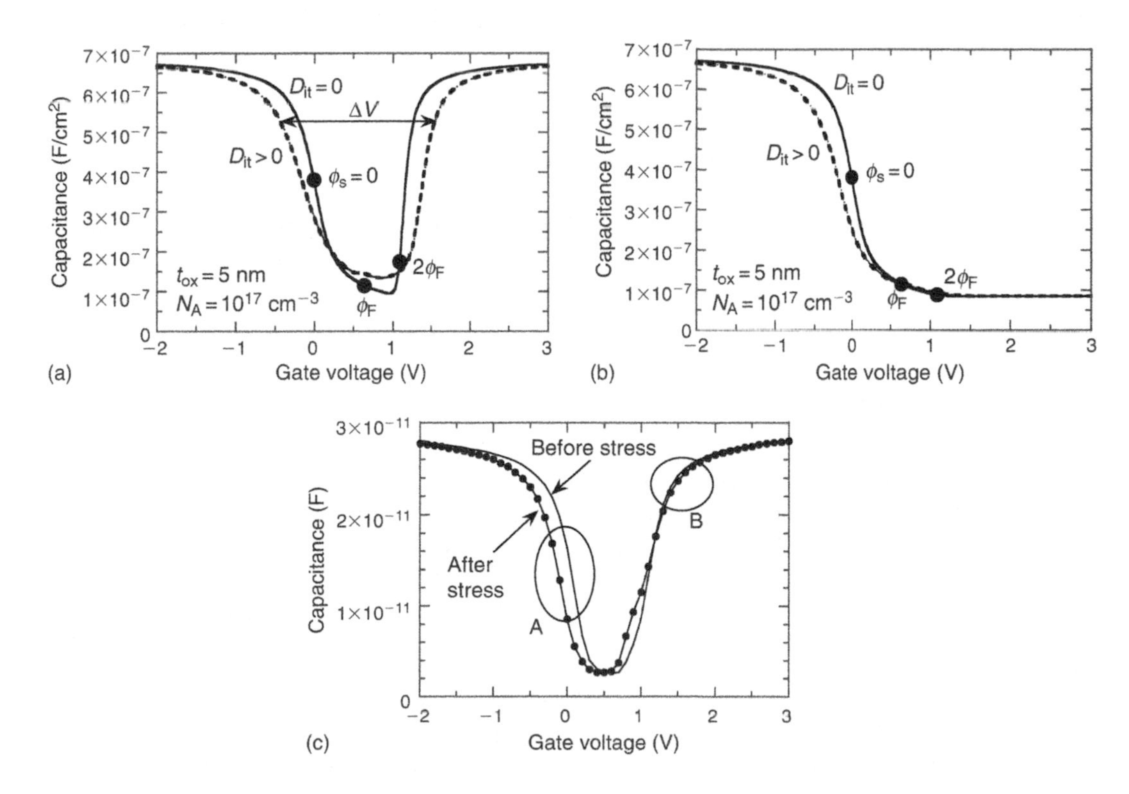

FIGURE 5.13
Theoretical (a) low-frequency, (b) high-frequency C–V_G ($D_{it,min} = 2.4 \times 10^{11}/\text{cm}^2$ eV), and (c) experimental data. φ_s is the surface potential.

Equation 5.23 is for interface traps with a single energy level in the band gap. Interface traps at the SiO_2/Si interface, however, are continuously distributed in energy throughout the Si band gap leading to a time constant dispersion and giving the normalized conductance as

$$\frac{G_P}{\omega} = \frac{qD_{it}}{2\omega\tau_{it}} \ln\left[1 + (\omega\tau_{it})^2\right] \quad (5.24)$$

The conductance is measured as a function of frequency and plotted as G_P/ω versus ω. G_P/ω has a maximum at $\omega = 1/\tau_{it}$ and at that maximum $D_{it} \approx 2.9G_P/q\omega$.

For thin oxides, there may be appreciable oxide leakage current. In addition, the device has series resistance leading to [58]:

$$\frac{G_P}{\omega} = \frac{\omega(G_c - G_t)C_{ox}^2}{G_c^2 + \omega^2(C_{ox} - C_c)^2} \quad (5.25)$$

where

$$C_c = \frac{C_m}{(1 - r_s G_m)^2 + (\omega r_s C_m)^2}, \quad G_c = \frac{\omega^2 r_s C_m C_c - G_m}{r_s G_m - 1} \quad (5.26)$$

where

G_t is the tunnel conductance
r_s is the series resistance
C_m and G_m are the measured capacitance and conductance

5.3.2.2 Border Traps

Absent tunneling to border traps, the G_p/ω–ω peak is symmetrical about the maximum peak frequency. However, a low-frequency conductance ledge is sometimes observed for insulators containing oxide traps. This was first reported by Eaton and Sah who observed a marked asymmetry in their conductance spectrum and attributed this to long time constants arising from carriers tunneling from interface traps into oxide traps [59]. The conductance spectrum in Figure 5.15 shows such a low-frequency ledge [60]. Such

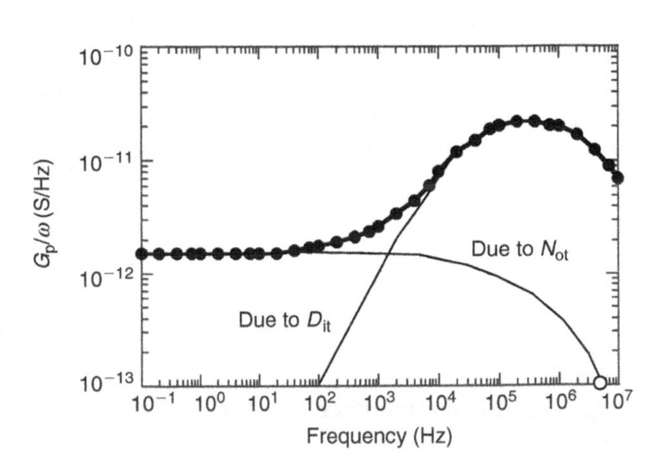

FIGURE 5.15
G_p/ω versus f for a MOS-C exhibiting a D_{it} peak and N_{ox} ledge. The lines indicate the contributions from $D_{it} = 1.46 \times 10^{11}/\text{cm}^2$ eV and $N_{ox} = 1.3 \times 10^{17}/\text{cm}^3$ eV. (Reused from Uren, M.J., Collins, S., and Kirton, M.J., *Appl. Phys. Lett.*, 54, 1448, 1989. With permission.)

ledges are usually not observed when measurements are confined to frequencies higher than 0.1–1 kHz.

5.3.3 Bias-Temperature Stress

In the bias-temperature stress (BTS) method the MOS device is heated to 150°C–250°C, and a gate bias to produce an oxide electric field of around 10^6 V/cm is applied for 5–10 min for any mobile oxide charge to drift to one oxide interface. The device is then cooled to room temperature under bias and a C–V_G curve is measured. The procedure is then repeated with the opposite bias polarity. The mobile charge is determined from the flatband voltage shift, according to the equation

$$Q_m = -C_{ox}\Delta V_{FB} \tag{5.27}$$

To distinguish between oxide-trapped charge and mobile charge, a BTS test is done with positive gate voltage. For oxide electric fields around 1 MV/cm mobile charge drifts, but the electric field is insufficient for appreciate charge injection. If the C–V_G curve shifts after BTS, it is due to positive mobile charge. For higher gate voltages, there is a good chance that electrons and/or holes can be injected into the oxide and mobile charge may also drift, making that measurement less definitive. For positive gate voltage, mobile charge drift leads to negative flatband voltage shifts while charge injection leads to positive shifts.

5.3.4 Triangular Voltage Sweep

In the triangular voltage sweep (TVS) method the MOS device is held at an elevated, constant temperature of 200°C–300°C and the low-frequency C–V_G curve is obtained by measuring the current in response to a slowly varying ramp gate voltage [61]. TVS is based on measuring the charge flow through the oxide at an elevated temperature in response to an applied time-varying voltage. If the ramp rate is sufficiently low, the measured current is the sum of the displacement and conduction current due to the mobile charge. If, at $-V_{G1}$, all mobile charges are located at the gate–oxide interface and at $+V_{G2}$ all mobile charges are located at the semiconductor–oxide interface, the mobile charge is determined by the area under the lf curve according to

$$\int_{-V_{G1}}^{V_{G2}} (I/C_{lf} - \alpha)C_{ox}\,dV_G = \alpha Q_m \tag{5.28}$$

The hf and lf C–V curves coincide at high temperatures except for the lf hump, due to mobile charge drifting though the oxide and illustrated in Figure 5.16 [62]. Figure 5.16a illustrates the effect of different mobile charge densities, and Figure 5.16b the effect of temperature. Clearly the temperature for this sample must be at least 120°C for all of the mobile charge to move.

Sometimes two peaks are observed in I–V_G curves at different gate voltages. These have been attributed to mobile ions with different mobilities. For an appropriate temperature and sweep rate, high-mobility ions (e.g., Na^+) drift at lower oxide electric fields than low-mobility ions (e.g., K^+) [63]. Hence, the Na peak occurs at lower gate voltages than the K peak, illustrated in Figure 5.17. Such discrimination between different types of mobile impurities is not possible with the bias-temperature method. This also explains why sometimes the total number of impurities determined by the BTS and the TVS

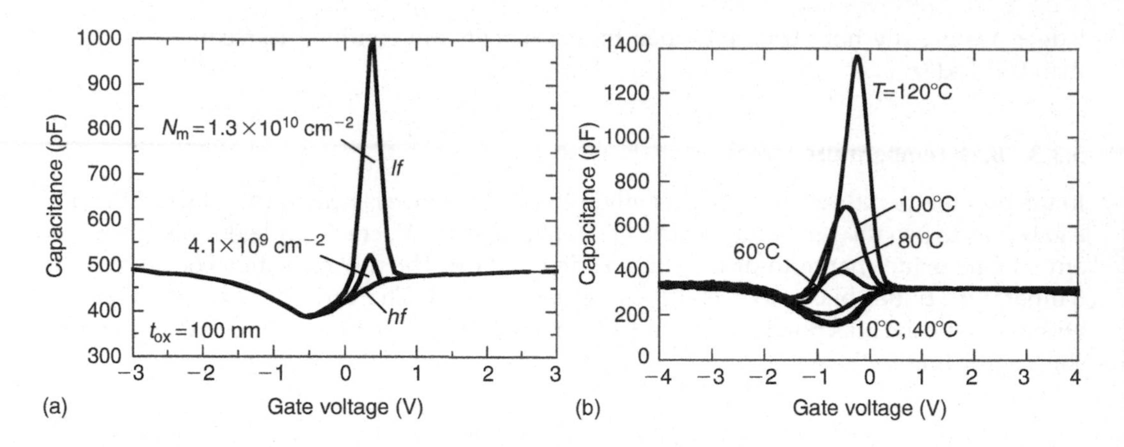

FIGURE 5.16
C_{lf} and C_{hf} measured at (a) $T = 250°C$ and (b) various temperatures. The mobile charge density is determined from the area between the two curves. (Reprinted from Schroder, D.K., *Semiconductor Material and Device Characterization*, 3rd edn., Wiley-Interscience, New York, 2006. With permission.)

methods differ. In the BTS method one usually waits long enough for all the mobile charge to drift through the oxide. If in the TVS method the temperature is too low or the gate ramp rate is too high, it is possible that only one type of charge is detected. For example, it is conceivable that high-mobility Na^+ drifts but low-mobility K^+ does not. The TVS method also lends itself to mobile charge determination in interlevel dielectrics, since a current or charge is measured instead of a capacitance.

5.3.5 Deep-Level Transient Spectroscopy

5.3.5.1 Bulk Traps

Deep-level transient spectroscopy (DLTS), introduced by Lang in 1972 [64], is commonly used to determine densities, energies, and capture cross sections of bulk and interface traps in semiconductors and sometimes for border traps. The device capacitance, current, or charge is measured as a function of time after pulsing the device between zero/forward and reverse bias. The traps capture electrons or holes, which they subsequently emit, leading to the time-dependent capacitance

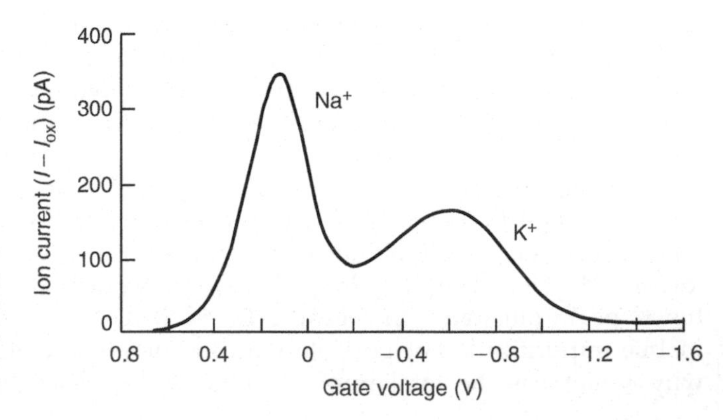

FIGURE 5.17
Ion current normalized by oxide current versus gate voltage. The ion moves to the gate. $T = 423°C$, $\alpha = 0.513$ V/s. (Reused from Hillen, M.W., Greeuw, G., and Verweij, J.F., *J. Appl. Phys.*, 50, 4834, 1979. With permission.)

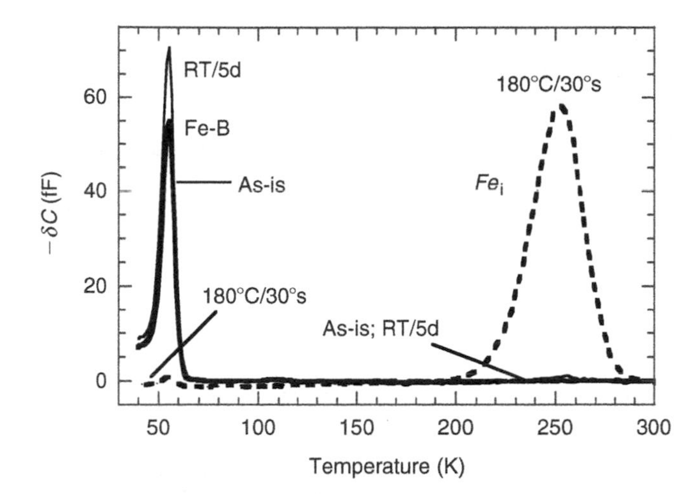

FIGURE 5.18 DLTS spectra for iron-contaminated Si wafer; "As-is," after 180°C/30 s dissociation anneal, and room temperature storage for 5 days. (Reprinted after Choi, B.D., et al., *Jap. J. Appl. Phys.*, 40, L915, 2001. With permission.)

$$C(t) = C_0\left[1 - \frac{N_T}{2N_D}\exp\left(-\frac{t}{\tau_e}\right)\right] \tag{5.29}$$

with the electron emission time constant τ_e depending on temperature as

$$\tau_e = \frac{\exp\left((E_c - E_T)/kT\right)}{\gamma_n \sigma_n T^2} \tag{5.30}$$

where γ_n is a constant for a given semiconductor. Determining the time constant at various temperatures allows the energy level E_T, density N_T, and capture cross section σ_n to be determined [65].

Example DLTS spectra of iron-contaminated Si are shown in Figure 5.18 [66]. Iron forms Fe–B pairs in boron-doped p-type Si with a DLTS peak at $T \approx 50$ K. When the sample is heated at 180°C–200°C for a few minutes, the Fe–B pairs dissociate into interstitial iron Fe_i and substitutional boron, and the DLTS peak for Fe_i occurs around $T \approx 250$ K. After a few days the interstitial iron again forms Fe–B pairs and the "Fe–B" peak returns while the "Fe_i" peak shrinks as shown in Figure 5.18.

5.3.5.2 Interface Traps

The instrumentation for interface-trapped charge DLTS is identical to that for bulk deep-level DLTS, but the data interpretation differs because interface traps are distributed in energy through the band gap. We illustrate the interface-trapped charge majority carrier DLTS concept for the MOS-C in Figure 5.19a. For a positive gate voltage most interface traps are occupied by majority electrons for n-substrates (Figure 5.19b). A negative gate voltage drives the device into deep depletion, causing electrons to be emitted from interface traps (Figure 5.19c). Although electrons are emitted over a broad energy spectrum, emission from interface traps in the upper half of the band gap dominates. DLTS is very sensitive, allowing interface trap density determination in the mid $10^9/\text{cm}^2$ eV range.

Interface trap characterization by DLTS was first implemented with MOSFETs [67]. Being three-terminal devices, they have an advantage over MOS-Cs. By reverse biasing the source/drain (S/D) and pulsing the gate, majority electrons are captured and emitted without interference from minority holes that are collected by the SD. This allows interface trap majority carrier characterization in the upper half of the band gap.

level E_F for simplicity. Of course, the device being in nonequilibrium during the CP measurement should be represented by two quasi-Fermi levels.

When the gate pulse falls from its high to its low value during its finite transition time, most channel electrons drift to source and drain and those electrons on interface traps near the conduction band are thermally emitted into the conduction band (Figure 5.21b) and also drift to source and drain. Those electrons on interface traps deeper within the band gap do not have sufficient time to be emitted and remain trapped. Once the hole barrier is reduced (Figure 5.21c), holes flow to the surface where some are captured by those interface traps still occupied by electrons and all interface traps are occupied by holes. As the device reverts back to inversion, some of the holes near the valence band are emitted to drift to the substrate (not shown), while those remaining on interface traps are annihilated by electron capture (Figure 5.21a). Hence, only a fraction of interface traps, those toward the center of the band gap, are not emitted and participate in capture or recombination processes and lead to CP current. This energy interval depends on the rise and fall times of the CP waveform. The waveforms can be constant base voltage in accumulation and pulsing with varying voltage amplitude ΔV into inversion or varying the base voltage from inversion to accumulation keeping ΔV constant. The current saturates for the former, while for the latter it reaches a maximum and then decreases.

From Shockley–Read–Hall recombination statistics the occupancy of interface traps is determined by carrier capture and emission. Considering capture processes, the variation of the occupancy during one charge pumping cycle is [73,74]:

$$\Delta F = \frac{[1 - \exp(-c_n/2f)][1 - \exp(-c_p/2f)]}{1 - \exp[-(c_n/2f) - (c_p/2f)]} \tag{5.33}$$

where

c_n and c_p are the capture coefficients for electrons and holes ($c_{n,p} = \sigma_{n,p}v_{th}$), $\sigma_{n,p}$ are the capture cross sections and v_{th} is the thermal velocity

f is the CP frequency

The charge pumping current then becomes

$$I_{cp} = qAf \int_{E_{low}}^{E_{high}} \Delta F D_{it}(E)\mathrm{d}E \tag{5.34}$$

where E_{high} and E_{low} are the Fermi energies for high and low gate bias.

The basic charge pumping technique gives an average value of D_{it} over the energy interval ΔE. Various refinements have been proposed to obtain energy-dependent interface trap distributions. For a sawtooth waveform, the recombined charge per cycle, $Q_{cp} = I_{cp}/f$, is given by [75]:

$$Q_{cp} = 2qkT\overline{D}_{it}fA_G \ln\left(v_{th}n_i\sqrt{\sigma_n\sigma_p}\sqrt{\zeta(1-\zeta)}\frac{|V_{FB} - V_T|}{|\Delta V_{GS}|f}\right) \tag{5.35}$$

where

$\overline{D}_{it}$ is the average interface trap density

ΔV_{GS} is the gate pulse peak–peak amplitude

ζ is the gate pulse duty cycle

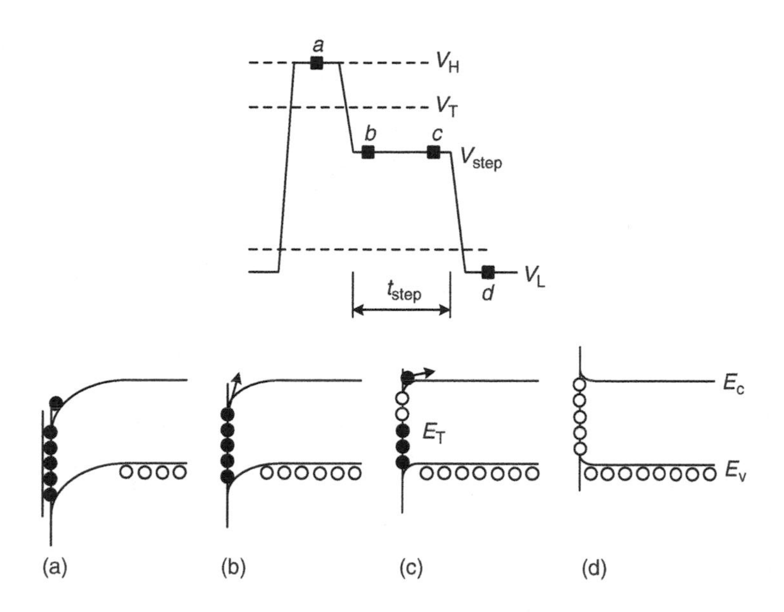

FIGURE 5.22
Trilevel charge pumping waveform and corresponding band diagrams. (Reprinted from Schroder, D.K., *Semiconductor Material and Device Characterization*, 3rd edn., Wiley-Interscience, New York, 2006. With permission.)

The slope of a Q_{cp} versus log(f) plot yields D_{it}, and the intercept on the log(f) axis yields $(\sigma_n\sigma_p)^{1/2}$. By varying the gate waveform rise and fall times, one obtains the interface trap energy distribution. Han et al., using this technique for SiON/HfO_2 insulators, found higher D_{it} in the upper half than in the lower half of the band gap [76]. This also agreed with V_T, mobility, and subthreshold slope data, e.g., the V_T shift of n-MOSFETs was higher than the flatband voltage shift for these HfO_2 samples.

The interface trap distribution through the band gap and capture cross sections can be determined with the trilevel waveform with an intermediate voltage level V_{step} [77], illustrated in Figure 5.22. At point (a), the device is in strong inversion with interface traps filled with electrons. As the waveform changes to (b), electrons are emitted from interface traps. The gate voltage remains constant to point (c). For $t_{step} \gg \tau_e$, where τ_e is the emission time constant of interface traps being probed, all traps above E_T have emitted their electrons and only those below E_T are available for recombination when holes come in to recombine with the electrons at point (d) on the waveform. This gives a charge pumping current that saturates as t_{step} increases. For $t_{step} < \tau_e$, fewer electrons have time to be emitted and more are available for hole recombination giving a correspondingly higher charge pumping current.

A typical I_{cp} versus t_{step} plot in Figure 5.23 shows the I_{cp} saturation and the $t_{step} = \tau_e$ break point. By varying V_{step} one can probe interface traps through the band gap. D_{it} is determined from the slope of the I_{cp} versus t_{step} curve according to the expression [78]:

$$D_{it} = -\frac{1}{qkTA_G f}\frac{dI_{cp}}{d\ln t_{step}} \tag{5.36}$$

The charge pumping current is assumed to be due to electron–hole pair recombination at interface traps. For thin oxides, gate current adds to the charge pumping current. The gate oxide leakage current can exceed I_{cp}. Figure 5.24 shows the effects of gate oxide leakage current on I_{cp} [79]. At sufficiently low frequencies, the gate leakage current dominates and can be subtracted from the total current.

Being unable to return to S/D from where they originated, they are injected into the substrate to recombine. This component is maximized by using long-channel MOSFETs and fast rise/fall times [84]. Under these conditions the long channel is rapidly pinched-off, forcing a large fraction of the inversion charge to recombine in the Si where it is measured as a substrate current. By grounding both source and drain, the inversion charge density is $N_{inv,S/D}$; by grounding the source and floating the drain it is $N_{inv,S}$, given by

$$N_{inv,S/D} = N_{inv} - 2N_{S/D}, \quad N_{inv,S} = N_{inv} - N_{S/D} \tag{5.41}$$

where N_{inv} is the true inversion charge density and $N_{S/D}$ is the geometric component. From Equation 5.41

$$N_{inv} = 2N_{inv,S} - N_{inv,S/D} \tag{5.42}$$

yielding the true N_{inv}. This technique was used to determine the mobility for SiO_2/HfO_2 gate insulators, eliminating trapping effects [84].

5.3.7 MOSFET Subthreshold Slope

5.3.7.1 *Interface Traps*

The effects of D_{it} on the MOSFET subthreshold I_D–V_G characteristics are illustrated in Figure 5.26. Similar to the broadening of the C–V_G curves in Figure 5.13, the subthreshold characteristic also broadens and only the surface potential region between about midgap ($\varphi_s = \varphi_F$) and inversion can be probed. The drain current of a MOSFET in the subthreshold regime for drain voltages higher than about $4kT/q$ is [85]:

$$I_D = \frac{W\mu_{eff}}{L}\left(\frac{kT}{q}\right)^2 \sqrt{\frac{qK_s\varepsilon_0 N_A}{4\phi_F}} \exp\left(\frac{q(V_{GS} - V_T)}{nkT}\right) \tag{5.43}$$

where

$n = 1 + (C_b + C_{it})/C_{ox}$

C_b is the bulk (substrate) capacitance

The usual subthreshold plot is $\log(I_D)$ versus V_G with subthreshold swing S being that gate voltage necessary to change the drain current by one decade, given by

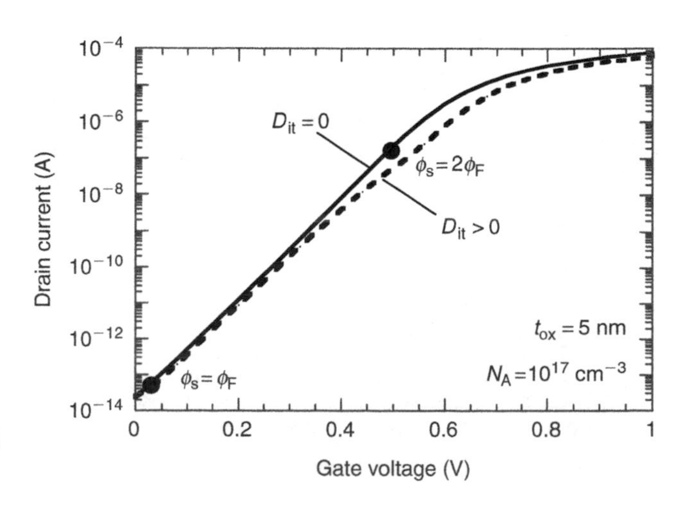

FIGURE 5.26
Theoretical I_D–V_G curves for $D_{it} = 0$ and $D_{it,min} = 2.7 \times 10^{10}/\text{cm}^2$ eV.

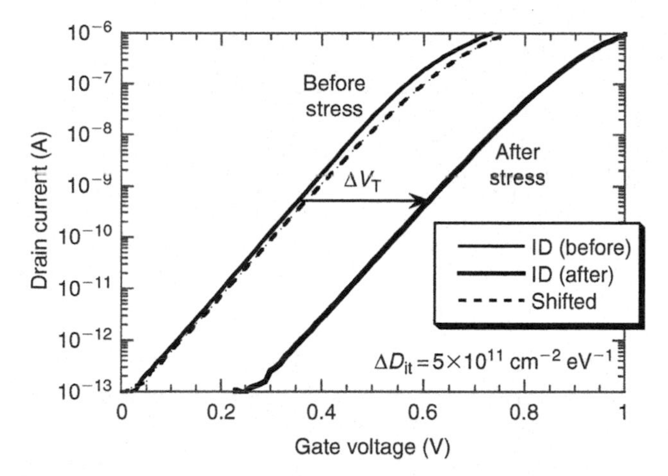

FIGURE 5.27
MOSFET subthreshold characteristics before and after MOSFET stress. The change in slope results in a stress-generated $\Delta D_{it} = 5 \times 10^{11}/\text{cm}^2$ eV. The dashed curve is the after stress curve shifted to the left to coincide with the before stress curve at $I_D = 10^{-13}$ A to bring out the slope change. (Reprinted from Schroder, D.K., *Semiconductor Material and Device Characterization*, 3rd edn., Wiley-Interscience, New York, 2006. With permission.)

$$S = \frac{\ln(10)nkT}{q} \approx \frac{60nT}{300} \text{ mV/decade} \quad (5.44)$$

with T in Kelvin. The interface trap density is

$$D_{it} = \frac{C_{ox}}{q^2}\left(\frac{qS}{\ln(10)kT} - 1\right) - \frac{C_b}{q^2} \quad (5.45)$$

requiring an accurate knowledge of C_{ox} and C_b. The slope also depends on surface potential fluctuations. This is the reason that this method is usually used as a comparative technique in which the subthreshold swing is measured, then the device is degraded and remeasured. The D_{it} change is given by

$$\Delta D_{it} = \frac{C_{ox}}{\ln(10)q^2kT}(S_{after} - S_{before}) \quad (5.46)$$

The subthreshold MOSFET curves are shown in Figure 5.27 before and after stress, causing a threshold voltage shift and a slope change.

5.3.7.2 Oxide Traps

Subthreshold measurements are also made to determine oxide charge densities. When the surface potential coincides with the Fermi level, as shown in Figure 5.28a by $\varphi_s = \varphi_F$, interface traps in the upper and lower half in the band gap are neutral, and neither contributes to a gate voltage shift. The corresponding gate voltage is V_{mg}, which is typically the gate voltage at $I_D \approx 0.1$–1 pA. Increasing the gate voltage from V_{mg} to V_T fills interface traps in the upper half of the band gap with electrons (Figure 5.28b). ΔE usually covers the range from midgap to strong inversion. Since at midgap the interface traps do not contribute any voltage shift, a shift of V_{mg} must be due to oxide-trapped charge, where the threshold voltage due to oxide-trapped charge and the change in oxide trap charge density as projected to the Si/SiO_2 interface may be estimated according to [86]:

$$\Delta V_{ot} = V_{mg2} - V_{mg1} \quad \text{and} \quad \Delta N_{ot} = \frac{\Delta V_{ot} C_{ox}}{q} \quad (5.47)$$

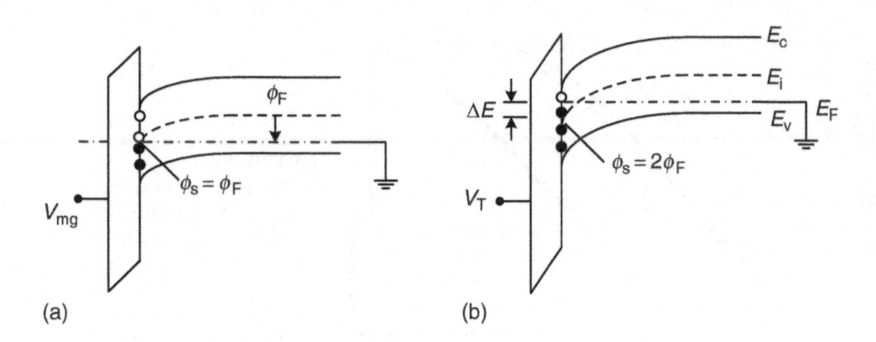

FIGURE 5.28
Band diagrams for midgap and threshold voltages. (Reprinted from Schroder, D.K., *Semiconductor Material and Device Characterization*, 3rd edn., Wiley-Interscience, New York, 2006. With permission.)

5.3.7.3 *Border Traps*

The threshold voltage of high-K dielectric MOSFETs frequently exhibits an instability that has been explained by charging preexisting defects in the HfO_2 in HfO_2/SiO_2 dual dielectrics with a strong dependence on gate bias and charging (discharging) time [87]. The measured drive current decay in n-MOSFETs is caused by the continuous increase in V_T due to a buildup of negative charge, caused by trapping in the HfO_2. Comparing the pulsed technique with the dc measurement methods shows the conventional dc techniques to underestimate the charging effects in SiO_2/HfO_2 dual layer gate dielectrics. An example of such dielectric charging is shown in Figure 5.29, clearly illustrating the hysteresis for the transient measurements due to charge trapping/detrapping. Information about border traps can be gained from such measurements. The pulsed techniques was recently used to estimate the capture cross sections in HfO_2 as 10^{-16}–10^{-14} cm^2 [88].

5.3.8 DC-IV

The DC-IV method is a dc technique to determine D_{it} illustrated in Figure 5.30a [89]. With the source S forward biased, electrons injected into the p-well diffuse to the drain to be

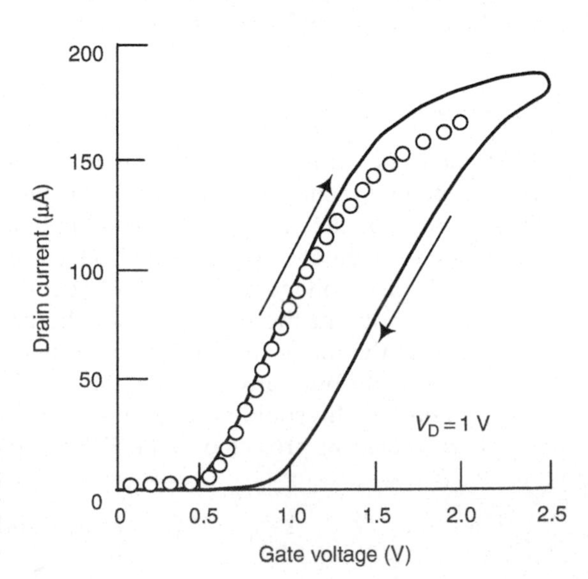

FIGURE 5.29
I_D–V_G dc (open circles) and transient (100 μs pulse width, rise and fall times) characteristics showing higher transient current and hysteresis due to carrier trapping/detrapping. (From Kerber, A. et al., *IEEE Intl. Rel. Phys. Symp.*, 41, 41, 2003. With permission.)

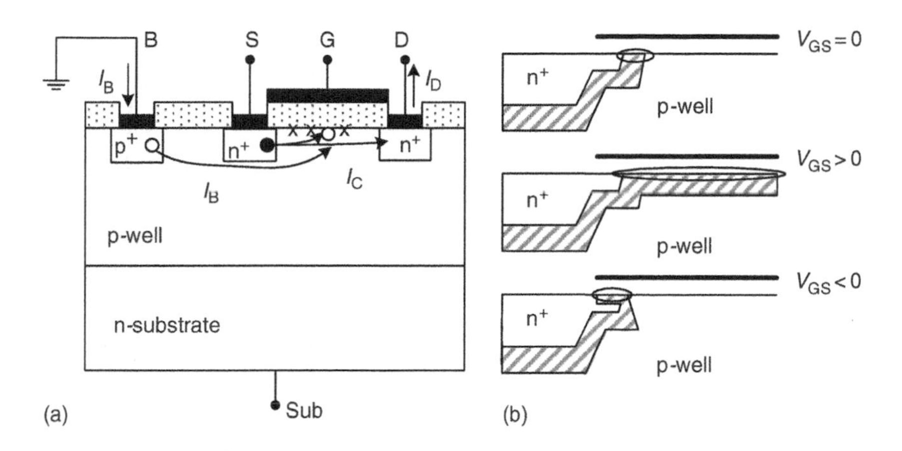

FIGURE 5.30
(a) MOSFET configuration for DC-IV measurements and (b) cross sections showing the space-charge regions and the encircled surface generation regions. (Reprinted from Schroder, D.K., *Semiconductor Material and Device Characterization*, 3rd edn., Wiley-Interscience, 2006. With permission.)

collected and measured as drain current I_D. Some electrons recombine with holes in the p-well bulk (not shown) and some recombine with holes at the surface below the gate with only the surface-recombining electrons controlled by the gate voltage. The recombined holes are replaced by holes from the body contact leading to body current I_B.

The electron–hole pair surface recombination rate depends on the surface condition. With the surface in strong inversion or accumulation, the recombination rate is low. The rate is highest with the surface in depletion [90]. The body current is given by

$$\Delta I_B = qA_G n_i s_r \exp(qV_{BS}/2kT) \tag{5.48}$$

where s_r is the surface recombination velocity given by

$$s_r = (\pi/2)\sigma_0 v_{th} \Delta N_{it} \tag{5.49}$$

with σ_0 the capture cross section (assuming $\sigma_n \approx \sigma_p \approx \sigma_0$).

When the gate voltage exceeds the flatband voltage, a channel forms between S and D and the drain current will increase significantly. For $V_{GB} = V_T$, the $I_D - V_{GB}$ curve saturates. If charge is injected into the oxide, leading to a V_T shift, the drain current will also shift. It is this shift that can be used to determine oxide charge. The interface trap density determined with the subthreshold slope method samples the band gap between midgap and strong inversion while the DC-IV body current samples the band gap between subthreshold and weak accumulation, i.e., surface depletion. By varying the gate voltage, different regions of the device are depleted (Figure 5.30b) and those regions can be characterized, allowing spatial D_{it} profiling. Experimental DC-IV data are shown in Figure 5.31 for a MOSFET before and after gate current stress [91]. A clear peak is observed at maximum surface recombination around $V_{GB} = 0$.

5.3.9 Stress-Induced Leakage Current

An effect frequently observed in thin electric field-stressed oxides is an enhanced gate oxide current—SILC, defined as the increase of oxide leakage current after high-field stress (≈10–12 MV/cm) compared to before stress and first reported in 1982 [92]. It is typically

grounded substrate. When the substrate is reverse biased with respect to S/D, the channel disappears. It is possible, however, for thermally generated electrons accelerated toward the Si/SiO_2 interface, to have sufficient energy to surmount the barrier φ_B at the interface, as shown in Figure 5.33a and c. These electrons have moderate energy and there is a good chance that they become trapped in the oxide. A problem with the arrangement in Figure 5.33a is the low density of electrons for injection. The electrons are thermally generated, i.e., the leakage current of the gate-induced space-charge region/substrate junction. Ning proposed to use a MOSFET with the p-substrate replaced by a p-epitaxial layer grown on an n-substrate. The resulting np junction can then be forward biased to inject electrons into the p-layer. Such a structure is a natural portion of MOSFETs fabricated in a p-well on an n-substrate [99]. It is also possible to form an np junction beside the source or drain and forward bias it, shown in Figure 5.33b [100], or surrounding the MOSFET with an n-type implant. Either configuration allows independent control of the oxide electric field and the electron injection current.

5.3.11 Constant Gate Voltage Oxide Stress

Gate oxide integrity (GOI) characterization determines the quality of the gate oxide and does provide trap densities. It is mentioned briefly here, because it is important to determine the gate oxide quality. GOI is measured by one of several methods. In the time-zero method, the gate voltage is swept and the gate current is measured until the oxide breaks down. In the time-dependent methods, the gate voltage is held constant (constant voltage source, CVS) and the gate current is measured, or the gate current is held constant (constant current source, CCS) and the gate voltage is measured, or the gate current is stepped and the gate voltage is measured. For thin oxides it has been shown that the time-to-breakdown, t_{BD}, increases with decreasing oxide thickness for constant current stress while it decreases for constant voltage stress [101]. For CCS, the oxide voltage changes during the measurement and the defect generation rate and critical defect density required for breakdown change. This makes the CVS technique the more appropriate method for GOI determination.

5.3.12 1/*f* Noise

Noise measurements can be used to characterize semiconductors. The recent review papers by Wong [102] and Claeys et al. [103] give a good overview of the present state of noise theory and measurements. At high frequencies, thermal noise and shot noise dominate. Both of these noises are fundamental in nature, forming an intrinsic lower noise limit. At low frequencies, flicker or $1/f$ noise dominates. Generation–recombination (G–R) noise can also occur in this frequency range. It is characterized by a Lorentzian spectrum with a constant plateau at $f < f_c$ and a $1/f^2$ roll-off beyond the characteristic frequency f_c. In contrast to the fundamental thermal and shot noise, $1/f$ and G–R noise depend on material and semiconductor processing. Only $1/f$ noise is discussed here since it is sensitive to interface and border traps.

Low-frequency or flicker noise, first observed in vacuum tubes over 80 years ago [104], dominates the noise spectrum at low frequencies. It gets its name from the anomalous plate current "flicker." Flicker noise is also called $1/f$ noise, because the noise spectrum varies as $1/f^n$, where the exponent n is very close to unity. Fluctuations with a $1/f$ power law have been observed in many electronic materials and devices, including homogenous semiconductors, junction devices, metal films, liquid metals, electrolytic solutions, Josephson junctions, and even in mechanical, biological, geological, and even musical systems. Two competing models have been proposed to explain flicker noise: the McWhorter number fluctuation theory [105] and the Hooge mobility fluctuation theory [106] with experimental evidence to support both models. Christensson et al. were the first to apply the McWhorter theory to MOSFETs, using the assumption that the necessary time constants are caused by the tunneling of carriers from the channel into traps located within the oxide (Chapter 7) [107]. Popcorn noise, sometimes called burst noise or random-telegraph-signal (RTS) noise, is a discrete modulation of the channel current caused by the capture and emission of as little as a single channel carrier [108].

RTS presents a serious problem to analog and digital devices as devices are scaled down. For example, in floating gate nonvolatile memories the capture/emission of even a single electron affects the threshold voltage and drain current. RTS-induced drain current produces significant variation in read current [109]. The threshold voltage change $\Delta V_T = -q/AC_{ox}$ increases as the area, A, decreases. The trap properties leading to RTS can be studied by statistical analysis. One analysis procedure is based on the difference in the statistical properties of discrete Marcovian telegraph fluctuations and Gaussian background noise. The average statistical lifetimes and amplitudes of the telegraph signal are then determined in an iterative way, allowing for analyzing noisy random telegraph signals with low ratio between the signal amplitude and the intensity of the background noise that cannot be analyzed by the classical approach. Separation of the time record enables an in-depth analysis of the spectral properties of the background noise observed together with the telegraph fluctuations [110].

The MOSFET current is proportional to the product of mobility μ_{eff} times the charge carrier density or number N_s. Low-frequency fluctuations in charge transport are caused by stochastic changes in either of these parameters, which can be independent (uncorrelated) or dependent (correlated). In most cases, fluctuations in the current, or more specifically in the $\mu_{eff} \times N_s$ product are monitored, which does not allow the separation of mobility from number effects and therefore obscures the identification of the dominant $1/f$ noise source.

The voltage noise spectrum density is [111]:

$$S_V(f) = \frac{q^2 kT\lambda}{\alpha WLC_{ox}^2 f}(1 + \sigma\mu_{eff}N_s)^2 N_{bt} \tag{5.50}$$

where

λ is the tunneling parameter
μ_{eff} is the effective carrier mobility
α is the Coulombic scattering parameter
N_s is the density of channel carriers
C_{ox} is the gate oxide capacitance/unit area
WL is the gate area
N_{bt} is the effective border trap density ($1/\text{cm}^3$ eV) near the interface

In weak inversion, the channel carrier density N_s is very low (10^7–$10^{11}/\text{cm}^2$), so the mobility fluctuation contribution becomes negligible and the second term in the bracket can be neglected.

It is assumed that the free carriers tunnel to traps in the oxide with a tunneling time constant, which varies with distance x from the interface. The tunneling parameter and time constant are given in Equation 5.6. For a trap at 1 nm from the semiconductor interface, $\tau_T \approx 0.5$ s or $f = 1/2\pi\tau_T \approx 0.3$ Hz. Hence, a distribution of traps in the oxide gives rise to a wide range of frequencies and can explain the $1/f$ dependence. $1/f$ noise shows sensitivity to the wafer orientation, which correlates with interface trap density. Figure 5.34 shows an example of low-frequency noise before and after annealing with the

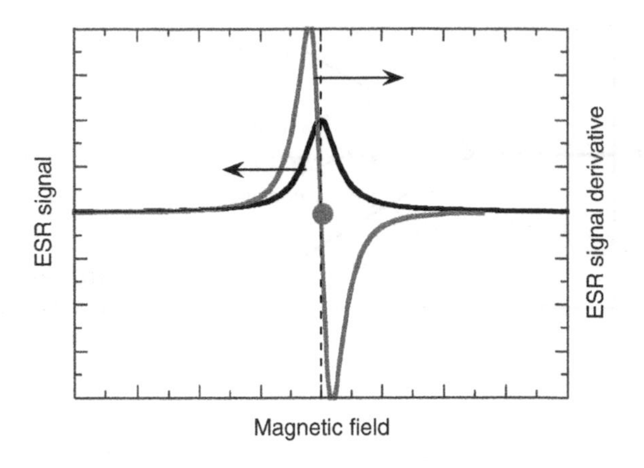

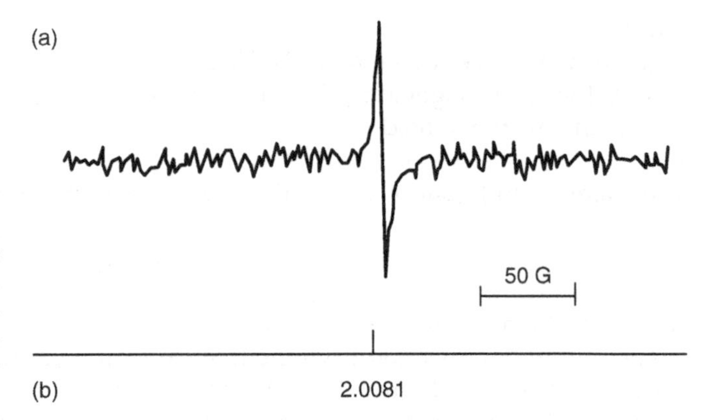

FIGURE 5.36
(a) ESR signal versus magnetic field and derivative of ESR signal versus magnetic field and (b) ESR signals from oxidized (111) silicon wafer. (Reused from Caplan, P.J., Poindexter, E.H., Deal, B.E., and Razouk, R.R., *J. Appl. Phys.*, 50, 5847, 1979. With permission.)

constant until resonance is achieved in either case. The latter is preferred. The ESR signal looks qualitatively like that in Figure 5.36a, showing a peak at a certain magnetic field. To determine the relevant magnetic field at resonance it is easier to plot the differentiated ESR signal versus magnetic field, as also illustrated in Figure 5.36a, because the signal is detected with a lock-in amplifier. Instead of the magnetic field at the peak, one simply determines the zero-point crossing, indicated by the point.

The ESR spectrometer consists of a frequency-stable radiation source provided by a Gunn diode and a magnet, which is slowly swept to record a spectrum. By far the most widely used frequency is X-band, about 9.5 GHz, due to its relatively moderate magnetic fields and inexpensive Gunn diode microwave sources. Higher frequencies require higher magnetic fields, leading to higher sensitivity. The most common higher frequency is 35 GHz with a wavelength of 8 mm. Microwave components and samples are thus correspondingly much smaller than for X-band measurements [115]. The sample is located in a resonant cavity with dimensions matching the wavelength of the radiation so that a standing wave pattern is set up in it. Varying the magnetic applied field to bring the sample to resonance leads to microwave power absorption by the sample. Magnetic field modulation is essential in ESR to allow for lock-in detection with sensitivities in the mid $10^{10}/cm^2$ range. The magnetic field is modulated at typically 100 kHz.

ESR has been extensively used to characterize defects in insulators and other materials. As mentioned earlier, it led to the identification of the nature of interface traps at the SiO_2/Si interface. Early irradiation studies used ESR to identify various radiation-induced defects (Chapter 6) [116]. Defect spectra in SiO_2 are easily separated from those due to defects in Si since the resonance line shapes of SiO_2 defects are independent of magnetic field direction, whereas those in Si are not. The magnetic field orientation is due to the physical defect location at the interface (interface of a crystalline material, hence repeated oriented lattice structure) or in the oxide (randomly oriented amorphous material). Negative bias-temperature instability (NBTI) is a significant reliability issue in p-MOSFETs. It manifests itself as a threshold voltage and drain current degradation and interface trap generation is one of the culprits. ESR was used early on during NBTI recognition to follow P_b center generation by negative gate voltage stress [117].

Corona charging is used routinely in semiconductor characterization [118]. The method consists of depositing corona charge on the surface of the sample and measuring a variety of material/device parameters. The corona charge replaces the metal/poly-Si gate. On oxidized samples, it can generate oxide electric fields as high as 10–15 MV/cm. Since the deposited ions are thermalized by their passage through a long air path, corona charging would seem to constitute an almost ideal nondamaging biasing tool. For positive biasing, the charging process involves hydrogenic species that can interact with the Si/SiO_2 system. The impact of room temperature corona charging biasing on thermal (100) and (111) Si/SiO_2 has been probed by ESR [119]. Both structures fully passivated in H_2 as well as exhaustively H-depleted ones were investigated. At least five types of ESR-active defects were generated, most likely due to the interaction of atomic hydrogen released during the corona process with the Si/SiO_2 system. However, for lower corona-induced oxided electric fields, this biasing technique does not damage the sample [120].

A variation of ESR is spin-dependent recombination (SDR) [121]. In conventional Shockley–Read–Hall recombination, electrons and holes are captured by deep-level impurities during a recombination event. In SDR, the semiconductor device is placed in a strong magnetic field that polarizes the spins of the dangling bond electrons as well as the conduction electrons and the holes. Consider a silicon dangling bond defect with an unpaired electron when it is electrically neutral. Let the SRH recombination process begin with electron capture by the defect. If both the dangling bond and electron have the same spin orientation, the conduction electron cannot be captured, because two electrons may not occupy the same orbital with the same spin quantum number. Thus, placing the semiconductor sample in the magnetic field modulates the trap capture cross section by a combination of magnetic field and microwaves.

5.4 Summary

This chapter has provided an overview of some of the more common defects observed in MOS devices, divided into oxide, border, and interface traps. While the defects in pure SiO_2 are quite well understood, that is not the case for high-*K*/oxide insulators. Not only are high-*K* dielectrics more complex, there is an additional interface in such insulators and the interface and bulk defects are usually higher and less well understood (Chapters 9 through 12). A comprehensive overview has also been provided of a fairly complete set of electrical characterization methods. In most cases, these techniques, first developed for SiO_2, were extended to more complex dielectrics. For some of the newer techniques a brief theoretical discussion was provided, but for the well-known methods, the reader is directed to the appropriate references. Wherever possible, applications are provided to interface oxide and border trap measurements. Other techniques may well be developed in the future, but for the most part existing techniques, some of which have been around since the 1960s, have served the IC community well and will continue to do so. Some of the techniques must be used with caution for advanced ICs because gate oxide leakage current and other nonidealities make data interpretation more difficult.

50. Berglund, C.N., Surface states at steam-grown silicon–silicon dioxide interfaces, *IEEE Trans. Electron Dev.* ED-13, 701, 1966.
51. Castagné, R. and Vapaille, A., Description of the SiO_2–Si interface properties by means of very low frequency MOS capacitance measurements, *Surf. Sci.* 28, 157, 1971.
52. Cohen, N.L., Paulsen, R.E., and White, M.H., Observation and characterization of near-interface oxide traps with *C–V* techniques, *IEEE Trans. Electron Dev.* 42, 2004, 1995.
53. Wu, W.H. et al., Spatial and energetic distribution of border traps in the dual-layer HfO_2/SiO_2 high-*k* gate stack by low-frequency capacitance–voltage measurement, *Appl. Phys. Lett.* 89, 162911, 2006.
54. Schroder, D.K., *Semiconductor Material and Device Characterization*, 3rd edn., Wiley-Interscience, New York, 2006, p. 432.
55. Hall, S., Buiu, O., and Lu, Y., Direct observation of anomalous positive charge and electron-trapping dynamics in high-*k* films using pulsed MOS-capacitor measurements, *IEEE Trans. Electron Dev.* 54, 272, 2007.
56. Nicollian, E.H. and Goetzberger, A., The Si–SiO_2 interface—electrical properties as determined by the metal-insulator-silicon conductance technique, *Bell Syst. Tech. J.* 46, 1055, 1967.
57. Nicollian, E.H. and Brews, J.R., *MOS (Metal Oxide Semiconductor) Physics and Technology*, Wiley, New York, 1982, p. 180.
58. Vogel, E.M. et al., Limitations of conductance to the measurement of the interface state density of MOS capacitors with tunneling gate dielectrics, *IEEE Trans. Electron Dev.* 47, 601, 2000; Ma, T.P. and Barker, R.C., Surface-state spectra from thick-oxide MOS tunnel junctions, *Solid State Electron.* 17, 913, 1974.
59. Eaton, D.H. and Sah, C.T., Frequency response of Si–SiO_2 interface states on thin oxide MOS capacitors, *Phys. Stat. Sol.* 12(a), 95, 1972.
60. Uren, M.J., Collins, S., and Kirton, M.J., Observation of "slow" states in conductance measurements on silicon metal-oxide-semiconductor capacitors, *Appl. Phys. Lett.* 54, 1448, 1989.
61. Kuhn, M. and Silversmith, D.J., Ionic contamination and transport of mobile ions in MOS structures, *J. Electrochem. Soc.* 118, 966, 1971; Hillen, M.W. and Verwey, J.F., Mobile ions in SiO_2 layers on Si, in *Instabilities in Silicon Devices: Silicon Passivation and Related Instabilities*, G. Barbottin and A. Vapaille (eds.), Elsevier, Amsterdam, 1986, p. 403.
62. Stauffer, L. et al., Mobile ion monitoring by triangular voltage sweep, *Solid State Technol.* 38, S3, 1995.
63. Hillen, M.W., Greeuw, G., and Verweij, J.F., On the mobility of potassium ions in SiO_2, *J. Appl. Phys.* 50, 4834, 1979.
64. Lang, D.V., Deep-level transient spectroscopy: a new method to characterize traps in semiconductors, *J. Appl. Phys.* 45, 3023, 1974; Lang, D.V., Fast capacitance transient apparatus: Application to ZnO and O centers in GaP p–n junctions, *J. Appl. Phys.* 45, 3014, 1974.
65. Schroder, D.K., *Semiconductor Material and Device Characterization*, 3rd edn., Wiley-Interscience, New York, 2006, p. 270.
66. Choi, B.D. et al., Latent iron in silicon, *Jpn. J. Appl. Phys.* 40, L915, 2001.
67. Wang, K.L. and Evwaraye, A.O., Determination of interface and bulk-trap states of IGFET's using deep-level transient spectroscopy, *J. Appl. Phys.* 47, 4574, 1976.
68. Schulz, M. and Johnson, N.M., Transient capacitance measurements of hole emission from interface states in MOS structures, *Appl. Phys. Lett.* 31, 622, 1977; Tredwell, T.J. and Viswanathan, C.R., Determination of interface-state parameters in a MOS capacitor by DLTS, *Solid State Electron.* 23, 1171, 1980; Yamasaki, K., Yoshida, M., and Sugano, T., Deep level transient spectroscopy of bulk traps and interface states in Si MOS diodes, *Jpn. J. Appl. Phys.* 18, 113, 1979.
69. Johnson, N.M., Measurement of semiconductor–insulator interface states by constant-capacitance, deep-level transient spectroscopy, *J. Vac. Sci. Technol.* 21, 303, 1982.
70. Yamasaki, K., Yoshida, M., and Sugano, T., Deep level transient spectroscopy of bulk traps and interface states in Si MOS diodes, *Jpn. J. Appl. Phys.* 18, 113, 1979.
71. Lakhdari, H., Vuillaume, D., and Bourgoin, J.C., Spatial and energetic distribution of Si–SiO_2 near-interface states, *Phys. Rev.* B38, 13124, 1988.
72. Brugler, J.S. and Jespers, P.G.A., Charge pumping in MOS devices, *IEEE Trans. Electron Dev.* ED-16, 297, 1969.

73. Bauza, D. and Ghibaudo, G., Analytical study of the contribution of fast and slow oxide traps to the charge pumping current in MOS structures, *Solid State Electron.* 39, 563, 1996.
74. Weintraub, C.E. et al., Study of low-frequency charge pumping on thin stacked dielectrics, *IEEE Trans. Electron Dev.* 48, 2754, 2001.
75. Groeseneken, G. et al., A reliable approach to charge-pumping measurements in MOS transistors, *IEEE Trans. Electron Dev.* ED-31, 42, 1984.
76. Han, J.P. et al., Asymmetric energy distribution of interface traps in n- and p-MOSFETs with HfO_2 gate dielectric on ultrathin SiON buffer layer, *IEEE Electron Dev. Lett.* 25, 126, 2004.
77. Tseng, W.L., A new charge pumping method of measuring Si–SiO_2 interface states, *J. Appl. Phys.* 62, 591, 1987; Hofmann, F. and Krautschneider, W.H., A simple technique for determining the interface-trap distribution of submicron metal-oxide-semiconductor transistors by the charge pumping method, *J. Appl. Phys.* 65, 1360, 1989.
78. Saks, N.S. and Ancona, M.G., Determination of interface trap capture cross sections using three-level charge pumping, *IEEE Electron Dev. Lett.* 11, 339, 1990; Siergiej, R.R., White, M.H., and Saks, M.S., Theory and measurement of quantization effects on Si–SiO_2 interface trap modeling, *Solid State Electron.* 35, 843, 1992.
79. Bauza, D., Extraction of Si–SiO_2 interface trap densities in MOS structures with ultrathin oxides, *IEEE Electron Dev. Lett.* 23, 658, 2002; Bauza, D., Electrical properties of Si–SiO_2 interface traps and evolution with oxide thickness in MOSFET's with oxides from 2.3 to 1.2 nm thick, *Solid State Electron.* 47, 1677, 2003; Masson, P., Autran, J.-L. and Brini, J., On the tunneling component of charge pumping current in ultrathin gate oxide MOSFET's, *IEEE Electron Dev. Lett.* 20, 92, 1999.
80. Paulsen, R.E. and White, M.H., Theory and application of charge pumping for the characterization of Si–SiO_2 interface and near-interface oxide traps, *IEEE Trans. Electron Dev.* 41, 1213, 1994.
81. Heiman, F.P. and Warfield, G., The effects of oxide traps on the MOS capacitance, *IEEE Trans. Electron Dev.* ED-12, 167, 1965.
82. Jakschik, S. et al., Influence of Al_2O_3 dielectrics on the trap-depth profiles in MOS devices investigated by the charge-pumping method, *IEEE Trans. Electron Dev.* 51, 2252, 2004.
83. Cartier, E. et al., Fundamental understanding and optimization of PBTI in nFETs with SiO_2/HfO_2 gate stack, *IEEE Int. Electron Dev. Meet.* 321, 2006.
84. Kerber, A. et al., Direct measurement of the inversion charge in MOSFETs: Application to mobility extraction in alternative gate dielectrics, *IEEE VLSI Symp. Tech. Dig.* 159, 2003.
85. Taur, Y. and Ning, T.H., *Fundamentals of Modern VLSI Devices*, Cambridge University Press, Cambridge, 1998.
86. McWhorter, P.J. and Winokur, P.S., Simple technique for separating the effects of interface traps and trapped-oxide charge in metal-oxide-semiconductor transistors, *Appl. Phys. Lett.* 48, 133, 1986.
87. Kerber, A. et al., Characterization of the V_T-instability in SiO_2/HfO_2 gate dielectrics, *IEEE Int. Rel. Phys. Symp.* 41, 41, 2003.
88. Zhao, C.Z. et al., Determination of capture cross sections for as-grown electron traps in HfO_2/HfSiO stacks, *J. Appl. Phys.* 100, 093716, 2006.
89. Neugroschel, A. et al., Direct-current measurements of oxide and interface traps on oxidized silicon, *IEEE Trans. Electron Dev.* 42, 1657, 1995.
90. Fitzgerald, D.J. and Grove, A.S., Surface recombination in semiconductors, *Surf. Sci.* 9, 347, 1968.
91. Guan, H. et al., Nondestructive DCIV method to evaluate plasma charging in ultrathin gate oxides, *IEEE Electron Dev. Lett.* 20, 238, 1999.
92. Maserijian, J. and Zamani, N., Behavior of the Si/SiO_2 interface observed by Fowler–Nordheim tunneling, *J. Appl. Phys.* 53, 559, 1982.
93. DiMaria, D.J. and Cartier, E., Mechanism for stress-induced leakage currents in thin silicon dioxide films, *J. Appl. Phys.* 78, 3883, 1995.
94. DiMaria, D.J., Defect generation in field-effect transistors under channel-hot-electron stress, *J. Appl. Phys.* 87, 8707, 2000.
95. Petit, C. et al., Low voltage SILC and P- and N-MOSFET gate oxide reliability, *Microelectron. Reliab.* 45, 479, 2005.
96. Ghetti, A. et al., Tunneling into interface states as reliability monitor for ultrathin oxides, *IEEE Trans. Electron Dev.* 47, 2358, 2000.

6.1 Introduction and Background

Two types of silicon dangling bond centers play dominating roles in metal/oxide/silicon (MOS) device limitations. Silicon dangling bonds defects at the Si/SiO_2 boundary, P_b centers dominate Si/SiO_2 interface trapping. Silicon dangling bond defects called E' centers dominate deep levels in the oxide. E' centers can also play dominating roles in the electronic properties of the near Si/SiO_2 interface dielectrics, serving as switching traps or border traps (Chapter 7). Under some circumstances, some E' centers also act as interface traps. Analysis of a combination of electron spin resonance (ESR) and electrical measurements provides a fairly detailed understanding of the relationship between the structure and electronic properties of the P_b interface centers and E' oxide centers. Straightforward application of the fundamental principles of statistical mechanics provides a partial explanation of the frequently noted correlation between oxide deep level defects and Si/SiO_2 interface trap defects.

This chapter reviews the spectroscopy, electronic properties, and device relevance of these two families of defects. Since the understanding of the defects has come from ESR, the chapter begins with a brief introduction to this technique. It is important to note that P_b and E' centers are not necessarily the only electrically active point defects in MOS devices. Although many studies have established the dominating roles of these two families of silicon dangling bond centers, the limited precision of ESR measurements as well as that of the relevant electrical measurements precludes concluding with certainty that either defect is entirely responsible for any device limitation. However, they are clearly responsible for the majority of the deep levels in most and perhaps all the phenomena discussed in this chapter. Furthermore, under some circumstances, it is clear that other defects are playing important roles. For example, the addition of nitrogen introduces greater complexity in plasma-nitrided gate oxides. Not surprisingly, several very different defects clearly play important roles in gate stacks based on hafnium oxide/silicon dioxide systems.

Several decades of ESR [1] studies of the MOS system have established the roles of these two families of silicon dangling bond defects in MOS device instabilities. At the Si/SiO_2 boundary the P_b center family dominates interface trapping under many circumstances [2–32]. Largely due to the simplicity of the ESR spectroscopy, most earlier studies focused upon the (111) Si/SiO_2 interface, where only one P_b defect, called simply P_b, appears [2–15].

At the technologically important (100) Si/SiO_2 interface, two P_b center variants are important: the P_{b0} and the P_{b1} centers [5,14,17,19,22–32]. The P_{b0} center has been reasonably well understood for quite some time because it is essentially identical to the (111) Si/SiO_2 silicon dangling bond center called P_b. The relative simplicity of the (111) Si/SiO_2 boundary not only makes sensitive ESR measurement relatively easy; the interface also makes the analysis particularly easy. This ease of measurement and analysis led to a fairly complete first-order understanding of the (111) P_b by the late 1980s. The (111) P_b and the (100) P_{b0} centers are both silicon dangling bond centers in which the central silicon is back bonded to three other silicons [3–5,9,17,22,25,27–29]. Both centers' dangling bond orbitals are directed along ⟨111⟩ symmetry axes. The centers have quite similar ESR parameters. The P_{b1} center has been less well understood, with gross conflicts appearing in the literature with regard to its electronic properties [14,28–30,32].

Although several dozen paramagnetic (ESR active) centers have been observed in MOS oxides, the E' center family [33–37] plays a dominating role in a wide variety of circumstances [10,11,16,18–22,24,38–43]. The centers clearly dominate oxide hole trapping [10,11,19,21,38,39] and almost certainly play important roles in instabilities resulting from the presence of very high electric fields such as stress-induced leakage currents [16,42,43].

Both the P_b and E' centers involve unpaired electrons primarily residing in high p-character wave function orbitals on silicon atoms [9,17,22,25,27–29,34–37,41]. The P_b centers involve silicons back bonded to three silicons. The E' centers involve silicons back bonded to oxygens.

6.2 ESR

In ESR measurements [1], the sample under study is exposed to a large slowly varying magnetic field and a microwave frequency magnetic field oriented perpendicularly to the applied field [1]. Usually the measurements are made at X band: a microwave frequency $\nu \approx 9.5$ GHz. An unpaired electron has two possible orientations in the large applied field and thus two possible orientation-dependent energies. (From classical electricity and magnetism, the energy of a magnetic moment $\boldsymbol{\mu}$ in a magnetic field $\mathbf{H}$ is $-\boldsymbol{\mu} \cdot \mathbf{H}$.) Magnetic resonance occurs when the energy difference between the two electron orientations is equal to the Planck constant, h, times the microwave frequency. For the very simple case of an isolated electron, the resonance requirement may be expressed as

$$h\nu = g_0 \beta_e H \tag{6.1}$$

where $g_0 = 2.002319$ and β_e is the Bohr magneton, $eh/4\pi m_e$, where e is electronic charge and m_e is the electron mass. The Bohr magneton is 9.274015×10^{-28} J/G.

Equation 6.1 describes the resonance condition for an electron that does not otherwise interact with its surroundings. The structural information provided by ESR is due to deviations from this simple equation. For the relatively simple trapping centers studied in MOS systems, these deviations are due to spin–orbit coupling and electron–nuclear hyperfine interactions.

6.2.1 Spin–Orbit Coupling

The deviations from Equation 6.1 due to spin–orbit coupling come about because a charged particle, the electron, traveling in an electric field due to the nuclear charge, experiences a magnetic field $\mathbf{B} = \mathbf{E} \times \mathbf{v}/c^2$, where $\mathbf{E}$ is the electric field, $\mathbf{v}$ is the velocity, and c is the speed of light [1]. The spin–orbit interaction may be understood qualitatively (and only qualitatively) in terms of the Bohr picture: an electron moves about the nucleus in a circular orbit. It would appear to an observer on the electron that the positively charged nucleus is in a circular orbit about the electron. (Similarly, it appears to an unsophisticated observer on Earth that the Sun is in a circular orbit about the Earth.) The nucleus thus generates a local magnetic field that would scale with the electron's orbital angular momentum, $\mathbf{r} \times \mathbf{p}$, and with the nuclear charge. One would thus correctly surmise that spin–orbit coupling interactions increase with increasing atomic number and orbital angular momentum quantum number.

In solids, the spin–orbit interaction is quenched, but a second-order effect appears from a coupling to excited states through spin–orbit coupling generated by the application of a large magnetic field. This effect scales with the applied magnetic field and depends on the orientation of the paramagnetic defect in the applied magnetic field. The spin–orbit coupling may be included in the ESR resonance condition by replacing the constant g_0 of Equation 6.1 with a g matrix with elements g_{ij}. The symmetry of this matrix reflects the symmetry of the paramagnetic center. Under some circumstances, the symmetry of the matrix alone allows identification of the defect under study. (The P_b and P_{b0} centers were fairly convincingly identified on the basis of their g matrix [3–5].)

Perturbation theory allows calculation (with modest accuracy) of the g matrix elements for both P_b and E' centers [1]. The components of the g matrix are given by

$$g_{ij} = g_0\delta_{ij} - 2\lambda \sum_k \frac{\langle \alpha|L_i|k\rangle\langle k|L_j|\alpha\rangle}{(E_k - E_\alpha)} \tag{6.2}$$

where

g_0 is the free electron value 2.0023193
λ is the atomic spin–orbit coupling constant
L_i and L_j are angular momentum operators appropriate for the x-, y-, or z-directions, and the summation is over all states k

State $|\alpha\rangle$ and energy E_α correspond to the paramagnetic state for which the g matrix is calculated. Equation 6.2 provides some very useful (qualitative and semiquantitative) physical insight in understanding the ESR spectra of both P_b and E' centers. As mentioned previously, the centers involve electrons residing in high p-character orbitals on silicon atoms. The angular momentum operator corresponding to the symmetry axis direction of a p-orbital yields a zero when operating on that orbital [44]. Therefore, the g value corresponding to the magnetic field parallel to the dangling bond orbital, $g_{\parallel}$, should be about $g_0 \cong 2.002$. The g value corresponding to the magnetic field perpendicular to this axis, $g_{\perp}$, should exhibit the largest deviation from this value [44]. Furthermore, since both E' and P_b levels appear within a band gap, one would intuitively expect that the denominator of the far right term of Equation 6.2 would result in larger deviations of $g_{\perp}$ from g_0 for the wide band gap ($E_g(SiO_2) \cong 9$ eV) E' centers than the narrower band gap ($E_g(Si) \cong 1.1$ eV) P_b centers. This is also the case. For an axially symmetric defect, we expect the g corresponding to an angle θ between the applied field and the defect symmetry axis to be:

$$g = \left(g_{\parallel}^2 \cos^2\theta + g_{\perp}^2 \sin^2\theta\right)^{1/2} \tag{6.3}$$

For the most general case of lower symmetry

$$g = \left(g_1^2 \cos^2\alpha + g_2^2 \cos^2\beta + g_3^2 \cos^2\gamma\right)^{1/2} \tag{6.4}$$

where α, β, and γ correspond to the field angle with respect to the principal axis directions of the g matrix.

6.2.2 Electron–Nuclear Hyperfine Interactions

The other important source of deviation from Equation 6.1 is the hyperfine interaction of the unpaired electron with nearby nuclei [1,45]. Certain nuclei have magnetic moments; in metal/silicon dioxide/silicon systems, some examples are ^{29}Si (spin 1/2), ^{1}H (spin 1/2), ^{31}P (spin 1/2), and ^{14}N (spin 1). A spin 1/2 nucleus has two possible orientations in the large applied field; a spin 1 nucleus has three possible orientations. Each nuclear moment orientation corresponds to one local nuclear moment field distribution.

One may envision the nuclear moment interacting with an unpaired electron residing in a wave function that is a linear combination of atomic orbitals (LCAOs). For all varieties of P_b and E' centers, a combination of s- and p-type wave functions provide a reasonable approximation to reality. The LCAO for an unpaired electron can therefore be written as

$$|\alpha\rangle = \sum_{n} a_n \{c_s|s\rangle + c_p|p\rangle\} \tag{6.5}$$

where

$|s\rangle$ and $|p\rangle$ represent the appropriate atomic orbitals corresponding to the nth site
a_n^2 represents the localization on the nth site
c_s^2 and c_p^2 represent, respectively, the amount of s- and p-character of the wave function on the nth atomic site

With possibly one exception, $a_1^2 \cong 1$ for all of the P_b and E' centers discussed in this chapter; that is, the defects' unpaired electron is reasonably well localized at a single nuclear site $(0.6 < a_2^1 < 1)$. To first order then, we can nearly always interpret the P_b and E' ESR spectra in terms of s/p hybridized atomic orbitals localized at a central site. We can understand the electron–nuclear hyperfine interaction of an electron in an sp orbital by considering the p and s orbital contributions individually and then add the contributions together. The electron–nuclear interaction of an electron in a p orbital is anisotropic: it corresponds to an essentially classical magnetic dipole interaction as schematically illustrated in Figure 6.1.

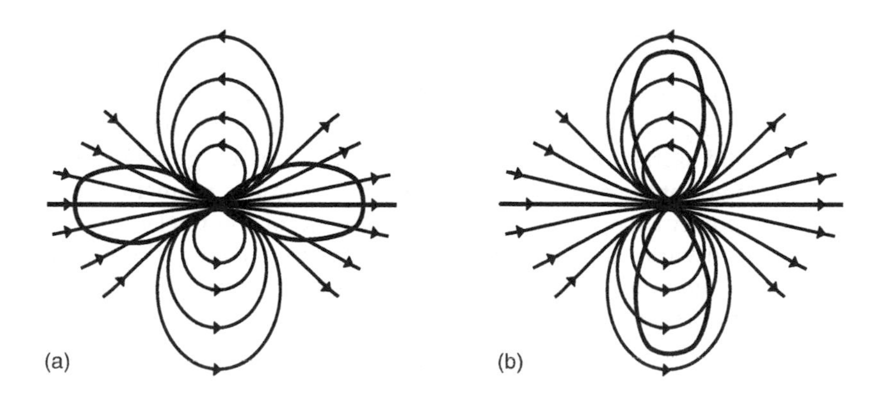

FIGURE 6.1
Schematic illustration of an electron in a p orbital interacting with a magnetic nucleus (a) for the nuclear moment parallel to the symmetry axis and (b) perpendicular to the symmetry axis. (Reprinted from Lenahan, P.M. and Conley, J.F., *J. Vac. Sci. Technol. B*, 16, 2134, 1998. With permission.)

where α, β, and γ correspond to the angles between the magnetic field and the principal axes of the hyperfine matrix.

Equations 6.2, 6.3, and 6.9 through 6.15 provide a very straightforward basis for analyzing ESR results for axially symmetric defects with a specific orientation with respect to the applied magnetic field. The dangling bond centers at the Si/SiO_2 boundary (P_b and P_{b0}) yield spectra readily described by Equations 6.2, 6.3, and 6.9 through 6.17 since the crystallinity of the silicon substrate provides a fixed relationship between the applied field and defect orientation. (The P_{b1} spectrum is slightly more complex due to its lower symmetry; it has three different values for both the g and hyperfine matrices.) One important qualitative point to be gleaned from the discussion of the hyperfine interactions is that, for an electron in an sp hybridized Si dangling bond orbital, the splitting due to the hyperfine interaction will be largest when the magnetic field points along the direction of a dangling bond direction and smallest when the field is perpendicular to the dangling bond direction. Similarly simple first-order logic (discussed previously) indicates that the g value corresponding to the dangling bond direction would be expected to be closer to the free electron value ($g_0 \cong 2.002$) and the g would be expected to deviate the most when the field is perpendicular to this direction.

The description of ESR spectra of defects within an amorphous film is more complex than is the case for defects within a crystalline environment. Within the amorphous matrix, all defect orientations are equally likely and, due to the lack of long range order, slight differences in local defect geometry may be anticipated. The presence of defects at all orientations leads to the continuous distribution of both g and A values from $g_{\|}$ and $A_{\|}$ to $g_{\perp}$ and $A_{\perp}$. The differences in local geometry lead to slight defect to defect variations in $g_{\|}$, $g_{\perp}$, $A_{\|}$, and $A_{\perp}$. (Additional complexity, in the form of three g and A terms are required for defects with lower than axial symmetry.)

These complications are relatively easy to deal with. The random distribution of defect orientation can be dealt with easy in terms of analytical expressions found in most ESR textbooks [1]. (For axially symmetric centers, the anticipated ESR patterns are intuitively obvious. Far fewer centers will have the symmetry axis parallel to the applied field than perpendicular to it; thus, the ESR spectrum intensity will be far stronger at the $A_{\perp}$ and $g_{\perp}$ values than at $A_{\|}$ and $g_{\|}$.) The slight defect-to-defect variations in g and A values lead to broadening of the line shapes anticipated for unbroadened tensor components. (The analysis is schematically illustrated in Figure 6.3.)

The evaluation of ESR hyperfine matrix components usually allows for a reliable and moderately precise identification of the unpaired electron's wave function. This is the case for silicon dangling bond centers like P_b and E' defects. For reasonably clean MOS oxides, we anticipate significant concentrations of only silicon, oxygen, hydrogen, and under certain circumstances, nitrogen, phosphorous, and boron. The nuclear moments of these atoms are all quite different [1,45]. Over 99.9% of oxygen atoms have nuclear spin zero; over 99.9% of phosphorous, nitrogen, boron, and hydrogen nuclei are magnetic. Silicon is readily distinguishable; 95% of silicon atoms have nuclear spin zero but 5% of the silicons, those with ^{29}Si nuclei, have a nuclear spin of one-half. This 95% spin zero/5% spin one-half ratio is unique among all the elements in the periodic table. Thus, a three-line pattern with the two side peaks each corresponding to about 2.5% the integrated intensity of the much more intense center line, can be convincingly linked to an unpaired electron associated with a silicon atom.

Having identified the nuclear species involved, a first-order analysis of the unpaired electron wave function is extremely straightforward in terms of the LCAO picture. For defects in a crystalline environment, Equation 6.16 can fit to the ESR spectrum for several values of θ. For defects in an amorphous (or polycrystalline) environment, one may fit the appropriately broadened analytical expressions to the ESR spectra to yield $A_{\perp}$ and

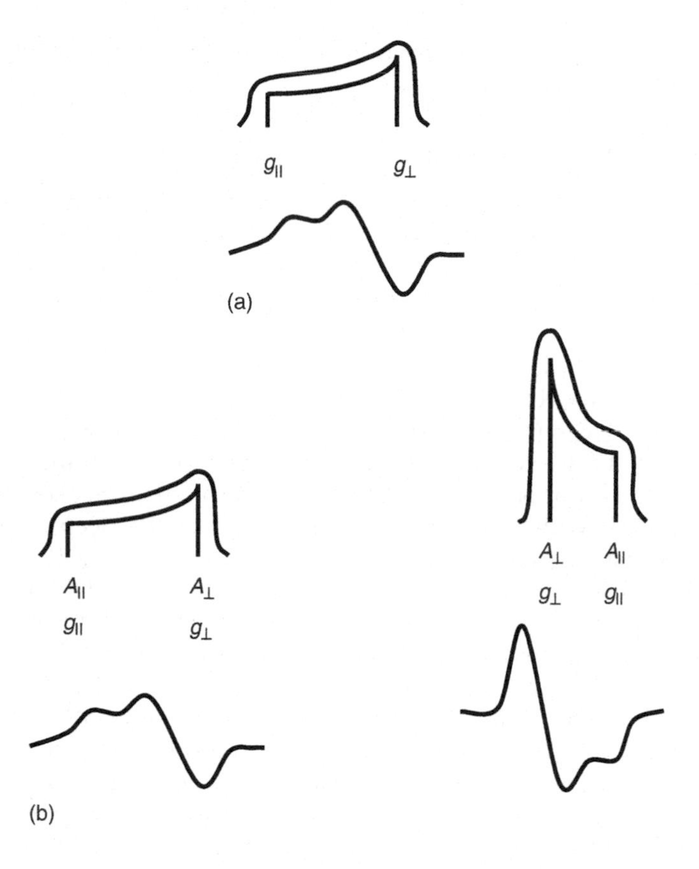

FIGURE 6.3
Schematic sketches of ESR absorption (a) and derivative (b) spectra for centers with axial symmetry in an amorphous matrix. The $g_{\perp}$, $g_{\|}$, $A_{\perp}$, and $A_{\|}$ values correspond to the resonance condition when the magnetic field is parallel ($\|$) or perpendicular ($\perp$) to the defect symmetry axes. The sketches in (a) correspond to the single line anticipated for defects that are not associated with a magnetic nucleus. The sketches in (b) correspond to the two lines anticipated for defects associated with a spin 1/2 nucleus. (The spin 1/2 nucleus has two possible orientations in the large applied field thus providing two different hyperfine contributions to the resonance condition.)

$A_{\|}$, as illustrated in Figure 6.3. Using Equations 6.9 through 6.13, one then obtains the isotropic and anisotropic coupling constants A_{iso} and A_{aniso}. Tabulated values [1] of A_{iso}, a_0, and A_{aniso}, b_0, calculated for 100% occupation probability can then be utilized to determine the hybridization and localization of the electronic wave functions. Although the crude analysis just discussed is not extremely precise it is, to first order, quite reliable for the relatively simple P_b and E' defects. One should realize that the isolated atomic values obtained for a_0 and b_0 are themselves only moderately accurate and that placing a silicon atom in an oxide matrix will inevitably alter the constants somewhat. Nevertheless, a straightforward analysis of hyperfine parameters provides moderately accurate measurement of hybridization and localization.

6.2.3 Quantitative Analysis (Spin Counting)

In conventional ESR detection, resonance is observed through changes in the quality factor of a microwave cavity induced by the resonance [1]. Conventional ESR sensitivity is about 10^{10} spins. (The precise number depends upon A and g and the coupling of the spins to the surrounding lattice.) Conventional ESR detection provides a count of paramagnetic defects within a sample to a precision of about ±10% in relative number and an absolute precision of a little better than a factor of 2.

6.2.4 Spin-Dependent Recombination

In semiconductor device studies, an approach very different from conventional ESR detection is quite powerful: spin-dependent recombination (SDR) [20,22,23,25,28,29,46,47]. SDR is an ESR technique in which the spin-dependent nature of recombination events is exploited so that one may observe ESR through measurement of recombination currents. A qualitative explanation of SDR, first proposed by Lepine [46], will be sufficient for the purposes of this chapter. The Lepine model combines the Shockley–Read–Hall (SRH) [48,49] recombination model and the Pauli exclusion principle.

In the SRH model [48,49], recombination of electrons and holes takes place through deep level defects in the semiconductor band gap. An electron is trapped at a deep level defect; then a hole is trapped at the same deep level defect. (The sequence could obviously be reversed, with a hole being trapped first at the deep level, and then an electron.) Suppose that the deep level defect is a silicon dangling bond. The dangling bond has an unpaired electron when it is electrically neutral. In SDR, as in conventional ESR, the semiconductor device is placed in a strong magnetic field that polarizes the spins of the dangling bond electrons as well as the conduction electrons and the holes. (That is, the electron spins tend to orient with the applied magnetic field.) The SRH recombination process begins with electron capture (or hole capture) at the deep level site. If both the dangling bond and conduction electrons have the same spin orientation, the conduction electron cannot be captured at the site because two electrons cannot occupy the same orbital with the same spin quantum number. Thus, placing the semiconductor sample in the magnetic field reduces the average trap capture cross section of the traps. In ESR, electron spins are flipped from one spin orientation to the other when the resonance condition is satisfied. The spin flipping at the trap site increases probability of oppositely oriented traps and conduction electron spins. This process increases the average trap capture cross section. Thus, if a semiconductor device is placed in a strong slowly varying magnetic field and is simultaneously subjected to microwave irradiation at frequency ν, the recombination current will increase when ν and H satisfy the resonance condition. The spin flipping event induced by ESR increases the possibility of oppositely oriented conduction electron and deep level defect spins, increasing the likelihood of recombination events. (A more sophisticated approach to SDR has been developed by Kaplan, Solomon, and Mott [47] (KSM). The KSM model is not precisely applicable to SDR of either E' or P_b centers. However, it more nearly approximates reality than the Lepine model. A detailed discussion of SDR models and the relevance to MOS defects has been published elsewhere [50].)

The observation of SDR thus, essentially by definition, demonstrates that the defect in question has levels deep in the band gap. Furthermore, with specific knowledge of the device in question and the various voltages and currents involved, one may generally, though only very crudely, identify the physical location and energy levels of the defect centers observed in SDR.

6.3 Defect Structure

6.3.1 P_b Center Structure

The chemical and structural nature of P_b centers has been established by several independent, consistent, and mutually corroborating studies. Three P_b variants have been consistently

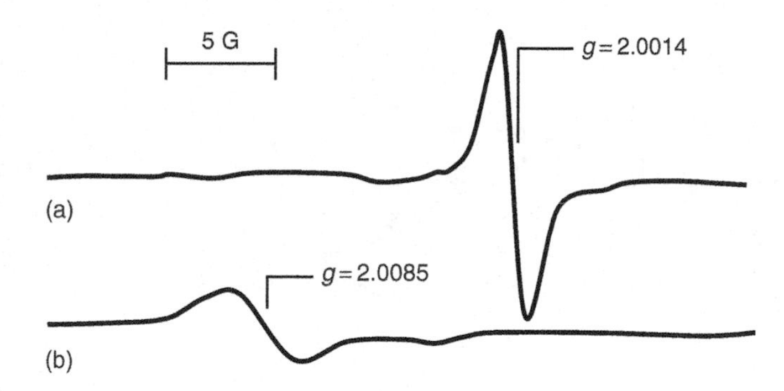

FIGURE 6.4
ESR traces of the P_b center for the magnetic field (a) parallel and (b) perpendicular to the (111) symmetry axis. (Reprinted from Lenahan, P.M. and Conley, J.F., *J. Vac. Sci. Technol. B*, 16, 2134, 1998. With permission.)

observed: at (111) Si/SiO_2 interfaces a defect called simply P_b [2–13,22], and at (100) Si/SiO_2 interfaces, two defects called P_{b0} and P_{b1} [5,14,19,22,25–31]. The structure of both the (111) Si/SiO_2 P_b and the (100) Si/SiO_2 P_{b0} are reasonably well understood; a more rudimentary understanding of P_{b1} exists at this time.

P_b centers were first observed by Nishi and coworkers [2,3]. Their work focused primarily on the (111) Si/SiO_2 system. They showed that the P_b centers were at or very near to the Si/SiO_2 interface and that they possess an axially symmetric g matrix with $g_{\|} \cong 2.000$ and $g_{\perp} \cong 2.01$ with the symmetry axis corresponding to the (111) direction. Figure 6.4 shows representative P_b spectra taken with the field perpendicular and parallel to the (111) surface normal. On the basis of this information, Nishi et al. concluded that P_b centers are trivalent silicon centers at or very near the Si/SiO_2 boundary. The structure of the (111) P_b center is illustrated in Figure 6.5.

Later, Caplan et al. [4,5] obtained more precise g matrix parameters. For the (111) Si/SiO_2 P_b, they found $g_{\|} \cong 2.0014$ and $g_{\perp} \cong 2.0081$ with, as Nishi had observed previously, the symmetry axis corresponding to the Si (111) direction. For the (100) Si/SiO_2 system, they found two P_b variants; P_{b0} and P_{b1} [4,5]. Although hampered by overlapping spectra, Caplan et al. were able to show that the P_{b0} g tensor was quite similar to that of the (111) Si/SiO_2 P_b; they found $g_1 \cong g_{\|} = 2.0015$ and $g_{\perp} \cong g_2 \cong g_3$, with $g_2 = 2.0087$ and $g_3 = 2.0080$. Later, more precise measurements by Kim and Lenahan demonstrated that the P_b g matrix, like that of the (111) P_b center, is also axially symmetric, with $g_{\|} = 2.0015 \pm 0.0002$ and $g_{\perp} = 2.0080 \pm 0.0002$ [19]. A g-map from that work is shown in Figure 6.6. [19] (The symmetry axes for the P_{b0} centers are also [111] directions.) This result was later confirmed by Stesmans and Afanas'ev who, within experimental error, found the same values: $g_{\|} = 2.0018 \pm 0.0003$ and $g_{\perp} = 2.0085 \pm 0.0003$ [26]. These studies strongly indicate that the (111) P_b and the (100) P_{b0} are essentially identical. (As discussed later in this chapter several

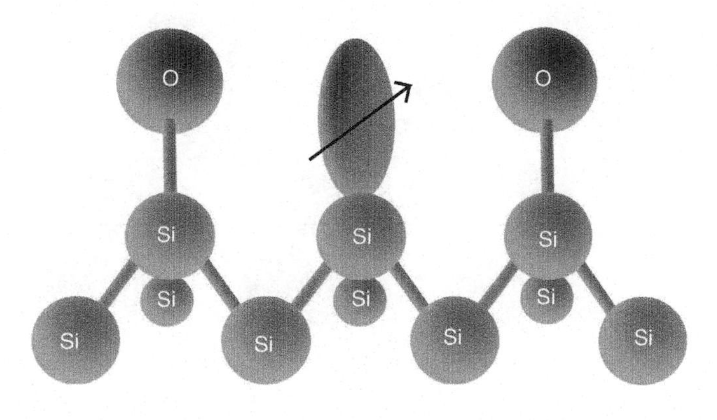

FIGURE 6.5
Schematic illustration of a P_b center at the (111) Si/SiO_2 interface.

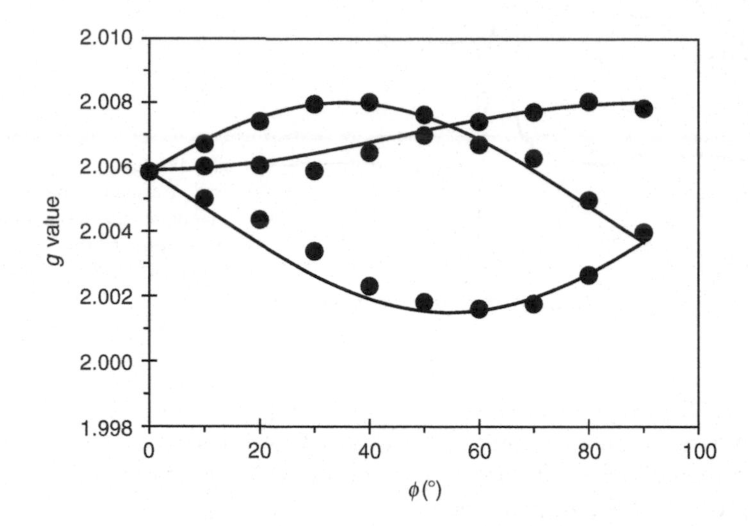

FIGURE 6.6
ESR g value map for P_{b0} centers on (100) silicon substrates. The data points correspond to various angles φ with respect to the (100) interface normal. The sample is rotated about the [011] direction. The solid lines are given by the relation $g = \left(g_{\parallel}^2 \cos^2\theta + g_{\perp}^2 \sin^2\theta\right)^{1/2}$ with $g_{\parallel} = 2.0015$ and $g_{\perp} = 2.0080$, where θ is the angle between the magnetic field and the [111] directions. Note the two different angles involved. The angle θ refers to the angle between the multiple <111> directions in which the P_{b0} centers can point and the magnetic field. The angle φ is the angle between the (100) interface normal and the magnetic field. (Reprinted from Kim, Y.Y. and Lenahan, P.M., *J. Appl. Phys.*, 64, 3551, 1988. With permission.)

other observations also support this conclusion.) The structure of the P_{b0} is illustrated in Figure 6.7.

Poindexter et al. found that the P_{b1} center exhibited lower symmetry, with $g_1 \cong 2.0076$, $g_2 \cong 2.0052$, and $g_3 \cong 2.0012$ [5]. Poindexter et al. reported that the axes corresponding to g_1 are approximately [111] directions, g_2 corresponds to a direction between the [100] and [111] directions, and g_3 approximately corresponds to [011]. Later, more precise measurements by Stesmans and Afanas'ev refined the original Poindexter et al. results yielding $g_1 = 2.0058$, $g_2 = 2.0074$, and $g_3 = 2.0022$ [26]. Stesmans et al. found g_1 to correspond to $[0\bar{1}\bar{1}]$, g_2 to near $[1\bar{1}\bar{1}]$, and g_3 to approximately [211].

The close correspondence between the P_{b0} and P_b g matrix components led Poindexter et al. to propose a near equivalence between the (111) P_b and the (100) P_{b0}, a conclusion that is clearly correct. The fact that the centers' matrices have (111) axial symmetry strongly

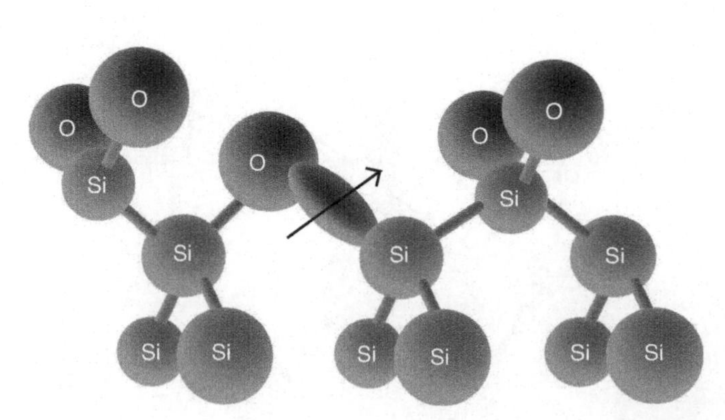

FIGURE 6.7
Schematic illustration of a P_{b0} center at the (100) Si/SiO_2 interface.

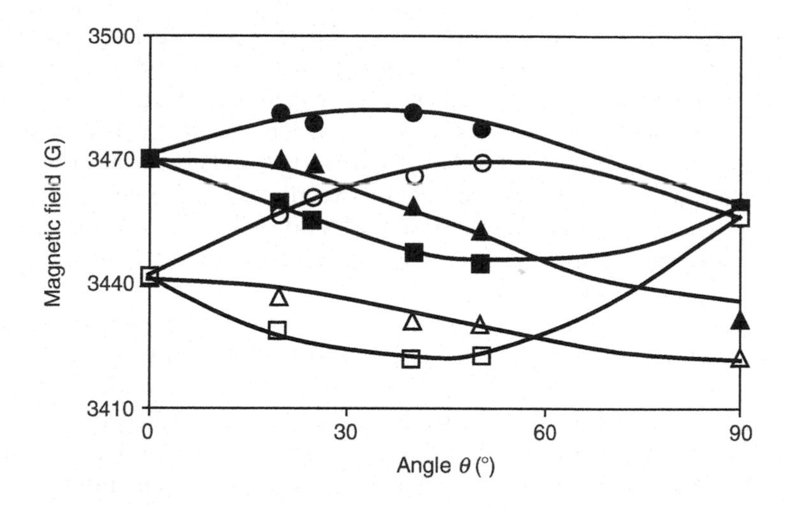

FIGURE 6.8
Plot of both P_{b0} (open data points) and P_{b1} (solid data points) hyperfine lines versus magnetic field orientation. The sample was rotated in the (110) plane. Angle θ is the angle between the [001] direction of the sample and the applied magnetic field. Note that multiple lines appear because multiple orientations for both P_{b0} and P_{b1} dangling bonds occur on the (100) interface, the (111) and approximately (211) directions, respectively. (Reprinted from Mishima, T.D., Lenahan, P.M., and Weber, W., *Appl. Phys. Lett.*, 78, 1453, 2001. With permission.)

supports the idea that they are interfacial silicon dangling bonds. The differences in the g matrix of P_{b1} with respect to P_{b0} and P_b led Poindexter et al. to propose that the defects have different structures. Although the specific P_{b1} model proposed by Poindexter et al., a silicon bonded to two silicons and an oxygen [5] is almost certainly incorrect, their conclusions that the P_{b1} is a somewhat different defect is certainly the case.

Any doubt about the basic structure of the (111) P_b and (100) P_{b0} centers has been resolved by measurements of the hyperfine interactions with ^{29}Si nuclei. Hyperfine results have also helped to refine our understanding of P_{b1} structure. Representative hyperfine results for both P_{b0} and P_{b1} are shown in Figure 6.8. The (111) P_b hyperfine matrix was first measured by Brower [9], who found $A_{\|} \cong 152$ G and $A_{\perp} \cong 89$ G. Using hyperfine parameter calculations tabulated by Weil et al. [1] indicating that $|a_0| = 1693.3$ G and $|b_0| = 40.75$ G, Equations 6.7 and 6.8 indicate that 12% s-character, 88% p-character, and 58% localization. Brower's (111) results were confirmed by Jupina and Lenahan [22]. The (100) P_{b0} hyperfine tensor was first measured by Gabrys et al. [25] and more recently by Stesmans and Afanas'ev [27]. The P_{b0} hyperfine parameters reported by Gabrys et al. are $A_{\perp} = 72 \pm 3$ G and $A_{\|} = 144 \pm 3$ G. Stesmans et al. reported $A_{\perp} = 75 \pm 5$ G and $A_{\|} = 149 \pm 4$ G. Utilizing hyperfine parameter calculations tabulated by Weil et al., these parameters indicate (both studies) 9% s-character, 91% p-character, and (Gabrys) 65% localization or (Stesmans) 67% localization on the central silicon. So, both studies quite clearly indicate that the P_{b0}, like the (111) P_b, involves a silicon dangling bond in which the wave function has very high p-character and is primarily localized on the central silicon. It should be mentioned that the 5%/95% ratio of side (^{29}Si) peaks to center (^{28}Si) peaks was reported in all of these studies. The ≅ 5%/95% ratio of hyperfine lines to center line intensity for both (111) P_b and (100) P_{b0} unequivocally establishes that both centers are silicon dangling bonds. This is so because, among all possible elements within the system, only silicon exhibits a 4.7% abundant magnetic nucleus with a nuclear spin of ½. The two possible orientations of the magnetic silicon nuclei produce two weak side lines with a total integrated intensity corresponding to the relative abundance of the magnetic nuclei, that is 4.67%, or about 5%. The (111) P_b and (100) P_{b0} have very similar hyperfine matrices that demonstrate, as had been indicated by earlier g tensor results, that P_b and P_{b0} are essentially identical defects. Of course quite subtle differences inevitably exist; the (111) P_b center's unpaired electron is in an orbital directed along the (111) surface normal. The (100) P_{b0} is also directed along a (111) direction but this direction is obviously not normal to the (100) surface. The very high p-character would tend to flatten the bonding arrangement with the three back-bonded silicons. It is easy to understand why this should be so. If the dangling bond wave function were purely p-character, all the s-character would be absorbed by the back bonding, making the back bonding pure planar sp^2 like graphite. If the dangling bond wave function were sp^3, the back bonding would be essentially tetrahedral.

The detailed structure of the P_{b1} center is not so well understood; although it too is clearly a silicon dangling bond at the interface, in which the central silicon atom is back bonded to three other silicons. The zero-order P_{b1} bonding arrangement was established in a study by Brower [17]. Brower observed P_{b1} (as well as P_{b0} and P_b) in oxides grown on both (111) and (100) silicon substrates in an ^{17}O enriched atmosphere. Since ^{17}O possesses a nuclear magnetic moment [1,17], if the P_b center silicon were bonded to an oxygen, the ^{17}O nuclear moment would greatly broaden the spectrum. None of the three P_b variants are broadened enough to indicate nearest neighbor oxygens. However, all are slightly broadened, indicating that they are all at the interface. Since nitrogen and hydrogen also possess magnetic moments, we know that the P_{b0} and P_{b1} silicons could not be bonded to them either. Since no other atoms are consistently present in numbers sufficient to account for the P_b centers we may conclude that all three varieties of P_b centers are silicons back bonded to silicons at the respective Si/SiO_2 boundaries.

The P_{b1} centers hyperfine interactions with central ^{29}Si nuclei were first reported by Stesmans and Afanas'ev [27] and later by Mishima and Lenahan [28] and Mishima et al. [29]. The Stesmans et al. results and Mishima et al. results differ, but only slightly beyond the stated error margins of each study. Stesmans et al. reported three hyperfine parameters: A_{xx} $[1\bar{1}0]$, A_{yy} $[\bar{1}\bar{1}1]$, and A_{zz} [112] to be respectively 102, 112, and 167 G. Mishima et al. reported the three hyperfine parameters: A_{xx} $[1\bar{1}0]$, A_{yy} $[\bar{1}\bar{1}1]$, and A_{zz} [112] to be respectively 90, 115, and 167 G. In both studies, the authors estimated that the parameters could be in error by up to ± 3 G, so only the A_{xx} parameters differ by more than the stated experimental error, and only slightly beyond those error estimates. Again, utilizing the hyperfine parameter calculations tabulated by Weil et al. [1] these parameters indicate 13% s-character (Mishima) or 14% s-character (Stesmans) and 87% p-character (Mishima) or 85% p-character (Stesmans) and 60% localization on the central silicon (Mishima) or 57% localization on the central silicon (Stesmans). In view of the previously discussed limits in the precision of this analysis, the conclusions of the two studies are essentially undistinguishable. The P_{b1} is thus also a silicon dangling bond interface trap with a high p-character wave function. Unlike the P_{b0}, its dangling bond points (approximately) along a [112] direction, not a [111] direction. The P_{b1} center wave function is slightly more delocalized than that of the P_{b0} wave function. It also has somewhat higher s-character. The paramagnetic (ESR active) status of these defects corresponds to the electrically neutral case for each center. Since the s-wave function is lower in energy than the p-wave function, the hyperfine results suggest that the neutral P_{b1} level would tend to be lower in the silicon band gap than the P_{b0} levels. Since the P_{b1} wave function is more delocalized than the P_{b0} wave function, one might expect the electron correlation energy of the P_{b1} centers to be smaller than that of the P_b centers. (As discussed later in the chapter, both of these intuitive arguments turn out to be correct.) The spectroscopic parameters of the P_b centers are summarized in Tables 6.1 and 6.2.

TABLE 6.1

^{29}Si Hyperfine Parameters for P_b and E' Centers

(111) P_b	$A_\perp$ **(G)**	$A_\parallel$ **(G)**	**% s**	**% p**	**% Loc.**	
Brower [9]	89 ± 3	152 ± 3	12	88	58	
(100) P_{b0}	$A_\perp$ **(G)**	$A_\parallel$ **(G)**	**% s**	**% p**	**% Loc.**	
Gabrys et al. [25]	72 ± 3	144 ± 3	9	91	65	
Stesmans and Afanas'ev [27]	75 ± 5	149 ± 4	9	91	67	
Mishima and Lenahan [28] and Mishima et al. [29]	72 ± 3	144 ± 3	9	91	65	
(100) P_{b1}	A_{xx} **[1$\bar{1}$0]**	A_{yy} **[$\bar{1}\bar{1}$1]**	A_{zz} **[112]**	**% s**	**% p**	**% Loc.**
Stesmans and Afanas'ev [27]	102 ± 3	112 ± 3	167 ± 3	14	86	57
Mishima and Lenahan [28] and Mishima et al. [29]	90 ± 3	115 ± 3	167 ± 3	13	87	60
E'	$A_\perp$ **(G)**	$A_\parallel$ **(G)**	**% s**	**% p**	**% Loc.**	
Figure 3 (crudely)	417	483	32	68	80	
Griscom et al. [36] (more precisely)	398	464	31	68	79	

6.3.2 *E′* Center Structure

6.3.2.1 Structure of Conventional E′ *Centers*

The most important oxide trapping centers are E' centers [10,11,18–22,38,39], which involve an unpaired electron localized on a silicon back bonded to three oxygens [34]. Often, though not always, the paramagnetic silicon site is coupled to a positively charged diamagnetic silicon [10,11,18–21,35] as shown in Figure 6.9. A single dangling bond E' is shown in Figure 6.10. As discussed later in this chapter, the cartoon figures of Figures 6.9 and 6.10 are at least reasonable first-order approximations, but may not precisely describe the most important E' centers in MOS oxides. The earliest studies of E' centers involve large volume (approximately cubic centimeter) samples of both crystalline quartz and amorphous SiO_2 [33–37]. The most commonly reported E' line shape found in the large volume amorphous SiO_2 samples as well as the most commonly observed E' line shapes in thin films closely resembles that of a randomly oriented array of crystalline quartz E'_1 centers. This line shape is often referred to as E'_γ. The crystalline quartz E'_1 center has a g matrix that is almost axially symmetric $g_1 \cong g_\parallel = 2.00176$, $g_2 = 2.00049$, $g_3 = 2.00029$; $g_2 \cong g_3 \cong g_\perp$. The amorphous SiO_2 E' line shape usually observed to dominate corresponds to a randomly

TABLE 6.2

g Matrix Parameters for P_b and E' Centers

(111) P_b	$g_\parallel$ **(111)**	$g_\parallel$	
Caplan et al. [4] and Poindexter et al. [5]	2.0014	2.0081	
(100) P_b	$g_\parallel$ **(111)**	$g_\perp$	
Kim and Lenahan [19]	2.0015 ± 0.0002	2.0080 ± 0.0002	
Stesmans and Afanas'ev [26]	2.0018 ± 0.0003	2.0085 ± 0.0003	
(100) P_{b1}	g_{xx} **[1$\bar{1}$0]**	g_{yy} **[$\bar{1}\bar{1}$1]**	g_{zz} **[112]**
Stesmans and Afanas'ev [26]	2.0058	2.0074	2.0022
E' **(amorphous SiO_2)**	$\cong g_\parallel$	$\cong g_\perp$	
Griscom et al. [36]	2.0018	2.0004	

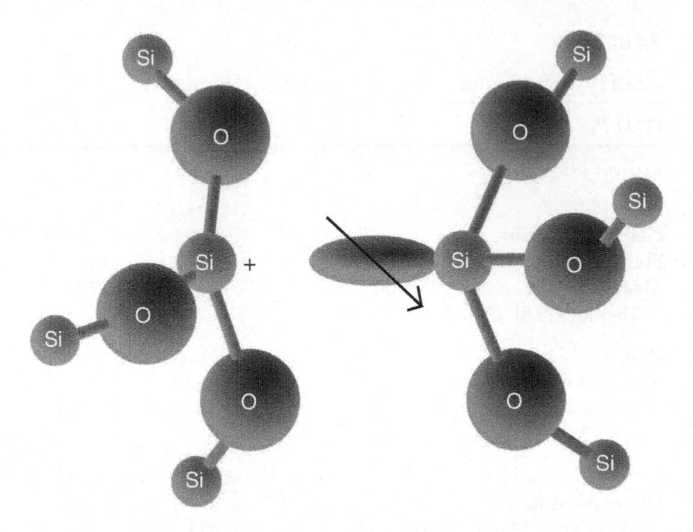

FIGURE 6.9
Schematic diagram of a positively charged E' center, a hole trapped at an oxygen vacancy.

oriented array of defects with $g_{\parallel} \cong 2.0018$ and $g_{\perp} \cong 2.0004$. The $g_{\parallel} \cong 2.0018$ is very close to the free electron $g_0 = 2.002319$. It is worth noting that this sort of g matrix is consistent with the predictions of Equation 6.2 for a dangling bond defect. An E' center line spectrum, measured on an MOS oxide is illustrated in Figure 6.11.

In large volume samples, it is quite easy to measure the hyperfine interactions of the unpaired electron with the single silicon atom on which it primarily resides [34,36,37]. Although the magnetic ^{29}Si nuclei are only $\cong$5% abundant, the number of centers present in large volume samples (typically $\sim 10^{16}$–$10^{17}/cm^3$ for irradiated SiO_2) is more than sufficient to generate quite strong ^{29}Si spectra. An ESR spectrum taken on such a large volume amorphous sample is shown in Figure 6.12. As discussed previously one may, by inspection, obtain a rough estimate of the hybridization and localization of the electron from this spectrum, as schematically illustrated in Figure 6.3. Also, as discussed previously, a second integration of the two side peaks (corresponding to spin 1/2 nuclei) yields a combined intensity of about 5% of the center line, unambiguously identifying the center as an unpaired spin localized on a silicon atom. An application of the analysis discussed previously indicates that $A_{iso} \cong 439$ G and $A_{aniso} \cong 22$ G. Recall that the isotropic and anisotropic coupling constants for an electron 100% localization in a silicon s and p orbital

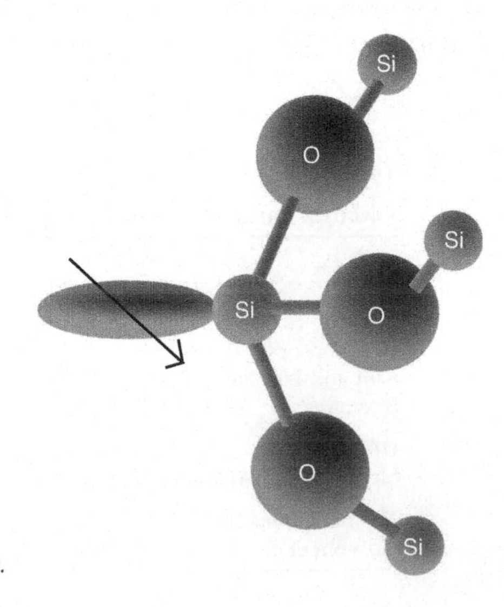

FIGURE 6.10
Schematic illustration of a neutral E' center.

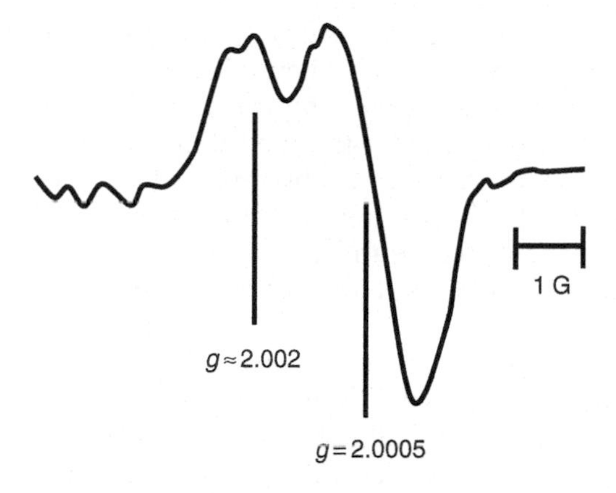

FIGURE 6.11
An 8 G wide ESR scan of a VUV irradiated MOS oxide, with the spectrometer settings optimized for the *E′* center signal. (Reprinted from Lenahan, P.M. and Mele, J.J., *J. Vac. Sci. Technol. B*, 18, 2169, 2000. With permission.)

are, respectively, $a_0 = 1639.3$ G and $b_0 = 40.75$ G. So, if the electron were 100% localized in a silicon s orbital one would expect an isotropic coupling constant of $a_0 = 1639.3$ G. Figure 6.4 indicates an isotropic splitting of about 439 G; thus, the orbital has about $439/1639 \cong 27\%$ s. If the electron were 100% localized in a silicon p orbital we would expect $A_{aniso} = b_0 = 40.75$ G. Figure 6.4 indicates about 22 G; thus, the orbital has about $22/40.75 \cong 54\%$ p. The analysis indicates a localization on the center silicon $(54 + 27) = 81\%$. Again, this analysis is crude, but, to first order, quite reliable. The reasonableness of this estimate is supported by comparison with a representative calculation of the wave function anticipated for these *E′* centers that estimates 32% s-character, 52% p-character, and 84% localization [51]. It has previously been mentioned that *E′* centers are quite frequently positively charged oxygen vacancies [10,11,18,19,21,38,39]. The connection between a positively charged oxygen vacancy and the just completed analysis is not obvious. (In fact, the *E′* spectrum is not always associated with the positively charged vacancy [40,43,52,53].) However, a very plausible explanation for a link between the *E′* spectrum and the positively charged vacancy was provided before direct experimental evidence linking the center to trapped holes was available.

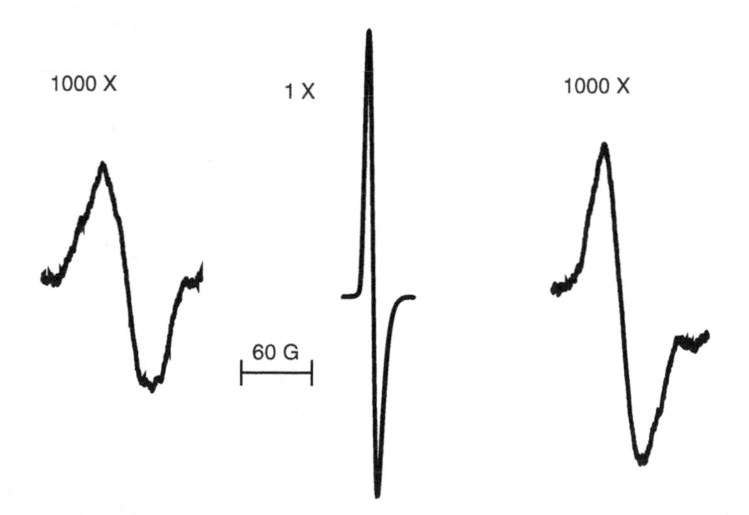

FIGURE 6.12
ESR trace of the *E′* center in a large volume amorphous SiO_2 sample showing a very narrow center line trace corresponding to the ≅95% abundant spin zero ^{28}Si nuclei and two broad lines corresponding to the ≅5% abundant spin 1/2 ^{29}Si nuclei.

In crystalline SiO_2, Feigl et al. [35] argued that E' centers are holes trapped in oxygen vacancies. In the Feigl et al. model, the unpaired electron resides on a neutral silicon on one side of the vacancy. The silicon on the other side of the vacancy is positively charged. Bonded to just three other atoms, the positive-side silicon adjusts its position to a flat-planar arrangement (expected for sp^2 bonding) with its three neighboring oxygens. In amorphous SiO_2, E' centers can be positively charged or neutral when paramagnetic. The Feigl et al. hypothesis explains how an ESR pattern consistent with an electron localized on a single silicon could correspond to a positively charged oxygen vacancy, which would obviously involve two silicons. The electron spends all of its time on one of them and is physically separated from the other due to the relaxation involved at the positively charged side of the vacancy. Quite a number of studies indicate that E' centers are dominating trapped hole centers in high-quality thermally grown oxide films [10,11,18–21,24,38,39]. These E' centers are very likely, for the most part, essentially the positively charged Feigl, Fowler, and Yip E' defects shown in Figure 6.9. It should probably be mentioned that models slightly different from a simple oxygen vacancy have been proposed for the positively charged E' center in amorphous SiO_2. An example is an E' model involving a neutral silicon dangling bond in which the silicon is back bonded to three oxygens near but not oriented opposite to a positively charged diamagnetic silicon [54,55].

6.3.2.2 Structure of Hydrogen Complexed E′ Centers

Two hydrogen complexed E' centers have been observed in thin oxide films [18,39,56–58]. They are known as the 74 G doublet and the 10.4 G doublet. Spectra of both hydrogen complexed E' centers are illustrated in Figure 6.13. The term doublet refers to the fact that both centers have two-line ESR spectra. The two lines arise because the hydrogen nucleus has a spin of 1/2, and thus can generate two local fields at the defect site [1]. One opposes

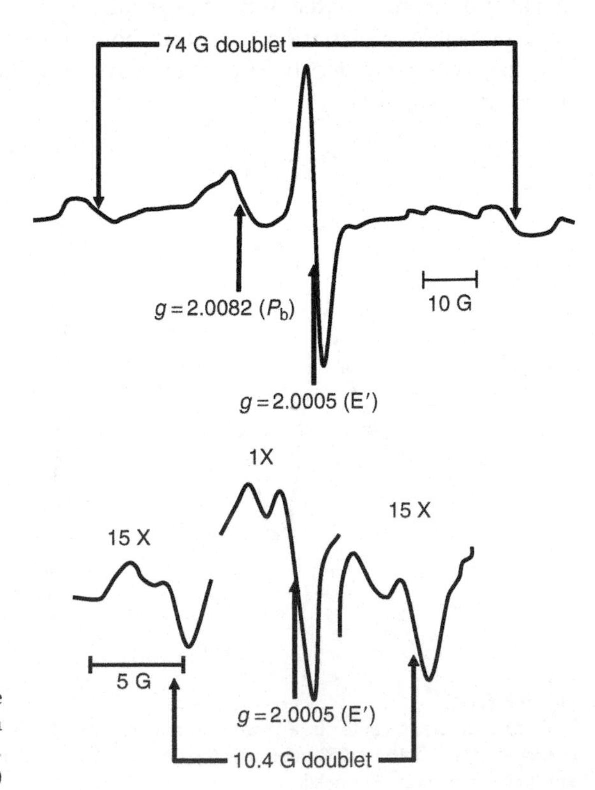

FIGURE 6.13
ESR traces of (a) the 74 G doublet and (b) the 10.4 G doublet centers. (Reprinted from Lenahan, P.M. and Conley, J.F., *J. Vac. Sci. Technol. B*, 16, 2134, 1998. With permission.)

the large applied field; the other adds to it [18,39,56–58]. These hydrogen coupled E' centers have also been observed in large volume (~1 cm^3) SiO_2 samples.

The 74 G doublet spectrum was conclusively linked to an E'/hydrogen complex by Vitko, who studied large volume (~1 cm^3) samples [59]. Vitko convincingly established this link by comparing heavily irradiated SiO_2 samples containing hydrogen and deuterium. The chemistry of ordinary hydrogen and deuterium are virtually identical; however, the one proton hydrogen nucleus and the proton and neutron deuterium nucleus have very different magnetic properties. Deuterium has a nuclear spin of 1; hydrogen has a nuclear spin of 1/2. The deuterium nuclear moment is much smaller than the magnetic moment of the hydrogen nucleus [1]. Vitko demonstrated that the deuterated SiO_2 samples exhibited an ESR spectrum precisely identical to what would be expected with a deuterium substituting for hydrogen. The reduction in the deuterated sample line spacing, 22.9 versus 74 G, scales perfectly with the anticipated difference due to the different moments of the two nuclei.

A later study, also involving large volume SiO_2, by Tsai and Griscom [60] extended and confirmed the Vitko results. Tsai and Griscom were able to observe the ^{29}Si hyperfine interactions of the central silicon of the hydrogen complexed defect. The results of Tsai and Griscom and Vitko indicate that the 74 G doublet center is a silicon dangling bond defect in which the central silicon is back bonded to two oxygens and a hydrogen. (A more detailed description of the defect became available with its observation in Si/SiO_2 systems. This will be discussed later in the chapter.)

A second hydrogen complexed center, called the 10.4 G doublet, has also been observed in both large volume samples [60] and in thin films [56,57]. The much smaller splitting (10.4 vs. 74 G) strongly suggests that the coupled hydrogen is more distant in the 10.4 G doublet case. Tsai and Griscom [60] have proposed that this center involves a silicon dangling bond center in which the central paramagnetic silicon is back bonded to three oxygen; one of the three oxygens is bonded to a hydrogen atom.

6.3.2.3 Structure of E'_δ (or EP) Center

The E'_δ (or EP) variant of the E' center has been observed in separation by implanted oxygen (SIMOX) buried oxides, bond and etch back oxides, and thermally grown oxides [52,61–67]. The spectrum is quite narrow and is centered on $g \cong 2.002$. In this film, it is nearly always superimposed on a much stronger conventional E' spectrum [61–64]. For example, Vanheusden and Stesmans [64] estimate that it accounts for about 20% of the E' centers in SIMOX buried oxides. It is also much less stable than the conventional E' centers; its intensity drops to a small fraction of its initial amplitude in a time of order 1 day at room temperature [62,63].

A nearly identical E'_δ (or EP) spectrum has been observed in large volume SiO_2 samples. Griscom and Friebele observed an ESR spectrum very similar to the oxide film E'_δ/EP pattern in large volume SiO_2 samples [68]. Their initial model was of an unpaired electron delocalized on a cavity of Cl capped Si atoms [68]. Thin films studies by Vanheusden and Stesmans [64] as well as Conley et al. [52] effectively rule out the role of chlorine. Vanheusden and Stesmans [64] as well as Warren et al. [65] proposed that the center involved a microcluster of four to five silicon atoms but did not offer specific spectroscopic evidence for this assignment.

On the basis of a close similarity between the conventional E' and E'_δ/EP response to molecular hydrogen, as well as on the E'_δ/EP signal response to microwave power, Conley and Lenahan [66] proposed that the defect must be simpler and more nearly related to the conventional E' centers involving an oxygen-vacancy-like site. ESR measurements by Zhang and Leisure [69] on large volume samples are sensitive enough to detect the

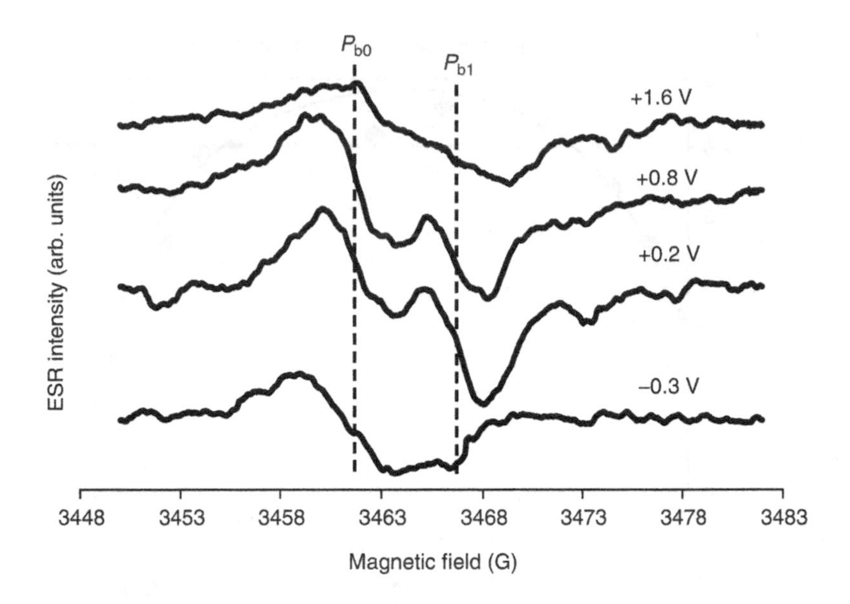

FIGURE 6.15
Four ESR traces taken with different applied corona biases on a 3.3 nm oxide on a (100) Si substrate. The P_{b0} and P_{b1} signal amplitudes clearly change with applied corona bias. (Reprinted from Campbell, J.P. and Lenahan, P.M., *Appl. Phys. Lett.*, 80, 1945, 2002. With permission.)

appears over a narrow range of bias over which the Si/SiO_2 Fermi level is fairly near the middle of the silicon band gap. The P_{b1} amplitude curve is quite a bit narrower than that of P_{b0}. The P_{b1} curve is also shifted negatively with respect to the P_{b0} curve. The results of Figure 6.16 lead to several conclusions regarding the P_{b1} density of states. (Although these conclusions are definitive, they are only semiquantitative.)

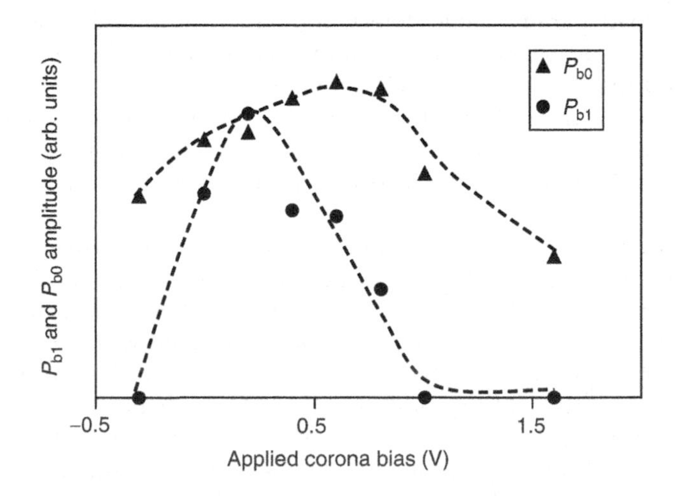

FIGURE 6.16
Plot of P_{b0} and P_{b1} signal amplitudes as a function of applied corona bias on a 3.3 nm oxide on a (100) Si substrate. The plot indicates that the P_{b0} correlation energy is much larger than the P_{b1} and that the P_{b1} energy distribution is skewed below midgap. (Reprinted from Campbell, J.P. and Lenahan, P.M., *Appl. Phys. Lett.*, 80, 1945, 2002. With permission.)

Earlier P_b/corona bias studies, such as those illustrated in Figures 6.15 and 6.16, utilizing much thicker oxides, allowed for a straightforward comparison between *CV* determination of the relationship between the Si/SiO_2 interface Fermi energy and the applied voltage [7,8,10,11,14,19]. Although such a *CV*/corona ion voltage comparison is, in principle, possible for thin (3.3 nm) oxides utilized in this case, the precision of both the measurements and the analysis is lower than had been the case in earlier studies on thicker oxides. However, reasonably precise knowledge of the P_{b0} defect's response to the Si/SiO_2 interface Fermi energy already exists; this knowledge can be used to crudely calibrate the P_{b1} response.

In examining the results of Figures 6.15 and 6.16, note first that both P_{b0} and P_{b1} amplitudes grow, and then shrink as the voltage is varied from the most negative to the most positive polarity; that is, as the Si/SiO_2 Fermi level is advanced from near the valence band edge, through the Si band gap, to near the conduction band edge. (The right and left sides of the voltage scale correspond to strong inversion and strong accumulation.) This response indicates that, as has been established earlier for the (111) P_b and (100) P_{b0}, P_{b1} defects are amphoteric [7,8,10,11,14,19]. Second, note that the P_{b1} curve is much narrower than that of P_{b0}, and that the P_{b1} maximum is shifted negatively with respect to the P_{b0} curve. The half-maximum width of the P_{b0} curve corresponds roughly to the distance between the peaks of the $+/0$ charge and $0/-$ charge transitions. In band bending (or relative Fermi energy positions) this corresponds to $\approx$0.6–0.7 eV [10,11]. This 0.6–0.7 eV is approximately the P_{b0} electron correlation energy. Since the P_{b1} curve has only about one-half the P_{b0} curve width, the P_{b1} correlation energy should be roughly half this size $\approx$0.3 eV. This result is qualitatively consistent with the earlier findings of Gerardi and coworkers [7,8,10,11,14,19]. However, note that the P_{b1} curve is clearly shifted negatively with respect to the P_{b0} curve. This indicates that the P_{b1} density of states is skewed below the P_{b0} density of states. The P_{b0} density of states is approximately centered about midgap. The results thus indicate rather clearly that the P_{b1} density of states is shifted (by several tenths of an electron volt) below midgap. Although a precise number is beyond the precision of these data, note that the P_{b1} peak is about halfway between the P_{b0} peak and the half amplitude value for the P_{b0} curve. This indicates that the P_{b1} density of states shifted about $\approx$0.7 eV/4 $\approx$ 0.2 eV downward relative to P_{b0}. This result is not consistent with the results of Gerardi and coworkers who argued that the P_{b1} levels were skewed toward the upper part of the silicon band gap [7,8,10,11,14,19]. The much narrower P_{b1} amplitude versus bias result means that the P_{b1} electron correlation energy is much smaller than the $\cong$0.7 eV value found for P_{b0}. The P_{b1} electron correlation energy is thus only a few tenths of an electron volt. This asymmetric distribution of P_{b1} levels results in the occupation of a significant fraction, perhaps one- to two-thirds of the sites, by net negative charge with the Si/SiO_2 Fermi level at midgap.

As mentioned previously, the SDR detection technique offers great sensitivity advantages. SDR measurements have also been made comparing the gate bias response of P_{b0} and P_{b1} centers. The measurements utilized metal–oxide semiconductor field-effect transistors (MOSFETs) configured as gate-controlled diodes [28,29,53]. The sensitivity of the measurements was so high that the ^{29}Si hyperfine side peaks of P_{b0} and P_{b1} could be monitored as a function of gate potential. These side peaks do not overlap, so the separate responses can be clearly separated [28,29,53]. The transistor gate devices were 10 nm thick. As Fitzgerald and Grove [76] demonstrated long ago, if a MOS device is configured as a gate-controlled diode, and the gate and drain to substrate diode is forward biased to a voltage V_J, one may probe Si/SiO_2 interface trap recombination by varying the gate potential. When the diode current is maximized, the electron and hole quasi-Fermi levels are symmetric about the interface intrinsic level and a range of interface traps in the middle of the gap about eV_J wide is surveyed. A shift in gate bias shifts the window

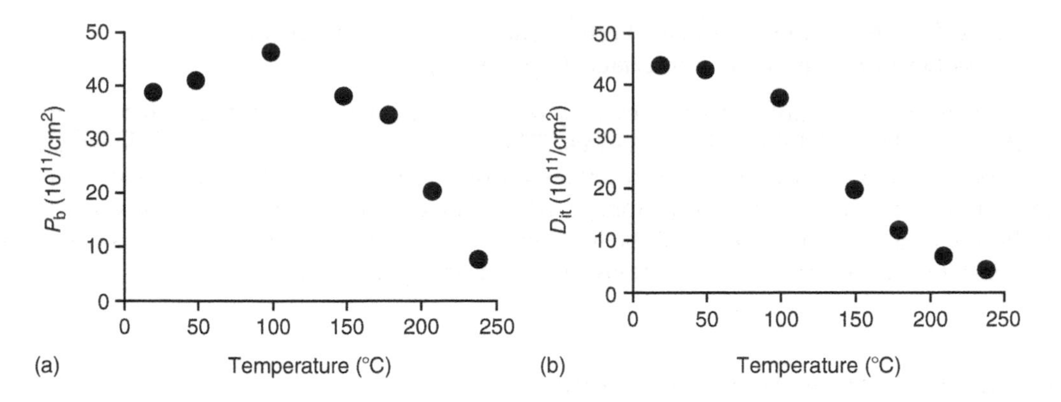

FIGURE 6.20
(a) Density of paramagnetic radiation-induced P_b centers as a function of isochronal annealing temperature. (b) Average radiation-induced interface trap density in the middle half of the band gap D_{it} as a function of isochronal annealing temperature. (Reprinted from Lenahan, P.M. and Dressendorfer, P.V., *J. Appl. Phys.*, 54, 1457, 1983. With permission.)

produced low yields of radiation-induced interface trap densities. Processing yielding higher P_b densities resulted in proportionately larger interface trap densities. A later study by Kim and Lenahan [19] extended these results to the more technologically important (100) Si/SiO$_2$ system. Some results from these studies are shown in Figures 6.21 and 6.22. As mentioned previously, an interesting aspect of the Kim and Lenahan study was the observation that the radiation-induced P_b centers were primarily (although not exclusively) P_{b0} defects. (Weaker P_{b1} spectra were also generated [19].) The results of these studies have been confirmed and extended by many other groups.

Miki et al. [21] compared the response of ultradry and steam-grown oxides to ionizing radiation. They found higher densities of radiation-induced interface traps and higher densities of radiation-induced P_b centers in the steam-grown oxides. They also found a rough numerical correspondence between the ratios of induced P_b and interface traps as well as in the absolute numbers of P_b centers and interface traps. However, since the ESR measurements were made on soft x-ray irradiated devices and the electrical measurements were made on devices in which holes were avalanche injected into the oxide (to simulate the radiation), precise numerical comparisons were not possible.

Awazu et al. [24] have also studied the role of processing parameters on the generation of P_b centers by ionizing radiation. Among the oxide processing parameters investigated were those used in the first study of Lenahan et al. [6]. For these oxides, Awazu et al. obtained the same P_b versus dose curve reported in the original 1981 study.

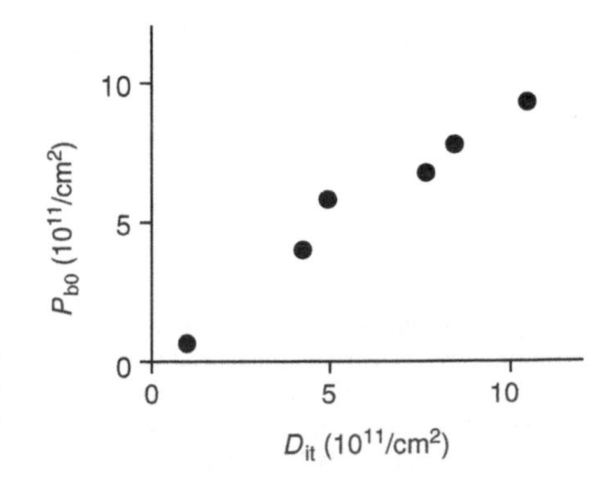

FIGURE 6.21
Density of radiation-induced P_{b0} centers versus mid-half band gap interface trap density in MOS devices subjected to ^{60}Co gamma irradiation. (Reprinted from Kim, Y.Y. and Lenahan, P.M., *J. Appl. Phys.*, 64, 3551, 1988. With permission.)

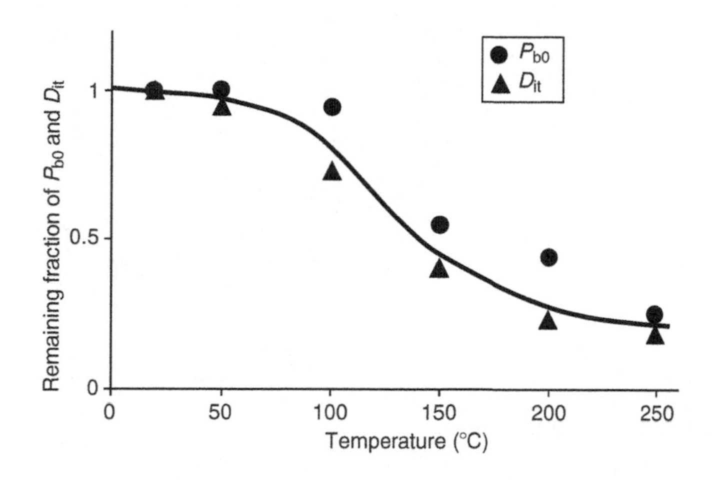

FIGURE 6.22
Plot of remaining fractions of P_{b0} and D_{it} as a function of 1 h isochronal annealing temperature. (Reprinted from Kim, Y.Y. and Lenahan, P.M., *J. Appl. Phys.*, 64, 3551, 1988. With permission.)

Awazu et al. [24] studied oxides grown on both (111) and (100) substrates. They found that if an as-processed interface had low P_b density then technologically relevant irradiation levels (10 Mrad(SiO_2)) generated very large ($\sim 10^{12}/cm^2$) P_b densities. However, if the as-processed P_b densities were extremely high ($\sim$2–3 $\times 10^{12}/cm^2$) the P_b density was reduced. They argued that this result should be expected from elementary reaction theory.

Vranch et al. [20] have also investigated the effects of ionizing radiation on the silicon–silicon dioxide interface. They showed that dose levels of several Mrad(SiO_2) could generate $\sim 10^{13}$ P_b centers/cm^2 as measured in a conventional ESR measurement. Their studies included conventional ESR as well as SDR. They found (as did several other groups) a strong preponderance of P_{b0} centers generated by radiation stressing.

6.4.2.3 P_b Centers Generated by High and Low Oxide Field Electron Injection

In addition to ionizing radiation, several other oxide stressing mechanisms have been investigated by ESR. Mikawa and Lenahan [13,15] showed that P_b centers can also be generated by injecting electrons into an oxide at low field by internal photoemission. Warren and Lenahan [16] showed that P_b centers can also be generated by high-field stressing oxides. In both studies, a rough (about one-to-one) correspondence was observed between the densities of P_b centers generated and the densities of interface traps in the middle half of the band gap. Results are shown in Figures 6.23 and 6.24.

6.4.2.4 P_b Centers Generated by Hot Carrier Stress of Short-Channel MOSFETs

Krick et al. [23] and Gabrys et al. [25] have used SDR to study hot carrier damage centers created near the drain of short-channel MOSFETs. These studies involved the highest sensitivity ESR measurements ever made in condensed matter up to that time. Both Krick et al. and Gabrys et al. generated strong P_b SDR signals by hot hole stressing the drain regions of the transistors. Krick et al. reported that the P_{b0} SDR signal scaled with increasing interface trap density as measured by charge pumping. The Gabrys measurements were sufficiently sensitive to allow for the first detection of the ^{29}Si hyperfine side peaks as well as an evaluation of the P_{b0} ^{29}Si hyperfine tensor.

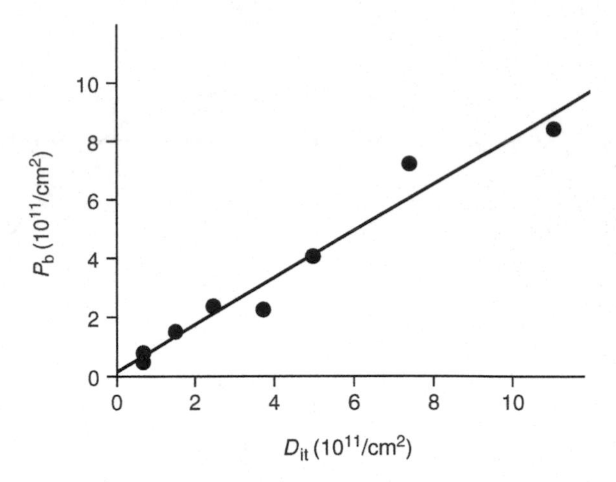

FIGURE 6.23
Density of trivalent silicon defects at the Si/SiO_2 interface, P_b, versus interface trap density, D_{it}, in the middle one-half eV of the silicon band gap. (Reprinted from Mikawa, R.E. and Lenahan, P.M., *Appl. Phys. Lett.*, 46, 550, 1985. With permission.)

6.4.2.5 P_b Centers Generated in the Negative Bias Temperature Instability

The negative bias temperature instability (NBTI) is perhaps the most important reliability problem of early twenty-first century complementary MOS (CMOS) integrated circuitry. NBTI involves a negative threshold voltage shift and a loss in drive current when pMOSFETs are subjected to a modest negative bias at moderately elevated (~120°C) temperature [77–79]. Fujieda et al. [80] utilized conventional ESR measurements on NBTI stressed MOS capacitors with p-type substrates to show that strong P_{b0} and P_{b1} spectra are generated when the structures are subjected to technologically relevant NBTI stressing conditions. Campbell et al. [81] utilized SDR measurements on NBTI stressed pMOSFETs to show that strong P_{b0} and P_{b1} SDR spectra are also generated when the transistors are subjected to technologically relevant NBTI stressing [81]. Campbell et al. [81,82] further demonstrated a strong correlation between P_{b0}/P_{b1} generation and NBTI-induced interface trap density (per unit energy) D_{it} generation.

In a rare exception to the rule Campbell et al. found that in plasma-nitrided pMOSFETs, P_b centers do not dominate NBTI-induced D_{it} generation [83]. They found that a different silicon dangling bond center, a silicon back bonded to three nitrogens [83], a defect called the K center or K_N center, dominates the NBTI-induced interface trap generation [83,84].

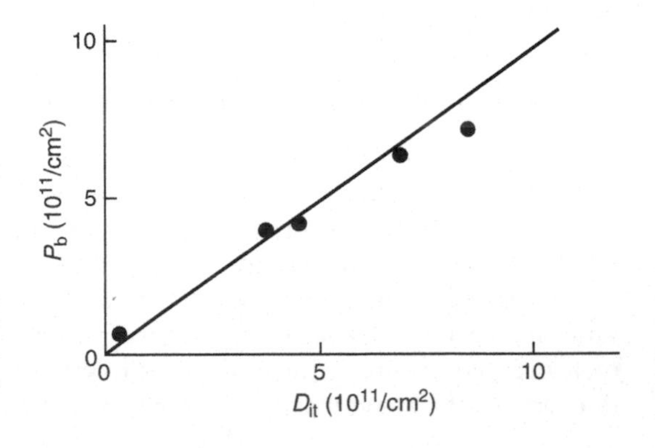

FIGURE 6.24
Density of trivalent silicon defects at the Si/SiO_2 interface versus interface trap density, D_{it}, in the middle half of the silicon band gap. (Reprinted from Warren, W.L. and Lenahan, P.M., *Appl. Phys. Lett.*, 49, 1296, 1986. With permission.)

6.4.2.6 *Conclusion Regarding* P_b *Centers and* Si/SiO_2 *Instabilities*

At least six groups [6–8,10,11,13,15,18–25,28] have reported essentially incontrovertible measurements on many oxides demonstrating a strong generation of P_b centers in device stressing. For radiation-induced instabilities, all five groups concluded that P_b centers play a dominating role. Although these studies have been most extensive for ionizing radiation, one may reasonably conclude that P_b centers do indeed play dominating roles in several technologically important instabilities. The earlier studies of Nishi et al. [2,3] and Poindexter et al. [5] established that P_b centers play dominating roles in as-processed Si/SiO_2 structures with relatively poor interfaces.

6.4.3 Roles of *E′* Centers in MOS Devices

The roles of *E′* centers in high-quality thermally grown oxides are fairly well but not yet completely understood. The electronic behavior of *E′* centers is summarized in the following list:

1. They can act as hole traps.
2. They certainly play a significant role in trap-assisted tunneling phenomena in electrically stressed oxides; for example, stress-induced leakage currents.
3. It is clear that both positively charged and neutral *E′* centers can be generated in SiO_2 films.
4. Positively charged and compensated *E′* centers can be generated under some circumstances.
5. Repeated electron–hole flooding can, at least under some circumstances, alter *E′* center structure.
6. It is also clear that several *E′* variants can be generated in thin films.
7. Several variants involve hydrogen complexed *E′* centers.
8. Detailed structure of several of the variants has yet to be identified.

6.4.3.1 E′ *Centers as Oxide Hole Traps*

E′ centers were probably first detected in thermally grown oxides by Marquardt and Sigel [85] who studied quite thick (up to 11,000 Å) oxide films subjected to quite high (up to 220 Mrad(SiO_2)) doses of ionizing radiation. They observed weak signals in these films which they attributed to *E′* centers. Although they did not report results of electrical measurements, they proposed (correctly) that *E′* centers can be thermal oxide hole traps.

The electronic properties of *E′* centers as hole traps and their significance in MOS device operation were demonstrated by Lenahan and Dressendorfer [10,11,38], who made a series of observations establishing that *E′* centers are dominating hole trap centers in a variety of MOS oxides. The role of *E′* centers in hole trapping was established through several observations, which are as follows:

1. There is a rough one-to-one correspondence between *E′* density and the density of hole traps in relatively hard and relatively soft oxides grown in both steam and dry oxygen (see Figure 6.25).
2. There is a rough one-to-one correspondence between *E′* density and trapped hole density in oxides irradiated under positive gate bias over a technologically meaningful range of ionizing radiation dose (see Figure 6.26).

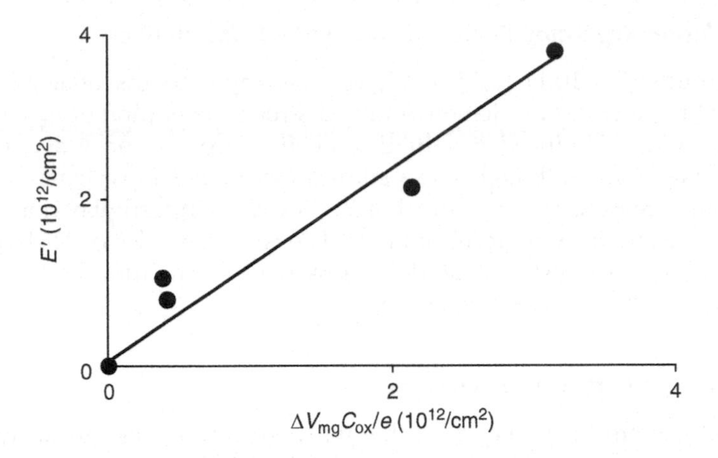

FIGURE 6.25
Concentration of E' centers plotted versus trapped hole density in the oxide ($\Delta V_{mg}C_{ox}/e$) for MOS structures subjected to four sets of processing parameters. All structures were irradiated to 10 Mrad(SiO_2); thus, differences in E' concentration here are all due to processing variations. (Reprinted from Lenahan, P.M. and Dressendorfer, P.V., *J. Appl. Phys.*, 55, 3495, 1984. With permission.)

3. MOS oxide E' centers and oxide trapped holes have the same annealing response in air (see Figure 6.27).
4. Distribution of E' centers and trapped holes are virtually identical in oxides subjected to ionizing irradiation under positive gate bias; both are close to the Si/SiO_2 boundary (see Figure 6.28).

The results of those early studies have been confirmed and extended in quite a few later studies.

1. Takahashi and coworkers [39] also reported an approximately one-to-one correspondence between E' centers and trapped holes; they also reported that the distribution of E' centers and trapped holes were the same in their irradiated oxides.
2. Lipkin et al. [86] also measured an approximately one-to-one correspondence between E' density and the density of trapped holes generated in oxides subjected to 10–20 Mrad(SiO_2) of gamma radiation.

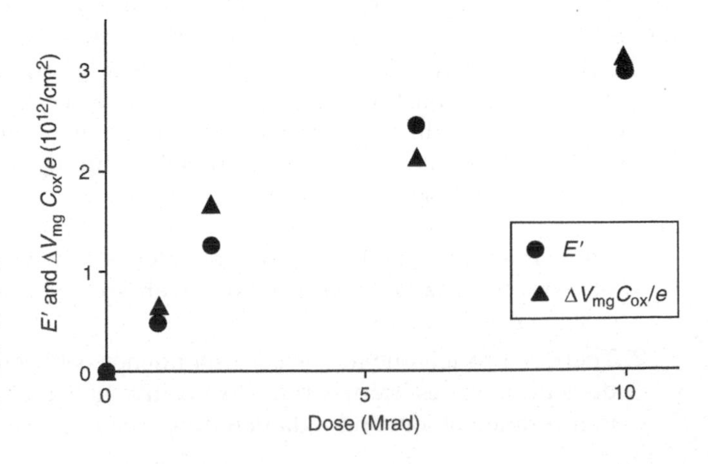

FIGURE 6.26
Distributions of E' and $\Delta V_{mg}C_{ox}/e$ versus irradiation dose for MOS structures with oxides grown in dry oxygen and subjected to a nitrogen anneal. (Reprinted from Lenahan, P.M. and Dressendorfer, P.V., *J. Appl. Phys.*, 55, 3495, 1984. With permission.)

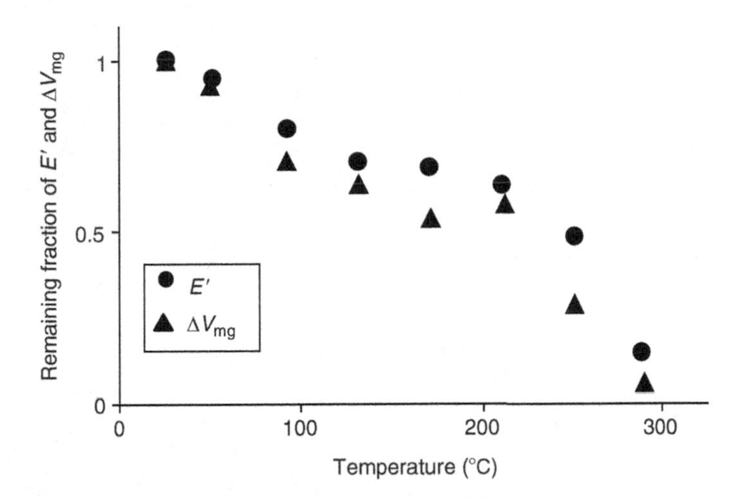

FIGURE 6.27
Plot of remaining fractions of ΔV_{mg} and E' as a function of isochronal annealing time. (Reprinted from Lenahan, P.M. and Dressendorfer, P.V., *J. Appl. Phys.*, 55, 3495, 1984. With permission.)

3. Miki et al. [21] compared both E' generation and trapped hole generation in ultradry and steam-grown oxides. They found that their ultradry oxides contained twice as many E' centers as the steam-grown oxides and that the ultradry oxides also had twice as many trapped holes as the steam-grown oxides. In addition, they found a rough numerical correspondence between the E' densities and trapped hole densities in the samples investigated. However, since Miki et al. [21] made electrical measurements on oxides subjected to avalanche injection of holes and ESR measurements on x-ray irradiated oxides, a precise numerical comparison between E' density and trapped hole density was not possible.
4. Awazu et al. [24] have explored the role of processing parameters on E' generation. They noted that, as Lenahan and Dressendorfer [10,11,38], and Miki et al. [21] had

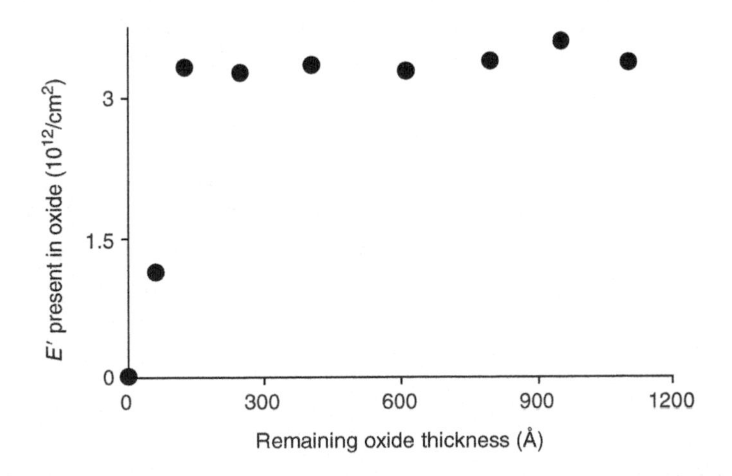

FIGURE 6.28
Distribution of E' density in an irradiated oxide. (Reprinted from Lenahan, P.M. and Dressendorfer, P.V., *J. Appl. Phys.*, 55, 3495, 1984. With permission.)

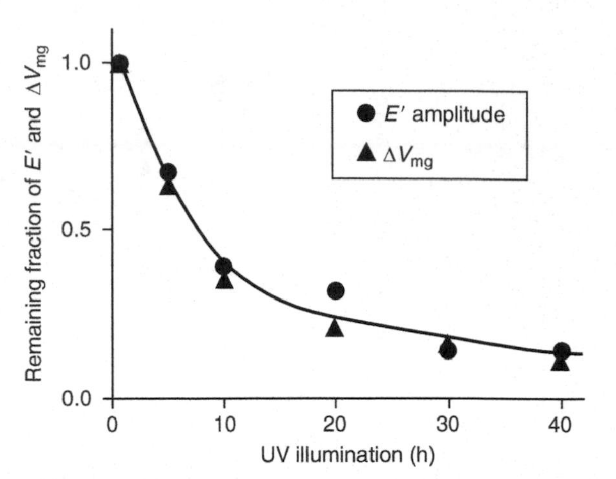

FIGURE 6.29
Plot of remaining fractions of E' and ΔV_{mg} as a function of UV illumination. (Reprinted from Kim, Y.Y. and Lenahan, P.M., *J. Appl. Phys.*, 64, 3551, 1988. With permission.)

previously reported, the densities of E' centers are strongly processing dependent. A point of particular interest in the Awazu study is their observation that the cooling rate after high-temperature processing strongly affects E' generation. Awazu et al. concluded that the E' centers in their oxides were holes trapped in oxygen vacancies $O_3 \equiv Si + \cdot Si \equiv O_3$. Since they observed $\sim 1-3 \times 10^{12}$ E' centers/cm^2 after modest ($\sim$1$-$11 Mrad(SiO_2)) doses of ionizing radiation, these centers would inevitably be the dominant hole trap centers in the oxides of their study.

A particularly extensive series of measurements on E' centers and oxide trapped holes was carried out by Kim and Lenahan [19]. They exposed (100) Si/SiO_2 poly-gate MOS structures to various levels of gamma irradiation while the MOS structures were under positive gate bias. Their study combined *CV* and ESR measurements. They found a very close, near one-to-one, correlation between E' density and the density of trapped holes as a function of total ionizing radiation dose. They subjected some irradiated devices to internal photoemission of electrons into the oxide. The E' centers and holes were annihilated at the same rate by the internal photoemission, also with an approximately one-to-one correspondence. This result is shown in Figure 6.29. In addition they found that a series of anneals in air removed both E' centers and trapped holes with, within experimental error, a one-to-one correspondence. This result is illustrated in Figure 6.30. These measurements clearly demonstrated the primary role of E' centers in hole trapping in these oxides.

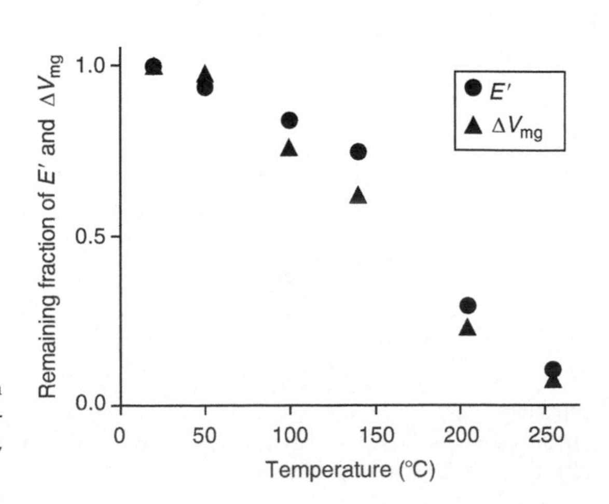

FIGURE 6.30
Plot of remaining fractions of E' and ΔV_{mg} as a function of isochronal annealing (1 h in air) temperature. (Reprinted from Kim, Y.Y. and Lenahan, P.M., *J. Appl. Phys.*, 64, 3551, 1988. With permission.)

The results of multiple independent ESR studies all indicate a dominant role for *E′* centers in oxide hole trapping in thermally grown oxide films on silicon. On the basis of this mutually corroborating work one may conclude that *E′* centers dominate oxide hole trapping in a wide variety of thermally grown oxide films on silicon.

6.4.3.2 *Neutral Paramagnetic* E′ *Centers*

Neutral *E′* centers can be generated in SiO_2 films under a number of circumstances. Conley et al. were able to generate neutral *E′* centers in thermally grown SiO_2 on silicon by exposing the oxides to extended periods of vacuum ultraviolet (VUV) light ($hc/\lambda <$ 10 eV) from a deuterium lamp [52]. Exposing the oxides to the equivalent of about 500 Mrad(SiO_2), they generated an ESR line shape within experimental error identical to that of the *E′* spectrum associated with hole trapping. Flooding the oxide with 5×10^{13} electrons/cm^2 did little to the ESR amplitude; if the defects were positively charged, the extremely large Coulombic capture cross section would have resulted in the annihilation of most or all of the *E′* amplitude. (Subsequent hole injection significantly altered the ESR spectrum generating an *E′* variant that will be discussed later in this chapter.)

Warren et al. generated neutral *E′* centers by internally photoemitting electrons into plasma-enhanced chemical vapor deposited oxide films on silicon [41]. The ultraviolet light utilized in the study ($hc/\lambda < 5.5$ eV) allowed photoemission of electrons from the silicon into the oxide but was not sufficiently energetic to generate electron–hole pairs in the (9.0 eV band gap) SiO_2. *CV* measurements (Figure 6.31) showed that the photoinjected electrons eliminated trapped holes initially present in the oxide. ESR measurements indicated a strong generation of *E′* centers as illustrated in Figure 6.32. A plot of *E′* generation versus the fluence of injected electrons, shown in Figure 6.33, indicates a very effective mechanism for neutral *E′* center generation in these films.

The generation of neutral *E′* centers in these high hydrogen content oxides would most plausibly be explained in terms of an *E′* precursor that is not an oxygen vacancy but a silicon atom back bonded to three oxygens and a hydrogen. (A hydrogen–silicon bond breaking event would leave behind a single neutral paramagnetic dangling bond.)

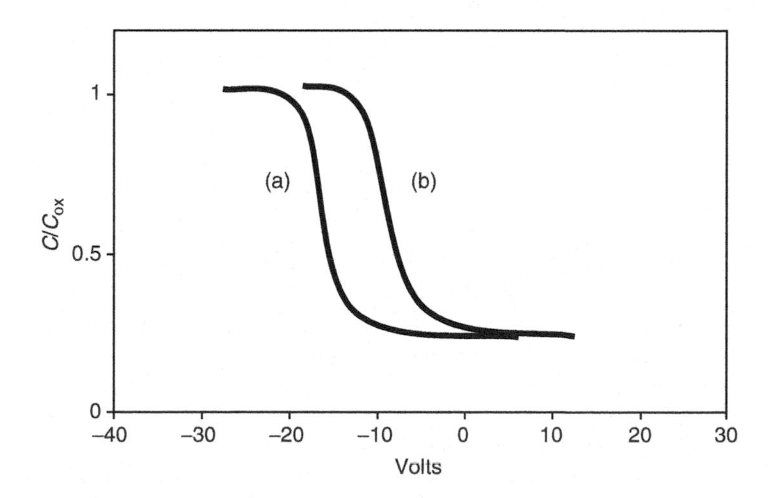

FIGURE 6.31
CV measurements. *C–V* curve (a) was taken before electron photoinjection. *C–V* curve (b) was taken after electron photoinjection. The photoinjected electrons eliminated oxide positive charge. (Reprinted from Warren, W.L. et al., *Appl. Phys. Lett.*, 53, 482, 1988. With permission.)

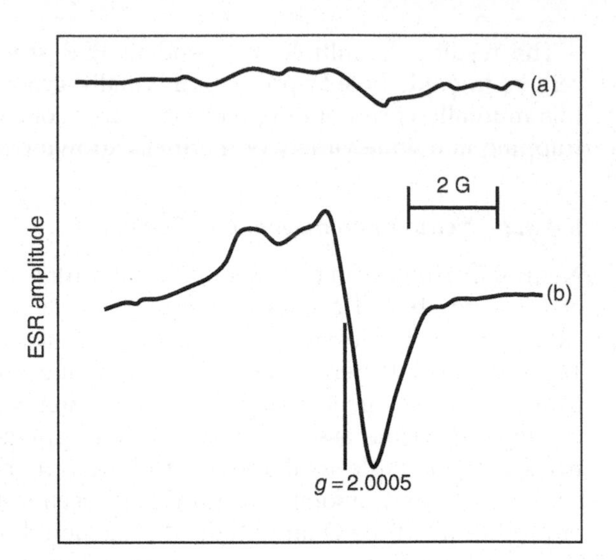

FIGURE 6.32
ESR measurements. ESR trace (a) was taken before electron photoinjection. ESR trace (b) illustrates an E' resonance after electron photoinjection. Electron injection substantially increased the E' center density. (Reprinted from Warren, W.L. et al., *Appl. Phys. Lett.*, 53, 482, 1988. With permission.)

6.4.3.3 *Positive but Compensated* E′ *Centers*

Conley et al. [62,63] reported the generation of positive but compensated E' centers in thick (400 nm) SIMOX buried oxides. These oxides had previously been subjected to an extended anneal in an inert ambient at a very high temperature (1325°C). The oxide exhibited a high density of E' centers when subjected to hole flooding. However, after gamma irradiation with the oxides unbiased, the oxides exhibited very high E' density with essentially no net positive charge. This result could be interpreted to mean either that the E' centers were neutral or positively charged and compensated. Conley et al. used internal photoemission to inject electrons into the oxides. The electron injection resulted in the annihilation of a large fraction of the E' centers with a Coulombic capture cross section clearly indicating that they had been positively charged and compensated. It should be mentioned that Walters and Reisman [87] have proposed that neutral electron traps may involve electron trapping at compensated E' centers. The Walters–Reisman proposal is generally consistent with these observations of Conley et al. [62,63].

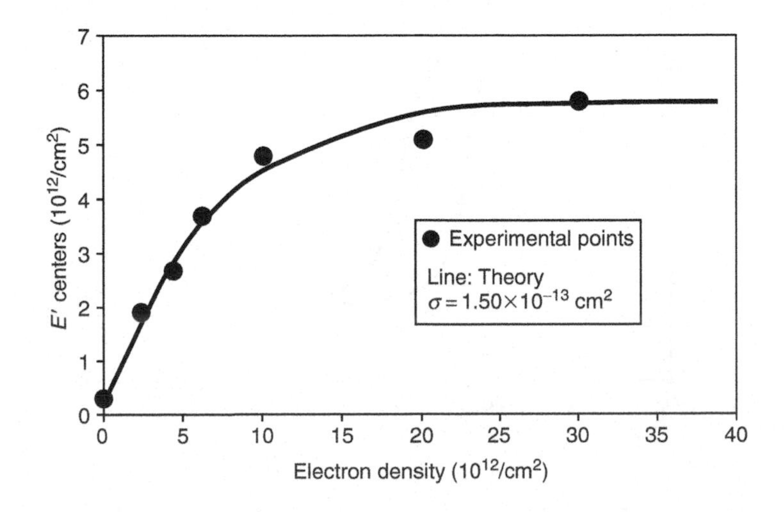

FIGURE 6.33
Plot of the E' density versus injected electron charge density. The circles are the experimental points and the line was plotted assuming $\sigma = 1.5 \times 10^{-13}$ cm^2. (Reprinted from Warren, W.L. et al., *Appl. Phys. Lett.*, 53, 482, 1988. With permission.)

6.4.3.4 *Structural Changes at E′ Sites Resulting from Repeated Charge Cycling*

Conley et al. [62,63] also noted that repeated cycles of hole and electron flooding could grossly alter the balance between *E′* density and trapped hole density. They noted an initial close correspondence between trapped hole and *E′* center density but with repeated electron–hole charge injection cycling noted a gradual increase in *E′* density that was not accompanied by positive charge. This response is illustrated in Figure 6.34. Such a result is not surprising if many of the *E′* center hole traps are essentially the Feigl–Fowler–Yip like oxygen vacancy centers illustrated in Figure 6.9. These defects would experience a large structural change upon hole capture. With subsequent electron capture, they might not all be able to return to the initial simple oxygen vacancy configuration.

6.4.3.5 *Near Si/SiO₂ Interface E′ Centers: Electronic Properties*

One could reasonably divide electrically active MOS defects into three categories: interface traps, which can communicate readily with charge carriers in the silicon, oxide traps which do not communicate with charge in the silicon, and very near interface traps which can, on fairly long timescales, communicate with charge carriers in the silicon. The near Si/SiO_2

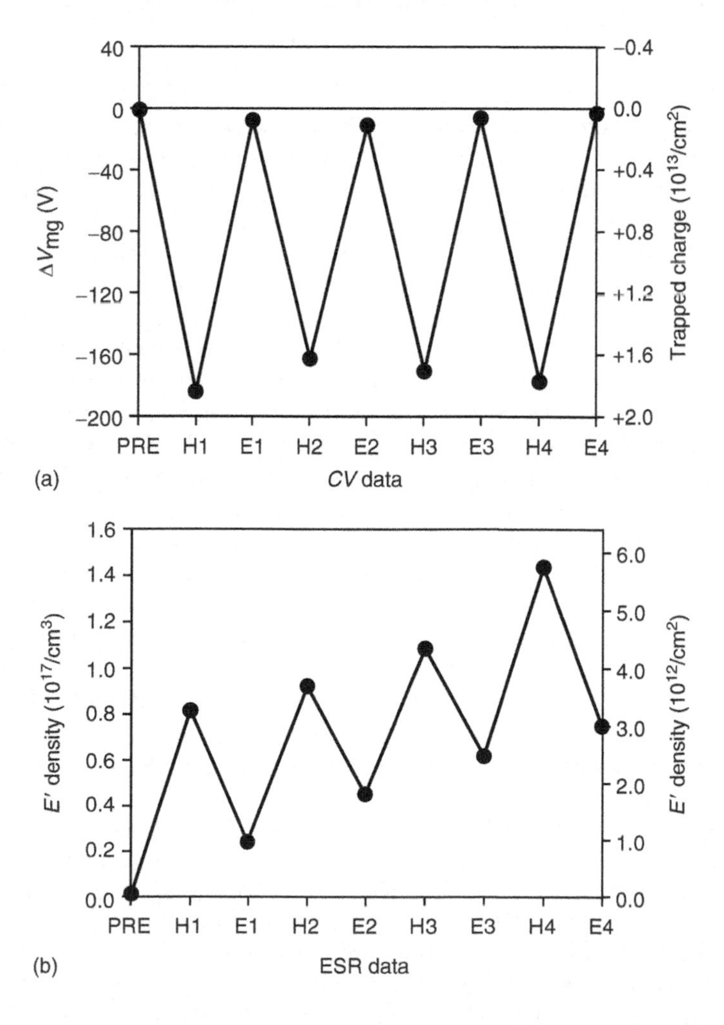

FIGURE 6.34
Effects of charge injection on (a) *CV* and (b) ESR measurements of unilluminated buried oxides. (Reprinted from Conley, J.F., Lenahan, P.M., and Roitman, P., *IEEE Trans. Nucl. Sci.*, 39, 2114, 1992. With permission.)

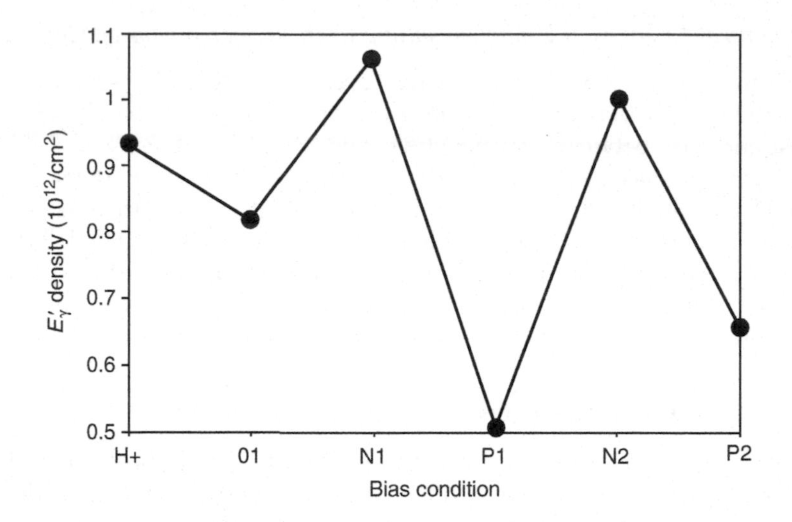

FIGURE 6.35
E' Density versus various 24 h bias sequences. (Reprinted from Campbell, J.P. et al., *Appl. Phys. Lett.*, 90, 123502, 2007. With permission.)

interface traps go by many names: slow states, border traps [88], switching traps [89], etc. (See Chapter 7.) It is possible, even likely, that more than one type of near Si/SiO_2 interface trap can exist in certain oxides under certain conditions. One type of near Si/SiO_2 interface trap has been directly identified via ESR, E' centers. Many studies (involving only electrical measurements) show that, when some MOS devices are subjected to ionizing radiation, *CV* and current versus voltage characteristics experience a negative voltage shift, ΔV, indicating the capture of positive charge in the oxide [90,91]. However, if a positive gate bias is applied, the magnitude of ΔV decreases logarithmically in time, indicating the annihilation or compensation of some of the positive charge via electron tunneling. If the applied bias is reversed from positive to negative some of the previously compensated charge returns [88,89,92]. The charge that returns is said to be in switching (border) traps. An ESR study by Conley et al. [93,94] clearly demonstrates that some E' centers can act as switching traps. These centers are presumably very close to the Si/SiO_2 boundary. Results from the Conley et al. [93,94] study are shown in Figure 6.35. The bias voltages and bias switching times approximately match those of earlier purely electronic measurements. The first point (H+) indicates the E' density initially after holes were injected into the oxide. The second point (01) was taken after 10^5 s with no bias across the oxide. Point (N1) was taken after a negative gate voltage corresponding to an average oxide field of 3.5 MV/cm was applied for 24 h. The negative bias increased the number of paramagnetic E' sites. Point (P1) was taken on the same sample after an additional 24 h under positive gate bias, also corresponding to an average oxide field of 3.5 MV/cm. The positive bias substantially decreases the density of paramagnetic E' centers. Two additional biasing points indicate the repeatability of this process. Clearly, the spin state and thus charge state of these E' centers can be repeatedly switched with bias; thus, E' centers can act as oxide switching traps. The results of Conley et al. [93,94] are consistent with, and clearly confirm, the basic premise of the switching trap model proposed earlier by Lelis and Oldman [89] and Lelis et al. [95]. After hole capture, subsequent electron capture does not always return the E' site involved to its original condition. This irreversibility leads to the switching behavior. The results of Conley et al. do not, of course, preclude the possibility that defects other than E' centers may act as switching traps. The Conley et al. [93,94] results extend earlier results generated by Jupina and Lenahan who reported the

SDR detection of E' centers [22]. Since SDR can only detect defects that in some way communicate with Si/SiO_2 interface charge carriers, their results strongly indicated that some near Si/SiO_2 interface E' centers did indeed behave in this way. The Conley et al. study is also consistent with an ESR study by Warren et al. [96] which suggested, but did not demonstrate, that E' centers may act as switching traps. The experimentally demonstrated role of E' centers as switching traps or border traps also is consistent with more recent purely theoretical work [54,55,97].

6.4.3.6 Hydrogen Complexed E′ Centers

Two hydrogen complexed E' variants may play important roles in device reliability. As discussed earlier in the chapter, the hydrogen complexed centers are called the 74 G doublet and the 10.4 G doublet. ESR traces of both defects are shown in Figure 6.13. Both defects were first observed in cubic centimeter size samples [59,60], and both centers clearly involve an E' hydrogen complex. As discussed previously, the 74 G doublet defect very likely involves an unpaired electron on a silicon back bonded to two oxygens and one hydrogen. The 10.4 G doublet defect very likely involves an unpaired electron on a silicon back bonded to three oxygens with one of the oxygens bonded to a hydrogen [59,60]. The 74 G doublet was first observed in thin oxide films by Takahashi et al. [39] who generated them at somewhat elevated temperatures (≅100°C) in irradiated oxides. More recently, Conley and Lenahan [57,58] observed the room temperature generation of both the 74 G doublet centers and 10.4 G doublet centers in oxides subjected to either VUV ($hc/\lambda \leq 10.2$ eV) or gamma irradiation. Takahashi et al. [39] suggested that the hydrogen complex E' defects might play an important role in Si/SiO_2 interface trap generation. Conley and Lenahan [57,58] provided strong circumstantial evidence linking E' hydrogen coupled centers to interface trap generation. Several (purely electrical measurement) studies [98,99] had shown that a molecular-hydrogen containing ambient leads to an enhancement in radiation-induced interface trap generation. Conley and Lenahan [57,58] showed that exposing an oxide previously flooded with holes (to generate E' centers) to an H_2/N_2 ambient leads to a conversion of conventional E' centers to 74 G doublet centers and other hydrogen complexed E' centers as well as to the generation of interface traps. They found that the number of E' centers converted to hydrogen complexed centers is approximately equal to the number of interface traps generated, with the time period involved in interface trap formation approximately equal to the time required to saturate the E'/hydrogen complexing process.

6.4.3.7 Electronic Properties of E'_δ (or EP)

Conley et al. [52,61–63,66] investigated the electronic properties of E'_δ (EP) centers in a variety of oxides. (Conley et al. chose to call the centers EP since they felt that the assignment as an E' variant was provisional (thus the *P*) at that time.) They flooded the oxides with holes and electrons. From plots of ESR intensity versus injected charge fluence, they were able to extract approximate capture cross sections and to determine the active sizes of the capture cross section and density of precursors for E'_δ (or EP) with respect to the conventional E' center [52]. From etch back measurements after hole injection, they were also able to determine the spatial density of the centers, finding them to be highly concentrated near the Si/dielectric boundary [52].

Figure 6.36 illustrates a plot of conventional E' (called E'_γ here) and E'_δ (referred to as EP here) versus fluence [52]. The conventional E' center has the smaller capture cross section for holes, but its precursor is clearly present in higher densities in the oxide under study. The E'_δ (or EP here) amplitude actually drops at higher hole fluence, indicating that it is much less stable than the conventional E' centers. Figure 6.37 illustrates a plot of both E'

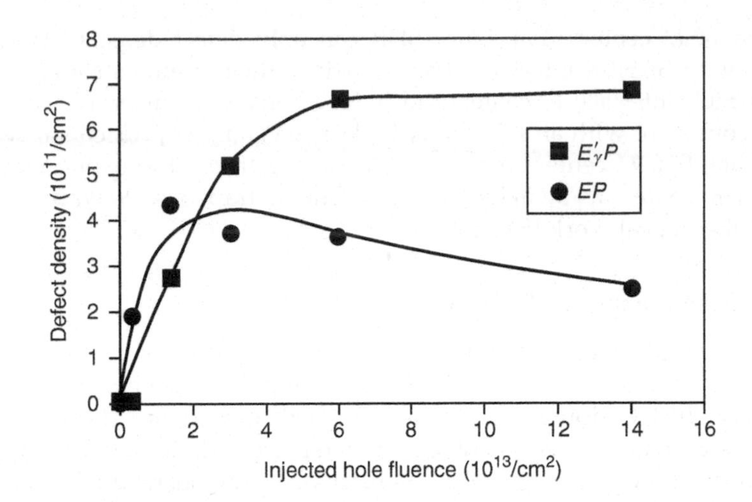

FIGURE 6.36
$E'_{\gamma p}$ and EP defect density versus hole injection fluence (the lines are drawn only as a guide to the eye). (Reprinted from Conley, J.F. et al., *J. Appl. Phys.*, 76, 2872, 1994. With permission.)

densities in response to electron injection [52]. The E'_{δ} (EP) center has about an order of magnitude larger capture cross section than the conventional E' center (E'_{γ} here).

6.4.3.8 Role of E′ Centers in Oxide Leakage Currents

Oxide leakage currents caused by various types of stressing are a fundamental concern. An unstressed oxide exhibits current density versus voltage characteristics consistent with the tunneling current density anticipated from an ideal band diagram. However, if an oxide is subjected to stressing, for example, by a high electric field across the oxide, damage caused by the stressing can cause leakage currents to appear at relatively low oxide fields. These leakage currents have been extensively investigated in many experimental studies of a purely electronic nature [100–102]. Several investigators have suggested a link between E' centers and these leakage currents [100,101,103]. The leakage currents are likely caused by inelastic tunneling of conduction band electrons to defect centers in the oxide near the Si/SiO_2 boundary. Recent work by Takagi et al. [100] suggests that E' centers are good candidates for the centers involved in the inelastic tunneling process.

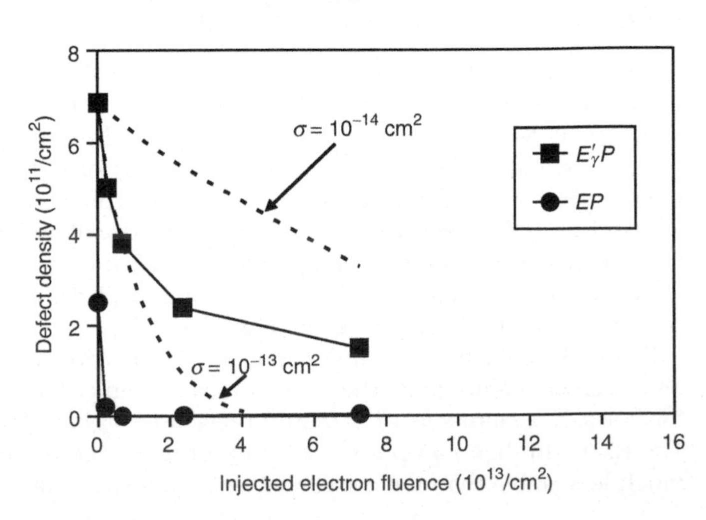

FIGURE 6.37
$E'_{\gamma p}$ and EP defect density versus injected electron fluence after hole injection (the lines are drawn only as a guide to the eye). (Reprinted from Conley, J.F. et al., *J. Appl. Phys.*, 76, 2872, 1994. With permission.)

Several earlier studies have shown that E' centers are generated in oxides subjected to high electric field [42,104,105]. Lenahan and Mele [43] showed that the generation of electrically neutral E' centers are accompanied by a large (several orders of magnitude) monotonic increase in leakage currents and that the disappearance of these E' centers via low-temperature (200°C) annealing is accompanied by a monotonic decrease in leakage currents. Recent calculations by Takagi et al. [100] indicate that defect densities in the $\sim 10^{18}$–$10^{19}/\text{cm}^3$ range in the vicinity of the Si/SiO_2 interface ($\sim 10^{12}$–$10^{13}/\text{cm}^2$) would lead to leakage current densities similar to those measured in the study.

In the Lenahan and Mele study, the leakage current generated with the corona ions bias of the oxide surface was measured using a Kelvin probe electrostatic voltmeter [43]. This approach, the measurement of oxide currents via corona ion decay, was pioneered many years ago by Weinberg et al. [106]. Taking the derivative of potential versus time, they determined current versus voltage dependence: $C(dV/dt) = dQ/dt = i$. Here C is the oxide capacitance, $\varepsilon_0 \kappa_{SiO_2}/t_{ox}$, where ε_0 is the permittivity of free space, κ_{SiO_2} is the relative dielectric constant of SiO_2, and t_{ox} is the oxide thickness; Q is charge; t is time; and i is current [106]. Although this method is quite crude, it provided reasonable semiquantitative electrical results to compare to the ESR measurements.

The central results are illustrated in Figures 6.38 and 6.39. Figure 6.38 illustrates pre-VUV, post-VUV, and post-VUV-postannealing ESR traces taken in the vicinity of the free electron $g \cong 2.002$. The magnetic field is swept over a range that is appropriate for both P_b and E' centers. The spectrometer settings used to obtain the data in Figure 6.38 are less than optimum for both E' and P_b spectra. The settings represent a compromise, chosen so that both spectra would be simultaneously visible. Before VUV illumination, no ESR signals are visible. After illumination, strong E' and P_b signals are present. An anneal of 15 min at 200°C eliminates most of the E' centers but does little to P_b intensity. Figure 6.11 illustrates a post-VUV stressing E' spectrum taken at a modulation amplitude and microwave power significantly lower than that of Figure 6.38. These spectrometer settings are optimized to detect the E' center defects.

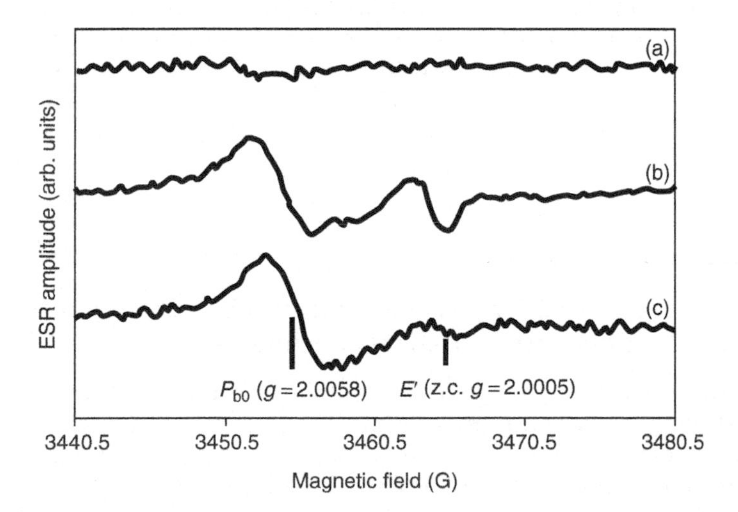

FIGURE 6.38
(a) Pre-VUV, (b) post-VUV, and (c) post-VUV postannealing ESR traces (40 G wide) taken on the 45 nm oxide films on silicon; the spectrometer settings were chosen so both the Si/SiO_2 interface P_b centers; and the E' centers are clearly visible. (Reprinted from Lenahan, P.M. and Mele, J.J., *J. Vac. Sci. Technol. B*, 18, 2169, 2000. With permission.)

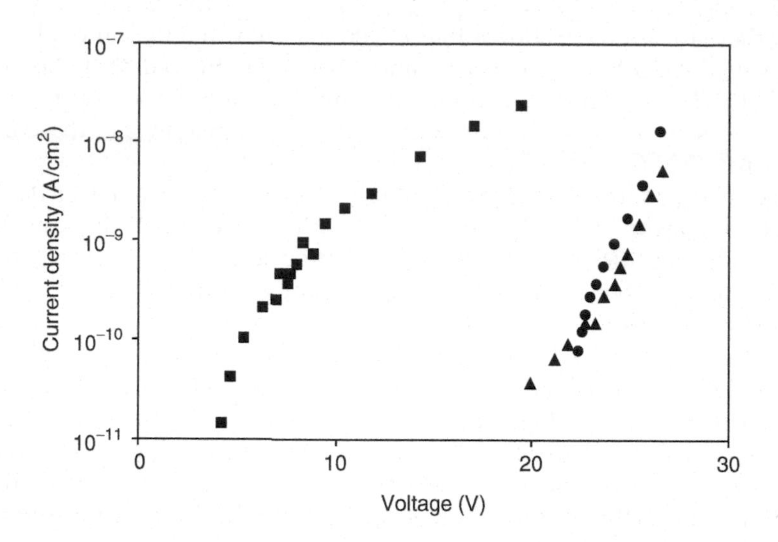

FIGURE 6.39
Current density versus voltage measurements on the 45 nm oxides before VUV illumination (as-processed), post-VUV illumination, and post-VUV postanneal at 200°C. Circles, as-processed; squares, post-VUV illumination; and triangles, post-anneal. (Reprinted from Lenahan, P.M. and Mele, J.J., *J. Vac. Sci. Technol. B*, 18, 2169, 2000. With permission.)

Figure 6.39 shows pre-VUV illumination, post-VUV illumination, and the postillumination, poststressing current–voltage curves for the 45 nm oxides. The VUV illumination greatly increases leakage currents. These currents were greatly reduced by the subsequent anneal. Pre- and post-VUV illumination *CV* curves indicated that essentially no oxide space charge is created by the VUV illumination; thus, the very large changes in current versus voltage characteristics illustrated in Figure 6.39 are not due to space charge in the oxides.

Figure 6.40 directly compares E' densities and leakage currents in the 45 nm oxides as a function of VUV illumination. Note the close correspondence between E' center generation and leakage current generation. These results clearly demonstrate a strong qualitative correlation between neutral E' center density and oxide leakage currents in several sets of different oxides. The results provide extremely strong, albeit circumstantial, evidence that E' defects, as suggested by the work of Takagi et al. [100], are largely responsible for the leakage currents due to inelastic tunneling of silicon conduction band electrons through their levels.

6.5 Intrinsic Defects and Device Reliability: Physically Based Predictive Models and Statistical Mechanics

Although, many ESR studies of MOS systems are of some general interest as physics, chemistry, and materials science, their ultimate significance must relate to their utility: can these studies help design better, more reliable, integrated circuits? The answer to this question is almost certainly yes, if the results can be utilized to predict and manipulate defect densities. It is clear that two families of point defects, E' centers and P_b centers, play dominating roles in a number of MOS reliability problems. Materials scientists and engineers have well-developed and widely verified methods of manipulating intrinsic point defect populations. These methods are based upon the fundamental principles of the statistical mechanics of solids as well as on basic principles of physical chemistry

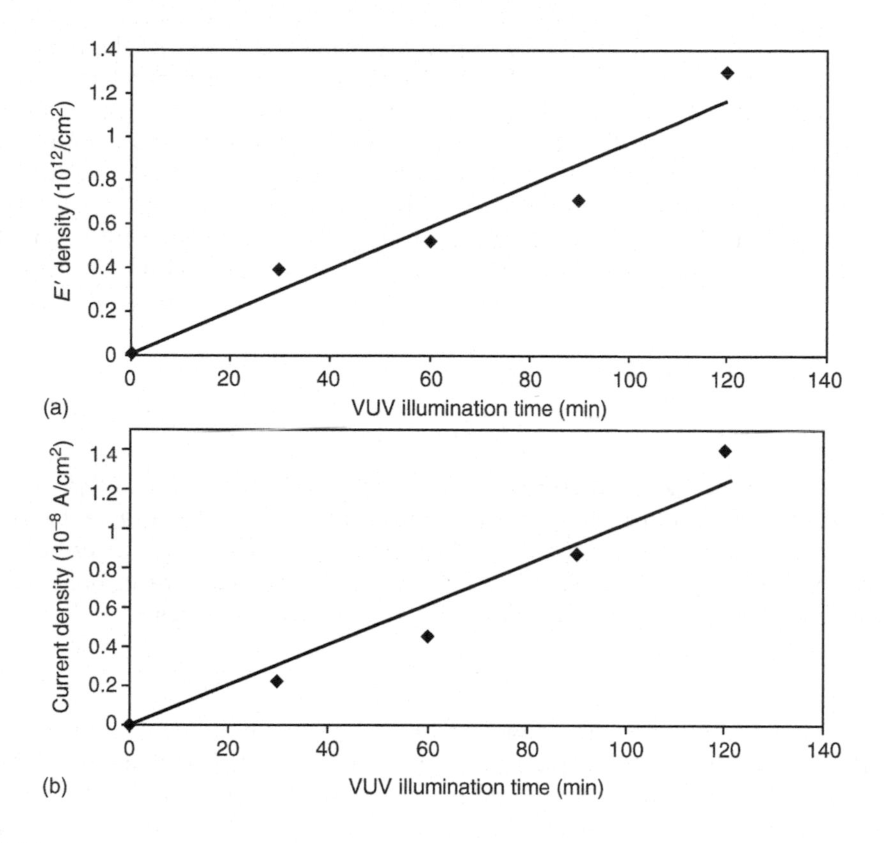

FIGURE 6.40
(a) E' density versus VUV illumination time in the 45 nm oxides. (b) Current density plotted at 17 V versus VUV illumination time for the same oxides. The lines are drawn as a guide for the eye. (Reprinted from Lenahan, P.M. and Mele, J.J., *J. Vac. Sci. Technol. B*, 18, 2169, 2000. With permission.)

[107,108]. One should be able to ameliorate device reliability problems by applying these well-established principles to E' and P_b centers.

6.5.1 Predicting Oxide Hole Trapping

Lenahan and Conley [109–113] used the standard approach of statistical mechanics [107,108] to calculate the density of oxygen vacancies in MOSFET oxides, calibrated the parameters of the expression with ESR measurements, and then tested the validity of calibrated (quantitative) expression on several oxide films. They found a close correspondence between the calibrated expression and experimental results. A consideration of the basic principles of statistical thermodynamics tells us that equilibrium occurs when the Gibbs free energy, G, of a solid is minimized [108]. It can be shown that, for the simplest cases, the minimization of Gibbs free energy leads to an equilibrium density of vacancy sites given by

$$n = Ne^{\Delta S_f/k - \Delta H_f/kT} \tag{6.18}$$

where

ΔS_f represents the nonconfigurational entropy contribution per defect site
ΔH_f represents the enthalpy of formation of a defect site
k is the Boltzmann constant
N represents the density of available sites

For the purposes of this discussion, the important points here are that the nonconfigurational entropy contribution is large and essentially temperature-independent, and the ΔH_f essentially represents the increase in system energy caused by vacancy creation of an unstressed lattice site minus the strain energy lost by removal from a compressed SiO_2 matrix. (This reduction in ΔH_f would be a strain energy $\sim P\,dV$ caused by the effective volume change resulting from the removal of the atom from its particular location.) As pointed out by Ohmameuda et al. [114] this strain energy reduction will be greatest for sites near the Si/SiO_2 boundary; this energy contribution should amount to several tenths of an electron volt [109–113]. One thus expects and finds [10,11,19] that the E' centers are primarily located close to the Si/SiO_2 boundary.

Anticipating then an oxygen vacancy/E' precursor density of the form

$$n = \alpha e^{-\beta/T} \tag{6.19}$$

where the temperature-independent constant α is given by $\alpha = N e^{\Delta S_f/k}$ and $\beta = \Delta H_f/k$, we may evaluate the relevant thermodynamic constants by making measurements on devices exposed to various high-temperature anneals. With a knowledge of E' center hole capture cross section [109–113] and the standard analysis of charge capture in oxide films, we would anticipate that, for a given fluence of holes through the oxides

$$N_{th} = \alpha e^{-\beta/T}(1 - e^{-\sigma\eta}) \tag{6.20}$$

where N_{th} is the density of trapped holes and h is the fluence of holes through the oxide. With α, β, and σ evaluated from spin resonance measurements, the expression provides a nonadjustable parameter prediction of oxide hole trapping. (However, due to the modest absolute precision of ESR measurements, the value of α as determined strictly from ESR could be in error by almost a factor of 2.) The potential validity of Equation 6.20 was assessed [109–113] through a series of measurements on MOS oxides subjected to anneals at 875°C, 950°C, 1025°C, and 1100°C. The oxides were all grown at 825°C and then a polycrystalline silicon gate was deposited. After gate deposition the anneals were carried out for 30 min in a dry N_2 atmosphere. After the anneals the capacitors were rapidly pulled from the furnace to quench in the defect densities at the annealing temperatures. The poly-Si gates were removed and two sets of measurements were made on the samples, both after subjecting the oxides to hole flooding. To evaluate the E' precursor enthalpy of creation, oxides of the three higher temperature annealing samples were each flooded with approximately 2×10^{13} holes/cm^2. The enthalpy was determined from the slope of a plot of the natural logarithm of E' density versus reciprocal temperature, shown in Figure 6.41; the activation enthalpy is approximately 1.5 ± 0.1 eV. To test the predictive capability of Equation 6.20, holes were injected into samples subjected to each of the four annealing steps; midgap CV shifts, ΔV_{mg}, were plotted versus injected hole fluence. Using Equation 6.20 and taking the trapped holes to be close to the Si/SiO_2 boundary, the simple model [109–113] predicts midgap shifts of

$$\Delta V_{mg} = q\alpha e^{-\beta/T} C_{ox}(1 - e^{\sigma\eta}) \tag{6.21}$$

where

q is the electronic charge
C_{ox} is the oxide capacitance
all other parameters are as previously defined

Figure 6.42 compares the experimental results and the predictions of Equation 6.16. The correspondence between prediction and experiment is quite close. It clearly demonstrates

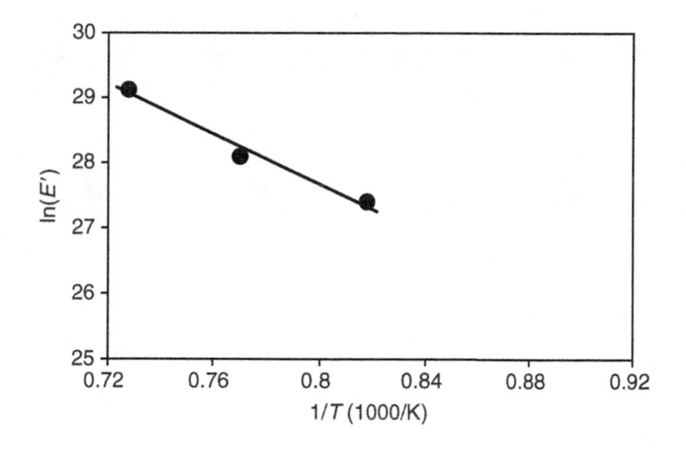

FIGURE 6.41
Plot of the natural logarithm of E' density versus the reciprocal of annealing temperature. Slope of the line yields an activation energy of about 1.5 eV. (Reprinted from Lenahan, P.M. and Conley, J.F., *J. Vac. Sci. Technol. B*, 16, 2134, 1998. With permission.)

that it is possible to predict the response of an oxide from an essentially nonadjustable parameter fit of a physically based model. The empirical results of Lenahan and Conley are consistent with more recent theoretical estimates of the enthalpy of E' formation. Their measurement of 1.5 ± 0.1 eV is close to calculated estimates of 1.3 eV by Boureau et al. [115] and Capron et al. [116]. As demonstrated by Conley et al. [112] the anticipated trap densities depend on both thermodynamics and kinetics.

6.5.2 Predicting Interface Trap Formation

Straightforward concepts from the equilibrium thermodynamics of chemical reactions also allow one to make some predictions about interface trap formation [110,113,117,118]. As discussed previously, at least five groups have demonstrated that significant (greater than

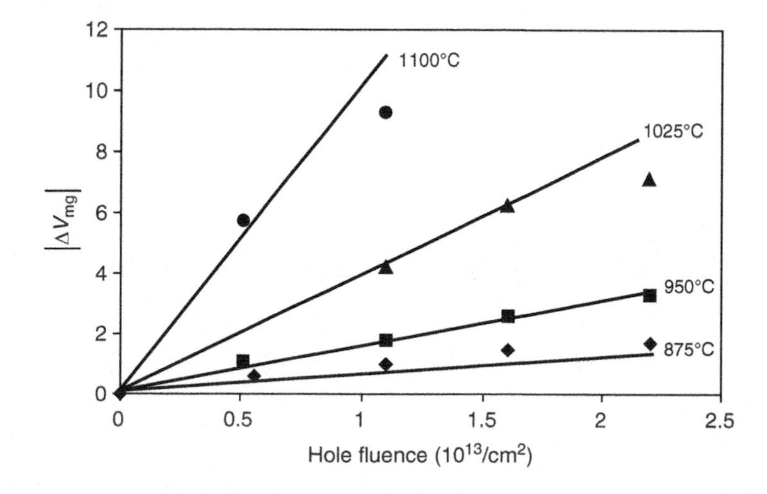

FIGURE 6.42
Solid lines represent Equation 6.21 evaluated for the various indicated temperatures. Dots represent experimental results. (Reprinted from Lenahan, P.M. and Conley, J.F., *J. Vac. Sci. Technol. B*, 16, 2134, 1998. With permission.)

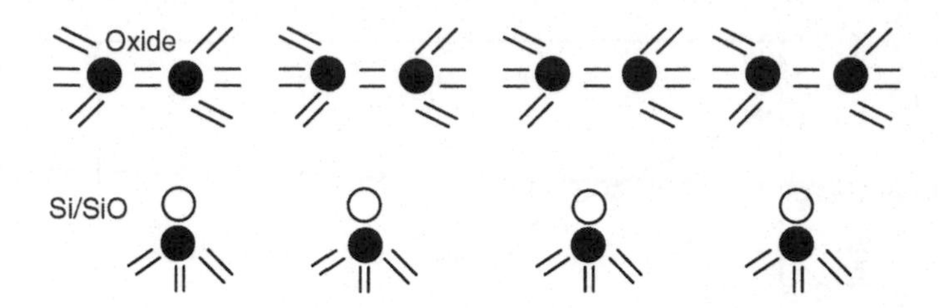

FIGURE 6.43
A schematic illustration of defect–defect precursor structure before stressing. The filled circles indicate silicon atoms; the small open circles indicate hydrogen atoms. (Reprinted from Lenahan, P.M., *Microelectron. Eng.*, 69, 173, 2003. With permission.)

or approximately equal to $10^{12}/\text{cm}^2$) generation of P_b centers occurs when MOS oxides are subjected to technologically relevant levels of ionizing radiation. More limited results indicate P_b generation resulting from high or low oxide electric field injection of electrons as well as the injection of hot holes into the oxide from the near drain region of short-channel MOSFETs. The previously discussed studies of Conley and Lenahan [56–58] show that E' centers rapidly react with molecular hydrogen at room temperature and that this reaction is accompanied by the simultaneous generation of Si/SiO_2 interface traps (these are P_b centers).

It is well established that $\sim 5 \times 10^{12}$ cm^2 Si/SiO_2 silicon dangling bonds sites are passivated by hydrogen [2–5]. As pointed out by Conley and Lenahan [110,113,118], in an unstressed oxide, one would expect to find the circumstances schematically illustrated in Figure 6.43. A large number of hydrogen dangling bond sites (P_bH) exist at the Si/SiO_2 boundary while a large number of E' precursor sites exist in the oxide. Essentially no unpassivated silicon dangling bonds exist at the E' precursor sites. For purposes of illustration, the precursor sites are taken to be oxygen vacancy sites. The stressing process (ionizing radiation, hot carrier injection, high-field stressing, etc.), for whatever reason, generates significant densities of E' center silicon dangling bonds in the oxide. This situation is schematically illustrated in Figure 6.44. Note the presence of both a large number of hydrogen passivated silicon dangling bond sites at the Si/SiO_2 boundary and a large number of unpassivated silicon dangling bond sites in the oxide. Such a combination of defects centers cannot be in thermodynamic equilibrium. Thermodynamics requires that a system in equilibrium will reach the lowest Gibbs free energy G [108]:

$$G = H - TS \tag{6.22}$$

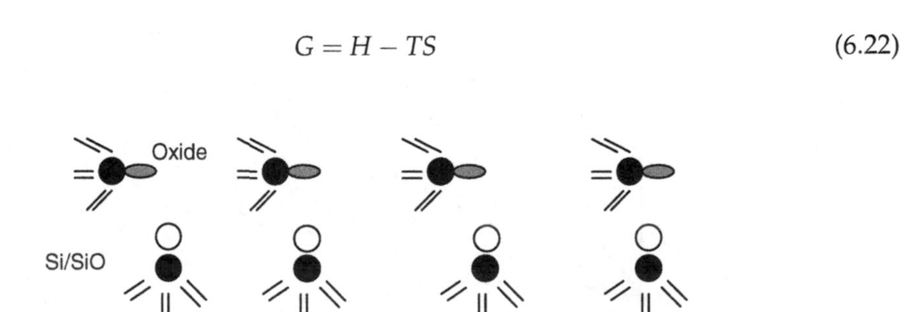

FIGURE 6.44
A schematic and perhaps somewhat artificial illustration indicating the creation of silicon dangling bonds (dangling bonds indicated by the gray ovals) in the oxide in the presence of passivated silicon atoms at the Si/SiO_2 boundary. The filled black circles indicate silicon atoms; the smaller open circles indicate hydrogen atoms. (Reprinted from Lenahan, P.M., *Microelectron. Eng.*, 69, 173, 2003. With permission.)

where
  H is the enthalpy
  T is the absolute temperature
  S is the entropy
  Enthalpy H is the sum of energy plus pressure times volume

The bond enthalpy of a silicon hydrogen bond (P_bH) at the Si/SiO_2 boundary must be very roughly equal to the enthalpy of a silicon hydrogen bond in the oxide ($E'H$) [110,113,118]. Thus, the exchange of an interface dangling bond hydrogen with an oxide dangling bond would cost very little enthalpy. The entropy of a system is defined as [108]:

$$S = k \ln \Omega \tag{6.23}$$

where
  k is the Boltzmann constant
  Ω is the number of microscopic configurations responsible for the macroscopic system

Consider a simplified situation in which all the M Si/SiO_2 interface dangling bond sites are hydrogen passivated (P_bH). The configurational entropy of this system is simply $k \ln 1 = 0$. Remove one hydrogen from any one of the M sites. The configurational entropy becomes $k \ln M$, a very large increase. (The removal of a second hydrogen would lead to a configurational entropy of $k \ln[(M)(M-1)/2]$, etc.)

Consider a simplified situation in the oxide in which each of the N E' dangling bond sites is unpassivated. The configurational entropy of these sites is also simply $k \ln 1 = 0$. Add one hydrogen to any one of the N sites. The configurational entropy contribution increases to $k \ln N$, again quite a large contribution. (The addition of a second hydrogen would similary lead to a configuration entropy of $k \ln[(N)(N-1)/2]$, etc.) This process is schematically illustrated in Figures 6.45 and 6.46.

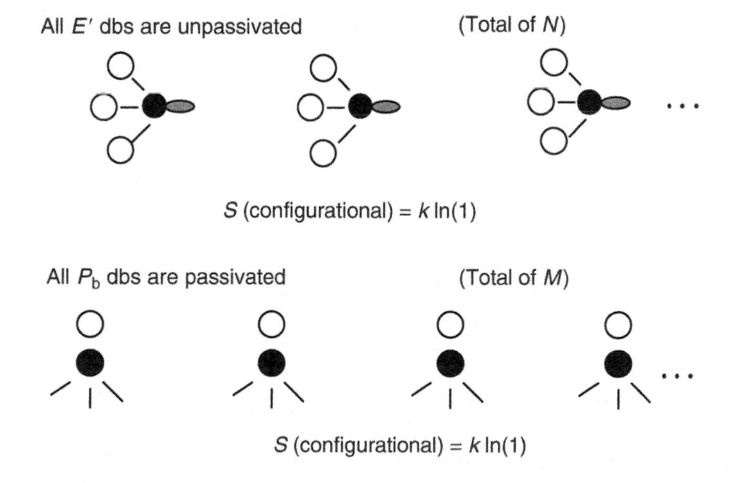

FIGURE 6.45
A schematic illustration of an oxide with N E' sites all unpassivated and M P_b sites all of which are passivated. In both the unpassivated E' dangling bond and the passivated P_b dangling bond cases, the configurational entropy, S, is given by $k \ln(1)$. The filled black circles indicate silicon atoms, the larger open circles indicate oxygen atoms, the smaller open circles indicate hydrogen atoms, and the gray ovals indicate silicon dangling bonds. (Reprinted from Lenahan, P.M., *Microelectron. Eng.*, 69, 173, 2003. With permission.)

Remove one H from the (M) PbH sites;
ΔS (configurational) = k ln(M)

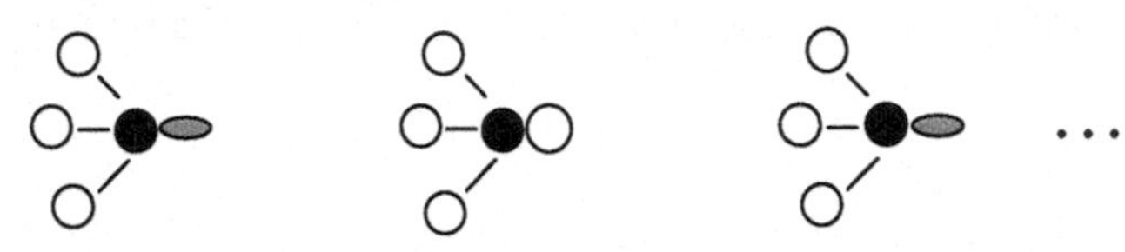
Add one H to the (N) E′ sites;
ΔS (configurational) = k ln(N)

FIGURE 6.46
A schematic illustration of the effect of the transfer of one hydrogen atom from a passivated interface dangling bond to an unpassivated oxide dangling bond. Note the large increase in configurational entropy in both cases. The filled black circles indicate silicon atoms, the larger open circles indicate oxygen atoms, the smaller open circles indicate hydrogen atoms, and the gray ovals indicate silicon dangling bonds. (Reprinted from Lenahan, P.M., *Microelectron. Eng.*, 69, 173, 2003. With permission.)

Thus, very basic statistical thermodynamics arguments indicate that the exchange of a large number of hydrogens from interface dangling bonds centers to oxide dangling bond centers is thermodynamically favored. This thermodynamic equiliblium can be expressed in the following equations:

$$P_bH + E' \leftrightarrow P_b + E'H \tag{6.24}$$

and

$$\frac{[P_b][E'H]}{[P_bH][E']} = K \tag{6.25}$$

In Equation 6.25, the constant K (not the SiO_2 dielectric constant) is defined as follows [108]:

$$K = \exp\left(\frac{-\Delta g_s}{kT}\right) \tag{6.26}$$

where Δg_s represents a Gibbs free energy per site [108]. As discussed above, the enthalpy change is near zero.

On the basis of the above discussion $K \approx 1$. Thus, if the system can reach some sort of quasiequilibrium condition, one would, inevitably, observe a fairly close correspondence between interface trap generation and oxide leakage currents. As noted previously, these E' defects are largely responsible for oxide hole trapping under many circumstances and are likely also responsible for oxide stress-induced leakage currents. As noted previously, the P_b centers are quite clearly the dominating source of interface traps under many circumstances. Thus, the often noted correspondence between oxide defect generation or population and interface trap generation [119–121] makes physical sense in terms of the chemistry and energy levels of these defects.

6.6 Relevance of P_b and E' Centers to High Dielectric Constant (High-K) MOSFETs

As the equivalent oxide thickness of MOS gate stacks approaches fundamental physical limits, high dielectric constant (high-K) oxide devices are under development [122,123]. The most promising high-K dielectrics are based on HfO_2 and have typically been deposited by the atomic layer deposition (ALD) technique [124]. The P_b and E' families may remain as important defects in these new device structures because they typically involve HfO_2 on a very thin (~1 nm) silicon dioxide interfacial layer on the silicon substrate [124]. Not surprisingly, P_b or P_b-like centers may be present at the Si/dielectric boundary of these new devices. E' centers may also be present.

Kang et al. [125,126] reported conventional ESR measurements of ALD HfO_2 films on both (111) and (100) silicon substrates. They detected the presence of P_b-like interface traps, nearly like (111) P_b centers and (100) P_{b0} centers, but with small shifts in the g matrix elements toward higher values ($g_{\parallel} = 2.0018$ and $g_{\perp} = 2.0094$ in contrast to the Si/SiO_2 values of $g_{\parallel} = 2.0013$ and $g_{\perp} = 2.008$). Triplett et al. [127] have also reported observation of P_b like centers, also with matrix elements shifted to slightly higher values in HfO_2-based structures. Stesmans and Afanas'ev [128] observed both P_{b0} and P_{b1} centers in HfO_2-based device structures. Unlike Kang et al. [125,126] and Triplett et al. [127] they did not observe any shifts in g matrix elements from the values reported for conventional Si/SiO_2 interfaces. Both Kang et al. [125,126] and Stesmans and Afanas'ev [128] reported that forming gas anneals (H_2/N_2) were effective in reducing the density of the P_b or P_b like centers.

E' centers may also play a significant role in HfO_2-based devices. Ryan et al. [129] recently reported the presence of quite high densities of two types of E' centers in the interfacial SiO_2 layer of HfO_2-based MOS structures (up to $10^{19}/cm^3$). They found high densities in device structures before any sort of stressing. The E' densities depend quite strongly upon the ALD deposition conditions, and the presence of a metal gate greatly enhances E' generation resulting from postdeposition anneals.

6.7 Summary

Two families of silicon dangling bond defects, P_b centers and E' centers, play extremely important roles in MOS device limitations. Although a comprehensive understanding of MOS limitations as well as of these defects has yet to be established, their overall importance as the dominating interface traps (P_b) and dominating oxide deep level centers (E') is quite clear. Under nearly all technologically reasonable stressing conditions, P_b family defects dominate interface trap generation. The E' family plays a more complex role in device operation. The defects clearly dominate oxide hole trapping when stressing conditions are not particularly severe. However, in very heavily stressed oxides or in oxides subjected to repeated sequences of electron–hole flooding, complex structural changes clearly take place at the E' sites. Since the E' defects have levels around the middle of the oxide band gap, they inevitably play an important role in oxide leakage current phenomena.

References

1. Weil, J.A., Bolton, J.R., and Wertz, J.E., *Electron Paramagnetic Resonance: Elementary Theory and Practical Applications*, John Wiley & Sons, New York, NY, 1994.
2. Nishi, Y., Study of silicon-silicon dioxide structure by electron spin resonance 1, *Jpn. J. Appl. Phys.*, 10, 52, 1971.
3. Nishi, Y., Tanaka, K., and Ohwada, A., Study of silicon–silicon dioxide structures by electron spin resonance 2, *Jpn. J. Appl. Phys.*, 11, 85, 1972.
4. Caplan, P.J. et al., ESR centers, interface states, and oxide fixed charge in thermally oxidized silicon wafers, *J. Appl. Phys.*, 50, 5847, 1979.
5. Poindexter, E.H. et al., Interface states and electron spin resonance centers in thermally oxidized (111) and (100) silicon wafers, *J. Appl. Phys.*, 52, 879, 1981.
6. Lenahan, P.M. et al., Radiation-induced trivalent silicon defect buildup at the Si–SiO_2 interface in MOS structures, *IEEE Trans. Nucl. Sci.*, 28, 4105, 1981.
7. Lenahan, P.M. and Dressendorfer, P.V., Effect of bias on radiation-induced paramagnetic defects at the silicon–silicon dioxide interface, *Appl. Phys. Lett.*, 41, 542, 1982.
8. Lenahan, P.M. and Dressendorfer, P.V., An electron spin resonance study of radiation-induced electrically active paramagnetic centers at the Si/SiO_2 interface, *J. Appl. Phys.*, 54, 1457, 1983.
9. Brower, K.L., Si-29 hyperfine-structure of unpaired spins at the Si/SiO_2 interface, *Appl. Phys. Lett.*, 43, 1111, 1983.
10. Lenahan, P.M. and Dressendorfer, P.V., Paramagnetic trivalent silicon centers in gamma-irradiated metal-oxide-silicon structures, *Appl. Phys. Lett.*, 44, 96, 1984.
11. Lenahan, P.M. and Dressendorfer, P.V., Hole traps and trivalent silicon centers in metal/oxide/silicon devices, *J. Appl. Phys.*, 55, 3495, 1984.
12. Poindexter, E.H. et al., Electronic traps and P_b centers at the Si/SiO_2 interface: band-gap energy distribution, *J. Appl. Phys.*, 56, 2844, 1984.
13. Mikawa, R.E. and Lenahan, P.M., Structural damage at the Si/SiO_2 interface resulting from electron injection in metal-oxide-semiconductor devices, *Appl. Phys. Lett.*, 46, 550, 1985.
14. Gerardi, G.J. et al., Interface traps and P_b centers in oxidized (100) silicon-wafers, *Appl. Phys. Lett.*, 49, 348, 1986.
15. Mikawa, R.E. and Lenahan, P.M., Electron-spin-resonance study of interface states induced by electron injection in metal-oxide-semiconductor devices, *J. Appl. Phys.*, 59, 2054, 1986.
16. Warren, W.L. and Lenahan, P.M., Electron-spin-resonance study of high-field stressing in metal-oxide-silicon device oxides, *Appl. Phys. Lett.*, 49, 1296, 1986.
17. Brower, K.L., Structural features at the Si/SiO_2 interface, *Zeitschrift Fur Physikalische Chemie Neue Folge, Z. Phys. Chem.*, Neue Folge, 151, 177, 1987.
18. Triplett, B.B., Takahashi, T., and Sugano, T., Electron-spin-resonance observation of defects in device oxides damaged by soft x-rays, *Appl. Phys. Lett.*, 50, 1663, 1987.
19. Kim, Y.Y. and Lenahan, P.M., Electron-spin-resonance study of radiation-induced paramagnetic defects in oxides grown on (100) silicon substrates, *J. Appl. Phys.*, 64, 3551, 1988.
20. Vranch, R.L., Henderson, B., and Pepper, M., Spin-dependent recombination in irradiated Si/SiO_2 device structures, *Appl. Phys. Lett.*, 52, 1161, 1988.
21. Miki, H. et al., Electron and hole traps in SiO_2-films thermally grown on Si substrates in ultra-dry oxygen, *IEEE Trans. Electron Devices*, 35, 2245, 1988.
22. Jupina, M.A. and Lenahan, P.M., Spin dependent recombination—a ^{29}Si hyperfine study of radiation-induced P_b centers at the Si/SiO_2 interface, *IEEE Trans. Nucl. Sci.*, 37, 1650, 1990.
23. Krick, J.T., Lenahan, P.M., and Dunn, G.J., Direct observation of interfacial point-defects generated by channel hot hole injection in n-channel metal-oxide silicon field-effect transistors, *Appl. Phys. Lett.*, 59, 3437, 1991.
24. Awazu, K., Watanabe, K., and Kawazoe, H., Generation mechanisms of paramagnetic centers by gamma-ray irradiation at and near the Si/SiO_2 interface, *J. Appl. Phys.*, 73, 8519, 1993.
25. Gabrys, J.W., Lenahan, P.M., and Weber, W., High-resolution spin-dependent recombination study of hot-carrier damage in short-channel MOSFETs—Si^{29} hyperfine spectra, *Microelectron. Eng.*, 22, 273, 1993.

26. Stesmans, A. and Afanas'ev, V.V., Electron spin resonance features of interface defects in thermal (100) Si/SiO_2, *J. Appl. Phys.*, 83, 2449, 1998.
27. Stesmans, A. and Afanas'ev, V.V., Electrical activity of interfacial paramagnetic defects in thermal (100) Si/SiO_2, *Phys. Rev. B*, 57, 10030, 1998.
28. Mishima, T.D. and Lenahan, P.M., A spin-dependent recombination study of radiation-induced P_{b1} centers at the (001) Si/SiO_2 interface, *IEEE Trans. Nucl. Sci.*, 47, 2249, 2000.
29. Mishima, T.D., Lenahan, P.M., and Weber, W., Response to comment on do P_{b1} centers have levels in the Si band gap? Spin-dependent recombination study of the P_{b1} hyperfine spectrum, *Appl. Phys. Lett.*, 78, 1453, 2001.
30. Campbell, J.P. and Lenahan, P.M., Density of states of P_{b1} Si/SiO_2 interface trap centers, *Appl. Phys. Lett.*, 80, 1945, 2002.
31. Lenahan, P.M. and Conley, J.F., What can electron paramagnetic resonance tell us about the Si/SiO_2 system? *J. Vac. Sci. Technol. B*, 16, 2134, 1998.
32. Stesmans, A. and Afanas'ev, V.V., Comment on do P_{b1} centers have levels in the Si band gap? Spin-dependent recombination study of the P_{b1} hyperfine spectrum, *Appl. Phys. Lett.*, 78, 1451, 2001.
33. Weeks, R.A., Paramagnetic resonance of lattice defects in irradiated quartz, *J. Appl. Phys.*, 27, 1376, 1956.
34. Silsbee, R.H., Electron spin resonance in neutron-irradiated quartz, *J. Appl. Phys.*, 32, 1459, 1961.
35. Feigl, F.J., Fowler, W.B., and Yip, K.L., Oxygen vacancy model for E_1' center in SiO_2, *Solid State Commun.*, 14, 225, 1974.
36. Griscom, D.L., Friebele, E.J., and Sigel, G.H., Observation and analysis of primary Si-29 hyperfine-structure of E′ center in non-crystalline SiO_2, *Solid State Commun.*, 15, 479, 1974.
37. Griscom, D.L., E′ center in glassy SiO_2—microwave saturation properties and confirmation of the primary Si-29 hyperfine-structure, *Phys. Rev. B*, 20, 1823, 1979.
38. Lenahan, P.M. and Dressendorfer, P.V., Microstructural variations in radiation hard and soft oxides observed through electron-spin resonance, *IEEE Trans. Nucl. Sci.*, 30, 4602, 1983.
39. Takahashi, T. et al., Electron-spin-resonance observation of the creation, annihilation, and charge state of the 74-gauss doublet in device oxides damaged by soft x-rays, *Appl. Phys. Lett.*, 51, 1334, 1987.
40. Warren, W.L. et al., Neutral *E*′ centers in microwave downstream plasma-enhanced chemical-vapor-deposited silicon dioxide, *Appl. Phys. Lett.*, 53, 482, 1988.
41. Warren, W.L. and Lenahan, P.M., Si-29 hyperfine spectra and structure of *E*′ dangling-bond defects in plasma-enhanced chemical-vapor deposited silicon dioxide films on silicon, *J. Appl. Phys.*, 66, 5488, 1989.
42. Hazama, H. et al., *Proceedings of the Workshop on Ultra Thin Oxides, Jpn. Soc. Appl. Phys.*, Tokyo, 207, 1998.
43. Lenahan, P.M. and Mele, J.J., *E*′ centers and leakage currents in the gate oxides of metal oxide silicon devices, *J. Vac. Sci. Technol. B*, 18, 2169, 2000.
44. Atkins, P., *Molecular Quantum Mechanics*, Chap. 14, 2nd edn., Oxford University Press, Oxford, 1983.
45. Slichter, C.P., *Principles of Magnetic Resonance*, 2nd edn., Springer-Verlag, Berlin, 1978.
46. Lepine, D.J., Spin-dependent recombination on silicon surface, *Phys. Rev. B*, 6, 436, 1972.
47. Kaplan, D., Solomon, I., and Mott, N.F., Explanation of large spin-dependent recombination effect in semiconductors, *J. Phys. Lett.*, 39, L51, 1978.
48. Hall, R.N., Electron-hole recombination in germanium, *Phys. Rev.*, 87, 387, 1952.
49. Shockley, W. and Read, W.T., Statistics of the recombinations of holes and electrons, *Phys. Rev.*, 87, 835, 1952.
50. Lenahan, P.M. and Jupina, M.A., Spin dependent recombination at the silicon-silicon dioxide interface, *Colloids Surf.*, 45, 191, 1990.
51. O'Reilly, E.P. and Robertson, J., Theory of defects in vitreous silicon dioxide, *Phys. Rev. B*, 27, 3780, 1983.
52. Conley, J.F. et al., Observation and electronic characterization of new *E*′ center defects in technologically relevant thermal SiO_2 on Si: an additional complexity in oxide charge trapping, *J. Appl. Phys.*, 76, 2872, 1994.

53. Mishima, T.D., Lenahan, P.M., and Weber, W., Do P_{b1} centers have levels in the Si band gap? Spin-dependent recombination study of the P_{b1} hyperfine spectrum, *Appl. Phys. Lett.*, 76, 3771, 2000.
54. Uchino, T. and Yoko, T., Density functional theory of structural transformations of oxygen-deficient centers in amorphous silica during hole trapping: structure and formation mechanism of the E' (gamma) center, *Phys. Rev. B*, 74, 2006.
55. Alemany, M.M.G. and Chelikowsky, J.R., Edge-sharing tetrahedra: precursors of the E-gamma′ defects in amorphous silica, *Phys. Rev. B*, 68, 2003.
56. Conley, J.F. and Lenahan, P.M., Room-temperature reactions involving silicon dangling bond centers and molecular-hydrogen in amorphous SiO_2 thin-films on silicon, *IEEE Trans. Nucl. Sci.*, 39, 2186, 1992.
57. Conley, J.F. and Lenahan, P.M., Molecular-hydrogen, E' center hole traps, and radiation-induced interface traps in MOS devices, *IEEE Trans. Nucl. Sci.*, 40, 1335, 1993.
58. Conley, J.F. and Lenahan, P.M., Radiation-induced interface states and ESR evidence for room-temperature interactions between molecular-hydrogen and silicon dangling bonds in amorphous SiO_2-films on Si, *Microelectron. Eng.*, 22, 215, 1993.
59. Vitko, J., ESR studies of hydrogen hyperfine spectra in irradiated vitreous silica, *J. Appl. Phys.*, 49, 5530, 1978.
60. Tsai, T.E. and Griscom, D.L., On the structures of hydrogen-associated defect centers in irradiated high-purity a-SiO_2, *J. Non-Cryst. Solids*, 91, 170, 1987.
61. Conley, J.F., Lenahan, P.M., and Roitman, P., *Proceedings of the IEEE SOI Technology Conference*, IEEE, New York, 12, 1991.
62. Conley, J.F., Lenahan, P.M., and Roitman, P., Electron-spin-resonance of separation by implanted oxygen oxides—evidence for structural-change and a deep electron trap, *Appl. Phys. Lett.*, 60, 2889, 1992.
63. Conley, J.F., Lenahan, P.M., and Roitman, P., Evidence for a deep electron trap and charge compensation in separation by implanted oxygen oxides, *IEEE Trans. Nucl. Sci.*, 39, 2114, 1992.
64. Vanheusden, K. and Stesmans, A., Characterization and depth profiling of E'-defects in buried SiO_2, *J. Appl. Phys.*, 74, 275, 1993.
65. Warren, W.L. et al., Paramagnetic defect centers in BESOI and SIMOX buried oxides, *IEEE Trans. Nucl. Sci.*, 40, 1755, 1993.
66. Conley, J.F. and Lenahan, P.M., Hydrogen complexed EP (E' delta) centers and EP/H_2 interactions—implications for EP structure, *Microelectron. Eng.*, 28, 35, 1995.
67. Conley, J.F. and Lenahan, P.M., Electron spin resonance analysis of EP center interactions with H_2: evidence for a localized EP center structure, *IEEE Trans. Nucl. Sci.*, 42, 1740, 1995.
68. Griscom, D.L. and Friebele, E.J., Fundamental radiation-induced defect centers in synthetic fused silicas—atomic chlorine, delocalized E′ centers, and a triplet-state, *Phys. Rev. B*, 34, 7524, 1986.
69. Zhang, L. and Leisure, R.G., The E'(delta) and triplet-state centers in x-irradiated high-purity amorphous SiO_2, *J. Appl. Phys.*, 80, 3744, 1996.
70. Chavez, J.R., et al., Microscopic structure of the E′(delta) center in amorphous SiO_2: a first principles quantum mechanical investigation, *IEEE Trans. Nucl. Sci.*, 44, 1799, 1997.
71. Buscarino, G., Agnello, S., and Gelardi, F.M., Investigation on the microscopic structure of E-delta′ center in amorphous silicon dioxide by electron paramagnetic resonance spectroscopy, *Mod. Phys. Lett. B*, 20, 451, 2006.
72. Stesmans, A., New intrinsic defect in as-grown thermal SiO_2 on (111) Si, *Phys. Rev. B*, 45, 9501, 1992.
73. Warren, W.L. et al., Hydrogen interactions with delocalized spin centers in buried SiO_2 thin-films, *Appl. Phys. Lett.*, 62, 1661, 1993.
74. Devine, R.A.B. et al., Point-defect generation during high-temperature annealing of the Si/SiO_2 interface, *Appl. Phys. Lett.*, 63, 2926, 1993.
75. Griscom, D.L., Characterization of 3 E'-center variants in x-irradiated and gamma-irradiated high-purity a-SiO_2, *Nucl. Instrum. Meth. Phys. Res. Sect. B*, 229, 481, 1984.
76. Fitzgerald, D.J. and Grove, A.S., Surface recombination in semiconductors, *Surf. Sci.*, 9, 347, 1968.
77. Alam, M.A. and Mahapatra, S., A comprehensive model of pMOS NBTI degradation, *Microelectron. Reliab.*, 45, 71, 2005.

78. Chakravarthi, S. et al., A comprehensive framework for predictive modeling of negative bias temperature instability, in *Proceeding of the IEEE International Reliability Physics Symposium*, 2004, p. 273.
79. Huard, V., Denais, M., and Parthasarathy, C., NBTI degradation: from physical mechanisms to modeling, *Microelectron. Reliab.*, 46, 1, 2006.
80. Fujieda, S. et al., Interface defects responsible for negative-bias temperature instability in plasma-nitrided SiON/Si(100) systems, *Appl. Phys. Lett.*, 82, 3677, 2003.
81. Campbell, J.P. et al., Direct observation of the structure of defect centers involved in the negative bias temperature instability, *Appl. Phys. Lett.*, 87, 204106, 2005.
82. Campbell, J.P. et al., Observations of NBTI-induced atomic-scale defects, *IEEE Trans. Device Mater. Reliab.*, 6, 117, 2006.
83. Campbell, J.P. et al., Location, structure, and density of states of NBTI-induced defects in plasma-nitrided pMOSFETs, in *IEEE International Reliability Physics Symposium*, Phoenix, AZ, 2007.
84. Campbell, J.P. et al., Identification of atomic-scale defect structure involved in the negative bias temperature instability in plasma-nitrided pMOSFETs, *Appl. Phys. Lett.*, 90, 123502, 2007.
85. Marquardt, C.L. and Sigel, G.H., Radiation-induced defect centers in thermally grown oxide films, *IEEE Trans. Nucl. Sci.*, 22, 2234, 1975.
86. Lipkin, L. et al., Correlation of fixed positive charge and *E*′gamma centers as measured via electron injection and electron-paramagnetic resonance techniques, *J. Electrochem. Soc.*, 138, 2050, 1991.
87. Walters, M. and Reisman, A., Radiation induced neutral electron trap generation in electrically biased insulated gate field effect transistor gate insulators, *J. Electrochem. Soc.*, 138, 2756, 1991.
88. Fleetwood, D.M. et al., Border traps—issues for MOS radiation response and long-term reliability, *Microelectron. Reliab.*, 35, 403, 1995.
89. Lelis, A.J. and Oldham, T.R., Time-dependence of switching oxide traps, *IEEE Trans. Nucl. Sci.*, 41, 1835, 1994.
90. Ma, T.P. and Dressendorfer, P.V., *Ionizing Radiation Effects in MOS Devices and Circuits*, John Wiley & Sons, New York, 1989.
91. Oldham, T.R., *Ionizing Radiation Effects in MOS Oxides*, World Scientific, Singapore, 1999.
92. Fleetwood, D.M. et al., The role of border traps in MOS high-temperature postirradiation annealing response, *IEEE Trans. Nucl. Sci.*, 40(6), 1323, 1993.
93. Conley, J.F. et al., Electron-spin-resonance evidence for the structure of a switching oxide trap—long-term structural-change at silicon dangling bond sites in SiO_2, *Appl. Phys. Lett.*, 67, 2179, 1995.
94. Conley, J.F. et al., Electron spin resonance evidence that *E*′(gamma) centers can behave as switching oxide traps, *IEEE Trans. Nucl. Sci.*, 42, 1744, 1995.
95. Lelis, A.J. et al., The nature of the trapped hole annealing process, *IEEE Trans. Nucl. Sci.*, 36, 1808, 1989.
96. Warren, W.L. et al., Microscopic nature of border traps in MOS oxides, *IEEE Trans. Nucl. Sci.*, 41, 1817, 1994.
97. Karna, S.P. et al., Electronic structure theory and mechanisms of the oxide trapped hole annealing process, *IEEE Trans. Nucl. Sci.*, 47, 2316, 2000.
98. Kohler, R.A., Kushner, R.A., and Lee, K.H., Total dose radiation hardness of MOS devices in hermetic ceramic packages, *IEEE Trans. Nucl. Sci.*, 35, 1492, 1988.
99. Stahlbush, R.E., Mrstik, B.J., and Lawrence, R.K., Postirradiation behavior of the interface state density and the trapped positive charge, *IEEE Trans. Nucl. Sci.*, 37, 1641, 1990.
100. Takagi, S., Yasuda, N., and Toriumi, A., A new I-V model for stress-induced leakage current including inelastic tunneling, *IEEE Trans. Electron Dev.*, 46, 348, 1999.
101. McPherson, J.W. and Mogul, H.C., Underlying physics of the thermochemical E model in describing low-field time-dependent dielectric breakdown in SiO_2 thin films, *J. Appl. Phys.*, 84, 1513, 1998.
102. Rosenbaum, E. and Register, L.F., Mechanism of stress-induced leakage current in MOS capacitors, *IEEE Trans. Electron Devices*, 44, 317, 1997.
103. Dumin, D.J. and Maddux, J.R., Correlation of stress-induced leakage current in thin oxides with trap generation inside the oxides, *IEEE Trans. Electron Devices*, 40, 986, 1993.

104. Warren, W.L. and Lenahan, P.M., Fundamental differences between thick and thin oxides subjected to high electric-fields, *J. Appl. Phys.*, 62, 4305, 1987.
105. Warren, W.L. and Lenahan, P.M., A comparison of positive charge generation in high-field stressing and ionizing-radiation on MOS structures, *IEEE Trans. Nucl. Sci.*, 34, 1355, 1987.
106. Weinberg, Z.A., Johnson, W.C., and Lampert, M.A., Determination of sign of carrier transported across SiO_2-films on Si, *Appl. Phys. Lett.*, 25, 42, 1974.
107. Henderson, B., *Defects in Crystalline Solids*, Crane-Russek, New York, 1972.
108. Chiang, Y.M., Birnie, D., and Kingery, W.D., *Physical Ceramics*, John Wiley & Sons, New York, 1997.
109. Lenahan, P.M., Conley, J.F., and Wallace, B.D., A model of hole trapping in SiO_2 films on silicon, *J. Appl. Phys.*, 81, 6822, 1997.
110. Lenahan, P.M. and Conley, J.F., A comprehensive physically based predictive model for radiation damage in MOS systems, *IEEE Trans. Nucl. Sci.*, 45, 2413, 1998.
111. Conley, J.F. et al., Quantitative model of radiation induced charge trapping in SiO_2, *IEEE Trans. Nucl. Sci.*, 44, 1804, 1997.
112. Conley, J.F., Lenahan, P.M., and McArthur, W.F., Preliminary investigation of the kinetics of postoxidation rapid thermal anneal induced hole-trap-precursor formation in microelectronic SiO_2 films, *Appl. Phys. Lett.*, 73, 2188, 1998.
113. Lenahan, P.M. et al., Predicting radiation response from process parameters: verification of a physically based predictive model, *IEEE Trans. Nucl. Sci.*, 46, 1534, 1999.
114. Ohmameuda, T. et al., Thermodynamical calculation and experimental confirmation of the density of hole traps in SiO_2-films, *Jpn. J. Appl. Phys., Part 2*, 30, L1993, 1991.
115. Boureau, G. et al., Thermodynamic analysis of hole trapping in SiO_2 films on silicon, *J. Appl. Phys.*, 89, 165, 2001.
116. Capron, N. et al., Thermodynamic properties of the Si/SiO_2 system, *J. Chem. Phys.*, 117, 1843, 2002.
117. Lenahan, P.M. and Conley, J.F., A physically based predictive model of Si/SiO_2 interface trap generation resulting from the presence of holes in the SiO_2, *Appl. Phys. Lett.*, 71, 3126, 1997.
118. Lenahan, P.M., Atomic scale defects involved in MOS reliability problems, *Microelectron. Eng.*, 69, 173, 2003.
119. Hu, G.J. and Johnson, W.C., Relationship between x-ray-produced holes and interface states in metal-oxide-semiconductor capacitors, *J. Appl. Phys.*, 54, 1441, 1983.
120. Dimaria, D.J. and Cartier, E., Mechanism for stress-induced leakage currents in thin silicon dioxide films, *J. Appl. Phys.*, 78, 3883, 1995.
121. Rofan, R. and Hu, C.M., Stress-induced oxide leakage, *IEEE Electron Device Lett.*, 12, 632, 1991.
122. Wilk, G.D., Wallace, R.M., and Anthony, J.M., High-kappa gate dielectrics: current status and materials properties considerations, *J. Appl. Phys.*, 89, 5243, 2001.
123. Green, M.L. et al., Ultrathin (<4 nm) SiO_2 and Si–O–N gate dielectric layers for silicon microelectronics: understanding the processing, structure, and physical and electrical limits, *J. Appl. Phys.*, 90, 2057, 2001.
124. Kirsch, P.D. et al., Nucleation and growth study of atomic layer deposited HfO_2 gate dielectrics resulting in improved scaling and electron mobility, *J. Appl. Phys.*, 99, 2006.
125. Kang, A.Y. et al., Electron spin resonance study of interface defects in atomic layer deposited hafnium oxide on Si, *Appl. Phys. Lett.*, 81, 1128, 2002.
126. Kang, A.Y., Lenahan, P.M., and Conley, J.F., The radiation response of the high dielectric-constant hafnium oxide/silicon system, *IEEE Trans. Nucl. Sci.*, 49, 2636, 2002.
127. Triplett, B.B. et al., Electron spin resonance study of as-deposited and annealed $(HfO_2)_x(SiO_2)_{(1-x)}$ high-kappa dielectrics on Si, *J. Appl. Phys.*, 101, 013703, 2007.
128. Stesmans, A. and Afanas'ev, V.V., Si dangling-bond-type defects at the interface of (100) Si with ultrathin HfO_2, *Appl. Phys. Lett.*, 82, 4074, 2003.
129. Ryan, J.T. et al., Electron spin resonance observations of oxygen deficient silicon atoms in the interfacial layer of hafnium oxide based metal-oxide-silicon structures, *Appl. Phys. Lett.*, 90, 173513, 2007.

7

Oxide Traps, Border Traps, and Interface Traps in SiO_2

Daniel M. Fleetwood, Sokrates T. Pantelides, and Ronald D. Schrimpf

CONTENTS

7.1 Introduction

In this chapter, we discuss defects in the critical bulk and near-interfacial SiO_2 regions of a metal-oxide-semiconductor (MOS) device or integrated circuit (IC). This discussion is derived mostly from experience in evaluating MOS radiation response, and therefore applies most directly to the performance and reliability of electronics in radiation environments. However, the defects that limit the radiation response of a device also can significantly affect its reliability outside a radiation environment. Hence, radiation exposure can be a very effective tool in MOS defect analysis, and the lessons learned from systems that must survive harsh radiation environments often enable insight into the defects that also limit MOS performance and long-term reliability.

When a MOS device or IC is exposed to ionizing radiation, electron–hole pairs are created in the transistor gate oxide, and in other (parasitic) insulating layers of the devices. This process is illustrated schematically in Figure 7.1. Under positive gate bias at room temperature, radiation-induced electrons rapidly transport to the gate and leave the oxide, while holes transport slowly toward the Si. A fraction of these holes is trapped near the Si/SiO_2 interface, leading to a shift in the threshold voltage of the transistor [1]. During the hole transport and trapping processes, hydrogen is released within the oxide, and under suitable bias conditions may transport to the interface and react with Si dangling bonds, forming interface traps [2–5]. Interface traps shift transistor threshold voltages and degrade channel carrier mobilities. Some positive charge that is trapped near the interface can induce compensating electron traps, which are often called border traps [6–9]. Faster border traps are sometimes mistaken for interface traps in studies of MOS performance, reliability, and radiation response [7–9].

In the MOS defect literature, there are wide varieties of nomenclatures used to characterize defects in materials, devices, and ICs. The different terms that are often used to describe defects that are similar or even identical in microstructure can vary with the method of characterization, the effect of the defect on the device of interest, the background of the investigator, the convention of the particular technical community, and many other factors. In Figure 7.2, a representative sampling is provided of many present and historical terms that are used to describe defects in the Si/SiO_2 system. The reader will see a variety of these terms used in this chapter and book that reflects the diversity of opinion and usage in the modern literature. In Figure 7.2, the defects are grouped schematically with their physical location in the Si bulk, at the Si/SiO_2 interface, in the near-interfacial SiO_2, in the SiO_2 bulk, or (in highly scaled devices), at the gate/SiO_2 interface. For a more extensive discussion of MOS defect nomenclature, please see Refs. [6–8, and references therein].

In this chapter, we first discuss briefly in Section 7.2 the measurement techniques that are used to distinguish the effects of MOS oxide, border, and interface-trap charge. We focus primarily on thermally stimulated current (TSC) methods to estimate MOS

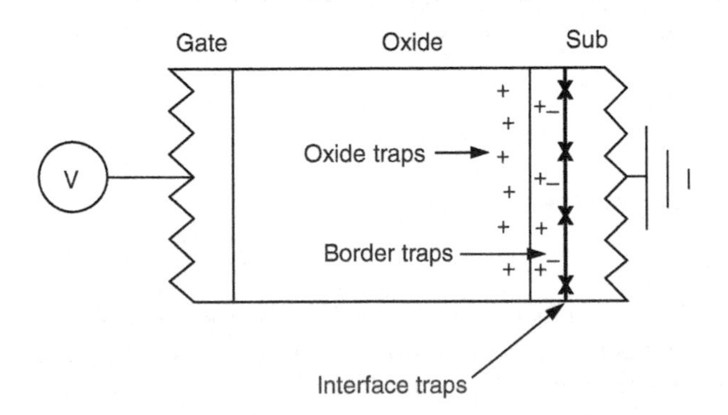

FIGURE 7.1
Schematic illustration of ionizing radiation effects on a MOS transistor. Irradiation under positive bias leads to the generation of electron–hole pairs. The electrons transport rapidly to the gate; the holes transport via a random walk process toward the Si/SiO_2 interface. During hole transport, protons can be liberated which can also transport to the interface and react with Si–H bonds to form interface traps. A subset of the trapped holes induce near-interfacial oxide-trap charge (border traps), which are defects that lie within the oxide, but are physically close enough to the Si that they can exchange charge with the channel on the timescale of the measurements being performed. (From Fleetwood, D.M., *IEEE Trans. Nucl. Sci.*, 39, 269, 1992. With permission.)

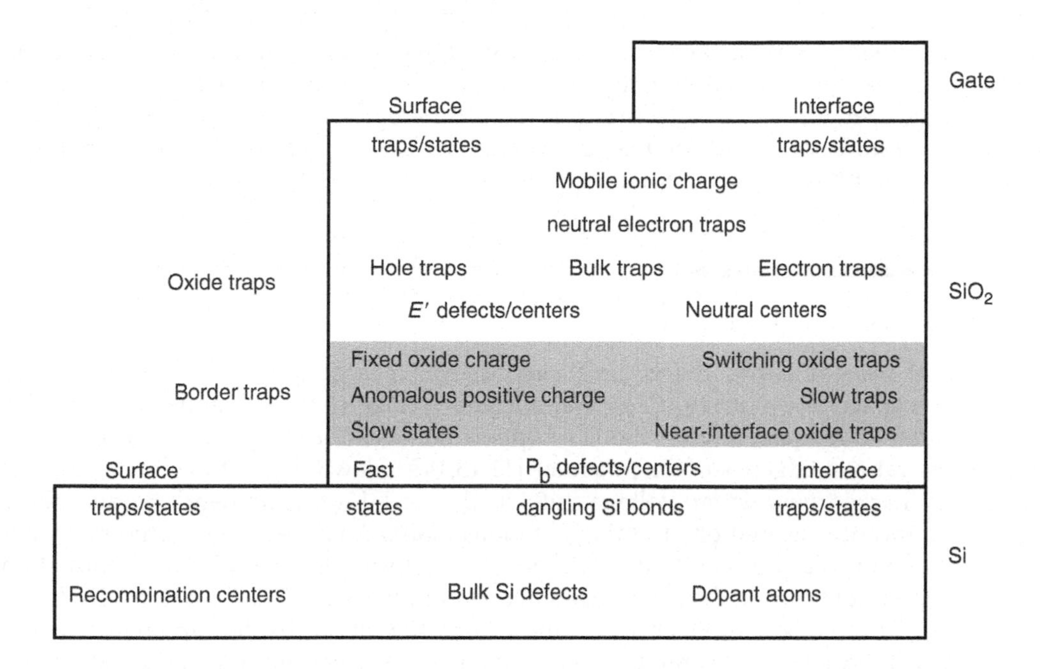

FIGURE 7.2
Schematic diagram that associates common names applied to defects in MOS devices with SiO_2 gate dielectrics with their approximate spatial locations. The diagram is not to scale. Although surface traps and states usually are reserved for cases in which there is truly a surface, in contrast to an interface, these terms are often used synonymously with interface traps and states. There have been efforts to standardize defect nomenclature with varying amounts of success. But one still finds all of these terms, and more, in the MOS defect literature.

oxide-trap charge densities and energy distributions, the results of which are then discussed in detail in Section 7.3. We also discuss the microstructures of defects in the near-interfacial SiO_2 that contribute to MOS oxide-trap charge and low-frequency noise. We discuss the nature of dipoles in the near-interfacial SiO_2 in Section 7.4; some of these dipoles are able to exchange charge easily with the underlying Si, and others are much more stable against both bias- and temperature-dependent annealing. In Section 7.5, we briefly review the effects of hydrogen on the buildup of MOS interface-trap charge. Finally, in Section 7.6, we discuss aging and device scaling effects on the radiation response and long-term reliability of MOS devices and ICs. Experimentally observed changes in MOS radiation response and/or inferred long-term reliability with time are attributed to changes in distributions of defects and/or impurities (especially hydrogenous species) with time and/or device thermal history. We conclude that the inferred radiation response or long-term reliability of devices with mobile impurities (e.g., hydrogen) can change significantly with time. Hence, efforts to manage MOS performance, reliability, and/or radiation response via defect engineering should be viewed with caution and checked for long-term stability. The same kinds of defects are observed in highly scaled devices, e.g., with ultrathin oxides, as are observed in previous generations of devices that have been studied extensively. However, their effects on device response can differ—for example, instead of oxide-trap charge shifting the device threshold voltage in a highly scaled circuit or device, it instead may increase gate leakage current or device noise. In addition, the total dose response of many deep submicron ICs is dominated by source-to-drain or device-to-device leakage currents, rather than charge trapping in ultrathin gate oxides. In these ICs, charge trapping and defect buildup in field oxides and shallow trench isolation (STI) regions is important; these isolation oxides frequently exhibit degradation characteristics that are similar to thick gate oxides from older technologies. Hence, the body of knowledge that has been developed over the past 40 or more years on defects in MOS systems is still quite relevant and applicable to modern and future MOS circuits and devices; however, the same defects can lead to different kinds of responses, and new materials introduce new kinds of defects and impurities that can also affect MOS performance, long-term reliability, and radiation response.

7.2 Measurement Techniques

Densities of net oxide-trap charge and interface-trap charge are most often estimated via simple capacitance–voltage (*C–V*) and current–voltage (*I–V*) techniques [10–12]. The most popular techniques rely on the assumption that interface traps are approximately charge neutral at midgap surface potential [11,13,14], allowing one to estimate the net oxide-trap charge from midgap voltage shifts in *C–V* or *I–V* measurements. Because there can be significant contributions to midgap voltage shifts from negative charge in border traps [6–9], and because standard *C–V* and *I–V* analysis do not provide information about trapped hole energy distributions, other techniques (e.g., TSC [15,16], as discussed in Section 7.3) must be employed to obtain information about oxide-trap charge that is more easily interpreted in a straightforward manner. Charge-pumping and conductance methods are among other techniques used to estimate densities and energy distributions of interface traps [10–12,17,18]. Techniques that are more amenable to rapid (e.g., less than ~1 ms) measurement times are more easily able to discriminate the effects of interface traps (which typically exchange charge with the underlying Si faster than border traps, which by definition lie within the oxide, and therefore are not in direct electrical contact with carriers in the Si channel [6–9]) than are techniques that require longer measuring times. This difference has been exploited to develop methods to (at least first order) separate the effects of interface and border traps [8,19,20], as discussed further below.

TSC and related techniques are often used to study the defects and impurities that limit the long-term reliability and radiation tolerance of electronic materials and devices [15,16,21–56]. In MOS capacitors with high-quality SiO_2 insulators that are thicker than about 7 nm, it is often possible to use TSC to estimate both the density and energy distribution of trapped positive charge for devices exposed to ionizing radiation or high-field electrical stress [7,14–16,24–27,29–55]. This is possible primarily because the dominant charge traps in thermal oxides tend to be at or near the Si/SiO_2 interface [4,5,30] and the TSC in these materials tends to be almost entirely due to trapped hole emission and transport [21,22,25–27,30,32,33]. For oxides thinner than ~7 nm, difficulties in obtaining high enough densities of trapped oxide charge and low enough baseline currents in the absence of trapped oxide charge, owing to the naturally high tunnel currents in ultrathin oxides, make it impossible to use this technique and less direct methods to estimate the oxide-trap charge must be used.

Figure 7.3 schematically illustrates the application of TSC to a MOS capacitor that has been exposed to ionizing radiation or high-field stress. Under suitable bias conditions, radiation- or stress-induced trapped positive charge is trapped infinitesimally close to the Si/SiO_2 interface. Then, ramped temperature measurements at a sufficiently large negative bias allow the density of trapped positive charge to be estimated by integrating the current due to charge transport across the oxide, and the trapped-charge energy distribution can be estimated from the temperature dependence of the current. Because interface traps exchange charge with the Si without displacement current, charge released from interface traps is not sensed by TSC measurement [27]. Thus, TSC is not useful for characterizing interface-trap charge. When one exposes MOS capacitors to ionizing radiation or high-field

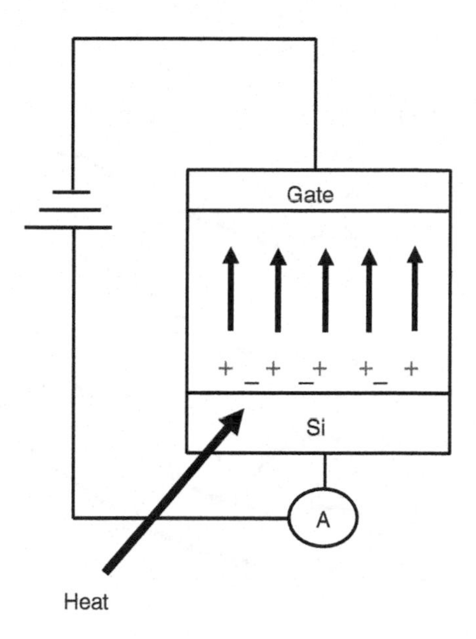

FIGURE 7.3
Schematic illustration of the TSC technique to estimate the density and energy of MOS oxide-trap charge. Irradiation or high-field stress can lead to hole trapping in the oxide, near the Si/SiO_2 interface, under appropriate bias conditions (e.g., positive-bias irradiation). When the capacitor is heated under negative bias, holes are emitted and transport across the oxide. Electrons in border traps do not contribute significantly to the measured TSC, allowing one to differentiate the density of trapped holes from the net oxide-trap charge. (After Fleetwood, D.M. et al., *Microelectron. Reliab.* 39, 1323, 1999. With permission.)

stress, both positive and negative charges are trapped in the SiO_2, although the dominant species are almost always positive, and usually dominated by trapped holes [1,4,5,13,15,16,29–34,37,38,44–46,49]. The currents for typical TSC tend to be quite small, on the order of 0.1–10 pA for typical devices and irradiation or stress conditions [15,16,30]. Thus, there are significant practical difficulties one must overcome to be able to accurately measure and interpret TSC in SiO_2, but in many cases these difficulties have been overcome and accurate estimates of MOS oxide-trap charge energy distributions and densities have been obtained [15,16,32].

7.3 Oxide-Trap Charge Energies

7.3.1 Trapped-Hole Energy Distributions

Oxide traps are distributed broadly in space and energy, as expected for an amorphous dielectric layer [1,32]. Figure 7.4 illustrates trapped-hole energy distributions as estimated by TSC for several kinds of oxides that were processed in different ways. The integrated TSC provides an estimate of the density of radiation-induced trapped positive charge, which is a function of device processing, radiation dose, bias, dose rate, and several other factors, as discussed in detail in Refs. [1,4,5,32]. Regardless of the trapped charge densities, the overall shapes of the inferred trapped-charge energy distributions are similar, consistent with the idea that the dominant defect types are similar in the different types of oxides (in most thermal oxides, the dominant hole trap is known to be an O vacancy in SiO_2 [13,40], as discussed further below), at least to first order [13,32]. The energy scale on the upper x-axis is derived from the current–temperature measurements under the assumption

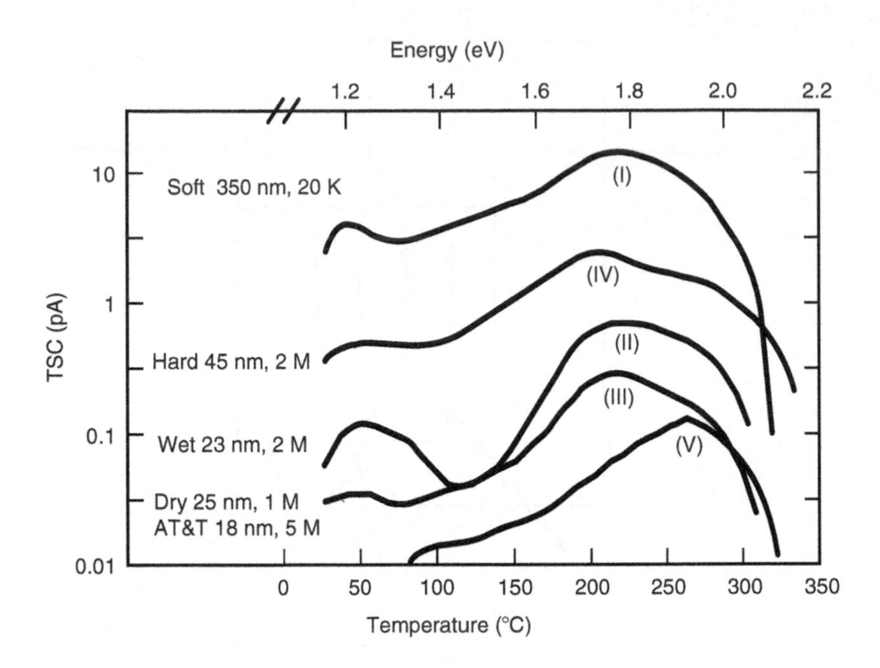

FIGURE 7.4
TSC as a function of oxide processing and radiation dose. Capacitors of different areas from different processes were irradiated to different doses [rad(SiO_2)] at electric fields of 1 or 2 MV/cm with 10 keV x-rays to obtain these data. All TSC spectra were measured at an electric field of −2 MV/cm at a heating rate of ~0.11 K/s. (After Fleetwood, D.M. et al., *IEEE Trans. Nucl. Sci.*, 39, 2192, 1992. With permission.)

that the excess TSC above the background observed in unirradiated devices is caused by simple hole emission from oxide traps that are distributed in energy and subsequent hole transport without retrapping across the SiO_2. With this assumption (which will be revisited below), a generalized activation model has been developed [32,33] to estimate effective trapped hole energies for arbitrary heating rates, and including the effects of electric field-induced barrier lowering. Briefly, the TSC emission process is modeled as Schottky emission of the trapped hole over a fixed energy barrier, which corresponds to the effective trap depth in the SiO_2. The rate at which traps are emptied is described by

$$\partial n(\varphi,t)/\partial t = -n(\varphi,t)\ F(t) \exp\left[-(\varphi - R(t))/A^*(t)\right] \tag{7.1}$$

where
- $n(\varphi,t)$ is the number of holes that remain trapped per unit area per unit energy
- φ is the effective trapped hole energy
- t is the annealing time
- $A^*(t) = kT(t)/q$, where k is the Boltzmann constant and T is the absolute temperature
- $R(t) = \beta\,(E_{ox})^{1/2}$, where $\beta = 0.5\,(q/\pi\varepsilon_{ox})^{1/2}$ for Schottky emission, $-q$ is the electronic charge, ε_{ox} is the SiO_2 dielectric constant, and E_{ox} is the oxide electric field applied during the TSC measurement
- $F(t) = a^* T(t)^2$ is analogous to the attempt-to-escape parameter ν that is typically used to characterize thermally activated processes that exhibit Arrhenius (e.g., simple activated emission over a single barrier) response [32,33]

In Equation 7.1, the value of a^* is given by [32,33,57]:

$$a^* = 2\sigma_t g(3k/m_h)^{1/2}(2\pi m_h k/h^2)^{3/2} \tag{7.2}$$

where

σ_t is the capture cross section of the dominant hole trap in SiO_2
g is the number of equivalent states into which a hole may be omitted
m_h is the effective mass of a hole in SiO_2
h is the Planck constant

To obtain an order-of-magnitude estimate for a^*, let $\sigma_t \approx 10^{-13}$ cm^2 be the hole capture cross section [1,58,59], $g \approx 1$, and $m_h \approx m$, where m is the free electron mass. With these simplifying assumptions, we estimate $a^* \approx 3 \times 10^8$ K^{-2} s^{-1}. This value of a^* corresponds to a value of $F(t)$ of $\sim 3 \times 10^{13}$ Hz at room temperature and $\sim 7.5 \times 10^{13}$ Hz at $\sim$500 K, which is the approximate temperature at which the TSC curves peak (i.e., the energy at which the highest trapped hole emission rates are observed) in Figure 7.4. To within the uncertainties in the assumptions underlying this analysis and the uncertainties in estimating $F(t)$ or ν experimentally, this value is consistent with experimental estimates of ν based on systematic variation of the TSC peak position with changes in the heating rate [32,33,50], which emphasizes the self-consistency of the effective trapped-hole energy scale in Figure 7.4.

7.3.2 Comparison of Radiation Response and 1/*f* Noise in SiO_2

The fairly straightforward model presented in Section 7.3.1 is at best only a first-order description of the defects in SiO_2 and their trapping and emission properties, as demonstrated in recent work that compares the radiation response, TSC, and low-frequency noise of MOS devices with thermal SiO_2 gate oxides. This comparison provides significant insight into oxide-trap microstructure and the dynamics of charge trapping and emission. In particular, several studies have compared the $1/f$ noise and radiation response of MOS devices [7,19,55,59–69]. An empirically observed correlation between the normalized $1/f$ noise K of MOS transistors before irradiation and threshold voltage shifts due to net positive radiation-induced oxide-trap charge ΔV_{ot} is shown in Figure 7.5 [59,60]. Here $K = S_V f(V_{GS} - V_{TH})^2/V_{DS}^2$; f is the frequency, S_V is the excess voltage-noise power spectral density (after correction for background leakage), V_{TH} is the threshold voltage, and V_{GS} and V_{DS} are the gate and drain voltages during the noise measurements [61,62]. Threshold voltage shifts due to oxide and interface trap charge were estimated via the midgap method of Winokur et al. [11].

The correlation in Figure 7.5 can be described numerically via the following expression derived from a simple, first-order model of the noise that attributes the excess low-frequency noise primarily to fluctuations in the number of channel carriers owing to trapping and emission events associated with near-interface oxide (border) traps [59–62]:

$$K = \frac{q^2 kT f_{ot} t_{ox}^2}{LW\ \sigma_t \varepsilon_{ox}^2 E_g\ \ln(t_{max}/t_{min})} \tag{7.3}$$

where

f_{ot} is the SiO_2 hole-trapping efficiency (that is, the probability that a given hole created by ionizing radiation exposure is trapped)
t_{ox} is the oxide thickness
L is the channel length
W is the channel width
E_g is the SiO_2 band gap
t_{max} and t_{min} are the presumed cutoff times for the noise process [59–62]

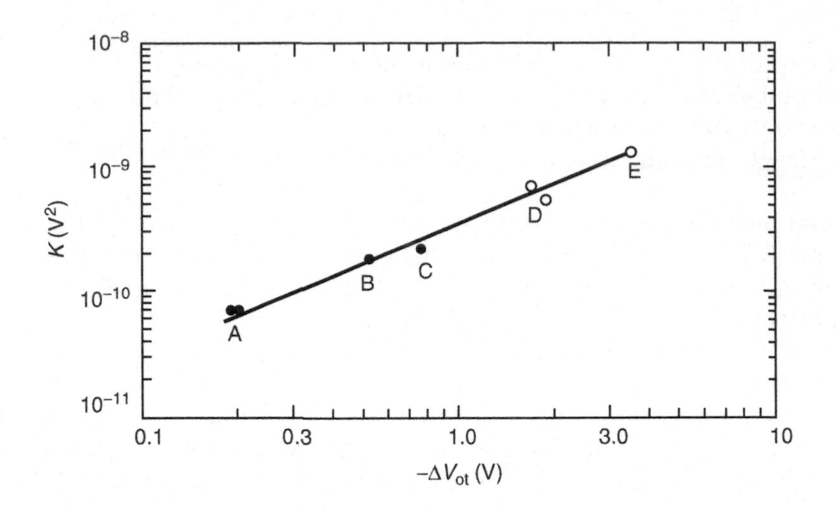

FIGURE 7.5
Normalized noise magnitude K as a function of threshold-voltage shifts due to radiation-induced oxide-trap charge ΔV_{ot} for 3×16 μm, *n*MOS transistors with gate oxides of different thickness (A, D: 32 nm; B, E: 48 nm; C: 60 nm) and radiation hardness (A–C hard; D, E soft) processed in the same lot. Noise measurements were performed in the linear region of device operation; values of ΔV_{ot} were obtained from room-temperature irradiation to 100 krad (SiO_2) in a Co-60 source at a dose rate of ~278 rad (SiO_2)/s at an oxide electric field of ~3 MV/cm. (From Fleetwood, D.M. and Scofield, J.H., *Phys. Rev. Lett.*, 64, 579, 1990. With permission.)

Equation 7.3 assumes (1) defects with similar average, effective capture cross sections σ_t are responsible for both $1/f$ noise and radiation-induced-hole trapping, (2) the preirradiation noise S_V is proportional to the density of oxide traps, which is in turn proportional to f_{ot}/σ_t, (3) oxide traps near the Si/SiO_2 interface are distributed approximately uniformly in space and energy, (4) carrier number fluctuations are the dominant cause of noise, and (5) that border traps have similar capture and emission cross sections to the bulk oxide traps that are located deeper within the SiO_2 layer [55,59–62]. A significant body of work in the literature suggests strongly that O vacancies play a key role in this process [5,13,40,55,63].

7.3.3 Defect Microstructure

The crucial significance of O vacancy-related defects to MOS oxide-trap charge has been recognized for a long time [1,13,40,55]. As an example of the significant impact that O vacancies can have on MOS oxide-trap charge, Figure 7.6 compares electrical estimates of MOS oxide-trap charge [70] with calculations using a simple Fick's law diffusion model of O vacancy formation [71], based on the idea that excess O vacancy creation occurs in these devices primarily as a result of O atoms out-diffusing from the SiO_2 during high-temperature process steps after gate oxidation. While other mechanisms of O vacancy formation in SiO_2 exist, the correlation between the experimental data and the calculations in Figure 7.6 provide striking confirmation of the importance of O vacancies to MOS oxide-trap charge, and consequently to MOS $1/f$ noise [59–63].

In Figure 7.7, we show four types of electronic processes of relevance to the correlation between low-frequency noise and MOS radiation response. Process 1 is an electron neutralizing-trapped positive charge under the proper electric field conditions, which is known to have a capture cross section that is on the order of $\sim 10^{-13}$ cm^2, consistent with the above assumptions [1,72,73]. Process 2 in Figure 7.7 depicts hole capture (i.e., valence

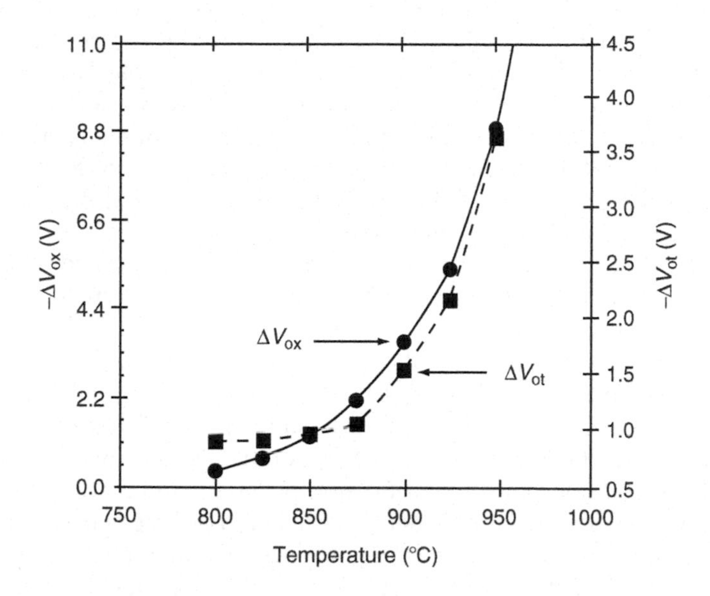

FIGURE 7.6
Estimated magnitudes of midgap voltage shifts ΔV_{ot} (squares, right-hand side y-axis) and predicted shifts ΔV_{ox} (circles, left-hand side y-axis), due to O vacancies in thermal SiO_2 films that were given postgate, 30 min anneals in N_2 at different temperatures (x-axis). The experimental data are obtained from 1 Mrad (SiO_2) x-ray irradiations of capacitors with 45 nm oxides at an electric field of ~2 MV/cm, and the model predictions are derived from a Fick's value diffusion model. (From Warren, W.L. et al., *Appl. Phys. Lett.*, 64, 3452, 1994. With permission.)

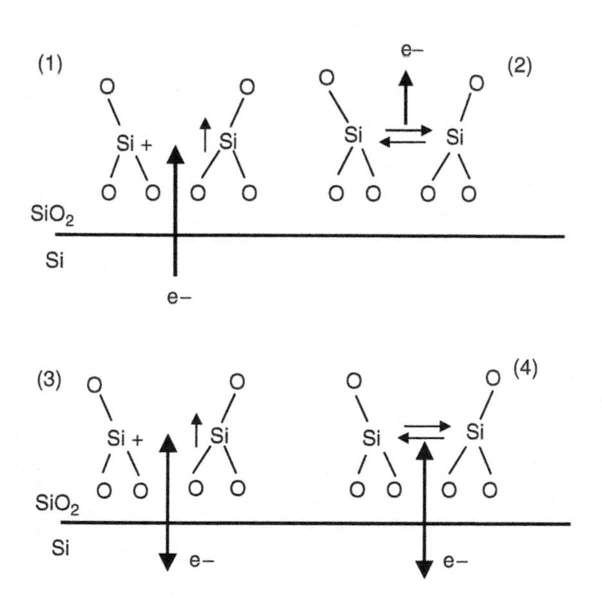

FIGURE 7.7
Schematic illustrations of (1) an electron from the Si neutralizing an E′ center (specific type not specified here); (2) an electron from an E′ precursor emitted from the O vacancy center, which will leave behind the positively charged defect shown in (1); (3) noise due to electron exchange between the Si and a positively charged E′ defect; and (4) noise due to electron exchange between the Si and a neutral E′ precursor (O vacancy in SiO_2). All of these processes can have similar effective cross sections. (From Fleetwood, D.M. et al., *IEEE Trans. Nucl. Sci.*, 49, 2674, 2002. With permission.)

electron emission), which has a similar cross section [1,58,73]. Processes 1 and 2 constitute a full capture and emission cycle for a defect near the Si/SiO_2 interface; hence, it is quite reasonable that these kinds of defects, located at suitable positions and energy levels, may play a key role in MOS $1/f$ noise, as long as similar kinds of charge capture and emission events can originate within the Si channel. A complete noise cycle that starts with a positively charged defect is illustrated by process 3 of Figure 7.7; process 4 is a capture and emission sequence involving an initially neutral site [55].

At least two plausible candidates for the O vacancy centers associated with processes 1–4 in Figure 7.7 have been identified; E'_γ and E'_δ defects are known to be high capture-cross-section hole traps in SiO_2 [1,40,74]. In several studies, the E'_δ has been associated with the dimer O vacancy defect that is illustrated in Figure 7.8a [55,75–80]; however, this association has been qualified and/or questioned as a result of subsequent embedded cluster and density functional theory (DFT) calculations [81,82] and recent Si^{29} hyperfine data [83,84]. The E'_γ defect has been associated with the defects shown in Figure 7.8b and c [78–82,85–89]. Final and conclusive identification of the E'_γ and E'_δ defect microstructures awaits future work; nevertheless, the dimer and puckered defects depicted in Figure 7.8 are found via DFT calculations to have several characteristics that emphasize their significance to comparative studies of MOS radiation response and low-frequency noise.

Recent DFT calculations [78,79] of the energetics of hole and/or electron capture by neutral and/or positively charged O vacancy centers show that: (1) the vast majority of O vacancy sites (as many as ~90%) in the supercells we have examined are dimers (Figure 7.8a) after hole capture [79]. This percentage may vary with device processing, but it suggests that the dimer configuration of the O vacancy center is the more common vacancy-related defect in SiO_2. (2) An initially neutral dimer can efficiently capture a hole, but the energy levels associated with this kind of hole trap typically are relatively shallow ($\leq$~1.0 eV) compared to those of either of the two puckered configurations (Figure 7.8b and c, $\geq$~3 eV) [79]. An exception is a stretched dimer (~2 eV) with no O available to form a puckered E'_γ; such a defect would not be observed in quartz-like SiO_2, but may well be present in regions of the oxide of low density [89] and/or near voids. (3) Electron paramagnetic resonance (EPR) signals from the two puckered configurations in Figure 7.8b

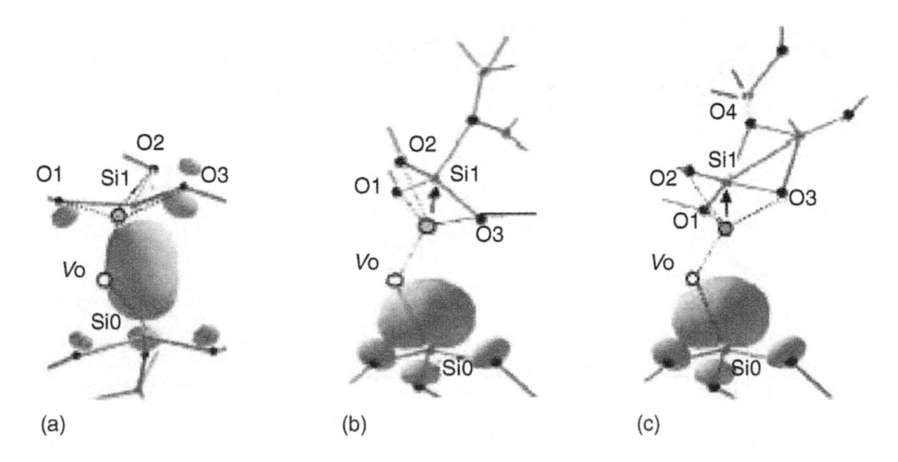

FIGURE 7.8
Schematic illustrations of the unpaired electron densities (gray regions) and atomic configurations of (a) a dimer O vacancy center, (b) a relaxed O vacancy center that has been associated with the E'_γ defect (the $E'_{\gamma 4}$), and (c) a second type of O vacancy center that apparently is also associated with the E'_γ (the $E'_{\gamma 5}$). The differences between defects (b) and (c) are the coordination of the atom denoted by the arrows and marked "Si1," with fourfold coordination in (b) and fivefold coordination in (c). (From Lu, Z.Y. et al., *Phys. Rev. Lett.*, 89, 285505, 2002. With permission.)

and c are likely quite similar, since the only difference in defect configuration is the coordination of the Si atom facing (not holding) the unpaired electron that is EPR active. But the fourfold coordinated, puckered Si vacancy center (Figure 7.8b) forms a stable dipole upon electron capture, while the fivefold coordinated, puckered (Figure 7.8c) does not form a dipole [79]. Whether an O vacancy center forms a fourfold or fivefold puckered configuration is determined largely by the local atomic spacing and bond angles; these determine whether O4 has a nearest neighbor Si close enough to bond with the puckered Si1.

DFT calculations also demonstrate that the probability of capturing an electron (e.g., from the Si) increases with increasing separation of the Si1–Si0 bond at the center of the complex shown in Figure 7.8a, as shown schematically in Figure 7.9. At an equilibrium spacing of ~0.25–0.30 nm in the bulk SiO_2, the electron trapping level for the neutral dimer in Figure 7.8a is near the SiO_2 conduction band. However, if one significantly stretches the Si–Si bond to ~0.35–0.4 nm, near-midgap states can open up that may at least metastably capture an electron, with the energetics for capture becoming more favorable with increasing Si–Si spacing. These kinds of stretched Si–Si bonds are likely exist near the Si/SiO_2 interface [79,81,82], in which a significant amount of strain must be accommodated [30,70,89–93]. Moreover, it has been demonstrated that the amount of strain near the Si/SiO_2 interface can change with radiation exposure [93].

Several kinds of charge exchanges between the Si and O vacancy-related defects such as those depicted in Figure 7.8 can lead to $1/f$ noise. (1) In a *p*MOS transistor, the simple capture and reemission of a hole from a dimer O vacancy defect near the Si/SiO_2 interface is an excellent candidate for $1/f$ noise. This defect has a high effective capture cross section, a modest barrier for reemission, and thus can contribute to *p*MOS noise before or after irradiation [8,55,65]. The charge transfer may occur via simple tunneling, trap-assisted tunneling, and/or thermal activation. Indeed, this may well be the dominant source of *p*MOS noise, although similar transitions with puckered O vacancy defects involving network relaxation may also play a significant role in *p*MOS noise. (2) In an irradiated *n*MOS transistor, electron exchange with fourfold-coordinated, puckered O vacancy defects is a possible source of noise. However, dipolar defects in irradiated SiO_2 can be quite stable, and electrons in some dipolar defects associated with trapped holes are found in TSC studies not to be reemitted to the Si even when the devices are heated to ~100°C under negative bias [52]. So this is a potential source of noise, but simple capture and emission of an electron from a fourfold-coordinated, puckered O vacancy defect may not

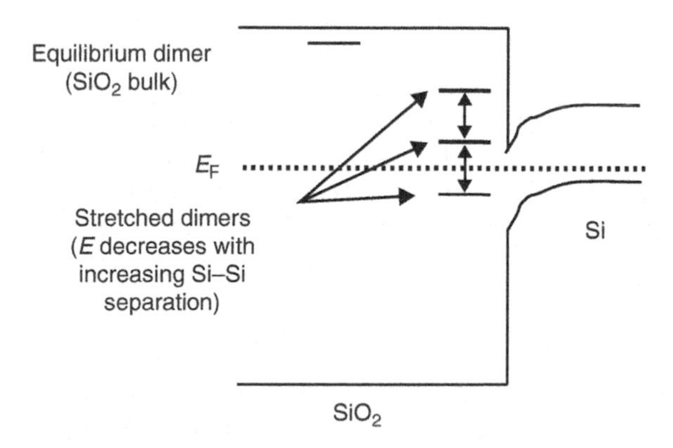

FIGURE 7.9
Schematic illustration of the energy levels of dimer O vacancies in bulk SiO_2 with equilibrium Si–Si atomic spacing of ~0.25–0.30 nm and near-interface dimers with stretched Si–Si spacing of ~0.35–0.40 nm and a distribution of energy levels near midgap. (From Fleetwood, D.M. et al., *IEEE Trans. Nucl. Sci.*, 49, 2674, 2002. With permission.)

be the only source of noise in these devices. (3) In an *n*MOS transistor, one must consider the possible interconversion of fourfold and fivefold-coordinated, puckered O vacancy defects due to thermal vibrations. Thermal energy (i.e., phonons) can cause atoms to stretch and rotate more strongly with increasing temperature. If one pictures the Si1–O4 atoms bending and stretching in Figure 7.8b, this site may switch from a configuration that locally favors fourfold coordination of the Si1 atom into one that favors fivefold coordination, and vice versa. This local change in atomic spacing and bond angle leads to an electron trap level in one configuration, but not in the other. Thus, it would be energetically favorable for a fourfold-coordinated, puckered O vacancy defect to release an electron when the O4–Si nearest neighbors bend toward and move closer to the Si1 atom in Figure 7.8b to form a fivefold-coordinated, puckered O vacancy defect, due to network vibrations and/or variations in the local electric field or strain. Similarly, it becomes more favorable to recapture the electron when the O4–Si atoms bend and move away from the Si1. Here, it is the thermally activated atomic reconfiguration that triggers the capture and release of the electron. Stretched fivefold-coordinated, puckered O vacancy defects may also exhibit this kind of behavior near a void, or in a low-density region where puckered defect configurations are not geometrically allowed [79,89]. (4) Finally, it is also possible that *n*MOS transistor noise before or after irradiation may involve thermally assisted capture and emission of an electron by a dimer O vacancy. This again must be mediated by network relaxation, since there is no electron trap state available when this defect has its equilibrium Si1–Si0 bond spacing. However, especially near the Si/SiO_2 interface, there likely exist highly strained dimer configurations, shown via DFT calculation to have near-midgap electron trap levels, as illustrated in Figure 7.9. When an electron is captured, the Si–Si spacing decreases due to the increased electron density between the two atoms. This leads to a rise in trap energy level and electron reemission. Although the dimers would have to be configured spatially and energetically so as to enable these transitions to occur, the timescale for the atomic relaxation process again is the rate-limiting step in the noise process [55]. In addition to the O vacancy-related defects considered in Figure 7.8, several hydrogen-related defects may also contribute to MOS $1/f$ noise.

7.3.4 Energy Scale for $1/f$ Noise

After illustrating the link between MOS radiation response and low-frequency noise, we now consider how the energy scales that are derived for $1/f$ noise measurements compare to those expected from DFT calculations for O vacancies, and also to energy scales derived from TSC measurements. Ultimately, since the defect energies should not be sensitive to the precise method of measurement, one expects the results from TSC and noise measurements to be self-consistent. In evaluating the conditions required for the energy scales to be consistent, additional insight into the underlying physical processes is obtained [55].

The frequency dependence of the measured $1/f$ noise is defined as

$$\alpha = -\frac{\partial \ln S_V}{\partial \ln f} \tag{7.4}$$

For devices in which the $1/f$ noise is due to a random thermally activated process having a broad distribution of energies relative to kT, Dutta and Horn have shown that the frequency and temperature dependences of the noise are related via [94]:

$$\alpha(\omega,T) = 1 - \frac{1}{\ln(\omega\tau_0)}\left(\frac{\partial \ln S_V(T)}{\partial \ln T} - 1\right) \tag{7.5}$$

where
$\omega = 2\pi f$
τ_0 is the attempt to escape frequency for the defect

For noise that is demonstrated to satisfy Equation 7.5, one can infer the shape of the defect energy distribution $D(E_0)$ from noise measurements versus temperature via:

$$D(E_0) \propto \frac{\omega}{k_B T} S_V(\omega,T) \tag{7.6}$$

where the defect energy is related to the temperature and frequency through the simple expression [94]:

$$E_0 \approx -kT \ln(\omega\tau_0) \tag{7.7}$$

To within a factor of 2π inside the logarithm (which does not significantly affect the results below, and arises only if one considers the radial frequency ω as opposed to f as per Dutta and Horn [94]; alternative developments omit this factor [95]), Equation 7.7 essentially restates the condition that the individual characteristic times of the noise processes τ are related to the activation energies E_0 via the expected [55]:

$$\tau = \tau_0 \exp(-E_0/kT) \tag{7.8}$$

Physically, if one presumes the noise is the result of thermally activated processes, E_0 is the barrier that the system must overcome for the system to move from one configurational state to the other [95]. Hence, Equation 7.6 provides a measure of the energy required not for the electronic transitions (e.g., the carrier motions depicted in Figure 7.8) that are associated with $1/f$ noise, but instead provides insight into the energetics of the thermally activated SiO_2 network relaxation processes.

Xiong et al. [96,97] have demonstrated that the $1/f$ noise of the transistors of Figure 7.5 clearly satisfy the Dutta–Horn frequency–magnitude correlation criterion of Equation 7.5. This makes it possible to use Equation 7.7 to extract defect energy distributions for the O vacancy-related defects causing the noise. This energy distribution is shown in Figure 7.10 for MOS transistors with 32 nm oxides from the same process lot as those in Figure. 7.5: (a) before irradiation, (b) after 500 krad (SiO_2) x-ray irradiation, and (c) after 200°C postirradiation anneal at 0 V. This energy scale assumes $f \cong 1$ Hz, and $\tau_0 \cong 1.8 \times 10^{-15}$s [55]—with the former choice for convenience, and the latter to facilitate comparison to TSC. The temperature dependence of the attempt-to-escape frequency [32,33] is neglected in Figure 7.10, as this has only second-order effects on inferred energy scales. Also, this particular value of τ_0 corresponds to the value of ν extracted from [50], in contrast to the slightly lower values inferred in Refs. [32,33], but this difference does not affect the results.

After annealing, the energy distribution of the defects in Figure 7.10 changes remarkably from preirradiation values. Evidently the irradiation and anneal have modified the defect energy distributions, perhaps consistent with work in the literature showing a relaxation of stress near the Si/SiO_2 interface with irradiation and/or annealing [93,98]. There is a broad enough range in energy distributions for the dimer and puckered O vacancy-related defects in DFT calculations (from <0.7 to >3.0 eV [79], with energy spreads for each defect associated with small variations in configuration) that these O vacancy-related defects may well be the primary cause for the observed noise [55]. Higher-energy processes tend to reflect dramatic changes in configuration from one defect type to another (e.g., puckered

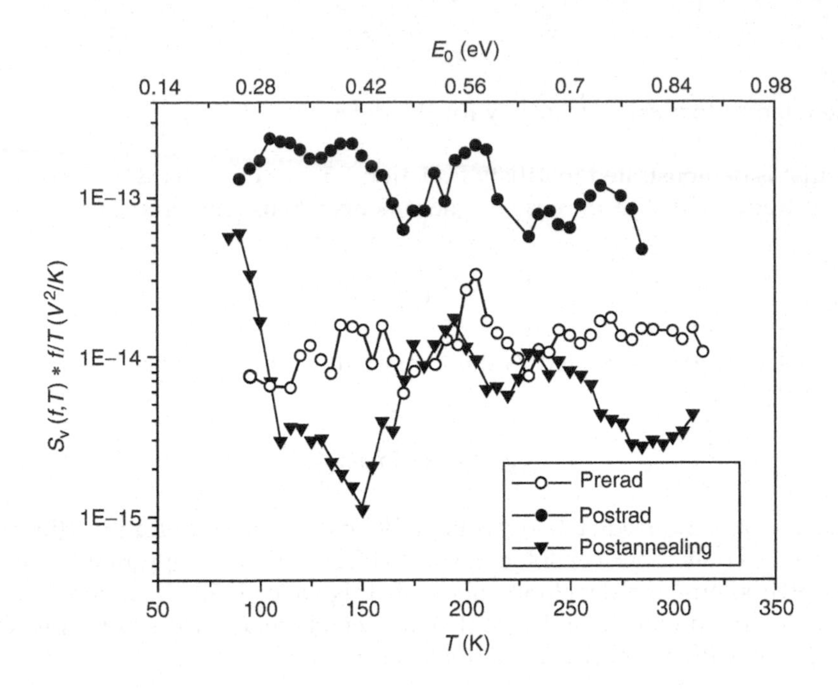

FIGURE 7.10
$S_V f/T$ (proportional to the defect energy distribution) versus T and E_0 for 3×16 μm nMOS transistors with 32 nm oxides (○) before irradiation, (•) after 500 krad (SiO_2) Co-60 irradiation at 6 V bias, and (▼) after 24 h of 0 V annealing at 473 K. The energy scale inferred from the Dutta–Horn model is given on the upper x-axis for $f = 1$ Hz and $\tau_0 \cong 1.8 \times 10^{-15}$ s. (After Xiong, H.D. et al., *IEEE Trans. Nucl. Sci.*, 49, 2718, 2002. With permission.)

to unpuckered), and smaller changes are associated with more subtle atomic rearrangement within a single defect type [55,79].

That the energy scale for the noise measurements in Figure 7.10 is consistent with the energy scale for the TSC measurements in Figure 7.4 is shown in detail in Ref. [55]. For these two energy scales to be consistent requires a commonality of capture and emission processes associated with MOS oxide traps and border traps. This implies [55] that the emission of trapped holes during TSC is not simple emission over a barrier, but is instead triggered by the kinds of SiO_2 network relaxation that lead to $1/f$ noise [55,95]. DFT calculations demonstrate that the energy levels of the dimer and puckered O vacancy-related defects shown in Figure 7.8 typically differ by ~1.0–2.5 eV [79]. This suggests that trapped hole emission may occur at the point during a TSC measurement when a puckered O vacancy-related center acquires enough thermal energy to convert to a dimer, after which the trapped hole is spontaneously emitted. The trigger for this process appears to the high vibrational rate inferred for the Si atom puckered through the plane in the E'_γ defect [7,50], with hole emission and rebonding of the Si–Si atoms at a temperature that is high enough to overcome the barrier to transform the puckered Si in the deeper hole trap back into its original plane in the shallower dimer level. Defects will not always reconfigure into their original (i.e., pre-stress) conditions, accounting for the changes in defect energy distributions with irradiation and annealing in Figure 7.10. The O vacancy centers in Figure 7.8a and b therefore form a simple two-level system, with both $1/f$ noise and TSC measurements sensing changes in atomic configurations that lead to changes in defect energy levels with time [55].

The above considerations help to resolve the long-standing discrepancies between thermal (<2 eV) and optical (>3 eV) energies inferred for trapped-hole annealing in the literature [1,32,33]. In optical measurements at room temperature, there is not enough thermal energy to enable defect reconfiguration, so one measures the deep barrier for trapped-hole emission from an E'_γ defect (>3 eV, as verified by DFT calculation [79]). At elevated temperature, the defect reconfigures before this emission occurs, so it likely is atomic reconfiguration that is the rate-limiting step in emission in this case [55]. This emphasizes the importance of matching the effective energy scale that is most appropriate to the application when attempting to compare experimental and theoretical results.

7.4 Dipolar Defects near the Si/SiO_2 Interface

7.4.1 Switched-Bias Annealing

In the discussion of defect energies, we discussed briefly two kinds of O vacancy-related defects that can function as border traps if located near the Si/SiO_2 interface. We now discuss other properties of border traps and related dipolar defects in the near-interfacial SiO_2. The presence of a significant density of dipoles in irradiated SiO_2 was first demonstrated by Pepper [99]; however, it was not studied intensively until later Schwank et al. [100] showed that trapped positive charge that was previously thought to be removed by positive-bias annealing could be restored during negative-bias annealing, and Nissan-Cohen et al. [101] found a similar reversibility of net trapped positive charge after high-field electrical stress. Dipoles associated with O vacancy-related defects were studied in detail by Lelis and coworkers [90,102], who proposed that the reversibility of the positive-charge annealing was caused by an electron tunneling into and out of a Si atom adjacent to a second, positively charged Si atom, associated with an E'_γ defect in the near-interfacial SiO_2 [90], similar to the processes considered in Figure 7.8.

The reversibility in MOS oxide-trap charge when bias is switched after irradiation or high-field stress can be significant in magnitude. In Figure 7.11, we show the irradiation and annealing response of devices that were exposed to Co-60 gamma rays under positive

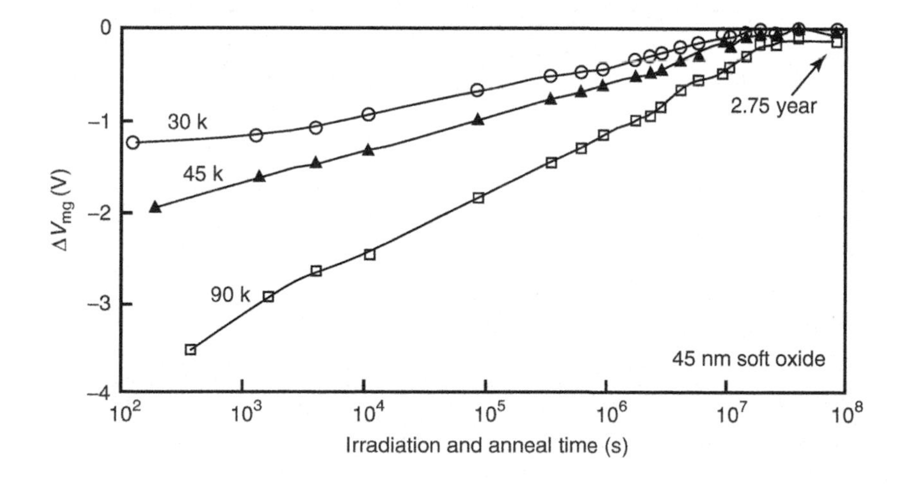

FIGURE 7.11
Midgap voltage shift as a function of irradiation and annealing time for *n*MOS transistors with 45 nm oxides that were intentionally softened with a postoxidation 30 min, 1100°C anneal to introduce a large number of O vacancies. The devices were irradiated at 6 V gate bias with Co-60 gamma rays at 240 rad (SiO_2)/s; the annealing temperature was 100°C. (From Fleetwood, D.M., *Appl. Phys. Lett.*, 60, 2883, 1992. With permission.)

bias, and then annealed for 2.75 years at positive bias at 100°C [14]. This long, elevated-temperature anneal completely neutralized the initially high density of positive oxide-trap charge, but did not reveal any net negative charge trapping even at very long times at negative bias at room temperature and 100°C. These results strongly reinforce the conclusion that interface traps are charge neutral at approximately midgap [2,13,14], confirming that the midgap voltage shift is approximately equal to the threshold voltage shift due to oxide-trap charge [11]. The lack of significant net electron trapping in these dielectric layers is typical of most experience with thermal SiO_2 [1,4,32], showing that the only locations for electrons to be trapped in these oxides with relatively high capture cross sections are at sites that compensate trapped positive charge.

Figure 7.12 shows that, even after such a long annealing period, the positive and negative charges have not recombined, so the defects have not truly annealed [14,90,100–103]. Indeed, when the temperature is restored to 100°C at negative bias in Figure 7.12, the net trapped positive charge in the SiO_2 increases dramatically. There are two points that are especially interesting about the reversibility of the net positive oxide-trap charge in Figure 7.12. First, the amount of net trapped positive charge in the two cases in Figure 7.12 is nearly independent of dose, despite the large difference in initial levels of net positive charge for the devices irradiated to 30 and 90 krad (SiO_2) in Figure 7.11 [34]. Second, the net positive oxide-trap charge after negative-bias annealing for the 30 krad (SiO_2) case in Figure 7.12 is greater than after initial irradiation in Figure 7.11. The latter result is consistent with the expectation that a significant amount of trapped positive charge already had been neutralized during the irradiation period [34]. This result is also consistent with the results of TSC experiments [30,32,44,49], as we discuss further below, and is also consistent with the presence of dipoles similar to those depicted in Figure 7.8.

Additional support for the idea that O vacancy-related defects may lead to the kinds of oxide-trap charge reversibility that is depicted in Figures 7.11 and 7.12 is illustrated by the reversibility in densities with changing bias observed in EPR studies by Conley et al. [104]. The key result of this study is shown in Figure 7.13. The initial E'_γ density in this study was generated via the vacuum ultraviolet (UV) injection of $\sim 5 \times 10^{13}$ holes/cm^2 into

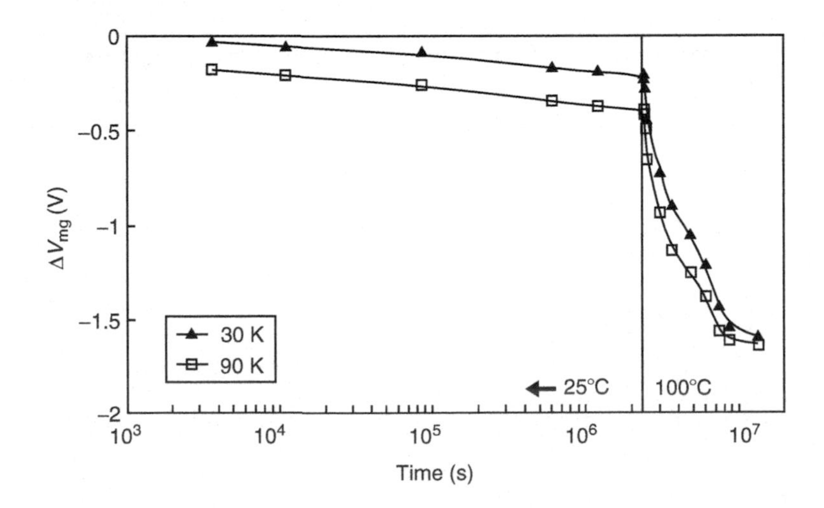

FIGURE 7.12
Midgap voltage shift versus negative-bias anneal time and temperature for the devices of Figure 7.9. The 2.75 years, 6 V, 100°C anneal in Figure 7.9 was followed by the room temperature and 100°C anneals at −6 V shown here. Results for the 45 krad (SiO_2) irradiations and anneals were similar. (From Fleetwood, D.M. et al., *IEEE Trans. Nucl. Sci.*, 40(6), 1323, 1993. With permission.)

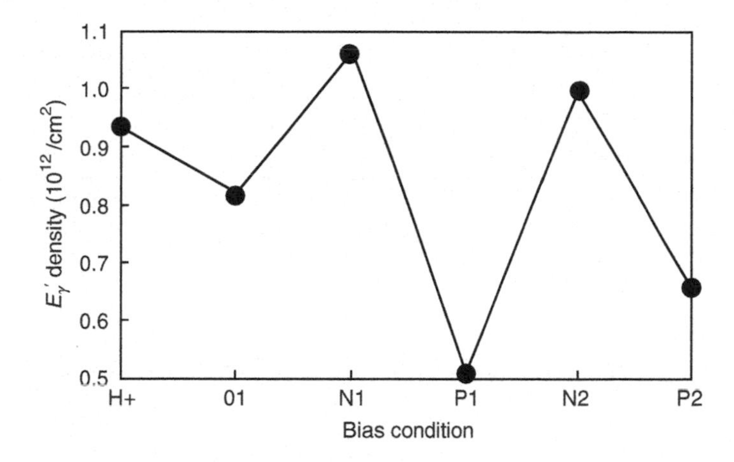

FIGURE 7.13
E'_γ defect densities measured via EPR after vacuum UV injection of $\sim 5 \times 10^{13}$ holes/cm^2 into 120 nm dry oxides, followed by two cycles of corona ion biases at alternating biases. (After Conley, J.H. Jr., *IEEE Trans. Nucl. Sci.*, 42, 1744, 1995. With permission.)

120 nm dry oxides grown at 1000°C. After hole injection, bias was applied onto these oxides using corona ions, establishing electric fields of alternating polarity, with magnitudes of $\sim$3.5 MV/cm. Point N1 in Figure 7.13 is the first negative corona ion bias, which leads to an increase in E'_γ density, followed by a large decrease when the bias is changed to positive (P1), with this pattern repeating for subsequent negative (N2) and positive (P2) corona bias. The striking correlation between the response of the E'_γ density in Figure 7.13 and the electrical response of MOS devices under similarly switched-bias conditions [5,34,90,100–102] is compelling evidence for O vacancy-related dipoles in the near-interfacial SiO_2.

Reversibility in MOS oxide-trap charge after irradiation or exposure to high-field stress is not limited to SiO_2 gate oxides. Even larger reversibility owing to shallow electron traps has been observed in silicon-on-insulator (SOI) buried oxides [105], in silicon nitride [57], and in some kinds of high-*K* dielectrics [106,107]. The magnitude and significance of these effects can be minimized with improved processing techniques for high-*K* dielectrics; however, the best one can apparently do is to obtain response similar to thermal SiO_2 [108].

7.4.2 Capacitance–Voltage Hysteresis

Similar results to the above switched-bias annealing response often can be observed on a somewhat more rapid timescale in the hysteresis of MOS capacitance–voltage (*C–V*) curves. Figure 7.14 illustrates this hysteresis after irradiation; Figure 7.15 displays similar hysteresis after high field stress, suggesting similar concentrations of border traps are caused by the two different stresses [8,44]. Consistent with the interpretation that the *C–V* hysteresis is caused by border traps, Figure 7.16 shows that the amount of measured hysteresis increases logarithmically with increasing ramp time during the *C–V* measurement, as expected for electrons tunneling into (at positive bias) and out of (at negative bias) border traps. As was the case for postirradiation switched-bias annealing, an excellent correlation has been observed between electrical estimates of border-trap densities (proportional to ΔV_{bt}, which is the difference in midgap voltages for forward and reverse *C–V* measurements) from *C–V* hysteresis and measurements of E'_γ densities via EPR during a room-temperature annealing sequence for devices with high O vacancy densities [40]. This is

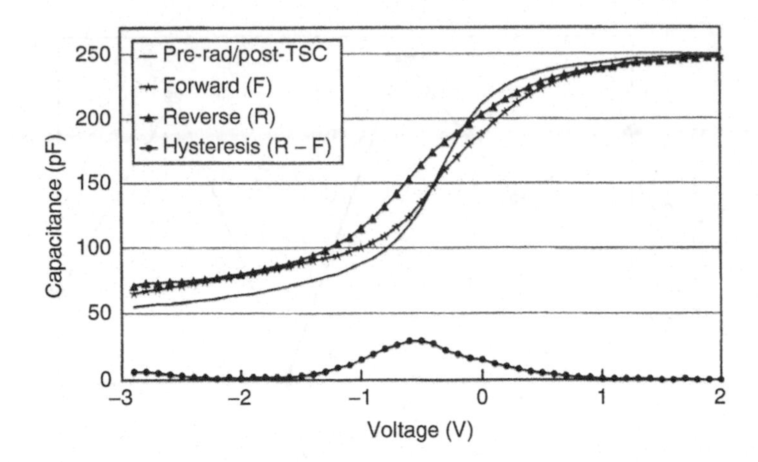

FIGURE 7.14
High-frequency (1 MHz) capacitance–voltage hysteresis curves for *n*-substrate capacitors with 17 nm thermal oxides irradiated to 2 Mrad (SiO_2) at 4 V bias with 10 keV x-rays at a dose rate of 3333 rad (SiO_2)/s. (From Fleetwood, D.M. and Saks, N.S., *J. Appl. Phys.*, 79, 1583, 1996. With permission.)

illustrated in Figure 7.17 for 410 nm thermal oxides grown in dry O_2, which were then given a 1320°C anneal to simulate the high-temperature processing that is used in the manufacturing of SOI buried oxides via the separation via implantation of oxygen (SIMOX) process [40,109]. The increase in both the E'_γ and border-trap densities with annealing time in Figure 7.17 is uncommon, and is attributed in these devices to an unusually high density of E'_δ centers, due to the extremely high-temperature anneal [40]. The presence of a large number of E'_δ defects can greatly retard hole transport [37,40,55,78,79]. The correlation in growth rates of the border traps and E'_γ defect densities in Figure 7.17 is striking. A similar correlation between trapped holes in SiO_2 and dipole formation at the Si/SiO_2 interface was noted by Lai in 1981 in studies of hole trapping and electron injection [110]. We note that the overall density of the E'_γ defects in these devices is more than an order of magnitude larger

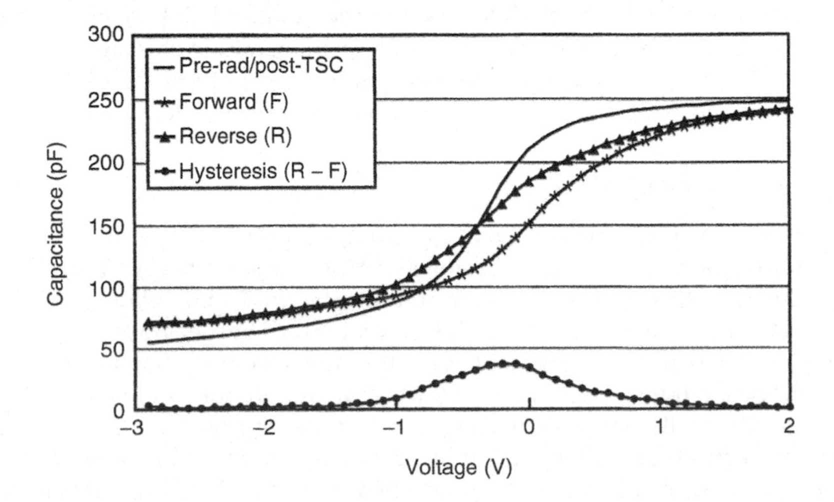

FIGURE 7.15
High-frequency (1 MHz) capacitance–voltage hysteresis curves for *n*-substrate capacitors with 17 nm thermal oxides subjected to 10 mC/cm² Fowler–Nordheim stress. (From Fleetwood, D.M. and Saks, N.S., *J. Appl. Phys.*, 79, 1583, 1996. With permission.)

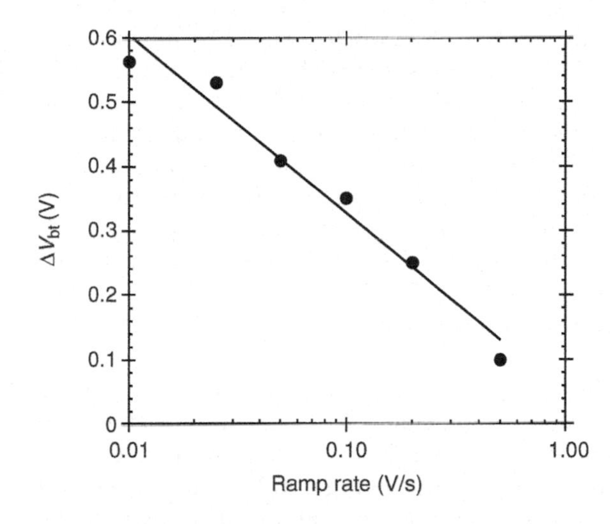

FIGURE 7.16
High-frequency *C–V* hysteresis as a function of voltage ramp rate for MOS capacitors with 45 nm oxides grown in dry O_2. These devices received a 1000°C N_2 anneal after oxidation to increase their O vacancy density. (From Fleetwood, D.M. et al., *Microelectron. Reliab.*, 35, 403, 1995. With permission.)

than the density of border traps, consistent with etch back studies that showed a large amount of bulk positive-charge trapping in these oxides [40]. So the interpretation of this result is simply that E'_γ centers that are not close enough to the Si/SiO_2 interface to exchange charge with the Si do not function as border traps, but act as bulk oxide-trap charge [40,111].

7.4.3 Stable Dipoles in the Near-Interfacial SiO_2

We now present evidence from TSC experiments that strongly suggest there is a second type of dipole in irradiated or stressed SiO_2. This defect is apparently distinct from the O vacancy-related defects discussed above, at least in position, and likely also in microstruc-

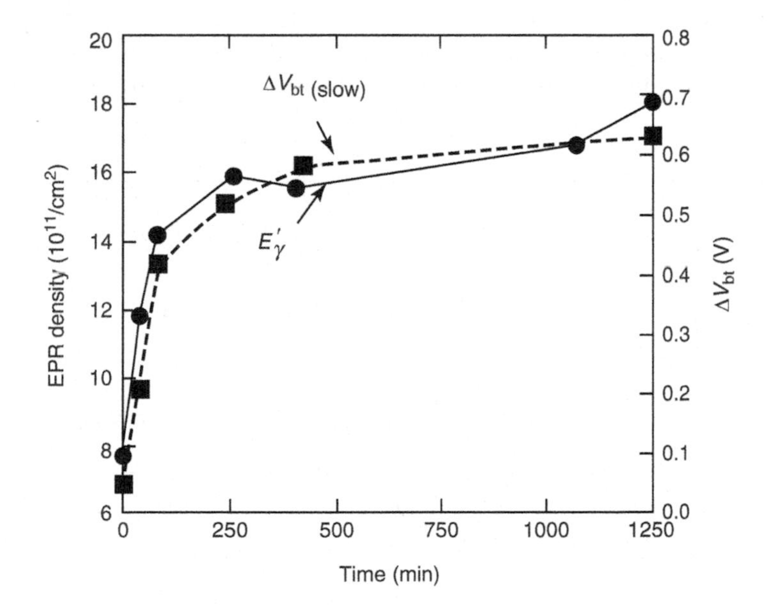

FIGURE 7.17
Comparison of E'_γ defect densities measured via EPR and threshold voltage shifts due to border traps estimated from capacitance–voltage hysteresis for 410 nm oxides annealed at 1320°C after gate oxidation and polycrystalline Si deposition. These devices emulate the thermal history of SIMOX buried oxides. (From Warren, W.L. et al., *IEEE Trans. Nucl. Sci.*, 41, 1817, 1994. With permission.)

ture. It has been observed commonly in TSC studies of irradiated or electrically stressed SiO_2 that the total amount of trapped positive charge estimated by integrating the TSC can greatly exceed the net positive oxide-trap charge estimated from the midgap voltage in C–V measurements, $Q_{CV} = -C_{ox}\Delta V_{mg}$, where C_{ox} is the oxide capacitance [24–27, 29–32,34,37,44–46,49–52]. An example is shown in Figure 7.18, for radiation-hardened oxides with a low O vacancy density [30,31]. The dashed line represents Q_{CV} ($\sim 2.3 \times 10^{12}$ cm^{-2}), which is ~56% of the total positive charge one infers from TSC measurements at large negative bias ($\sim 4.1 \times 10^{12}$ cm^{-2}). The reduction in charge collected at low electric field is a result of space charge effects [15,29,30].

That there is no significant TSC for positive bias in Figure 7.18 from trapped positive charge is easy to understand from Figure 7.3, since under positive bias, the trapped positive charge moves only an infinitesimal distance before reaching the Si. It requires charge moving over a finite distance to produce significant TSC [15,24–27,30]. However, the lack of TSC at positive bias from the negative charge inferred to be in the oxide provides significant insight into the nature of the negative charge. Indeed, results similar to those in Figure 7.18 strongly suggest that the negative charge that must exist in the near-interfacial SiO_2 after radiation exposure or high-field stress to account for the difference between TSC and C–V estimates of MOS oxide-trap charge must be unable to transport across the oxide [29–32,44,46,49,52].

Another interesting property of dipoles in the near-interfacial SiO_2 is illustrated in Figure 7.19, where we show inferred densities of trapped positive and negative charge from TSC and C–V measurements. The trapped negative charge is categorized in the figure as easily exchanged with the Si (shallow electrons), or difficult to exchange with the Si (deep electrons). In Figure 7.19, there is very little change in the inferred deep electron trap density until the postirradiation annealing temperature exceeds 115°C. In contrast, the inferred density of shallow electrons decreases by more than 30%. Larger differences in the annealing characteristics of the deep and shallow electrons are noted after isothermal annealing at negative bias at ~115°C [112]. It is possible that the results in Figure 7.19 can be explained by an O vacancy-related defect with a single basic microstructure (e.g., the E'_{γ}) that is distributed in space and energy in such a way as to alter its

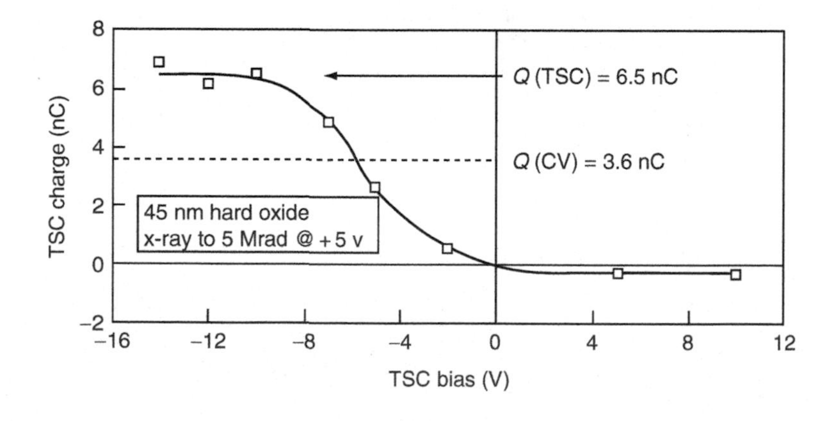

FIGURE 7.18
Integrated TSC, corrected for background leakage, as a function of bias for n-substrate capacitors with 45 nm oxides grown at 1000°C in dry O_2, after which they received a 900°C forming gas postoxidation anneal (before polycrystalline Si deposition). The devices were irradiated to 5 Mrad (SiO_2) with 10 keV x-rays at a dose rate of 5550 rad (SiO_2)/s at a bias of 5 V. The total positive charge Q_p estimated from TSC is compared to the net positive charge Q_{CV} estimated from midgap C–V measurements. (After Fleetwood, D.M., Reber, R.A. Jr., and Winokur, P.S., *IEEE Trans. Nucl. Sci.*, 38, 1066, 1991. With permission.)

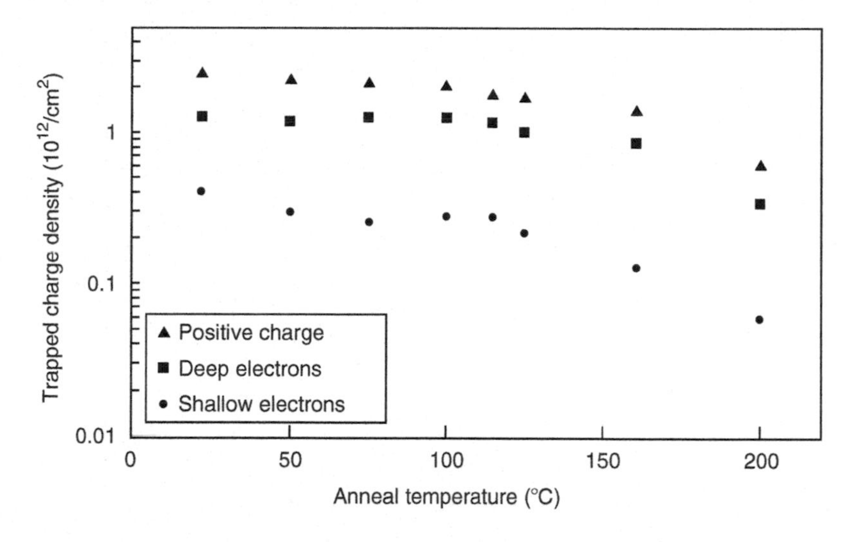

FIGURE 7.19
Trapped positive charge estimated from the integrated TSC (triangles), estimated density of electrons in border traps from *C–V* hysteresis (dots; shallow electrons), and differences between Q_p estimated from the TSC and Q_{CV} ($Q_n = Q_p - Q_{CV}$) as a function of isochronal annealing temperature. To estimate Q_n (squares, deep electrons), *C–V* curves were swept from negative-to-positive bias to exclude contributions of electrons in border traps that contribute to *C–V* hysteresis. Capacitors with 45 nm oxides were irradiated to 2 Mrad (SiO_2) at a dose rate of 1100 rad (SiO_2)/s at 10 V, after which they were annealed at −10 V for 15 min for the temperatures shown between irradiation and TSC measurement at −12 V bias. (From Fleetwood, D.M. et al., *Appl. Phys. Lett.*, 74, 2969, 1999. With permission.)

annealing properties in the manner required to account for the observed response differences in Figure 7.19. However, it is also quite possible that hydrogen-related defects may also be responsible for the stable dipoles in the near-interfacial SiO_2 [110,111]. Hydrogen- related defects have also been identified that can contribute to postirradiation or post-electrical stress reversibility in observed oxide-trap charge [101,113–119], but these switching defects are obviously not a candidate for the fixed negative charge observed in the TSC experiments [32,111]. The microstructure of these defects remains an open and interesting question.

The relative rates of reversibility are compared among oxide, interface, and border traps in Ref. [20]. Interestingly, in those devices and in recent work on high-*K* dielectrics [107], there is a close correspondence between the reversibility in bulk oxide traps and interface traps, which are dangling Si bonds at the Si/SiO_2 interface. An exchange of protons between an O vacancy-related defect in the near-interfacial SiO_2 and the interface-trap site can explain this reversibility in a natural way [107]. Thus, both hydrogenous species and O vacancy-related defects may play significant roles in the reversibility of MOS oxide-trap charge in ways that are inextricably linked and affect MOS long-term reliability and radiation response in significant and fundamental ways that are still under intense investigation [119–126].

7.5 Interface Traps

7.5.1 Two-Stage Interface-Trap Buildup Model

For the last 30 years, it has been recognized that hydrogenous impurities and Si dangling bond defects are critical to the formation of interface traps in MOS radiation response

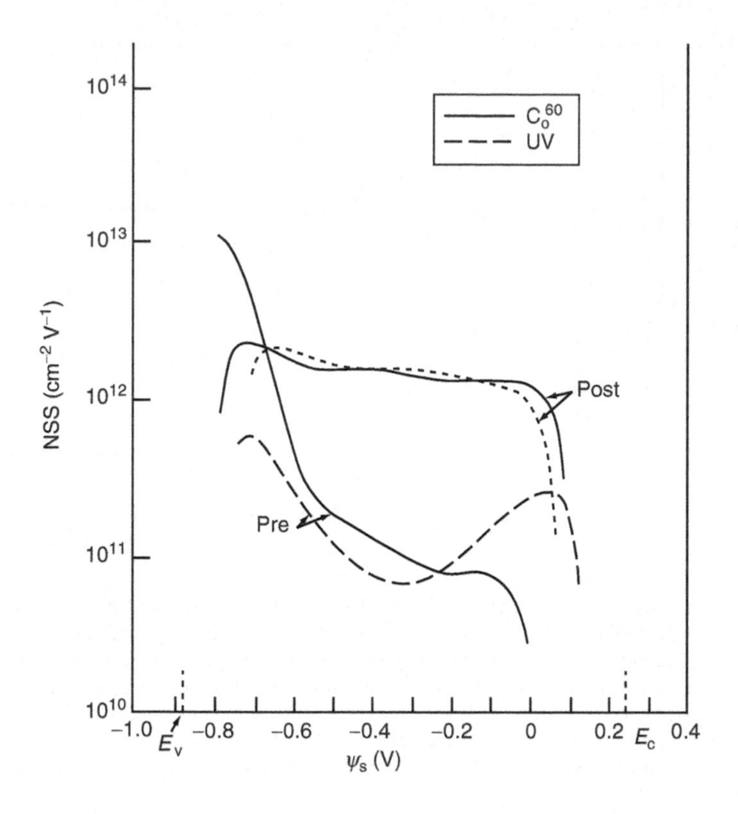

FIGURE 7.20
Interface-trap charge densities as a function of surface potential for 83 nm dry oxides with 5000 nm Al gates before and after ~100 krad (SiO_2) Co-60 irradiation at 10 V, and 90 nm dry oxides with 50 nm Au gates before and after 90 min of ~10 eV UV irradiation at 10 V. (From Winokur, P.S. and Sokoloski, M.M., *Appl. Phys. Lett.*, 28, 627, 1976. With permission.)

[2,4,5,118,127–135]. Figure 7.20 illustrates a pivotal experiment that compares interface-trap densities estimated from high- and low-frequency *C–V* methods [10,12] after Co-60 gamma (~1 MeV) and ~10 eV UV irradiation [130]. Energy from the Co-60 irradiation is absorbed uniformly throughout the SiO_2 and surrounding device layers. In contrast, the nonpenetrating 10 eV photons are absorbed only at the top surface of the SiO_2. Hence, the similarities in estimated densities and energy distributions indicate that interface traps mostly are not created by the direct interaction of highly penetrating radiation with weak bonds at the Si/SiO_2 interface [130]. Instead, interface-trap formation usually occurs in radiation effects studies (and many types of high-field stress experiments as well [44]) as a result of a series of processes initiated by the creation of electron–hole (e–h) pairs in the SiO_2, and/or the subsequent transport of the holes to the Si/SiO_2 interface [28,132]. This process is consistent with the schematic diagram that was shown in Figure 7.1. Follow-on studies by Winokur and coworkers evaluate the dose, temperature, processing, radiation bias, and annealing time and bias dependences of radiation-induced interface-trap buildup in metal gate capacitors [2,130,132–134]. From these and other related studies, it was inferred that the activation energy for MOS interface-trap buildup in metal gate capacitors is ~0.8 eV [133]. Moreover, it was also demonstrated that interface-trap formation is a two-stage process [2,130,132–135]. The first stage of the process is associated with hole transport through the SiO_2 and/or trapping near the Si/SiO_2 interface, with the accompanying release of a positive ion.

The second stage involves the transport of this ion to the Si/SiO_2 interface and its subsequent interaction, leading to the formation of an interface trap. It was argued persuasively in 1980 by McLean that this ion was H^+ [135]. This semiempirical model of H^+ transport and reactions captures many of the essential features of interface-trap formation, and forms a foundation for present models of interface-trap formation in MOS devices. McLean also demonstrated the consistency of the hydrogen model of radiation-induced interface-trap formation with experimental results on the temperature dependence of interface-trap formation during high-field stress obtained by Hu and Johnson [135,136].

Despite the successes of the two-stage hydrogen model, alternative explanations for interface-trap buildup to the hydrogen model of McLean et al. have been proposed. Many studies have observed the apparent conversion of trapped positive charge to interface traps [110,136–144]. Because both H^+ ions and trapped holes are charged positively when transporting or trapped in SiO_2, the interpretation of some studies is open to alternative explanation. Moreover, EPR studies found no link [145] between the reduction of holes trapped in E' centers (as noted in Section 7.3, this is the most common type of hole trap in SiO_2 [1,13,85–88]) and the buildup of P_b interface defects. The P_b defect has been identified as a dangling Si bond at the SiO_2 interface, as is known to be the dominant radiation-induced interface trap at the Si/SiO_2 interface [13,87,146–149]. In contrast, studies of the time and electric field dependences of interface-trap formation in polycrystalline Si-gate devices appeared at least initially to be more consistent with trapped-hole models than hydrogen models [150]. However, a very careful set of time, electric field, bias, and temperature dependence experiments by Boesch [151] and by Saks et al. [152–159] have clearly demonstrated that hydrogen release in the SiO_2, transport to the SiO_2 interface, and subsequent reaction at or near the interface play a central, enabling role in radiation-induced interface-trap buildup. We now examine some of the key results of this work.

Figure 7.21 shows the interface-trap buildup, measured via the charge-pumping technique, for MOS transistors with 42 nm oxides [153]. For one set of devices, a constant 8.4 V bias was applied throughout the irradiation and annealing period. The remainder of the devices were irradiated and annealed at a bias of −8.4 V, with the applied gate bias then switched to 8.4 V at varying times after the radiation pulse was delivered. In all switched-bias cases, the bias remained negative for times much longer than that calculated for holes to transport out of the oxide and into the gate ($\sim$10 μs) for this oxide thickness and electric field [153,160]. Clearly, there is significant interface-trap formation in all of the switched bias cases in Figure 7.21. This indicates that neither the direct interaction of radiation at the Si/SiO_2 interface nor the direct conversion of transporting or trapped holes to interface traps is the rate-limiting step in interface-trap formation in these devices. Instead, the results of these studies are quite consistent with the two-stage buildup of interface traps in metal gate devices [2,130,132–134], as described by the McLean model [135].

The magnitude and the time dependence of the interface-trap buildup in the devices of Figure 7.21 have been modeled successfully by Brown and Saks [157] by assuming that the rate-limiting step for the majority of radiation-induced interface trap buildup is the time it takes for H^+ to transport from its point of release in the SiO_2 to its point of reaction at the Si/SiO_2 interface. Significantly, a similar activation energy is reported for interface-trap buildup in polycrystalline Si gate devices ($\sim$0.8 eV [152,153]) as for metal gate devices [2,132–134]. An isotope effect in interface-trap formation also has been observed in radiation effects studies, with interface-trap formation occurring on a slower timescale for deuterium-treated oxides than it does for hydrogen-treated oxides [159]. This is consistent with the heavier D^+ taking more time to transport to the interface after release during a radiation pulse than the lighter H^+, once again showing the significance of hydrogenous species to interface-trap formation.

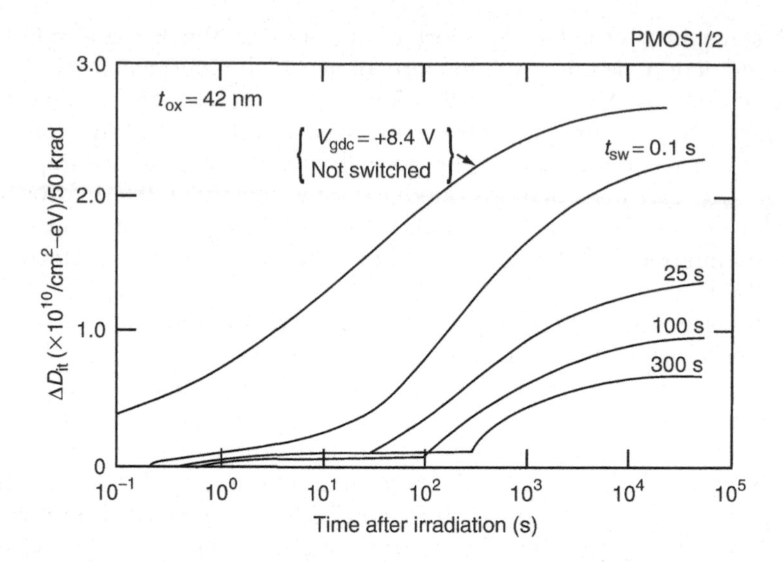

FIGURE 7.21
Interface-trap charge densities per unit energy as a function of postirradiation annealing time after a 50 krad (SiO_2), 1.5 μs radiation pulse from a 40 MeV electron linear accelerator. These devices were polycrystalline Si-gate MOS transistors with 42 nm oxides, irradiated at either a constant 2 MV/cm oxide electric field, or with a −2 MV/cm electric field that was switched to 2 MV/cm during the postirradiation anneal at the times indicated in the figure. (From Saks, N.S. and Brown, D.B., *IEEE Trans. Nucl. Sci.*, 35, 1168, 1988. With permission.)

Saks et al. [154] have used isochronal annealing measurements after low-temperature irradiation at positive bias to further investigate the kinetics of radiation-induced interface-trap formation, as shown in Figure 7.22. Here, we observe two distinct regimes of interface-trap formation. Below 150 K, the interface-trap formation is roughly independent of annealing bias (positive, negative, or zero), and is attributed to the diffusion of neutral H. Above 180 K, the interface-trap formation is strongly affected by annealing bias (enhanced for positive or zero bias, and suppressed for negative bias), and is attributed to proton drift. In these experiments, the total fraction of interface-trap formation associated with the low-temperature diffusion mechanism only corresponds to about 1%–2% of the portion of the interface-trap density associated with proton drift to the interface. This affirms that proton drift dominates over neutral hydrogen diffusion in room-temperature interface-trap formation [154].

Substantial interface-trap buildup is also reported in oxides that are first irradiated or subjected to high-field stress, and then are exposed to hydrogen after the radiation exposure has been stopped [114,116,161–166]. In this case, the experimental evidence strongly suggests that neutral hydrogen diffuses into the oxide, cracks at a defect center, and creates atomic hydrogen or a proton, which then transports to and reacts at the interface. Further, significant interface-trap buildup has been observed in oxides exposed to high fluences of atomic hydrogen, even in the absence of ionizing radiation exposure or high-field stress [167,168]. Hence, although alternative modes of interface-trap formation are possible, there is a wealth of evidence that suggests hydrogen plays a key role in interface-trap formation.

7.5.2 Hydrogen Transport and Reactions

Although there remains little doubt that hydrogen plays a central role in radiation-induced interface-trap formation, the details of the hydrogen release, transport, and reaction

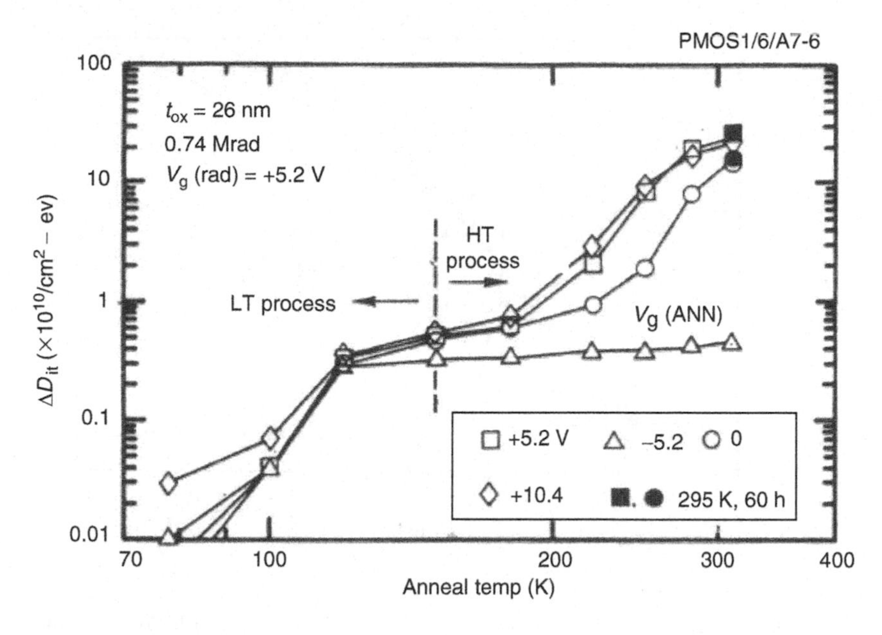

FIGURE 7.22
Interface-trap charge densities per unit energy inferred via the charge pumping technique as a function of postirradiation annealing temperature and bias for *p*MOS transistors with 26 nm oxides irradiated to 740 krad (SiO_2) at 78 K with Co-60 gamma rays. The annealing time at each temperature was 1200 s, and the devices were recooled to 78 K for all measurements. (From Saks, N.S., Klein, R.B., and Griscom, D.L., *IEEE Trans. Nucl. Sci.*, 34, 1234, 1987. With permission.)

processes remain somewhat uncertain. It is often assumed that hydrogen release occurs throughout the bulk of the oxide as a response to hole transport [2,132–135,152–159]. Indeed, such bulk release is consistent with the dependence on oxide thickness of the time required for radiation-induced interface trap buildup in the work of Saks et al., as illustrated in Figure 7.23 [153]. The thicker the oxide, the longer it takes for interface traps to build up following a radiation pulse for the devices of Figure 7.23. This is consistent with protons being released in the bulk of the SiO_2 for these kinds of oxides. However, Shaneyfelt et al. have performed a similar study on different oxides. They found minimal oxide thickness dependence, as illustrated in Figure 7.24, for dry oxides [169]. A slightly greater thickness dependence for the interface-trap buildup rate is observed for wet oxides by Shaneyfelt and coworkers [169], but still the dependence of the trap buildup rate on oxide thickness is neither as pronounced nor as systematic as that observed by Saks et al. [153]. These results, coupled with the strong similarities of the electric field dependences of trapped-positive-charge buildup and interface-trap formation in polycrystalline Si-gate devices [150,170], led Shaneyfelt et al. to suggest that most of the hydrogen that ultimately leads to interface-trap buildup in the oxides they have examined is released when a hole is captured at or released from an O vacancy site near an interface (Si or gate), as opposed to more distributed release throughout the bulk of the SiO_2 [169,170]. This "hole-trapping/hydrogen transport" model is quite similar to the McLean model, only with a different release point for the protons. The dominant release mechanism in a particular device depends on the differences in densities of hydrogen contained within the bulk SiO_2 network versus the densities that are attached to defect sites (e.g., O vacancies) near the Si/SiO_2 interface, as well as the dynamics between the transporting holes and hydrogen species bonded at such different sites.

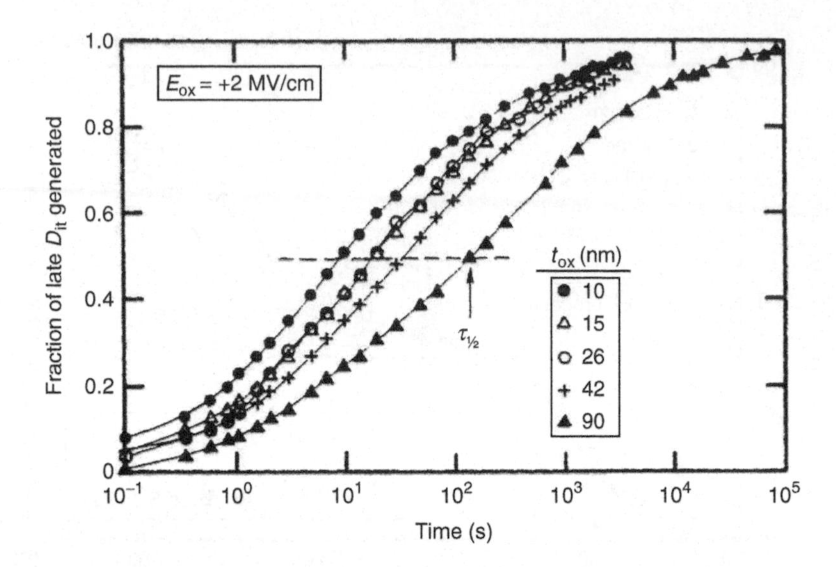

FIGURE 7.23
Normalized interface-trap charge densities per unit energy as a function of postirradiation annealing time after a 50 krad (SiO_2) radiation pulse from a 40 MeV electron linear accelerator. A small early component of interface-trap formation has been subtracted from these results to focus on the two-stage interface-trap formation mechanisms. These devices were polycrystalline Si-gate MOS transistors with 10–90 nm oxides from a single process lot, irradiated and annealed at a constant 2 MV/cm oxide electric field. (From Saks, N.S. and Brown, D.B., *IEEE Trans. Nucl. Sci.*, 35, 1168, 1988. With permission.)

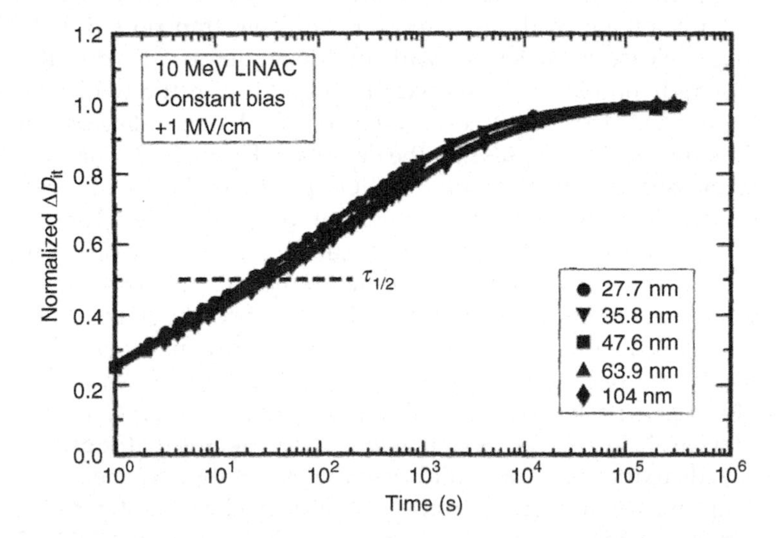

FIGURE 7.24
Normalized interface-trap charge densities per unit energy as a function of postirradiation annealing time after radiation pulses from a 10 MeV electron linear accelerator. These devices were polycrystalline Si-gate MOS transistors with 27.7–104 nm dry oxides from a single process lot, irradiated and annealed at a constant 1 MV/cm oxide electric field. Doses were 175, 95, 70, 50, and 35 krad (SiO_2) for the 27.7, 35.8, 47.6, 63.9, and 104 nm oxides, respectively. (From Shaneyfelt, M.R. et al., *IEEE Trans. Nucl. Sci.*, 39, 2244, 1992. With permission.)

In past studies of interface-trap formation mechanisms, it has been often assumed that, as a transporting proton approached the Si/SiO_2 interface, it is neutralized by an electron from the Si, diffuses until it is near a Si–H bond, and then reacts via [171–174]:

$$Si - H + H = Si - + H_2 \tag{7.9}$$

While intuitively plausible, this reaction process poses several practical difficulties. First, atomic hydrogen is highly chemically reactive in SiO_2 and therefore less stable than H^+ during transport [173]. So there are several competing reactions, including simple dimerization ($H + H = H_2$), that compete with Equation 7.9 both in the bulk of the oxide and near the Si/SiO_2 interface. Second, it has also been argued theoretically that H^+ is a lower energy state than H in SiO_2 [122,123,175–177]. This suggests that, even if the atomic H does not react immediately, it would donate an electron and become H^+ again as soon as kinetically feasible. Consistent with this interpretation, both Afanase'ev et al. [178] and Vanheusden et al. [179] have obtained evidence that H^+ has an approximately 100-times lower cross section for capturing an electron in SiO_2 than does a trapped hole. Finally, it has always been conceptually difficult to understand why interface-trap formation would be as efficient as commonly observed under positive bias during and/or after radiation exposure if the drifting proton were neutralized, thereby transforming its transport from field-directed drift to a diffusion process [118]. Indeed, evidence of excess hydrogen at or near the Si/SiO_2 interface [180,181] might even be likely to favor hydrogen motion away from the interface toward areas of lower concentration in the bulk, in the latter case.

The theoretical and conceptual difficulties associated with Equation 7.9 have been addressed by Rashkeev and coworkers. DFT calculations [177,182] strongly suggest that protons interact directly at the Si/SiO_2 interface via the simple reaction:

$$H^+ + Si - H = Si^+ + H_2 \tag{7.10}$$

Theoretical calculations suggest that two electrons leave the Si–H almost simultaneously when a proton approaches the Si/SiO_2 interface in the vicinity of a passivated dangling Si bond, forming a H_2 molecule and leaving the resulting dangling bond positively charged, at least momentarily, as per Equation 7.10. The positive charge on the trivalent Si at the interface will quickly be transformed to a negative charge following double electron capture from the inversion charge layer [177,182] when devices are irradiated or subjected to high-field electrical stress under positive bias. Interface-trap formation under negative bias requires hydrogenous species to transport from the substrate [183], if positively charged, or to be neutral [152,154] or negatively charged. The form of the reaction in Equation 7.10 avoids the above conceptual and theoretical difficulties associated with electron capture prior to reaction.

Similar kinds of mechanisms to those discussed above have also often been invoked to explain interface-trap buildup in MOS devices during hot carrier testing and/or high-field stress [113,174,184–190]. In these experiments a very high electric field is applied, leading to hot-carrier injection. These hot carriers can deposit enough energy to release hydrogen from the oxide, or in very thin oxides, perhaps even from the polycrystalline Si gate region. This release could either be a result of direct interaction, or triggered during anode hole injection and/or subsequent hole transport, for example [189–192]. After its release, it is presumed that the subsequent hydrogen transport and reactions at or near the Si/SiO_2 interface occur very much like that described above for radiation effects experiments.

7.6 Aging and Scaling Effects

7.6.1 Aging Effects

As illustrated in Figure 7.25, MOS oxide-trap charge is typically positive (thereby causing a negative threshold voltage shift) and recovers after its creation due to radiation exposure or high-field stress [1,100,193,194]. Consistent with the result of Sections 7.3 and 7.4, this recovery is due more to the neutralization of trapped positive charge by compensating electrons than to the true annealing of trapped holes [31,32,52]. Hence, the effects of oxide-trap charge tend to be mitigated with increasing time, in the absence of continuing irradiation or high-field stress. In contrast, interface traps can continue to build up with increasing time after irradiation or electrical stress, owing to the continuing transport and reactions of hydrogenous species under appropriate bias conditions, as illustrated in Figure 7.26 [2,100,193,194]. Border traps are known to be less significant in these devices than bulk oxide traps or interface traps. In nMOS devices, interface traps shift MOS transistor threshold voltages positively [2,13]; in pMOS devices, the threshold voltage shifts are negative, owing to the amphoteric nature of interface-trap charge in the Si band gap [2,13]. Interface traps also degrade the transistor channel mobility much more strongly than oxide-trap charge, owing to the higher efficiency in scattering rates [195,196]. These combinations of threshold voltage shifts and mobility degradation make device and IC testing a challenging and complex issue, although test methods have been developed to assess and assure the suitability of MOS parts for use in high radiation environments [193,194,197].

In Figures 7.25 and 7.26, the oxide-trap and interface-trap densities at long annealing times after pulsed irradiation exposure are similar to those during low-dose-rate irradiations to the same dose over the same elapsed time. This similarity shows that (1) there are no true dose-rate effects in these MOS devices under these irradiation and annealing conditions [194], and (2) there are no time-dependent aging effects that change the

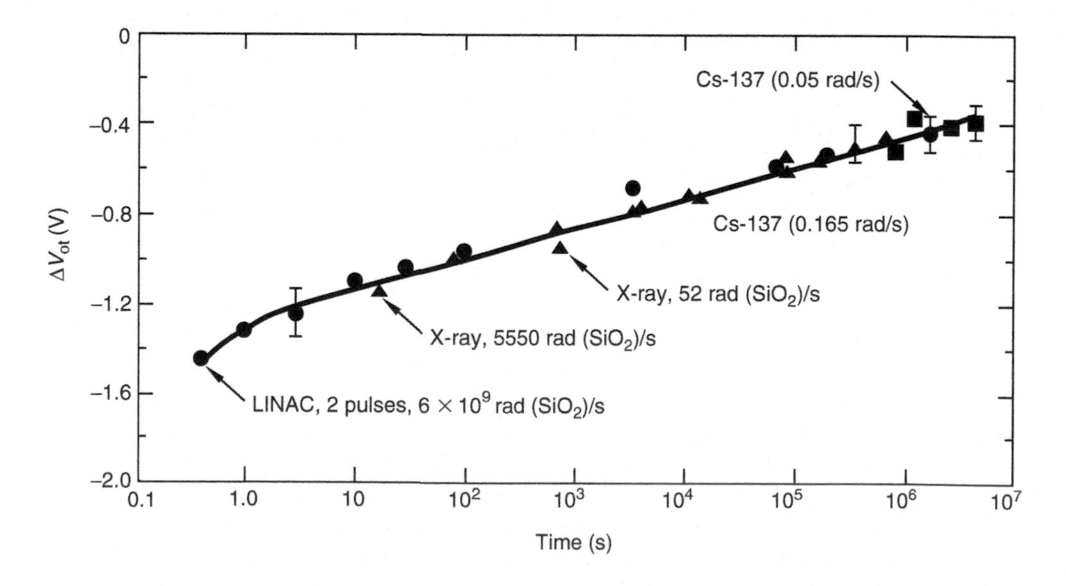

FIGURE 7.25
Threshold voltage shifts due to oxide-trap charge as a function of irradiation and annealing time for MOS transistors with 60 nm oxides irradiated to 100 krad (SiO_2) and annealed at room temperature at an applied electric field of 1 MV/cm. (From Fleetwood, D.M., Winokur, P.S., and Schwank, J.R., *IEEE Trans. Nucl. Sci.*, 35, 1497, 1988. With permission.)

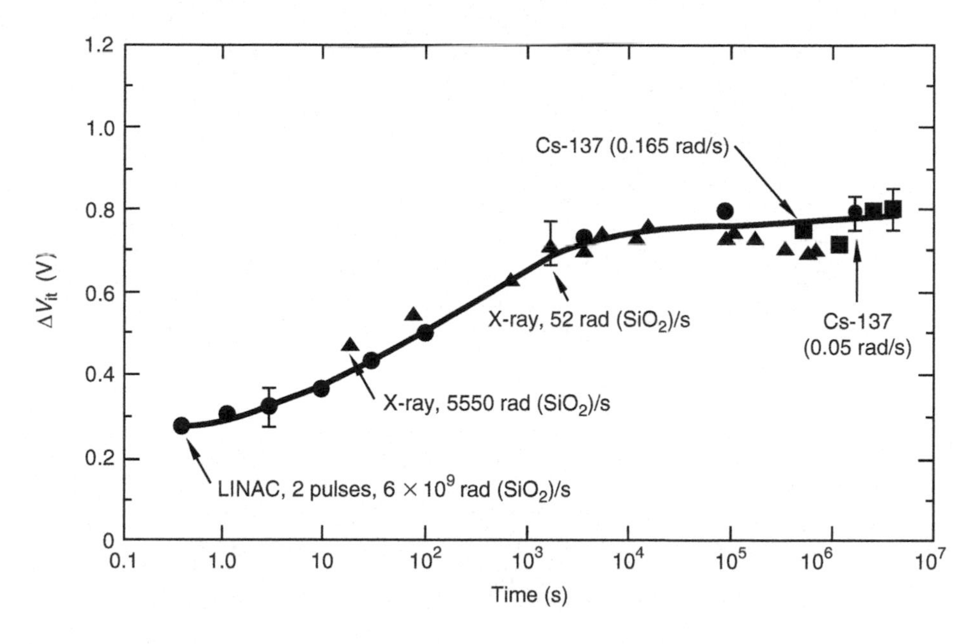

FIGURE 7.26
Threshold voltage shifts due to interface-trap charge as a function of irradiation and annealing time for MOS transistors with 60 nm oxides irradiated to 100 krad (SiO_2) and annealed at room temperature at an applied electric field of 1 MV/cm. (From Fleetwood, D.M., Winokur, P.S., and Schwank, J.R., *IEEE Trans. Nucl. Sci.*, 35, 1497, 1988. With permission.)

underlying radiation hardness of the devices during the testing. The latter point is often implicitly assumed to be true, but is often not the case. Significant changes in MOS oxide-trap charge and interface-trap charge have been observed with differences in aging times and/or thermal histories before irradiation. For example, Figure 7.27 shows a comparison of oxide-trap charge density for MOS capacitors with 33.4 nm oxides irradiated in 1986, compared to devices from the same wafer irradiated 15 years later [198]. In this case, less oxide-trap charge is observed. However, a subsequent heat treatment of these capacitors (processed with no passivation layers, and given a high-temperature anneal to increase their O vacancy density [199]) before irradiation restores the original radiation response. In this case it is assumed that the absorption of moisture passivated some hole trapping sites, but baking the device removed the passivating species, likely a hydrogen complex [198].

Aging has also been shown to degrade MOS radiation response. Figure 7.28 shows experiments that were performed on fully processed MOS devices with P-glass (3%

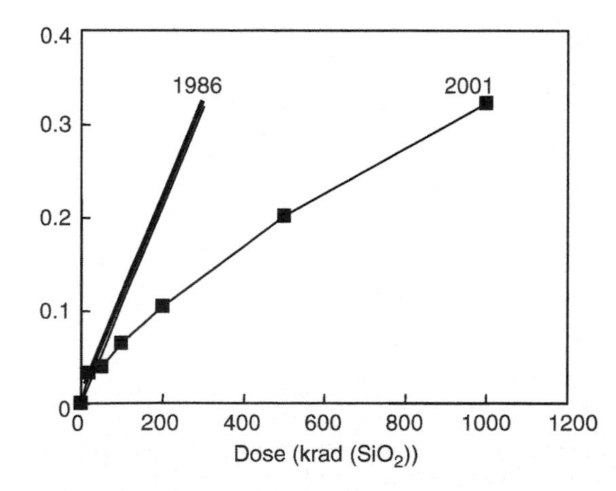

FIGURE 7.27
Oxide trap charge densities for Al-gate MOS capacitors irradiated with 10 keV x-rays at an applied electric field of ~1.5 MV/cm and dose rate of ~1 krad (SiO_2)/min. (From Karmarkar, A. et al., *IEEE Trans. Nucl. Sci.*, 48, 2158, 2001. With permission.)

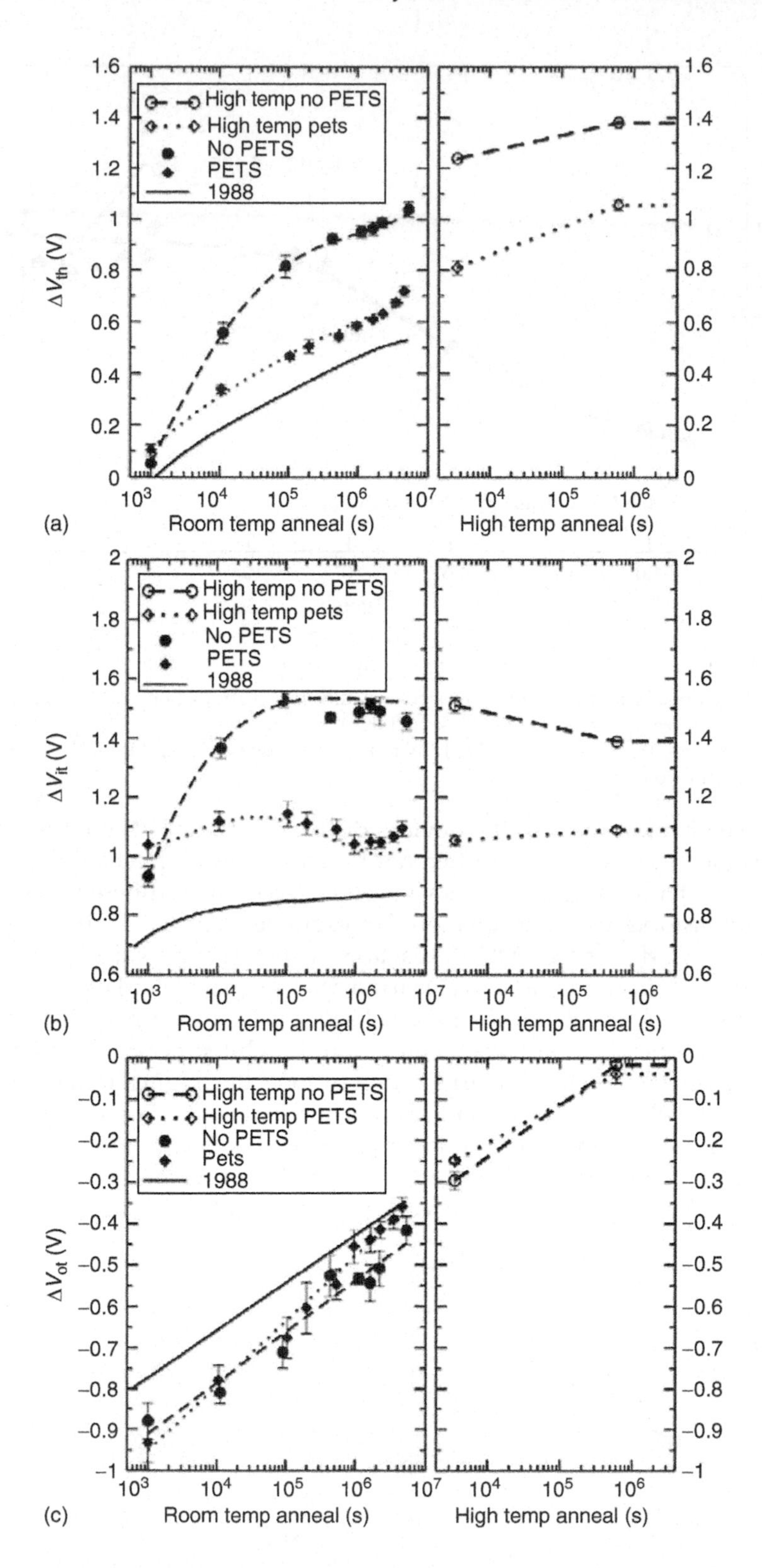

FIGURE 7.28
(a) ΔV_{th}, (b) ΔV_{it}, and (c) ΔV_{ot} for 60 nm gate oxide *n*MOS transistors stored hermetically since 1987 versus postirradiation anneal time for exposures to 100 krad (SiO_2). The irradiation and anneal bias was 6 V. (The 1988 data are from Fleetwood, D.M., Winokur, P.S., and Schwank, J.R., *IEEE Trans. Nucl. Sci.*, 35, 1497, 1988.) (From Rodgers, M.P. et al., *IEEE Trans. Nucl. Sci.*, 52, 2652, 2005. With permission.)

P-doped SiO_2) passivation layers [200]. Devices tested for aging effects in Ref. [200] include nMOS transistors processed at Sandia National Laboratories in 1984, with polycrystalline Si gates and oxide thicknesses of 32 or 60 nm. All irradiations were performed with a 10 keV x-ray irradiator at room temperature at 6 V bias. Some devices were baked before irradiation. Devices were annealed at room temperature and then 100°C at 6 V bias. Figure 7.28a compares threshold voltage shifts for nMOS transistors for 32 nm oxide parts irradiated in 1988 [194] and in 2005 [200]; the latter exposures were performed with and without exposure to preirradiation elevated temperature stress (PETS). The threshold voltage rebound for parts irradiated in 2005 is much larger than for parts irradiated in 1988 (Figure 7.28a). Figure 7.28b shows the estimated threshold voltage shifts due to interface-trap charge ΔV_{it} for the 32 nm parts. By the end of the postirradiation room-temperature anneal, the values of ΔV_{it} for parts not exposed to PETS are ~67% greater than the maximum ΔV_{it} experienced by these parts in 1988. However, when parts are exposed to PETS prior to irradiation, these shifts in magnitude decrease substantially, but are still greater than the 1988 values [194,200,201]. Figure 7.28c shows the shifts in threshold voltage due to oxide-trap charge ΔV_{ot} for the 32 nm gate oxide transistors [200]. Much less change during irradiation and annealing is observed for ΔV_{ot} with aging or baking than for ΔV_{it}. These changes in ΔV_{it} have been attributed to moisture absorption and the subsequent transport and reactions of water in the sensitive gate oxide region [200,201].

7.6.2 Scaling Effects and Emerging Materials

The total ionizing dose response of MOS devices has been increasingly less affected in recent years due to the scaling down of gate oxide thicknesses [1,4,202–204]. For thicker oxides, threshold voltage shifts due to oxide and interface-trap charge tend to decrease as $\sim 1/t_{ox}^2$ (for otherwise identical gate oxides) [1,4,5]. For thinner oxides (less than ~6 nm), the neutralization of trapped positive oxide-trap charge via tunneling further reduces radiation-induced threshold voltage shifts [202,203]. Interface-trap formation follows similar trends with decreasing oxide thickness as oxide-trap charge [2,4,204,205]. The reduction in interface trap density also leads to a decrease in mobility degradation with radiation exposure, and a lessening of the requirements to test for interface-trap related failures at very long times during radiation hardness assurance testing that can be masked in shorter-term tests due to the effects of oxide-trap charge [197,206]. However, we note that highly scaled microelectronics devices and ICs are increasingly affected by negative-bias temperature instabilities, which may well involve hydrogen release within the Si, followed by migration to and reaction at the Si/SiO_2 interface [207–212]. There is also evidence, at least in alternative dielectrics to SiO_2, that this process can be enhanced by radiation exposure [106,107]. So one should not neglect the effects of interface traps and trapped positive charge on threshold voltage shifts entirely in ultrathin gate oxides—alternative mechanisms may become important, especially as operating voltages decrease and operating speeds increase. These make some devices and ICs susceptible to even very small threshold voltage shifts.

In addition to threshold voltage shifts, radiation exposure can also lead to enhanced leakage in ultrathin MOS gate oxides. This is shown in Figure 7.29, for example, for 8 MeV electron irradiation of 6 nm oxides [213]. For these devices, the primary effect of the radiation exposure was to increase the gate oxide leakage. This is usually attributed to trap-assisted tunnelling across the gate oxide, and has also been observed for heavy ion irradiation [214]. In the case of heavy ion exposure, one can see catastrophic failure for even light fluences of high-LET ion exposure at high enough electric fields [214,215], but for lower electric fields, one sees radiation-induced leakage current. More generally, stress-induced leakage current (for applications outside a radiation environment) [175,216–218],

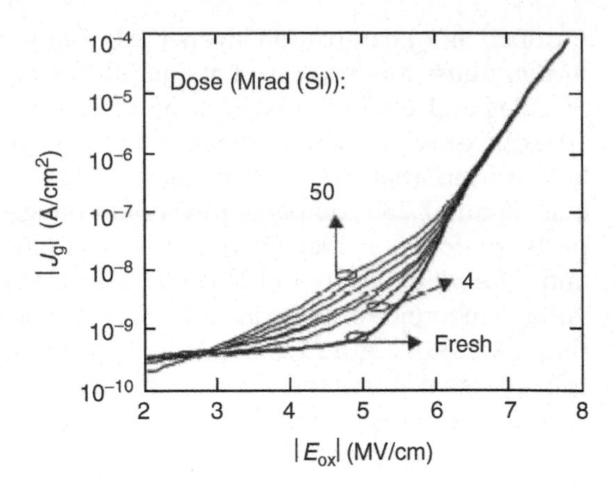

FIGURE 7.29
Current–voltage measurements for capacitors with 6 nm oxides irradiated from 0 to 50 Mrad (Si) with 8 MeV electrons. (From Ceschia, M. et al., *IEEE Trans. Nucl. Sci.*, 45, 2375, 1998. With permission.)

enhanced noise [62,219,220], and oxide breakdown [190,216,221–223] are also extremely important issues associated with oxide and interface defects in highly scaled MOS devices.

To avoid high leakage currents in highly scaled MOS gate oxides, many electronics manufacturers are exploring the use of high-*K* dielectrics as potential alternatives to thermal SiO_2, which can show unacceptably high leakage currents for oxide thicknesses below ~1.5–2 nm [224,225]. The use of high-*K* materials would enable device dielectric scaling to continue over ranges in which SiO_2 becomes unmanageably thin to manufacture and/or integrate reliably, by replacing it with a material that is physically thicker, but equivalent to ultrathin SiO_2 electrically. The radiation responses of high-*K* dielectrics have been investigated, and typically are inferior to high-quality thermal SiO_2, but are improving with increasing high-*K* material quality [106–108,226,227], and should be expected to continue to do so as high-*K* dielectrics are adopted in commercial manufacturing processes.

For modern commercial microelectronics, the most sensitive regions of the device for total ionizing dose are the relatively thick isolation oxides. These isolate devices effectively

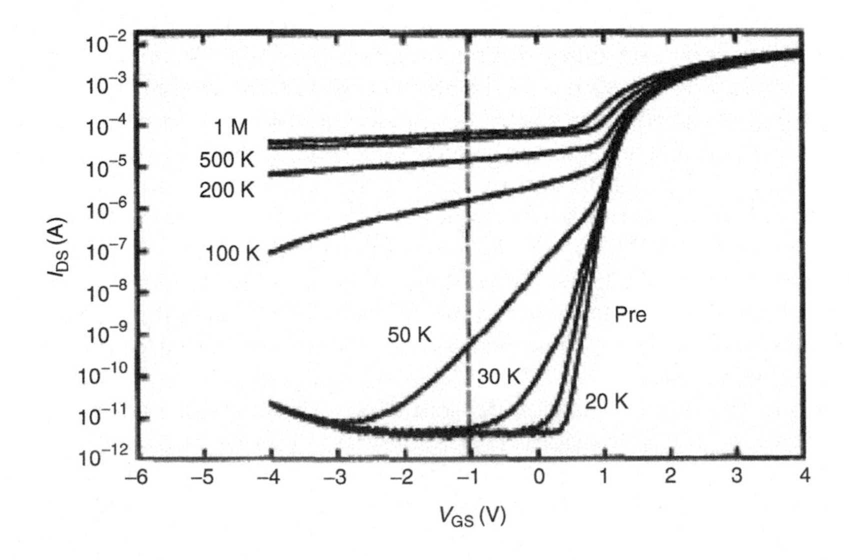

FIGURE 7.30
Current as a function of voltage for nMOS transistors with unhardened, shallow trench isolation (STI), irradiated with 10 keV x-rays at 167 rad (SiO_2)/s. (After Shaneyfelt, M.R. et al., *IEEE Trans. Nucl. Sci.*, 45, 2584, 1998. With permission.)

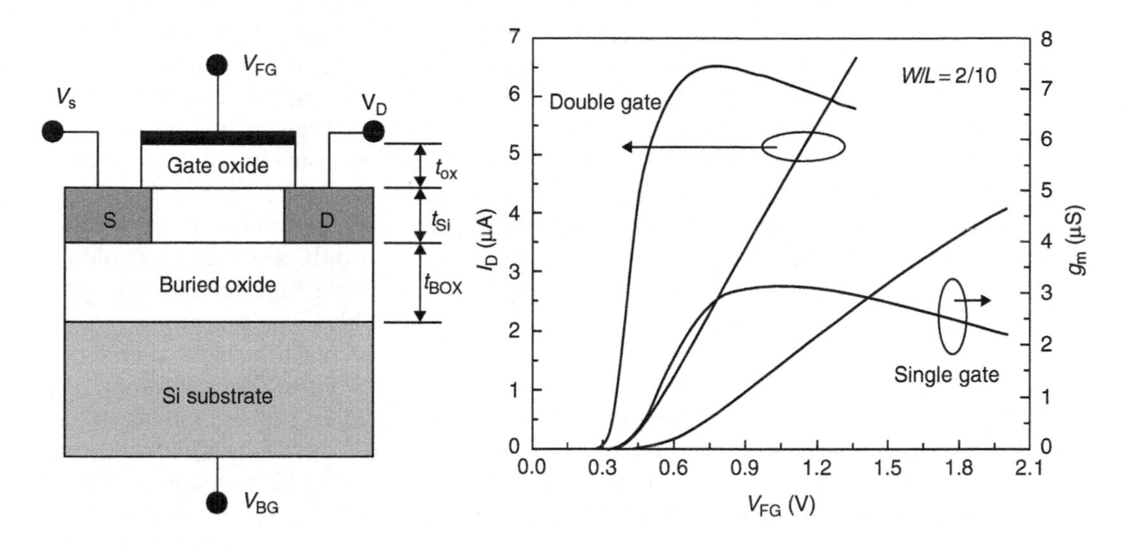

FIGURE 7.31
(Left) Schematic cross-sectional diagram of a fully depleted SOI MOSFET. (Right) Current–voltage and transconductance characteristics for fully depleted nMOS transistors operated in double gate mode. Front to back-gate coupling of the electric fields can lead to improved device mobility and potentially enhanced radiation response in double gate mode of device operation. (After Jun, B. et al., *IEEE Trans. Nucl. Sci.*, 51, 3767, 2004. With permission.)

for applications that do not include radiation exposure, but often trap enough charge to cause failures at relatively low doses when users attempt to employ these in radiation environments. An example of this for unhardened trench isolation is seen in Figure 7.30 [228]. These devices would only be suitable for very low-dose radiation environments. However, other commercial microelectronics technologies with STI have shown more acceptable hardness for space environments [229]. Similar issues are observed for the buried oxides of commercial SOI technologies [230]. For SOI technologies with hardened buried oxides, though, opportunities to operate the devices in double gate mode, as illustrated in Figure 7.31 [231], and other novel device structures make SOI of continuing interest for future commercial and radiation-tolerant microelectronics.

There are considerable hardness assurance challenges (lot-to-lot variability, process changes made without consideration for radiation effects response, etc.) associated with using commercial microelectronics technologies, as opposed to radiation hardened or tolerant devices. But the lower the dose of the application and the less critical the part, the more benefits some commercial parts may offer in terms of increased functionality and reduced purchase expense. However, often qualification costs can overwhelm purchase price savings, so the costs to the full system must be considered as well. For this reason, hardening by design is an increasingly popular approach to developing radiation-hardened technologies [229,232] and likely will become an even more popular approach for advanced defense and space systems with high radiation requirements.

7.7 Summary and Conclusions

Significant insights have been obtained into the nature of the defects that limit MOS performance, reliability, and radiation response. Oxide-trap charge in high-quality thermal

SiO_2 tends to be dominated by O vacancy-related defects. Interface traps are formed primarily through the interaction of hydrogenous species, liberated by irradiation or high-field stress, with Si–H bonds at the Si/SiO_2 interface, leaving a dangling Si bond. Border traps are near-interfacial oxide traps that can exchange charge with the underlying Si on the timescale of measurements of interest. These defects can trap charge on timescales ranging from nanoseconds to years, the effects of which can often be difficult to describe in a simple and straightforward fashion. Hence, the most reliable method of ensuring that devices are not limited by defects is to limit their densities through processing techniques that limit the number of O vacancies in SiO_2 (especially reducing high-temperature processing steps) and the amount of hydrogen in the devices. With these precautions, MOS devices have been successfully built and integrated into systems that have functioned successfully for many years in commercial and defense applications. The integration of a wide variety of new materials into the microelectronics fabrication sequence ensures that new challenges will continue to arise in future generations of technologies, leading to a continuing need to develop a deeper understanding of defects in the MOS system and their sometimes surprising effects on device and circuit response.

Acknowledgments

Portions of this work have been supported by the Air Force Office of Scientific Research, the U.S. Navy, the Defense Threat Reduction Agency, and the U.S. Department of Energy. We wish to thank our many collaborators that have contributed strongly to the efforts described above. The insights obtained are due to their dedicated efforts.

References

1. McLean, F.B., Boesch, H.E. Jr., and Oldham, T.R., Electron–hole generation, transport, and trapping in SiO_2, in *Ionizing Radiation Effects in MOS Devices & Circuits*, Ma, T.P. and Dressendorfer, P.V., Eds., Wiley, New York, 1989, Chapter 3.
2. Winokur, P.S., Radiation-induced interface traps, in *Ionizing Radiation Effects in MOS Devices & Circuits*, Ma, T.P. and Dressendorfer, P.V., Eds., Wiley, New York, 1989, Chapter 4.
3. Holmes-Seidle, A. and Adams, L., *Handbook of Radiation Effects*, 2nd edn., Oxford University Press, Oxford, 2002, Chapter 4.
4. Schwank, J.R., Total dose effects in MOS devices, in *2002 IEEE Nuclear and Space Radiation Effects Conference Short Course Notebook*, Dressendorfer, P.V., Ed., IEEE, Piscataway, New Jersey, available at nsrec.com, 2002, Chapter 3.
5. Oldham, T.R. and McLean, F.B., Total ionizing dose effects in MOS oxides and devices, *IEEE Trans. Nucl. Sci.*, 50, 483, 2003.
6. Fleetwood, D.M., Border traps in MOS devices, *IEEE Trans. Nucl. Sci.*, 39, 269, 1992.
7. Fleetwood, D.M. et al., Effects of oxide traps, interface traps, and "border traps" on MOS Devices, *J. Appl. Phys.*, 73, 5058, 1993.
8. Fleetwood, D.M. et al., Border traps: Issues for MOS radiation response and long-term reliability, *Microelectron. Reliab.*, 35, 403, 1995.
9. Fleetwood, D.M., Fast and slow border traps in MOS devices, *IEEE Trans. Nucl. Sci.*, 43, 779, 1996.
10. Nicollian, E.H. and Brews, J.R., *MOS (Metal Oxide Semiconductor) Physics and Technology*, Wiley, New York, 1982.

11. Winokur, P.S. et al., Correlating the radiation response of MOS capacitors and transistors, *IEEE Trans. Nucl. Sci.*, 31, 1453–1460, 1984.
12. Schroder, D.K., *Semiconductor Material and Device Characterization*, 3rd edn. Wiley, New York, 2006.
13. Lenahan, P.M. and Dressendorfer, P.V., Hole traps and trivalent Si centers in MOS devices, *J. Appl. Phys.*, 55, 3495, 1984.
14. Fleetwood, D.M., Long-term annealing study of midgap interface-trap charge neutrality, *Appl. Phys. Lett.*, 60, 2883, 1992.
15. Reber, R.A. Jr. and Fleetwood, D.M., Thermally-stimulated-current measurements of SiO_2 defect density and energy in irradiated MOS capacitors, *Rev. Sci. Instrum.*, 63, 5714, 1992.
16. Fleetwood, D.M. et al., Thermally stimulated current in SiO_2, *Microelectron. Reliab.*, 39, 1323, 1999.
17. Groeseneken, G. et al., A reliable approach to charge-pumping measurements in MOS transistors, *IEEE Trans. Electron Dev.*, 31, 42, 1984.
18. McWhorter, P.J., Winokur, P.S., and Pastorek, R.A., Donor/acceptor nature of radiation-induced interface traps, *IEEE Trans. Nucl. Sci.*, 35, 1154, 1988.
19. Fleetwood, D.M., Shaneyfelt, M.R., and Schwank, J.R., Estimating oxide-trap, interface-trap, and border-trap charge densities in MOS transistors, *Appl. Phys. Lett.*, 64, 1965, 1994.
20. Fleetwood, D.M. et al., Effects of interface traps and border traps on MOS postirradiation annealing response, *IEEE Trans. Nucl. Sci.*, 42, 1698, 1995.
21. Simmons, J.G. and Taylor, G.W., High-field isothermal currents and thermally stimulated currents in insulators having discrete trapping levels, *Phys. Rev. B*, 5, 1619, 1972.
22. Simmons, J.G., Taylor, G.W., and Tam, M.C., Thermally stimulated currents in semiconductors and insulators having arbitrary trap distributions, *Phys. Rev. B*, 7, 3714, 1973.
23. Esqueda, P.D. and Das, M.B., Characterization of surface states in HCl grown oxides using MOS transient currents, *Solid State Electron.*, 23, 365, 1980.
24. Shanfield, Z., Thermally stimulated current measurements on irradiated MOS capacitors, *IEEE Trans. Nucl. Sci.*, 30, 4064, 1983.
25. Shanfield, Z. and Moriwaki, M.M., Characteristics of hole traps in dry and pyrogenic gate oxides, *IEEE Trans. Nucl. Sci.*, 31, 1242, 1984.
26. Shanfield, Z. and Moriwaki, M.M., Radiation-induced hole trapping and interface state characteristics of Al-gate and poly-Si gate MOS capacitors, *IEEE Trans. Nucl. Sci.*, 31, 3929, 1985.
27. Shanfield, Z. and Moriwaki, M.M., Critical evaluation of the midgap-voltage-shift method for determining oxide trapped charge in irradiated MOS devices, *IEEE Trans. Nucl. Sci.*, 34, 1159, 1987.
28. Antonenko, V.I., Zhdan, A.G., and Sul'zhenko, P.S., Identification of spatial localization of surface states in experiments involving thermally stimulated discharge of a metal-insulator-semiconductor capacitor, *Sov. Phys. Semicond.*, 22, 473, 1988 [*Fiz. Tekh. Poluprovodn.*, 20, 758, 1988].
29. Shanfield, Z. et al., A new MOS radiation-induced charge: Negative fixed interface charge, *IEEE Trans. Nucl. Sci.*, 39, 303, 1992 [*J. Radiat. Effects Res. Eng.*, 8, 1, 1990].
30. Fleetwood, D.M., Reber, R.A. Jr., and Winokur, P.S., Effect of bias on thermally stimulated current (TSC) in irradiated MOS devices, *IEEE Trans. Nucl. Sci.*, 1066, 38, 1991.
31. Fleetwood, D.M., Reber, R.A. Jr., and Winokur, P.S., Trapped-hole annealing and electron trapping in MOS devices, *Appl. Phys. Lett.*, 60, 2008, 1992.
32. Fleetwood, D.M. et al., New insights into radiation-induced oxide-trap charge through thermally-stimulated-current measurement and analysis, *IEEE Trans. Nucl. Sci.*, 39, 2192, 1992.
33. Miller, S.L., Fleetwood, D.M., and McWhorter, P.J., Determining the energy distribution of traps in insulating thin films using the thermally stimulated current technique, *Phys. Rev. Lett.*, 69, 820, 1992.
34. Fleetwood, D.M. et al., The role of border traps in MOS high-temperature postirradiation annealing response, *IEEE Trans. Nucl. Sci.*, 40(6), 1323, 1993.
35. Miller, S.L. et al., A general centroid determination methodology, with application to multilayer dielectric structures and thermally stimulated current measurements, *J. Appl. Phys.*, 74, 5068, 1993.

36. Raynaud, C. et al., Electrical characterization of instabilities in 6H SiC MOS capacitors, *J. Appl. Phys.*, 76, 993, 1994.
37. Fleetwood, D.M. et al., Physical mechanisms contributing to enhanced bipolar gain degradation at low dose rates, *IEEE Trans. Nucl. Sci.*, 41, 1871, 1994.
38. Saks, N.S. et al., Radiation effects in oxynitrides grown in N_2O, *IEEE Trans. Nucl. Sci.*, 41, 1854, 1994.
39. Shaneyfelt, M.R. et al., Effects of burn-in on radiation hardness, *IEEE Trans. Nucl. Sci.*, 41, 2550, 1994.
40. Warren, W.L. et al., Microscopic nature of border traps in MOS oxides, *IEEE Trans. Nucl. Sci.*, 41, 1817, 1994.
41. Lu, Y. and Sah, C.T., Thermal emission of trapped holes in thin SiO_2 films, *J. Appl. Phys.*, 78, 3156, 1995.
42. Paillet, P. et al., Trapping-detrapping properties of irradiated ultra-thin SIMOX buried oxides, *IEEE Trans. Nucl. Sci.*, 42, 2108, 1995.
43. Warren, W.L. et al., Electron and hole trapping in doped oxides, *IEEE Trans. Nucl. Sci.*, 42, 1731, 1995.
44. Fleetwood, D.M. and Saks, N.S., Oxide, interface, and border traps in thermal, N_2O, and N_2O-nitrided oxides, *J. Appl. Phys.*, 79, 1583, 1996.
45. Fleetwood, D.M. et al., Radiation effects at low electric fields in thermal, SIMOX, and bipolar-base oxides, *IEEE Trans. Nucl. Sci.*, 43, 2537, 1996.
46. Fleetwood, D.M., Revised model of thermally stimulated current in MOS capacitors, *IEEE Trans. Nucl. Sci.*, 44, 1826, 1997.
47. Cai, J. and Sah, C.T., Theory of thermally stimulated charges in MOS gate oxide, *J. Appl. Phys.*, 83, 851, 1998.
48. Dmitriev, S.G. and Markin, Yu.V., Macroscopic ion traps at the silicon-oxide interface, *Semiconductors*, 32, 1284, 1998 [*Fiz. Tekh. Poluprovodn.*, 32, 1439, 1998].
49. Fleetwood, D.M. et al., Bulk oxide and border traps in MOS capacitors, *J. Appl. Phys.*, 84, 6141, 1998.
50. Fleetwood, D.M. et al., Effects of isochronal annealing and irradiation temperature on radiation-induced trapped charge, *IEEE Trans. Nucl. Sci.*, 45, 2366, 1998.
51. Paillet, P. et al., Simulation of multi-level radiation-induced charge trapping and thermally, activated phenomena in SiO_2, *IEEE Trans. Nucl. Sci.*, 45, 1379, 1998.
52. Fleetwood, D.M. et al., The role of electron transport and trapping in MOS total-dose modeling, *IEEE Trans. Nucl. Sci.*, 46, 1519, 1999.
53. Fleetwood, D.M. et al., Dielectric breakdown of thin oxides during ramped current-temperature stress, *IEEE Trans. Nucl. Sci.*, 47, 2305, 2000.
54. Felix, J.A. et al., Bias and frequency dependence of radiation-induced charge trapping in MOS devices, *IEEE Trans. Nucl. Sci.*, 48, 2114, 2001.
55. Fleetwood, D.M. et al., Unified model of hole trapping, $1/f$ noise, and thermally stimulated current in MOS devices, *IEEE Trans. Nucl. Sci.*, 49, 2674, 2002.
56. Kaschieva, S., Todorova, Z., and Dmitriev, S.N., Radiation defects induced by 20 MeV electrons in MOS structures, *Vacuum*, 76, 307, 2004.
57. McWhorter, P.J. et al., Modeling the memory retention characteristics of SNOS nonvolatile transistors in a varying thermal environment, *J. Appl. Phys.*, 68, 1902, 1990.
58. Tzou, J.J., Sun, J.Y.C., and Sah, C.T., Field dependence of two large hole capture cross sections in thermal oxide on silicon, *Appl. Phys. Lett.*, 43, 861, 1983.
59. Fleetwood, D.M. and Scofield, J.H., Evidence that similar point defects cause $1/f$ noise and radiation-induced-hole trapping in MOS transistors, *Phys. Rev. Lett.*, 64, 579, 1990.
60. Scofield, J.H., Doerr, T.P., and Fleetwood, D.M., Correlation of preirradiation $1/f$ noise and postirradiation oxide trapped charge in MOS transistors, *IEEE Trans. Nucl. Sci.*, 36, 1946, 1989.
61. Scofield, J.H. and Fleetwood, D.M., Physical basis for nondestructive tests of MOS radiation hardness, *IEEE Trans. Nucl. Sci.*, 38, 1567, 1991.
62. Fleetwood, D.M., Meisenheimer, T.L., and Scofield, J.H., $1/f$ noise and radiation effects in MOS devices, *IEEE Trans. Electron Dev.*, 41, 1953, 1994.

63. Fleetwood, D.M. et al., Enhanced MOS $1/f$ noise due to near-interfacial oxygen deficiency, *J. Non-Cryst. Solids*, 187, 199, 1995.
64. Meisenheimer, T.L. and Fleetwood, D.M., Effect of radiation-induced charge on $1/f$ noise in MOS devices, *IEEE Trans. Nucl. Sci.*, 37, 1696, 1990.
65. Meisenheimer, T.L. et al., $1/f$ noise in n- and p-channel MOS devices through irradiation and annealing, *IEEE Trans. Nucl. Sci.*, 38, 1297, 1991.
66. Babcock, J.A. et al., Effects of ionizing radiation on the noise properties of DMOS power transistors, *IEEE Trans. Nucl. Sci.*, 38, 1304, 1991.
67. Simoen, E. et al., DC and low frequency noise characteristics of gamma-irradiated gate-all-around SOI MOS transistors, *Solid State Electron.*, 38, 1, 1995.
68. Simoen, E. and Claeys, C., The low-frequency noise behavior of silicon-on-insulator technologies, *Solid State Electron.*, 39, 949, 1996.
69. Simoen, E. et al., Short-channel radiation effect in 60 MeV proton irradiated 0.13 μm CMOS transistors, *IEEE Trans. Nucl. Sci.*, 50, 2426, 2003.
70. Schwank, J.R. and Fleetwood, D.M., The effect of post-oxidation anneal temperature on radiation-induced charge trapping in poly-crystalline Si gate MOS devices, *Appl. Phys. Lett.*, 53, 770, 1988.
71. Warren, W.L. et al., Links between oxide, interface, and border traps in high-temperature annealed Si/SiO_2 systems, *Appl. Phys. Lett.*, 64, 3452, 1994.
72. Ning, T.H., High-field capture of electrons by Coulomb-attractive centers in SiO_2, *J. Appl. Phys.*, 47, 3203, 1976.
73. Fleetwood, D.M., Radiation-induced charge neutralization and interface-trap buildup in MOS devices, *J. Appl. Phys.*, 67, 580, 1990.
74. Conley, J.F. Jr., et al., Observation and electrical characterization of new E′ center defects in technologically relevant thermal SiO_2 on Si: An additional complexity in oxide charge trapping, *J. Appl. Phys.*, 76, 2872, 1994.
75. Chavez, J.R. et al., Microscopic structure of the E_δ' center in amorphous SiO_2: A first principles quantum mechanical investigation, *IEEE Trans. Nucl. Sci.*, 44, 1799, 1997.
76. Karna, S.P. et al., Electronic structure theory and mechanisms of the oxide trapped hole annealing process, *IEEE Trans. Nucl. Sci.*, 47, 2316, 2000.
77. Pineda, A.C. and Karna, S.P., Effect of hole trapping on the microscopic structure of oxygen vacancy sites in a-SiO_2, *J. Phys. Chem. A*, 104, 4699, 2000.
78. Lu, Z.Y. et al., Structure, properties, and dynamics of oxygen vacancies in amorphous SiO_2, *Phys. Rev. Lett.*, 89, 285505-1, 2002.
79. Nicklaw, C.J. et al., The structure, properties, and dynamics of oxygen vacancies in amorphous SiO_2, *IEEE Trans. Nucl. Sci.*, 49, 2667, 2002.
80. Stirling, A. and Pasquarello, A., First-principles modeling of paramagnetic Si dangling-bond defects in amorphous SiO_2, *Phys. Rev. B*, 66, 245201, 2002.
81. Mukhopadhyay, S. et al., Modeling of the structure and properties of oxygen vacancies in amorphous silica, *Phys. Rev. B*, 70, 195203, 2004.
82. Uchino, T. and Yoko, T., Density functional theory of structural transformations of oxygen-deficient centers in amorphous silica during hole trapping: Structure and formation mechanism of the E_γ' center, *Phys. Rev. B*, 74, 125203, 2006.
83. Buscarino, G., Agnello, S., and Gelardi, F.M., Delocalized nature of the E_δ' center in amorphous SiO_2, *Phys. Rev. Lett.*, 94, 125501, 2005.
84. Buscarino, G., Agnello, S., and Gelardi, F.M., Hyperfine structure of the E_δ' center in amorphous SiO_2, *J. Phys. Condens. Matter*, 18, 5213, 2006.
85. Feigl, F.J., Fowler, W.B., and Yip, K.L., Oxygen vacancy model for E_1' center in SiO_2, *Solid State Commun.*, 14, 225, 1974.
86. O'Reilly, E.P. and Robertson, J., Theory of defects in vitreous SiO_2, *Phys. Rev. B*, 27, 3780, 1983.
87. Warren, W.L. et al., Paramagnetic point defects in amorphous SiO_2 and amorphous Si_3N_4 thin films, I. *a*-SiO_2, *J. Electrochem. Soc.*, 139, 872, 1992.
88. Rudra, J.K. and Fowler, W. B., Oxygen vacancy and the E_1' center in crystalline SiO_2, *Phys. Rev. B*, 35, 8223, 1987.

89. Mrstik, B.J. et al., A study of the radiation sensitivity of non-crystalline SiO_2 films using spectroscopic ellipsometry, *IEEE Trans. Nucl. Sci.*, 45, 2450, 1998.
90. Lelis, A.J. et al., The nature of the trapped hole annealing process, *IEEE Trans. Nucl. Sci.*, 36, 1808, 1989.
91. Zekeriya, V. and Ma, T.P., Effect of stress relaxation on the generation of radiation-induced interface traps in post-metal-annealed Al–SiO_2–Si devices, *Appl. Phys. Lett.*, 45, 249, 1984.
92. Kasama, K. et al., Mechanical stress dependence of radiation effects in MOS structures, *IEEE Trans. Nucl. Sci.*, 33, 1210, 1985.
93. Witczak, S.C. et al., Relaxation of the Si–SiO_2 interfacial stress in bipolar screen oxides due to ionizing radiation, *IEEE Trans. Nucl. Sci.*, 42, 1689, 1996.
94. Dutta, P. and Horn, P.M., Low-frequency fluctuations in solids: $1/f$ noise, *Rev. Mod. Phys.*, 53, 497, 1981.
95. Weissman, M.B., $1/f$ noise and other slow, non-exponential kinetics in condensed matter, *Rev. Mod. Phys.*, 60, 537, 1988.
96. Xiong, H.D. et al., Temperature dependence and irradiation response of $1/f$ noise in MOSFETs, *IEEE Trans. Nucl. Sci.*, 49, 2718, 2002.
97. Xiong, H.D., Low-frequency noise and radiation response of buried oxides in SOI nMOS transistors, *IEEE Proc. Circ. Dev. Syst.*, 151, 118, 2004.
98. Shu, K. et al., Role of stress in irradiation-then-anneal technique for improving radiation hardness of metal–insulator–semiconductor devices, *Appl. Phys. Lett.*, 61, 675, 1992.
99. Pepper, M., Inversion layer transport and the radiation hardness of the Si–SiO_2 interface, *IEEE Trans. Nucl. Sci.*, 25, 1283, 1978.
100. Schwank, J.R. et al., Physical mechanisms contributing to device "rebound," *IEEE Trans. Nucl. Sci.*, 31, 1434, 1984.
101. Nissan-Cohen, Y., Shappir, J., and Frohman-Bentchkowsky, D., Dynamic model of trapping–detrapping in SiO_2, *J. Appl. Phys.*, 57, 2252, 1985.
102. Lelis, A.J. et al., Reversibility of trapped hole annealing, *IEEE Trans. Nucl. Sci.*, 35, 1186, 1988.
103. Roh, Y., Trombetta, L., and DiMaria, D.J., Interface traps induced by hole trapping in MOS devices, *J. Non-Cryst. Solids*, 187, 165, 1995.
104. Conley, J.H. Jr. et al., Electron spin resonance evidence that E_{γ}' centers can behave as switching oxide traps, *IEEE Trans. Nucl. Sci.*, 42, 1744, 1995.
105. Schwank, J.R. et al., Generation of metastable electron traps in the near interfacial region of SOI buried oxides by ion implantation and their effect on device properties, *Microelectron. Eng.*, 72, 362, 2004.
106. Zhou, X.J. et al., Bias-temperature instabilities and radiation effects in MOS devices, *IEEE Trans. Nucl. Sci.*, 52, 2231, 2005.
107. Zhou, X.J. et al., Effects of switched-bias annealing on charge trapping in HfO_2 gate dielectrics, *IEEE Trans. Nucl. Sci.*, 53, 3636, 2006.
108. Lucovsky, G. et al., Differences between charge trapping states in irradiated nano-crystalline HfO_2 and non-crystalline Hf silicates, *IEEE Trans. Nucl. Sci.*, 53, 3644, 2006.
109. Devine, R.A.B. et al., Point defect generation and oxide degradation during annealing of the Si/SiO_2 interface, *Appl. Phys. Lett.*, 63, 2926, 1993.
110. Lai, S.K., Two-carrier nature of interface-state generation in hole trapping and radiation damage, *Appl. Phys. Lett.*, 39, 58, 1981.
111. Fleetwood, D.M. et al., Dipoles in SiO_2: Border traps or not? in *PV 2003–02 Silicon Nitride and Silicon Dioxide Thin Insulating Films*, 7th edn., Sah, R.E. et al. Eds., The Electrochemical Society, Pennington, NJ, 2003, p. 291.
112. Fleetwood, D.M. et al., Stability of trapped electrons in SiO_2, *Appl. Phys. Lett.*, 74, 2969, 1999.
113. Trombetta, L.P. et al., An EPR study of electron injected oxides in MOS capacitors, *J. Appl. Phys.*, 64, 2434, 1988.
114. Stahlbush, R.E., Mrstik, B.J., and Lawrence, R., Post-irradiation behavior of the interface state density and the trapped positive charge, *IEEE Trans. Nucl. Sci.*, 37, 1641, 1990.
115. Trombetta, L.P., Feigl, F.J., and Zeto, R.J., Positive charge generation in MOS capacitors, *J. Appl. Phys.*, 69, 2512, 1991.

116. Stahlbush, R.E. et al., Post-irradiation cracking of H_2 and formation of interface states in irradiated MOSFETs, *J. Appl. Phys.*, 73, 658, 1993.
117. Freitag, R.K., Brown, D.B., and Dozier, C.M., Evidence for two types of radiation-induced trapped positive charge, *IEEE Trans. Nucl. Sci.*, 41, 1828, 1994.
118. Fleetwood, D.M., Effects of hydrogen transport and reactions on microelectronics radiation response and reliability, *Microelectron. Reliab.*, 42, 523, 2002.
119. Vanheusden, K. et al., Nonvolatile memory device based on mobile protons in SiO_2 thin films, *Nature*, 386, 587, 1997.
120. Vanheusden, K. et al., Irradiation response of mobile protons in buried SiO_2 films, *IEEE Trans. Nucl. Sci.*, 44, 2087, 1997.
121. Fleetwood, D.M. et al., $1/f$ noise, hydrogen transport, and latent interface-trap buildup in irradiated MOS devices, *IEEE Trans. Nucl. Sci.*, 44, 1810, 1997.
122. Pantelides, S.T. et al., Reactions of hydrogen with Si–SiO_2 interfaces, *IEEE Trans. Nucl. Sci.*, 47, 2262, 2000.
123. Bunson, P.E. et al., Hydrogen-related defects in irradiated SiO_2, *IEEE Trans. Nucl. Sci.*, 47, 2289, 2000.
124. Rashkeev, S.N. et al., Statistical modeling of radiation-induced proton transport in Si: Deactivation of dopant acceptors in bipolar devices, *IEEE Trans. Nucl. Sci.*, 50, 1896, 2003.
125. Shaneyfelt, M.R. et al., Annealing behavior of linear bipolar devices with enhanced low-dose-rate sensitivity, *IEEE Trans. Nucl. Sci.*, 51, 3172, 2004.
126. Rashkeev, S.N. et al., Effects of hydrogen motion on interface trap formation and annealing, *IEEE Trans. Nucl. Sci.*, 51, 3158, 2004.
127. Revesz, A.G., Defect structure and irradiation behavior of noncrystalline SiO_2, *IEEE Trans. Nucl. Sci.*, 18, 113, 1971.
128. Revesz, A.G., Chemical and structural aspects of the irradiation behavior of SiO_2 films on silicon, *IEEE Trans. Nucl. Sci.*, 2102, 24, 1977.
129. Svensson, C.M., The defect structure of the Si–SiO_2 interface, a model based on trivalent silicon and its hydrogen compounds, in *The Physics of SiO_2 and its Interfaces*, Pantelides, S.T. Ed., Pergamon, New York, 1978, pp. 328–332.
130. Winokur, P.S. and Sokoloski, M.M., Comparison of interface-state buildup in MOS capacitors subjected to penetrating and nonpenetrating radiation, *Appl. Phys. Lett.*, 28, 627, 1976.
131. Ma, T.P., Scoggan, G., and Leone, R., Comparison of interface-state generation by 25-keV electron beam irradiation in *p*-type and *n*-type MOS capacitors, *Appl. Phys. Lett.*, 27, 61, 1975.
132. Winokur, P.S., McGarrity, J.M., and Boesch, Jr., H.E., Dependence of interface-state buildup on hole generation and transport in irradiated MOS capacitors, *IEEE Trans. Nucl. Sci.*, 23, 1580, 1976.
133. Winokur, P.S. et al., Field- and time-dependent radiation effects at the SiO_2/Si interface of hardened MOS capacitors, *IEEE Trans. Nucl. Sci.*, 24, 2113, 1977.
134. Winokur, P.S. et al., Two-stage process for buildup of radiation-induced interface states, *J. Appl. Phys.*, 50, 3492, 1979.
135. McLean, F.B., A framework for understanding radiation-induced interface states in SiO_2 MOS structures, *IEEE Trans. Nucl. Sci.*, 27, 1651, 1980.
136. Hu, G. and Johnson. W.C., Relationship between trapped holes and interface states in MOS capacitors, *Appl. Phys. Lett.*, 36, 590, 1980.
137. Sah, C.T., Models and experiments on degradation of oxidized silicon, *Solid-St. Electron*, 33, 147, 1990.
138. Grunthaner, F.J., Grunthaner, P.J., and Maserjian, J., Radiation-induced defects in SiO_2 as determined with XPS, *IEEE Trans. Nucl. Sci.*, 29, 1462, 1982.
139. Lai, S.K., Interface trap generation in SiO_2 when electrons are captured by trapped holes, *J. Appl. Phys.*, 54, 2540, 1983.
140. Sabnis, A.G., Characterization of annealing of Co-60 gamma-ray damage at the Si/SiO_2 interface, *IEEE Trans. Nucl. Sci.*, 30, 4094, 1983.
141. Sabnis, A.G., Process dependent buildup of interface states in irradiated n-channel MOSFETs, *IEEE Trans. Nucl. Sci.*, 32, 3905, 1985.
142. Grunthaner, F.J. and Gruthaner, P.J., Chemical and electronic structure of the SiO_2/Si interface, *Mater. Sci. Rep.*, 1, 65, 1986.

143. Kenkare, P.U. and Lyon, S.A., Relationship between trapped holes, positive ions, and interface states in irradiated Si–SiO_2 structures, *Appl. Phys. Lett.*, 55, 2328, 1989.
144. Pershenkov, V.S. et al., Three-point method of prediction of MOS device response in space environments, *IEEE Trans. Nucl. Sci.*, 40(6), 1714, 1993.
145. Witham, H.S. and Lenahan, P.M., The nature of the deep hole trap in MOS oxides, *IEEE Trans. Nucl. Sci.*, 34, 1147, 1987.
146. Nishi, Y., Study of Si–SiO_2 structure by electron spin resonance I, *Jpn. J. Appl. Phys.*, 10, 52, 1971.
147. Poindexter, E.H. et al., Interface states and electron spin resonance centers in thermally oxidized (111) and (100) Si wafers, *J. Appl. Phys.*, 52, 879, 1981.
148. Brower, K. L., Kinetics of H_2 passivation of P_b centers at the (111) Si–SiO_2 interface, *Phys. Rev. B*, 38, 9657, 1988.
149. Kim, Y.Y. and Lenahan, P.M., Electron-spin-resonance study of radiation-induced paramagnetic defects in oxides grown on (100) Si substrates, *J. Appl. Phys.*, 64, 3551, 1988.
150. Schwank, J.R. et al., Radiation induced interface-state generation in MOS devices, *IEEE Trans. Nucl. Sci.*, 33, 1178, 1986.
151. Boesch, Jr., H.E., Time-dependent interface trap effects in MOS devices, *IEEE Trans. Nucl. Sci.*, 35, 1160, 1988.
152. Saks, N.S. and Ancona, M.G., Generation of interface states by ionizing radiation at 80 K measured by charge pumping and subthreshold slope techniques, *IEEE Trans. Nucl. Sci.*, 34, 1348, 1987.
153. Saks, N.S., Dozier, C.M., and Brown, D.B., Time dependence of interface trap formation in MOSFETs following pulsed irradiation, *IEEE Trans. Nucl. Sci.*, 35, 1168, 1988.
154. Saks, N.S., Klein, R.B., and Griscom, D.L., Formation of interface traps in MOSFETs during annealing following low temperature irradiation, *IEEE Trans. Nucl. Sci.*, 35, 1234, 1988.
155. Saks, N.S. and Brown, D.B., Interface trap formation via the two-stage H^+ process, *IEEE Trans. Nucl. Sci.*, 36, 1848, 1989.
156. Saks, N.S. and Brown, D.B., Observation of H^+ motion during interface trap formation, *IEEE Trans. Nucl. Sci.*, 37, 1624, 1990.
157. Brown, D.B. and Saks, N.S., Time dependence of radiation-induced interface trap formation in MOS devices as a function of oxide thickness and applied field, J. *Appl. Phys.*, 70, 3734, 1991.
158. Saks, N.S., Brown, D.B., and Rendell, R.W., Effects of switched gate bias on radiation-induced interface-trap formation, *IEEE Trans. Nucl. Sci.*, 38, 1130, 1991.
159. Saks, N.S. and Rendell, R.W., The time-dependence of post-irradiation interface trap buildup in deuterium annealed oxides, *IEEE Trans. Nucl. Sci.*, 39, 2220, 1992.
160. McLean, F.B., Boesch, Jr., H.E., and McGarrity, J.M., Hole transport and recovery characteristics of SiO_2 gate insulators, *IEEE Trans. Nucl. Sci.*, 23, 1506, 1976.
161. Kohler, R.A., Kushner. R.A., and Lee, K.H., Total dose radiation hardness of MOS devices in hermetic ceramic packages, *IEEE Trans. Nucl. Sci.*, 35, 1492, 1988.
162. Mrstik, B.J. and Rendell, R.W., Model for Si–SiO_2 interface state formation during irradiation and during post-irradiation exposure to hydrogen environment, *Appl. Phys. Lett.*, 59, 3012, 1991.
163. Mrstik, B.J. and Rendell, R.W., Si–SiO_2 interface state generation during x-ray irradiation and during post-irradiation exposure to a hydrogen ambient, *IEEE Trans. Nucl. Sci.*, 38, 1101, 1991.
164. Conley, J.F. and Lenahan, P.M., Room temperature reactions involving Si dangling bond centers and molecular hydrogen in *a*-SiO_2 thin films on Si, *IEEE Trans. Nucl. Sci.*, 39, 2186, 1992.
165. Conley, J.F. Jr. and Lenahan, PM., Molecular hydrogen, E′ center hole traps, and radiation induced interface traps in MOS devices, *IEEE Trans. Nucl. Sci.*, 40, 1335, 1993.
166. Saks, N.S. et al., Effects of post-stress hydrogen annealing on MOS oxides after Co-60 irradiation or Fowler-Nordheim injection, *IEEE Trans. Nucl. Sci.*, 40(6), 1341, 1993.
167. Cartier, E., Stathis, J.H., and Buchanan, D.A., Passivation and depassivation of Si dangling bonds at the Si/SiO_2 interface by atomic hydrogen, *Appl. Phys. Lett.*, 63, 1510, 1993.
168. Stahlbush, R.E. and Cartier, E., Interface defect formation in MOSFETs by atomic hydrogen exposure, *IEEE Trans. Nucl. Sci.*, 41, 1844, 1994.
169. Shaneyfelt, M.R. et al., Interface-trap buildup rates in wet and dry oxides, *IEEE Trans. Nucl. Sci.*, 39, 2244, 1992.

170. Shaneyfelt, M.R. et al., Field dependence of interface-trap buildup in polysilicon and metal gate MOS devices, *IEEE Trans. Nucl. Sci.*, 37, 1632, 1990.
171. Griscom, D.L., Diffusion of radiolytic molecular hydrogen as a mechanism for the post-irradiation buildup of interface states in SiO_2-on-Si structures, *J. Appl. Phys.*, 58, 2524, 1985.
172. Brown, D.B., The time dependence of interface state production, *IEEE Trans. Nucl. Sci.*, 32, 3900, 1985.
173. Griscom, D.L., Brown, D.B., and Saks, N.S., Nature of radiation-induced point defects in amorphous SiO_2 and their role in SiO_2-on-Si structures, in *The Physics and Chemistry of SiO_2 and the Si–SiO_2 Interface*, Helms, C.R. and Deal, B.E., Eds., Plenum Press, New York, 1988, pp. 287–297.
174. Zhang, W.D., et al., On the interface states generated under different stress conditions, *Appl. Phys. Lett.*, 79, 3092, 2001.
175. Blöchl, P.E. and Stathis, J.H., Hydrogen electrochemistry and stress-induced leakage current in silica, *Phys. Rev. Lett.*, 83, 372, 1999.
176. Bunson, P.E. et al., Ab initio calculations of H^+ energetics in SiO_2: Implications for transport, *IEEE Trans. Nucl. Sci.*, 46, 1568, 1999.
177. Rashkeev, S.N. et al., Defect generation by hydrogen at the Si–SiO_2 interface, *Phys. Rev. Lett.*, 87, 165506, 2001.
178. Afanase'ev, V.V., de Nijs, J.M.M., and Balk, P., SiO_2 hole traps with small cross section, *Appl. Phys. Lett.*, 66, 1738–1740, 1995.
179. Vanheusden, K. et al., Thermally activated electron capture by mobile protons in SiO_2 thin films, *Appl. Phys. Lett.*, 72, 28, 1998.
180. Gale, R. et al., Hydrogen migration under avalanche injection of electrons in Si MOS capacitors, *J. Appl. Phys.*, 54, 6938, 1983.
181. Chen, P.J. and Wallace, R.M., Examination of deuterium transport through device structures, *Appl. Phys. Lett.*, 73, 3441, 1998.
182. Rashkeev, S.N. et al., Proton-induced defect generation at the Si–SiO_2 interface, *IEEE Trans. Nucl. Sci.*, 48, 2086, 2001.
183. Tsetseris, L. et al., Common origin for enhanced low-dose-rate sensitivity and bias temperature instability under negative bias, *IEEE Trans. Nucl. Sci.*, 52, 2265, 2005.
184. Fischetti, M.V., Generation of positive charge in SiO_2 during avalanche and tunnel electron injection, *J. Appl. Phys.*, 57, 2860, 1985.
185. Do Thanh, L. and Balk, P., Elimination and generation of Si–SiO_2 interface traps by low temperature hydrogen annealing, *J. Electrochem. Soc.*, 135, 1797, 1988.
186. Stojadinovic, N. and Dimitrijev, S., Instabilities in MOS transistors, *Microelectron. Reliab.*, 29, 371, 1989.
187. Buchanan, D.A. and DiMaria, D.J., Interface and bulk trap generation in MOS capacitors, *J. Appl. Phys.*, 67, 7439, 1990.
188. DiMaria, D.J., Temperature dependence of trap creation in SiO_2, *J. Appl. Phys.*, 68, 5234, 1990.
189. Stathis, J.H. and DiMaria, D.J., Identification of an interface defect generated by hot electrons in SiO_2, *Appl. Phys. Lett.*, 61, 2887, 1992.
190. DiMaria, D.J., Cartier, E., and Arnold, D., Impact ionization, trap creation, degradation, and breakdown in SiO_2 films on Si, *J. Appl. Phys.*, 73, 3367, 1993.
191. Al-Kofahi, I.S. et al., Continuing degradation of the SiO_2/Si interface after hot hole stress, *J. Appl. Phys.*, 81, 2686, 1997.
192. Degraeve, R., Kaczer, B., and Groeseneken, G., Degradation and breakdown in thin oxide layers: Mechanisms, models, and reliability prediction, *Microelectron. Reliab.*, 39, 1445, 1999.
193. Winokur, P.S. et al., Total-dose radiation and annealing studies: Implications for hardness assurance testing, *IEEE Trans. Nucl. Sci.*, 33, 1343, 1986.
194. Fleetwood, D.M., Winokur, P.S., and Schwank, J.R., Using laboratory x-ray and cobalt-60 irradiations to predict CMOS device response in strategic and space environments, *IEEE Trans. Nucl. Sci.*, 35, 1497, 1988.
195. Sexton, F.W. and Schwank, J.R., Correlation of radiation effects in transistors and integrated circuits, *IEEE Trans. Nucl. Sci.*, 32, 3975, 1985.
196. McLean, F.B. and Boesch, H.E. Jr., Time-dependent degradation of MOSFET channel mobility following pulsed irradiation, *IEEE Trans. Nucl. Sci.*, 36, 1772, 1989.

197. Fleetwood, D.M. and Eisen, H.A., Total-dose radiation hardness assurance, *IEEE Trans. Nucl. Sci.*, 50, 552, 2003.
198. Karmarkar, A.P. et al., Aging and baking effects on the radiation hardness of MOS capacitors, *IEEE Trans. Nucl. Sci.*, 48, 2158, 2001.
199. Fleetwood, D.M. et al., The response of MOS devices to dose-enhanced low-energy MOS radiation, *IEEE Trans. Nucl. Sci.*, 33, 1245, 1986.
200. Rodgers, M.P. et al., The effects of aging on MOS irradiation and annealing response, *IEEE Trans. Nucl. Sci.*, 52, 2642, 2005.
201. Batyrev, I.G. et al., Effects of water on the aging and radiation response of MOS devices, *IEEE Trans. Nucl. Sci.*, 53, 3629, 2006.
202. Saks, N.S., Ancona, M.G., and Modolo, J.A., Radiation effects in MOS capacitors with very thin oxides at 80 K, *IEEE Trans. Nucl. Sci.*, 31, 1249, 1984.
203. Benedetto, J.M. et al., Hole removal in thin-gate oxide MOSFETs by tunneling, *IEEE Trans. Nucl. Sci.*, 32, 3916, 1985.
204. Saks, N.S., Ancona, M.G., and Modolo, J.A., Generation of interface states by ionizing radiation in very thin oxides, *IEEE Trans. Nucl. Sci.*, 33, 1185, 1986.
205. Dressendorfer, P.V., Radiation effects on MOS devices and circuits, in *Ionizing Radiation Effects in MOS Devices & Circuits*, Ma, T.P. and Dressendorfer, P.V., Eds., Wiley, New York, 1989, Chapter 5.
206. Fleetwood, D.M., Winokur, P.S., and Meisenheimer, T.L., Hardness assurance for low-dose space applications, *IEEE Trans. Nucl. Sci.*, 38, 1552, 1991.
207. Jeppson, K.O. and Svensson, C.M., Negative bias stress of MOS devices at high electric-fields and degradation of MNOS devices, *J. Appl. Phys.*, 48, 2004, 1977.
208. Blat, C.E., Nicollian, E.H., and Poindexter, E.H., Mechanism of negative-bias-temperature instability, *J. Appl. Phys.*, 69, 1712, 1991.
209. Ogawa, S. and Shiono, N., Generalized diffusion-reaction model for the low-field charge-buildup instability at the Si/SiO_2 interface, *Phys. Rev. B*, 51, 4218, 1995.
210. Schroder, D.K. and Babcock, J.A., Negative bias temperature instability: Road to cross in deep submicron semiconductor manufacturing, *J. Appl. Phys.*, 94, 1, 2003.
211. Zhou, X.J. et al., Negative bias-temperature instabilities in MOS devices with SiO_2 and SiO_xN_y/HfO_2 gate dielectrics, *Appl. Phys. Lett.*, 84, 4394, 2004.
212. Tsetseris, L. et al., Physical mechanisms of negative-bias temperature instability, *Appl. Phys. Lett.*, 86, 142103, 2005.
213. Ceschia, M. et al., Radiation induced leakage current and stress induced leakage current in ultra-thin gate oxides, *IEEE Trans. Nucl. Sci.*, 45, 2375, 1998.
214. Sexton, F.W. et al., Precursor ion damage and angular dependence of single event gate rupture in thin oxides, *IEEE Trans. Nucl. Sci.*, 45, 2509, 1998.
215. Sexton, F.W. et al., Single event gate rupture in thin gate oxides, *IEEE Trans. Nucl. Sci.*, 44, 2345, 1997.
216. DiMaria, D.J. and Cartier, E., Mechanism for stress-induced leakage currents in thin SiO_2 films, *J. Appl. Phys.*, 78, 3883, 1995.
217. Depas, M., Nigam, T., and Heyns, M.M., Soft breakdown of ultra-thin gate oxide layers, *IEEE Trans. Electron Dev.*, 43, 1499, 1996.
218. Ricco, B., Gozzi, G., and Lanzoni, M., Modeling and simulation of stress-induced leakage current in ultrathin SiO_2 films, *IEEE Trans. Nucl. Sci.*, 45, 1554, 1998.
219. Ralls, K.S. et al., Discrete resistance switching in submicrometer Si inversion layers: Individual interface traps and low-frequency ($1/f$?) noise, *Phys. Rev. Lett.*, 52, 228, 1984.
220. Kirton, M.J. and Uren, M.J., Noise in solid-state microstructures—a new perspective on individual defects, interface states and low-frequency ($1/f$) noise, *Adv. Phys.*, 38, 367, 1989.
221. Schuegraf, K.F. and Hu, C.M., Hole injection SiO_2 breakdown model for very-low voltage lifetime extrapolation, *IEEE Trans. Electron Dev.*, 41, 761, 1994.
222. Degraeve, R. et al., New insights in the relation between electron trap generation and the statistical properties of oxide breakdown, *IEEE Trans. Electron Dev.*, 45, 904, 1998.
223. Stathis, J., Percolation models for gate oxide breakdown, *J. Appl. Phys.*, 86, 5757, 1999.

224. Buchanan, D.A., Scaling the gate dielectric: Materials, integration, and reliability, *IBM J. Res. Dev.*, 43, 245, 1999.
225. Wilk, G.D., Wallace, R.M., and Anthony, J.M., High-*K* gate dielectrics: Current status and materials properties considerations, *J. Appl. Phys.*, 89, 5243, 2001.
226. Felix, J.A. et al., Charge trapping and annealing in high-κ gate dielectrics, *IEEE Trans. Nucl. Sci.*, 51, 3143, 2004.
227. Felix, J.A. et al., Effects of radiation and charge trapping on the reliability of high-κ gate dielectrics, *Microelectron. Reliab.*, 44, 563, 2004.
228. Shaneyfelt, M.R. et al., Challenges in hardening technologies using shallow-trench isolation, *IEEE Trans. Nucl. Sci.*, 45, 2584, 1998.
229. Lacoe, R.C. et al., Application of hardness-by-design methodology to radiation-tolerant ASIC technologies, *IEEE Trans. Nucl. Sci.*, 47, 2334, 2000.
230. Schwank, J.R. et al., Radiation effects in SOI technologies, *IEEE Trans. Nucl. Sci.*, 50, 522, 2003.
231. Jun, B. et al., Total dose effects on double gate fully depleted SOI MOSFETs, *IEEE Trans. Nucl. Sci.*, 51, 3767, 2004.
232. Nowlin, N. et al., A total-dose hardening-by-design approach for high-speed mixed-signal CMOS integrated circuits, in *Radiation Effects and Soft Errors in Integrated Circuits and Electronic Devices*, Schrimpf, R.D. and Fleetwood, D.M. Eds., World Scientific, Singapore, 2004, p. 83.

8

From 3D Imaging of Atoms to Macroscopic Device Properties

S.J. Pennycook, M.F. Chisholm, K. van Benthem, A.G. Marinopoulos, and Sokrates T. Pantelides

CONTENTS

8.1 Introduction

The quest for smaller, faster semiconductor devices is rapidly approaching the so-called "end of the roadmap" for silicon-based devices, in which the thickness of the gate oxide is becoming too small to maintain the macroscopic properties of that mainstay of the semiconductor industry, SiO_2. As the thickness approaches the size of the intrinsic ring structure of SiO_2, the electronic integrity of the gate oxide becomes compromised, and undesirable characteristics such as leakage and dielectric breakdown cannot be managed effectively. This situation is the driver for the search for alternative, high dielectric constant (high-k) gate structures. The introduction of new materials to augment a thin SiO_2 interlayer raises the possibility of new kinds of impurities entering either SiO_2 or the Si substrate, and that dopant atoms may enter new configurations, adversely affecting macroscopic device properties. It therefore becomes very valuable to be able to characterize these nanometer scale device structures with probes that are sensitive to individual atoms, ideally in three dimensions. At the same time, the ability to determine local electronic structure with similar resolution and sensitivity would reveal any local defect states introduced into the band gap as a result of these impurity or dopant atoms. In conjunction with theoretical calculations of electronic structure, it would then become possible to link the atomic-scale characterization of actual device structures to their macroscopic characteristics, and thus to find the true atomic origins of, for example, high leakage current or low mobility.

In this chapter, we describe the latest advances in aberration-corrected scanning transmission electron microscopy (STEM), which have in the last few years brought this vision to a practical possibility. The ability to correct for the primary aberrations of electron lenses represents a revolution in the field of electron microscopy. The rate of instrumental advance today is faster than at any time since the invention of the electron microscope in the 1930s, as shown in Figure 8.1 [1]. Aberrations occur inevitably in round lenses, as a result of physics, and not only from imperfections, as was recognized by Scherzer over 70 years ago [2]. The spherical aberration of the simple round lens has limited electron microscope resolution for most of its history, as was appreciated by Feynman in his 1959 lecture, [3] "There's Plenty of Room at the Bottom," where he explicitly called for a 100-fold improvement in resolution by overcoming spherical aberration, "why must the field be symmetric?" Designs for aberration correctors have existed for over 60 years [4], but they require multipole lenses, which makes such schemes enormously complex. It was not until the era of the fast computer and the efficient detection of the charge-coupled device that it has become practically possible to measure aberrations with the necessary sensitivity, $\sim$1 part in 10^7, and to control all the multipoles to the required accuracy. Focusing, or tuning, the aberration corrector involves optimization in a space of 40 or more dimensions, which has proved to be beyond human capability. Therefore, it is not so surprising that the successful correction of aberrations in electron microscopy has had to wait for the computer age.

Now, aberration correction has been successfully achieved both in the conventional transmission electron microscope (TEM) [5] and in the STEM [6]. In the STEM, direct imaging of a crystal lattice has been achieved at subangstrom resolution; see Figure 8.2 [7]. But aberration correction brings more than just a factor of 2 or 3 in resolution. As shown, in the STEM it also brings a greatly improved signal-to-noise ratio, resulting in the ability to image individual Hf atoms inside the nanometer-wide gate oxide of an advanced dielectric device structure [8]. Furthermore, since the lens aperture is much wider after

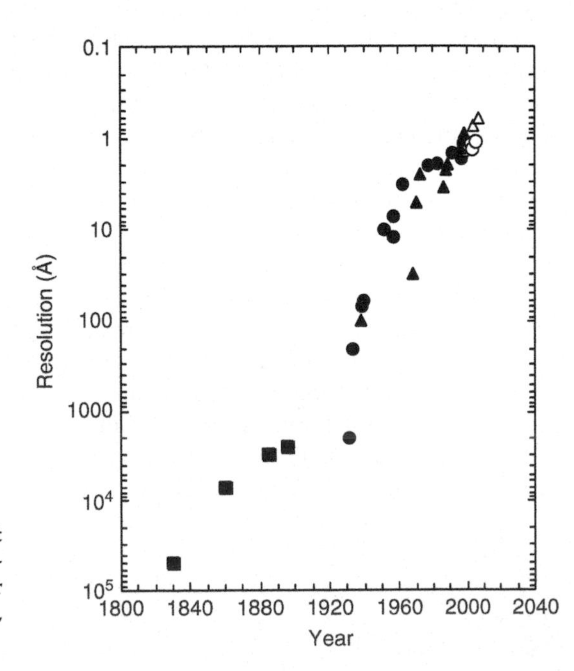

FIGURE 8.1
Evolution of resolution; squares represent light microscopy, circles TEM, triangles STEM, solid symbols before aberration correction, open symbols after correction. (After Rose, H., *Ultramicroscopy*, 56, 11, 1994. With permission.)

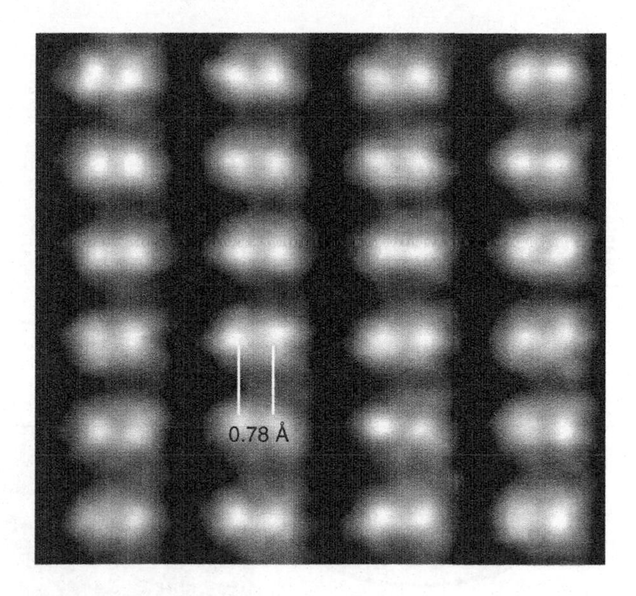

FIGURE 8.2
Direct imaging of a Si crystal in the <112> projection showing resolution of pairs of Si columns just 0.78 Å apart, obtained with a VG Microscopes HB603U dedicated STEM equipped with a Nion aberration corrector. (Adapted from Nellist, P.D. et al., *Science*, 305, 1741, 2004. With permission.)

aberration correction, the depth of field is much smaller, and it becomes possible to locate atoms in three dimensions to a precision better than 1 nm in depth and 0.1 nm in lateral position [9,10]. Three-dimensional mapping of heavy atoms inside SiO_2 has become possible, and in principle, three-dimensional spectroscopy should be feasible. With this level of sensitivity, new insights are available into the microscopic defects in device structures and their influence on device characteristics. From such knowledge, properties can be calculated from first principles, enabling the determination of the atomic-scale defects responsible for macroscopic device properties such as leakage and mobility, leading to improved device design.

8.2 Scanning Transmission Electron Microscopy

Figure 8.3 shows a schematic diagram of a STEM. As in any scanning microscope, the image is built up sequentially, pixel by pixel, as the probe is scanned across a specimen, and a variety of signals can be detected to form an image. The two most important imaging signals are the transmitted electron signal, which gives a bright-field image, and the scattered electron signal, which gives a dark-field signal. These two images are complementary, giving quite different information about the sample. The bright-field image can be arranged to give a high-resolution phase contrast image just as seen in a conventional TEM [11]. A phase contrast image is a coherent image formed by interference between diffracted beams, and fringes in the image can be black or white depending on whether the interference is constructive or destructive, which depends sensitively on parameters such as specimen thickness and microscope focus. Consequently there can be significant ambiguity on where the atoms actually are, and direct interpretation of atomic structure from the image is generally not feasible. Instead, simulations of trial structures are usually performed.

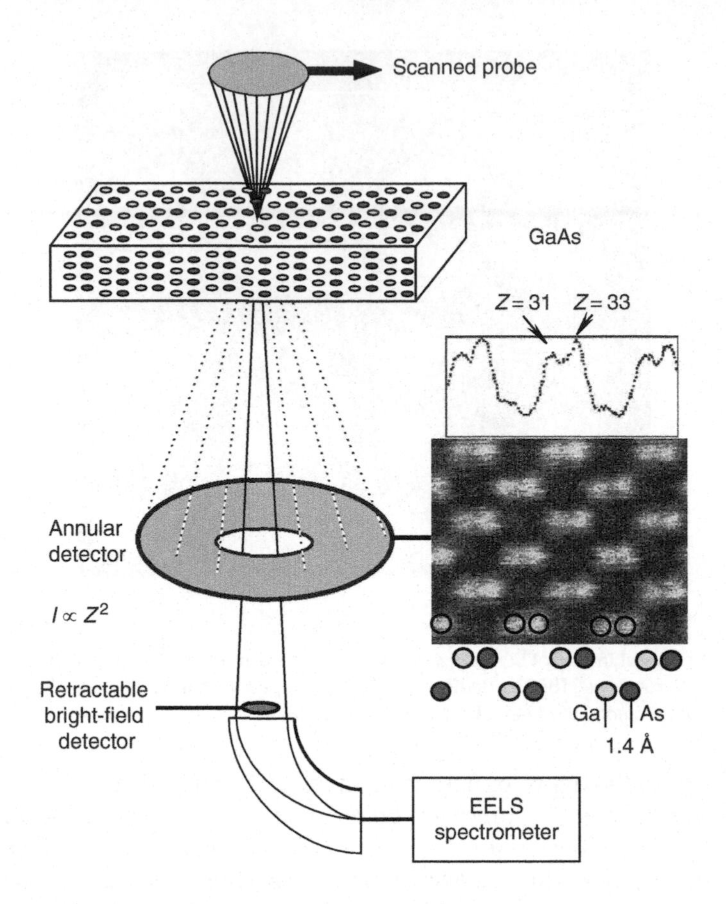

FIGURE 8.3
Schematic of a STEM. A finely focused probe is scanned over a specimen and an image is built up point-by-point. Simultaneous bright-field and dark-field images are possible or the bright-field detector may be replaced with an electron spectrometer to provide atomically resolved spectroscopic imaging. (Adapted from Pennycook, S.J. et al., *Philos. Trans. R. Soc. London, Ser. A*, 354, 2619, 1996. With permission).

The annular (or high-angle annular) detector gives an image of quite different character, which is essentially an atomic-resolution map of the specimen's scattering power [12,13]. The intensity of electrons scattered by a specimen is approximately proportional to its mean square atomic number (Z), and also its thickness, at least up to a few tens of nanometers, by which time most of the beam has been scattered and the image saturates. The annular dark-field (ADF) image is therefore an incoherent image, which has a directly interpretable relationship between object and image, just like the image formed with an optical camera or projector. Because of these characteristics, the image is often referred to as a *Z*-contrast image. High-*Z* atoms can be directly distinguished in the image, and unexpected configurations can be immediately apparent, properties that have been very useful in identifying many unexpectedly complex defect arrangements at semiconductor interfaces [14–20]. Until recently, however, *Z*-contrast images have been substantially noisier than the phase contrast images obtained in conventional TEM. This is because only a small current can be focused into a small spot, which is then scanned sequentially over the field of view. Conventional TEM images are obtained with a broad, high-current beam, with all image points being imaged in parallel.

In recent years major improvements in STEM instrumentation have occurred, including the successful incorporation of high-resolution STEM capabilities into commercial

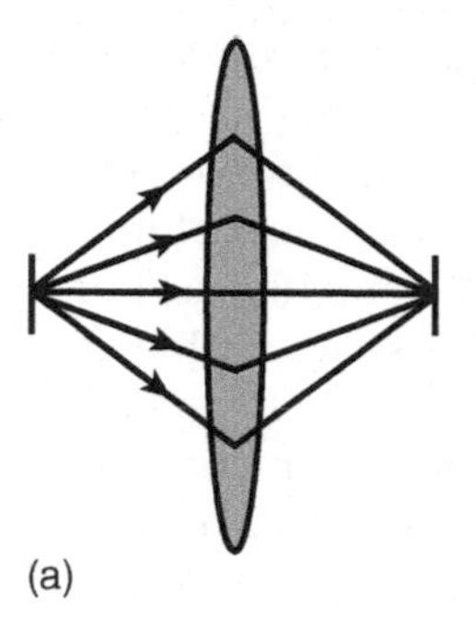

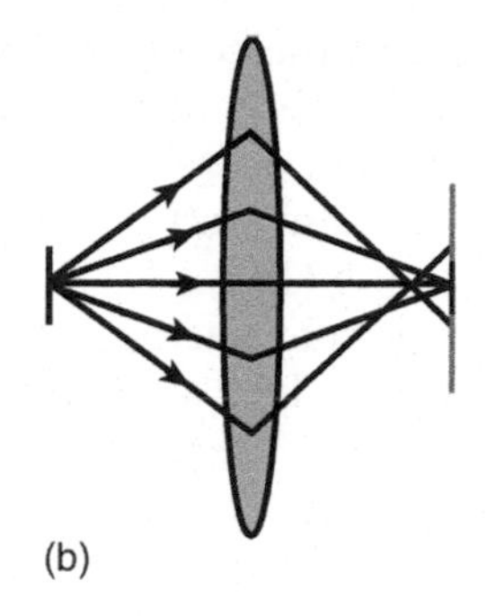

FIGURE 8.4
Schematic showing the action of spherical aberration in overfocusing high-angle rays. (Adapted from Varela, M. et al., *Annu. Rev. Mater. Res.*, 539, 2005, With permission.)

microscope columns with efficient detectors [21]. In particular, the development of aberration correctors has dramatically improved STEM performance [22,23]. For around 60 years, spherical aberration has been the major limitation to microscope resolution. As shown in Figure 8.4, spherical aberration is the tendency for high-angle rays passing through the lens to be overfocused, meaning that beyond a certain maximum aperture they are not usefully contributing to the focused spot but produce an extended tail in the image. Thus, spherical aberration results in a maximum effective aperture. Microscope resolution is proportional to λ/θ, where λ is electron wavelength and θ is the semiangle of the objective lens. By correcting for the lowest-order aberrations of the objective lens, the aperture can be opened up by a significant factor of between 2 and 3, which translates directly into improved resolution. However, not only has reducing the probe size enhanced the resolution, but also, the same current can now be focused into a smaller probe, giving a vastly higher peak intensity in the narrower beam. As a result, the signal-to-noise ratio in aberration-corrected images is greatly improved compared to uncorrected images, and the sensitivity to individual high-Z atoms is enormously increased [24–36].

Another key advantage of the STEM is that multiple detectors can be used to obtain complementary images with pixel-to-pixel correlation. Simultaneously acquired bright-field and dark-field images can be directly compared, although the optimum focus for each is slightly different so that a focal series of images may be necessary for quantitative work [37]. Figure 8.5 shows simultaneously acquired, aberration-corrected phase- and

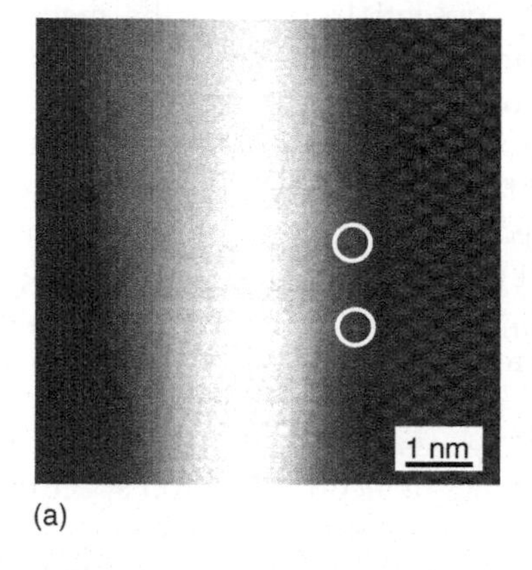

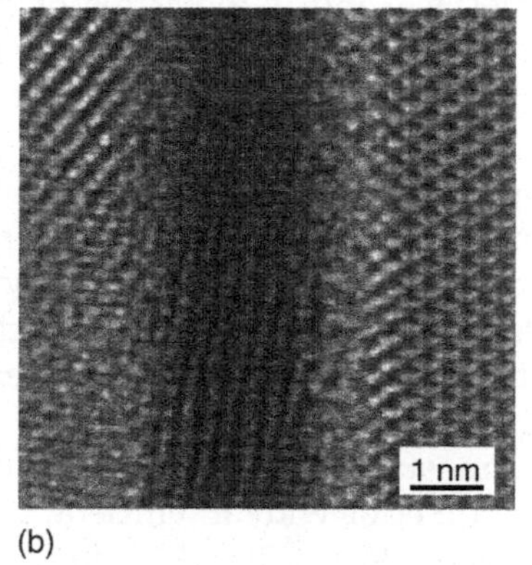

FIGURE 8.5
Simultaneously acquired (a) Z-contrast and (b) phase contrast images of a poly-Si/HfO_2/SiO_2/Si gate dielectric structure obtained with the aberration-corrected HB603U. (Adapted from Rashkeev, S.N. et al., *Microelectron. Eng.*, 80, 416, 2005; van Benthem, K., Rashkeev, S.N., and Pennycook, S.J., in *Characterization and Metrology for ULSI Technology*, American Institute of Physics, Richardson, TX, 2005, 79. With permission.)

Z-contrast images of a poly-Si/HfO_2/SiO_2/Si gate dielectric structure in which the different characteristics are very clear [38,39]. In the phase contrast image, the SiO_2 shows the speckle pattern typical of an amorphous material, caused by random interference effects within the structure, and the crystalline regions show lattice fringes. The Z-contrast image is dominated by the high scattering power of the HfO_2 ($Z = 72$ for Hf compared to 14 for Si and 8 for O). In the thin gate oxide layer there are small bright spots visible that represent single stray Hf atoms [8]. These are not detectable in the phase contrast image.

The exact position of the Si/SiO_2 interface is difficult to determine from the phase contrast image without extensive simulations, and it has therefore been proposed that microscopic measurements of dielectric layer thickness be determined from intensity traces across the Z-contrast images [40]. The incoherent nature of the Z-contrast image makes it reasonably immune to the influence of specimen thickness and interface roughness. An incoherent image is given by a mathematical convolution of a specimen object function with a resolution function of the optical instrument. In the case of ADF STEM, the resolution function is the intensity profile of the scanning probe, which is convoluted with the specimen scattering power. A useful concept is that of an effective probe, which is the probe profile integrated through the thickness of the specimen. In a crystal aligned to a major zone axis, the probe can channel along the columns. At the same time some intensity is scattered to high angles leading to probe broadening. In the amorphous oxide there is no channeling, but beam broadening remains. Because of the change in beam channeling, the effective probe will change somewhat as it is scanned across the Si/SiO_2 interface. However, these effects are generally small and normally ignored. The method proposed for thickness determination assumes the same effective probe in the crystal and in the oxide. For any radially symmetric probe profile, a convolution with a step function (for example, the abrupt change in intensity from Si to SiO_2) will lead to a blurring of the image intensity, with the point of inflection locating the interface. Hence, differentiating the intensity trace will regenerate the form of the effective probe. If now the interface has structural roughness, this will further broaden the intensity profile. However, the point of inflection will remain as the mean interface position, which is why the method is relatively immune to details of the specimen and microscope parameters.

The above procedure is illustrated in Figure 8.6. Differentiating the ADF intensity profile gives a peak at the interface position. Note, however, that the derivative trace is somewhat asymmetric. This illustrates the different channeling conditions on either side of the interface, and is indicative of the potential errors in the method. Also, the width of the derivative peak, ~5 Å, is much broader than the width of the probe, which is definitely of the order of 1.3 Å, since it is resolving the Si dumbbell separation at 1.26 Å. This is probably mostly due to interfacial roughness; however, even abrupt interfaces typically show a transition in intensity that is broader than expected from a simple convolution model. Such behavior is not predicted even by sophisticated image simulations. The problem is an unexplained background intensity in the images, which is higher for higher-Z materials, and the background intensity does not change abruptly across the interface. This issue is often referred to as the Stobbs factor, and is yet to be resolved [41].

Perhaps the most important simultaneous signals in STEM are obtained by replacing the bright-field detector with an efficient, parallel detection, electron spectrometer. Then, it becomes possible to use the ADF image to locate atomic planes or columns and to perform spectroscopy from specific columns selected from the image. The first atomically resolved spectroscopic analysis was demonstrated in 1993, in which a characteristic Co edge was measured plane-by-plane across a $CoSi_2$/Si interface [42]. The drop in measured Co concentration in moving from the last plane of the silicide to the first plane of the Si well exceeded the 50% points required to demonstrate atomic resolution. Later,

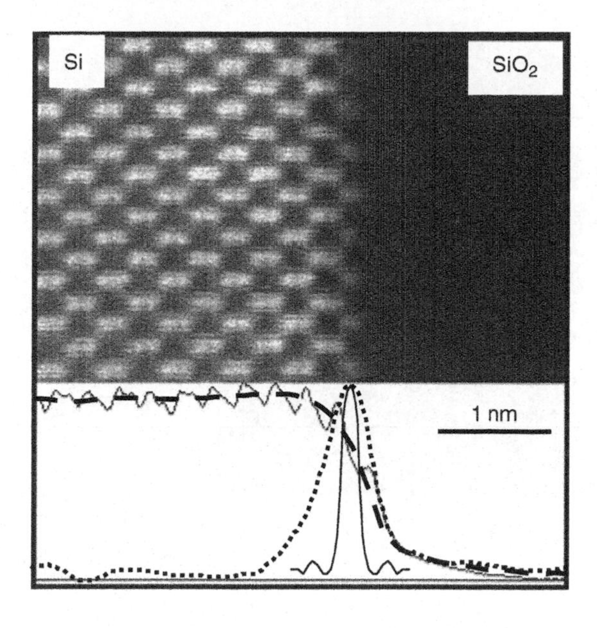

FIGURE 8.6
Z-contrast image of an Si–SiO_2 interface obtained with the uncorrected HB603U STEM. Its probe size of about 0.13 nm is still sufficiently small to resolve the Si dumbbells. A vertically averaged line trace (grey) is fitted with a smooth curve (dashed line), differentiated (dotted line), and the width can then be compared to the theoretically calculated probe profile (solid line). Reasons for the increased interface width can be interfacial roughness, and also probe broadening effects. (Reproduced from Diebold, A.C. et al., *Microsc. Microanal.*, 9, 493, 2003. With permission.)

column-by-column spectroscopy became common [43–48], but recently, with the incorporation of aberration correctors into STEM columns, the improvements have become even more spectacular. First was the spectroscopic identification of an individual atom inside a bulk solid as shown in Figure 8.7, [49] and second, the recent achievement of true, two-dimensional atomic-resolution maps [50], as opposed to point spectra obtained from columns chosen from the Z-contrast image, both achievements due to the higher signal levels available after aberration correction.

Not only does electron energy loss spectroscopy (EELS) provide compositional information from the energies of characteristic absorption edges, but also the fine structure on those edges gives information on local electronic structure in a similar manner to x-ray absorption spectroscopy (XAS). Figure 8.8 shows spectra characteristic of bulk Si and SiO_2, and of the (rough) interface between them which shows additional peaks due to suboxide bonding [51]. STEM and EELS have developed into a powerful tool for the atomic and electronic characterization of interfaces in device structures [52–59]. Compared with XAS, EELS has the benefit of spatial resolution at the atomic level, and can probe changes that occur as the probe is moved across an interface, whereas XAS provides an average measurement. The problem, as always, is that EELS has poorer count statistics and so quantitative analysis is always limited by signal-to-noise.

Whereas the motivation for aberration correction was primarily the resolution, implicitly the lateral resolution, a somewhat unanticipated advantage has been the gain in depth resolution, which is given by $2\lambda/\theta^2$. So a factor of 2 or 3 improvement in the transverse resolution brings also a factor of 4–9 in depth resolution. We have moved from a situation where the depth of field of the microscope is typically greater than the specimen thickness (the image is a projection of the specimen) to a situation where the depth of field is usually

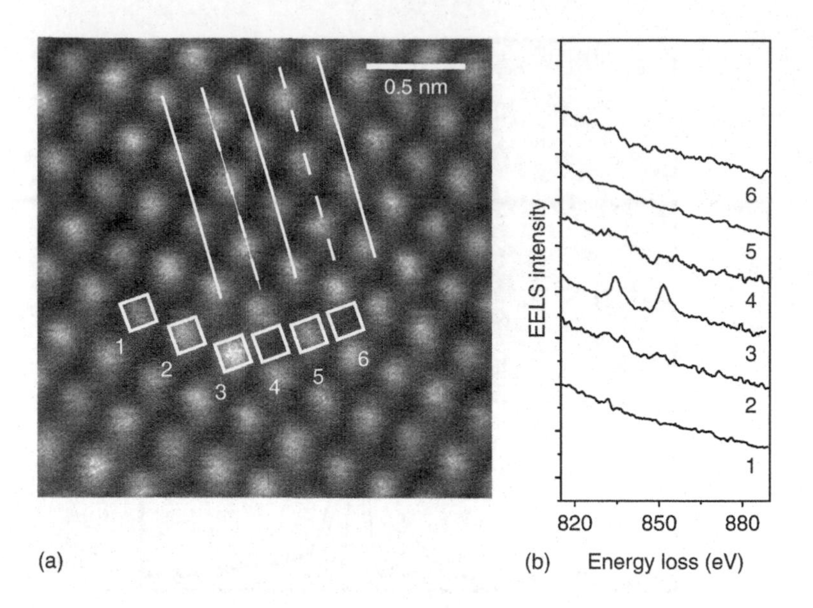

FIGURE 8.7
Spectroscopic identification of a single La atom in a column of Ca in the perovskite $CaTiO_3$. (a) *Z*-contrast image of $CaTiO_3$ showing traces of the CaO and TiO_2 {100} planes as solid and dashed lines, respectively. A single La dopant atom in column 3 causes this column to be slightly brighter than other Ca columns, and EELS shows a clear La $M_{4,5}$ signal (b). Moving the probe to adjacent columns gives reduced or undetectable signals. Results obtained using a VG Microscopes HB501UX STEM equipped with a Nion aberration corrector. (Adapted from Varela, M. et al., *Phys. Rev. Lett.*, 92, 095502, 2004. With permission.)

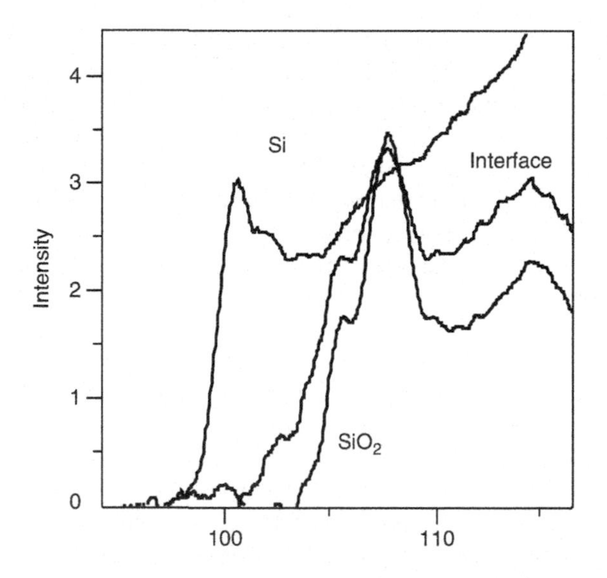

FIGURE 8.8
Si L EELS edges obtained from an Si/SiO_2 interface compared to spectra from Si and SiO_2 far from the interface. Due to the high band gap in SiO_2 the edge onset is shifted higher in energy. The spectrum from the interface shows the presence of states in the band gap of SiO_2 indicating suboxide bonds. (Reproduced from Pennycook, S.J. et al., *Encyclopedia of Materials: Science and Technology*, Elsevier Science Ltd., Kidlington, Oxford, 2001 p. 1. With permission.)

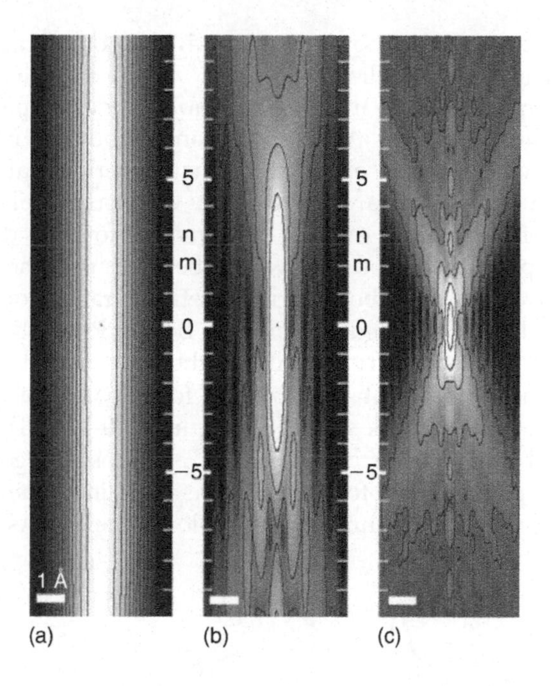

FIGURE 8.9
Probe intensity distributions for three generations of STEM, (a) an uncorrected VG Microscopes HB501UX 100 kV STEM with depth of field ~45 nm, (b) the aberration-corrected HB603U 300 kV STEM with depth of field ~3.7 nm, and (c) for a next generation fifth-order aberration-corrected 200 kV microscope with predicted depth of field ~1 nm. Intensities are shown on a logarithmic scale and curves have been shifted vertically to locate maxima at zero defocus. (Reproduced from Borisevich, A.Y., Lupini, A.R., and Pennycook, S.J., *Proc. Nat. Acad. Sci. USA*, 103, 3044, 2006. With permission.)

smaller than the specimen thickness, as illustrated in Figure 8.9. This means that we are able to optically section the specimen, rather as in a confocal optical microscope. A through-focal series of images becomes a depth sequence of images, which can be reconstructed to obtain useful information on the three-dimensional structure of the object [9,10]. The sectioning is opposed by the tendency of a crystal to channel the electrons along atomic columns [60], and so the method works best in amorphous materials or, if aligned crystals are used, the columns should be weakly channeling, i.e., low Z.

For more detailed discussions on STEM and the benefits and applications of aberration correction the reader is referred to several recent reviews [32,34,37,61–63].

8.3 Theoretical Microscope

Density-functional total-energy calculations provide valuable complementary insight into atomic-scale structure and the only quantitative link to macroscopic properties. Theory can validate impurity or defect configurations seen in the microscope as stable, low-energy structures, and provide formation and migration energies, predict local electronic structure for comparison with EELS data, and, as we show later, even predict transport properties. This joint approach to understand structure and property relationships has proved fruitful in many semiconductor systems. In some cases theoretical predictions stimulate experimental observations, while in other cases experiment comes first and theory provides the rationalization.

As an example of the additional insights available from complementary theory and experiment, we present the case of As-doped grain boundaries in Si. This began as a theoretical study motivated by the observation that experimental segregation energies for As in Si grain boundaries are around 0.5 eV; whereas, available theory predicted segregation energies of around 0.1 eV [64]. The reason for this discrepancy was soon found to be the assumption that impurities would be present in the boundary as isolated atoms. It was well known that grain boundaries in tetrahedrally coordinated semiconductors were fully rebonded, i.e., contained no dangling bonds. Priory theory assumed that As is incorporated in the grain boundary just like in the bulk, i.e., at isolated substitutional sites. The new hypothesis was that As dopants form pairs, each atom obtaining its preferred threefold configuration, as shown in Figure 8.10 [65]. The segregation energy per As atom increased threefold for certain sites. Z-contrast images of As in a Si grain boundary (see below) confirmed the formation of As pairs in certain sites, as predicted. It is the environment of the grain boundary that allows the As–As distance to increase,

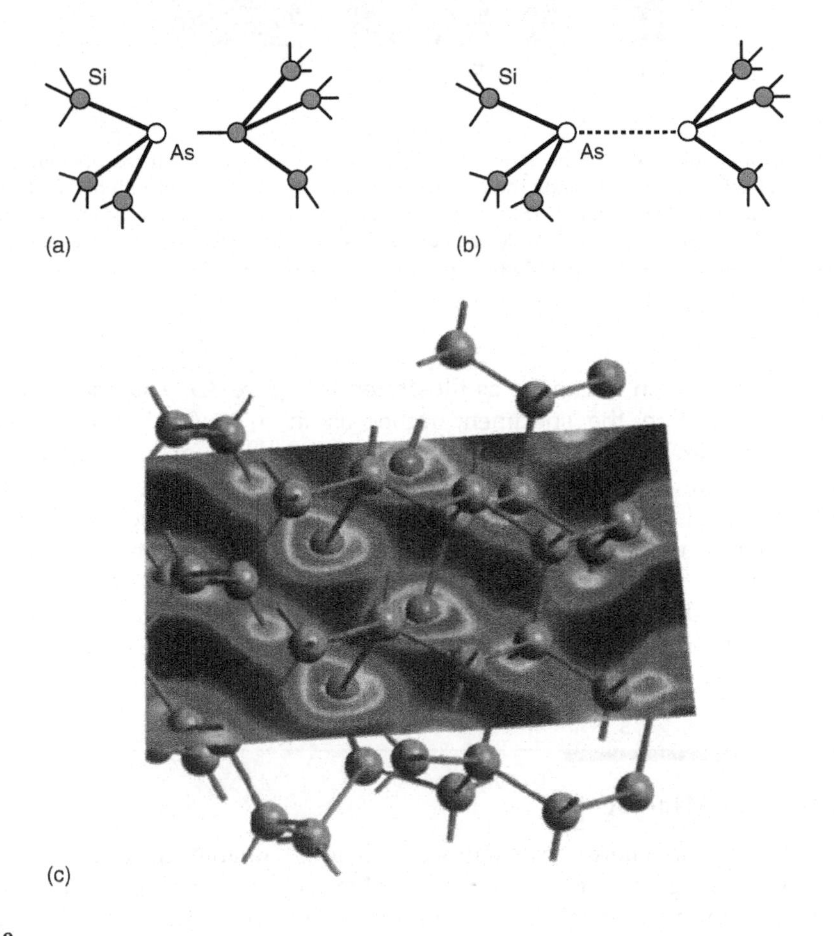

FIGURE 8.10
(a) Arsenic (As) prefers threefold coordination, so isolated As dopants create Si dangling bonds. (b) As dimers, however, can lower their energy by relaxing away from each other leaving no dangling bonds. This can only be accomplished in the grain boundary environment. (c) Plot showing the reduced charge density between the As dimers compared to between Si atoms. (Adapted from Maiti, A. et al., *Phys. Rev. Lett.*, 77, 1306, 1996. With permission.)

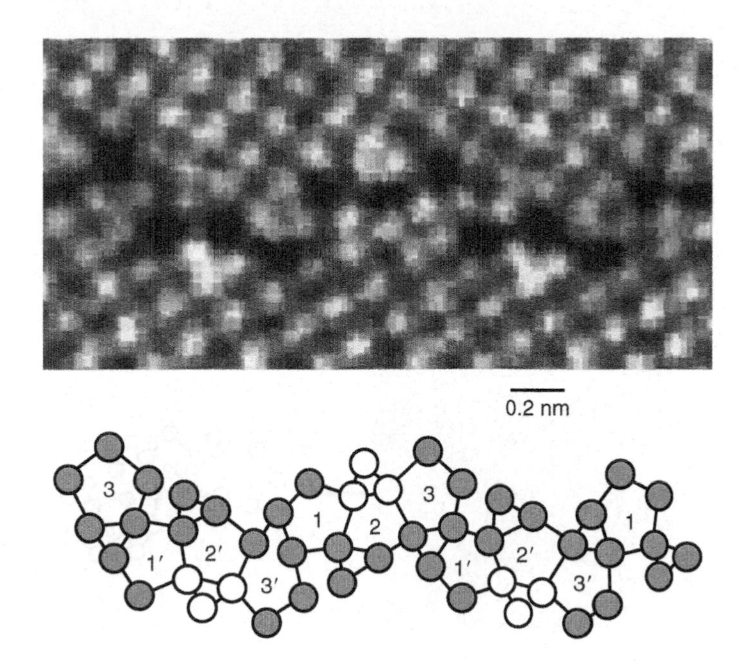

FIGURE 8.11
Z-contrast image of a 23° $\Sigma = 13\{510\}$ <001> tilt grain boundary in Si, with schematic diagram showing the boundary to comprise three dislocation cores which are then mirrored across the boundary plane. As is seen to segregate only to one of the three dislocation cores, seen as bright triangles in the image. Results obtained with the uncorrected HB603U. (Reproduced from Chisholm, M.F. et al., *Phys. Rev. Lett.*, 81, 132, 1998. With permission.)

allowing the repulsive energy of near neighbor As atoms to be avoided, an example of cooperative chemical rebonding. In the bulk crystal there is no such elastic relaxation. In fact, the energy of a dimer is slightly higher than that of two isolated atoms. Figure 8.10c shows the calculated charge density between the two atoms of an As dimer, showing it to be significantly reduced compared to that in the Si–Si bond.

The theory was performed on the lowest period grain boundary to minimize computational cost, a symmetric $\Sigma = 5\{310\}$ <001> tilt grain boundary, which has a misorientation of 36°. Later, experimental observations were made on a lower-angle grain boundary [66], a symmetric 23° $\Sigma = 13\{510\}$ <001> tilt grain boundary. The structure of this boundary was found to be surprisingly complex, as shown in Figure 8.11, comprising a contiguous array of six dislocation cores, a perfect edge dislocation (1), and two perfect mixed type dislocations (1,2) arranged as a dipole (i.e., their dislocation content cancels and they could be replaced by perfect crystal without changing the misorientation of the grain boundary). The same sequence is then mirrored across the boundary plane, denoted by primed numbers. The presence of As is revealed as a higher intensity in one of the two mixed type dislocation cores, repeated periodically along the boundary structure. Clearly, only two of the six cores forming the boundary structure contain As. The extra intensity in the image was about 18%, indicating an As concentration of about 5% per column, or about two As atoms in the specimen thickness. This image was obtained before aberration correction, and so single atom sensitivity was not possible for atoms of such low Z (33 for As compared to 14 for Si).

Theory was then performed on this more complex 23° boundary structure, but as with the 36° boundary, segregation energies for isolated As atoms never exceeded 0.1 eV. Although calculations for truly isolated As dimers could not be carried out because the required supercell would be too large, calculations for As chains (a continuous line of As

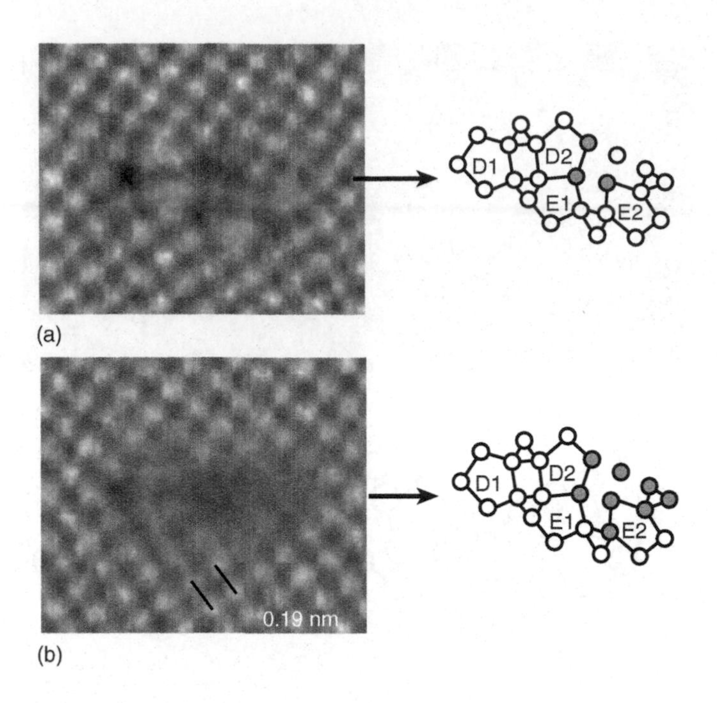

FIGURE 8.12
Z-contrast image of a 16° $\Sigma = 25\{710\}$ <001> tilt grain boundary in Si and derived structures with increasing electron irradiation. (a) With little electron irradiation only the three columns shaded show reduced intensity due to damage. (b) With increasing irradiation the damage spreads. Results obtained with the uncorrected HB603U. (Adapted from Maiti, A. et al., *Appl. Phys. Lett.*, 75, 2380, 1999. With permission.)

dimers) showed that only in core 2 was there significant segregation energy, in full accord with experiment. The theory showed further the reason for this. In cores 1 and 3, the relaxation of a dimer caused rotation toward the tilt axis, pushing the As atoms toward the neighboring dimer. In core 2, the relaxation rotated the dimer away from the tilt axis, allowing the As–As separation to increase, thereby lowering the energy. Thus the theory provides a perfect rationalization of the experimental observations, showing how it is the fine details of the boundary geometry that controls segregation, and specific sites can be favored to a significant degree.

Another example where theory rationalized microscopic observation is in the generation of vacancies by the fast electron beam. The observations were made on a lower-angle 16° $\Sigma = 25\{710\}$ <001> tilt grain boundary where the dislocations formed isolated, but still complex arrangements, as shown in Figure 8.12 [67]. Again, cores D1 and D2 represent a dipole, their dislocation content canceling, while cores E1 and E2 are mixed dislocations. With electron irradiation the intensity of specific columns shaded in the schematic decreased, indicating atomic displacements, the onset of amorphization. It was clear that the damage initiated in the vicinity of core E1. To explain these observations vacancy and interstitial formation energies were computed for various configurations. In the case of vacancies, all configurations in the boundary had much lower formation energies than in the bulk, as expected, but in particular, all the low-energy defect structures indeed occurred in the vicinity of the region seen dark in the micrograph. On the contrary, with interstitials, no correlation was found to the dark region of the image. It is clear therefore that the irradiation induces stable vacancy complexes; whereas, the ejected interstitials diffuse rapidly away.

8.4 Structure and Properties of $Si/SiO_2/HfO_2$ Gate Stacks by Aberration-Corrected STEM and Theory

One of the more promising routes to overcome the potential "end of the roadmap" for shrinking Si-based devices is to move to alternative gate dielectrics with higher dielectric constant, the so-called high-*k* dielectrics, of which HfO_2 is perhaps the most widely studied material (Chapters 9 through 14). The high-*k* gives a smaller effective dielectric thickness, but introduces new concerns. Key issues include whether the Si/SiO_2 interface remains well-passivated or whether Hf atoms would segregate to the interface and degrade its characteristics through introduction of localized states, either directly or through introduction of dangling bonds. Localized states could act as scattering centers for electrons moving in the channel, reducing mobility, or, at a sufficiently high density, might enable leakage through the thin SiO_2 layer. To investigate these effects, aberration-corrected STEM combined with theory is emerging as a powerful combination, providing the ability to locate single Hf atoms in three dimensions within the gate oxide, and through theory, to predict macroscopic properties.

One of the most important advantages of the short depth of focus in the aberration-corrected microscope is with the probe focused inside the specimen; atoms on the top and bottom surfaces are out of focus, and hence invisible. Therefore, a through focal series ensures that Hf atoms inside the actual device structure are indeed being examined, with no possibility of the atoms being part of any surface layers produced during specimen preparation. For the present study, HfO_2 films approximately 3 nm thick were deposited on silicon substrates by atomic layer deposition, and then rapid thermally annealed at 950°C for 30 s in N_2. The films were then capped with undoped polycrystalline Si. Samples were prepared purely by mechanical polishing to avoid damaging the thinnest regions by ion milling. Part of a through focal series of images is shown in Figure 8.13, in which a bright spot is clearly seen at a defocus of zero and -5 Å, but fades at lower or higher defocus. A line trace across such an image is shown in Figure 8.14, revealing that the excess intensity of the Hf atom is comparable to that of an entire column of Si atoms, which is in agreement with image simulations [8,10]. This rules out the possibility that the spots could be due to a lighter element such as Zr, a known possible impurity in these systems. It is also apparent that the intensity across the oxide is highly nonuniform, which represents the out of focus contributions of other Hf atoms and particularly the HfO_2 at other depths. Because of this out of focus background, the atoms are only visible when in sharp focus. Figure 8.15 shows plots of the excess intensity extracted from such line traces for a few representative atoms as a function of focus, or atom depth. The plot clearly shows how the depth of the Hf atoms can be determined to a precision of much better than ±0.5 nm, just by locating the position of the peak intensity. For atoms that are widely separated, as in the present example, the depth precision therefore significantly exceeds the expected depth resolution. The specimen thickness is assumed to be approximately equal to the greatest difference in depth found between Hf atoms, in this case 6 nm.

A total of 65 atoms were analyzed in this focal series which were randomly distributed in depth. Interestingly, their lateral position was not found to be random. The lateral precision is an order of magnitude greater than the depth precision, at a level of 0.1 nm or higher limited primarily by specimen drift. Figure 8.16a shows an image with positions of some Hf atoms marked, along with a histogram of the lateral positions with respect to the interface plane. No Hf atoms are present right at the interface, and they prefer to sit at specific distances from the interface, around 2.5 and 4 Å.

Now it is obvious that a transition from crystalline Si to a truly disordered oxide cannot occur instantaneously, and some order is therefore expected right near to the last Si plane.

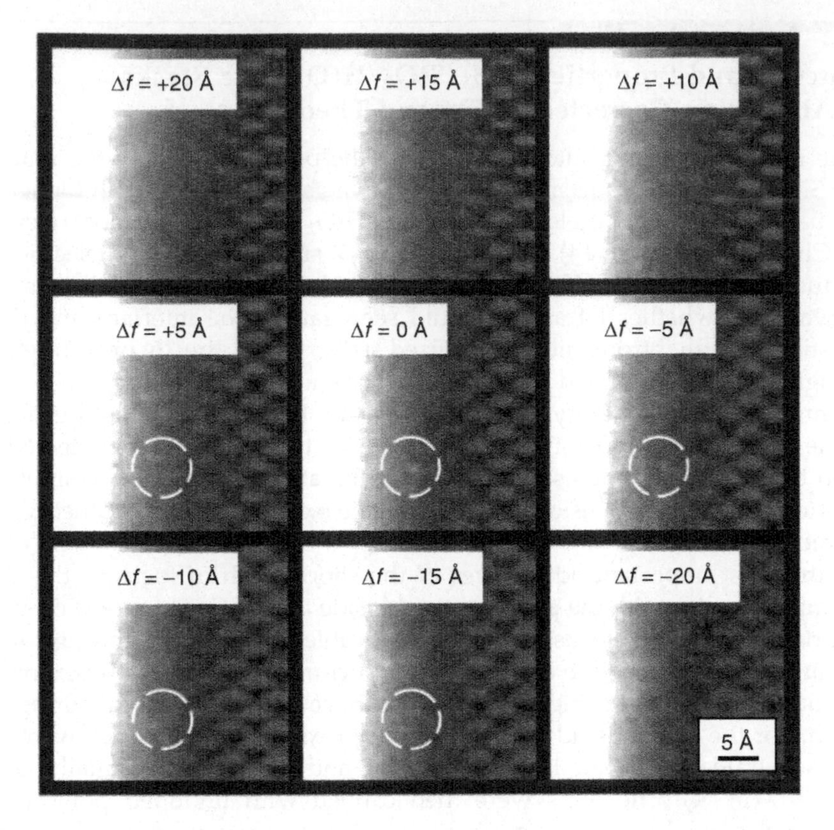

FIGURE 8.13
Sequence of frames from a through-focal series of Z-contrast images of an $Si/SiO_2/HfO_2$ high-*k* device structure showing an individual Hf atom coming in and out of focus (circled). Results obtained with the aberration-corrected HB603U. (Adapted from van Benthem, K. et al., *Appl. Phys. Lett.*, 87, 034104, 2005. With permission.)

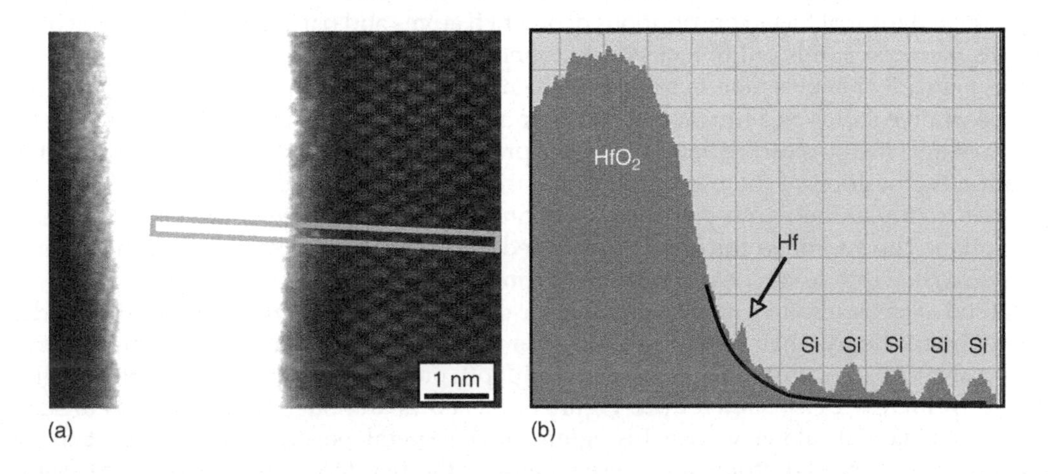

FIGURE 8.14
(a) One frame from the through-focal series showing a single atom with an intensity profile taken from the rectangular box (b) showing a large nonuniform background contribution. (Adapted from van Benthem, K. et al., *Ultramicroscopy*, 106, 1062, 2006. With permission.)

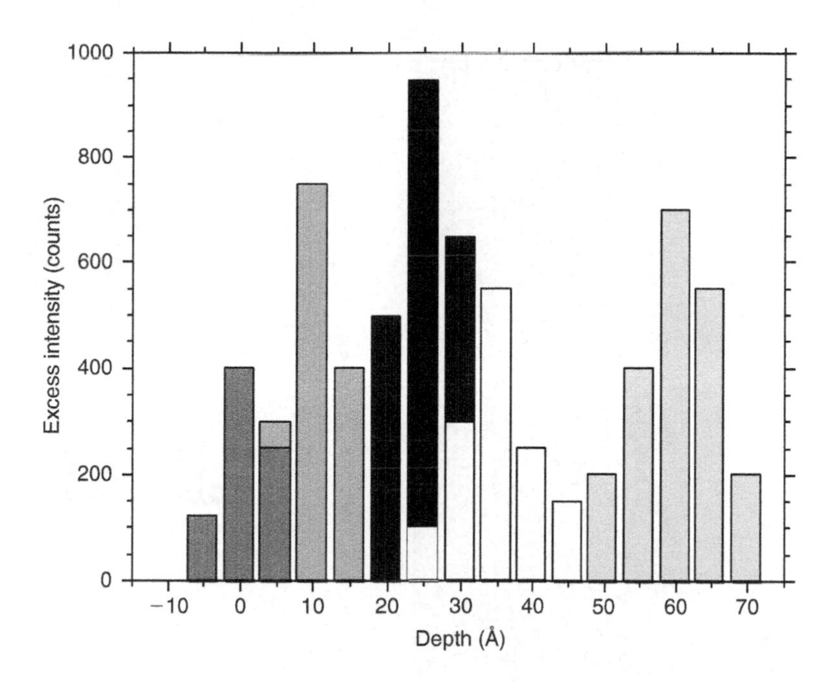

FIGURE 8.15
Plot showing the excess intensity from several individual Hf atoms as a function of depth. (Adapted from van Benthem, K. et al., *Ultramicroscopy*, 106, 1062, 2006. With permission.)

None is apparent in the Z-contrast image, however, and so we turn to theory. Several studies of the Si/SiO_2 interface have been carried out, and here we compare the experimental data to two studies, the first to a study without Hf using classical potentials [68] with a large number of atoms, and the second to density functional calculations including Hf, using by necessity a smaller number of atoms [69]. The classical potential calculations give a fair reproduction of the experimental Hf histogram (see Figure 8.16b and c), therefore suggesting that the Hf atoms substitute for Si in the oxide. This was tested explicitly by density functional theory using various crystalline models for the interface studied previously. In that work it was found energetically preferable to have abrupt interfaces, avoiding suboxide bonds [69]. The rationalization of this result was that Si–Si bonds are stiff, and the structural mismatch at the interface can be better accommodated by inserting an oxygen atom into any Si–Si bond that may protrude into the oxide. A Si–O–Si bond has a soft angle at the O atom that allows the bond angle to vary, as shown in Figure 8.17. Two particular oxide structures were found to have the lowest energy, tridymite-like bonding and quartz-like bonding. Quartz-like bonding results in tension in the oxide, in the plane of the interface, whereas tridymite-like bonding results in compression. It was therefore concluded that the actual interface structure would most likely consist of an intimate phase mixture of these two types of bonding to avoid any long-range elastic strains. We therefore looked at the two models to extract the distance from the interface plane to the first Si atoms, and found 2.3 and 4.4 Å for the quartz-like structure, and longer distances for the crystobalilte-like structure, 2.9 and 5.4 Å. Inserting these distances onto a histogram, we find that the result is also in reasonable agreement with the data (Figure 8.16d). Furthermore, the theory directly confirmed that substitutional Hf is preferred over interstitial Hf, by at least 4 eV [70]. Interstitial Hf results in disruption of the oxide lattice, because the Hf atoms steal O atoms from the Si leading to the creation of dangling bonds. Comparing the energy of substitutional Hf in different

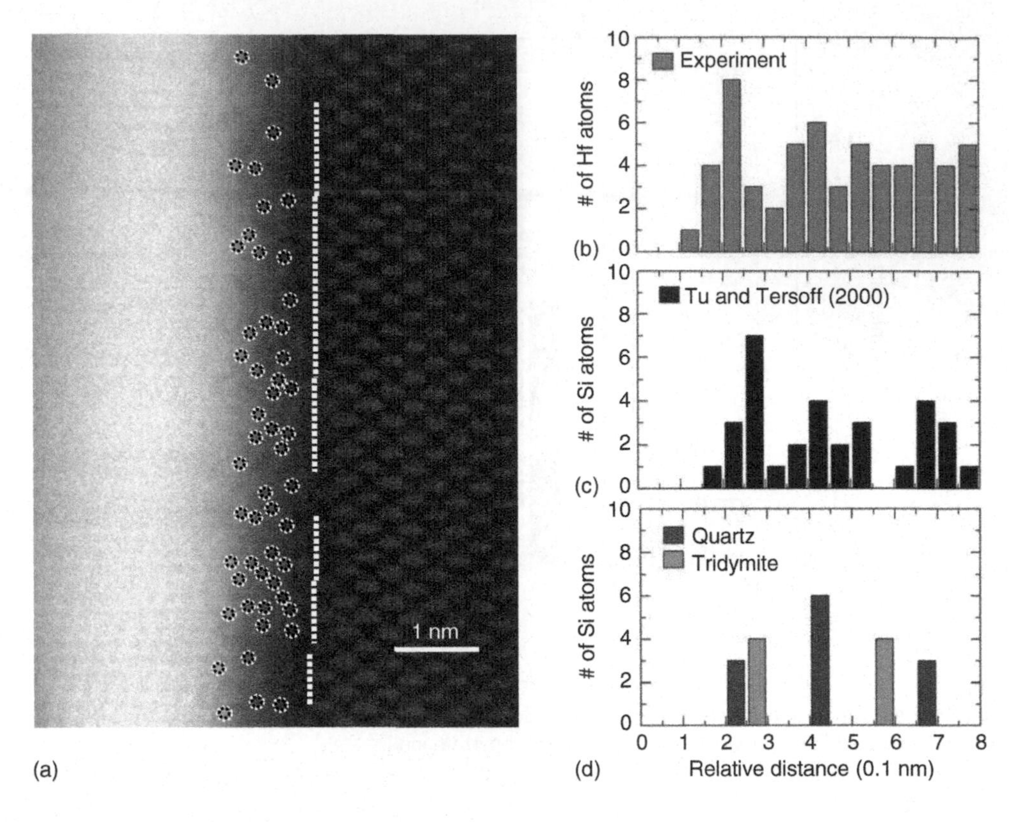

FIGURE 8.16
(a) Image of the device structure with Hf atom positions superimposed. No Hf atoms are found at the Si/SiO_2 interface. (b) Histogram showing the distance of Hf atoms from the Si/SiO_2 interface compared to (c) Si positions obtained by classical potential calculations (courtesy J. Tersoff) and (d) first-principles calculations of model abrupt interfaces to crystalline SiO_2. (Adapted from Marinopoulos, A.G. et al., *Phys. Rev. B77*, 195317, 2008. With permission.)

sites also explains why the Hf atoms keep away from the interface. Figure 8.18 shows a plot of the energy of formation for substitutional Hf, taking the zero to be in bulk Si, which shows how the Hf prefers the more open oxide structure, where it can have O neighbors, and even how it prefers to stay away from the interface plane by a significant energy, 1.4 eV. Also, since the source of the Hf atoms is the HfO_2, we would not expect Hf to migrate into the Si substrate. Thus, the theory completely explains all the

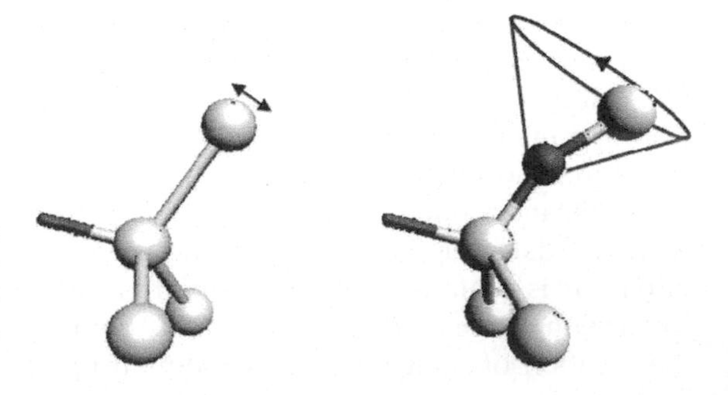

FIGURE 8.17
Schematic diagram showing the additional degree of freedom obtained with an Si–O–Si bond compared to the stiff Si–Si bond. (Reproduced from Buczko, R., Pennycook, S.J., and Pantelides, S.T., *Phys. Rev. Lett.*, 84, 943, 2000. With permission.)

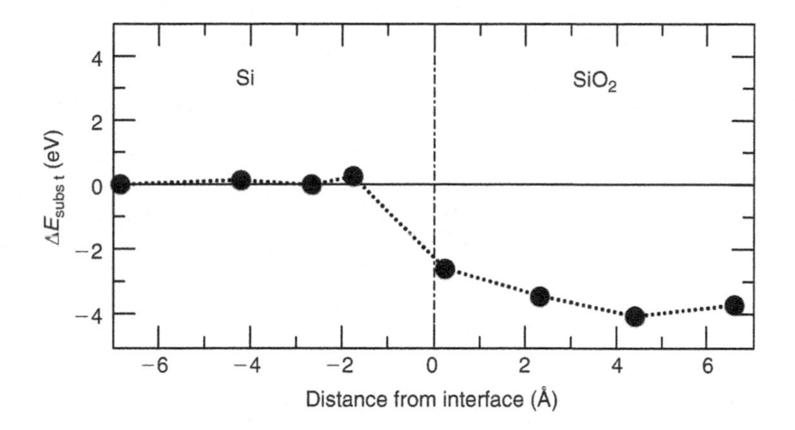

FIGURE 8.18
Plot showing the formation energy for substitutional Hf atoms across the interface. Hf atoms at the interface have 1.4 eV higher energy than those in the oxide, explaining why none were found in that position. (Adapted from Marinopoulos, A.G. et al., *Phys. Rev. B77*, 195317, 2008. With permission.)

experimental observations leading to a detailed microscopic explanation of the structure and energetics of the system.

Theory also gives insight into the electronic properties of the material. In Figure 8.19, we plot the densities of states around two interstitial configurations of Hf showing how they overlap the Si band gap and are therefore electrically active. Also shown is the wave function of the 4d localized state, along with the wave function of the bottom of the Si conduction band, showing a significant lateral extent overlapping the neighboring O and Si atoms. Substitutional Hf showed similar behavior. From the data shown in Figure 8.16, knowing the coordinates of all the atoms it is simple to deduce their density, 1.4 nm^{-3} (1.7×10^{14} cm^{-2}), and hence determine their average separation to be about 7 Å. It is clear

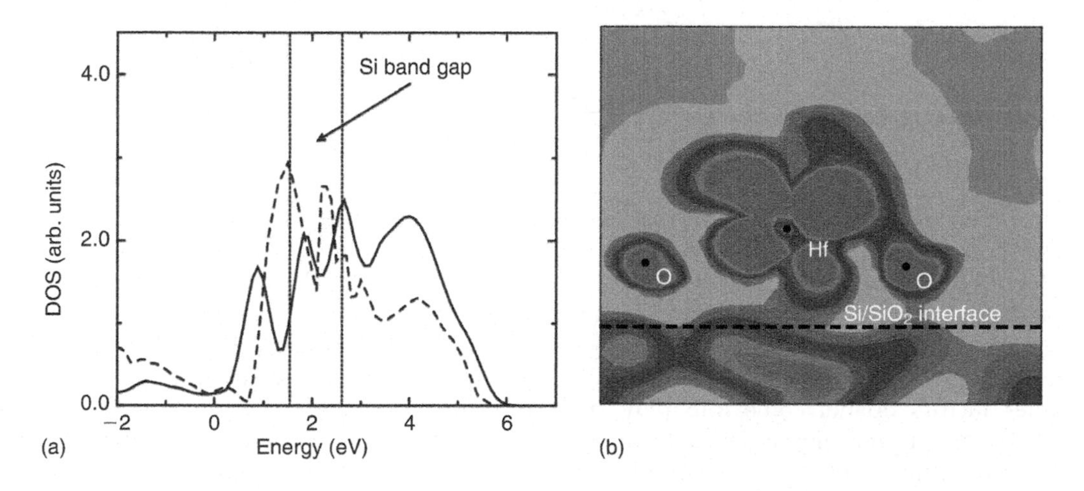

FIGURE 8.19
(a) Densities of states around two interstitial Hf atoms overlap the Si band gap. (b) Plot of the Hf 4d localized state, and the wave function of the bottom of the Si conduction band, showing a significant lateral extent. (Adapted from Pantelides, S.T. et al., *NATO Advanced Research Workshop on Defects in High-K Dielectrics and Nano-Electronic Semiconductor Devices*, Russia, 2005. With permission.)

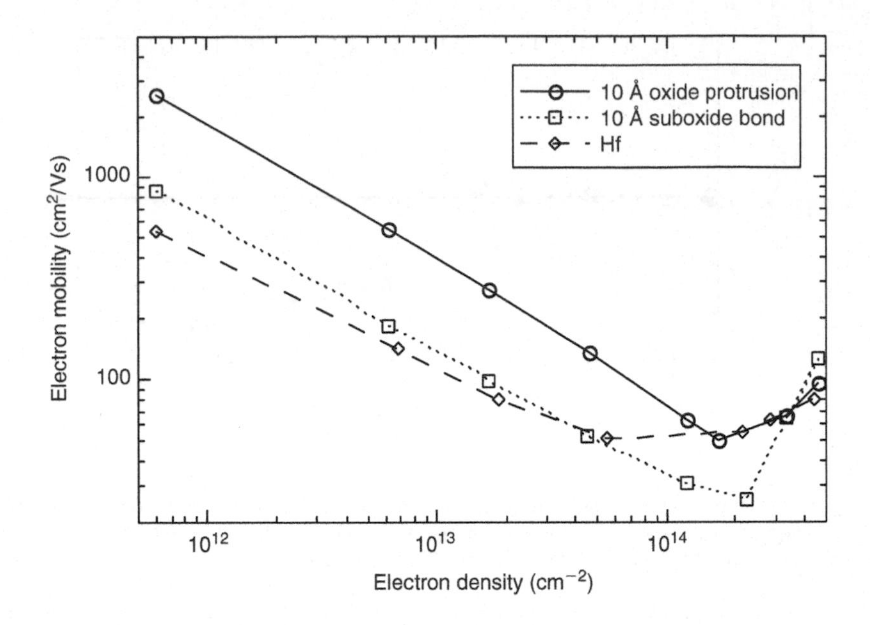

FIGURE 8.20
Electron mobilities for a variety of defects at an Si/SiO_2 interface calculated by first principles. (Data taken from Evans, M.H. et al., *Phys. Rev. Lett.*, 95, 106802, 2005; Evans, M.H. et al., *IEDM Tech. Dig.*, 597, 2005. With permission.)

that at this density their wave functions could overlap significantly and explain the high leakage observed in this particular sample.

Mobilities can also be determined from first principles, by calculating a scattering potential as the difference between a perfect interface and one containing the defect of interest, a suboxide bond, an oxide protrusion, and a Hf atom [71]. Results for these defects are compared in Figure 8.20, and it is clear that the Hf atoms do represent a large potential contribution to mobility reduction in these thin gate oxide structures [72].

8.5 Future Directions

The aberration-corrected results presented here were all using third-order correctors, which correct for geometric aberrations up to third order, leaving the resolution-limiting aberrations to be of fifth order. However, correctors are in development that will correct the fifth-order aberrations, leaving resolution-limiting aberrations to be of seventh order, which would allow roughly another factor of 2 in aperture angle [73,74]. We can look forward to further improved resolution (in principle, another factor of 2, assuming that other factors do not come into play such as instabilities) and a factor of 4 in depth resolution, approaching a depth resolution of 1 nm [9]. Again, we can expect sensitivities for individual atoms to improve, and all benefits will apply not only to the image but also to EELS. In principle, we should be able to probe electronic structure around individual Hf atoms with a depth resolution that is not much larger than the extent of the wave function seen in Figure 8.19, and a through depth sequence of EELS data could be used to isolate the contribution from the region of the Hf atom. An indication of the possibilities is seen in

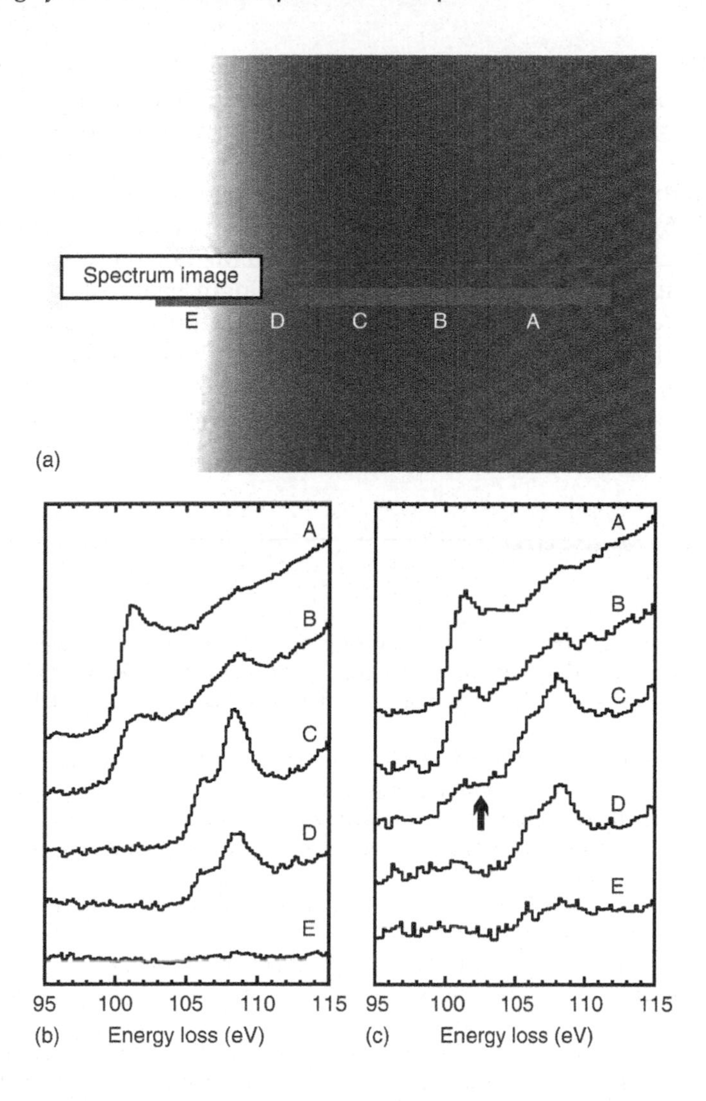

FIGURE 8.21
(a) Z-contrast image of an $Si/SiO_2/HfO_2$ high-*k* device structure showing regions analyzed by EELS (b) before and (c) after rapid thermal annealing, which introduced stray Hf atoms and a significant density of states in the region of the Si band gap (arrowed).

Figure 8.21, which shows EELS spectra from various regions of the gate stack before and after rapid thermal annealing. Before annealing, no Hf atoms were seen in the SiO_2, but they were present after annealing at a density of 0.3 Hf nm^{-3} in this case. The EELS data show that the density of states within the Si band gap increased substantially after annealing, suggesting a correlation with the presence of Hf. Such ideas could be tested directly with the improved three-dimensional resolution available with next generation STEM instruments.

Acknowledgments

The authors are grateful to many colleagues for the collaborations mentioned in this article, including A.R. Lupini, A.Y. Borisevich, M. Varela, P.D. Nellist, O.L. Krivanek, N. Dellby,

M.F. Murfitt, Z.S. Szilagyi, S.N. Rashkeev, S.F. Findlay, M.P. Oxley, L.J. Allen, H.M. Christen, G. Duscher, A. Maiti, M. Kim, H.S. Baik, R. Buczko, S.N. Rashkeev, M.H. Evans, W.H. Sides, and J.T. Luck. Work at Oak Ridge National Laboratory (ORNL) was supported by the Division of Materials Science and Engineering (S.J.P., M.F.C., K.v.B., S.T.P.), at Vanderbilt by the Air Force Office of Scientific Research under a MURI grant (FA9550-05-1-0306), the National Science Foundation grant ECS-0524655, and by the McMinn Endowment at Vanderbilt University (A.G.M., S.T.P.), also by appointments to the ORNL postdoctoral program (K.v.B.) and distinguished visiting scientist program (S.T.P.) administered jointly by ORNL and Oak Ridge Institute for Science and Education, and by a fellowship from the Alexander-von-Humboldt Foundation (K.v.B.). Theoretical work was supported in part by a grant of computer time from the DoD High Performance Computing Modernization Program at Maui High Performance Computer Center and U.S. Army Engineer Research and Development Center.

References

1. Rose, H., Correction of aberrations, a promising means for improving the spatial and energy resolution of energy-filtering electron microscopes, *Ultramicroscopy* 56, 11, 1994.
2. Scherzer, O., Das theoretisch erreichbare Auflösungsvermögen des Elektronenmikroskops, *Z. Physik* 114, 427, 1939.
3. Feynman, R.P., There's plenty of room at the bottom, available at Caltech, 1959, available at http://www.its.caltech.edu/~feynman/plenty.html
4. Scherzer, O., Sparische und chromatische Korrektur von Electronen-Linsen, *Optik* 2, 114, 1947.
5. Haider, M. et al., A spherical-aberration-corrected 200 kV transmission electron microscope, *Ultramicroscopy* 75, 53, 1998.
6. Batson, P.E., Dellby, N., and Krivanek, O.L., Sub-angstrom resolution using aberration corrected electron optics, *Nature* 418, 617, 2002.
7. Nellist, P.D. et al., Direct sub-angstrom imaging of a crystal lattice, *Science* 305, 1741, 2004.
8. van Benthem, K. et al., Three-dimensional imaging of individual hafnium atoms inside a semiconductor device, *Appl. Phys. Lett.* 87, 034104, 2005.
9. Borisevich, A.Y., Lupini, A.R., and Pennycook, S.J., Depth sectioning with the aberration-corrected scanning transmission electron microscope, *Proc. Nat. Acad. Sci. USA* 103, 3044, 2006.
10. van Benthem, K. et al., Three-dimensional ADF imaging of individual atoms by through-focal series scanning transmission electron microscopy, *Ultramicroscopy* 106, 1062, 2006.
11. Cowley, J.M., Image contrast in a transmission scanning electron microscope, *Appl. Phys. Lett.* 15, 58, 1969.
12. Pennycook, S.J. and Jesson, D.E., High-resolution incoherent imaging of crystals, *Phys. Rev. Lett.* 64, 938, 1990.
13. Pennycook, S.J. and Jesson, D.E., High-resolution Z-contrast imaging of crystals, *Ultramicroscopy* 37, 14, 1991.
14. Jesson, D.E., Pennycook, S.J., and Baribeau, J.M., Direct imaging of interfacial ordering in ultrathin $(Si_mGe_n)_P$ superlattices, *Phys. Rev. Lett.* 66, 750, 1991.
15. Pennycook, S.J. and Jesson, D.E., Atomic resolution Z-contrast imaging of interfaces, *Acta Metall. Mater.* 40, S149, 1992.
16. Pennycook, S.J. et al., Atomic-resolution imaging and spectroscopy of semiconductor interfaces, *Appl. Phys. A* 57, 385, 1993.
17. Chisholm, M.F. et al., Z-contrast investigation of the ordered atomic interface of $CoSi_2$/Si(001) layers, *Appl. Phys. Lett.* 64, 3608, 1994.
18. Chisholm, M.F. et al., New interface structure for a-type $CoSi_2$/Si(111), *Appl. Phys. Lett.* 64, 2409, 1994.
19. McGibbon, A.J., Pennycook, S.J., and Angelo, J.E., Direct observation of dislocation core structures in CdTe/GaAs(001), *Science* 269, 519, 1995.

20. Pennycook, S.J. et al., Direct determination of interface structure and bonding with the scanning transmission electron microscope, *Philos. Trans. R. Soc. London Ser. A* 354, 2619, 1996.
21. James, E.M. and Browning, N.D., Practical aspects of atomic resolution imaging and analysis in STEM, *Ultramicroscopy* 78, 125, 1999.
22. Dellby, N. et al., Progress in aberration-corrected scanning transmission electron microscopy, *J. Electron Microsc.* 50, 177, 2001.
23. Krivanek, O.L., Dellby, N., and Lupini, A.R., Towards sub-angstrom electron beams, *Ultramicroscopy* 78, 1, 1999.
24. Abe, E., Pennycook, S.J., and Tsai, A.P., Direct observation of a local thermal vibration anomaly in a quasicrystal, *Nature* 421, 347, 2003.
25. Pennycook, S.J. et al., Aberration-corrected scanning transmission electron microscopy: The potential for nano- and interface science, *Z. Metalkd.* 94, 350, 2003.
26. Varela, M. et al., Nanoscale analysis of $YBa_2Cu_3O_{7-x}/La_{0.67}Ca_{0.33}MnO_3$ interfaces, *Solid State Electron.* 47, 2245, 2003.
27. McBride, J.R. et al., Aberration-corrected Z-contrast scanning transmission electron microscopy of CdSe nanocrystals, *Nano Lett.* 4, 1279, 2004.
28. Shibata, N. et al., Observation of rare-earth segregation in silicon nitride ceramics at subnanometre dimensions, *Nature* 428, 730, 2004.
29. Wang, S.W. et al., Dopants adsorbed as single atoms prevent degradation of catalysts, *Nat. Mater.* 3, 274, 2004.
30. Klie, R.F. et al., Enhanced current transport at grain boundaries in high-Tc superconductors, *Nature* 435, 475, 2005.
31. Prabhumirashi, P. et al., Atomic-scale manipulation of potential barriers at $SrTiO_3$ grain boundaries, *Appl. Phys. Lett.* 87, 121917, 2005.
32. Varela, M. et al., Materials characterization in the aberration-corrected scanning transmission electron microscope, *Annu. Rev. Mater. Res.* 35, 539, 2005.
33. Chisholm, M.F. and Pennycook, S.J., Direct imaging of dislocation core structures by Z-contrast STEM, *Philos. Mag.* 86, 4699, 2006.
34. Pennycook, S.J. et al., Materials advances through aberration-corrected electron microscopy, *MRS Bull.* 31, 36, 2006.
35. Varela, M. et al., Atomic scale characterization of complex oxide interfaces, *J. Mater. Sci.* 41, 4389, 2006.
36. Shibata, N. et al., Nonstoichiometric dislocation cores in alpha-alumina, *Science* 316, 82, 2007.
37. Pennycook, S.J., Microscopy: Transmission electron microscopy, in *Encyclopedia of Condensed Matter Physics*, Bassani, F., Liedl, J., and Wyder, P., Eds., Elsevier Science Ltd., Kidlington, Oxford, 2006, p. 240.
38. Rashkeev, S.N. et al., Single Hf atoms inside the ultrathin SiO_2 interlayer between a HfO_2 dielectric film and the Si substrate: How do they modify the interface? *Microelectron. Eng.* 80, 416, 2005.
39. van Benthem, K., Rashkeev, S.N., and Pennycook, S.J., Atomic and electronic structure investigations of $HfO_2/SiO_2/Si$ gate stacks using aberration-corrected STEM, in *Characterization and Metrology for ULSI Technology 2005*, Seiler, D.G., Diebold, A.C., McDonald, R., Ayre, C.R., Khosla, R.P., Zollner, S., and Secula, E.M., Eds., American Institute of Physics, Richardson, TX, 2005, p. 79.
40. Diebold, A.C. et al., Thin dielectric film thickness determination by advanced transmission electron microscopy, *Microsc. Microanal.* 9, 493, 2003.
41. Hytch, M.J. and Stobbs, W.M., Quantitative comparison of high-resolution TEM images with image simulations, *Ultramicroscopy* 53, 191, 1994.
42. Browning, N.D., Chisholm, M.F., and Pennycook, S.J., Atomic-resolution chemical-analysis using a scanning-transmission electron-microscope, *Nature* 366, 143, 1993.
43. Browning, N.D. and Pennycook, S.J., Atomic-resolution electron energy-loss spectroscopy in the scanning transmission electron microscope, *J. Microsc.-Oxford* 180, 230, 1995.
44. Browning, N.D. et al., EELS in the STEM: Determination of materials properties on the atomic scale, *Micron* 28, 333, 1997.
45. Dickey, E.C. et al., Structure and bonding at Ni–ZrO_2 (cubic) interfaces formed by the reduction of a NiO–ZrO_2 (cubic) composite, *Microsc. Microanal.* 3, 443, 1997.

9

Defect Energy Levels in HfO_2 and Related High-K Gate Oxides

J. Robertson, K. Xiong, and K. Tse

CONTENTS

9.1 Introduction

The scaling of complementary metal oxide semiconductor (CMOS) transistors is leading to the replacement of SiO_2 gate oxide by oxides of higher dielectric constant (K) such as HfO_2 to avoid excessive gate leakage currents. The leading candidates are presently HfO_2, Hf silicate, and their nitrogenated alloys [1–5]. They will be introduced at the 45 nm node. Their use has required a wide-ranging optimization of these materials.

There are three main problems with devices using high-K gate oxides. First, there is a large fast transient charge trapping, which causes a shift in the transistor threshold voltages, instability, and ultimately a loss of reliability [5–9]. The problem is a band of charge traps in the bulk oxide that lie at an energy near the Si conduction band edge [6,8]. The second problem is a degradation of the Si carrier mobility, which is lower than in the equivalent SiO_2 gate oxide devices, particularly for negative MOS (NMOS) [5,10–13]. This is partly due to remote phonon scattering by soft phonons [11], and partly due to a remote Coulombic scattering by trapped charge in the oxide. The third problem is controlling the threshold voltage of the gate electrode, whether it is polycrystalline Si or a metal gate [14]. This complex effect may be due to dipoles and due to charged vacancies [15]. This indicates a critical need to identify the nature of point defects present in the oxides and to find their energy levels.

The oxides of interest are oxides of early transition metals or rare earths, with fully ionized d shells, so that they have a large gap of order 6 eV. They were chosen to be thermodynamically stable in contact with Si and to have a large conduction band offset with Si [3,4]. In general, we consider HfO_2, ZrO_2, La_2O_3, Y_2O_3, and Gd_2O_3. The most common defect in such oxides is the oxygen vacancy. This class of oxides, especially CeO_2 and ZrO_2, is also of interest as fast ion conductors and as catalyst supports, such as in the three-way catalyst in car exhausts, where oxygen deficiency plays an important role. Thus, we focus particularly on the oxygen vacancy.

This chapter first reviews calculations of the electronic structure and energy levels of the defects. It then considers experimental data on the energy levels from optical and electrical studies, which support the importance of the oxygen vacancy. The interesting fact is that the traps are electron traps, meaning that an electron becomes trapped at an O vacancy. This is counterintuitive, as the O vacancy is expected to be positively charged, in compounds of such electropositive metals. The chapter then describes the effects of the oxygen vacancy in trapping, mobility, and the interaction with gate electrodes. One question is the concentration of defects, and why such large concentrations can occur for defects with a large formation energy.

9.2 Ionic Oxides and Defect Densities

The oxides of interest are ionic, and can be nanocrystalline or amorphous. Amorphous is preferable, but often to obtain the highest K value, we accept nanocrystalline oxides. Because they are ionic, their defects have a fundamentally different character than those in SiO_2. In SiO_2, the main defects are broken or dangling bonds, either Si dangling bonds or oxygen dangling bonds. The other possible type is like atom bonds, Si–Si bonds or O–O bonds. The key point is that SiO_2 is a random network with a low average coordination number of 2.67. Thus, if a local stress or bombardment creates a broken bond, the network can usually relax to reform that bond, and remove the defect. This means that the defect density in bulk SiO_2 or at the SiO_2/Si interface is very low. Also, any O vacancy formed can rebond as a Si–Si bond.

In contrast, HfO_2, ZrO_2, La_2O_3, etc. are ionic oxides. They have high coordination numbers; the average is 5.3. If a process creates a defect in this structure, it is not easily able to relax to remove the defect. Thus, the defect density is naturally higher [16,17].

9.3 Bulk Band Structures

In bulk cubic HfO_2, the Hf is surrounded by eight oxygen ions and the oxygen is surrounded by four Hf ions. Figure 9.1 shows the band structure of HfO_2. The band gap is 5.8 eV and is direct at the X point [18,19]. Figure 9.2 shows the bulk density of states (DOS) on the Hf and O sites. The valence band (VB) consists mainly of O p states and the conduction band mainly of Hf d states, followed by Hf s states at 7.5 eV and above. This is typical of an ionic oxide consisting of Hf^{4+} and O^{2-} ions. The valence bandwidth of 6.5 eV is typical of a fairly ionic oxide.

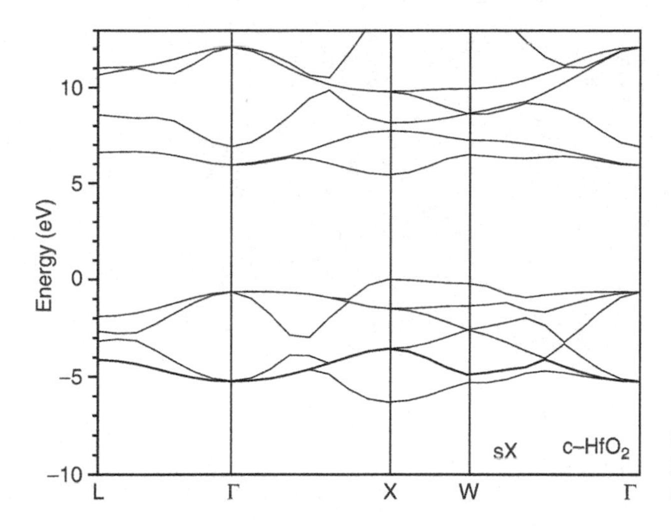

FIGURE 9.1
Band structure of cubic HfO_2 by the sX functional.

9.4 Defect Types and Their Formation Energies

In binary ionic oxides, there are four possible defects: metal vacancy, metal interstitial, oxygen vacancy, and oxygen interstitial. These correspond to metal or oxygen excess. The formation energy for all four types has been calculated for HfO_2 and ZrO_2 by Foster et al. [20,21]. They find that the metal site defects have a higher formation energy. Thus, for a given case of oxygen or metal excess, we need only consider oxygen site defects. Figure 9.3 shows a diagram of the neutral configurations of the oxygen vacancy (V) and interstitial (I) in cubic HfO_2. Note the O–O bond of the interstitial [22], which we will return to.

The formation energies of the neutral defects are plotted against oxygen chemical potential μ_O, as shown in Figure 9.4. The case of $\mu_O = 0$ corresponds to oxygen as molecular O_2. The formation energy of V_O is 9.35 eV and that of the interstitial is

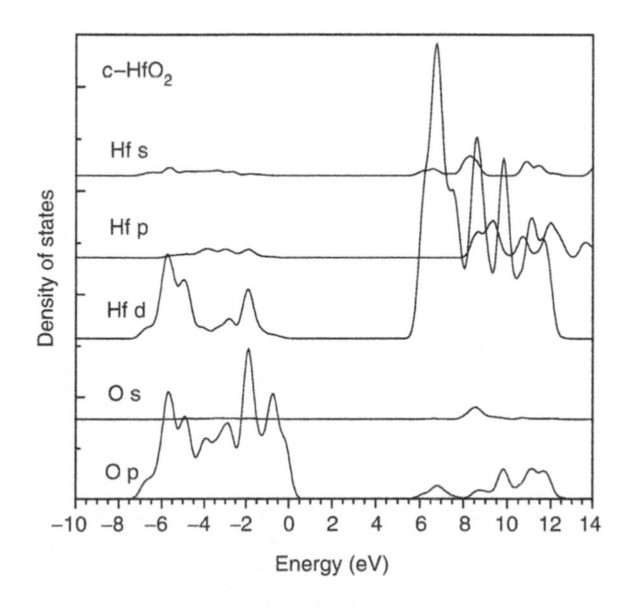

FIGURE 9.2
Partial DOS of cubic HfO_2.

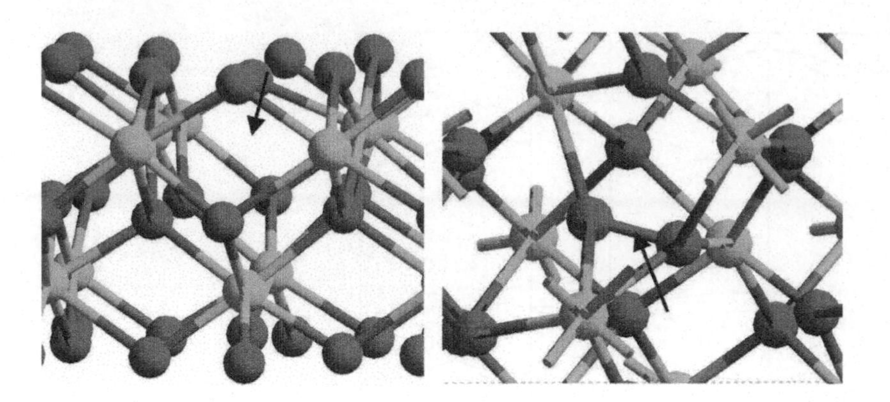

FIGURE 9.3
The oxygen vacancy and neutral oxygen interstitial in HfO_2.

-1.35 eV at $\mu_O = 0$. The calculated formation energies of Foster et al. [20,21], Ikeda et al. [23], and Scopel et al. [24] are all consistent. Some reaction energies are given below:

$$Hf + O_2 \rightarrow HfO_2 \quad -5.85 \text{ eV (exothermic)}$$
$$0 \rightarrow V^0 + I^0 \quad 8.0 \text{ eV}$$
$$0 \rightarrow V^{2+} + I^{2-} \quad 5.8 \text{ eV}$$

9.5 Defect Energy Levels

9.5.1 Oxygen Vacancy

There have now been a number of calculations of defect energies and energy levels in HfO_2 and ZrO_2 [21,24–28]. The calculations generally used the local density approximation

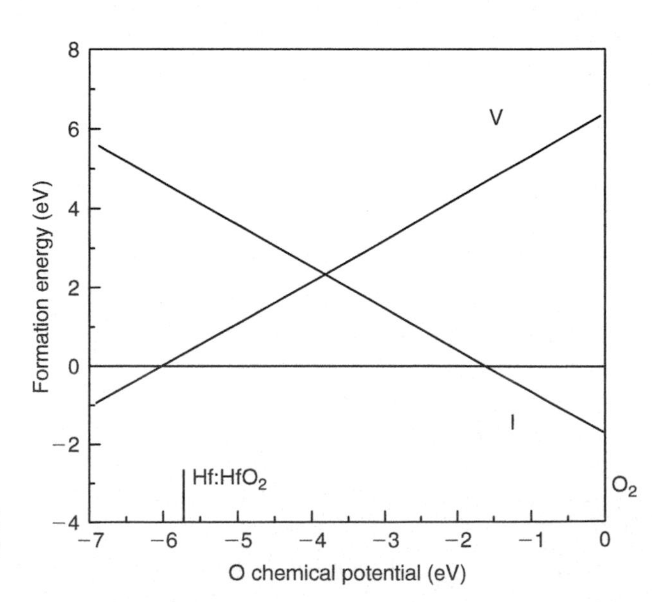

FIGURE 9.4
Formation energy of neutral O vacancy and interstitial versus O chemical potential for HfO_2.

(LDA) or the generalized gradient approximation (GGA). LDA and GGA are well known to give good values for the ground state structures, but to underestimate the band gap of semiconductors and insulators by 30%–50%. This error can be corrected for bulk materials by a rigid upward shift of the conduction bands to fit the experimental gap—the so-called scissors operator. However, for defects, it is unclear how to apply this correction. In Foster et al. [21], both the O vacancy and interstitial levels are placed below the Si band gap in the Si:HfO_2 system. In contrast, other LDA calculations leave the levels high in the gap [25,26]. This is a critical factor, as it controls the trapping characteristics.

As the LDA and GGA methods underestimate band gaps, there are a number of methods that can give improved band gaps, such as the GW approximation [28], B3LYP, self-interaction correction, and LDA plus U. Here, we use LDA-based calculation methods, the screened exchange (sX) [29], and the weighted density approximation (WDA) [30]. These methods are summarized earlier [19].

We carried out calculations of the oxygen vacancy and interstitial using the GGA, sX, and WDA functionals. The calculations used the total energy plane wave pseudopotential code CASTEP [31]. For sX calculations, norm-conserving pseudopotentials must be used with a plane wave cutoff of 600 eV. For GGA and WDA calculations, Vanderbilt ultrasoft pseudopotentials are used with a cutoff of 400 eV [32,33].

For cubic HfO_2 the GGA band gap is found to be 3.6 eV. This increases to 6.1 eV in WDA, which is slightly above the experimental value of 6 eV, while sX gives a gap of 5.5 eV, which is a slight underestimate [19].

The defects are created by making a periodic supercell of about 48 atoms of the original cell and removing or adding an oxygen atom [32,33]. The structures are first relaxed in GGA. The energy levels are then calculated in sX or WDA. Usually, the size of the defect wave function determines the convergence with respect to supercell size. However, for these ionic oxides, the defect wave functions are very localized. However, the supercells must be large enough to allow the considerable atomic relaxations around the defects, without them interfering between cells. This requires at least 48 atoms.

The oxygen vacancy creates a single gap state. This state has the fully symmetric A_1 symmetry, and it is strongly localized on the d orbitals of the adjacent Hf ions [32]. It is occupied by two electrons for the neutral vacancy, V^0. The defect state eigenvalue of the ideal (unrelaxed) V^0 lies at 3.0 eV above the oxide VB maximum in GGA. The defect state was then calculated in sX to be 3.75 eV. The vacancy is relaxed in GGA, and Hf atoms move outward only slightly from their ideal positions, so the Hf–Hf distance is 3.59 Å compared to 3.54 Å in bulk HfO_2. This causes the level as calculated in sX to move up by 0.05 to 3.8 eV (Figure 9.5).

The A_1 defect state is singly degenerate and it is filled with two electrons for V^0. For the positively charged vacancy V^+, one electron is removed from this state. This state lies at

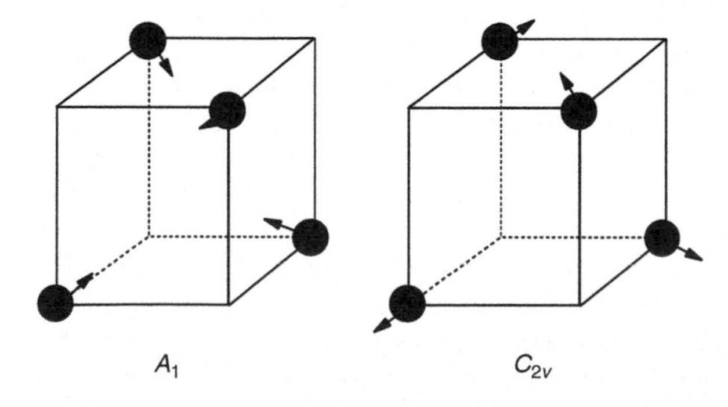

FIGURE 9.5
(a) A_1 symmetry breathing mode relaxation of the O vacancy, (b) C_{2v} relaxation of the negatively charged vacancies.

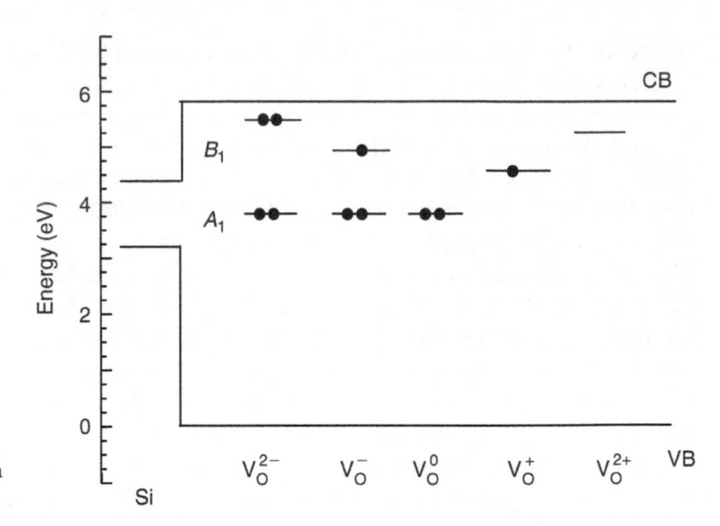

FIGURE 9.6
Summary of vacancy energy levels in HfO_2.

3.2 eV in the unrelaxed configuration. Unlike at V^0, the net charge of V^+ causes the adjacent Hf ions to have a substantial outward relaxation from the vacancy site. The relaxation is an A_1 symmetric breathing relaxation, as shown in Figure 9.6a. The Hf–Hf spacing is now 3.74 Å compared to 3.54 Å in the bulk. This relaxation causes the defect state to move upward from 3.2 to 4.7 eV, as shown in Figure 9.5.

At the doubly positive vacancy V^{2+}, the A_1 gap state is now empty. The state lies at 2.6 eV for the unrelaxed vacancy. The greater positive charge causes a further outward relaxation of the Hfs, increasing the Hf–Hf separation to 3.90 Å. The gap state moves up to 5.2 eV after the relaxation (Figure 9.5). This is a very large shift. For $V^0 - V^{2+}$ charge states, there are no other empty gap states above the A_1 state.

As oxygen is the second most electronegative element, it is natural to expect removing O to create positively charged states like $V^0 - V^{2+}$. Interestingly and perhaps unexpectedly, the O vacancy in HfO_2 can also trap one or two electrons [31]. A trapped electron causes the O vacancy to undergo a distortion of C_{2v} symmetry (Figure 9.6), in which adjacent Hf ions move away from the vacancy, but one pair comes together. This distortion pulls down a new, singly degenerate B_1 state from the conduction band, in addition to the existing, filled A_1 state. This B_1 state can be occupied by one or two electrons. For V^-, it lies at about 5.0 or 0.8 eV below the conduction band edge in sX (Figure 9.5). For the V^{2-} center, the C_{2v} distortion is stronger, and the B_1 level is now occupied by two electrons.

These energy levels for the various charge states are summarized in Figure 9.5. The vacancy levels lie in the upper part of the HfO_2 band gap. The levels can be aligned to the bands of the underlying Si by using the band offset of HfO_2 on Si of ~3.3 eV [4,34]. This places the V^0 level within the Si gap and the V^+ level just above the Si gap.

The nature of the defect levels can also be seen by plotting the charge density of the defect state. For V^0, the defect state is shown in Figure 9.7. It has A_1 symmetry, and it lies symmetrically between its four neighboring Hf ions, three of which are shown here in the (111) plane. The plot emphasizes its localization. This state has similar character at the V^+ and V^{2+} centers. At V^-, this A_1 state is fully occupied and C_{2v} distortion moves the state slightly toward two of the four Hf neighbors (Figure 9.7a). The B_1 state wave function has a node plane in the plane of the two close-lying Hf neighbors (Figure 9.7b).

We can compare the O vacancy with that in other semiconductors or oxides. In a polar semiconductor like GaAs, an anion vacancy (like V_O) would be expected to give rise to gap states in the upper gap, because the states are localized on the cation (Ga), which forms the main part of the conduction band. The states would be delocalized over many neighbors

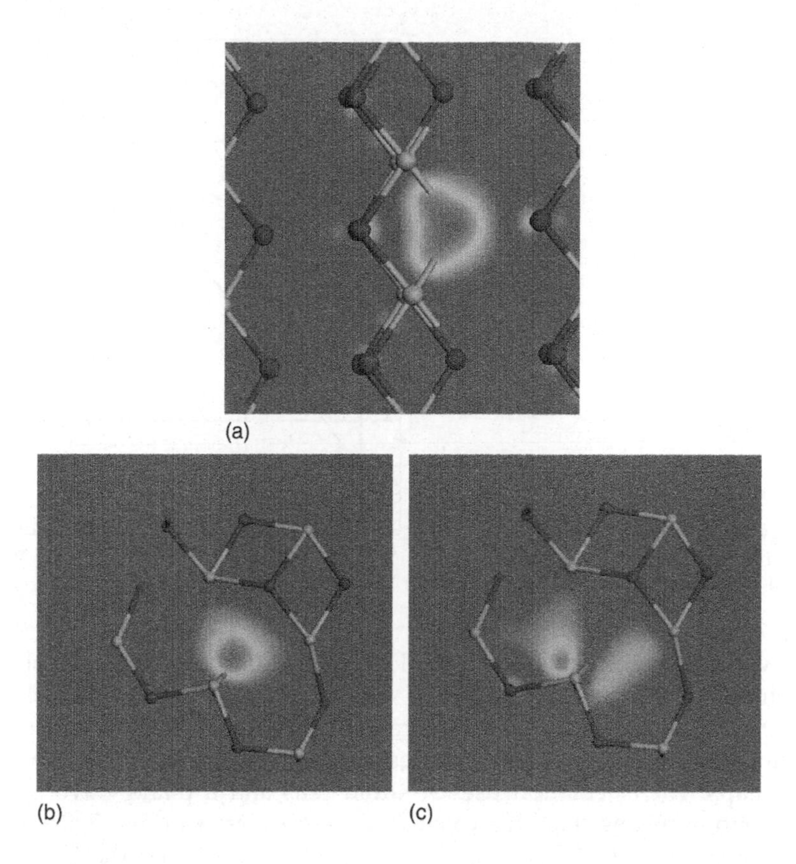

FIGURE 9.7
(a) Charge density of the A_1 defect state for the V^0 center in a (110). (b, c) Charge densities of the A_1 and B_1 states for the V^- center.

because the effective mass is small and screening is large. In an ionic oxide like MgO, the gap is much larger, the screening lower and effective masses are higher, so defect states are very deep. The vacancy state is now localized largely inside the vacancy itself. In some cases, it is not even localized on the adjacent Mg ions; it is a vacancy-site wave function. In this case, the vacancy levels tend toward midgap [35].

We have also derived a formation energy diagram for the O vacancy in the various charge states (Figure 9.8). The formation energy of V^0 is taken from Foster et al. [21], and that of the charged states from our calculated energy levels. We take the free energy of O as zero in its standard state of molecular O_2. The formation energy is 1.8 eV lower, if the free O atom is used as the reference state. The calculated formation energies of Foster et al. [21], Ikeda et al. [23], and Scopel et al. [24] are all consistent. We see that the O vacancy has a very large formation energy, of order 9.3 eV for V^0.

We see that V_O possesses a small negative correlation energy or "U" [21]. The single positive state V^+ is unstable with respect to V^0 and V^{2+}. The V^- also has a negative U. A positive U is the normal situation, where electrons in a defect repel each other due to their charge. A negative U is an effective electron–electron attraction, which arises when there is a significant electron–lattice coupling, so that the ions relax as the defect charge state changes.

A number of other groups have calculated the oxygen vacancy. Feng et al. [27] noted the negative U character of the states in their calculation, and suggested that this was important for bias stress instability. Gavartin et al. [36] used the B3LYP functional to find the defect levels. The calculation uses 96 atoms, which is good, but the basis set is less good

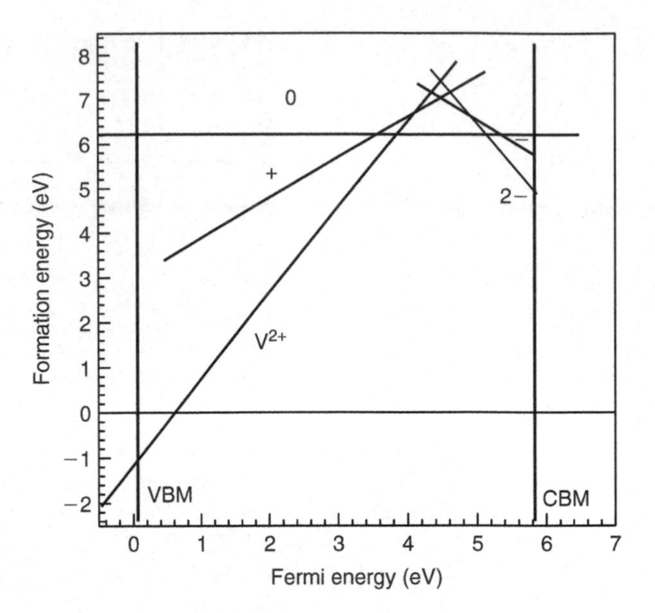

FIGURE 9.8
Formation energy of the O vacancy in HfO_2 as a function of Fermi level energy. Zero of the oxygen chemical potential is the O_2 molecule.

than a plane wave basis set. Their levels are lower in the gap than we find, but overall the results are similar.

The most recent calculation of the O vacancy by Broqvist and Pasquarello [37] uses the PBE0 functional for both total energies and energy levels. This functional gives correct band gaps. The calculation uses 96 atom cells and a plane wave basis set. This overall combination means that this calculation is the best to date. They find that the 1−/2− transition is just above the conduction band edge, so that the 2− state is never stable for Fermi energies within the gap. They also find that the states have a slight positive U, unlike other groups. This means negative U is probably not essential to model bias stress instability.

9.5.2 Oxygen Interstitial

Now consider the oxygen interstitial. The O^{2-} interstitial ion forms a closed shell ion just like a bulk O^{2-} ion, whose energy states all lie in the VB. It adds more states to the VB of the bulk material, with some states repelled slightly above the VB edge.

The O^- interstitial is similar to the O^{2-}. It is separated from the other O^{2-} ions. It now lies about 2.0 Å away from the nearest O^{2-} ion [32]. Figure 9.9 shows the calculated local DOS of the O^- interstitial and on a nearby bulk O^{2-} ion. This is calculated in the spin-polarized state to allow for the partial occupancy. The structure was relaxed in the GGA and the DOS was calculated in the WDA. The energy zero is referenced to the VB maximum of the bulk oxide. We see that there is now a state 1 eV above the oxide VB maximum. This is partially occupied and the Fermi level lies in this state.

Figure 9.3b shows the calculated configuration of the neutral interstitial. There is now a clear O–O bond of bond length 1.49 Å. It is a split interstitial with the O–O axis along {100}. This dumbbell structure is the typical configuration of a neutral O interstitial in ionic oxides, and is also found in MgO and similar materials. The DOS is calculated in WDA. This structure gives rise to an empty state in the upper band gap, which lies at 4.6 eV, the antibonding or σ^* state of the O–O bond. Its bonding σ partner lies just below the main VB at −5.5 eV. There are also filled π and π^* states lying inside the VB and just above the VB top. This configuration is sometimes called the peroxyl anion. The Fermi level in this case lies near midgap.

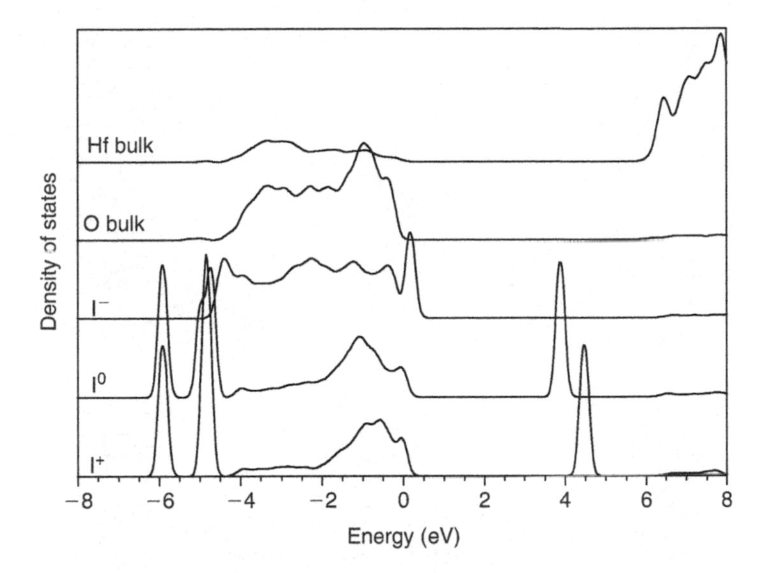

FIGURE 9.9
Calculated local DOS at the oxygen interstitial site, for its different charge states: −, 0, and +.

For the positive O^+ interstitial, sometimes called the peroxyl radical, the local DOS is shown in Figure 9.9. The O^+ also forms an O–O bond, but the positive charge causes the bond length to reduce to 1.39 Å. This repels the σ^* state toward the conduction band. The σ state falls deeper in the VB at −6.5 eV. There are still π and π^* states, and there is a hole in the π^* state above the VB. The Fermi level lies in the π^* state. The states of the various interstitial configurations are summarized in Figure 9.10.

9.5.3 Defects in Hf Silicate

Here, we model the structure of amorphous Hf silicate by the crystalline $HfSiO_4$ structure. In this lattice in Figure 9.11, each Hf is eightfold coordinated, each Si is fourfold

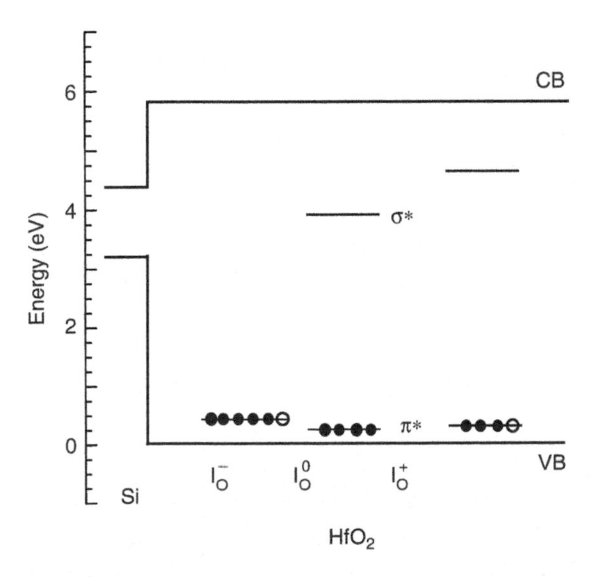

FIGURE 9.10
Summary of the energy levels of the interstitial defects in HfO_2.

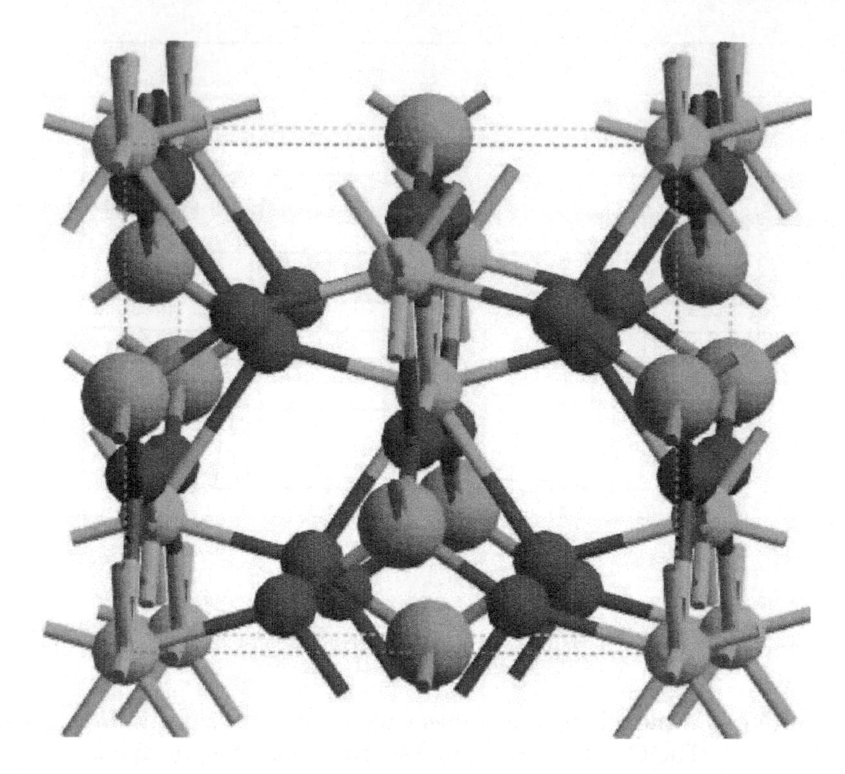

FIGURE 9.11
Structure of tetragonal $HfSiO_4$.

coordinated, and each oxygen is threefold coordinated to two Hfs and one Si. The GGA gap of $HfSiO_4$ is 4.5 and 6.5 eV in WDA. The states of an O vacancy are calculated for a 48 atom unit cell in the WDA, after structural relaxation in GGA [38]. The upper VB is nonbonding oxygen states. The lowest conduction band is due to Hf d states, followed by Si states (Figure 9.12).

The total energies of defects in the related crystal $ZrSiO_4$ have been calculated by Pruneda and Artacho [39] as a function of the charge state and O chemical potential.

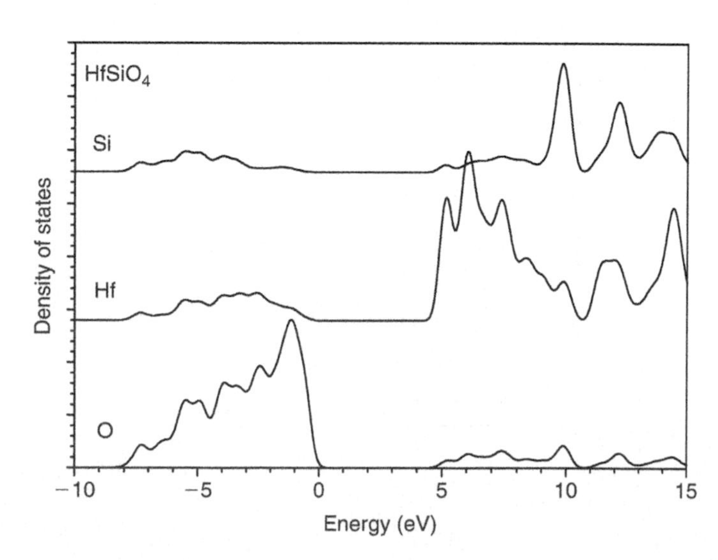

FIGURE 9.12
Partial DOS of bulk $HfSiO_4$.

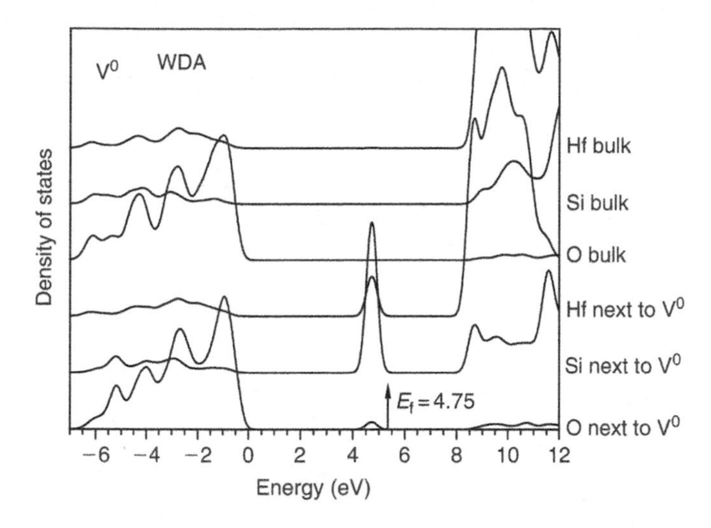

FIGURE 9.13
Local DOS of the neutral oxygen vacancy in $HfSiO_4$.

In WDA, the neutral O vacancy V^0 has a A_1 state at 4.75 eV occupied by two electrons, Figure 9.13. It is mainly localized on the adjacent Si atom (Figure 9.14). The Si relaxes toward the vacancy but the Hfs relax away from it slightly. V^0 has a second, empty state that lies just below the conduction band edge, and is localized mainly on the adjacent two Hf ions (Figure 9.14).

At V^+ configuration, the adjacent Si atom moves back to its original site, and the Hfs are repelled further away. It gives a singly occupied state at 5.1 eV, again mainly localized on the Si atom. There is a second empty state lying just below the conduction band edge that is mainly localized on the adjacent Hf ions, similar to Figure 9.14. At V^{2+}, this gives an empty state at 6.7 eV, in the upper gap. The Si has relaxed away from the vacancy to give a nearly planar Si site. The Hf ions are also repelled. The defect state is strongly localized on the Hfs. These states are summarized in Figure 9.8.

The O vacancy also has negative charged configurations, as in HfO_2. The V^- creates an extra state from the conduction band. The A_1 state at 4.55 eV is filled and this new state is singly occupied. The energy levels of the O vacancy are summarized in Figure 9.15. The defect states of HfO_2, $HfSiO_4$, and SiO_2 are all aligned in Figure 9.16.

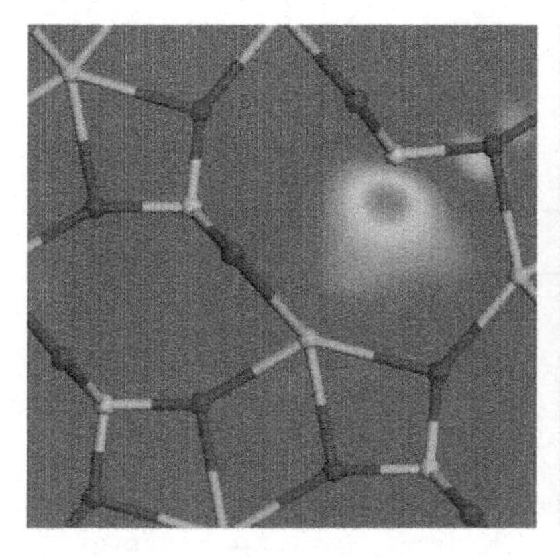
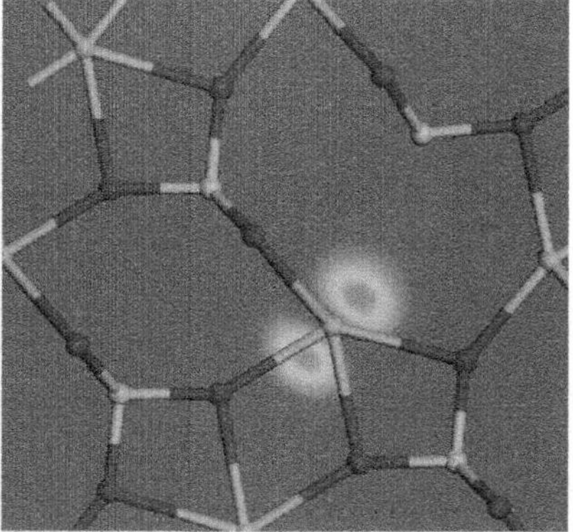

FIGURE 9.14
Charge density maps of the defect states of the neutral oxygen vacancy in $HfSiO_4$.

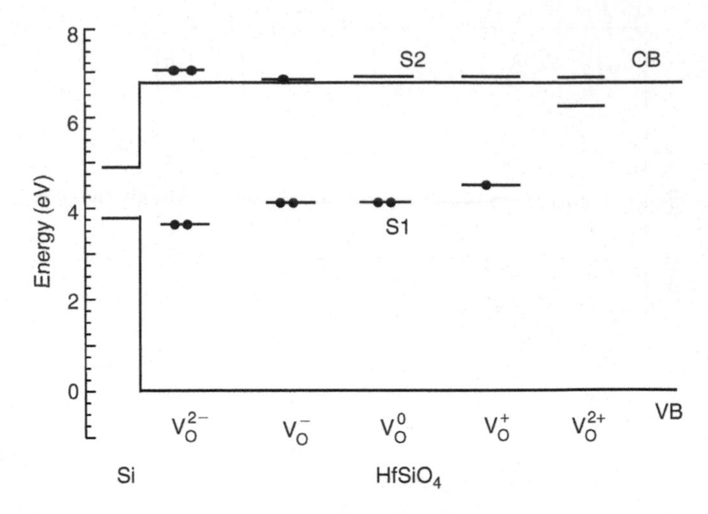

FIGURE 9.15
Summary of energy levels of the oxygen vacancy in $HfSiO_4$.

9.6 Discussion

The trapped charge in high-*K* oxides is generally attributed to trapping at existing defect levels rather than the creation of new defects [6]. The trapping causes hysteresis between the up and down ramps of gate voltage cycles (Figure 9.17). This is interpreted as fast trapping and detrapping of electrons in HfO_2. Higher field stressing will create extra states; this has been reviewed by Zafar et al. [9].

We first compare the calculated energy levels with optical data. Takeuchi et al. [40] used spectroscopic ellipsometry on HfO_2 films to identify an absorption band within the band gap at about 4.5 eV. They attributed this to the oxygen vacancy because the absorption increased with increasing oxygen deficiency. Our work is consistent with this interpretation.

Afanas'ev et al. [41] and Nguyen et al. [42] observed an optical absorption peak below the HfO_2 conduction band. This would be consistent with an oxygen vacancy level. Note, however, that an exciton absorption is also expected (it is a direct gap material), but this

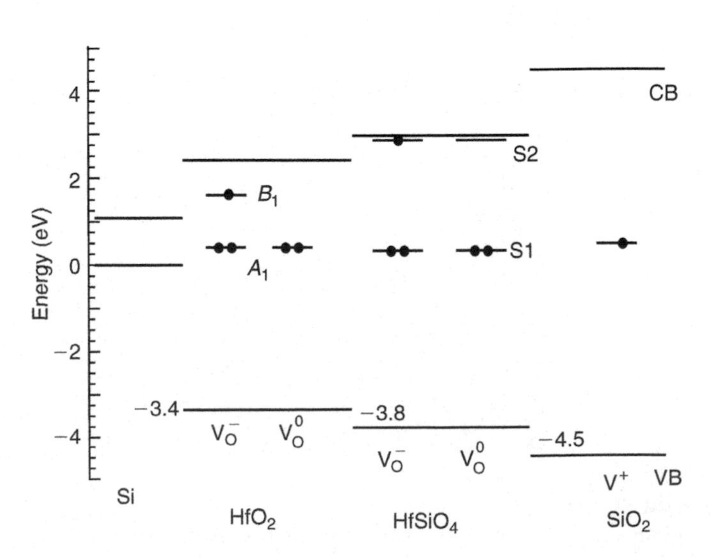

FIGURE 9.16
Alignment of the defect energy levels of HfO_2, $HfSiO_4$, and SiO_2.

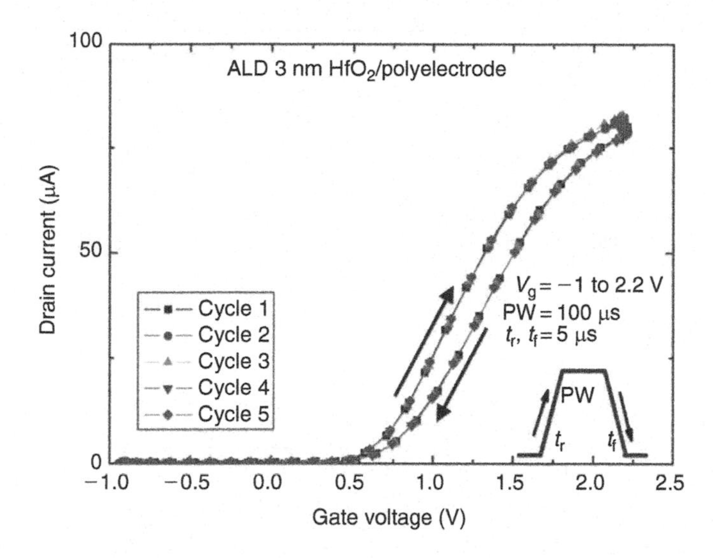

FIGURE 9.17
Hysteresis due to transient electron trapping in HfO_2.

may be buried under the absorption tail. Lucovsky [43] observed peaks in x-ray absorption spectra. Walsh et al. [44] observed a number of peaks in cathodo-luminescence, one around 4.2 eV on HfO_2 thin films. These peaks correspond reasonably well to the calculated vacancy levels. It is possible that absorptions at 4.2 eV could be due to the E2 peak of the underlying Si wafer. However, this is less likely for the cathodo-luminescence case.

We can also compare calculation results with electrical data. Kerber et al. [6] noted that their trapping data in HfO_2 films on n-type Si wafers were consistent with an electrical level lying just above the Si conduction band edge. Zhao et al. [45] noted that the defect is a trapped electron, meaning the V^0 to V^- transition. Bersuker et al. [7] carried out trapping and release experiments, and analyzed their data to find two levels; they place the upper level rather shallow at ~0.3 eV. Mitard et al. [46] analyzed the trapping data over a wide temperature range and derived two trap levels at 0.8 and 2.1 eV below the oxide conduction band. Cartier et al. [47] studied defects in HfO_2 films on Si using charge pumping spectroscopy. The resulting spectrum is consistent with the oxygen vacancy. Moreover, the peak intensities are proportional to oxygen deficiency, although the exact values are proprietary.

The consistency of our results with trapping data implies that the O vacancy is the dominant defect in HfO_2 gate oxide films. Only the O vacancy can give an electron trap in this energy range; the interstitial gives an electron trap near the VB. It means that conduction in HfO_2 films is by electrons, by Poole–Frenkel hopping or trap-assisted tunneling [48]. The trap depth is therefore of order 1–1.5 eV, that of V^- below the oxide conduction band edge. Thus, electrons are likely to be the dominant carriers in both HfO_2 and ZrO_2.

We see that the O vacancy has a very large formation energy for V^0. It is therefore not clear why vacancies should be present in such high concentrations, to be seen so clearly by optical absorption, as in Refs. [42,43]. The partial answer is that the vacancy is usually measured in an electronic system after metallization, which tends to create vacancies. Gusev et al. [49] and Wen [50] recently found that the degree of charge trapping in HfO_2 depended on the nature of the gate electrode material. The conventional poly-Si gates cause strong trapping while tungsten gates gave much less trapping. This suggests that O vacancies are the source of traps and could be related to the faster O diffusion in HfO_2.

Charged defects are responsible for some of the reductions of carrier mobility seen in devices using high-*K* oxides (Figure 9.18). Mobility degradation has been one of the key problems in implementing high-*K* dielectrics in manufacturing. The degradation is larger

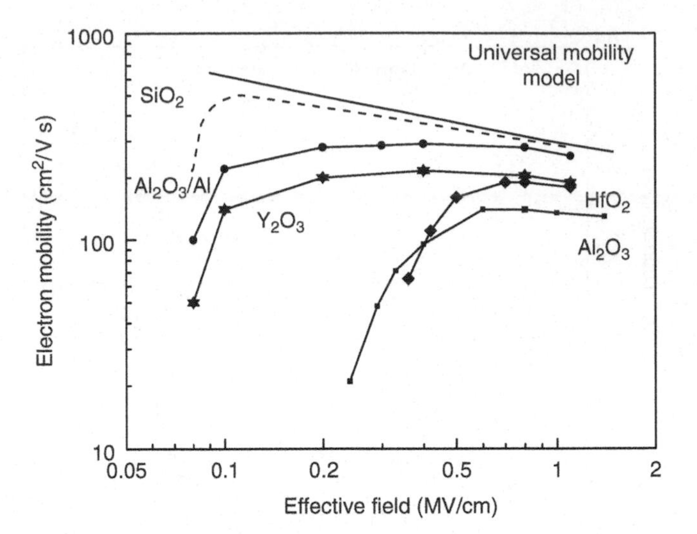

FIGURE 9.18
Reduction in electron mobility due to various unoptimized high-*K* gate oxides. (After Gusev, E.P. et al., *Tech Digest—Int. Electron Devices Meeting*, 455, 2001. With permission.)

for electrons than for holes [5], because the electron mobility is higher in Si. The mobility reduction has been shown to be due to remote scattering of carriers by effects in the high-*K* layer (Chapter 4). The high-*K* layer is separated from the Si channel by a SiO_2 layer, intentional or otherwise. By varying the thickness of this layer, and keeping the HfO_2 layer constant, the change in mobility was found to decrease exponentially with SiO_2 thickness [51], as expected for remote scattering.

There are two mechanisms that can give remote scattering, either Coulombic scattering by charged defects, or remote phonon scattering by low energy soft polar modes. The latter effect arises because high-*K* oxides are incipient ferroelectrics; this is where their high-*K* comes from. Unfortunately, such polar modes are very efficient carrier scatterers. This problem was a potential show stopper for high-*K* gate oxides [11]. The contributions of each mechanism can be derived from the temperature dependence, and by the dependence on gate field (carrier density) [12,13,51–53]. The inverse mobility can be separated into three contributions, Coulomb scattering, phonon scattering, and interface roughness scattering, according to Matthiessen's rule

$$\frac{1}{\mu} = \frac{1}{\mu_C} + \frac{1}{\mu_{PH}} + \frac{1}{\mu_{SR}}$$

where

C is the Coulombic scattering
PH is the phonon scattering
sr is the surface roughness

At low fields, mobility is limited by Coulombic scattering by trapped charges in the oxide or channel or the gate electrode interface; at moderate fields, it is limited by phonon scattering; and at high fields by scattering by surface roughness (Figure 9.19). Phonon scattering is the only one with a significant temperature dependence. Ren et al. [12] and Chau et al. [13] concluded that phonon scattering was more important from the temperature dependence. On the other hand, Casse et al. [51] deduced that Coulombic scattering was dominant in HfO_2 layers from the field dependence.

Recently, a number of groups, including IBM [54,55] and Sematech [56–58] have optimized their HfO_2 layers to such an extent that the mobility degradation due to both

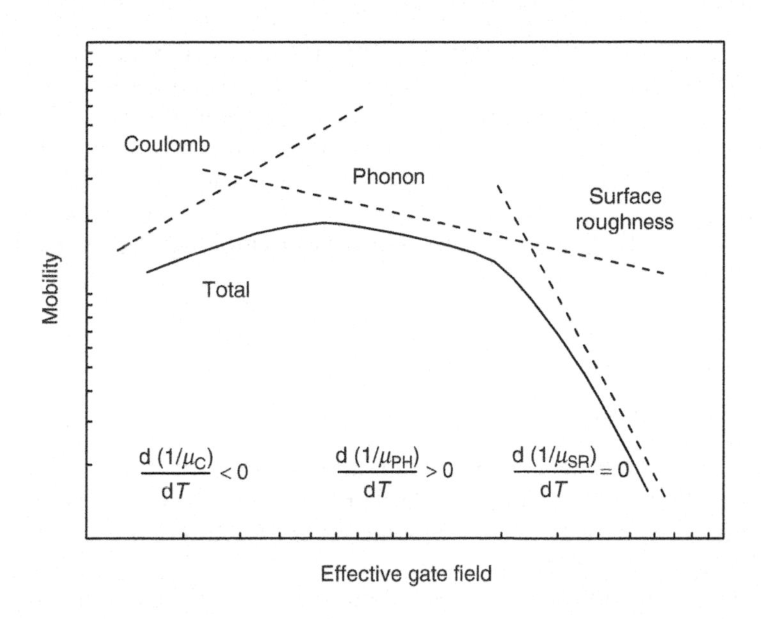

FIGURE 9.19
Contributions to scattering mechanisms and mobility versus gate field in an inversion layer.

mechanisms is now much less than 5 years ago; see Figure 9.20. It still occurs at low effective oxide thickness, so the problem is only postponed. One insight was to ensure that the HfO_2 layer is thin enough for electrons to be able to tunnel out of the trap states, and thus avoid becoming charged [58].

9.7 Interface Defects

The oxygen vacancy has often been identified as the cause of the problems in HfO_2-based dielectrics, such as trapping, or Fermi level pinning. However, the large concentrations of

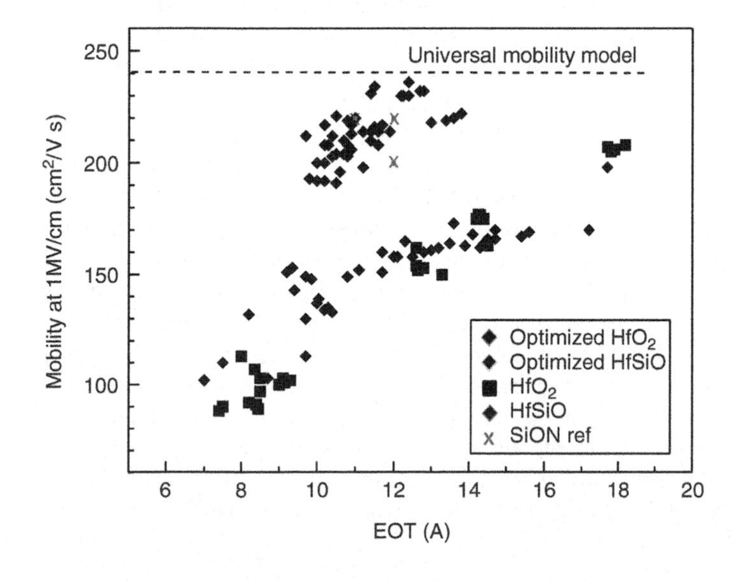

FIGURE 9.20
Electron mobility versus effective oxide thickness for recent optimized HfO_2 gate stacks. (After Lee, B.H. et al., *Mater. Today*, 9(6), 32, 2006. With permission.)

the vacancy needed to cause this contrasts with the large formation energy of the defect, of order 6.3 eV for the neutral defect referred to the oxygen molecule. The reason for the large defect concentrations has recently become apparent. It is caused by the presence of a Si substrate or a metal gate electrode. The oxygen atom liberated by the vacancy can form a MO unit with a metal atom in the gate. Or it can form half a SiO_2 with a Si atom from the Si substrate, or from the SiO_2 interfacial layer. The metal case was discussed by Demkov [59], while the Si case was discussed by Akasaka et al. [60] and Robertson et al. [61].

9.8 Defect Passivation

The high defect concentrations typical of HfO_2 and other high-*K* oxides mean that it is also important to be able to passivate defects. Defect passivation is an important feature in SiO_2. Even though the defect density at SiO_2/Si interface is naturally quite low, it can be further reduced by the action of hydrogen.

Most important semiconductors are covalent, e.g., Si, a–Si, etc., and their defects are unsatisfied or dangling bonds. These defects can be passivated by monovalent hydrogen. This also applies to SiO_2 and the Si/SiO_2 interface. The main interface defect is the P_b center, a Si dangling bond on the Si side, which can be passivated by H to form a Si–H bond. While the Si dangling bond creates a state in the midgap of Si, the Si–H bond only has states well away from the gap. This absence of gap states means there are no longer any trap states near the Si gap, to trap charges. This is passivation. The second aspect is that the Si–H bond is moderately strong, so the reaction to form Si–H is exothermic. Unfortunately, it can be reversed by electron bombardment or hydrogen reactions, so that passivation is never completely irreversible. This leads to phenomena such as bias stress instability in conventional CMOS.

In contrast to Si, HfO_2 and related oxides are ionic, and have more defects. Is there an effective passivation mechanism? The first passivation method considered was nitrogen. Nitrogen addition has been used to fix up all sorts of problems with high-*K* oxides. Mainly this has been to lower atomic diffusion rates and thereby raise crystallization temperatures. In these empirical studies, nitrogen was also found to have a beneficial role in reducing charge trapping, when applied in the correct process sequence. The mechanism of this improvement was not understood. Gavartin et al. [62] considered a wide range of configurations of N in HfO_2 without and with the presence of hydrogen. This was because N is often introduced in the form of ammonia, NH_3. However, all the configurations considered by Gavartin still possessed gap states, so they were not passivated.

The first idea of how nitrogen passivation could occur was by Umezawa et al. [63]. They noted that two nitrogens and one O vacancy (as in Figure 9.21) is actually a closed shell system. It turns out that this configuration does not possess gap states. The Ns have one less electron than an oxygen, so the two electrons of a neutral O vacancy will fall down from the vacancy level into the two holes in the VB due to the nitrogens (Figure 9.22). In addition, the state originally due to the vacancy has been repelled out of the gap into the conduction band (or at least close to the conduction band edge), and there is no longer a trap state in the gap (Figure 9.22). In detail, the vacancy itself has become positively charged, and so the adjacent Hf ions relax outward [64]. This relaxation raises the vacancy level, as in the simple V^{2+}. It turns out that the relaxation here raises the state even more. The combination of VN_2 as a closed shell configuration is also recognized from the work of Shang et al. [65]. The passivation of vacancies by nitrogen is

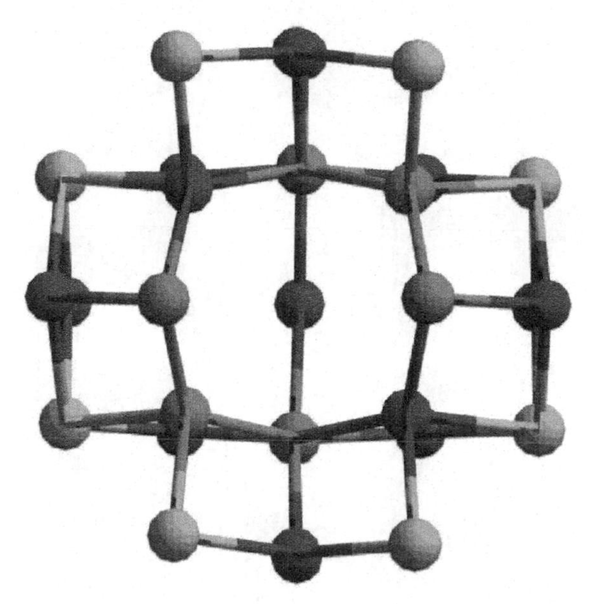

FIGURE 9.21
Two nitrogens adjacent to an oxygen vacancy in HfO_2.

moderately effective. The binding energy of N for the vacancy is rather weak, only of order 0.4 eV [63].

A second passivation mechanism is to implant fluorine. It appears that implanted fluorine ions strongly reduce trap densities [66–68]. However, this method needs care as excess can cause formation of a SiOF material of low-*K*, defeating the objective of high-*K*. We have calculated the electronic structure of F substituting at an O vacancy [69] (Figure 9.23). The local DOS is shown in Figure 9.24. This gives no gap states for both F^+ and F^0. The calculations are carried out in GGA and WDA. Thus, F has removed all gap states, as summarized in Figure 9.25. F has one more electron than O, so the F^+ is the more usual situation, and the Fermi level will lie near midgap. The fluorine is strongly attracted to the O vacancy compared to nitrogen. The binding energy is of the order of 2.8 eV. F is the most effective passivant because it strongly repels all antibonding states out of the gap, and it is more electronegative than O so that any π states lie below the VB maximum, whereas Cl will not do this.

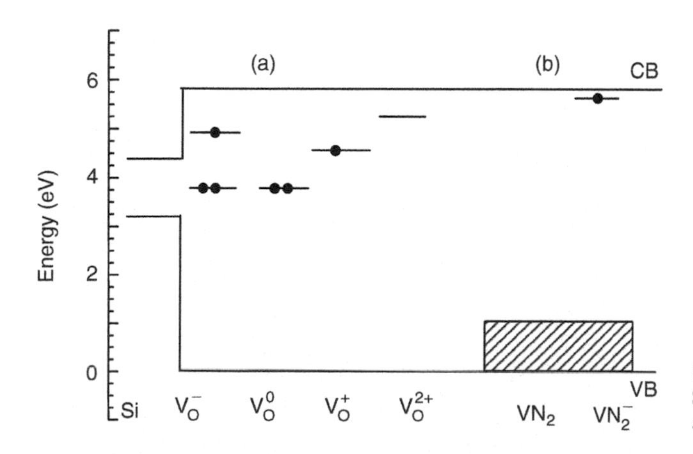

FIGURE 9.22
Summary of energy levels of N configurations in HfO_2.

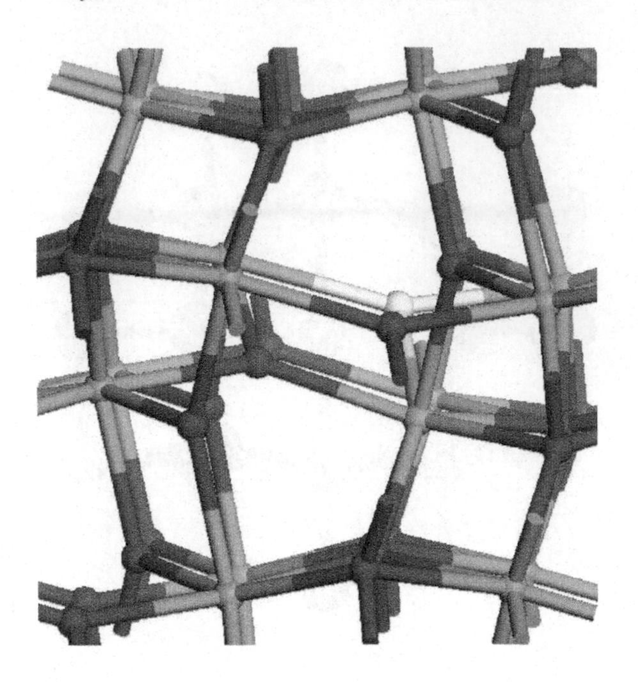

FIGURE 9.23
Substitutional F at an O vacancy.

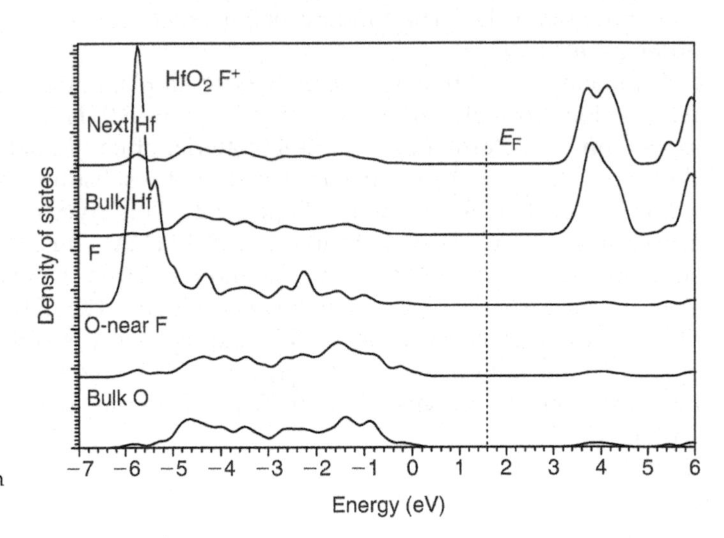

FIGURE 9.24
Local DOS of F^+ at the O vacancy in HfO_2.

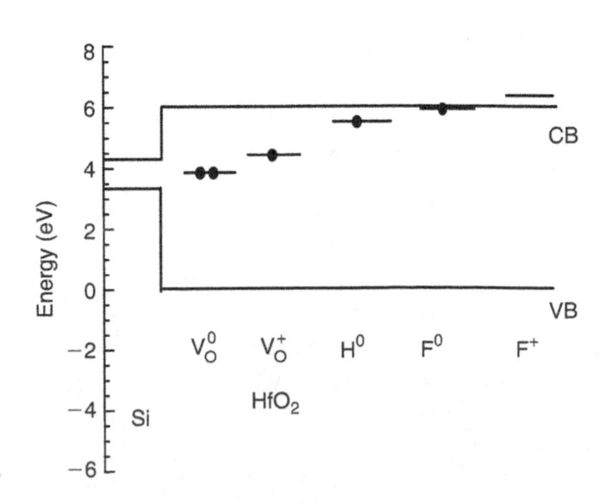

FIGURE 9.25
Summary of energy levels of H and F in HfO_2.

9.9 Summary

In brief summary of the key points, the O vacancy is the dominant defect in HfO_2. It can adopt five charge states, 2+, +, 0, −, and 2−. The 2+ to 0 states create a single A_1 state in the gap, which lies in the upper gap close to the Si CB edge. The vacancy undergoes a large relaxation in the + and 2+ states. For the - and 2− states, an additional level of B_1 symmetry is pulled out of the oxide CB, and this lies at about 0.8 eV below the oxide CB. This is the key trapping level responsible for V_T instability. Fluorine is the most effective passivant of defects in HfO_2.

References

1. Wilk, G.D., Wallace, R.M., and Anthony, J.M., High-K dielectrics: current status and materials properties considerations, *J. Appl. Phys.*, 89, 5243, 2001.
2. Robertson, J., High dielectric constant oxides, *Eur. Phys. J. Appl. Phys.*, 28, 265, 2004.
3. Hubbard, K.J. and Schlom, D.G., Thermodynamic stability of binary oxides in contact with Si, *J. Mater. Res.*, 11, 2757, 1996.
4. Robertson, J., Band offsets of wide-band-gap oxides and implications for future electronic devices, *J. Vac. Sci. Technol.*, B18, 1785, 2000.
5. Gusev, E.P. et al., Ultrathin high-K gate stacks for advanced CMOS devices, *Tech Digest—Int. Electron Devices Meeting*, 455, 2001.
6. Kerber, A. et al., On the thermal stability of atomic layer deposited TiN as gate electrode in MOS devices, *IEEE Electron Device Lett.*, 24, 87, 2003.
7. Bersuker, G. et al., Dielectrics for future transistors, *Mater. Today*, 7(1), 26, 2004.
8. Pantisano, L. et al., Dynamics of threshold voltage instability in stacked high-K dielectrics: role of the interfacial oxide, *VLSI Symp. Tech. Dig.*, 163, 2003.
9. Zafar, S. et al., Charge trapping related threshold voltage instabilities in high permittivity gate dielectric stacks, *J. Appl. Phys.*, 93, 9298, 2003.
10. Ragnarsson, L.A. et al., Molecular-beam-deposited yttrium-oxide dielectrics in aluminium-gated MOSFETs: effective electron mobility, *Appl. Phys. Lett.*, 78, 4169, 2001.
11. Fischetti, M.V., Neumayer, D.A., and Cartier, E.A., Effective electron mobility in Si inversion layers in MOS systems with a high-K insulator: The role of remote phonon scattering, *J. Appl. Phys.*, 90, 4587, 2001.
12. Ren, Z. et al., Inversion channel mobility in high-K high performance MOSFETs, *Tech. Digest IEEE IEDM*, 793, 2003.
13. Chau, R. et al., High-K/metal-gate stack and its MOSFET applications, *IEEE Electron Device Lett.*, 25, 408, 2004.
14. Hobbs, C. et al., Fermi-level pinning at the poly-Si/metal oxide interface, *VLSI Symp. Tech. Dig.*, 9, 2003.
15. Shiraishi, K. et al., Oxygen vacancy induced substantial threshold voltage shifts in the Hf-based high-K MISFET with p+ poly gates—a theoretical approach, *Jpn. J. Appl. Phys.*, 43, L1413, 2004.
16. Lucovsky, G., Transition from thermally grown gate dielectrics to deposited gate dielectrics for advanced silicon devices: A classification scheme based on bond ionicity, *J. Vac. Sci. Technol.*, A19, 1553, 2001.
17. Robertson, J., Interfaces and defects of high-K oxides on silicon, *Solid-State Electron.*, 49, 283, 2005.
18. Peacock, P.W. and Robertson, J., Band offsets and Schottky barrier heights of high dielectric constant oxides, *J. Appl. Phys.*, 92, 4712, 2002.
19. Robertson, J., Xiong, K., and Clark, S.J., Band structure of functional oxides by screened exchange and weighted density approximation, *Phys. Stat. Solidi. B.*, 243, 2054, 2006.
20. Foster, A.S. et al., Structure and electrical levels of point defects in monoclinic zirconia, *Phys. Rev. B*, 64, 224108, 2001.

21. Foster, A.S. et al., Vacancy and interstitial defects in hafnia, *Phys. Rev. B*, 65, 174117, 2002.
22. Kang, A.Y., Lenahan, P.M., and Conley, J.F., Electron spin resonance observation of trapped electron centers in atomic-layer-deposited hafnium oxide on Si, *Appl. Phys. Lett.*, 83, 3407, 2003.
23. Ikeda, M. et al., First principles molecular dynamics simulations for amorphous HfO_2 and $HfSiO_2$, *Mater. Res. Soc. Symp. Proc.*, 786 E5.4, 2004.
24. Scopel, W.L. et al., Comparative study of defect energetics in HfO_2 and SiO_2, *Appl. Phys. Lett.*, 84, 1492, 2004.
25. Torii, K. et al., Physical model of BTI, TDDB, and SILC in HfO_2-based high-K gate dielectrics, *Tech. Digest IEEE IEDM*, 129, 2004.
26. Shen, C. et al., Negative U traps in HfO_2 gate dielectrics and frequency dependence of dynamic BTI in MOSFETs, *Tech. Digest IEDM*, 733, 2004.
27. Feng, Y.P., Lim, A.T.L., and Li, M.F., Negative U property of oxygen vacancy in cubic HfO_2, *Appl. Phys. Lett.*, 87, 062105, 2005.
28. Kralik, B., Chang, E.K., and Louie, S.G., Structural properties and quasiparticle band structure of zirconia, *Phys. Rev. B*, 57, 7027, 1998.
29. Bylander, D.M. and Kleinman, L., Good semiconductor band gaps with a modified local density approximation, *Phys. Rev. B*, 41, 7868, 1990.
30. Rushton, P.P., Tozer, D.J., and Clark, S.J., Nonlocal density functional description of exchange and correlation in silicon, *Phys. Rev. B*, 65, 235203, 2002.
31. Segall, M.D. et al., First principles simulation: Ideas, illustrations, and the CASTEP code, *J. Phys.: Condens. Matter*, 14, 2717, 2002.
32. Xiong, K. et al., Defect energy levels in HfO_2 high-dielectric-constant oxide, *Appl. Phys. Lett.*, 87, 183505, 2005.
33. Xiong, K., Robertson, J., and Clark, S.J., Defect energy states in high-K gate oxides, *Phys. Status Solidi B*, 243, 2071, 2006.
34. Sayan, S., Garfunkel, E., and Suzer, S., Soft x-ray photoemission studies of the $HfO_2/SiO_2/Si$ system, *Appl. Phys. Lett.*, 80, 2135, 2002.
35. Klein, B.M. et al., Structural properties of ordered high-melting temperature intermetallic alloys from 1st-principles total-energy calculations, *Phys. Rev. B*, 41, 10311, 1990; Kantorovich, L.N., Holender, J.M., and Gillan, M.J., The energetics and electronic structure of defective and irregular surfaces on MgO, *Surf. Sci.*, 343, 221, 1995.
36. Gavartin, J.L. et al., Negative oxygen vacancies in HfO_2 as charge traps in high-K stacks, *Appl. Phys. Lett.*, 89, 082908, 2006.
37. Broqvist, P. and Pasquarello, A., Oxygen vacancy in monoclinic HfO_2: A consistent interpretation of trap assisted conduction, direct electron injection, and optical absorption experiments, *Appl. Phys. Lett.*, 89, 262904, 2006.
38. Xiong, K. et al., Defect states in the high-dielectric-constant gate oxide $HfSiO_4$, *J. Appl. Phys.*, 101, 024101, 2007.
39. Pruneda, J.M. and Artacho, E., Energetics of intrinsic point defects in $ZrSiO_4$, *Phys. Rev. B*, 71, 094113, 2005.
40. Takeuchi, H., Ha, D., and King, T.J., Observation of bulk HfO_2 defects by spectroscopic ellipsometry, *J. Vac. Sci. Technol.*, A22, 1337, 2004.
41. Afanas'ev, V.V. et al., Band alignment between (100) Si and complex rate earth/transition metal oxides, *Appl. Phys. Lett.*, 85, 5917, 2004.
42. Nguyen, N.V. et al., Sub-bandgap defect states in polycrystalline hafnium oxide and their suppression by admixture of silicon, *Appl. Phys. Lett.*, 87, 192903, 2005.
43. Lucovsky, G., Band edge electronic structure of transition metal/rare earth oxide dielectrics, *Appl. Surf. Sci.*, 253, 311, 2006.
44. Walsh, S. et al., Process-dependent defects in $Si/HfO_2/Mo$ gate oxide heterostructures, *Appl. Phys. Lett.*, 90, 052901, 2007.
45. Zhao, C.Z. et al., Properties and dynamic behavior of electron traps in HfO_2/SiO_2 stacks, *Microelectron. Eng.*, 80, 366, 2005.
46. Mitard, J. et al., Characterization and modeling of defects in high-K layers through fast electrical transient measurements, in *Defects in High-K Gate Dielectric Stacks*, Gusev, E. (Ed.), Springer, Berlin, 2005, p. 73.

47. Cartier, E. et al., Fundamental understanding and optimization of PBTI in nFETs with SiO_2/HfO_2 gate stack, *Tech. Digest IEEE IEDM*, 321, 2006.
48. Houssa, M. et al., Trap assisted tunnelling in high permittivity gate dielectric stacks, *J. Appl. Phys.*, 87, 8615, 2000.
49. Gusev, E.P. et al., Advanced gate stacks with fully silicided (FUSI) gates and high-K dielectrics: Enhanced performance at reduced gate leakage, *Tech. Digest IEEE IEDM*, 729, 2004.
50. Wen, H.C. et al., Oxygen deficiency and fast transient charge-trapping effects in high-K dielectrics, *IEEE Electron Dev. Lett.*, 27, 984, 2006.
51. Casse, M. et al., Carrier transport in HfO_2/metal gate MOSFETs: Physical insight into critical parameters, *IEEE Trans. Electron Dev.*, 53, 759, 2006.
52. Zhu, W.J., Han, J.P., and Ma, T.P., Mobility measurement and degradation mechanisms of MOSFETs made with ultrathin high-K dielectrics, *IEEE Trans. Electron Devices*, 51, 98, 2004.
53. Ragnarsson, L.A. et al., Hall mobility in hafnium oxide based MOSFETs: charge effects, *IEEE Electron Device Lett.*, 24, 689, 2003.
54. Callegari, A. et al., Interface engineering for enhanced electron mobilities in W/HfO_2 gate stacks, *Tech. Digest IEEE IEDM*, 825, 2004.
55. Narayanan, V. et al., Process optimization for high electron mobility in nMOSFETs with aggressively scaled HfO_2 metal stacks, *IEEE Electron Device Lett.*, 27, 591, 2006; Ioeng, M. et al., Transistor scaling with novel materials, *Mater. Today*, 9(6), 26, 2006.
56. Kirsch, P.D. et al., Nucleation and growth study of atomic layer deposited HfO_2 gate dielectrics resulting in improved scaling and electron mobility, *J. Appl. Phys.*, 99, 023508, 2006.
57. Kirsch, P.D. et al., Mobility and charge trapping comparison for crystalline and amorphous HfON and HfSiON gate dielectrics, *Appl. Phys. Lett.*, 89, 242909, 2006.
58. Lee, B.H. et al., Gate stack technology for nanoscale devices, *Mater. Today*, 9(6), 32, 2006.
59. Demkov, A.A., Thermodynamic stability and band alignment at a metal-high-K dielectric interface, *Phys. Rev. B*, 74, 085310, 2006.
60. Akasaka, Y. et al., Modified oxygen vacancy induced Fermi level pinning model extendable to p-metal pinning, *Jpn. J. Appl. Phys.*, 45, L1289, 2006.
61. Robertson, J., Sharia, O., and Demkov, A., Fermi level pinning by defects in HfO_2-metal gate stacks, *Appl. Phys. Lett.*, 91, 132912, 2007.
62. Gavartin, J.L. et al., The role of nitrogen-related defects in high-K dielectric oxides: Density functional studies, *J. Appl. Phys.*, 97, 053704, 2005.
63. Umezawa, N. et al., First principles studies of the intrinsic effect of nitrogen atoms on reduction in gate leakage current through Hf-based high-K gate dielectrics, *Appl. Phys. Lett.*, 86, 143507, 2005.
64. Xiong, K., Robertson, J., and Clark, S.J., Passivation of oxygen vacancy states in HfO_2 by nitrogen, *J. Appl. Phys.*, 99, 044105, 2006.
65. Shang, G., Peacock, P.W., and Robertson, J., Stability and band offsets of nitrogenated high-dielectric-constant gate oxides, *Appl. Phys. Lett.*, 84, 106, 2004.
66. Tseng, H.H. et al., Defect passivation with fluorine in a Ta_xC_y/high-K gate stack for enhanced device threshold voltage stability and performance, *Tech. Digest IEEE IEDM*, 713, 2005.
67. Inoue, M. et al., Fluorine incorporation into HfSiON dielectric for Vth control and its impact on reliability for poly-Si gate pFET, *Tech. Digest IEEE IEDM*, 425, 2005.
68. Seo, K.I. et al., Improvement in high-K (HfO_2/SiO_2) reliability by incorporation of fluorine, *Tech. Digest IEEE IEDM*, 429, 2005.
69. Tse, K. and Robertson, J., Defect passivation in HfO_2 gate oxide by fluorine, *Appl. Phys. Lett.*, 89, 142914, 2006.

10

*Spectroscopic Studies of Electrically Active Defects in High-*K *Gate Dielectrics*

Gerald Lucovsky

CONTENTS

10.1 Introduction

This chapter deals with intrinsic defects in high-K dielectrics attributed to local bonding arrangements that differ from the intrinsic bonding arrangements of the host material independent of whether or not the particular thin-film dielectric is nanocrystalline or noncrystalline. As such this chapter addresses three different areas of experimental and theoretical research: (1) spectroscopic studies of the intrinsic electronic structure of the high-*K* dielectrics, as well as intrinsic bonding defects in these same dielectrics, (2) a theoretical framework for describing electronic structure and the intrinsic bonding defects, and (3) relationships between the defect properties identified spectroscopically, and defect studies on test device structures addressing (a) preexisting defects in as-deposited films, and also (b) noting the importance of defects generated by subjecting the most promising high-*K* dielectrics for radiation hard survivable electronics to ionizing radiations such as x-rays and γ-rays.

The field of high-*K* dielectrics now has been narrowed through about 10 years of worldwide research and technology studies, and the materials that have received the most attention are Hf-based dielectrics [1], including (1) ultrathin films of HfO_2 with a physical thickness of less than about 2 nm [2], and (2) noncrystalline Hf-based alloys, generally characterized as HfSiON [3], and more recently including specific pseudoternary alloy compositions such as $(HfO_2)_x(SiO_2)_y(Si_3N_4)_{1-x-y}$, with $SiO_2 \sim HfO_2 \approx 0.3$, and $Si_3N_4 \approx 0.4$ [4]. There has recently been renewed interest in Zr-based high-*K* dielectrics, especially in the context of using semiconductor substrates other than Si, strained Si, or Si, Ge alloys for advanced devices [1]. These nonsilicon substrates include Ge, SiC, and III–V semiconductors, such as GaAs [5,6].

There are several excellent published reviews that include discussions of high-*K* dielectrics; the one that is most timely with respect to the subject matter addressed in this chapter is a collection of articles edited by M. Houssa and published by the Institute of Physics in 2004 [1]. There are also many annual conferences sponsored by a wide range of technical societies that include a strong focus on high-*K* dielectrics as well.

10.1.1 Differences between High-*K*, SiO_2, and Si Oxynitride Gate Dielectrics

The driving force for introducing high-*K* dielectrics into advanced Si devices is the reduction of direct tunneling current through the gate dielectric; this contribution to the leakage current in a metal–oxide–semiconductor field effect transistor (MOSFET) is not significant for SiO_2 devices when the gate oxide physical thickness (t_{phys}) or equivalently, the equivalent oxide thickness (EOT) is more than 4 nm [7]. For other dielectrics, EOT is defined by

$$\mathrm{EOT} = \frac{k(\mathrm{SiO_2})}{k(\mathrm{dielectric})} t_{\mathrm{phys}} \tag{10.1}$$

Here $k(SiO_2)$ is nominally 3.9, and the designation high-*K* has been generally been applied to dielectrics with $k(\mathrm{dielectric}) > 15$, and more frequently >20–25. Thus, a high-*K* dielectric with dielectric constant of ~ 20 would yield approximately the same EOT as an SiO_2 dielectric that is about five times thinner. This was anticipated to reduce direct tunneling currents in ultrathin dielectrics by many orders of magnitude, greater than at least 5–10, and based on this criterion, to thereby extend EOT scaling to <1 nm, and perhaps as low as 0.5 nm.

Direct tunneling through an SiO_2 gate dielectric becomes a significant issue for $t_{phys}(SiO_2) = \mathrm{EOT} < \sim 3$ nm. The initial estimates of the end of the road for SiO_2 dielectrics were based on an assumption that tunneling leakage in excess of 1 A cm^{-2} would be detrimental to the operation of MOSFETs, particularly in complementary MOS (CMOS) circuit arrangements. For device quality thermally grown SiO_2 dielectrics, this level of direct tunneling occurs for an EOT of ~ 1.5 nm [7]. A rule of thumb for tunnel current scaling indicated that this current is increased by approximately one order of magnitude for each 0.2 nm (or 2 Å) decrease in $t_{phys}(SiO_2)$ below about 2–5 to 3 nm. An experimentally determined reference for establishing whether SiO_2 prepared by deposition or thermal oxidation is of device quality, is a direct tunneling current of approximately 10^{-2} A cm^{-2}

for $t_{phys}(SiO_2) = EOT = 2$ nm at a gate bias of 1 V in excess of the flat band voltage for an MOS capacitor (MOSCAP), or MOSFET.

There is a first generation alternative gate dielectric technology for MOSFETs that has extended Moore's law scaling for several generations before requiring the introduction of high-*K* dielectrics with $K > 15$–25 [1]. This approach was initially predicated on the use of silicon nitride (Si_3N_4), or more generally Si oxynitride dielectrics, as a replacement dielectric for thermally grown SiO_2 [8]. Si oxynitrides, by definition, are pseudobinary alloys of SiO_2 and Si_3N_4, $(Si_3N_4)_x(SiO_2)_{1-x}$, with only Si–O and Si–N bonds, and no Si–Si or O–N bonds. These alloys have dielectric constants that scale approximately linearly between those of SiO_2, ~3.9, and Si_3N_4, ~7.6. Since the direct tunneling current can be approximated by an exponential function of physical thickness, a factor of 2 in the dielectric constant, for same effective tunneling mass (m_e*) and tunneling barrier height (E_b) as in SiO_2, would then be expected to produce more than a 10 order of magnitude decrease in direct tunneling leakage for an Si_3N_4 dielectric with an EOT of 2 nm. However, since the effective mass is reduced by more than a factor of 2, from $\sim 0.55m_o$ in SiO_2 to $0.25m_o$ in Si_3N_4, where m_o is the free electron mass, and the tunneling barrier is reduced by about 1 eV from ~3.1 to 2.1 eV, the reduction in the direct tunneling current for an EOT = 2 nm is only a few orders of magnitude, consistent with what has been found experimentally [9,10]. This estimate is based on the following figure of merit scaling parameter (f_{dt}) for the direct tunneling current:

$$f_{dt} = k\sqrt{m_e{}^* E_b} \tag{10.2}$$

Here $f_{dt} = 4.96$ for SiO_2, increasing to 5.5 for Si_3N_4, accounting for only a modest decrease in the direct tunneling current. In the absence of the reductions in m_e^* and E_b for Si_3N_4 with respect to their values in SiO_2, f_{dt} for Si_3N_4 would have been a factor of 2 larger, 9.92, than the value for SiO_2. Since the direct tunneling is an exponential function of the value of f_{dt}, with a large negative prefactor, increases of 2 for f_{dt} correspond to very large tunneling current reduction, e.g., greater than approximately six orders of magnitude. Assuming that m_e^* and E_b vary linearly with alloy composition for Si oxynitride alloys, the tunneling current for a 50–50 Si oxynitride alloy was estimated to be about one order of magnitude less than for Si_3N_4, corresponding to $f_{dt} \approx 6.0$. This difference in direct tunneling has been confirmed experimentally [9,10], and has resulted in Si oxynitride becoming the dielectric of choice for commercial devices with EOT extending to about 1.1 nm, and with direct tunneling leakage currents of about 5 A cm^{-2} [11,12].

If the effective tunneling masses and tunneling barriers for high-*K* dielectrics such as HfO_2 were the same as that for SiO_2, then the f_{dt} would be increased by the ratio of the dielectric constant of HfO_2, ~20–25, to SiO_2, or by factor of ~5.1–6.4, with tunneling currents much reduced with respect to Si oxynitride alloys, making it relative easy to achieve EOTs < 1 nm. This same conclusion would apply even if the product of electron tunneling mass and tunneling barrier heights were reduced by a factor of 2 with respect to the empirical values determined for Si_3N_4. For this case, the tunneling current reductions with respect to SiO_2 would still be more than four orders of magnitude, even for EOTs ~1 nm [10].

These estimates for tunneling current reductions in HfO_2, and in nitrided Hf silicate alloys have been realized for EOTs ~1 nm, and have been the driving force for the worldwide research and technology investments in extending Moore's law scaling through the introduction of high-*K* dielectrics.

This anticipated result has been further realized in a recent announcement from Intel Corporation for the introduction of CMOS circuits with a high-*K* Hf alloy dielectric [11,12]. This heralds a new era of scaled devices that will sustain Moore's law scaling for at least another two and possibly three device generations.

10.1.2 Roadmap for This Chapter

This chapter identifies three different types of Hf-based medium ($K \sim 10$–15) and high-K dielectrics ($k > 20$) that can be classified by different thin-film morphologies: (1) nanocrystalline HfO_2 with a nanograin size of >4 nm, (2) nanocrystalline HfO_2 with a nanograin size of <2 nm, and (3) noncrystalline Hf silicate and Si oxynitride alloys. The electronic structures of these three different Hf-based dielectrics are differentiated by combinations of spectroscopic and high-resolution transmission electron microscopy imaging as well (this is not emphasized in this chapter). Their respective electronic structures can be compared by the application of an ab initio chemical bonding theory based on small terminated and embedded clusters [4,13] (Chapter 9). Intrinsic electronically active defects in these three different classes of dielectrics differ significantly, and have been addressed by spectroscopic methods, and by a local chemical bonding theory as well.

Section 10.2 presents spectroscopic studies for thick (>4 nm) high-K nanocrystalline dielectrics. Section 10.3 presents the theory of O monovacancy defects, illustrating the necessity to consider extended or multi-O-vacancy, rather than only monovacancies. Section 10.4 addresses defect reductions associated with dimensional, kinetic, and alloy effects. Section 10.5 discusses the interfacial issues for Ge CMOS devices, illustrating significant band gap reductions between (1) GeO_2 and Ge_3N_4, and (2) SiO_2 and Si_3N_4, respectively. Section 10.6 presents the results of electrical measurements, which together with the spectroscopic studies identify dielectric options that must be studied in more detail for applications in radiation hard survivable electronic circuits and systems. Section 10.7 summarizes the results addressed above.

10.2 Spectroscopic Studies of Thick (>4 nm) High-K Nanocrystalline Dielectrics

The distinctions between two different types of nanocrystalline transition metal/rare earth (TM/RE) atom elemental and complex oxides is a recent discovery, and is something that had not been anticipated by the behavior of other nanocrystalline thin-film materials [4,14,15]. However, in retrospect, it provides some important insights to other nanocrystalline solids with a length scale for bonding that separates the total regime of nanocrystallinity, e.g., from 2 to >100 nm into two different regimes with significant differences in electronic structure and defect properties. Since the TM elemental oxide dielectrics and TM Si oxynitride are clearly the high-K materials of choice for extending the current CMOS technology into the nano-CMOS regime of feature sizes, or generation nodes below 32 nm, the next several sections will focus on the electronic structure and intrinsic defects as determined spectroscopically, and correlate these with studies with electrical measurements for test devices with dielectrics with different compositions, physical thickness, and on Ge, as well as Si substrates. This section describes spectroscopic studies for TM elemental oxides with a physical thickness of at least 4 nm [4,15]. The next section addresses differences in defect states in two difference regimes of nanocrystallinity with length scales >4 nm as discussed immediately below, and $< \sim 2.5$ nm. The spectroscopic approach of this section is also extended to noncrystalline Ti, Zr, and Hf silicates, and more importantly to Ti, Zr, and Hf Si oxynitrides, leading to the discovery of a new class of low defect density and thermally stable noncrystalline gate TM Si oxynitride dielectrics, but with a very narrow compositional range.

10.2.1 Spectroscopic Studies of Intrinsic Electronic Structure in Thick TM Oxides

Even though the asymmetric trapping of electrons and holes has been identified as an important issue in HfO_2 and ZrO_2 dielectric films >4 nm in physical thickness [16–18], and these films have not been considered as candidates for scaled CMOS devices beyond the 45 and 32 nm process nodes, it is informative to develop a microscopic description of the electronic structure and defect bonding arrangements. This is first addressed for TiO_2, and then used as a basis for addressing the similarities and differences between the spectroscopic signatures of intrinsic bonding states and defects in HfO_2 and ZrO_2 [19]. In addition to identifying the local bonding of intrinsic O-atom vacancy defect states in TiO_2, this approach provides significant insights into the microscopic origin of qualitatively similar intrinsic defects in Zr and Hf elemental oxides as well, and additionally has played a significant role in better understanding the unique and special properties of the Hf and Zr Si oxynitride alloys.

All of the thin film results in this section are for samples deposited at low temperature, 300°C, by remote plasma-enhanced chemical vapor deposition (RPECVD). These films have been subsequently annealed at temperatures between 700°C and 900°C. Films prepared by atomic layer deposition (ALD) and reactive evaporation (RE) display essentially the same electronic states and defects as well. The initial focus is on the distinction between (1) the crystal field (C–F) splitting of d-states into threefold degenerate, T_{2g}, and twofold degenerate, E_g, groups as determined by the coordination and local symmetry of the TM atoms with respect to their O-atom neighbors, and then (2) the removal of these two- and threefold degeneracies by Jahn–Teller (J–T) bonding distortions that are accompanied by symmetry reductions at the TM atom bonding sites [12,20]. See Figure 10.1 for a schematic

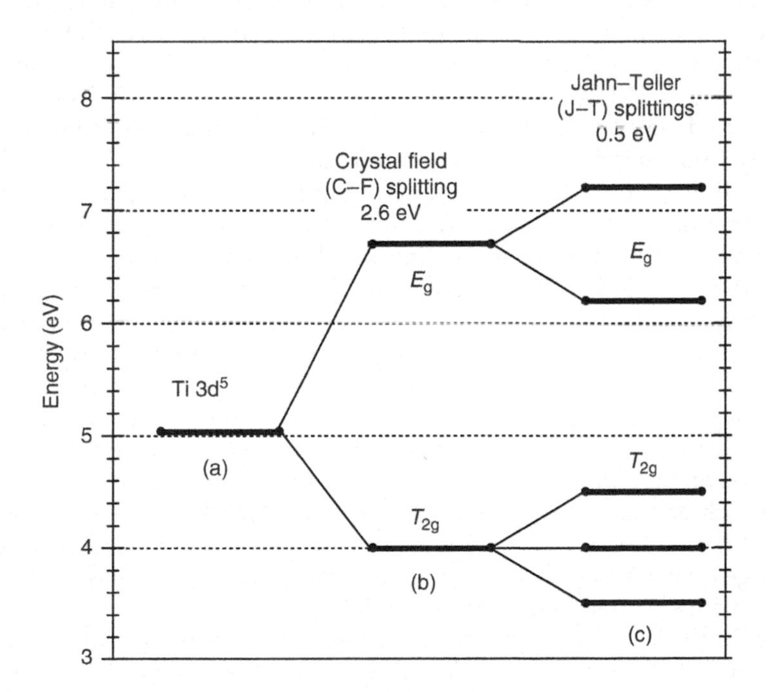

FIGURE 10.1
Schematic representation of C–F and Jahn–Teller splittings for Ti relative to the fivefold degenerate atomic 3d state: (a) the atomic Ti 3d state; (b) in an ideal octahedral field; and (c) in a distorted octahedral field. The energy scale is arbitrary, i.e., the energy of the triply degenerate T_{2g} state has been set to 4 eV.

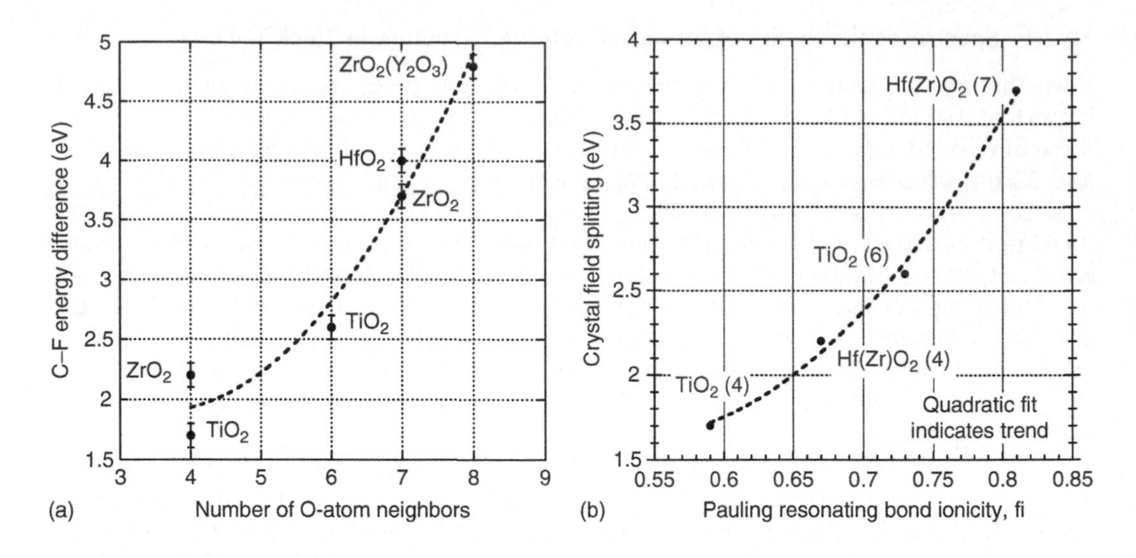

FIGURE 10.2
Average C–F energy differences determined from NEXAS O K_1 spectra as a function (a) the number of O-atom ligands for Ti, Zr, Zr(Y), and Hf elemental oxides, and (b) the Pauling resonating bond ionicity. (After Lucovsky, G., *J. Mol. Struct.*, 838, 187, 2007. With permission.)

representation of the C–F and J–T splittings for a sixfold coordinated TM, such as TiO_2. The C–F splitting of the atomic d-states of a TM or RE atom scales monotonically, but not linearly, with the number of O-atom nearest neighbors [20]. This scaling holds rigorously in a molecule or molecular ion, and in solids is influenced to a lesser extent by additional shells of negatively charged second nearest-neighbor O-atoms that are screened by a shell of positively charged TM/RE atoms. Departures from linearity are also associated with differences in bond ionicity [20]. Figure 10.2 contains plots of C–F splittings as determined experimentally from spectroscopic studies presented in Refs. [4,15,19]. The plot in Figure 10.2a is C–F splitting versus the coordination of the TM atom (the number of negatively charged O-atoms that make the dominant contribution to the C–F), and in Figure 10.2b, it is C–F splitting versus the Pauling resonating bond ionicity (which is determined by the atomic properties of the TM atom, as well as the bonding coordination) [4,15,19]. The relationship between atomic d-states and C–F and J–T splittings of these d-states in TM elemental oxides is addressed below. This approach is supported by an analogy between electronic states and bonding constraints, which is based on a paper published in 1973 that demonstrated a linear scaling between the electronic contribution to the dielectric constant that determines the index of refraction in the transparent region of insulators and semiconductors, and the infrared effective charges that determine the strength of infrared absorption by vibrational modes that contribute significantly to the static dielectric constant [21]. In the transition dichalcogenides, and the elemental oxides of this review, these effective charges have a significant dynamic component that is derived from dynamic charge transfer during phonon displacements [22,23].

This approach to electronic structure and defect states introduces an inherent relationship between atomic states and molecular orbital states. The atomic states must first be cast in symmetries consistent with the solid state bonding, i.e., the local site symmetry, and these representations are then combined into symmetry-adapted linear combinations (SALCs) of atomic states that make up the filled molecular orbital states of valence filled, and empty molecular orbital states of the conduction band. These are the symmetry-determined basis states, derived from TM and O atomic states, which are used as a basis for assigning features in the respective valence and conduction band spectra addressed below [4,15]. This relationship between atomic states and electronic spectra is essentially the same as the relationship between atomic motions in symmetric diatomic molecules and the normal mode motions of the solid state that are functionally equivalent to the SALCs of TM and O-atom states that contribute to final states molecular orbital states [24,25]. Additionally and equally important are the changes in π-state coupling between TM- and O-atom nearest neighbors in the two nanoscale regions with different grain size length scales (t_{grain}). This coupling is incoherent in the first nanoscale regime in which $t_{grain} < 2.5$ nm, and coherent in the second regime in which $t_{grain} > 3$ nm [4,15]. This distinction will become more transparent in later sections of this chapter.

Fundamental electronic structure and defect state features in O K_1, vacuum ultraviolet spectroscopic ellipsometry (VUV-SE), and soft x-ray photoelectron spectroscopy (SXPS) spectra are qualitatively different in these two regimes of grain size within an extended domain of nanocrystallinity between 1 and to more than >100 nm. The marked changes in electronic structure and defects, to be addressed later in this chapter, are associated with these different length scale regimes, $t_{grain} < 2.5$ nm, and $t_{grain} > 3$ nm, and have many of the same spectral signatures as the changes in Raman spectra and heat flow in the transition between the intermediate and stressed rigid phases in continous random network (CNRs) in noncrystalline dielectrics [24,25].

Figure 10.3a and b display, respectively, O K_1 (more often designated as O K, even though O K_1 is more consistent with subscript labeling, for other shells) and Ti L_3 core level near-edge x-ray absorption spectroscopy (NEXAS) results for nanocrystalline TiO_2. As noted earlier, unless otherwise indicated, all spectra displayed in this section are for nanocrystalline thin films that have been annealed in an inert nonoxidizing ambient at a temperature between 700°C and 900°C [4,15]. The transitions for the O K_1 spectra terminate in final states that are derived primarily from O 2p antibonding states mixed in SALC molecular orbitals with empty/antibonding Ti-atom 3d, 4s, and 4p states. The lowest energy states are π-state, and all other features are associated predominantly with σ-states. These final states display the same d-state average C–F energy splittings, as well the same J–T term splittings that are present in optical transitions between the valence and the conduction bands as addressed by SE [4]. The relative d-state energies are the same in

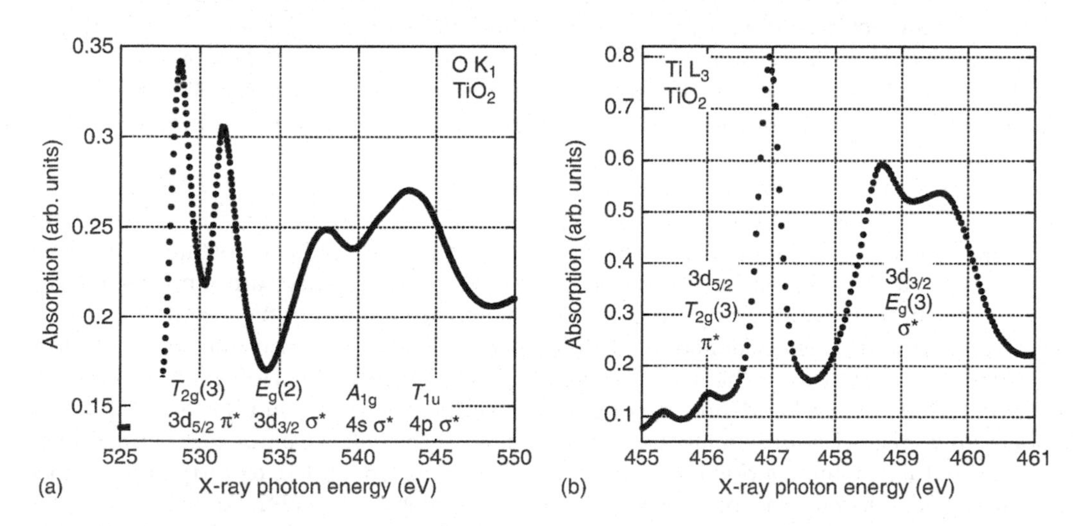

FIGURE 10.3
NEXAS spectra for nanocrystalline TiO_2: (a) O K_1 edge, and (b) Ti L_3 edge. (After Lucovsky, G. et al., *Jpn. J. Appl. Phys.*, 46, 1899, 2007. With permission.)

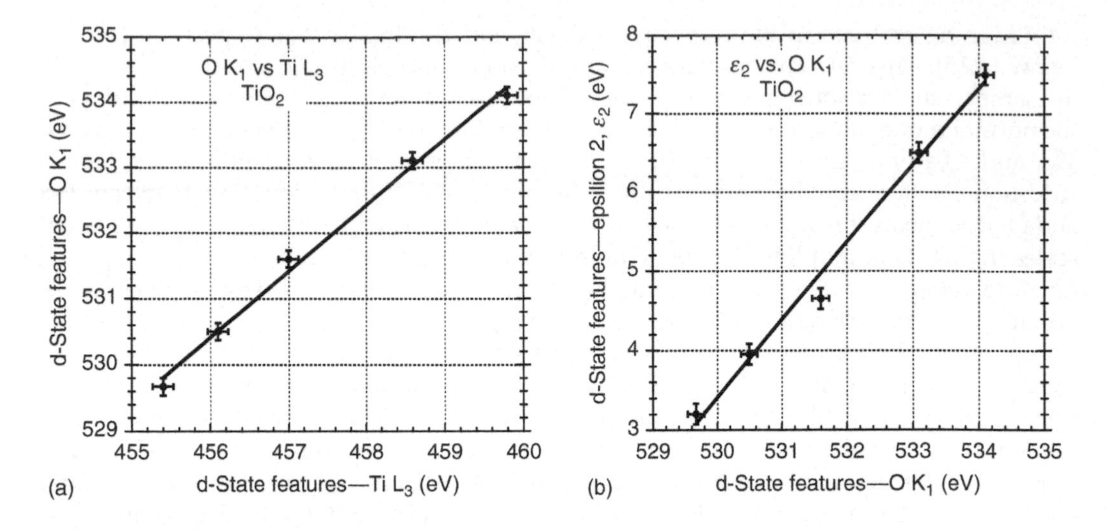

FIGURE 10.4
Photon energies of Ti d-state features: (a) O K_1 edge versus Ti L_3 edge, and (b) ε_2 versus O K_1 edge. (After Lucovsky, G. et al., *Jpn. J. Appl. Phys.*, 46, 1899, 2007. With permission.)

the O K_1 and Ti L_3 spectra, while the final state features are qualitatively markedly different due to different matrix elements or transition probabilities. As already noted above, the final states for O K_1 transitions are to antibonding O 2p states on the same O-atom that are mixed antibonding Ti d-states of the neighboring Ti-atoms, while the final states for the Ti L_3 transitions are from spin–orbit split Ti 2p core states to antibonding Ti d-states localized on the same Ti-atoms. The photon energies that are identified by the minima for second derivatives of these spectra identify the five d-state features, which are presented in Figure 10.4a. The *x*-axis values in Figure 10.4a are from the Ti L_3 differentiation, and the *y*-axis values are from the O K_1 differentiation. The five spectral features in each of the spectra are assigned to J–T d-state splittings, i.e., removal of three- and twofold degeneracies, respectively, from the C–F split T_{2g} and E_g d-states [4,15]. The energy spacing between the average values of these transitions yields a C–F field splitting of 2.6 ± 0.2 eV. The J–T splittings for the T_{2g} and E_g states are $\sim 0.7 \pm 0.2$ eV, and smaller by about a factor of about 4 than the average C–F energy difference. The linearity of the relationship in Figure 10.4a and the slope ~ 1 support this interpretation of the C–F and J–T splittings.

Figure 10.5a and b display, respectively, (a) the epsilon 2, ε_2, spectrum for TiO_2, obtained from analysis of visible-VUV-SE (vis-VUV-SE) measurements, and (b) the valence band spectrum for TiO_2 is obtained from SXPS measurements [4,15]. The d-states have been extracted from the minima in the second derivative of the ε_2 spectrum, and the relative energies in these are compared in Figure 10.4b with the energies of the d-state features in the O K_1 spectrum, also obtained by differentiation. This plot is also linear with a slope approximately equal to 1, the same as the slope in Figure 10.4a. However and equally important, the valence band, and ε_2 conduction band edge spectra in Figures 10.5b and a, respectively, display additional features as a doublet at the valence band edge, and a broad spectral feature at the conduction band edge. These are attributed to intrinsic occupied bonding, and unoccupied antibonding states, respectively, of O-atom defects. For TiO_2, these defect states are consistent with the electronic states of Ti^{3+} atomic species that are associated with Ti_2O_3 bonding, or equivalently extended O-atom vacancies, pinned and clustered at the grain boundaries of these nanocrystalline films [4,5,15].

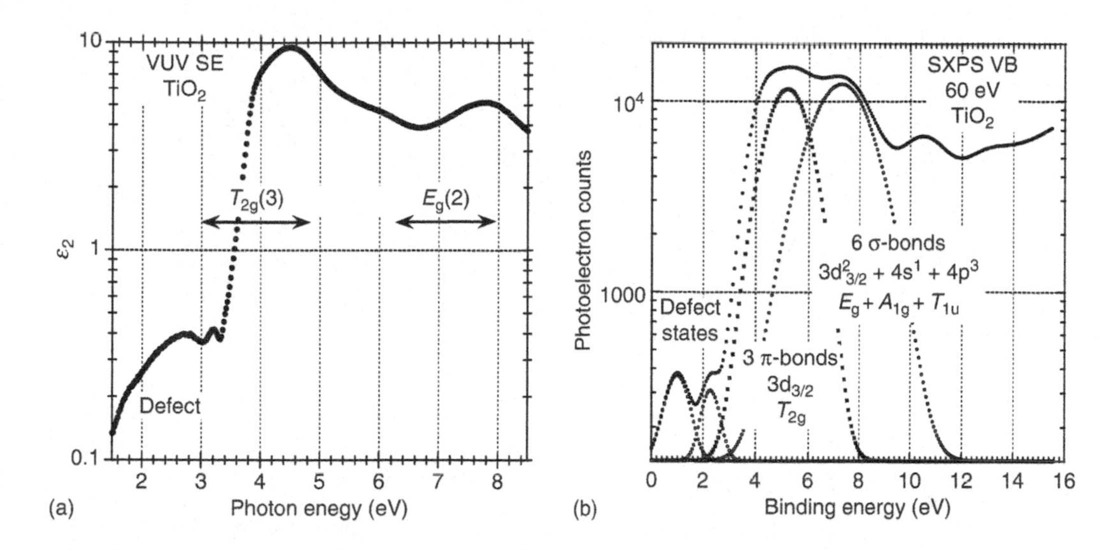

FIGURE 10.5
(a) Vis-VUV-SE ε_2 spectrum and (b) SXPS VB structure versus photon energy for TiO_2. (After Lucovsky, G. et al., *Jpn. J. Appl. Phys.*, 46, 1899, 2007. With permission.)

Before continuing to the focus on intrinsic electronic structure, and defects, and presenting spectra for ZrO_2 and HfO_2, there are two quantitative aspects of the spectra for TiO_2 that are markedly different for the corresponding spectra of ZrO_2 or HfO_2. The extraction of five d-state features directly from the Ti L_3 (Ti $2p_{3/2}$ core level to Ti $3_{d5/2}$ empty state) spectrum is possible for TiO_2 because the core lifetime broadening is smaller than the energy separation between the J–T spectral components, <0.2 eV compared with 0.7 eV on average [26]. However, for the Zr M_3 (Zr 3p to Zr 4d) and Hf N_3 (Hf 4p to 5d) transitions, the respective Zr and Hf core hole lifetimes (τ_{core}) are significantly shorter, and via the uncertainly principle, $\Delta\tau_{core}\Delta E_\lambda > 2\pi h$, have significantly increased line-widths (ΔE_λ); $\Delta E_\lambda \sim 2.5$ eV for ZrO_2, and $\Delta E_\lambda > 5$ eV for HfO_2 [27]. This makes it possible to extract only the average C–F splitting of ~3.5 eV for the Zr 4d-states from the Zr M_3 spectrum of ZrO_2, but impossible to extract the corresponding C–F splitting of ~4 eV for Hf 5d states in HfO_2 [28]. On the other hand, the average C–F and J–T splittings are readily obtained for the 4d states in ZrO_2, and 5d states in HfO_2 from second derivative spectra of the respective O K_1 edges [4,15].

Figure 10.6a presents the O K_1 edge spectra for HfO_2 and ZrO_2; each of these spectra show d-state, s-state, and p-state features, with an asymmetry in the lowest energy d-state feature that is associated with a cooperative J–T effect splitting [4,15]. Figure 10.6b presents the SXPS valence band spectrum for HfO_2, indicating occupied defect states above the valence band edge [4,15]. Figure 10.6c displays the VUV-SE absorption edge spectrum, α, for HfO_2, displaying defect features ~1–3 eV below the intrinsic d-state conduction band edge [4,15]. The α spectrum contains d-state features at approximately the same relative energies as the ε_2 spectrum, and more importantly, the energy differences between d-state features in each of these band edge spectra are the same to within an experimental uncertainty of ~0.2 eV. These spectral features have been used to create the electron energy level diagram for HfO_2 in Figure 10.7a. SXPS valence band, and VUV-SE α and ε_2 spectra for ZrO_2 (not shown) are essentially the same as those in Figure 10.6a and b for HfO_2, and therefore the energy level diagram in Figure 10.7a applies to ZrO_2 as well, with only small decreases of ~0.2 eV in the band gap energy, and the average splitting of the defect states at the respective band edges. Photoconductivity and cathodoluminescence (CLS) spectra support the intrinsic character of these defect states in the nanocrystalline elemental oxides

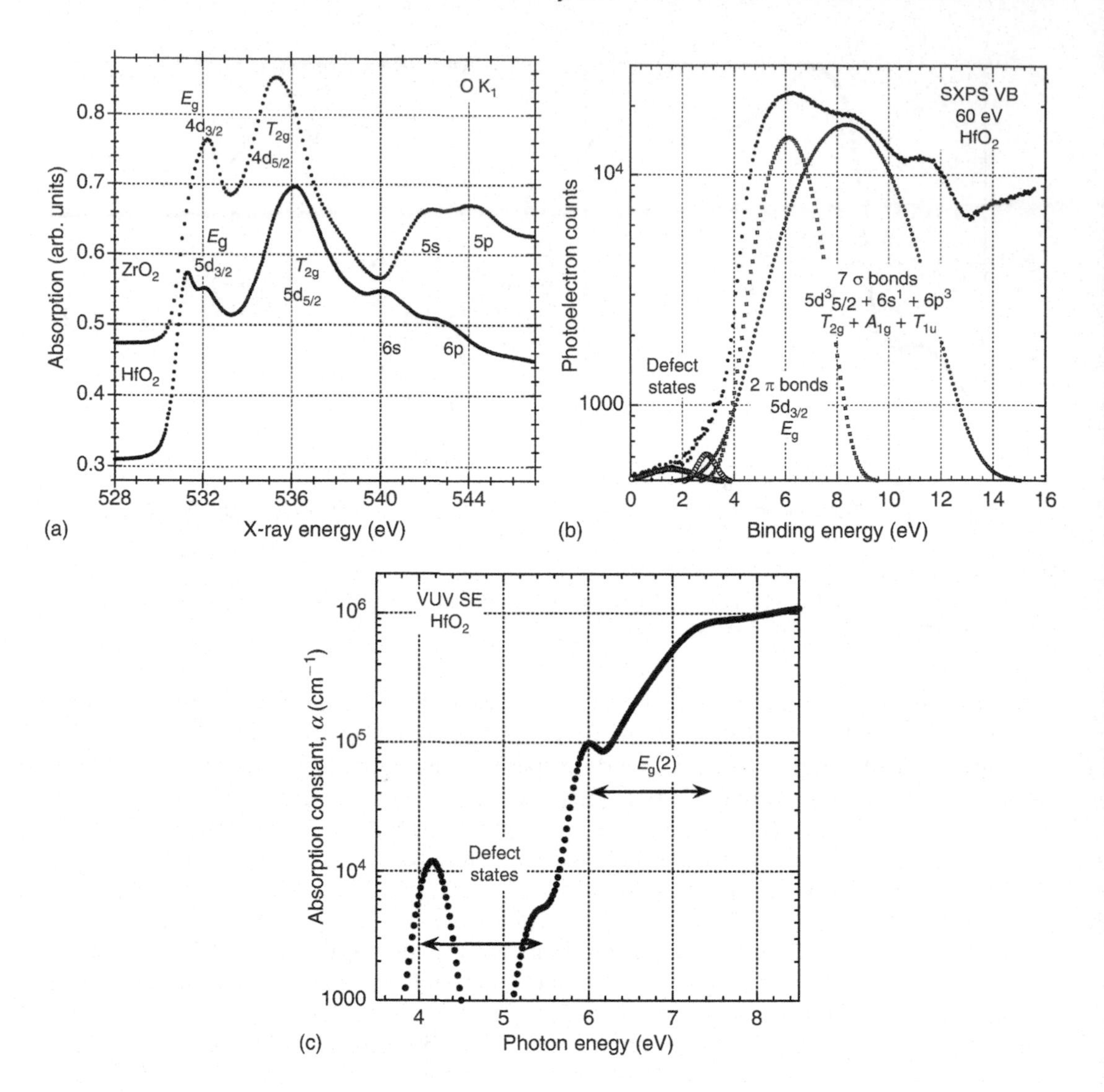

FIGURE 10.6
(a) O K_1 edge spectra for HfO_2 and ZrO_2 showing d-, s-, and p-state features, (b) SXPS VB structure versus photon energy for HfO_2, and (c) Vis-VUV-SE ε_2 spectrum. (After Lucovsky, G. et al., *Jpn. J. Appl. Phys.*, 46, 1899, 2007. With permission.)

[4,15]. Figure 10.7b is a similar energy level diagram derived from the vis-VUV ε_2 conduction band and SXPS valence band spectra of TiO_2. There are important qualitative differences between the energy level schemes in Figure 10.7a and b. There are four distinct and different defect states for HfO_2, two at each band edge, whereas the same two defect state features for TiO_2 appear in both the SXPS valence band, and vis-VUV-SE ε_2 (or α) spectra. These aspects of the energy band schemes are discussed in more detail in Section 10.3.

Figure 10.8a presents the CLS spectrum for low-energy electron-excited nanoscale (LEEN) luminescence spectroscopy, also designated as depth-resolved CLS (DRCLS) [29,30]. This has been used to probe the (1) defect states in as-deposited relatively thick (~100 nm) HfO_2 films grown by RPECVD, as well as (2) changes in these states after high temperature (900°C) annealing [30]. Figure 10.8a presents the DRCLS spectrum for a 100 nm thick HfO_2 film in which the electron beam energy (E_b) has been adjusted to provide excitation of energetic electron–hole pairs through the volume of the film. This spectrum

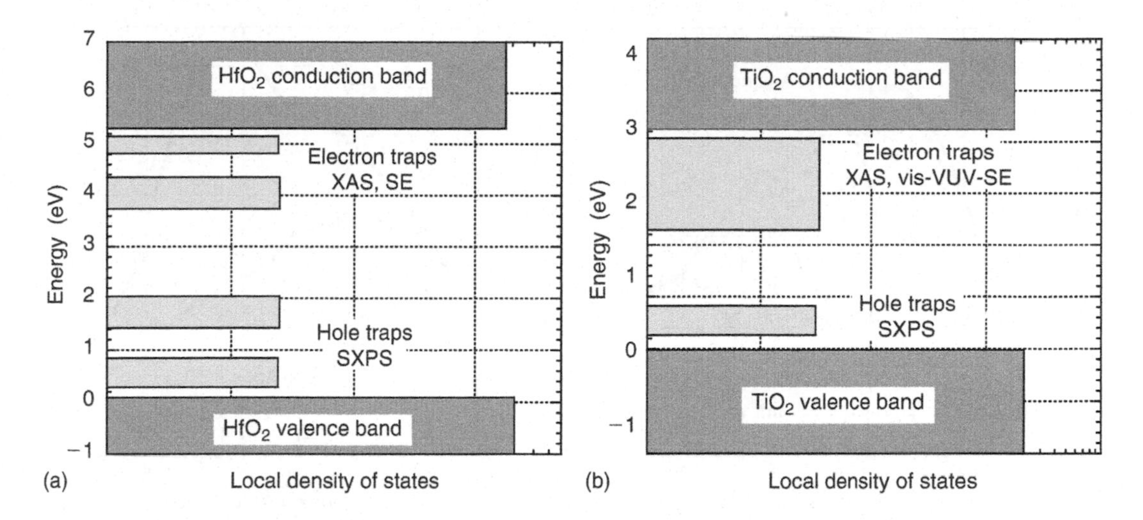

FIGURE 10.7
Band edge and preexisting defect energy level diagrams from spectroscopic studies for (a) HfO_2 and (b) TiO_2. (After Lucovsky, G., *J. Mol. Struct.*, 838, 187, 2007. With permission.)

has been deconvolved into four Gaussian bands that indicate four different optical transition energies in the 900°C annealed HfO_2 films that are assigned to defect-associated radiative transition energies, respectively, of 2.7, 3.4, 4.2, and 5.5 eV. These transition energies are in excellent agreement with the energy level diagram in Figure 10.7a for HfO_2 created from NEXAS, SE, and SXPS studies of the electronic structure of thick (>4 nm) HfO_2 films that have been annealed at temperatures between 700°C and 900°C. This diagram has been expanded in Figure 10.8b to indicate seven transition energies originating in the defect states below and at the conduction band edge [30]. Four of these are intradefect state transitions. Two of the remaining three are between defect states and terminate at the valence band edge, and the final transition is a band gap transition from

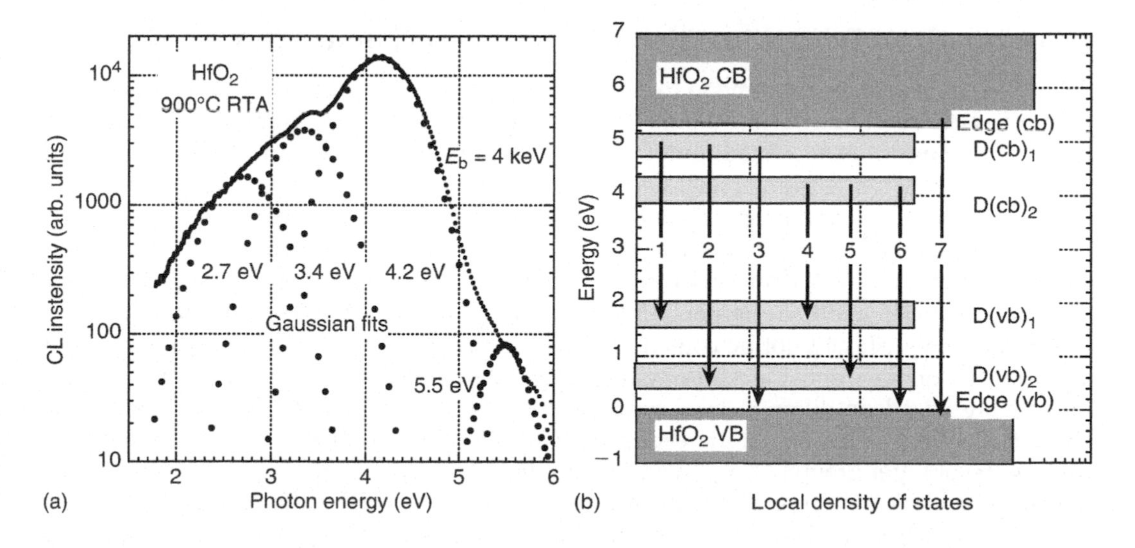

FIGURE 10.8
(a) Gaussian fits to four transition bands in the DRCLS spectrum for $E_b = 4$ keV from a 100 nm thick HfO_2 film after a 900°C anneal, and (b) electron energy level diagram from Figure 10.7a indicating radiative transitions contributing to Gaussian fit in (a).

(continued)

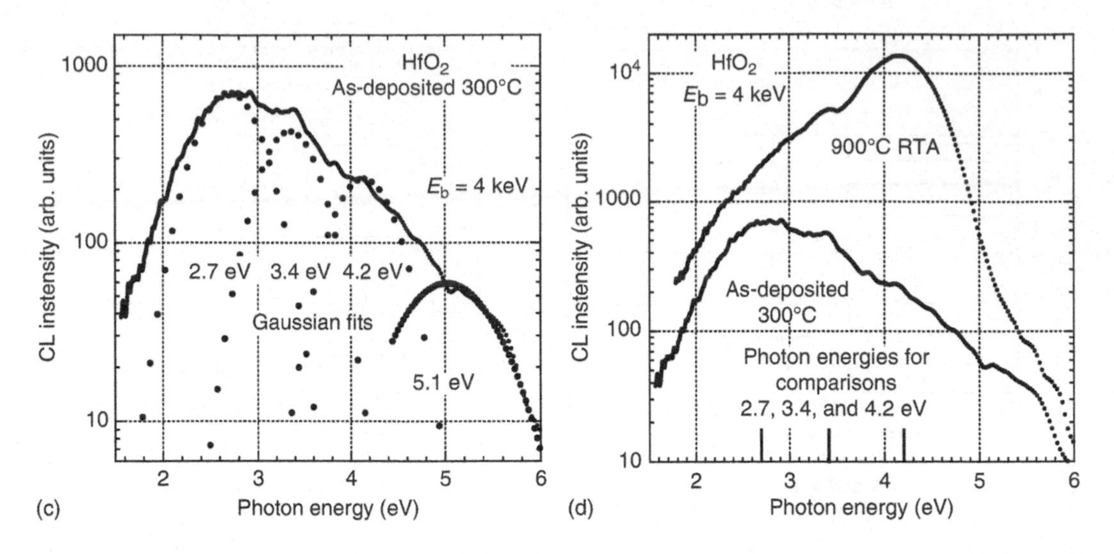

FIGURE 10.8 (continued)
(c) Gaussian fits to four transition bands in the DRCLS spectrum for $E_b = 4$ keV from a 100 nm thick HfO_2 film as-deposited at 300°C, and (d) compares the spectra in (a) and (c). (After Strzhemechy, Y.M. et al., *J. Vac. Sci. Technol. B*, 26, 232, 2008. With permission.)

states at the bottom of the conduction to top of the valence band. In addition, three of these transitions are at essentially the same energies, 3.4 ± 0.4 eV, 4.25 ± 0.4 eV, and 5.2 ± 0.4 eV. The four transition energies estimated from the energy level diagram in Figure 10.4a are 2.7 ± 0.4 eV, 3.4 ± 0.4 eV, 4.25 ± 0.4 eV, and 5.5 ± 0.4 eV, in excellent agreement with the deconvolution of the DRCLS spectrum into Gaussian bands [30]. DRCLS, LEEN spectra have also been obtained for the as-deposited HfO_2; these spectra are consistent with the changes in defect spectra discussed below between as-deposited films (RPECVD at 300°C) and HfO_2 annealed at 900°C.

10.3 Theory of Defects in Thick HfO_2 Nanocrystalline Films

Several proposals have been made to describe the microscopic origins of the electron and hole traps in ZrO_2 and HfO_2. A distinction has been made between (1) preexisting defects, those occurring in as-deposited and/or thermally annealed films, and (2) stress-induced defects, those resulting from accelerated electrical stress, or exposure to ionizing radiation [2,4,31]. Some of the first proposals for the origins of preexisting defects are based on the calculations of the Robertson and Shluger groups [32–34]. These groups have addressed electronic states within the forbidden energy gap associated with different charge states for O-atom vacancies and O-atom and molecule interstitials [15]. If these are present in the same films, they should not be considered to be a Frenkel defect pair, since they are not assumed to derive from the same geminate creation process. The calculations of these groups are based on different local bonding models: (1) the ideal CaF_2 structure by the Robertson group [32], and (2) the experimentally determined $Hf(Zr)O_2$ monoclinic structure by the Shluger group [33,34]. Even though the Hf(Zr) atoms are sevenfold coordinated in the monoclinic structure and eightfold coordinated as in CaF_2, the energies of defects associated with different charge states and electron occupancies, e.g., neutral, or negatively or positively charged vacancies and interstitials are approximately the same with respect to the conduction and valence band edges. O-atom vacancy defect states with different charge states are in the upper half of the forbidden band gap, while the O-atom interstitial defect states are within about 1–2 eV of the valence band edge. In addition, a DRCLS, LEEN spectrum has also been obtained for the as-deposited HfO_2 film, which is displayed in Figure 10.8c [30]. This spectrum is qualitatively different than the DRCLS, LEEN spectrum for the annealed film in Figure 10.8a, peaking at the lowest photon energies, in contrast to the reverse dependence on photon energy in Figure 10.8a. These differences are compared in Figure 10.8d, and will be addressed in more detail in Section 10.4.

In the context of the Robertson and Shluger model calculations [32–34], the trapping/transport state below the conduction band edge is associated with either an empty, or partially occupied O-atom vacancy [30,35]. The conduction band edge defects in ε_2 and absorption, α, spectra derived <0.5 eV below the respective conduction band edges in Hf(Zr)O_2, and ~1.8–2 eV below the spectral peak, the lowest d-state E_g spectral features, are assigned to transitions from the valence band edge to unoccupied states of O-vacancies. The CLS feature in HfO_2 at ~4.2 eV [30,35] has been assigned to a transition from the deeper O-atom vacancy state to the valence band edge. In the spirit of these same calculations, occupied, or partially occupied states above the valence band edge must be assigned to O-atom interstitials, and these may contribute to hole trapping [16–18,31,35].

However, there is a potentially significant problem with these calculations; they utilize TM d-states constructed from orthogonalized plane wave basis sets [32,33]. While this approach is valid for O 2s, Zr 5s and 5p, and Hf 6s and 6p states, because of sufficiently rapid convergence, it is known that the convergence is not fast enough for TM 3d, 4d, and 5d states in oxides; this issue has been resolved for NiO in the 1974 seminal paper of Koiller and Falicov [36], who point that atomic TM d-states must be used in the basis set, rather than an orthogonalized plane wave basis. In a recent publication, Gavartin and coworkers [34], used an atomic d-state basis set, and their results for HfO_2 should then compared with the Xiong et al. paper from the Robertson group which uses a plane wave d-state basis [33].

The primary difference between these two calculations is that the occupied states for the neutral and negatively charged O-atom vacancies are just below midgap for the atomic d-state calculation of Ref. [34], and above midgap in Ref. [33]. In neither of these calculations are the O-vacancy occupied states close to the valence band edge, as they are in the SXPS valence band spectrum for HfO_2 in Figure 10.6b. This suggests that calculations for isolated O-vacancies, even if they are done in the spirit of Ref. [36] with atomic d-states for the TM atom, cannot explain what is observed experimentally, without invoking defects states associated with O-atom interstitials as well as vacancies. In particular, to account for all four spectral features in Figure 10.7a, it would be necessary to include charged O-atom vacancy and interstitial sites, and the combinations used would be subject to a condition of charge neutrality, on the one hand, and not be limited by vacancy–interstitial recombination.

A second model proposed by the Lucovsky research group and addressed in Refs. [4,37] attributes the relatively high density of preexisting trapping states, $\sim 10^{12}$ defects cm^{-2}, or equivalently $\sim 10^{18}$ defects cm^{-3}, to grain-boundary defects that are intrinsic to nanocrystalline oxygen-deficient TM oxides. The grain-boundary description is supported by the SE studies of the Berkeley group that show that defect state spectral features decrease with increasing annealing temperature [35]. This decrease is in the spectral amplitude is consistent with increases in crystallite size as a function of postdeposition annealing at increasingly higher temperatures. A microscopic basis for the grain-boundary model is initially proposed for TiO_2, and then extended to HfO_2 and ZrO_2 as well. Even though the spectroscopic data for TiO_2 are qualitatively similar to those for HfO_2 and ZrO_2, the relative energies of defect features in the respective forbidden band gaps in Figure 10.7a and b for HfO_2 and TiO_2 are quantitatively different. In HfO_2 in Figure 10.7a, the higher lying occupied state is less than 2 eV above the top of the valence band edge, and the lower defect state in the upper half of the band gap is at least 4 eV above the valence band edge.

In contrast, for TiO_2 in Figure 10.7b, the higher lying valence band edge state and broad feature below the conduction band edge are at essentially the same energy, at 2.3 ± 0.2 eV above the valence band edge, suggesting that these features are associated with a partially occupied state of the same defect. The broad feature in the ε_2 spectrum for TiO_2 with a spectral peak $\sim 2.6 \pm 0.1$ eV, is essentially the same as the charge transfer absorption feature of Ti^{3+} in the $Ti(H_2O)_6{}^{3+}$ complex and in Ti_2O_3 [20,26]. This suggests that the local atomic structure of the intrinsic Ti-atom defect in TiO_2 also has a formal valence of 3+, Ti^{3+}. In the grain-boundary model, the Ti^{3+} ions are in grain-boundary bonding arrangements with a Ti_2O_3 stoichiometry. Based on the electronic structure for the $Ti(H_2O)_6{}^{3+}$ complex in Ref. [26], the lowest energy ground state for Ti^{3+} is a nonbonding T_{2g} state, and the higher lying state is an E_g state. The energy difference between these states is approximately equal to average C–F splitting for a sixfold coordinated Ti-atom [13,26]. This average C–F splitting can then be obtained from Figure 10.4a and b as 2.6 ± 0.1 eV.

A similar model is proposed for the valence and conduction band paired defects in Figure 10.7a for HfO_2. The average C–F splittings between T_{2g} and E_g features in the O K_1 edge spectra of monoclinic nanocrystalline HfO_2 and ZrO_2 are essentially the same, and equal to 3.9 ± 0.1 eV. This average energy separation between the conduction and valence band edge defects for HfO_2 in Figure 10.7a is 3.6 ± 0.2 eV, approximately the same as the C–F splitting in the XAS spectrum. The construction of a similar energy level diagram for ZrO_2 (not shown in this chapter) also yields essentially the same energy level difference of 3.6 ± 0.2 eV. These comparisons are consistent with the defects in HfO_2, as well as ZrO_2, being associated, respectively, with Hf^{3+} and Zr^{3+} bonding in suboxide, or O-deficient bonding arrangements clustered on grain boundaries. As already noted earlier, these defects are assigned to O-atom extended vacancies (e.g., divacancies, trivacancies, etc.) clustered and pinned at internal grain boundaries. As such they can also be described in terms of Ti^{3+}, Zr^{3+}, and Hf^{3+} electronic states [4,37]. This assignment was recently confirmed by studies of electron spin resonance (ESR) [38]. These studies rule out the ESR defects being at isolated vacancies as proposed initially [32–34], but instead are consistent with oxygen atom vacancy complexes.

Based on differences in absorption constant alone, the densities of electrons in these band edge defect states would be ~0.5 to more than 3% of the occupied density of valence states, and about 50–100 times larger than defect state densities determined from electrical measurements. This difference is explained by noting that matrix elements for spatially localized defect–dopant states in semiconductors are larger by one to two orders of magnitude than those for direct valence to conduction band transitions [39]; e.g., this difference can be estimated using the N-sum rule as applied to GaAs band edge defect features [40]. The effective density of valence band atomic states in III–V direct band gap semiconductors is counted by absorption that extends from the band edge, ~1.5 eV for GaAs to >30 eV, whereas band edge defects $\sim 10^{18}$ cm^{-3} yield measurable and comparable absorption in a spectral range within 0.5 eV of the band gap at 1.4 eV [40].

The spectra in Ref. [40] for doped GaAs display band edge tails that are very similar to those in Figure 10.6 of this chapter. Applying the same approximation to band edge absorption in other TM/RE oxides gives a volume density of band edge defects $\sim 10^{-4}$ in good agreement with electrical measurements [2,4].

This model, including the effect of matrix element enhancement, can be quantified by describing O-deficient $TiO_{2-\delta}$, $ZrO_{2-\delta}$, and $HfO_{2-\delta}$ as a mixture that contains Ti_2O_3, Zr_2O_3, and Hf_2O_3 fractions, or equivalently, local bonding arrangements that include the trivalent ions, Ti^{3+}, Zr^{3+}, and Hf^{3+}. If δ is the relative concentration of defects, $\sim 10^{18}$ defects cm^{-3}/3×10^{22} states cm^{-3} or $\sim 3.3 \times 10^{-5}$, then for any of these oxides

$$HfO_{2-\delta} = (1 - 2\delta)HfO_2 + 2\delta Hf_2O_3 \tag{10.3}$$

The values of δ are consistent with ~0.006% deviations from the HfO_2 elemental oxide stoichiometry.

The multiplicity of defect energy states for HfO_2 (and ZrO_2), as displayed in Figure 10.7a, is also supported by qualitative differences between radiation-induced charge trapping states in nanocrystalline HfO_2 and noncrystalline Hf silicate alloys that have been integrated into test devices with high-*K* gate stacks [41–44]. Similar to the measurements reported in Refs. [16–18,31,34], these results also indicate asymmetric trapping of electrons and holes. Comparisons have also been made between radiation-induced defects in MOS devices with SiO_2 gate dielectrics and those with high-*K* dielectrics. These have established qualitatively different trapping efficiencies for electrons and holes in the Hf-based alternative high-*K* dielectrics subjected to x-ray and γ-ray irradiation [41–44]. The mechanism proposed for radiation-induced charged defect generation and trapping in devices with Hf silicate gate dielectrics is essentially the same as that proposed for devices with SiO_2 gate dielectrics [41]. For an SiO_2 MOS device with a metal gate, Si substrate, and irradiation with x-rays or γ-rays under positive gate bias, electrons are swept out at the gate electrode, while holes are transported to the Si/SiO_2 interface [43]. A fraction of these holes are trapped within the oxide, typically giving rise to a net positive oxide-trap charge density that increases linearly with the γ-ray integrated dose.

In another set of experiments [42], it was shown that, for HfO_2 gate dielectric stacks, the density of trapped charge in the oxide depends strongly on the film thickness and processing conditions. The midgap voltage shifts, extracted from an analysis of capacitance–voltage (*C–V*) curves, are a nonlinear function of dose, being relatively small up to a critical concentration at which the voltage shift increases as a function of dose with a slope that exceeds the linear changes in Hf silicate alloys and SiO_2. In this regime, the fixed charge is also positive, but hysteresis in the *C–V* trace is inversely proportional to the magnitude of the radiation-induced voltage shift, indicating compensating, but unequal hole and electron trapping.

Other tests of the O-atom vacancy–interstitial nanocrystalline grain-boundary defect combined model of this chapter additionally offer possible engineering options for incorporation of medium- and high-*K* dielectrics into advanced Si devices. These engineering solutions/device options are based on spectroscopic and electrical studies of (1) noncrystalline IVB TM Si oxynitride alloys [4,35,44], (2) ultrathin HfO_2 and ZrO_2 dielectrics, and (3) ultrathin phase-separated Hf and Zr silicate alloys with relatively high TM oxide content, 50%–80% [4].

10.4 Defect Reductions Associated with Dimensional, Kinetic, and Alloy Effects

10.4.1 Noncrystalline Films

It has been well documented that Zr and Hf binary silicate alloys such $(ZrO_2)_x(SiO_2)_{1-x}$, are unstable at all compositions, x, with respect to chemical phase separation (CPS) into SiO_2 and the respective TM oxides, e.g., (ZrO_2) or (HfO_2) [45,46]. Quaternary alloys composed of Si, Hf and Zr, N and O, e.g., commonly designated as ZrSiON and HfSiON, respectively, have been shown stable against CPS, but only for a limited range of compositions; see Refs. [47,48]. This stability also includes Ti Si oxynitride alloys [48]. This section addresses quaternary alloys of Hf, Zr, Ti, and Si, O and N, and on compositions that are better characterized as pseudoternary alloys of the respective TM oxides, and SiO_2 and Si_3N_4 [4,48]. The stability with respect to CPS of high Si_3N_4 content TM Si oxynitrides, $(SiO_2)_x(Ti/Zr/HfO_2)_y$ and $(Si_3N_4)_{1-x-y}$, has been established by spectroscopic studies [4,47,48]. The compositions that are of interest generally lie on a tie line from Si_3N_4 to the

TM silicate composition with equal concentrations of SiO_2 and the TM oxide, i.e., $(SiO_2)_x(Ti/Zr/HfO_2)_x(Si_3N_4)_{1-2x}$, and the compositions that are stable with respect to CPS have x ~0.30 ± 0.02, and therefore Si_3N_4 concentrations of 0.40 ± 0.04 [48]. The techniques emphasized in this part of the chapter include (1) derivative x-ray photoelectron spectroscopy (DXPS), (2) NEXAS, and (3) vis-VUV-SE.

TM Si oxynitride alloys were deposited at 300°C by RPECVD and annealed in Ar at temperatures up to 1100°C. Alloys with approximately equal SiO_2 and $Ti/Zr/HfO_2$ content, 37%–43%, and a smaller Si_3N_4 content, <15%–25%, display CPS into noncrystalline SiO_2 and nanocrystalline $Ti/Zr/HfO_2$ after annealing to 900°C, while alloys with higher Si_3N_4 content, ~40%, and approximately equal SiO_2 and TM oxide content, ~30%, do not display spectral features indicative of CPS [4,47,48]. As noted above, the behavior with respect to CPS is qualitatively different than for Zr and Hf silicate alloys [45,46]. These alloys are unstable with respect to CPS at ~900°C, regardless of the alloy composition, i.e., the relative SiO_2, and ZrO_2 or HfO_2 alloys fractions. Since the stable IVB TM Si oxynitride alloys are noncrystalline, they do not include grain boundaries at which O-vacancies cluster, and therefore represent a possible engineering solution for introducing medium-K dielectrics with low defect densities into advanced Si devices.

The research issues addressed are twofold: (1) is there spectroscopic evidence for band edge defects other than the discrete O-atom or clustered vacancy defect states? and (2) is there a reduced density of preexisting electron and hole traps as determined from electrical measurements on test devices?

The CPS issues are most easily addressed in the Ti Si oxynitrides for two reasons: (1) as noted in Section 10.3, the J–T splitting of the band edge T_{2g} and E_g final states is evident in the Ti L_3 edge spectrum for transitions from the Ti $2p_{3/2}$ to the Ti 3d spin–orbit split T_{2g} triplet and E_g doublet states, and (2) the same Ti 3d T_{2g} and E_g states are at photon energies in the O K_1 edge spectra are below those for excitation into Si–O final states in the Ti silicate and Ti Si oxynitride alloys as well. However, before any discussion of these spectra, the spectroscopic detection of CPS in TM Si oxynitrides is addressed for the Ti Si oxynitrides [48]. As examples of the CPS detection in these noncrystalline alloys, the first derivatives of O 1s XPS spectra for low and high Si_3N_4 content Ti Si oxynitride alloys are displayed respectively in Figures 10.9a for an alloy with 19% Si_3N_4 and Figure 10.9b for an alloy with 40% Si_3N_4. After the 900°C rapid thermal anneal (RTA), the spectrum in Figure 10.9a indicates two features, indicative of CPS into two different bonding arrangements, one with Si–O bonds, and a second with Ti–O bonds. These are also evident after the 1100°C anneal; there is an indication of some separation as-deposited, but the spectral features are clearly not as definitive as those after the 900°C anneal. In contrast, the spectrum in Figure 10.9b displays a single strong spectral feature indicative of a chemical environment comprising only one Si–O–Ti bonding arrangement. There is, however, a shoulder at ~530 cm^{-1} that is also evident in the as-deposited low Si_3N_4 content alloy.

Similar derivative O 1s XPS spectra for low and high Si_3N_4 content Zr Si oxynitrides are shown in Figure 10.9c and d [49]. The occurrence of two spectral features in Figure 10.9c indicates CPS similar to that reported for Zr silicates [47], while the occurrence of a single spectral feature in Figure 10.9d indicates the suppression of CPS, and therefore indicates a single phase Zr Si oxynitride alloy. Qualitative similar spectra have been obtained for low and high Si_3N_4 Hf Si oxynitride films as shown in Figure 10.9e and f, respectively [48].

Figures 10.10a and b compare, respectively, the Ti L_3 NEXAS spectra for as-deposited and annealed (900°C) low and high Si_3N_4 content Ti Si oxynitride alloys with 16% and 40% Si_3N_4, respectively [48]. The relative magnitudes of the C–F and J–T splittings in Figure 10.10a are consistent with fourfold coordination as-deposited, and sixfold coordination with CPS following the 900°C anneal, while the coordination is fourfold in Figure 10.10b both as-deposited and after annealing to 900°C.

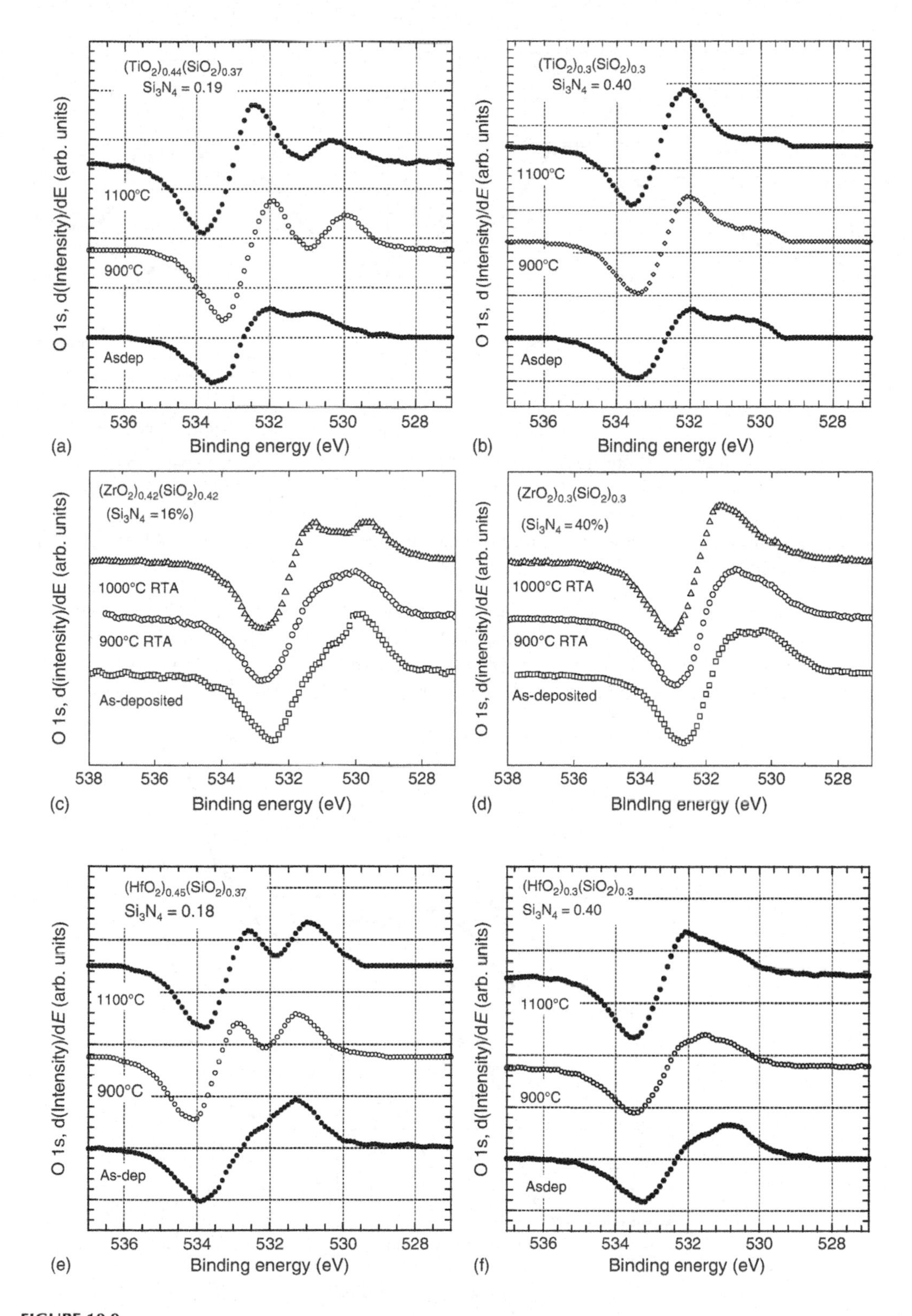

FIGURE 10.9
Derivative O 1s XPS spectra for (a) low and (b) high Si_3N_4 content Ti Si oxynitrides, (c) low and (d) high Si_3N_4 content Zr Si oxynitrides, and (e) low and (f) high Si_3N_4 content Hf Si oxynitrides. (After Lucovsky, G. et al., *Jpn. J. Appl. Phys.*, 46, 1899, 2007. With permission.)

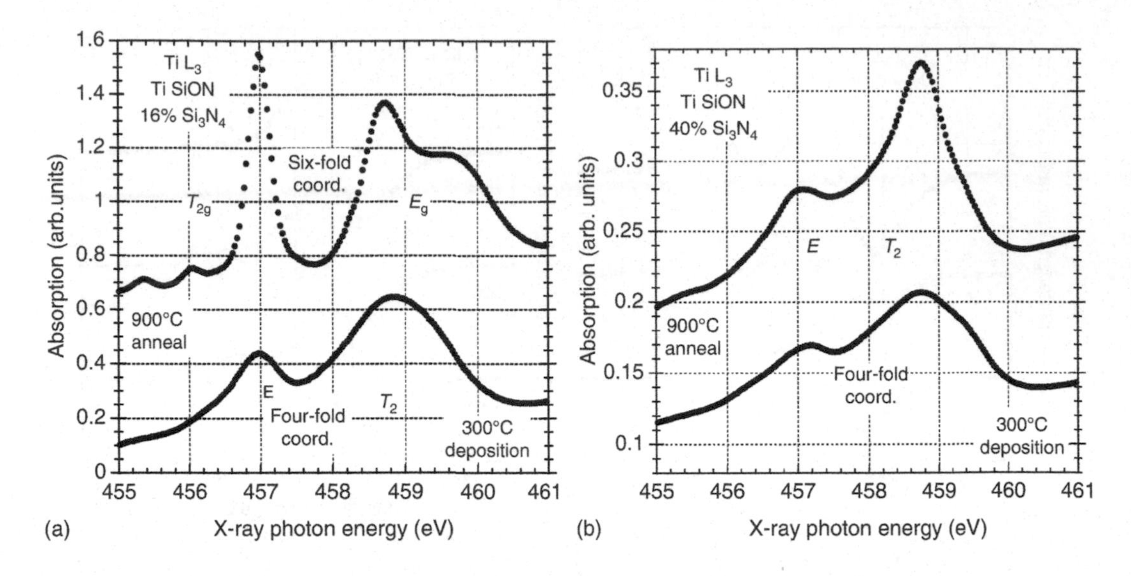

FIGURE 10.10
Ti L_3 spectra: (a) low and (b) high Si_3N_4 content Ti Si oxynitrides. (After Lucovsky, G. et al., *Jpn. J. Appl. Phys.*, 46, 1899, 2007. With permission.)

Figure 10.11 compares O K_1 NEXAS and band edge ε_2 Ti Si oxynitride alloys, confirming CPS for the low Si_3N_4 content alloy following the 900°C anneal. The comparison between the ε_2 spectrum derived from vis-VUV-SE measurements and O K_1 edge spectra for the 40% Si_3N_4 alloy indicates no spectroscopic evidence for a coordination change associated

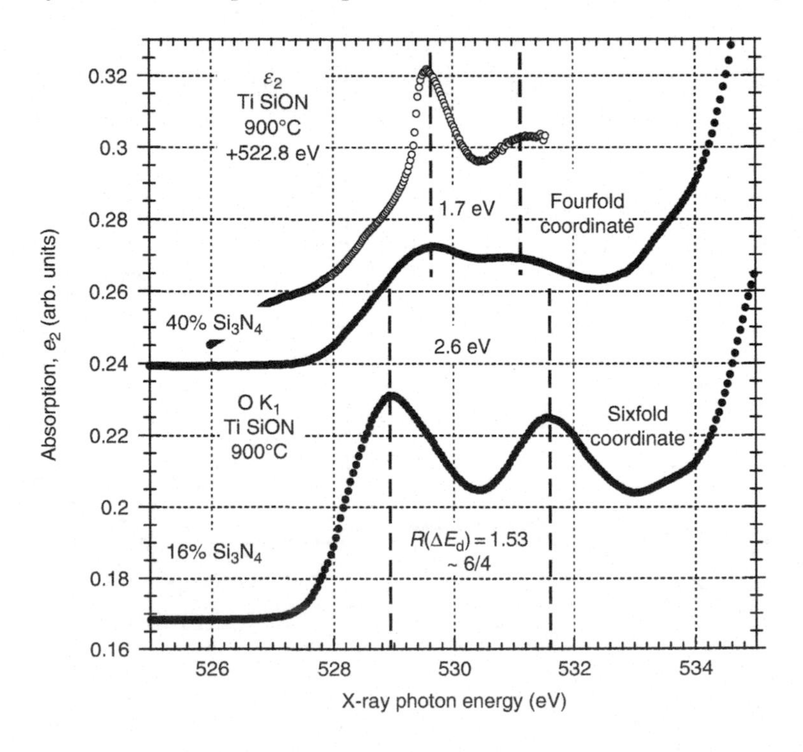

FIGURE 10.11
O K_1 and ε_2 spectra for Ti Si oxynitride alloys. (After Lucovsky, G. et al., *Jpn. J. Appl. Phys.*, 46, 1899, 2007. With permission.)

with CPS, i.e., the C–F splitting indicates fourfold coordinated Ti as-deposited and after the 900°C anneal. The ε_2 spectrum has been projected onto the x-ray photon energy scale of the O K_1 edge spectrum by matching the energy of the low-lying E states at the band edge. The splitting of the C–F states is 1.7 ± 0.1 eV, and is equal to the E–T_2 splitting in the Ti L_3 spectrum in Figure 10.10b. Additionally, and consistent with the corresponding behavior for TiO_2 noted above, the spectra in Figure 10.11 establish that the d-state features in O K_1 spectrum terminate in the same antibonding d-states as in the ε_2 spectrum; specifically, the states in each spectrum exhibit the same C–F splitting. However, it should be noted that each of the spectra in Figure 10.11 display a significant tailing of states into the forbidden band gap, suggesting intrinsic defects associated with the noncrystalline character of these films. Since these states are at energies less than 5.5 eV, they may be associated with the Si–N bonding of these alloys, and in particular the characteristic band tail defects [50].

The O K_1 spectra in Figure 10.12 for a high Si_3N_4 content Zr Si oxynitride indicate no changes in the Zr E_g feature, consistent with no CPS [49]. Only one completely resolved 4d state feature is evident in the O K_1 edge. An analysis of d-splitting, $\Delta(T_g - E_g)$ based on extraction of E_g and T_{2g} features from second derivative NEXAS provides information relative to the bonding coordination of Zr. For the low Si_3N_4 content alloy, $\Delta(T_g - E_g)$ increases from ~2.2 eV in as-deposited films to >3 eV after a 900°C anneal, consistent with fourfold coordinated Zr in as-deposited films, and an increase to sevenfold coordination after CPS. NEXAS spectra for low and high Si_3N_4 content Hf Si oxynitrides are similar to those presented above for Zr Si oxynitrides, with differentiation of the O K_1 edge demonstrating the different stabilities of low and high Si_3N_4 content films.

Measurements of Zr Si oxynitride dielectrics on Si substrates indicate EOTs as low as ~0.7–0.8 nm, consistent with the best results reported for ultrathin HfO_2 dielectrics on Si [47,49]. Reduced tunneling in TM Si oxynitrides relative to HfO_2 derives from higher tunneling barriers and larger tunneling electron masses, which more than compensate reduced values of K ~12 for the Hf Si oxynitrides [44], compared to ~20 for HfO_2. Similar results have been obtained for the ultrathin HfO_2 and chemically phase-separated silicates. In addition, the ultrathin HfO_2 and chemically phase-separated Hf silicates also display

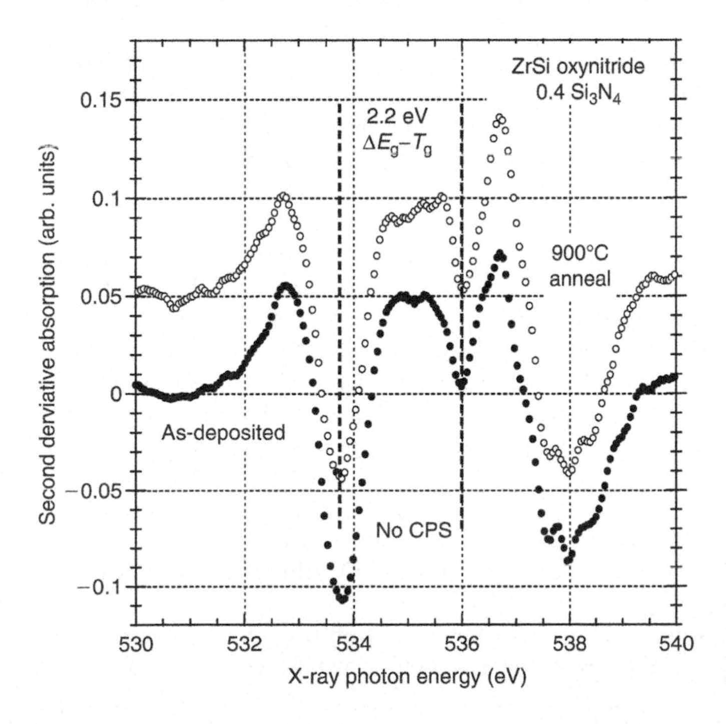

FIGURE 10.12
Derivative O K_1 for high Si_3N_4 content Zr Si oxynitride after 900°C anneal. (After Lucovsky, G. et al., *Jpn. J. Appl. Phys.*, 46, 1899, 2007. With permission.)

reduced levels of traps, consistent with the assignment of these traps to grain-boundary defects [4,15].

Finally, the total-dose x-ray irradiation response of low- and high-Si_3N_4 content Hf Si oxynitride, ~15%–20% HfO_2 has been compared with SiO_2 devices, and other devices with Hf-based dielectrics [41–44]. The interfaces for these devices received a postremote plasma oxidation nitridation treatment to suppress interfacial reactions during film deposition. Significant electron trapping, similar to what was found in HfO_2 and Hf silicates in Ref. [43] was also found in the low-Si_3N_4 content devices, but not in the high-Si_3N_4 content devices [44]. Furthermore, by comparing the radiation response and *C–V* results of the low- and high-Si_3N_4 content devices, we conclude that electron trap sites originate from the grain-boundary induced defect states in the HfO_2 nanograins in the chemically phase-separated low-Si_3N_4 film. Finally, the level of hole trap generation in the high-Si content Hf Si oxynitride films was comparable to what was found in SiO_2 devices, confirming the validity of the bond constraint theory calculations that indicate a chemical bond self-organization with broken bond-bending constraints that reduce the average number of bonding constraints per atom, C_{av}, to values where defect densities are anticipated to comparable to those in SiO_2 devices [51]. A similar occurrence is found in a strain-free intermediate phase of the second kind, wherein there is a confluence of broken bond-bending constraints reducing $C_{av} = 3$, and a percolation of reduced strain associated with a bonding arrangement in which the Hf occupy 16% of the fourfold coordinated bonding sites (the other fourfold coordinated atoms are all Si). These are spatially extended to include four O-atom nearest neighbors that are bonded to four next nearest neighbor $-SiN_3$ groups [38].

10.4.2 As-Deposited and Annealed HfO_2 Films >4 nm Thick

Figure 10.13a presents O K_1 edge spectra for 6 nm thick HfO_2 films (1) as-deposited by RPECVD at 300°C, and (2) annealed in Ar for 1 min at temperatures of (a) 500°C, (b) 700°C, and (c) 900°C. Four Hf-related MO final state features are indicated in this figure: (1) the band edge antibonding π-symmetry $5d_{3/2}$ or doublet E_g feature, (2) the antibonding σ-symmetry $5d_{5/2}$ or triplet T_{2g} feature, and (3) the antibonding σ-symmetry 6s and 6p features. There are significant qualitative differences between the (1) 300°C and 500°C spectra, and (2) the 700°C and 900°C spectra. All four spectral features in the films annealed at 700°C and 900°C are sharper, and therefore better resolved than in the films processed at 300°C and 500°C. Most importantly, there is an asymmetry and partially resolved doublet structure in the band edge $5d_{3/2}$ or E_g feature that indicates final J–T splitting in the films annealed at 700°C and 900°C, whereas in the films processed at 300°C and 500°C, the same feature is broad with no indication of a doublet structure; these differences are also readily evident in first and second derivative spectra (not shown). These aspects of the spectra are also evident in Figure 10.13b, which presents four expanded-scale band edge O K_1 spectra for the 6 nm thick films in the spectral range between 530 and 535 eV. There are three features to note with respect to differences between the films processed at 300°C and 500°C and annealed at 700°C and 900°C: (1) the qualitative differences in the $5d_{3/2}$, E_g π-state band edge feature between the 300°C–500°C and 700°C–900°C processing, (2) in particular, the doublet structure is clearly evident with a J–T splitting of about 1 eV, and additionally, (3) the band edge slope is greater for the 700°C–900°C processing, 1.4, compared with 1.0 for the 300°C–500°C processing. The softer band edge slope, and absence of a J–T splitting is similar to what we have previously found for ultrathin HfO_2 films, both as-deposited and after 700°C–900°C annealing.

This has been attributed in Refs. [4,15] to incoherent π-bonding between strings of Hf–O–Hf–O–Hf–O, etc., which extend between nearest-neighbor Hf-atoms interconnected

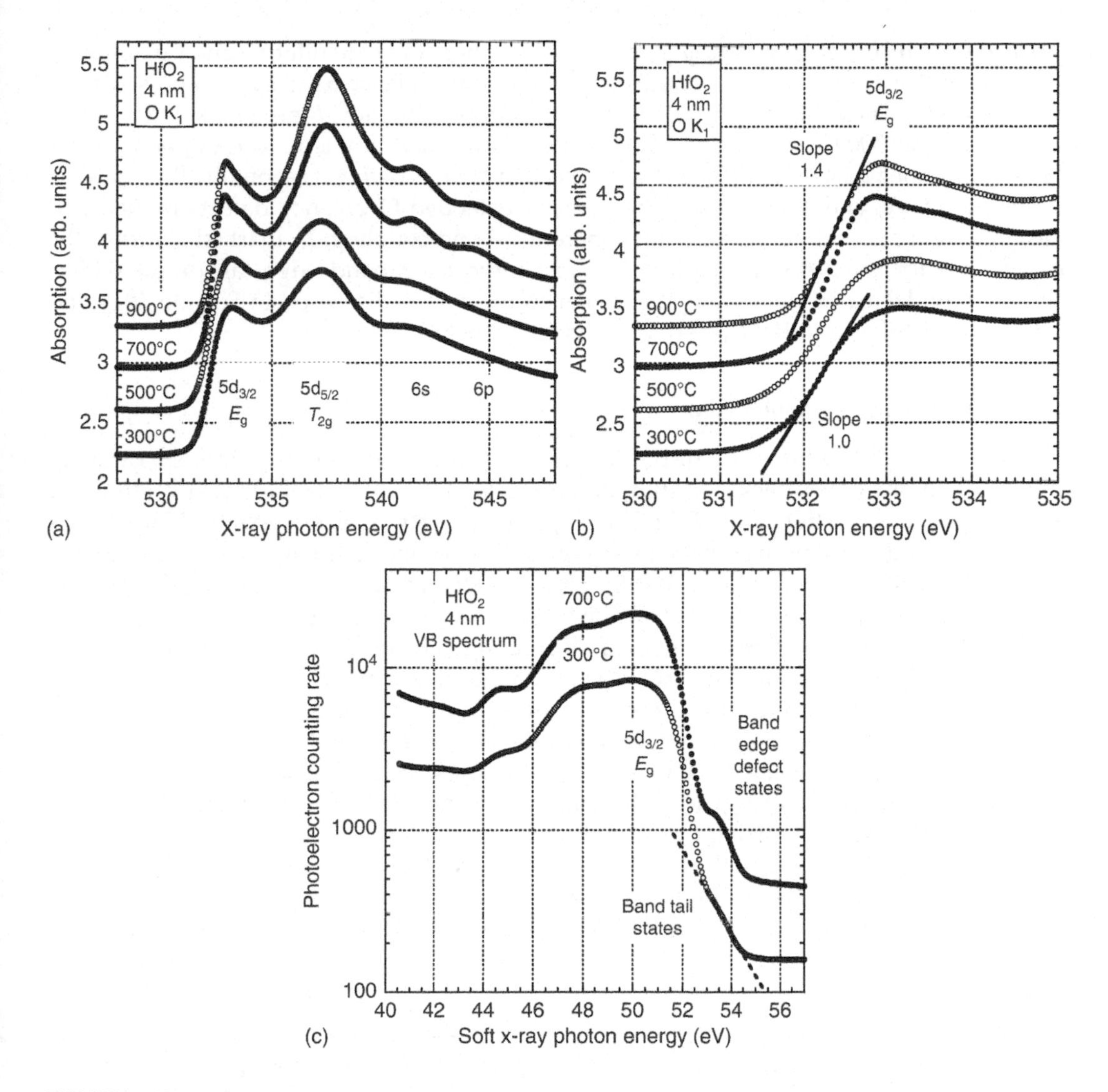

FIGURE 10.13
(a) O K_1 edge spectra for 4 nm thick HfO_2 films (1) as-deposited by RPECVD at 300°C, and (2) annealed in Ar for 1 min at temperatures of (a) 500°C, (b) 700°C, and (c) 900°C, (b) expanded-scale band edge O K_1 spectra for the 4 nm thick films of (a) in the spectral range between 530 and 546 eV, and (c) SXPS valence band spectra for 4 nm thick HfO_2 films, as-deposited at 300°C, and after a 700°C anneal.

in at least five primitive cells with a characteristic length scale of ~2–2.5 nm. There is also a direct correlation between discrete band edge defects associated with extended O-vacancies, e.g., divacancies, which are pinned and clustered at grain boundaries. Stated differently, extended grain boundaries are a necessary requisite for the formation of divacancy defects with J–T splittings. The results presented above indicate that a length scale of <2.5 nm is then too small for discrete band edge defects; coherent grain boundaries with pinned extended vacancy defects can only occur at grain sizes >3 nm where coherent π-bond coupling is present. This length scale for π-state coupling has been demonstrated in model calculations [52], and is similar to antiferromagnetic superexchange coupling between Mn atoms separated by O-atoms as in cubic MnO [53,54].

Comparing the differences in the O K_1 spectra in Figure 10.13a and b between the 300°C and 500°C processed films, with those processed at 700°C and 900°C, these results are also consistent with change in π-bonding coupling, and therefore a qualitative and quantitative change in defects as well. This is confirmed by the SXPS valence band spectra presented in Figure 10.13c. These are SXPS valence band edge spectra for 4 nm thick HfO_2 films, as-deposited at 300°C, and after a 700°C anneal in an inert ambient. The upper trace for the 700°C annealed film indicates a discrete band edge defect, essentially the same as discussed above, and assigned to multiatom or extended O-vacancy defects pinned and clustered along grain boundaries with coherent π-bond coupling. In marked contrast, the lower trace for the 300°C as-deposited defect indicates a band edge tail defect with a significantly decreased density, again similar to the results discussed in Section 10.3, and the spectra displayed in Refs. [4,34], and discussed below in Section 10.4.3.

10.4.3 Thin Nanocrystalline Films

The other examples of films in which electronic and defect states are qualitatively different than those in physically thicker (≥4nm) and annealed HfO_2 films include (1) ultrathin (~2 nm) nanocrystalline HfO_2, and (2) high-HfO_2 content (~80%) phase-separated Hf silicates [5]. Spectra for these films are displayed in Figure 10.14a and b. For HfO_2 films that are 2 nm thick with no Hf–N bonding, the J–T splitting of the Hf 6d E_g band edge state is suppressed and appears as a single spectral feature, whereas the J–T splitting in a doublet is clearly evident in the 3 and 4 nm films with no Hf–N bonding in Figure 10.14a. Similar results have been obtained for ZrO_2 films. For ZrO_2, the O K_1 edge spectrum of a thick film, e.g., 10 nm, does not show distinct doublet features. However, in this case the J–T E_g term splitting is unambiguously extracted by differentiation of the O K_1 spectrum. For ZrO_2 films, changes in the E_g band edge doublet with changes in film thickness are evident as a systematic shift in the E_g band edge feature energy (not shown). These changes display a sigmoidal behavior, with the total change occurring between 2 and 5 nm (20 and 50 Å) being approximately one-half of the J–T splitting of the band edge E_g state as determined by differentiation.

In a complementary way, the band edge Hf E_g J–T splitting in phase-separated Hf silicates is suppressed in films that are 2, 3, and 4 nm thick in Figure 10.14b, but these

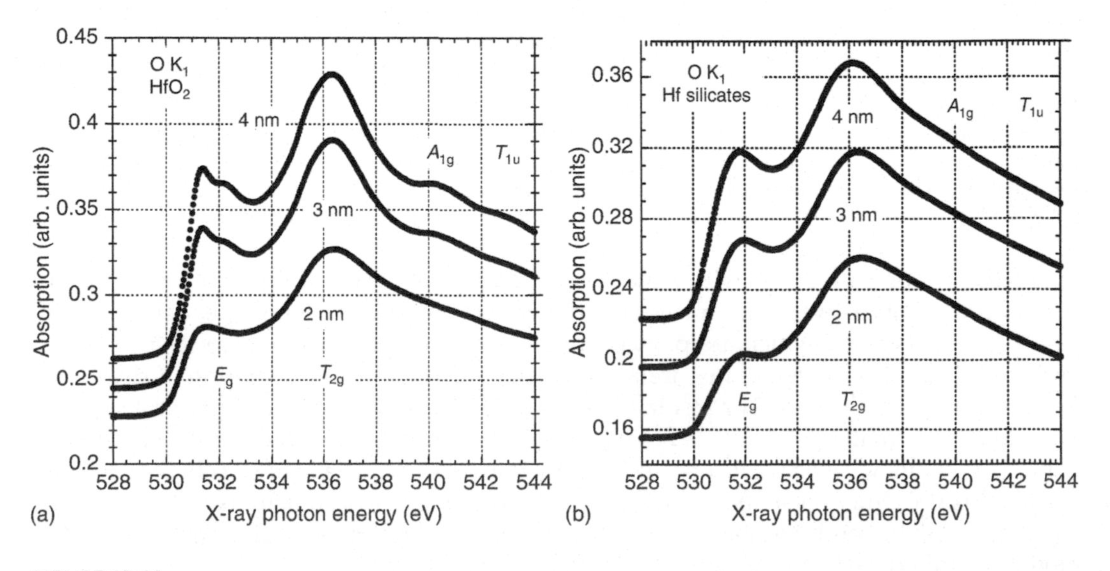

FIGURE 10.14
O K_1 spectra for 2, 3, and 4 nm films: (a) HfO_2 and (b) phase-separated Hf silicate. (After Lucovsky, G., *J. Mol. Struct.*, 838, 187, 2007. With permission.)

spectra display an increase in the relative amplitude of the Hf E_g feature in the thicker films, consistent with anisotropic grain growth, i.e., larger crystallite size in the direction of the film growth. Similar results have been obtained for phase-separated Zr silicate films.

There are two mechanisms that contribute to the small grain size in HfO_2 and ZrO_2 films: (1) total film thickness, as for the ultrathin 2 nm thick HfO_2 and ZrO_2 films discussed above, and (2) inclusion of noncrystalline SiO_2 as a component in phase-separated Hf and Zr silicates. These SiO_2 regions comprise a significant fraction of the total film volume, ~16%–20% for the Hf silicates; these large volume fractions limit the grain sizes and eliminate coherent π-bonding, and hence discrete band edge defects. However, as already noted above, in 4 nm thick films, anisotropic grain growth in the direction of the film growth promotes increased π-bonding in this direction as evident in the differences in NEXAS spectra between the 2 and 4 nm thick phase-separated Hf silicate alloys in Figure 10.14b.

Turning now to SXPS valence band spectra, a valence band edge defect state is readily observable in Figure 10.15a as a distinct shoulder in the 4 nm thick HfO_2 film. In contrast, for a 2 nm film, there is not a well-defined band edge defect that can be fit with two distinct spectral features, but instead is a weaker band edge feature corresponding to an approximately 10-fold reduction in the integrated defect state density.

10.4.4 HfO_2 Films with Bonded Nitrogen

A final approach to suppress defects in HfO_2 dielectrics involves postdeposition annealing in N_2 and NH_3 at temperatures of 700°C. Films annealed in N_2 show no spectroscopic evidence for N-incorporation, i.e., Hf–N bonding in N K_1 NEXAS spectra, while films annealed in NH_3 display Hf–N bonding. Figure 10.15b compares the SXPS valence band spectra of 4 nm thick HfO_2 films. As in Figure 10.15a, the valence band spectrum for the HfO_2 film with no NH_3 anneal and no Hf–N bonding indicates a distinct spectrally resolved band defect, while the film annealed in NH_3 with Hf–N bonding in Figure 10.15b does not show this feature, but instead displays a band edge tail that is significantly

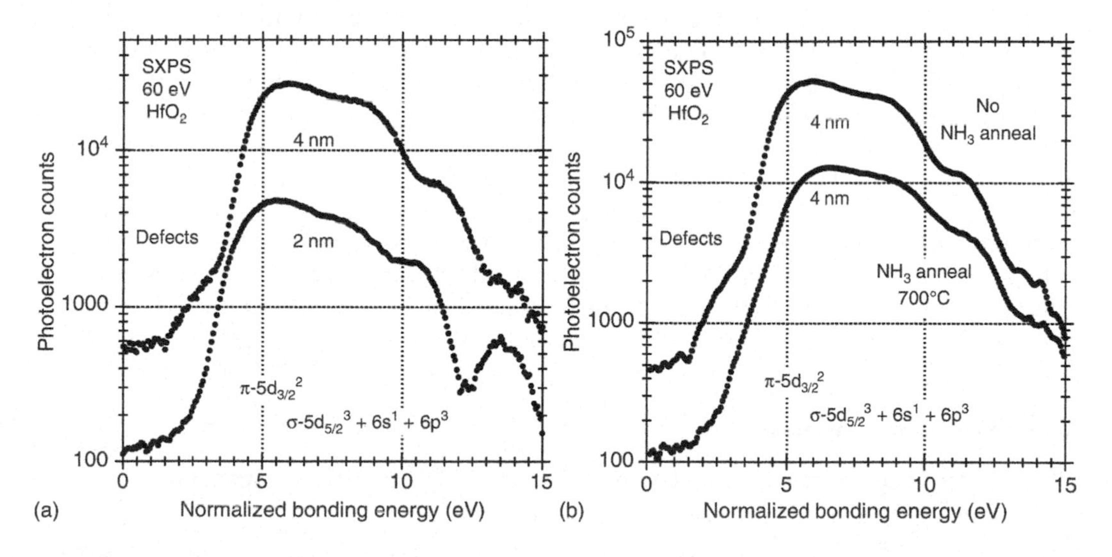

FIGURE 10.15
SXPS VB spectra for (a) 2 and 4 nm HfO_2, and (b) 4 nm HfO_2 with and without a 700°C NH_3 postdeposition anneal. (After Lucovsky, G., *J. Mol. Struct.*, 838, 187, 2007. With permission.)

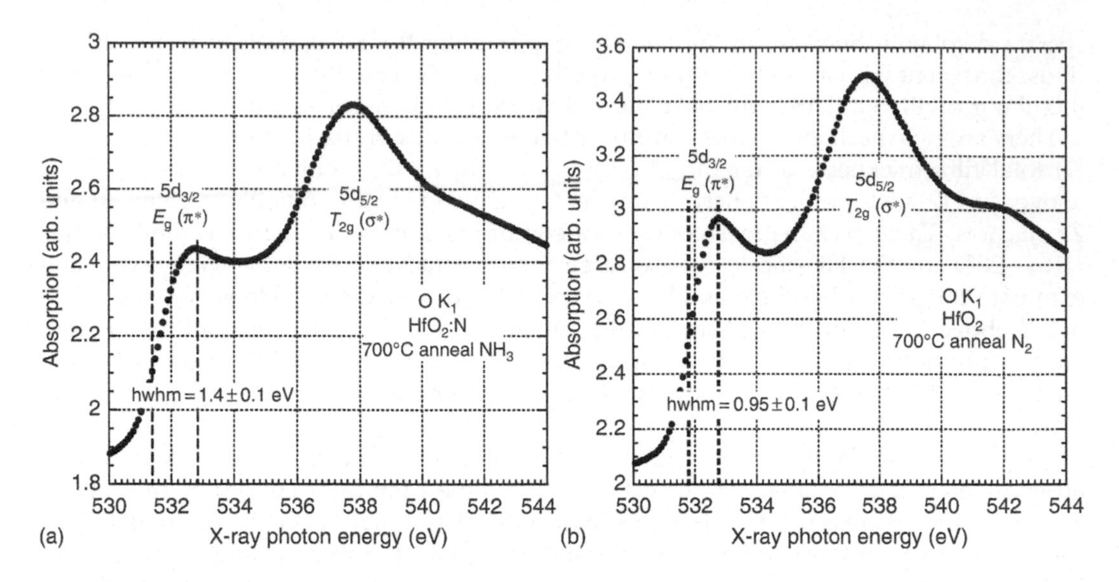

FIGURE 10.16
O K_1 spectra for 2 nm thick films, postdeposition annealed at 700°C, respectively, in NH_3 in (a) and in N_2 in (b).

softer than in the 4.0 nm films with no Hf–N bonding in Figure 10.15a and b. This softer edge is assigned to band tail defects, and the specific bonding mechanism giving rise to the band tail softening and defects in Figure 10.15b is addressed below.

Similar band edge tailing is also present at the conduction band edge and evident in the O K_1 spectrum in Figure 10.16a for a 2 nm thick film annealed in NH_3. The low-energy half-width at half-maximum (hwhm) of the band edge 5d E_g feature is about 20% larger than that displayed in Figure 10.16b for a HfO_2 film of the same physical thickness annealed in N_2. The film in Figure 10.16b does not display any spectroscopic evidence for Hf–N bonding, even though the film has been annealed in N_2 at 700°C. It is proposed that the results for defect reduction in Figure 10.16a for the 2 nm thick HfO_2 with no Hf–N bonding, and for the softer valence and conduction band edges, respectively, in Figure 10.16b and a for the 4 and 2 nm films with Hf–N bonding, are due to a similar microscopic mechanism. Discrete valence band edge intrinsic and defect states are associated with d-state π-bonding with a scale of order of more than 2 nm as the 4 nm thick films without Hf–N bonds in Figure 10.16a and b. However, for the 4 nm thick film in Figure 10.19b and the 2 nm film in Figure 10.19a with Hf–N bonding, randomized interactions between π-bonding MOs on different pairs of Hf-atoms with both O- and N-neighbors give rise to increased band tailing at both the valence and conduction band edges rather than discrete defect states. This explanation for films with Hf–N bonding, as well as the thickness dependence in films without Hf–N bonding is supported by the model calculations in Ref. [52].

10.5 Interface Properties between Si and SiO_2 and Si_3N_4, and Ge between GeO_2 and Ge_3N_4

High Si_3N_4 content noncrystalline Hf Si oxynitride pseudoternary alloy films, $(HfO_2)_{0.3}(SiO_2)_{0.3}(Si_3N_4)_{0.4}$, were deposited at 300°C by RPECVD onto remote plasma-nitrided Ge substrates [55], and compared with the same dielectric films deposited onto Si substrates

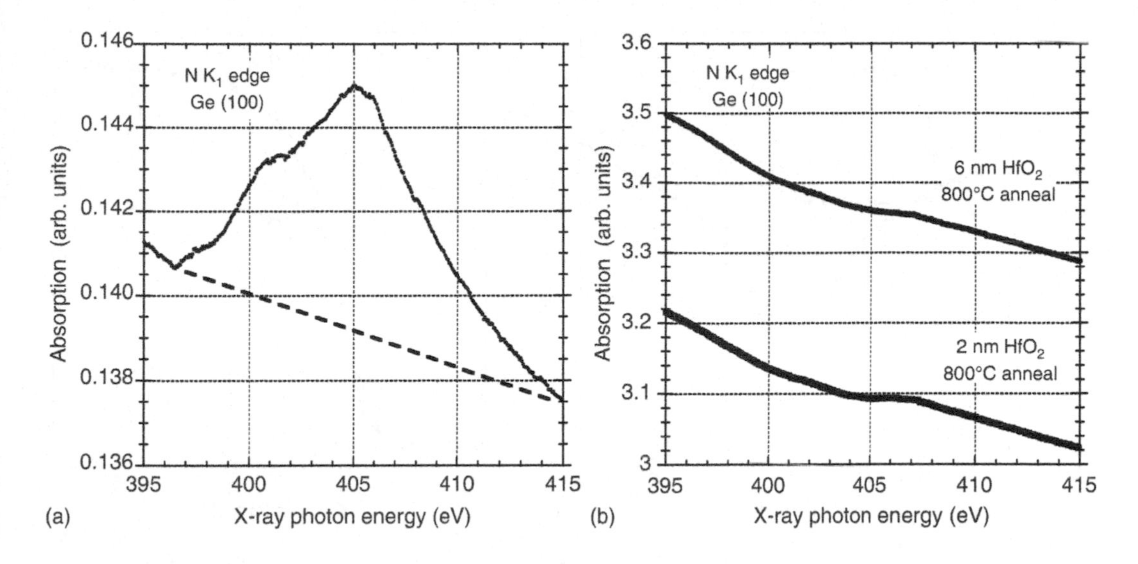

FIGURE 10.17
(a) N K_1 edge spectrum of a nitrided Ge substrate before deposition of HfO_2 or Hf Si oxynitride. (b) N K_1 spectrum for a HfO_2–Ge gate stack after an anneal in Ar for about 1 min at 800°C. (After Lee, S. et al., *Microelectron. Reliab.* 48, 364, 2008. With permission.)

with ~0.6–0.8 nm thick SiON interfacial layers [56]. NEXAS measurements were performed at the Stanford Synchrotron Research Laboratory (SSRL) primarily in the spectral regimes of the O K_1 and N K_1 edges, 520–550 and 390–430 eV, respectively [4,57,58]. Figure 10.17a indicates an N K_1 edge spectrum of a nitrided Ge substrate before deposition of a HfO_2 film, and Figure 10.17b is an N K_1 spectrum after an anneal in Ar for about 1 min at 800°C. There is a detectable Ge–N interfacial layer at the buried interface in Figure 10.17a, but the spectrum in Figure 10.17b is simply the background for this spectral regime. This approach for studying buried Ge–N layers is possible because there are essentially no x-ray absorption in the N K_1 spectral regime between 390 and 420 eV for any of the constituent atoms in the HfO_2 dielectric [57,58]. Since Ge–N bonding in the intentionally nitrided Ge–N passivation layer has been eliminated, we assume that a similar situation prevails for deposition of the Hf Si oxynitride alloy; however, we cannot rule out substrate Ge bonding with N-atoms of the Hf Si oxynitride alloy.

Figure 10.18a and b indicate the *I–V* and *C–V* characteristics for a 4 nm high-Si_3N_4 content (~40%) film deposited directly onto nitrided Ge (100), and then subjected to a 1 min 800°C anneal in Ar. The direct tunneling currents through this film for substrate and gate injection in Figure 10.18a are essentially the same for $V = \pm 1.0$ V. The small reduction in current for gate injection is due to the voltage drop across the depletion region in the n-type substrate when a positive bias is applied to the n-type Ge. When corrected for this voltage shift, the near equality of the currents indicates that the dielectric does not have traps that can provide a pathway for a trap-assisted tunneling process, as has been discussed for trap-assisted tunneling in HfO_2 films with a nanograin size >3.5 nm [4,57].

Analysis of *C–V* traces in Figure 10.18b yields an accumulation capacitance of 129 pF, and an EOT of 1.22 nm, consistent with a dielectric constant of ~12, essentially the same as that obtained for the same Hf Si oxynitride films deposited on Si substrates, where a correction was made for the interfacial SiON layer [59]. There is no measurable frequency dispersion for the *C–V* trace, and the stretch-out is small as well; i.e., the *C–V* characteristic is near-ideal. This in turn means that the interface trap density D_{it} is low; our best and conservative estimate is that $D_{it} < 10^{11}$ cm^{-2} [59].

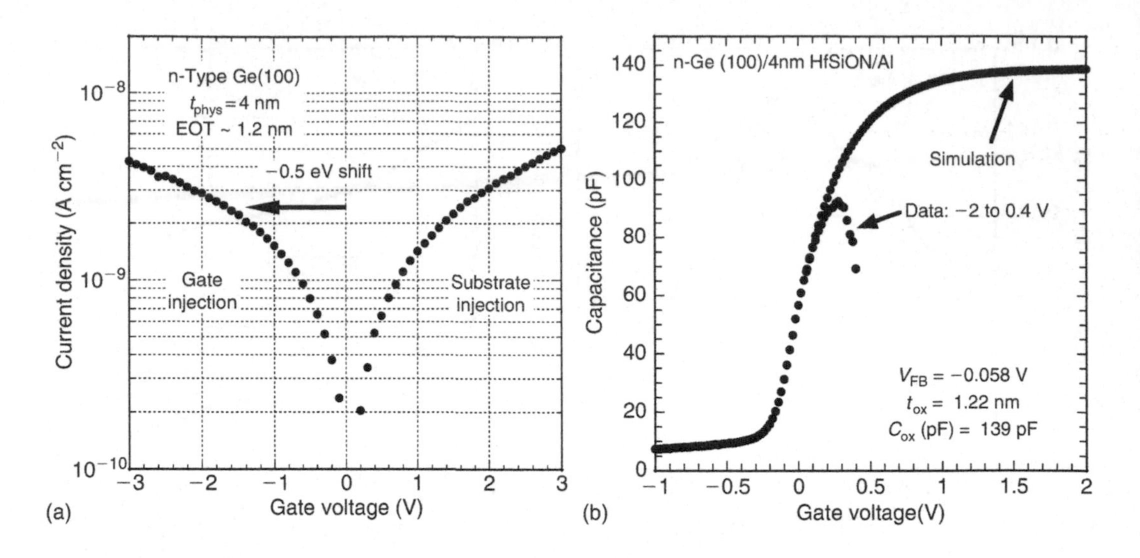

FIGURE 10.18
(a) *I–V* and (b) *C–V* characteristics for a 4 nm high-Si_3N_4 content (~40%) film deposited directly onto nitrided Ge (100), and subjected to a 1 min 800°C anneal in Ar. The *C–V* characteristics have been corrected for a series resistance effecting following the method in Ref. [9].

To understand why defect densities at interfaces between n-type Ge and GeO_2 displayed high defect densities $>10^{12}$ cm^{-2} [5,6], differences between the band edge absorption threshold differences for (1) plasma-deposited SiO_2 on Si substrates, (2) plasma-oxidized Si, resulting in SiO_2 layers, and (3) plasma-oxidized Ge, resulting in GeO_2 layers were determined by O K_2 NEXAS measurements. Plasma-deposited SiO_2 and SiO_2 layers formed by plasma-oxidation of Si display essentially the same O K_1 edge absorption edge features, to ~0.2 eV. The respective thresholds for absorption and other band edge features are shifted to lower photon energies for GeO_2 with respect to SiO_2; these differences are presented in Figure 10.19a. Figure 10.19b displays differences in the first two edge features

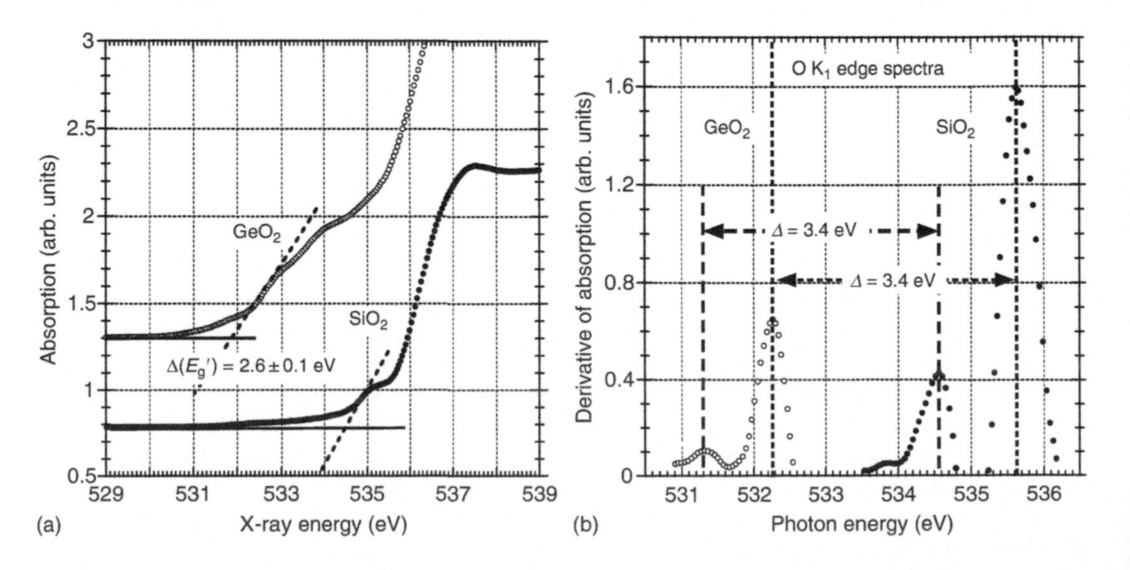

FIGURE 10.19
(a) O K_1 edges of GeO_2 and SiO_2. (b) Numerical differentiation of spectra in panel (a).

that have been determined by plotting the first derivatives of the respective absorption edges. These GeO_2 features in the derivative plots are redshifted by 3.4 ± 0.2 eV with respect to the SiO_2 spectral features. This gives a band gap for GeO_2 of $\sim 5.6 \pm 0.2$ eV in excellent agreement with the band gap of 5.6 eV obtained from a dispersion analysis of ultraviolet reflectivity spectra, and the extraction of ε_2 spectra [60]. The band gap of GeO_2 will be assumed to be $\sim$5.6 eV in the discussion of conduction band offset energies that is addressed below. Based on a similar analysis of N K_1 edge spectra, the spectral features of Ge_3N_4, formed by plasma nitridation of Ge, are redshifted by 1.0 ± 0.2 eV with respect to the corresponding spectral features of Si_3N_4 formed by plasma nitridation of Si, suggesting a band gap of 4.3 ± 0.2 eV for noncrystalline Ge_3N_4. This is approximately the same band gap of $\sim 4.5 \pm 0.2$ eV as determined experimentally, as well as from the calculations presented in Refs. [61,62]. A band gap 4.4 eV for Ge_3N_4 will be used for the comparisons of conduction band offset energies in the schematic band edge energy diagrams that are displayed below.

Figure 10.20a displays the band alignments between Ge, GeO_2 and HfO_2, and Ge, Ge_3N_4 and HfO_2. These diagrams have incorporated the GeO_2 and Ge_3N_4 band gaps identified

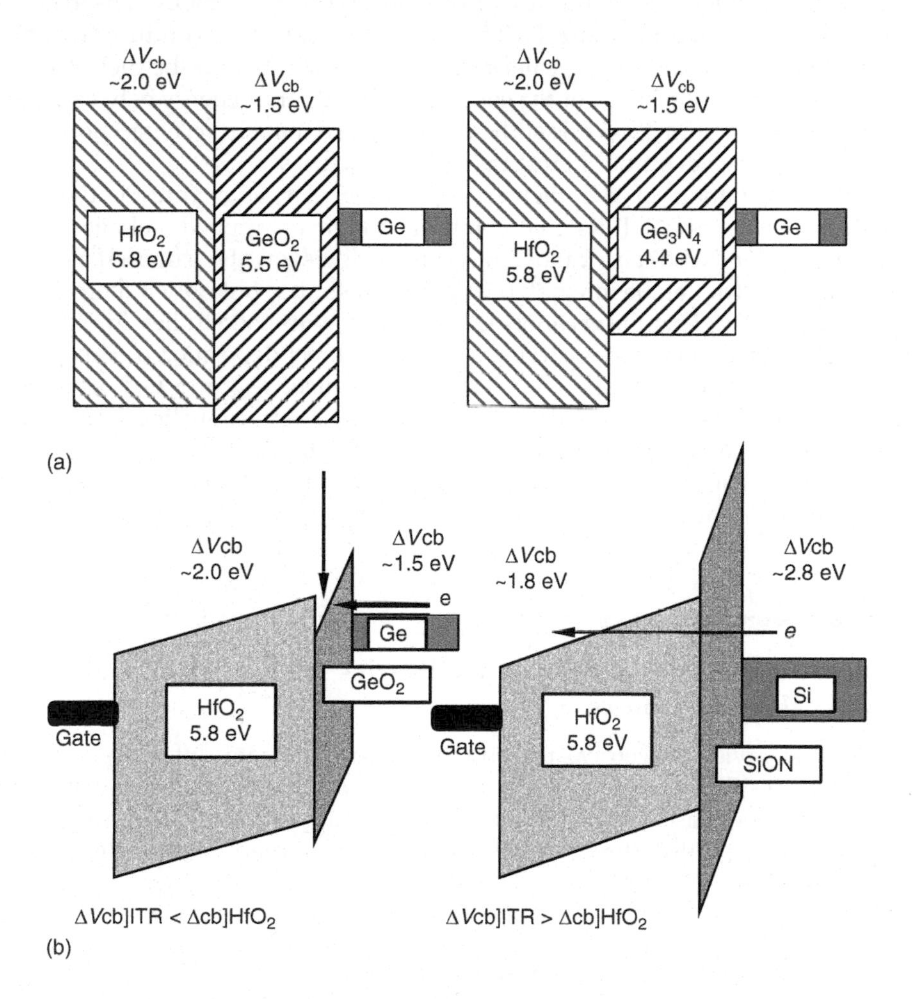

FIGURE 10.20
(a) Band alignments between n-type Ge, GeO_2, and HfO_2, and nm-type Ge, Ge_3N_4, and HfO_2. (b) Substrate electron injection for the n-type Ge, GeO_2, and HfO_2 stack in (a), in which the conduction band offset energy between Ge and GeO_2 ($\sim$1.5 eV) which is less than that between Ge and HfO_2 ($\sim$2.0 eV) and Si and SiON ($\sim$2.8 eV) which is greater than that between Si and HfO_2 ($\sim$1.8 eV).

above and demonstrate that, when these layers are used as interfacial transition regions (ITRs) between Ge and a high-*K* dielectric, the band gaps of these respective ITRs are less than that of the high-*K* dielectric film, and also SiO_2. Figure 10.20b compares injection from an n-type Ge substrate for a Ge/GeO_2/HfO_2 gate stack, in which the conduction band offset energy between Ge and the ITR (Δcb]ITR) is less than the conduction band offset energy between Ge and the high-*K* dielectric (Δcb]HfO_2), with that from an n-type Si substrate for an Si/SiON/HfO_2 gate stack, in which the conduction band offset energy between Si and the SiON ITR (Δcb]ITR) is greater than the conduction band offset energy between Si and SiO_2 (Δcb]HfO_2). Under these bias conditions, there is a potential well in the GeO_2 film between the n-type Ge substrate and HfO_2. The same condition would apply if the HfO_2 film were replaced by SiO_2 as in Ref. [63], in which the fixed or trapped charge density is $>10^{13}$ cm^{-2}, as determined by *C–V* measurements for an n-Ge/GeO_2/SiO_2 gate stack structure. Similar results were reported in Refs. [5,6] for nGe MOSCAPS, and nGe MOSFETs for GeO_2, Ge_3N_4, and GeON interfacial transitions consistent with the band alignments illustrated in Figure 10.20a. The situation for dielectrics, either high-*K*, SiO_2 or Si oxynitride on Si, using either SiO_x or SiON ITRs in gate stacks with n-Si is also illustrated in Figure 10.20b. In this case either the conduction band offset energy between the ITR and Si is greater than that between Si and the dielectric, e.g., for high-*K* dielectrics including HfO_2, or the conduction band offset energy between n-Si, or an inverted p-type Si in an nMOSFET, is sufficiently large so that electron injection into a well does not occur for bias levels of interest in device structures.

Finally, the integration of the high Si_3N_4 content Hf Si oxynitride dielectrics on Si substrates requires SiON ITRs that limit the range of applied voltages before the threshold for Fowler–Nordheim tunneling occurs. However, the same Hf Si oxynitride noncrystalline dielectric can be integrated onto Ge substrates, but only if there is no GeO_2, Ge_3N_4, or GeON ITR. This observation is consistent with the studies in Refs. [5,6]. In marked contrast, it has been demonstrated above that direct deposition of Si_3N_4-rich (40%) Hf Si oxynitrides on Ge (100) and Ge (111) have excellent *I–V* and *C–V* characteristics; e.g., for a physical thickness of ~4 nm and an EOT ~1.2 nm, tunneling leakage is in the 10^{-9} A cm^{-2} range, and defect densities are $<10^{11}$ cm^{-2}. By extrapolation, EOTs as low as 0.6 nm are possible with direct tunneling less than at most 10^{-4} A cm^{-2}; this is compatible with feature sizes <10 nm.

10.6 Electrical Measurements and Dielectrics for Radiation Hard Survivable Electronics

This section first reviews electrical measurements that indicate differences between the current–voltage (*I–V*) characteristics of (1) SiO_2 and (2) representative Hf-based gate dielectrics, and then based on these results that support the differences in defects identified spectroscopically, three different options for radiation hard survivable MOS devices are identified.

10.6.1 Electrical Measurements on Hf-Based Dielectrics

The substrate-injected gate tunneling leakage current for an Al/SiO_2/n-Si MOSCAP is measured at 25°C and 125°C for an SiO_2 EOT of ~2.5 nm; the *I–V* data are shown in Figure 10.21a. There are no significant differences between the two traces at 25°C and 125°C. The tunneling barrier is composed of the SiO_2 film, as well as significantly thinner

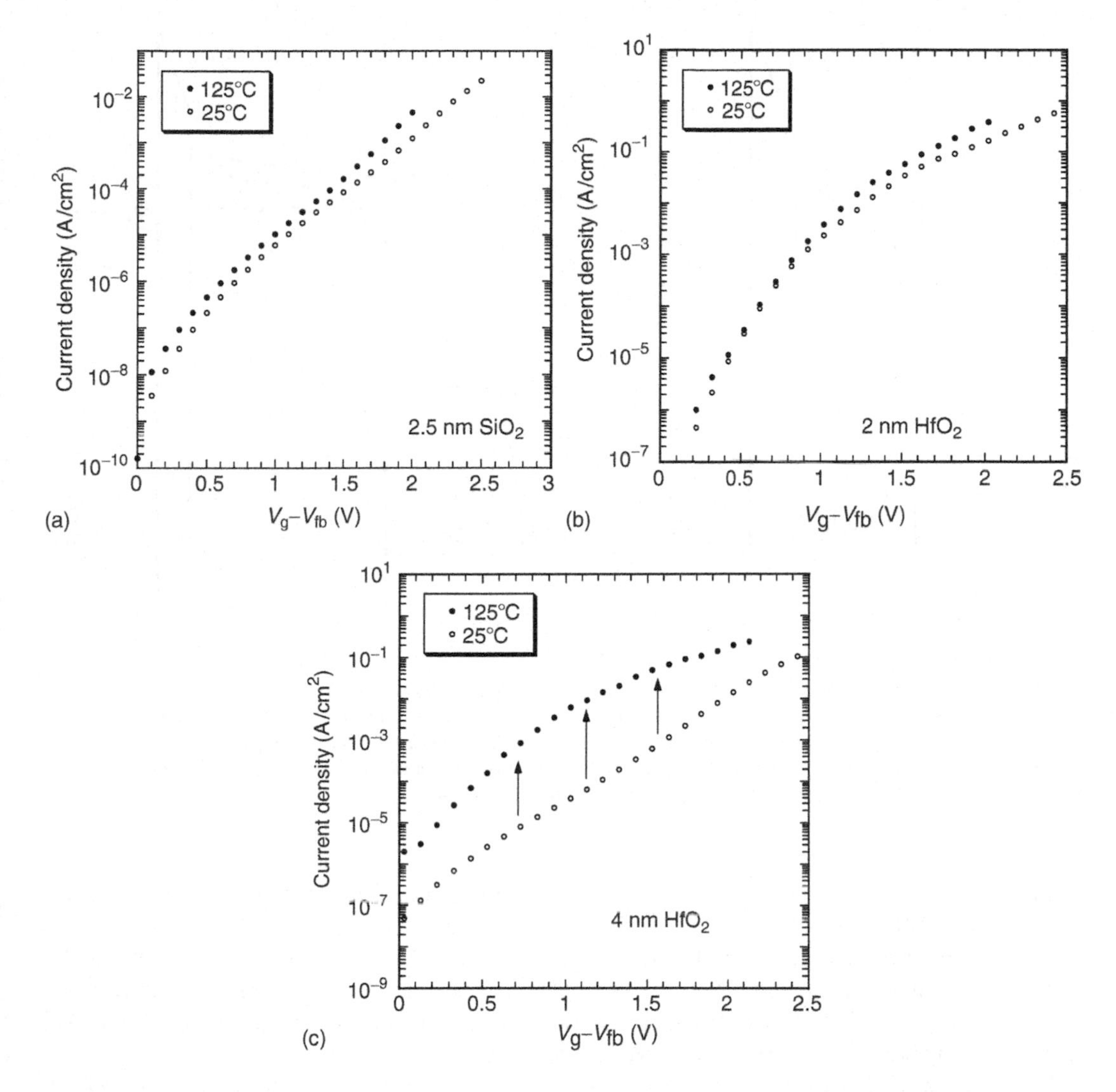

FIGURE 10.21
Substrate-injected gate tunneling leakage current measured at 25°C and 125°C. (a) Al/SiO_2/n-Si MOSCAP with an EOT of ~2.5 nm. (b) MOSCAP with a 2 nm HfO_2 film annealed to 900°C. (c) MOSCAP with a 4 nm HfO_2 film annealed to 900°C measured at 25°C and 125°C.

~0.6–0.8 nm thick SiON ITR. The tunneling current through a 2 nm HfO_2 film, annealed to 900°C is qualitatively similar to that of the SiO_2 film, including the absence of any significant temperature dependence, which is displayed in Figure 10.21b. As shown schematically in Figure 10.22a, the electrons are transported the trapezoidal HfO_2 barrier and SiON interface by a direct tunneling process. A qualitatively similar barrier applies to the test device with the SiO_2/SiON barrier, with the only difference being the division of the applied tunneling voltage between the SiON substrate and the respective dielectrics. This is determined by the ratio of the dielectric constants, $K(SiO_2)$ ~3.9 and $K(HfO_2) \sim 20$, of the respective dielectrics, and that of the ITR K(SiON) ~5.5, through the electric field, e.g., $E(SiO_2)K(SiO_2) = E(SiON)K(SiON)$.

There is, however, a significant difference in the temperature dependence of I–V characteristics of the 4 nm HfO_2 film that is readily evident in Figure 10.21c. The room

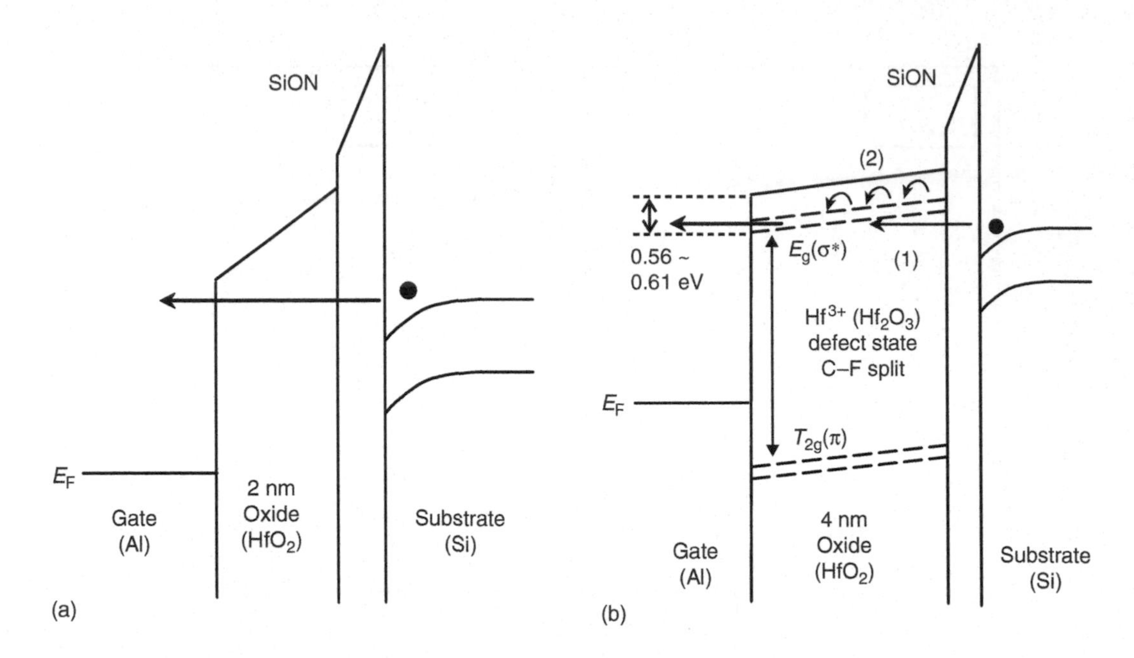

FIGURE 10.22
Energy band alignment between an n-type Si substrate a gate dielectric stack and Al metal gate electrode that lines up with the Fermi level in the n-type Si. (a) Direct tunneling barrier for an SiON ITR, and 2 nm thick HfO_2 annealed film in which the conduction band edge defects are included in a band tail regime, and at integrated density of $\sim 10^{11}$ cm^{-2}. (b) Trap-assisted tunneling/Poole–Frenkel transport by injection into discrete band edge traps on average 0.5–0.6 eV below the conduction band edge of HfO_2.

temperature trace is qualitatively similar to that of the 2 nm film, but the 125°C trace shows a markedly different behavior. Analysis of the increased current for the 125°C trace is explained by the band diagram in Figure 10.22b, which includes localized defects at the conduction band edge of HfO_2 [4]. Analysis of the enhanced current requires a continuous and smooth transition between a trap-assisted tunneling mechanism and Poole–Frenkel transport, similar to that previously proposed in Ref. [16].

Pseudobinary Hf silicate alloys, $(HfO_2)_x(SiO_2)_{1-x}$, $0 < x < 1$, are unstable with respect to CPS when annealed to ~800°C–900°C in nonoxidizing ambients (Ar or N_2) [46]. Alloys with $0.10 < x < 0.4$ phase separate into nanograins of nanocrystalline HfO_2 that are completely encapsulated by noncrystalline SiO_2, whereas for $60 < x < 90$, the reverse situation applies. Nanocrystalline clusters of SiO_2, ~2 nm in size, are completely encapsulated in nanograins of HfO_2, also ~2 nm in size. These alloys display band tail defects similar to HfO_2 for a film thickness of ~2 nm. This is reflected in *I–V* characteristics that are quantitatively similar to those in Figure 10.21b.

There is a relatively narrow composition range for Ti, Zr, Hf Si oxynitride alloys that are stable against CPS to temperatures >1000°C [4,48]. However, the Hf Si oxynitrides are particularly interesting for applications as gate dielectrics on either Si or Ge substrates. Analysis of electrical measurements of *C–V*, and *I–V* measurements performed on films with a physical thickness of approximately 4 nm and incorporated on n-type and p-type Si and Ge substrates in conventional MOS test structures have indicated (1) static dielectric constants, K_{static}, ~12 (three to four times higher than SiO_2), (2) defect levels and reliability comparable to SiO_2 [44], and (3) tunneling leakage currents $<10^{-6}$ A cm^2 for 1 V bias, with an EOT of ~1.3 nm, as normalized to the dielectric constant of SiO_2. These properties are correlated with a pseudoternary alloy composition of $(HfO_2)_{0.3}(SiO_2)_{0.3}(Si_3N_4)_{0.4}$, which we have characterized as an Si_3N_4-rich composition (uncertainty in composition, $\delta \sim 0.02$). In marked contrast, alloys with reduced Si_3N_4 content between 15% and 25% Si_3N_4 and equal concentrations of SiO_2 and HfO_2, are not stable, and chemically phase separate into SiO_2 and HfO_2, as detected by DXPS [48].

Figure 10.23a and b, respectively, compare (1) the tunneling currents through 4 nm thick high-Si_3N_4 content Hf Si oxynitride films on Si, with an SiON interface, and on Ge (100) with no ITR, and (2) schematic diagrams of the band alignments in these two MOS gate stacks. The processing steps for eliminating Ge–N and Ge–O bonding at Ge-dielectric interfaces, and the regrowth of an epitaxial Ge surface layer after annealing at 800°C are addressed in the Section 10.6.2. The doping densities of the n-type Si and Ge substrates have been chosen to provide an interface alignment to the Al gate electrode. The physical thickness of these dielectrics are comparable, ~4 nm, on the Si and Ge substrates. The trace for the MOSCAP with the Si substrate with an SiON ITR displays a transition from direct to Fowler–Nordheim tunneling current [11]. In contrast, the trace for the same dielectric directly on Ge displays only a direct tunneling characteristic. This marked difference is

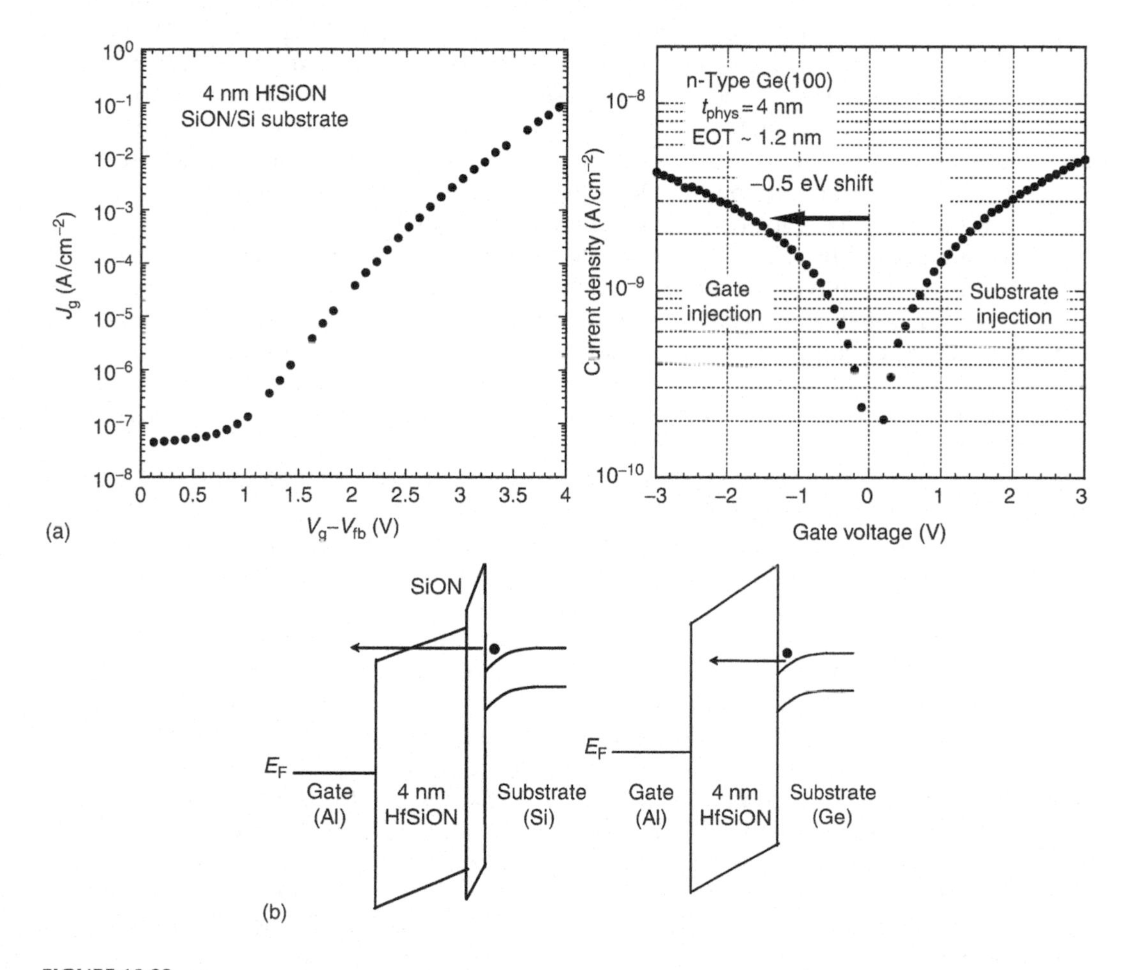

FIGURE 10.23
(a) Tunneling currents through Si_3N_4-rich (40%) Hf Si oxynitride dielectrics on (left) SiON interfacial layers on n-type Si (100) and annealed in Ar at 900°C, and (right) in direct contact with n-type Ge (111) and annealed in Ar at 800°C. (b) Energy level alignment under substrate injection (left) transition from direct to Fowler–Nordheim tunneling, and (right) direct tunneling.

readily accounted for by the band alignment profiles. The field distribution in the Hf Si oxynitride with an SiON interface on Si allows injection into the conduction band of the dielectrics, while the field profile for direct contact with Ge does not.

None of the Hf Si oxynitride compositions are associated with a noncrystalline phase, and the compositions that are stable occupy only a limited portion of a ternary alloy phase diagram. This behavior is explained by an extension of bond constraint theory [49] that is beyond the scope of this discussion [24,25]. These alloys are effectively strain free with respect to local bond strain, as well as a macroscopic strain reduction associated with the fourfold coordination of Hf. To apply these concepts to the Hf Si oxynitride alloys, it is first important to distinguish between two classes of intermediate phase or IP widows with an approximately 5–10 times difference in their compositional window widths. In the first type, locally compliant bonding, and extended compliant molecular groups are competitive with locally rigid bonding, and extended rigid molecular groups, in defining the transitions that define IP windows. These IPs are designated by us as competitive double-percolation IP windows, which are defined by two well-separated transitions [24]. For example, with increasing cross-linking of chalcogenide atoms, the first transition is soft (floppy) to hard (rigid), while the second transition at the termination of the IP window is a more abrupt transition from a rigid to a stressed rigid state. Within the IP window, there are chemical bonding self-organizations involving competitive locally compliant and locally rigid bonding arrangements that minimize bond- and macroscopic-strain through the encapsulation of locally rigid clusters by a skin of locally compliant bonding arrangements. This unstressed IP state proceeds up to an alloy composition at which the locally compliant bonding drops below a percolation limit and the rigid locally rigid bonding arrangements lock the alloy into a stressed-rigid state.

A second class of IP windows, designated by us as confluent double percolation IP windows, is based on two independent contributions to strain reduction that add coherently, one from locally compliant bonding arrangements above the percolation limit, and the second from a compliant extended molecular group [25]. Compositional windows for this second class of IPs are typically ~1%–2%, in contrast to the 8%–12% windows of the first class of IP phases. The high Si_3N_4 content TM Si oxynitride alloys belong to this second class of IP windows.

As noted above, but restated from comparisons with the spectra presented in this section, Figures 10.18a and b present the *I–V* and *C–V* traces for a 4 nm thick 40% Si_3N_4 noncrystalline Hf Si oxynitride pseudoternary alloy dielectric on a Ge (100) surface. Figure 10.18a shows the *I–V* traces for both substrate and gate injection, and Figure 10.18b shows the *C–V* trace, and a fit to the trace that includes the accumulation capacitance for gate voltages beyond the reduction in the parallel capacitance response that derives from the series resistance [58]. It is important to note that the direct tunneling currents through this film for substrate and gate injection are essentially the same for $I = \pm 1.0$ V. The small reduction in current for gate injection is due to the voltage drop across the depletion region in the n-type substrate when a positive bias is applied to the n-type Ge. When corrected for this voltage shift, the near equality of the current indicates that the dielectric does not have traps that can provide a pathway for a trap-assisted tunneling process, as was discussed above for HfO_2 with a nanograin size >3.5 nm. Spectroscopic studies [64,65] and the symmetry of the *I–V* trace in Figure 10.18a are consistent with direct bonding between the n-Ge substrate, and the Hf Si oxynitride nitride, layer, i.e., the absence of a low-*K*, GeO, GeON, or GeN ITR.

The *C–V* trace yields an accumulation capacitance of 139 pF and an EOT of 1.22 nm, consistent with dielectric constant of ~12, essentially the same as that obtained for the same Hf Si oxynitride films deposited on Si substrates, where a correction was made for the interfacial SiON layer. There is no measurable frequency dispersion for the *C–V* trace, and the stretch-out is small well; i.e., the C–V characteristic is near-ideal. This in turn means that D_{it} is low; our best and conservative estimate is that $D_{it} < 10^{11}$ cm^{-2}.

10.6.2 Options for Hf-Based Radiation Hard Devices for Survivable Electronics

The results presented above have identified three different dielectrics for use on Si substrates with SiON ITRs that have sufficiently low defect densities to be considered for applications in commercial electronics, and more importantly in radiation hard devices for survivable electronics. These are (1) ultrathin nanocrystalline HfO_2, $t_{phys} < 2$ nm, (2) chemically phase-separated 84%–80% HfO_2 and 16%–20% SiO_2 with nanocrystalline grain size <2 nm, and (3) Hf Si oxynitride pseudoternary alloys with high Si_3N_4 content, ~40%: $(HfO_2)_{0.3}(SiO_2)_{0.3}(Si_3N_4)_{0.4}$. Studies are in progress to determine the relative radiation hardness of these dielectrics. These studies will involve comparisons with state-of-the-art SiO_2 dielectrics subjected to x-ray and γ-ray irradiation.

10.7 Summary

The spectroscopic data in this chapter have established that: (1) relative energies of d-state C–F and/or J–T term split features in NEXAS O K_1 edge spectra are the same as those determined from band gap transitions: either the absorption constant, α, or imaginary part of the dielectric constant, ε_2, and as determined from vis-VUV-SE measurements; (2) discrete defect state features in films >4 nm thick are at the valence band edge in TiO_2, ZrO_2, and HfO_2 by SXPS, and below the conduction band edge by NEXAS, vis-VUV-SE, PC; (3) based on comparisons with intrinsic d-state features, the defect densities are ~10^{12} defects cm^{-2} (~10^{18} defects cm^{-3}); (4) defect states in ZrO_2 and HfO_2 are qualitatively different than those in TiO_2; (5) the defects are assigned to Ti^{3+} bonding in TiO_2, as Zr^{3+} and Hf^{3+} bonding in ZrO_2 and HfO_2, in suboxide bonding arrangements clustered at nanocrystalline grain boundaries; (6) a connection between intrinsic bonding arrangements and grain-boundary defects is supported by studies on (a) noncrystalline IVB TM Si oxynitrides, (b) ultrathin, <2 nm physical thickness, nanocrystalline ZrO_2 and HfO_2 in which the suppression of J–T splittings by π-bonded states suppress grain-boundary formation; and finally (7) even though O-atom vacancy and interstitial defect state features are not found in thin film dielectrics identified above in point (6), these dielectrics display band edge tailing, indicative of preexisting defects at an order of magnitude lower densities than the grain-boundary associated defects in the nanocrystalline IVB TM oxides that are ~4 nm thick.

In the opinion of this author, and supported by the recent report of Intel Corporation announcing the introduction of Hf alloy dielectric in the 45 nm mode devices [9], the most promising TM gate dielectrics for Si-based nano-CMOS, are two in number: (1) ultrathin HfO_2 deposited on SiON interfacial layers, and (2) high-Si_3N_4 content Hf Si oxynitrides, also deposited on SiON interfacial layers. The most likely successor to Si, strained Si, and Si, Ge as a successor the Si-based CMOS is Ge; however, this will depend critically on the extent to which defects at Ge–dielectric interfaces can be reduced. The limiting factors are (1) the limitation of EOT imposed by SiON layers that utilize thin Si grown on Ge as a template for subsequent high-K TM dielectric deposition, and (2) the extent to which TM dielectrics can be deposited directly onto Ge with formation of Ge–O bonding. Very recent results that address this issue more thoroughly in the context of process-induced defects are addressed in Ref. [66]. These include electrical measurements that complement and update Figures 10.18 and 10.23.

Acknowledgments

This research has been supported in part by ONR, SRC, the SRC/Sematech Front End Processes Center (FEP), and AFOSR. The author acknowledges informative discussions with (1) Jan Lüning at SSRL, (2) Sanghyun Lee, Hyungtak Seo, Chris Hinkle, Les Fleming, Jerry Whitten, and Jack Rowe at NC State University, (3) Marc Ulrich at NC State University and the Army Research Office, (4) Len Brillson at the Ohio State University, (5) Gennadi Bersuker and Pat Lysaght at Sematech, (6) Alex Shluger and Jacob Gavartin at University College London, and (7) John Robertson at the University of Cambridge. Finally, the author has benefited significantly from interactions with Ron Schrimpf and Dan Fleetwood of Vanderbilt University, and Hap Hughes of the Naval Research Laboratory, for informative discussions and collaborations relating to radiation hardness testing of Hf-based dielectrics prepared by my students at NC State University, as well as testing of some of these devices. The author also acknowledges the contributions of many of his graduate students and postdoctoral fellows whose papers are referenced in this chapter.

References

1. Houssa, M. and Heyns, M.M. High-K dielectrics: Why do we need them? in *High-K Gate Dielectrics*, Houssa, M. (Ed.), Institute of Physics, Bristol, 2004, Chap. 1.1.
2. Bersuker, G. et al., Mechanism of electron trapping and characteristics of traps in HfO_2 gate stacks, *IEEE Trans. Dev. Mater. Reliab.*, 7, 138, 2007.
3. Visokay, M.R. et al., Application of HfSiON as a gate dielectric material, *Appl. Phys. Lett.*, 85, 4460, 2002.
4. Lucovsky, G. et al., Intrinsic electronically-active defects in transition metal elemental oxides, *Jpn. J. Appl. Phys.*, 46, 1899, 2007.
5. Krishnamohan, T. et al., High performance, uniaxially-strained, silicon and germanium, double-gate p-MOSFETs, *Microelectron. Eng.*, 84, 2063, 2007.
6. Takagi, S. et al., Gate dielectric formation and MIS interface characterization on Ge, *Microelectron. Eng.*, 84, 2314, 2007.
7. Lo, S.-H., Buchanan, D.A., and Taur, Y., Modeling and characterization of quantization, polysilicon depletion, and direct tunneling effects in MOSFETs with ultrathin oxides, *IBM J. Res. Dev.*, 43, 327, 1999.
8. Lucovsky, G., Ultrathin nitrided gate dielectrics: Plasma processing, chemical characterization, performance, and reliability, *IBM J. Res. Dev.*, 43, 301, 1999.
9. Wu, Y.D., Lee, Y.M., and Lucovsky, G., 1.6 nm oxide equivalent gate dielectrics using nitride/oxide (N/O) composites prepared by RPECVD/oxidation process, *Electron Dev. Lett.*, 22, 116, 2000.
10. Hinkle, C.L. et al., A novel approach for determining the effective tunneling mass of electrons in HfO_2 and other high-K alternative gate dielectrics for advanced CMOS devices, *Microelectron. Eng.*, 72, 257, 2004.
11. "Meet the world's first 45 nm transistors" available at: http://intel.feedroom.com 2007.
12. McFadden, R.M. et al., Intel Corp., Plasma nitridation for reduced leakage gate dielectric layers, US Patent 6,610,615, Issued August 25, 2003.
13. Cotton, F.A., *Chemical Applications of Group Theory*, 2nd edn., Wiley-Interscience, New York, 1971, Chap. 8.
14. Lucovsky, G., Transition from thermally grown gate dielectrics to deposited gate dielectrics for advanced silicon devices: A classification scheme based on bond ionicity, *J. Vac. Sci. Technol. A*, 19, 1553, 2001.
15. Lucovsky, G., Jahn–Teller d-state term splittings in Ti, Zr, and Hf elemental oxides: intrinsic bonding/anti-bonding states and conduction/valence band edge intrinsic defects, *J. Mol. Struct.*, 838, 187, 2007.

16. Autran, J-L., Munteanu, D., and Houssa, M., Electrical characterization, modelling and simulation of MOS structures with high-K gate stacks, in *High-K Gate Dielectrics*, Houssa, M. (Ed.), Institute of Physics, Bristol, 2004, Chap. 3.4.
17. Houssa, M. et al., Effect of O_2 post-deposition anneals on the properties of ultra-thin SiO_x/ZrO_2 gate dielectric stacks, *Semicond. Sci. Technol.*, 16, 31, 2001.
18. Xu, Z. et al., Polarity effect on the temperature dependence of leakage current through HfO/SiO gate dielectric stacks, *Appl. Phys. Lett.*, 80, 1975, 2002.
19. Lucovsky, G. et al., Conduction band-edge states associated with the removal of d-state degeneracies by the Jahn-Teller effect, *IEEE Trans. Dev. Reliab.*, 5, 65, 2005.
20. Cox, P.A., *Transition Metal Oxides*, Clarendon, Oxford, 1992, Chaps. 2, 3, and 5.
21. Lucovsky, G. and White, R.M., The effects of resonance bonding on the properties of crystalline and amorphous semiconductors, *Phys. Rev. B*, 8, 660, 1973.
22. Burstein, E., Brodsky, M.H., and Lucovsky, G., The dynamic ionic charge of zincblende-type crystals, *Int. J. Quantum Chem.*, 1s, 759, 1967.
23. Lucovsky, G. et al., Infrared reflectance spectra of layered group IV and group VI transition metal dichalcogenides, *Phys. Rev. B*, 7, 3855, 1973.
24. Lucovsky, G. and Phillips, J.C., Intermediate phases in binary and ternary alloys: A new perspective on semi-empirical bond constraint theory, *J. Phys.: Condens. Matter*, 19, 455218, 2007.
25. Lucovsky, G. and Phillips, J.C., A new class of intermediate phases in non-crystalline films based on a confluent double percolation mechanism, *J. Phys.: Condens. Matter*, 19, 455219, 2007.
26. Gray, H.B., *Electrons and Chemical Bonding*, Benjamin, New York, 1965, Chap. IX.
27. Lucovsky, G. et al., Studies of bonding defects, and defect state suppression in HfO_2 by soft X-ray absorption and photoelectron spectroscopies, *Surf. Sci.*, 601, 4236, 2007.
28. Lucovsky, G. et al., Electronic structure of high-K transition metal oxides and their silicate and aluminate alloys, *J. Vac. Sci. Technol. B*, 20, 1739, 2002.
29. Brillson, L.F. et al., Depth-resolved detection and process dependence of traps at ultrathin plasma-oxidized and deposited SiO_2/Si interfaces, *J. Vac. Sci. Technol. B*, 18, 1737, 1999.
30. Strzhemechy, Y.M. et al., Low energy electron-excited nanoscale luminescence spectroscopy studies of intrinsic defects in HfO_2 and SiO_2-HfO_2-SiO_2-Si stacks, *J. Vac. Sci. Technol. B*, 26, 232, 2008.
31. Lucovsky, G. et al., Intrinsic band edge traps in nano-crystalline HfO_2 gate dielectrics, *Microelectron. Eng.*, 80, 110, 2005.
32. Xiong, K. et al., Defect energy levels in HfO_2 high-dielectric-constant gate oxide, *Appl. Phys. Lett.*, 87, 183505, 2005.
33. Foster, A.S. et al., Structure and electrical levels of point defects in monoclinic zirconia, *Phys. Rev. B*, 64, 224108, 2001; Vacancy and interstitial defects in hafnia, *Phys. Rev. B*, 65, 174117, 2002.
34. Gavartin, J.L. et al., Negative oxygen vacancies in HfO_2 as charge traps in high-K stacks, *Appl. Phys. Lett.*, 89, 082908, 2006.
35. Takeuchi, H., Ha, D., and King, T.J., Observation of bulk HfO_2 defects by spectroscopic ellipsometry, *J. Vac. Sci. Technol. A*, 22, 1337, 2004.
36. Koiller, B. and Falicov, L.M., Electronic structure of the transition-metal monoxides, *J. Phys. C: Solid State Phys.*, 7, 299, 1974.
37. Lucovsky, G. et al., Defect reduction by suppression of π-bonding coupling in nano- and non-crystalline high-(medium)-κ gate dielectrics, *Microelectron. Eng.*, 84, 2350, 2007.
38. Wright, S., Feeney, S., and Barklie, R.C., EPR study of defects in as-received, γ-irradiated and annealed monoclinic HfO_2 powder, *Microelectron. Eng.*, 84, 2378, 2007.
39. Philipp, H.R. and Ehrenreich, H., Optical properties of semiconductors, *Phys. Rev.*, 129, 1550, 1963.
40. Lucovsky, G., Absorption edge measurements in compensated GaAs, *Appl. Phys. Lett.*, 5, 37, 1964.
41. Felix, J.A. et al., Total-dose radiation response of hafnium-silicate capacitors, *IEEE Trans. Nucl. Sci.*, 49, 3191, 2002.
42. Felix, J.A. et al., Effects of radiation and charge trapping on the reliability of high-κ gate dielectrics, *Microelectron. Reliab.*, 44, 563, 2004.
43. Lucovsky, G. et al., Differences between charge trapping states in irradiated nano-crystalline HfO_2 and non-crystalline Hf silicates, *IEEE Trans. Nucl. Sci.*, 53, 3644, 2006.

44. Chen, D.K. et al., Total dose and bias temperature stress effects for HfSiON on Si MOS capacitors, *IEEE Trans. Nucl. Sci.*, 54, 1931, 2007.
45. Rayner, G.B., Kang, D., and Lucovsky, G., Spectroscopic study of chemical phase separation in zirconium silicate alloys, *J. Vac. Sci. Technol. B*, 21, 1783, 2003.
46. Maria, J.P. et al., Crystallization in SiO_2-metal oxide alloys, *J. Mater. Res.*, 17, 1571, 2002.
47. Ju, B., Properties of Zr-Si oxynitride dielectric alloys, PhD thesis, NC State University, Raleigh, 2000.
48. Lee, S., Lucovsky, G., and Luning, J., Suppression of chemical phase separation in Ti and Hf Si oxynitride alloys with high silicon nitride content, *J. Vac. Sci. Technol. A*, Unpublished.
49. Lucovsky, G., A spectroscopic study distinguishing between chemical phase separation with different degrees of crystallinity in Hf(Zr) silicate alloys, *Appl. Surf. Sci.*, 234, 439, 2004.
50. Williams, M.J. et al., Hydrogenated amorphous silicon-nitrogen alloys, a-Si,N:H: a candidate alloy for the wide band gap photo-active material in tandem photovoltaic (PV) devices, *J. Non-Cryst. Solids*, 164–166, 67, 1993.
51. Lucovsky, G. and Phillips, J.C., Reduction of bulk and interface defects by network self-organizations in gate dielectrics for silicon thin film and field effect transistors (TFTs and FETs, respectively), *J. Non-Cryst. Solids*, 352, 42, 2006.
52. Wheeler, R.A. et al., Symmetric vs. Asymmetric linear M—X-M linkages in molecules polymers and extended networks, *J. Am. Chem. Soc.*, 108, 2222, 1986.
53. White, R.M. and Geballe, T.H., *Long Range Order in Solids, Solid State Physics Supplement 15*, Academic Press, New York, 1979, Chap. IV.
54. Anderson, P.W., Antiferromagnetism—theory of superexchange interaction, *Phys. Rev.*, 79, 350, 1950.
55. Hattangady, S.V., Niimi, H., and Lucovsky, G., Controlled nitrogen incorporation at the gate oxide surface, *Appl. Phys. Lett.*, 66, 3495, 1995.
56. Niimi, H. and Lucovsky, G., Monolayer-level controlled incorporation of nitrogen at Si–SiO_2 interfaces using remote plasma processing, *J. Vac. Sci. Technol. B*, 17, 3185, 1999.
57. Lucovsky, G. et al., Spectroscopic studies of O-vacancy defects in transition metal oxides, *J. Mater. Sci.: Mater. Electron.*, 18, 263, 2007.
58. Lee, S. et al., Suppression of Ge-O and Ge-N bonding at Ge-HfO_2 and Ge-TiO_2 interfaces by deposition onto plasma-nitrided passivated Ge substrates; integration issues for Ge gates stacks into advanced devices, *Microelectron. Reliab.*, 48, 364, 2008.
59. Henson, W.K. et al., Estimating oxide thickness of tunnel oxides down to 1.4 nm using conventional capacitance-voltage measurements on MOS capacitors, *IEEE Electron. Dev. Lett.*, 20, 179, 1999.
60. Pasajova, L., Optical properties of GeO_2 in the ultraviolet region, *Czech. J. Phys. B*, 19, 1265, 1969.
61. Ren, S.-Y. and Ching, W.Y., Electronic structures of β-and α-silicon nitride, *Phys. Rev. B*, 23, 5454, 1981.
62. Dong, J. et al., Theoretical study of β-Ge_3N_4 and its high-pressure spinel γ phase, *Phys. Rev. B*, 61, 11979, 2000.
63. Johnson, R.S., Niimi, H. and Lucovsky, G., New approach for the fabrication of device-quality Ge/GeO_2/SiO_2 interfaces using low temperature remote plasma processing, *J. Vac. Sci. Technol. A*, 18, 1230, 2000.
64. Chung, K-B., unpublished results.
65. Vasic, R., unpublished results.
66. Lucovsky, G., Elimination of native Ge dielectrics at Ge/High-K dielectric interfaces for Ge MoS devices, *Proceedings of ECS Conference*, Honolulu, Hawaii, 2008.

11

Defects in CMOS Gate Dielectrics

Eric Garfunkel, Jacob Gavartin, and Gennadi Bersuker

CONTENTS

11.1 Introduction

The electrical and optical behavior of semiconducting devices is often dominated by the quantity, energy, and physical location of defects. Five decades of research on Si-based devices have led to a reasonable (although not definitive) consensus concerning the nature of defects in complementary metal-oxide semiconductor (CMOS) gate stack dielectrics [1–3]. This understanding has resulted from a continuous interplay between theoretical computation of model structures and experimental measurements of films and devices using a variety of methods. Some of the defects involve changes in local structure or stoichiometry. For example, a slight excess of Si atoms in an otherwise perfect SiO_2 film will result in the appearance of Si–Si bonds. These bonds result in electronic states in the SiO_2 band gap that can become charged under certain conditions. Other defects involve dangling bonds, either at the Si/SiO_2 interface or in the bulk of the SiO_2 film. A third class of defect involves changes in local coordination (Si becoming three- or fivefold coordinated, or O becoming threefold coordinated). Yet another class involves impurity atoms in the film, hydrogen being predominant. Although most impurities degrade device performance, hydrogen can also improve device properties when present at appropriate concentrations and in the appropriate location (usually by bonding with uncoordinated/dangling bonds). The role of nitrogen incorporation into these films as industry has moved from SiO_2 to SiON dielectrics has been extensively studied [2]. Finally, the role of radiation damage in dielectrics has also received much attention over the past few decades, especially for space and some military applications.

Over the past few years, a new class of dielectrics with higher permittivity than SiO_2 or SiON has appeared in the gate stack of CMOS devices. The first new gate dielectric is hafnium oxide based, as hafnia has an optimal combination of properties, including being quite stable against reduction when integrated onto a silicon channel-based platform [4–6].

During the exploration of HfO_2 and other high-permittivity metal oxide gate dielectrics over the past decade, the critical issue of the nature, quantity, energy, and electrical behavior of defects has come to the fore. As with SiO_2-based dielectrics, defects again are thought to involve nonideal stoichiometry, structural imperfections, and impurities. Current thinking by many experts is that O vacancies are a predominant cause of defects. We spend some time in this review outlining our understanding of this class of defects, including presenting recent results of ab initio based calculations of the energies of the various defects.

In most cases currently being explored, the gate stacks formed using Hf oxide and Hf silicate thin films represent multilayer structures that include a SiO_2 layer intentionally or spontaneously formed at the interface with the substrate. This interfacial SiO_2 layer may be nonuniform across its thickness due to its interaction with the high-*K* film [7]. Defects in both high-*K* and interfacial layers are believed to affect device electrical characteristics, although the nature and concentration of each set is still a matter of debate. Hydrogen has also been implicated in both enhancing and passivating defects in high-*K* oxides. Finally, nitrogen may appear in high-*K* gate stacks primarily as a structural/thermal/chemical-stabilizing species, rather than directly enhancing the device electrical properties (as it does in current generation SiON structures).

As industry and the research community progress in their understanding of high-*K* integration, control of the various defects and mitigation of the negative electrical consequence of defects will continue to be aggressively pursued. The understanding that is coming from high-*K* on Si studies also helps the field as it moves from Si to higher mobility substrates such as Ge and III–V materials. Given the higher relative concentration of electrical defects in metal oxides that have been observed in most studies to date, there is also concern that radiation effects could become an even more serious problem with high-*K* oxide gate stacks than similar SiO_2-based ones.

In this chapter, we briefly review selected experimental measurements and computational calculations of defects in HfO_2-based CMOS gate stacks. From one perspective, defects in the dielectric can be viewed as structural and/or compositional anomalies (vacancies, impurities, multiple phases, etc.). There are several classes of possible defects in the HfO_2/SiO_2 stack, for example: (1) vacancies—O, Hf, or Si; (2) impurities—H, Si, Ti, Zr, Hf, N, C, Cl, F, etc.; (3) interstitials—O, Hf, and Si; and (4) various combinations of (1)–(3). At the same time, the defects can be classified by their origins: thermodynamic defects whose concentrations are defined solely by their free energies of formation, and kinetic defects, whose concentration and distribution are metastable, process dependent and, to some extent controllable. Moreover, as the thermodynamic conditions in the dielectric stack vary during the device operation, the concentration and distribution of thermodynamic and kinetic defects may also evolve reversibly or irreversibly.

We also discuss the electronic structure of proposed defects, including their densities and charge state (if any) in the dielectrics (and at the interfaces). And from yet another vantage, we discuss how defects affect various electrical properties of the gate stack, such as leakage current, band alignment, charge trapping, threshold voltage, mobility, reliability, etc.

The chapter is organized as follows. In Section 11.2, we present theoretical models of defects. In Section 11.3, we review some of the experimental results of physical measurements (most involving different forms of spectroscopy). In Section 11.4, we evaluate electrical measurements that have been performed in our labs and elsewhere. There are still some uncertainties in our understanding of defects; however, the agreements between theory and experiment are tantalizing. Finally, we speculate on how these defects will respond to radiation based on initial results and understanding in the field.

11.2 Theoretical Approaches to Understand Dielectric Defects

With regard to defects in a gate stack, atomistic theories have a dual aim:

1. *Explorational*. Given the materials composition, identify most likely (energetically favorable) structural defects and evaluate their thermodynamic properties such as free energies of formation, migration and aggregation, ionization energies, and electron affinities.
2. *Metrological*. Calculate measurable characteristics of defects such as local structure, vibrational frequencies and their activities, electronic energy levels and corresponding optical absorption spectra, electronic spin resonance (ESR) parameters, effective cross sections for charge trapping, chemical shifts for x-ray spectroscopy, etc.

Although present state-of-the-art ab initio calculations allow for evaluation of many properties listed above, none of the existing methods are capable of doing this with high accuracy for all of the properties. The atomistic modeling approaches can be classified by their three main components: the Hamiltonian (Hartree–Fock and post Hartree–Fock, various approximations to density functional theory (DFT), etc.), the basis set (plane waves, atomic orbitals, numeric), and the boundary conditions (periodic, finite cluster, or embedded cluster). These three components may be realized in various combinations. The merits and limitations of various techniques are discussed in many textbooks and reviews (see Ref. [8]).

The most widely used approach is the plane wave density functional (PW-DFT) method utilizing various pseudopotential approaches to treat core electrons and the Kohn–Sham approximation to DFT, and periodic boundary conditions. The PW-DFT method is implemented in many software packages such as VASP [9], CASTEP [10], ABINIT [11], and CPMD [12] to name a few. The PW-DFT approach combines relative simplicity and accuracy and has been proven effective for reliable structural and vibrational information. However, its main well-documented limitation is a severe underestimation of the single-particle band gaps of insulators and semiconductors, and defect level splittings from the band edges. This seriously limits the reliability and accuracy of direct calculations of band alignment at the dielectric/semiconductor and dielectric/metal interfaces [13–15] and optical and electrical properties of defects [16,17]. The band gap problem can be resolved by lifting the local or semilocal approximation to the exchange energy functional in local (LDA) and generalised gradient (GGA) corrected functionals. Several existing methods such as the GW approximation [18], time-dependent DFT [19], screened exchange (sX) [20], and self-interaction correction (SIC) [21] have been used, but their merits and limitations still must be validated. The complexity of these approaches severely restricts the size of the systems that can be calculated and the possibility of structure optimization. An alternative, more economical, but also more empirical approach is to use hybrid functionals in conjunction with plane waves (PBE0 functional [22] and CPMD package [12]) or local orbitals (B3LYP functional [23] and CRYSTAL03 package [24]). The applicability of hybrid functionals has been tested on numerous oxide systems [25].

As noted above, a likely culprit behind one or more of the dielectric defects is the oxygen vacancy. We suspect this for several reasons: among other things, oxygen and hydrogen anneals have a significant effect on the charge in the dielectric that is measured electrically, and perovskite oxides are notorious for having oxygen vacancies (quite useful for oxygen transport in this class of materials). So, what is the argument making oxygen vacancies a most likely defect? First, consider an oxygen vacancy–interstitial pair of defects (Frenkel pair V^q–O_i^{-q}) in monoclinic hafnia (m-HfO_2). The calculated formation energies $E_{Fr}(q)$ are

8.1, 7.0, and 5.6 eV for charges $q = 0$, 1, 2, respectively, strongly suggesting the V^{2+}–O^{2-} combination to be the most stable [16,26]. Once created, oxygen interstitials are found to be extremely mobile even at room temperature [27]. Therefore, oxygen diffusion out of the dielectric into either the metal or the substrate is possible. Reaction of the interstitial with the substrate or metal is also quite exothermic (for most relevant materials). For example, oxidation of the Si–Si bond in the substrate is favorable with respect to oxygen interstitials in m-HfO_2 by ~3–4 eV [28]. Thus, the extended Frenkel pair of a vacancy in HfO_2 and oxygen in Si has a formation energy of only 1.6–2.6 eV. Similar studies of oxygen exchange between HfO_2 and various metal gates [29] also found reduction of formation energies in the extended Frenkel pairs. Therefore, hafnium-based thin films have a propensity for oxygen to escape to a semiconducting substrate or a metal gate, leaving excess oxygen vacancies in the dielectric. However, the sizable positive defect formation energies suggest that this process is technically suppressible (e.g., by interface nitridation).

Numerous earlier calculations of isolated oxygen vacancies in HfO_2 were based on PW DFT and Kohn–Sham approaches using LDA or GGA approximations (see Refs. [8,17,26]). These calculations predicted three stable charge states for oxygen vacancies in various polymorphs of HfO_2—V^{2+}, V^{+}, and V^{0}. Calculated electrical levels associated with thermodynamic charge switching between various charge states of the vacancy were predicted in the mid gap region of HfO_2, and thus, also in the Si gap region in the Si/HfO_2 stack. However, this result was in conflict with electrical measurements placing the energies of electron traps in the conduction band offset region of Si/HfO_2. This discrepancy prompted suggestions of alternative shallow trap candidates. For example, various forms of oxygen interstitials were considered [15,16,26,27], and impurity-induced states from H and Zr were discussed by Shluger et al. [8] and Bersuker et al. [7]. Nevertheless, the almost universally observed sensitivity of trap density to oxygen exposure during and after high-*K* film deposition, and the much weaker dependency on the metallic precursors, strongly points to oxygen deficiency related defects.

This controversy was resolved in recent atomistic studies that revealed that one key vacancy in HfO_2 is an amphoteric defect, i.e., it may accept negative as well as positive charge. Although the existence of negative oxygen vacancies was first suggested from the GGA calculations [7], the obtained trapping energies were far too small. More refined treatment of the electron exchange interaction was necessary for accurate predictions. Negative oxygen vacancies were later predicted using sX approximations [15,30], hybrid density functional [31–33], and the SIC approximation [34].

The results of these different calculations are qualitatively similar, although the numerical differences and accuracies of these approaches warrant separate comparative analysis, as seen below. Here, we follow results [32,33] obtained for monoclinic HfO_2 using atomic basis sets and nonlocal B3LYP density functional [23] as implemented in the CRYSTAL03 package [24]. The single-particle energy diagram for various charge states of a four-coordinated oxygen vacancy is schematically depicted in Figure 11.1. The calculated single-particle band gap of the monoclinic (m)-HfO_2 is about 6.1 eV, i.e., slightly larger than is experimentally observed, 5.6–5.9 eV [35], although there remains some uncertainty in the experimental value as well. As discussed before, oxygen vacancies in m-HfO_2 may exist in five charge states +2, +1, 0, −1, −2, with up to four extra electrons in the vicinity of the vacant O^{2-} site. This property is truly surprising: an ability of a defect to act both as a donor and an acceptor is well established in nonpolar covalent systems (e.g., a single vacancy in Si [36]), but is unexpected in polar wide-gap dielectrics. Interestingly, the preferred oxygen vacancy site in m-HfO_2 depends on the charge state: V^{2+} and V^{+} are more stable in threefold coordinated sites, whereas the V^{0}, V^{-}, and V^{2-} states are energetically more favorable at the fourfold coordinated sites. However, the difference in

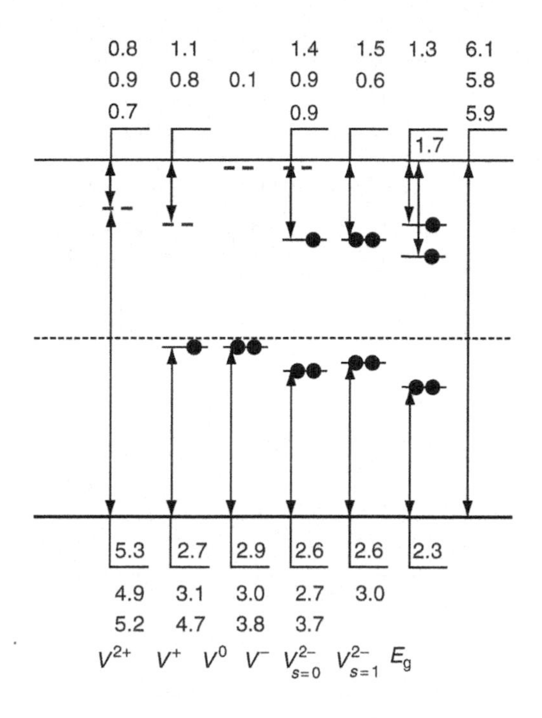

FIGURE 11.1
One-electron level diagram for oxygen vacancies in monoclinic and cubic HfO_2 calculated using different approaches. At the top and bottom of the figure, the first row refers to periodic calculations with an atomic basis set, a B3LYP density functional and a 96 atoms supercell (From Gavartin, J.L. et al., *Appl. Phys. Lett.*, 89, 082908, 2006; Ramo, D.M. et al., *Phys. Rev. B*, 75, 205336, 2007); the second row refers to PW-DFT calculations with an RPBE0 density functional and a similar supercell (From Broquist, P. and Pasquarello, A., *Appl. Phys. Lett.*, 89, 262904, 2006); the third row refers to PW-DFT calculations with an sX density functional and a cubic supercell of 54 atoms. (From Robertson, J., *Rep. Prog. Phys.*, 69, 327, 2006; Xiong, K. et al., *Appl. Phys. Lett.*, 87, 183505, 2005.)

formation energies for doubly negatively charged vacancies between three- and four-coordinated sites is only about 0.2 eV.

The change of the charge state of the vacancy (trapping and detrapping) is associated with a large local lattice relaxation associated with polarization energies in the range of 0.5–1.0 eV. The direct consequences are an asymmetry in charge trapping and detrapping kinetics and differences in characteristic energies as observed in optical (reflection ellipsometry, adsorption, and photoluminescence) and electrical (*I–V* and *C–V*) probes. The difference between optical and thermal ionization of defects has been extensively discussed in the literature (see Ref. [37]), and can be qualitatively explained by the potential surface diagram depicted in Figure 11.2. Optical ionization occurs at time scales much faster than characteristic lattice vibrations. Thus, an electronic transition from the trap of charge q (state A) onto the conduction band minimum (CBM) (state B*) can be considered in a frozen lattice, which corresponds to a Frank–Condon type vertical transition E_{opt}. However, the ionized trap induces lattice polarization, which in the case of the oxygen vacancy is mainly associated with the outwards displacement of the nearest neighbor Hf ions. This relaxation coordinate is denoted as Q in the diagram. The minimum energy state B corresponds to a fully relaxed trap state $q-1$, and the electron is delocalized at the CBM. Thermal ionization of the trap is a slow phonon-assisted process with the activation energy E_{th}.

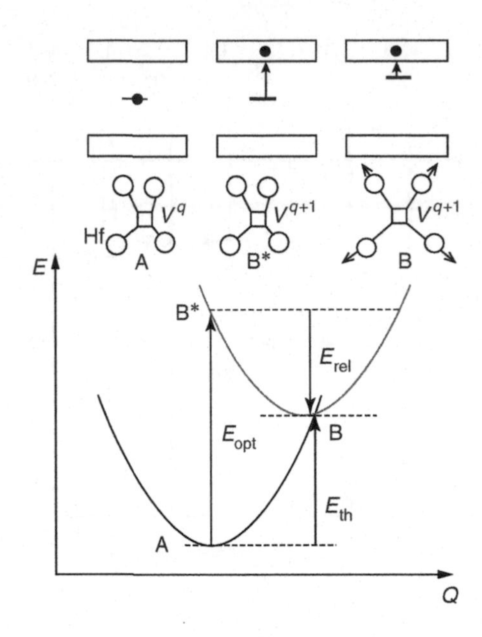

FIGURE 11.2
Schematic potential energy surface for optical and thermal detrapping processes. Generalized displacement of the nearest neighbor Hf ions, *Q*, is shown, and the energy scale is chosen to represent ionization of V^{2-} vacancy. The band diagram schematically illustrates optical and thermal ionization processes, respectively, without (B*) and with (B) lattice relaxation. (After Gavartin, J.L. et al., *Appl. Phys. Lett.*, 89, 082908, 2006. With permission.)

By construction of Figure 11.2, E_{th} may be approximated as

$$E_{th} = E_{opt} - E_{rel}. \tag{11.1}$$

Optical and thermal ionization energies can be deduced as a combination of the calculated total energies of different systems as follows:

$$E_{opt}(V^q) = E_q(V^{q+1}) - E_q(V^q) + E^- - E^0, \tag{11.2}$$

$$E_{rel}(V^{q+1}) = E_{q+1}(V^{q+1}) - E_q(V^{q+1}), \tag{11.3}$$

where
- E^0 is the total energy of the perfect HfO_2 crystal
- E^- is the total energy of the perfect HfO_2 crystal with an electron at the bottom of the conduction band
- $E_q(V^q)$ is the total energy of HfO_2 with the vacancy in a charge state q ($q = +2, +1, 0, -1, -2$) in the optimized geometry
- $E_q(V^{q+1})$ is the total energy of the vacancy in the charge state $q+1$ but at the equilibrium geometry corresponding to the vacancy in the charge state q

The calculated optical excitation energies from the defect states into the CBM and the thermal detrapping energies of oxygen vacancies are summarized in Table 11.1. The defect energies, especially for the shallow defects, critically depend on the nonlocal exchange interaction method used. Most of the existing approaches are based on so-called hybrid functionals in which the local or GGA exchange functional is mixed with the Hartree–Fock exchange using one or another empirical scheme. Although these

TABLE 11.1

Optical Excitation Energies, E_{opt}, Relaxation Energies, E_{rel}, and Thermal Activation Energies, E_{th}, for Oxygen Vacancies in m-HfO_2 Calculated Using Equations 11.1 through 11.3

Charge	E_{opt} (eV)	E_{rel} (eV)	E_{th} (eV)
V^+	3.33	1.01	2.32
V^0	3.13	0.80	2.33
V^-	1.24	0.48	0.76
V^{2-}	0.99	0.43	0.56

Source: After Gavartin, J.L. et al., *Appl. Phys. Lett.*, 89, 082908, 2006. With permission.

approaches seem relatively accurate in general, the error margins associated with specific systems are essentially unknown. Therefore, it is instructive to compare the results obtained using different methods.

Comparisons of the B3LYP calculations [32,33] with PBE0 [31] and sX [15,30] indicate an agreement within 0.3–0.4 eV in most of the single-particle energies with respect to the band edges (Figure 11.1). Most of the discrepancies may be attributed to different single-particle band gap energies predicted in these approaches. As noted above, the B3LYP method predicts the largest one-electron band gap with the largest deviation from E_g measured by optical absorption (~5.6 eV [34]). However, single-electron estimates of band gaps neglect excitonic effects and the effects of phonon broadening, which may reduce the measured optical band gap of HfO_2 by up to 0.8 eV [38]. Similar arguments apply to the adsorption energies associated with optical transitions between the valence band and V^{2+} level. An apparent close agreement of the calculations [31] with the reflection ellipsometry measurements [39] is somewhat deceptive. Single-particle calculations do not include electron–hole interaction, which is significant for such transitions. When taken into account (say using the TDDFT method and M3LYP functional), this interaction reduces the transition energy to 4.9 eV, that is, ~0.5 eV lower than the single-particle energy difference [33], bringing it closer to experiment. Apparently, the positive and negative oxygen vacancy interplay is not unique for HfO_2. Recent atomistic studies predict similar behavior in tetragonal hafnium and zirconium silicates [40–42], and even in silica [43]. However, no experimental confirmation has yet been reported for negative vacancy states in either silica, hafnon ($HfSiO_4$), or zircon ($ZrSiO_4$) to the best of our knowledge.

An important feature of the optical signal associated with the oxygen deficiency in HfO_2 is that its width strongly increases with vacancy concentration [39]. This issue has been studied theoretically using the PW-DFT approach and a GGA energy functional [44]. First, it was concluded that oxygen divacancies, although not energetically favorable, may form under certain conditions (with near zero-binding energy). Second, the case of multiple vacancies was considered in a large supercell of 324 atoms. It was demonstrated that, by a 5% vacancy concentration, the vacancy signal significantly broadens. This effect is illustrated in the time evolution of the one-electron levels calculated using Born–Oppenheimer molecular dynamics (Figure 11.3). Remarkably, even at room temperature the dynamical fluctuations of localized defect and band edge levels occur on a scale much larger than the thermal energy. This gives rise to a broad (>1.5 eV) defect band and long band tails, which are also observed in reflection ellipsometry of oxygen-deficient films [39]. The static lattice relaxation near the isolated neutral vacancy is negligible, so the strong interaction between the separated vacancies is yet another manifestation of a strong electron–lattice coupling in high-*K* systems [45]. Another phenomenon associated with strong electron–phonon coupling is the possibility of small polaron formation; these are electronic states localized by

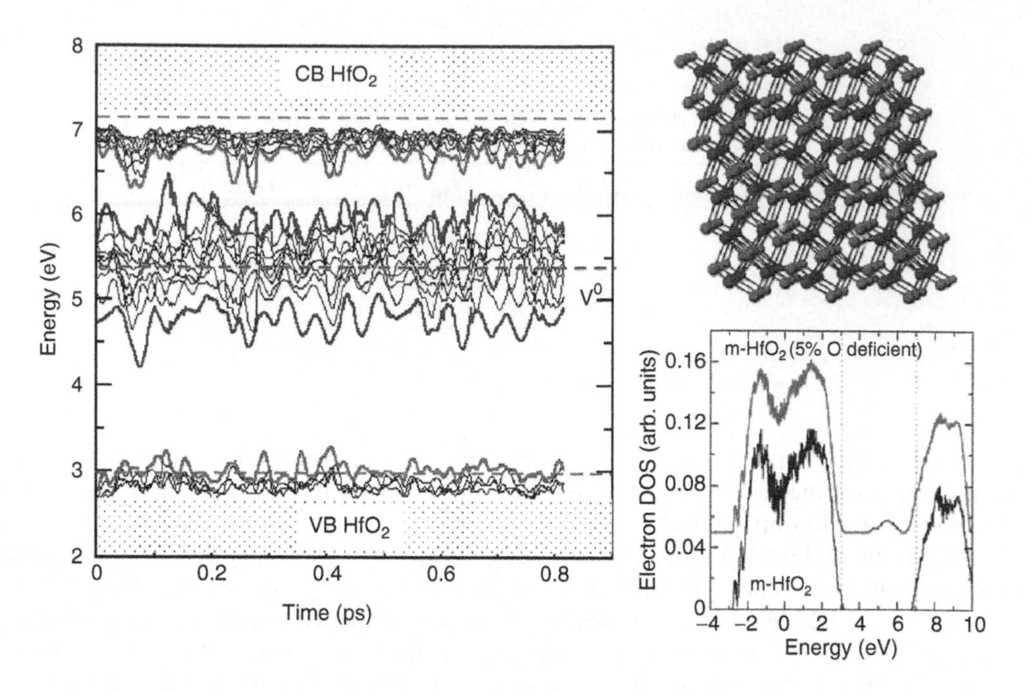

FIGURE 11.3
Dynamics of the one-electron levels of m-HfO_2 with ~5% neutral oxygen vacancies calculated in a supercell of 324 ions using Born–Oppenheimer MD at $T = 300$ K. The positions of the band edges and a single vacancy level in a static calculation are shown as dashed lines. The resulting electron DOS is shown in comparison with a perfect m-HfO_2 at the same temperature. (After Gavartin, J.L., et al., *Physics and Technology of High-K Dielectrics*, 4, Kar, S., et al., Eds., *Electro Chemical Society*, Pennington, New Jersey, 2006, p. 227. With permission.)

lattice polarization, induced by the carriers themselves. The negative oxygen vacancy is one example of a small polaron [46]. The possibility of electron and hole self-trapping was also suggested in amorphous HfO_2 films [44,46] and even in perfect monoclinic hafnia [47,48]. Small polarons in high-*K* dielectrics may have a significant contribution to the trapping kinetics, leakage currents, defect formation, and film degradation. It is also a challenging problem for atomistic modeling, and currently is a subject of intensive research.

11.3 Physical Methods of Defect Detection in Dielectrics

In this section, we briefly review two classes of "nonelectrical" experimental dielectric defect analysis methods—one microscopic and the other spectroscopic. We note ESR work in the final "electrical methods" section. There are other tools that have been used to help elucidate defect behavior that we have not reviewed here, including Fourier transform infrared (FTIR) [49–51] and cathodoluminescence [52].

Imaging defects in HfO_2 dielectrics has proven a difficult task. Surface defects, as can be seen in scanning probe microscopy, are unlikely to be the same as defects in the bulk. Electron microscopy is also difficult at the single defect level; it is virtually impossible to imagine seeing a single oxygen vacancy in a SiO_2 or HfO_2 film. One exception is that of imaging heavy *Z* atoms (e.g., Au or Hf) in a light *Z* matrix (e.g., C or even SiO_2), which makes seeing a Hf atom in an SiO_2 interfacial layer possible [53]. On the other hand, oxygen deficiencies in the interfacial SiO_2 layer may manifest themselves in other

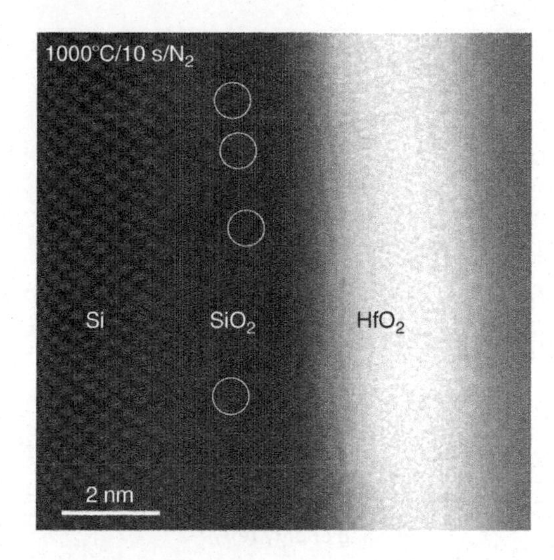

FIGURE 11.4
ADF TEM image of a 3 nm HfO_2/2 nm SiO_2/Si stack after a 1000°C anneal. Stray Hf atoms in the IL are circled. (From Bersuker, G., et al., *J. Appl. Phys.*, 100, 09418, 2006. With permission.)

spectroscopic ways, including via their spectral features corresponding to under-coordinated Si atoms as observed in the Si $L_{2,3}$ electron energy loss spectra (EELS). In Figure 11.4, we show a transmission electron microscopy (TEM) image of a high-K gate stack fabricated with a 2 nm thermal SiO_2 interfacial layer followed by 3 nm ALD HfO_2 film. After the stack was exposed to a 1000°C/10 s anneal, isolated Hf atoms, in the density range of 10^{12} to 10^{13} cm^{-2}, were found in the SiO_2 layer. At the same time, the high-temperature anneal resulted in significant oxygen deficiency of this interfacial layer, as demonstrated by the rise of the shoulder associated with the under-oxidized Si atoms in the EELS spectra (Figure 11.5).

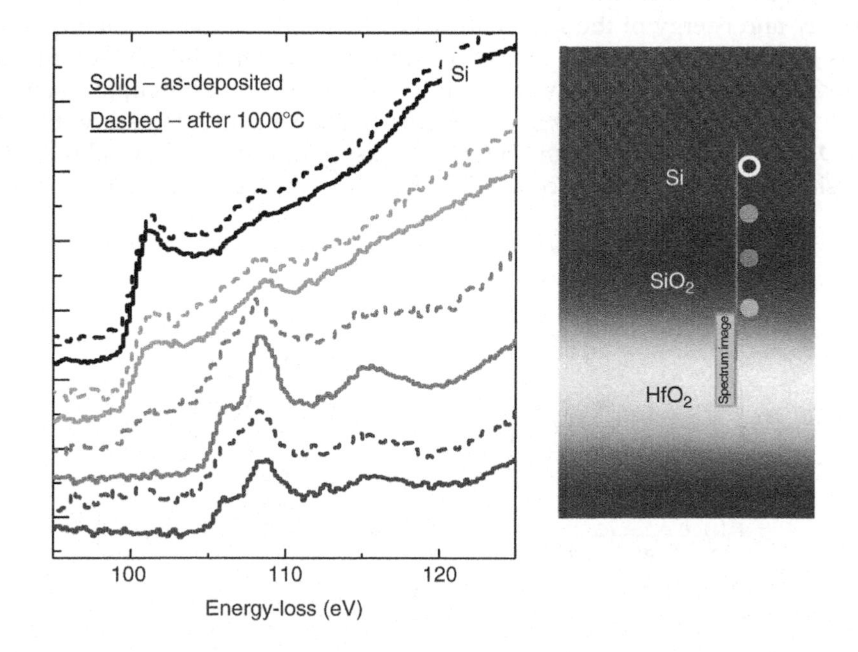

FIGURE 11.5
ADF TEM image and Si $L_{2,3}$ edge EELS taken at the identified positions in the 3 nm HfO_2/2 nm SiO_2 stack before (a) and after (b) anneal. (From Bersuker, G., et al., *J. Appl. Phys.*, 100, 09418, 2006. With permission.)

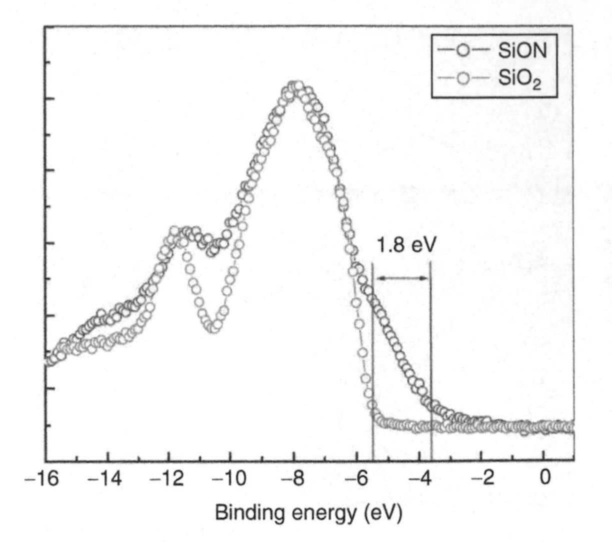

FIGURE 11.6
PES (taken at Brookhaven) of HfSiO/Si and HfSiNO/Si ultrathin films. Note especially the difference in top of the valence band. (From Sayan, S., et al., *Appl. Phys. Lett.*, 87, 212905, 2005. With permission.)

Variants of photoemission spectroscopy (PES) have also been employed to examine defects in dielectrics. One set of photoemission methods focuses on developing an accurate picture of the density of states (DOS) in the valence band. The spectra of two $HfSiO_x$ systems, one with and one without nitrogen present, are shown in Figure 11.6 [54]. The incorporation of nitrogen leads to a significant DOS at the top of the valence band, which in the pure oxides (hafnium and/or silicon oxide) is predominantly of O 2p character. Whether nitrogen can be called a defect or not is mainly a semantic issue; it is clear that it results in a significant DOS at the bottom of the gap (at the top of the valence band). A second, related method is called inverse photoemission (InvPES). In InvPES, electrons with an energy of order 20 eV impinge upon a surface, fall into unoccupied levels, and emit a photon with an energy given by the difference between the incident kinetic energy and energy of the state into which they fall. As the probability for photon emission is directly related to the density of unoccupied states into which the electrons fall, the InvPES spectrum roughly maps out the density of unoccupied states within the first 10 eV of the Fermi energy. In Figure 11.7, we present an InvPES spectrum from a HfO_2/Si sample. In the same figure, we also plot out an ultraviolet light excited photoemission (UPS) spectrum (occupied DOS) taken on the same sample in the same

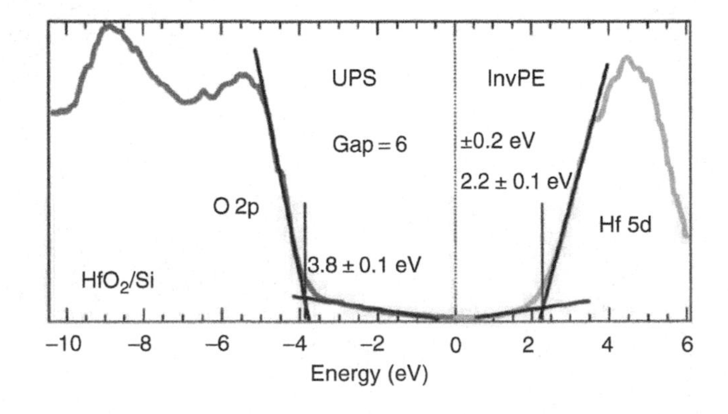

FIGURE 11.7
UPS and InvPES of a 2 nm HfO_2 film on Si. The densities of states relative to the Fermi level ($E = 0$) and the gap can readily be observed. Defects and band tail states are also sometimes observed. (From Bartynski, Garfunkel, et al., previously unpublished data.)

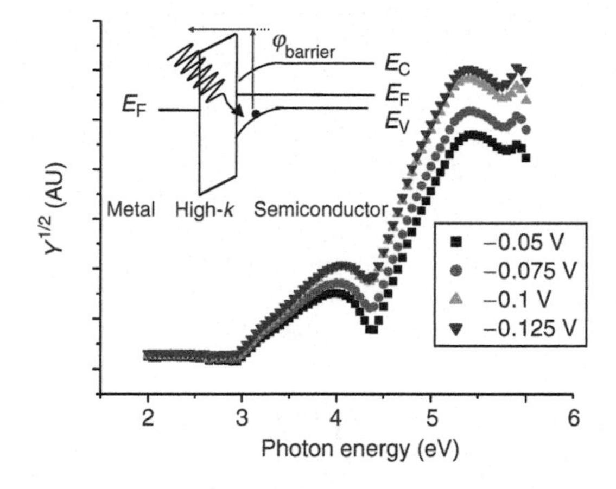

FIGURE 11.8
IntPES spectra from a Al/HfO_2/Ge gate stack. HfO_2 (conduction band) to Ge (valence band) energy offset measured to be ~3.1 eV. (From Garfunkel and Celik, previously unpublished data.) Energies listed in box refer to bias across MIS structure.

chamber on the same day. Although very useful for getting a big picture of the major components of the system's electronic structure, PES and InvPES are not particularly useful for determining the electronic structure of defects that may be present in the system at a level that is much less than 1% of the total atomic density.

Two more variants of photoemission that yield very helpful information about the system's electronic structure are internal photoemission (IntPES—see especially the extensive work of Afanas'ev and Stesmans [55,56] and x-ray photoelectron (XPS) energy loss spectroscopy, and total photoelectron yield spectroscopy (especially the work of Miyazaki, Hirose, et al. [57,58]). IntPES is performed by shining monochromatic light onto an ultrathin capacitor structure (metal–insulator–metal [MIM] or metal–insulator–semiconductor [MIS]) and measuring the photoinduced current as a function of photon energy, bias across the insulator, sample and processing conditions, etc. Some believe that this is an accurate way to both determine band edge energies and to observe states in the gap. As the photocurrent yield axis is not linear (following standard practice in the field), low density defects are enhanced. In Figure 11.8, we present one set of our IntPES results from a study of various substrates and metal gates sandwiching a HfO_2 dielectric. One result relevant to defect characterization is the appearance of band tail states near the top of the gap on samples that also show a high electrical defect concentration.

Another set of experimental measurements that yields information about defects comes from optical spectroscopy. The idea is that the cross section for optical absorption comes only when there is an unoccupied state present that can accept an electron. There are various ways to examine this, one being ellipsometry [51,59]. In ellipsometry a plot of the imaginary part of the dielectric function versus energy is loosely related to the density of unoccupied states. In Figure 11.9, we show the imaginary part of the dielectric function for two hafnium silicates (the same sample set as presented in Figure 11.6) [54]. Clearly the gap narrows when nitrogen is included in the dielectric. It is not clear from these data alone whether that narrowing arises because of states at the top or bottom of the gap—here PES and InvPES offer a more complete picture of the DOS changes. The optical-based methods of characterization such as IntPES, ellipsometry, and cathodoluminescence are usually more sensitive to lower defect concentrations.

11.4 Electrical Methods of Defect Characterization

Electrical characterization allows one to identify electrically active defects relevant to device performance with exceptional sensitivity, on the order of 10^{10} cm^{-2}. This level

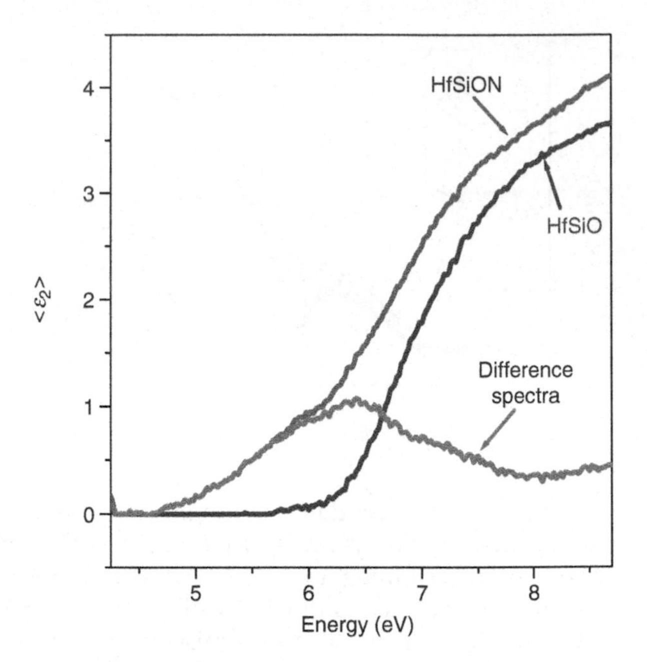

FIGURE 11.9
The imaginary part of the dielectric function for two hafnium silicates, and the difference spectrum, showing how N narrows the effective gap. (After Sayan, S., et al., *Appl. Phys. Lett.*, 87, 212905, 2005. With permission.)

of sensitivity, developed over several decades as devices have scaled from the micron to nanometer regime, is much higher than can be realized with most "physical" methods (e.g., PES, TEM, FTIR, ion scattering, ellipsometry, etc.). Gate stacks formed using hafnium oxide and hafnium silicate thin films currently being considered for gate dielectric applications represent multilayer structures that include an SiO_2 layer formed at the interface with the substrate [60]. The high density of the as-grown electron/hole traps found to be present in the high-*K* dielectric stacks may dramatically degrade transistor performance due to very fast charge trapping processes (in the μs range) to which they can contribute. Specific electrical techniques with good spatial (especially depth) and time resolution are then required to probe electrically active defects either in the high-*K* or interfacial SiO_2 layers. In particular, during pulsed I_d–V_g measurements on nMOSFETs [61], some electrons injected from the transistor inversion layer into the dielectric during nanosecond–microsecond pulses applied to the gate become trapped in as-grown defects in the HfO_2 film. This type of measurement results in well-localized (from the dielectric thickness standpoint) trapped charges, since trapping of the tunneling electrons proceeds most effectively via a resonant process [62], while short pulse time prevents redistribution of the trapped charges between the adjacent traps. By subtracting contributions due to the fast resonant trapping from the total threshold voltage shift after long-term stress, one can obtain an electrical signature of the slow charge redistribution process (process B in Figure 11.10), and use it for extraction of the trap energies. Fitting results for all voltages in the whole temperature range yields trap energies on the order of 0.35 and 0.45 eV below the conduction band edge. This value is consistent with calculated energies of the single and double charged negative oxygen vacancies in monoclinic hafnia [31,32]. Measured and modeled trapping effects on V_t are shown in Figure 11.11.

Interfacial SiO_2 layers in high-*K* gate stacks can be profiled using frequency-dependent charge pumping techniques. The frequency dependence can be converted into a distance (across the dielectric) dependence [63]. Trap generation in the interlayer (IL), ΔN_{it}, is higher close to the high-*K* layer, while in the gate stacks with different IL thicknesses, $N_{it}(x)$ values in the IL are similar at the same distance x from the high-*K* film, Figure 11.12 (here N_{it} is plotted against the distance from the HfO_2/SiO_2 interface) [64]. At the same time, the rate

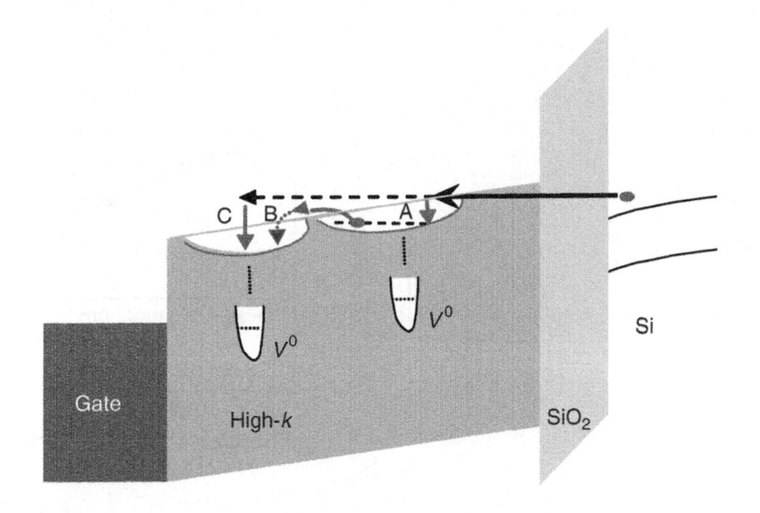

FIGURE 11.10
Schematic diagram of the electron trapping processes: A—electrons injection from the conduction band and trapping at a shallow defect (presumably neutral O-vacancy (V^0)), which leads to the formation of a negative O-vacancy accompanied by lattice relaxation (down arrow); B—a temperature-assisted electron detrapping/retrapping; Process C is ineffective due to the mismatch of the electron and trap energies. (From Bersuker, G., et al., *Proc. IEEE Intl. Reliab. Phys. Sympos.*, 2006, p. 179. With permission.)

of trap generation in the IL, $R_{it}(t) = \Delta N_{it}(t + \Delta t)/\Delta t$, decreases with the stress time, t, along with the stress-induced leakage current (SILC) (Chapters 15 through 17) generation rate, $R_{SILC}(t)$, in Figure 11.13.

Similar growth rates of SILC and N_{it} suggest that their increase is most likely driven by the same defects within the IL. The decrease in the degradation rate with time, Figure 11.13, indicates that the electron traps are generated in the IL at preexisting, precursor defect sites, whose density increases closer to the high-K film. That they exist prior to electrical stressing suggests that these precursor defects could be induced by the interaction of the IL with the high-K dielectric. Recent physical results (scanning tunneling electron microscopy (STEM)/EELS, ESR, XPS, ion scattering) strongly suggest that the Hf-based high-K films modify the stoichiometry of the underlying SiO_2 layer by rendering

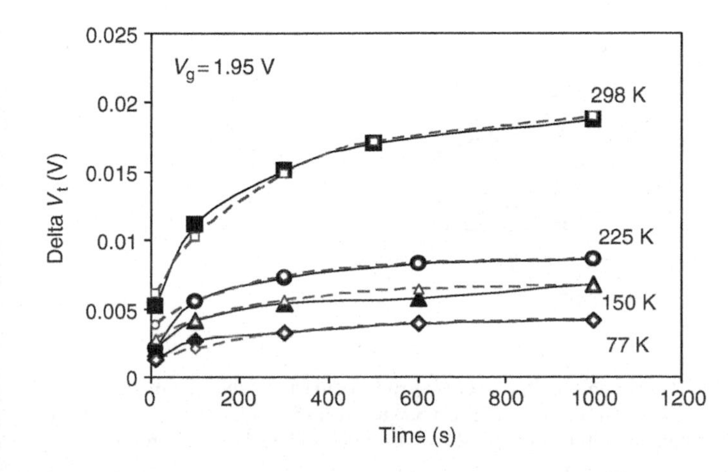

FIGURE 11.11
Measured and modeled (open symbols) V_t shift associated with the slow trapping process (A) of Figure 11.10 during the $V_g = 1.95$ V stress at different temperatures. (From Bersuker, G., et al., *Proc. IEEE Intl. Reliab. Phys. Sympos.*, 2006, p. 179. With permission.)

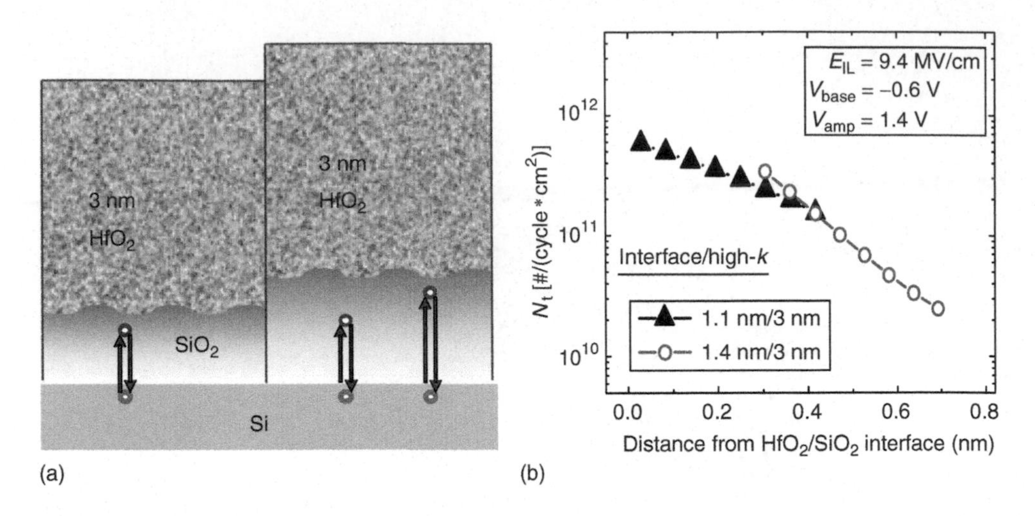

FIGURE 11.12
Simple schematic diagram (a) of the electron probing of the traps during charge pumping measurements, and (b) N_{it} plotted versus the distance from the HfO_2/SiO_2 interface. (After Young, C.D., et al., *IEEE Trans. Dev. Mater. Reliab.*, 6, 123, 2006. With permission.)

it oxygen-deficient. Precursor defects associated with the oxygen vacancies (for instance, Si–Si defects) can be converted into electron traps during stress (by breaking the Si–Si bonds), giving rise to SILC.

The electrically detected ESR technique called spin-dependent recombination (SDR), has several orders of magnitude greater sensitivity than "conventional" electron spin resonance (Chapter 6) and allows measurements in fully processed transistors [65].

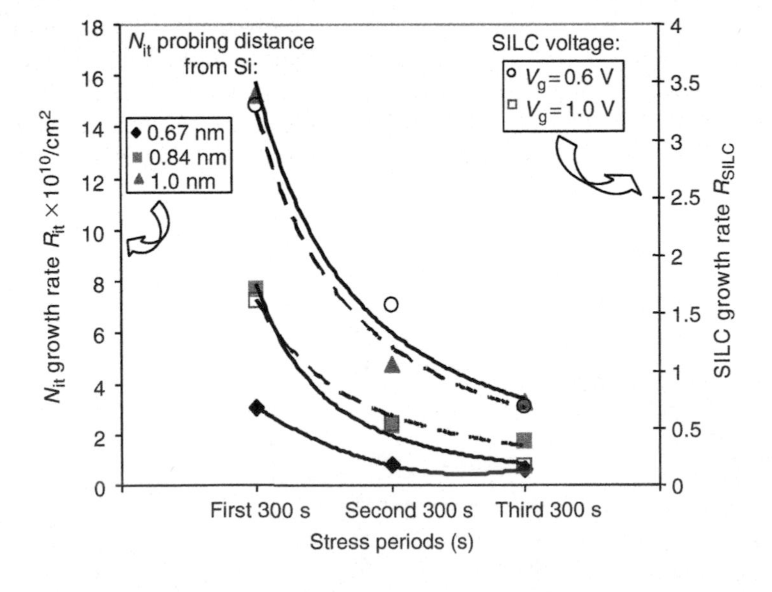

FIGURE 11.13
SILC and trap generation rates obtained by averaging the SILC and N_{it} growth over the subsequent 300 s stress intervals at different distances from the Si substrate and different SILC voltages, respectively, under a low-voltage CVS ($V_g = 2.4$ V). Trends for the trap generation and SILC rates are shown by solid and broken lines, respectively.

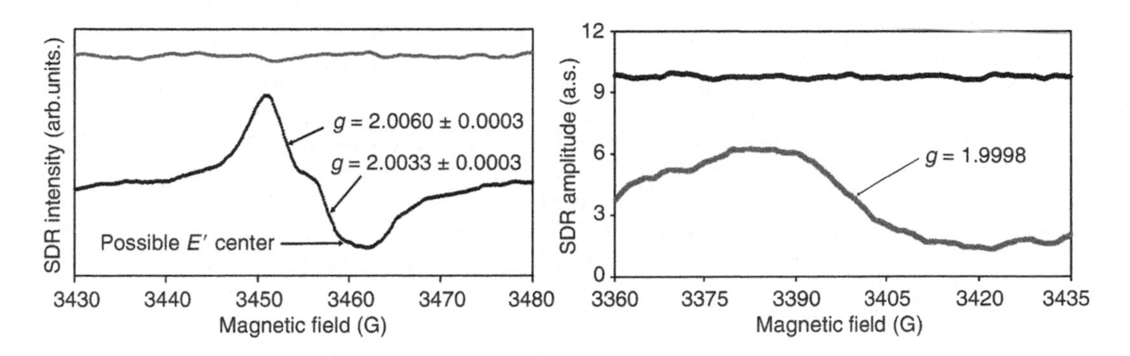

FIGURE 11.14
SDR signal measured on a $HfO_2/SiO_2/Si$ stack. (After Neugroschel, A., et al., *IEEE IEDM*, 2006, p. 137. With permission.)

By simultaneously exposing a device, appropriately biased (for maximum recombination), to a large magnetic field and microwave irradiation, one can observe the magnetic resonance response of deep level recombination centers by monitoring the response of the recombination current versus magnetic field. SDR measurements performed on high-*K* transistors show a new signal after the initial few seconds of stress, which is not observed in the SiO_2 stack, Figure 11.14 [66]. Due to the broadness of this signal, indicative of the presence of d-type electronic states, the additional defects can tentatively be assigned to Hf atoms, which were found to be present in low concentrations ($\sim 10^{13}/cm^2$) in the interfacial SiO_2 layer (Figure 11.4).

Research on the effects of radiation on high-*K* devices is still in its initial phase [67–76]. In an ionizing radiation environment, SiO_2-based MOSFETs are known to predominantly trap holes with some electron trapping being observed due to the existence of neutral electron traps or new electron trap creation in the oxides. The presence of the interfacial SiO_2 layer in the high-*K* gate stack suggests that one may expect an electrical response similar to that of a very thin SiO_2 film, although significantly amplified by the SiO_2 interaction with the high-*K* film. Studies of radiation effects in high-*K* stacks are ongoing.

References

1. Buchanan, D.A., Scaling the gate dielectric: Materials, integration, and reliability, *IBM J. Res. Dev.*, 43, 245, 1999.
2. Green, M.L. et al., Ultrathin (<4 nm) SiO_2 and Si–O–N gate dielectric layers for silicon microelectronics: Understanding the processing, structure, and physical and electrical limits, *J. Appl. Phys.*, 90, 2121, 2001.
3. Gusev, E.P., Narayanan, V., and Frank, M.M., Advanced high-*k* dielectric stacks with poly-Si and metal gates: Recent progress and current challenges, *IBM J. Res. Dev.*, 50, 387, 2006.
4. Wallace, R.M. and Wilk, G.D., High-*k* dielectric materials for microelectronics, *Crit. Rev. Solid State Mater. Sci.*, 28, 231, 2003.
5. Wilk, G.D., Wallace, R.M., and Anthony, J.M., High-*k* gate dielectrics: Current status and materials properties considerations, *J. Appl. Phys.*, 89, 5243, 2001.
6. Gusev, E., *Defects in Advanced High-k Dielectric Nano-Electronic Semiconductor Devices*, Springer, Dordrecht, the Netherlands, 2006.
7. Bersuker, G. et al., Mechanism of charge trapping reduction in scaled high-*k* gate stacks, in *Defects in Advanced High-k Dielectric Nano-Electronic Semiconductor Devices*, Gusev, E., Ed. Springer, Dordrecht, the Netherlands, 2006, p. 227.

8. Shluger, A.L. et al., *Models of Defects in Wide-Gap Oxides: Perspective and Challenges*, Elsevier, Amsterdam, 2003, p. 151.
9. Kresse, G. and Furthmüller, J., Efficiency of ab-initio total energy calculations for metals and semiconductors using a plane-wave basis set, *Comput. Mater. Sci.*, 6, 15, 1996.
10. Segall, M.D. et al., First-principles simulation: Ideas, illustrations and the CASTEP code, *J. Phys. Condens. Matter*, 14, 2717, 2002.
11. ABINIT software code. Available at: www.abinit.org.
12. CPMD software code. Available at: www.cpmd.org.
13. Demkov, A.A. et al., Density functional theory of high-*k* dielectric gate stacks, *Microelectron. Reliab.*, 47, 686, 2007.
14. Gavartin, J.L. and Shluger, A.L., Modeling $HfO_2/SiO_2/Si$ interface, *Microelectron. Eng.*, 84, 2412, 2007.
15. Robertson, J., High dielectric constant gate oxides for metal oxide Si transitors, *Rep. Prog. Phys.*, 69, 327, 2006.
16. Gavartin, J.L. et al., The role of nitrogen-related defects in high-*k* dielectric oxides: Density-functional studies, *J. Appl. Phys.*, 97, 053704, 2005.
17. Xiong, K. and Robertson, J., Point defects in HfO_2 high *K* gate oxide, *Microelectron. Eng.*, 80, 408, 2005.
18. Onida, G., Reining, L., and Rubio, A., Electronic excitations: Density-functional versus many-body Green's-function approaches, *Rev. Mod. Phys.*, 74, 601, 2002.
19. Botti, S. et al., Time-dependent density-functional theory for extended systems, *Rep. Prog. Phys.*, 70, 357, 2007.
20. Rushton, P.P., Tozer, D.J., and Clark, S.J., Nonlocal density-functional description of exchange and correlation in silicon, *Phys. Rev. B*, 65, 235203, 2002.
21. Chu, S.I., Recent development of self-interaction-free time-dependent density-functional theory for nonperturbative treatment of atomic and molecular multiphoton processes in intense laser fields, *J. Chem. Phys.*, 123, 062207, 2005.
22. Todorova, T. et al., Molecular dynamics simulation of liquid water: Hybrid density functionals, *J. Phys. Chem. B*, 110, 3685, 2006.
23. Becke, A.D., A new mixing of Hartree–Fock and local density-functional theories, *J. Chem. Phys.*, 98, 1372, 1993.
24. Saunders, V. et al., *CRYSTAL2003 User's Manual*, University of Torino, Torino, 2003.
25. Cora, F. et al., The performance of hybrid density functionals in solid state chemistry, in *Principles and Applications of Density Functional Theory in Inorganic Chemistry II*, Springer, Berlin/Heidelberg, 2004, Vol. 113, p. 171.
26. Foster, A.S. et al., Vacancy and interstitial defects in hafnia, *Phys. Rev. B*, 65, 174117, 2002.
27. Foster, A.S., Shluger, A.L., and Nieminen, R.M., Mechanism of interstitial oxygen diffusion in hafnia, *Phys. Rev. Lett.*, 89, 225901, 2002.
28. Gavartin, J.L. et al., Ab initio modeling of structure and defects at the HfO_2/Si interface, *Microelectron. Eng.*, 80, 412, 2005.
29. Demkov, A.A., Thermodynamic stability and band alignment at a metal-high-*k* dielectric interface, *Phys. Rev. B*, 74, 085310, 2006.
30. Xiong, K. et al., Defect energy levels in HfO_2 high-dielectric-constant gate oxide, *Appl. Phys. Lett.*, 87, 183505, 2005.
31. Broqvist, P. and Pasquarello, A., Oxygen vacancy in monoclinic HfO_2: A consistent interpretation of trap assisted conduction, direct electron injection, and optical absorption experiments, *Appl. Phys. Lett.*, 89, 262904, 2006.
32. Gavartin, J.L. et al., Negative oxygen vacancies in HfO_2 as charge traps in high-*k* stacks, *Appl. Phys. Lett.*, 89, 082908, 2006.
33. Ramo, D.M. et al., Spectroscopic properties of oxygen vacancies in monoclinic HfO_2 calculated with periodic and embedded cluster density functional theory, *Phys. Rev. B*, 75, 205336, 2007.
34. Filippetti, A., Fiorentini, V., and Lopez, G.M., Electronic structure of defects in dielectrics with electronic correlation, *Electrochem. Soc. Trans.*, 3, 267, 2006.
35. Houssa, M., *High-k Dielectrics*, Institute of Physics Publishing, Bristol, 2004, p. 597.

36. Baraff, G.A., Kane, E.O., and Schlüter, M., Theory of the silicon vacancy: An Anderson negative-*U* system, *Phys. Rev. B*, 21, 5662, 1980.
37. Van de Walle, C.G. and Neugebauer, J., First-principles calculations for defects and impurities: Applications to III-nitrides, *J. Appl. Phys.*, 95, 327, 2006.
38. Sayan, S. et al., Band alignment issues related to HfO_2/SiO_2/p-Si gate stacks, *J. Appl. Phys.*, 96, 7485, 2004.
39. Takeuchi, H., Ha, D., and King, T.-K., Observation of bulk HfO_2 defects by spectroscopic ellipsometry, *J. Vac. Sci. Technol. Sci. Technol. A*, 22, 1337, 2004.
40. Pruneda, J.M. and Artacho, E., Energetics of intrinsic point defects in $ZrSiO_4$, *Phys. Rev. B*, 71, 094113, 2005.
41. Stoneham, M. et al., The vulnerable nanoscale dielectric, *Phys. Status Solidi (a)*, 204, 653, 2007.
42. Xiong, K. et al., Defect states in the high-dielectric-constant gate oxide $HfSiO_4$, *J. Appl. Phys.*, 101, 024101, 2007.
43. Sushko, P.V. et al., Oxygen vacancies in amorphous silica: Structure and distribution of properties, *Microelectron. Eng.*, 80, 292, 2005.
44. Gavartin, J.L. et al., Polaron-like charge trapping in oxygen deficient and disordered HfO_2: Theoretic insight, in Physics and Technology of High-*k* Dielectrics 4, Kar, S. et al., Eds. Electrochemical Society, Pennington, NJ, 2006, p. 277.
45. Gavartin, J.L. and Shluger, A.L., Ab initio modeling of electron–phonon coupling in high-*k* dielectrics, *Phys. Status Solidi C*, 3, 3382, 2006.
46. Stoneham, A.M. et al., Trapping, self-trapping and the polaron family, *J. Phys. Condens. Matter*, 19, 255208, 2007.
47. Ramo, D.M. et al., Intrinsic and defect-assisted trapping of electrons and holes in HfO_2: An ab initio study, *Microelectron. Eng.*, 84, 2362, 2007.
48. Ramo, D.M. et al., Theoretical prediction of intrinsic self-trapping of electrons and holes in the monoclinic HfO_2, *Phys. Rev. Lett.*, 99, 155504, 2007.
49. Diebold, A.C. et al., Characterization and production metrology of thin transistor gate oxide films, *Mater. Sci. Semicond. Process.*, 2, 103, 1999.
50. Frank, M.M. et al., HfO_2 and Al_2O_3 gate dielectrics on GaAs grown by atomic layer deposition, *Appl. Phys. Lett.*, 86, 152904, 2005.
51. Nguyen, N.V. et al., Sub-bandgap defect states in polycrystalline hafnium oxide and their suppression by admixture of silicon, *Appl. Phys. Lett.*, 87, 192903, 2005.
52. Walsh, S. et al., Process-dependent defects in Si/HfO_2/Mo gate oxide heterostructures, *Appl. Phys. Lett.*, 90, 052903, 2007.
53. Bersuker, G. et al., The effect of interfacial layer properties on the performance of Hf-based gate stack devices, *J. Appl. Phys.*, 100, 094108, 2006.
54. Sayan, S. et al., Effect of nitrogen on band alignment in HfSiON gate dielectrics, *Appl. Phys. Lett.*, 87, 212905, 2005.
55. Afanas'ev, V.V. and Stesmans, A., Spectroscopy of electron states at interfaces of (100) Ge with high-kappa insulators, *Mater. Sci. Semicond. Process.*, 9, 764, 2006.
56. Afanas'ev, V.V. et al., Band alignment between (100) Si and Hf-based complex metal oxides, *Microelectron. Eng.*, 80, 102, 2005.
57. Miyazaki, S. et al., Chemical and electronic structure of ultrathin zirconium oxide films on silicon as determined by photoelectron spectroscopy, *Solid State Electron.*, 46, 1679, 2002.
58. Yamaoka, M., Murakami, H., and Miyazaki, S., Diffusion and incorporation of Zr into thermally grown SiO_2 on Si(100), *Appl. Surf. Sci.*, 216, 223, 2003.
59. Cho, Y.J. et al., Spectroscopic ellipsometry characterization of high-*k* dielectric HfO_2 thin films and the high-temperature annealing effects on their optical properties, *Appl. Phys. Lett.*, 80, 1249, 2002.
60. Kirsch, P.D. et al., Nucleation and growth study of atomic layer deposited HfO_2 gate dielectrics resulting in improved scaling and electron mobility, *J. Appl. Phys.*, 99, 023508, 2006.
61. Kerber, A. et al., Origin of the threshold voltage instability in SiO_2/HfO_2 dual layer gate dielectrics, *IEEE Electron Dev. Lett.*, 24, 87, 2003.

62. Bersuker, G. et al., Intrinsic threshold voltage instability of the HfO_2 NMOS transistors, *Proc. IEEE Intl. Reliab. Phys. Sympos.*, San Jose, CA, 2006, p. 179.
63. Heh, D. et al., Spatial distributions of trapping centers in HfO_2/SiO_2 gate stack, *IEEE Trans. Electron Dev.*, 54, 1338, 2007.
64. Young, C.D. et al., Electron trap generation in high-*k* gate stacks by constant voltage stress, *IEEE Trans. Dev. Mater. Reliab.*, 6, 123, 2006.
65. Campbell, J.P. et al., Observations of NBTI-induced atomic-scale defects, *IEEE Trans. Dev. Mater. Reliab.*, 6, 117, 2006.
66. Neugroschel, A. et al., An accurate lifetime analysis methodology incorporating governing NBTI mechanisms in high-k/SiO_2 gate stacks, *IEEE IEDM*, San Francisco, CA, 2006, p. 317.
67. Felix, J.A. et al., Total-dose radiation response of hafnium-silicate capacitors, *IEEE Trans. Nucl. Sci.*, 49, 3191, 2002.
68. Felix, J.A. et al., Effects of radiation and charge trapping on the reliability of high-kappa gate dielectrics, *Microelectron. Reliab.*, 44, 563, 2004.
69. Felix, J.A. et al., Radiation-induced charge trapping in thin $Al_2O_3/SiO_xN_y/Si(100)$ gate dielectric stacks, *IEEE Trans. Nucl. Sci.*, 50, 1910, 2003.
70. Kang, A.Y., Lenahan, P.M., and Conley, J.F., The radiation response of the high dielectric-constant hafnium oxide/silicon system, *IEEE Trans. Nucl. Sci.*, 49, 2636, 2002.
71. Kang, A.Y., Lenahan, P.M., and Conley, J.F., Electron spin resonance observation of trapped electron centers in atomic-layer-deposited hafnium oxide on Si, *Appl. Phys. Lett.*, 83, 3407, 2003.
72. Lucovsky, G. et al., Differences between charge trapping states in irradiated nano-crystalline HfO_2 and non-crystalline Hf silicates, *IEEE Trans. Nucl. Sci.*, 53, 3644, 2006.
73. Ryan, J.T. et al., Identification of the atomic scale defects involved in radiation damage in HfO_2 based MOS devices, *IEEE Trans. Nucl. Sci.*, 52, 2272, 2005.
74. Zhou, X.J. et al., Bias-temperature instabilities and radiation effects in MOS devices, *IEEE Trans. Nucl. Sci.*, 52, 2231, 2005.
75. Zhou, X.J. et al., Effects of switched-bias annealing on charge trapping in HfO_2 gate dielectrics, *IEEE Trans. Nucl. Sci.*, 53, 3636, 2006.
76. Charge trapping in ultrathin hafnium dioxide based MOSFETs, *IEEE Trans. Nucl. Sci.*, 54, 1883, 2007.

12

Negative Bias Temperature Instabilities in High-κ Gate Dielectrics

M. Houssa, M. Aoulaiche, S. De Gendt, G. Groeseneken, and M.M. Heyns

CONTENTS

12.1 Introduction

High-κ gate dielectrics, in combination with metal gates, are currently under investigation for the replacement of the conventional SiON/polycrystalline-Si gate stack in future generations of metal-oxide-semiconductor field effect transistors (MOSFETs) [1,2]. One of the major reasons is that the leakage current flowing through SiON layers thinner than 1 nm exceeds 100 A/cm^2 at operating voltage (around 1 V), leading to unacceptable power dissipation in the circuits. For a given technology, determined, e.g., by the transistor gate length and equivalent electrical thickness of the gate insulator, the use of high-κ gate dielectrics would allow the use of physically thicker layers, aiming at reducing the leakage current flowing through the devices. In that respect, HfSiO(N) [3–5] and HfO_2 [6–8] are considered as potential candidates, due to their high dielectric constants and good thermal stability in contact with Si, allowing them to sustain the high thermal budgets required for the fabrication of advanced metal-oxide semiconductor (MOS) devices. In addition, the Si/HfSiO(N) and Si/HfO_2 conduction and valence band offsets are sufficiently high, about 1.5–2 eV [2], preventing thermoionic emission over these energetic barriers.

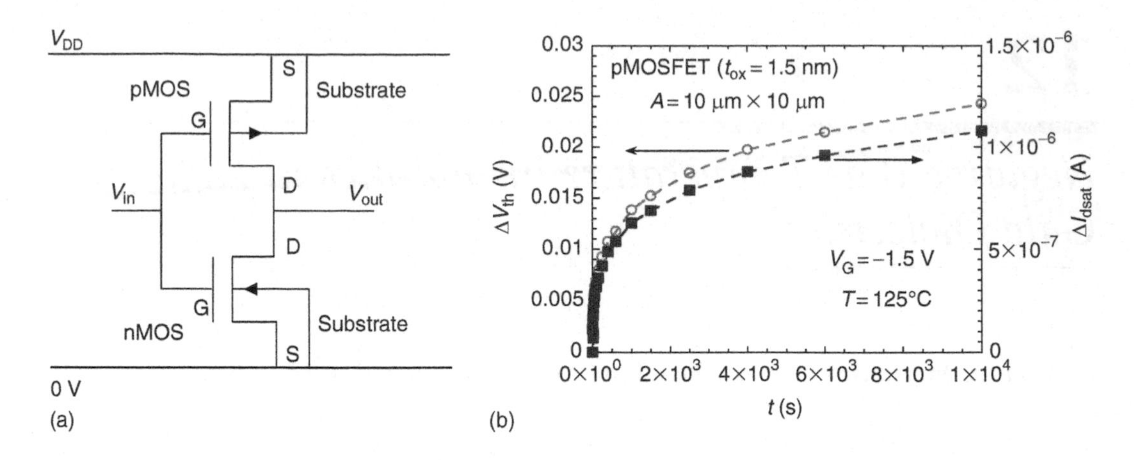

FIGURE 12.1
(a) Schematic illustration of a complementary MOS inverter. (b) Drain current reduction and V_{th} shift as a function of time, observed during negative bias temperature stress of a pMOSFET.

On the other hand, negative bias temperature instabilities (NBTI) in pMOSFETs are considered as major reliability issues in advanced analog and digital integrated circuits [9]. NBTI occur in p-channel MOSFETs under inversion conditions. Let us consider the simple example of a complementary MOS inverter, schematically illustrated in Figure 12.1a. When the p-channel transistor is turned on at elevated temperature (typically between 100°C and 150°C), i.e., when the gate is negatively biased with respect to the substrate, defects are generated in the device, resulting in threshold voltage (V_{th}) shifts and reduction of the drive current (I_{dsat}) of the devices, as illustrated in Figure 12.1b. The degradation of these device parameters can then lead to the failure of integrated circuits [10].

In SiO_2-based devices, the most important defects involved in NBTI are P_{b0} centers (Chapter 6), i.e., Si trivalent dangling bonds ($Si{\equiv}Si^{\bullet}$) at the (100) Si/SiO_2 interface [11–13]. During the negative bias temperature (NBT) stress, holes attracted to the Si/SiO_2 interface can induce the depassivation of P_{b0} centers (initially passivated by hydrogen during the forming gas anneal of the device), followed by the diffusion of hydrogen species away from the interface. The rate limiting step for P_{b0} centers generation is the diffusion process [12,13]; see Figure 12.2. This is the so-called reaction–diffusion model of NBTI,

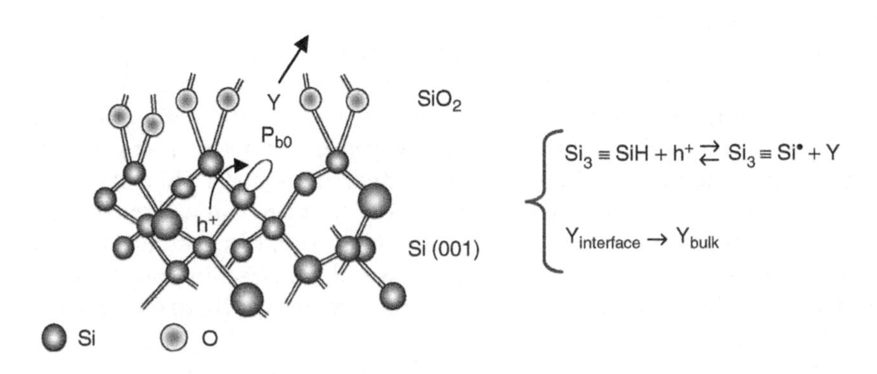

FIGURE 12.2
Schematic illustration of interface trap generation during NBT stress.

which can explain the time, temperature, and oxide electric field (E_{ox}) dependence of V_{th} shifts in SiO_2-based devices, which can be phenomenologically described by the expression:

$$\Delta V_{th} \propto E_{ox}^{m} \exp\left(\frac{-E_a}{k_B T}\right) t^{\alpha} \tag{12.1}$$

where $m \approx 3$–4, $E_a \sim 0.1$–0.2 eV, and $\alpha \approx 0.2$–0.25.

In this chapter, we review NBTI in HfSiO(N)/metal gate stacks. By combining V_{th} extractions with charge pumping measurements, the contribution of fast interface traps (interface traps) and slow bulk states (border traps) are discriminated. It is found that bulk states have a significant contribution to device degradation in these gate stacks. The discharging of these defects, when the gate bias is removed or switched positive, is mainly responsible for the recovery of V_{th} in these devices. The impact of different process steps on NBTI, like H_2/D_2 anneals, nitrogen incorporation, either by NH_3 anneals or plasma exposure, and Si substrate orientation are now discussed in detail.

12.2 Experimental Details

HfSiO(N)-based pMOSFETs with metal gates were fabricated with a conventional self-aligned transistor flow. Two nanometer thick HfSiO layers with ~50 atomic% Hf were deposited by metal organic chemical vapor deposition (MOCVD) at 600°C with tetrakis-diethylaminohafnium (TDEAH), tetrakis-dimethylaminosilicon (TDMAS), and O_2 as precursors, or by atomic layer deposition (ALD) at 300°C with H_2O and $HfCl_4$ as sources. The high-κ layers were deposited on a 1 nm thick chemical SiO_x interlayer. TaN or TiN metal gates were next deposited by physical vapor deposition (PVD) or ALD, respectively, and defined by photolithography. The dopant activation was performed during a spike anneal at 1030°C. Forming gas annealing was performed at 520°C after silicidation of the source and drain junctions for 20 min, and at 420°C after metal 1 (Al) deposition for 20 min. This process allows efficient passivation of interface traps [14].

The conventional method used to characterize NBTI consists of measuring the I_D–V_G characteristics of the device periodically during the electrical stress, performed under constant negative gate voltage, at temperatures typically ranging between 100°C and 150°C. The (stress) time dependence of the threshold voltage is extracted from these measurements, and the device lifetime is extrapolated, assuming that device failure occurs when a given shift in V_{th} is reached. Typical failure criteria are 10% or 30 mV shift in V_{th}.

As will be discussed in more detail in Section 12.3.3, a substantial recovery in V_{th} is observed when the electrical stress is interrupted [15,16], as illustrated in Figure 12.3. Consequently, the conventional stress-sense method has been modified to reduce the recovery effect as much as possible. This alternative characterization method consists of recording the I_D–V_G characteristic of the device before and after stress. Between these two I_D–V_G measurements, a pulse-like stress is applied to the gate of the transistor and a second pulse is applied to the drain, as shown in Figure 12.4a and b [17,18]. The transistor is stressed under NBT when the gate voltage is at $V_{G,stress}$ and drain voltage at 0 V. The evolution of the drain current is then recorded during the electrical stress when the gate voltage is switched to $V_{G,sense} \sim V_{th}$ and $V_D = -50$ mV. The delay in switching between $V_{G,stress}$ and $V_{G,sense}$ is minimized (typically below 500 ms) to reduce the V_{th} recovery effect. The measured $I_D(t)$ trace is next converted to a $V_{th}(t)$ trace by interpolating the subthreshold slope of the fresh I_D–V_G characteristic. The second I_D–V_G characteristic measured at the end of the stress is used to control and eventually correct for the

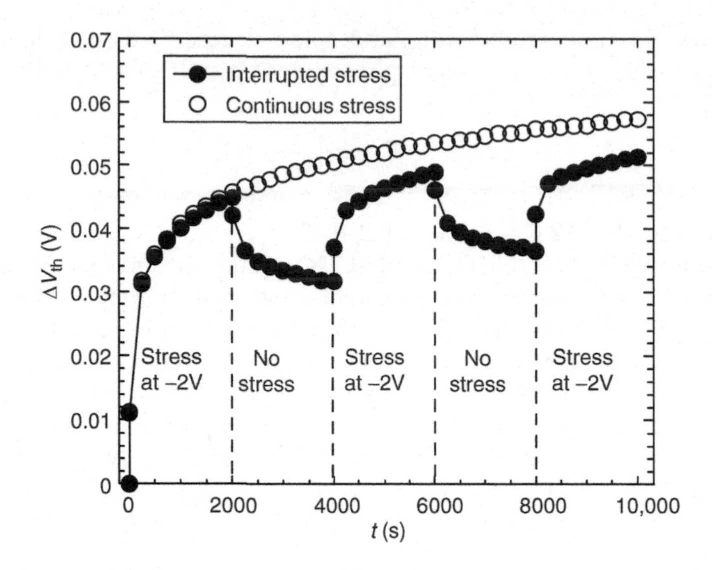

FIGURE 12.3
ΔV_{th} as a function of stress time, showing the threshold voltage degradation during the stress and the partial recovery when the gate bias is switched to 0 V.

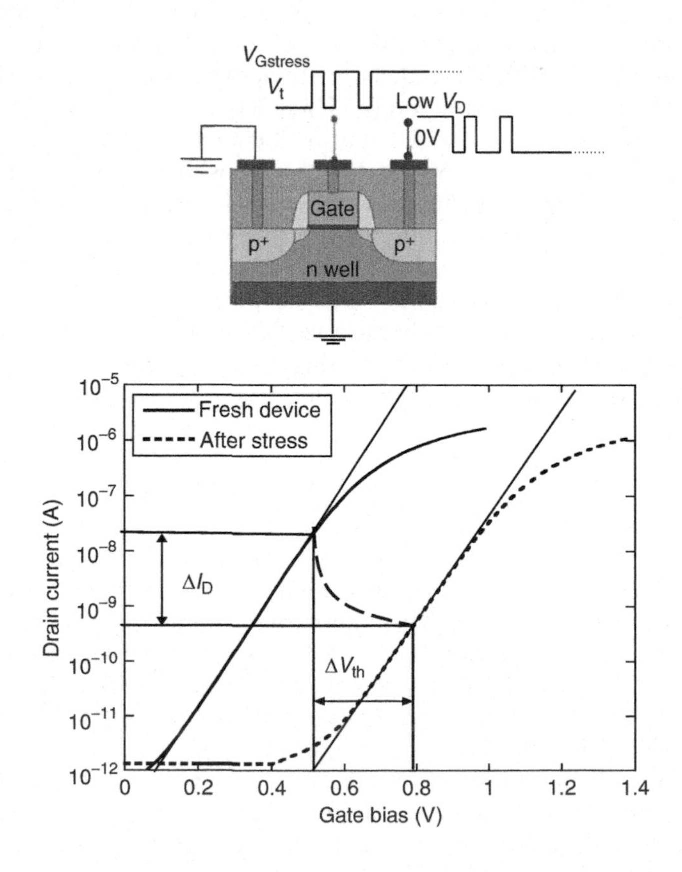

FIGURE 12.4
Schematic illustration of the pulsed stress method (top). The decrease in the drain current during NBT stress is converted to the threshold voltage shift using the subthreshold slope interpolation (bottom).

subthreshold slope change. NBT stress and charge pumping current (I_{CP}) measurements are performed on (10 × 10) and (10 × 1) μm^2 transistors at a typical temperature of 125°C (unless otherwise specified).

12.3 Results and Discussion

As in SiO_2 and SiON-based devices, shifts in the threshold voltage of high-κ-based transistors are observed during negative bias temperature stress. It has been reported that recently that NBTI could be a potential reliability issue in these devices [19–23]. In this section, we first discuss the generation of fast (interface) states and slow (bulk) states during NBT stress; these two contributions are separated by combining V_{th} extraction with charge pumping measurements. Next, the impact of different process steps on NBTI, like H_2/D_2 postmetallization anneals, nitridation of the gate stack, and the Si substrate orientation are reviewed.

12.3.1 Fast and Slow State Generation

The V_{th} shifts of SiO(N)/HfSiON/TaN pMOSFETs stressed at different negative gate biases at 125°C are shown in Figure 12.5 as a function of (stress) time [24]. Nitrogen was incorporated in the gate stack during a 60 s anneal in NH_3 at 800°C. The characteristic power-law time dependence of ΔV_{th} is clearly observed, with a power-law time exponent ~0.2. The threshold voltage shift can be written as $\Delta V_{th}(t) = C_{ox}(\Delta N_{it}(t) + \Delta N_{ot}(t))/q$, where C_{ox} is the gate dielectric capacitance per unit area, N_{it} is the fast state density (interface traps), N_{ot} is the slow state density, which are most likely bulk defects located in the SiON/HfSiON gate stack [25,26] (Figure 12.6), and $-q$ is the electron charge.

To separate the contributions of N_{it} and N_{ot}, charge pumping current during NBT stress was periodically measured as a function of base level sweep at 3 MHz; at this frequency, only fast states located at or very near the Si/SiON interface are supposed to be able to respond to the signal applied on the gate. Only a few points of the I_{CP}–V_{base} curve, near

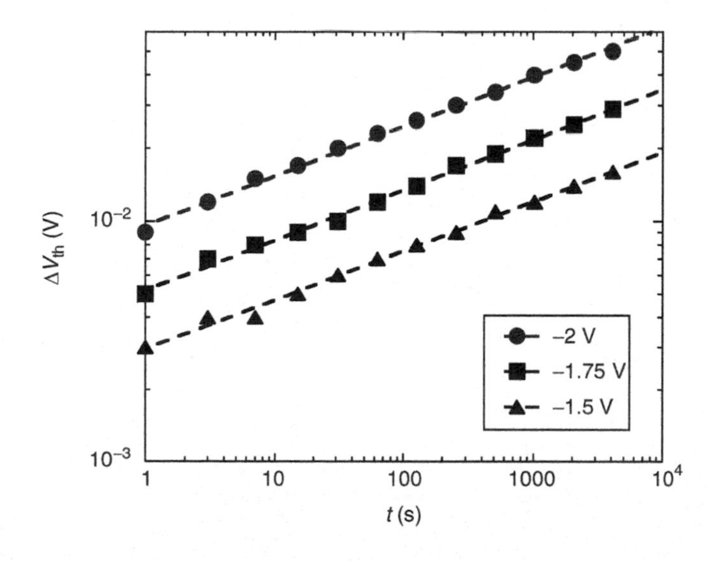

FIGURE 12.5
Threshold voltage shift as a function of NBT stress time for different gate voltages at 125°C.

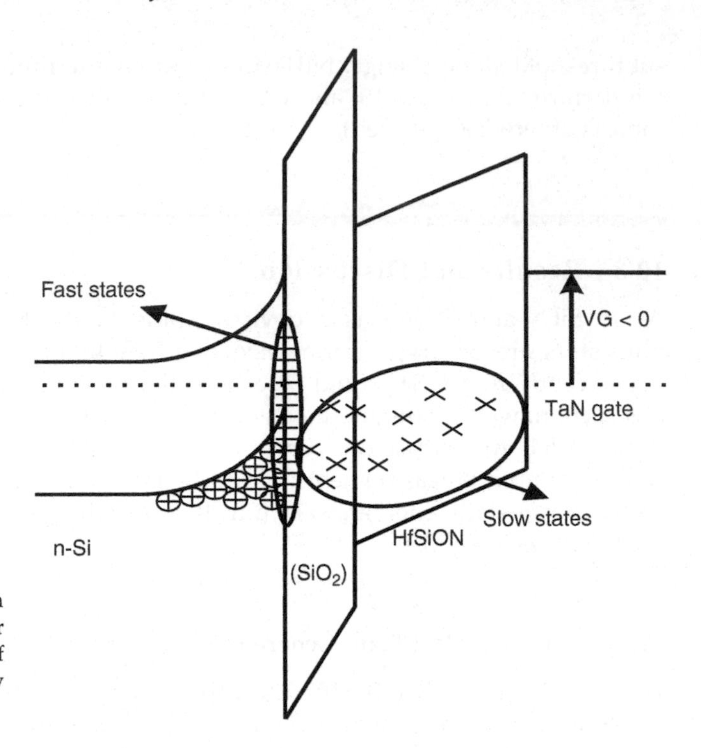

FIGURE 12.6
Schematic energy band diagram of a n-Si/SiO_x/HfSiON/TaN gate stack under NBT stress, illustrating the generation of fast states at the Si/SiO_x interface and slow states in the SiO_x/HfSiON stack.

$I_{CP,max}$, were recorded, in order to minimize recovery effects as much as possible (typically less than 2 s).

The increase in the maximum charge pumping current (in absolute value) observed after NBT stress (Figure 12.7a) suggests the generation of fast states in the gate stack. The maximum charge pumping current is related to the density of fast states N_{it} according to the expression:

$$I_{cp,max}(t) = qAfN_{it}(t) \quad (12.2)$$

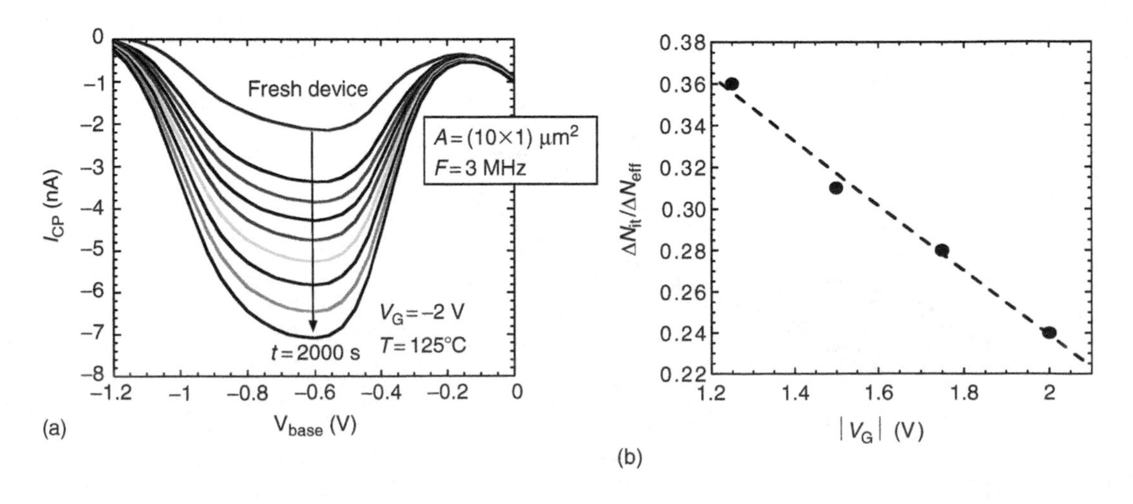

FIGURE 12.7
(a) Charge pumping current measured as a function of base sweep during NBT stress. (b) Ratio between N_{it} and N_{eff} as a function of stress voltage. (From Houssa, M., et al., *Microelectron. Reliab.*, 47, 880, 2007. With permission.)

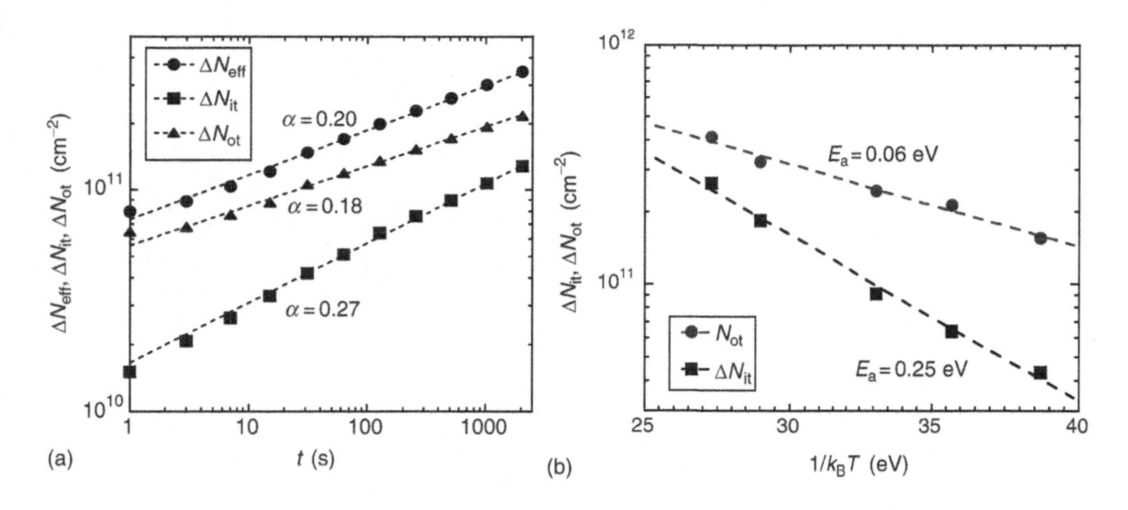

FIGURE 12.8
(a) Fast state and slow state densities as a function of stress time. (b) Arrhenius plot showing fast state and slow state densities as a function of inverse temperature.

where

- $-q$ is the electron charge
- A is the device area
- f is the frequency of the (trapezoidal) pulse applied to the gate [27]

Note that N_{it} is the average interface trap density in the Si band gap scanned during the charge pumping measurement, which is typically ±0.3 eV around midgap.

The fraction of fast interface traps to the total effective density of defects produced, $N_{eff} = N_{it} + N_{ot}$ is shown in Figure 12.7b as a function of the stress voltage [25]. The N_{it} contribution ranges between about 20% and 40%, depending on the gate bias. Consequently, a substantial fraction of the V_{th} shift can be attributed to the generation of slow states. The typical time and temperature dependences of the fast and slow states generated during NBT stress are shown in Figure 12.8a and b, respectively. Fast states increase with time like a power law with an exponent ~0.25 and are thermally activated, with an apparent activation energy of about 0.25 eV. These time and temperature dependences are characteristics of the generation of P_{b0} centers, as predicted by the reaction–diffusion model. On the other hand, the slow states are characterized by a reduced power-law time exponent ($\alpha \sim 0.16$–0.18) and are almost independent of temperature. This suggests that the mechanism of slow state generation is different from the fast interface traps generation. A possible explanation for the slow state generation is the trapping of holes injected from the Si substrate inversion layer during the electrical stress, resulting in the generation of positively charged defects in the gate dielectric stack [26], as will be discussed in more detail in Section 3.3.

12.3.2 H_2/D_2 Isotopic Effect

The substitution of hydrogen by deuterium at the Si/SiO_2 interface and/or in the SiO_2 layer has been shown to improve the immunity of MOSFETs against hot carrier degradation [28,29], irradiation [30], and NBTI [9,31]. As far as irradiation-induced degradation is concerned, the isotopic effect has been attributed to the slower transport of D_2 (or D^+), compared to H_2 (or H^+) in the gate stack [30,32].

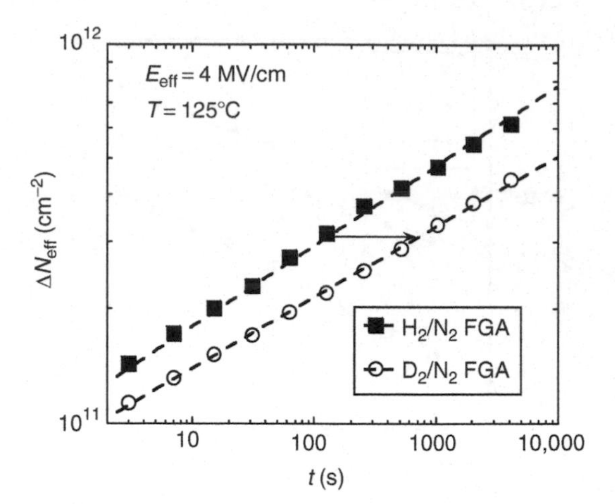

FIGURE 12.9
Effective density of positive charges, ΔN_{eff}, generated during negative bias temperature stress at an electric field of 4 MV/cm and 125°C, as a function of stress time, for devices annealed in H_2/N_2 or D_2/N_2. (From Houssa, M., et al., *Electrochem. Solid State Lett.*, 9, G10, 2006. With permission.)

In this section, we study the H_2/D_2 isotopic effect on NBTI in HfSiON/TaN gate stacks. A retardation effect on the device degradation is clearly revealed at low electric field (below about 5 MV/cm), while the isotopic effect is much reduced at higher electric field. These results suggest that fast interface traps, which substantially contribute to device degradation at low electric field, are related to the transport of hydrogen species, while slow bulk states, which dominate device degradation at high electric fields, are most probably not related to hydrogen transport/trapping.

The kinetics of ΔN_{eff} are shown in Figure 12.9 for devices annealed in H_2 or D_2, and stressed under a constant average electric field $E_{eff} = 4$ MV/cm, where $E_{eff} = (V_G - V_{th})/t_{phys}$ and t_{phys} is the physical thickness of the SiO_x/HfSiON stack (3 nm). ΔN_{eff} increases with time like a power law, with an exponent about 0.2–0.18 [24]. A clear retardation effect on ΔN_{eff} is observed for the D_2-annealed device, characteristic of the H_2/D_2 isotopic mass effect [32]; the time to reach a given defect density is increased by about a factor of 5 in D_2-annealed devices compared to those annealed in H_2.

ΔN_{eff} is presented in Figure 12.10, for devices annealed in H_2 or D_2, as a function of E_{eff}, for stress time fixed at 1000 s. The isotopic effect is clearly observed for E_{eff} below about 5 MV/cm; this effect is much reduced at higher E_{eff}. A possible interpretation of these results is given below.

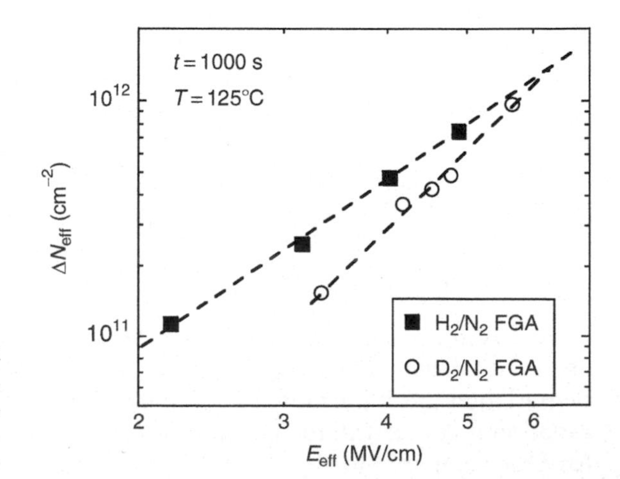

FIGURE 12.10
ΔN_{eff} as a function of E_{eff}, for devices annealed in H_2/N_2 or D_2/N_2. Stress time was fixed at 1000 s, and temperature at 125°C. (From Houssa, M., et al., *Electrochem. Solid State Lett.*, 9, G10, 2006. With permission.)

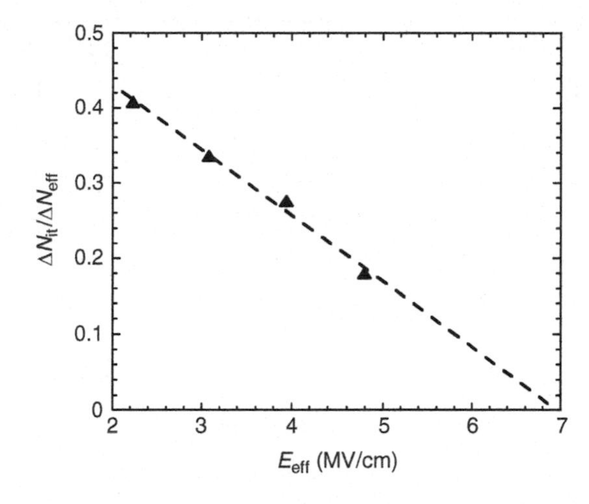

FIGURE 12.11
Ratio between the fast interface traps, N_{it}, and effective density of positive charges, N_{eff}, as a function of E_{eff}.

During the NBT stress, fast interface traps, N_{it}, and slow bulk traps, N_{ot}, are generated in the gate stack, as discussed in Section 12.3.1. The N_{it}/N_{eff} ratio is shown in Figure 12.11 as a function of E_{eff}. The contribution of fast interface traps is found to decrease from about 40% to below 10% when E_{eff} increases from 2 to 6 MV/cm; as the electric field is increased, a larger fraction of slow bulk traps is generated in the gate stack, with the N_{ot} contribution being the dominant one at high electric fields.

The E_{eff} dependence of the isotopic effect can be explained as follows. At low electric fields, fast interface traps are substantially contributing to ΔN_{eff}, and the generation mechanism of these defects is most probably related to the transport of hydrogen species, possibly H^+ [33]. On the other hand, at high electric fields, device degradation is mainly due to the generation of bulk slow states. Assuming that these defects are not related to the transport/trapping of hydrogen species, the strong reduction of the isotopic effect at large electric field can be explained. These latter defects are possibly induced by the trapping of holes injected from the channel during the electrical stress, as discussed in Section 12.3.3 [35].

To further highlight this interpretation, the H_2/D_2 isotopic effect on the generation of fast interface traps has been simulated, using the reaction-dispersive H^+ transport model [33]. In this model, one assumes that $Si_3{\equiv}SiH$ centers at the (100) Si/SiO_x interface are depassivated during the NBT stress, with the rate-limiting step for interface trap generation being the dispersive transport of H^+ in the gate dielectric stack, as discussed in more detail in Ref. [34]. The isotopic effect was taken into account by assuming that the average displacement per H^+ or D^+ hop, μ, depends on the effective mass m of the transported species like, $\mu \propto 1/\sqrt{m}$, as expected for the transport of small polarons [34,35]. Consequently, the ratio between the average displacement per hop for D^+ and H^+ is given by

$$\mu_{D^+}/\mu_{H^+} = \sqrt{m_{H^+}/m_{D^+}} = 1/\sqrt{2}. \tag{12.3}$$

The calculated kinetics of interface trap ($Si_3{\equiv}Si^\bullet$ dangling bonds) generation is shown in Figure 12.12a, for H^+ and D^+ transport. The reduced average displacement per hop for D^+ leads to slower transport of this species in the gate stack, hence to a delay in defect production, defined as $\tau_{D+}/\tau_{H+} \sim 3$, where τ is the time corresponding to the generation of a fixed amount of interface traps. This delay is in reasonable agreement with the data shown in Figure 12.9 (about 5). Within the dispersive transport model, τ is related to μ according to the expression [36]:

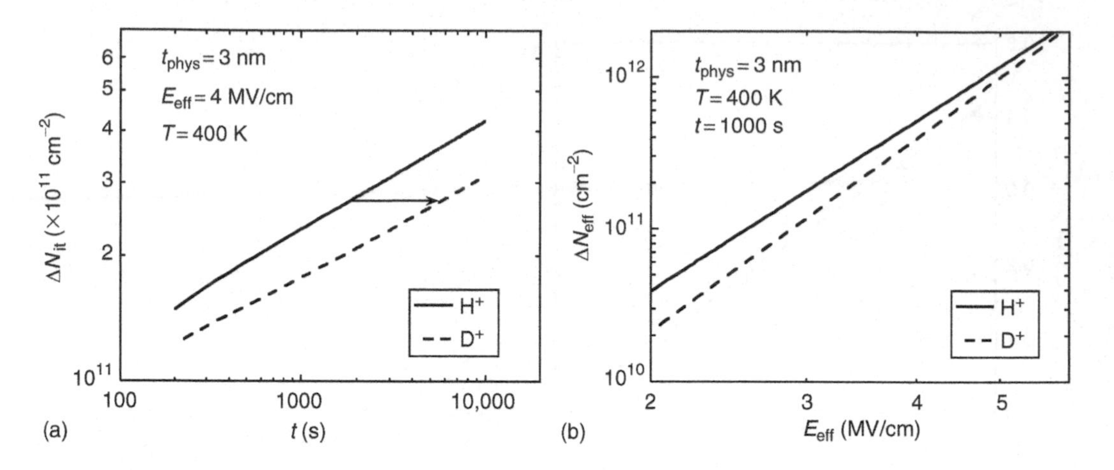

FIGURE 12.12
(a) Simulated kinetics of interface trap generation, considering H^+ or D^+ dispersive transport in a 3 nm gate dielectric layer, for oxide electric field and temperature fixed at 4 MV/cm and 400 K, respectively. (b) Simulated effective density of positive charges as a function of the electric field, considering H^+ or D^+ dispersive transport. The stress time and temperature were fixed at 1000 s and 400 K, respectively. (From Houssa, M., et al., *Electrochem. Solid State Lett.*, 9, G10, 2006. With permission.)

$$\tau \propto \mu^{-1/a} \tag{12.4}$$

where $a \sim 0.3$ is the dispersive transport exponent [33]. From Equation 12.3, this leads to $\tau_{D+}/\tau_{H+} \sim 3.2$, in very good agreement with the simulations.

To simulate the electric field dependence of the isotopic effect, the contribution of N_{ot} was added to N_{it}. This contribution was calculated such that the fraction N_{it}/N_{eff} agrees with the data shown in Figure 12.11, and N_{ot} was assumed to be unaffected by the H_2/D_2 isotopic exchange, considering that the slow states are not related to the transport/trapping of H^+/D^+ (i.e., the same amount of N_{ot} was added to N_{it} for the H_2 and D_2-annealed devices, for a given electric field). The simulated $\Delta N_{eff} = \Delta N_{it} + \Delta N_{ot}$ is shown in Figure 12.12b vs E_{eff}. A noticeable isotopic effect is observed at low E_{eff} (typically below 5 MV/cm), where a substantial fraction of N_{it} contributes to N_{eff}. On the other hand, at higher electric fields, where device degradation is mainly caused by N_{ot} generation, the isotopic effect tends to disappear. These simulations are in good qualitative agreement with the experimental results shown in Figure 12.10.

12.3.3 Impact of Nitrogen Incorporation

Nitrogen incorporation has allowed prolonging the use of SiO_2 dielectric by reducing leakage current and improving devices characteristics. Indeed SiO_2 nitridation has been reported to achieve advantages such as reduction of penetration of dopants and other impurities, dielectric constant increase, resistance enhancement to high electric field and radiation hardness [37]. The nitridation of high-κ layers, like HfSiO, is usually performed to improve the thermal stability of the material, by increasing the temperature corresponding to the SiO_2 and HfO_2 phase separation [3]. Besides, nitrogen incorporation also increases the dielectric constant of the material [38]. In this section, we investigate the impact on NBTI of N incorporation in HfSiO layers, either by NH_3 or decoupled plasma nitridation (DPN).

TABLE 12.1

Capacitance Equivalent Thickness (CET), and Initial Threshold Voltage $V_{th}(0)$ of the SiO_x/HfSiO Gate Stacks Exposed to Different Postdeposition Anneals

PDA	CET at V_{DD} (nm)	V_{th} (V)
As deposited	2.41	−0.633
N_2, 800°C	2.32	−0.590
O_2, 800°C	2.35	−0.630
NH_3, 800°C	2.03	−0.550
DPN + O_2, 800°C	2.15	−0.620
DPN + N_2, 900°C	2.07	−0.600

12.3.3.1 Experimental Results

The shifts of the threshold voltage of pMOSFETs ($L \times W = 10 \times 10\ \mu m^2$) with nitrided and nonnitrided SiO_x/HfSiO stacks (Table 12.1) are shown in Figure 12.13 as a function of the stress time, for an electric field across the interfacial layer $E_{ox} = (V_G - V_{th})/\text{EOT}$ (EOT is the equivalent oxide thickness of the gate stack) fixed at 7 MV/cm and $T = 125°C$. One observes that ΔV_{th} increases with time like a power law, with an exponent $\alpha \approx 0.25$ and $\alpha \approx 0.2$ for the nonnitrided stacks (annealed in N_2, O_2 or as deposited) and for the nitrided stacks (annealed in NH_3 or exposed to DPN), respectively. It appears also that the nitrided stacks present a much larger V_{th} shift as compared to the stacks annealed in N_2 or O_2.

The effective charge density N_{eff} for the different stacks are compared in Figure 12.14 as a function of the electric field across the interfacial layer; comparing ΔN_{eff} at fixed E_{ox} allows us to account for the different initial $V_{th}(0)$ and EOT of the different gate stacks (Table 12.1). One observes that NBTI is much enhanced in the nitrided stacks, as compared to the stacks annealed in N_2 and O_2 (which behave very similarly). Like in SiON-based devices [39,40], the incorporation of nitrogen in the SiO_x/HfSiO gate stack leads to much enhanced NBT-induced degradation.

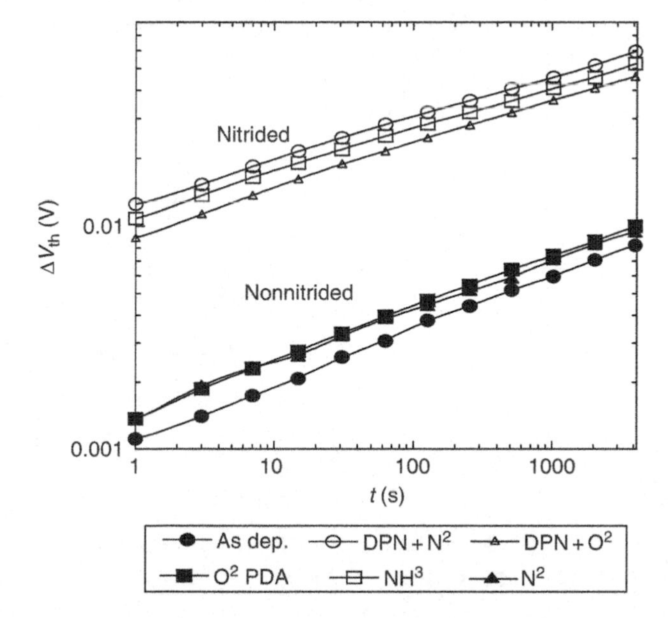

FIGURE 12.13
Threshold voltage shift as a function of stress time of pMOSFETs with HfSiO(N)/TaN gate stacks, stressed at 7 MV/cm and $T = 125°C$. Solid lines are power-law fits to the data. (From Aoulaiche, M., et al., *IEEE Trans. Dev. Mater. Reliab.*, 7, 146, 2007. With permission.)

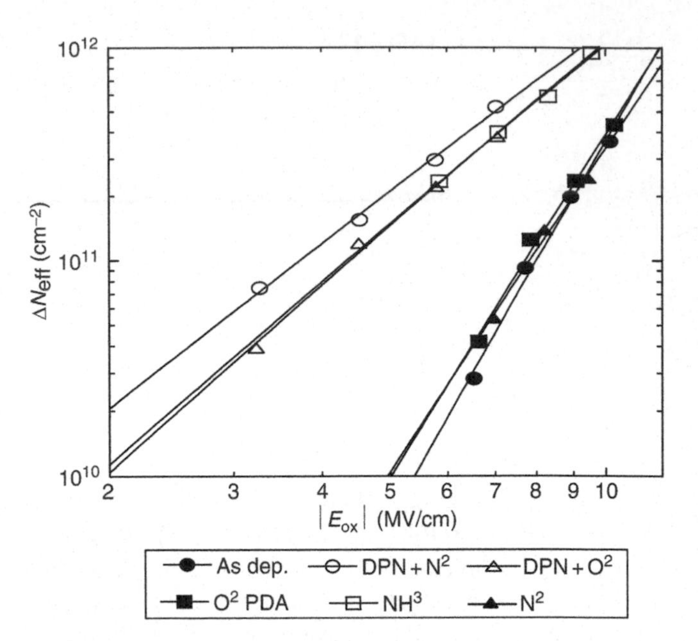

FIGURE 12.14
Effective density of defects N_{eff} as a function of electric field across the SiO_x interlayer, for stressed time fixed at 1000 s and $T = 125°C$. (From Aoulaiche, M., et al., *IEEE Trans. Dev. Mater. Reliab.*, 7, 146, 2007. With permission.)

The NBTI-lifetime of the devices, extrapolated from the kinetics of ΔV_{th} and corresponding to $\Delta V_{th} = 30$ mV, is presented in Figure 12.15 as a function of the gate overdrive $V_G - V_{th}(0)$. Note that lifetime is presented as a function of gate overdrive on a log–log scale, due to the power-law electric field dependence of ΔV_{th} observed experimentally on similar devices [24]. As expected from the previous results, the lifetime of the nonnitrided stacks is superior to the nitride ones, with a corresponding extrapolated gate overdrive at 10 years of about 1.1 V, compared to $V_G - V_{th}(0) \approx 0.4$–0.6 V for the nitrided stacks.

The density of fast interface traps generated during NBT stress is presented in Figure 12.16a as a function of the generated slow states density (E_{ox} fixed at 7 MV/cm). One observes almost a one-to-one correlation between N_{it} and N_{ot} for the nonnitrided

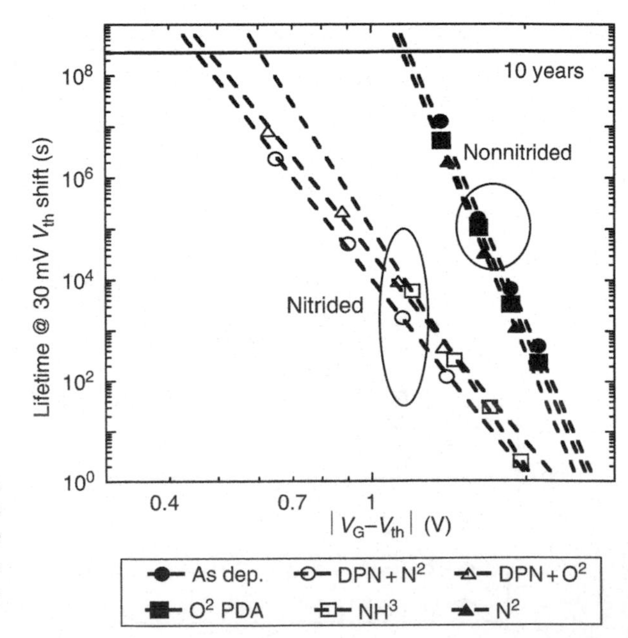

FIGURE 12.15
Lifetime corresponding to 30 mV V_{th} shift as a function of gate overdrive for gate stacks exposed to different PDA. Solid lines are power-law fits to the data. (From Aoulaiche, M., et al., *IEEE Trans. Dev. Mater. Reliab.*, 7, 146, 2007. With permission.)

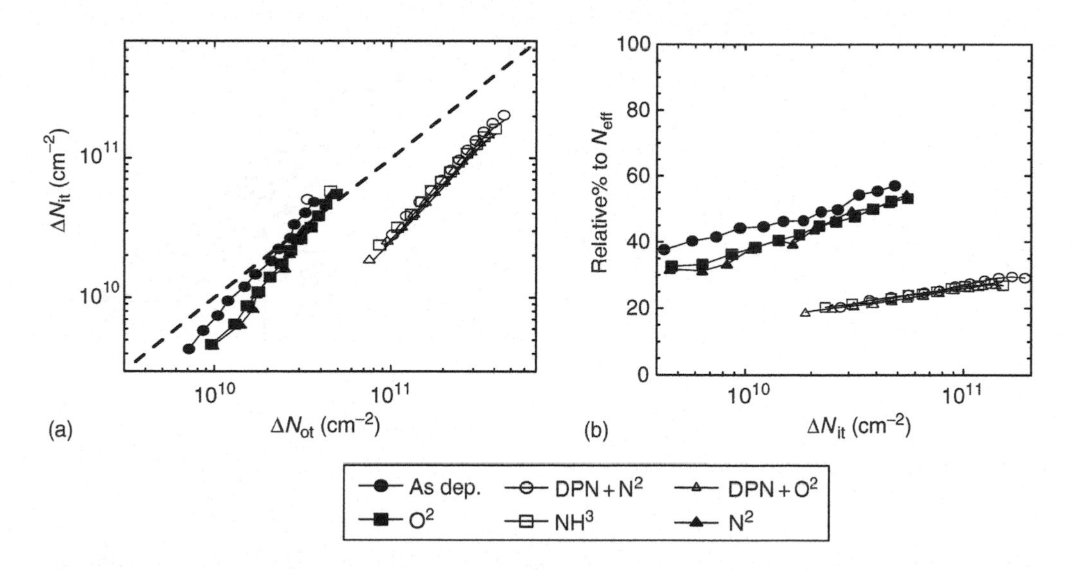

FIGURE 12.16
(a) Fast interface trap density as a function of the density of slow states produced during NBT stress of HfSiON/TaN pMOSFETs. (b) Fraction of interface traps generated during NBT stress.

stacks, indicating a substantial contribution of fast states to the device degradation. On the other hand, the N_{it}/N_{ot} ratio is smaller than 30% in the nitrided stacks, indicating that a substantial density of slow (bulk) states is produced in these devices, as also illustrated in Figure 12.16b.

As discussed below, the bulk defect generation is likely related to the tunneling of holes through the gate stack. The bulk defect generation probability, P_{gen}, with respect to the density of holes injected from the channel, $N_{h,inj}$, is then calculated from

$$P_{gen} = \frac{\Delta N_{ot}}{N_{h,inj}} \tag{12.5}$$

The slow state generation probability is shown in Figure 12.17 as a function of $N_{h.inj}$, as extracted from carrier-separation measurements [41]. One observes that for a fixed $N_{h,inj}$, P_{gen} increases with the stress bias. This indicates that the injection of holes from the channel results in the generation of new positively charged bulk defects in the HfSiON gate stacks, like in SiO_2-based devices [41]. One possible defect creation mechanism could be hole-induced bond breaking in the SiON and/or the HfSiON layer (see below).

12.3.3.2 Model for Slow States Generation

It has been reported by Lenahan and coworkers [42,43] that in Si_3N_4 and SiON layers annealed in NH_3 or in N_2O, Si_2–N• dangling bonds (N centers) and N_3–Si• dangling bonds (K centers) are potentially efficient hole trapping centers. By analogy, a large fraction of the nitrogen-related slow states observed here are possibly the counterparts of the N and K centers in the SiON/HfSiON stack.

To simulate the N_{ot} generation, the following electrochemical model, adapted from the work of McPherson et al. [44], is considered. The precursor defect is tentatively attributed here to be the N center, passivated by H [42], as illustrated in Figure 12.18. When an electric

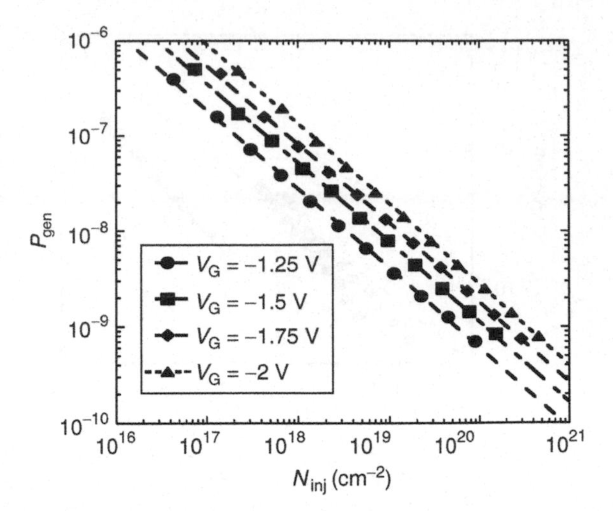

FIGURE 12.17
Probability of defect generation, P_{gen}, as a function of the density of holes injected during NBT stress, for devices subjected to different stress voltages. (From Houssa, M., et al., *Microelectron. Reliab.*, 47, 880, 2007. With permission.)

field is applied to the structure, the polarization of the $N^{\delta-}$–$H^{\delta+}$ bond leads to the distortion of these bonds, namely the increase (reduction) of bond lengths for bonds with dipolar moment parallel (antiparallel) to the applied electric field. This results in a decrease of the N–H bonding energy, given by [44]:

$$\varepsilon_{di} = \varepsilon_{di}(0) - \vec{p}.\vec{E}_{loc} \tag{12.6}$$

where

- $\varepsilon_{di}(0)$ is the N–H bonding energy in absence of an external electric field
- $\vec{p}$ is the dipolar moment of the bonds
- $\vec{E}_{loc} = (1 + \chi/3)\vec{E}_{ox}$ is the local electric field, where χ is the electric susceptibility of the material

If a hole, injected from the channel, is trapped at such a distorted bond, bond breaking can be induced, leading to the generation of a positively charged defect, resulting from the localization of the hole at the N-dangling bond, as shown in Figure 12.18. The equation describing the kinetics of N_{ot} generation within this electrochemical model is given by [44]:

$$N_{ot}(t) = N_0\left(1 - \frac{1}{\sqrt{2\pi\sigma}}\int_{-\infty}^{+\infty} \exp\left(\frac{\varepsilon_d - \varepsilon_{d,i}}{2\sigma^2}\mathrm{d}\varepsilon_d\right)\left[\frac{k_2\exp(-k_1 t) - k_1\exp(-k_2 t)}{(k_2 - k_1)}\right]\right) \tag{12.7}$$

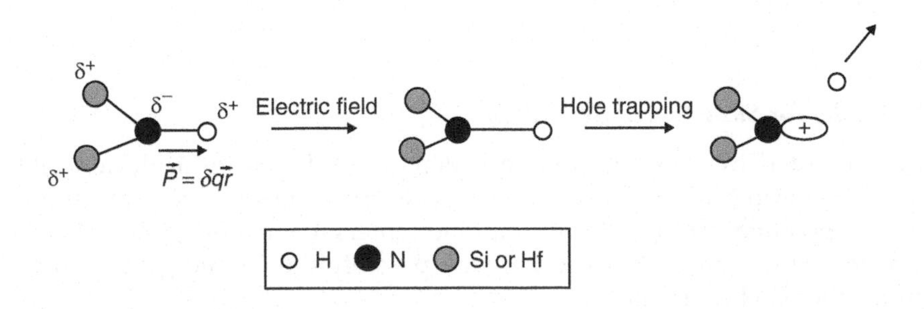

FIGURE 12.18
Schematic illustration of the electric field and hole-induced Si_2N—H or Hf_2N—H bond breaking, resulting in the generation of a positively charged =N• dangling bond (N-center).

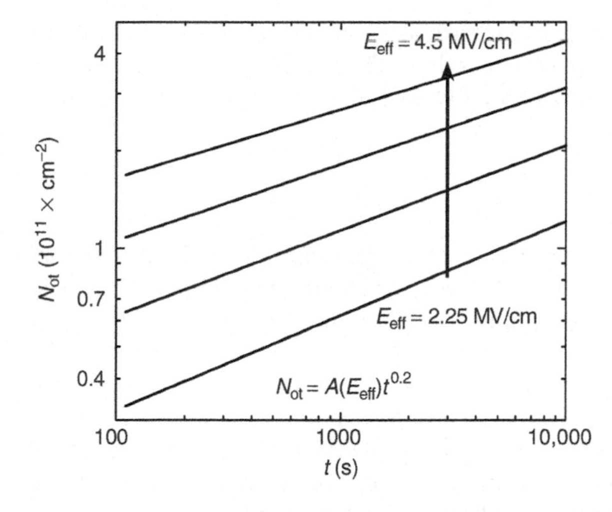

FIGURE 12.19
Calculated density of slow states as a function of stress time, for different values of the electric field across the SiO_x/HfSiON stack. (From Houssa, M., et al., *Microelectron. Reliab.*, 47, 880, 2007. With permission.)

where N_0 is the initial density of Si_2N–H or Hf_2N–H centers. Note that a Gaussian distribution of activation energies for the defect generation was considered here to account for the amorphous (i.e., disordered) nature of the gate dielectric stack [45,46].

The kinetics of N_{ot} generation, simulated from Equation 12.7, is shown in Figure 12.19 for different values of the effective electric field E_{eff}. The following values of the parameters were fixed in these simulations: $N_0 = 10^{12}$ cm^{-2}, $\sigma = 0.2$ eV, $\varepsilon_{di}(0) = 3$ eV, $\mathbf{I}\,\vec{p}\,\mathbf{I} = 1.3 \times 10^{-29}$ cm, and $\chi = 9$, corresponding to a mean dielectric constant of 10 for the SiON/HfSiON stack. One observes that N_{ot} increases with time like a power law, with an exponent ~0.2, in reasonable agreement with the experimental results (Figure 12.8a); this power-law time dependence arises from the Gaussian distribution of activation energies [45,46].

The simulated bulk defect generation probability is shown in Figure 12.20a as a function of the injected hole density for different values of E_{eff}. These results are in good qualitative agreement with the data shown in Figure 12.17. P_{gen} is found to increase with the effective

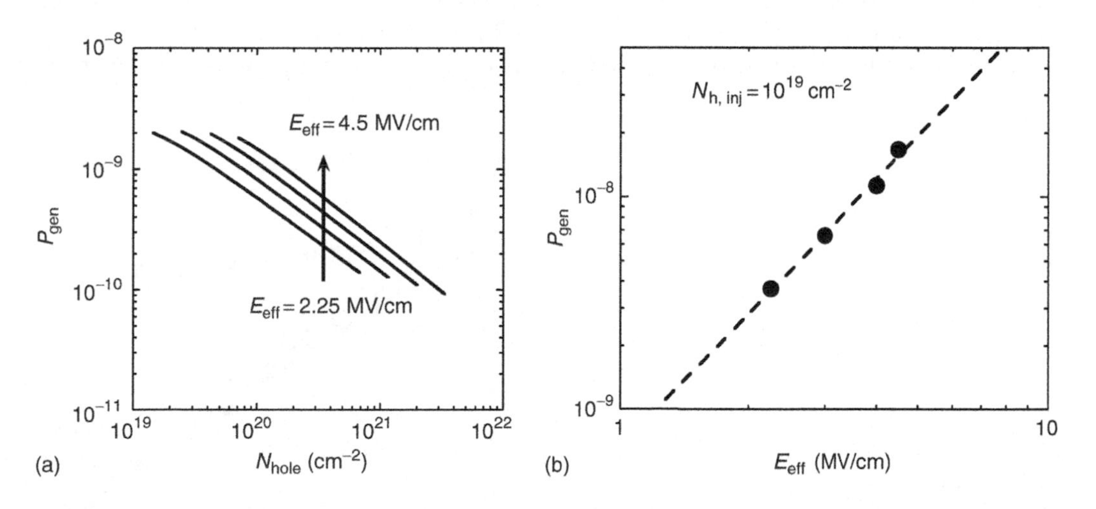

FIGURE 12.20
(a) Calculated generation probability of slow states, as a function of the injected hole density in the gate stack, for different values of the electric field E_{eff} across the SiO_x/HfSiON stack. (b) Calculated P_{gen} as a function of E_{eff}. The dashed line is a power-law fit to the calculated results and the full circles are the experimental results. (From Houssa, M., et al., *Microelectron. Reliab.*, 47, 880, 2007. With permission.)

electric field like a power law, with an exponent ~2, as shown in Figure 12.20b, in excellent agreement with the experimental results, represented by the symbols in Figure 12.20b. The electric field dependence of P_{gen} is solely related to the dependence of ε_{di} on the local electric field, as shown in Equation 12.6.

12.3.3.3 V_{th} Recovery

As already mentioned in Section 12.1, the threshold voltage partially recovers to its initial value when the gate bias is switched to 0 V after NBT stress, as illustrated in Figure 12.21 [26], where the V_{th} recovery of nitrided and nonnitrided SiO_x/HfSiO stacks is compared. The ΔV_{th} recovery is logarithmically time-dependent, like in the SiON-based devices [18]. Interestingly, the V_{th} recovery is enhanced in the nitrided SiON/HfSiON stacks, as compared to the N_2-annealed stack. A correlation between degradation and recovery is thus clearly observed; in nitrided stacks, enhanced NBTI also leads to enhanced recovery. This suggests that a hole trapped in a nitrogen-related defects during stress can tunnel out or be neutralized when the stress is removed; i.e., recovery results mainly from hole detrapping/neutralization.

The threshold voltage shift of SiON/HfSiON stacks annealed in NH_3 is presented as a function of stress time in Figure 12.22. The device was stressed under NBT at −2 V during 2000 s (first stress phase). Then the stress was removed ($V_G = 0$ V) for 2000 s (recovery phase), and the device was again exposed to NBT stress at −2 V for 2000 s (second stress phase). Note that the temperature was kept at 125°C during the stress and recovery phases. As illustrated in Figure 12.22, the initial increase of ΔV_{th} is much faster (by a factor of ~3.5) during the second stress phase compared to the first stress phase. Indeed, the time Δt between $\Delta V_{th} \approx 32$ mV (starting point of the second stress phase) and $\Delta V_{th} \approx 45$ mV (end-point of the first stress phase) is found to be ~1750 s and 500 s for the first and second stress phases, respectively. After the initial faster increase of ΔV_{th} observed on the second stress phase, the ΔV_{th} kinetics are very similar for the two phases, as also shown in Figure 12.22.

The above results can be interpreted as follows. During the first stress phase, nitrogen-related defects (possibly N dangling bonds) are generated in the HfSiON stack, possibly by dissociation of N—H bonds, as discussed in Section 12.3.3.2. During the recovery phase, (part of) the holes trapped at these defects are released, leading to partial V_{th} relaxation. During the second stress phase, holes injected in the gate stack are initially filling again the N-dangling bonds generated during the first stress phase, leading to a fast ΔV_{th} increase.

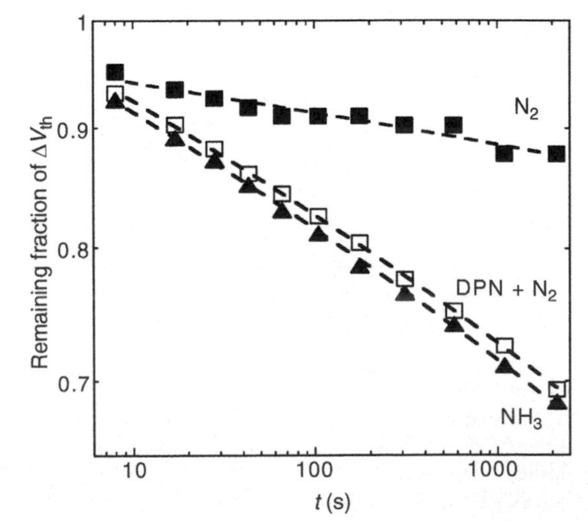

FIGURE 12.21
Remaining fraction of V_{th} shift as a function of time during the recovery phase ($V_G = 0$ V, 125°C), for devices previously subjected to NBT stress at −2 V for 1000 s. Dashed lines are logarithmic fits to the data. (From Houssa, M., et al., *Microelectron. Reliab.*, 47, 880, 2007. With permission.)

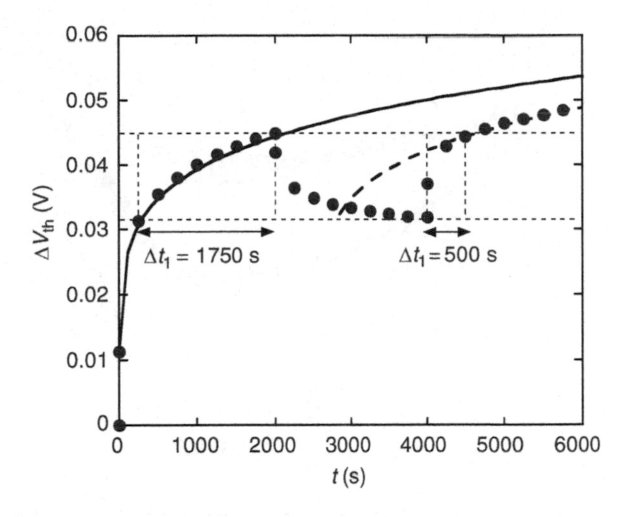

FIGURE 12.22
V_{th} shift as a function of time of HfSiON/TaN stacks alternatively exposed to NBT stress (−2 V, 125°C) and recovery phase (0 V, 125°C). The solid line is a power-law fit to the data recorded during the first stress phase. The dashed line corresponds to a translation of the solid line to the data recorded during the second stress phase. (From Houssa, M., et al., *Microelectron. Reliab.*, 47, 880, 2007. With permission.)

Additional N-dangling bonds are subsequently generated, leading to ΔV_{th} kinetics very similar to the first stress phase.

12.3.4 Si-Substrate Orientation Effect

It has been reported that hole mobility is higher on (110) Si-substrate surface orientation while electron mobility is higher on (100) surface orientation [47–49]. The hybrid orientation (110) for pMOSFETs and (100) for nMOSFETs is thus a promising technology for future high-performance CMOS devices [50]. However, the density of Si–H bonds at the Si/dielectric interface is potentially higher on (110) Si surfaces [50], which can have a negative impact on NBTI robustness of the device, as discussed below.

The effective density of positively charged defects, N_{eff}, generated during NBT stress of both (100) and (110) Si substrates with SiO_x/HfSiON/TiN gate stacks (HfSiON deposited by ALD and exposed to soft plasma nitridation) is shown in Figure 12.23 as a function of

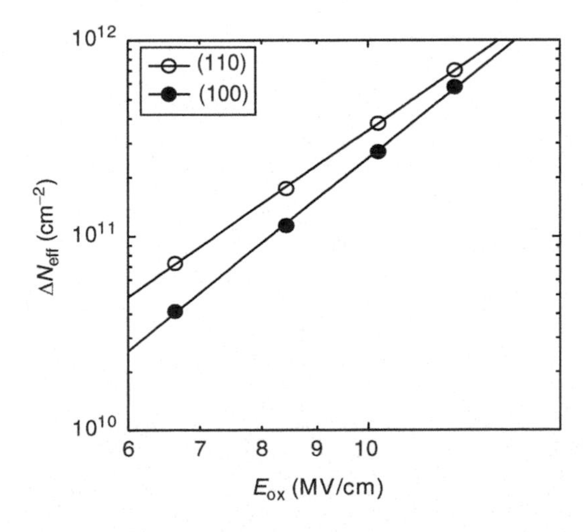

FIGURE 12.23
ΔN_{eff} as a function of the electric field across the interfacial layer for the (100) and (110) Si surface orientations (stress time fixed at 1024 s).

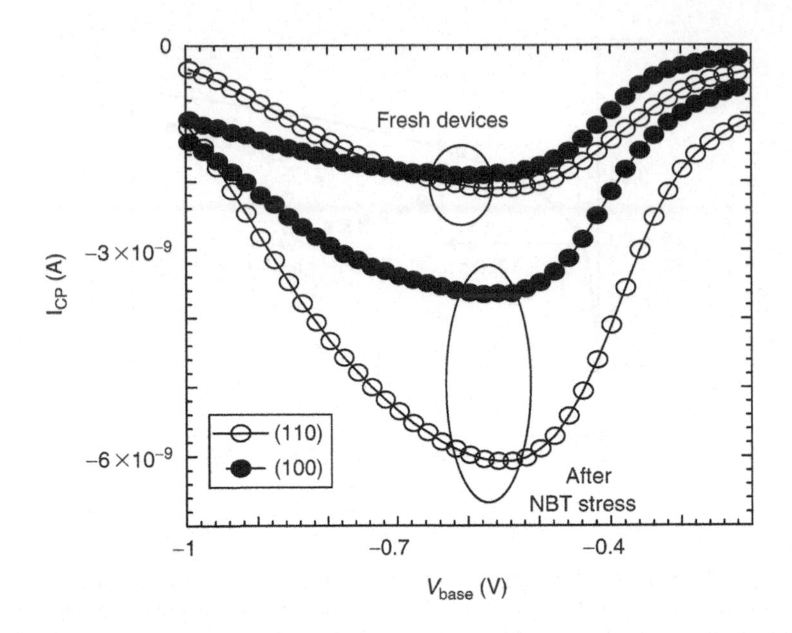

FIGURE 12.24
Charge pumping current as a function of base level sweep, measured at 3 MHz on fresh and NBT-stressed pMOSFETs with (100) and (110) Si surfaces.

the electric field across the interfacial layer. More NBT degradation is observed on (110) Si-surface orientation than on (100). Charge pumping current measured at high frequency on fresh (unstressed) devices with (100) and (110) Si-surface orientation shows about the same density of interface traps as shown in Figure 12.24. However, the generation of interface traps during NBT stress is higher on (110) than on (100), as evidenced from the data shown in Figures 12.24 and 12.25.

Consequently, the contribution of N_{it} to the V_{th} shift is larger on (110) Si surface orientation, as shown in Figure 12.26. The enhanced generation of interface traps on (110) surfaces is likely related to a higher density of hydrogen-passivated dangling bonds at the interface [47] (i.e., interface defect precursors), predisposed to be activated during the electrical stress. Thus, the (110) surface orientation is likely to show a

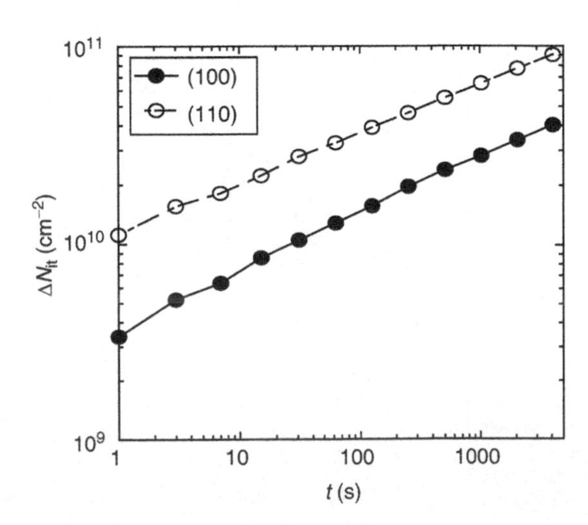

FIGURE 12.25
Kinetics of interface trap generation during NBT stress of pMOSFETs with (100) and (110) Si surfaces.

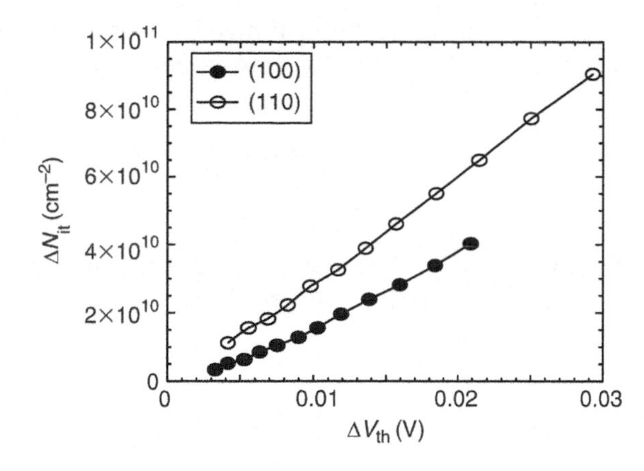

FIGURE 12.26
Interface trap density as a function of V_{th} shift for (100) and (110) Si surface orientations.

reduced device lifetime under NBT-stress compared to its (100) counterpart, especially at low electric fields.

12.4 Summary

Negative bias temperature instabilities in SiO_x/HfSiO(N)/metal gate stacks were reviewed. The contributions of fast (interface) states and slow (bulk) states to device degradation were separated by combining fast V_{th} extraction and high-frequency charge pumping measurements. In general, a substantial contribution of the fast states to the device degradation was observed at low electric field and high temperature, while slow state generation dominated at low temperature and/or high electric field.

The kinetics and temperature dependence of fast states were consistent with the reaction–diffusion model (or the reaction-dispersive transport model), suggesting that these defects were P_{b0} centers at the (100) Si/SiO_x interface, with the rate-limiting step for defect generation being the transport of hydrogen species away from this interface. The H_2/D_2 isotopic effect observed at low electric field, i.e., the retardation of defect generation observed in devices annealed in N_2/D_2, was also consistent with the reaction-dispersive transport model. In addition, the enhanced interface defect generation observed on (110) Si surfaces could be explained by the higher density of $P_{b0}H$ defect precursors present at this interface.

On the other hand, the different observed time and temperature behavior of slow states suggest another defect generation mechanism. These defects were also correlated to the presence of nitrogen in the gate stack. We have proposed that the slow states precursors were likely N-related dangling bonds passivated by hydrogen. The distortion of the polar N–H bonds induced by the electric field and the trapping of holes at these bonds were assumed to result in the generation of positively charged N-related centers in the gate stack. Besides, the revealed correlation between enhanced NBTI and enhanced V_{th} recovery observed in nitrided stacks indicated that the recovery resulted most likely from the hole detrapping/neutralization when the NBT stress was interrupted.

From a reliability point of view, the presence of nitrogen in the SiO_x/HfSiO gate stacks was shown to reduce the NBTI robustness of the MOSFETs; 10 years of lifetime were usually not extrapolated at operating gate overdrive in these devices. If the incorporation of nitrogen in high-κ gate stacks is necessary, e.g., to improve their thermal stability with respect to advanced complementary metal-oxide semiconductor (CMOS) device processing, the concentration and profile of nitrogen has to be optimized, in order to reduce the concentration of N-related defects near the channel of the transistors.

Acknowledgments

The authors are grateful to Professors A. Stesmans and V.V Afanas'ev (K.U. Leuven) for fruitful discussions. This work has been financially supported by the IMEC Industrial Affiliation Program on high-κ dielectrics and metal gates.

References

1. Wilk, G.D., Wallace, R.M., and Anthony, J.M., Hafnium and zirconium silicates for advanced gate dielectrics, *J. Appl. Phys.*, 87, 484, 2000.
2. Houssa, M. (Ed.), *High-K Gate Dielectrics*, IOP, London, 2003.
3. Visokay, M.R. et al., Application of HfSiON as a gate dielectric material, *Appl. Phys. Lett.*, 80, 3183, 2002.
4. Koyama, M. et al., Careful examination on the asymmetric V_{fb} shift problem for poly-Si/HfSiON gate stack and its solution by the Hf concentration control in the dielectric near the poly-Si interface with small EOT expense, *IEDM Tech. Dig.*, IEEE, New Jersey, 2004, 499.
5. Watanabe, T. et al., Impact of Hf concentration on performance and reliability for HfSiON-CMOSFET, *IEDM Tech. Dig.*, IEEE, New Jersey, 2004, 507.
6. Tseng, H.H. et al., Improved short channel device characteristics with stress relieved pre-oxide (SRPO) and a novel tantalum carbon alloy metal gate/HfO_2 stack, *IEDM Tech. Dig.*, IEEE, New Jersey, 2004, 821.
7. Callegari, A. et al., Interface engineering for enhanced electron mobilities in W/HfO_2 gate stack, *IEDM Tech. Dig.*, IEEE, New Jersey, 2004, 825.
8. Lebedinskii, Y. et al., In situ investigation of growth and thermal stability of ultrathin Si layers on the HfO2/Si (100) high-κ dielectric system, *Appl. Phys. Lett.*, 86, 191904, 2005.
9. Schroder, D.K. and Babcock, J.A., Negative bias temperature instability: Road to cross in deep submicron silicon semiconductor manufacturing, *J. Appl. Phys.*, 94, 1, 2003.
10. Krishnan, A.T. et al., NBTI impact on transistor and circuit: Models, mechanisms and sclaing effects, *IEDM Tech. Dig.*, IEEE, New Jersey, 2003, 349.
11. Helms, C.R. and Poindexter, E.H., The silicon-silicon-dioxide system: Its microstructure and imperfections, *Rep. Prog. Phys.*, 57, 791, 1994.
12. Ogawa, S. and Shiono, N., Generalized diffusion–reaction model for low field charge buildup instability at the Si–SiO_2 interface, *Phys. Rev. B*, 51, 4218, 1995.
13. Alam, M.A., A critical examination of the mechanics of dynamic NBTI for pMOSFETs, *IEDM Tech. Dig.*, IEEE, New Jersey, 2003, 346.
14. Callegari, A. et al., Physical and electrical characterization of hafnium oxide and hafnium silicate sputtered films, *J. Appl. Phys.*, 90, 6466, 2001.
15. Ershov, M. et al., Dynamic recovery of negative bias temperature instability in *p*-type metal-oxide-semiconductor field-effect transistors, *Appl. Phys. Lett.*, 83, 1647, 2003.
16. Chen, G. et al., Dynamic NBTI of pMOS transistors and its impact on MOSFET scaling, *IEEE Electron Dev. Lett.*, 23, 734, 2002.
17. Kaczer, B. et al., Disorder-controlled kinetics model for negative bias temperature instability and its experimental verification, in *Proc. Int. Reliab. Phys. Symp.*, IEEE, New Jersey, 2005, 381.
18. Denais, M. et al., On-the-fly characterization of NBTI in ultra-thin gate oxide PMSOFETs, *IEDM Tech. Dig.*, IEEE, New Jersey, 2004, 109.
19. Onishi, K. et al., Bias-temperature instabilities of polysilicon gate HfO_2 MOSFETs, *IEEE Trans. Electron Dev.*, 50, 1517, 2003.

20. Zhou, X.J. et al., Negative bias-temperature instabilities in metal-oxide-silicon devices with SiO_2 and SiO_xN_y/HfO_2 gate dielectrics, *Appl. Phys. Lett.*, 84, 4394, 2004.
21. Zafar, S. et al., A model for negative bias temperature instability in oxide and high-k pFETs, *VLSI Symp.*, IEEE, New Jersey, 2004, 208.
22. Houssa, M. et al., Role of hydrogen on negative bias temperature instability in HfO_2-based hole channel field-effect transistors, *Appl. Phys. Lett.*, 85, 2101, 2004.
23. Houssa, M. et al., Detrimental impact of hydrogen on negative bias temperature instabilities in HfO_2-based pMOSFETs, *VLSI Symp.*, IEEE, New Jersey, 2004, 212.
24. Houssa, M. et al., Negative bias temperature instabilities in HfSiON/TaN-based pMOSFETs, *IEDM Tech. Dig.*, IEEE, New Jersey, 2004, 121.
25. Aoulaiche, M. et al., Contribution of fast and slow states to negative bias temperature instabilities in $Hf_xSi_{(1-x)}ON$/TaN based pMOSFETs, *Microelectron. Eng.*, 80, 134, 2005.
26. Aoulaiche, M. et al., Impact of nitrogen incorporation in SiO_x/HfSiO stacks on negative bias temperature instability, in *Proc. Int. Reliab. Phys. Symp.*, IEEE, New Jersey, 2006, 317.
27. Groeseneken, G. et al., A reliable approach to charge pumping measurements in MOS transistors, *IEEE Trans. Electron Dev.*, 31, 42, 1984.
28. Lyding, J.W., Hess, K., and Kizilyalli, I.C., Reduction of hot electron degradation in metal oxide semiconductor transistors by deuterium processing, *Appl. Phys. Lett.*, 68, 2526, 1996.
29. Hess, K., Kizilyalli, I.C., and Lyding, J.W., Giant isotope effect in hot electron degradation of metal oxide silicon devices, *IEEE Trans. Electron Dev.*, 45, 406, 1998.
30. Saks, N.S. and Rendell, R.W., Time-dependence of the interface trap buildup in deuterium-annealed oxides after irradiation, *Appl. Phys. Lett.*, 61, 3014, 1992.
31. Kimizuka, N. et al., NBTI enhancement by nitrogen incorporation into ultrathin gate oxide for 0.10-μm gate CMOS generation, *VLSI Symp.*, IEEE, New Jersey, 2000, 92.
32. Saks, N.S. and Rendell, R.W., The time-dependence of post-irradiation interface trap buildup in deuterium-annealed oxides, *IEEE Trans. Nucl. Sci.*, 39, 2220, 1992.
33. Houssa, M. et al., Reaction-dispersive proton transport model for negative bias temperature instabilities, *Appl. Phys. Lett.*, 86, 093506, 2005.
34. Stoneham, M., Electronic and defect processes in oxides: The polaron in action, *IEEE Trans. Diel. Electr. Insul.*, 4, 604, 1997.
35. Rendell, R.W., Role of dynamical cooperativity for an enhanced isotope effect during transport, *J. Appl. Phys.*, 75, 7626, 1994.
36. Brown, D.B. and Saks, N.S., Time dependence of radiation-induced interface trap formation in metal-oxide-semiconductor devices as a function of oxide thickness and applied field, *J. Appl. Phys.*, 70, 3734, 1991.
37. Hori, T., *Gate Dielectrics and MOS ULSIs*, Springer, Berlin, 1997.
38. Koyama, M. et al., Effects of nitrogen in HfSiON gate dielectric on the electrical and thermal characteristics, *IEDM Tech. Dig.*, IEEE, New Jersey, 2002, 849.
39. Kushida-Abdelghafar, K. et al., Effect of nitrogen at SiO_2/Si interface on reliability issues: Negative-bias-temperature instability and Fowler–Nordheim-stress degradation, *Appl. Phys. Lett.*, 81, 4362, 2002.
40. Huard, V. and Denais, M., Hole trapping effect on methodology for DC and AC negative bias temperature instability measurements in pMOS transistors, in *Proc. Inter. Reliab. Phys. Symp.*, IEEE, New Jersey, 2004, 40.
41. DiMaria, D.J. and Stathis, J.H., Anode hole injection, defect generation, and breakdown in ultrathin silicon dioxide films, *J. Appl. Phys.*, 89, 5015, 2001.
42. Yount, J.T., Lenahan, P.M., and Krick, T.J., Comparison of defect structure in N_2O- and NH_3-nitrided oxide dielectrics, *J. Appl. Phys.*, 76, 1754, 1994.
43. Warren, W.L., Lenahan, P.M., and Curry, S.E., First observation of paramagnetic nitrogen dangling-bond centers in silicon nitride, *Phys. Rev. Lett.*, 65, 207, 1990.
44. McPherson, J.W. et al., Complementary model for intrinsic time-dependent dielectric breakdown in SiO_2 dielectrics, *J. Appl. Phys.*, 88, 5351, 2000.
45. Hess, K. et al., Simulation of Si–SiO_2 defect generation in CMOS chips: From atomistic structure to chip failure rates, *IEDM Tech. Dig.*, IEEE, New Jersey, 2000, 93.

46. Houssa, M. et al., Model for defect generation at the (100) Si/SiO_2 interface during electron injection in MOS structures, *Appl. Surf. Sci.*, 212–213, 749, 2003.
47. Yang, M. et al., Hybrid-orientation technology (HOT): Opportunities and challenges, *IEEE Trans. Electron Dev.*, 53, 965, 2006.
48. Sato, T. et al., Effects of crystallographic orientation on mobility, surface state density, and noise in p-type inversion layers on oxidized silicon surfaces, *Japan. J. Appl. Phys.*, 8, 588, 1969.
49. Zafar, S. et al., A comparative study of NBTI as a function of Si substrate orientation and gate dielectrics (SiON and SiON/HfO_2), in *Intl. Symp. on VLSI Technology (VLSI-TSA)*, IEEE, New Jersey, 2005, 128.
50. Maeda, S. et al., Negative bias temperature instability in triple gate transistors, in *Proc. Int. Reliab. Phys. Symp.*, IEEE, New Jersey, 2004, 8.

13

Defect Formation and Annihilation in Electronic Devices and the Role of Hydrogen

Leonidas Tsetseris, Daniel M. Fleetwood, Ronald D. Schrimpf, and Sokrates T. Pantelides

CONTENTS

13.1 Introduction

The control of concentrations and properties of intrinsic defects and impurities has been an integral part of the microelectronics revolution. Overcoordinated and undercoordinated atoms, extrinsic species, and defect complexes have been the subjects of numerous studies, both theoretical and experimental. Deviations from ideal crystallinity in the bulk of electronic materials and their interfaces have been identified as the key culprits of important reliability phenomena. Nevertheless, despite the mass of accumulated information on defect formation in traditional materials, especially in silicon and silicon-based devices, key questions remain under debate. Moreover, with the current drive to replace silicon-based materials in various applications, including designs with nanostructures and organic systems, new challenges emerge for the characterization and control of novel defect configurations.

Defects and impurities may appear in electronic materials in all stages, from the initial growth of materials to aging of fully developed devices. For example, oxygen is a common impurity in silicon, especially Czochralski-grown Si, incorporated into the bulk material during growth from the melt. Oxygen of course can appear in various forms, including defect configurations during the later stage of thermal oxidation of silicon. Finally, oxygen can migrate to various parts of a device and participate in defect formation during the long-term operation of a device. Similar scenarios are possible for several other impurities and defects. One common device impurity that stands out in significance and complexity of behavior is hydrogen. The presence of hydrogen may be either intentional, for example during the postoxidation passivation of dangling bonds of the Si–SiO_2 system, or inadvertent when it is provided by extraneous or internal sources.

In this chapter, we review results from theoretical studies on several of the most important effects of device defects and impurities, especially hydrogen. Emphasis is placed on first-principles approaches (also known as ab initio approaches), in particular, quantum mechanical calculations within density functional theory (DFT) [1,2]. In recent years, first-principles approaches have become an integral part of studies of electronic devices, because of the availability of increased computational power. The strengths and limitations of these approaches are presented with examples of combinations with other theoretical methods. We discuss defect formation and dynamics in a silicon substrate, in amorphous SiO_2, and at the Si–SiO_2 interface. We point out the relevance of impurities for notable reliability phenomena, such as bias-temperature instability (BTI) [3] and radiation damage [4], and we discuss the importance of defects in novel materials, namely pentacene [5].

13.2 Theoretical Approaches

The majority of the results presented below were obtained with so-called supercell studies in the framework of DFT. In the supercell approach, isolated defects are approximated by periodic arrays of defect configurations. The distance between periodic images is made as large as possible, minimizing the impact of the artificial periodicity on the conclusions about individual defects. The quantum mechanical problem for valence electrons was solved in a plane wave basis and their interactions with ionic cores were described through either ultrasoft pseudopotentials [6] or projector augmented waves [7], as implemented in the VASP code [8,9]. Convergence of results is checked with respect to relevant quantities, namely the size of supercells, the number of *k*-points for Brillouin zone sampling [10], and the energy cutoff for the plane wave basis.

The calculations provide the total energies of supercells; the relative stability of various defect configurations is obtained by evaluating the energy differences between possible structures. Moreover, an assessment can be made on the activation of relevant defect transformations, including diffusion. In particular, reaction and migration barriers can be obtained with the elastic band (EB) method [11,12]. Figure 13.1 contains a schematic of the way that activation energies are calculated within the EB approach. The barriers can be combined with typical attempt frequencies to provide an estimate of the time scale of events at certain temperatures. For example, for an attempt frequency of 10^{13} Hz, barriers of 0.5, 0.9, and 1.5 eV result at room temperature in time scales of 48 μs, 431 s, and 10^{13} s, respectively. Based on the energy differences and barriers, we can determine whether certain processes are activated in a particular temperature range.

First-principles calculations are a unique tool in determining atomic-scale details that are not often accessible with other theoretical methods and experiments. However, there are also notable limitations that should be taken into account in the interpretation of results. For example, DFT in the commonly used local-density approximation for exchange-correlation potentials underestimates the energy band gap of nonmetallic solids. More generally, the DFT description of unoccupied states is not as accurate as the information on valence states. Furthermore, approximations are often employed to decrease the system of study to sizes appropriate for the computationally demanding ab initio calculations. With respect to kinetics, the error in activation energies obtained with first-principles calculations is typically

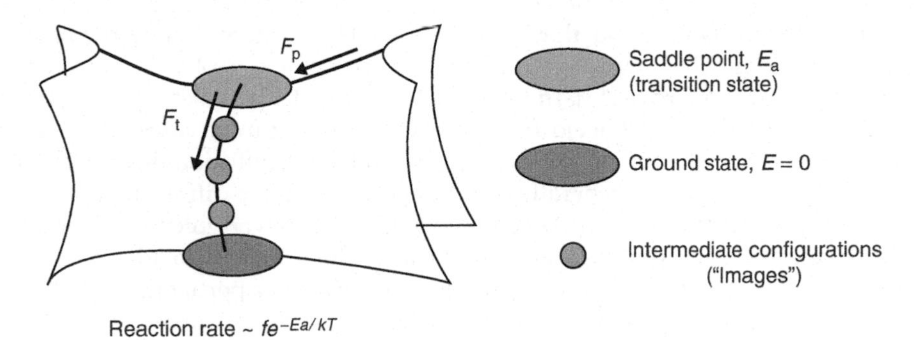

FIGURE 13.1
Schematic diagram of the calculation of activation energies E_a with the EB method. A transition pathway is modeled by a series of intermediate configurations (images). The chain is relaxed so that at each stage the tangential to the path quantum mechanical force is replaced by a spring force between neighboring images.

less than 0.1–0.2 eV. Overall, first-principles calculations can give information of exceptional quality about structural properties, such as bond lengths or lattice constants, and they normally describe correctly physical trends in terms of energy differences. In this respect, even though they may not always provide highly accurate values for rates of processes, they typically elucidate issues of relative stability and activation of transformations.

The bulk of theoretical approaches in defect physics and engineering have traditionally relied on more efficient approaches, which do not match the accuracy of ab initio tools, but they allow for the treatment of larger-scale phenomena. One large class of studies comprises the use of rate equations to describe processes, such as chemical reactions or diffusion. The equations contain parameters such as diffusion barriers or reaction barriers that can be chosen by fitting to experiments, or to a higher-level theoretical approach. Another approach, which can be regarded as intermediate between rate-equations and first-principles, employs Monte Carlo or molecular dynamics simulations to describe the formation and dynamics of defects. Once again, the parameters used, such as pair interaction energies, are selected either by fitting to experiments or to first-principles calculations. A systematic approach that uses parameters from high-level calculations in small systems and for short times to perform lower-level calculations in larger scale is called a sequential multiscale approach. A more sophisticated method involves the simultaneous use of high-level and low-level tools to describe both the details of a small area around the system of interest and the effects of larger-scale phenomena, such as long-range stress. These so-called concurrent multiscale approaches are significantly more demanding in terms of computational power, but they are gaining an ever increasing popularity in defect physics.

13.3 Core Boundary of Microelectronics: Si–SiO_2 Interface

Let us start with a discussion of one of the most important elements of today's silicon-based electronic devices: the Si–SiO_2 interface. It is well known that optimal conditions for growth of smooth interfaces are achieved with thermal oxidation at temperatures T_{th} typically higher than 800°C. The thermally grown interface has been the subject of many studies that have identified several key features. First, the rate of oxide growth is generally described within reaction–diffusion theory. O_2 molecules enter SiO_2, they diffuse in the amorphous silica network, and they reach the interface where they participate in silicon oxidation reactions. The original quantitative model of these physical mechanisms was proposed by Deal and Grove [13]. In the Deal–Grove model two distinct activation energies (E_a) of 2 and 1.2 eV were extracted from measured oxide growth rates. These values were attributed [13] to an initial reaction-limited phase and a late diffusion-limited phase, respectively. The 1.2 eV process has generally been connected with the diffusion of O_2 in silica. The former 2 eV process, however, has been the subject of renewed interest lately, and studies [14–18] have questioned the original assignment to the interfacial oxidation reaction.

First-principles calculations [19], including calculations of pertinent activation barriers, led to a comprehensive picture of thermal oxidation of Si as a process analogous to the growth of thin films on substrates, with oxidized islands that nucleate, grow, and coalesce in a smooth layer-by-layer mode. As is known, a particular mode of thin-film growth depends on the competition of a limited number of processes, namely deposition, relaxation, surface diffusion, and possibly desorption [20]. Thermal oxidation can be treated as an effective deposition problem, and one can calculate binding energies and reaction and diffusion barriers for processes that are reminiscent of the surface analogs described above.

The key findings that support a smooth layer-by-layer propagation of oxidation fronts with small critical roughness exponents are the following: (1) there is an effective barrier of about 2 eV for O_2 molecules to cross the near interfacial region and arrive at the substrate. This barrier accounts for the observed 2 eV activation energy for the reaction step in the Deal–Grove model of thermal oxidation. Figure 13.2 includes distinct stages of the

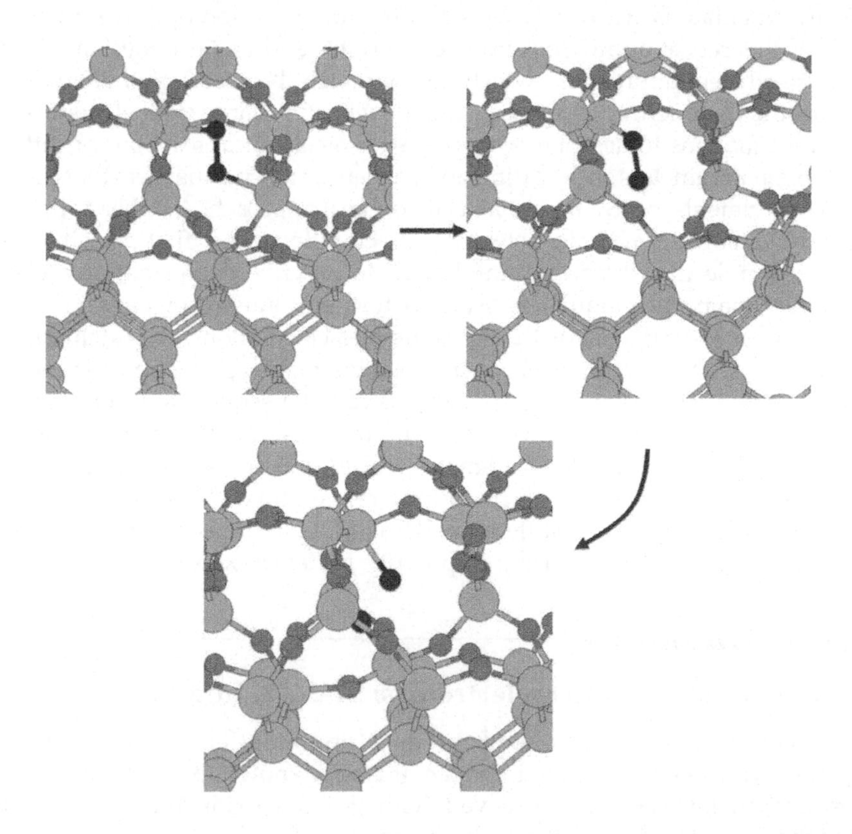

FIGURE 13.2
Migration of an O_2 molecule in the near-interface layer of the Si–SiO_2 boundary. The layer is denser than bulk silica. The diffusion barrier is 2.0 eV, accounting for the reaction rate-limiting step of the Deal–Grove model.

migration of an oxygen molecule inside the near-interface layer. (2) The oxidation reaction of Si–Si bonds through dissociation of O_2 molecules is almost barrier-less and strongly exothermic. This fact by itself points to a random deposition (RD) problem with large roughness, contrary to the established smoothness of the Si–SiO_2 interface [21]. (3) Diffusion of O atoms between Si–Si bonds along the oxidation front is activated at thermal oxidation temperatures T_{th} due to relatively small barriers of 2–2.3 eV. (4) Si–O–Si bridges prefer to agglomerate with binding energies of about 0.3–0.4 eV per O atom. Findings (3) and (4) are the basic conditions that favor nucleation and island growth mode. (5) Migration of O in bulk Si is also activated at T_{th}, which raises the question why O does not simply dissolve in Si. This puzzle is resolved by showing that O species that arrive from the substrate get trapped at the interface and eventually relax in configurations consistent with a smooth front. However, occasional desorption back into the bulk Si decreases the long-range roughness of the interface. Thus, the processes described in (3)–(5) resolve the quandary posed in (2) and cancel the RD character of the evolution of the oxidation front, providing for overall smoothness, as observed.

The above scenario corroborates suggestions made in earlier studies [15] that both the 1.2 and 2 eV rate-limiting steps relate to diffusion of O_2. The former value is related to O_2 migration in the bulk of silica, and the latter corresponds to diffusion in the denser near-interface SiO_2 layer. In addition to the above features, many key effects have been unraveled and debated in theoretical studies. Monte Carlo studies showed that so-called canonical interfaces [22,23] with abrupt interfaces are terminated by oxygen at interfacial Si dimers. Other studies have explored the varying oxidation states at the near-interface layer and found that substrate Si atoms close to the interface are displaced [24] from their original crystal positions.

The above set of physical mechanisms suggests that postoxidation annealing at temperatures lower than T_{th} may ameliorate partly the disorder in the near-interface layers, as long as O migration at the oxidation front is activated. Moreover, the set of mechanisms described here is key for an understanding of the morphology of novel interface systems, involving for example high-*k* dielectrics as gate oxides or SiC as the substrate. In general, one can expect that growth of oxides at temperatures significantly lower than the temperatures needed for interfacial O migration and reorganization results in rougher interfaces.

13.4 Defects at the Si–Dielectric Interface

Long efforts have achieved growth of thermal oxides with outstanding interface smoothness and very low densities of defects. However, defects are always present at the interface, even in limited numbers. The most common imperfection is a threefold coordinated Si, whose unpaired electron corresponds to a so-called dangling bond. These defects are known as P_b centers [25]. Elimination of the defect is achieved by passivating the dangling bond with, for example, hydrogen, as is shown in Figure 13.3. Si–SiO_2 interfaces grown in the (111) direction possess a single class of P_b centers. In contrast, two distinct centers, the so-called P_{b0} and P_{b1}, are encountered at (100) interfaces. Various geometries have been proposed for the structure around a P_b center: (1) the geometry proposed by Stirling et al. [26], pertaining to a Si back-bonded to three other Si atoms with three oxygen atoms as second nearest neighbors (n.n.). First-principles calculations found good agreement between the electron paramagnetic resonance (EPR) signal for this structure and experimental data [26]. (2) A similar geometry of a normal dimer configuration with only one oxygen as second n.n., and (3) a singly oxidized threefold coordinated Si, i.e., a Si_2O geometry as originally proposed by Poindexter and coworkers (Chapter 6) [27].

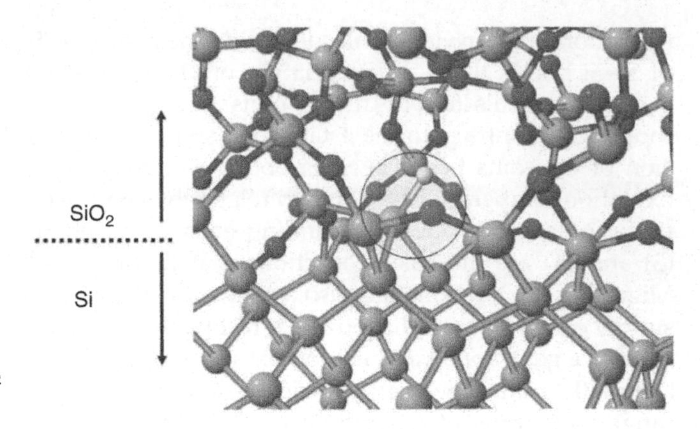

FIGURE 13.3
Passivated Si–H bond at the Si–SiO_2 interface.

The kinetics of passivation in a H_2-rich environment has been studied experimentally first by Brower [28] and Brower and Myers [29], who monitored the diminishing EPR signal of the dangling bonds of a (111) Si–SiO_2 interface during passivation. They reported an activation energy of 1.66 eV for the overall process and demonstrated that the rate-limiting step is associated with the passivation reaction itself. A subsequent series of experiments by Stesmans [30,31] refined this activation energy to about 1.5 eV for the (111) grown interface, and in the range of 1.51 to 1.57 eV for the (100) interface.

The Si–H bond is a very stable entity that has no levels in the energy band gap of silicon. However, there are processes that may result in the removal of H from the Si–H bond and the reappearance of dangling bonds that can act as carrier traps. One such process is the so-called depassivation reaction [32,33]:

$$\mathrm{Si-H + H^+} \leftrightarrow P_b^+ + \mathrm{H_2} \tag{13.1}$$

The reverse reaction describes the passivation of a dangling bond by a H_2 molecule (Figure 13.4). This reaction can also be viewed as a mechanism for the cracking or

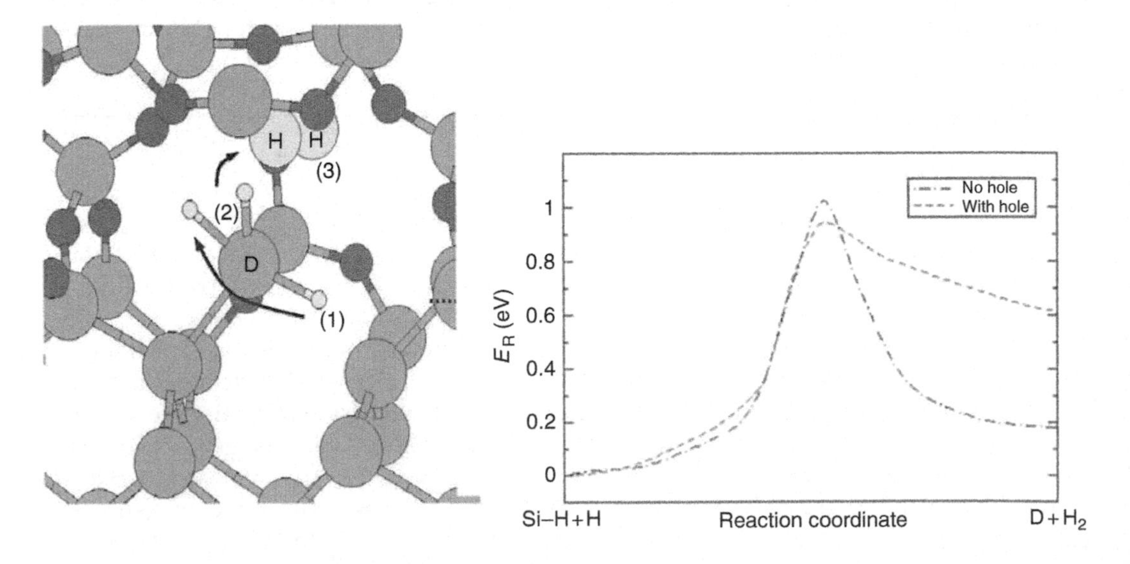

FIGURE 13.4
Depassivation reaction at the Si–SiO_2 interface. A schematic diagram of the reaction is shown on the left part. The results for the variation of energy along the reaction pathway are shown on the right. The reverse reaction describes the passivation of an Si dangling bond by hydrogen.

dissociation of molecular hydrogen at the interface. In the presence of a hole, the reverse of Equation 13.1 has a relatively small barrier of about 0.5 eV. The forward depassivation reaction has a barrier of 1.0 eV, indicating that the depassivation reaction can reach quasi-equilibrium quickly at even moderately elevated temperatures, a condition fulfilled for the reliability phenomenon of BTI we discuss below. If dangling bonds are passivated with fluorine, the depassivation reaction is suppressed in the presence of holes [34]. Therefore, a limited amount of fluorine can improve resistance to degradation. However, F plays also a detrimental role at the Si–SiO_2 interface because excess F creates new dangling bonds.

Interface traps in the form of Si dangling bonds can be created through another process:

$$\mathrm{Si-H} \leftrightarrow P_b + \mathrm{H} \tag{13.2}$$

which we term the dissociation reaction. There are several different dissociation pathways with the H atom either hopping backward to bind with an O atom, hopping across an interface void around the Si–H bond, or wagging sideways to a vicinal O atom of the interfacial SiO_2 network. All these pathways are strongly endothermic reactions that lead to an increase of energy by more than 2 eV. The corresponding activation energies are even higher, more than 2.3 eV [35], indicating that dissociation reactions are activated only at very high temperatures. Indeed, experiments [31,36,37] on Si–H dissociation found activation energies in the range of 2.5–2.8 eV; they documented breakup events of Si–H bonds at temperatures of more than 400°C.

Water molecules are other impurities that can be found in the vicinity of the Si–SiO_2 interface, where they can participate in various reactions. It has long been known that water molecules react at the Si–SiO_2 interface to further oxidize the Si substrate, but the atomic-scale details remain largely unknown. Two possible oxidation reactions are shown in Figure 13.5. One path has an activation energy [38] of 1.8 eV, in excellent agreement with the observed [13] oxidation activation energy of 2 eV. The reaction is exothermic with reaction energy of 0.8 eV. Clearly, this reaction is suppressed when the temperature is lowered and any water molecules that lurk in interstitial sites near the interface would remain dormant. The other oxidation process has an activation energy of only 1.5–1.6 eV in

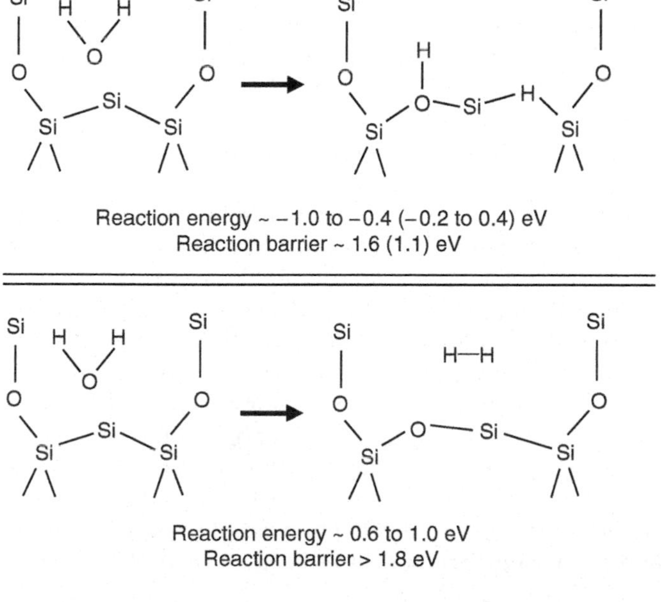

Reaction energy ~ −1.0 to −0.4 (−0.2 to 0.4) eV
Reaction barrier ~ 1.6 (1.1) eV

Reaction energy ~ 0.6 to 1.0 eV
Reaction barrier > 1.8 eV

FIGURE 13.5
Oxidation reactions of an H_2O molecule at the Si–SiO_2 interface.

the absence of holes [34]. This reaction, however, is endothermic with reaction energy of −0.6 eV so that it is not competitive with the normal exothermic reaction that operates at high temperatures. However, the reaction activation energy drops to 1.1–1.3 eV in the presence of excess holes, and the reaction energy shifts to a lower range −0.2 to +0.4 eV. The forward reaction is then favorable. It is in fact further enhanced because one of the end products, namely H^+, is highly mobile, reducing the probability of the reverse reaction.

The hole-enhanced local oxidation reaction leads to the degradation of the interface and the oxide in three distinct ways: (1) it causes interface roughening (Si–O–Si protrusions on the Si side of the interface); (2) the released H^+ is confined in a potential trough in the interface plane, where it migrates easily and can depassivate dangling bonds as in Equation 13.1, and (3) H^+ can overcome a barrier of 1 eV to enter the oxide where it can form positively charged defects, e.g., get trapped at O vacancies (a negative bias enhances migration into the oxide).

13.5 Defects in the Substrate

Because of the importance of silicon as the material of choice for electronic applications, its defects have attracted extensive attention over a long period of time. The cases include the so-called native point defects of Si vacancies and self-interstitials and their complexes, for example Frenkel pairs. Extrinsic species in silicon have also been studied extensively. Oxygen [39] and hydrogen [40] are among the most important Si impurities. Other noted impurities that are encountered in varying numbers in silicon include metal atoms that may diffuse from other parts of a device, carbon, as well as defect complexes between impurities and dopants [41–43] or native defects. Similar defects and defect complexes are found in other electronic materials, like germanium [44] and silicon carbide [45].

As we discussed above, hydrogen can be introduced in various stages of growth and operation of an electronic device. When it migrates to silicon, it can appear in many different forms. An isolated free hydrogen atom induces [46] levels in the energy band gap of silicon. The positions of these defect levels depend on the charge state of hydrogen, and hydrogen can act both as a donor and as an acceptor. The charge state of the species determines [47] also the structural details of hydrogen in Si. In particular, whereas positive (H^+) and neutral (H^0) hydrogen reside between two silicon atoms, the ground state position of negatively charged H (H^-) is the tetrahedral position inside a void of the Si crystal. Moreover, the diffusivity varies significantly between H^-, H^0, and H^+. First-principles calculations indicated that H^0 is highly mobile with a diffusion barrier [47] in the range of 0.1–0.2 eV, whereas H^+ and H^- are less mobile with migration activation energies [46] of about 0.48 and 0.7 eV, respectively.

Another interesting property of isolated hydrogen in silicon relates [47] to its so-called negative-U character. Normally, the Coulomb repulsion between carriers of the same sign that occupy a certain level raises the energy of the system, rendering the state unfavorable. However, when the system, for example an atom, is embedded in a crystal, the interaction with the surrounding electronic density may alter the energetics considerably. Specifically, the charging of a species can lead to significant relaxation so that the relative stability of different charge states is not determined only by the on-site Coulomb repulsion. This is the case for hydrogen, for which, as we noted above, different charge states have completely different lowest energy configurations (Si–H–Si bridge for H^0 and H^+, tetrahedral interstitial for H^-).

Hydrogen appears in many other configurations in silicon. For example, complexes of H with dopants in Si have been known for a long time and have been studied [41–43] both

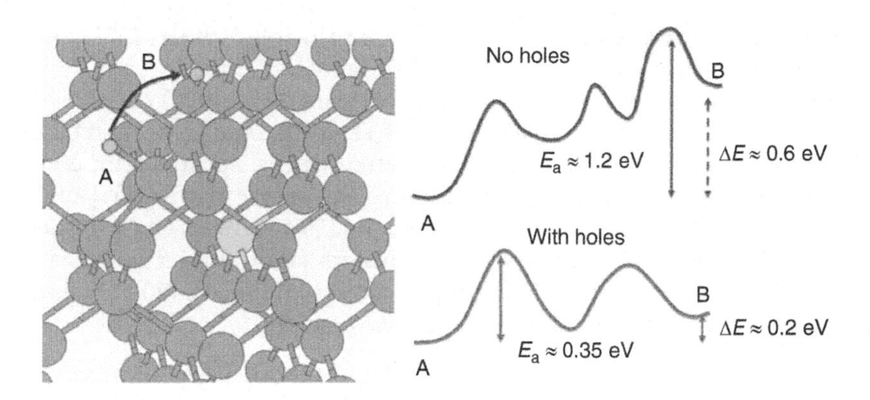

FIGURE 13.6
Release of hydrogen from phosphorus dopants in silicon.

experimentally and theoretically. It is known [48] that H binds in the antibonding site of a P dopant in *n*-type silicon, as shown on the left-hand side of Figure 13.6. Formally, the binding can be described as

$$D^+ + H^- \rightarrow D^+H^- \tag{13.3}$$

where D is a dopant atom. The calculated [48,49] H–P binding energy is 0.65 eV. However, the fundamentals of H–P trapping change significantly in the presence of holes. Experiments [41,43] find that the release of hydrogen from P dopants is enhanced significantly when minority carriers are introduced in the sample. Indeed, first-principles calculations of the binding energy of the H–P complex in the presence of holes found [49] a reduced value of 0.23 eV. Combined with the difference between the H^- (0.7 eV [46]) and H^0 (0.1–0.2 eV [47]) migration barriers in Si, the activation energy for H release drops substantially in the presence of holes. Similar effects of holes on the dopant-H binding energies have been calculated [50] for other typical dopants, namely As, Sb, and B. If H captures another hole, as can happen, for example, in the inversion layer of an *n*-type substrate, the resulting proton H^+ no longer binds to a P dopant.

Hydrogen can interact also with oxygen in silicon, and in particular with oxygen clusters known [51–54] as thermal donors (TD). TD are electrically active structures with shallow energy levels below the conduction band of Si. A family of such defects with a varying number of oxygen atoms has been detected experimentally, and theoretical studies have investigated the details of TD formation and dynamics.

Two cases can be distinguished for the formation of TD in silicon. First, one can consider the case that large numbers of O clusters preexist in Si. In this case, TD formation is rate-limited by the transformation of inactive staggered chains of O atoms to TD configurations with ring structures and threefold coordinated oxygen atoms. The calculated [55] activation energy of this transformation is 1.15 eV, in agreement with the lower measured [56] activation energies of 1.25 eV for small TD formation. The efficiency of this process depends on the position of the Fermi level with respect to the TD energy levels.

In the second case, there are no preexisting O clusters; therefore, O migration and agglomeration must precede the formation of TD. One of the interesting features of O in silicon is that oxygen dimers are a lot more mobile in silicon than isolated oxygen atoms. The migration of the former has an activation energy [55] of 1.55 eV, whereas the diffusion barrier of isolated O is about 2.5 eV. The process of O cluster migration is enhanced significantly when H is present in the vicinity. First-principles studies [57] confirmed that O–H codiffusion is greatly enhanced compared to isolated O migration.

In addition to migration, however, direct interactions of H with O clusters also play an integral role in understanding the enhancement of TD formation. Hydrogen binds strongly to O clusters and transforms them to TD-like configurations. For low temperatures, these sites are very efficient H-traps with a high hydrogen-release barrier [55] of 1.9 eV. For temperatures typical of TD formation, the release of H is activated, and repeated cycles of trapping and release of hydrogen from such clusters amount to a hydrogen-catalyzed creation of TD. Even small amounts of hydrogen can lead to the creation of a large number of TD, an effect that has been measured [58] experimentally.

13.6 Defects in Gate Oxides

A missing oxygen atom in silica corresponds to an oxygen vacancy, one of the most common defects in thermally grown oxides. Oxygen vacancies have distinct EPR signals, and they are classified as E' centers in SiO_2. One of their most interesting characteristics is that certain oxygen vacancies [59] are bi-stable, undergoing large relaxation upon capture of a carrier. In particular, whereas neutral O vacancies correspond normally to a Si–Si dimer, when a hole is captured the dimer is broken and one of the silicon atoms forms a back-bond with another O atom of the network. The effect is also known as puckering. Even though first-principles studies show that only a small fraction of oxygen vacancies of amorphous SiO_2 exhibit this bi-stability, the effect facilitates capture of carriers and renders oxygen vacancies as candidates for oxide-fixed charge traps.

Positively charged hydrogen H^+ migrates in silica by hopping between O atoms of the network. Because of amorphicity, there is a variation between about 0.6 and 0.8 eV in the migration activation energy. Thus, it migrates readily at room temperature, and in the presence of a positive bias, the migration toward the interface is enhanced by the electric field. Negatively charged hydrogen binds to electropositive silicon atoms, whereas neutral H^0 does not form strong bonds with Si or O. When the migrating hydrogen atoms encounter defects such as nonbridging oxygen atoms or oxygen vacancies, they passivate them by forming very stable silanol Si-OH groups. Passivation of oxygen vacancies can be achieved also with molecular hydrogen through the formation of a vicinal pair of Si-OH entities.

Water-related complexes are among the most pervasive defects for a wide spectrum of materials. Insertion and reactions of water in a solid, either during processing or after long-term operation, typically result in significant modifications of physical properties. Amorphous SiO_2, because of its immense significance for electronic and optoelectronic devices is a prototype system to study the role of disorder on H_2O reactions with a network. Infrared (IR) spectroscopy measurements typically [60] find water molecules and silanol groups (SiOH) inside a-SiO_2. Therefore, IR data provide unambiguous evidence for water incorporation in silica, but they are limited in identifying the atomic-scale details in the vicinity of the impurity. Theoretical investigations, especially first-principles calculations, are well suited to elucidate these details; these have indeed been used [61–63] to probe reaction energies and diffusion barriers of H_2O in a-SiO_2. In particular, Bakos et al. [61] investigated the energetics of formation of silanol groups. They found pertinent structures with energies higher than nondissociated water molecules in a-SiO_2 voids. Recent calculations [62,63] found energetically favorable configurations of silanol groups in silica that result from water dissociation. The stability of the silanol group configurations depends on the local structure, and H_2O dissociation is an exothermic reaction only as hydrolysis of Si–O–Si bonds that are members of small rings, i.e., rings that contain less than six Si–O–Si bonds. The smaller the rings are, the more favorable is the hydrolysis and formation of SiOH groups. The stable SiOH configurations are dormant species under equilibrium conditions at room temperature due to high barriers. However, further transformations that result in release of H can be activated under nonequilibrium conditions. For example, irradiation introduces holes, which can in turn trigger [63] the breakup of SiOH groups and the release of H atoms.

The above processes indicate that water is part of the hydrogen dynamics in a Si–SiO_2 system, initiating hydrogen-induced degradation reactions. Another aspect that links the evolution of water and hydrogen in SiO_2 is their mutual interaction. When a hydrogen atom that migrates in SiO_2 arrives in the vicinity of a water molecule, it can hop [64] to H_2O and form hydronium (H_3O) species. The interaction between H and H_2O depends strongly on the charge state of H. In particular, the binding energy of H^+ with a water molecule is considerable and equal to about 1 eV. Therefore, it is clearly energetically favorable for H^+ to hop away from the network O atoms and bind with a H_2O molecule. The formation of stable hydronium species is important for the interpretation of IR spectra of silica. In addition, it can change the dynamics of hydrogen in silica by altering the mobility of the species involved. Experiments [65] found that hydrogen diffusion in SiO_2 is enhanced significantly when traces of water are present in the amorphous network. The details on the role of H_3O on trapping, release, and diffusion of hydrogen remain under investigation.

13.7 Reliability Phenomena and the Role of Hydrogen and Other Defects

Reactions of hydrogen and other impurities in various parts of a device often manifest themselves as changes in the characteristics of electronic devices. Important reliability phenomena can trace their origins to the formation and dynamics of defects under specific conditions. Examples include the aging of materials and devices, the displacement damage induced by irradiation with energetic particles or photons, dielectric breakdown and leakage current, change of device characteristics under stress and annealing, and so on.

BTI [3] is one of those processes for which hydrogen is known [49,66–76] to play a key role. In BTI experiments on metal-oxide semiconductor (MOS) capacitors and metal-oxide-semiconductor field effect transistors (MOSFETs), strong electric fields are applied across the gate oxide and at elevated temperatures. The electric field is typically in the range of 2–6 MV/cm and typical BTI temperatures are between 100°C and 200°C. Under these conditions, modification of threshold voltage, transconductance, and other key device characteristics is observed. The underlying phenomena relate to creation of interface traps and oxide trapped charges (Chapters 6, 12, and 14). The effect is often more pronounced for negative BTI (NBTI) and for *n*-type silicon substrates, even though similar degradation has been reported for positive BTI (PBTI) and for *p*-type Si.

Even though BTI has been known [66] for more than 30 years, there is a recent resurgence of interest in the phenomenon and BTI is now regarded as one of the most important reliability issues affecting advanced complementary metal-oxide semiconductor (CMOS) devices. That hydrogen plays a central role in BTI has received strong support from experiments and theory. However, several details of the phenomena remain under debate, and further systematic studies on the details of the phenomenon are being pursued in the context of several different models.

One set of mechanisms [49] has been proposed based on first-principles calculations for hydrogen trapping, release, and reactions in Si, in SiO_2, and at the Si–SiO_2 interface. In the case of NBTI, the negative gate bias drives the *n*-type substrate of *p*-channel MOSFETs into inversion. Under these conditions, H is released easily from dopants in the depletion region, as discussed above. Some of the released hydrogen atoms enter the inversion layer and capture holes to become H^+. The positively charged protons are then driven to the interface by the applied field. Protons that arrive at the interface can migrate into the oxide where they can be trapped and contribute to the buildup of oxide charge. Another possibility for them is to diffuse laterally along the interface. In this case, protons can get trapped [49] in the vicinity of an interfacial Si–H bond, where they can participate in depassivation [33,49] reactions (Equation 13.1) of Section 13.4 and create dangling bonds.

For small stress times t_s the increase of the number N_{it} of interface traps is reaction-limited with an exponential growth with time. As the concentrations of H_2 products grow, the hydrogen molecules diffuse in SiO_2 and a fraction of them comes back to repassivate the newly created dangling bonds in the reverse reactions of Equation 13.1 (Section 13.4). Reaction–diffusion theory shows [66,67] then that the increase of N_{it} is limited by the diffusion of the product H_2 molecules in the oxide, and the increase of N_{it} obeys a power law with respect to t_s,:

$$N_{it}(t_s) \sim \sqrt{\frac{k_f N_D}{k_r}} (D_H t_s)^{1/4} \quad (13.4)$$

where

D_H is the diffusivity of H_2 in SiO_2

N_D is the initial number of passivated Si–H bonds

k_f and k_r are the forward and reverse rates of reaction of Equation 13.1, respectively.

Based on Equation 13.4, the apparent activation energy E^{BTI} is given by [66]:

$$E^{BTI} = \frac{\Delta E}{2} + \frac{\phi_D}{4} \quad (13.5)$$

where ΔE is the reaction energy of process (Equation 13.1), and φ_D is the diffusion barrier of H_2 in SiO_2. Using the calculated value $\Delta E \approx 0.5$ eV and the experimental [67] φ_D value of 0.45 eV, the apparent activation energy of E^{BTI} is found [49] to be about 0.35 eV. The above ΔE value corresponds to the case that H_2 resides in an interfacial void next to the dangling bond. When the molecule migrates deeper in silica, the energy decreases, and so does the reaction energy. Given this fact, and the approximations involved in these first-principles calculations, the extracted activation energy is in very good agreement with typical experimental NBTI activation energies.

As noted above, other BTI models place emphasis on different hydrogen-related reactions and effects in devices. Several studies have focused on the dissociation reaction (Equation 13.2) (Section 13.4) as the main process [3,68] that creates interface traps under BTI stress conditions. The dispersive character of hydrogen migration [70] and additional hydrogen reactions [72,73] have also been suggested as key factors, especially in the case that the measured time exponents deviate from the classical reaction–diffusion theory value of 0.25. Ongoing and future studies will of course address those features that are proven to be truly robust and systematic in BTI degradation. Chapter 14 also includes a detailed review of advances in BTI modeling.

Degradation can occur also under different conditions, for example irradiation or Fowler–Nordheim injection, or under a combination [77,78] of BTI conditions with these phenomena. In the case of irradiation, hydrogen has been identified as the main culprit for several effects, including the so-called [79,80] enhanced-low-dose-rate sensitivity (ELDRS). ELDRS refers to increased degradation of bipolar devices during low dose rate irradiation.

Its origin can be traced to hydrogen reactions either in the gate oxide or in the substrate and the interface. Hydrogen can be released in either the oxide or substrate of irradiated devices, and it can then migrate to participate in interfacial depassivation reactions and cause device degradation. Under positive or positive bias, the dependence of degradation on the dose-rate can be linked either to second-order reactions [81] of recombination of atomic H into H_2, or to the buildup at high-rates of positive space charge that hinders [82] the arrival of H^+ at the interface, and hence the ensuing degradation. Under negative bias, the conditions are similar [50] to NBTI, with irradiation taking the place of annealing in the release of H from substrate sources (Figure 13.7).

The substitution of hydrogen (H) by deuterium (D) in thermally annealed Si–SiO_2 devices has been shown [83] to lead to a large increase of the immunity of the device against hot-carrier stressing. This large isotope effect has been attributed to the strong coupling of interfacial Si–D vibrational modes to the modes of the substrate that make Si–D bonds resistant to breakup in the presence of hot carriers. Experiments [84] of annealing in H_2 and D_2 ambients found an activation energy of 1.84 eV for isotope exchange at the Si–SiO_2 interface. This value has been confirmed by first-principles calculations [85] for exchange reactions at the Si–SiO_2 boundary. The exchange can either occur directly from the molecular state, or it can comprise first the breakup of the molecule, the formation of Si–H (or Si–D) bonds in the vicinity of passivated P_b centers, and the reformation of the molecule. The corresponding theoretical activation energies are 1.94 and 1.97 eV, in agreement with the experimental findings.

Let us finally describe briefly theoretical findings on the role of hydrogen on stress-induced leakage current in silica gate oxides and dopant deactivation in a silicon substrate. Leakage current and eventual dielectric breakdown has its origin in the defect-assisted tunneling of carriers from the gate electrode of a MOS transistor to the substrate through the gate oxide. First-principles calculations showed [86] that a hydrogen atom bridging

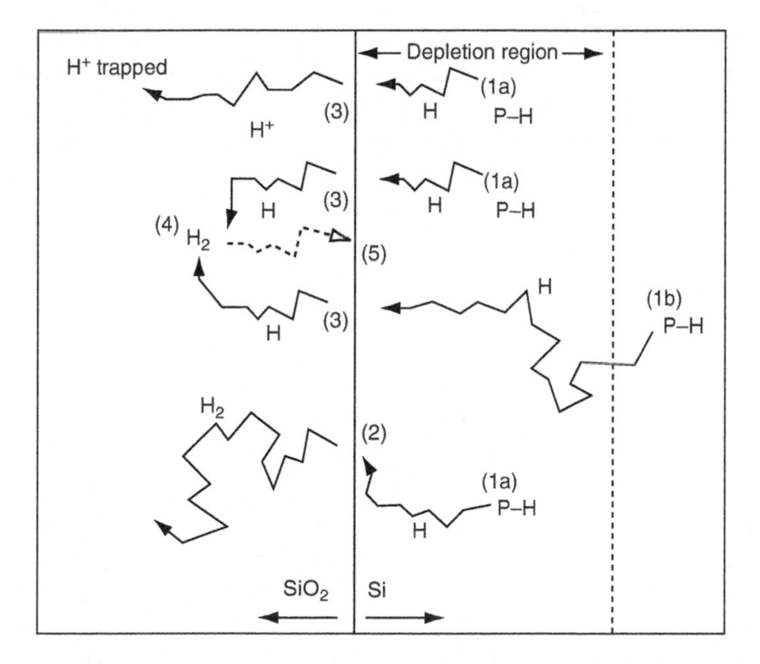

FIGURE 13.7
Main processes that give rise to BTI or ELDRS. (From Tsetseris, L. et al., *IEEE Trans. Nucl. Sci.*, 52, 2265, 2005. With permission.)

two silicon atoms at an oxygen vacancy site in SiO_2 fulfills the conditions of such a defect. In particular, this hydrogen bridge has a defect level close to the Fermi level of the electrode and the level has only small energy shifts upon charging. In the case [87] of dopant deactivation, we already mentioned above the H makes stable complexes with all typical dopants in silicon. The direct deactivation of boron impurities by H^+ in the presence of bias and radiation was also modeled with Monte Carlo simulations. The calculations found good agreement with experimental data on numbers of deactivated boron atoms as a function of irradiation bias, attributing the deactivation mainly to drift and diffusion of H^+ in the depletion region.

13.8 Defects in Pentacene Devices

Even though the quantitative details are expected to vary, most of the issues discussed above on the formation and dynamics of defects and their effects on physical properties of devices are relevant also for novel devices where the substrate material or the gate oxide is replaced with a system other than silicon and SiO_2. For example, bias-stress has been shown to result in the appearance of defect levels [88] in the energy band gap of pentacene, a prototype system of organic electronics. Selected effects of hydrogen and of other defects in pentacene are discussed in this section.

Pentacene films can grow [89] on a variety of substrates, and they are normally characterized [5] by relatively high carrier mobility. The molecules are packed with noncovalent bonds into a crystal with a herringbone structure inside successive layers. The noncovalent character of cohesion underlies the flexibility of the structure. However, the existence of space between molecules allows impurities to enter the crystal and alter its physical properties. Oxygen, water, and hydrogen are among the most common defect culprits whose insertion into a pentacene crystal have been shown [90–92] to be energetically favorable.

Oxygen and water have [92] different behaviors in pentacene. Water molecules stay nonbonded between the layers of the molecular crystal. They do not induce any defect levels in the energy band gap and they can only limit carrier mobility by acting as scattering centers. In contrast, oxygen has a more complex behavior. Both individual O defects and

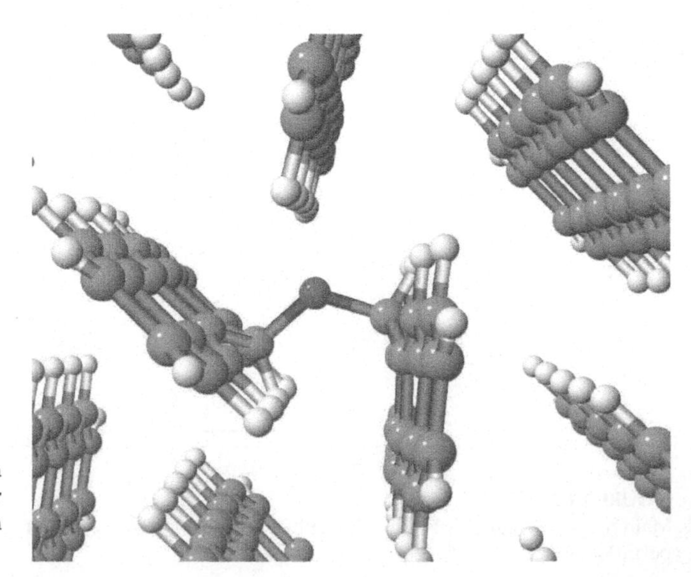

FIGURE 13.8
Oxygen defect in a pentacene crystal. An oxygen atom forms an intermolecular bridge and it induces defect levels in the energy band gap of pentacene.

O defect complexes are energetically favorable. The most stable O_2 configurations correspond to two atom O-complexes that result after dissociation of an O_2 molecule inside a pentacene layer. One of the stable configurations includes an intermolecular bridge with a single O atom. An isolated O atom in such a bridge configuration is shown in Figure 13.8.

The most interesting feature of the defect structure of Figure 13.8 is that it has a pair of levels in the pentacene energy band gap. The levels are located about 0.36 eV above the valence band maximum, in excellent agreement with experiments. Similar to oxygen, hydrogen has also been shown to induce levels in the band gap of pentacene. Therefore, there is a need for a systematic study of the conditions that lead to the appearance of the above defect configurations. Similar to the case of silicon-based devices, issues of trapping, release, and migration of impurities are of critical importance for the operation of organic-based devices.

13.9 Summary

Theoretical modeling has always been a valuable tool for the characterization of defects and defect dynamics in electronic devices. With increased available computational power, first-principles calculations in particular now play a key role in understanding the details of defects and the mechanisms that may limit defect concentration. The above sections described representative examples with key findings of ab initio studies on the effect of hydrogen and other impurities in silicon, in SiO_2 and their interface, as well as in pentacene-based devices. The result of such studies, in cases combined with more efficient lower-level approaches like Monte Carlo modeling, can provide the framework of understanding important device phenomena such as BTI, sensitivity to radiation, leakage currents, and others.

Acknowledgments

We acknowledge support by the Air Force Office of Scientific Research through the MURI program, the U.S. Navy, and the McMinn Endowment at Vanderbilt University.

References

1. Hohenberg, P. and Kohn, W., Inhomogeneous electron gas, *Phys. Rev.*, 136, B864, 1964.
2. Kohn, W. and Sham, L.J., Self-consistent equations including exchange and correlation effects, *Phys. Rev.*, 140, A1133, 1965.
3. Stathis, J.H. and Zafar, S., The negative bias temperature instability in MOS devices: A review, *Microelectron. Reliab.*, 46, 270, 2006.
4. Schrimpf, R.D. and Fleetwood, D.M., *Radiation Effects and Soft Errors in Integrated Circuits and Electronic Devices*, World Scientific, Singapore, 2004.
5. Dimitrakopoulos, C.D. et al., Low-voltage organic transistors on plastic comprising high-dielectric constant gate insulators, *Science*, 283, 822, 1999.
6. Vanderbilt, D., Soft self-consistent pseudopotentials in a generalized eigenvalue formalism, *Phys. Rev. B*, 41, 7892, 1990.
7. Blöchl, P.E., Projector augmented-wave method, *Phys. Rev. B*, 50, 17953, 1994.

8. Kresse, G. and Furthmuller, J., Efficient iterative schemes for ab initio total-energy calculations using a plane-wave basis set, *Phys. Rev. B*, 54, 11169, 1996.
9. Kresse, G. and Joubert, D., From ultrasoft pseudopotentials to the projector augmented wave method, *Phys. Rev. B*, 59, 1758, 1999.
10. Chadi, D.J. and Cohen, M.L., Special points in Brillouin zone, *Phys. Rev. B*, 8, 5747, 1973.
11. Mills, G., Jonsson, H., and Schenter, G.K., Reversible work transition-state theory, *Surf. Sci.*, 324, 305, 1995.
12. Johnson, H., Mills, G., and Jacobsen, K.W., in *Classical and Quantum Dynamics in Condensed Phase Simulations*, Berne, B.J., Cicotti, G., and Coker, D.F., (Eds.), World Scientific, Singapore, 1998.
13. Deal, B.E. and Grove, A.S., General relationship for the thermal oxidation of silicon, *J. Appl. Phys.*, 36, 3770, 1965.
14. de Almeida, R.M.C. et al., Dynamics of thermal growth of silicon oxide films on Si, *Phys. Rev. B*, 61, 12992, 2000.
15. Bongiorno, A. and Pasquarello, A., Reaction of the oxygen molecule at the Si–SiO_2 interface during thermal oxidation, *Phys. Rev. Lett.*, 93, 086102, 2004.
16. Bongiorno, A. and Pasquarello, A., Atomic-scale modeling of kinetic processes occurring during thermal oxidation, *J. Phys. Condens. Matter*, 17, 2051, 2005.
17. Akiyama, T. and Kageshima, H., Reaction mechanism of oxygen at SiO_2/Si(100) interface, *Surf. Sci.*, 576, L65, 2005.
18. Watanabe, T., Tatsumura, K., and Ohdomari, I., New linear-parabolic rate equation for thermal oxidation of silicon, *Phys. Rev. Lett.*, 96, 169102, 2006.
19. Tsetseris, L. and Pantelides, S.T., Oxygen migration, agglomeration and trapping: Key factors for the morphology of the Si–SiO_2 interface, *Phys. Rev. Lett.*, 97, 116101, 2006.
20. Barabasi, A.L. and Stanley H.E., *Fractal Concepts in Surface Growth*, Cambridge University Press, Cambridge, 1995.
21. Cho, E.C. et al., Atomistic structure of SiO_2/Si/SiO_2 quantum wells with an apparently crystalline silicon oxide, *J. Appl. Phys.*, 96, 3211, 2004.
22. Tu, Y. and Tersoff, J., Structure and energetics of the Si–SiO_2 interface, *Phys. Rev. Lett.*, 84, 4393, 2000.
23. Tu, Y. and Tersoff, J., Microscopic dynamics of silicon oxidation, *Phys. Rev. Lett.*, 89, 086102, 2002.
24. Bongiorno, A. et al., Transition structure at the Si–SiO_2 interface, *Phys. Rev. Lett.*, 90, 186101, 2004.
25. Lenahan, P.M. and Dressendorfer, P.V., Hole traps and trivalent silicon centers in metal-oxide silicon devices, *J. Appl. Phys.*, 55, 3495, 1984.
26. Stirling, A. et al., Dangling bond defects at Si–SiO_2 interfaces: Atomic structure of the P_{b1} center, *Phys. Rev. Lett.*, 85, 2773, 2000.
27. Poindexter, E.H. et al., Interface states and electron-spin resonance centers in thermally oxidized (111) and (100) silicon wafers, *J. Appl. Phys.*, 52, 879, 1981.
28. Brower, K.L., Kinetics of H_2 passivation of P_b centers at (111) Si–SiO_2 interfaces, *Phys. Rev. B*, 38, 9657, 1988.
29. Brower, K.L. and Myers, S.M., Chemical kinetics of hydrogen and (111) Si–SiO_2 interface defects, *Appl. Phys. Lett.*, 57, 162, 1990.
30. Stesmans, A., Influence of interface relaxation in passivation kinetics in H_2 of P_b defects, *J. Appl. Phys.*, 92, 1317, 2002.
31. Stesmans, A., Passivation of P_{b0} and P_{b1} interface defects in thermal (100) Si–SiO_2 with molecular hydrogen, *Appl. Phys. Lett.*, 68, 2076, 1996.
32. Stathis, J.H. and Cartier, E., Atomic hydrogen reactions with P_b centers at the (100) Si–SiO_2 interface, *Phys. Rev. Lett.*, 72, 2745, 1994.
33. Tsetseris, L. and Pantelides, S.T., Migration, incorporation and passivation reactions of molecular hydrogen at the Si–SiO_2 interface, *Phys. Rev. B*, 70, 245320, 2004.
34. Tsetseris, L. et al., Dual role of fluorine at the Si–SiO_2 interface, *Appl. Phys. Lett.*, 85, 4950, 2004.
35. Tsetseris, L. et al., Hydrogen-related instabilities under bias-temperatures stress, *IEEE Trans. Dev. Mater. Reliab.*, 7, 502, 2007.
36. Khatri, R. et al., Kinetics of hydrogen interaction with SiO_2–Si interface trap centers, *Appl. Phys. Lett.*, 65, 330, 1994.
37. Stathis, J.H., Dissociation kinetics of hydrogen-passivated (100) Si–SiO_2 interface defects, *J. Appl. Phys. Lett.*, 77, 6205, 1995.

38. Tsetseris, L. et al., Hole-enhanced reactions of water at the Si–SiO_2 interface, *Mater. Res. Soc. Proc.*, 786, 171, 2004.
39. Pritchard, R.E. et al., Interactions of hydrogen molecules with bond-centered interstitial oxygen and another defect center in silicon, *Phys. Rev. B*, 56, 13118, 1997.
40. Estreicher, S.K., Hydrogen-related defects in crystalline semiconductors, *Mater. Sci. Eng.*, R14, 319, 1995.
41. Johnson, N.M. and Herring, C., Kinetics of minority-carrier enhanced dissociation of hydrogen-dopant complexes in semiconductors, *Phys. Rev. B*, 46, 11379, 1992.
42. Pearton, S.J. and Lopata, J., Dissociation of P-H, As-H, and Sb-H complexes in silicon, *Appl. Phys. Lett.*, 59, 2841, 1991.
43. Seager, C.H. and Anderson, R.A., Minority carrier-induced release of hydrogen from donors in silicon, *Phys. Rev. B*, 52, 1708, 1995.
44. Coutinho, J. et al., Oxygen and di-oxygen centers in Si and Ge: Density-functional calculations, *Phys. Rev. B*, 62, 10824, 2000.
45. Casady, J.B. and Johnson, R.P., Status of silicon carbide (SiC) as a wide band-gap semiconductor for high-temperature applications: A review, *Solid State Electron.*, 39, 1409, 1996.
46. Herring, C., Johnson, N.M., and van de Walle, C.G., Energy levels of isolated hydrogen levels in silicon, *Phys. Rev. B*, 64, 125209, 2001.
47. van de Walle, C.G. et al., Theory of hydrogen diffusion and reactions in crystalline silicon, *Phys. Rev. B*, 39, 10791, 1989.
48. Denteneer, P.J.H., van de Walle, C.G., and Pantelides, S.T., Microscopic structure of the hydrogen–phosphorus complex in crystalline silicon, *Phys. Rev. B*, 41, 3885, 1990.
49. Tsetseris, L. et al., Physical mechanisms of negative-bias temperature instability, *Appl. Phys. Lett.*, 86, 142103, 2005.
50. Tsetseris, L. et al., Common origin for enhanced low-dose-rate sensitivity and bias temperature instability under negative bias, *IEEE Trans. Nucl. Sci.*, 52, 2265, 2005.
51. Fuller, C.S. et al., Resistivity charges in silicon induced by heat treatment, *Phys. Rev.*, 96, 833, 1954.
52. Gotz, W., Pensl, G., and Zulehner, W., Observation of five additional thermal donor species TD12 to TD16 and of regrowth of thermal donors at initial stages of the new oxygen donor formation in Czochralski-grown silicon, *Phys. Rev. B*, 46, 4312, 1992.
53. Pesola, M. et al., Structure of thermal double donors in silicon, *Phys. Rev. Lett.*, 84, 5343, 2000.
54. Coutinho, J. et al., Thermal double donors and quantum dots, *Phys. Rev. Lett.*, 87, 235501, 2001.
55. Tsetseris, L., Wang, S.W., and Pantelides, S.T., Thermal donor formation processes in silicon and the catalytic role of hydrogen, *Appl. Phys. Lett.*, 88, 051916, 2006.
56. Hallberg, T. and Lindstrom, J.L., Activation energies for the formation of oxygen clusters related to the thermal donors in silicon, *Mater. Sci. Eng. B*, 36, 13, 1996.
57. Capaz, R.B. et al., Mechanism for hydrogen-enhanced oxygen diffusion in silicon, *Phys. Rev. B*, 59, 4898, 1999.
58. Stein, H.J. and Hahn, S., Hydrogen introduction and hydrogen-enhanced thermal donor formation in silicon, *J. Appl. Phys.*, 75, 3477, 1994.
59. Lu, Z.Y. et al., Structure, properties, and dynamics of oxygen vacancies in amorphous SiO_2, *Phys. Rev. Lett.*, 89, 285505, 2002.
60. Davis, K.M. and Tomozawa, M., An infrared spectroscopic study of water-related species in silica glasses, *J. Non-Cryst. Solids*, 201, 177, 1996.
61. Bakos, T., Rashkeev, S.N., and Pantelides, S.T., Reactions and diffusion of water and oxygen molecules in amorphous SiO_2, *Phys. Rev. Lett.*, 91, 226402, 2003.
62. van Ginhoven, R.M. et al., Hydrogen release in SiO_2: Source sites and release mechanisms, *J. Phys. Chem. B*, 109, 10936, 2005.
63. Batyrev, I.G. et al., Effects of water on the aging and radiation response of MOS devices, *IEEE Trans. Nucl. Sci.*, 53, 3629, 2006.
64. Tsetseris, L. et al., Hydrogen effects in MOS devices, *Microelectron. Eng.*, 84, 2348, 2007.
65. Nogami, M. and Abe, Y., Evidence for water-cooperative proton conduction in silica glasses, *Phys. Rev. B*, 55, 12108, 1997.
66. Jeppson, K.O. and Svensson, C.M., Negative bias stress of MOS devices at high electric fields and degradation of MNOS devices, *J. Appl. Phys.*, 48, 2004, 1977.

67. Ogawa, S. and Shiono, N., Generalized diffusion–reaction model for the low-field charge buildup instability at the Si–SiO_2 interface, *Phys. Rev. B*, 51, 4218, 1995.
68. Schroder, D.K. and Babcock, J.A., Negative bias temperature instability: Road to cross in deep submicron silicon semiconductor manufacturing, *J. Appl. Phys.*, 93, 1, 2003.
69. Zhou, X.J. et al., Negative bias-temperature instabilities in metal-oxide-silicon devices with SiO_2 and SiO_xN_y/HfO_2 gate dielectrics, *Appl. Phys. Lett.*, 84, 4394, 2004.
70. Campbell, J.P. et al., Direct observation of the structure of defect centers involved in the negative bias temperature instability, *Appl. Phys. Lett.*, 87, 204106, 2005.
71. Houssa, M. et al., Insights on the physical mechanism behind negative bias temperature instabilities, *Appl. Phys. Lett.*, 90, 043505, 2007.
72. Krishnan, A.T. et al., Negative bias temperature instability mechanism: The role of molecular hydrogen, *Appl. Phys. Lett.*, 88, 153518, 2006.
73. Kufluoglou, H. and Alam, M.A., Theory of interface-trap-induced NBTI degradation for reduced cross section MOSFETs, *IEEE Trans. Electron Dev.*, 53, 1120, 2006.
74. Houssa, M. et al., Reaction-dispersive proton transport model for negative bias temperature instabilities, *Appl. Phys. Lett.*, 86, 093506, 2005.
75. Blat, C.E., Nicollian, E.H., and Poindexter, E.H., Mechanism of negative-bias-temperature instability, *J. Appl. Phys.*, 69, 1712, 1991.
76. Zhang, J.F. and Eccleston, W., Positive bias temperature instability in MOSFETs, *IEEE Trans. Electron Devices*, 45, 116, 1998.
77. Zhou, X.J. et al., Bias-temperature instabilities and radiation effects in MOS devices, *IEEE Trans. Nucl. Sci.*, 52, 2231, 2005.
78. Busani, T., Devine, R.A.B., and Hughes, H.L., Negative bias temperature instability and Fowler–Nordheim injection in silicon oxynitride insulators, *Appl. Phys. Lett.*, 90, 163512, 2007.
79. Enlow, E.W. et al., Response of advanced bipolar processes to ionizing radiation, *IEEE Trans. Nucl. Sci.*, 38, 1342, 1991.
80. Pease, R.L. et al., A proposed hardness assurance test methodology for bipolar linear circuits and devices in a space ionizing radiation environment, *IEEE Trans. Nucl. Sci.*, 44, 1981, 1997.
81. Hjalmarson, H.P. et al., Mechanisms for radiation dose-rate sensitivity of bipolar transistors, *IEEE Trans. Nucl. Sci.*, 50, 1901, 2003.
82. Rashkeev, S.N. et al., Physical model for enhanced interface-trap formation at low dose rates, *IEEE Trans. Nucl. Sci.*, 49, 2650, 2002.
83. Hess, K., Kizilyalli, I.Z., and Lyding, J.W., Giant isotope effect in hot electron degradation of metal oxide silicon devices, *IEEE Trans. Electron Dev.*, 45, 406, 1998.
84. Cheng, K., Hess, K., and Lyding, J.W., A new technique to quantify deuterium passivation of interface traps in MOS devices, *IEEE Electron Dev. Lett.*, 22, 203, 2001.
85. Tsetseris, L. and Pantelides, S.T., Hydrogenation-deuteration of the Si–SiO_2 interface: Atomic-scale mechanisms and limitations, *Appl. Phys. Lett.*, 86, 112107, 2005.
86. Blöchl, P.E. and Stathis, J.H., Hydrogen electrochemistry and stress-induced leakage current in silica, *Phys. Rev. Lett.*, 83, 372, 1999.
87. Rashkeev, S.N. et al., Radiation-induced acceptor deactivation in bipolar devices: Effects of electric field, *Appl. Phys. Lett.*, 83, 4646, 2003.
88. Lang, D.V. et al., Bias-dependent generation and quenching of defects in pentacene, *Phys. Rev. Lett.*, 93, 076601, 2004.
89. Ruiz, R. et al., Structure of pentacene thin films, *Chem. Mater.*, 16, 4497, 2004.
90. Jurchescu, O.D., Baas, O.D., and Palstra, T.T., Electronic transport properties of pentacene single crystals upon exposure to air, *Appl. Phys. Lett.*, 87, 052102, 2005.
91. Zhu, Z.T. et al., Humidity sensors based on pentacene thin-films transistors, *Appl. Phys. Lett.*, 2002, 81, 4643.
92. Tsetseris, L. and Pantelides, S.T., Intercalation of oxygen and water molecules in pentacene crystals, *Phys. Rev. B*, 75, 153202, 2007.

14

Toward Engineering Modeling of Negative Bias Temperature Instability

Tibor Grasser, Wolfgang Goes, and Ben Kaczer

CONTENTS

14.1 Introduction

Negative bias temperature instability (NBTI) has been known for 40 years [1] and is attracting an ever growing industrial and scientific attention as one of the most important reliability issues in modern complementary metal-oxide semiconductor (CMOS) technology. It affects mostly p-metal-oxide-semiconductor field-effect transistors (pMOSFETs) at elevated temperatures with a large negative voltage applied to the gate. While the typical NBT setup requires the other terminals to be grounded, an application of a larger voltage at the drain creates interesting mixed patterns with hot-carrier degradation (HCI) and a large voltage at the bulk contact can be used to investigate the dependence of NBTI on hot or cold holes. Altogether, as a result of this stress condition, a shift in the threshold voltage is observed [2,3]. In addition to this threshold voltage shift, other crucial transistor parameters degrade as well, such as the drain current, the transconductance, the subthreshold slope, the gate capacitance, and the mobility [2,3].

The evolution of the threshold voltage during stress is commonly described by a power law of the form:

$$\Delta V_{\mathrm{th}}(t) = A(T,E_{\mathrm{ox}})\, t^{n}, \tag{14.1}$$

with the prefactor A strongly depending on the temperature and the electric field. The actual dependencies of the power-law exponent n are still not fully clarified with some groups [4,5] reporting a temperature- and technology-independent value around $n \approx 0.15$, while recent publications show considerably smaller values [6,7]. Alternatively, some groups have reported a log-like dependency [6,8,9], for instance of the form:

$$\Delta V_{\mathrm{th}}(t) = A(T,E_{\mathrm{ox}}) \log(1 + t/\tau), \tag{14.2}$$

at least at early times. A typical scenario is depicted in Figure 14.1 where the same data are shown once on a lin–log and on a log–log plot. Depending on the accuracy of the initial threshold voltage determination or, in that example, the initial drain current in the linear regime, different interpretations seem possible [6].

The detailed microscopic physics behind NBTI are not yet fully understood [10–14] but the creation of interface states seems to be a universally acknowledged feature of NBTI [2,15]. A growing number of recent publications, however, attribute at least a part of the degradation to positive charge generation in the oxide bulk [11,13,16,17]. Possible positive charges that have been suggested include holes trapped in either preexisting traps [11,16] or in traps generated by the hydrogen species released during the creation of the interface states [13].

Other potential contributions to a threshold voltage shift like mobile charges are commonly assumed to be negligible for NBTI [2] and the total threshold voltage shift is thus given by

$$\Delta V_{\mathrm{th}}(t) = -\frac{\Delta Q_{\mathrm{it}}(t) + \Delta Q_{\mathrm{ox}}(t)}{C_{\mathrm{ox}}}, \tag{14.3}$$

with ΔQ_{it} and ΔQ_{ox} being the effective charges due to interface and oxide states and C_{ox} the gate capacitance per area.

The fundamental problem in the context of NBTI is given by the fact that the degradation created during the stress phase begins to recover immediately once the stress is removed. This makes the classic measurement technique where the stress is interrupted during the

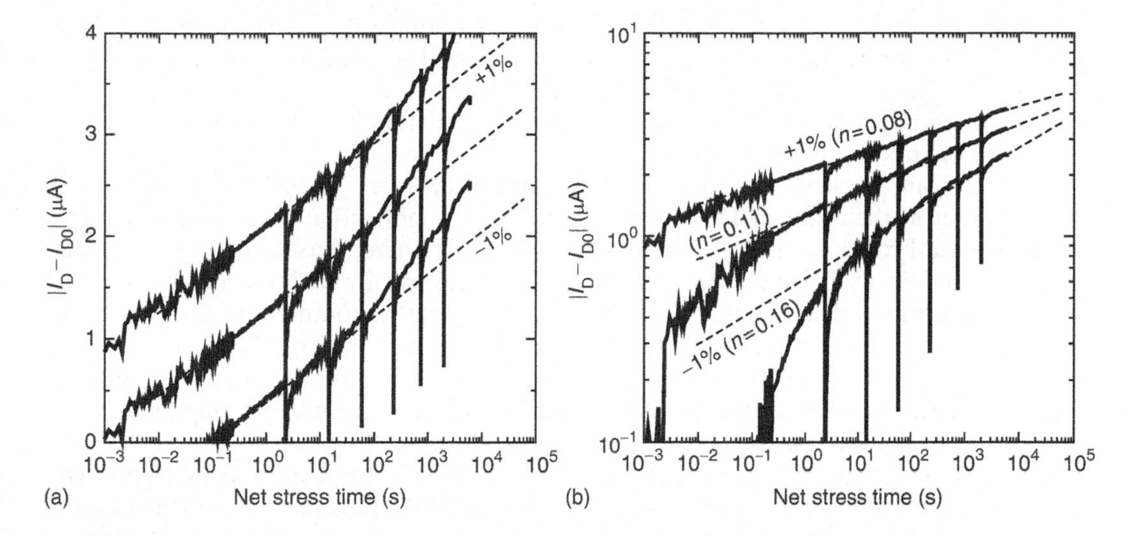

FIGURE 14.1
(a) Degradation of the drain current in the linear regime during stress ($V_G = V_G^{stress}$). The measurement is periodically interrupted to record the drain current around $V_G = V_{th}$. The initial degradation appears to be linear on a lin–log scale and any uncertainty in the initial drain current results in a shift of the whole curve. (b): Same data on a log–log plot. When the drain current measured at $t_s = 1$ ms is directly used for I_{D0}, a power-law exponent of $n = 0.11$ is obtained. An uncertainty in I_{D0} of ±1% changes the slope to 0.08 and 0.16, thereby making the interpretation of the data extremely difficult.

extraction of the threshold voltage problematic [9,18]. In particular, the value of the extracted power-law exponent depends significantly on the delay introduced during the measurement [5,14,19]. Experimental results obtained with delayed measurements show a linear increase of the exponent with temperature [5,8,14] with values around 0.2–0.3. In contrast, temperature-independent exponents in the range 0.07–0.2 have been extracted from recent delay-free measurements [4,6,20].

Of particular interest is the question related to the origin of this extremely fast relaxation [9,14,21]. While some authors assume that hole trapping is negligible and both degradation as well as relaxation are determined by the temporal change of the interface state density and an associated back- and forth-diffusion of hydrogen [5], others acknowledge at least partial importance of trapped charges [6,8,13,22]. In the latter case it has been assumed that trapped charges either form the fast component of NBTI relaxation superimposed onto some interface defect relaxation [6,22] or are solely responsible for any recovery while created interface defects do not recover at all [8,13].

14.2 Interface States

The most commonly and earliest reported effect related to NBTI is the creation of defects at the fundamentally important Si/SiO_2 interface. These interface states are often assumed to be P_b centers [23–25] which are known to have electrically active levels within the silicon band gap. In particular, for industrially relevant samples with (100) surfaces, two variants of P_b centers have been identified [24], the P_{b0} and the P_{b1} center. Both defects are silicon dangling bonds, with the Si atom backbonded to three other Si atoms [26]. While some researchers argue that only the P_{b0} is electrically active [27], others have observed additional electrically active peaks which were claimed to originate from P_{b1} centers [26].

A recent study suggests that in nitrided oxides the role of the P_b center is taken over by K centers, which are silicon dangling bonds backbonded to three nitrogen atoms [28]. K centers are located inside the nitrided oxide, rather than at the interface as P_b centers. As such, a model relating NBTI to K centers could be different from the available theories, a question open to future research.

P_b centers are present in a considerable number at every Si/SiO_2 interface with a concentration in the order of 10^{12} cm^{-2}. During device fabrication these defects have to be passivated through some sort of hydrogen anneal [2], thereby reducing the electrically active trap levels to a value below 10^{10} cm^{-2}. The electrically active trap levels are of amphoteric nature, meaning that each interface state can accommodate two electrons. Possible transitions are from the positive to the neutral state (+/0), which appears as a donor-like trap level in the lower half of the silicon band gap, and the neutral to negative charge state (0/−) which is commonly assumed to act as an acceptor-like trap level in the upper half of the band gap.

Although the P_bH bonds obtained after the passivation step are relatively stable, they can be broken at elevated temperatures and higher electric fields, thus reactivating the electrically active trap levels. In our analysis, we will denote the time-dependent density of interface states as $N_{it}(t) = [P_b^\bullet]$. Depending on the trap occupancy, the initial value of $N_{it0} = N_{it}(t_0)$ is inherently visible in the reference threshold voltage $V_{th}(t_0)$ and the change in the density of interface states is given through $\Delta N_{it}(t) = N_{it}(t) - N_{it0}$. It is normally assumed that charging and discharging of these interface states is very fast, and consequently that the positive charge in these interface states immediately follows the Fermi-level via

$$\Delta Q_{it}(t) = q \int \Delta D_{it}(E_t,t) f_{it}(E_F,E_t,t) \mathrm{d}E_t. \tag{14.4}$$

Here, ΔD_{it} is the time-dependent density of interface states in the units of cm^{-2} eV^{-1}, which is by a still to be quantified relation [29] directly linked to $\Delta N_{it}(t)$, and $f_{it}(E_t)$ their occupancy with holes. In addition to the exponential band-tail states of a passivated Si/SiO_2 interface, the P_b centers create Gaussian peaks in the Si band gap where the broadening is probably linked to the disorder at the interface [30]. As an example, the measured concentration of P_{b0} centers as obtained by Ragnarsson and Lundgren [31] is shown in Figure 14.2 for an initially unpassivated interface and after a short hydrogen passivation step. This may correspond to the inverse process occurring during NBT stress, that is, the relaxation part which we have argued to be of fundamental importance for the understanding of NBTI [32]. The measurement data can be nicely fitted by two Gaussian peaks or by using a Fermi-derivative function [33] (which can be analytically integrated):

$$g_P(E_t,E_P,\sigma) = \frac{1}{\sigma} \frac{\exp\left(\frac{E_P - E_t}{\sigma}\right)}{\left(1 + \exp\left(\frac{E_P - E_t}{\sigma}\right)\right)^2} \tag{14.5}$$

as

$$\Delta D_{it}(E_t) = N_{it}\left(g_P(E_t,E_{P1},\sigma_1) + g_P(E_t,E_{P2},\sigma_2)\right). \tag{14.6}$$

Note that in order to fit the data of Ragnarsson and Lundgren, the variances of the two peaks have to evolve differently in time, with the acceptor-like peaks becoming narrower sooner (Figure 14.2). In contrast, other groups have reported a similar time evolution of both peaks [30]. This disorder-induced broadening of the electrical active levels is

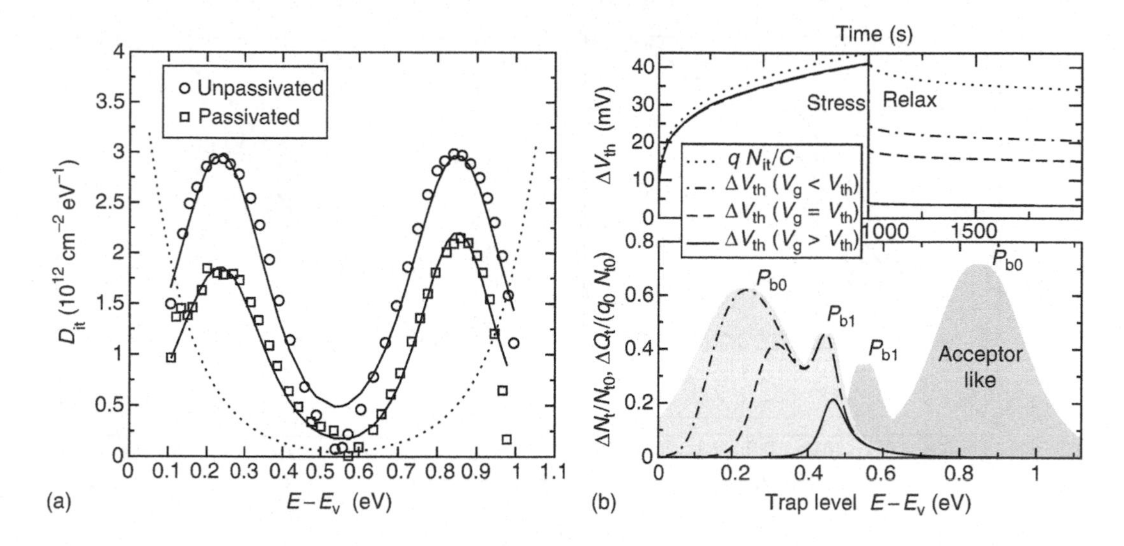

FIGURE 14.2
(a) Measured density of interface states supposedly related to P_{b0} centers before and after a short hydrogen passivation process [31]. The symbols are the measurement data, the solid lines give the analytic fit, while the exponential band-tail states are schematically represented by the dotted lines. The donor-like peak is located ≈0.24 eV above the valence band edge, while the acceptor-like peak is at ≈0.85 eV. Note that the variance of the unpassivated sample is $\sigma_1 = \sigma_2 = 0.085$ eV^2 while after the passivation step one obtains $\sigma_1 = 0.074$ eV^2 and $\sigma_2 = 0.062$ eV^2, a fact to be included into a model. (b) Influence of the interface state occupancy on the observed threshold voltage shift using on-the-fly measurements. During stress, nearly all interface traps are positively charged. When a different gate voltage is used during relaxation, only a fraction of the traps are visible which must be separated from the real relaxation. Schematically shown is the density-of-states typically associated with P_{b0} and P_{b1} centers [26]. (From Grasser, T., et al., *Proc. IRPS*, 268, 2007. With permission.)

suspected to be closely related to a disorder-induced Gaussian variation of the binding energies of the Si–H bonds at the interface [30,34,35].

Nevertheless, during NBT stress, the Fermi-level E_F is close to the valence band edge and $f_{it}(E_t) \approx 1$ throughout the silicon band gap. Thus, under the assumption that P_b centers introduce states only within the silicon band gap, see Ref. [13] for a different interpretation, all newly generated interface states ΔN_{it} are positively charged and one obtains $\Delta Q_{it}(t) \approx q\Delta N_{it}(t)$, independently of the exact form of the density-of-states. This is the usual assumption employed for instance in the widely used reaction–diffusion (RD) model and quite reasonable during the stress phase. However, in order to measure the degradation, the stress is often interrupted and the various forms of degradation are assessed using different possible techniques. Regardless of the actual measurement technique employed, be it a complete or partial $I_D V_G$ sweep, single point V_{th} determination [9,14], ultrafast pulse $I_D V_G$ [16], capacitance–voltage (CV), DCIV [36], or charge-pumping (CP) [37] measurements, the trap occupancy changes significantly because a different fraction of the traps is charged during stress and measurement. Furthermore, this Fermi-level dependence causes a change in the subthreshold slope during $I_D V_G$ measurements and humps in the CV characteristics, in contrast to constant shifts induced by fixed positive charges, see Figure 14.3 for a qualitative description.

Alternatively, in the model of Zafar [13], a different interpretation is introduced. Zafar assumed that a large number of dangling bonds always exists but that only a fraction can be observed in electrical measurements, while the majority is too close to the band-edges to contribute. During NBT stress the total number of interface states is increased and only this increase is visible during measurements. To properly account for this partial contribution

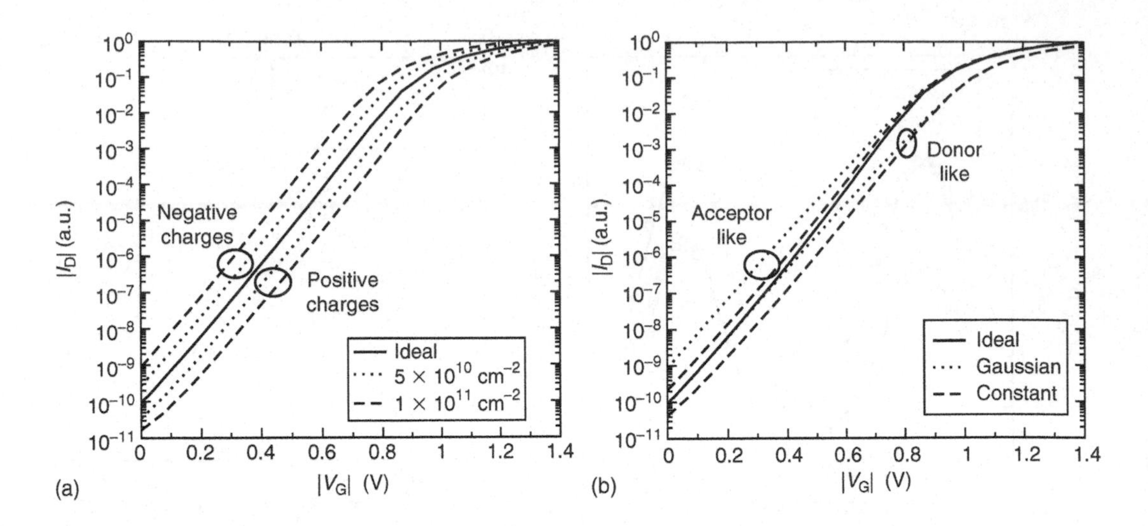

FIGURE 14.3
(a) Simulated influence of fixed charges on the $I_D V_G$ characteristics of a PMOS transistor. (b) Simulated influence of interface states on the $I_D V_G$ characteristics of a PMOS transistor. Depending on the form of the trap density-of-states, a different shift in ΔV_{th} and a different change in the subthreshold slope are obtained.

of the generated interface states to the observable threshold voltage shift, the occupancy of the interface states as a function of the Fermi-level position has to be introduced.

14.3 Oxide Charges

On the top of generated interface defects, charge may be stored in existing or newly created oxide traps. Although some of these traps may still be considered fast, they are more difficult to charge and discharge, that is, have larger time constants than interface states due to their location inside the oxide bulk. It has also been suggested that holes trapped in energetically deep levels give rise to practically permanent charge contributions which can only be neutralized through the application of unusual bias conditions [38]. Altogether, the occupancy of oxide traps cannot follow the Fermi-level immediately and $\Delta Q_{ox}(t)$ will be governed by different dynamics. The contribution of the oxide charges to the threshold voltage shift is formally written as

$$\Delta Q_{ox}(t) = q \int\int \Delta D_{ox}(x,E_t,t) f_{ox}(x,E_t,t)(1 - x/t_{ox})\mathrm{d}x\,\mathrm{d}E_t, \tag{14.7}$$

where
ΔD_{ox} is the spatially and energy-dependent density-of-states in the oxide
f_{ox} is the hole occupancy of these traps
t_{ox} is the oxide thickness

Note that the issue of whether oxide charges are important during NBTI or not is currently one of the most debated ones [5,11,13,16]. Also, the question whether ΔD_{ox} consists mainly of preexisting traps [11,16] or traps that are created during stress [13] remains to be answered.

14.4 Measurement Issues

The understanding and characterization of NBTI is considerably hampered by the difficulties arising during measurement. Currently, two techniques are often used to characterize NBTI: the classic measurement/stress/measurement (MSM) technique, which is handicapped by undesired relaxation, and on-the-fly (OTF) measurements, which avoid any relaxation by maintaining a high stress level throughout the measurement and directly monitor the drain current in the linear regime, ΔI_{Dlin}. Since the threshold voltage shift ΔV_{th} is more suitable to study the creation of charges, ΔI_{Dlin} has to be converted to ΔV_{th} which is commonly done using the simple SPICE Level-1 model [8] or an empirical formalism [39]. The applicability of the OTF technique is particularly troublesome when one switches from stress to relaxation. When V_{G} is left at $V_{\mathrm{G}}^{\mathrm{relax}}$, the interface trap occupancy is considerably lower than during the stress phase [29], resulting in spurious additional relaxation (Figure 14.2). Conversely [4], when V_{G} is brought back to $V_{\mathrm{G}}^{\mathrm{stress}}$, one faces the opposite problem one is trying to avoid during the stress phase, since now additional uncontrolled stress is introduced during the measurement cycles. Even more important is the fact that the initial value of I_{Dlin} is extremely difficult to determine as it is already obtained at the stress voltage. Conventionally, the time required for this is in the milliseconds range where already significant degradation can be observed [6] but any uncertainty in I_{D0} modifies the time exponent (the slope) of ΔV_{th} on a log–log plot in a somewhat arbitrary manner, see also Figure 14.1. This may render many results obtained by the OTF technique questionable.

In contrast, the MSM technique probes the interface under comparable conditions during both the stress and relaxation phase. In addition, the voltage applied to the gate can be kept close to the threshold voltage where only negligible degradation can be expected. However, as has been pointed out [7,32,40], it is probably very difficult to minimize the measurement delay in such a way that the true degradation is observed.

14.5 Characterization of Relaxation

In order to properly understand and characterize NBTI it is mandatory to take a close look at the relaxation behavior. Particularly noteworthy are the long tails of logarithmic-like nature that may cover more than 12 decades in time [9,14,42,43]. In order to formalize the description, we use the term $S(t_{\mathrm{s}}) = \Delta V_{\mathrm{th}}(t_{\mathrm{s}})$ for the real degradation accumulated during the stress phase. As soon as the stress voltage is removed, relaxation sets in as a function of the accumulated stress time t_{s} and the relaxation time $t_{\mathrm{r}} = t - t_{\mathrm{s}}$. In the following, we will assume that the accumulated degradation $S(t_{\mathrm{s}})$ consists of a recoverable component $R(t_{\mathrm{s}})$ and a permanent component [8,38] $P(t_{\mathrm{s}})$ as

$$S(t_{\mathrm{s}}) = R(t_{\mathrm{s}}) + P(t_{\mathrm{s}}). \tag{14.8}$$

As the recoverable component depends on the recovery time t_{r}, any measurement conducted with a certain delay $t_{\mathrm{r}} = t_{\mathrm{M}}$ observes only

$$S_{\mathrm{M}}(t_{\mathrm{s}},t_{\mathrm{M}}) = R_{\mathrm{M}}(t_{\mathrm{s}},t_{\mathrm{M}}) + P(t_{\mathrm{s}}) \leq S(t_{\mathrm{s}}), \tag{14.9}$$

with the subscript M indicating quantities observed in a measurement. Of course, $S_{\mathrm{M}}(t_{\mathrm{s}},0) = S(t_{\mathrm{s}})$ and $R_{\mathrm{M}}(t_{\mathrm{s}},0) = R(t_{\mathrm{s}})$ hold.

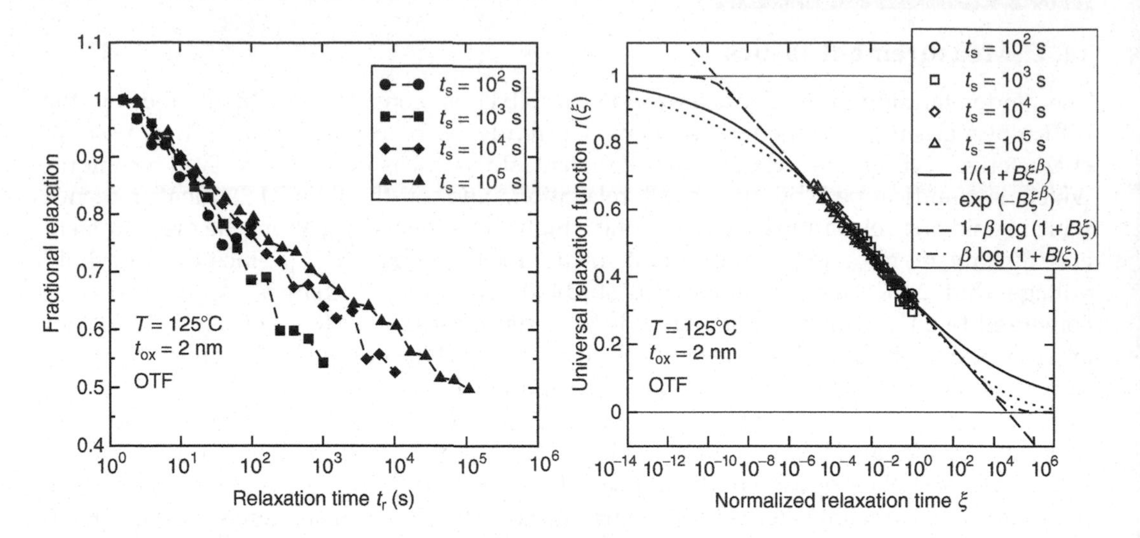

FIGURE 14.4
Demonstration of universal recovery for the OTF data of Denais et al [41]. The left figure shows a conventional view of the fractional recovery as a function of the relaxation time t_r. Apparently, data obtained after longer stress times seem to relax more slowly than data obtained at shorter times. The right figure, on the other hand demonstrates the universality of relaxation when the relaxation data are normalized to the last stress value and plotted over the ratio $\xi = t_r/t_s$ [41]. Also shown are some possible empirical expressions which can be fit to the data. (From Grasser, T., et al., *Proc. IRPS*, 268, 2007. With permission.)

Due to the onset of recovery which may occur at timescales possibly shorter than nano- or even picoseconds [7,9], a rigorous characterization of the relaxation phase is extremely challenging [4,21,40,41]. Typically, the relaxation data $R(t_s,t_r)$ recorded at different stress times t_s have been normalized to the first measurement point t_M as

$$r_f(t_s,t_r) = \frac{S_M(t_s,t_r)}{S_M(t_s,t_M)}, \tag{14.10}$$

giving the fractional measurable recovery, and aligned as a function of the absolute relaxation time t_r [9,14,21,40], see Figure 14.4. The functional form of the relaxation remains elusive in such a plot.

Instead, it has been demonstrated that it is highly advantageous to study the recoverable component in its universal representation [32] which is based on the observation that all individual relaxation curves obtained at different stress times $t_{s,i}$ can be represented by a single universal curve when [41]

- Relaxation data are normalized to the last stress value $S(t_{s,i}) = S_M(t_{s,i},0)$ rather than the first measurement point $S_M(t_{s,i},t_M)$
- Relaxation time t_r is normalized to the last stress time $t_{s,i}$ as $\xi = t_r/t_{s,i}$

The above results in the definition of the universal relaxation function as [32]:

$$r(\xi) = \frac{R_M(t_s,t_r)}{R(t_s)} = \frac{S_M(t_s,t_r) - P(t_s)}{S(t_s) - P(t_s)}, \tag{14.11}$$

which is a function of ξ only. For the special case of a negligible permanent component, note the relationship between the universal recovery function and the fractional recovery given by $r_f(t_s,t_r) = r(\xi)/r(\xi_M)$ with $\xi_M = t_M/t_s$. This concept is visualized in Figure 14.5.

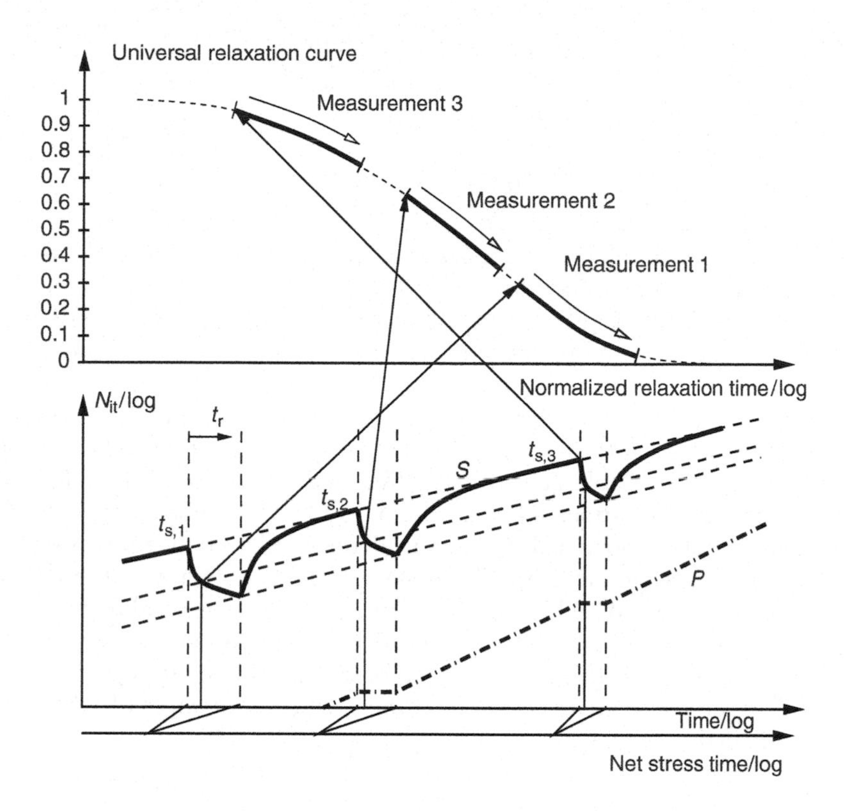

FIGURE 14.5
Schematic view of universal relaxation. The stress is interrupted three times to record relaxation data on the relative time scale $t_r = t - t_s$. For stress intervals considerably larger than the relaxation intervals, the device forgets the interruption. Note how for larger stress times the relaxation data move to smaller normalized relaxation times $\xi = t_r/t_s$ and how the relative recovery becomes less significant. Also indicated is a possible permanent/slowly relaxing component P.

Because the bulk of relaxation data available in the literature do not allow a definite identification of the permanent component, which requires very detailed data [7], we will consider the permanent component to be negligible in the following and assume $P(t_s) = 0$. Recent studies have shown how more detailed data allow for a clear identification of a permanent component and the extension of the method presented here [7,43].

14.5.1 Functional Form of the Universal Relaxation Function

Lacking a universally accepted and valid theory for NBTI, the exact form of the universal relaxation function $r(\xi)$ remains illusive at this point and empirical functions have to be used. So far, excellent results have been obtained with the power-law-like expression:

$$r(\xi) = (1 + B\xi^{\beta})^{-1}, \tag{14.12}$$

where the parameters B and β are in the range $B \approx 0.3$–3 and $\beta \approx 0.15$–0.2 for most of the data available. Of particular interest is the relaxation predicted by the RD model which is well described by Equation 14.12 using $B = 1$ and $\beta = 1/2$. However, we have to point out that the available data are not conclusive at the time being, making alternative expressions such as the logarithmic dependence suggested by Denais et al. [41]:

$$r(\xi) = 1 - \beta \log(1 + B\xi), \tag{14.13}$$

or the frequently used stretched-exponential [44]:

$$r(\xi) = \exp(-B\xi^{\beta}). \tag{14.14}$$

viable alternatives as well [32]. We remark that Equation 14.13 is nonphysical at larger times, making a reformulation mandatory.

One difficulty in determining the correct choice of the empirical function is the fact that relaxation may occur over more than 12 decades in time [9]. As the delay times in conventional measurements are around 1 ms and relaxation data are not normally recorded for $t_r > 10^5$ s, only eight decades in time is commonly available. By employing fast measurements which start at $t_r = 1$ μs 12 decades has been reported [42]. Interestingly, the measurement data available behave logarithm-like over most of the recorded regime and excellent fits with Equations 14.12 through 14.14 can be obtained [32]. Only more detailed relaxation data and a solid theoretical description will allow to differentiate between possible expressions which differ mostly in the behavior at extremely short and long times. This is illustrated in Figure 14.4 where possible empirical expressions for the universal relaxation function are compared. All expressions can be fit to the measurement data and give fits of practically the same accuracy. However, they result in different extrapolations for large and small relaxation times, the consequences of which need to be carefully investigated.

14.6 Characterization of MSM Data

Although more delicate to apply, universal relaxation is of particular interest for data obtained by the MSM technique. For the normalization needed in Equation 14.11 one has to keep in mind that the value of $S(t_s) = R_M(t_s, 0)$ is essentially unknown, one only knows $R_M(t_s, t_M)$ determined at the first measurement point available after a short but probably nonnegligible relaxation period t_M. However, making use of the universal relaxation expression 14.11 and assuming for the time being that $r(\xi)$ is known, $S(t_s) = R_M(t_s, 0)$ can be obtained as

$$S(t_s) = R(t_s) = \frac{R_M(t_s, t_M)}{r(t_M/t_s)}. \tag{14.15}$$

Inserting the above into the universal relaxation relation 14.11 we obtain

$$\frac{r(\xi)}{r(\xi_M)} = \frac{R_M(t_s, t_r)}{R_M(t_s, t_M)}. \tag{14.16}$$

From Equation 14.16 the as of yet unknown parameters B and β can be easily determined from a measured sequence of relaxation data $R(t_{s,i}, t_r)$ obtained after N stress intervals $t_{s,i}$, by minimizing for instance

$$\varepsilon_t = \sum_{i=1}^{N} \int \left(\frac{r(t_r/t_{s,i})}{r(t_M/t_{s,i})} - \frac{S_M(t_{s,i}, t_r)}{S_M(t_{s,i}, t_M)} \right)^2 \mathrm{d}\log(\xi_i). \tag{14.17}$$

Note that the parameter extraction is independent of the functional form of R and that the final form of R is directly related to the measurement data through the universal relaxation relation as $R(t_s) = R_M(t_s, t_M)/r(t_M/t_s)$ [32].

Naturally, in contrast to data obtained by OTF measurements where $R_M(t_s, 0)$ is known, the analytical expression determines the final value of $R_M(t_s, 0)$ through the extrapolation given by Equation 14.16. This results in a floating behavior of $r(\xi_M)$ which reflects the uncertainty of this approach [32].

A particularly intriguing feature of Equation 14.16 is that it can be applied to a whole sequence of stress and relaxation sequences as typically encountered during MSM measurements. This is because during MSM sequences the duration of the stress intervals usually grows exponentially while the measurement interval t_M is short and of constant duration. This implies that after a certain stress time, which we determined empirically to be of the order $t_s \gtrsim 10 \times t_M$, the relaxation during the measurement does not significantly alter the degradation at the end of each stress phase, meaning that the degradation relaxed during each measurement interval is mostly restored during the next stress phase. This is in agreement with the reports of Reisinger et al. [9] who report that "the sample completely forgets the effect of the interruption" provided the stress phase following the interruption is by a factor of 100 longer than the interruption.

Consequently, Equation 14.15 holds for every stress point t_s, where t_s is now the accumulated net stress time. The applicability of the procedure outlined above to the detailed relaxation data published by Reisinger et al. [9] and for the IMEC data otherwise published in Ref. [14] is outlined in Figure 14.6. For the IMEC data the universality is also shown at three different temperatures, 50°C, 125°C, and 200°C.

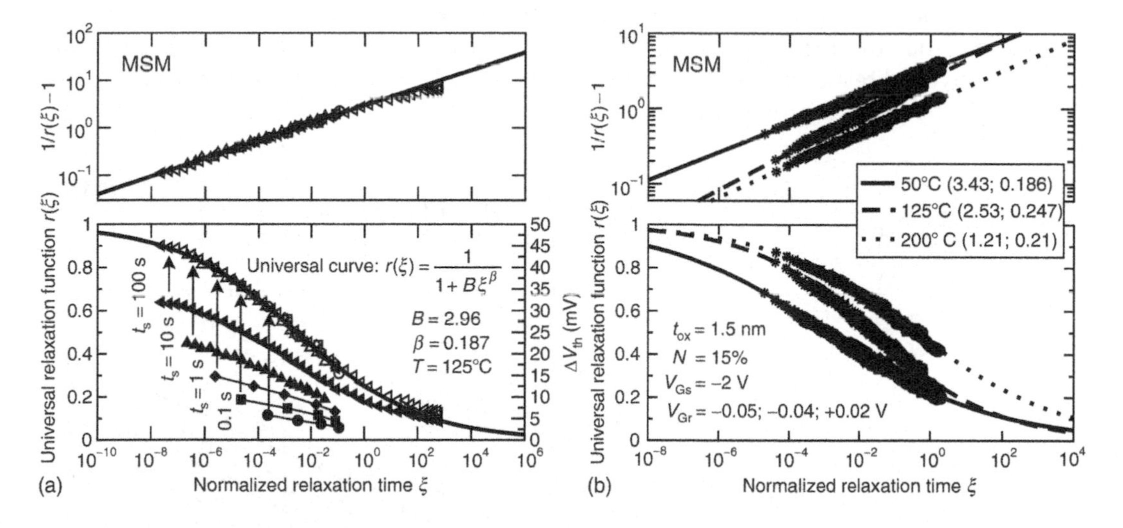

FIGURE 14.6
(a) Application of universal relaxation to the fast MSM data obtained by Reisinger et al. [9]. Depending on the choice of the universal relaxation function, the individual data points can be mapped onto the respective universal curve, in this case Equation 14.12. Note the linear behavior of $1/r - 1$ shown in the upper plot. The slight deviation for $\xi > 10^2$ is introduced by a permanent component $P(t_s)$, see Ref. [43]. (b) Same as (a) but with data from IMEC [14]. Relaxation data of three devices stressed in a single MSM sequence were recorded at 10 different stress times in the interval 10–10^4 s at three different temperatures. The values of B and β (given in parenthesis) depend on the temperature, β even in a nonmonotonic manner which may indicate the existence of two different processes with different temperature dependencies. (From Reisinger, H. et al., *Proc. IRPS*, 2006.)

Universal relaxation thus results in the interesting possibility to reconstruct the true (undelayed) measurement curve from delayed data sets. This suggests a novel measurement technique:

(1A) Determine $R_M(t_s,t_M)$ using a single delay time and add a long relaxation period at the end. In case a permanent component is present, multiple devices can be subjected to different stress intervals for an accurate determination of the time-dependence of P [7].

(1B) Alternatively, one may determine $R_M(t_s,t_M)$ using different delay times. This approach probably only works for stress cases were negligible permanent degradation is created [7].

(2) From that data determine B and β.

(3) Finally, calculate the true degradation using Equation 14.15.

Variant A, where B and β have been obtained from detailed relaxation data, has already been demonstrated in Figure 14.6. However, the method also works for MSM data obtained with different delay times where no relaxation data are available (Variant B). In that case the parameters A, n, B, and β can be directly extracted through fitting of Equation 14.18. This is demonstrated in Figure 14.7 for the data published by Li et al. [45]. Again, the extracted parameter values agree very well with the cases where we had access to the full relaxation data.

14.6.1 Influence of Measurement-Delay on the Power-Law Parameters

Next, we show that the universal relaxation expression naturally connects individual stress curves obtained using the MSM technique with different delay times. For simplicity, we assume that the true degradation behavior follows a power law as $S(t_s) = At_s^n$ and that the universal relaxation is given by Equation 14.12. Due to the measurement delay one observes instead of the power law

$$S_M(t_s,t_M) = S(t_s)r(t_s,t_M) = \frac{At_s^n}{1 + B(t_M/t_s)^\beta}. \quad (14.18)$$

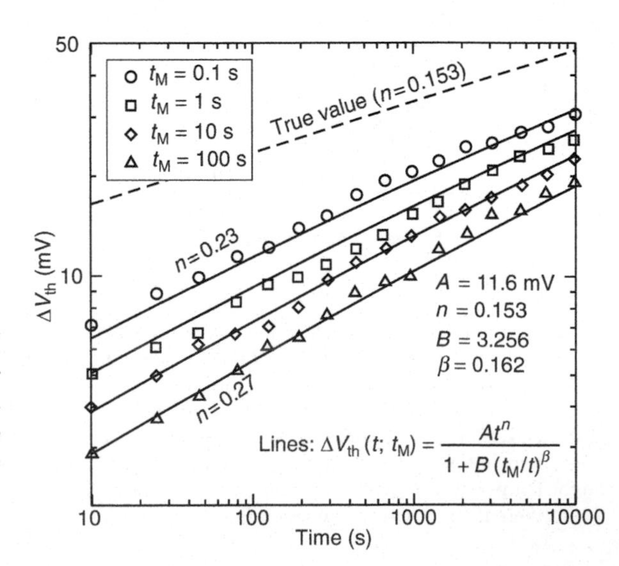

FIGURE 14.7
Reconstruction of the true degradation from MSM data obtained by Li et al. [45] with four different delay times without the knowledge of the detailed relaxation behavior. Again, a corrected slope of about $n \approx 0.15$ is obtained. Note that even at $t_s = 10^4$ s the lines do not merge and the impact of the delay is still clearly visible. (From Grasser, T., et al., *Proc. IRPS*, 268, 2007. With permission.)

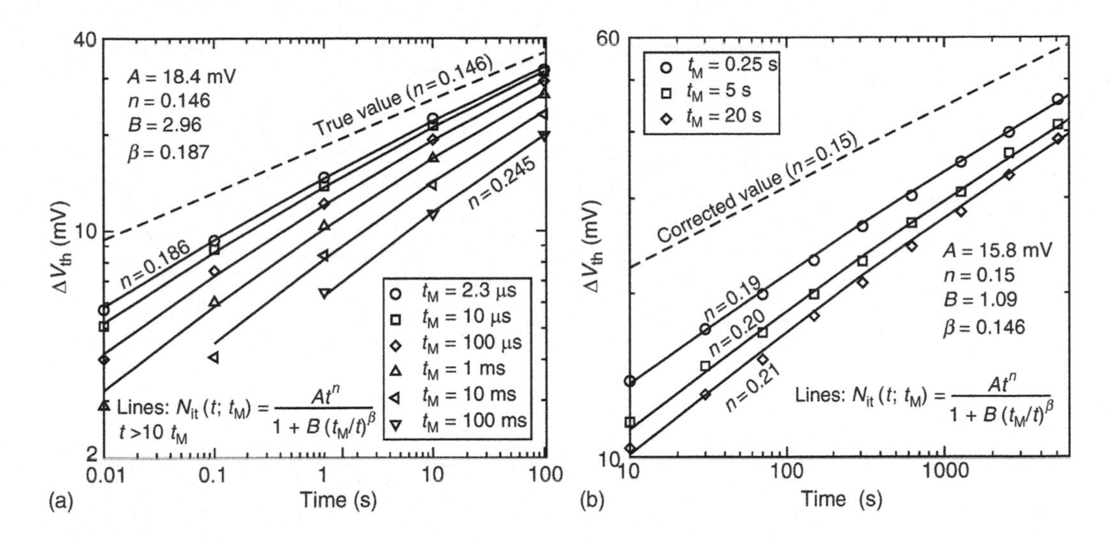

FIGURE 14.8
Comparison of the analytic model for MSM measurements based on the universal relaxation to the data of Reisinger et al. [9] (a) and Kaczer et al. [14] (b). Excellent accuracy of the analytic model is obtained for all available delay times. In addition, the true NBTI degradation can be recovered by extrapolating to $t_M = 0$ s. (From Grasser, T., et al., *Proc. IRPS*, 268, 2007. With permission.)

Equation 14.18 is validated against the Infineon and IMEC data in Figure 14.8 where the parameters B and β are given by the universal relaxation law. The analytic expression 14.18 exactly reproduces the delayed measurement results for various delay times t_M and thereby convincingly confirms our assumptions stated above.

As a consequence of the measurement delay, the observed power-law exponent n_M will be time-dependent and given through Equation 14.18 as

$$n_M(t_s,t_M) = n - r'(t_M/t_s)\frac{t_M/t_s}{r} = n + \frac{\beta B}{B + (t_s/t_M)^{\beta}}, \tag{14.19}$$

with $r'(\xi) = \partial r(\xi)/\partial \xi$.

It is worthwhile to point out that although many groups report a constant measured power-law exponent over three to four decades which varies as a function of the temperature and delay time, this can of course only be approximately correct. The fact that all curves obtained with different delay times have to merge at larger times, makes a time-dependent slope a necessity. However, depending on the actual values of B and β this time-dependence will be more-or-less visible in a log–log plot. In general, the smaller β, the less visible the time-dependence will be. A comparison of measured power-law exponents as a function of the delay time t_M and temperature is given in Figure 14.9. Most of the data show an apparently constant power-law exponent (within the measurement accuracy) over three to four decades. Clearly, the measured power-law exponents, and consequently B and/or β (see Table 14.1), depend on temperature, on the particular technology, and/or the measurement technique.

14.7 Modeling of NBTI

As has been detailed in the previous sections, the fundamental dilemma encountered during the development of NBTI models is the question of what exactly should be modeled. While conventional models for semiconductor device simulation can rely on a

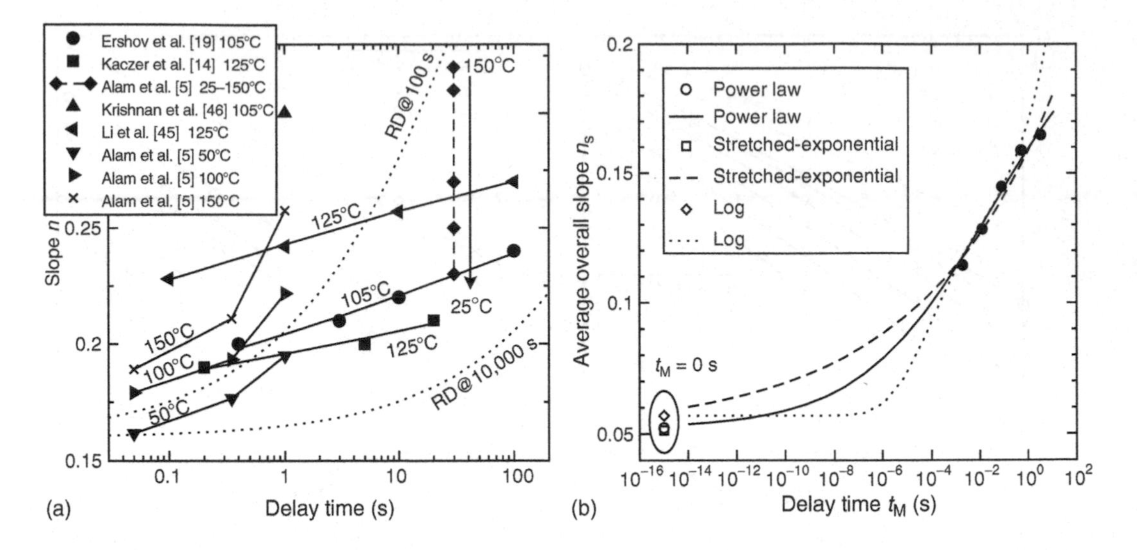

FIGURE 14.9
(a) Influence of the measurement delay on the measured slope as reported by various groups [5,10,14,19,45,46]. The solid lines are given by a fit to Equation 14.19 using the parameters in Table 14.1. Note that the strong temperature-dependence of the reported slopes and that the slopes were found to be constant over three to four decades in many measurements. Clearly, there is a large spread in the measurement data indicating a technology dependence. The dotted lines show the slopes predicted by the RD model at $t_s = 100$ s and $t_s = 10{,}000$ s. Note that the RD slope changes considerably within two decades, is per construction temperature independent, and cannot be adjusted to the technology. (b) Observed slope in a delayed measurement as a function of the measurement delay [7]. The symbols are the measurement data while the lines give the extrapolation to the true slope using our algorithm. Note that according to the power-law and stretched-exponential model only for delay times in the picosecond range the true slope could be measured. (From Grasser, T., et al., *Proc. IRPS*, 268, 2007. With permission.)

rather robust set of measurement data which need to be captured by the models, the modeling of NBTI has been plagued by a still open definition of what NBTI degradation actually is. The question of whether the model should result in a power law or a logarithmic evolution of the threshold voltage, whether it should predict a temperature-dependent slope, or whether it should relax universally is of fundamental importance to any modeling attempt.

In contrast to previous model validation attempts that have focused almost exclusively on the stress phase, we put a special emphasis on the analysis of the model prediction

TABLE 14.1
Parameters for Equation 14.19 Used to Fit the Data in Figure 14.9 Assuming $t_s = 1000$ s

Reference	*T*	*N*	*B*	*β*
Ershov et al. [19]	105	0.15 (fixed)	1.49	0.179
Kaczer et al. [14]	125	0.15 (fixed)	1.29	0.136
Li et al. [45]	125	0.15 (fixed)	4.08	0.163
Alam et al. [5]	50	0.155	4.79	0.611
Alam et al. [5]	100	0.177	40.23	0.973
Alam et al. [5]	150	0.186	102.2	1.048

Source: From Grasser, et al., *Proc. IPRS.*, 2007, With permission.
Note: The fit was obtained using a fixed $n = 0.15$ with a simple least-square algorithm. However, in order to fit the data of Ref. [5], which are somewhat different from the other sources considered in this study, n had to be included as a free parameter. Interestingly, this results in a significant temperature-dependence of the zero-delay slope, well described by a linear relationship for n reported in Refs. [8,14] for delayed measurements. Keep in mind that these values should be taken with care, since they were extracted by a fit to three or four rather inaccurate slope values using two/three free parameters. The inaccuracy of the slope values is a result of both the measurement uncertainty as well as the time-dependence of the slope.

during the relaxation phase. Thereby two features are of interest, namely the large distribution of timescales and the universal behavior. For the analysis, the models under consideration have been implemented into a partial-differential-equation solver and solved numerically in order to rule out any uncertainties related to approximate analytic expressions. Since most of the published NBTI models can be derived from a generalized RD formalism [7], a short review of the assumptions employed in this model is given.

14.7.1 Reaction-Diffusion Models

RD-like models consist basically of an electrochemical reaction at the semiconductor–oxide interface which is coupled to a transport equation in the oxide bulk. We remark that the questions whether the depassivation process is field-driven [10,47], why holes at the interface are required and how they influence the reaction [48], and in which charge state, neutral or positive, the created trap and the released hydrogen species are, are highly controversial and are put aside for the moment. Nevertheless, for the discussion of the basic properties it is instructive to write the electrochemical reaction at the interface, which creates a dangling bond $Si^{\bullet}$ from a passivated interface defect SiH, as

$$\mathrm{Si-H} \rightleftharpoons \mathrm{Si}^{\bullet} + \mathrm{H_c} + \mathrm{H_t}. \tag{14.20}$$

Thereby we differentiate between hydrogen in a conduction/mobile state, H_c, and trapped hydrogen [49], H_t. Such a distinction is important, since in dispersive transport models most hydrogen becomes trapped quickly and might not be available for the reverse reaction. We also note that a large background concentration of hydrogen may exist in the vicinity of the interface, possibly in the order of 10^{19} cm^{-3} [50], which, if assumed to be freely available, could dominate the reverse reaction and completely compensate the forward reaction in a standard RD model.

It has been claimed that the binding energies of the Si–H bonds display a Gaussian broadening [8,35]. Previously published dispersive NBTI models consider either a dispersion in the forward rate [8] or a dispersion in the transport properties [13,14,51]. Models based on these assumptions will be discussed in Section 14.7.4. In particular, the variations in the energy barrier for the reverse reaction is important for the investigation of dispersive transport.

14.7.1.1 Standard RD Model

In the standard RD formulation the dissociation barrier is considered to be single valued (dispersion-free) and $H_t = 0$, meaning that all released hydrogen remains in the conduction state. The kinetic equation describing the interface reaction is commonly assumed to be of the form [46,52,53]:

$$\frac{\partial N_{\mathrm{it}}}{\partial t} = k_{\mathrm{f}}(N_0 - N_{\mathrm{it}}) - k_{\mathrm{r}} N_{\mathrm{it}} H_{\mathrm{it}}^{1/a}, \tag{14.21}$$

where $N_{it} = [Si^{\bullet}]$ is the interface state density, $N_0 = [Si\text{–}H]_0$ is the initial density of passivated interface defects, H_{it} is the hydrogen concentration at the semiconductor–oxide interface, k_f and k_r are the temperature and possibly field-dependent rate coefficients, while a gives the order of the reaction (1 for H^0 and H^+, 2 for H_2, assuming an instantaneous conversion of H^0 to H_2, cf. Ref. [5,46,54]). In our context it is important to recall that the usual assumptions are that $N_{it0} = N_{it}(t_0) = 0$ at the beginning of the stress period and that all generated N_{it} contribute equally to the threshold voltage shift. A somewhat unappreciated feature of the RD equations is, as will be shown in Section 14.7.4.2, that by allowing for a larger number of initial interface defects, a completely different behavior is obtained.

Hydrogen motion is assumed to be controlled by conventional drift–diffusion [53]:

$$\frac{\partial H_c}{\partial t} = -\nabla \cdot F_c + G_c, \tag{14.22}$$

$$F_c = -D_c\left(\nabla H_c - Z\frac{E_{ox}}{V_T}H_c\right), \tag{14.23}$$

with the (possibly unrealistic) assumption of a negligible initial hydrogen concentration, $H_c(\mathbf{x},0) = 0$. Hydrogen transport is postulated to occur on a single energy level, which will be referred to as the conduction state, with H_c, D_c, and G_c the hydrogen concentration, diffusivity, and generation rate in the conduction state, F_c the particle flux, Z the charge state of the particle, $V_T = k_B T_L/q$ the thermal voltage, T_L the lattice temperature, and E_{ox} the electric field inside the oxide.

The generation rate G_c is given by the interface reaction and reads for the usually considered one-dimensional problem:

$$G_c(x,t) = \frac{1}{a}\frac{\partial N_{it}}{\partial t}\delta(x) \tag{14.24}$$

with the interface assumed to be located at $x = 0$.

For the calculation of the time-dependent density of interface states, N_{it}, Equations 14.21 and 14.22 can be solved numerically on an arbitrary geometry. Although the solution of the RD model depends on the underlying geometry [55], it is commonly assumed that NBTI is a one-dimensional problem. In particular, for some special cases analytical approximations can be given [46,56,57] which are helpful for understanding the basic kinetics.

Depending on the parameter values and boundary conditions, different phases are observed which are shown in Figure 14.10 for the three most commonly used species

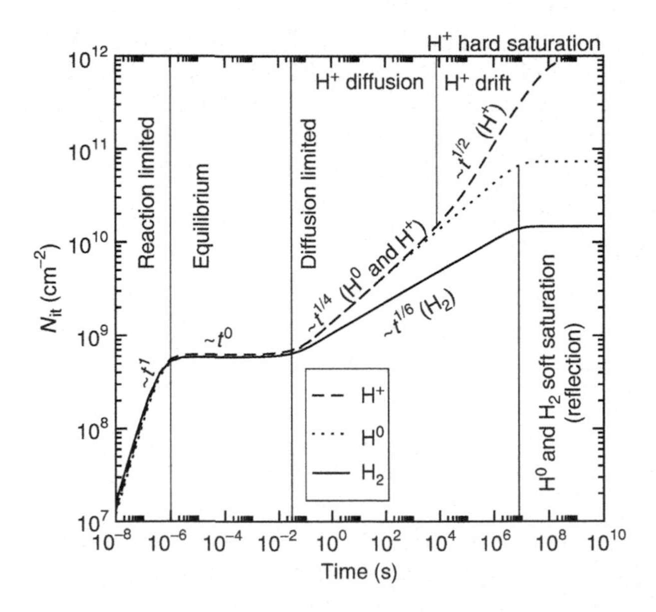

FIGURE 14.10
Five phases of the standard RD model obtained from a numerical solution of Equations 14.21 and 14.22 on a 2 nm oxide, using the parameters $N_0 = 10^{12}$ cm^{-2} and $N_{it0} = 0$. Shown are the results for the three species H^+, H^0, and H_2. The time exponent $n = 1$ is the signature of the reaction-limited phase while $n = 1/4 \ldots 1/2$, $n = 1/4$, and $n = 1/6$ are observed for the three species in the diffusion-limited phase. At the beginning of the diffusion-limited phase H^+ behaves like H^0. Furthermore, in the nonself-consistent simulation, where the feedback of the charges on the field distribution is neglected, H^+ does not show a soft saturation since all hydrogen is pulled away from the interface. (From Grasser, T., et al., *Trans. Dev. Mater. Reliab.*, 8(1), 79, 2008. With permission.)

H^+, H^0, and H_2: (1) The reaction-dominated regime with an exponent $n=1$, where the reverse rate is negligible due to the lack of available H_{it}. (2) Depending on the parameter values, a transition regime where $\partial N_{it}/\partial t=0$ which gives an exponent $n=0$. (3) The quasiequilibrium regime where $\partial N_{it}/\partial t$ is much smaller than the generation and passivation terms. This is assumed to be the dominant regime and displays the characteristic time exponent depending on the created species. (4) A saturation regime which could for instance be a soft saturation due to a reflecting boundary condition or a hard saturation resulting from the depassivation of all passivated interface states [58].

The RD model assumes the quasiequilibrium of the interface reaction ($\partial N_{it}/\partial t \approx 0$) to be the dominant regime [10,13,14]. Consequently, we obtain from Equation 14.21 together with the assumption $\Delta N_{it}(t) \gg N_{it0}$ the standard RD model as

$$\Delta N_{it}(t) = A_{RD} C^{1/(1+a)}(t), \tag{14.25}$$

with the species-dependent prefactor

$$A_{RD} = \left(a \left(\frac{k_f}{k_r} \Delta N_{it,\,max} \right)^a \right)^{1/(1+a)} \tag{14.26}$$

and the maximum value of ΔN_{it} given by $\Delta N_{it,max} = N_0 - N_{it0}$. For nondispersive transport $C(t) = D_c E_{ox} t / V_T$ for the proton case while $C(t) \approx \sqrt{D_c t}$ for the neutral species H^0 and H_2 [10,59]. This results in the well-known exponents 1/2, 1/4, and 1/6 for proton, atomic, and molecular hydrogen transport, respectively. These exponents do not depend on temperature nor is it possible to include process dependencies. We recall that such an exponent of 1/2 obtained for H^+ transport is not observed experimentally which led researchers to discard the possibility of drifting protons.

14.7.1.2 Pre-RD Regime

Interestingly, by allowing a relatively large initial concentration of interface states N_{it0} and by assuming $\Delta N_{it}(t) \ll N_{it0}$, a completely different solution is obtained [59],

$$\Delta N_{it}(t) = A_P C(t), \tag{14.27}$$

with the prefactor

$$A_P = a \left(\frac{k_f}{k_r} \frac{\Delta N_{it,\,max}}{N_{it0}} \right)^a = A_{RD}^{1+a} N_{it0}^{-a}. \tag{14.28}$$

This regime is termed pre-RD regime [59], because for intermediate concentrations of N_{it0} the number of created interface states ΔN_{it} will eventually become larger than N_{it0}, changing the overall behavior to that of the standard RD model. This is demonstrated in Figure 14.11 for the H_2-RD model.

In the pre-RD regime the exponents have the values 1, 1/2, and 1/2 for proton, atomic, and molecular hydrogen transport, respectively. Note that these exponents do not depend on the kinetic exponent a as in the standard RD model. For classic drift–diffusion, these resulting exponents are not compatible with measurements. However, as has been shown [13,59], the introduction of dispersive transport can bring the exponents within the observed ranges.

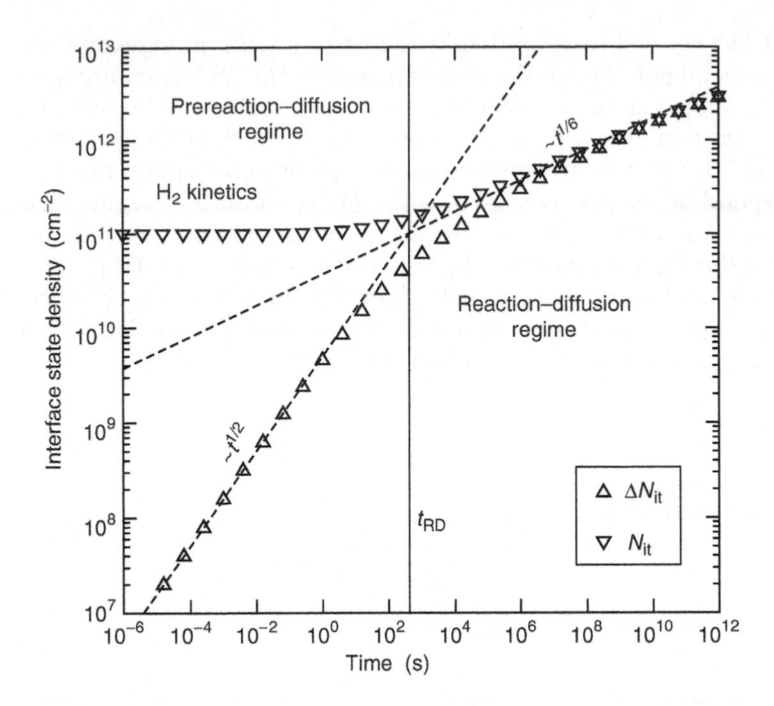

FIGURE 14.11
Two different regimes for a medium number of initial interface defects $N_{\mathrm{it0}} = 10^{11}\ \mathrm{cm}^{-2}$. The transition between the pre-RD regime and the standard RD regime can be clearly observed. A value of $N_0 = 10^{13}\ \mathrm{cm}^{-2}$ was used in the simulations. (From Grasser, T., et al., *Trans. Dev. Mater. Reliab.*, 8(1), 79, 2008. With permission.)

The assumption $\Delta N_{\mathrm{it}}(t) \ll N_{\mathrm{it0}}$ has originally been introduced by Zafar [13]. This is based on the (actually mandatory) notion that the occupancy of the interface states depends on the position of the Fermi-level and that not all interface states are electrically active. In this context N_0 is now the maximum number of hydrogen binding sites rather than the maximum number of electrically observable interfaces states in a completely depassivated sample.

14.7.1.3 Relaxation as Predicted by the RD Model

As soon as the stress condition is removed, the forward-rate of the RD model is assumed to be negligible. Just like during the stress phase, the reaction is in quasiequilibrium, resulting in the left-hand side of Equation 14.21 to become very small. With $k_{\mathrm{f}} \approx 0$, the actual values of k_{r} and D_{c} become irrelevant, except for a very short and insignificant reaction-limited initial phase. In addition, the species type has no influence on the relaxation and the overall behavior is again diffusion-limited. Consequently, the RD model predicts a universal relaxation practically independent of the species (H and H_2) as

$$r(\xi) = 1/(1 + \xi^{1/2}). \qquad (14.29)$$

This analytic expression is compared to the numerical results for both species in Figure 14.12. Also shown is the measurement data of Reisinger et al. [9]. It is worthwhile to realize that the relaxation predicted by the RD model does not depend on any model parameters. Consequently, it must be clearly emphasized that since the relaxation predicted by the RD

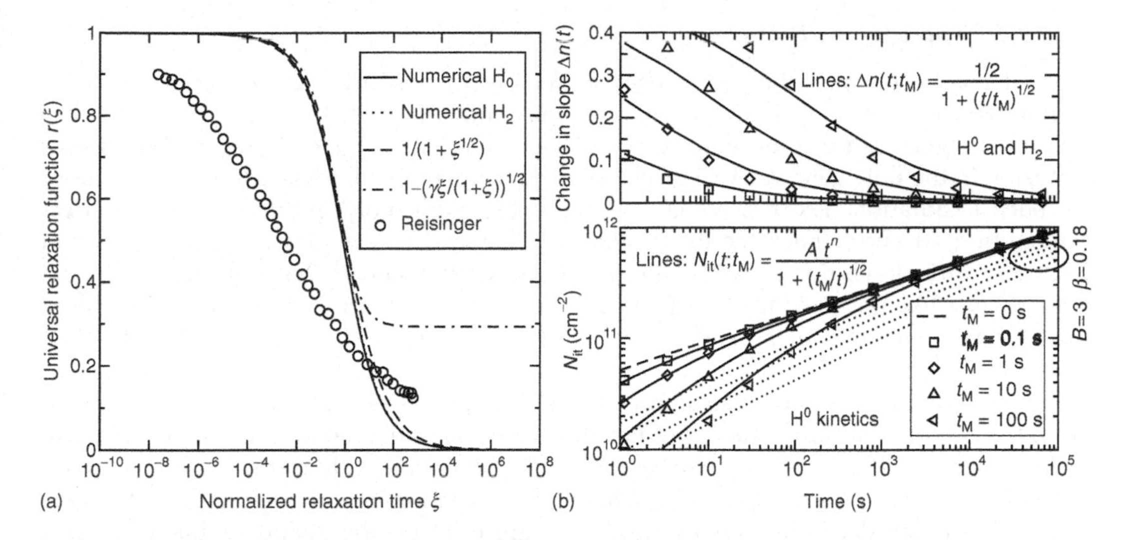

FIGURE 14.12
(a) Comparison of two analytic expressions for the RD relaxation behavior with numerical results obtained for H and H_2 kinetics. The power-law-like expression is accurate for all relaxation times and will be used as reference throughout this work. Also shown is the analytic expression derived in Ref. [56]. Due to the lack of parameters there is no way to fit the measurement data with the RD model. (b) Influence of the measurement delay t_M as predicted by the RD model. Comparison of the analytic model (lines) with the numerical solution (symbols) proves the excellent accuracy of the analytic model for $t > t_M$. Note that the RD model predicts a very small influence of delay for longer stress times, in contrast to Figure 14.8. For the sake of comparison, a more realistic influence of the measurement delay is given by the dotted lines, obtained with typical parameter values $B = 3$ and $\beta = 0.18$. There the individual curves obtained with different delay times remain clearly separated even after $t_s > 10^3$ s. (From Grasser, T., et al., *Proc. IRPS*, 268, 2007. With permission.)

model cannot be made to depend on gate bias, temperature, and process conditions, in stark contradiction to Figure 14.9, there is no way to fit the available measurement data. In particular, $\beta = 1/2$ is much larger than observed experimentally, leading to a relaxation which is too slow in the beginning and too fast in the end. This is clearly visible in Figure 14.12 where most of the relaxation occurs within three to four decades whereas the measurements show relaxation over 12 decades. Consequences of this erroneous relaxation prediction are a heavily time-dependent but temperature-independent slope in the RD simulated delayed measurements, and a vanishing influence of the delay on the measurement result for $t_s \gtrsim 10 \times t_M$ (Figure 14.12), in contradiction to measurements [14,19], see also Ref. [6]. The only way to move the relaxation curve to shorter relaxation times is to bring the forward reaction into the quasisaturation regime where hydrogen has already piled up considerably in the oxide (assuming for instance a reflecting boundary condition). However, in addition to the fact that this behavior is not universal, the slope during the stress phase approaches zero.

14.7.2 Extended Classical RD Models

As the standard form of the RD model has been found to have also limitations during the stress phase [4,54,60], extended versions have been introduced. However, the question of whether these extended models are better able to describe the relaxation behavior has so far only been qualitatively assessed and a rigorous statement is missing. This will be done in the following.

14.7.2.1 Two-Region RD Model

First, it has been noted that due to the extremely thin oxides used in modern CMOS technology, the diffusing hydrogen species may quickly reach the oxide/poly interface [4]. As a consequence, the degradation will be dominated by the presumably slower diffusion in the poly gate. We will discuss two variants of RD models extended to account for such a situation. The first variant assumes the oxide/poly interface to be a perfect transmitter. At short times the oxide will be filled with H_2. At later times, the overall hydrogen diffusion is dominated by the slower diffusion inside the polygate and the model behaves just like the standard H_2-RD model. One might suspect that the hydrogen stored inside the oxide, where the diffusivity has been assumed to be larger, modifies the relaxation behavior. Under certain conditions this is indeed the case, with undesired properties, though, as shown in Figure 14.13. For large stress times, most hydrogen is stored in the poly and the model predicts the same relaxation as the RD model. Thus, in order to see the influence of the two regions we have to look at shorter stress times, in our particular case $t_s = 10$ s and $t_s = 100$ s, where the population in both regions is of the same order of magnitude. However, as show in Figure 14.13, the shape of the relaxation curve does not agree with measurement data. Furthermore, the shape depends on the ratio of both reservoir occupancies, which changes with time and consequently results in a nonuniversal relaxation. We also remark that the assumptions underlying this model are in contradiction to a study which did not show a dependence of NBTI on the gate material [61].

14.7.2.2 Two-Interface RD Model

Next, we discuss a two-interface model which can be considered a refined form of the two-region model. It assumes that atomic hydrogen is released from the silicon/oxide interface which then diffuses through the thin oxide and depassivates defects at the oxide/poly interface [4,62]. The creation of defects at the opposite interface is supported by SILC

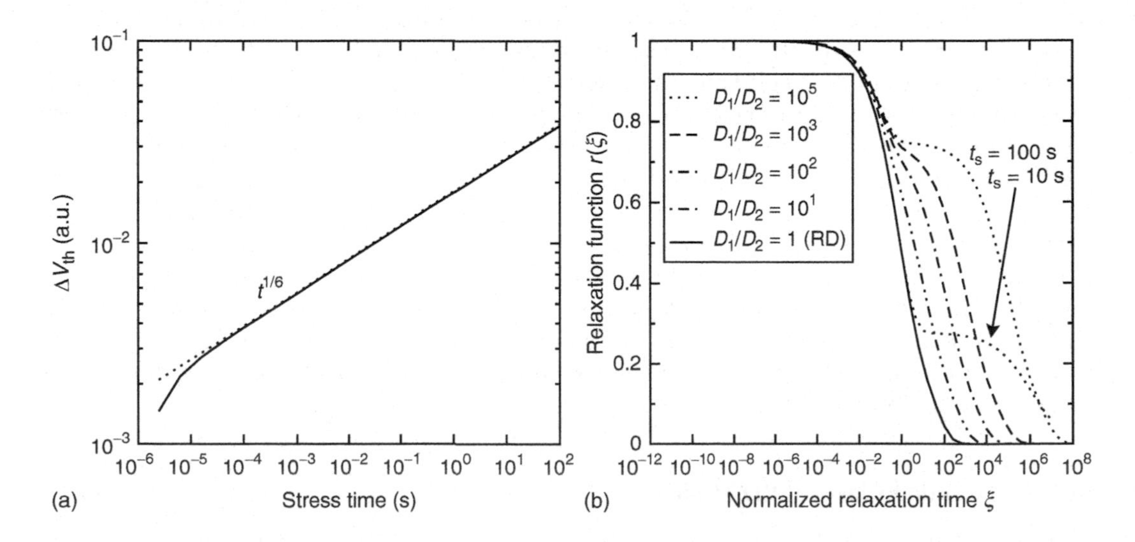

FIGURE 14.13
Numerical simulation of a generalized RD model with two different diffusion coefficients in the oxide and poly. For this particular set of parameters and the small stress time required to bring out this effects, no difference is visible during the stress phase (a), while the relaxation behavior slows down and displays nonuniversal humps (b). (From Grasser, T., et al., *Proc. IRPS*, 268, 2007. With permission.)

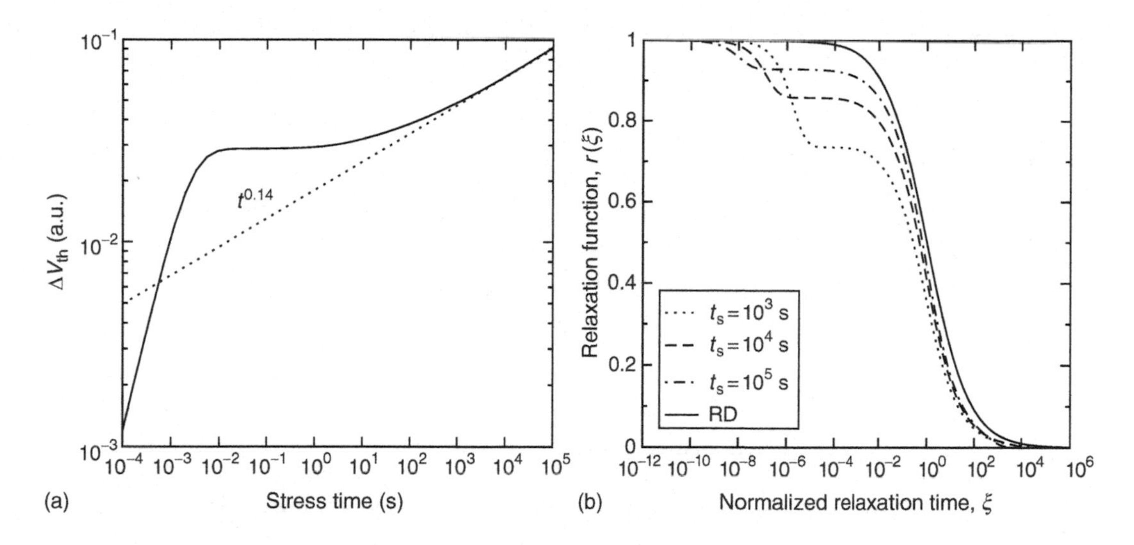

FIGURE 14.14
Numerical simulation of the two-interface RD model stress (a) and relaxation (b) behavior. Although faster relaxation than with the standard RD model is possible, the relaxation is not universal and the shape of the relaxation curve starkly disagrees with measurement data. In order to obtain a visible influence on the relaxation behavior, the hydrogen stored in both regions has to be of the same order of magnitude, which considerably lowers the power-law slope during the stress phase. (From Grasser, T., et al., *Proc. IRPS*, 268, 2007. With permission.)

measurements [4]. The hydrogen from the oxide and the released hydrogen at the oxide/poly interface diffuse as H_2 through the poly and result in an overall power-law exponent of 1/6 at large times. It has been suggested that such a two-interface model may predict a faster recovery compared to the standard RD model [62]. For this to be the case, the amount of fast hydrogen stored in the oxide must be of the same order of magnitude compared to the slow hydrogen stored in the poly. As in the case of the two-region RD model, it is again possible to modify the relaxation behavior to a certain extent, see Figure 14.14. In this case the relaxation can be made faster than with the standard RD model because the fast hydrogen concentration inside the oxide is saturated, resulting in a shift to smaller normalized relaxation times ξ on the universal plot. However, just as with the two-region model, the resulting relaxation is not universal, as the ratio of these two hydrogen storage areas changes with time, see Figure 14.14.

14.7.2.3 *Explicit H–H_2 Conversion RD Model*

Another variant of the classic RD model aims at improving the model prediction at early times [54]. This is based on the suggestions that measurements might display a power-law exponent of 1/3 during the initial stress phase [5,40], which is incompatible with the standard RD model. This has been explained by an extended RD model which explicitly accounts for the dimerization of H into H_2 [5,54],

$$\frac{\partial[\mathrm{H}]}{\partial t} = k_{\mathrm{H}_2}[\mathrm{H}_2] - k_{\mathrm{H}}[\mathrm{H}]^2, \tag{14.30}$$

rather than assuming an instantaneous transition, in addition to the diffusion of both hydrogen species. Depending on the values of k_{H_2} and k_{H}, either pure H or H_2 kinetics can be observed. In addition, a regime with the aforementioned transitional power-law exponent of 1/3, which eventually changes to 1/6, is possible. Since recent measurements

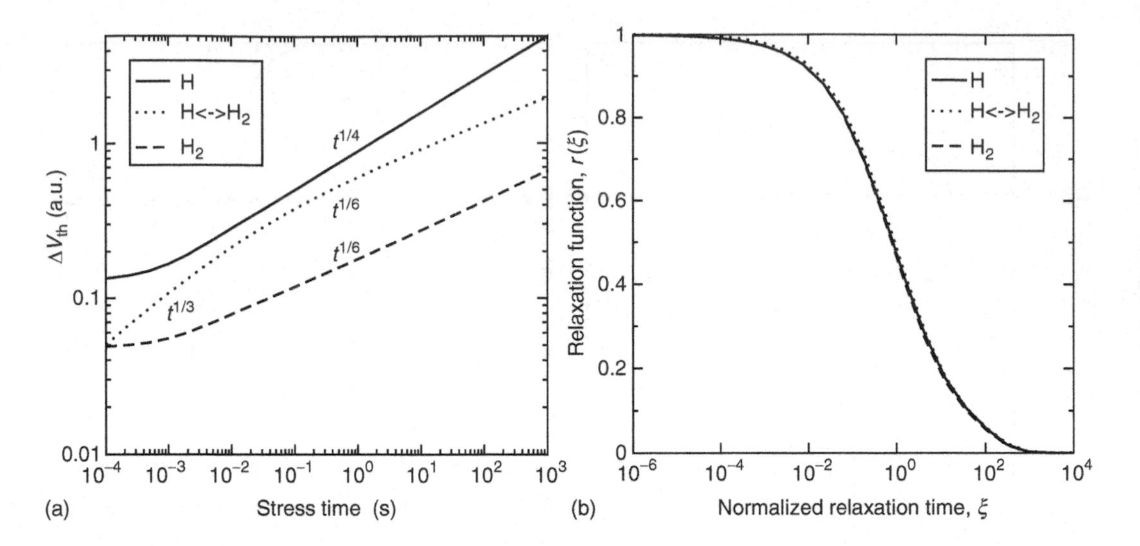

FIGURE 14.15
Numerical simulation of a generalized RD model with explicit H to H_2 conversion. (a) Depending on the choice of parameters, the model gives power-law exponents known from the H and the H_2 models during the stress phase, in addition to a transitional region with $n = 1/3$. (b) Since the model remains within the boundaries set by the pure H^0 and H_2 model (which are equal during relaxation), the overall relaxation behavior cannot be influenced by any of the available parameters. (From Grasser, T., et al., *Proc. IRPS*, 268, 2007. With permission.)

give a long-term exponent closer to 1/6 than to 1/4, the parameters have to be chosen in such a way that the total amount of stored [H_2] is much larger than [H]. One might conclude from this that the two distinct reservoirs of H and H_2 may allow for a modified relaxation behavior. However, this is not the case for the simple reason that the model stays within the limits set by pure H and H_2 behavior, just as during the stress phase. Since the relaxation of both species is practically equivalent, no influence on the relaxation behavior is obtained from such a model, see Figure 14.15.

14.7.2.4 Vanderbilt Model

By employing first-principles calculations, Tsetseris et al. [48] investigated the electrochemical reaction (Equation 14.20), which is one of the foundations of the RD model. They found an activation energy of about 2.4 eV, in agreement with measurement data [63]. Such a barrier is way too large to allow the bond to be broken during typical bias temperature conditions. Although the presence of holes lowers the activation energy to values around 2.1 eV, this value is still too high for a relevant contribution. Consequently, they suggested an alternative reaction triggered by protons supplied from the semiconductor bulk

$$H^+ + Si - H \rightleftharpoons Si^+ + H_2. \qquad (14.31)$$

Provided H^+ is readily supplied from the bulk, the differential equations resulting from Equation 14.31 combined with the standard diffusion Equation 14.22 are from a mathematical point of view equivalent to the equations resulting from the standard RD model for atomic hydrogen diffusion [62]. Consequently, the model predicts a slope of 1/4 and the same relaxation as the RD model and can therefore not be used in this form to explain NBTI.

14.7.3 Final Notes on RD Models

We have shown that irrespective of the extensions applied to the RD model, the recovery behavior observed during measurement cannot be described with the published RD variants in their present form. The fact that some OTF measurements and the corrected MSM measurements give exponents of around $n = 0.15$, which is close to the value predicted by the H_2-based RD model ($n = 1/6$), should not let one arrive at the conclusion that the RD model is consequently reasonable. In particular, we think one has to be extremely cautious with a point of view that the RD model correctly covers the stress part while only the relaxation part needs to be refined. The point to make here is that the 1/6 exponent during the RD stress phase is a result of a delicate interplay between the forward and backward reaction [5]. Without the backward reaction, which dominates the time evolution by inserting the diffusion-limited component into the RD model, the forward reaction alone would result in $n = 1$. It is only during relaxation, where the forward rate is suppressed, that the poor performance of the RD reverse reaction becomes visible. Consequently, we do not see any reason to believe the very same reverse reaction to be valid during the stress phase to constructively change the reaction-limited exponent of $n = 1$ to the proposed diffusion-limited value of $n = 1/6$.

14.7.4 Dispersive NBTI Models

It has been clearly shown in the previous sections that the RD model predicts 80% of the relaxation to occur within three to four decades, while in reality relaxation is observed to span more than 12 decades [9,14,21]. This indicates some form of dispersion in the underlying physical mechanism(s). Various forms of dispersion have already been introduced into NBTI models based on either (1) diffusion [12–14], (2) hole tunneling from/into states in the oxide [16], and (3) reaction rates at the interface [11,64]. The models suggested to capture these mechanisms will be benchmarked in the following using the universality as a metric.

14.7.4.1 Reaction-Dispersive–Diffusion (RDD) Models

First, we consider generalized RD models based on dispersive transport of the hydrogen species [14,65]. These models are obtained by replacing the classic drift–diffusion Equation 14.22 in the RD model by its dispersive counterpart. Several different variants of dispersive transport equations have been used for the formulation of NBTI models [10,13,14]. It can be shown that the transport models themselves give practically identical results and that the differences in the final model prediction can be traced back to different assumptions used for the boundary and initial conditions [59].

These differences are best studied using the dispersive multiple-trapping (MT) transport equations [66–68]. In the MT model the total hydrogen concentration H consists of hydrogen in the conduction states H_c and trapped hydrogen as

$$H(\mathbf{x},t) = H_c(\mathbf{x},t) + \int \rho(\mathbf{x},E_t,t)\mathrm{d}E_t, \qquad (14.32)$$

with $\rho(\mathbf{x},E_t,t)$ being the trapped hydrogen density ($\mathrm{cm}^{-3}\,\mathrm{eV}^{-1}$) at the trap level E_t. Transport is governed by the continuity equation and the corresponding flux relation:

$$\frac{\partial H}{\partial t} = -\nabla \cdot \mathbf{F}_c + G_c. \qquad (14.33)$$

Note that in contrast to Equation 14.22 the time derivative of the total hydrogen concentration is used in Equation 14.33, which also accounts for the exchange of particles with the trap levels. The occupancy of the trap levels is governed by balance equations which have to be solved for each trap level

$$\frac{\partial \rho(E_t)}{\partial t} = \frac{\nu}{N_c}(g(E_t) - \rho(E_t))H_c - \nu \exp\left(-\frac{E_c - E_t}{k_B T_L}\right)\rho(E_t), \tag{14.34}$$

with ν being the attempt frequency, N_c the effective density-of-states in the conduction band, and E_c the conduction band edge. An exponential trap density-of-states is commonly used [14,68]:

$$g(E_t) = \frac{N_t}{E_0}\exp\left(-\frac{E_c - E_t}{E_0}\right), \tag{14.35}$$

which, in this particular context, results in a power law [14,67] for the time-dependence of ΔN_{it}. It is also worth recalling that the transport will only be dispersive for $E_0 > k_B T_L$, that is, for sufficiently deep trap distributions [68].

As the MT equations are rather complex and can in general only be solved numerically, simplified equations have been derived by Arkhipov and Rudenko [68,69]. Their approximate solution relies on the existence of the demarcation energy,

$$E_d(t) = E_c - k_B T_L \log(\nu t), \tag{14.36}$$

separating shallow from deep traps and was derived to describe the broadening of an initial particle distribution in the conduction band, see Figure 14.16. This is not the case during NBT stress, however, where we have to deal with a continuous generation of particles at the interface during the stress phase. An extended model suitable for NBTI

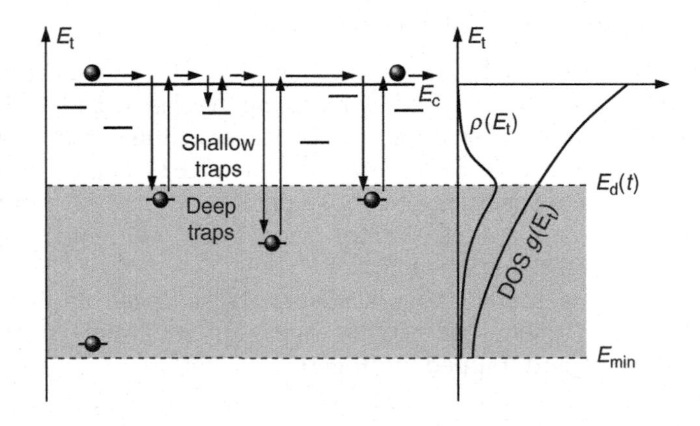

FIGURE 14.16
Schematic illustration of MT dispersive transport. Particles from the conduction band fall into the traps and are thermally reemitted into the conduction band. Reemission is more likely for shallow traps. The time-dependent demarcation energy E_d separates shallow from deep traps. With time, the demarcation energy becomes more negative, until the bottom of the trap distribution is reached ($E_d \rightarrow E_{min}$) and equilibrium is obtained. Before equilibrium, the motion of the particle packet slows down with time. Note how the individual trap levels, which microscopically correspond to the different energy levels of hydrogen in an amorphous material, are approximated by a macroscopic density-of-states. (From Grasser, T., et al., *Trans. Dev. Mater. Reliab.*, 8(1), 79, 2008. With permission.)

has been derived in Ref. [59] which, in the highly nonequilibrium regime, describes the overall motion as

$$H(\mathbf{x},t) - H_0(\mathbf{x}) = -\nabla \cdot F(\mathbf{x},t) + \int_{t_0}^{t} G_c(\mathbf{x},t')\mathrm{d}t', \tag{14.37}$$

with the effective flux of the total concentration of the species H given by

$$F(\mathbf{x},t) = -D_c\tau(t)\left(\nabla H - Z\frac{E_{ox}}{V_T}H\right). \tag{14.38}$$

Note that Equation 14.37 is basically the time integral of Equation 14.33 and thus does not contain a time derivative anymore. This is a consequence of the fact that the dynamics of the system can be incorporated solely into $\tau(t)$, which directly depends on the hydrogen trap density-of-states and the demarcation energy as

$$\frac{1}{\tau(t)} = \frac{\nu}{N_c}\int_{-\infty}^{E_d(t)} g(E_t)\mathrm{d}E_t. \tag{14.39}$$

This is a characteristic feature of any adiabatic process, where the time-dependence of the whole system is determined by the slowest process, in our case, the thermal equilibration of hydrogen [70]. For the particular case of an exponential trap density-of-states as in Equation 14.35, τ can be evaluated in closed form to

$$\tau(t) = \frac{1}{\nu}\frac{N_c}{N_t}(\nu t)^{\alpha}, \tag{14.40}$$

with the dispersion parameter

$$\alpha = k_B T_L / E_0. \tag{14.41}$$

It is also worth pointing out that by introducing a time-dependent diffusivity $D(t) = D_c\tau(t)$, under certain circumstances a link to empirical dispersive transport models can be established [59].

Of particular interest for the derivation of NBTI models is the concentration of the mobile hydrogen H_c, which is directly linked to the total hydrogen concentration H as

$$H_c(\mathbf{x},t) = \frac{\partial \tau(t)H(\mathbf{x},t)}{\partial t}. \tag{14.42}$$

This relation will be used for the formulation of the different NBTI boundary conditions.

In order to obtain an NBTI model, the dispersive transport equation has to be coupled to the electrochemical reaction assumed to take place at the interface. For the present analysis we remark that the macroscopic hydrogen trap density-of-states is derived for an amorphous bulk material and is unlikely to be valid close to an interface. In that context, the physical mechanisms justifying the conduction band concept in conjunction with hydrogen hopping next to the interface need to be evaluated and justified. Published dispersive NBTI models, however, are based on the validity of this concept, and the different interpretations explain the discrepancies in these models. As in the RD model, the kinetic equation describing the interface reaction is assumed to be of the form Equation 14.21. Also, the interface reaction is assumed to be in quasiequilibrium.

A crucial question for the formulation of dispersive NBTI models is how to link it to the interfacial hydrogen concentration H_{it}. In the following, we will consider two different models. The first assumes that H_{it} is given by the total hydrogen concentration H. Thus, in the RD regime one obtains for neutral particles [14,59]:

$$\Delta N_{\text{it}}(t) = A_{\text{RD}} \left(\frac{D_{\text{c}}}{\nu} \frac{N_{\text{c}}}{N_{\text{t}}} \right)^{1/(2+2a)} (\nu t)^{\alpha/(2+2a)}, \tag{14.43}$$

with the same prefactor A_{RD} as in the RD model given through Equation 14.26. For atomic hydrogen ($a = 1$) the exponent is given through $n = \alpha/4$ while molecular hydrogen ($a = 2$) gives $n = \alpha/6$ (Figure 14.17).

For the proton, one can show that [14,59]

$$\Delta N_{\text{it}}(t) = A_{\text{RD}} \left(\frac{D_{\text{c}}}{\nu} \frac{N_{\text{c}}}{N_{\text{t}}} \frac{E_{\text{ox}}}{V_{\text{T}}} \right)^{1/2} (\nu t)^{\alpha/2} \tag{14.44}$$

holds. Note that the numerical solution for H^+ may contain a transitional regime with $n = \alpha/4$, where the diffusive component still dominates.

Since α equals 1 in the diffusive limit and 0 in the extremely dispersive case, Equations 14.44 and 14.43 imply that with dispersive transport an exponent smaller than the RD exponents of 1/2, 1/4, and 1/6 can be obtained. Also, for increasing trap density N_{t}, the total amount of degradation decreases.

A qualitative explanation for the reduction of the exponent can be given by noting that dispersive transport results in most particles being trapped close to the interface, yielding a steeper profile compared to classic diffusion. As all hydrogen is available for the reverse rate in Equation 14.21, the net interface state generation is suppressed, resulting in a smaller exponent.

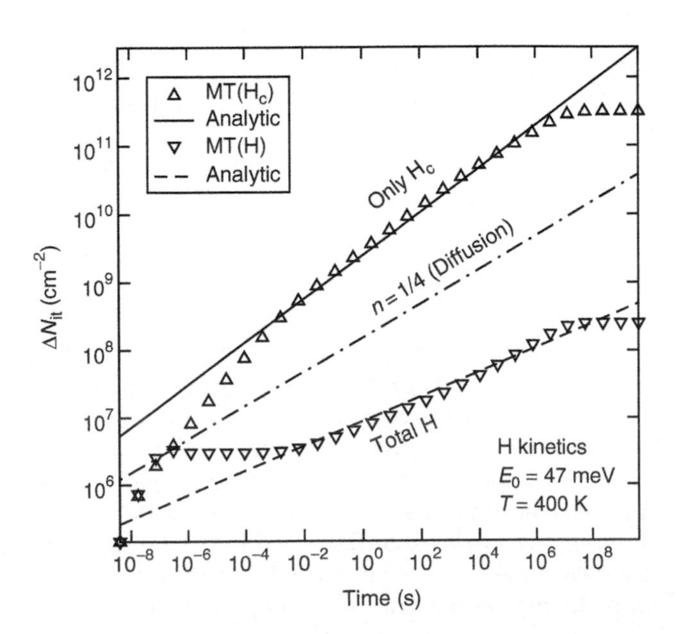

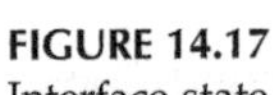

FIGURE 14.17
Interface state density as a function of the boundary condition calculated numerically by solving the MT equations in comparison to the analytic expressions (Equations 14.43 and 14.46) for $N_{\text{it0}} = 0$. Good agreement between the numerical and analytical solution is obtained for both boundary conditions. (From Grasser, T., et al., *Trans. Dev. Mater. Reliab.*, 8(1), 79, 2008. With permission.)

Since the dispersion parameter α depends linearly on the temperature, a linear temperature-dependence of the exponent is obtained as [14]:

$$n_1 = \frac{\alpha}{2 + 2a} = \frac{k_B T_L}{2E_0(1 + a)}. \tag{14.45}$$

This is consistent with experimental results obtained with delayed measurements [8,14].

The previous model was based on the assumption that all hydrogen, mobile and trapped, can participate in the NBTI reverse rate. In contrast, if we now assume that only the mobile hydrogen can participate in the reverse rate, that is, $H_{it} = H_c(0)$, which appears to be the more natural boundary condition for the MT model [71], one obtains for neutral particles

$$\Delta N_{it}(t) = A_{RD}\left(\frac{D_c}{\nu}\frac{N_t}{N_c}\right)^{1/(2+2a)}\left(\frac{1+a}{1+a\alpha/2}\right)^{1/(1+a)}(\nu t)^{(1-\alpha/2)/(1+a)}. \tag{14.46}$$

For atomic hydrogen, the exponent $n = 1/2 - \alpha/4$ is obtained while H_2 results in $n = 1/3 - \alpha/6$. Hence, for increased dispersion the exponents become now larger than their RD equivalents. Furthermore, when the trap density is increased, the degradation increases. This is in agreement with the previously stated result that the inclusion of traps into a standard RD model increases the exponent [29,72].

Interestingly, for H^+ one obtains [59]:

$$\Delta N_{it}(t) = \left(2\frac{k_f N_0}{k_r} D_c \frac{E_{ox}}{V_T}\right)^{1/2} t^{1/2}, \tag{14.47}$$

with an exponent $n = 1/2$. This is equal to the result obtained by the standard RD model [59].

Again, qualitatively, in this model the newly released hydrogen quickly falls into the traps, but for times larger than $1/\nu$ most hydrogen resides in deep traps and is therefore not as easily available for the reverse rate in Equation 14.21. This suppresses the reverse reaction and consequently enhances the net interface state generation and results in a larger exponent.

In contrast to the total hydrogen boundary condition, now the exponent decreases with increasing temperature through

$$n_2 = \frac{1 - \alpha/2}{1 + a} = \frac{2E_0 - k_B T_L}{2E_0(1 + a)} = \frac{1}{1 + a} - n_1. \tag{14.48}$$

This is in contradiction to currently available observations [8,14,20]. Note, however, that this particular temperature-dependence is a consequence of the exponential trap density-of-states and a hardly noticeable temperature-dependence has been reported [29] using a Gaussian distribution on top of the exponential density-of-states.

14.7.4.2 Dispersive Pre-RD Regime

For the case that a large initial concentration of interface states is allowed, the pre-RD result (Equation 14.27) can be directly transferred to the dispersive case and one obtains for the total-hydrogen-boundary-condition [59]:

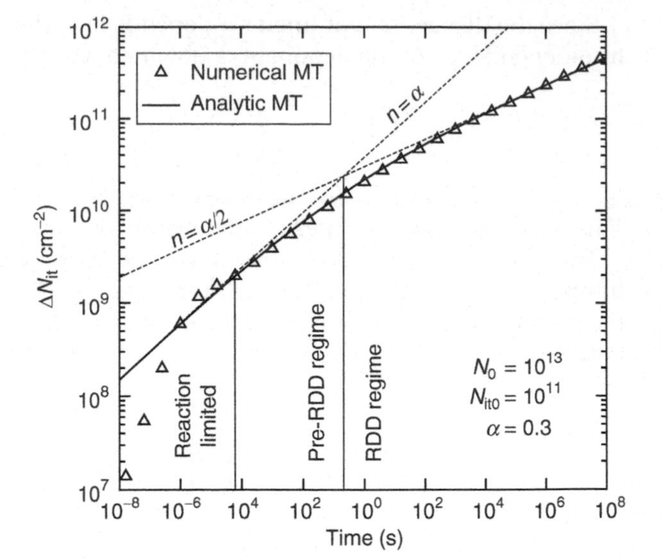

FIGURE 14.18
Example simulation showing the transition between the pre-RDD regime and the RDD regime which could be used to explain a different initial exponent compared to the long-time exponent. Shown are the numerical and analytical solutions of the MT equations. Note that the analytic solution is only valid after the reaction-limited phase. (From Grasser, T., et al., *Trans. Dev. Mater. Reliab.*, 8(1), 79, 2008. With permission.)

$$\Delta N_{\mathrm{it}}(t) = A_{\mathrm{P}}\left(\frac{D_{\mathrm{c}}}{\nu}\frac{N_{\mathrm{c}}}{N_{\mathrm{t}}}\right)^{1/2}(\nu t)^{\alpha/2}. \tag{14.49}$$

For the proton one obtains

$$\Delta N_{\mathrm{it}}(t) = A_{\mathrm{P}}\left(\frac{D_{\mathrm{c}}}{\nu}\frac{N_{\mathrm{c}}}{N_{\mathrm{t}}}\frac{E_{\mathrm{ox}}}{V_{\mathrm{T}}}\right)(\nu t)^{\alpha}. \tag{14.50}$$

As before, the exponents $n = \alpha$, $\alpha/2$, and $\alpha/2$ for H^+, H^0, and H_2 reduce to their pre-RD equivalents 1, 1/2, and 1/2 for $\alpha = 1$. Also, the exponent increases linearly with temperature similarly to Equation 14.45, and the same compatibility to measurements is given.

Interestingly, it can be shown that Equation 14.50 is equivalent to the Zafar model, which also has a slope $n = \alpha$ rather than $n = \alpha/2$ as obtained in the RD regime [59]. For an intermediate concentration of interface states, the transition between the pre-RD and the RD regime is shown in Figure 14.18.

14.7.4.3 Relaxation as Predicted by the RDD Models

A previous analysis of the relaxation behavior predicted by dispersive transport equations [14] was based on various assumptions (such as pulse-like excitation [65], uncertainties in the boundary conditions [65], and a neglected history of previously trapped hydrogen atoms during relaxation) which led to only approximative solutions. As it turned out, a more rigorous analytic derivation is rather involved. An approximation for $\xi < 1$ ($t_r < t_s$, as normally encountered during typical MSM measurements), is given by Equation 14.11 with B and β depending on the boundary condition and the dispersion coefficient α. Interestingly, for $\xi > 1$ the behavior changes and different values for B and β have to be used (cf. Figure 14.19).

In order to avoid any uncertainties inherent in approximate analytical solutions, we numerically solve the full time-dependent multiple trapping model [66] to allow for an accurate description of both the stress and the relaxation phase. The results shown in Figure 14.19 display a much broader range of possible relaxation characteristics compared to classic diffusion. Nevertheless, the dispersive transport models in their present form are

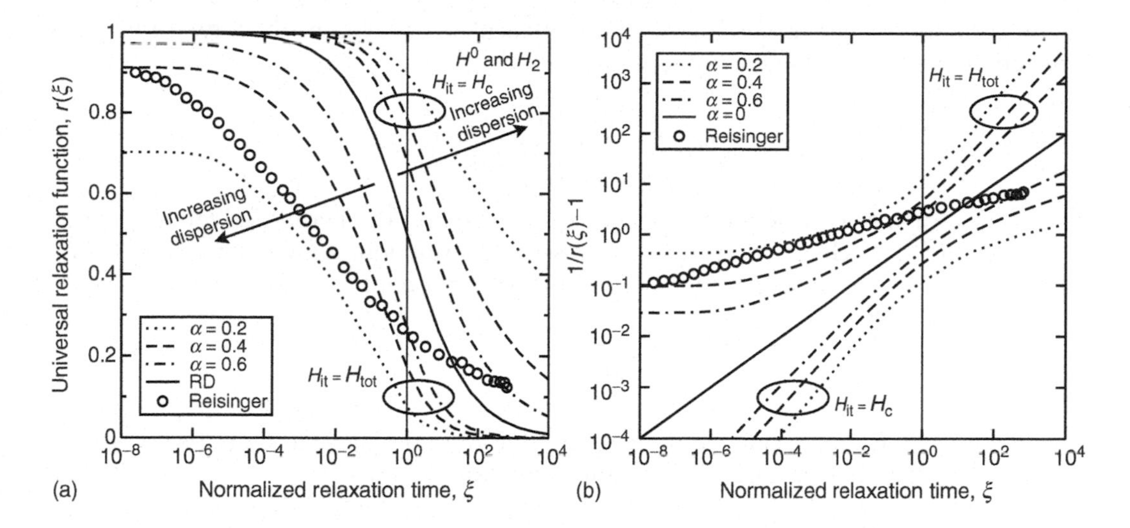

FIGURE 14.19
Relaxation as predicted by the full numerical solution of the dispersive transport models for various values of the dispersion parameter α. (a) Shows the universal relaxation function r(ξ) while (b) shows $1/r(\xi)-1$ which should be close to a straight line according to Equation 14.12. Also shown are the data from Reisinger as a reference. The model with $H_{it}=H_{tot}$ always stays below the diffusive (RD) limit, while the model with $H_{it}=H_c$ always stays above. The diffusive limit is like a watershed which cannot be crossed by either model. Also note that the $H_{it}=H_{tot}$ model appears to have a limit different from unity for $\xi=0$, which is a result of an extremely fast relaxation triggered by the hydrogen stored right at the interface. Note that this component is not universal. (From Grasser, T., et al., *Proc. IRPS*, 268, 2007. With permission.)

not able to fully explain the experimentally observed relaxation on their own. They might, however, be combined with a hole trapping model in order to account for a slow component during relaxation. Note that the standard RD model is an unlikely candidate for this slow component since any contribution would be negligible for large relaxation times.

Also note that the $H_{it}=H_{tot}$ model appears to have a limit different from unity for $r(\xi\rightarrow 0)$. This is a result of the extremely fast relaxation triggered by the hydrogen stored directly at the interface. The exact shape of this initial hump (not shown) depends on the stress time and the width of the interfacial layer, thereby rendering this model nonuniversal.

14.7.4.4 Dispersive-Rate Coefficients

Next we consider reaction-limited models using a dispersion in the rate coefficients [11,64]. This is based on the observation of Stesmans et al. [35] who could best describe the dissociation kinetics of hydrogen-passivated P_b centers at the interface using first-order kinetics and a Gaussian distribution of interface states. A similar observation was made regarding the passivation of P_{b0} and P_{b1} interface defects [34]. Huard et al. [11] base their permanent component on such a dispersive forward rate, assuming that the generated interface states do not relax at all, or at least not at shorter and medium relaxation times [73].

The model derivation uses the RD interface reaction given in Equation 14.21. In contrast to the RD model, however, it is assumed that the generated interface states are permanent. Thus, k_r can be set to zero and Equation 14.21 has the solution:

$$N_{it}=N_0\big(1-\exp\left(-k_f(E_d)t\right)\big). \qquad (14.51)$$

By invoking the Arrhenius law for the rate coefficient k_f, one obtains the forward rate as a function of the dissociation energy E_d as

$$k_f(E_d) = k_{f0} \exp\left(-\frac{E_d}{k_B T_L}\right). \tag{14.52}$$

Assuming a distribution of dissociation energies given by the Fermi-derivative function (Equation 14.5), which, in contrast to a Gaussian distribution, allows for a closed form solution, one obtains by summing the individual contributions

$$\frac{\Delta N_{it}}{N_0} = \int g_P(E_d, E_{dm})(1 - \exp(-k_f(E_d)t))\,dE_d. \tag{14.53}$$

This integral can be approximated by realizing that $N_{it}(E_d)$ is close to unity below $E^*(t) = k_B T_L \ln(k_{f0} t)$ and zero otherwise. One can thus approximately write

$$\frac{\Delta N_{it}}{N_0} \approx \int_0^{E^*(t)} g_P(E_d)\,dE_d = \frac{1}{1 + \left(\frac{t}{\tau}\right)^{-\alpha}}, \tag{14.54}$$

with $\tau = k_{f0}^{-1} \exp(E_{dm}(E_{ox})/k_B T_L)$ and $\alpha = k_B T_L/\sigma_f$. Note the similarity with the relaxation expression (Equation 14.12) and the correspondence between E^* and the demarcation energy in the dispersive multiple trapping equations. The median dissociation energy E_{dm} was assumed to depend on the oxide electric field in order to accommodate for the reported field dependence. For short stress times, the above simplifies to a power law

$$\Delta N_{it}(t) = \Delta N_{it,\max}\left(\frac{t}{\tau}\right)^{\alpha}. \tag{14.55}$$

Again, as with the dispersive transport model, a temperature-dependent slope is obtained. The analytic solution (Equation 14.54) is compared to the numerical solution in Figure 14.20, where excellent accuracy is obtained for $\sigma_f > 0.12$, which corresponds to $\alpha < 0.21$, and is thus well within the required regime.

In the above model the backward rate was assumed to be negligible, resulting in an unrecoverable degradation of ΔN_{it}. In order to generalize this model to allow at least for some recovery, one has to account for the reverse rate in Equation 14.21. In contrast to the RD model, however, where the diffusion of the hydrogen species eventually limits the reverse rate, it is now assumed that hydrogen at the interface is readily available. Formally, this may be done by setting H_{it} constant in Equation 14.21, equivalent to a large background hydrogen concentration. The solution of Equation 14.21 with $k_r \neq 0$ is readily obtained as

$$N_{it}(t, E_d, E_a) = N_0 \frac{k_d(E_d)}{k_d(E_d) + k_a(E_a)H_{it}}(1 - \exp(-k_d(E_d)t - k_a(E_a)H_{it}t)), \tag{14.56}$$

with the overall time evolution of N_{it} given through

$$N_{it}(t) = \int dE_d \int dE_a N_{it}(t, E_d, E_a) g_P(E_d, E_{dm}) g_P(E_a, E_{am}). \tag{14.57}$$

A numerical solution of Equation 14.57 is given in Figure 14.20 for varying parameters σ_f and σ_r. Obviously, the introduced reverse rate strongly influences the stress phase.

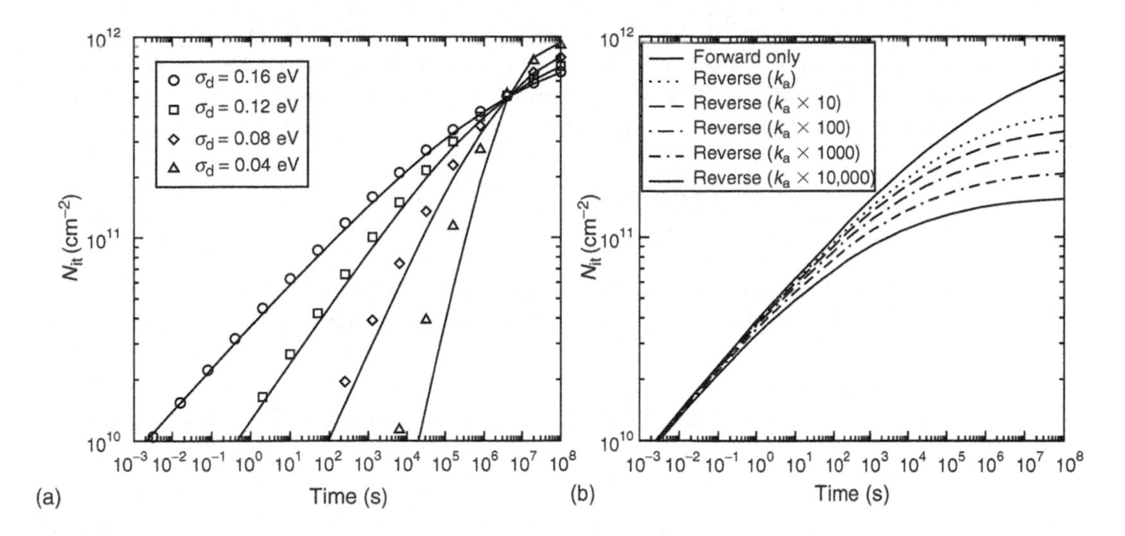

FIGURE 14.20
(a) Comparison of the analytic result of Equation 14.54 (lines) with the numerical solution of Equation 14.53 (symbols). Good agreement is obtained for larger dispersion σ_f. (b) As soon as the reverse rate is taken into account, an additional curvature in the slope is introduced which increases with the reverse rate.

During the relaxation phase an apparently very flexible model behavior is observed and the model can be nicely fit to a single relaxation curve. Unfortunately, however, the excellent fit during the single relaxation phase adversely affects the quasi-power-law exponent during the stress phase which reduces to very small values ($n \approx 0.03$). Furthermore, the model does not scale universally as demonstrated in Figure 14.21.

14.7.4.5 Simple Dispersive Hole Trapping Model

In addition to the creation of interface states, trapped charges have been made responsible for the observed threshold voltage shift during NBT stress. In particular, it has been argued

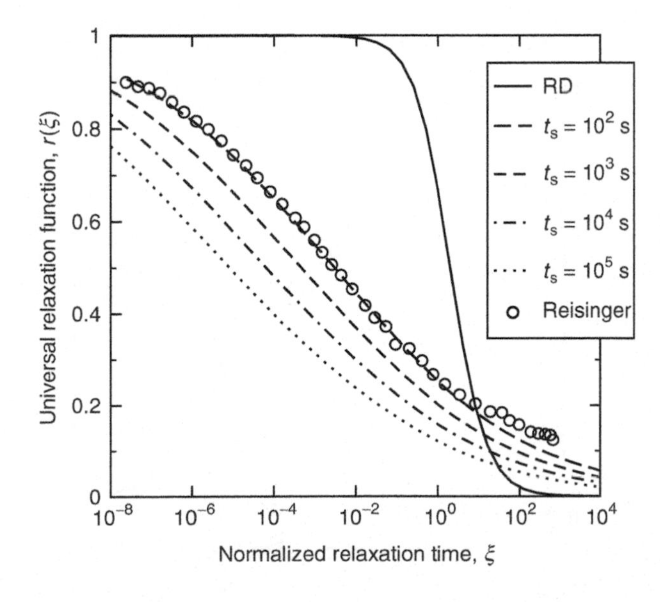

FIGURE 14.21
The dispersive rate model can be fit to an individual relaxation curve, $t_s = 100$ s in this case, but does not scale universally. In addition, the calibrated model gives a rather strong curvature during the stress phase with a too small power-law exponent $n \approx 0.03$, the variance of the rate coefficients had to be set to a value considerably larger than reported ($\sigma_A = 0.211$ eV and $\sigma_D = 0.264$ eV, compare Refs. [34,35]), and in general the model cannot be fit to both the stress and relaxation phase.

that these charges are responsible for the fast component observed both during stress and relaxation [11,16]. A simple phenomenological hole trapping model has been used by Yang et al. [16] based on a broad distribution of trapping times. It is assumed that hole trapping occurs over a broad spectrum of capture and release times following first-order kinetics

$$\frac{\partial_{\mathrm{p}}(t,\tau_{\mathrm{c}},\tau_{\mathrm{e}})}{\partial t} = \frac{1}{\tau_{\mathrm{c}}}(N_{\mathrm{ot}} - p) - \frac{p}{\tau_{\mathrm{e}}}. \tag{14.58}$$

Here, p is the hole concentration of a trap with capture and emissions times τ_c and τ_e, while N_{ot} is the trap density. In order to fit their measurement data, Yang et al. coupled the two time-constants via $\tau_e = k\tau_c$ and employed different capture and release times during stress and relaxation. The overall time evolution of all trapped charges is obtained by weighing all contributions using a probability density function in a manner similar to the rate-limited model (Equation 14.53):

$$p(t) = \int \mathrm{d}\tau\ p(t,\tau)f(\tau). \tag{14.59}$$

The probability density function for the relaxation times was assumed to be given by a log normal distribution:

$$f(\tau) = \frac{1}{\sqrt{2\pi}\tau\sigma} \exp\left(-\frac{1}{2}\left(\frac{\log(\tau) - \mu}{\sigma}\right)^2\right). \tag{14.60}$$

The numerical solution of Equation 14.59 is shown in Figure 14.22. Although a fast component can indeed be formed and a single relaxation curve can be nicely fit, no universality is observed. Note that this was to be expected due to the mathematical similarity of Equation 14.59 with Equation 14.53. We remark that this is not in disagreement with the good agreement to the measurement data reported in Ref. [16], but possibly a consequence of the narrow range of stress times employed in that study.

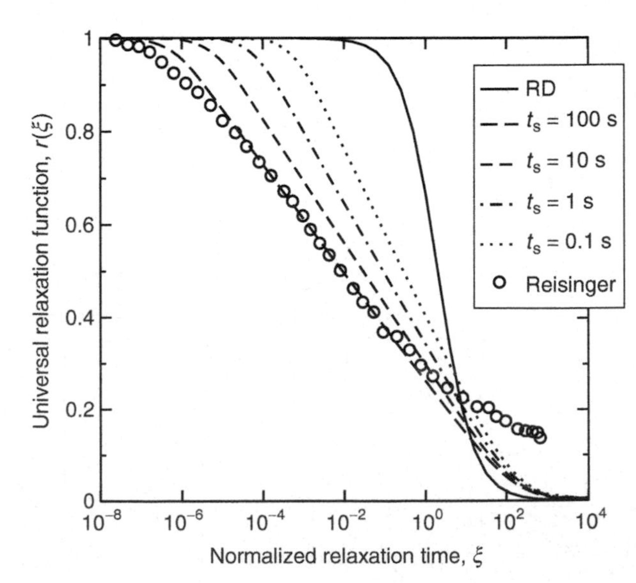

FIGURE 14.22
(a) Simple dispersive hole trapping model used in Ref. [16] can be fit to an individual relaxation curve but does not scale universally and gives a very small slope during the stress phase. (From Grasser, T., et al., *Proc. IRPS*, 268, 2007. With permission.)

14.7.4.6 Detailed Dispersive Hole Trapping Model

A more detailed hole trapping model has been derived by Tewksbury [74] based on a number of possible transitions from conduction, valence, and interface states into bulk oxide traps. Its use for NBTI has been suggested by Huard et al. [75] to cover the recoverable part of the degradation. For the following discussion we limit ourselves to the component of the model which results from charge transfer from an interface state into an oxide trap and back to the interface state, the other suggested mechanisms behave similarly [74] and follow analogously. During stress, the threshold voltage shift due to trapped bulk charge accumulated via transfer from the interface states can be given as $S(t_s) \approx A \ln(t_s/\tau_{0s})$ while the absolute relaxation is given by $R_M(t_s,t_r) \approx A \ln(t_s\tau_{0r}/(t_r\tau_{0s}))$ and depends (at least in this approximate form) universally on t_r/t_s. However, using the previous two relations, the relative relaxation function is given by

$$r(t_s,t_r) \approx 1 - \ln\left(\frac{t_r}{\tau_{0r}}\right) \ln^{-1}\left(\frac{t_s}{\tau_{0s}}\right), \tag{14.61}$$

which cannot be written as a function of t_s/t_r and is consequently not universal in our sense. The full numerical solution of the Tewksbury model is given in Figure 14.23 together with an excellent fit for a single relaxation curve. However, in order to obtain such a fit, the logarithmic behavior of the hole trapping component results in a slope close to zero during the stress phase. Also shown in Figure 14.23 is a permanent component modeled by a numerical solution of a dispersive forward rate only, as suggested by Huard et al. [75]. Note however, that after a certain stress time the degradation will be dominated by ΔQ_{it} and the observed relaxation given only through ΔQ_{ox} will be minimal. This is also not compatible with the data at hand where even at large stress times considerable relaxation can be observed.

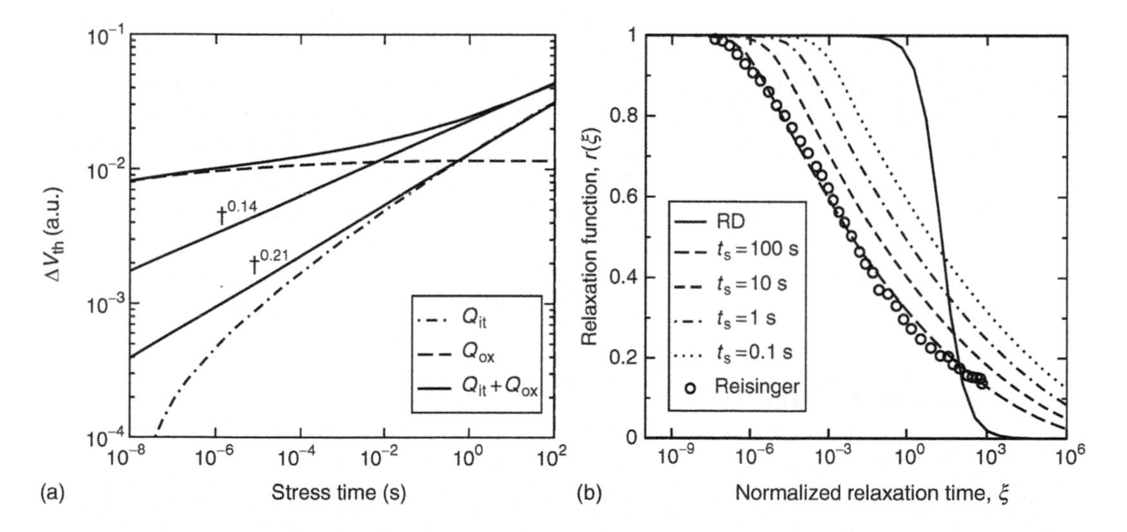

FIGURE 14.23
(a) Behavior during stress as predicted by the Tewksbury model on top of the dispersive rate (Equation 14.54). (b) Relaxation predicted by the Tewksbury model. The model can be fit to the data for an individual relaxation curve, here again at 100 s, but does not scale universally. Also, the excellent fit comes at the price of a very small power-law exponent at early times during the stress phase, in contradiction to Figure 14.4 of Ref. [9]. We were not able to fit both the stress and relaxation phases with the same set of parameters.

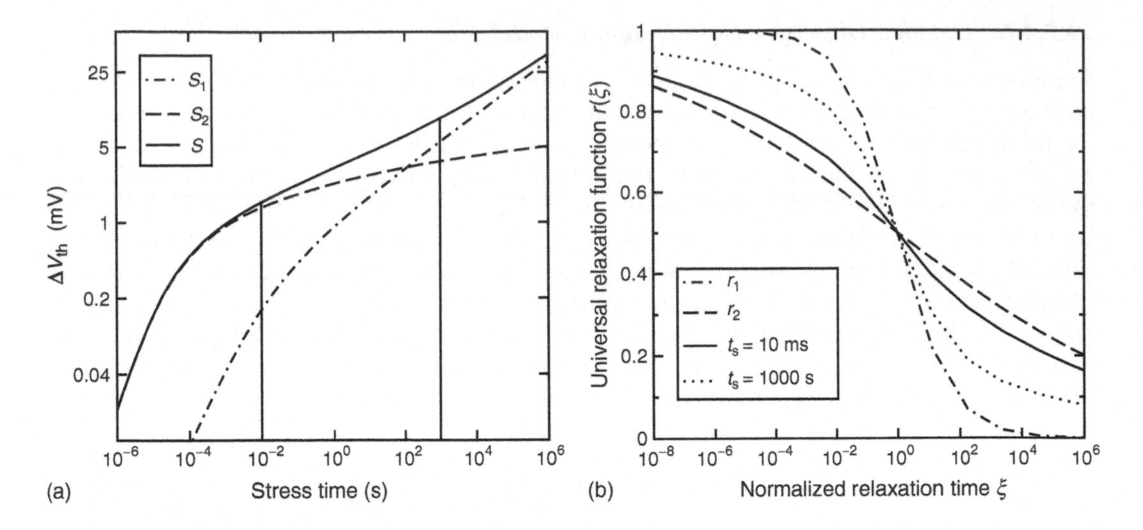

FIGURE 14.24
Schematic illustration of the contribution of two processes to the overall NBTI stress and relaxation characteristics. Process 1 could be the RD model while process 2 might be a simple hole-trapping process. If both processes are of the same order of magnitude, as is frequently assumed, the relaxation recorded at short stress times will be considerably different from a relaxation characteristic recorded at later times, in contrast to measurement data which are universal for usually employed stress times.

14.7.5 Multiple Mechanisms

Since none of the studied mechanisms can fully capture the universal relaxation, we have considered various combinations in our numerical framework. In order to obtain a universal behavior, some points need to be considered. Consider the case that the total observed threshold voltage shift is the result of two independent mechanisms, that is, $S = S_1 + S_2$. During relaxation one observes $R = S_1 r_1 + S_2 r_2$ and the normalized relaxation function is given by $r = \rho r_1 + (1 - \rho) r_2$ with $\rho = S_1/(S_1 + S_2)$. If the two degradation mechanisms progress differently with time, ρ will be a function of t_s and r cannot be universal, see Figure 14.24. We thus conclude that for the relaxation to be universal, the two mechanisms need to be tightly coupled, that is, $S_1/S_2 = \text{const}$, or at least roughly constant within the range of measured ξ and within the measurement tolerance. Alternatively, both mechanisms could relax equally, $r_1 \approx r_2$, or one process could be dominant for the range of recorded stress times. Finally, one process could be permanent or slowly relaxing, forming the permanent component identified in Ref. [59] which would have to subtracted from the total relaxation data in order to study the universally recovering component.

14.8 Conclusions

We have thoroughly analyzed the relaxation of NBT stress-induced degradation using data from various groups. The observed universal relaxation behavior has been quantified and modeled using possible empirical expressions. It has been demonstrated that data obtained via conventional MSM sequences can be analytically described as a function of the delay introduced during the measurement. In particular, this analytic expression allows one to reconstruct a corrected degradation curve. Using this corrected curve, it might be possible to more accurately estimate the time-to-failure.

We have then used the relaxation behavior and in particular the universality as a benchmark for existing NBTI models. There we have found that none of the existing models are capable of reproducing both the stress and the relaxation phase with the same set of parameters. While the classic RD model scales universally, it predicts relaxation to occur mainly during three to four decades, in contradiction to detailed relaxation measurements available in literature which span more than 12 decades. No improvement could be found in extended RD models using two regions, a second interface, or an explicit transition from atomic to molecular hydrogen. Models based on an extension of the RD model with dispersive transport somewhat improve on the situation but are still not able to cover the whole relaxation regime. Other dispersive models, like dispersive forward and backward rates or dispersive hole-trapping models allow one to fit an individual relaxation curve only but are not universal. In addition, we were not able to describe both the stress and relaxation phase with the same set of parameters. This indicates a significant gap in our current understanding of NBTI.

We particularly wish to point out that it is of utmost importance not to consider the inaccuracies of existing models during the relaxation phase of secondary importance compared to the stress phase. The reason for this is only partially related to the frequently quoted fact that continuous DC stress is rarely observed in a circuit and that duty-cycle dependent corrections have to be applied. The more important point we want to make here is that during the stress phase the relaxation mechanism in existing models always interacts with the degradation mechanism, dominating the overall time behavior during the stress phase. It is only during the relaxation phase, where the degradation mechanism is more or less absent, that the relaxation mechanism can be studied in full detail, despite the difficulties arising during measurements. We therefore stipulate that a more complete NBTI model needs to focus on the relaxation phase first before attempting to cover the stress phase as well.

It is also important to stress that the observed discrepancies in the available models with measurement data do not necessarily indicate that the physical processes involved in NBTI, predominantly hole trapping and interface state generation, have been wrongly identified. The main finding of this study is that NBTI relaxation, and consequently the stress phase, are strongly influenced by physical mechanisms that are not yet fully understood and require a refined set of models.

References

1. Miura, Y. and Matukura, Y., Investigation of silicon–silicon dioxide interface using MOS structure, *Jpn. J. Appl. Phys.*, 5, 180, 1966.
2. Schroder, D.K. and Babcock, J.A., Negative bias temperature instability: Road to cross in deep submicron silicon semiconductor manufacturing, *J. Appl. Phys.*, 94(1), 1–18, 2003.
3. Schroder, D.K., Negative bias temperature instability: What do we understand? *Microelectron. Reliab.*, 47(6), 841–852, 2007.
4. Krishnan, A.T. et al., Material dependence of hydrogen diffusion: Implications for NBTI degradation, in *Proc. IEDM*, Washington, Washington D.C., 2005, pp. 688–691.
5. Alam, M.A. et al., A comprehensive model for PMOS NBTI degradation: Recent progress, *Microelectron. Reliab.*, 47(6), 853–862, 2007.
6. Shen, C. et al., Characterization and physical origin of fast V_{th} transient in NBTI of p-MOSFETs with SiON dielectric, in *Proc. IEDM*, San Francisco, CA, 2006, pp. 333–336.
7. Grasser, T. and Kaczer, B., Negative bias temperature instability: Recoverable versus permanent degradation, in *Proc. ESSDERC*, Munich, Germany, 2007, pp. 127–130.

8. Huard, V., Denais, M., and Parthasarathy, C., NBTI degradation: From physical mechanisms to modelling, *Microelectron. Reliab.*, 46(1), 1–23, 2006.
9. Reisinger, H. et al., Analysis of NBTI degradation- and recovery-behavior based on ultra fast V_{th}-measurements, in *Proc. IRPS*, San Jose, CA, 2006, pp. 448–453.
10. Alam, M.A. and Mahapatra S., A comprehensive model of PMOS NBTI degradation, *Microelectron. Reliab.*, 45(1), 71–81, 2005.
11. Huard, V. et al., A thorough investigation of MOSFETs NBTI degradation, *Microelectron. Reliab.*, 45(1), 83–98, 2005.
12. Houssa, M., Modelling negative bias temperature instabilities in advanced p-MOSFETs, *Microelectron. Reliab.*, 45(1), 3–12, 2005.
13. Zafar, S., Statistical mechanics based model for negative bias temperature instability induced degradation, *J. Appl. Phys.*, 97(10), 1–9, 2005.
14. Kaczer, B. et al., Disorder-controlled-kinetics model for negative bias temperature instability and its experimental verification, in *Proc. IRPS*, San Jose, CA, 2005, pp. 381–387.
15. Stathis, J.H. and Zafar, S., The negative bias temperature instability in MOS devices: A review, *Microelectron. Reliab.*, 46(2–4), 270–286, 2006.
16. Yang, T. et al., Fast DNBTI components in p-MOSFET with SiON dielectric, *IEEE Electron Dev. Lett.*, 26(11), 826–828, 2005.
17. Mahapatra, S. et al., On the physical mechanism of NBTI in silicon oxynitride p-MOSFETs: Can differences in insulator processing conditions resolve the interface trap generation versus hole trapping controversy? in *Proc. IRPS*, Phoenix, AZ, 2007, pp. 1–9.
18. Denais, M. et al., Interface trap generation and hole trapping under NBTI and PBTI in advanced CMOS technology with a 2-nm gate oxide, *IEEE Trans. Dev. Mater. Reliab.*, 4(4), 715–722, 2004.
19. Ershov, M. et al., Transient effects and characterization methodology of negative bias temperature instability in pMOS transistors, in *Proc. IRPS*, Dallas, TX, 2003, pp. 606–607.
20. Varghese, D. et al., On the dispersive versus Arrhenius temperature activation of NBTI time evolution in plasma nitrided gate oxides: Measurements, theory, and implications, in *Proc. IEDM*, Washington, Washington D.C., Dec. 2005, pp. 1–4.
21. Denais, M. et al., On-the-fly characterization of NBTI in ultra-thin gate oxide p-MOSFETs, in *Proc. IEDM*, San Francisco, CA, 2004, pp. 109–112.
22. Li, M.F. et al., Dynamic bias-temperature instability in ultrathin SiO_2 and HfO_2 metal-oxide-semiconductor field effect transistors and its impact on device lifetime, *Jpn. J. Appl. Phys.*, 43(11B), 7807–7814, 2004.
23. Poindexter, E.H. et al., Electronic traps and P_b centers at the Si/SiO_2 interface: Band-gap energy distribution, *J. Appl. Phys.*, 56(10), 2844–2849, 1984.
24. Lenahan, P.M. and Conley, J.F. Jr., What can electron paramagnetic resonance tell us about the Si/SiO_2 system? *J. Vac. Sci. Technol. B*, 16(4), 2134–2153, 1998.
25. Campbell, J.P. et al., Direct observation of the structure of defect centers involved in the negative bias temperature instability, *Appl. Phys. Lett.*, 87(20), 1–3, 2005.
26. Campbell, J.P. et al., Observations of NBTI-induced atomic-scale defects, *IEEE Trans. Dev. Mater. Reliab.*, 6(2), 117–122, 2006.
27. Stesmans, A. and Afanas'ev, V.V., Electrical activity of interfacial paramagnetic defects in thermal (100) Si/SiO_2, *Phys. Rev. B*, 57(16), 10030–10034, 1998.
28. Campbell, J.P. et al., Density of states and structure of NBTI-induced defects in plasma-nitrided pMOSFETs, in *Proc. IRPS*, Phoenix, AZ, 2007, pp. 503–510.
29. Grasser, T. et al., TCAD modeling of negative bias temperature instability, in *Proc. SISPAD*, Monterey, USA, Sept. 2006, pp. 330–333.
30. Cartier, E. and Stathis, J.H., Hot-electron induced passivation of silicon dangling bonds at the Si (111)/SiO_2 interface, *Appl. Phys. Lett.*, 69(1), 103–105, 1996.
31. Ragnarsson, L.-A. and Lundgren, P., Electrical characterization of P_b centers in (100)Si/SiO_2 structures: The influence of surface potential on passivation during post metallization anneal, *J. Appl. Phys.*, 88(2), 938–942, 2000.

32. Grasser, T. et al., The universality of NBTI relaxation and its implications for modeling and characterization, in *Proc. IRPS*, Phoenix, AZ, 2007, pp. 268–280.
33. Haggag, A. et al., High-performance chip reliability from short-time-tests, in *Proc. IRPS*, Orlando, FL, 2001, pp. 271–279.
34. Stesmans, A., Passivation of P_{b0} and P_{b1} interface defects in thermal (100) Si/SiO_2 with molecular hydrogen, *Appl. Phys. Lett.*, 68(15), 2076–2078, 1996.
35. Stesmans, A., Dissociation kinetics of hydrogen-passivated P_b defects at the (111)Si/SiO_2 interface, *Phys. Rev. B*, 61(12), 8393–8403, 2000.
36. Neugroschel, A. et al., Direct-current measurements of oxide and interface traps on oxidized silicon, *IEEE Trans. Electron. Dev.*, 42(9), 1657–1662, 1995.
37. Groeseneken, G. et al., A reliable approach to charge-pumping measurements in MOS transistors, *IEEE Trans. Electron. Dev.*, 31(1), 42–53, 1984.
38. Zhang, J.F. et al., Hole traps in silicon dioxides—Part I: Properties, *IEEE Trans. Electron. Dev.*, 51(8), 1267–1273, 2004.
39. Parthasarathy, C.R. et al., New insights into recovery characteristics post NBTI stress, in *Proc. IRPS*, San Jose, CA, 2006, pp. 471–477.
40. Rangan, S., Mielke, N., and Yeh, E.C.C., Universal recovery behavior of negative bias temperature instability, in *Proc. IEDM*, Washington, Washington D.C., 2003, pp. 341–344.
41. Denais, M. et al., Paradigm shift for NBTI characterization in ultra-scaled CMOS technologies, in *Proc. IRPS*, San Jose, CA, 2006, pp. 735–736.
42. Reisinger, H. et al., A comparison of very fast to very slow components in degradation and recovery due to NBTI and bulk hole trapping to existing physical models, *IEEE Trans. Dev. Mater. Reliab.*, 7(1), 119–129, 2007.
43. Grasser, T. et al., Simultaneous extraction of recoverable and permanent components contributing to bias-temperature instability, in *Proc. IEDM*, 2007, pp. 801–804.
44. Kakalios, J., Street, R.A., and Jackson, W.B., Stretched-exponential relaxation arising from dispersive diffusion of hydrogen in amorphous silicon, *Phys. Rev. Lett.*, 59(9), 1037–1040, 1987.
45. Li, J.-S. et al., Effects of delay time and AC factors on negative bias temperature instability of PMOSFETs, in *IIRW Final Rep.*, Lake Tahoe, CA, 2006, pp. 16–19.
46. Krishnan, A.T. et al., Negative bias temperature instability mechanism: The role of molecular hydrogen, *Appl. Phys. Lett.*, 88(15), 1–3, 2006.
47. Ushio, J., Okuyama, Y., and Maruizumi, T., Electric-field dependence of negative-bias temperature instability, *J. Appl. Phys.*, 97(8), 1–3, 2005.
48. Tsetseris, L. et al., Physical mechanisms of negative-bias temperature instability, *Appl. Phys. Lett.*, 86(14), 1–3, 2005.
49. Van De Walle, C.G., Stretched-exponential relaxation modeled without invoking statistical distributions, *Phys. Rev. B*, 53(17), 11292–11295, 1996.
50. Nickel, N.H., Yin, A., and Fonash, S.J., Influence of hydrogen and oxygen plasma treatments on grain-boundary defects in polycrystalline silicon, *Appl. Phys. Lett.*, 65(24), 3099–3101, 1994.
51. Houssa, M., Reaction-dispersive proton transport model for negative bias temperature instabilities, *Appl. Phys. Lett.*, 86(9), 1–3, 2005.
52. Ogawa, S., Shimaya, M., and Shiono, N., Interface-trap generation at ultrathin SiO_2 (4 nm–6 nm)–Si interfaces during negative-bias temperature aging, *J. Appl. Phys.*, 77(3), 1137–1148, 1995.
53. Ogawa, S. and Shiono, N., Generalized diffusion-reaction model for the low-field charge build up instability at the Si/SiO_2 interface, *Phys. Rev. B*, 51(7), 4218–4230, 1995.
54. Islam, A.E. et al., Temperature dependence of the negative bias temperature instability in the framework of dispersive transport, *Appl. Phys. Lett.*, 90(1), 083505-1–083505-3, 2007.
55. Kufluoglu, H. and Alam, M.A., Theory of interface-trap-induced NBTI degradation for reduced cross section MOSFETs, *IEEE Trans. Electron Dev.*, 53(5), 1120–1130, 2006.
56. Alam, M.A., A critical examination of the mechanics of dynamic NBTI for p-MOSFETs, in *Proc. IEDM*, Washington, Washington D.C., 2003, pp. 345–348.
57. Yang, J.B. et al., Analytical reaction–diffusion model and the modeling of nitrogen-enhanced negative bias temperature instability, *Appl. Phys. Lett.*, 88(17), 1–3, 2006.

58. Alam, M.A. and Kufluoglu, H., On quasi-saturation of negative bias temperature degradation, *ECS Trans.*, 1(1), 139–145, 2005.
59. Grasser, T., Gös, W., and Kaczer, B., Dispersive transport and negative bias temperature instability: Boundary conditions, initial conditions, and transport models, *IEEE Trans. Dev. Mater. Reliab.*, 8(1), 79–97, 2008.
60. Alam, M.A., NBTI: A simple view of a complex phenomena, in *Proc. IRPS*, San Jose, CA, 2006, (Tutorial).
61. Zafar, S. et al., A comparative study of NBTI and PBTI (charge trapping) in SiO_2/HfO_2 stacks with FUSI, TiN, Re gates), in *Proc. VLSI Symp.*, Honolulu, HI, 2006, pp. 23–25.
62. Chakravarthi, S. et al., Probing negative bias temperature instability using a continuum numerical framework: Physics to real world operation, *Microelectron. Reliab.*, 47(6), 863–872, 2007.
63. Stathis, J.H., Dissociation kinetics of hydrogen-passivated (100) Si/SiO_2 interface defects, *J. Appl. Phys.*, 77(12), 6205–6207, 1995.
64. Houssa, M. et al., Modeling negative bias temperature instabilities in hole channel metal-oxide-semiconductor field effect transistors with ultrathin gate oxide layers, *J. Appl. Phys.*, 95(5), 2786–2791, 2004.
65. Grasser, T., Gös, W., and Kaczer, B., Modeling of dispersive transport in the context of negative bias temperature instability, in *IIRW Final Rep.*, Lake Tahoe, CA, 2006, pp. 5–10.
66. Noolandi, J., Multiple-trapping model of anomalous transit-time dispersion in *a*-Se, *Phys. Rev. B*, 16(10), 4466–4473, 1977.
67. Orenstein, J., Kastner, M.A., and Vaninov, V., Transient photoconductivity and photo-induced optical absorption in amorphous semiconductors., *Philos. Mag. B*, 46(1), 23–62, 1982.
68. Arkhipov, V.I. and Rudenko, A.I., Drift and diffusion in materials with traps, *Philos. Mag. B*, 45(2), 189–207, 1982.
69. Rudenko, A.I. and Arkhipov, V.I., Drift and diffusion in materials with traps: Quasi-equilibrium transport regime, *Philos. Mag. B*, 45(2), 177–187, 1982.
70. Arkhipov, V.I. and Bässler, H., An adiabatic model of dispersive hopping transport, *Philos. Mag. B*, 68(4), 425–435, 1993.
71. Kaczer, B. et al., Temperature dependence of the negative bias temperature instability in the framework of dispersive transport, *Appl. Phys. Lett.*, 86(14), 1–3, 2005.
72. Chakravarthi, S. et al., A comprehensive framework for predictive modeling of negative bias temperature instability, in *Proc. IRPS*, Phoenix, AZ, 2004, pp. 273–282.
73. Huard, V., Private communications.
74. Tewksbury, T.L. and Lee, H.-S., Characterization, modeling, and minimization of transient threshold voltage shifts in MOSFETs, *IEEE J. Solid State Circ.*, 29(3), 239–252, 1994.
75. Huard, V. et al., Physical modeling of negative bias temperature instabilities for predictive extrapolation, in *Proc. IRPS*, San Jose, CA, 2006, pp. 733–734.

15

Wear-Out and Time-Dependent Dielectric Breakdown in Silicon Oxides

John S. Suehle

CONTENTS

15.1 Definitions

The following definitions of terms apply to this chapter:

SILC Stress-induced leakage current (A). Excess leakage current through a gate dielectric generally caused by trap-assisted tunneling.

N_{it} Number of interface traps that exist at or close to the silicon–silicon dioxide interface; these can change occupancy depending on the position of the Fermi level at the interface.

N_{ot} Number of fixed oxide traps that can exist throughout the volume of a dielectric film. Their charge state generally remains fixed and can affect the flatband voltage of the device.
N_{BD} Number of defects required to be generated in an oxide film before dielectric breakdown occurs.
P_g Defect generation rate. The rate by which oxide traps or defects are generated during high-voltage stress.
Q_{BD} Charge-to-breakdown (C). The amount of charge that has tunneled through a gate dielectric up until dielectric breakdown occurs.
J_g Tunneling current density through a gate dielectric (A/cm^2).
t_{BD} Time to dielectric breakdown (s).
E_{ox} Stress electric field across the oxide film (MV/cm).
γ Electric field acceleration factor for dielectric breakdown (decades/MV).
E_a Thermal activation energy for dielectric breakdown.
q_p Injected hole charge density to dielectric breakdown (C/cm^2).
SHEI Substrate hot electron injection. This is a technique using an *n*-channel metal-oxide-semiconductor field effect transistor (MOSFET) and a separate injector to inject electrons into the gate dielectric independent of the applied gate voltage.
SHHI Substrate hot hole injection. This is a technique using a *p*-channel MOSFET and a separate injector to inject holes into the gate dielectric independent of the applied gate voltage.
AHR Anode hydrogen release. This is a mechanism where atomic hydrogen is released via an energy transfer of tunneling electrons in the anode region of a thin gate dielectric.
AHI Anode hole injection. This is a mechanism where holes in the anode region gain energy via tunneling electrons and are injected into a thin gate dielectric.

15.2 Introduction

Time-dependent dielectric breakdown (TDDB) or "wear-out" of thin silicon dioxide films is a process where defects are created throughout the volume of the dielectric by the application of a voltage across the film. These defects continue to accumulate with time until a localized conductance path is formed and shorts the anode to the cathode producing a large current surge through the path. High joule heating from the localized power dissipation produces permanent structural damage and the film loses its insulating properties, essentially shorting the cathode to the anode. Even though TDDB has been studied for over three decades, the exact physical mechanism remains elusive. What is known is that the process is driven by voltage and temperature.

The continued scaling of MOSFET dimensions in advanced microelectronics to achieve higher performance and circuit density requires the thinning of the gate dielectric to control short channel effects. Unfortunately, dielectric thinning not only increases leakage current and standby power dissipation, but also compromises the intrinsic reliability of the dielectric layer. The driving forces for TDDB appear to be increasing as devices are scaled. Circuits are operating at higher temperatures, and the operating voltage is not scaling as quickly as device dimensions. It was projected at one time that devices would not have sufficient reliability if their dielectrics were scaled thinner than 2.2 nm [1]. Extensive research studies were conducted over the last decade to understand the reliability consequences of dielectric thinning and how to accurately project device lifetime. Of particular interest was the occurrence of soft breakdown in ultrathin dielectric films. Because of the reduced power dissipation at the instant of dielectric breakdown in ultrathin films, the thermal transient is significantly reduced. The degree of dielectric structural damage was limited, and the conductance through the broken-down film is much smaller than that in a thicker dielectric. In fact, it was observed that some logic circuits continued to operate even after gate dielectrics in the constituent devices failed [2,3]. Device lifetime may be different than circuit or product lifetime and projection of both require different procedures. Accurate device or circuit life projection requires correct physical models for voltage and temperature acceleration, correct statistics for modeling failure distributions, and correct circuit models for devices that have limited dielectric damage due to soft breakdown.

This chapter will discuss the various physical mechanisms proposed for defect generation and dielectric breakdown in thin silicon dioxide films. The present understanding of the driving forces for defect generation will be presented, as well as statistical models for describing the failure distributions observed after TDDB experiments. Finally, recent developments in ultrathin dielectric breakdown, including soft breakdown, the power-law voltage acceleration model, and breakdown in advanced gate dielectric stacks will be discussed.

15.3 Physical Models for Dielectric Breakdown

15.3.1 Defect Generation Model

The generally accepted process of time-dependent dielectric breakdown is illustrated in Figure 15.1. The left panel represents a typical plot of gate current versus stress time when a voltage is applied across a thin gate dielectric film. The right panel is a schematic representation of three phases of defect generation in a dielectric where shaded circles are defects that have electrically connected to another defect or to the anode or cathode of the device. In phase 1 wear-out defects begin to form randomly in the early stages of

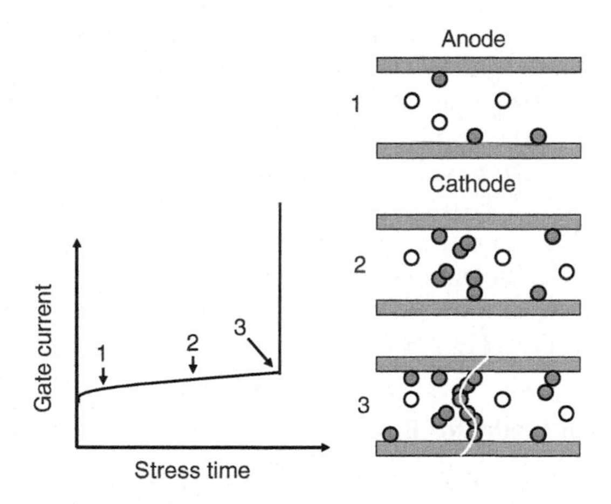

FIGURE 15.1
Schematic representation illustrating defect generation and dielectric breakdown in a thin dielectric film. The figure shows the sequence of defect generation (step 1), defect coalescence (step 2), and percolation filament formation (step 3) in the right panel. The corresponding change in the gate leakage current as a function of stress time is shown in the left panel. The dielectric breaks down when a filament connects the anode to the cathode producing an electrical short.

wear-out. These defects may be correlated to measurable electrically active defects such as stress-induced leakage current (SILC), interface traps (N_{it}), or fixed oxide charge (N_{ot}). The left panel illustrates a gradual increase in leakage current as SILC defects accumulate; however, in thick dielectric films the leakage current could decrease in time due to electron trapping. In phase 2, wear-out defects begin to coalesce and form conduction or percolation paths. Finally, in phase 3 a percolation path connects the cathode to the anode, and a large surge of current flows through a localized region in the dielectric causing permanent structural damage and loss of dielectric properties.

The concept of a defect generation model where defects accumulate until a critical number of defects are produced (N_{BD}) to cause dielectric breakdown was formalized by DiMaria and coworkers [4–7]. The defect generation model relates the charge-to-breakdown (Q_{BD}) to N_{BD} divided by the defect generation rate (P_g) or

$$Q_{BD} = \frac{qN_{BD}}{P_g} \tag{15.1}$$

Q_{BD} is obtained by simply integrating the tunneling current until the oxide breaks down or

$$Q_{BD} = \int_{t=0}^{t=t_{BD}} i_{ox}\,dt \tag{15.2}$$

The time-to-breakdown (t_{BD}) is equal to the gate tunneling current (J_g) divided by Q_{BD} or

$$t_{BD} = \frac{qN_{BD}}{J_g P_g}. \tag{15.3}$$

Figure 15.2 shows that P_g of electrically measurable defects including SILC, N_{it}, and fixed oxide charges (N_{ot}) depends exponentially on gate voltage [6]. It is interesting to note that the voltage dependences of P_g fall on the same line independent of what defect was measured. The deviation from the exponential dependence observed at higher gate voltages ($V_g > 6.0$ V) has been explained as the transport mechanism of the tunneling carriers changing from a ballistic to a quasi-steady-state process in thicker oxides [7]. The model provides a convenient oxide projection methodology, where one can measure the generation rate of SILC or N_{it} and extrapolate when an oxide breakdown will occur. N_{BD} would need to be independent of stress voltage for such a projection to be valid. Indeed, this has been demonstrated in Refs. [5,8–10]. However, it was shown that at very long stress times N_{BD} was larger than values measured at shorter stress times [11], and the value was larger under alternating bias conditions when compared to those measured under dc conditions with the same effective stress time [12]. N_{BD} was also observed to decrease for decreasing stress voltage in other studies [1,13–14].

15.3.2 Thermochemical Model for Breakdown

The exact physical mechanism by which defects are created is still subject to debate. Defects that can be probed (i.e., can be measured by electrical means) may not necessarily be correlated to defects that form percolation paths leading to breakdown. This was illustrated in those studies mentioned above where N_{BD} was not constant. Early physical models for defect generation were developed empirically by observations of the functional dependence of dielectric breakdown data with stress voltage, field, and temperature. Two different models emerged based on very different physical mechanisms for defect

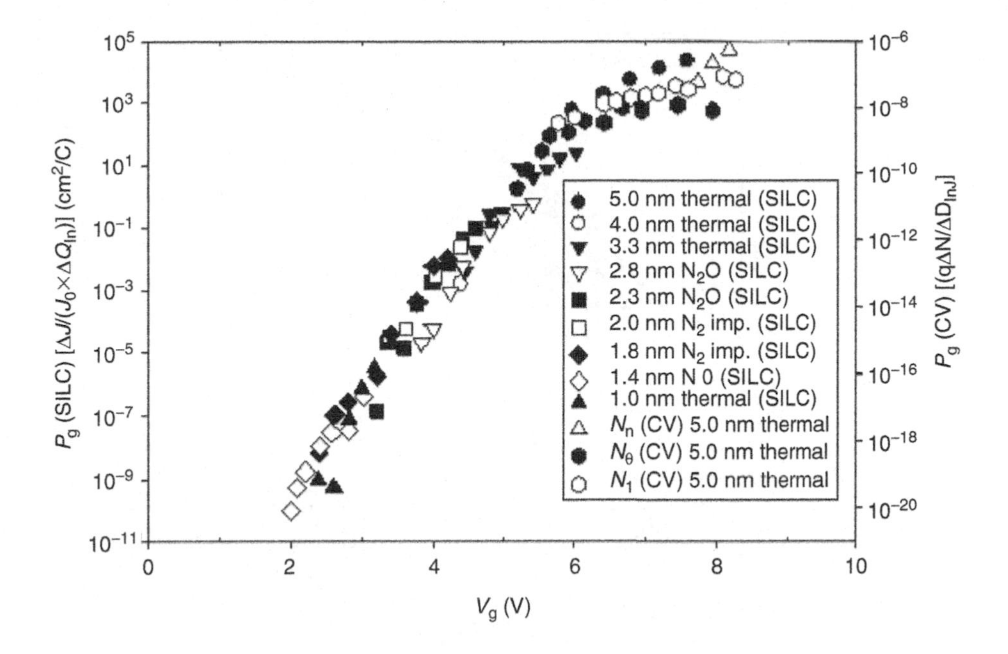

FIGURE 15.2
Plot showing that the defect generation rate, P_g, for SILC, N_{it}, N_h (trapped holes obtained from CV), and N_n (trapped electrons from CV) depends exponentially on the gate voltage over a wide range of oxide processing conditions and thickness. The data are collected over the direct tunneling regime ($V_g < 3$ V) through the Fowler–Nordheim regime ($V_g > 3$ V), and indicate that P_g is related to the energy of the electrons arriving at the anode. The saturation shown above $V_g \approx 7.5$ V for N_h, N_n, and N_{it} reflects the threshold for the AHI process. (Reprinted from Stathis, J.H., *IEEE Trans. Dev. Mater. Reliab.*, 1, 43, 2001. With permission.)

generation. The first such model was developed by observing that time-to-breakdown was exponentially related to stress electric field or:

$$t_{BD} \sim \exp(-\gamma E_{ox}) \tag{15.4}$$

Here E_{ox} is the stress electric field across the oxide film and γ is the field acceleration parameter [15–18]. Known as the "E" model, a straight line would be observed if the log (t_{BD}) was plotted as a function of the applied electric field. Figure 15.3 shows typical TDDB data for several different oxide thicknesses plotted as a function of E_{ox} and $1/E_{ox}$ [19]. The physical model leading to a $1/E_{ox}$ dependence will be discussed in Section 15.3.3. Note that the data fit a straight line much better when plotted as a function of E_{ox} than $1/E_{ox}$. It is also important to observe from the plots that the data appear to fit E_{ox} and $1/E_{ox}$ equally well for stress electric fields above 8–9 MV/cm. The deviation from the $1/E_{ox}$ becomes evident for lower electric fields.

A thermochemical model based on electric field-induced bond breakage was developed by McPherson et al. [18,20,21], which correctly characterized data from a variety of studies. The kinetics of the model are illustrated in Figure 15.4. Amorphous silicon dioxide is composed of molecules that have a range of bonding angles between the silicon and oxygen atoms as shown in illustration A. Bond angles deviating from the mean angle of 150° are severely weak bonds, which can break upon oxide formation, leading to an oxygen vacancy structure as shown in illustration B. This structure is also referred to as an E' center and is paramagnetic [22]. The applied electric field can break the Si–Si bond associated with the oxygen vacancy, and the molecule collapses in the hybridization shown in illustration C. Such a structure behaves as a hole trap [23]. The thermochemical model

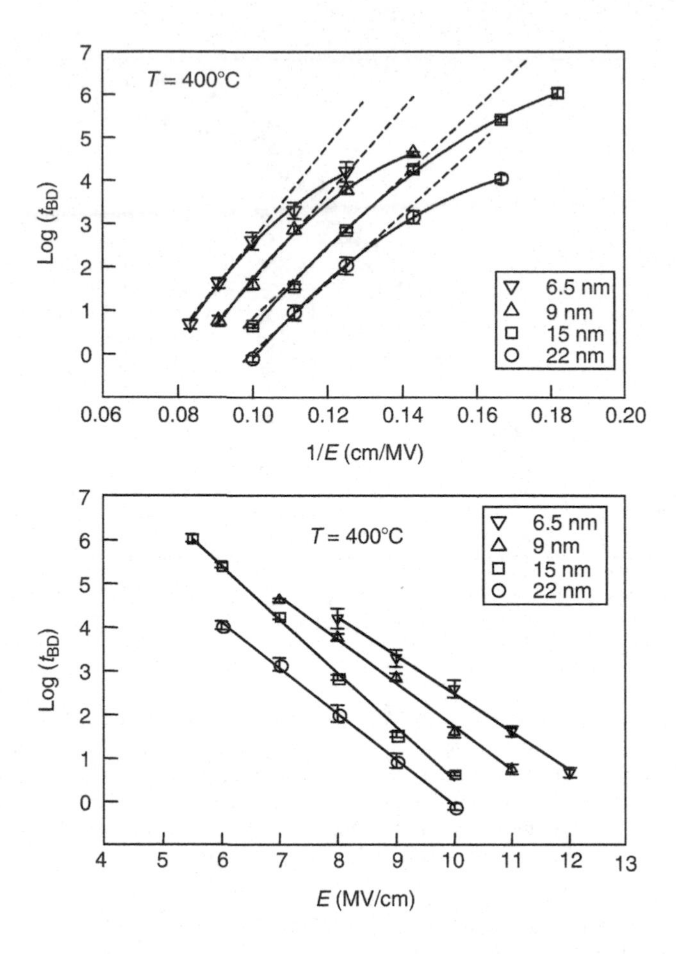

FIGURE 15.3
Plots showing log (t_{BD}) versus the inverse of the electric field (top plot) and electric field (bottom plot). Note that the data deviate away from the straight line at stress fields below ~7 MV/cm in the inverse electric field plot. All of the data fit a straight line well in the bottom plot where the data are plotted as a function of electric field. Such plots were typically used to infer the physical model for dielectric breakdown. (Reprinted from Suehle, J.S. and Chaparala, P., *IEEE Trans. Electron Dev.*, 44, 801, 1997. With permission.)

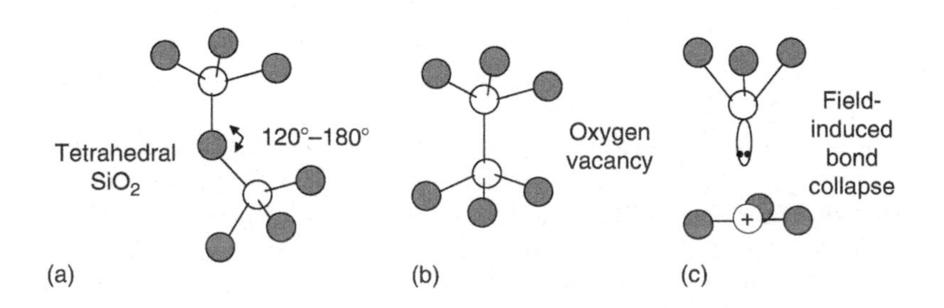

FIGURE 15.4
Illustrations showing the kinetics of field-induced bond breakage in the thermochemical dielectric breakdown model. (a) The range of Si–O–Si bond angles in a network of amorphous silicon dioxide can vary between 120° and 180°. (b) Bond angles outside the optimal range are very weak and can break resulting in an oxygen vacancy defect structure. (c) The applied electric field can break the Si–Si bond resulting in a hole trap.

includes a thermal activation energy, E_a, which decreases with increasing E_{ox}, and a field acceleration parameter, γ, which decreases with increasing temperature. The complete equation for the model becomes:

$$t_{BD} \sim \exp[\gamma(T)E_{ox}] \exp[E_a(E_{ox})/k_B T] \quad (15.5)$$

It is important to note that the thermochemical breakdown mechanism does not require tunneling electrons to induce bond breakage, and only the stress electric field and temperature are the driving forces. Experimental evidence presented in later sections of this chapter will show that energetic carriers associated with tunneling current play a significant role in oxide breakdown.

15.3.3 Anode Hole Injection Breakdown Mechanism

Anode hole injection (AHI) is a mechanism in which electrons tunneling from the cathode transfer energy to holes in the anode region. These energetic holes can then be injected into the oxide and produce wear-out defects. The initial formulation of the mechanism is depicted schematically in Figure 15.5. The model was based on impact ionization occurring inside the oxide induced by Fowler–Nordheim tunneling carriers [24–26]. Energetic holes are injected into the oxide and become trapped. The net positive trapped charge lowers the potential barrier near the cathode (Figure 15.5). Tunneling current begins to increase as a result of the lower barrier. As a result, more holes are injected via a positive feedback. Eventually, the oxide breaks down. The $1/E_{ox}$ dependence of breakdown time is due to the reciprocal electric field dependence of the Fowler–Nordheim tunneling process. The model has the form:

$$t_{BD} \sim \exp[(B + H)/E_{ox}]. \quad (15.6)$$

The field acceleration parameter $B + H$ is related to the prefactor of the Fowler–Nordheim current equation and the hole generation rate. Injected hole current density, j_{SUB}, can be

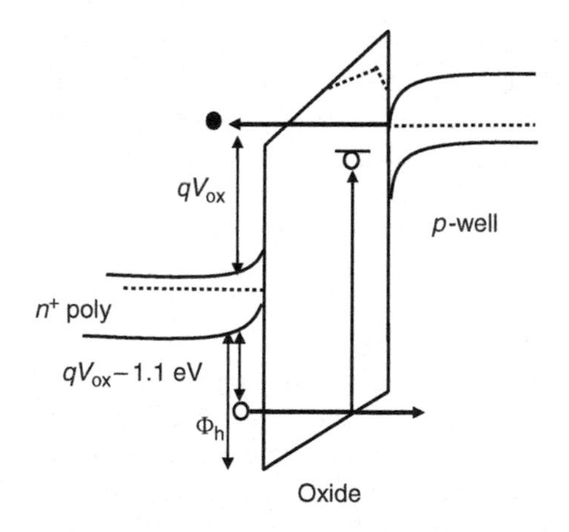

FIGURE 15.5
Band diagram showing the AHI process by which tunneling carriers create holes through impact ionization ($E \sim 1.1$ eV) and can be injected into the oxide. The initial formulation of the model was based on holes accumulating near the cathode where the local barrier height is lowered. The lower barrier favors additional injection resulting in more injected holes. Eventually a feedback results causing dielectric breakdown. For thinner oxides, the applied gate voltage is too low to provide enough energy for holes to surmount the hole barrier (Φ_h) of 4.7 eV. Other processes such as minority ionization would allow hole injection at lower applied gate voltages.

measured as a substrate current and integrated to obtain a critical hole charge density to breakdown:

$$q_{\mathrm{P}} = \int_{t=0}^{t=t_{\mathrm{BD}}} j_{\mathrm{SUB}} \mathrm{d}t \tag{15.7}$$

It was observed that q_{p} was independent of stress voltage and decreased as oxide thickness is decreased [27]. It was observed that q_{p} has a value of approximately 0.1 $\mathrm{C/cm^2}$, and q_{p} was also reported to be temperature dependent [28].

It was proposed that AHI could take place via a surface plasmon process that would require a gate voltage of about 7 V [29,30]. This high-energy process would appear to preclude AHI to be operative at lower voltages typical of circuit operating conditions. Another mechanism involving minority ionization was proposed where tunneling electrons could transfer energy to holes in *p*-type material or lightly doped inverted *n*-type material by losing energy in the silicon conductance band or silicon valence band [31,32]. This mechanism would allow AHI to be viable at lower voltages.

Another difficulty with AHI is the value and origin of Q_{P}. As mentioned above, Q_{P} is obtained by integrating the substrate current during stress until oxide breakdown. Later studies showed that substrate current could be the result of mechanisms other than holes tunneling into the oxide film, such as generation–recombination processes [30] and photoexcitation mechanisms [33]. One study showed that, at low gate voltages, the hole current is many orders of magnitude lower than the electron current during constant voltage stress, and that the number of oxide defects generated per injected hole were similar to those generated per injected electron as long as the energy of the carrier was the same [6]. The same study also showed that Q_{p} was over eight orders of magnitude larger than the reported value of 0.1 $\mathrm{C/cm^2}$. Another study used substrate hot hole injection (SHHI) [34] to inject holes in a controlled manner independent of the applied gate voltage to determine if the additional holes would shorten the dielectric lifetime under constant voltage stress. The experimental setup is shown in the left panel of Figure 15.6. Holes can be injected via an additional $p+$ diffusion contact in close proximity to a channel region of a *p*-channel MOSFET. The gate voltage (V_{g}) is defined by the potential between the gate electrode and the inversion region in the device channel, and is accomplished by connecting the source and drain to the ground reference of the gate power supply. Hole current is injected by forward biasing the $p+$ diffusion and is adjusted by varying V_{inj}. Injected holes drift toward the space charge region beneath the inverted channel region. They are accelerated through the space charge region and are injected into the gate dielectric. The substrate voltage, V_{b}, controls the energy acquired by holes as they traverse the space charge region. Both the number and energy of the injected holes can be varied by adjusting V_{inj} and V_{b}. The plot in the right panel of Figure 15.6 shows the Weibull TDDB failure distributions of test capacitors with an oxide thickness of 3.0 nm. One set shown by the solid symbols was stressed until breakdown with a contact voltage of 5.0 V and were not subjected to SHHI before the stress commenced. The second sample set shown as open symbols was subjected to SHHI, where 4 $\mathrm{C/cm^2}$ of holes were injected before the constant voltage stress. Note that statistically both failure distributions are similar, indicating that the injected holes were very inefficient in generating those defects that lead to dielectric breakdown. Note that the q_{p} injected during the SHHI phase of this experiment is much larger than the reported value of 0.1 $\mathrm{C/cm^2}$ and should have quickly broken down the oxide films even before the constant voltage stress of the experiment was performed.

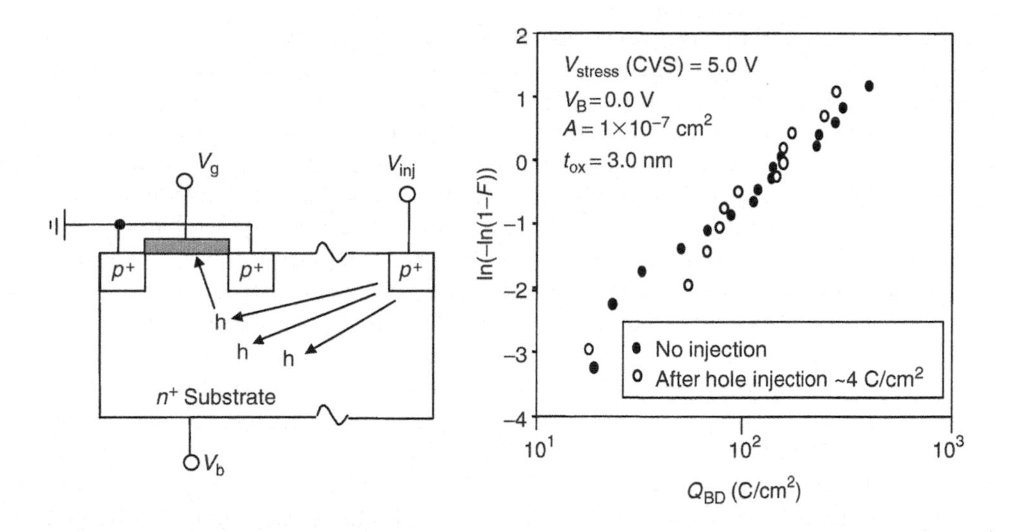

FIGURE 15.6
Plot showing the experimental setup for SHHI (left panel) and t_{BD} Weibull distributions for the case with and without injection of additional substrate holes (right panel). The injection of ~4 C/cm^2 did not appreciably change t_{BD} under constant voltage stress conditions indicating that holes are not very effective in causing wear-out defects. (Reprinted from Vogel, E.M., Edelstein, M.D., and Suehle, J.S., *J. Appl. Phys.*, 90, 1, 2001. With permission.)

It should be mentioned that a recent experimental study by Nicollian [35] showed devices with a lightly doped $n+$ polysilicon gate electrode exhibited a reduced Q_{BD} when biased to invert the polycrystalline silicon to favor the injection of holes.

15.3.4 Anode Hydrogen Release Breakdown Mechanism

Quantum mechanical tunneling precludes the ability of carriers to interact with the dielectric layer as they tunnel through. Defect generation then must be connected to mechanisms in the anode as electrons dissipate their energy. In Section 15.3.3, it was discussed how holes could be injected into the dielectric by gaining energy from tunneling carriers arriving at the anode. In a similar way, a hydrogen-related species could be released at the anode and drift into the dielectric film. This process termed "Anode Hydrogen Release," AHR, was proposed by DiMaria and coworkers [7]. The process is shown schematically in Figure 15.7. Tunneling electrons release energy in the anode and break silicon–hydrogen bonds releasing hydrogen to drift into the gate dielectric. The energy required to disassociate the Si–H structure was determined to be around 2 eV, requiring a gate voltage of about 5 V [7], although it has

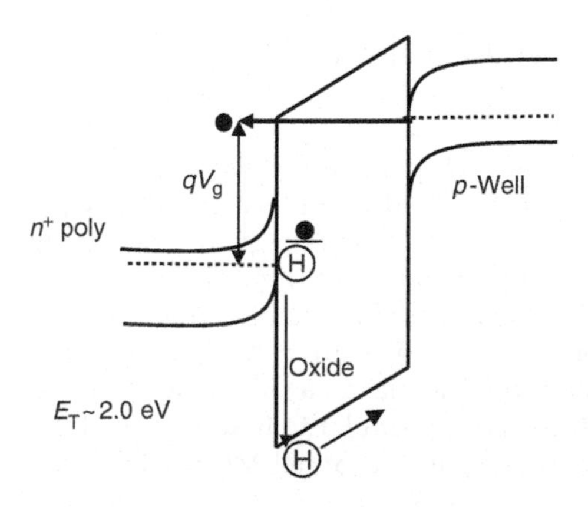

FIGURE 15.7
Band diagram showing the AHR mechanism. Tunneling carriers arrive at the anode with energy about equal to the gate voltage. If the energy exceeds ~2 eV, hydrogen can be disassociated at the interface and diffuse into the dielectric, causing defects, and eventually leading to dielectric breakdown.

been shown that the AHI process can be operative at voltages as low as 1.2 V [5,36]. It has been known for some time that hydrogen can induce a number of defects in silicon dioxide by exposing devices to hydrogen processing [37–42]. For example, increased SILC was observed in n-channel MOSFETs with a 4.5 nm thick gate dielectric after subsequent exposures to a hydrogen plasma [40]. There are several mechanisms proposed to explain how tunneling electrons can excite Si–H bonds and release hydrogen. These mechanisms include multiple-excitation of the Si–H bond by inelastic resonant tunneling electrons [43,44] and by coherent excitation of the bond by a single resonant tunneling electron [45,46].

One of the arguments against AHI being responsible for generating wear-out defects is the lack of experimental evidence showing the effect of deuterium annealing on oxide breakdown. Experimental studies have shown that silicon–deuterium bonds are much more difficult to break than silicon–hydrogen bonds. Also, theoretical models have predicted that electronic or vibrational excitation processes have a much lower desorption yield for deuterium [47], and it should be more difficult for tunneling electrons to release this specie. It has been shown that deuterium-annealed dielectrics exhibited less generation of SILC when stressed than films annealed with H_2 [48]. However, deuterium-treated SiO_2 films did not show a longer time-to-breakdown when compared to identical films annealed with forming gas, although fewer channel hot-electron-induced defects were generated [49].

15.3.5 Power Law Model for Breakdown

As TDDB data were being collected for oxides thinner than 3.0 nm, it was noticed that the voltage acceleration parameter, γ, was increasing as oxide thickness decreased [32]. For example, γ increased from 5.2 decades/V for a 2.4 nm thick oxide to 6.8 decades/V for a 1.4 nm thick oxide. Wu et al. later realized that γ was increasing as the stress voltage decreased, not as oxide thickness decreased [50]. Wu formulated a model known as the "power law" model, which predicted that t_{BD} was proportional to V^{-n} not $\exp(-\gamma V)$ [51]. Other studies also validated the power law dependence of t_{BD} and observed that the value of n varied between 33 and 50, depending on the type of MOSFET tested and polarity of stress voltage [52]. The power law dependence was explained by hydrogen being released by resonantly enhanced incoherent vibration excitations [46]. Inelastic tunneling carriers provide multiple excitations of the Si–H bond, and eventually break it, releasing hydrogen. Defect generation in the oxide film is related to the carrier density and their energy (applied gate voltage). The voltage power law was observed in experiments measuring the desorption yield of hydrogen, and the value of n was similar to those reported in TDDB experiments [46].

15.4 Driving Forces for Defect Generation and Breakdown

15.4.1 Fluence-Driven Breakdown

In Section 15.3 several physical mechanisms for defect generation were proposed. The thermochemical model requires only a stress electric field to induce defects, while the AHI and AHR models require energetic carriers to release energy at the anode where some species diffuse or a particle is injected into the dielectric film. The validity of the thermochemical model, where the electric field is the driving force, comes into question when the results of substrate hot electron injection (SHEI) studies are considered [53–56]. SHEI allows the injection of electrons independent of the gate bias in a similar way as was shown in Figure 15.6 for SHHI. The experimental setup is very similar to that used in SHHI and is shown in the left panel of Figure 15.8. In this case, an n-channel MOSFET is used

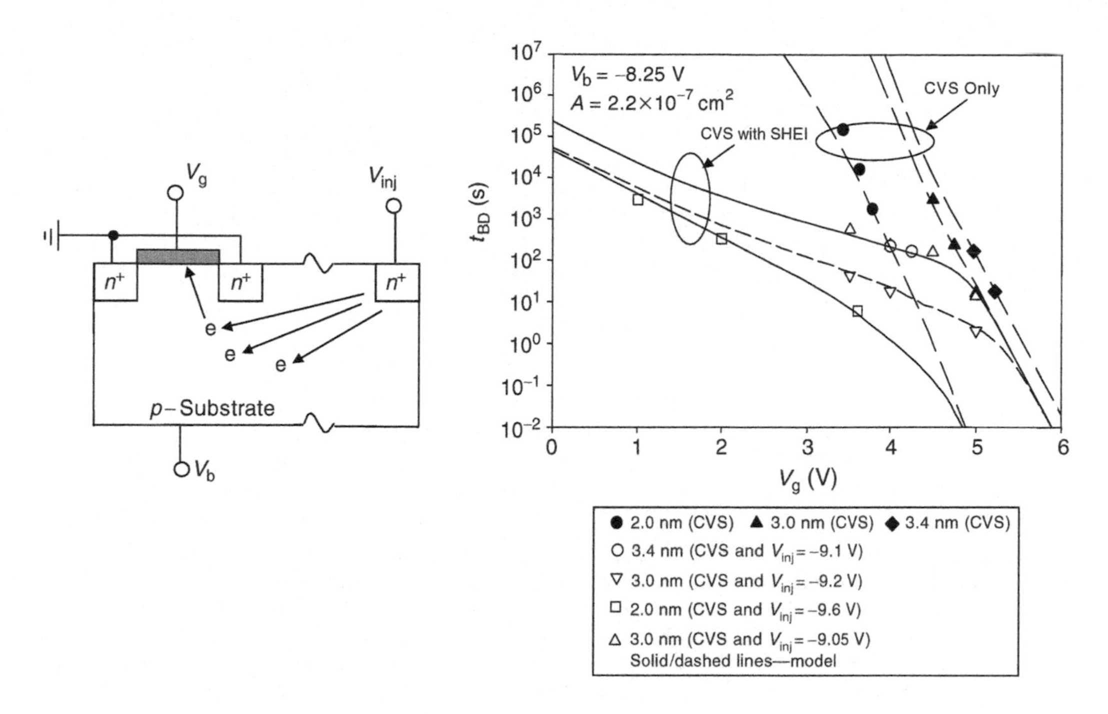

FIGURE 15.8
Plot showing the experimental setup for SHEI (left panel) and t_{BD} versus gate voltage for the case with and without injection of additional substrate electrons (right panel). Note that constant voltage stress with substrate hot electron injection results in a considerable reduction in t_{BD}. The solid and dashed lines show a model prediction where t_{BD} depends inversely on the total current tunneling through the oxide (current due to the gate voltage plus the additional current injected by SHEI). (Reprinted from Vogel, E.M. et al., *IEEE Trans. Electron Dev.*, 47, 1183, 2000. With permission.)

where an additional $n+$ diffusion region is fabricated close to the channel region of the device. Electrons are injected by forward biasing an $n+p$ diode formed by the $n+$ diffusion region and the p-type substrate. The amount of injected current is controlled by adjusting V_{inj}. Electrons diffuse toward the space charge region under the inverted channel where they are accelerated toward the SiO_2 gate dielectric. V_b can be adjusted to control the energy of the injected electrons. Note that the gate voltage is determined by the potential difference between the gate electrode and inverted channel connected via the source and drain contacts and is independent of V_b and V_{inj}.

The right panel shows the time-to-failure as a function of gate voltage under constant voltage stress (CVS) for thin oxides with an oxide thickness ranging between 2.0 and 3.4 nm [55]. The solid symbols are devices that were subjected to CVS with no additional current injected ($V_{inj} = 0.0$ V). The open symbols are for devices that received additional injected carriers by forward biasing the $n+p$ diode. Note that the time-to-failure is significantly decreased. An important result of the experiment is that the time-to-failure is inversely proportional to the injected current density, $t_{BD} \sim 1/J_T$, where J_T is the total current injected. J_T includes the tunneling current and the injected current. The model is shown by the solid and dashed lines in the plot.

15.4.2 Gate Voltage-Driven Breakdown

For ultrathin oxide films direct (or ballistic) transport is the dominant mechanism for quantum mechanical tunneling. In this regime electrons can tunnel directly from the cathode

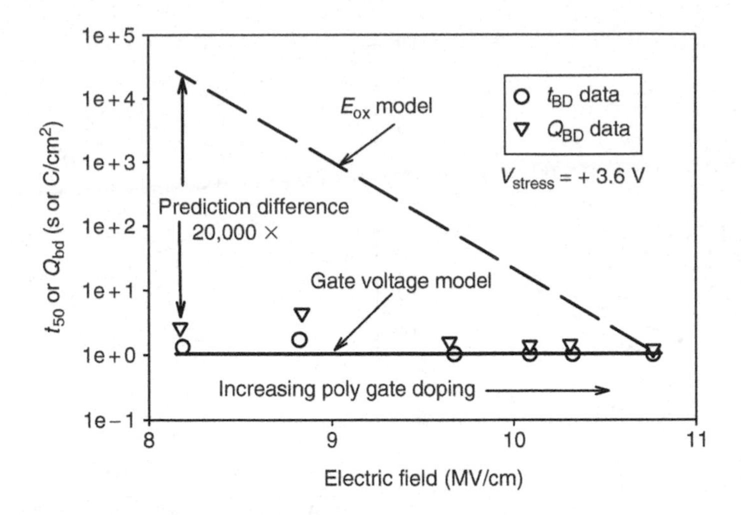

FIGURE 15.9
Plot showing t_{BD} as a function of electric field for devices with varying polycrystalline silicon gate electrode doping. The gate voltage was kept constant during time-dependent dielectric breakdown experiments. The plot shows that t_{BD} remained unchanged although the oxide electric field varied by 3 MV/cm due to the doping of the polycrystalline silicon gate electrode. The data verify that breakdown is driven by gate voltage, not oxide electric field. (Reprinted from Nicollian, P.E., Hunter, W.R., and Hu, J.C., *Proc. Intl. Reliab. Phys. Symp.*, 38, 7, 2000. With permission.)

to the anode elastically when the applied gate voltage is less than the electron barrier height. The energy of the electrons arriving at the anode can be approximated by the magnitude of the gate voltage [31]. AHR and AHI mechanisms are highly dependent on the energy released by the tunneling carrier at the anode. If energetic carriers are the primary driving force in defect generation, not only their numbers but also their energies must also be considered. Electric field-induced breakdown mechanisms would only depend on the total electric field across the dielectric film, which is different than the energy of the tunneling carriers. To resolve the issue whether the electric field or the carrier energy was the driving force in dielectric breakdown, Nicollian [35] devised an experiment where the oxide electric field could be modulated by several MV/cm, while keeping the gate voltage constant. This was achieved by changing the doping in the polycrystalline silicon gate electrode. The result is shown in Figure 15.9. The figure plots time-to-fail or charge-to-breakdown versus oxide electric field for a constant gate voltage of 3.6 V. The time-to-breakdown (circles) and charge-to-breakdown (inverted triangles) do not change as a function of oxide electric field. If breakdown were field dependent, it should fall on the dashed line for the E_{ox} model as shown. This study is additional evidence that dielectric breakdown is fluence and energy driven, not electric field driven. Another study by McKenna et al. [57] modulated the polycrystalline silicon doping of the gate electrode to change the fluence of carriers injected by the gate for the same electric field. The lower doped polycrystalline silicon electrode resulted in over an order of magnitude increase in the tunneling current over the highly doped electrode. The time-to-breakdown for the lower doped electrode (with the higher tunneling current) was shorter by over an order of magnitude compared to the higher doping, even though the oxide electric field was the same in both cases.

15.4.3 Temperature

The studies discussed above showing that tunneling carriers and their energy drive defect generation and dielectric breakdown formulate the physical picture that is generally

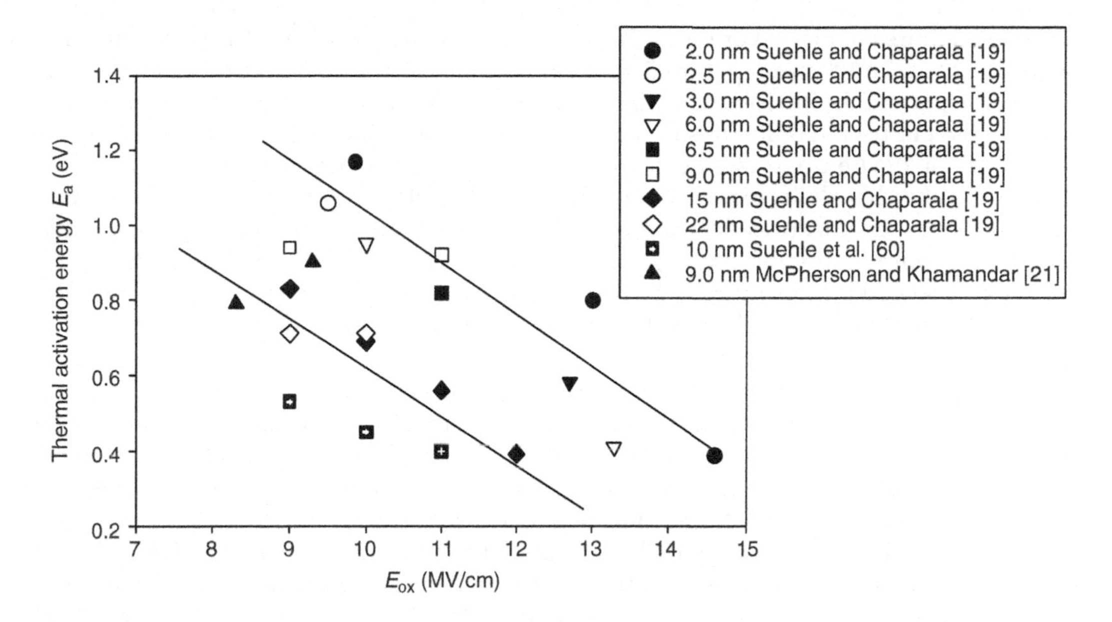

FIGURE 15.10
Thermal activation (E_a) energy of TDDB plotted as a function of stress electric field. The plot shows that E_a increases as electric field is decreased. Thinner oxides also appear to have larger E_a values than thicker oxides. (Reprinted from Suehle, J.S., et al., *Proc. Intl. Reliab. Phys. Symp.*, 10, 38, 2000. With permission.)

accepted for ultrathin gate dielectric breakdown. The exact details of defect generation are still unknown; however, it is widely believed that hydrogen plays some role. Temperature has been observed to accelerate the breakdown process. The mechanism of temperature acceleration is not well understood; however, it must be related to the defect generation process, since tunneling current depends only weakly on temperature. Indeed, it has been shown that N_{BD} determined by SILC and N_{ot} decreased significantly with temperature [58]. TDDB studies of thicker oxide films showed that t_{BD} is related to temperature through an Arrhenius relationship, $t_{BD} \sim \exp(E_a)/kT$. E_a is the thermal activation energy and was found to decrease with increasing electric field [19,59]. Figure 15.10 plots E_a as a function of E_{ox} for the oxide thickness range between 2.0 and 22 nm from four different studies [19,21,57,59,60]. It is apparent from the plot that E_a is a strong function of the applied electric field. The plot also shows a trend where the thinner dielectrics have a larger E_a than the thicker dielectrics for the same electric field. This trend was also observed in other studies [58,61]. Later studies of ultrathin gate oxides (<5 nm) indicated that the relationship was not Arrhenius, because E_a was observed to change as a function of temperature [58,60–62]. Wu et al. [63] later showed the stronger temperature dependence for the thinner oxides was due to the lower voltages used in the TDDB experiments, not to the oxide thickness. More experimental and theoretical work is required to fully understand the origin of the temperature dependence in defect generation and dielectric breakdown.

15.5 Statistics for Dielectric Breakdown

The accurate projection of oxide lifetime not only requires correct models for the voltage and temperature dependence, but also requires the proper statistical model for the failure distribution of the population of devices. Of critical importance is the accurate description of the low failure population or the early "tail" of the failure population. Typical reliability projection is at the 100 parts per million (ppm) level, i.e., for those devices that fail at the very early part of the distribution. Early studies of TDDB used lognormal statistics to describe the population of failures in the experiment. Many studies have shown that t_{BD} is a function of device area and decreases significantly as device area increases. Recent studies of ultrathin dielectric films showed that the slope of the failure distribution decreased as oxide thickness decreased [9]. The problem with lognormal statistics is that they do not comprehend area scaling of the failure distribution or the change of slope with thickness observed in experiments. Lognormal statistics also do not describe the critically important early tail of the failure distribution as shown by Wu et al. [64] for a very large sample size greater than 4000 devices. The Weibull distribution was shown to provide a much better description of the statistics of failure, especially for ultrathin films [9,65].

Suñé et al. [66] developed one of the first statistical models that describe a process where defects are generated randomly throughout a two-dimensional array of cells in the oxide. Breakdown occurs when a critical number of defects are generated in one of the cells. The model comprehends both the area and slope dependence of experimentally observed failure distributions. Dumin et al. [67] incorporated the concepts of the model to describe experimental failure distributions obtained for thin oxides. Percolation theory was later used by Degraeve et al. [9] to describe the statistical breakdown process. The model describes the breakdown process as defects of a certain radius accumulating randomly throughout the volume of the oxide film until a critical density, N_{BD}, is reached to form a percolation path between the anode and cathode. N_{BD} decreases as oxide thickness decreases and the slope of the Weibull distribution, β, also decreases. When the oxide becomes thinner than the defect radius, then only one defect is required to cause breakdown, and β is equal to unity. Figure 15.11 illustrates the statistical process of percolation

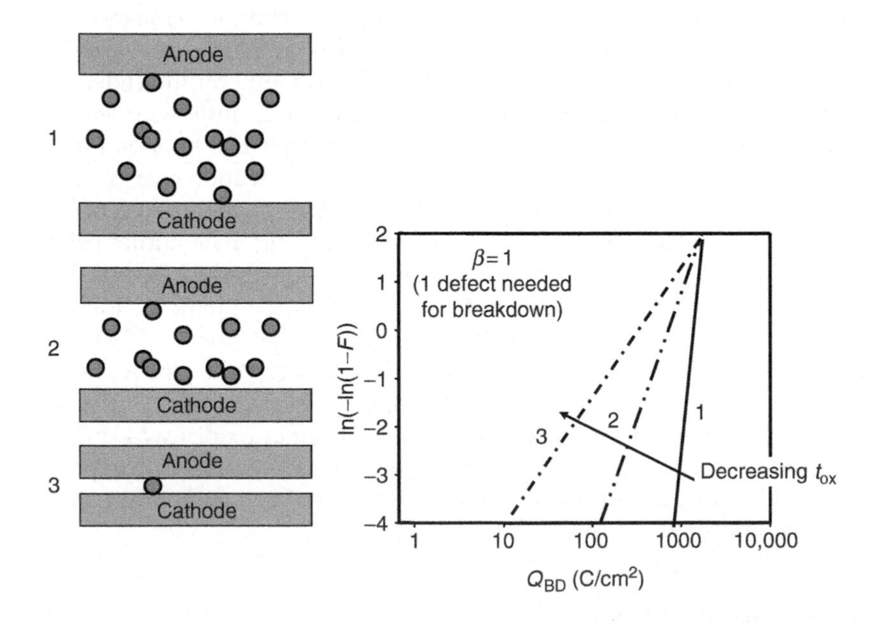

FIGURE 15.11
Illustration showing how the Weibull slope changes as oxide thickness is decreased. The schematic illustration in the left panel shows that fewer defects are required to cause dielectric breakdown as the anode and cathode of the device become closer together. The variability of the percolation path formation increases as the oxide thickness is decreased, resulting in a smaller Weibull slope as shown in the right panel. A Weibull slope of unity indicates that only one defect is required to bridge the anode and the cathode resulting in dielectric breakdown.

theory and how β decreases with oxide thickness. The left panel shows defects accumulating in an oxide between the anode and cathode for a thick oxide in schematic 1. The defects represented by the shaded circles are of a fixed radius. As the oxide becomes thinner (schematics 2 and 3), the cathode and anode become closer and fewer defects are required to "connect" to form a breakdown path. The extreme case is shown in schematic 3 where only one defect is required to connect the cathode and anode. The corresponding Weibull distributions are shown in the right panel. A property of percolation theory indicates that β becomes smaller as oxide thickness decreases and approaches unity where only one defect produces breakdown [5,65,68]. The gate oxide area is related to β by [65,68,69]:

$$\frac{t_{\mathrm{BD1}}}{t_{\mathrm{BD2}}} = \left[\frac{A_2}{A_1}\right]^{1/\beta} \tag{15.8}$$

Here t_{BD1} is the time-to-fail for a device with area A_1 and t_{BD2} is the time-to-fail for a device with area A_2. Usually test capacitors used to obtain TDDB data are much smaller than the total gate area in an actual product. Test results must then be projected to the larger area. A decreasing value of β presents an increasing reliability hazard as oxide thickness is scaled down since the failure distribution becomes more dispersed and t_{BD} becomes more sensitive to device area as predicted by Equation 15.8. β is relatively independent of voltage [50] and exhibits a slight increase as stress temperature increases [61].

It is generally very difficult to obtain a statistically meaningful value of β from a failure distribution if the sample size is less than 1000 units [64]. Equation 15.8 can be used to obtain a measure of β but requires devices with areas spanning several orders of magnitude [50]; however, the number of samples tested for each area can be less. Other factors have been identified that can affect the extracted values of β, including test structure series resistance and process-induced nonuniformity in oxide thickness across the samples [50].

15.6 Soft Breakdown and Projecting Circuit Failure

15.6.1 Introduction and Physical Explanation

"Soft" or "quasi" breakdown was first observed in sub-5 nm thick oxide films [70]. Unlike "hard" breakdown, where a sudden surge of gate current exceeded the tunneling current by many orders of magnitude, soft breakdown is accompanied by only a slight increase in the gate current that exhibited noisy or erratic fluctuations. Figure 15.12 illustrates the signature of soft breakdown in the current versus time characteristics shown in the left panel. The corresponding current–voltage characteristics are shown in the right panel. Note that immediately after soft breakdown the leakage current is only slightly larger than the prestress tunneling characteristic. After some time the leakage current can continue to increase, finally resulting in a hard breakdown. There were several explanations for soft breakdown, including localized damage producing an effective thinner oxide [71]. The noise was explained as a result of electrons trapping and detrapping from the damaged region. It was also explained that the current fluctuations were due to variable range hopping [72] and the reduced softer breakdown was due to a locally modified energy barrier [73]. Bude [31] described soft breakdown as a result of inadequate energy transfer from tunneling electrons to holes that were injected at the anode. Suñé [74] modeled the post breakdown conductance as tunneling through a quantum point contact

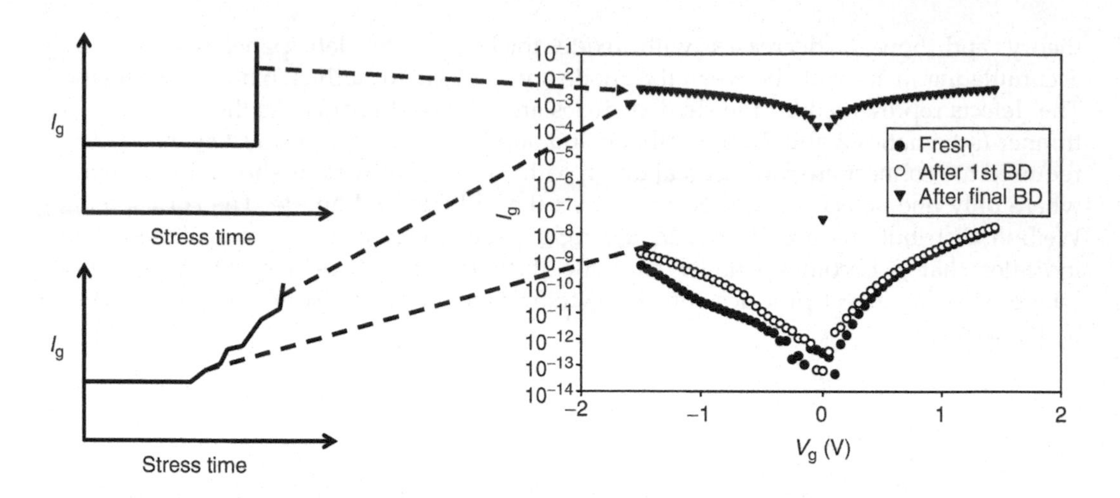

FIGURE 15.12
Gate current versus time for devices that experienced hard breakdown (left panel, top illustration) and soft breakdown (left panel, bottom illustration). The corresponding gate current versus gate voltage is shown in the right panel. Hard breakdown results in a substantial increase in gate leakage current, while soft breakdown produces only a slight increase in current.

(QPC). The model was able to accurately fit post soft breakdown current–voltage characteristics similar to those shown in Figure 15.12 (right panel) with two fitting parameters, the energy barrier height and a barrier thickness parameter.

The dominance of soft breakdown behavior in ultrathin oxides was explained by Alam [75] as a reduction in the power dissipation at the onset of breakdown. As discussed in Section 15.4, defects accumulate in the oxide until a critical number of traps form a percolation path that connects the cathode and anode. Current begins to flow through the path and the instantaneous power dissipation is equal to the percolation path current times the voltage across it. The amount of heat and structural damage in the oxide is related to the power dissipation at the onset of breakdown. If the voltage is reduced or series resistance in the structure limits the percolation path current, the power dissipation is reduced limiting thermal damage in the structure. The applied stress voltages in TDDB experiments are less for ultrathin oxides and narrow interconnect features in scaled technologies increase the series resistance of the device.

One of the real difficulties in characterizing the reliability of ultrathin dielectrics is detecting the breakdown event. The current surge when the percolation path forms could be much less than the tunneling current flowing through the oxide due to the applied stress voltage, especially for large area devices. This complication is illustrated in Figure 15.13, which shows current versus time (normalized to the time when 2.0 mA flows through the oxide) for four devices with different areas. Breakdown is quite easily observed for the smallest area device with an area of 4×10^{-6} cm^2. However, the instant of breakdown is quite difficult to detect for the larger area devices where the percolation path current is masked by the tunneling current. As shown in the figure, the largest area device (4×10^{-4} cm^2) exhibits a gradual increasing level of leakage current; a more abrupt increase in current is observed later in time. A robust soft breakdown detection technique that utilizes the measurement of current noise was reported by Weir et al. [76]. More sophisticated techniques for detecting soft breakdown were developed later that were able to take into account gradual leakage current increases due to SILC and transient current fluctuations due to random telegraph signals (RTS) [77,78].

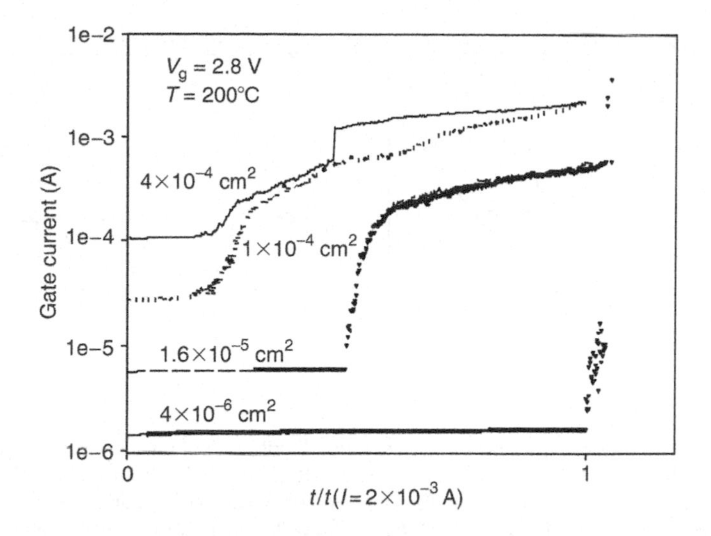

FIGURE 15.13
Gate current versus time (normalized to the time for the current to reach 2 mA) for four different device areas. The figure shows that the instant of dielectric breakdown becomes very difficult to detect as the device area increases. This is due to the percolation path current being masked by the tunneling current for the larger area devices.

15.6.2 Prevalence and Successive Breakdown

There have been several soft breakdown modes observed by which the low conductance leakage path develops into a higher conductance hard breakdown. One mode is a result of the percolation path that contributes to the observable soft breakdown behavior remaining stable as the oxide continues to age. A second percolation path causing a more damaging hard breakdown will eventually form that is statistically independent temporally and spatially to the first percolation path. Figure 15.14 illustrates the process schematically. The top illustration shows three percolation paths forming between the anode and cathode in the oxide film and the corresponding features in the gate current versus stress time. It is assumed that the percolation paths form randomly in the oxide film, and the formation of one path does not influence the development of subsequent paths. In this case, the current

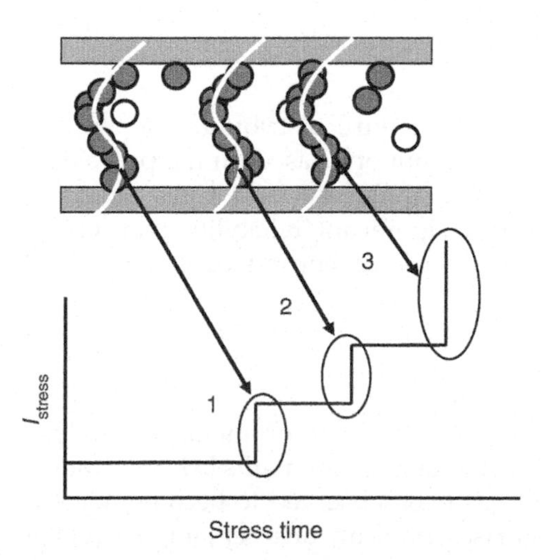

FIGURE 15.14
Illustration showing successive soft breakdown events in the gate current versus stress time characteristic where the steps in gate current are due to independent percolation paths. The formation of each percolation path results in a step in the gate current.

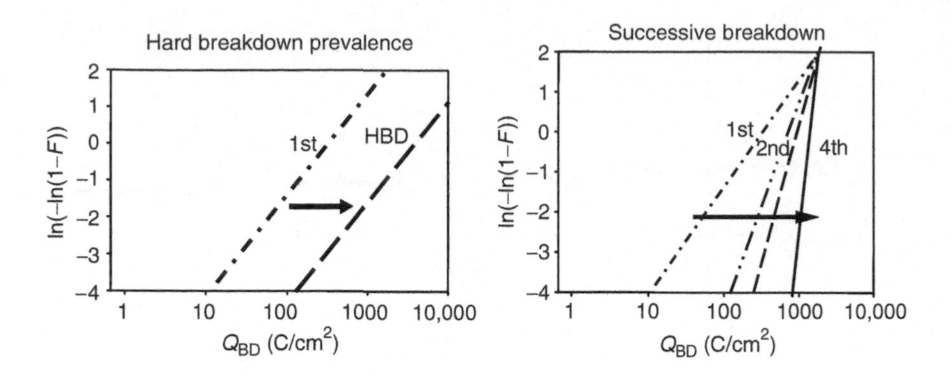

FIGURE 15.15
(Left panel) Illustration showing the prediction of the failure distribution of hard breakdown from the initial soft breakdown failure distribution using the prevalence method. (Right panel) Illustration showing the prediction of k successive breakdown failure distributions from the initial soft breakdown distribution using the method reported in Ref. [83]. Both prediction methods assume that the percolation filament formed during soft breakdown remains stable and subsequent breakdowns are statistically uncorrelated.

versus stress time characteristic appears as a series of steps. The first percolation path (labeled 1 in Figure 15.14) may be either a hard or soft breakdown depending on the magnitude of the power dissipation during the breakdown transient. If the first breakdown is soft and the second breakdown is hard, i.e., the percolation path (labeled 2 in Figure 15.14) is of sufficient size to cause significant oxide damage, the failure distribution of the second "hard" breakdown can be predicted by statistical methods. Termed as the hard breakdown prevalence method [79,80], the final hard breakdown failure distribution can be predicted by shifting the first breakdown distribution by ln (α_{HBD}) [80]. α_{HBD} is defined as the cumulative hard breakdown prevalence ratio. If the second percolation path is not sufficient to cause a hard breakdown, then the statistical distribution of the third percolation path (labeled 3 in Figure 15.14) can be predicted by successive breakdown theory [81,82]. If the third percolation path also does not result in a hard breakdown then the t_{BD} Weibull distribution of the kth percolation path that results in a hard breakdown can be predicted by [83]:

$$W_k \approx k\beta \ln\left(\frac{t_{\mathrm{BD}}}{t_{63\%}}\right) - \ln(k!) \tag{15.9}$$

Here $t_{63\%}$ and β are the scale factor and slope of the failure distribution of the first breakdown. The resulting projection of the Weibull distributions based on the prevalence method and successive breakdown theory are shown in the left and right panels of Figure 15.15, respectively. Note that in both cases a significant reliability margin can be attained if the circuit can tolerate the amount of leakage current caused by the k percolation paths.

15.6.3 Progressive Breakdown

The post soft breakdown leakage current also has been observed to continually degrade as stress continues without any indication of steps or stable percolation paths [84–86]. Known as "progressive breakdown," gate leakage current has been observed to become noisy at the instant of soft breakdown and gradually increases until an unacceptable current is

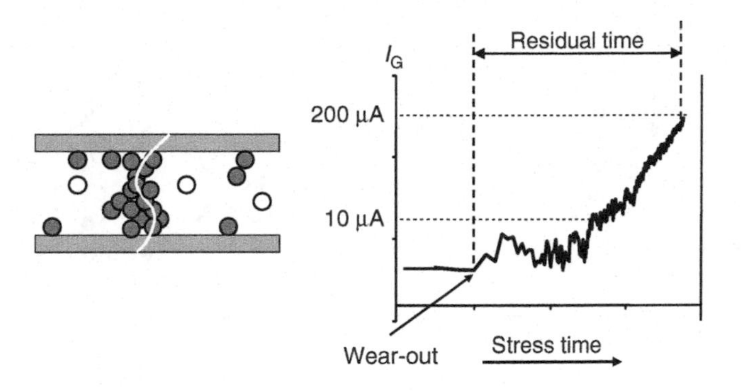

FIGURE 15.16
(Right panel) Illustration showing gate current versus stress time after soft breakdown where the filament is unstable and progressive breakdown takes place. The residual time is the time when oxide wear-out is first detected and the time when the gate current reaches a predetermined level. The DR defined in Ref. [84] was defined as the slope of the current increase from a level of 10 μA to a level of 100 μA. (Left panel) Illustration showing the percolation path that forms during the soft breakdown continues to enlarge due to enhanced local defect generation.

reached where circuit function is expected to cease. The process of progressive breakdown is illustrated in Figure 15.16. A typical gate current versus time characteristic is shown in the right panel. The development of the first percolation path or ''wear-out'' is coincident with the observation of current noise as shown. In this case no clear sudden current jumps or steps are observed. A schematic of the first percolation path is shown in the left panel and is depicted as becoming larger in cross section and conductance as stress continues. Although the remaining part of the oxide can still form additional percolation paths as stress continues, the vicinity of the original path experiences enhanced defect generation due to greater carrier flux (through the original path) and a possibly larger local temperature. The oxide will ultimately fail due to the enlargement of the first percolation path as defects continue to ''connect'' to the original path before additional independent paths form. The time between oxide wear-out and hard breakdown is referred to as the ''residual time'' [87]. The assignment of the gate leakage current to define hard breakdown depends on the circuit architecture and other factors. The figure shows an arbitrarily value of 200 μA to define hard breakdown.

The kinetics of progressive breakdown are not entirely known. Monsieur et al. [86] showed that the process was driven by the generation of defects by noting that the progressive breakdown could be accelerated by the application of a substrate voltage which increased the energy at the anode. It was also shown that additional electrons injected via SHEI decreased the progressive breakdown time [88]. In the case of progressive breakdown a new set of acceleration parameters are required to project the residual time. Linder et al. [84] showed that the degradation rate (DR), defined as the slope of the post soft breakdown leakage current characteristic between 10 and 100 μA (Figure 15.16) was exponentially related to the stress voltage. An additional reliability margin of greater than 10 years after the first breakdown could be obtained if a circuit could tolerate a gate leakage current of 10 μA before it malfunctioned.

Figure 15.17 shows the gate voltage and temperature dependence of the residual time, defined as the time between the detection of oxide wear-out and the time at which the gate leakage current reaches 200 μA [88]. It is interesting to note that the voltage acceleration is independent of temperature, and the temperature acceleration is non-Arrhenius and independent of voltage. The voltage acceleration parameter for the data in the left panel of

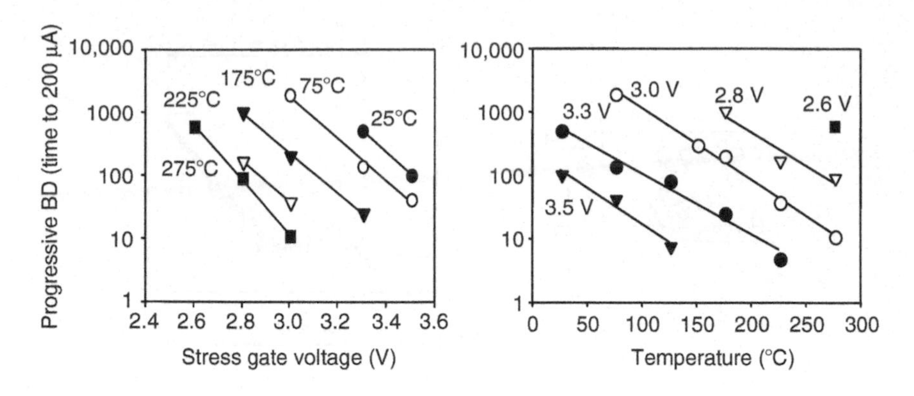

FIGURE 15.17
Acceleration parameters for progressive breakdown. The left plot shows the residual time (defined from the instant of oxide wear-out to the time the gate current exceeds 200 μA) as a function of gate voltage for five temperatures. The voltage acceleration parameter is observed to be independent of temperature. The right plot shows the residual time as a function of temperature for five different gate voltages. The temperature dependence is not Arrhenius and the slope does not change with gate voltage. (Reprinted from Suehle, J.S. et al., *Proc. Intl. Reliab. Phys. Symp.*, 42, 96, 2004. With permission.)

Figure 15.17 is about 3.2 decades/V, which is smaller than the 5 decades/V reported in Ref. [84]. The physical mechanism of progressive breakdown is still a subject of research, especially to identify the mechanism leading to the observed voltage and temperature dependence. Pey et al. [89] were able to use high-resolution tunneling electron microscopy (TEM) to image breakdown regions in MOSFET devices, and observed epitaxial Si that grew from the anode and extended into the cathode. Dielectric breakdown-induced epitaxy (DBIE) forms under high levels of power dissipation at the onset of breakdown. This behaves as a stable breakdown filament due to a negative feedback effect that limits current flowing through the structure [90].

As discussed above, the method used to project the residual time after soft breakdown depends on whether the filament formed at the onset of breakdown remains stable or unstable. Wu et al. [87] reported that there could be a mixture of breakdown modes in a population of devices depending on gate voltage, temperature, and oxide thickness. It was observed that breakdown filaments became more stable as temperature decreased and MOSFET channel length increased. Regardless of the nature of the breakdown filament, a significant reliability margin can be realized if soft breakdown is considered as a means to extend circuit lifetime. This advantage may diminish for future technologies that utilize alternative gate dielectric materials and metal electrodes.

15.7 Dielectric Breakdown in Advanced Gate Dielectric Stacks

As device dimensions continue to scale down, short channel effects (the loss of gate control) and punch-through must be mitigated by minimizing the extension of the depletion width of the source and drain junctions (x_j) into the channel. The channel doping must be increased to minimize x_j and consequently the threshold voltage, V_t, must increase to achieve the same inversion charge in the channel. During scaling, the inversion charge must be maintained at a level to achieve the channel current required to meet the performance specifications of the device. The threshold voltage can only be increased to a level where the externally applied operation voltage can adequately turn on the channel, i.e., $V_g - V_t$, where V_g is the applied gate voltage. Note that it is desirable to also scale the operation voltage down as the device dimensions decrease. The channel inversion charge can also be increased by increasing the device capacitance, since $Q_{inv} = C_{ox}(V_g - V_t)$. The gate capacitance, C_{ox}, is inversely proportional to the dielectric thickness, t_{ox}; therefore, the dielectric thickness is usually scaled down to increase the capacitance. Because tunneling current increases exponentially with decreasing oxide thickness, SiO_2-based dielectrics are near their limit in scalability.

Recently, dielectrics with a larger permittivity are being investigated as a replacement for SiO_2 since the gate capacitance can be increased by increasing the dielectric permittivity:

$$C_{ox} \sim \kappa \varepsilon_0 A / t_{ox} \tag{15.10}$$

where

κ is the permittivity of the dielectric material
ε_0 is the permittivity of free space needs carriage return
A is the area of the gate

The high-κ dielectric material can be thicker that a SiO_2 film by the ratio of its dielectric constant with that of SiO_2:

$$t_{\text{high-}\kappa} = \left[\frac{\kappa_{\text{high-}\kappa}}{\kappa_{SiO_2}}\right] t_{SiO_2} \tag{15.11}$$

Figure 15.18 shows a schematic representation of the energy band diagram of a dielectric stack consisting of HfO_2 and a thin interfacial layer at the Si interface. The interfacial layer is usually comprised of SiO_2 or SiO_x. The square symbols represent traps that can exist in the high-κ layer or in the interfacial layer. The traps in the high-κ layer have been attributed to significant slow and fast trapping events leading to a recoverable threshold voltage instability [91]. Ultrafast current–voltage characterization is required to determine transistor parameters without the interference associated with transient electron trapping by the traps in the high-κ layer.

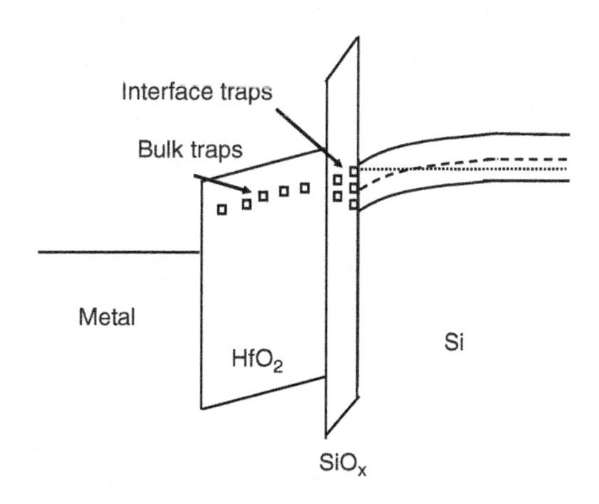

FIGURE 15.18
Band diagram of a metal gate, high-κ gate dielectric stack. The material system usually has an unavoidable interfacial layer at the Si interface that is primarily comprised of SiO_x. The figure depicts traps or defects in the interfacial layer and in the high-κ bulk layer. Such traps are responsible for transient threshold voltage instability observed in MOSFET devices.

Note that the band diagram is not symmetric. Electrons injected from the Si substrate experience a different barrier than electrons injected from the gate electrode. The tunneling currents can be very different depending on the injection polarity [92]. This is expected to result in a polarity-dependent reliability of the stack. It was reported that the Weibull slopes were different depending on injection polarity for TiN/Al_2O_3/SiO_2 gate stacks [93]. The Weibull slope was much larger in the case of substrate injection than in gate injection. As discussed in Section 15.4, the Weibull slope was observed to increase as the insulator thickness increased. The larger slope observed for substrate injection was explained as bulk defect generation in the thicker Al_2O_3 layer determining the reliability of the stack. In the case of gate injection, the reliability of the stack was determined by defect generation in the SiO_x/Si interfacial layer. This observation is consistent with defect generation mechanisms discussed previously in this chapter that suggest that the material in the anode region of the dielectric is damaged or some species is released by energetic tunneling carriers. Weibull slopes observed for HfO_2 layers are generally lower than SiO_2 films of similar thickness [94]. Based on percolation theory, this observation was explained as the defects responsible for wear-out in the HfO_2 layer were larger than in the SiO_2 case.

There has been considerable discussion in the literature whether defect generation in the high-κ layer or the interfacial layer is responsible for dielectric breakdown in gate stacks. Trap profiling techniques were used to show that defect generation only occurred in the transition region between the high-κ layer and the interfacial layer during high-field stress [95]. It was argued that traps are not produced in the bulk high-κ layer because the dielectric metal–oxygen bond is formed by d-shell electrons and are very stable against breakage from high-field stress or tunneling electrons. Other reports showed that permanent defects in the high-κ layer are created after long stress times [96,97]. The defects were believed to be oxygen vacancy defects located in the high-κ bulk 1.3 eV above the Si intrinsic Fermi level [96].

Soft breakdown also occurs in high-κ gate dielectric stacks. It was shown that traps are created in the bulk high-κ layer and the voltage acceleration of leakage path creation and defect generation are similar. The progressive wear-out phase is much longer than intrinsic wear-out but the dielectric becomes very leaky [96]. Another study showed that progressive breakdown in high-κ/metal gate stacks was triggered by trap generation in the SiO_2 interfacial layer as evidenced by a correlation of interface trap density and stress-induced leakage current [98]. The gradual increase in leakage current observed in $HfAlO_x$/SiO_2 stacks was attributed to multiple soft breakdown filaments, not SILC caused by generated traps [99]. In this case the breakdown of the stack occurs when the SiO_2 layer breaks down.

The increased reliability margin due to soft breakdown discussed in Section 15.5 may not be obtainable with a metal gate electrode technology. It was shown that breakdown is less "soft" when the device has a metal gate electrode [100]. It was explained that a polycrystalline silicon gate electrode, unlike a metal electrode, behaves like a ballasting resistor to limit the transient power dissipation during dielectric breakdown.

15.8 Conclusion

The exact physical mechanism of defect generation and dielectric breakdown still remains elusive after several decades of studying the silicon–silicon dioxide system. However, there has been much progress made toward the understanding of dielectric breakdown recently. Major contributors to this progress were studies that showed that electron fluence (current) and energy (voltage) are the driving factors for defect generation and eventual breakdown. Accurate lifetime projection demands the correct statistics for the failure distribution and

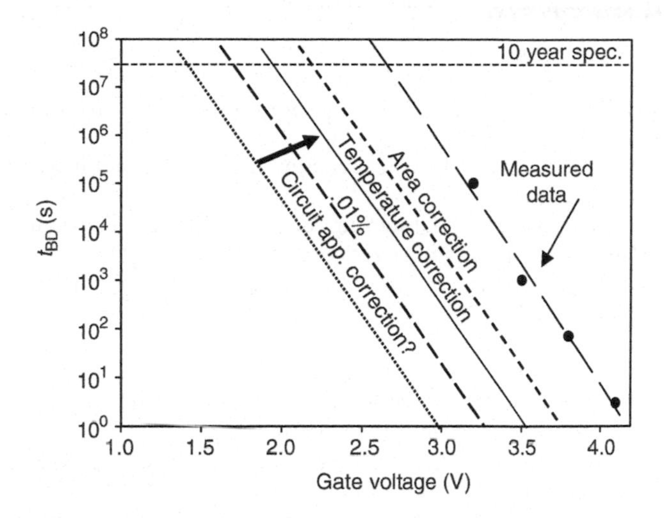

FIGURE 15.19
Illustration showing that correct physical models for voltage acceleration, temperature dependence, and post soft breakdown leakage current increases are required to accurately project product lifetime from test data. It is also essential that the proper statistical model be used.

correct physical models for voltage and temperature acceleration. Figure 15.19 shows an example of the methodology for projecting the lifetime of a product based on time-dependent dielectric breakdown data collected on test samples. Usually, large voltages are used to accelerate breakdown such that data (shown as solid circles) can be collected in a reasonable time. A model that characterizes the correct voltage acceleration is required first. Since the product gate area is much larger than that of the test samples, the voltage acceleration curve must be derated and moved to lower voltages, as shown in the example. Another change is required when considering that the product typically operates at a higher temperature than test data collected at room temperature. Once again the voltage acceleration is shifted to lower voltages. It is essential that the proper statistical model be used to shift the voltage acceleration curve again to lower voltages to project the tail of the distribution, typically 100 ppm. It was shown that Weibull statistics correctly characterize the failure distribution of ultrathin dielectrics, including the thickness dependence of the Weibull slope. The distinction between device life and circuit life can move the voltage acceleration back to higher voltages if a circuit architecture can still function after a soft breakdown event. Care must be used in using soft breakdown as a means to extend product reliability through the residual time. Competing mechanisms resulting in stable and nonstable breakdown paths complicate post soft breakdown projection methodology. The residual time for stable post breakdown filaments can be determined by prevalence and successive breakdown techniques. Unstable breakdown filaments resulting in progressive breakdown are more complicated, and estimating the residual time may require an experimentally determined set of acceleration parameters.

Many challenges in reliability characterization remain as advanced technology nodes begin to utilize alternative gate dielectrics such as high-κ dielectrics and dielectric stacks. Dielectric stacks are no longer symmetrical with respect to the electron-injecting electrode; differences in wear-out mechanisms have been observed, depending on the polarity of the gate voltage. The interfacial layer between the high-κ dielectric and the silicon oxide grown on the silicon substrate is of particular concern as a region susceptible for defect generation. It is also expected that new wear-out and failure mechanisms will be observed as device dimensions approach the nanoscale regime.

References

1. Stathis, J.H. and DiMaria, D.J., Reliability projection for ultra-thin oxides at low voltage, *IEDM Tech. Dig.*, 167, 1998.
2. Kaczer, B. et al., Impact of MOSFET oxide breakdown on digital circuit operation and reliability, *IEDM Tech. Dig.*, 553, 2000.
3. Linder, B.P. et al., Gate oxide breakdown under current limited constant voltage stress, *IEEE Symposium on VLSI Technology, Digest of Technical Papers*, 214, 2000.
4. DiMaria, D.J., Defect generation in ultrathin silicon dioxide films produced by anode hole injection, *Appl. Phys. Lett.*, 77, 2716, 2000.
5. DiMaria, D.J. and Stathis, J.H., Explanation for the oxide thickness dependence of breakdown characteristics of metal-oxide-semiconductor structures, *Appl. Phys. Lett.*, 70, 2708, 1997.
6. Stathis, J.H., Physical and predictive models of ultrathin oxide reliability in CMOS devices and circuits, *IEEE Trans. Dev. Mater. Reliab.*, 1, 43, 2001.
7. DiMaria, D.J. and Stasiak, J.W., Trap creation in silicon dioxide produced by hot electrons, *J. Appl. Phys.*, 65, 2342, 1989.
8. Vogel, E.M. et al., Reliability of ultra-thin silicon dioxide under combined substrate hot-electron and constant voltage tunneling stress, *IEEE Trans. Electron Dev.*, 47, 1183, 2000.
9. Degraeve, R. et al., New insights in the relation between electron trap generation and the statistical properties of oxide breakdown, *IEEE Trans. Electron Dev.*, 45, 904, 1998.
10. Vincent, E. et al., Dielectric reliability in deep submicron technologies: From thin to ultra-thin oxides, *Microelectron. Reliab.*, 37, 1499, 1997.
11. Stathis, J.H. et al., Breakdown measurements of ultra-thin SiO_2 at low voltage, *VLSI Technology Digest of Technical Papers*, 94, 2000.
12. Wang, B. et al., Time-dependent breakdown of ultra-thin SiO_2 gate dielectrics under pulsed biased stress, *IEEE Electron Dev. Lett.*, 22, 224, 2001.
13. Cheung, K.P. et al., Field dependent critical trap density for thin gate oxide breakdown, *Proc. Int. Reliab. Phys. Symp.*, 37, 52, 1999.
14. Okada, K., Analysis of the relationship between defect site generation and dielectric breakdown utilizing A-mode stress induced leakage current, *IEEE Trans. Electron Dev.*, 47, 1225, 2000.
15. Anolick, E. and Nelson, G., Low field time dependent dielectric integrity, *Proceedings of the International Reliability Physics Symposium*, 17, 8, 1979.
16. Crook, D., Method of determining reliability screens for time dependent dielectric breakdown, *Proceedings of the International Reliability Physics Symposium*, 17, 1, 1979.
17. Berman, A., Time-zero dielectric reliability test by a ramp method, *Proc. Int. Reliab. Phys. Symp.*, 19, 204, 1981.
18. McPherson, J.W. and Baglee, D.A., Acceleration factors for thin gate oxide stressing, *Proc. Int. Reliab. Phys. Symp.*, 23, 1, 1985.
19. Suehle, J.S. and Chaparala, P., Low electric field breakdown of thin SiO_2 films under static and dynamic Stress, *IEEE Trans. Electron Dev.*, 44, 801, 1997.
20. McPherson, J.W. and Khamankar, R.B., Molecular model for intrinsic time-dependent dielectric breakdown in SiO_2 dielectrics and the reliability implications for hyper thin gate oxide, *Semicond. Sci. Technol*, 15, 462, 2000.
21. McPherson, J.W. and Khamandar, R.B., Disturbed bonding states in SiO_2 thin-films and their impact on time-dependent dielectric breakdown, *Proc. Int. Reliab. Phys. Symp.*, 36, 47, 1998.
22. Lenahan, P.M. and Conley, J.F., What can electron paramagnetic resonance tell us about the Si/SiO_2 system? *J. Vac. Sci Technol. B*, 16, 2134, 1998.
23. Lenahan, P.M. and Dressendorfer, P.V., Hole traps and trivalent silicon centers in metal/oxide/silicon devices, *J. Appl. Phys.*, 55, 3495, 1984.
24. Chen, I.C. et al., A quantitative physical model for time-dependent breakdown in SiO_2, *Proc. Int. Reliab. Phys. Symp.*, 23, 24, 1985.
25. Chen, I.C. et al., Substrate hole current and oxide breakdown, *Appl. Phys. Lett.*, 49, 669, 1986.

26. Schuegraf, K.F. and Hu, C., Hole injection SiO_2 breakdown model for very low voltage lifetime extrapolation, *IEEE Trans. Electron Dev.*, 41, 761, 1994.
27. Schuegraf, K.F. and Hu, C., Effects of temperature and defects on breakdown lifetime of thin SiO_2 at very low voltages, *Proc. Int. Reliab. Phys. Symp.*, 7, 126, 1994.
28. Satake, H. and Toriumi, A., Temperature dependent hole fluence to breakdown in thin gate oxides under Fowler–Nordheim electron tunneling injection, *Appl. Phys. Lett.*, 66, 3516, 1995.
29. Fischetti, M.V., Model for the generation of positive charge at the Si–SiO_2 interface based on hot-hole injection from the interface, *Phys. Rev. B*, 31, 2099, 1985.
30. DiMaria, D.J., Anode hole injection and trapping in silicon dioxide, *J. Appl. Phys.*, 80, 304, 1996.
31. Bude, J.D. et al., Explanation of stress-induced damage in thin oxides, *IEDM Tech. Dig.*, 179, 1998.
32. Weir, B.E. et al., Gate oxide reliability projection to the sub-2 nm regime, *Semicond. Sci. Technol.*, 15, 455, 2000.
33. Rasras, M. et al., Photo-carrier generation as the origin of Fowler–Nordheim-induced substrate hole current in thin oxides, *IEDM Tech. Dig.*, 465, 1999.
34. Vogel, E.M., Edelstein, M.D., and Suehle, J.S., Defect generation and breakdown of ultrathin silicon dioxide induced by substrate hole injection, *J. Appl. Phys.*, 90, 1, 2001.
35. Nicollian, P.E., Hunter, W.R., and Hu, J.C., Experimental evidence for voltage driven breakdown models in ultrathin gate oxides, *Proc. Int. Reliab. Phys. Symp.*, 38, 7, 2000.
36. DiMaria, D.J., Electron energy dependence of metal-oxide-semiconductor degradation, *Appl. Phys. Lett.*, 75, 2427, 1999.
37. Cartier, E., Stathis, J.H., and Buchanan, D.A., Passivation and depassivation of silicon dangling bonds at the Si/SiO_2 interface by atomic hydrogen, *Appl. Phys. Lett.*, 63, 1510, 1993.
38. Stahlbush, R.E. and Cartier, E., Interface defect formation in MOSFETs by atomic hydrogen exposure, *IEEE Trans. Nucl. Sci.*, 41, 1844, 1994.
39. Stathis, J.H. and Cartier, E., Atomic hydrogen reactions with P_b centers at the <100>Si/SiO_2 interface, *Phys. Rev. Lett.*, 72, 2745, 1994.
40. DiMaria, D.J. and Cartier, E., Mechanism for stress-induced leakage currents in thin silicon dioxide films, *J. Appl. Phys.*, 78, 3883, 1995.
41. Nissan-Cohen, Y. and Gorczyca, T., The effect of hydrogen on trap generation, positive charge trapping, and time-dependent dielectric breakdown of gate oxides, *IEEE Electron Dev. Lett.*, 6, 287, 1988.
42. Pompl, T. et al., Change of acceleration behavior of time-dependent dielectric breakdown by the BEOL process: Indications for hydrogen induced transition in dominant degradation mechanism, *Proc. Int. Reliab. Phys. Symp.*, 43, 388, 2005.
43. Walkup, R.E., Newns, D.M., and Avouris, Ph., Role of multiple inelastic transitions in atom transfer with the scanning tunneling microscope, *Phys. Rev. B*, 48, 1858, 1993.
44. McMahon, W., Haggag, A., and Hess, K., Reliability scaling issues for nanoscale devices, *IEEE Trans. Nanotechnol.*, 2, 33, 2000.
45. Salam, G.P., Persson, M., and Palmer, R.E., Possibility of coherent multiple excitation in atom transfer with a scanning tunneling microscope, *Phys. Rev. B*, 49, 10655, 1994.
46. Wu, E.Y. and Suñé, J., Hydrogen-release mechanisms in the breakdown of thin SiO_2 films, *Phys. Rev. Lett.*, 92, 87601, 2004.
47. Hess, K. et al., Giant isotope effect in hot electron degradation on metal oxide silicon devices, *IEEE Trans. Electron Dev.*, 45, 406, 1998.
48. Mitani, Y., Satake, H., and Toriumi, A., Experimental evidence of hydrogen-related SILC generation in thin gate oxide, *IEDM Tech. Dig.*, 2, 2001.
49. Wu, J. et al., Anode hole injection versus hydrogen release: The mechanism for gate oxide breakdown, *Proc. Int. Reliab. Phys. Symp.*, 38, 27, 2000.
50. Wu, E.Y. et al., Voltage-dependent voltage-acceleration of oxide breakdown for ultra-thin oxides, *IEDM Tech. Dig.*, 541, 2000.
51. Wu, E.Y. et al., Experimental evidence of T_{BD} power-law for voltage dependence of oxide breakdown in ultrathin gate oxides, *IEEE Trans. Electron Dev.*, 49, 2244, 2002.

52. Ohgata, K. et al., Universality of power-law voltage dependence for TDDB lifetime in thin gate oxide PMOSFETs, *Proc. Int. Reliab. Phys. Symp.*, 43, 372, 2005.
53. Degraeve, R. et al., Oxide and interface degradation and breakdown under medium and high field injection conditions: A correlation study, *Microelectron. Eng.*, 28, 265, 1995.
54. Umeda, K., Tomita, T., and Taniguchi, K., Silicon dioxide breakdown induced by SHE (substrate hot electron) injection, *Electron. Commun. Jpn. Part 2*, 80, 11, 1997.
55. Vogel, E.M. et al., Combined substrate hot-electron and constant voltage tunneling stress, *IEEE Trans. Electron Dev.*, 47, 1183, 2000.
56. DiMaria, D.J., Defect generation under substrate-hot-electron injection into ultrathin silicon dioxide layers, *J. Appl. Phys.*, 86, 2100, 1999.
57. McKenna, J.M., Wu, E.Y., and Lo, S.-H., Tunneling current characteristics and oxide breakdown in P + poly gate PFET capacitors, *Proc. Int. Reliab. Phys. Symp.*, 16, 38, 2000.
58. DiMaria, D.J. and Stathis, J.H., Non-Arrhenius temperature dependence of reliability in ultrathin silicon dioxide films, *Appl. Phys. Lett.*, 74, 1752, 1999.
59. Kimura, M., Oxide breakdown mechanism and quantum physical chemistry for time-dependent dielectric breakdown, *Proc. Int. Reliab. Phys. Symp.*, 35, 190, 1997.
60. Suehle, J.S. et al., Temperature dependence of soft breakdown and wear-out in sub-3 nm SiO_2 films, *Proc. Int. Reliab. Phys. Symp.*, 10, 38, 2000.
61. Kaczer, B. et al., The influence of elevated temperature on degradation and lifetime prediction of thin silicon-dioxide films, *IEEE Trans. Electron Dev.*, 47, 1514, 2000.
62. Pompl, T. et al., Investigation of ultra-thin gate oxide reliability behavior by separate characterization of soft breakdown and hard breakdown, *Proc. Int. Reliab. Phys. Symp.*, 38, 10, 2000.
63. Wu, E.Y. et al., Interplay of voltage and temperature acceleration of oxide breakdown for ultra-thin gate oxides, *Solid State Electron.*, 46, 1787, 2002.
64. Wu, E.Y. et al., Challenges for accurate reliability projections in the ultra-thin oxide regime, *Proc. Int. Reliab. Phys. Symp.*, 37, 57, 1999.
65. Degraeve, R., Consistent model for the thickness dependence of intrinsic breakdown in ultra-thin oxides, *IEDM Tech. Dig.*, 866, 1995.
66. Suñé, J. et al., On the breakdown statistics of very thin SiO_2 films, *Thin Solid Films*, 185, 347, 1990.
67. Dumin, D.J. et al., High field related thin oxide wear-out and breakdown, *IEEE Trans. Electron Dev.*, 42, 760, 1995.
68. Stathis, J.H., Percolation models for gate oxide breakdown, *J. Appl. Phys.*, 86, 5757, 1999.
69. Nigam, T. et al., A fast and simple methodology for lifetime predictions in thin oxides, *Proc. Int. Reliab. Phys. Symp.*, 37, 381, 1999.
70. Depas, M., Nigam, T., and Heynes, M.H., Soft breakdown of ultra-thin gate oxide layers, *IEEE Trans. Electron Dev.*, 43, 1499, 1996.
71. Lee, S.-H. et al., Quasi-breakdown of ultrathin gate oxide under high field stress, *IEDM Tech. Dig.*, 605, 1994.
72. Okada, K. and Taniguchi, K., Electrical stress-induced variable range hopping conduction in ultrathin silicon dioxides, *Appl. Phys. Lett.*, 70, 351, 1997.
73. Halimaouia, A., Brièrea, O., and Ghibaudo, G., Quasi-breakdown in ultrathin gate dielectrics, *Microelectron. Eng.*, 36, 157, 1997.
74. Suñé, J. et al., Point contact conduction at the oxide breakdown of MOS devices, *IEDM Tech. Dig.*, 191, 1998.
75. Alam, M.A. et al., Explanation of soft and hard breakdown and its consequences for area scaling, *IEDM Tech. Dig.*, 449, 1999.
76. Weir, B.E. et al., Ultra-thin gate dielectrics: They break down, but do they fail? *IEDM Tech. Dig.*, 73, 1997.
77. Schmitz, J., Soft breakdown triggers for large area capacitors under constant voltage stress, *Proc. Int. Reliab. Phys. Symp.*, 39, 393, 2001.
78. Roussel, P., Accurate and robust noise-based trigger algorithm for soft breakdown detection in ultra thin oxides, *Proc. Int. Reliab. Phys. Symp.*, 39, 386, 2001.
79. Alam, M. et al., The statistical distribution of percolation resistance as a probe into the mechanics of ultra-thin oxides breakdown, *IEDM Tech. Dig.*, 529, 2000.

80. Suñé, J. et al., Understanding soft and hard breakdown statistics, prevalence ratios and energy dissipation during breakdown runaway, *IEDM Tech. Dig.*, 117, 2001.
81. Suñé, J. and Wu, E., Statistics of successive breakdown events for ultra-thin gate oxides, *IEDM Tech. Dig.*, 47, 2002.
82. Alam, M. et al., Statistically independent soft-breakdowns redefine oxide reliability specifications, *IEDM Tech. Dig.*, 151, 2002.
83. Suñé, J. and Wu, E.Y., Statistics of successive breakdown in gate oxides, *IEEE Electron Dev. Lett.*, 24, 272, 2003.
84. Linder, B.P. et al., Growth and scaling of oxide conduction after breakdown, *Proc. Int. Reliab. Phys. Symp.*, 41, 402, 2003.
85. Monsieur, F. et al., A thorough investigation of progressive breakdown in ultra-thin oxides, physical understanding and application for industrial assessment, *Proc. Int. Reliab. Phys. Symp.*, 40, 45, 2002.
86. Monsieur, F. et al., Evidence for defect-generation-driven wear-out of breakdown conduction path in ultra-thin oxides, *Proc. Int. Reliab. Phys. Symp.*, 41, 424, 2003.
87. Wu, E.Y. et al., Critical assessment of soft breakdown stability time and the implementation of new post-breakdown methodology for ultra-thin gate oxides, *IEDM Tech. Dig.*, 919, 2003.
88. Suehle, J.S. et al., Acceleration factors and mechanistic study of progressive breakdown in small area ultra-thin gate oxides, *Proc. Int. Reliab. Phys. Symp.*, 42, 96, 2004.
89. Pey, K.L. et al., Size difference in dielectric-breakdown-induced epitaxy in narrow *n*- and *p*-metal oxide semiconductor field effect transistors, *Appl. Phys. Lett.*, 83, 2940, 2003.
90. Tung, C.-H. et al., Percolation path and dielectric-breakdown-induced-epitaxy evolution during ultrathin gate dielectric breakdown transient, *Appl. Phys. Lett.*, 83, 2223, 2003.
91. Young, C.D. et al., Interfacial layer dependence of $HfSi_xO_y$ gate stacks on V_T instability and charge trapping using ultra-short pulse *I–V* characterization, *Proc. Int. Reliab. Phys. Symp.*, 43, 75, 2005.
92. Vogel, E.M. et al., Modeled tunnel currents for high dielectric constant dielectrics, *IEEE Trans. Electron Dev.*, 45, 1350, 1998.
93. Kerber, A., Charge trapping and dielectric reliability of SiO_2/Al_2O_3 gate stacks with TiN electrodes, *IEEE Trans. Electron Dev.*, 50, 1261, 2003.
94. Kim, Y.-H. et al., Thickness dependence of Weibull slopes of HfO_2 gate dielectrics, *IEEE Electron Dev. Lett.*, 24, 40, 2003.
95. Young, C. et al., Detection of electron trap generation due to constant voltage stress on high-*k* gate stacks, *Proc. Int. Reliab. Phys. Symp.*, 44, 169, 2006.
96. Mitard, J. et al., Large-scale time characterization and analysis of PBTI in HfO_2/metal gate stacks, *Proc. Int. Reliab. Phys. Symp.*, 44, 174 2006.
97. Degraeve, R. et al., Degradation and breakdown of 0.9 nm EOT SiO_2 ALD HfO_2/metal gate stacks under positive constant voltage stress, *IEDM Tech. Dig.*, 408, 2005.
98. Bersuker, G. et al., Progressive breakdown characteristics of high-k/metal gate stacks, *Proc. Int. Reliab. Phys. Symp.*, 45, 49, 2007.
99. Okada, K. et al., Mechanism of gradual increase of gate current in high-*k* gate dielectrics and its application to reliability assessment, *Proc. Int. Reliab. Phys. Symp.*, 44, 189, 2006.
100. Kauerauf, T. et al., Abrupt breakdown in dielectric/metal gate stacks: A potential reliability limitation? *IEEE Electron Dev. Lett.*, 26, 773, 2005.

16

Defects Associated with Dielectric Breakdown in SiO_2-Based Gate Dielectrics

Jordi Suñé and Ernest Y. Wu

CONTENTS

16.1 Introduction

The focus of this chapter is the breakdown (BD) of the thin SiO_2 films used as gate insulators in metal–oxide–semiconductor (MOS) field-effect transistors (FETs). This includes the nitrided silicon oxides that have replaced pure SiO_2 for ultrathin gate oxide applications because the incorporation of nitrogen allows a reduction in the gate leakage current. Although the main concepts presented in this chapter might also (at least partially) apply to the BD of the high-K dielectrics required to further proceed with MOS technology downscaling, we will not explicitly deal with these types of dielectrics. In the process of reducing the transistor dimensions, the gate oxide thickness (T_{OX}) has been reduced from hundreds of nanometers to roughly 1 nm. Here, we will deal with oxides thinner than 10 nm, with emphasis on ultrathin oxides with $T_{OX} < 3$ nm. Specification of the thickness range is important because the phenomenology associated with defect generation and BD changes with T_{OX}.

When an oxide is subjected to a constant-voltage (CV) stress, leakage current flows through the oxide by tunneling. If the stress continues for a certain time, the current is found to evolve due to the continuous degradation of the oxide (generation of defects in the oxide bulk and the interfaces), and finally a current jump reveals the occurrence of the

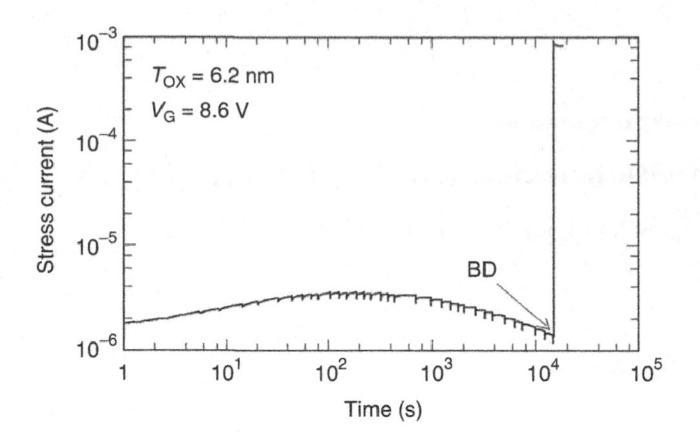

FIGURE 16.1
Evolution of the current that flows through a MOS structure with an oxide of 6.2 nm subjected to a CV stress at 8.6 V. The BD is detected as an abrupt current jump after more than 10^4 s of stress. A turnaround effect is observed in the pre-BD current. This indicates that there is an initial transient of positive charge trapping, later followed by the dominating electron trapping, which causes a continuous reduction of the tunneling current until the occurrence of the BD.

dielectric BD. This is illustrated in Figure 16.1 for a 6.2 nm oxide. Analogously, if the stress is performed under constant current conditions, the BD signature is an abrupt voltage drop. The abruptness of the BD event tends to disappear in the case of ultrathin dielectrics, and this introduces serious difficulties not only for the BD detection but also for its definition [1,2]. Figure 16.2 shows an example of a nonabrupt BD event measured in a p-channel FET (PFET) with $T_{OX} = 1.25$ nm. In general, we can define the BD as a local loss of the insulating properties of the dielectric. Whether this loss is abrupt or progressive depends mainly on the oxide thickness and the stress conditions.

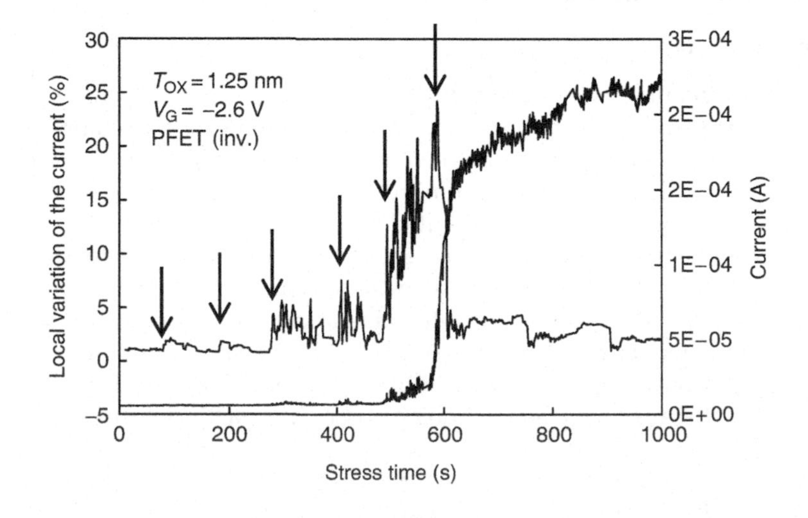

FIGURE 16.2
Evolution of the stress current during a CV stress experiment ($V_G = -2.6$ V; $T = 400$°C) performed on a 1.25 nm gate oxide PFET showing the progressive growth of the BD current. The time evolution of a BD current trigger function $\Delta I/I$ is shown to reveal relevant changes in the electrical properties long before the $I(t)$ curve is affected significantly. The vertical arrows indicate possible time locations of the first BD event. This example shows the difficulty of defining dielectric BD in ultrathin oxides and, consequently, to determine the time to BD.

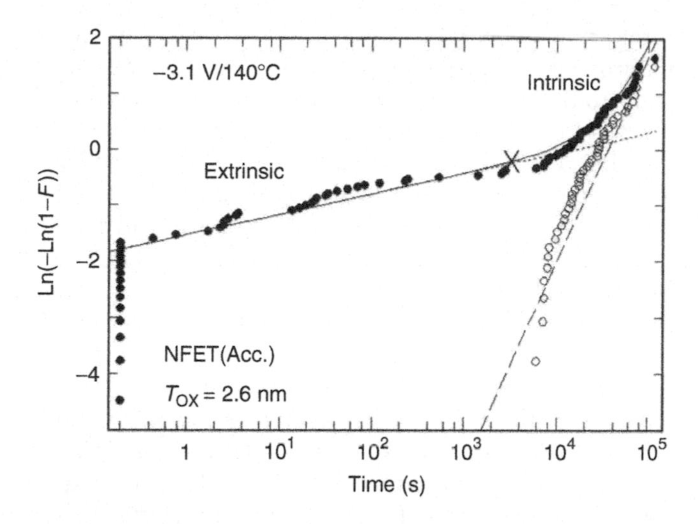

FIGURE 16.3
Cumulative BD distribution, F, represented in a Weibull plot versus time to BD. The devices under test, NFETs with $T_{OX} = 2.6$ nm, were subjected to CV stress ($V_G = -3.1$ V and $T = 140°C$). The dashed (dotted) line represents the distribution of intrinsic (extrinsic) BD. The solid line is the combined bimodal BD distribution. The solid circles are the raw experimental BD data points and the open circles correspond to only those samples that show intrinsic BD, which in this case are those with BD times >3000 s (cross-point). The separation between extrinsic and intrinsic samples is arbitrary but, in this case, it is supported by the examination of the distribution of initial current in the as-fabricated devices. (From Wu, E.Y. and Suñé, J. in *Proceedings of the International Reliability Physics Symposium*, The Institute of Electrical and Electronics Engineers, Piscataway, NY, 2006, 36. With permission.)

While the pre-BD degradation of the oxide is roughly uniform across the oxide area, the BD is a local phenomenon that is triggered in a region of about 10–100 nm^2 [3–5]. Oxide BD is a stochastic phenomenon, which means that a number of nominally identical structures subjected to identical stress conditions do not show the same time to BD. On the contrary, any BD variable shows a certain distribution of results (see Figure 16.3). Hence, the BD always requires a statistical description. Related to the stochastic nature and to the extreme localization of the BD is the weakest link character of the phenomenon, a property that determines the scaling of the BD statistical distribution with the oxide area [6,7]. Since the BD is triggered in an area much smaller than the device area, we may consider each device to be equivalent to a system of structures (spots) connected in parallel, each with equal (much smaller) area. From the point of view of system reliability, this parallel combination of structures is a series combination of elements, because the failure of a single element causes the failure of the whole system. The weakest spot is the one that causes the device BD. In the same way, the failure of a chip can be modeled as that of a system with N devices in parallel, and its BD coincides with that of the device that breaks down first.

As stated above, the weakest link property determines the way the BD distribution changes with area. If the cumulative probability of failure at time t of a device of area A_1 is known to be $F_1(t)$, the weakest link property can be used to calculate the cumulative probability of failure for a structure of arbitrary area. Without any loss of generality let us consider a device with area $A_2 > A_1$ (or a chip with total active oxide area A_2). Assuming that the device of area A_2 is divided into N identical structures of area A_1, the weakest link property tells us that the survival of the complete structure requires the survival of each of its elements, so that the cumulative survival probabilities are related by $1 - F_2(t) = (1 - F_1(t))^N$. This property (widely known as the Poisson area scaling property) is routinely used to calculate the cumulative BD distribution of large chips by scaling the distribution measured in stress experiments performed on small area test devices. The Poisson area scaling is a basic property of general validity provided that the degradation that finally causes the BD is uniform across the device area. In Figure 16.3, the cumulative distribution function (F) of the time to BD (t_{BD}) is depicted in the Weibull plot, which is a representation of $\text{Ln}(-\text{Ln}(1-F))$ versus $\text{Ln}(t_{BD})$. The choice of this plot is usual when dealing with oxide reliability because the BD distribution is found to be well described by a Weibull distribution model, which yields a straight line in the Weibull plot. The Poisson area scaling property yields a vertical shift of the cumulative distribution in the Weibull plot, i.e., if $A_2 = NA_1$ it immediately follows that $\text{Ln}(-\text{Ln}(1-F_2(t))) = \text{Ln}(N) + \text{Ln}(-\text{Ln}(1-F_1(t)))$. However, although the Weibull distribution is the most usual choice, the lognormal distribution has also been used by some authors to fit the experimental BD data. It has been argued that both Weibull and lognormal distributions are able to fit the data in a limited failure percentile window [8,9]. Although some controversy still exists about which distribution is better suited, the Weibull distribution is found to provide a much better description of the thin-oxide BD data when sufficiently large sample sizes are considered or when the area scaling property is used to artificially enlarge the experimental percentile window [10,11]. Other important pieces of evidence supporting the Weibull model are found in the percolation model of oxide BD [12] and that, as a particular case of extreme-value distribution, the Weibull model is invariant to area scaling while the lognormal distribution is not [13]. In other words, using the Poisson area scaling law, it is evident that, if the BD distribution is a Weibull distribution for a particular oxide area, it is also Weibull for structure of any area. On the contrary, if the BD distribution is lognormal for structures of one area, it cannot be lognormal for any other area.

The fact that the BD distribution is well described by the Weibull model is not evident in Figure 16.3 because the cumulative distribution is bimodal in this case. This bimodality corresponds to the combination of the so-called intrinsic and extrinsic BD modes. Both types of BDs are related to defects. However, extrinsic BD is caused by native defects related to imperfections of the fabrication process, whereas intrinsic BD is triggered by defects created during operation or accelerated stress. In this regard, intrinsic BD represents the BD of a nominally perfect oxide film. Extrinsic BD is responsible for early failures and shows a very wide distribution. On the contrary, intrinsic BD occurs after much longer stress times and its distribution is much narrower. In the time range dominated by extrinsic BD, the failure rate decreases with time (a typical indication of infant mortality related to fabrication defects). In the intrinsic BD regime, the failure rate increases with time, which is an unequivocal signature of a wear-out process. Both BD modes are important for reliability. Intrinsic BD determines the maximum reliability of the devices, hence it is the relevant mechanism for the qualification of the fabrication technology. On the other hand, extrinsic BD is the mechanism of most interest for product reliability. Since the design of integrated circuits (ICs) is performed under the constraints imposed on the technology by the intrinsic failure mechanisms, IC reliability is usually limited by the density of extrinsic defects. Although it is possible to separate the statistical distribution of intrinsic and extrinsic BD modes (see Figure 16.3), the use of small area structures is very convenient to avoid extrinsic defects and measure a single-mode (intrinsic) statistical distribution. In this chapter, we focus mainly on intrinsic BD and its relation to defect generation.

16.2 Defect Generation and Dielectric BD

Oxides are usually driven to dielectric BD by electrical stress experiments under strongly accelerated conditions. This means that the voltage applied during stress, and sometimes also the temperature, are much higher than expected for the operating conditions of ICs.

During these experiments the oxide wears out due to the generation of defects in the oxide bulk and at the interfaces. Some of these defects are related to oxide reliability because the BD is triggered when their density locally reaches a critical value. Their atomic structure is not yet well determined and they are generically referred to as traps. In this regard, our description is mostly phenomenological and focused on the impact of these traps on the electrical properties of the insulating film.

16.2.1 Monitoring Defect Generation during Electrical Stress Experiments

The degradation of the oxide is usually monitored by measuring electron trapping in newly generated bulk traps, the increase of the density of interface traps at the Si/SiO_2 interface, or the increase of the leakage current at low fields.

The generation of interface traps, for example, has been measured by means of capacitance–voltage characterization of MOS structures and also by the charge pumping technique in transistors [14–18]. These experiments revealed several important results: (1) the average density of defects required to trigger the BD decreases with the oxide thickness and is roughly independent of the applied voltage; (2) the width of the statistical distribution of the BD critical density of defects increases when the oxide thickness is reduced; and (3) the defect generation rate decreases very strongly when the applied voltage decreases. Although these experiments have demonstrated that there is a certain correlation between oxide BD and interface trap generation, the formation of a local conduction path through the oxide is inherent to the dielectric BD phenomenon, and this requires the generation of bulk rather than interfacial defects. Thus, although the generation of interface traps might be an indicator of oxide degradation, the relation between interface traps and BD is probably not causal. The case of ultrathin oxides might be an exception because the separation of bulk and interface defects is not so evident in such films. Moreover, recent results reported by Nicollian suggest that interface traps control the BD of gate oxides below $V_G \sim 2.7$ V and $T_{OX} \sim 1.5$ nm [19].

Other well-known effects of oxide degradation are an initial increase of positive charge trapping [20–22] and a long-lasting process of generation of neutral electron traps [4,23–28]. In oxides thicker than about 5 nm, the tunnel current significantly evolves during stress due to the partial occupation of native and generated bulk traps. Under constant voltage (CV) stress conditions, for instance, Figure 16.1 shows that, after first increasing for a short while before the turnaround point, the current continuously decreases until the BD. This current reduction has been related to electron trapping. Although sometimes it has been claimed that the positive charge trapping plays a fundamental role in the triggering of oxide BD [20–22], experiments demonstrate that it is only dominant during the early stages of stress, and that the net charge trapping is negative when the BD occurs. Moreover, while positive charge trapping saturates well before BD, the negative trapping shows a nonsaturating behavior that has been linked to the generation of neutral electron traps [4,23–28]. As in the case of interface traps, the BD is observed to occur when the density of neutral electron traps reaches on the average a critical (thickness dependent) value [25].

In oxides thinner than about 5 nm, electrons can easily tunnel from the traps to the electrodes so that trapping effects decrease significantly. On the contrary, the traps can efficiently act as stepping stones for electrons, thus giving rise to an inelastic trap-assisted tunneling (TAT) component that adds to the background tunneling current [29–36]. This is the stress-induced leakage current (SILC) which, reported for oxides thinner than about 8 nm, has also been used as a monitor of oxide degradation in thin oxides [17,37–42]. Figure 16.4 shows how the current–voltage characteristics change with stress for thin oxides with T_{OX} ranging from 1.2 to 7 nm [41]. The most usual quantitative indicator of SILC is the relative current increase $\Delta I/I_0$ with respect to the initial current, I_0, usually

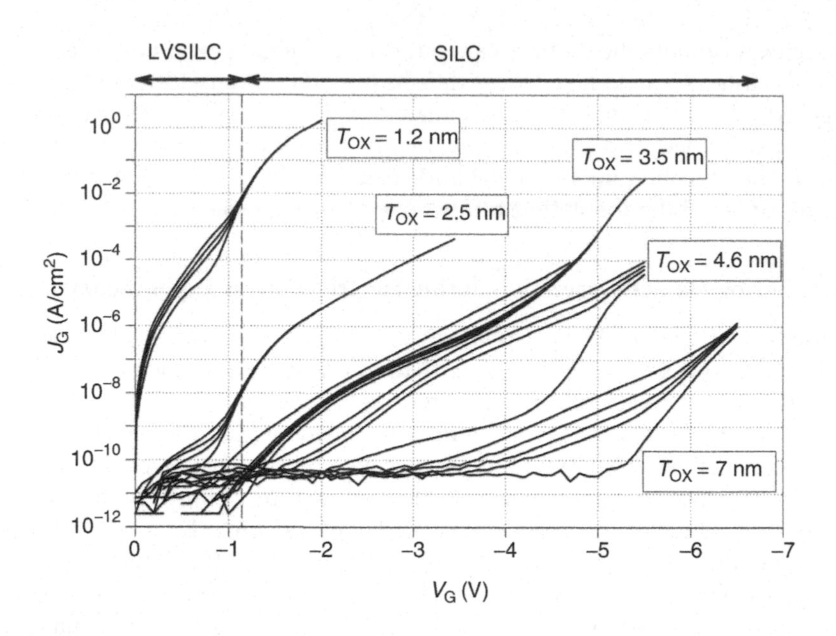

FIGURE 16.4
Typical J_G versus V_G curves in n-MOSFETs subjected to electrical stress under negative bias conditions. For each thickness, several $J_G(V_G)$ characteristics are shown which correspond to increasing stress times. SILC increase is shown to be important in oxides of thickness 3.5–7 nm. Below $T_{OX} = 2.5$ nm, the conventional SILC increase is completely masked by the background direct tunneling current, but there is a noticeable effect of the stress when the semiconductor is biased in depletion ($-1 < V_G < 0$). This increase of the low-voltage current is known as LVSILC. (From Petit, C. and Zander, D., *Microelectron. Reliab.*, 47, 401, 2007. With permission.)

measured at a low sense voltage because the SILC is relatively more important at low voltages. Therefore, the high-voltage stresses used to cause the device BD within a reasonable time frame are periodically stopped to measure the evolution of the current at the sense voltage. As shown in Figure 16.5, the evolution of the SILC is usually found to depend on time as a power law $\Delta I/I_0 = at^{\alpha}$ with $\alpha < 1$ [41]. From the time derivative of the SILC degradation monitor, a measure of the defect generation rate has been obtained [38]. On the other hand, the value of $\Delta I/I_0$ measured immediately before the BD has been interpreted by some authors as a measure of the average density of defects required to trigger the BD, N_{BD} [43]. The defect generation rate measured as a function of the stress voltage [44] and the critical defect density measured as a function of T_{OX} have been used to forecast chip reliability as a function of gate bias down to operation conditions [44]. However, the detailed analysis of the SILC statistics [45], the fact that different annealing kinetics have been found for SILC and BD [46], and other issues related to the SILC definition and voltage extrapolation [47] have called into question the use of SILC for quantitative reliability prediction. In ultrathin oxides with T_{OX} below 2.5 nm, the relative importance of the conventional SILC is smaller due to the strong increase of the background current due to direct tunneling. However, a noticeable increase of the current has been reported at very low sense voltages (see Figure 16.4) [41,48–50]. This low-voltage SILC (LVSILC) has also been suggested to be a good indicator of oxide degradation leading to BD [41]. Although some doubts about the quantitative measure of oxide degradation always remain, linear correlations between the measured SILC increase, the interface trap density buildup, and the density of generated neutral electron traps have been reported (see Figure 16.6 [17]). This justifies the alternative use of each of these different measurements as relevant monitors for the degradation of the oxide.

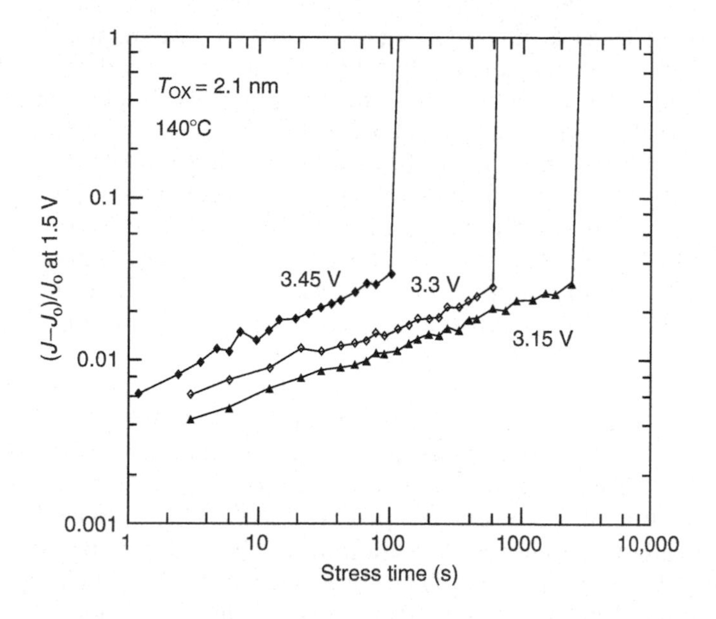

FIGURE 16.5
Time evolution of the relative SILC increase measured at a sensing gate voltage of 1.5 V during three different stress experiments performed on NFETs with $T_{OX} = 2.1$ nm. Stresses were performed in inversion at three different stress voltages and at 140°C. The time dependence of SILC increase is well described by a power law with exponent $\alpha < 1$. The final abrupt current jump corresponds to oxide BD. (From Lai, W.L. et al., in *Proceedings of the International Reliability Physics Symposium*, The Institute of Electrical and Electronics Engineers, Piscataway, NY, 2004, 102. With permission.)

Recently, Degraeve and coworkers directly measured the evolution of the number of defects during stress by periodically recording the *I*(*V*) characteristic in small area ($\sim 5 \times 10^{-10}$ cm^2) MOS structures with ultrathin (~1.8 nm physical thickness) SiON films

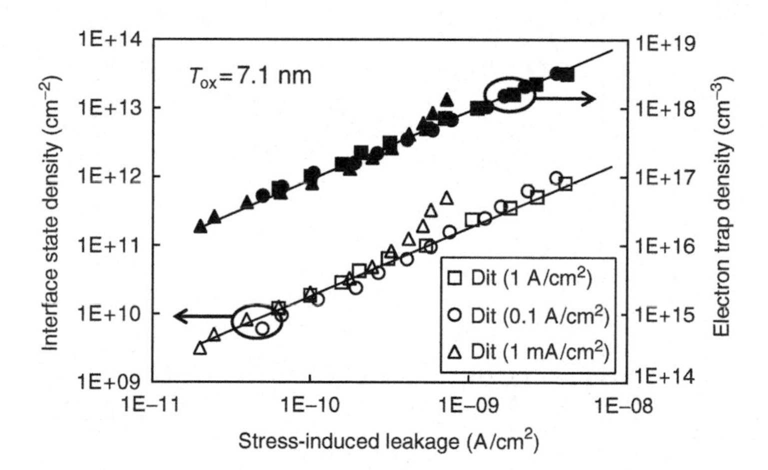

FIGURE 16.6
Correlation between SILC, interface trap density, and electron trap density generated in a 7.1 nm oxide subjected to constant current stresses. Squares correspond to a stress current density $J_{stress} = 1$ A/cm^2, circles to $J_{stress} = 0.1$ A/cm^2 and triangles to $J_{stress} = 1$ mA/cm^2. Interface trap densities (empty symbols) were measured using the charge pumping technique, electron trap densities (filled symbols) by CV flatband voltage shift (after filling the traps by substrate hot electron [SHE] injection), and SILC is the excess leakage measured at an oxide field of 5 MV/cm. The lines correspond to linear correlations (slope = 1). (From De Blauwe, J. et al., *IEEE Trans. Electron Devices*, 45, 1745, 1998. With permission.)

[51]. Their procedure consists of comparing successive measurements of $I(V)$ to isolate the contribution of each single generated defect. In this way, not only do they obtain a defect-by-defect measurement of the evolution of the defect density, but also statistical information about the properties (energy and position) of the generated traps. These measurements are only possible in very small area structures and ultrathin oxides. Ultrathin oxides are required to enhance the importance of the trap-assisted current provided by a single defect. The small area is needed to avoid averaging effects due to the generation of multiple defects between successive $I(V)$ measurements and to reduce the background current. The reported results support the power-law time evolution of the defect density (see Figure 16.7), but a model accounting for the random telegraph fluctuations of the generated defects is required to reconcile the measured exponent ($\alpha \sim 0.38$) with that ($\alpha \sim 0.26$) obtained for the average evolution of SILC. The results reported by Degraeve reveal the difficulty to detect and even to define BD in ultrathin oxides [51]. The BD event has usually been defined as an abrupt current jump (see Figure 16.1) related to the creation of a local conduction path that suddenly connects the electrodes. In ultrathin oxides, the growth of the BD current is not abrupt but progressive [52,53], and even the generation of a single defect can produce an abrupt current jump. The combination of these two phenomena complicates the definition (and hence, the detection) of the BD in ultrathin oxides.

Since the BD current growth is not abrupt and one desires to detect the BD event just when it is triggered, one must be able to detect small BD-related changes in the current time evolution. Hence, the test structures must have a small area to avoid a large background direct tunneling (DT) current that could mask the signature of the BD. However, if the oxide is ultrathin and the structure area is small, the generation of a single trap can give rise to a current jump that might be wrongly interpreted as the BD. An example of this difficulty is found in Figure 16.2, which shows the BD current transient (for a PFET with $T_{OX} = 1.25$ nm) and the evolution of $\Delta I/I$, which has been considered as a useful BD trigger for ultrathin oxides [1,54]. In this case, $\Delta I(t)/I(t)$ shows noisy bursts long before the final transition of the BD current. The arrows shown in Figure 16.2 represent different possible locations of the first BD according to the evolution of the $\Delta I/I$. Whether these bursts are

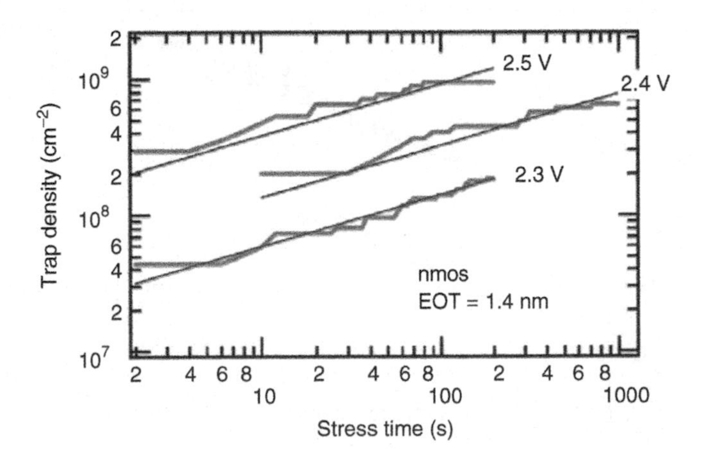

FIGURE 16.7
Evolution of the trap surface density, N_S, measured by Degraeve, using a technique that apparently allows one to identify the defects one by one as they are generated during stress. The time evolution is nicely described by a power law $N_S = N_{S1}\ t^{\alpha}$ with $\alpha \sim 0.38$. The devices are small area (6.25×10^{-10} cm^2) transistors with SiON dielectric with a physical thickness of about 1.8 nm. (From Degraeve, R. et al., in *Proceedings of the International Reliability Physics Symposium*, The Institute of Electrical and Electronics Engineers, Piscataway, NY, 2005, 360. With permission.)

pre-BD noise due to fluctuations in trapped charge density or the signature of the initial stages of the BD is a question that has no clear answer on a single-sample basis. An early study of the pre-BD noise in MOS structures of interest for microelectronics was performed by Neri et al. [55]. These authors reported the observation of large current fluctuations during a relatively short period of time before the BD current jump. However, this interpretation is not unique, and other authors consider that these fluctuations are not a precursor of the dielectric BD, but instead are evidence that the BD path is already formed [56]. Following this interpretation, several groups have suggested the use of noise-based triggers to detect BD in ultrathin oxides. Recently, a method has been proposed to bypass the problem of BD definition by recording the whole $I(t)$ transient during the stress experiment, and analyzing of the statistical data for different values of the BD trigger ΔI [2]. Since it is very difficult (or impossible) to identify the critical value of $\Delta I/I$ that corresponds to the BD in the evolution of the stress current of an individual sample, this choice must be made on the basis of the statistical analysis of the data corresponding to many samples [1,2,57]. The analysis of experimental BD data using this procedure has been recently shown to provide very interesting information about ultrathin oxide reliability in both PFETs and n-type field-effect transistors (NFETs) [1,2,57].

The main problem of using oxide degradation monitors for BD reliability forecasting is not being sure that the measured defects are actually those that are involved in the formation of the BD path. To avoid this uncertainty, a completely different approach has been proposed which is based on the direct analysis of the BD data [58,59]. The method consists of measuring the BD statistics to obtain information about defect generation by interpreting the BD data in the framework of the percolation model [3,11,25,60]. The details of this method will be discussed in Section 16.3, after first presenting the percolation model Section 16.2.3.

16.2.2 Physics of Defect Generation and BD

After 50 years of intense research on the degradation and BD of SiO_2 gate oxides, no definitive conclusion about the nature of the defects that trigger the BD, or about the mechanisms involved in the generation of these defects, has been reached. This is due at least in part to the BD phenomenology, and also because the mechanisms involved in the generation of oxide defects change with the oxide thickness and stress conditions. In this period of time, T_{OX} has been scaled from hundreds of nanometers to the limit of 1.2 nm found in the 90 and 65 nm complementary MOS (CMOS) technology nodes, while the gate voltage has been scaled accordingly. With a focus on oxides thinner than 10 nm, we will first discuss the role of the oxide field, the impact of carrier injection, and the influence of the energy of the injected carriers on the process of defect generation. Then, we will briefly review three relevant BD models. In this section, we will consider the anode hole injection (AHI) model and the thermochemical model of BD. The hydrogen release (HR) model will be considered in more detail in Section 16.3.

The oxide electric field has always been considered to have an important role in the dielectric BD process. Early work focused on determining the BD field (BD strength) of the oxide, with the underlying idea that this was an intrinsic property of the material [61]. Nowadays, the simplistic idea of a maximum BD field has been abandoned, since it is known that the BD is related to the previous generation of defects in the oxide (at least in the case of oxides thinner than 10 nm). However, the oxide field still plays an important role. If one plots the gate voltage required to keep the mean time to BD fixed as a function of the oxide thickness, a linear relation is found [62]. Reliability engineers typically assume on the basis of experience that keeping an oxide field of about 10 MV/cm provides a zeroth-order estimation of the gate voltage that can be applied to keep the time to BD within a reasonable time frame for wafer-level stress experiments. Given these experimental

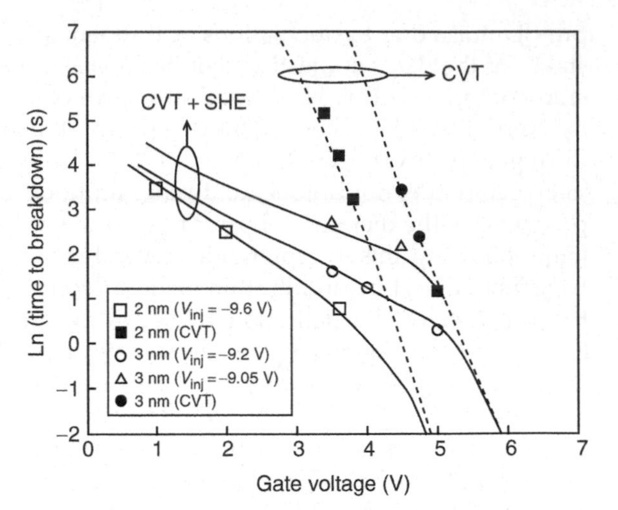

FIGURE 16.8
Time to BD versus gate voltage for CV tunneling (CVT) stress, and for the combined CVT and substrate hot electron (SHE) injection experiments as reported by Vogel and coworkers. Lines correspond to models of Vogel et al. (From Vogel, E.M. et al., *IEEE Trans. Electron Devices*, 47, 1183, 2000. With permission.)

results, the most direct conclusion is that the BD is controlled by the oxide field. Hence, a popular model of oxide BD is based on the assumption of a thermally activated field-assisted bond-breaking mechanism [63,64]. However, although the electric field is certainly important, there is strong experimental evidence indicating that electron injection through the oxide and the energy of the electrons that reach the anode SiO_2/Si interface play the most relevant role in the oxide BD process [65–67]. As claimed by Selmi et al. [67], the transition toward ballistic transport in ultrathin oxides, associated with voltage downscaling, has made carrier energy emerge as the driving force for oxide wear-out, and also for other transistor failure mechanisms such as hot-carrier degradation or negative bias temperature instability (NBTI). The roles of current and energy have been studied by means of hot electron injection experiments. This technique allows one to separately change the oxide field, the stress current, and the energy distribution of the injected carriers. As shown in Figure 16.8, the time to BD is strongly influenced by the injection of hot carriers under fixed oxide field conditions. In this way, it has been conclusively demonstrated that the oxide field alone cannot explain the complexity of the BD process. On the contrary, the injection of carriers and their energy distribution at the anode are relevant variables that significantly alter the dynamics of defect generation and BD.

Given these results, defect generation models based on the injection of carriers and on the dissipation of energy at the anode interface have been proposed. This electron energy-driven approach has two main versions: the HR model [24,44,58,68–70], and the AHI model [20–22,71–73]. In these models, defect generation is described as a two-step process: (1) injected electrons gain energy from the oxide field and release some positively charged species from the anode, and (2) the released species travel backward toward the cathode, and generate defects by reacting with some precursor sites in the oxide.

16.2.2.1 Thermochemical Model of Oxide BD

In this model, a process of dipole–field interaction is assumed to explain bond breakage, defect generation, and oxide BD. The dipole moment is related to oxygen vacancies. It is assumed that, when an oxide field (F_{OX}) is applied to this dipole moment, the activation energy for bond breakage is reduced, thus leading to enhanced defect generation. This is a fundamental degradation mechanism since it is known that there are always many oxygen vacancies in SiO_2 films. Recently, the model has been extended to incorporate a distribution of different bond states (oxygen vacancies, distorted Si–Si bonds, Si–H bonds, and many others) [64], and further refinements which are related to the theoretical description of the molecular dipole moment [74]. The main success of the thermochemical model was to provide a physics-based justification for the $\exp(-\gamma F_{OX})$ dependence of the time to BD on the oxide field that had been traditionally used for voltage extrapolation [75,76]. McPherson and coworkers stressed 9 nm thick oxide modules for 3 years at low fields, trying to determine the functional form of voltage acceleration; their results were found to agree with the $\exp(-\gamma F_{OX})$ model. Moreover, in that oxide thickness range, the derived field acceleration factor γ was found to be in good accordance with the experimental data, provided that the involved dipole moment is in the reasonable range from ~7 to 15 eV.

In spite of this partial success, there has been controversy about the actual functional form that best fits the T_{BD} dependence on the oxide field. In particular, in the mid-1980s, a much more aggressive extrapolation model based on an exponential dependence on the inverse of the electric field $T_{BD} \propto \exp(-F_{OX}/G)$ gained support from the AHI model developed by the group led by Hu [20–22]. Whether the field dependence of T_{BD} is better represented by $T_{BD} \sim \exp(-\gamma F_{OX})$ or by $T_{BD} \sim \exp(B/F_{OX})$ has been the subject of extensive debate in recent years. Extrapolation of the results of accelerated stress performed at high voltages to low-voltage operation conditions is a key element of any reliability practical methodology. These two models (commonly known as the E-model and the $1/E$-model, respectively) both provide a reasonable fit of the high-voltage stress experimental data. However, the required extrapolation spans many orders of magnitude in time [77], and the extrapolated IC lifetime (under operation conditions) is significantly different for each of the models. Thus, the choice of one model or the other has a significant impact on the performance-reliability trade-off (see Chapter 15). On the other hand, in the ultrathin oxides of present interest, the time to BD is found to depend on the gate voltage rather than on the oxide field [78,79], and the acceleration factor is found to depend on the applied voltage [73,80–82]. These features cannot be explained by the thermochemical model, and extrapolations based on a constant value of γ are largely pessimistic [62]. In any case, the most important limitation of the thermochemical model is that it assumes that the BD is controlled by the oxide field, while there is compelling experimental evidence that both electron fluence and electron energy play a crucial role [65–67].

16.2.2.2 AHI Model

In the mid-1980s, Hu and coworkers developed a positive feedback BD model based on the assumption that holes generated in the oxide become trapped in localized areas near the cathode and cause an increase of the local current density which further enhances hole generation and trapping [20,21]. However, it was not until 1994 that the AHI model was presented in its complete form [71,83]. According to this model, electrons are injected through the oxide, gain energy from the electric field and generate electron–hole pairs in the anode with a hole generation efficiency, α_p. The generated hot holes have a certain probability Θ_p of being reinjected into the oxide by tunneling through the anode Si/SiO_2 hole barrier. Once in the oxide, the holes travel toward the cathode, accelerated by the oxide field, and some of them become trapped. Finally, the trapped holes generate (in cooperation with injected electrons [84]) the defects that trigger oxide BD. The probability for an injected hole to generate one defect, k_p, is usually considered to be temperature dependent, but field or voltage independent. While a reasonable Fowler–Nordheim model was considered for the field-dependent tunneling probability Θ_p in this early version of the AHI model, a phenomenological energy relaxation model was used to calculate the energy of the electrons at the anode [85], and the efficiencies for anode hole generation α_p and hole-assisted defect generation k_p were used as fitting parameters. The main result supporting the AHI model was the finding of a rather constant total hole fluence to BD, as measured by integrating the substrate current in carrier separation experiments. Subsequently, the AHI model received considerable support from physics-based Monte Carlo (MC) simulations of the transport of electrons, including a detailed consideration of electron–hole pair generation processes in the silicon anode [72,73]. Using these MC simulations, the role of minority carrier impact ionization was revealed [72], which explained why the electron energy must be referred to the anode Fermi level instead of the anode silicon conduction band [78]. However, serious doubts about the actual involvement of holes in the process of defect generation arise from recent experiments [86–90]. As discussed above, substrate currents historically have been used to measure the total hole fluence and to demonstrate the existence of a critical hole fluence to BD. However, mechanisms different from the generation of holes in the anode have been suggested recently as alternative explanations for the substrate current. In particular, generation–recombination processes in the substrate [86] and the creation of hole–electron pairs by photons generated by energetic electrons in the gate [87] have been proposed. On the other hand, serious doubts recently were raised by the publication of results concerning the efficiency of defect generation (k_p) by holes injected through thin oxides [88–90]. Using PFETs stressed at low voltages, in which the hole tunneling current becomes comparable to the electron current, it was shown that only when the hole current reaches at least the magnitude of the electron current do the effects of hole injection on additional defect buildup become observable [88]. Hence, under conventional NFET inversion conditions, the AHI mechanism, which relies on a tiny anode hole current that is orders of magnitude smaller than the injected electron current, cannot be the dominant defect generation mechanism. On the other hand, using the substrate hot hole (SHH) injection technique, it was shown that different amounts of injected holes have no impact on subsequent time-dependent dielectric background (TDDB) measurements [90]. Surprisingly enough, while many efforts have been dedicated during the last 25 years to improve the understanding of the generation and injection of holes from the anode, the main criticisms of the AHI picture arise from results concerning the efficiency of the holes to react with precursors and generate defects in the oxide, a process that has not been theoretically studied in detail. Given these serious criticisms of the AHI picture, the multiple experimental evidence that supports the involvement of hydrogen in the wear-out of oxides [65], and the proposal of a power-law voltage acceleration empirical model for oxide BD [80], several groups have recently developed a quantitative oxide degradation model based on the release of hydrogen species. This HR picture will be explained in detail in Section 16.3.

16.2.3 Relationship between Defect Generation and Breakdown Statistics

Although the relation between the generated electron traps and oxide BD had already been suggested in a number of earlier publications [23,24,27,91], the relation between defect generation and BD statistics was established directly in 1990 by Suñé et al. [3]. In that work, the oxide area was divided in a two-dimensional array of cells, and the BD was considered to occur when a critical number of defects, n_{BD}, had been generated in one of those cells. Assuming a uniform distribution of the defects across the oxide area, the statistical distribution of the number of defects in the cells could be modeled by the Poisson distribution, and the cumulative BD distribution was computed as the probability of finding one or more cells with a number of defects equal or larger to n_{BD}. Subsequent work by Dumin demonstrated that the model was adequate to describe the experimental BD distributions and gave the first estimation of the area ($\sim 10^{-14}$ to 10^{-12} cm^2) of the region where the BD is triggered [4,92]. This model introduced the important idea that a critical density of defects triggers the BD but, being a two-dimensional picture, it failed to

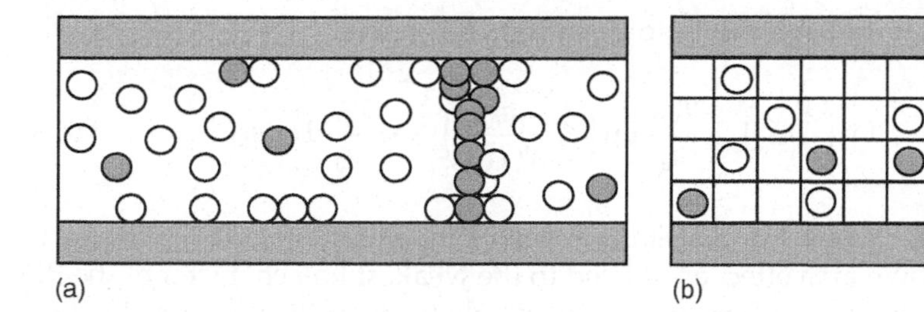

FIGURE 16.9
Schematic cross-sectional representation of defect generation in the oxide as considered in percolation models. (a) Defects modeled by spheres generated at random as considered by Degraeve et al. [25]. (b) Cell-based model as considered by Suñé [12]. (From Suñé, J., *IEEE Electron Device Lett.*, 22, 296, 2001. With permission.)

consider the impact of the oxide thickness. To capture the scaling of the BD distribution with T_{OX}, the concept of percolation path introduced by Degraeve is crucial [25]. In the Degraeve's version of the percolation model, defects are modeled as spheres generated at random positions in the oxide film, and the BD is triggered by the first complete path of overlapping defects, as shown schematically in Figure 16.9. Slightly different versions of the percolation model have been developed recently [12,93,94], but the main concepts are the same as in Degraeve's picture. The percolation model is usually implemented through MC simulation, which considers all possible types of overlapping defect paths and which is a very convenient approach to incorporate second-order effects such as nonuniformity or correlation [60]. However, for practical reliability assessment methodologies, simpler analytical pictures are highly desirable. With this goal, a simple three-dimensional cell-based model was developed that only considers vertical paths, yet still reproducing all the important results of the percolation picture [12]. This model has been recently revisited by Krishnan and Nicollian, who incorporated nonvertical paths to make it completely equivalent to the full percolation model [94].

Taking the cell-based version of the percolation model as the simplest reference framework, we now discuss the model properties in more detail. The oxide is considered to be divided into a lattice of cubic cells of side a_0 [12]. The generation of defects is modeled as a transition of a cell to a defective state. The BD occurs when a vertical column of cells that covers the whole oxide thickness is fully defective. Hence, if the oxide thickness is T_{OX}, the minimum number of defects required to trigger the BD is $n_{BD} = T_{OX}/a_0$. Assuming a uniform and uncorrelated generation of defects, straightforward calculations yield an analytical equation for the cumulative distribution of BD F_{BD}, which in the standard Weibull representation:

$$W_{BD} \equiv \mathrm{Ln}(1 - \mathrm{Ln}(1 - F_{BD})) = \mathrm{Ln}\left(\frac{A_{OX}}{a_0^2}\right) + \frac{T_{OX}}{a_0}\mathrm{Ln}\left(N_S \frac{a_0^3}{T_{OX}}\right) \tag{16.1}$$

Here N_S is the density of generated defects per unit area. However, for comparison with the experimental results, we introduce the relation between N_S and the usual BD variables, t_{BD} or the charge injected to BD (q_{BD}). The density of generated defects is experimentally found to follow a power law of the fluence q_{INJ}:

$$N_S = \zeta_\alpha \left(\frac{q_{INJ}}{q}\right)^\alpha \tag{16.2}$$

Here ζ_α and α are constants. Substitution into Equation 16.1 yields:

$$\mathrm{Ln}(-\mathrm{Ln}(1-F_{\mathrm{BD}})) = \mathrm{Ln}\left(\frac{A_{\mathrm{OX}}}{a_0^2}\right) + \frac{T_{\mathrm{OX}}}{a_0}\mathrm{Ln}\left(\zeta_\alpha \frac{a_0^3}{q^\alpha T_{\mathrm{OX}}}\right) + \alpha\frac{T_{\mathrm{OX}}}{a_0}\mathrm{Ln}(q_{\mathrm{BD}}) \tag{16.3}$$

This equation gives the BD cumulative distribution versus the measurable variable q_{BD}. In this equation, the well-known area effect associated to the weakest link character of the BD and to the uniform distribution of defects appears explicitly through the term $\mathrm{Ln}(A_{\mathrm{OX}}/a_0^2)$. Since the plot of W_{BD} versus $\mathrm{Ln}(q_{\mathrm{BD}})$ is a straight line, it is confirmed that this is a Weibull distribution with Weibull slope

$$\beta = \alpha\frac{T_{\mathrm{OX}}}{a_0} \tag{16.4}$$

and Weibull scale factor

$$Q_{\mathrm{BD}} = \frac{qT_{\mathrm{OX}}^{1/\alpha}}{a_0^{3/\alpha}\zeta_\alpha^{1/\alpha}}\exp\left[-\frac{a_0}{\alpha T_{\mathrm{OX}}}\mathrm{Ln}\left(\frac{A_{\mathrm{OX}}}{a_0^2}\right)\right] \tag{16.5}$$

where Q_{BD} is the characteristic injected charge required to cause the BD of ~63.2% of the samples. These two equations provide β and Q_{BD} as a function of T_{OX} and A_{OX}. Experimentally, the Weibull slope is found to decrease linearly with T_{OX}, as predicted by Equation 16.4 (see Figures 16.10 and 16.11). However, the experimental relation between β and T_{OX} does not always pass through the origin, but has a positive [94–96] or a negative Y-intercept [25]. In the approach of Krishnan and Nicollian, who have extended the cell-based model to full percolation, the nonzero Y-intercepts were explained in terms of native defects, multiple types of traps, and misalignment [94]. These authors found that the linear scaling of the Weibull slope is a robust result that does not depend on the kind of defects or types of percolation paths implemented in the model. However, they found that the relation between Q_{BD} and T_{OX} is more sensitive to the model implementation details. In particular, the more types of paths that are considered to trigger the BD, the lower is the value predicted for Q_{BD} because the BD becomes more probable for the same density of generated defects.

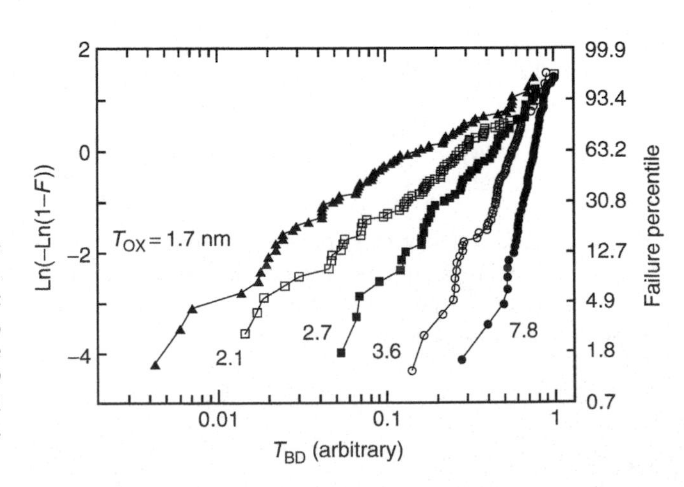

FIGURE 16.10
Normalized T_{BD} distributions for oxide thickness from 1.7 to 7.8 nm at 140°C. The data were collected using different stress voltages. The behavior of the Q_{BD} distributions is fully analogous, with the Weibull slope being the same for both BD variables. (From Wu, E.Y., Suñé, J., and Lai, W.L., *IEEE Trans. Electron Devices*, 49, 2141, 2002. With permission.)

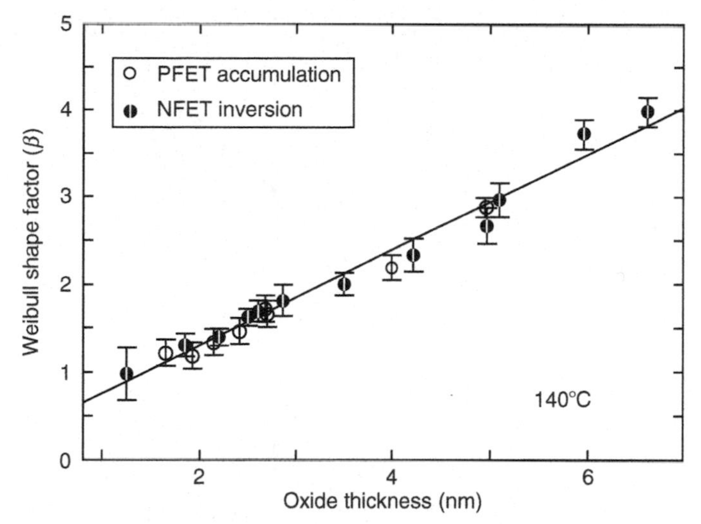

FIGURE 16.11
Weibull shape factor (Weibull slope) as a function of oxide thickness.

The percolation model also can be used to describe the statistics of successive BD events. In the case of thin and ultrathin oxides, where the post-BD current is sometimes not very high, the occurrence of multiple BD events has been reported when the stress experiment continues after the detection of the first BD (Figure 16.12). Within the percolation approach, the cumulative distribution of the *K*th BD event can be readily calculated [97–100]. Agreement between model and experiment is excellent (see Figure 16.13), thus giving further support to the main assumptions of the percolation approach. In some cases, however, correlation effects have been observed between successive events, and slight modifications of the theory are required to fully capture these experimental details.

Although the percolation model has represented one of the most significant advances in the understanding of the oxide BD statistics, it is a very simple geometrical picture that has some important limitations that become more evident as the oxide thickness is reduced and approaches the size of the defects. To address these limitations, the concept of BD path efficiency has been introduced [60] to match the critical defect densities measured at BD [12]. The concept of path efficiency recognizes that some paths predicted by the percolation model are not able to effectively trigger the BD. In the case of ultrathin oxides, moreover, it

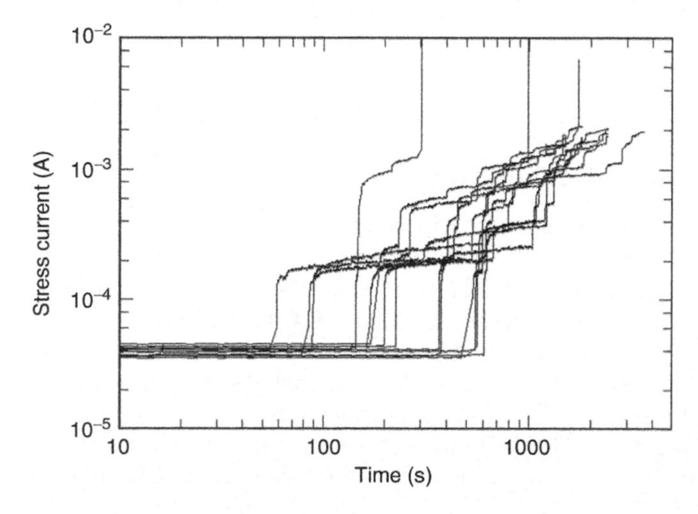

FIGURE 16.12
Detection of multiple BD events during CV stress experiments performed at 140°C and $V_G = 3.7$ V on NFET capacitors with $T_{OX} = 2.6$ nm and $A_{OX} = 6.2 \times 10^{-4}$ cm^2. Each line corresponds to a different sample, and each current jump to a BD event.

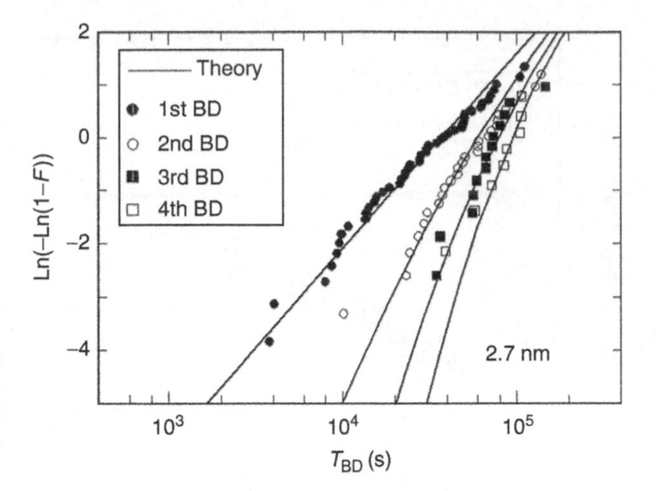

FIGURE 16.13
Statistical cumulative distributions of the first, second, and third BD event as measured (marks) in 2.7 nm oxide samples stressed at 4.2 V and 30°C. Lines correspond to the percolation model predictions. (From Suñé, J. and Wu, E.Y., *IEEE Electron Device Lett.*, 24, 272, 2003. With permission.)

is difficult to relate the geometrical definition of the BD path to the actual BD detection criteria used in experiments. As we have already discussed, in the thin-oxide limit, the BD current growth is progressive [52,53], and the generation of a single defect can sometimes cause a measurable current jump [51]. Hence, the simple BD experimental criterion of a large current jump, which had been used without any difficulty in the case of thicker oxides, is no longer applicable to ultrathin oxides. In this limit, the BD distribution has been found to depend on the magnitude of the current threshold $(\Delta I/I)|_{BD}$ used to detect the BD, which is not considered by the traditional percolation model. The limitations of the percolation approach are not unexpected, since this is a very simple geometrical abstraction of physical reality. Although it captures first-order effects such as the thickness scaling of the Weibull slope, if we want to understand second-order effects and establish a closer relation to the conditions of BD experiments, a more physics-based approach is required. Some key steps in this direction were done by Degraeve and coworkers with the goal of modeling the distribution of leakage current in nonvolatile memory cells [36]. These authors abandoned the geometric description to consider a physics-based model for the TAT current, coupled with a statistical model for the position of the defects in the oxide [35,36]. Recently, we have revisited their approach and proposed an improved nongeometrical picture for oxide BD that is able to recover all the results of the early versions of the percolation model and, at the same time, establish a much better connection to the experimental conditions of BD detection [57].

16.3 HR Model of Defect Generation and BD

There is plenty of hydrogen in MOS devices due to its intentional introduction during processing, and hydrogen is known to form many different defects in silicon dioxide (Chapter 7). One of them, the hydrogen bridge, has been recently identified as a trap with all the properties required to explain the characteristics of SILC [101]. Moreover, it is known that hydrogen has a significant impact on defect generation and oxide BD, as has been reported by different authors. For example, it is known that the introduction of hydrogen during high-temperature anneals or in the deposition of Si_3N_4 causes a strong reduction of Q_{BD} during high-voltage stress [102]. On the other hand, the rate of generation of the traps involved in SILC nicely correlates with the hydrogen dose in remote hydrogen plasma exposure experiments [38]. Recently, it was also shown that excessive hydrogen can lead to degraded T_{BD} and Q_{BD} and to a change of the voltage acceleration factors [103,104]. In summary, there is a lot of experimental evidence suggesting that hydrogen is involved in oxide degradation and BD. However, the so-called HR model of oxide BD has remained only a qualitative proposal for years. Only very recently, and given the difficulties found by the AHI model (see Section 16.2.2.2), several groups have worked out quantitative formulations of a model based on the involvement of hydrogen [58,59,69,70]. The main trends of the proposed picture are common to all these formulations, which are based on the release of hydrogen from bonding sites at the anode interface, and on the ulterior reaction of the released species with precursors in the oxide. These proposals are quite recent, and require some time for critical assessment. However, promising results have been obtained by comparing oxide BD results with those of scanning tunneling microscope (STM) experiments of hydrogen desorption from H-passivated silicon surfaces [105,106].

As in the AHI model, the HR approach assumes that the defect generation rate is determined by two processes occurring in series, namely the release of species from the anode interface and the resulting reaction of the released species with defect precursors in the oxide. As a first-order approximation, it is assumed that the reaction rate is thermally activated (and hence temperature dependent) but independent of the gate voltage. Hence, the relation between defect generation efficiency and gate voltage (i.e., related to the energy of the carriers when they reach the anode interface) is assumed to be fully determined by the H release process. Before dealing with the actual mechanisms involved in the release of hydrogen from the anode interface, we will first discuss the details of the experiments employed and the procedure used to calculate the defect generation efficiency from the raw BD statistical data.

NFETs with T_{OX} ranging from ~1.4 to ~6.8 nm were stressed to BD under CV stress conditions at 140°C. Figure 16.14 shows the experimental Q_{BD} data versus the stress voltage (V_G). In all cases, the number of samples used to obtain the BD distribution was large enough to ensure that the error of the Q_{BD} data points is comparable to the size of the symbols. From these data, the goal is to determine the relation between the defect generation efficiency and the energy of carriers at the anode. In this way, i.e., not using other indirect monitors of defect generation such as those discussed above, we ensure that the defects we are considering are only those finally involved in oxide BD. In particular, Equation 16.5 relates Q_{BD} to ζ_α, a parameter that is directly linked to the defect generation efficiency ζ, which is defined as the probability that one injected electron generates one defect. Although the experimental time evolution of the density of defects during stress

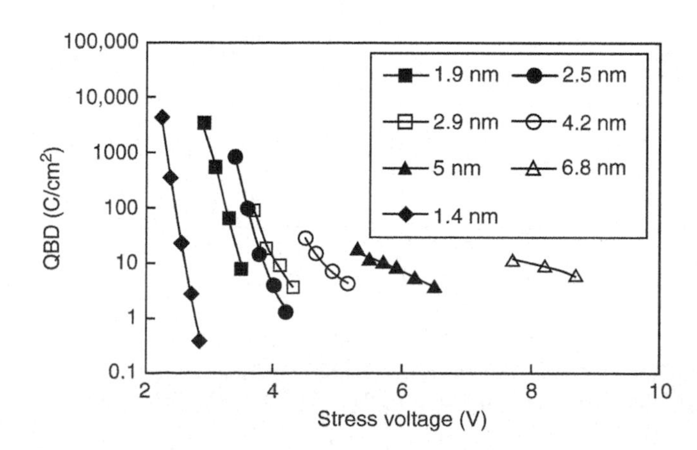

FIGURE 16.14
Mean charge to BD versus stress voltage for NFETs with different T_{OX} stressed under CVS at 140°C. (From Suñé, J. and Wu, E.Y., in *International Electron Device Meeting Digest Technical Papers*, The Institute of Electrical and Electronics Engineers, Piscataway, NY, 2005, 388. With permission.)

experiments has been found to follow the power law given in Equation 16.4 with values of α usually smaller than 1 ($\alpha = 1$ [68]; $\alpha = 0.38$ [25]; and $\alpha = 0.26$ [19]), some authors have assumed $\alpha = 1$ [68,95], for simplicity. It can be demonstrated that this assumption provides consistent results for the defect generation efficiency and its voltage dependence [107]. Hence, the defect generation efficiency ζ can be derived from Q_{BD} data using Equation 16.5 with $\alpha = 1$:

$$\zeta = \frac{qT_{OX}}{a_0^3 Q_{BD}} \exp\left\{-\frac{1}{\beta}\mathrm{Ln}\left(\frac{A_{OX}}{a_0^2}\right)\right\} \tag{16.6}$$

In this equation there is only one free parameter, a_0, which is independently determined from the relation between the Weibull slope β and T_{OX} (as shown in Section 16.2.3). Applying this equation for Q_{BD} to experimental data obtained from devices with different T_{OX} and A_{OX}, so as to cover a wide range of V_G, we can obtain ζ as a function of the maximum energy that the electrons can release at the anode (E_{max}). In Ref. [108], the quantitative relation between E_{max} and V_G was considered at a phenomenological level. Under ballistic direct tunneling injection, $E_{max} = V_G$ in the accumulation injection mode (both for NFETs and PFETs), and $E_{max} = V_G - V_{poly}$ for NFETs stressed in inversion, where V_{poly} is the voltage drop in the depleted n^+-poly gate. For thicker oxides stressed under FN conditions, the model of Chang et al. [85] was considered. This is a phenomenological approach that includes electron energy dissipation in the SiO_2 conduction band using a mean free path as the single parameter. Figure 16.15 shows the defect generation efficiency versus E_{max}, as obtained from the Q_{BD} data of Figure 16.14. It is remarkable that the different curves of Figure 16.14, which correspond to different T_{OX} values, now fall onto a single universal curve. This is a strong indication that the thickness scaling properties of the percolation model of BD (implicit in Equation 16.6) correctly capture the experimental behavior, and that ζ depends strongly on E_{max} but is independent of T_{OX} to first order. Within the HR model this is a reasonable result, since the energy dependence of the defect generation efficiency is determined by an interface phenomenon. As shown in Figure 16.15,

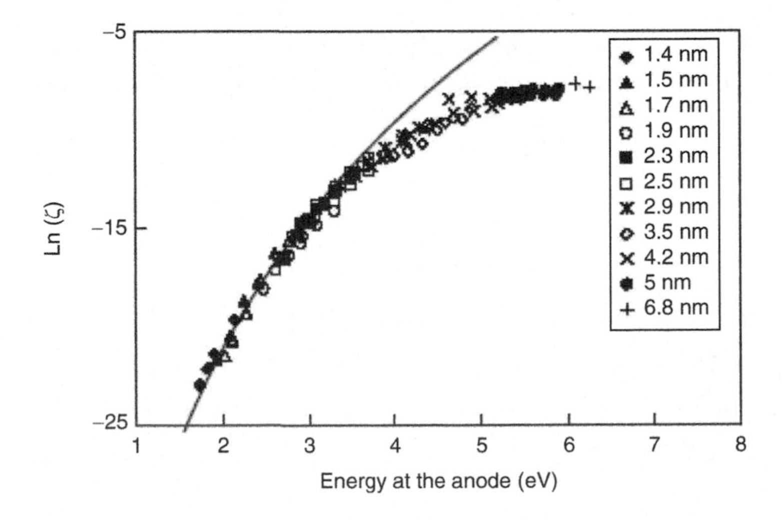

FIGURE 16.15
Defect generation efficiency calculated from raw Q_{BD} data. Symbols correspond to experimental results and the solid line is a fit to a power law $\zeta \propto (E_{max})^{38}$. (From Suñé, J. and Wu, E.Y., *Phys. Rev. Lett.*, 92, 87601, 2004. With permission.)

a power law $\zeta \propto (E_{max})^m$, with $m=38$ nicely fits the ζ data at low E_{max}. This result is consistent with the previously reported $T_{BD} \sim V_G^{-n}$ (with $n \sim 44$) voltage acceleration relationship, provided that the V_G dependence of the tunneling current is accounted for correctly [10].

We now discuss the ζ versus E_{max} relation in terms of the physical mechanisms involved in the release of hydrogen from Si–H bonds. To this purpose we will compare the BD results with those obtained in experiments of STM desorption of hydrogen from H-passivated silicon surfaces. In these experiments, the desorption yield is found to be independent of the electron energy above an energy threshold of ~7 eV [105,106]. This is consistent with an electronic excitation (EE) mechanism in which an incident carrier causes the transition of one electron from the bonding 6σ state to the 6σ* antibonding state [106]. The desorption yield is temperature independent (from 11 to 300 K) [109], and a very large isotope effect has been reported when the silicon surface is passivated with deuterium [110]. This isotope effect has been explained in terms of the dynamics of the bond in the excited state. When the electron is excited in the antibonding state, the bond is unstable and the system wave packet evolves toward bond breakage but, during this evolution, the bond can be quenched back to the ground state. The desorption rate depends on how far the wave packet moves from the equilibrium bond distance before being relaxed to the ground state, and this is different for hydrogen and deuterium due to the different isotope masses. Our $\zeta(E_{max})$ data do not actually reach full saturation at high E_{max}, but the general trend is fully compatible with that of STM experiments as shown in Figure 16.16.

The reason why saturation is not reached in our BD experiments is that the electron transport in the oxide becomes dispersive at high values of V_G so that the relation between V_G and E_{max} is sublinear [59]. Moreover, given a particular value of T_{OX}, the range of V_G that will cause oxide BD within a reasonable time frame is very narrow (few tenths of a volt), so to explore a wide voltage range we need to use a wide range of oxide thickness (see Figure 16.14). Thus, the $\zeta(E_{max})$ data reported at high energies were obtained from thick oxide stress experiments, where the transport is no longer ballistic but dispersive. Although the gate voltage was larger than 10 V in the thicker oxides ($T_{OX}=6.8$ nm), the maximum energy did not actually reach the 7 eV threshold expected for EE of the Si–H bonds. In any case, the trend toward saturation observed in Figure 16.15 can be explained in terms of approaching the EE regime, as we will further discuss below. For electron energies below the $E(6\sigma \rightarrow 6\sigma^*)$ threshold, the probability of EE decays very steeply in STM experiments, and the excitation of the Si–H vibrational degrees of freedom has been

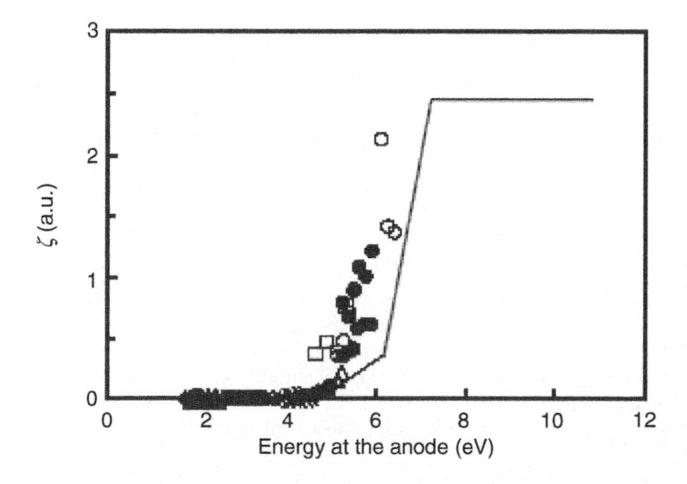

FIGURE 16.16
Experimental defect generation efficiency as a function of the electron energy at the anode. The solid line is a schematic representation of the results obtained in STM desorption experiments in Refs. [105,106]. (From Suñé, J. and Wu, E.Y., *Phys. Rev. Lett.*, 92, 87601, 2004. With permission.)

suggested to play a relevant role [105,110,111]. The vibrational excitation (VE) of the Si–H bond involves the transient resonant trapping of the tunneling electron in the 6σ* orbital until it decays into the available anode states and leaves the Si–H bond in an excited vibrational state. Two mechanisms have been theoretically considered to explain HR by VE: the multiple carrier (incoherent) thermal heating process [106,110], and the single carrier-induced (coherent) multiple VE [111]. A truncated harmonic potential with N eigenstates has been considered to model the Si–H bond vibrations and to study the voltage and current dependences of the involved transitions [110,111]. A schematic representation of the incoherent [110] and coherent mechanisms [111] is shown in Figure 16.17.

Determining whether HR takes place by coherent or incoherent VE of the Si–H bonds cannot be based on the bias dependence, since both mechanisms have a strongly nonlinear (power law) dependence on the applied voltage [111]. However, the dependence on the current is completely different in both cases. Following the notation of Salam et al. [111], the desorption rates corresponding to the coherent and incoherent mechanisms can be written as

$$R_{\text{incoh}} = (N-1)\gamma\left|\frac{I_0}{e\gamma}\right|^N\left(\frac{I_1}{I_0}\right)^N \tag{16.7}$$

$$R_{\text{coh}} = N!\left|\frac{I_0}{e}\right|\left(\frac{I_1}{I_0}\right)^N \tag{16.8}$$

where

N is the number of levels in the truncated harmonic oscillator that describes the Si–H bond vibrations (when the bond is excited to the Nth level, the hydrogen atom in released)

$n\gamma$ is the lifetime of the nth excited vibrational state

I_0 is the elastic current flowing through the Si–H bond resonance

I_1 is the inelastic current corresponding to the excitation of the bond to the first vibrational level

Since the most probable component of inelastic tunneling corresponds to the excitation to the first excited level and since $I_0 \gg I_1$, the ratio I_1/I_0 roughly represents the inelastic

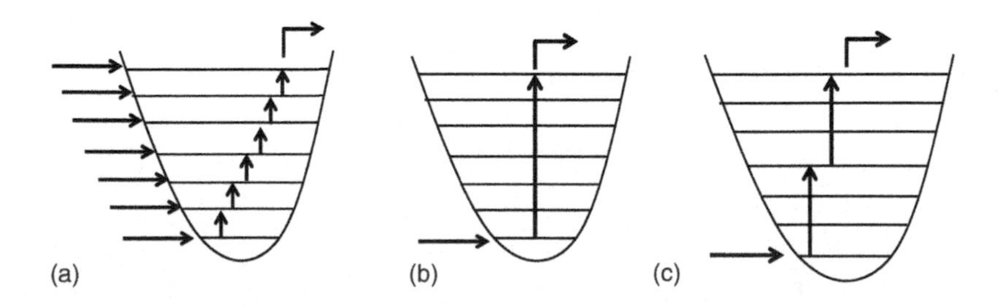

FIGURE 16.17
Representation of HR by excitation of the local vibration modes of the Si–H bond: (a) the incoherent thermal heating mechanism which involves the action of several electrons (arrows) that induce one-level transitions between adjacent vibrational states [105,106,110]; (b) the coherent excitation mechanism involves a single carrier to excite the bond vibrations to the highest energy state [111]; and (c) more general situations that involve several electrons and also multiple level transitions [112]. (From Suñé, J. and Wu, E.Y., in *International Electron Device Meeting Digest Technical Papers*, The Institute of Electrical and Electronics Engineers, Piscataway, NY, 2005, 388. With permission.)

tunneling fraction. The empirical fitting of Shen's data yields the voltage dependence for this ratio, $I_1/I_0 \sim V^4$. If $N = 10$, both VE mechanisms justify a power-law dependence of $\zeta \sim R/I_0$ on the voltage (i.e., carrier energy) with an exponent of the order of 40, which is fully consistent with the low-voltage power law with $m = 38$ reported in Figure 16.15. However, the desorption efficiency due to the incoherent thermal heating mechanism also depends strongly on the current $(\zeta \propto I_0^{N-1})$, while that of the coherent mechanism is current independent. This is because the former involves N electrons while the latter is a single-electron mechanism. In general, processes involving K transitions $(1 < K < N)$ are also possible (Figure 16.17c), and would yield intermediate current dependencies $\zeta \propto I_0^{K-1}$ [112]. As claimed in Ref. [112], the experimental current dependence should allow a determination of the dominant mechanism. However, while in STM experiments, current and voltage can be varied independently, in BD experiments, the current depends on V_G, so it is more difficult to decouple the dependencies on these two variables. Nevertheless, since the current also depends on T_{OX} at fixed V_G, we can examine the T_{OX} dependence of ζ (Figure 16.18) as a way to indirectly check the current dependence. Based on the results in Figure 16.18, we conclude that ζ does not significantly depend on the current, so that, contrary to what happens in STM desorption experiments, the coherent single-electron process is more likely to explain HR in BD experiments. This thickness independence is consistent with the results of Figure 16.15, where the ζ data corresponding to devices with different T_{OX} fall onto a universal $\zeta(E_{max})$ curve. However, the data obtained at the lowest voltage (2.3 V) suggest a possible current dependence at low voltages. This is not surprising since the coherent VE mechanism can only explain HR if the energy of the involved electrons is larger than the total energy required for Si–H bond breakage (potential barrier of the Si–H bond potential). For energies below the Si–H bond barrier height, multiple-electron VE mechanisms become more likely because the single-electron coherent HR is no longer possible. Although the Si–H binding energy is about 3.6 eV, many authors have reported significantly lower theoretical values (2.5–3 eV) for this barrier in oxide [113,114]. Moreover, Brower measured a barrier of 2.56 eV for thermally activated desorption of hydrogen from the (111) Si–SiO_2 interface [115], and Stesmans reported a value of 2.83 eV for the dissociation of H-passivated Pb centers in vacuum [116].

In conclusion, we should expect that below ~2.5 eV at least two electrons are involved in the HR process. This is consistent with the T_{OX} dependence of ζ at 2.3 V (Figure 16.18), as explored in further detail in Ref. [59]. To this purpose, we must examine the dependence of ζ on E_{max} at the lowest voltages. If ζ does not depend on the current, we expect that the $\zeta(E_{max})$ data corresponding to different T_{OX} should fall on a universal curve. If, on the contrary, the relevant HR mechanism involves K electrons $(1 < K < N)$, a dependence $\zeta \propto J^{K-1}$ (J being the current density) should be found, and the magnitude giving a

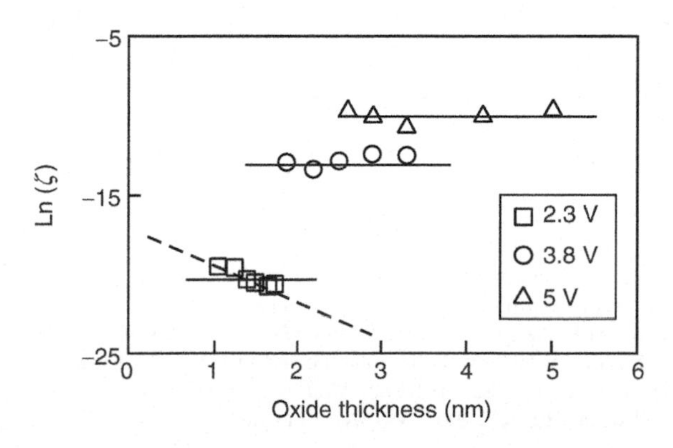

FIGURE 16.18
Dependence of defect generation rate on T_{OX} at fixed V_G (4). The red dashed line reveals a possible T_{OX} dependence at low V_G. (From Suñé, J. and Wu, E.Y., *Phys. Rev. Lett.*, 92, 87601, 2004. With permission.)

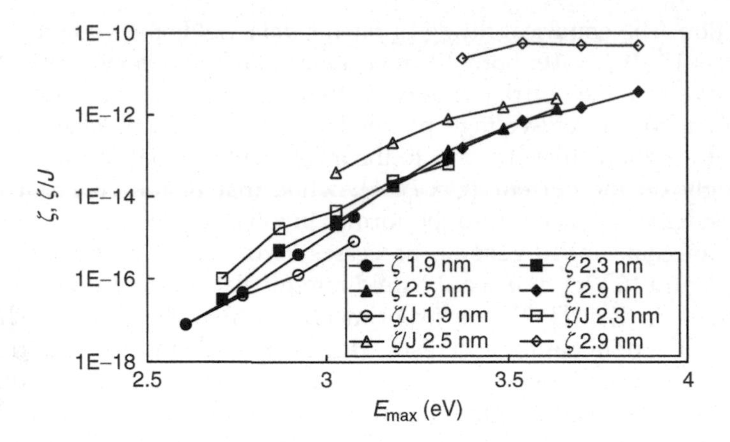

FIGURE 16.19
Representation of ζ (solid symbols) and ζ/J (empty symbols) versus E_{max} for devices with different T_{OX} such that the E_{max} range is above 2.5 eV. The data points correspond to gate oxides with 1.9 nm $< T_{OX} <$ 2.9 nm. While $\zeta(E_{max})$ appears as T_{OX} independent, $\zeta/J(E_{max})$ shows a clear dependence on T_{OX}. This indicates that the HR mechanism involves a single-electron process, which is likely the coherent VE mechanism [111]. (From Suñé, J. and Wu, E.Y., in *International Electron Device Meeting Digest Technical Papers*, The Institute of Electrical and Electronics Engineers, Piscataway, NY, 2005, 388. With permission.)

universal curve should be ζ/J^{K-1} rather than ζ. Figure 16.19 shows that above 2.5 eV $\zeta(E_{max})$ is thickness independent, while $\zeta/J(E_{max})$ changes with T_{OX}. On the contrary, Figure 16.20 shows that the situation is different for E_{max} below 2.5 eV, since $\zeta(E_{max})$ depends on T_{OX} while $\zeta/J(E_{max})$ does not. This suggests that below $E_{max} \sim 2.5$ eV two

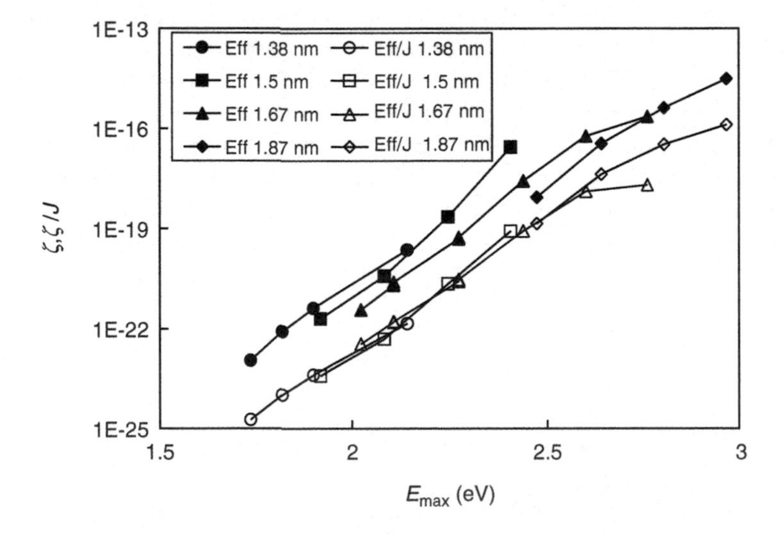

FIGURE 16.20
Representation of ζ (solid symbols) and ζ/J (empty symbols) versus E_{max} for ultrathin oxides in the E_{max} range below 3 eV. The data points correspond to gate oxides with 1.4 nm $< T_{OX} <$ 1.9 nm. While $\zeta(E_{max})$ appears as T_{OX} dependent below ~2.5 eV, $\zeta/J(E_{max})$ becomes independent of T_{OX}, in contrast to what is observed at higher E_{max} in Figure 16.19. Notice that between 2.5 and 3 eV, a transition toward the behavior reported in Figure 16.19 is also evident in this plot. (From Suñé, J. and Wu, E.Y., in *International Electron Device Meeting Digest Technical Papers*, The Institute of Electrical and Electronics Engineers, Piscataway, NY, 2005, 388. With permission.)

electrons cooperate in the HR process. This is consistent with the barrier of Si–H bond rupture being ~2.5 eV. Since the Si–H stretching mode frequency is well known to be ~0.25 eV [117–119], this means that the number of excited levels in the truncated harmonic potential is $N \sim 10$. Hence, as previously suggested, the exponent expected for the $\zeta(E_{max})$ power law is $4N \sim 40$, the factor 4 coming from a $\sim V_G^4$ dependence of the inelastic tunneling fraction [105], in reasonable agreement with the value of 38 found in Figure 16.15. In conclusion, the results presented in Figures 16.19 and 16.20 suggest that there is a transition from purely coherent single-electron coherent HR by VE to a multiple-electron heating mechanism [112] when the energy of the carriers is lower than the Si–H bond barrier. Below 2.5 eV we have found that two electrons are likely involved and we expect that below 1.25 eV (half the Si–H barrier), at least three electrons should be involved. This result is very important for reliability extrapolation to the low operating voltages of ICs. Ignoring the transition from single- to multiple-electron release mechanisms may cause significant errors in the forecasted IC reliability.

Up to now we have discussed the two extremes of the $\zeta = \zeta(E_{max})$ relation of Figure 16.15. In particular, we have claimed that the trend toward saturation observed at high values of E_{max} corresponds to the EE mechanism. On the other hand, we have discussed the low E_{max} power-law behavior in terms of VE mechanisms, having concluded that above the Si–H bond desorption barrier of ~2.5 eV, the coherent excitation mechanism dominates [111], while two electron processes (a mixture between coherent excitation and incoherent heating mechanisms [112]) are required when the carrier energy is below this barrier. Between these two extremes, there is a transition region in which we suggest that there is a cooperation of VE and EE mechanisms. In this regard, the excitation of the Si–H vibrations leads to an enlargement of the Si–H bond distance. At the same time, the energy required for the transition between the 6σ and the 6σ* states diminishes with the bond distance [106,110]. Hence, there is a natural path for the cooperation between VE and EE. One electron excites the Si–H bond to the *n*th excited vibrational state, and then a second electron with an energy smaller than the EE threshold can more easily cause the Si–H bond rupture by EE. According to Avouris' results [106], the $E(6\sigma \rightarrow 6\sigma^*)$ threshold decreases approximately linearly with the Si–H bond distance as shown in Figure 16.21. On the other hand, the increase of the average bond distance for excited vibrational states is not captured by the truncated harmonic potential, and this requires a more realistic model

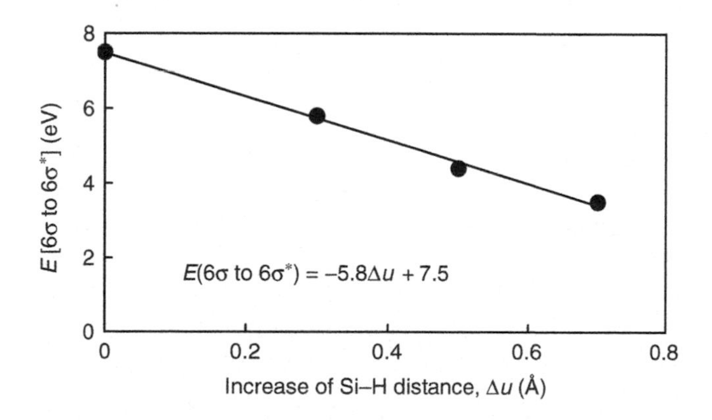

FIGURE 16.21
Energy required for the electronic transition $6\sigma \rightarrow 6\sigma^*$ in an Si–H bond as a function of the increase of the Si–H bond distance with respect to that of equilibrium. The points have been calculated from the results of Avouris et al. [106]. (From Suñé, J. and Wu, E.Y., in *International Electron Device Meeting Digest Technical Papers*, The Institute of Electrical and Electronics Engineers, Piscataway, NY, 2005, 388. With permission.)

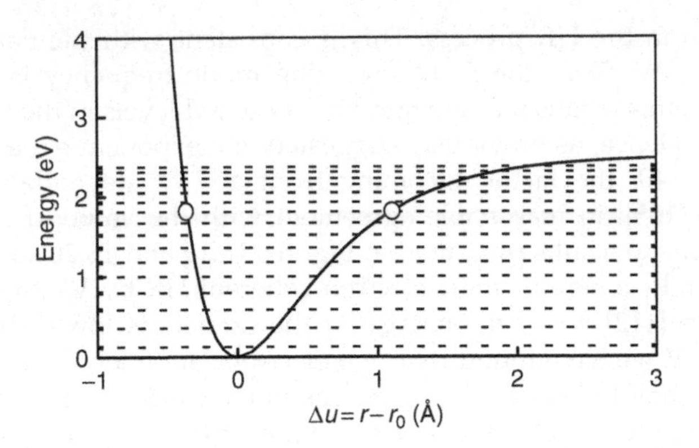

FIGURE 16.22
Morse potential representing the Si–H bond [119,120]. The barrier height (depth of potential well) is 2.5 eV. Dashed lines represent the position of the energy eigenstates. The separation between the first excited state and ground state is 0.24 eV, in agreement with the Si–H stretching mode frequency. The circles represent the classical turning points of one intermediate excited state. (From Suñé, J. and Wu, E.Y., in *International Electron Device Meeting Digest Technical Papers*, The Institute of Electrical and Electronics Engineers, Piscataway, NY, 2005, 388. With permission.)

such as the Morse potential shown in Figure 16.22 [119,120]. In this asymmetric potential, the average Si–H distance increases linearly with the eigenvalue n $(1 < n < N)$ as shown in Figure 16.23. Combining these two results, we can calculate that the energy required for the EE process is $E(6\sigma \rightarrow 6\sigma^*) \approx (6.45 - 0.35n)$ eV, if the bond is in the nth excited vibrational state. Thus, we suggest that, in the range between ~2.5 and ~6.5 eV, HR takes place when one electron excites the bond vibration to the nth state and then another one causes HR by EE. Assuming that the EE process is only possible if $E_{max} \geq E(6\sigma \rightarrow 6\sigma^*)$, we can calculate to which state (n) the bond must be excited to allow a subsequent electron

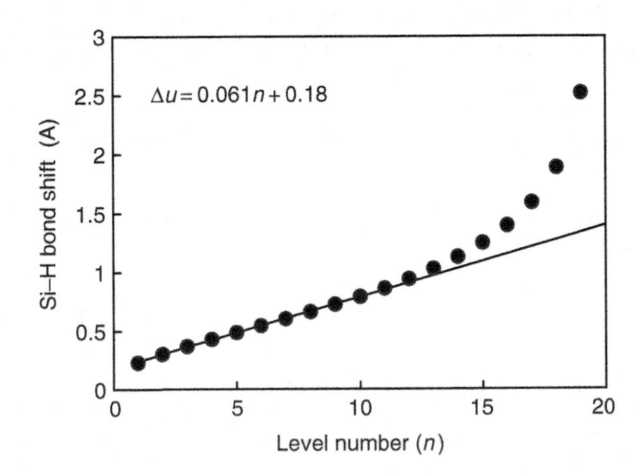

FIGURE 16.23
Shift of the average Si–H distance with respect to that of the ground state (equilibrium) associated with the eigenstates of the Morse potential considered in Figure 16.22. The linear fit to the first 10 states gives the equation given in the legend. Although it is known that the Morse potential is a better model than the truncated harmonic for the description of the bonds of diatomic molecules, it begins to fail when the bond length is stretched beyond 0.5–1 Å. (From Suñé, J. and Wu, E.Y., in *International Electron Device Meeting Digest Technical Papers*, The Institute of Electrical and Electronics Engineers, Piscataway, NY, 2005, 388. With permission.)

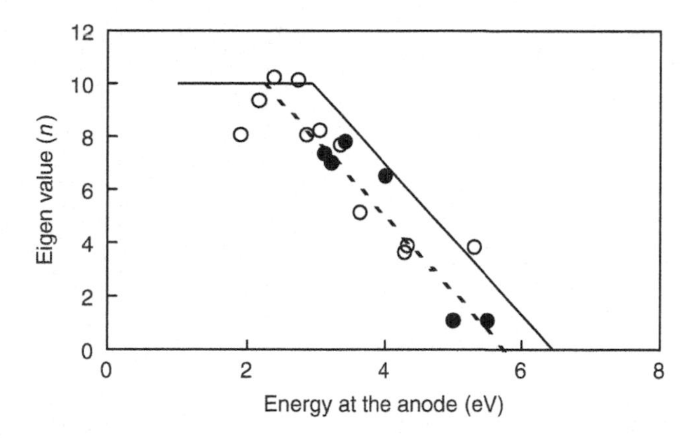

FIGURE 16.24
Symbols are calculated by dividing the experimental $\zeta(E_{max})$ power-law exponent by 4 (to calculate the eigenvalue of the vibrational state involved in the HR process). The solid line is obtained by combining the change of $E(6\sigma$ to $6\sigma^*)$ with Si–H distance (Figure 16.11) with the average Si–H bond distance when it is excited to the nth eigenstate. The dashed line better fits the data and is the result of an arbitrary shift of the EE threshold by 0.7 eV. (From Suñé, J. and Wu, E.Y., in *International Electron Device Meeting Digest Technical Papers*, The Institute of Electrical and Electronics Engineers, Piscataway, New York, 2005, 388. With permission.)

to cause HR by EE. Moreover, since the exponent of the $\zeta(E_{max})$ power law is $\sim 4n$, we can compare the calculated $n(E_{max})$ with the values obtained from experiments. This comparison is done in Figure 16.24, which shows excellent agreement between this simple model and the experimental results, thus providing quantitative support to the cooperation of VE and EE mechanisms in the intermediate E_{max} range. In general, one would expect that a bond breakage mechanism that involves several electrons should reveal a current-dependent release efficiency. In this particular case, however, although two electrons are required for the cooperation between the VE and the EE mechanisms, ζ shows no sign of current dependence in the energy range of interest. This is because the probability that a second electron causes the EE event (once the bond is in the excited vibrational state) is orders of magnitude larger than that of the first electrons to cause the VE. In other words, it is the first electron that primarily determines the HR rate.

Although the results presented in this section do not demonstrate the implication of hydrogen in the generation of the defects that finally trigger oxide BD, they indeed show that the HR model is a reasonable framework to interpret the BD results. Although other authors [121,122] have stressed the importance of multiple-electron incoherent thermal heating processes (which dominate hydrogen desorption in STM experiments), the final conclusions regarding the defect generation efficiency in the low carrier energy limit are quite similar to those presented here. Our discussion in support of the HR model has been fully based on the examination of the defect generation efficiency dependence on the carrier energy. However, other sources of evidence are also available. In particular, some advances in the understanding of the hydrogen role have been made by examining the dynamics of defect generation. In particular, recent experimental results by Nicollian and coworkers suggest the release of two different hydrogenic species, H^+ and H_0, may be involved in the generation of bulk and interface traps, respectively. Their conclusions are based on interpreting experimental results in terms of the reaction–diffusion theory [123]. The reaction–diffusion theory has been applied to model other degradation mechanisms that also involve hydrogenic species such as NBTI [122–125]. Application of this theory to the understanding of the generation of the bulk defects that trigger oxide BD suggests the interesting possibility of a unified theory of oxide degradation, eventually including oxide

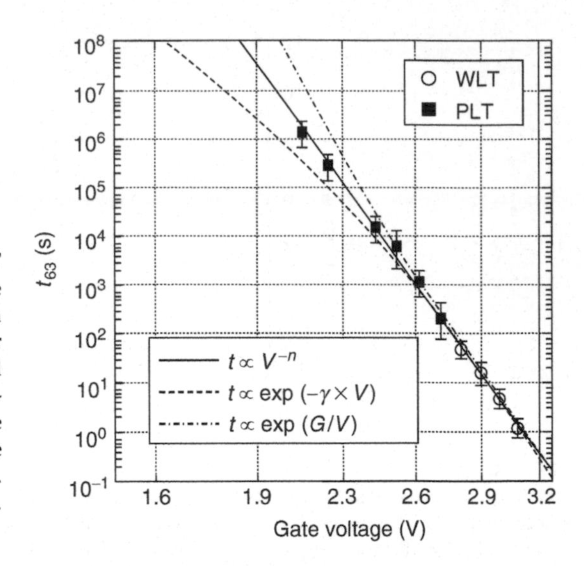

FIGURE 16.25
Double-logarithmic plot of T_{BD} versus V_G for PFET devices with $T_{OX}=1.6$ nm. The different voltage acceleration models were adjusted to the data obtained at relatively high values of V_G at wafer level by Pompl and Röhner [77]. The results obtained with packaged devices stressed at much lower voltages and for much longer times demonstrate that the best of the three considered acceleration models is the power-law model first proposed by Wu et al. [80]. (From Pompl, T. and Röhner, M., *Microelectron. Reliab.*, 45, 1835, 2005. With permission.)

BD, NBTI, and hot-carrier effects. However, other proposals have been made to explain the power-law time evolution of the defect density, such as those based on dispersive transport [126,127], or on a distribution of Si–H dissociation barriers [128]. Based on their results and theory, Nicollian and coworkers also claimed that the release of H^+ requires the involvement of hot holes, while electrons or holes may be involved in the release of H_0. As for the possible role of holes, a mechanism of VE through the Si–H 5σ hole resonance has also been studied both theoretically and by means of STM desorption experiments under negative sample bias [129]. If the VE excitation of Si–H bonds should finally be demonstrated to be hole-mediated, this could resolve the controversy between AHI and HR models, since both hole generation in the anode and HR from the anode interface would be involved in defect generation and oxide BD. However, this possibility still requires detailed investigation.

Finally, it is worth mentioning that the HR model based on VE mechanisms provides a reasonable theoretical basis for the use of the power-law model of voltage acceleration [80,130]. As previously stated, the gate voltage rather than the oxide field has been shown to be the proper electrical parameter to describe the degradation of ultrathin oxides [79]. In Figure 16.25, it is shown how the power-law model of Wu, which gives an intermediate prediction between more traditional exponential (i.e., $\exp(-\gamma_V V_G)$ and $\exp(-B_V/V_G)$) models, provides a best fit of the experimental results. Since the proposal of the power-law model in 2000, many different groups have confirmed the power-law dependence of time and charge to BD on the gate voltage and have even extended its application to thicker oxides [131]. Nowadays, this model is well accepted by many semiconductor companies (e.g., see the special section of the journal *Microelectronics Reliability* recently devoted to the power-law TDDB model [132]).

16.4 Conclusions

The relation between defect generation and BD has been reviewed. Although the exact nature of the involved defects is not well known yet, it has been argued that electrically active defects generated during stress finally cause the dielectric BD of gate oxides. The BD statistics have been shown to be well understood in the framework of the percolation model of BD. However, the geometric description that is implicit in this simple model does not apply when oxide thickness is scaled down to be comparable to the effective electrical size of the involved defects. Different models that aim at explaining the defect generation physics have also been reviewed, with emphasis on the recent quantitative formulation of the anode HR picture. In this regard, it has been shown that the inelastic interaction of injected electrons with Si–H bonds at the anode interface can explain the quite complex relation between defect generation efficiency and gate stress voltage.

Acknowledgments

J. Suñé acknowledges the funding support of IBM Microelectronics, the Spanish Ministry of Science and Education under contract TEC2006-13731-C02-01, and the Departament d'Universitats, Recerca i Societat de la Informació de la Generalitat de Catalunya.

References

1. Wu, E.Y. et al., A comprehensive investigation of gate oxide breakdown of p+poly/pFETs under inversion mode, in *IEEE International Electron Devices Meeting Technical Digest*, The Institute of Electrical and Electronics Engineers, Piscataway, NY, 2005, 407.
2. Kaczer, B. et al., Implications of progressive wear-out for lifetime extrapolations of ultra-thin (EOT ~1 nm) SiON films, in *IEEE International Electron Devices Meeting Technical Digest*, The Institute of Electrical and Electronics Engineers, Piscataway, NY, 2004, 713.
3. Suñé, J. et al., On the breakdown statistics of very thin SiO_2 films, *Thin Solid Films*, 185, 347, 1990.
4. Dumin, D.J. et al., A model relating wearout to breakdown in thin oxides, *IEEE Trans. Electron Devices*, 41, 1570, 1994.
5. Porti, M., Nafría, M., and Aymerich, X., Current limited stress of SiO_2 with conductive atomic force microscope, *IEEE Trans. Electron Devices*, 50, 933, 2003.
6. Wu, E.Y. and Suñé, J., Post-breakdown characteristics of extrinsic failure modes for ultra-thin gate oxides, in *Proceedings of the International Reliability Physics Symposium*, The Institute of Electrical and Electronics Engineers, Piscataway, NY, 2006, 36.
7. Wolters, D.R. and Van der Schoot, J.J., Dielectric breakdown in MOS devices. Part I. Defect related and intrinsic breakdown, *Philips J. Res.*, 40, 115, 1985.
8. McPherson, J.W. et al., Comparison of E and 1/E TDDB modes for SiO_2 under long-term/low-field test conditions, in *IEEE International Electron Devices Meeting Technical Digest*, The Institute of Electrical and Electronics, Piscataway, NY, 1998, 171.
9. Prendergast, J., O'Driscoll, E., and Mullen, E., Investigation into the correct statistical distribution for oxide breakdown over oxide thickness range, *Microelectron. Reliab.*, 45, 973, 2005.
10. Wu, E., Stathis, J.H., and Han, L.-K., Ultra-thin oxide reliability for ULSI applications, *Semicond. Sci. Technol.*, 15, 425, 2000.
11. Degraeve, R., Kaczer, B., and Groeseneken, G., Reliability: A possible showstopper for oxide thickness scaling? *Semicond. Sci. Technol.*, 15, 436, 2000.
12. Suñé, J., New physics-based analytic approach to the thin oxide breakdown statistics, *IEEE Electron Device Lett.*, 22, 296, 2001.
13. Hunter, W., The analysis of oxide reliability data, in *Integrated Reliability Workshop Final Report*, The Institute of Electrical and Electronics Engineers, Piscataway, NY, 1998, 114.
14. DiMaria, D.J., Cartier, E., and Arnold, D., Impact ionization, trap creation, degradation and breakdown in silicon dioxide films on silicon, *J. Appl. Phys.*, 73, 3367, 1993.
15. DiMaria, D.J. et al., Interface traps induced by the presence of trapped holes near the silicon–silicon-dioxide interface, *J. Appl. Phys.*, 77, 2032, 1995.
16. DiMaria, D.J. and Stathis, J.H., Explanation for the oxide thickness dependence of breakdown characteristics of metal-oxide-semiconductor structures, *Appl. Phys. Lett.*, 70, 2708, 1997.

17. De Blauwe, J. et al., SILC-related effects in flash E^2PROM's—part I: a quantitative model for steady-state SILC, *IEEE Trans. Electron Devices*, 45, 1745, 1998.
18. Ghetti, A. et al., Low voltage tunneling in ultra-thin oxides: a monitor for interface traps and degradation, in *IEEE International Electron Devices Meeting Technical Digest*, The Institute of Electrical and Electronics, Piscataway, NY, 1999, 731.
19. Nicollian, P.E. et al., The traps that cause breakdown in deeply scaled SiON dielectrics, in *IEEE International Electron Devices Meeting Technical Digest*, The Institute of Electrical and Electronics Engineers, Piscataway, NY, 2006, 743.
20. Chen, I.C., Holland, S., and Hu, C., Electrical breakdown of thin gate and tunneling oxides, *IEEE Trans. Electron Devices*, 32, 413, 1985.
21. Chen, I.C., Holland, S., and Hu, C., Hole trapping and breakdown in thin SiO_2, *IEEE Electron Device Lett.*, 7, 164, 1986.
22. Chen, I.C. et al., Substrate hole current and oxide breakdown, *Appl. Phys. Lett.*, 49, 669, 1986.
23. Nissan-Cohen, Y., Shappir, J., and Frohman-Bentchkowsky, D., Trap generation and occupation dynamics in SiO_2 under charge injection stress, *J. Appl. Phys.*, 60, 2024, 1986.
24. DiMaria, D.J. and Stasiak, J.W., Trap creation in silicon dioxide produced by hot electrons, *J. Appl. Phys.*, 65, 2342, 1989.
25. Degraeve, R. et al., New insights in the relation between electron trap generation and the statistiscal properties of oxide breakdown, *IEEE Trans. Electron Devices*, 45, 904, 1988.
26. Suñé, J. et al., Degradation and breakdown of gate oxides in VLSI devices, *Phys. Status Solidi A*, 111, 675, 1989.
27. Olivo, P., Riccò, B., and Sangiorgi, E., Electron trapping/detrapping within thin SiO_2 films in the high field tunneling regime, *J. Appl. Phys.*, 54, 5267, 1983.
28. Zhang, W.D. et al., Two types of neutral electron traps generated in the gate silicon dioxide, *IEEE Trans. Electron Devices*, 49, 1868, 2002.
29. Moazzami, R. and Hu, C., Stress-induced current in thin silicon dioxide films, in *IEEE International Electron Devices Meeting Technical Digest*, The Institute of Electrical and Electronics Engineers, Piscataway, NY, 1992, 139.
30. Rosenbaum, E. and Register, L.F., Mechanism of stress-induced leakage current in MOS capacitors, *IEEE Trans. Electron Devices*, 44, 317, 1997.
31. Riccó, B., Gozzi, G., and Lanzoni, M., Modeling and simulation of stress-induced leakage current in ultrathin SiO_2 films, *IEEE Trans. Electron Devices*, 45, 1554, 1998.
32. Takagi, S., Yasuda, N., and Toriumi, A., Experimental evidence of inelastic tunneling in stress-induced leakage current, *IEEE Trans. Electron Devices*, 46, 335, 1999.
33. Wu, J., Register, L.F., and Rosenbaum, E., Trap-assisted tunneling current through ultra-thin oxide, in *Proceedings of the International Reliability Physics Symposium*, The Institute of Electrical and Electronics Engineers, Piscataway, NY, 1999, 389.
34. Ielmini, D. et al., A detailed investigation of the quantum yield experiment, *IEEE Trans. Electron Devices*, 48, 1696, 2001.
35. Schuler, F. et al., Physical description of anomalous charge loss in floating gate based NVM's and identification of its dominant parameter, in *Proceedings of the International Reliability Physics Symposium*, The Institute of Electrical and Electronics Engineers, Piscataway, NY, 2002, 26.
36. Degraeve, R. et al., Analytical percolation model for predicting anomalous charge loss in flash memories, *IEEE Trans. Electron Devices*, 51, 1392, 2004.
37. Olivo, P., Nguyen, T.N., and Riccó, B., High-field induced degradation in ultra-thin SiO_2, *IEEE Trans. Electron Devices*, 20, 2259, 1988.
38. DiMaria, D.J. and Cartier, E., Mechanisms for stress-induced leakage currents in silicon dioxide films, *J. Appl. Phys.*, 78, 3883, 1995.
39. Okada, K. and Yoneda, K.A., Consistent model for time-dependent dielectric breakdown in ultrathin silicon dioxides, in *IEEE International Electron Devices Meeting Technical Digest*, The Institute of Electrical and Electronics Engineers, Piscataway, NY, 1999, 445.
40. Rodríguez, R. et al., Monitoring the degradation that causes the breakdown of ultrathin (<5 nm) SiO_2 gate oxides, *IEEE Electron Device Lett.*, 21, 251, 2000.
41. Petit, C. and Zander, D., Low voltage stress induced leakage current and time to breakdown in ultra-thin (1.2–2.3 nm) oxides, *Microelectron. Reliab.*, 47, 401, 2007.

42. Lai, W.L. et al., Impact of stress induced leakage current on power-consumption in ultra-thin gate oxides, in *Proceedings of the International Reliability Physics Symposium,* The Institute of Electrical and Electronics Engineers, Piscataway, NY, 2004, 102.
43. Buchanan, D.A. et al., On the relationship between stress induced leakage currents and catastophic breakdown in ultra-thin SiO_2 based dielectrics, *Microelectron. Eng.*, 36, 329, 1997.
44. Stathis, J.H. and DiMaria, D.J., Reliability projections for ultra-thin oxides at low voltage, in *IEEE International Electron Devices Meeting,* The Institute of Electrical and Electronics Engineers, Piscataway, NY, 1998, 167.
45. Wu, E.Y. et al., Weibull slopes, critical defect density, and the validity of stress-induced-leakage current (SILC) measurements, in *IEEE International Electron Device Meeting,* The Institute of Electrical and Electronics Engineers, Piscataway, NY, 2001, 125.
46. Pantisano, L. and Cheung, K.P., Stress-induced leakage current (SILC) and oxide breakdown: are they from the same oxide traps? *IEEE Trans. Device Mater. Reliab.*, 1, 109, 2001.
47. Alam, M.A., SILC as a measure of trap generation and predictor of T_{BD} in ultrathin oxides, *IEEE Trans. Electron Devices*, 49, 226, 2002.
48. Nicollian, P.E. et al., Low voltage stress-induced-leakage-current in ultrathin gate oxides, in *Proceedings of the International Reliability Physics Symposium,* The Institute of Electrical and Electronics Engineers, Piscataway, NY, 1999, 400.
49. Ielmini, D. et al., Modeling of SILC based on electron and hole tunneling—part II: Steady-state, *IEEE Trans. Electron Devices*, 47, 1266, 2000.
50. Ghetti, A. et al., Tunneling into interface traps as reliability monitor for ultrathin oxides, *IEEE Trans. Electron Devices*, 47, 2358, 2000.
51. Degraeve, R. et al., Measurement and statistical analysis of single-trap current-voltage characteristics in ultrathin SiON, in *Proceedings of the International Reliability Physics Symposium,* The Institute of Electrical and Electronics Engineers, Piscataway, NY, 2005, 360.
52. Monsieur, F. et al., A thorough investigation of progressive breakdown in ultra-thin oxides, physical understanding and application for industrial reliability assessment, in *Proceedings of the International Reliability Physics Symposium,* The Institute of Electrical and Electronics Engineers, Piscataway, NY, 2002, 45.
53. Linder, B.P. et al., Voltage dependence of hard breakdown growth and the reliability implications in thin dielectrics, *IEEE Electron Device Lett.*, 23, 661, 2002.
54. Schmitz, J. et al., Comparison of soft breakdown triggers for large-area capacitors under constant voltage stress, *IEEE Trans. Device Mater. Reliab.*, 1, 150, 2001.
55. Neri, B., Olivo, P., and Riccó, B., Low-frequency noise in silicon-gate metal-oxide-silicon capacitors before oxide breakdown, *Appl. Phys. Lett.*, 51, 2167, 1987.
56. Depas, M., Nigam, T., and Heyns, M.M., Definition of dielectric breakdown for ultra thin (<2 nm) gate oxides, *Solid-State Electron.*, 41, 725, 1997.
57. Suñé, J., Wu, E.Y., and Tous, S., A physics-based deconstruction of the oxide breakdown percolation picture, *Microelectron. Eng.*, 84, 1917, 2007.
58. Suñé, J. and Wu, E.Y., Hydrogen release mechanisms in the breakdown of thin SiO_2 films, *Phys. Rev. Lett.*, 92, 87601, 2004.
59. Suñé, J. and Wu, E.Y., Mechanisms of hydrogen release in the breakdown of SiO_2-based gate oxides, in *International Electron Device Meeting Digest Technical Papers,* The Institute of Electrical and Electronics Engineers, Piscataway, NY, 2005, 388.
60. Stathis, J.H., Percolation models for gate oxide breakdown, *J. Appl. Phys.*, 86, 5757, 1999.
61. O'Dwyer, J.J., Theory of high-field conduction in a dielectric, *J. Appl. Phys.*, 40, 3887, 1969.
62. Vollertsen, R.P. and Wu, E.Y., Voltage acceleration and $t_{63.2}$ of 1.6–10 nm gate oxides, *Microelectron. Reliab.*, 44, 906, 2004.
63. McPherson, J., Stress dependent activation energy, in *Proceedings of the International Reliability Physics Symposium Proceedings,* The Institute of Electrical and Electronics Engineers, Piscataway, NY, 1986, 12.
64. McPherson, J.W. and Mogul, H.C., Underlying physics of the thermo-chemical E model in describing low-field time-dependent dielectric breakdown in SiO_2 thin films, *J. Appl. Phys.*, 84, 1513, 1998.

65. DiMaria, D.J. and Stathis, J.H., Non-Arrhenius temperature dependence of reliability in ultrathin silicon dioxide films, *Appl. Phys. Lett.*, 74, 1752, 1999.
66. Vogel, E.M. et al., Reliability of ultrathin SiO_2 under combined substrate hot-electron and constant voltage tunneling stress, *IEEE Trans. Electron Devices*, 47, 1183, 2000.
67. Selmi, L., Esseni, D., and Palestri, P., Towards microscopic understanding of MOSFET reliability: the role of carrier energy and transport simulations, *International Electron Device Meeting Digest Technical Papers*, The Institute of Electrical and Electronics Engineers, Piscataway, NY, 2003, 333.
68. Stathis, J.H., Physical and predictive models of ultrathin oxide reliability in CMOS devices and circuits, *IEEE Trans. Device Mater. Reliab.*, 1, 43, 2001.
69. McMahon, W., Haggag, A., and Hess, K., Reliability scaling issues for nanoscale devices, *IEEE Trans. Nanotechnol.*, 2, 33, 2003.
70. Ribes, G. et al., Modeling charge-to-breakdown using hydrogen multivibrational excitation (thin SiO_2 and high-K dielectrics), in *IEEE International Integrated Reliability Workshop, Final Report*, The Institute of Electrical and Electronics Engineers, Piscataway, NY, 2004, 1.
71. Schuegraf, K.F. and Hu, C., Hole injection SiO_2 breakdown model for very low voltage lifetime extrapolation, *IEEE Trans. Electron Devices*, 41, 761, 1994.
72. Bude, J.D., Weir, B.E., and Silverman, P.J., Explanation of stress-induced damage in thin oxides, in *IEEE International Electron Devices Meeting Technical Digest*, The Institute of Electrical and Electronics Engineers, Piscataway, NY, 1998, 179.
73. Alam, M.A., Bude J., and Ghetti, A., Field acceleration for oxide breakdown—can an accurate anode hole injection model resolve the E vs. 1/E controversy? in *Proceedings of the International Reliability Physics Symposium*, The Institute of Electrical and Electronics Engineers, Piscataway, NY, 2000, 26.
74. McPherson, J.W., Determination of the nature of molecular bonding in silica from time-dependent dielectric breakdown data, *J. Appl. Phys.*, 95, 8101, 2004.
75. Anolik, E.S., Low field time-dependence dielectric integrity, in *Proceedings of the IEEE International Reliability Physics Symposium*, The Institute of Electrical and Electronics Engineers, Piscataway, NY, 1979, 8.
76. Crook, D.L., Method of determining reliability screens for time-dependent dielectric breakdown, in *Proceedings of the IEEE International Reliability Physics Symposium*, The Institute of Electrical and Electronics Engineers, Piscataway, NY, 1979, 1.
77. Pompl, T. and Röhner, M., Voltage acceleration of time-dependent breakdown of ultra-thin gate dielectrics, *Microelectron. Reliab.*, 45, 1835, 2005.
78. DiMaria, D.J., Explanation for the polarity dependence of breakdown in ultra-thin silicon dioxide films, *Appl. Phys. Lett.*, 68, 3004, 1996.
79. Nicollian, P.E., Hunter, W.R., and Hu, J.C., Experimental evidence for voltage driven breakdown models in ultrathin gate oxides, in *Proceedings of the International Reliability Physics Symposium*, The Institute of Electrical and Electronics Engineers, Piscataway, NY, 2000, 7.
80. Wu, E.Y. et al., Voltage-dependent voltage acceleration of oxide breakdown for ultra-thin oxides, in *IEEE International Electron Device Meeting*, The Institute of Electrical and Electronics Engineers, Piscataway, NY, 2000, 54.
81. Wu, E.Y. et al., Experimental evidence of T_{BD} power-law for voltage dependence of oxide breakdown in ultra-thin gate oxide, *IEEE Trans. Electron Devices*, 49, 2244, 2002.
82. Takayanagi, M., Takagi, S., and Toyoshima, Y., Experimental study of gate voltage scaling for TDDB under direct tunneling regime, in *Proceedings of the International Reliability Physics Symposium*, The Institute of Electrical and Electronics Engineers, Piscataway, NY, 2001, 380.
83. Schuegraf, K.F. and Hu, C., Effects of temperature and defects on breakdown lifetime of thin SiO_2 at very low voltages, *IEEE Trans. Electron Devices*, 41, 1227, 1994.
84. Chen, I.C., Holland, S., and Hu, C., Electron trap generation by recombination of electron and holes in SiO_2, *J. Appl. Phys.*, 61, 4544, 1987.
85. Chang, C., Hu, C., and Brodersen, R.W., Quantum yield of electron impact ionization in silicon, *J. Appl. Phys.*, 57, 302, 1985.
86. DiMaria, D.J., Cartier, E., and Buchanan, D.A., Anode hole injection and trapping in silicon dioxide, *J. Appl. Phys.*, 80, 304, 1996.

87. Rasras, M. et al., Photo-carrier generation as the origin of Fowler-Nordheim substrate hole current in thin oxides, *IEEE Trans. Electron Devices*, 48, 231, 2001.
88. DiMaria, D.J. and Stathis, J.H., Anode hole injection, defect generation, and breakdown in ultrathin silicon dioxide films, *J. Appl. Phys.*, 89, 5015, 2001.
89. Vogel, E.M., Edelstein, M.D., and Suehle, J.S., Defect generation and breakdown of ultra-thin silicon dioxide induced by substrate hot-hole injection, *J. Appl. Phys.*, 90, 2338, 2001.
90. Heh, D., Vogel, E.M., and Bernstein, J.B., Impact of substrate hot hole injection on ultra-thin silicon dioxide breakdown, *Appl. Phys. Lett.*, 82, 3242, 2003.
91. Harari, E., Dielectric breakdown in electrically stressed thin films of thermal SiO_2, *J. Appl. Phys.*, 49, 2478, 1978.
92. Subramonian, R., Scott, R.S., and Dumin, D. J., A statistical model of oxide breakdown based on the physical description of wearout, in *IEEE International Electron Device Meeting*, The Institute of Electrical and Electronics Engineers, Piscataway, NY, 1992, 285.
93. Stathis, J.H., Quantitative model of the thickness dependence of breakdown in ultra-thin oxides, *Microelectron. Eng.*, 36, 325, 1997.
94. Krishnan, A.T. and Nicollian, P.E., Analytic extension of the cell-based oxide breakdown model to full percolation and its implications, in *Proceedings of the International Reliability Physics Symposium*, The Institute of Electrical and Electronics Engineers, Piscataway, NY, 2007, 232.
95. Wu, E.Y. and Vollertsen, R.P., On the Weibull shape factor of intrinsic breakdown of dielectric films and its accurate experimental determination—part I: Theory, methodology, experimental techniques, *IEEE Trans. Electron Devices*, 49, 2131, 2002.
96. Wu, E.Y., Suñé, J., and Lai, W.L., On the Weibull shape factor of dielectric films and its accurate experimental determination, part II: Experimental results and the effects of stress conditions, *IEEE Trans. Electron Devices*, 49, 2141, 2002.
97. Alam, M.A., Statistically independent soft-breakdowns redefine oxide reliability specifications, in *International Electron Devices Meeting Technical Digest*, The Institute of Electrical and Electronics Engineers, Piscataway, NY, 2002, 151.
98. Suñé, J. and Wu, E.Y., Statistics of successive breakdown events in gate oxides, in *International Electron Devices Meeting Technical Digest*, The Institute of Electrical and Electronics Engineers, Piscataway, NY, 2002, 147.
99. Suñé, J. and Wu, E.Y., Statistics of successive breakdown events in gate oxides, *IEEE Electron Device Lett.*, 24, 272, 2003.
100. Wu, E.Y. and Suñé, J., Successive breakdown events and their relation with soft and hard breakdown modes, *IEEE Electron Device Lett.*, 24, 692, 2003.
101. Blöchl, P.E. and Stathis, J.H., Hydrogen electrochemistry and stress-induced leakage current in silica, *Phys. Rev. Lett.*, 83, 372, 1999.
102. Nissan-Cohen, Y. and Gorczyca, T., The effect of hydrogen on trap generation, positive charge trapping, and time-dependent dielectric breakdown of gate oxides, *IEEE Electron Device Lett.*, 9, 287, 1988.
103. Gelatos, C. et al., The effects of passivation and post-passivation anneal on the integrity of thin oxides, in *1997 International Symposium on VLSI Technology, Systems, and Applications, Proceedings of Technical Papers*, Taipei, Taiwan, 1997, 188.
104. Pompl, T. et al., Change of acceleration behavior of time-dependent dielectric breakdown by the BEOL process: indications for hydrogen induced transition in dominant degradation mechanism, in *Proceedings of the International Reliability Physics Symposium*, The Institute of Electrical and Electronics Engineers, Piscataway, NY, 2005, 388.
105. Shen, T.C. et al., Atomic-scale desorption through electronic and vibrational excitation mechanisms, *Science*, 268, 1590, 1995.
106. Avouris, Ph. et al., STM-induced H atom desorption from Si(100): Isotope effects and site selectivity, *Chem. Phys. Lett.*, 257, 148, 1996.
107. Suñé, J. and Wu, E.Y., Defect generation efficiency and time non-linearity in oxide breakdown experiments, unpublished.
108. Wu, E.Y. and Suñé, J., New insights in polarity-dependent oxide breakdown for ultra-thin gate oxide, *IEEE Electron Device Lett.*, 23, 494, 2002.

109. Foley, E.T. et al., Cryogenic UHV-STM study of hydrogen and deuterium desorption from Si(100), *Phys. Rev. Lett.*, 80, 1336, 1998.
110. Stokbro, K. et al., First-principles theory of inelastic currents in a scanning tunneling microscope, *Phys. Rev. B*, 58, 8038, 1998.
111. Salam, G.P., Persson, M., and Palmer, R.E., Possibility of coherent multiple excitation in atom transfer with a scanning tunneling microscope, *Phys. Rev. B*, 49, 10655, 1994.
112. Stipe, B.C. et al., Single-molecule dissociation by tunneling electrons, *Phys. Rev. Lett.*, 78, 4410, 1997.
113. Hess, K. et al., Magnitude of the threshold energy for hot electron damage in metal–oxide–semiconductor field effect transistors by hydrogen desorption, *Appl. Phys. Lett*, 75, 3147, 1999.
114. Tuttle, B. and Van de Walle, C.G., Structure, energetics, and vibrational properties of Si-H bond dissociation in silicon, *Phys. Rev. B*, 59, 12884, 1999.
115. Brower, K.L., Dissociation kinetics of hydrogen-passivated (111) Si-SiO_2 interface defects, *Phys. Rev. B*, 42, 3444, 1990.
116. Stesmans, A., Dissociation kinetics of hydrogen-passivated P_b defects at the (111)Si/SiO_2 interface, *Phys. Rev. B*, 61, 8393, 2000.
117. Chabal, Y.C. and Patel, C.K., Infrared absorption in a-Si:H: First observation of gaseous molecular H_2 and Si-H overtone, *Phys. Rev. Lett.*, 53, 210, 1984.
118. Li, X.P. and Vanderbilt, D., Calculation of phonon-phonon interactions and two-phonon bound states on the Si(111):H surface, *Phys. Rev. Lett.*, 69, 2543, 1992.
119. Honke, R., et al., Anharmonic adlayer vibrations on the Si(111):H structure, *Phys. Rev. B*, 59, 10996, 1999.
120. Morse, P.M., Diatomic molecules according to the wave mechanics. II. Vibrational levels, *Phys. Rev. B*, 34, 57, 1929.
121. Ribes, G. et al., Multi-vibrational hydrogen release: Physical origin of T_{BD} and Q_{BD} power-law voltage dependence of oxide breakdown in ultra-thin gate oxides, *Microelectron. Reliab.*, 45, 1842, 2005.
122. Haggag, A. et al., Physical model for the power-law voltage and current acceleration of TDDB, *Microelectron. Reliab.*, 45, 1855, 2005.
123. Jeppson, K.O. and Svensson, C.M., Negative bias stress of MOS devices at high electric fields and degradation of MOS devices, *J. Appl. Phys.*, 48, 2004, 1977.
124. Ogawa, S. and Shiono, N., Generalized diffusion-reaction model for the low-field charge build up instability at the Si-SiO_2 interface, *Phys. Rev. B*, 51, 4218, 1995.
125. Alam, M.A. and Mahapatra, S., A comprehensive model of PMOS NBTI degradation, *Microelectron. Reliab.*, 45, 71, 2005.
126. Zafar, S., Statistical mechanics based model for negative bias temperature instability induced degradation, *J. Appl. Phys.*, 97, 103709, 2005.
127. Houssa, M. et al., Stress-induced leakage current in ultrathin SiO_2 layers and the hydrogen dispersive transport model, *Appl. Phys. Lett.*, 78, 3289, 2001.
128. Hess, K. et al., Theory of channel hot-carrier degradation in MOSFETs, *Physica B*, 272, 527, 1999.
129. Stokbro, K. et al., STM-induced hydrogen desorption via a hole resonance, *Phys. Rev. Lett.*, 80, 2618, 1998.
130. Wu, E.Y. and Suñé, J., Power-law voltage acceleration: A key element for ultra-thin gate oxide reliability, *Microelectron. Reliab.*, 45, 1809, 2005.
131. Duschl, R. and Vollertsen, R.-P., Is the power-law applicable beyond the direct tunneling regime? *Microelectron. Reliab.*, 45, 1861, 2005.
132. Vollertsen, R.-P. and Miranda, E., The TDDB power-law model—physics and experimental evidences, *Microelectron. Reliab.*, 45, 1807, 2005.

17

Defects in Thin and Ultrathin Silicon Dioxides

Giorgio Cellere, Simone Gerardin, and Alessandro Paccagnella

CONTENTS

17.1 Introduction

The first studies on metal-oxide-semiconductor (MOS) structures date back to the 1960s, and several years were spent in research before the first complementary MOS (CMOS) was introduced in 1968 [1] by integrating both nMOS and pMOS in the same chip, thus allowing high-speed and very low power dissipation. Since then, silicon integrated circuits have penetrated into virtually every apparatus with electrical components. The key of the technology growth [2] has been the drive to smaller and smaller dimensions using the principle of scaling [3]. MOSFET scaling-down is based on the idea of changing device parameters and operating voltage together. Generally, in the conventional scaling, over the last 10–15 years the operating electric field has been maintained around $E = 5$ MV/cm in the oxide (constant-electric-field scaling), as dictated by transistor reliability, which is dominated by time-dependent dielectric breakdown (TDDB) [4,5] (Chapters 15 and 16), channel hot-carrier injection (CHC) [5,6], and other degradation phenomena such as negative bias temperature instability (NBTI) [7,8] (Chapters 12 through 14). Scaling the gate oxide and channel length by a factor k allows designers to increase the device density by a factor k^2 due to the smaller wiring and device dimensions, improves the speed thanks to a k gain in the transconductance-to-capacitance ratio (g_m/C), and reduces the power dissipation by a factor k^2 because of the reduced voltage and current in each device, while keeping control over short-channel effects. However, the voltage is actually scaling slower than device physical dimensions. More importantly, the oxide is becoming thinner and thinner, and reaching its ultimate physical limits. Now gate oxides in modern state-of-the-art technologies are so thin (~1–2 nm) that leakage current due to direct tunneling is a major issue [9]. New materials such as aluminum or hafnium oxides, having a larger dielectric constant in comparison to SiO_2, have been studied to replace silicon dioxide as the gate dielectric in the last several years [10] (Chapters 9 through 11), and are finally entering mass production [11], at least for high-end applications. Some high-κ dielectric materials are already in use in devices such as dynamic random-access memories (DRAMs) [12]. The advantage of having a high-κ material is that the same equivalent electrical thickness (i.e., the same capability to control short-channel effects) is obtained with a larger physical thickness (hence, lower tunnel current). Despite the advancement of such technologies, SiO_2 is still the workhorse gate dielectric of the semiconductor industry, and is expected to retain its role for several years. In fact, high-κ materials are difficult to integrate into the CMOS process flow [13], requiring complex process changes such as suitable gate replacements [11]. For this reason, even state-of-the-art technologies may be realized without using them at all [14].

In this chapter, we will review the reliability of thin and ultrathin oxides. We will discuss SiO_2 only, because of its ubiquity and the huge amount of literature on this subject. Silicon nitridation [15,16] is routinely used to prevent boron penetration in the gate oxide. The use of nitrided oxides (briefly, SiON) slightly enhances the dielectric constant of the dielectric, thus helping to reduce direct tunneling currents, but may also result in enhanced degradation of the oxides [8,17]. Since nitridation is performed in various measures in all modern technologies, we will not distinguish between nitrided or nonnitrided oxides, speaking simply of SiO_2 (unless otherwise stated). Furthermore, we will focus on the degradation of the insulating properties of the oxide due to two different causes. At first, we will consider high-field electrical stress, which is encountered during actual operation for some classes of devices (such as the tunnel oxide of nonvolatile memories) and used to simulate the long-term degradation for others (thin gate oxides in advanced logic devices). Then, we will review the loss of insulating properties due to ionizing radiation. Once again, this will be done for the two classes of oxide thicknesses previously mentioned. We will generally use "gate oxide" to refer to the gate dielectric of an aggressively scaled metal-oxide-semiconductor field-effect transistor (MOSFET), which varies with thickness and nitride concentration, depending on the technology node (a 5 nm oxide in a 0.25 μm long transistor actually was aggressively scaled just a few years ago), and "tunnel oxide" to refer to the dielectric separating the floating gate from the substrate in nonvolatile memories (as we will discuss later, the thickness of this oxide is not scaling at all with the technology due to reliability constrains).

In this chapter, we will not cover catastrophic phenomena induced by electrical stress such as breakdown or soft breakdown (SB) [18,19]; these topics are covered in Chapters 15 and 16. To be consistent, neither we will address important issues induced by ionizing radiation such as single event gate rupture (SEGR) [20,21], that is, the catastrophic breakdown following the impact of a single, high-energy ion on a biased oxide.

17.2 Stress-Induced Leakage Current

Hot carriers can produce defects in thin gate dielectrics by means of several mechanisms. In this section, we are going to discuss how trap generation can result in a leakage current in thin gate oxides, and how this can impact device reliability as a function of oxide thickness. In a first approximation, the degradation of the insulating properties of the oxide can be described as a sequence of different stages (see Figure 17.1). At the beginning of the stress, a gradual and continuous increase of the gate leakage current can be observed. Such current is referred to as stress-induced leakage current (SILC) [22]. SILC is characterized by a uniform conduction across the whole gate active area, due to neutral trap generation inside the insulating layer. Then, by further stressing the oxide, the gate current can abruptly increase, due to SB (initially called B-mode SILC [23]). SB is associated with a single or few localized conductive spots in the oxide, rather than to a uniform conduction across the whole gate area as for SILC. In small devices, intermediate events known as microbreakdown [24] can be detected as well. Finally, catastrophic oxide breakdown may be reached at high stress levels, and the gate current can approach 1 mA or more. Hard breakdown (HB) represents this last degradation step. The increasing complexity of devices is reflected in the increasing complexity of the reliability assessment: breakdown itself is becoming a progressively evolving phenomenon instead than the abrupt one to which we were accustomed [25,26].

Overall, many scientific works have been devoted to these issues, so this section should be intended as a guide to approach the problems, rather than a in-depth discussion of their characteristics and implications. We will focus on SILC only (with a slight mention of microbreakdown), since the other aspects are treated in more detail in Chapters 15 and 16.

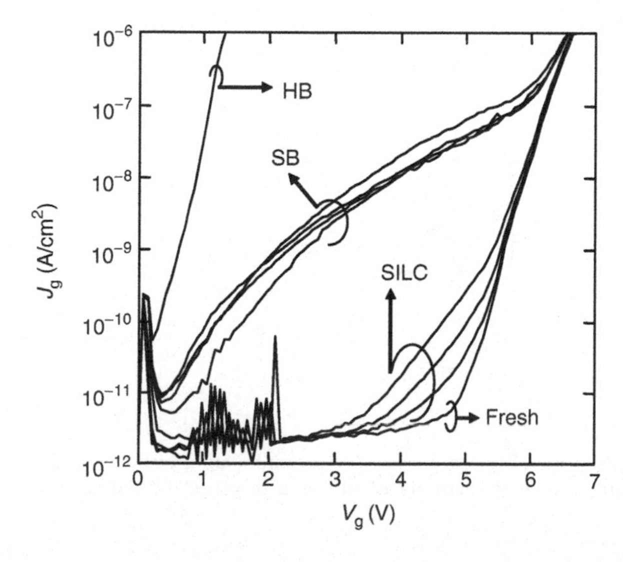

FIGURE 17.1
Evolution of the gate leakage current through a thin oxide during an accelerated electrical stress.

17.2.1 SILC Phenomenology

The first observations of leakage current in thin SiO_2 layers subjected to high-field electrical stresses date back to the 1980s [27], when a significant degradation of the oxide insulating properties, leading to enhanced tunneling current, was found after injection of 10^{17}–10^{18} electrons/cm^2. Later in the same decade it was proposed that this oxide leakage current originates from localized weak spots uniformly distributed across the whole oxide area [22]. The weak spots were related to some defect-rich regions produced by the electrical stress in the SiO_2 layer. The nature of SILC has been the subject of a large number of studies, which have demonstrated the presence of different contributions. In particular, these different contributions can be grouped into three families [28–31]:

1. Nonreproducible component, mainly deriving from the recombination and/or passivation of the oxide positive trapped charge. This component can be observed only once during the first positive gate voltage sweep after stress, and is associated with electrons tunneling into the positively charged site during the application of a positive electric field (see Figure 17.2a). Once electrons have tunneled into the positively charged site, the site is neutralized and annihilated.
2. Transient component, due to charging and discharging of oxide traps located close to the oxide interfaces. This transient current is strongly dependent on the polarity of gate voltage sweeps and on oxide thickness. In particular, it decreases with decreasing oxide thickness. The transient component includes several contributions, generally measurable even after repeated voltage scans and associated with electron tunneling into neutral traps (Figure 17.2b). Such components have been widely studied by many authors and shown to strongly decrease with oxide thicknesses below 6 nm [32], which are of major interest for virtually all modern logic applications. Among the transient contributions we also consider transient currents that are measurable whenever the polarity of the applied gate voltages is switched from negative to positive or vice versa. They were attributed to electron tunneling into electron traps with a repulsive potential barrier, whose energy level

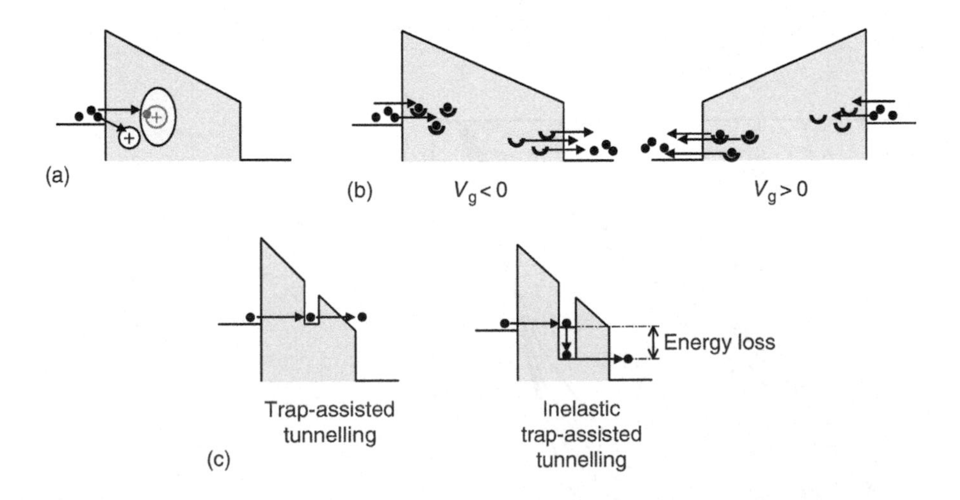

FIGURE 17.2
Different components contributing to SILC: (a) irreproducible component due to annihilating positive trapped charge; (b) transient component, due to charging and/or discharging of interface traps; (c) DC component, attributed to an elastic or inelastic TAT process.

is lower than the Fermi level of the silicon substrate [29]. The defect site is positively or negatively charged only during the first negative or positive voltage sweep, respectively. The oxide charge is maintained until a voltage scan with the opposite polarity is performed.

3. True DC component. This last contribution, which is the most difficult to measure and model, but also less dependent on the specific stress-and-test setup, has been the subject of several successful modeling efforts. The first observations supported the idea that the electrical stress generates weak spots in the oxide, where the electron conduction is easier than across the undamaged oxide regions. Starting from an empirical approach, where SILC was still described on the basis of the Fowler–Nordheim (FN) tunneling equation with a reduced barrier height [22], more refined models have been proposed. Lately, the model that has received general, if not universal, agreement was the trap-assisted tunneling (TAT) through oxide neutral traps, as shown in Figure 17.2c [30,32–36]. According to this model, SILC is due to inelastic TAT across neutral defects generated by the electrical stress and uniformly distributed across the whole oxide area. The tunneling process is accompanied with the trap energy relaxation. A possible atomic-scale model, which successfully describes the neutral traps involved in SILC conduction, is based on hydrogen-bridge-related defects [32]. The calculated energy-relaxation value of 1.71 eV is in substantial agreement with experimentally observed values [34,35] (roughly 1.5 eV). To make this picture even more complex, it has been demonstrated that the DC component is not stable over time, but tends to gradually decrease with time upon injection of low energy electrons [37], likely due to local reconstruction of the damaged lattice due to electron capture by the neutral traps.

Based on the TAT conduction mechanism, SILC is proportional to both the defect concentration and the TAT probability. When the oxide thickness is reduced, the distance between the neutral defects mediating SILC and the oxide interfaces decreases. This exponentially increases the probability that traps capture and/or emit electrons. In contrast, the transient component becomes practically undetectable in thin oxides, because of the enhanced emission probability. It has been experimentally verified by several authors that both the transient and the nonreproducible components practically disappear in 6 nm (or thinner) oxides, and the DC SILC component controls the leakage current intensity [32]. This is good news for the reliability of advanced devices, where 3 nm oxides have already been left behind the mainstream.

17.2.2 SILC Growth with Stress Level

Based on the TAT model, the leakage current linearly depends on the defect concentration inside the oxide. The SILC growth kinetics have been widely studied in the literature, and many works in the last decade have been devoted to the understanding of leakage current growth and defect generation kinetics. SILC gradually grows until the breakdown event. The growth kinetics of the excess current J_e (defined as the difference between the current measured after and before stress) follows a power-law [38–40] relationship:

$$\frac{\partial \ln(J_e)}{\partial N_{inj}} = K_e \cdot N_{inj}^{\nu} \qquad (17.1)$$

where N_{inj} is the injected charge. This kinetics law applies to all devices after constant current stress, regardless of the oxide thickness and technology, as shown in Figure 17.3a.

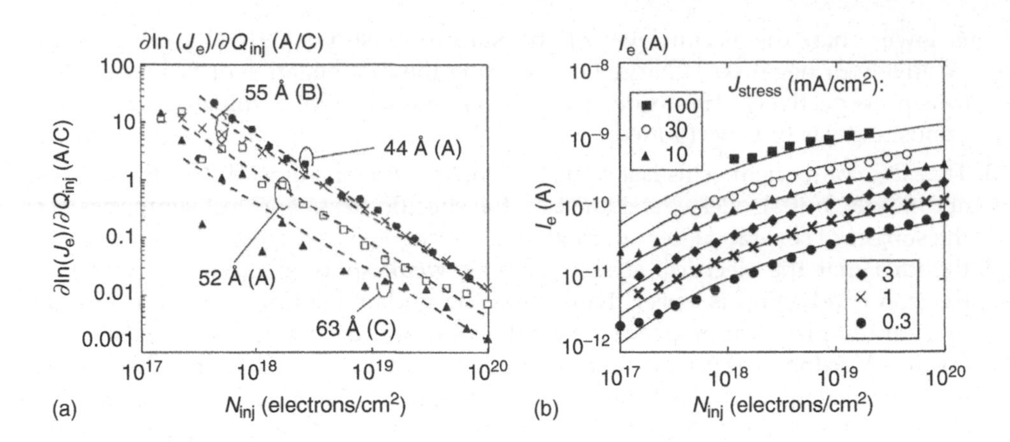

FIGURE 17.3
(a) SILC growth rate $d[\ln(J_e)/dN_{inj}]$, as a function of injected charge for different oxide thickness and technology. Symbols, experimental data; lines, fitting. A, steam grown oxide and N_2 annealing; B, dry grown oxide; C, steam grown oxide and N_2O annealing. (b) SILC in 4.4 nm oxide as a function of injected charge symbols = experimental data, lines = fitting. (Reprinted from Scarpa, A. et al., *Microelectron. Reliab.*, 40, 57, 2000. With permission.)

The slope of the curves ν ranges between the narrow limits −1.1 and −1.4 over several decades of injected charge N_{inj}, independent of stress intensity, gate voltage, and measurement polarities. The preexponential coefficient K_e weakly depends on the stress current density J_{str} following:

$$K_e = K \cdot J_{str}^{-\varepsilon} \tag{17.2}$$

where K only depends on the oxide thickness and technology.

From Equations 17.2 and 17.3 the following relation can be obtained for the excess current:

$$J_e = C \cdot J_{str}^{\beta} \cdot \exp\left(\frac{K}{\alpha} \cdot J_{str}^{-\varepsilon} \cdot N_{inj}^{\alpha}\right) \tag{17.3}$$

where the values of the parameters (α, ε, β, K, and C) depend on the oxide thickness. An example of the excellent matching between model predictions and experimental data are shown in Figure 17.3b [38]. This model predicts that SILC saturates at a maximum current value J_{sat}, (i.e., the first term in Equation 17.2), which increases with the stress current density J_{str} [41]. However, the experimental evidence of breakdown occurrence even after SILC saturation, excludes that trap generation kinetics actually saturates the same way: trap concentration should be increasing until breakdown occurs in some region of the oxide layer, even if SILC growth already saturated. This problem is still open and no definitive solution has yet been found. SILC growth kinetics, trap generation in bulk oxide, and interface states generation may be related, since a strong one-to-one correlation has been reported between SILC, interface defects density (D_{it}), and bulk oxide traps (D_{ot}) [42,43]:

$$J_{SILC} \sim D_{it} \sim D_{ot} \tag{17.4}$$

This indicates that SILC, interface traps, and bulk traps have associated origins. Nevertheless, after thermal annealing at 250°C, the interface trap density D_{it} substantially decreases, whereas SILC and D_{ot} are only partially annealed. In other words, the defects responsible for SILC do not (or only marginally) contribute to D_{it}. On the contrary, the strong correlation between SILC and D_{ot} may be a causal one.

17.2.3 Origin of Defects Leading to SILC

The origin of the neutral traps mediating SILC is still a matter of debate and none of the proposed models have encountered a general agreement. Since a deeper discussion of this topic would be outside the scope of this chapter, we briefly present the most important models without going into details.

In one of the first studies devoted to shedding light upon the nature and origin of neutral traps responsible for SILC [44], it was proposed that the increase in the leakage current is related to hydrogen-induced defects. This has been supported by the experimental evidence that exposure of thin oxides to atomic hydrogen from a remote plasma causes leakage currents similar to those observed after high-field stress. The atomic-scale model discussed above nicely fits with this approach.

A completely different approach is followed by some other authors, who identify anode hole injection (AHI) as responsible for defect generation (see, for instance, [45 and references therein]). Holes have been indicated as possible precursors of defects causing SILC by other authors [43], who studied the polarity dependence of the trap generation rate and charge-to-breakdown in nitrided oxides. They found that nitridation significantly affects the oxide degradation process as well as the stress polarity dependence. Nitridation was demonstrated to cause a large decrease of bulk oxide traps during gate injection, while trap creation is less affected by the nitridation during substrate injection. The asymmetrical distribution of incorporated nitrogen in the oxide was proposed as responsible for the different hole generation, transport, and trapping in nitrided and nonnitrided oxides. The atomic-scale model previously described explains creation starting from hole capture at an oxygen vacancy [46].

The two approaches can be joined into a model which accounts for both holes and hydrogenous species [47]. In this case, for both positive and negative stress polarity, holes are generated and injected into the oxide. Such holes can be trapped or react with the Si–H bonds in the silicon dioxide layer, leading to hydrogen release, according to one of the following reactions:

$$\mathrm{Si-H+h^+ \Rightarrow Si\bullet + H^+}$$
$$\mathrm{Si-H+h^+ \Rightarrow Si^+ + H}$$

The first process leads to the release of positively charged hydrogen ions, while the second one releases a neutral hydrogen atom. The mobile hydrogen can then drift and/or diffuse to the Si/SiO_2 interface, where it can break a Si–H bond and create an interface trap, according to one of the following reactions:

$$\mathrm{Si-H+H^+ + e^- \Rightarrow Si\bullet + H_2}$$
$$\mathrm{Si-H+H \Rightarrow Si\bullet + H_2}$$
$$\mathrm{Si-H+H^+ \Rightarrow Si^+ + H_2}$$

In the case of the latter reaction, electron capture occurs at the Si atom, as opposed to the proton, owing to the low probability with which a proton in SiO_2 can capture an electron, as discussed in Chapters 7 and 13.

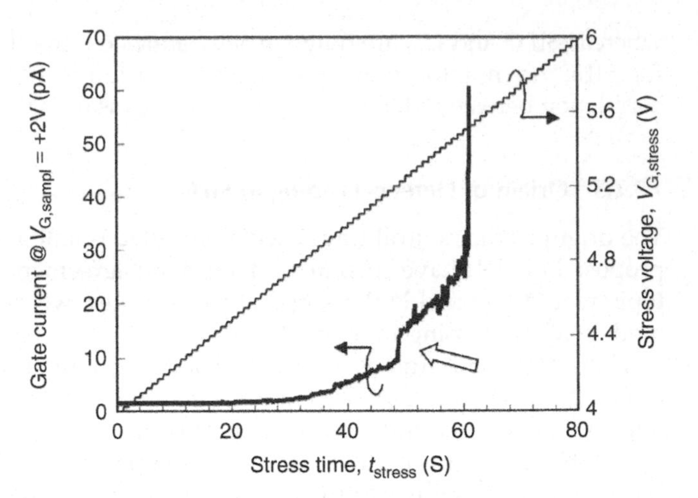

FIGURE 17.4
Gate current measured at $V_{G,sampl} = 2$ V during ramped voltage stress (bold line). The thin line (right-hand side Y-axis) is the stress voltage as a function of time. (Adapted from Cellere, G. et al., *IEEE Trans. Electron Dev.*, 49, 1367, 2002. With permission.)

17.2.4 Microbreakdown

Microbreakdown [24], or prebreakdown [48,49], can be defined as an abrupt but small (tens of pA) current jump, preceding SB in thin oxides. A typical gate current versus stress time in a microbroken device is shown in Figure 17.4. In this experiment, the gate stress voltage shown in the right-hand side Y-axis increases with time, but the gate leakage reported on the left-hand side Y-axis was measured at a constant bias during the whole experiment. Due to the tiny currents involved, these events are not detectable in large-area devices, but can be a crucial phase in degradation of actual devices, where most area is occupied by many small area devices.

The microbreakdown current has been found to be related to latent damage in the oxides [50], and is the consequence of a percolation path made of only two defects [24]. The two-defect nature (hence, the mid-way character between SILC and SB) is confirmed by experiments showing how microbreakdown recovers after a high-temperature anneal in the same way that SILC does [51,52]. Since microbreakdown spots develop only if the device is stressed at low electric field, they are thought to be a precursor for breakdown in actual operating conditions [24,50], even if a role in high-voltage experiments is all but excluded [48,49].

17.2.5 SILC in Tunnel Oxides

Studies on large-area capacitors have established a solid relationship between oxide defects and gate leakage, but they can provide only spatially averaged information about the conduction properties of a large number of defects. Here we present an alternative experimental approach to study SILC, based on the analysis of charge loss from tiny floating gates (FG) across a relatively thick oxide (7–10 nm). This method makes it possible to investigate leakage currents due to only one or two defects with intensities lower than 10^{-18} A. In addition to its implications for physical models, studying such relatively thick oxides also is of crucial technological importance since they are at the heart of Flash memories (Figure 17.5) [53]. This technology relies on the storage of charge in a conductive polycrystalline silicon region (the floating gate), which is insulated from the control gate and from the transistor channel by two dielectric layers: the interpoly dielectric and the tunnel oxide. To prevent the onset of any leakage path, the interpoly dielectric is usually realized using an oxide–nitride–oxide (ONO) sandwich. The tunnel oxide is the thinner and the more critical of the two, since it must permit the passage of carriers to charge

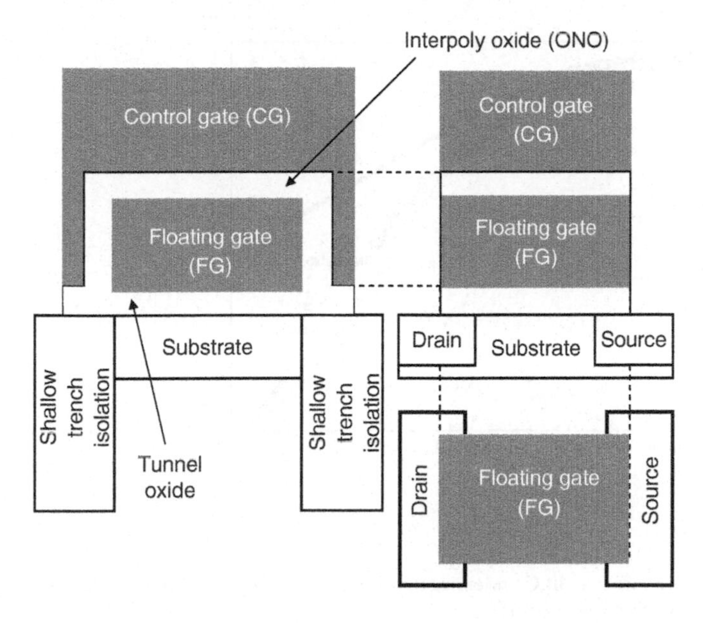

FIGURE 17.5
First-principles sketch of an industry-standard NOR flash cell.

(program) and discharge (erase) the floating gate, while avoiding any unwanted leakage path that may severely compromise the data retention characteristics of the FG cell. State-of-the-art FG cells are programmed with a few thousand electrons. This means that not even one electron per day is allowed to leak across the tunnel oxide if the information has to be retained for 10 years, as customarily required for these devices. SILC usually affects only a small number of storage nodes, giving rise to a tail in the threshold voltage distributions (tail bits), but it is exacerbated by the high electric fields required by program and erase operations, each constituting an electrical stress. Not only the operation of the FG cell, but also its data retention capability, has to be guaranteed after many (typically 100,000) write/erase cycles.

The extreme sensitivity of FG cells to charge loss is the key to the measurement of leakage current with unmatched accuracy. In fact, since the threshold voltage V_{TH} of an FG cell is determined by the number of electrons stored on the FG, the leakage current through the tunnel oxide can be monitored by measuring the V_{TH} of the FG transistors over extended periods of time (Figure 17.6). SILC studies in FG memories are often performed by applying a gate bias to increase the leakage current across the tunnel oxide. Accelerated tests can involve either data retention (charge loss from programmed cells) or gate-disturb (charge gain in erased cells), according to the polarity of the gate bias. TAT has been used to model SILC in large-area capacitors or MOSFETs, including inelastic tunneling in stressed oxides [54] and electron–hole recombination effects [55]. Yet, modeling charge loss in memories also requires anomalous effects to be considered [56]. Both a physical model for the conduction across the gate oxide and a statistical approach to calculate the probability of each defect configuration are needed.

Figure 17.7 shows the cumulative SILC distributions for three flash arrays with different tunnel oxide thicknesses. For thinner and thinner oxides, the statistical spreads [57] of the SILC distribution increase. The large spreads in leakage currents shown in Figure 17.7 are due to a number of reasons. SILC depends on the spatial position (in particular the distance from the interfaces), energy, and density of the defects. However, simulations have shown that the energy distribution of SILC-related defects is quite narrow and cannot account for the observed variations [55]. The total number of defects should play at least a small role as well, since Poisson statistics predicts a small standard deviation (discussed below), which

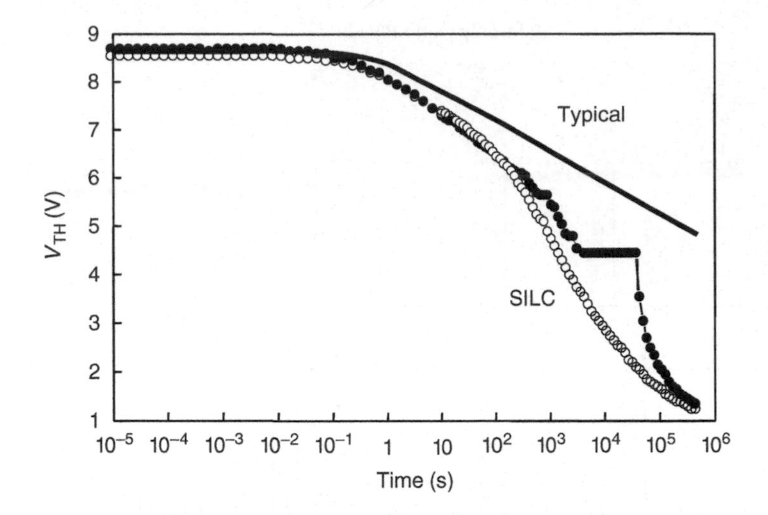

FIGURE 17.6
Evolution of V_{TH} for a typical FG and two FGs exhibiting SILC under negative gate stress. (Reprinted from Cellere, G. et al., *IEEE Trans. Nucl. Sci.*, 53, 3349, 2006. With permission.)

again cannot justify the measured spread. On the contrary, the generation of two or more defects close to each other can significantly enhance the local SILC [56]. Figure 17.8 shows the $I-V$ characteristics of two bits A and B belonging to the distribution of Figure 17.7. The $I-V$ curve of bit A can be modeled with a TAT mechanism through a single oxide defect. This is not the case for bit B, where a 2TAT mechanism, i.e., tunneling assisted by two traps, yields a much better fit.

In the general case, contributions due to TAT, 2TAT, and SILC at more than two cooperating defects must be considered. The 2TAT model has also been shown to be consistent with anomalous erratic SILC. For instance, unstable SILC activation and/or deactivation

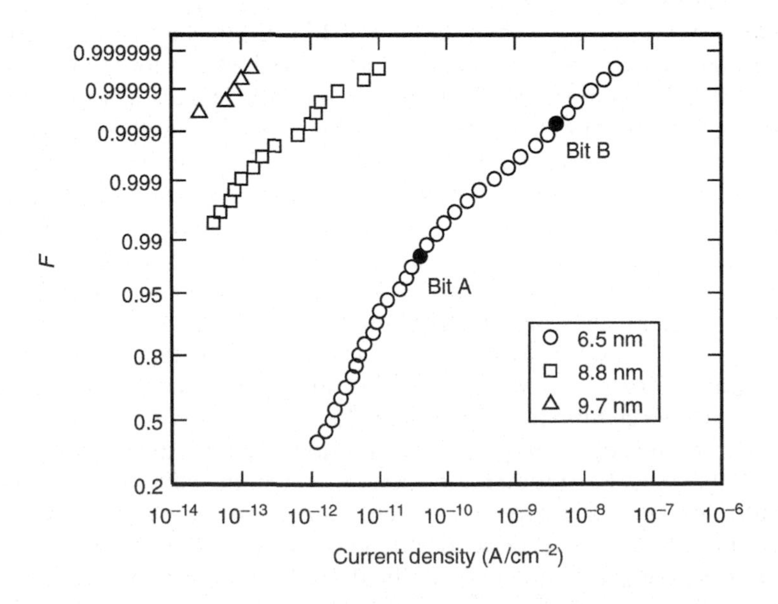

FIGURE 17.7
Cumulative SILC distribution for different oxide thicknesses ($V_{FG} = 4$ V). (Reprinted from Ielmini, D. et al., *IEEE Trans. Electron Dev.*, 49, 1955, 2002. With permission.)

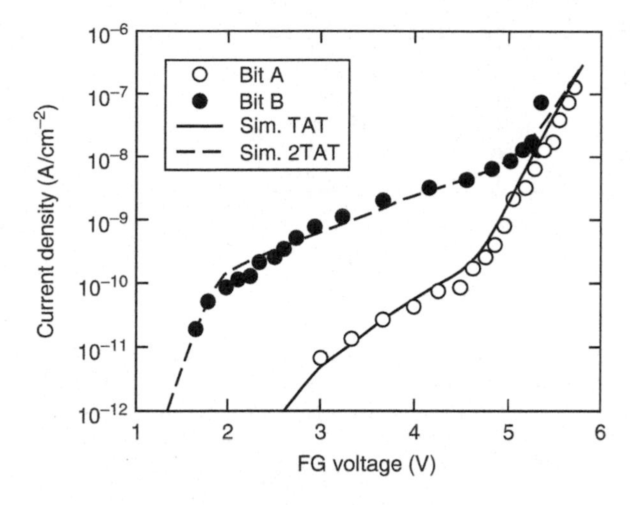

FIGURE 17.8
I–V curves of bits A and B in Figure 17.7. Simulated *I–V* curves have been obtained with TAT and 2TAT models. (Reprinted from Ielmini, D. et al., *IEEE Trans. Electron Dev.*, 49, 1955, 2002. With permission.)

(i.e., sudden and unpredictable switching of the anomalous leakage current) occurring in tail cells can be modeled as a transition between TAT and 2TAT conduction [58]. Figure 17.9 shows the probabilities (calculated according to Poisson statistics) of having clusters with different amount of defects. As shown, two-trap leakage effects are not negligible at all for defect densities larger than 10^{14} cm^{-3}, which is a relatively small defect concentration.

Physical models based on TAT and percolation have been proposed as well [59]. In these works, a single parameter, i.e., the trap density, is used to characterize the oxide, and the conductivity of the leakage path is determined by the longest trap-to-trap or trap-to-interface distance. Studies also have been made to measure the SILC trap energy using the unique properties of flash memories, namely the SILC roll-off regime [60]. The results of the work show that different types of traps at different energy positions contribute to TAT.

Concerning defect generation, as mentioned above, Poisson statistics is usually employed [61]. In other words, the defect creation process is considered to have no memory of past events, meaning that the position of a new defect is statistically independent from previous ones. Yet, evidence of the contrary, i.e., correlated defect generation, has been reported in Ref. [62]. Possible mechanisms for correlated generation also have been pointed out and are depicted in Figure 17.10.

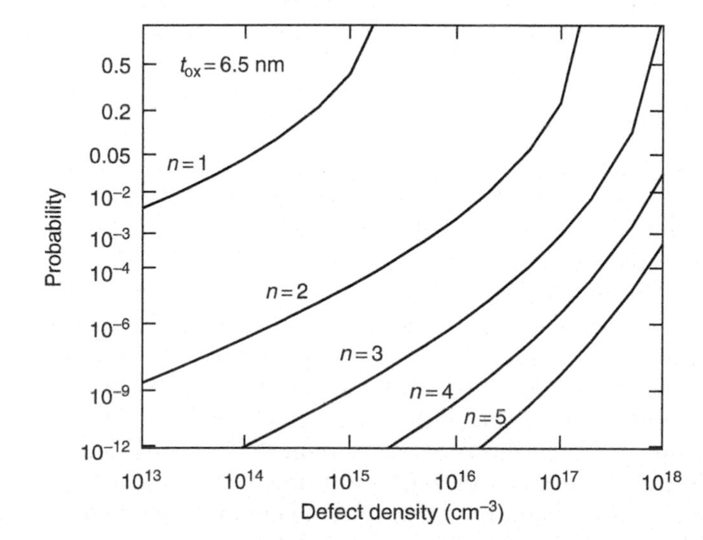

FIGURE 17.9
Probability of having clusters of order $n = 1$–5 for a flash array with $T_{OX} = 6.5$ nm. (Reprinted from Ielmini, D. et al., *IEEE Trans. Electron Dev.*, 49, 1955, 2002. With permission.)

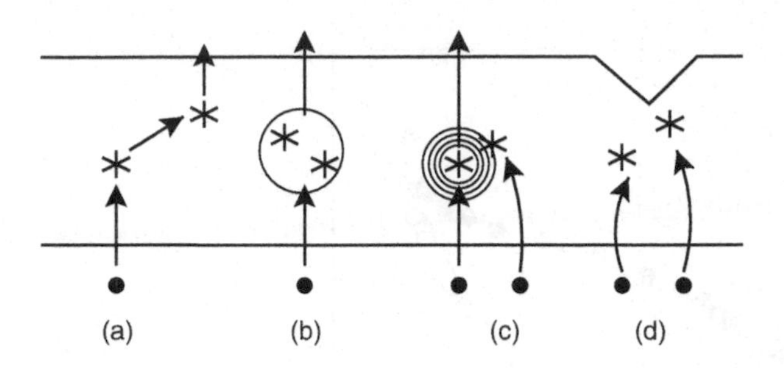

FIGURE 17.10
Possible microscopic mechanisms for correlated defect generation: (a) successive trap generation by a single injected electron/hole; (b) simultaneous generation of the two traps by a single injected electron/hole; (c) generation of a correlated trap near a preexisting precursor; and (d) generation of a correlated trap due to locally enhanced degradation induced by oxide thinning or trapped charge. (Reprinted from Ielmini, D. et al., *IEEE Trans. Electron Dev.*, 51, 1288, 2004. With permission.)

17.3 Ionizing Radiation Effects in Thin Oxides

We have seen that electrically active defects responsible for SILC are generated by electrical stress. This is not the only mechanism that can damage oxides. In particular, ionizing radiation can have major implications on CMOS device reliability. Interestingly, ionizing radiation offers the possibility of generating defects while the oxides are subjected to the electric field found during operating conditions, or even at zero electric fields. In contrast, in accelerated electrical stresses the oxide under study is subjected to a carrier density, a carrier energy, a temperature, or various combinations of the three, much larger than those it is designed to withstand. Hence, we will concentrate in Sections 17.3.1 and 17.3.2 on the degradation of insulating properties of thin gate oxides and of tunnel oxides, respectively. This is not only useful for investigating degradation mechanisms, but ionizing radiation effects also are important for the reliability of devices used in harsh environments or for safety critical or dependable applications, such as satellites, nuclear technology, high energy physics, and some medical applications [63]. Also, while ionizing radiation is often though not to be an issue at all in the majority of operating conditions, things are becoming worse and worse with novel technologies [64]; in the future, at least certain aspects of radiation-induced defects will need to be addressed even for mainstream technologies.

17.3.1 Radiation—A Concern for Harsh Environments Only

Before beginning our discussion, we will try to answer a fundamental question: is ionizing radiation an issue for applications in radiation harsh environments only? Further, how can we define a radiation harsh environment? We first consider the space and satellite industries. Charged particles coming from the Sun are trapped in the near-Earth environment (Van Allen particle belts) [65,66], and interstellar particles called galactic cosmic rays drift into the solar system [65,66]. The 11-year periodic activity of the Sun, and its explosive activities such as flares or coronal mass ejections [66] contribute to make the picture more complex. The Earth (ground level) environment is influenced by "space weather" more strongly than might be thought. First, electronics systems can be severely endangered by solar storms such as those of November 2003, which led to disruptions or system failures in some satellites, cellular communications, and power grids [67]. Further, high-energy protons entering the atmosphere generate a cascade of secondary particles resulting in a significant neutron flux at ground level [68,69], which increases by orders of magnitude at the cruise altitude for airliners [70]. While neutrons are nonionizing, that is, they cannot directly generate electron–hole pairs, they also interact with atomic nuclei, via recoil [71] or spallation reactions [72]. Also, thermal neutrons can be adsorbed by ^{10}B, which will then decay to ^{7}Li by emitting an alpha particle [73]. The by-products of neutron interaction with the chip materials are ionizing particles, which may be generated in the device active region. Finally, many man-made applications require the use of ionizing radiation. These include high energy physics experiments, and also energy production-related research, such as the forthcoming international thermonuclear experimental reactor (ITER), which will be characterized by an harsh radiation environment in proximity of the reactor [74]. Other less exotic but not less important applications are, for example, some medical equipment, known sources of neutrons [75]; food [76] and mail [77] sanitation; and security controls in airports, especially in recent years [78].

17.3.2 Two Broad Classes of Ionizing Radiation Effects

When looking more closely at the interactions of ionizing particles with matter, we need to distinguish two broad classes of phenomena, namely, total ionizing dose (TID) effects and single event effects (SEE) [79]. TID effects are due to the progressive buildup of defects and trapped charge in the device, such as in the gate and isolation [80,81] oxides in bulk MOS technologies and in the buried oxide in silicon-on-insulator (SOI) technologies [82]. This is the case for 10 keV x-rays, and 1.25 MeV γ-rays emitted by ^{60}Co, both standard radiation sources for TID tests, for high energy electrons, and, in space, for protons [66]. High-energy photons cross a device and generate electron–hole pairs, due to the photoelectric and/or the Compton effects, depending on their energies. The energy that should be delivered to the material to create an electron–hole pair is 17 ± 1 eV in SiO_2 [83], and 3.6 eV in bulk silicon [79]. Right after their generation, and depending on the electric field in the relevant region, these carriers tend to recombine in very short times, smaller than a picosecond [79]. Since a very low density of electron–hole pairs is created, typically each electron can recombine only with its hole partner, unless an external electric field separates the newly created electron–hole pair; this is the geminate recombination model [84,85]. Surviving electrons are quickly swept away by the electric field due to their high mobility [79]; holes are almost immobile in comparison, and are the direct or indirect cause of the subsequent oxide damage we discuss below [86].

The second broad category of ionizing radiation effects descends from the charge released by a single (heavy) ion, which results in a SEE; that is, in a macroscopic electrical phenomenon produced by a single microscopic and highly localized event. From a phenomenological point of view, a huge variety of different phenomena are grouped under the SEE acronym, such as single event upset (SEU) [87,88], single event functional interruption (SEFI) [89], multiple bit upset (MBU) [90], single event gate rupture (SEGR) [20,91], single event transient (SET) [92], single event latch-up (SEL) [21], and others. From a device physics perspective, the energy of the impinging ion is first transferred to lattice atoms and electrons, resulting in photons, phonons, and electron–hole pair generation. These carriers then thermalize to the lattice temperature and follow solid-state physics laws of transport [65]. For each ion passing through a material, the amount of energy lost in ionization processes per unit length is defined as linear energy transfer (LET), and is measured in $MeVcm^2/mg$. LET is directly proportional to the number of electron–hole pairs generated per unit length, N_0. What happens after the ion strike strongly depends on the material. In a SiO_2 layer, a very efficient columnar recombination occurs [93,94], while carriers are separated by two different mechanisms: diffusion and

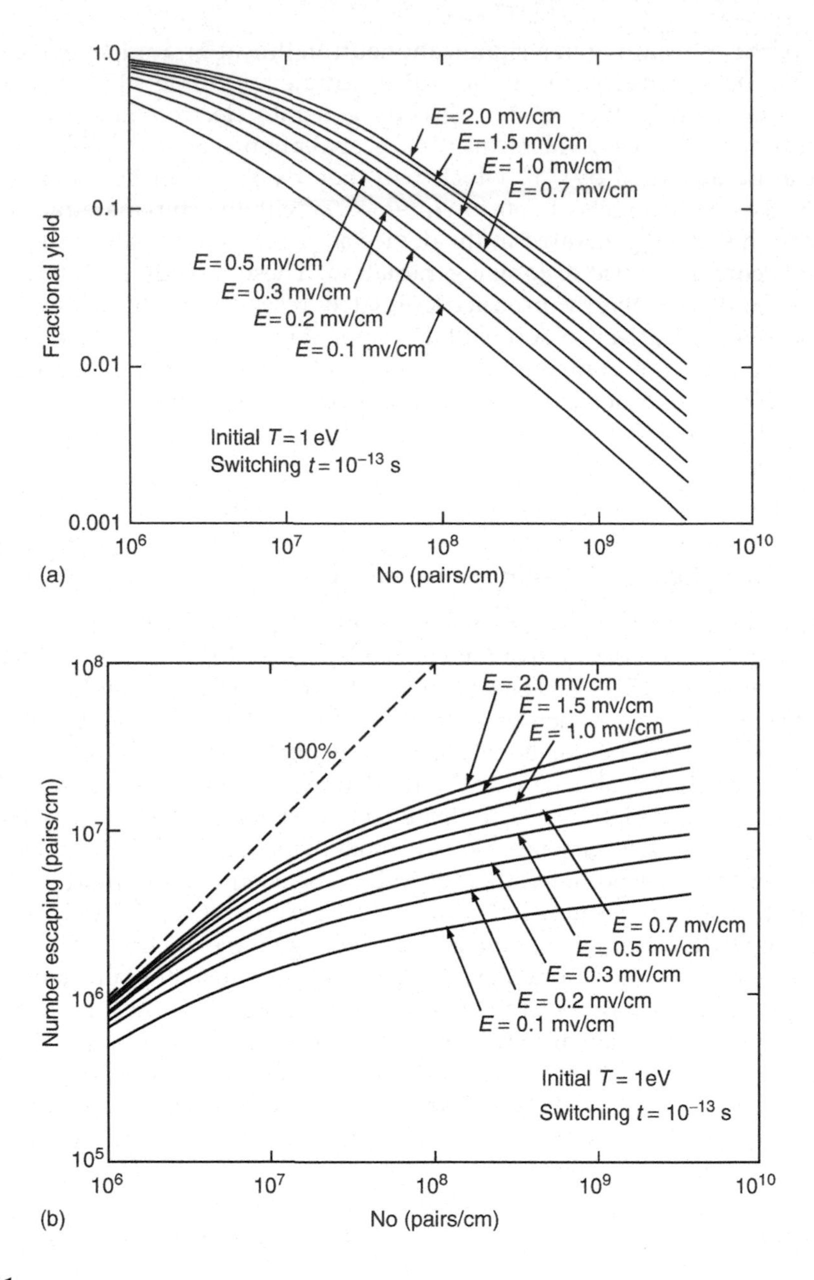

FIGURE 17.11
Charge yield (a) and number of carriers surviving recombination (b) in SiO_2, for ions having different LET (the number of pairs/cm reported on the X-axis is directly proportional to ion LET), at different electric field values. (Reprinted from Oldham, T.R., *J. Appl. Phys.*, 57, 2695, 1985. With permission.)

electric field. The dependence of charge yield on electric field is reported in Figure 17.11 for SiO_2, where N_0 is reported on the X-axis and the fraction (number) of surviving carriers is reported on the upper (lower) Y-axis.

Charges generated in the oxides are (directly or indirectly) sources for electrically active defects. Close to the Si/SiO_2 interface, the oxygen vacancies can act as trapping centers for holes. Often, trapping sites for holes are associated with those for electrons [95,96], resulting in an E' center [97]. In ultrathin gate oxides, with thicknesses smaller than 6–7 nm, similar defects are the origin of radiation-induced leakage current (RILC), similar to SILC [39,98], which will be the main subject of the remainder of this chapter. Finally, Si/SiO_2 interface traps are created after irradiation with a complex kinetic process [99–101], due to a two-step generation mechanism linked to hydrogen migration and bond breaking [102–104]. Here we will concentrate on the loss of isolation properties of tunnel and gate oxides.

17.4 RILC in Ultrathin Oxides

RILC can be defined as an increase in oxide leakage at low electric field after ionizing radiation exposure. As an example, the current density as a function of oxide electric field ($J_g - E_{OX}$) is shown in Figure 17.12 for a 6 nm oxide, before and after a high dose irradiation with 8 MeV electrons, produced by a pulsed LINAC accelerator. The main effect of irradiation is the increase of the low-field gate current, observed between $E_{OX} = 3$ MV/cm and $E_{OX} = 6$ MV/cm. Also note that the oxide trapped charge in the stressed capacitors is negligible, as deduced from the overlap of the high-field characteristics of irradiated and unirradiated devices, corresponding to the FN tunneling regime. The lack of charge trapping is expected for ultrathin oxides, holding true also after electrical stresses, on the basis of experiments in the 1980s that showed that, in the first ~3–4 nm from the interfaces, charge detrapping is almost instantaneous, even at very low temperatures [105]. RILC can be evaluated as the excess current J_e after stress:

$$J_e = J_g - J_0 \tag{17.5}$$

where J_g and J_0 are the gate current density after and before the stress, respectively.

17.4.1 RILC Model

As previously discussed, SILC is the result of TAT across the oxide, mediated by neutral traps created by the stress. Owing to the similar J_g–E_{OX} characteristics of RILC and SILC, RILC also can be correlated to TAT, where traps are generated by irradiation. The field dependence of the excess gate current has been studied to investigate the details of this conduction mechanism. As shown in Figure 17.13, the J_e slope suddenly increases at a

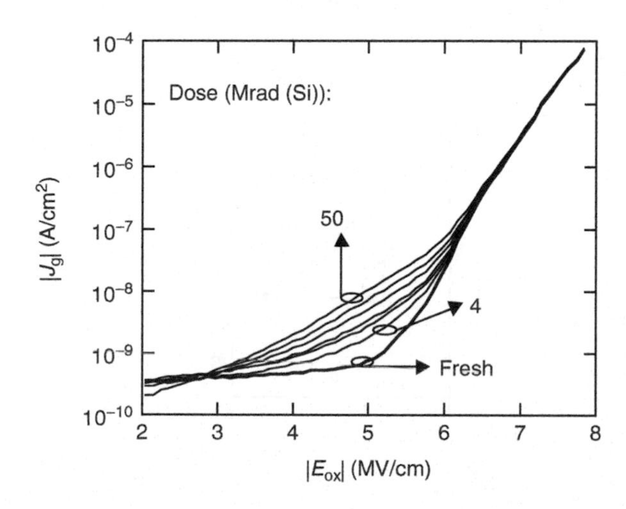

FIGURE 17.12
Negative $J_g - E_{OX}$ curves measured before (fresh) and after irradiation for various doses ranging from 4 to 50 Mrad (Si). (Reprinted from Ceschia, M. et al., *IEEE Trans. Nucl. Sci.*, 45, 2375, 1998. With permission.)

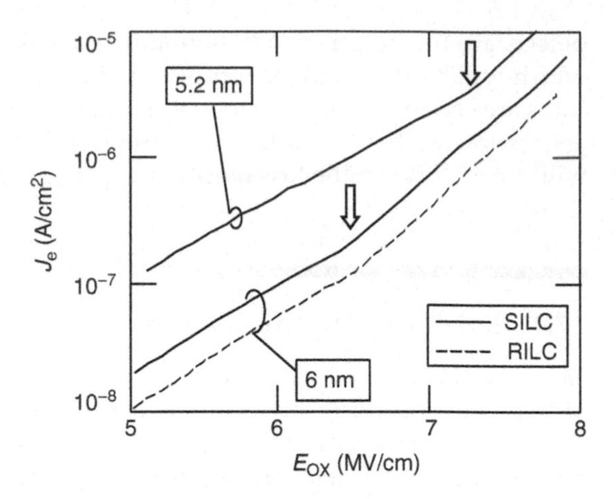

FIGURE 17.13
Excess current after radiation or electrical stress for two different oxide thickness (5.2 and 6 nm). (Reprinted from Ceschia, M. et al., *IEEE Trans. Nucl. Sci.*, 45, 2375, 1998. With permission.)

critical kink field E_k, which is characteristic of the oxide thickness. For a given oxide thickness, E_k is the same for both positive and negative excess currents, and increases when the oxide thickness is reduced. This kink appears to be associated with the modification of the barrier seen by the electron inside the oxide trap before being emitted toward the anode. The barrier is trapezoidal for $E_{OX} < E_k$, and becomes triangular for $E_{OX} > E_k$. This enhances the barrier transparency and the corresponding tunneling current, as shown in Figure 17.14 [98]. This allows us to exclude an elastic tunneling model for RILC. If no energy is lost by the tunneling electron during its transition through the trap, the current kink would occur when the oxide voltage drop equals the oxide barrier height at the cathode. Assuming a cathodic barrier height $\Phi_B \cong 3$ eV and an oxide thickness $T_{OX} = 6$ nm, the critical field is $E_k = \Phi_B/qT_{OX} \cong 5$ MV/cm. The experimental results (Figure 17.13) indicate that E_k is larger; for instance, $E_k \cong 6.4$–6.5 MV/cm, when $T_{OX} = 6$ nm for both RILC and SILC. Hence, an inelastic tunneling mechanism evidently is responsible for RILC (Figure 17.14) [29,106].

Based on this assumption, an analytical model of RILC has been developed for ultrathin oxides submitted to ionizing radiation, based on an analytical solution of the Schrödinger equation for a simplified oxide band structure [107]. Here RILC occurs through a two-step process: first, an electron tunnels into the oxide defect from the cathode conduction band edge. Then, the electron tunnels out of the trap after losing approximately 1.5 eV, in agreement with previous findings for SILC. Simulation results have shown that the most effective traps promoting RILC conduction are located close to the middle of the oxide and are energetically placed 1.3 eV below the oxide conduction band [107]. The RILC current density J_{RILC} can be then expressed as [107]:

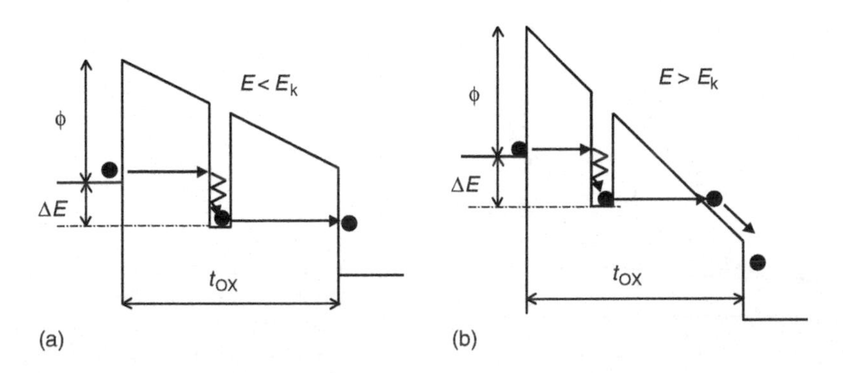

FIGURE 17.14
Inelastic TAT for (a) $E_{OX} < E_k$ and (b) $E_{OX} > E_k$.

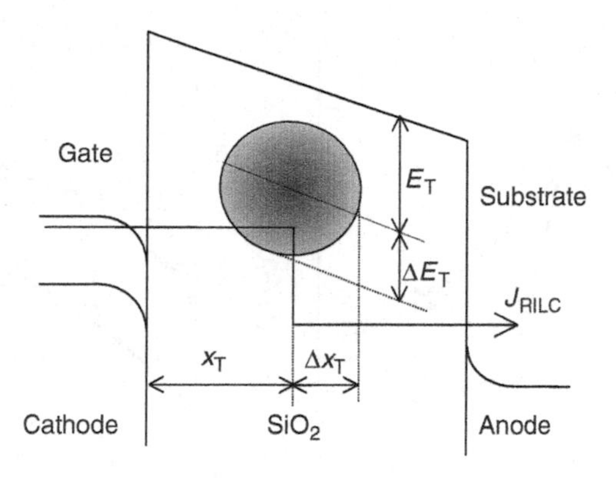

FIGURE 17.15
Schematic representation of the radiation-induced trap distribution in space and energy. x_T is the distance from the cathodic interface to the center of the trap distribution; E_T is the energy level of the center of the trap distribution; Δx_T and ΔE_T are the widths of the spatial and energetic Gaussian trap distributions, respectively. (Reprinted from Larcher, L., et al., *IEEE Trans. Nucl. Sci.*, 46, 1553, 1999. With permission.)

$$J_{RILC}(t_{OX},F_{OX},E_T) = J_{INC}\int_0^{t_{OX}} \frac{P_{IN}(x_T,F_{OX}) \times P_{OUT}(t_{OX} - x_T,F_{OX},E_T)}{P_{IN}(x_T,F_{OX}) + P_{OUT}(t_{OX} - x_T,F_{OX},E_T)} \times N_T(x_T,F_{OX}) \times \sigma \times \partial x_T \tag{17.6}$$

where (see also Figure 17.15)

P_{IN} and P_{OUT} are the tunneling probabilities into and out of the defects
N_T is the defect density
σ is the defect cross section
J_{INC} is the flux of electrons incident into the oxide
x_T is the defect position within the oxide

With appropriate fitting parameters, which depend on the oxide thickness, this model yields excellent fitting capabilities, as shown in Figure 17.16.

17.4.2 RILC Dependence on Applied Bias during Irradiation

The influence of bias on the distribution of neutral traps responsible for RILC is an important issue to assess device reliability. In Figure 17.17a ($T_{OX} = 4$ nm) and Figure 17.17b ($T_{OX} = 6$ nm)

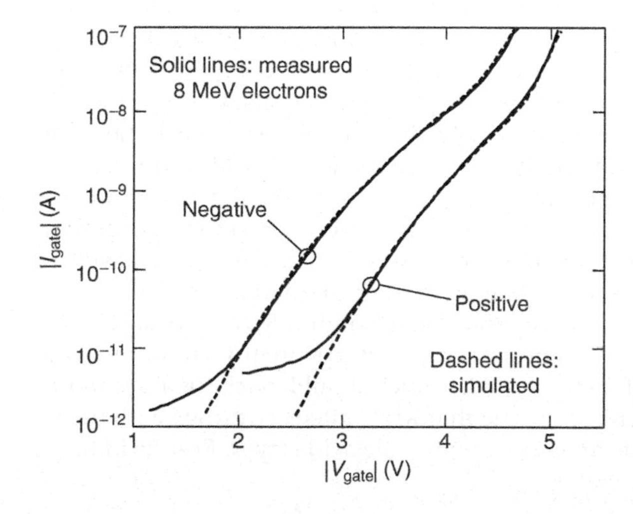

FIGURE 17.16
Comparison between simulation and measured RILC curves. (Reprinted from Blochl, P.E. and Stathis, J.H., *Phys. Rev. Lett.*, 83, 372, 1999. With permission.)

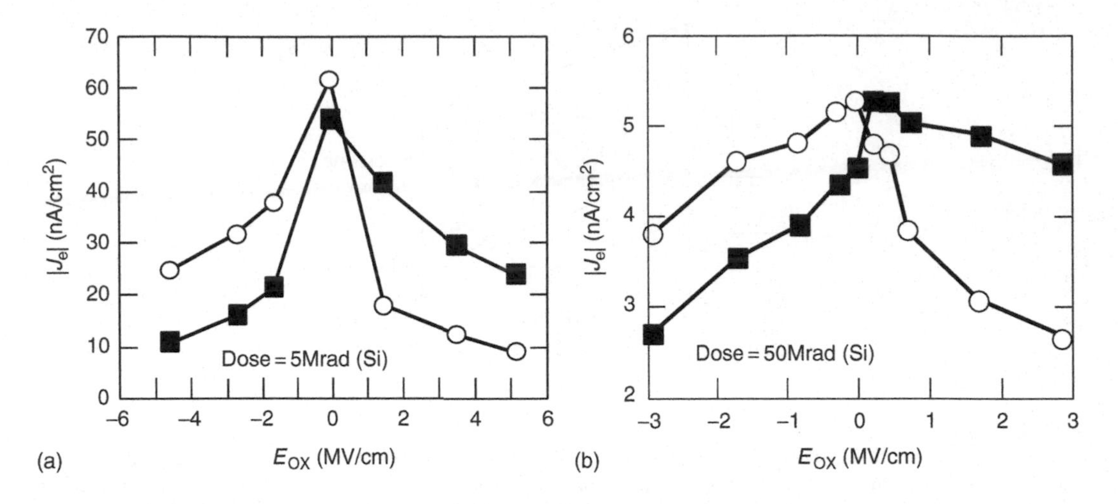

FIGURE 17.17
Excess current read at $|E_{OX}| = 5$ MV/cm for different gate voltages applied during 8 MeV electron irradiation for (a) $T_{OX} = 4$ nm and (b) $T_{OX} = 6$ nm (○, positive RILC; ■, negative RILC). (Reprinted from Ceschia, M. et al., *IEEE Trans. Nucl. Sci.*, 45, 2375, 1998. With permission.)

RILC is shown as a function of the oxide field applied during irradiation [98]. The 6 nm capacitors of Figure 17.17b have been subjected to a radiation dose 10 times higher than the 4 nm devices of Figure 17.17a, to enhance the measured current and the corresponding experimental precision. The obtained results are similar on both device types; positive RILC is higher than negative RILC in devices negatively biased during irradiation, and the opposite holds true for devices positively biased when irradiated. This difference disappears and RILC is at its maximum when the oxide field is close to zero during irradiation, while RILC decreases for increasing oxide field applied during irradiation.

These results show that, in agreement with the results for damage on thicker oxides, the trap distribution is controlled by the oxide field during irradiation. A strong correspondence between the generation of an oxygen-deficient, silicon-dangling-bond defect in the oxide and the appearance of oxide leakage current has been demonstrated [108]. Also, the generation of E' centers, as detected by the electron spin resonance technique, is accompanied by the appearance of oxide leakage current. Similarly, after a thermal annealing cycle at 200°C, both the E' center density and the oxide leakage current decrease (see Figure 17.18). Neutral defects could result from the neutralization of an E' center after hole capture [109]. A neutral electron trap could result from hole capture at a weak Si–Si covalent bond, which may relax into a Si^{-}:Si^{+} neutral amphoteric defect after the hole compensation. Even though holes appear as the first candidate for generating neutral defects, the possible role of hydrogen cannot be neglected [44]. In fact, ionizing radiation can easily break the weak bonds of bonded H atoms, which can either migrate under an electric field or diffuse and be trapped in different sites, possibly generating neutral defects in the oxide and/or leaving electron traps in the original site. A model based on a two-step generation mechanism linked to hydrogen migration and bond breaking [102–104] works well to describe the complex kinetics of interface trap generation after irradiation [99–101].

Returning to Figure 17.17, it may be surprising that the maximum RILC is observed for null field, when the charge yield is at a minimum [110]. This apparently contradicts the claim that neutral traps are a product of trapped holes, which should reach their minimum density at zero oxide field. However, while it is true that RILC intensity grows with defect density, it is also depends strongly on the defect position in the oxide layer. Results indicate

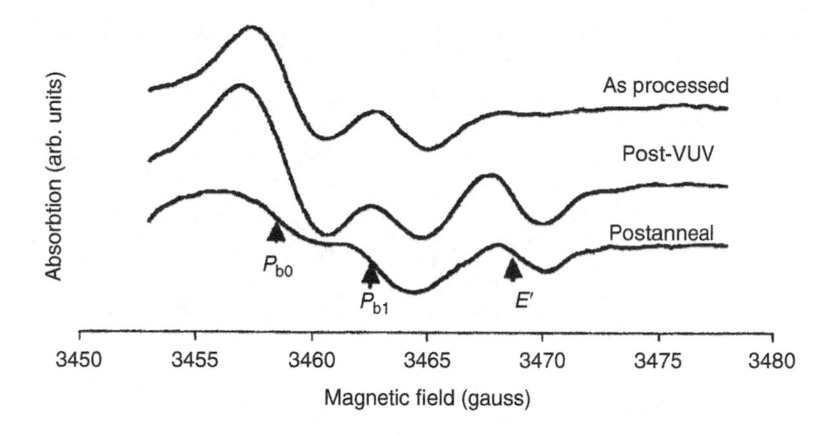

FIGURE 17.18
Electron spin resonance traces of As processed, post-VUV irradiation, and post 20 min annealing at 200°C taken on a 4 nm oxide film. (Reprinted from Lenahan, P.M. et al., *39th IEEE International Reliability Physics Symposium*, Orlando, 2001, p. 150. With permission.)

that RILC is a maximum for a homogeneous distribution of traps (obtained after zero-field irradiation), and decreases when trapped charge is located preferentially near one interface ($E_{OX} \neq 0$), even though the trap density is likely to be much higher in the latter case. The thermalization distance of an e–h pair is around 5–10 nm [79], which is comparable to the thickness of oxides we are discussing. Hence, there is still some net hole trapping during zero bias irradiation. On the other side, when an electric field is applied, it displaces holes surviving recombination, hence changing the distribution of defects in the oxide. As a consequence, differences between positive and negative RILC result from this asymmetric trap distribution, indicating that traps are less effective when close to the cathodic interface during measurement, in agreement with simulations that identify traps close to the middle of the oxide thickness as the most effective for RILC [107].

17.4.3 Radiation versus Electrical Stress

Radiation stress can be compared with constant current stress (CCS), as shown in Figure 17.19a and b for 4 nm and for 6 nm devices, respectively, as a function of the stress current. The FN stress current is an exponential function of the oxide field, and it has been reported on the *X*-axis. SILC is at its minimum for the lowest stress electric field; in fact, the energy of the injected electrons grows with the oxide field, and so does the oxide damage. When a negative CCS is performed (negative oxide field), the negative SILC is smaller than the positive one, and the opposite trend is observed for positive CCS. However, for positive CCS the negative–positive SILC difference is much larger than for negative CCS. This represents a striking difference with RILC, and it is probably related to the presence of an asymmetrically nitrided oxide. In fact, the nitrided layer has been proposed to act as a diffusion barrier against the diffusion of holes or hydrogen [43,44]. Hence, during positive stress, the positive species are generated at the gate/oxide interface by the high-energy FN electrons, and reinjected into the oxide creating traps across all the oxide, but more efficiently in the nitrided part of the oxide. Hence, negative SILC is higher than positive SILC, since traps accumulate toward the substrate. On the contrary, during negative CCS, holes (injected from the substrate) are easily blocked by the nitrided layer close to the injecting interface (i.e., in a position quite ineffective to produce SILC), with a reduced trap generation in the residual oxide thickness, thus producing lower SILC.

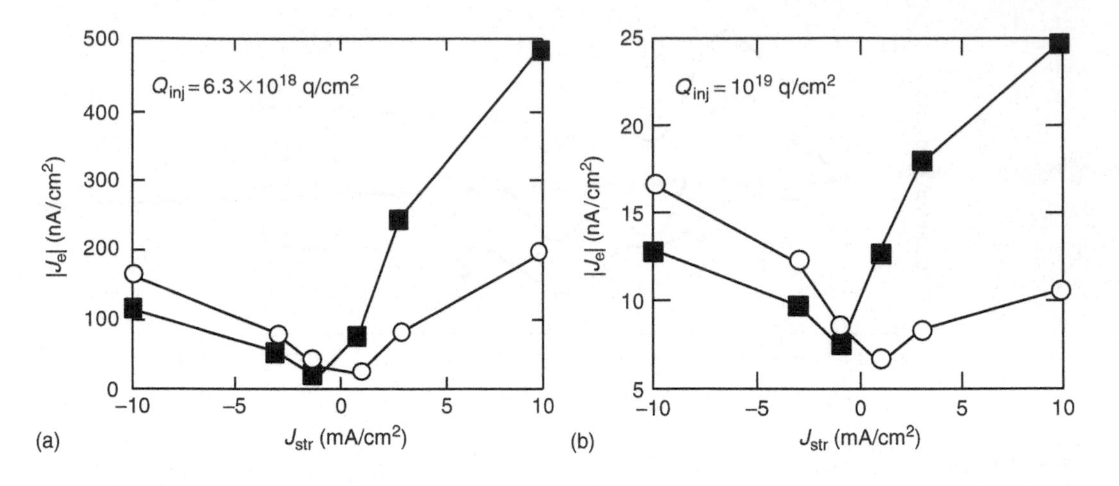

FIGURE 17.19
Excess current read at $|E_{OX}| = 5$ MV/cm as a function of constant current stress current density, for (a) $T_{OX} = 4$ nm and (b) $T_{OX} = 6$ nm (○ = positive SILC; ■ = negative SILC). (Reprinted from Ceschia, M. et al., *IEEE Trans. Nucl. Sci.*, 45, 2375, 1998. With permission.)

Further information about the differences between RILC and SILC can be obtained by looking at the growth kinetics of the two phenomena. Figure 17.20 compares the RILC growth with the accumulated total dose (Figure 17.20a) to SILC growth with the injected current. RILC has been measured at an oxide field $E_{OX} = 6$ MV/cm on both 4 and 6 nm oxides. For a given radiation dose, the excess current is much higher in the 4 nm oxides, owing to the higher tunneling probability across the barrier, which is thinner than in the 6 nm oxide. RILC data can be well fitted by using the following relation:

$$J_e = K_R \cdot \mathrm{Dose}^{\beta} \tag{17.7}$$

Here $\beta \cong 0.9$ for $T_{OX} = 6$ nm and $\beta \cong 0.94$ for $T_{OX} = 4$ nm, while K_R is a constant that depends on the read-out gate voltage and oxide thickness. RILC grows approximately

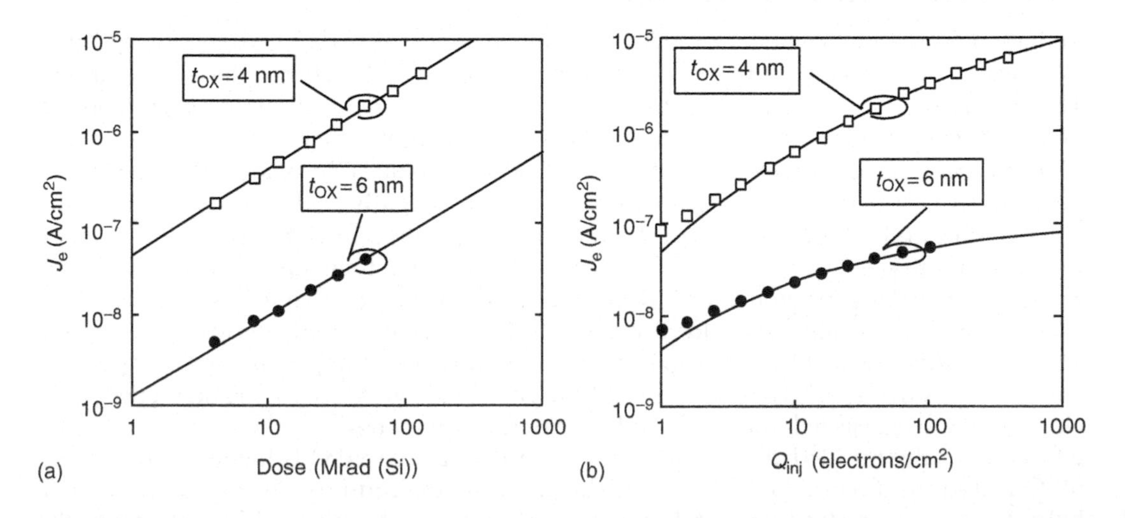

FIGURE 17.20
RILC (a) and SILC (b) kinetics (negative excess current) for $T_{OX} = 4$ nm (□) and $T_{OX} = 6$ nm (●). $V_g = 0$ V during 8 MeV electron irradiation. (Reprinted from Ceschia, M. et al., *IEEE Trans. Nucl. Sci.*, 45, 2375, 1998. With permission.)

linearly with the radiation dose, indicating that the oxide defect density follows a linear growth rate as well.

In Figure 17.20b, SILC is plotted versus the injected charge (Q_{inj}) during CCS, with $J_{str} = -10$ mA/cm^2 for both 4 and 6 nm devices. At least at high stress levels, SILC follows the empirical law presented in Section 17.3 [40]:

$$J_e = J_{sat} \cdot \exp\left(-\frac{DJ_{stress}^{\gamma}}{Q_{inj}^{\alpha}}\right) \qquad (17.8)$$

where

$\alpha \cong 0.31$ for $T_{OX} = 6$ nm
$\alpha \cong 0.18$ for $T_{OX} = 4$ nm
$\gamma \cong 0.04$
J_{sat} is the saturation SILC level
D is a constant depending on the oxide thickness and quality

The most important difference is that SILC shows a saturating behavior for current levels where RILC is still linearly growing. Moreover, α is lower in 4 nm than in 6 nm oxides, indicating that SILC saturates faster in thicker oxides. This basic difference between RILC and SILC may be related on one side to the defect distribution within the oxide layer, which results from uniform generation of holes across the oxide (RILC) or high-field injection from the anode (SILC). On the other hand, many more electrons are injected by electrical stress than by radiation for the same gate leakage level, suggesting that electrons may effectively passivate and/or anneal part of the stress-induced defects for electrically stressed devices.

17.5 RILC in Tunnel Oxides

To study the impact of a single, high-LET ion on the tunnel oxide of the FG it impacts, the first step is to identify all the FGs hit after irradiating an array, and them alone [111]. Luckily, this is rather easy if one has access to the threshold voltage V_{TH} of the single FG, instead of working on the digital ("0" or "1") output only [112,113]. In fact, in this case (and if the ion LET is large enough [114,115]), the hit FGs can be identified since they (and they alone) experience a large threshold voltage shift ΔV_{TH}. In Figure 17.21, an FG array with all FGs programmed at "1" has been irradiated with 2×10^7 I ions/cm^2 [112]. Each square represents a single cell featuring a gray level related to the V_{TH} shift (ΔV_{TH}); the darker the square, the larger the threshold variation, which is always negative (V_{TH} decreases because electrons are lost from the FG). Hit cells appear randomly distributed across the chip surface, but their number is in good agreement with the number expected based on purely geometrical considerations [112,116]. Actually, in most recent technologies even cells not directly hit can experience a nonnegligible charge loss [116]; however, the two populations ("hit" FGs and "not hit but still leaking charge" FGs) are clearly separated one from the other [116]. Interestingly, the charge loss following the ion hit is by far larger than that expected based on simple models, and depends on ion LET, electric field, and technology [112,113,115]. However, these aspects will be neglected here, since we will focus on what is going on after the ion strike.

Figure 17.21 allows one to easily identify hit cells, for example by selecting those whose $|\Delta V_{TH}|$ exceeds 50 mV. Subsequently, the whole array can be erased, programmed again at

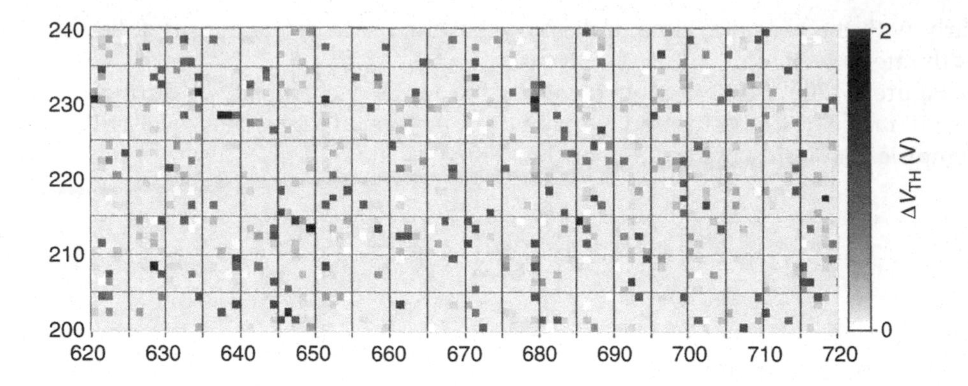

FIGURE 17.21
Spatial distribution of threshold voltage for a small subset of cells from a chip after 2×10^7 iodine ions/cm^2. The gray scale (right) indicates the amount of ΔV_{TH}; the two cells indicated by arrows have $\Delta V_{TH} > 2$ V. Each square represents an FG. (Reprinted from Cellere, G. et al., *IEEE Trans. Nucl. Sci.*, 48, 2222, 2001. With permission.)

high V_{TH}, and then the V_{TH} evolution can be monitored. For FG cells that did not experience V_{TH} shift after irradiation (i.e., cells not hit by ions), V_{TH} remained constant during that period, as expected for nonvolatile memories. For comparison, results for the "hit" cells of a device irradiated with I ions are shown in Figure 17.22. The "fresh" (i.e., before irradiation) distribution is very steep and centered around 7 V. The distribution after irradiation is much gentler and features a leftward shift with respect to the fresh one, since the data refer to the set of hit cells only. Then, after programming the memory array, an unexpected tail appears, extending down to almost 5 V. This tail is due to the time (half an hour) elapsed between the program and read operations, a time that is long enough for the most damaged FGs to get appreciably discharged. From this moment on, the tail always increases, and the worst bits eventually join the "after rad" distribution 481 h

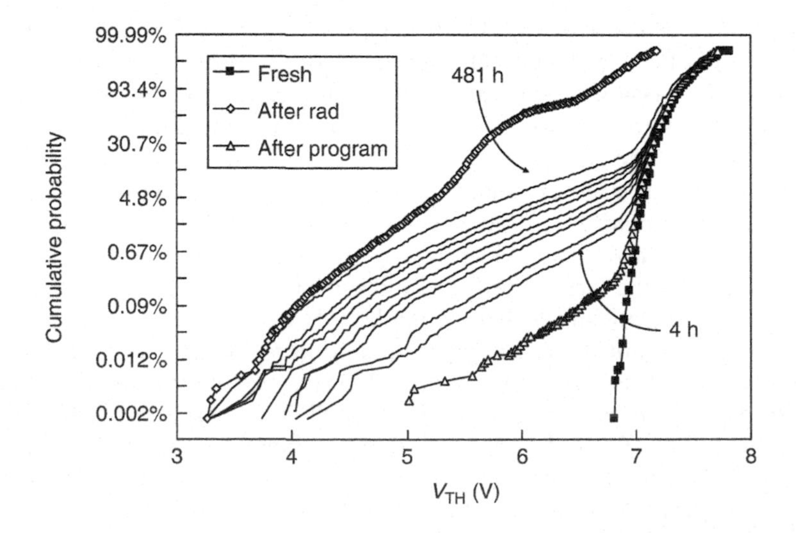

FIGURE 17.22
Cumulative probability plots of threshold voltages for hit cells only of an FG array irradiated with 2×10^7 I ions/cm^2. Curves indicated with thin lines were taken 4, 7, 21, 30, 46, 81, 126, 191, and 481 h after the program operation. (Reprinted from Larcher, L. et al., *IEEE Trans. Nucl. Sci.*, 50, 2176, 2003. With permission.)

after the programming operation. After 3 weeks, the worst bits of the lot reach the same threshold voltage measured immediately after irradiation. This discharge is due to a gate leakage current through the FG oxide that is too small to be detected by any MOSFET measurements, but is large enough to effectively discharge an FG. Only a few thousand electrons are stored in a programmed FG, so a leakage current of just 1 fA would completely discharge the FG in about 1 s, while we detect a charge loss of 100–1000 electrons over several days. This sub-fA current flows through a path created by the impinging ion, which generates a high-density electron–hole column in SiO_2 about 4 nm in radius [94]. Similar results were obtained after irradiation with other ion beams, but the extent of the tail decreases with ion LET. While this is clearly a different manifestation of the RILC described for ultrathin oxides in the previous section, there are noticeable differences that will emerge in the following.

17.5.1 Physical Characteristics of the RILC Path

The physical nature of the oxide defects involved in the discharge path, and their close affinity with electrical defects, such as those responsible for SILC or RILC in ultrathin oxides, can be studied by performing a forming gas anneal (FGA) after irradiation and before the electron retention experiment. In fact, the neutral traps responsible for phenomena such as SILC and RILC are known to anneal out at high temperatures [51,52,117–120]. Only one of the two devices reported in Figure 17.23 was subjected to FGA at 250°C for 24 h. After this treatment, it has been programmed again, and the V_{TH} evolution has been monitored. No V_{TH} shift can be detected in this device 72 h after programming, whereas a large tail is present in the unannealed device after only 48 h. The direct comparison between data for the annealed and nonannealed device in Figure 17.23 clearly demonstrates that the oxide defects are (at least partially) annealed, and consequently the multi-TAT path across the tunnel oxide is switched off by the thermal treatment.

By programming the FG arrays to different average V_{TH} values before irradiation, it is possible to evaluate the impact of electric field on the generation of defects. Quite surprisingly, if devices programmed in these conditions are irradiated with the same ion beam at the same fluence, and then programmed again, the RILC-induced V_{TH} shift is the same [121,122]. In other words, it appears that the generation of defects is not linked to the electric field in the tunnel oxide. This allows us to draw some interesting considerations on the actual mechanisms leading to oxide damage. The rapid current transient that discharges the FG in the first 10 fs after the ion impact [114,115] cannot be the cause of the

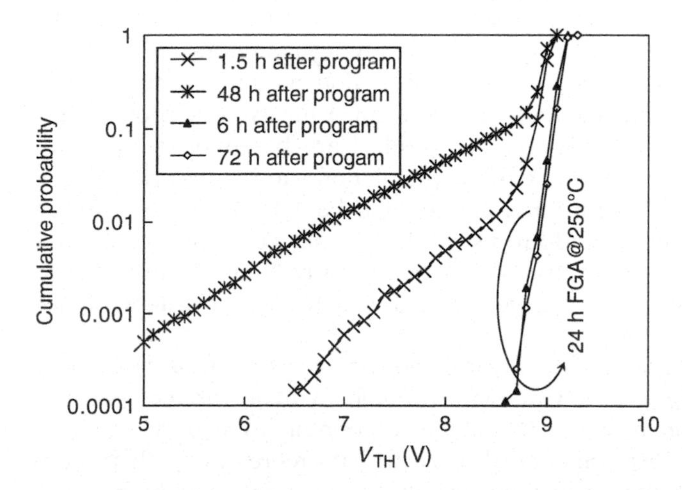

FIGURE 17.23
Cumulative probability of V_{TH} for two devices: the first was programmed after being irradiated, the second was at first subjected to forming gas anneal, and then programmed. (Reprinted from Larcher, L. et al., *IEEE Trans. Nucl. Sci.*, 50, 2176, 2003. With permission.)

oxide damage, since it does not flow when the electric field is close to zero. More interestingly, the oxide damage appears not to be generated by the motion of the holes surviving recombination. In fact, the charge yield after the impact of an ion with such a large LET as in this case is well below 1% even at relatively high E_{OX} [94]. An ion with LET = 85 MeVcm^2/mg generates about 10,000 holes in the tunnel oxide, of which fewer than 100 will survive recombination at high E_{OX}. This number will decrease more and more with smaller E_{OX}; for example, if E_{OX} reduces from 2 to 0.1 MV/cm, the charge yield decreases by roughly an order of magnitude [94], and virtually no hole survives recombination if $E_{OX} = 0$ MV/cm. If the oxide defects were generated by these holes, which survive recombination and move toward the FG, one should find a lower number of defects for the low-field irradiation case, which is not true, since RILC appears to be independent of the electric field during irradiation. On the contrary, for high-LET ions the percentage of recombined holes ranges from 99% at high electric fields to 100% at zero electric field—a difference too small to be detected.

According to the majority of reports in the literature [79,86,123], positive charges (hydrogen [103,123] or holes [45,79,86]) moving in the oxide are responsible for oxide damage. However, these results do no contradict previous reports, for at least two reasons. First, the FG memory cell is a very specific case, and it is well known that recombination in very thin layers does not follow the usual laws found in thicker oxides [105]. Second, a few reports are available on the evaluation of actual charge yield for heavy ions in SiO_2. For example, in Ref. [94] charge yield is calculated based on the assumption that the ion direction and the electric field are orthogonal; when the angle approaches 0°, the number of holes surviving recombination mainly depends on diffusion processes and only weakly depends on the electric field [94]. Since in the FG case the electric field is parallel to the ion direction, one may suppose that the percentage of holes surviving recombination under normal incidence in thin SiO_2 layers is a small but constant number, and these holes are the precursors for oxide damage. Unfortunately, while this can explain the independence of RILC on electric field on the high-to-moderate E_{OX} range, it appears very unlikely that the same number of holes will survive recombination when E_{OX} is close-to-zero, in a 10 nm oxide layer.

17.5.2 Model

The RILC model previously discussed can be improved to simulate the much more complex situation of a thicker oxide [121,125,126]. In this case, electrons are coupled to oxide phonons (PTAT, phonon trap-assisted tunneling), which results in a series of virtual states in the oxide energy gap broadening the trap energy level, E_T. In other words, electron transport between two adjacent traps may occur involving a phonon, and accounting for the energy and momentum difference. This conduction model can supply the current through the oxide driven by conductive paths comprising two or more traps, i.e., percolation paths, whereas most common TAT models are based on single trap [36,54,55,58,127] or 2TAT [128,129]. Also, the leakage calculation can be extended to the case of positively charged traps with a simplified mathematical apparatus. Once coupled with a statistical simulator generating a random distribution of defects in the considered volume (Figure 17.24), and by repeating the simulations many times (Monte Carlo approach), this model can be used to derive the gate current versus oxide field dependence shown in Figure 17.25.

Here, defects are generated considering a cylindrical coordinate system (r, θ, z); θ and z are uniformly distributed, whereas r is a Gaussian variable, with a variance ~4 nm [94,130,131]. A defect cross section $\sigma_T = 10^{-14}$ cm^2, corresponding to neutral traps, is considered. Symbols correspond to experimental data, with bars representing their standard deviations. From simulations, the current mean values, mean + σ (maximum), and

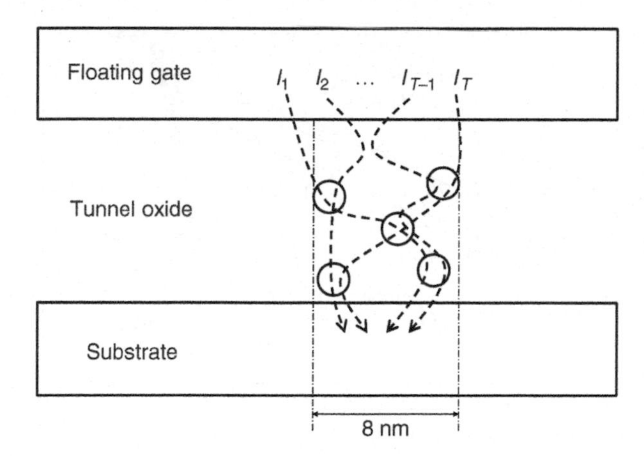

FIGURE 17.24
Schematic representation of the model: first defects are randomly distributed across the oxide (along the ion track), and then contributions by all T possible paths are calculated and finally added to determine the actual current. (Reprinted from Larcher, L. et al., *IEEE Trans. Nucl. Sci.*, 50, 2176, 2003. With permission.)

mean $- \sigma$ (minimum) have been obtained. As clearly shown in this figure, experimental currents are reproduced with high accuracy by the mean leakage current distribution calculated by averaging 10^4 simulation trials. Furthermore, the simulated standard deviation accounts well for statistical variations measured among different samples. Note that, as mentioned above, the currents involved are extremely small, down to 10^{-24} A. Obviously, such a small current, corresponding to about 1 electron/day, cannot be directly measured, no matter the precision of the instrumentation used. However, these devices store just a few thousand excess electrons, and the loss of hundreds of them can be clearly detected.

When coupled to an accurate electrostatic model of the cell, the current–electric field relationship can be translated into a V_{TH}–time relationship by using [132]:

$$I_{PTAT} = C_T \frac{dV_{FG}}{dt} \tag{17.9}$$

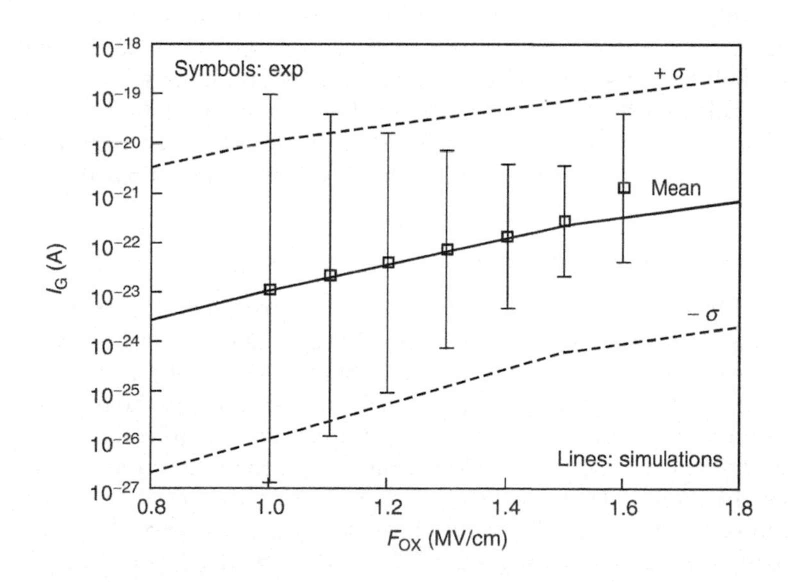

FIGURE 17.25
Average value of the experimental current and its standard deviation (symbols and bar), compared to simulated data. 10,000 simulations were carried out, with 20 traps ($\sigma_T = 10^{-14}$ cm^2) distributed along the ion track (see text for details). (Reprinted from Larcher, L. et al., *IEEE Trans. Nucl. Sci.*, 50, 2176, 2003. With permission.)

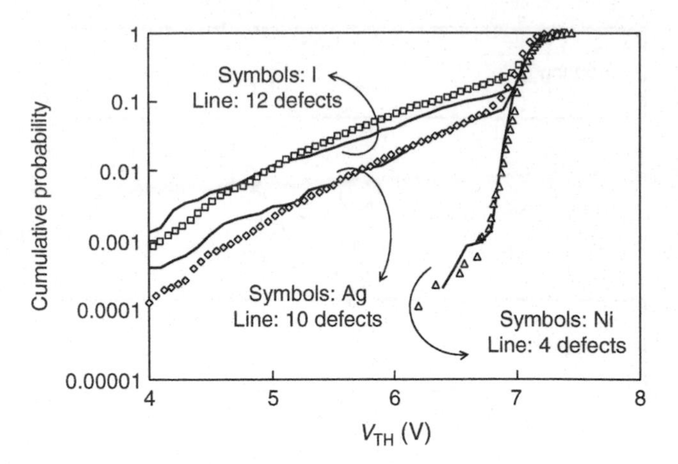

FIGURE 17.26
Cumulative probability of V_{TH} for *T2* FGs irradiated with different ions, 529 h after reprogramming: simulated (lines) versus experimental data (symbols). (Reprinted from Cellere, G. et al., *IEEE Trans. Nucl. Sci.*, 52, 2144, 2005. With permission.)

where I_{PTAT} is the percolation path current, C_T is the total capacitance between the FG and all other terminals (substrate, drain, source, and control gate), and V_{FG} is the floating gate voltage. The first step is to choose the number of defects involved in the path and the statistical model of their spatial and energetic distributions. Then, one can start to form an actual distribution of V_{TH}. For each hit FG it is now possible to generate a distribution of oxide defects, and to calculate its V_{TH} after a given retention time. When repeated on all hit FGs (whose number is known based on our previous discussion), this yields the distribution of V_{TH} after the chosen retention time [121]. Noticeably, the only free parameter in this process is the number of defects in the path. Some results are shown in Figure 17.26, where the good fitting capabilities of the approach can be seen for devices irradiated with different ions. The number of defects in the percolation path becomes clearly smaller for lower LET ions, as expected, since both the number of carriers that recombine and the number of holes surviving recombination (possibly leading to the subsequent creation of defects) decrease with LET. Note that the relatively large number of defects involved in these paths, if compared with the two or three defects involved in most SILC models [49,130], is reasonable, since the traps we are dealing with are randomly generated inside a given volume. Therefore, they are not perfectly aligned and are not disposed in positions that maximize the tunneling probability, as usually done when simulating anomalous SILC [58]. Also, this is not in contrast with the RILC measured after irradiation with high fluences of low-LET ions [98,133], since in that case the overall current derives from the superimposition of a large number of small contributions, each impossible to be individually measured.

17.5.3 RILC in Thin versus Thick Oxides

To assess the thickness scaling of the model, let us consider, for example, the results published in Ref. [134] and reported in Figure 17.26. Here the authors show the excess RILC current after irradiation at the same total dose (10 Mrad (Si)), measured at 6 MV/cm on 10^{-2} cm^2 capacitors with a 4 nm oxide. We will at first consider the point corresponding to irradiation with silver ions. Discussion is summarized in Table 17.1, where the upper part refers to data in Ref. [134] and the lower part refers to the discussion in Ref. [121]. The silver ions used in Ref. [134] have LET = 52.9 MeV cm^2/mg in silicon, leading to about 1.2×10^8 Ag ions that hit the device (bold). If Ag ions generate on average $N_{10} = 10$ defects in a 10 nm oxide, they generate on average $N_4 = 4$ defects in a 4 nm oxide. The model has been used to calculate how many paths, each comprising four randomly generated defects, which should be considered to obtain the 0.6 μA excess current corresponding to Ag. This number (3.3×10^8) is reported in bold at the end of the table, and is about a factor of

TABLE 17.1

Experimental versus Simulation Results

	Ag	Si	
LET in Si ($MeVcm^2/mg$)	52.9	9.3	
TID (Mrad (Si))	10	10	
Ion fluence ($ions/cm^2$)	1.2×10^{10}	7.2×10^{10}	
Ions impacting on oxide	$\mathbf{1.2 \times 10^8}$	$\mathbf{7.2 \times 10^8}$	
LET in SiO_2 ($MeVcm^2/mg$)	57.3	9.9	9.9
Fitting function	—	Linear	Power law
Number of defects generated in 10 nm, N_{10}	10	1.7	1
Number of defects generated in 4 nm, N_4	4	0.7	0.4
Excess current, I_{ex}	0.6 μA	5 nA	5 nA
Number of paths responsible for excess current, N_{ex}	3.3×10^8	10^8	10^8
Number of ions responsible for excess current, N_{ions}	$\mathbf{3.3 \times 10^8}$	$\mathbf{1.4 \times 10^8}$	$\mathbf{2.5 \times 10^8}$

Source: Reprinted from Cellere, G. et al., *IEEE Trans. Nucl. Sci.*, 52, 2144, 2005. With permission.

2 larger than the previously calculated number of Ag ions hitting the capacitor. Given the large uncertainties in all parameters, the tiny currents, and the large number of events we are dealing with, we consider this to be good agreement.

What happens when we scale to lower LET ions, which are by far more common in space? In relatively thick oxides RILC is evident only for high-LET ions, so that no data on RILC have been measured up to now for silicon ions [121]. However, the number of defects in the path as a function of LET can be approximated with both a linear relationship and a power law (Figure 17.28), with a slight preference toward the second because of the better correlation (R^2) coefficient. Based on these relationships, it is possible to calculate the number of defects generated in a 10 nm oxide by a silicon ion as $N_{10} = 1.7$ by using a linear fit, and as $N_{10} = 1$ by using a power law fit. Therefore, Si ions generate about $N_4 = 0.7$ (using the linear fit) or $N_4 = 0.4$ (using the power law fit) defects in a 4 nm oxide layer. $N_4 < 1$ means that not all impinging ions actually generate defects, and that, at most, an ion generates a single defect in the oxide. Therefore, the excess current is now the result of an unknown number of defects, N_{ex}, randomly generated in the oxide (i.e., defects are not generated in the mid-oxide position that maximizes tunneling current). About 10^8 defects are needed to carry the 5 nA excess current of Figure 17.27 after Si irradiation. The number of ions needed to generate these defects is given by $N_{ions} = N_{ex}/N_4$, and varies between 1.4×10^8 ions (linear-fit extrapolation) and 2.5×10^8 (power-law extrapolation). In Ref. [134] the device was hit with 7.2×10^8 ions to deliver 10 Mrad (Si). Therefore, even in this case, results from simulations are in good agreement with the calculated number of ions impinging on the capacitor, with a slight preference (once again) for the power-law extrapolation. In short, what all these calculations demonstrate is that the same model developed to describe results on 0.01 μm^2 oxides that are 10 nm thick is consistent with results obtained on a 10^{-2} cm^2 (10^6 μm^2) oxide that is 3 nm thick. As a consequence, the model predicts, for example, how silicon ions are not expected to cause major RILC problems in 10 nm oxides, and how the cumulative damage due to multiple hits in adjacent positions may be responsible for the damage due to ion implantation [135], where extremely high fluences, well above 10^{12} $ions/cm^2$, are used.

17.5.4 Erratic Behavior of RILC

The easiest way to study the sometimes erratic behavior of RILC is simply to repeat more than once the retention experiments described so far. For example, in Figure 17.29 this has

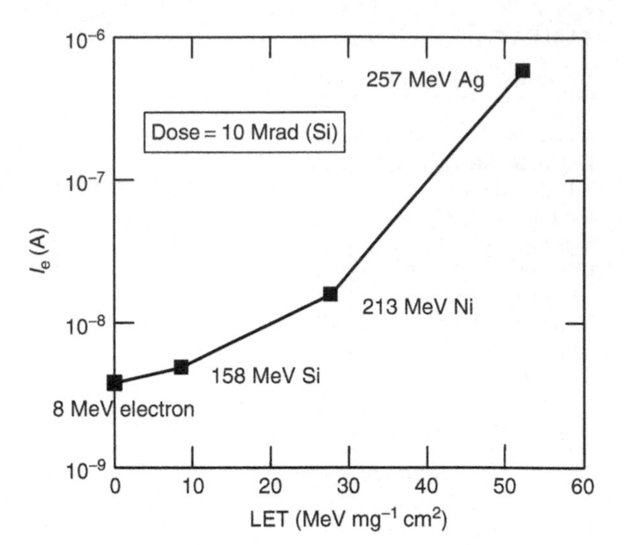

FIGURE 17.27
Excess current (RILC) as a function of ion LET, for capacitors with 4 nm thick oxide and 10^{-2} cm^2 area, measured at $E_{OX} = 6$ MV/cm. (Reprinted from Ceschia, M. et al., *IEEE Trans. Nucl. Sci.*, 3, 566, 2000. With permission.)

been done three times. Resulting distributions are very well described by the superposition of the main distribution where the majority of FGs lie, and a tail that basically hosts a fraction of hit FGs. The lack of any difference between the tails corresponding to the different retention experiments implies that the retention characteristics of the arrays as a whole do not change among different retention experiments. However, it is very interesting to study what happens during retention experiments when looking at a single FG instead of a distribution, possibly after many retention experiments (instead of just three). While possible in principle, it is also very time consuming. However, by applying a negative voltage to the control gate CG (this is the so-called "gate stress condition"), one can obtain similar results in a far shorter time. Also, a larger fluence helps by providing more hit FGs on the same device. In Figure 17.30 the V_{TH} of some selected FGs are shown during such experiments. Figure 17.30a shows the V_{TH} of four selected FGs right after programming, as a function of the experiment number (called "run #" in the X-axes). Only very small changes are seen, meaning that the E_{OX} was almost constant during all the

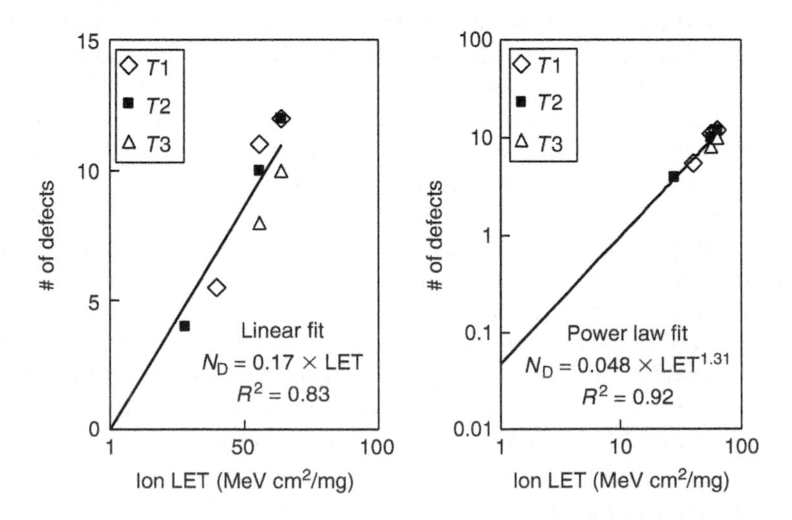

FIGURE 17.28
Number of oxide defects as a function of ion LET (symbols) and linear-(left) and power-law (right) fit. (Reprinted from Cellere, G. et al., *IEEE Trans. Nucl. Sci.*, 52, 2144, 2005. With permission.)

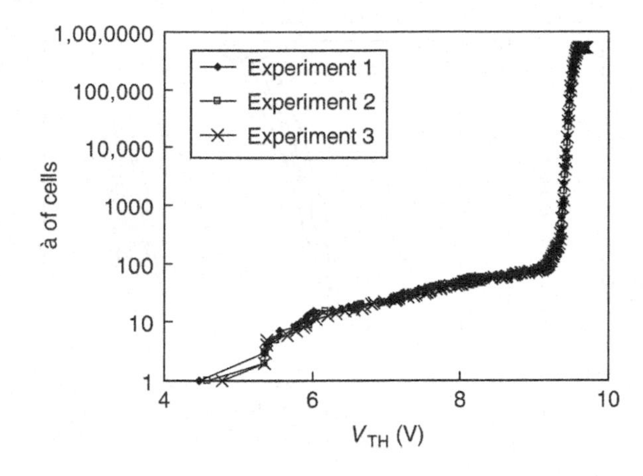

FIGURE 17.29
V_{TH} cumulative distribution for a sector irradiated with Au ions, after three retention experiments, each 166 h long. The sector was programmed at "low" V_{TH} before irradiation; holes stored in the FG. (Reprinted from Cellere, G. et al., *IEEE Trans. Nucl. Sci.*, 53, 3349, 2006. With permission.)

experiments. Such (very) small differences are due to small variations of the programming pulse due to the algorithm used. However, if one looks at V_{TH} after a gate stress of 10,000 s (Figure 17.30b), there are four different behaviors. The FG indicated with x's shows an almost constant V_{TH}. In contrast, the FG indicated with triangles has $V_{TH} \sim 7.2$ V for the first 10 experiments, and then moves to $V_{TH} \sim 8$ V for the remaining experiments.

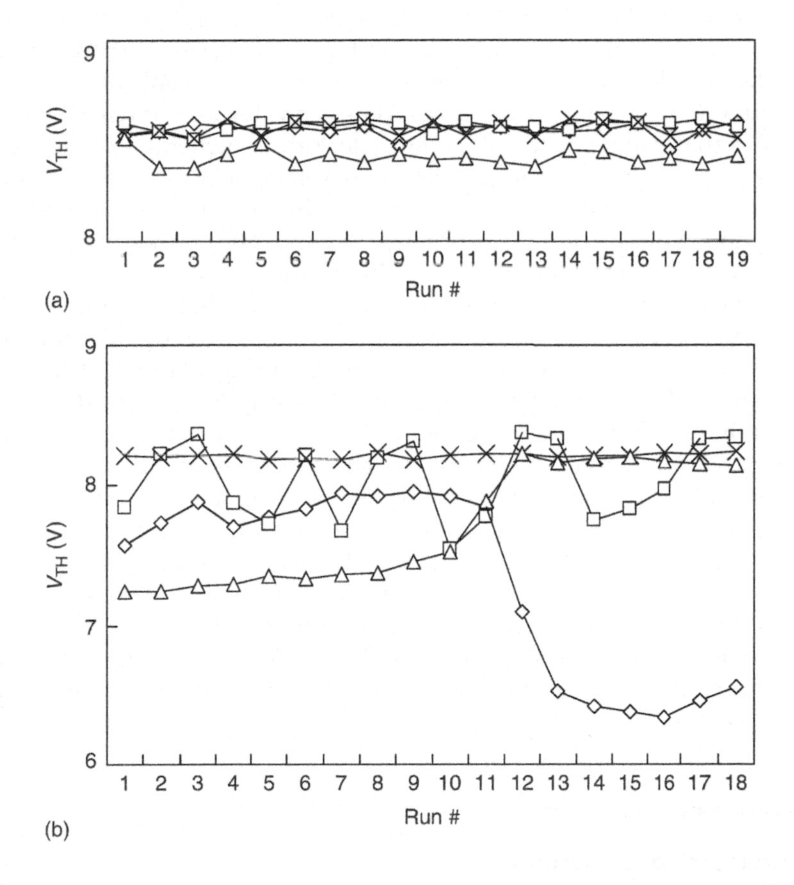

FIGURE 17.30
Evolution of V_{TH} measured as a function of the experimental run (i.e., the same experiment was repeated many times). (a) After programming, and (b) after a 10,000 s gate stress. Same symbols in the two graphs correspond to the same FG. (Reprinted from Cellere, G. et al., *IEEE Trans. Nucl. Sci.*, 53, 3349, 2006. With permission.)

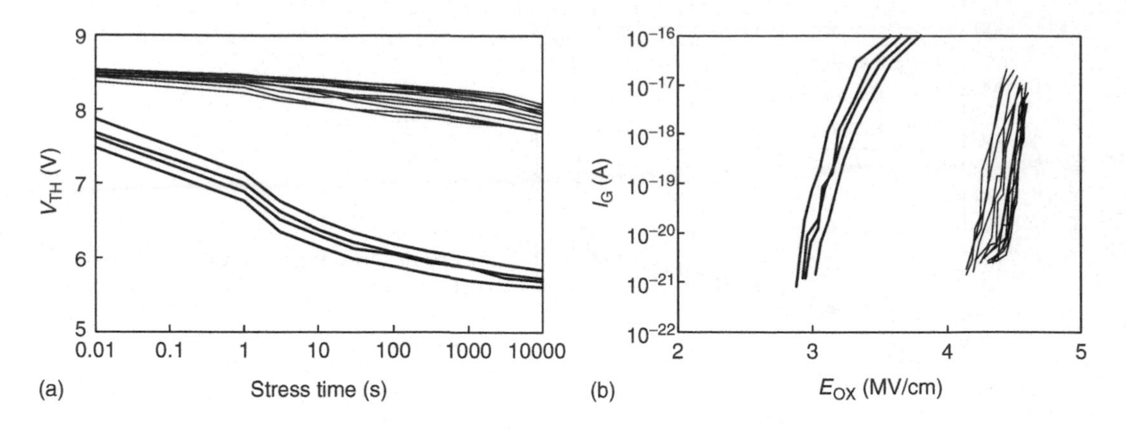

FIGURE 17.31
Evolution of V_{TH} as a function of time (a) and calculated gate current versus oxide electric field (b) for a selected FG. All curves are relative to different experimental runs on the same FG. (Reprinted from Cellere, G. et al., *IEEE Trans. Nucl. Sci.*, 53, 3349, 2006. With permission.)

This means that, between experiments 10 and 11, the percolation path has switched to a lower conductance, resulting in smaller V_{TH} degradation. The device shown with diamonds has the opposite behavior. From these data it is impossible to assess if the conduction states of the percolation paths after the last experiment are stable or metastable, that is, if the path conductivity will always be (around) its last value, or if it will change again. The FG shown with squares appears to follow this latter rule. In general, the change of the path conductance may be due to the change in occupancy state of border traps [136,137], whose (non)availability actively changes the electron tunneling probability. The occurrence probability of this phenomenon is hard to estimate, since the time constants of border traps are comprised between 1 μs and more than 1 s [138] (Chapter 7).

The random change in the path conductivity impacts the whole evolution of V_{TH} during each of the gate stresses, as can be seen in Figure 17.31. In Figure 17.31a, we show the evolution of V_{TH} as a function of time for the same FG, during all the repeated experiments. For this particular FG device, the path conductance abruptly changes during the experiment, resulting in the characteristics of Figure 17.31a, where it is clear a semipermanent change of the path conductivity has occurred. The corresponding $I_G - E_{OX}$ characteristics of Figure 17.31b show a change of several orders of magnitude in I_G for the same E_{OX} value, once the path switches on or off. This is reasonable since the current of a percolation path strongly decreases as soon as one of the defects involved in the path is no more available. Also, the characteristics are very steep, thus explaining the relatively small slope of the curves in Figure 17.30a; in fact, a steep slope in the I_G–E_{OX} curve means that, as soon as some charge leaves the FG and the electric field decreases, the current becomes much smaller. On the contrary, very steep V_{TH}-time characteristics would have led to very smooth I_G–E_{OX} curves.

17.6 Summary and Conclusions

Technological evolution, as synthesized by Moore's law, has led to thinner and thinner dielectric layers in advanced devices, which have reached the physical limits of a few atomic layers. While new dielectrics such as hafnium oxide, featuring a higher dielectric constant and lower direct tunnel current, are finally entering the market after having been discussed for years [11], SiO_2 is still and is expected to remain the workhorse of the industry for years to come.

In this review, we have shown the results of the loss of insulating properties that can be caused by the creation of defects due to electrical stress or ionizing radiation. We discussed two classes of oxides, namely, thin (<5 nm) oxides used as gate insulators and slightly thicker (~10 nm) oxides used as tunnel dielectrics in nonvolatile memories, focusing on SILC and RILC.

SILC is easily and widely studied in gate oxides, but for virtually all existing technologies today, it is not a primary reliability concern. In fact, SILC is not large enough to endanger the operation of the device. In addition, when oxides become thinner than 2–3 nm, direct tunneling current increases to a level that overwhelms the possible increase due to SILC. On the other hand, SILC is a serious problem in thicker tunnel oxides of nonvolatile memory devices, where it is smaller in magnitude and by far more difficult to measure, but extremely dangerous. In fact, the tiny amount of charge stored in modern nonvolatile memories makes them extremely sensitive even to very small leakage. This is exacerbated because manufacturers need to guarantee retention in a large number of devices per wafer, where each device can have billions of FG cells. We also have discussed how ionizing radiation can generate defects in the oxide, leading to a degradation of its insulating properties, known as RILC, which bears a close resemblance to SILC. RILC has been studied at first in ultrathin oxides and/or in large-area capacitors, since the current levels are too small to be measured in small devices with relatively thick oxides. With these kinds of oxides, RILC is very interesting, since the defects are generated at low-oxide field intensities during normal operating life. This is in contrast to accelerated stresses that use high electric fields to conduct SILC experiments in a reasonable time. While methods to translate the accelerated stress results into low field stresses do exist [139], this it is not always easy and the basic physics of defect generation can change depending, for example, on the gate voltage [123]. By using ionizing radiation, defects are generated in the operating conditions, i.e., at low fields, since the radiation dose rate and not the electric field the accelerating factor.

Once again small devices with relatively thick oxides (i.e., FG cells) are those where the reliability implications are more profound. While in these devices RILC and SILC are similar in many regards, they also have peculiar and distinctive features. SILC is effective only when $T_{OX} < 8$–9 nm, and it is almost totally suppressed for thicknesses of ~10 nm and is triggered by the device degradation during writing/erasing operations; for converse, RILC occurs for all thicknesses, provided that the ion LET is high enough, and depends on external factors such as the radiation environment. Defects responsible for SILC are distributed across the whole tunnel oxide volume, whereas those responsible for RILC are condensed in a tiny oxide volume; this could in principle lead to different erratic behavior. Finally, SILC is usually tested at high electric fields to accelerate the degradation, whereas, RILC may be observed at very low fields.

Acknowledgments

Authors are deeply indebted to the following contributors who have played an important role in this work and wish to gratefully acknowledge their help: A. Visconti, M. Bonanomi, S. Beltrami, and A. Modelli (STMicroelectronics, Agrate Brianza, Italy); L. Larcher (University of Modena and Reggio Emilia, Italy); A. Candelori (INFN, Padova, Italy); and A. Cester and M. Ceschia (DEI, University of Padova, Italy).

References

1. Richman, P. and Zloczower, W., P+π P+MOSFETS: A new approach to complementary integrated circuits, in *IEDM Tech. Digest*, Washington, 22, 1968.
2. Moore, G., Cramming more components into integrated circuits, *Electronics*, 38, 1965.
3. Dennard, R.H. et al., Design of ion-implanted MOSFET's with very small physical dimensions, *IEEE J. Solid State Circ.*, 9, 256, 1974.
4. Dumin, D.J. et al., High field related thin oxides wearout and breakdown, *IEEE Trans. Electron Dev.*, 42, 760, 1995.
5. Groeseneken, G. et al., Hot Carrier degradation and time-dependent dielectric breakdown in oxides, *Microelectron. Eng.*, 49, 27, 1999.
6. Kobayashi, K. et al., Electron trapping and excess current induced by hot-hole injection into thin SiO_2 films, *J. Electrochem. Soc*, 143, 3377, 1996.
7. Yamamoto, T. et al., Bias temperature instability in scaled p+ polysilicon gate p−MOSFET's, *IEEE Trans. Electron Dev.*, 921, 1999.
8. Mitani, Y. et al., NBTI mechanism in ultra thin gate dielectric-nitrogen-originated mechanism in SiON, in *IEDM Tech. Digest*, San Francisco, 509, 2002.
9. Choi, C.H. et al., Impact of gate direct tunneling current on circuit performance: A simulation study, *IEEE Trans. Electron Dev.*, 48, 2823, 2001.
10. *International Technology Roadmap for Semiconductors*, 2006 edition, Available on-line at www.itrs.net.
11. Mistry, K. et al., A 45 nm logic technology with high-k + metal gate transistors, strained silicon, 9 Cu interconnect layers, 193 nm dry patterning, and 100% Pb-free packaging, in *IEDM Tech. Digest*, 247, 2007.
12. Seidl, H. et al., A fully integrated Al_2O_3 trench capacitor DRAM for sub-100 nm technology, in *IEDM Tech. Digest*, 839, 2002.
13. Process integration, devices, and structures, in *International Technology Roadmap for Semiconductors, 2006 Update*; available on line at www.itrs.net.
14. Lee, S. et al., Record RF performance of 45-nm SOI CMOS technology, in *IEDM Tech. Digest*, 255, 2007.
15. Kobayashi, H. et al., Nitridation of silicon oxide layers by nitrogen plasma generated by low energy electron impact, *Appl. Phys. Lett.*, 71, 1978, 1997.
16. Kraft, R. et al., Surface nitridation of silicon dioxide with a high density nitrogen plasma, *J. Vac. Sci. Technol.B*, 15, 967, 1997.
17. Jha, N.K. and Rao, V.R., A new oxide trap assisted NBTI degradation model, *IEEE Electron Dev. Lett.*, 26, 687, 2005.
18. Miranda, E. et al., Soft breakdown conduction in ultrathin (3–5 nm) gate dielectrics, *IEEE Trans. Electron Dev.*, 47, 82, 2000.
19. Miranda, E. et al., Modeling of soft breakdown spots in silicon dioxide films as point contacts, *Appl. Phys. Lett.*, 74, 959, 1999.
20. Boruta, N. et al., A new physics-based model for understanding single event gate rupture in linear devices, *IEEE Trans. Nucl. Sci.*, 48, 1917, 2001.
21. Johnston, A. and Hughlock, B., Latchup in CMOS from single particles, *IEEE Trans. Nucl. Sci.*, 37, 1886, 1990.
22. Olivo, P. et al., High-field-induced degradation in ultra-thin SiO_2 films, *IEEE Trans. Electron Dev.*, 35, 2259, 1988.
23. Lee, S.-H. et al., Quasi-breakdown of ultrathin gate oxide under high field stress, in *IEDM Tech. Digest*, San Francisco, 605, 1994.
24. Cellere, G. et al., Micro breakdown in small-area ultra-thin gate oxide, *IEEE Trans. Electron Dev.*, 49, 1367, 2002.
25. Kerber, A. et al., Lifetime prediction for CMOS devices with ultra thin gate oxides based on progressive breakdown, in *45th IEEE International Reliability Physics Symposium* (IRPS), Phoenix, 217, 2007.
26. Wu, E. et al., On the progressive breakdown statistical distribution and its voltage acceleration, in *IEDM Tech. Digest*, Washington, 493, 2007.

27. Maserjian, J. and Zamani, N., Behavior of the Si/SiO_2 interface observed by Fowler–Nordheim tunneling, *J. Appl. Phys.*, 53, 559, 1982.
28. Sakakibara, K. et al., A quantitative analysis of stress induced excess current (SIEC) in SiO_2 films, in *40th IEEE International Reliability Physics Symposium (IRPS)*, Dallas, 100, 2002.
29. Sakakibara, K. et al., Identification of stress induced leakage current components and corresponding trap models in SiO_2 films, *IEEE Trans. Electron Dev.*, 44, 986, 1997.
30. Sakakibara, K. et al., Influence of holes on neutral trap generation, *IEEE Trans. Electron Dev.*, 44, 2274, 1997.
31. Sakakibara, K. et al., A quantitative analysis of time-decay reproducible stress induced leakage current in SiO_2 films, *IEEE Trans. Electron Dev.*, 44, 1002, 1997.
32. Runnion, E.F. et al., Thickness dependence of stress-induced leakage currents in silicon oxide, *IEEE Trans. Electron Dev.*, 44, 993, 1997.
33. Chou, A.I. et al., Modeling of stress-induced leakage current in ultrathin oxides with the trap-assisted tunneling mechanism, *Appl. Phys. Lett.*, 70, 3407, 1997.
34. Rosenbaum, E. and Register, L.F., Mechanism of stress induced leakage current in MOS capacitors, *IEEE Trans. Electron Dev.*, 44, 317, 1997.
35. Takagi, S. et al., A new *I–V* model for stress-induced leakage current including inelastic tunneling, *IEEE Trans. Electron Dev.*, 46, 348, 1999.
36. Larcher, L. et al., A model of the stress induced leakage current in gate oxides, *IEEE Trans. Electron Dev.*, 48, 285, 2001.
37. Cester, A. et al., Time stability of stress induced leakage current in ultra-thin gate oxides, *Solid State Electron.*, 45, 1345, 2001.
38. Scarpa, A. et al., Electrically and radiation induced leakage current in thin oxides, *Microelectron. Reliab.*, 40, 57, 2000.
39. Scarpa, A. et al., Reliability extrapolation model for stress induced leakage current in thin silicon oxides, *Electron. Lett.*, 33, 1342, 1997.
40. Scarpa, A. et al., Stress induced leakage current dependence on oxide thickness, technology and stress level, in *27th European Solid-State Device Research Conference (ESSDERC)*, Stuttgart, 592, 1997.
41. Rodriguez, R. et al., Relation between defect generation, SILC and soft breakdown in thin (<5 nm) oxides, *Microelectron. Reliab.*, 40, 707, 2000.
42. De Blauwe, J. et al., SILC-related effects in flash E^2PROM's-part I: A quantitative model for steady-state SILC, *IEEE Trans. Electron Dev.*, 45, 1745, 1998.
43. Degraeve, R. et al., A new polarity dependence of the reduced trap generation during high-field degradation of nitrided oxides, *in IEDM Tech. Digest*, San Francisco, 327, 1996.
44. DiMaria, D.J. and Cartier, E., Mechanism for stress-induced leakage currents in thin silicon dioxide films, *J. Appl. Phys*, 78, 3883, 1995.
45. Alam, M.A. et al., Field acceleration for oxide breakdown—can an accurate anode hole injection model resolve the *E* vs. $1/E$ controversy?, in *Intl. Reliab. Phys. Sympos. (IRPS)*, San Jose, 21, 2000.
46. Yokozawa, A. et al., Oxygen vacancy with large lattice distortion as an origin of leakage current in SiO_2, in *IEDM Tech. Digest*, Washington, 703, 1997.
47. Chen, T.C. et al., Post-stress interface trap generation induced by oxide-field stress with FN injection, *IEEE Trans. Electron Dev.*, 45, 1972, 1998.
48. Sakura, T. et al., A detailed study of soft- and pre-soft-breakdowns in small geometry MOS structures, in *IEDM Tech. Digest*, San Francisco, 183, 1998.
49. Degraeve, R. et al., Statistical model for stress induced leakage current and pre-breakdown currents jumps in ultra-thon oxide layers, in *IEDM Tech. Digest*, Washington, 125, 2001.
50. Cellere, G. et al., Plasma induced micro breakdown in small-area MOSFETs, *IEEE Trans. Electron Dev.*, 49, 1768, 2002.
51. Cellere, G. et al., Different nature of process-induced and stress-induced defects in thin SiO_2 layers, *IEEE Electron Dev. Letters*, 24, 393, 2003.
52. Cellere, G. et al., Forming gas anneal effect on plasma-induced damage: Beyond the appearances, *IEEE Trans. Electron Dev.*, 51, 302, 2004.
53. Cappelletti, P. et al., *Flash Memories*, Kluver Academic Press, Boston, 2000.

54. Takagi, S. et al., Experimental evidence of inelastic tunneling in stress-induced leakage current, *IEEE Trans. Electron Dev.*, 46, 335, 1999.
55. Ielmini, D. et al., Modeling of SILC based on electron and hole tunneling—part II: Steady-state, *IEEE Trans. Electron Dev.*, 47, 1266, 2000.
56. Ueno, S. et al., Impact of the two traps related leakage mechanism on the tail distribution of DRAM retention characteristics, in *IEDM Tech. Digest*, 37, 1999.
57. Yamada, S. et al., Nonuniform current flow through thin oxide after Fowler–Nordheim current stress, in *Proc. IRPS*, 108, 1996.
58. Ielmini, D. et al., A new two-trap tunneling model for the anomalous SILC in flash memories, *Microelectron. Eng.*, 59, 189, 2001.
59. Degraeve, R. et al., Analytical percolation model for predicting anomalous charge loss in flash memories, *IEEE Trans. Electron Dev.*, 51, 1392, 2004.
60. Ielmini, D. et al., Characterization of oxide trap energy by analysis of the SILC roll-off regime in flash memories, *IEEE Trans. Electron Dev.*, 53, 126, 2006.
61. Ielmini, D. et al., A statistical model for SILC in flash memories, *IEEE Trans. Electron Dev.*, 49, 1955, 2002.
62. Ielmini, D. et al., Defect generation statistics in thin gate oxides, *IEEE Trans. Electron Dev.*, 51, 1288, 2004.
63. Wilkinson, J. and Hareland, S., A cautionary tale of soft errors induced by SRAM packaging materials, *IEEE Trans. Dev. Mater. Reliab.*, 5, 428, 2005.
64. Baumann, R., Radiation induced soft errors in advanced semiconductor technologies, *IEEE Trans. Dev. Mater. Reliab.*, 5, 305, 2005.
65. Mazur, J., The radiation environment outside and inside a spacecraft, in *IEEE Nuclear and Space Radiation Effects Conference (NSREC) Short Course*, 2002.
66. Barth, J., Modeling space radiation environments, in *IEEE NSREC Short Course*, 1997.
67. Moore, S., Extreme solar storm strikes Earth, *IEEE Spectrum*, 40, 15, 2003.
68. Braley, G.S. et al., Modeling of secondary neutron production from space radiation interactions, *IEEE Trans. Nucl. Sci.*, 49, 2800, 2002.
69. Gordon, S.M. et al., Measurement of the flux and energy spectrum of cosmic-ray induce neutrons on the ground, *IEEE Trans. Nucl. Sci.*, 51, 3427, 2004.
70. Goldhagen, P. et al., Measurement of the energy spectrum of cosmic-ray induced neutrons aboard an ER-2 high altitude airplane, *Nucl. Instrum. Meth. Phys. Res. A*, 476, 42, 2002.
71. Wrobel, F. et al., Contribution of SiO_2 in neutron-induced SEU in SRAMs, *IEEE Trans. Nucl. Sci.*, 50, 2055, 2003.
72. Ziegler, J., *SER—History, Trends and Challenges*, Cypress semiconductor, San Jose, CA, 2004.
73. Maurer, R. et al., Neutron production from polyethylene and common spacecraft materials, *IEEE Trans. Nucl. Sci.*, 48, 2029, 2001.
74. Aymar, R. et al., ITER-FEAT—the future international burning plasma experiment, in *18th IAEA Fusion Energy Conference*, Sorrento, OV/1, 2000.
75. Wilkinson, J. et al., Cancer radiotherapy equipment as a cause of soft errors in electronic equipment, *IEEE Trans. Dev. Mater. Reliab.*, 5, 449, 2005.
76. Goldstein, H., Can electron beams and x-rays make our food safe? *IEEE Spectrum*, 40, 24, 2003.
77. Sexton, F.W. et al., Effects of E-beam mail sanitizing process on commercial electronics, in *40th IEEE International Reliability Physics Symposium (IRPS)*, Dallas, 2002.
78. Blish, R.C. et al., Filter optimization of x-ray inspection of surface-mounted ICs, in *40th IEEE International Reliability Physics Symposium (IRPS)*, Dallas, 377, 2002.
79. Ma, T.P. and Dressendorfer, P.V., *Ionizing Radiation Effects in MOS Devices and Circuits*, Wiley, New York, 1989.
80. Brisset, C. et al., Two-dimensional simulation of total dose effects on NMOSFET with lateral parasitic transistor, *IEEE Trans. Nucl. Sci.*, 43, 2651, 1996.
81. Sallagoity, P. et al., Analysis of width edge effects in advanced isolation schemes for deep submicron CMOS technologies, *IEEE Trans. Nucl. Sci.*, 43, 1900, 1996.
82. Musseau, O. and Ferlet-Cavrois, V., Silicon on insulator technologies: Radiation effects, in *IEEE NSREC Short Course*, Vancouver, 2001.

83. Benedetto, J.M. and Boesch, H.E., The relationship between ^{60}Co and 10 keV x-ray damage in MOS devices, *IEEE Trans. Nucl. Sci.*, 33, 1318, 1986.
84. Osanger, L., Initial recombination of ions, *Phys. Rev.*, 54, 554, 1938.
85. Ausman, G.A. and McLean, F.B., Electron–hole pair creation energy in SiO_2, *Appl. Phys. Letters*, 45, 173, 1975.
86. Schwank, J.R., Total dose effects in MOS devices, in *IEEE NSREC Short Course*, Phoenix, 2002.
87. O'Gorman, T.J., The effect of cosmic rays on the soft error rate of a DRAM at ground level, *IEEE Trans. Nucl. Sci.*, 4, 553, 1994.
88. Scheick, L.Z., Swift, G.M., and Guertin, S.M., SEE evaluation of SRAM memories for space application, in *IEEE Radiation Effects Data Workshop*, Reno, 61, 2000.
89. Hiemstra, D.M. and Baril, A., Single event upset characterization of the Pentium MMX and Celeron microprocessors, in *IEEE Radiation Effects Data Workshop*, Reno, 39, 2000.
90. Makihara, A. et al., Analysis of single-ion multiple bit upset in high density DRAM, *IEEE Trans. Nucl. Sci.*, 47, 2400, 2000.
91. Sexton, F. et al., Single event gate rupture in thin gate oxides, *IEEE Trans. Nucl. Sci.*, 44, 2345, 1997.
92. Johnston, A. et al., Single-event transients in high-speed comparators, *IEEE Trans. Nucl. Sci.*, 49, 3082, 2002.
93. Jaffe, G., Zur Theorie der Ionization in Kolonnen, *Ann. Phys.*, (Leipzig) 42, 353, 1913.
94. Oldham, T.R., Recombination along the tracks of heavy charged particles in SiO_2 films, *J. Appl. Phys.*, 57, 2695, 1985.
95. Lelis, A.J. et al., Reversibility of trapped hole charge, *IEEE Trans. Nucl. Sci.*, 35, 1186, 1988.
96. Lelis, A.J. et al., The Nature of the trapped hole annealing process, *IEEE Trans. Nucl. Sci.*, 36, 1808, 1989.
97. Warren, W.L. et al., Microscopic nature of border traps in MOS oxides, *IEEE Trans. Nucl. Sci.*, 41, 1817, 1994.
98. Ceschia, M. et al., Radiation-induced leakage current and stress induced leakage current in ultra-thin gate oxides, *IEEE Trans. Nucl. Sci.*, 45, 2375, 1998.
99. Saks, N.S., Dozier, C.M., and Brown, D.B., Time dependence of interface trap formation in MOSFETs following pulsed irradiation, *IEEE Trans. Nucl. Sci.*, 35, 1168, 1988.
100. Shaneyfelt, M.R. et al., Interface trap buildup rates in wet and dry oxides, *IEEE Trans. Nucl. Sci.*, 39, 2244, 1992.
101. Schwank, J.R. et al., Latent thermally activated interface-trap generation in MOS devices, *Electron Dev. Lett.*, 13, 203, 1992.
102. Svensson, C.M., The defect structure of the Si–SiO_2 interface, a model based on trivalent silicon and its hydrogen 'compounds,' in *The Physics of SiO_2 and Its Interfaces*, Pantelides, S.T. Ed., Pergamon Press, Elmsford, 328, 1978.
103. Schwank, J.R. et al., The role of hydrogen in radiation-induced defect formation in polysilicon gate MOS devices, *IEEE Trans. Nucl. Sci.*, 34, 1152, 1987.
104. Rashkeev, S.N. et al., Proton induced defect generation at the Si–SiO_2 interface, *IEEE Trans. Nucl. Sci.*, 48, 2086, 2001.
105. Saks, N. et al., Radiation effects in MOS capacitors with very thin oxides at 80 K, *IEEE Trans. Nucl. Sci.*, 31, 1249, 1984.
106. Takagi, S. et al., Experimental evidence of inelastic tunneling and new *I–V* model for stress-induced leakage current, in *IEDM Tech. Digest*, San Francisco, 703, 1996.
107. Larcher, L. et al., A model of radiation induced leakage current (RILC) in ultra-thin gate oxides, *IEEE Trans. Nucl. Sci.*, 46, 1553, 1999.
108. Lenahan, P.M. et al., Direct experimental evidence linking silicon dangling bond defects to oxide leakage currents, in *39th IEEE International Reliability Physics Symposium*, Orlando, 150, 2001.
109. Walters, M. and Reisman, A., Radiation-induced neutral electron trap generation in electrically biased insulated gate field effect transistor gate insulators, *J. Electrochem. Soc.*, 138, 2756, 1991.
110. Oldham, T.R. and McGarrity, J.M., Comparison of ^{60}Co response and 10 keV x-ray response in MOS capacitors, *IEEE Trans. Nucl. Sci.*, 30, 4377, 1983.

111. Cellere, G. and Paccagnella, A., A review of ionizing radiation effects in flash memories, *IEEE Trans. Dev. Mater. Reliab.*, 3, 359, 2004.
112. Cellere, G. et al., Radiation effects on floating-gate memory cells, *IEEE Trans. Nucl. Sci.*, 48, 2222, 2001.
113. Cellere, G. et al., Anomalous charge loss from floating-gate memory cells due to heavy ions irradiation, *IEEE Trans. Nucl. Sci.*, 49, 3051, 2002.
114. Cellere, G. et al., Transient conductive path induced by a single ion in 10 nm SiO_2 layers, *IEEE Trans. Nucl. Sci.*, 6, 3304, 2004.
115. Cellere, G. et al., Subpicosecond conduction through thin SiO_2 layers triggered by heavy ions, *J. Appl. Phys.*, 99, 074101, 2006.
116. Cellere, G. et al., Secondary effects of single ions on floating gate memory cells, *IEEE Trans. Nucl. Sci.*, 53, 3291, 2006.
117. Stesmans, A., Interaction of Pb defects at the (111) Si/SiO_2 interface with molecular hydrogen: Simultaneous action of passivation and dissociation, *J. Appl. Phys.*, 88, 489, 2000.
118. Pantisano, L. and Cheung, K.P., Stress induced leakage current (SILC) and oxide breakdown: Are they from the same oxide traps? *IEEE Trans. Dev. Mater. Reliab.*, 109, 2001.
119. Riess, P. et al., Reversibility of charge trapping and SILC creation in thin oxides after stress/anneal cycles, *Microelectron. Reliab.*, 38, 1057, 1998.
120. Nissan-Cohen, Y. and Gorczyca, T., The effect of hydrogen on trap generation, positive charge trapping, and time-dependent dielectric breakdown of gate oxides, *IEEE Electron Dev. Lett.*, 9, 287, 1988.
121. Cellere, G. et al., RILC in 10 nm SiO_2 layers, *IEEE Trans. Nucl. Sci.*, 52, 2144, 2005.
122. Cellere, G. et al., Variability in FG memories performance after irradiation, *IEEE Trans. Nucl. Sci.*, 53, 3349, 2006.
123. Stathis, J.H. and DiMaria, D.J., Reliability projection for ultra-thin oxides at low voltage, in *IEDM Tech. Digest*, San Francisco, 167, 1998.
124. Blochl, P.E. and Stathis, J.H., Hydrogen electrochemistry and stress-induced leakage current in silica, *Phys. Rev. Lett.*, 83, 372, 1999.
125. Cellere, G. et al., Sub-attoampere current induced by single ions in silicon oxide layers of nonvolatile memory cells, *Appl. Phys. Lett.*, 88, 192909, 2006.
126. Larcher, L. et al., Data retention after heavy ion exposure of floating gate memories: Analysis and simulation, *IEEE Trans. Nucl. Sci.*, 50, 2176, 2003.
127. Riccò, B. et al., Modeling and simulation of stress-induced leakage current in ultrathin SiO_2 films, *IEEE Trans. Electron Dev.*, 45, 1554, 1998.
128. Ielmini, D. et al., Statistical modeling of reliability and scaling projections for flash memories in *IEDM Tech. Digest*, Washington, 703, 2001.
129. Schuler, F. et al., Physical description of anomalous charge loss in floating gate based NVM's and identification of its dominant parameter, in *40th IEEE International Reliability Physics Symposium (IRPS)*, Dallas, 26, 2002.
130. Meftah, A. et al., Track formation in SiO_2 quartz and the thermal-spike mechanism, *Phys. Rev. B*, 49, 12457, 1994.
131. Toulemonde, M. et al., Transient thermal process after a high energy heavy ion irradiation of amorphous metals and semiconductors, *Phys. Rev. B*, 46, 14362, 1992.
132. Pavan, P. et al., Flash memory cells—an overview, *Proc. IEEE*, 85, 1248, 1997.
133. Candelori, A. et al., Thin oxide degradation after high energy ion irradiation, *IEEE Trans. Nucl. Sci.*, 48, 1735, 2001.
134. Ceschia, M. et al., Low field leakage current and soft breakdown in ultra-thin gate oxides after heavy ions, electron or x-ray irradiation, *IEEE Trans. Nucl. Sci.*, 3, 566, 2000.
135. Goguenheim, D. et al., Comparison of oxide leakage currents induced by ion implantation and high electric field stress, *Solid State Electron.*, 45, 1355, 2001.
136. Fleetwood, D.M., Border traps in MOS devices, *IEEE Trans. Nucl. Sci.*, 39, 269, 1992.
137. Fleetwood, D.M. et al., Effects of oxide traps, interface traps, and "border traps" on metal-oxide-semiconductor devices, *J. Appl. Phys.*, 73, 5058, 1993.
138. Fleetwood, D.M., Fast and slow border traps in MOS devices, *IEEE Trans. Nucl. Sci.*, 43, 779, 1996.
139. Cheung, K.P., *Plasma Charging Damage*, Springer-Verlag, New York, 2000.

18

Structural Defects in SiO_2–Si Caused by Ion Bombardment

Antoine D. Touboul, Aminata Carvalho, Mathias Marinoni, Frederic Saigne, Jacques Bonnet, and Jean Gasiot

CONTENTS

18.1 Introduction

A charged particle entering matter interacts both with the surrounding atoms and electrons. A part or all of its energy is then transferred to the electrons or converted into displacement damage, depending on the kinetic energy of the particle. Until recently, the fraction of energy transferred to the electrons was considered almost exclusively to be an ionization phenomenon that does not induce any atomic motion. However, some recent studies have pointed out that structural modifications could even occur at very high energy via electronic interactions [1,2].

From a historical point of view, experiments have already shown that damage may be induced through ionization processes. In 1912, Wilson observed alpha particle tracks using a cloud chamber [3]. The ionizing path of the ion was revealed by the formation of condensed droplets. The first photography of fission fragment tracks obtained by means of a Wilson chamber was reported in 1939 [4,5]. An important stage was reached in the middle of the twentieth century with the development of the electron microscope. With a spatial resolution about a few tens of nanometers, this tool allows the investigation of matter at the atomic scale. The first direct observation of damage tracks of fission fragments was published by Silk and Barnes in 1959 [6]. From that time, the understanding of such specific interactions became of major interest. Three physicists of the General Electric research laboratory, Fleischer, Price, and Walker, succeeded in quantifying the diameter of ^{235}U fission fragment tracks in mica. They concluded that the damaged region around the ion path could reach a diameter of 150 Å [7]. Consequently, they realized that observing such small tracks may quickly became a real technological challenge. For that reason, they decided to add to their experimental process a new stage comprising a chemical enhancement of the tracks [8]. They thereby showed that a single ion track could be enlarged by a chemical etching process to a diameter of 0.5 μm [9].

18.2 Swift Heavy Ion-Induced Structural Modifications: A Review of Some Phenomenological Models

Swift heavy ion effects on matter are still under investigation. A great number of studies devoted to this field of interest were developed during the 1960s. A significant increase in the pace of investigation of these interactions has been especially motivated by the observation of nanometric modifications occurring in some materials. For such reasons, nanotechnology sciences were the first ones to have some interest in this field. However, at present, the geometrical features of a single transistor in a commercial integrated circuit are of the same order of magnitude as the structural modifications that can be induced by a swift heavy ion. Moreover, such modifications are not the consequences of a uniform particle flux, as considered for displacement damage studies, but can be induced by a single swift heavy ion. The huge amount of energy lost by the ion can indeed lead to structural modifications through atomic motion in some particular materials (e.g., amorphous SiO_2). For the sake of reaching a better understanding of swift heavy ion effects on SiO_2–Si structures, the most accurate models in the frame of heavy ion interactions with matter are reviewed. For each model the domain of validity is described.

18.2.1 Displacement Spike

Brinkman initially developed this model in 1954 to describe the created defects in neutron-irradiated materials [10]. It is assumed that each incoming particle creates a nuclear collision cascade in the material. Based on the Kinchin–Pease theory, Brinkman assessed the free mean path of each ejected atom on the scale of a few angstroms. This implicitly leads to the assumption that the displacement damage is strongly localized along the ion path. Two main regions of interaction are defined in Brinkman's theory. The first zone begins in the area of the ejection site of the primary knocked-on atom (PKA). It corresponds to a high-energy interaction region in which some sporadic formation of Frenkel pairs may occur. The second one occurs at the end of the PKA range and corresponds to a highly disordered region, the so-called displacement spike. In this model, Brinkman considered that material modifications were basically caused by displacement damage. The displacement spike model leads naturally to the assumption that nuclear tracks could be formed independently from the material type. However, some studies have pointed out that each material can behave differently with regards to track formation [11,12].

18.2.2 Ion Explosion Spike

This theory was developed in 1965 by Fleischer et al. [13]. Historically, this model was the first one that could describe the track formation efficiently. From the dynamics point of view, we can describe the ion explosion spike in three main stages. The medium becomes positively charged within a few femtoseconds, leaving the atomic structure highly unstable. Afterward, during the next picoseconds, an electronic relaxation occurs that is followed by atomic motions along the ion track into interstitial sites. The region under stress is finally subjected to an elastic relaxation that creates a constraint on both sides of the ion track. This electrostatic model aims to describe the repulsive forces that should occur to overcome the bonding strength of the lattice. The electrostatic repulsive force between two charged entities, F, can be expressed as

$$F = \frac{1}{\varepsilon_m}\frac{n_1 n_2 e^2}{a_0{}^2}, \tag{18.1}$$

where

n_1 and n_2 are the units of charge for each ion
e is the fundamental charge
a_0 is the average atomic spacing
ε_m is the dielectric constant of the medium

We can also note that the theoretical strength of a material, σ, is described by the famous Orowan–Polanyi law reported in Equation 18.2 [14]:

$$\sigma = \left(\frac{Yg}{x_0}\right)^{1/2} \approx \frac{Y}{10}, \tag{18.2}$$

where

g is the free surface energy
Y is Young's modulus
x_0 is the equilibrium spacing between the two planes to be separated (a_0 in our case)

It follows directly that the electrostatic stress will be larger than the mechanical strength if Equation 18.3 is satisfied:

$$n_1 n_2 > \frac{Y\varepsilon_m a_0^4}{10e^2} \tag{18.3}$$

This condition implies that materials exhibiting a low mechanical strength, a low dielectric constant, and dense atomic network will be more sensitive to track formation. However, some studies have pointed out that the electrostatic force could not be large enough to induce any atomic displacement [15].

18.2.3 Thermal Spike Model

18.2.3.1 Philosophical Origins

Dessauer published the very first description of this theory in the early 1920s. At this time, he was working on the effects of electron irradiation on biological cells. He pointed out in particular that an electronic radiation of 11 eV on a molecule could induce a thermal increase up to 100°C. He described the temperature increase as a consequence of the energy transfer between the electronic and the vibrational system [16]. He initiated the thermal spike theory. Among the different models that have been developed, the thermal spike is so far the only one that allows one to quantify the geometrical features of the defected region. For this reason, this theory is described in more detail than the other ones.

18.2.3.2 Thermodynamical Model

This model was developed to describe the interaction of high-energy particles with matter, especially in the electronic excitation regime. When dealing with heavy ion energies greater than 1 MeV u^{-1} (u is the atomic mass unit), we can consider that the interaction is mainly related to the electron-cloud excitation. The ratio between the electronic and the nuclear energy loss, based on linear energy transfer (LET) calculations, is greater than 100. The thermal spike theory has many things in common with a thermal diffusion model. The key parameters are the thermal diffusivity, D, the thermal conductivity, K_{Th}, the density, ρ, and the specific heat, C. The thermal diffusivity can be classically expressed according to

$$D = \frac{K_{Th}}{C_\rho} \tag{18.4}$$

Considering a thermal energy quantity, Q, emitted from a point source, we can describe the temporal and spatial evolution of the temperature according to [17]:

$$T(r,t) = \frac{Q}{8\pi^{3/2}C\rho}\frac{1}{(Dt)^{3/2}}\exp\left(-\frac{r^2}{4Dt}\right) \tag{18.5}$$

Extending this equation to the specific study of heavy ion-induced effects, the following expression has been obtained by Seitz [18]:

$$T(r,t) = \frac{dE/dx}{4\pi K_{Th}t}\exp\left(-\frac{r^2}{4Dt}\right) \tag{18.6}$$

where dE/dx is the quantity of energy per unit path length transmitted to the electrons. It is then possible to represent the thermal spike phenomenon from a qualitative point of view in a SiO_2–Si structure as shown in Figure 18.1. ΔT_1 and ΔT_2 correspond to the maximum temperature increases occurring, respectively, in the oxide layer and in the

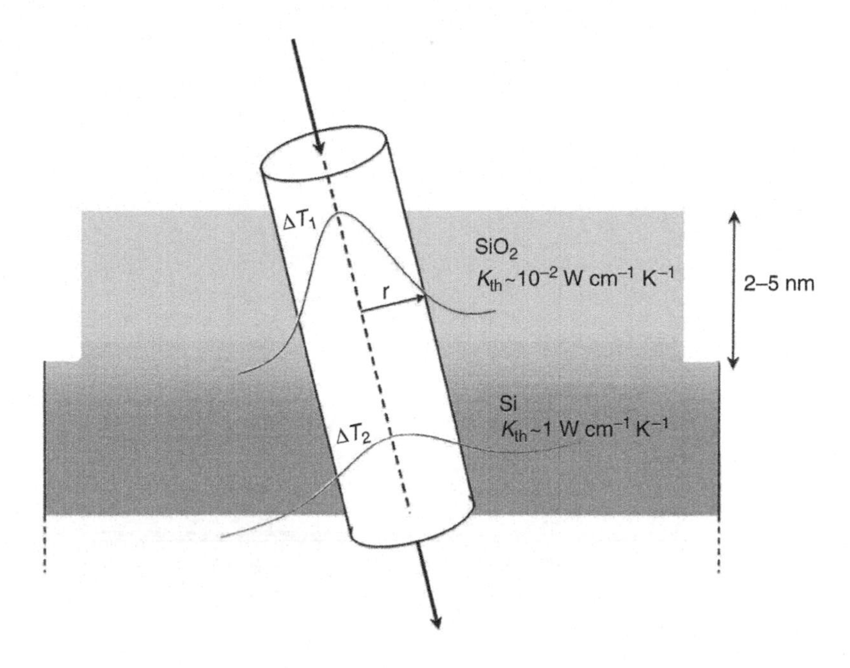

FIGURE 18.1
Qualitative illustration of the thermal spike phenomenon in an SiO_2–Si structure.

silicon substrate, and r is the direction in which the thermal diffusion occurs. Both Gaussian shapes represent the temperature increases in SiO_2 and Si. Since those two materials exhibit different thermal diffusion properties, different temperature evolutions occur in each layer. The silicon layer dissipates heat more easily and over a greater distance than any thermal contribution in the oxide.

18.2.3.3 What about the Phonon Contribution?

The first calculations dedicated to the quantification of the electronic relaxation were performed by Seitz and Koehler [18]. Regarding the ultimate geometrical features of the current elementary MOS devices, it becomes of paramount interest to study the effects of radiation at the atomic scale. Initially, an ion passes through the elementary SiO_2–Si structure. In a time less than a femtosecond, the ion has ionized the matter along its path. At this stage, most of the interactions are related to elastic collisions between excited electrons. To verify this assumption, the excited electron density, δ_e, can be estimated using [19]:

$$\delta_e = \frac{3Z}{4\pi r_\nu^3} \tag{18.7}$$

where Z is the number of valence electrons. The van der Waals radius, r_v, is estimated to be about 200 pm. Thus, the excited electron density can reach 10^{23} cm^{-3}. Hence, the high-energy electrons are going to be thermalized during some electron–electron interactions, leading to production of secondary electrons at lower energy. At this time, depending on the intrinsic properties of the material, part of the energy can be transferred to the lattice through electron–phonon interactions. A semiquantitative description is represented in Figure 18.2. The temporal dynamic response of the atomic system is slower than that of the

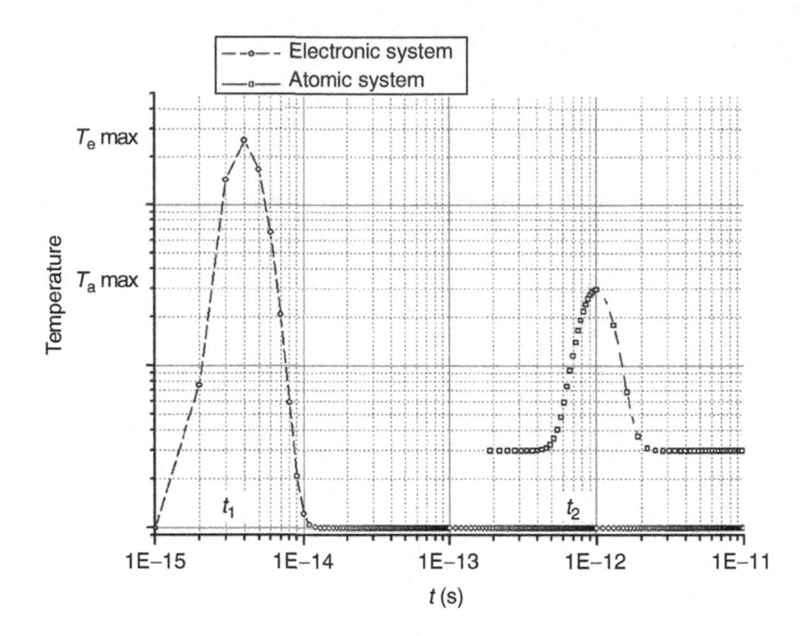

FIGURE 18.2
Semiquantitative description of the electronic and atomic relaxation during a thermal spike phenomenon.

electronic system. Moreover, the temperature increase occurring in the atomic system is strongly dependent on the coupling factor between the electrons and the lattice vibrations. To summarize, the electron–phonon coupling represents the energy transfer efficiency between the electrons and the lattice [20]. Some studies on this topic lead to the assumption that the coupling factor strongly depends on the ability of a material to be conductive [21–23]. It means that the less conductive the material is, the more degradation that may be induced.

18.2.3.4 Some Quantitative Development of the Thermal Spike

To evaluate the energy threshold for track formation, Toulemonde developed a code based on solving the heat flow equation. The space and time evolution of electronic system and lattice temperature, respectively T_e and T, are governed by a set of coupled nonlinear equations [24]:

$$\begin{aligned}\rho C_e(T_e)\frac{\partial T_e}{\partial t} &= \frac{\partial}{\partial r}\left(K_e(T_e)\frac{\partial T_e}{\partial r}\right) + \frac{K_e(T_e)}{r}\frac{\partial T_e}{\partial r} - g(T_e - T) + A(r)\\ \rho C(T)\frac{\partial T}{\partial t} &= \frac{\partial}{\partial r}\left(K(T)\frac{\partial T}{\partial r}\right) + \frac{K(T)}{r}\frac{\partial T}{\partial r} + g(T_e - T)\end{aligned} \tag{18.8}$$

where

C_e, C and K_e, and K are the specific heat and the thermal conductivity for the electronic and atomic systems, respectively
g is the electron–phonon coupling
$A(r)$ is the energy given to the electrons
r is the transverse axis of the ion path

Based on this calculation, the author has pointed out that the melting temperature of the material could be reached locally, inducing structural modifications along the ion track.

Also based on a thermal spike theory, another useful model was developed by Szenes to quantify the ion track radius within which local melting has occurred [25]. Since the calculated values are in agreement with the experimental ones reported in the literature, this calculation appears to be sufficiently accurate to evaluate the track radius that can be induced in an SiO_2 layer by a single swift heavy ion.

$$\begin{aligned} \mathrm{LET_{eth}} &= \frac{\rho\pi a(0)^2\Delta T}{g} \\ r_0^2 &= a(0)^2\ln\left(\frac{\mathrm{LET}}{\mathrm{LET_{eth}}}\right), \quad \text{for } 1 \leq \frac{\mathrm{LET}}{\mathrm{LET_{eth}}} \leq 2.7 \\ r_0^2 &= \frac{a(0)^2\mathrm{LET}}{2.7\mathrm{LET_{eth}}}, \quad \text{for } \frac{\mathrm{LET}}{\mathrm{LET_{eth}}} > 2.7 \end{aligned} \tag{18.9}$$

The parameters involved in this model are the temperature increase needed to reach the melting point, ΔT, the material density, ρ, the electron–phonon coupling efficiency, g, and a parameter related to the thermal diffusion, $a(0)$. Accurate values of $a(0)$ and g can be found in the literature [25,26]. We have calculated SiO_2 track radii that have been reported in Figure 18.3 versus the heavy ion energy loss. Both lines correspond to our calculated values. Note that they are of the same order of magnitude as the experimental data extracted from the literature (triangle symbols) [27].

18.2.4 Nonexhaustive List of Some Other Models

In addition to the models described above, there are additional models that provide insight into the mechanisms in certain situations. Depending on the kind of interaction (e.g., electronic interaction, displacement damage), the right model must used in the right

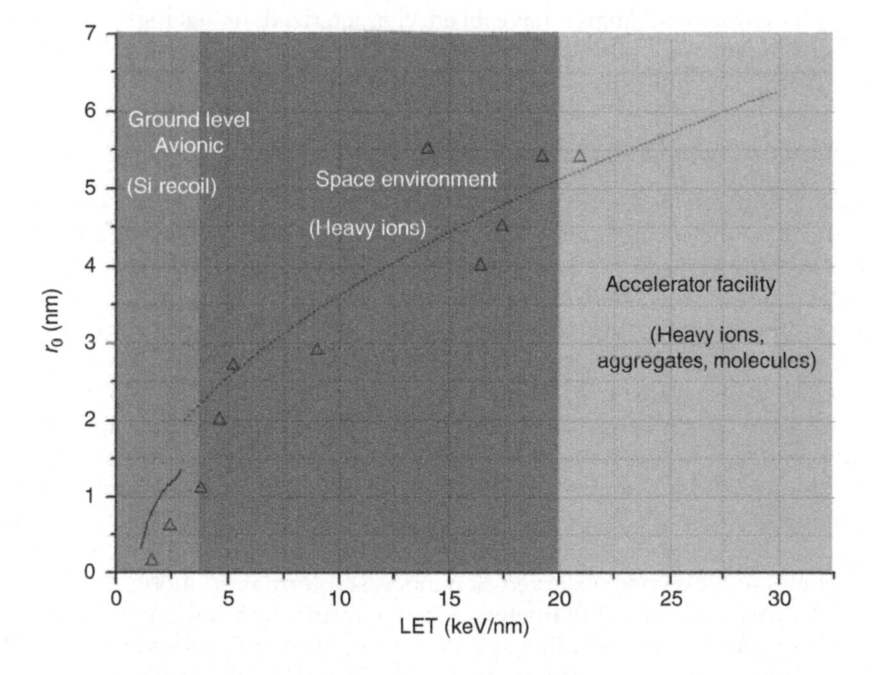

FIGURE 18.3
Track radii in SiO_2 are evaluated using Szenes model as a function of the electronic LET value. The environment of occurrence has also been integrated in the background.

place. The criteria to determine the best model for use in a given situation are strongly related to both target and incoming ion properties. In some specific cases (electronic sputtering, potential sputtering, etc.), additional phenomena should be considered. For instance, the reader who is interested in these areas should refer to some complementary theories, such as the core plasma model of charged particles in insulators developed by Ritchie and Claussen [28], and/or the excitonic model developed by Itoh [29,30].

18.3 Heavy Ion-Induced Phase Transition

A crucial question that must be answered in many situations is "What is the physical structure of the latent track?" Crystalline materials have been studied more frequently than amorphous ones after heavy ion irradiation. However, a few works have investigated ion irradiation effects in amorphous materials (e.g., porous silicon, amorphous silicon dioxide). These experiments have pointed out that either a high-fluence irradiation at low energy or even a single energetic ion might induce a phase transition effect in the target material. That can suggest either an amorphization or a crystallization effect. In Section 18.3.1, we will especially focus on single energetic ion effects, since low-energy ion irradiations have been extensively investigated and many documents on that specific topic are already available [31–34].

18.3.1 Amorphization under Swift Heavy Ion Irradiation

In the previous parts of this chapter, we have highlighted the intrinsic complexity of high-energy ion interaction with matter. Some preeminent works have built the basis of ion interaction with matter [7,13,18,35–37]. In the 1990s, the development of near-field microscopy allowed the first direct observation of structural modifications induced by energetic ions [38]. Latent tracks in quartz have been characterized using high-resolution transmission electronic microscopy (HRTEM) after high-energy irradiation [27]. The observed tracks appear to be cylinder-like and filled with amorphous matter.

18.3.2 Electronic Excitation-Induced Crystallization

The only way of understanding ion-induced phase transitions so far has been in terms of thermal processes that occur along the ion track. However, some recent studies have improved the comprehension of heavy ion interactions at the atomic scale [39]. From the experimental point of view, only a few data are available, although crystallization effects are of great importance. This is related to the inherent difficulty of characterizing nanometric structural modifications. Crystalline track formation in silicon and germanium after high-energy heavy ion irradiations has been reported [11,40]. However, it is also important to consider the behavior of insulators and especially silicon dioxide, regarding crystallization effects. Amorphous SiO films have been investigated after high-energy ion irradiation (Ni at 575 MeV and Pb at 863 MeV). Formation of Si nanocrystals was then observed along each ion track for fluences ranging from 10^{11} to 10^{13} cm^{-2} [41]. Using an identical set of fluences, Chaudhari et al. have irradiated SiO_X films with Ni ions at 100 MeV. They have likewise observed Si nanocrystal formation along the ion tracks [1]. They estimated the nanocrystal diameters to range from 2 to 5 nm. Those studies pointed out that SiO_X regions clearly exhibit a specific sensitivity to the crystallization effect. That could be due to the strong oxygen-atom deficit inherent to those films. This key point, which will be discussed in Section 18.4, is of major interest regarding the reliability of integrated metal-oxide semiconductor (MOS) devices.

18.4 Brief Summary of Experimental Results

18.4.1 Growth of Silicon Bumps at the SiO_2–Si Interface under Swift Heavy Ion Irradiation

The results described here were obtained by surface analysis using a Digital 3100 (Veeco Instruments) atomic force microscope (AFM) working in ambient air conditions at room temperature in tapping mode. Samples consist of 30 nm of thermally grown oxide on (111) *p*-type silicon substrates. Measured root-mean-square (RMS) oxide surface roughness is typically in the range of 0.2 nm. Irradiation was carried out at room temperature with a VIVITRON (Strasbourg) delivering 210 MeV Au+ ions under normal incidence, corresponding to electronic LET of 16.6 keV nm^{-1} in silicon and 18.3 keV nm^{-1} in SiO_2. The fluence was 2×10^9 Au+ ions cm^{-2}, corresponding to an average areal ion impact density of 20 μm^{-2}. This fluence was chosen to give a reasonable number of ion impacts in the area defined by well-resolved AFM images. The measured oxide surface roughness did not exhibit a significant change after irradiation. To observe ion tracks in the oxide and at the interface, hydrofluoric acid (HF) was used to etch the silicon oxide. Gradual chemical etching of the oxide was performed at room temperature in a 10% HF solution, immediately followed by rinsing in deionized water and drying with nitrogen. The etching rate of amorphous SiO_2 in a 10% HF bath is approximately 5 Å s^{-1}. More rapid etching occurs along the ion track in the oxide, due to structural modifications of the oxide along the ion path. Holes of growing diameter are revealed along the tracks as the etching time increases. The hole density after etching was 20 (±2) μm^{-2}, in agreement with the ion fluence. After 60 s of etching time, a nanodot was observed in the bottom of each etched ion track, as shown in Figure 18.4. Those nanodots are less than 1 nm high, and their tops are below the oxide surface level. Further etching allows us to remove the oxide layer completely and to uncover the nanodots further. Figure 18.5 represents an AFM image of the Au+ irradiated sample surface after the oxide layer is completely removed. It shows bumps around 8 nm high, easily distinguishable on the silicon surface. The bumps are distributed

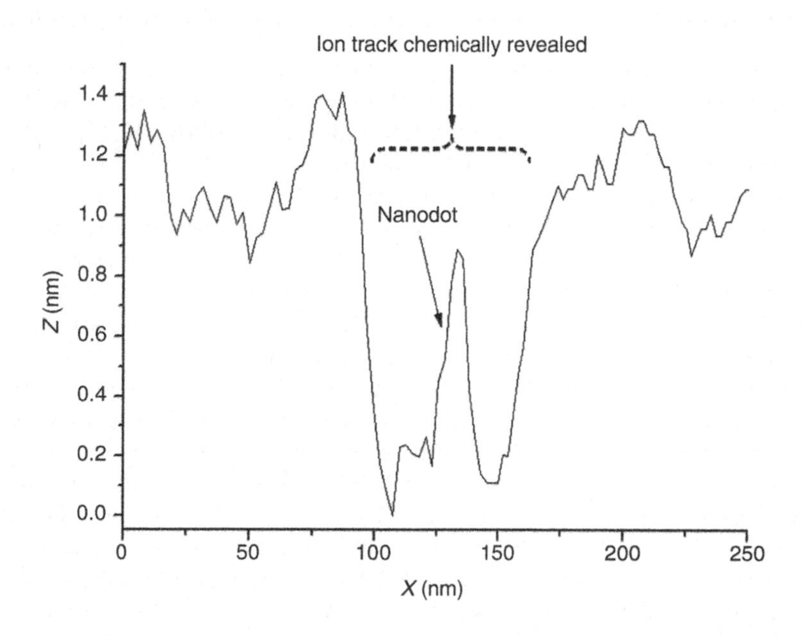

FIGURE 18.4
AFM height profile of a nanodot in an ion track. Scale X: 250 nm and Z: 1.4 nm. (After Carlotti, J.F. et al., *Appl. Phys. Lett.*, 88, 041906, 2006. With permission.)

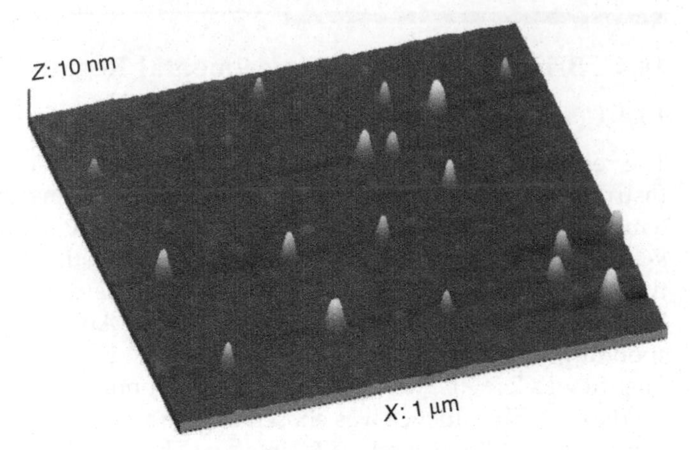

FIGURE 18.5
AFM topography of Si surface after complete etching of SiO_2 layer. Scale *X*: 1 μm, *Y*: 1 μm, *Z*: 10 nm. (After Carlotti, J.F. et al., *Appl. Phys. Lett.*, 88, 041906, 2006. With permission.)

randomly over the sample surface. In the typical example of Figure 18.5, 18 bumps μm^{-2} are shown. The number of bumps remains the same after long etching in a commercial 40% HF solution, thus supporting that the bumps are made of silicon. No holes, nanodots, or bumps were observed after similar etching of nonirradiated samples. The results suggest that each impinging ion induces a track in the silicon oxide that can be etched. Each hole corresponds to a nanodot at one stage of the etching that becomes a bump once the oxide layer is totally etched away. Moreover, the areal density of bumps coincides closely to the density of impinging ions. The Si-substrate structure and the near-interfacial silicon-rich oxide can provide favorable seeds for silicon nanostructure growth. Silicon oxide has a higher melting point than silicon, but is a worse thermal conductor. Thus most of the energy deposited by incoming ions in the oxide contributes to local melting and even to local plasma formation [12]. In silicon oxide, oxygen diffuses away from the ion tracks, leaving Si nanostructures wherever quenching of the melted zone is slower than oxygen diffusion [1].

18.4.2 Discontinuous Ion Tracks in SiO_2–Si after Grazing-Angle Heavy Ion Irradiation

Very low angle irradiations allow direct observation of the geometrical features of the track and are an easier way to get information about the interaction itself than the irradiations at normal incidence. Until now, except for a few bioorganic solids [42,43], only crystalline materials have been subjected to swift heavy ion irradiation at grazing incidence, and just a few of them have exhibited discontinuous tracks. Actually, since only layered media exhibited intermittent tracks, the origin of those tracks has been assumed to be related to the intrinsic atomic structure of the material [44–48]. Irradiations were carried out at GANIL (Caen, France) using a Pb-ion beam with an energy of 0.63 MeV u^{-1} delivered by the beam line IRRSUD, at an angle of 1° with respect to the SiO_2 surface. The corresponding electronic linear energy transfer (LET_{elec}) was about 15.5 keV nm^{-1} in SiO_2, according to the SRIM2003 code [49]. To avoid any spatial overlap of the surface tracks, the fluence of the beam was 5×10^{10} ions cm^{-2}, corresponding to an average areal ion impact density of 9 μm^{-2}. Irradiated samples consisted of an 11.8 nm oxide, thermally grown on a (100) *p*-type silicon substrate. All the experiments were performed at room temperature. Surface tracks were then probed by means of AFM using a Dimension 3100 AFM controlled by a Nanoscope IIIa quadrex electronic from Veeco Instruments working in ambient air conditions in tapping mode. After irradiation, extended surface tracks due to ion impacts are directly visible on the SiO_2 surface by AFM observation, as shown in Figure 18.6. In addition, we can note that the number of tracks is in agreement with the expected areal density. Then, focusing on the topography along each ion track, we can point out an apparent intermittence in the shape of the nanostructures (Figure 18.7).

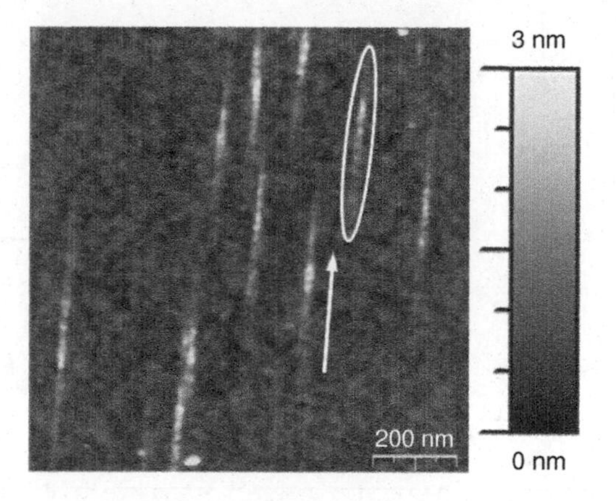

FIGURE 18.6
Top view of an AFM topographical measurement (tapping mode) of 0.63 MeV $^{208}Pb^{32+}$ ion tracks in Si crystals with a 11.8 nm SiO_2 layer irradiated at grazing incidence under 1° at 5×10^{10} ions cm^{-2}. Ions have traversed from bottom to top (white arrow). The number of tracks is in agreement with the expected average ion impact areal density of 9 μm^{-2}. Brightest regions correspond to the highest topographical areas and darkest zones represent lowest topographical edges. The outline around a track corresponds to the reference mark for the topographic height profile along the track and the undisturbed material near a surface track. (After Carvalho, A.M.J.F. et al., *Appl. Phys. Lett.*, 90, 073116, 2007. With permission.)

This observation is directly corroborated by a comparison between the height profile of one track and that of an unirradiated region, which displays a random structure, as shown in Figure 18.8. Each track exhibits a trail of intermittent nanodots, with an average periodicity of 30 nm determined by considering all the surface tracks produced in this experiment. The key result of this study is that a glancing angle heavy ion irradiation has induced intermittent surface modifications in SiO_2. Several different models have been developed to get a better understanding of swift heavy ion-induced tracks in materials. However, despite much work having been done on this topic [18,19,42,50–58], the energy deposition mechanisms at the nanometric scale are still under investigation. The critical point is that nearly

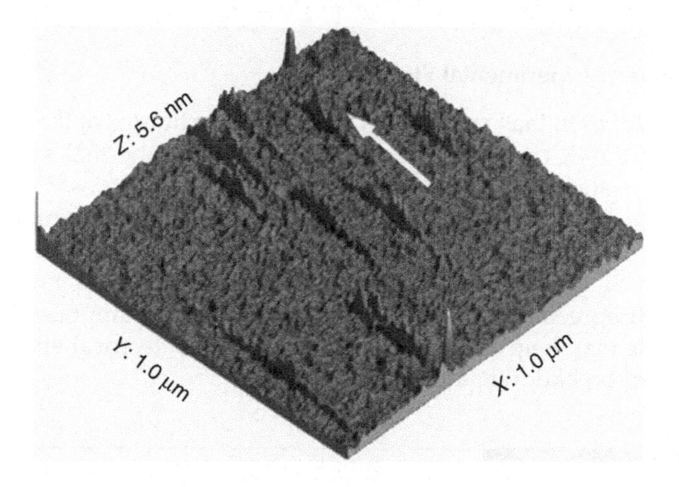

FIGURE 18.7
AFM three-dimensional view of the surface defects observed on 11.8 nm thick SiO_2–Si samples showing clearly an intermittence in the elongated surface tracks produced after 0.63 MeV $^{208}Pb^{32+}$ ions bombardment by surface-grazing. The discontinuous surface tracks seem to exhibit a periodic structure. The white arrow indicates the ion beam direction. (After Carvalho, A.M.J.F. et al., *Appl. Phys. Lett.*, 90, 073116, 2007. With permission.)

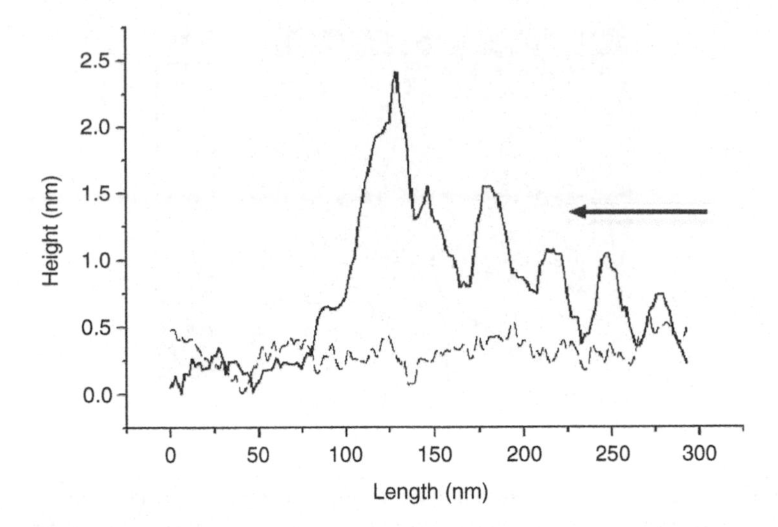

FIGURE 18.8
AFM height profile of an elongated surface track (encircled by the outline reference mark in Figure 18.1) formed after 0.63 MeV $^{208}Pb^{32+}$ ion irradiation at 1° (full line) and height profile of the undisturbed material from an unirradiated sample that displays a random structure (dashed line). The black arrow refers to the ion beam direction. All of the tracks exhibit a trail of humps, which have an intermittent structure with a noticeable periodicity about an average value of 30 nm, statistically determined considering all the surface tracks produced in this experiment. (After Carvalho, A.M.J.F. et al., *Appl. Phys. Lett.*, 90, 073116, 2007. With permission.)

all theories of these effects are mainly based on the LET_{elec}. Since the discontinuity is highlighted in an amorphous material, one can assume that the intermittent structure does not result from the layered organization of the target material, but should be attributed to other processes. Moreover, considering that the electron density distribution of the target material is not homogeneous [59], this feature of the tracks could be explained by the existence of several local melting points subsequent to transient thermal spikes along the ion path. These results thus offer an experimental look at the electronic energy transfer to the matter along the ion track.

18.4.3 Conclusions of Experimental Studies

Swift heavy ion irradiation may create defects in SiO_2 and bumps at the oxide–Si interface. The appearance of the Si bumps can be explained by the thermal spike model, and it corresponds to a drastic decrease of the effective oxide thickness. On the other hand, we observed that the SiO_2 material could exhibit elongated and intermittent surface tracks after swift heavy ion irradiation under grazing incidence. This result leads to the assumption that in SiO_2, the energy transfer itself along the incoming ion trajectory is discontinuous, in disagreement with the usual simplifying assumptions [42,51–54]. These experimental results may help to explain more fully some electrical effects like radiation-induced soft or hard breakdown [60–63].

18.5 Single Ion-Induced Structural Modifications and Consequences for Integrated MOS Device Reliability

The reliability of gate oxides has been of prime interest since the early days of MOS technology. Technological downscaling, leading to thinner oxides, makes it imperative to assess the intrinsic reliability of swift heavy ion-irradiated oxides. Earlier studies have pointed out that swift heavy ions may induce a latent reliability effect as a result of precursor ion damage [64]. This lifetime degradation was observed for devices that were unbiased during irradiation. It suggests that, even if no electrical field is applied, the intrinsic reliability of the oxide can drastically be altered after swift heavy ion irradiation. In recent work, we pointed out that the interface was a preferential region favoring the oxide crystallization along the swift heavy ion track. This theory is in agreement with the "thermal spike" model [1] and with earlier experimental results [11,12]. We concluded that those silicon nanostructures were embedded in the silicon oxide layer. Here, the consequences of such silicon growth in SiO_2 on the intrinsic reliability of ultrathin gate oxides are discussed. It is well known that thin oxides are susceptible to a variety of radiation effects such as radiation soft breakdown, radiation-induced leakage current, and single-event gate rupture (Chapter 17) [65–67]. But despite those recent works, the effects of radiation on ultrathin oxides are still not fully understood [62]. Concerning SEGR or even single-event burnout, it is usually assumed that these processes are field-assisted degradation modes along the heavy ion path. But some recent works have pointed out that latent oxide defects might be induced by swift heavy ions even when the device is unbiased [64]. Thus, it seems clear that a swift heavy ion induces precursor damage, revealed after irradiation by any electrical stress. In their paper, Suehle et al. do not propose a precise physical explanation concerning the oxide degradation mechanisms [64]. The experimental results described above, correlated with experiments related to swift heavy ion-induced nanostructure growth, may provide possible explanations [1,12,68,69]. Our analysis focuses on the initial defect induced by a single swift heavy ion as a potential risk for the oxide reliability. As the oxide degradation mechanisms seem to be consistent with a thermal spike theory, we can give a possible analysis of oxide degradation that occurs after swift heavy ion irradiation. We emphasize that the structural defect extension may vary with the energy density of the ion and with the intrinsic thermal and conductive characteristics of the target material. We have shown that this specific silicon growth process may preferentially occur in SiO_x regions close to the silicon-rich interface. Considering the created nanodot as a more conductive part of the oxide, it can be described as a local decrease of the equivalent oxide thickness, not unlike what is assume in purely electrical models of oxide breakdown (Chapter 15). We have defined a spatial extension ranging from 0.5 to 8 nm for the structural defects induced in SiO_2 by a single heavy ion. Such silicon growth in the oxide may then induce a localized increase of the electrical field applied across the remaining oxide, which could be responsible for oxide lifetime degradation under bias. The main reliability problems considered here for thin gate oxides are radiation-induced leakage current and radiation-induced soft or hard breakdown [62–65,68–70].

Recent studies have pointed out that radiation-induced soft breakdown may even occur for 4 nm thin oxides [71]. Some of the results concerning the oxide reliability are still under investigation [62], and recent relevant models have been developed such as the "quantum point" contact (QPC) model (Chapter 16) [72]. As the basic idea of the QPC model is a drastic localized increase of the conduction, swift heavy ion-induced silicon nanodot growth in the oxide bulk may be a possible physical explanation. In an earlier paper, Toulemonde has pinpointed the effective radii of ion tracks versus LET_{elec} in the case of a SiO_2 quartz target [27]. The track diameter can easily range up to 10 nm. In the case of soft breakdowns, some results indicate that the heavy ion irradiation substantially reduces the intrinsic oxide lifetime [62]. The leakage path is currently considered to be a permanent conductive region. If we consider that swift heavy ions may induce some local silicon or silica crystallization ranging from 1 to 10 nm in diameter, as observed in earlier studies [1,11,27], we can assume that such crystallization may be a possible physical origin for post-irradiation oxide breakdown.

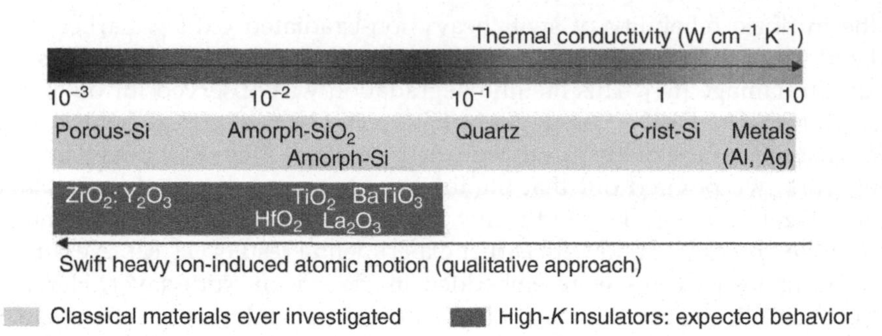

FIGURE 18.9
Correlation between the thermal conductivity of two types of insulating materials, plus Si and metals for comparison, and their relative sensitivity to swift heavy ion-induced atomic motion. (After Touboul, A.D. et al., *J. Non-Cryst. Solids*, 351, 3834, 2005. With permission.)

Scaling down the oxide thickness induces a diminution of the time to the first breakdown. It is also more and more frequent to observe leakage currents during operation, which induce a drastic increase of the device power consumption. For those reasons, alternate oxides are under investigation in the microelectronic industry. High-dielectric constant materials (high-*k*) may be in the future a potential alternative to SiO_2. In high-*k* devices, a very thin SiO_x layer is usually necessary at the interfaced with the Si substrate. In this case, an interface identical to the one described above is included in the device. As described before, one of the key parameters of the thermal spike model is the thermal conductivity. It is thus used to evaluate the sensitivity to swift heavy ion-induced degradation for some materials. In Figure 18.9, a qualitative approach of the possible sensitivity to swift heavy ion-induced atomic motion for classical materials and for several high-*k* insulators is presented. This parameter fits very well with the material sensitivity in the case of the classical materials already investigated [12,27,69], Concerning the high-*k* insulators, we have pinpointed their expected behavior in terms of their macroscopic thermal properties. It is also well known that from porous to amorphous silicon, ion tracks are easily induced by swift heavy ions. On the other hand, ion track creation is much more difficult to induce in metals. Following their thermal properties, high-*k* insulators might be as sensitive to swift heavy ions as materials ranging from porous to amorphous silicon. We want to emphasize that, according to the evaluated sensitivity reported in Figure 18.9, it will be clearly of interest to study the evolution of the electrical characteristics of high-*k* devices after swift heavy ion irradiation [73]. The intrinsic reliability of gate oxides is a major challenge in the short-term development of microelectronics. From the application point of view, in the case of a harsh radiation environment and even if the device is unbiased, the physical oxide degradation may be considered both as a latent defect and also as an acceleration factor of failures.

18.6 Conclusion

Although heavy ion interactions with matter have been extensively studied during the past 50 years, the understanding of structural defects created in matter has been mainly based on traditional displacement damage theories. However, it has been shown that many other theories may be useful as well and that "the right model has to be used at the right place." From the experimental point of view, some studies have shown that even a single swift heavy ion can induce structural modifications in thin SiO_2 layers [1,41,74]. Moreover, some recent experimental results have pointed out the urgent necessity to understand the electronic energy transfer along the ion track. It has been especially shown that the energy transfer itself is discontinuous at the nanometric scale [75]. This experimental overview is of prime interest regarding the reliability of advanced devices.

References

1. Chaudhari, P.S. et al., Swift heavy ion induced growth of nanocrystalline silicon in silicon oxide, *J. Appl. Phys.*, 93, 3486, 2003.
2. Touboul, A.D. et al., Growth of heavy ion-induced nanodots at the SiO_2–Si interface: Correlation with ultrathin gate oxide reliability, *J. Non-Cryst. Solids*, 351, 3834, 2005.
3. Wilson, C.T.R., On an expansion apparatus for making visible the tracks of ionizing particles in gases and some results obtained by its use, *Proc. Roy. Soc.*, 87, 277, 1912.
4. Joliot, F., Observation par la methode de Wilson des trajectoires de brouillard des produits de l'explosion des noyaux d'uranium, *CR Acad. Sci.*, 208, 647, 1939.
5. Corson, D.R. and Thornton, R.L., Disintegration of uranium, *Phys. Rev.*, 55, 509, 1939.
6. Silk, E.C.H. and Barnes, R.S., Examination of fission fragment tracks with an electron microscope, *Phil. Mag.*, 4, 970, 1959.
7. Price, P.B. and Walker, R.M., Electron microscope observation of a radiation-nucleated phase transformation in mica, *J. Appl. Phys.*, 33, 2625, 1962.
8. Fleischer, R.L. and Price, P.B., Tracks of charged particles in high polymers, *Science*, 140, 1221, 1963.
9. Price, P.B. and Walker, R.M., Chemical etching of charged particle tracks, *J. Appl. Phys.*, 33, 3407, 1962.
10. Brinkman, J.A., On the nature of radiation damage in metals, *J. Appl. Phys.*, 25, 961, 1954.
11. Izui, K. and Furuno, S., High energy heavy ion tracks in germanium and silicon, in *Proceedings of the XIth International Congress on Electron Microscopy*, Kyoto, Japan, 1986, pp. 1299–1300.
12. Toulemonde, M. et al., Track formation and fabrication of nanostructures with MeV-ion beams, *Nucl. Instrum. Meth. B*, 216, 1, 2004.
13. Fleischer, R.L., Price, P.B., and Walker, R.M., Ion explosion spike mechanism for formation of charged-particle tracks in solids, *J. Appl. Phys.*, 36, 3645, 1965.
14. Orowan, E., Fracture and strength of solids, *Rep. Prog. Phys.*, 12, 185, 1948.
15. Katz, R. and Kobetich, E.J., Formation of etchable tracks in dielectrics, *Phys. Rev.*, 170, 401, 1968.
16. Dessauer, F., Uber einige wirkungen von strahlen, *Zeit. Phys.*, 12, 38, 1922.
17. Smoluchowski, R., Mayer, J.E., and Weyl, W.A., in *Phase Transformation in Solids*, Smoluchowski, R., Mayer, J.E., and Weyl, W.A. (Eds.), John Wiley & Sons, New York, 1951.
18. Seitz, F., Displacement of atoms during irradiation, *Solid State Phys.*, 2, 307, 1956.
19. Izui, K., Fission fragment damage in semiconductors and ionic crystals, *J. Phys. Soc. Jpn.*, 20, 915, 1965.
20. Kaganov, M.I., Lifshitz, I.M., and Tanatarov, L.V., Relaxation between electrons and the crystalline lattice, *Sov. Phys.*, 4, 173, 1956.
21. Toulemonde, M. et al., Track creation in SiO_2 and $BaFe_{12}O_{19}$, *Nucl. Instrum. Meth. B*, 116, 37, 1996.
22. Baranov, I.A. et al., Inelastic sputtering of solids by ions, *Sov. Phys. Usp.*, 31, 1015, 1988.
23. Katin, V.V., Martinenko, Y.V., and Yavlinskii, Y.N., Low-temperature ionization wave, *Sov. Tech. Phys. Lett.*, 13, 276, 1987.
24. Toulemonde, M., Dufour, C., and Paumier, E., Transient thermal process after a high-energy heavy-ion irradiation of amorphous metals and semiconductors, *Phys. Rev. B*, 46, 14362, 1992.
25. Szenes, G., General feature of latent track formation in magnetic insulators irradiated with swift heavy ions, *Phys. Rev. B*, 51, 8026, 1995.
26. Szenes, G., Amorphous track formation in SiO_2, *Nucl. Instrum. Meth. B*, 122, 530, 1997.
27. Meftah, A. et al., Track formation in SiO_2 quartz and the thermal spike mechanism, *Phys. Rev. B*, 49, 12457, 1994.

28. Ritchie, R.H. and Clausen, C., A core plasma model of charged particle track formation in insulators, *Nucl. Instrum. Meth. B*, 198, 133, 1982.
29. Itoh, N. et al., The lattice relaxation energy associated with self-trapping of a positive hole and an exciton in alkali halides, *J. Phys. Condens. Matter*, 1, 3911, 1989.
30. Itoh, N. and Stoneham, A.M., Excitonic model of track registration of energetic heavy ions in insulator, *Nucl. Instrum. Meth. B*, 146, 362, 1998.
31. Shockley, W., Forming semiconductive devices by ionic bombardment, US Patent 2,787,564, 1954.
32. Maszara, W.P. and Rozgonyi, G.A., Kinetics of damage production in silicon during self-implantation, *J. Appl. Phys.*, 60, 2310, 1986.
33. Glaser, E. et al., Ion implantation and recrystallization of SiC, *Inst. Phys. Conf. Ser.*, 142, 557, 1996.
34. Morehead, F.F., Crowder, B.L., and Title, R.S., Formation of amorphous silicon by ion bombardment as a function of ion, temperature and dose, *J. Appl. Phys.*, 43, 1112, 1972.
35. Seitz, F., On the theory of electron multiplication in crystals, *Phys. Rev.*, 76, 1376, 1949.
36. Seitz, F., On the disordering of solids by action of fast massive particles, *Discuss. Faraday Soc.*, 30, 271, 1949.
37. Fleischer, R.L., Price, P.B., and Walker, R.M., *Nuclear Tracks in Solids, Principles and Applications*, University of California Press, Berkeley and Los-Angeles, 1975.
38. Thibaudau, F. et al., Atomic-force-microscopy observations of track induced by swift Kr ions in mica, *Phys. Rev. Lett.*, 67, 1582, 1991.
39. Walker, D.G. and Weller, R.A., Phonon production and nonequilibrium transport from ion strikes, *IEEE Trans. Nucl. Sci.*, 51, 3318, 2004.
40. Hyde, S.C.W. et al., Electronic excitation-enhanced crystallization of amorphous films, in *Proceedings of Quantum Electronics and Laser Science Conference*, Anaheim, California, 1996, p. 58.
41. Rodichev, D. et al., Formation of Si nanocrystals by heavy ion irradiation of amorphous SiO films, *Nucl. Instrum. Meth. B*, 107, 259, 1996.
42. Kopniczky, J. et al., MeV-ion-induced defects in organic crystals, *Radiat. Meas.*, 1–4, 47, 1995.
43. Barlo Daya, D.D.N. et al., Crater formation due to grazing incidence C_{60} cluster ion impacts on mica: A tapping-mode scanning force microscopy study, *Nucl. Instrum. Meth. Phys. Res. B*, 124, 484, 1997.
44. Geguzin, Ya and Vorobyova, I.V., Thermal stability of tracks formed by fragments of fissioned nuclei at the surface of non-metallic crystals, *Sov. Phys. Dokl.*, 17, 674, 1973.
45. Chadderton, L.T., On the anatomy of a fission fragment track, *Nucl. Tracks Radiat. Meas.*, 15, 11, 1988.
46. Chadderton, L.T., Biersack, J.P., and Koul, S.L., Discontinuous fission tracks in crystalline detectors, *Nucl. Tracks Radiat. Meas.*, 15, 31, 1988.
47. Vorobyova, I.V. and Kolesnikov, D.A., Discontinuous fission tracks on the surface of layered dielectrics, *Radiat. Meas.*, 25, 103, 1995.
48. Vorobyova, I.V. and Kolesnikov, D.A., Tracks of heavy multiply charged ions on the surface of anisotropic crystals, *Sov. Phys. Solid State*, 38, 1953, 1996.
49. Ziegler, J.F., Biersack, J.P., and Littmark, U. (Eds.), *The Stopping and Range of Ions in Solids*, Pergammon, NY, 1985; New edition in 2003. (SRIM is freely available at http://www.srim.org. The current version is 2003.26.).
50. Ronchi, C., The nature of surface fission tracks in UO_2, *J. Appl. Phys.*, 44, 3575, 1973.
51. Gol'danskii, V.I., Lantsburg, E.Ya, and Yampol'skii, P.A., Hydrodynamic effect in the passage of fission fragments through condensed matter, *JETP Lett.*, 21, 166, 1975.
52. Vorobyova, I.V. et al., Formation of surface tracks of heavy ions in solids by the shock wave mechanism, *Sov. Phys. Solid State*, 26, 1191, 1984.
53. Vorobyova, I.V., Geguzin, Ya, and Monastyrenko, V.E., Shock wave mechanism of formation of surface tracks of heavy ions in solids, *Sov. Phys. Solid State*, 28, 88, 1986.
54. Vorobyova, I.V., Geguzin, Ya, and Monastyrenko, V.E., Ion tracks in mica studied with scanning force microscopy using force modulation, *Sov. Phys. Solid State*, 28, 2402, 1986.
55. Toulemonde, M., Defect creation by swift heavy ions: Material modifications in the electronic stopping power regime, *Appl. Radiat. Isot.*, 46, 375, 1995.
56. Fenyö, D. and Jonhson, R.E., Computer experiments on molecular ejection from an amorphous solid: Comparison to an analytic continuum mechanical model, *Phys. Rev. B*, 46, 5090, 1992, and references therein.

57. GEANT4 is available at http://geant4.web.cern.ch.
58. Johnson, R.E. et al., Sputtering by fast ions based on a sum of impulses, *Phys. Rev. B*, 40, 49, 1989.
59. Sarnthein, J., Pasquarello, A., and Car, R., Structural and elelctronic properties of liquid and amorphous SiO_2: An ab initio molecular dynamics study, *Phys. Rev. Lett.*, 74, 4682, 1995.
60. Sexton, F.W. et al., Precursor ion damage and angular dependence of single event gate rupture in thin oxides, *IEEE Trans. Nucl. Sci.*, 45, 2509, 1998.
61. Ceschia, M. et al., Heavy ion irradiation of thin oxides, *IEEE Trans. Nucl. Sci.*, 47, 2648, 2000.
62. Conley, J.F. Jr. et al., Heavy-ion-induced soft breakdown of thin gate oxides, *IEEE Trans. Nucl. Sci.*, 48, 1913, 2001.
63. Choi, B.K. et al., Long-term reliability degradation of ultrathin dielectric films due to heavy-ion irradiation, *IEEE Trans. Nucl. Sci.*, 49, 3045, 2002.
64. Suehle, J.S. et al., Observation of latent reliability degradation in ultrathin oxides after heavy-ion irradiation, *Appl. Phys. Lett.*, 80, 1282, 2002.
65. Cester, A. et al., Post-radiation-induced soft breakdown conduction properties as a function of temperature, *Appl. Phys. Lett.*, 79, 1336, 2001.
66. Sexton, F.W. et al., Single event gate rupture in thin gate oxides, *IEEE Trans. Nucl. Sci.*, 44, 2345, 1997.
67. Johnston, A.H. et al., Breakdown of gate oxides during irradiation with heavy ions, *IEEE Trans. Nucl. Sci.*, 45, 2500, 1998.
68. Dienes, G.J., Radiation effects in solids, *Ann. Rev. Nucl. Sci.*, 2, 187, 1953.
69. Dartyge, E. et al., New model of nuclear particle tracks in dielectric minerals, *Phys. Rev. B*, 23, 5213, 1981.
70. Cester, A. et al., Accelerated wear-out of ultra-thin gate oxides after irradiation, *IEEE Trans. Nucl. Sci.*, 50, 729, 2003.
71. Ceshia, M. et al., Low field leakage current and soft breakdown in ultrathin gate oxides after heavy ions, electron or x-ray irradiations, *IEEE Trans. Nucl. Sci.*, 47, 566, 2000.
72. Sune, J. et al., Modeling the breakdown spots in silicon dioxide films as point contacts, *Appl. Phys. Lett.*, 75, 959, 1999.
73. Massengill, L.W. et al., Heavy-ion-induced breakdown in ultra-thin gate oxides and high-*k* dielectrics, *IEEE Trans. Nucl. Sci.*, 48, 1904, 2001.
74. Carlotti, J.F. et al., Growth of silicon bump induced by swift heavy ion at the silicon oxide–silicon interface, *Appl. Phys. Lett.*, 88, 041906–1, 2006.
75. Carvalho, A.M.J.F. et al., Discontinuous ion tracks on silicon oxide on silicon surfaces after grazing-angle heavy ion irradiation, *Appl. Phys. Lett.*, 90, 073116, 2007.

19

Impact of Radiation-Induced Defects on Bipolar Device Operation

Ronald D. Schrimpf, Daniel M. Fleetwood, Ronald L. Pease, Leonidas Tsetseris, and Sokrates T. Pantelides

CONTENTS

19.1 Introduction

Bipolar junction transistors (BJTs) play critical roles in modern integrated circuits (ICs), ranging from low-cost analog ICs to high-performance silicon–germanium (SiGe) bipolar complementary metal–oxide–semiconductor (BiCMOS) technologies. In applications such as space systems, defense systems, medical facilities, and high-energy particle accelerators these ICs may be exposed to radiation. The principal effect of total ionizing dose (TID) and displacement damage on bipolar transistors is an increase in the number of defects that participate in Shockley–Read–Hall (SRH) recombination (either at the Si/SiO_2 interface or in the Si bulk) [1,2]. These defects are manifested at the device level as increases in the base current, which lead directly to reductions of the current gain.

Bipolar and BiCMOS technologies are particularly important for analog, radio frequency (RF), and mixed-signal ICs. High-speed bipolar transistors ($f_T > 100$ GHz) based on SiGe are particularly important for the wireless communications market [3]. However, much of the market for BJTs remains in the traditional silicon-based analog IC area. Many of these analog ICs are fabricated in relatively old technologies with relaxed feature sizes compared to state-of-the-art CMOS ICs.

Irradiated bipolar transistors may exhibit long-term degradation of their electrical properties, possibly resulting in functional failure, or transient effects that occur due to collection of radiation-generated charge at device junctions. Transient radiation effects caused by individual ionizing particles (single-event effects) are a serious problem for electronics operated in space, and are becoming an increasingly important issue for advanced technologies at sea level. The charge deposited by a single ionizing particle can produce a wide range of effects, including single-event upset, single-event transients, single-event functional interrupt, single-event latchup, single-event dielectric rupture, and others. The physical structure of SiGe heterojunction bipolar transistors (HBTs), with a large reverse-biased collector junction surrounded by deep trench isolation, makes them particularly vulnerable to single-event effects [4]. While these transient effects are important for many applications, they are not considered further in this work.

The focus of this work is on defects produced by the cumulative amount of energy that goes into electron–hole pair creation. The effects of this radiation, typically called TID radiation, are manifested as semipermanent changes in the BJT current–voltage characteristics. The TID response of BJTs is dominated by defects created at the Si/SiO_2 interface over the base region or by charge trapping in the oxide that covers the base. In addition to TID effects, we also consider the effects of displacement damage in the Si substrate. This damage is usually related to the amount of nonionizing energy loss (NIEL), and it is much more significant for particle irradiation than for photon irradiation.

The oxides that affect the TID radiation response of bipolar transistors are generally field or isolation oxides that have properties significantly different from those of high-quality gate oxides. They are much more likely to see high temperatures during the fabrication sequence and to serve as masking layers during ion implantation. The density of defects in these oxides is frequently quite high and there is usually a great deal of hydrogen present. The radiation response of these oxides may be much more complicated than that observed in MOS-quality oxides; in particular, the amount of degradation that occurs at a given total dose may depend on the dose rate at which the energy is deposited.

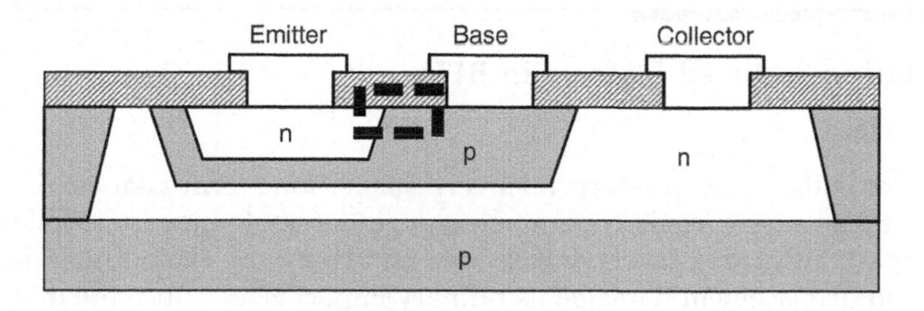

FIGURE 19.1
Cross-sectional view of a conventional npn BJT. The dashed line indicates where the base region intersects the overlying oxide, which is the portion of the device that is most sensitive to ionizing radiation. (After Schrimpf, R.D., *Int. J. High Speed Electron. Syst.*, 14, 503, 2004. With permission.)

19.2 Device Structures

19.2.1 Vertical npn BJTs

A cross-sectional view of a traditional vertical npn BJT is shown in Figure 19.1 [5]. The dashed line in this figure indicates the surface of the base region where it intersects the overlying oxide, which is the portion of the device that dominates the radiation response. Defects at the Si/SiO_2 interface over the base have the greatest impact on the radiation response. Charge trapped in the oxide over the base modulates the surface potential and affects the surface recombination.

19.2.2 Vertical, Lateral, and Substrate pnp BJTs

There are three basic types of pnp bipolar transistors that are used in ICs. Vertical pnp transistors are similar in structure to vertical npn devices. Lateral pnp (LPNP) transistors have the active region of the device at the silicon surface, and the current flows between emitter and collector regions that are both at the surface. Substrate pnp devices have a vertical current-flow pattern, but the substrate serves as the collector for the device. Figure 19.2 shows qualitative cross sections of lateral and substrate pnp transistors. The lightly doped n-type base is usually particularly sensitive to radiation, particularly at the oxide interface. An optional gate electrode is shown over the neutral base. While standard devices do not have this electrode, it is very useful in experiments to understand the mechanisms of radiation-induced damage. The gate allows independent control of the surface potential, which in turn affects the surface recombination rate.

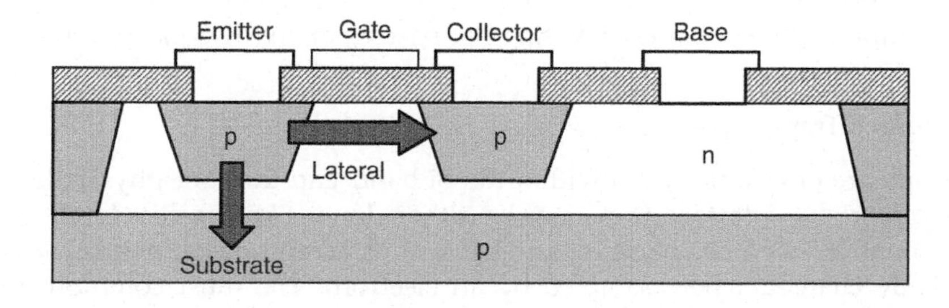

FIGURE 19.2
Representative cross sections of lateral and substrate pnp transistors. The arrows indicate the current-flow paths for each type of device. The lightly doped n-type base can be quite sensitive to radiation, particularly at the oxide interface. An optional gate electrode is shown over the neutral base; this electrode is not present in standard devices.

19.3 Radiation-Induced Defects in BJTs

19.3.1 Overview

Exposure to radiation may produce relatively stable, long-term changes in device and circuit characteristics that result in parametric degradation or functional failure. The TID primarily impacts insulating layers, which may trap charge or exhibit interfacial changes. NIEL results in displacement damage; its primary impact is to reduce the minority-carrier lifetime in the semiconductor regions.

Oxide-trapped charge (described by the areal number density, N_{ot}) refers to radiation-induced charges, typically net positive, that are relatively stable. In thin, high-quality gate oxides, the effects of oxide-trapped charge are minimal because of the small volume in which charge is generated and the ease with which it can tunnel from the oxide. In state-of-the-art MOS ICs, field oxides and isolation structures are usually much less radiation-tolerant than the active device regions [6]. Ionizing radiation also results in formation of interface traps (described by the areal number density, N_{it}) at semiconductor–insulator boundaries that exchange charge with the semiconductor on relatively short timescales. In MOS field effect transistors (MOSFETs), interface traps stretch out the subthreshold *I–V* characteristics and reduce the inversion-layer mobility. In BJTs, the current gain decreases with total dose due to increased surface recombination caused by interface-trap formation [5]. Border traps (Chapter 7) are defects that are similar to oxide traps in microstructure, but electrically behave like slow interface traps [7].

In the Si, some of the carriers generated by the radiation recombine and those that survive are transported by drift and diffusion. The fraction of carriers that survive initial recombination is a function of the carrier concentration and the electric field [8]. The transporting carriers result in transient currents that may affect circuit operation, but have no long-term effect on device characteristics.

19.3.2 Oxide-Trapped Charge

When radiation generates electron–hole pairs in the oxide, the electrons are relatively mobile and, under typical operating biases, leave the oxide under the influence of the electric field. The remaining holes move slowly by a process that involves defect sites in the oxide. These defects trap holes for an amount of time that depends on the trap energy level. While the holes initially are distributed throughout the oxide, the density of deep hole traps is typically highest near the Si/SiO_2 interface [8]. In a MOSFET, the electric field in the gate oxide is determined by the gate bias, but in a BJT, there is usually no gate over the sensitive base region, so the electric field in the oxide is relatively low. As a result of the low electric field and high defect concentrations, the charge yield and subsequent carrier transport may be significantly different in the oxides covering BJT bases than in MOS gate oxides.

19.3.3 Interface Traps

Interface traps are electronic states within the Si band gap that are physically located at the Si/SiO_2 interface. Interface traps can be either donor-like (positively charged when empty and neutral when occupied by an electron) or acceptor-like (neutral when empty and negatively charged when occupied by an electron). The most common process for interface-trap formation is believed to comprise the following steps [9]: (1) release of hydrogen trapped in the oxide by radiation-generated holes, (2) transport of the hydrogen (typically in the form of protons) to the Si/SiO_2 interface through drift or diffusion, (3) reaction of the protons with hydrogen-passivated defects at the interface, forming H_2 molecules and electrically active defects [10–12], and (4) transport of the H_2 molecules away from the interface. Charge in the interface traps affects device operation through electrostatic effects. The energy levels associated with the traps also increase surface recombination velocity by serving as recombination centers, as described below.

19.3.4 Displacement Damage

The nonionizing energy deposited by particle irradiation displaces atoms and creates defects, some of which are electrically active [13]. The amount of energy that goes to the process of creating displacement damage is defined in terms of the NIEL, which is typically given in units of MeV cm^2/g. Many important device parameters, such as leakage current and inverse lifetime, are proportional to NIEL for common devices and irradiation conditions. The proportionality between the radiation-induced increase in thermal generation rate per unit depletion-region volume (ΔG) and NIEL is illustrated in Figure 19.3 for various particle types and energies [14]. The fraction of the total energy loss that produces ionization (as opposed to the NIEL) is typically calculated using a theory originally described by Lindhard et al. [15]. When comparing different irradiation conditions, it is frequently useful to represent the particle fluence in terms of the fluence of 1 MeV neutrons that would produce equivalent damage [16].

The NIEL is partitioned into two general categories: production of isolated defects and production of damage clusters [14]. The electrical performance of devices irradiated with high-NIEL particles (e.g., 100 MeV protons) is dominated by the effects of clusters, while low-NIEL particles (e.g., 1 MeV electrons) produce only isolated defects [14]. The effect of clusters on recombination lifetime is generally greater than that produced by isolated defects.

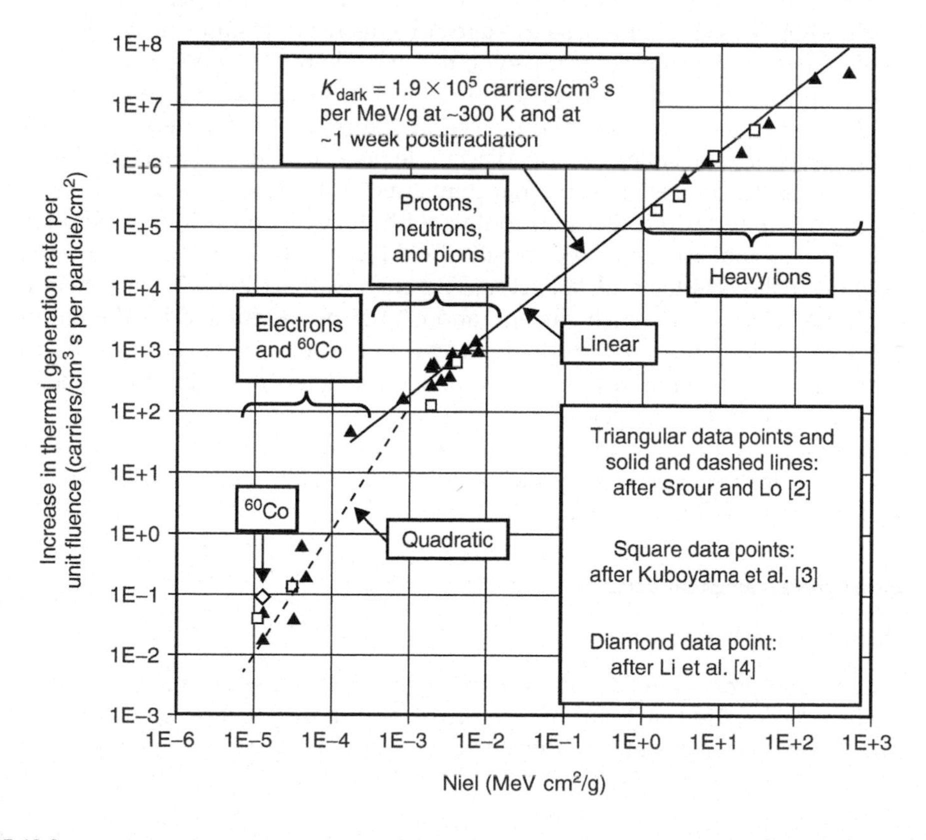

FIGURE 19.3
Radiation-induced increase in thermal generation rate per unit particle fluence versus NIEL in Si devices exposed to a variety of particle types. (After Srour, J.R., *IEEE Trans. Nucl. Sci.*, 53, 3610, 2006. With permission.)

One of the most common types of isolated defect is the Frenkel pair, comprising a single interstitial atom and a corresponding vacancy. The electrical effect of damage produced by low-NIEL particles differs for n- and p-type Si, but this difference largely disappears for high-NIEL irradiations. The relative stability of Frenkel pairs in n- and p-type Si may be responsible for some of this dependence on doping type [17].

The electrically active defects resulting from displacement damage reduce carrier lifetimes and mobilities, change carrier densities, and increase nonradiative transitions in optical devices, among other effects. Minority-carrier devices, including BJTs, are particularly susceptible to displacement damage. In the sections below, we focus on the impact of reduced carrier lifetimes on recombination, since that is the dominant degradation mechanism in irradiated BJTs. The defects create states within the Si band gap that reduce the lifetime and facilitate recombination.

19.3.5 Relationship between Defects and Device Degradation

The principal effect of TID and displacement damage on BJTs is an increase in the number of defects that participate in SRH recombination (either at the Si/SiO_2 interface or in the Si bulk), resulting in increased base current and decreased current gain [1,2]. This process is discussed in detail in Section 19.4. TID has more impact on oxides and interfaces than it does on bulk Si properties, so it is usually sufficient to consider its effect on surface recombination. However, displacement damage can affect recombination in both the bulk and at the Si/SiO_2 interface [5].

In addition to gain degradation, bipolar ICs also may suffer from device-to-device or collector-to-emitter leakage current, similar to the isolation-related leakage currents that may occur in MOSFETs [18–22]. These leakage currents result from positive oxide-trapped charge, which tends to invert adjacent p-type regions. There are two types of leakage that may occur in npn BJTs: (1) buried layer-to-buried layer leakage and (2) collector-to-emitter leakage [18]. In the first case, positive charge in the field oxide inverts the p-substrate between the buried layers (collectors) of adjacent devices. The second mechanism, which is a form of sidewall leakage, is an issue that may affect BJTs with walled emitters. In these devices, the n-type emitter is directly in contact with the isolation oxide. If the p-type base is inverted due to radiation-induced positive oxide charge, it will form a conducting channel between the emitter and collector. The current that flows in this channel is in parallel to the conventional device current, and cannot be modulated by the emitter-base bias in the same way as the conventional current. Either leakage-current mechanism will result in circuit failure if the current exceeds a level determined by power-dissipation issues or circuit-operation constraints.

In some bipolar circuits, the only requirement for the current gain is that it exceeds some minimum value. In these circuits, radiation-induced leakage current is likely to be the dominant failure mechanism.

19.4 SRH Recombination

The recombination process in Si BJTs is dominated by SRH recombination. Since Si is an indirect-gap material, the transition of electrons from the conduction band to the valence band is mediated by defect states within the band gap. The SRH recombination rate, R, is described by [23,24]:

$$R = \frac{pn - n_i^2}{\tau_{n0}(p + p_1) + \tau_{p0}(n + n_1)} \quad (19.1)$$

where
 n is the electron concentration
 p is the hole concentration
 n_i is the intrinsic concentration
 τ_{n0} is the electron lifetime
 τ_{p0} is the hole lifetime

p_1 and n_1 are given by

$$p_1 = n_i \exp\left(\frac{E_i - E_T}{kT}\right) \tag{19.2}$$

and

$$n_1 = n_i \exp\left(\frac{E_T - E_i}{kT}\right) \tag{19.3}$$

where
 E_i is the intrinsic energy
 E_T is the energy of the defects that dominate the recombination process
 k is the Boltzmann constant
 T is the absolute temperature

The electron and hole lifetimes are related to the defect density by

$$\tau_{n0} = \frac{1}{\sigma_n N_T v_t} \tag{19.4}$$

and

$$\tau_{p0} = \frac{1}{\sigma_p N_T v_t} \tag{19.5}$$

where
 σ_n is the electron capture cross section
 σ_p is the hole capture cross section
 N_T is the volumetric density of defects contributing to recombination
 v_t is the thermal velocity

Defects at the Si/SiO_2 interface over the base region typically dominate the TID response of BJTs. The defects at the interface are distributed in energy and the surface recombination rate, R_s, is obtained by integrating the recombination rate as a function of trap energy level over the Si band gap:

$$R_s = \int_{E_V}^{E_C} \frac{p_s n_s - n_i^2}{(p_s + p_{1s})/c_{ns} + (n_s + n_{1s})/c_{ps}} D_{it}(E) \mathrm{d}E \tag{19.6}$$

where
 subscript s indicates that the quantities are to be evaluated at the surface (interface)
 $D_{it}(E)$ is the areal density of interface traps per unit energy
 c_{ns} and c_{ps} are constants of proportionality

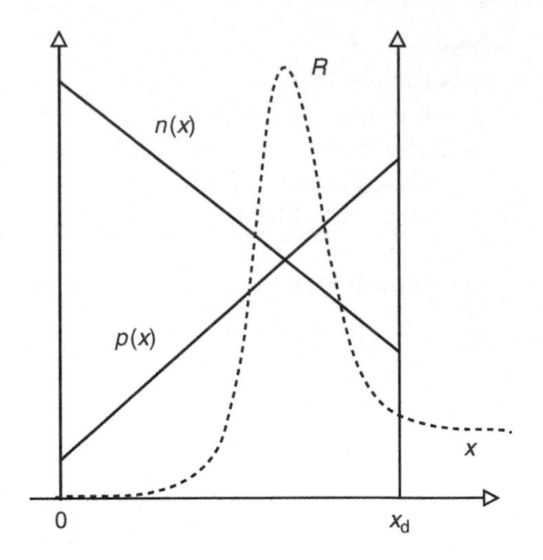

FIGURE 19.4
Electron concentration, hole concentration, and recombination rate versus position within a depletion region. The recombination rate is maximum where the electron and hole concentrations are approximately equal. (After Schrimpf, R.D., *Int. J. High Speed Electron. Syst.*, 14, 503, 2004. With permission.)

A simple approximation to this expression is obtained when the surface is quasineutral:

$$R_s = s_n \Delta n \tag{19.7}$$

where
s_n is the electron surface recombination velocity
Δn is the excess electron concentration

This equation is written for the specific case of excess electrons in p-type Si, corresponding to the base region of an npn BJT. An analogous equation applies for holes in n-type Si. If the recombination process is dominated by traps at a single energy level, the electron surface recombination velocity is related to the physical properties of the defects by

$$s_n = \sigma_n N_{sT} v_t \tag{19.8}$$

where N_{sT} is the areal density of interface traps.

The recombination process requires both electrons and holes. Thus, defects with energy levels located approximately in the middle of the Si band gap are most effective as recombination centers. If the electron and hole lifetimes are equal, these defects have equal probabilities of capturing electrons and holes. The recombination rate also varies with position, since n and p are functions of position. The maximum recombination rate occurs where the electron and hole concentrations are approximately equal, as illustrated in Figure 19.4 [5]. This condition occurs within a depletion region, so the recombination process in bipolar transistors is typically dominated by recombination in the emitter-base depletion region, rather than in the neutral base.

19.5 Bipolar Devices

19.5.1 Current Components

The collector current in an npn BJT operating in the forward-active region consists of electrons that are injected from the forward-biased emitter–base junction, diffuse across the neutral base, and are swept through the collector depletion region by the electric field. If the recombination in the base region is negligible (usually a good assumption), the collector current density is equal to the diffusion current density of electrons in the neutral base:

$$J_C = \frac{qD_{nB}n_{B0}}{W_B}\exp\left(\frac{qV_{BE}}{kT}\right) \tag{19.9}$$

where
- J_C is the collector current density
- D_{nB} is the electron diffusivity in the base
- n_{B0} is the equilibrium electron concentration in the base
- W_B is the neutral base width
- V_{BE} is the base–emitter voltage

The main components of base current in an npn BJT are back-injection of holes from the base to the emitter (J_{B1}), recombination in the emitter-base depletion region (J_{B2}), and recombination in the neutral base (J_{B3}). In unirradiated devices, the back-injection component of the base current typically dominates. However, for devices that have been exposed to ionizing radiation, both J_{B2} and J_{B3} increase, with J_{B2} normally dominating. When the degradation is dominated by displacement damage, J_{B2} and J_{B3} both may increase significantly. The back-injected hole-current density from the base to the emitter, J_{B1}, is a diffusion current:

$$J_{B1} = \frac{qD_{pE}p_{E0}}{L_{pE}}\exp\left(\frac{qV_{BE}}{kT}\right) \tag{19.10}$$

where
- D_{pE} is the hole diffusivity in the emitter
- p_{E0} is the equilibrium hole concentration in the emitter
- L_{pE} is the diffusion length for holes in the emitter

Analogous equations describe the operation of pnp transistors, with the roles of holes and electrons interchanged.

19.5.2 Recombination in the Emitter–Base Depletion Region

The primary effect of ionizing radiation on BJTs is usually an increase in the base current resulting from enhanced recombination in the emitter–base depletion region. The amount by which the base current increases above its preirradiation value is called the excess base current (defined as $\Delta I_B = I_B - I_{B0}$, where I_{B0} is the preirradiation base current). The recombination rate increase occurs mainly where the depletion region intersects the Si–SiO_2 interface, due to formation of interface traps that serve as recombination centers (the surface recombination velocity increases with the interface-trap density).

When a BJT is exposed to energetic particles, displacement damage occurs in the bulk Si. The resulting defects reduce the minority-carrier lifetime, which increases recombination in both the emitter–base depletion region and the neutral base [25].

The recombination rate is a function of position within the depletion region, exhibiting a strong peak where $n = p$, as illustrated in Figure 19.4. The ideality factor (defined as n_B in $\exp(qV/n_BkT)$) is 2 for recombination occurring at this peak and 1 for the ideal component of the base current. However, the recombination rate must be integrated throughout the depletion region to obtain the contribution of recombination to the base current. The excess base current due to surface recombination thus has an ideality factor between 1 and 2, which combines the effects of the different spatial locations at which the recombination takes place. However, it is a reasonable approximation in many cases to assume that the radiation-induced excess base current is dominated by the peak recombination rate, which leads to the results [2]:

$$J_{\mathrm{B2,surf}} \propto s_n \exp\left(\frac{qV}{2kT}\right) \tag{19.11}$$

and

$$J_{\mathrm{B2,bulk}} \propto \frac{1}{\tau_{\mathrm{d}}} \exp\left(\frac{qV}{2kT}\right) \tag{19.12}$$

where

$J_{\mathrm{B2,surf}}$ and $J_{\mathrm{B2,bulk}}$ are the base current densities due to recombination in the depletion region where it intersects the Si/SiO_2 interface and in the depletion region in the bulk Si, respectively

v_{surf} is the surface recombination velocity

τ_{d} is the minority-carrier lifetime in the depletion region

Since the base current due to recombination in the depletion region increases more slowly with voltage than the back-injected (J_{B1}) component of the base current, its effect is greatest at low emitter–base voltages.

19.5.3 Recombination in the Neutral Base

When minority carriers are injected from the emitter into the neutral base, some of them recombine before they reach the collector junction. This neutral–base recombination may take place in either the bulk Si or at the Si/SiO_2 interface (if the Si surface is neutral, as determined by the charge in the oxide or the bias on any electrodes lying over the base region). The fraction of carriers that recombine is small in most unirradiated devices, but it may become significant following irradiation.

If the average lifetime of minority carriers in the neutral base is τ_{B}, the base current density required to supply the recombination process is

$$J_{\mathrm{B3}} = \frac{Q_{\mathrm{B}}}{\tau_{\mathrm{B}}} = \frac{qW_{\mathrm{B}}n_{\mathrm{B0}} \exp\left(\frac{qV_{\mathrm{BE}}}{kT}\right)}{2\tau_{\mathrm{B}}} \tag{19.13}$$

Unlike recombination in the emitter–base depletion region, which exhibits an ideality factor close to 2, the ideality factor associated with recombination in the neutral base is approximately 1. As a consequence, the excess base current associated with neutral–base recombination decreases rapidly as the emitter–base voltage decreases. Recombination in the neutral base is typically more significant for particle irradiation than for photon irradiation since the displacement damage produced by high-energy photons is small compared to that produced by protons or neutrons. Proton irradiation causes both ionization- and displacement-related damage, so the resulting degradation includes all of the processes described above. Additional complications may arise because the charge produced in the oxide can modulate the degradation that occurs due to the bulk traps [26–28].

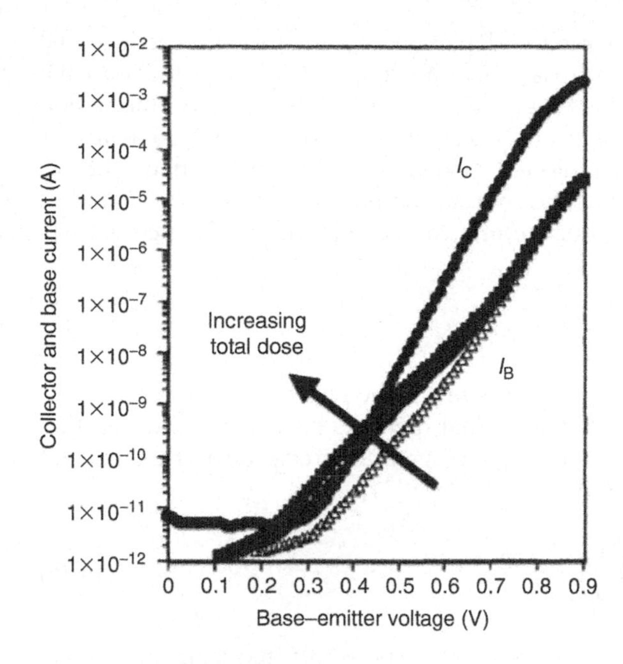

FIGURE 19.5
Collector and base current versus base–emitter voltage for an irradiated npn BJT. (After Schrimpf, R.D., *Int. J. High Speed Electron. Syst.*, 14, 503, 2004. With permission.)

19.5.4 Current Gain

The common-emitter current gain (β) is defined as the ratio of the collector current to the base current:

$$\beta = \frac{I_C}{I_B} \tag{19.14}$$

When a BJT is irradiated, the base current increases, but the collector current typically remains relatively constant, causing the current gain to decrease [29,30]. An example Gummel plot (log I_C and I_B vs. V_{BE}) is shown in Figure 19.5. For this device, the collector current remains virtually unchanged except at very low bias levels, while the base current increases significantly. This causes a large reduction in the current gain, especially at low bias levels where the base current increases most rapidly. The current gain of an irradiated npn BJT is plotted versus V_{BE} in Figure 19.6.

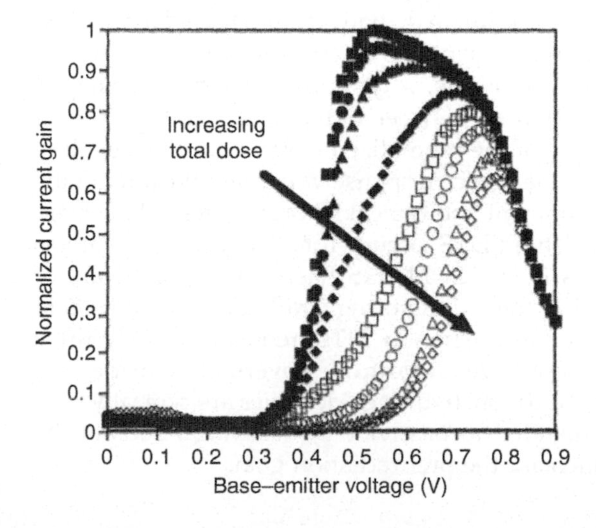

FIGURE 19.6
Normalized current gain versus base–emitter voltage for an npn BJT irradiated to various total doses. (After Schrimpf, R.D., *Int. J. High Speed Electron. Syst.*, 14, 503, 2004. With permission.)

Increased recombination in the emitter–base depletion region does not reduce the collector current at a given bias level because the number of carriers injected into the base depends only on the doping of the base and the applied bias. If recombination increases in the depletion region, the emitter and base currents increase, but the collector current remains constant. However, when injected carriers recombine in the neutral base, they do not reach the collector junction and the collector current decreases.

It has been shown that the minority-carrier lifetime varies with particle fluence as [31]:

$$\frac{1}{\tau_0} = \frac{1}{\tau_i} + \frac{\phi}{K} \tag{19.15}$$

for neutron fluences up to about $2.5 \times 10^{15}\ \mathrm{cm}^{-2}$, where τ_i is the preirradiation lifetime, ϕ is the fluence, and K is a constant of proportionality that depends on doping and injection level. Based on this relation, the radiation-induced change in the reciprocal of the current gain due to neutron irradiation is frequently described by [31]:

$$\Delta\left(\frac{1}{\beta}\right) = \frac{\phi}{2\pi f_t K} \tag{19.16}$$

where f_t is the intrinsic common-emitter cutoff frequency. The cutoff frequency appears in this equation because f_t is inversely proportional to the square of the active base width (W_B^2) and $\Delta(1/\beta)$ is proportional to W_B^2. This equation is valid only for devices in which the gain degradation is dominated by recombination in the neutral base, although it may work empirically in situations dominated by recombination in the emitter–base depletion region, since the neutral base width and the depletion-region width tend to be correlated. As discussed above, the ideality factor of the excess base current will be approximately unity when it is dominated by recombination in the neutral base. This condition is more likely to be satisfied at high current densities near the peak gain.

19.6 Enhanced Low-Dose-Rate Sensitivity (ELDRS)

19.6.1 Overview

The amount of degradation exhibited by many bipolar technologies at a given total dose depends on the rate at which the dose is accumulated, with more degradation often occurring at lower dose rates [32–41]. This phenomenon is referred to as ELDRS. When the amount of degradation is plotted versus dose rate, it typically exhibits an S-shaped curve, with the most rapid variation somewhere between 0.1 and 10 rad(SiO_2)/s, as illustrated in Figure 19.7. Figure 19.7 is a plot of the normalized current gain versus dose rate for a particular LPNP BJT, showing that the device response varies significantly in the range of dose rates between the space environment and typical laboratory tests. However, the specific range of dose rates over which ELDRS occurs (or if it occurs at all) is technology-dependent. Many of the oxides in which ELDRS has been reported are of relatively low quality, and the electric field in the oxide during irradiation that results in ELDRS is usually low. Oxides covering the base region of BJTs frequently satisfy these conditions because they may be thick, may not have an electrode covering the oxide, and may have been used as an implant mask. In addition, hydrogen densities are typically high in these oxides, owing to process steps intended to passivate Si dangling bonds at the Si/SiO_2 interface (Chapters 6 and 7) and increase the preirradiation gain.

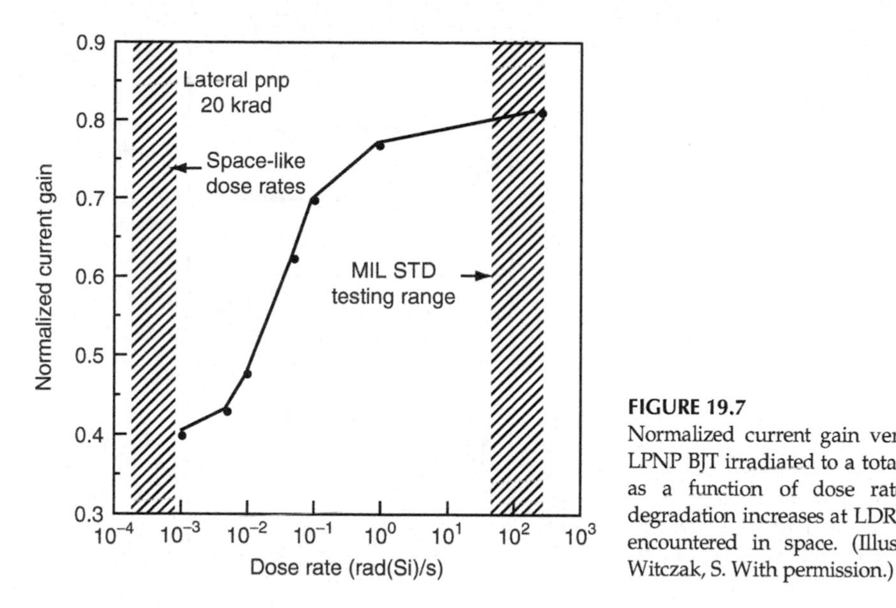

FIGURE 19.7
Normalized current gain versus dose rate for a LPNP BJT irradiated to a total dose of 20 krad(Si) as a function of dose rate. The amount of degradation increases at LDRs approaching those encountered in space. (Illustration courtesy of Witczak, S. With permission.)

At low dose rates (LDRs), the net positive oxide charge and the interface-trap density are both found to be greater in some bipolar base oxides [42]. Thermally stimulated current (TSC) measurements (Chapter 7) showed that the increased net positive oxide charge in these devices was due to increased electron compensation, rather than reduced hole trapping [42].

19.6.2 ELDRS Mechanisms

19.6.2.1 Physical Models

There are three general categories of explanations that have been proposed for ELDRS: (1) space charge-related effects [41–47], (2) bimolecular processes [48–51], and (3) binary reaction rate processes [52]. Each of these models agrees with a significant amount of experimental data, at least for some devices and irradiation conditions. Given the complexity of the phenomenon, it is possible that multiple mechanisms affect the occurrence of ELDRS, perhaps even simultaneously.

The first physical explanations proposed for ELDRS in BJTs were related to space-charge effects due to slowly transporting holes [41–45,53]. Hole transport is mediated by shallow traps in the oxide, while electrons in the oxide are highly mobile and disappear at short times. The space charge that exists at higher dose rates retards the transport of holes (and also protons) to the interface. The positive charge in the bulk of the oxide results in an electric field that increases with proximity to the interface. At high dose rates (HDRs) (large space-charge densities), the field changes sign near the middle of the oxide, becoming negative in the portion of the oxide far from the interface [44]. This field reversal phenomenon reduces the number of holes and protons that can reach the interface. This phenomenon is illustrated schematically in Figure 19.8, in which the low dose-rate case is shown in the top panel and the high dose-rate case in the bottom panel [45]. Note that, in the high dose-rate case, the holes at the interface limit the arrival of protons. The electric field corresponding to the conditions that exist during high dose-rate irradiation (a large amount of transporting positive charge) is illustrated in Figure 19.9 [44]. The holes and protons generated far from the interface encounter a field that moves them away from the interface.

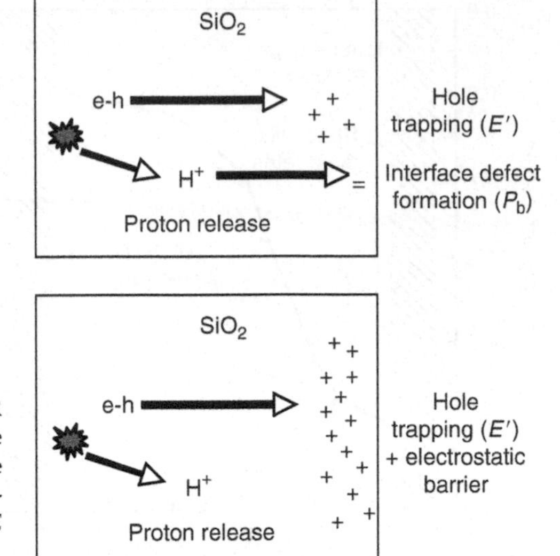

FIGURE 19.8
Schematic illustration of interface-trap buildup for LDR and HDR. For HDR, an electrostatic barrier near the Si/SiO_2 interface (located at the right-hand side of the image) exists due to trapped holes, which limits the interface-trap formation. (After Rashkeev, S.N. et al., *IEEE Trans. Nucl. Sci.*, 49, 2650, 2002. With permission.)

A critical dose rate exists at which the space charge has a significant effect on the electric field in the oxide. At dose rates far above or far below this, the charge buildup is nearly independent of dose rate. The transition between the two regions results in an S-shaped curve of current gain versus dose rate, as observed experimentally and illustrated in Figure 19.7. Since both holes and protons are affected by the space charge, this same process may affect the dose-rate dependence observed experimentally in the buildup of both oxide charge and interface traps [45].

The transport properties of holes and protons play critical roles in determining the dose-rate dependence of interface-trap buildup [45]. The effective mobility of protons in the oxide is usually much lower than that of holes (several orders of magnitude difference). At LDRs, many of the protons are able to reach the interface, release hydrogen, and form interface traps before enough holes have been trapped to reverse the field and prevent additional protons from reaching the interface. However, at HDRs, many holes reach the interface and are trapped there before a significant number of protons are able to arrive. These holes produce a component of the electric field that is directed away from the interface, which suppresses proton arrival and interface-trap formation.

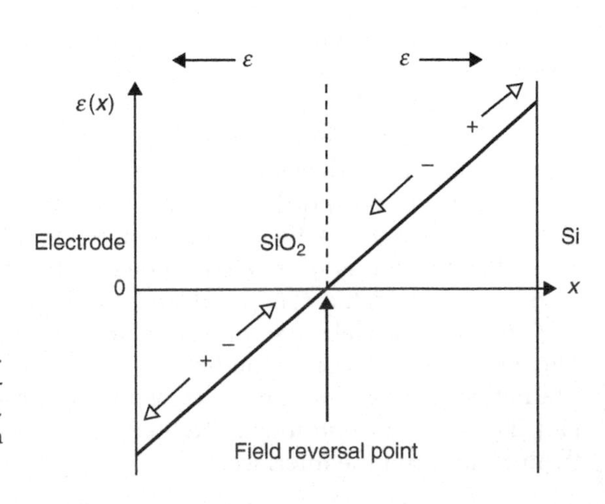

FIGURE 19.9
Electric field versus position in an irradiated oxide. The field reversal occurs due to space charge associated with transporting holes. (After Graves, R.J. et al., *IEEE Trans. Nucl. Sci.*, 45, 2352, 1998. With permission.)

Bimolecular explanations of ELDRS are based on processes that require two different entities for the process to take place. The processes that have been proposed include those involving hydrogen reactions or molecular hydrogen formation [49], and electron–hole recombination [48–51]. Five different possibilities are considered by Hjalmarson et al. [49]: (1) free electron-free hole recombination, (2) free electron–trapped hole recombination, (3) hydrogen release that occurs when a hole cracks a hydrogen molecule, (4) hydrogen dimerization, and (5) hydrogen retrapping. For models based on any of these processes, the defect formation at HDR may be suppressed relative to that at LDR.

In models of the ELDRS phenomenon that emphasize the critical role of electron–hole recombination, the key mechanism is believed to be competition between trapping and recombination processes [50,51]. Due to trapping, some of the radiation-generated electrons may remain in the oxide sufficiently long that they can recombine with transporting holes. The probability of electron trapping is much higher at low electric fields, so more of the electrons recombine under these conditions, reducing the charge yield. Since the carrier concentrations in the oxide are larger at HDRs, the fraction of the carriers that recombine is greater. The increased recombination is manifested as reduced defect formation at HDRs.

Another possibility has been proposed in which the interface-trap buildup results from the interaction of two species according to binary reaction rate theory [52]. In this work, the species were not identified, but they are most likely holes (reactant A) and hydrogen (reactant B). Reactant A is generated (or released) during irradiation and subsequently interacts with reactant B, which is already present at the interface, to form interface traps. The rate of interface-trap formation increases at short times, peaks, and then decreases with time as the preexisting quantity of reactant B at the interface is consumed. At later times, a new supply of reactant B that has been released by the radiation in the bulk of the oxide reaches the interface and the rate of interface-trap formation then becomes greater than the initial rate. The amount of degradation that occurs when the delayed species B reaches the interface is much greater if the part is being irradiated at the arrival time. Hence the much greater degradation at LDR, where the irradiation times are quite long compared to irradiation at HDR [52,54].

The ELDRS phenomenon also may depend on the thermal history of the devices [55] and the presence (or absence) of passivation layers on the die [56]. Both of these issues may be related to the packaging process. Although the physical mechanisms responsible for these effects have not been identified conclusively, hydrogen appears to play an important role.

19.6.2.2 Temperature Effects on ELDRS

Based on a space-charge explanation of ELDRS, it was suggested that a similar amount of enhanced degradation could be produced by a high dose-rate irradiation at elevated temperature [43]. It was explained that if ELDRS is related to the space charge produced by slowly transporting holes, higher degradation also would result if the hole transport is speeded up by increasing the temperature. This phenomenon was observed [43] and subsequent work based on this approach examined issues and limitations of elevated-temperature irradiation in detail [37,57–60]. An optimum temperature is also observed for the enhancement produced by high-temperature irradiation; at higher temperatures, increased annealing overtakes the enhancement produced by the reduced space charge [39,40]. The existence of this optimum temperature is illustrated in Figure 19.10, where it is also seen that the maximum degradation observed for elevated-temperature irradiation approaches, but does not reach, the degradation observed at LDR [40].

The other physical explanations of ELDRS summarized here also are consistent with increased degradation for high-temperature irradiation. If one considers a bimolecular process involving electron–hole recombination, the temperature dependence of the

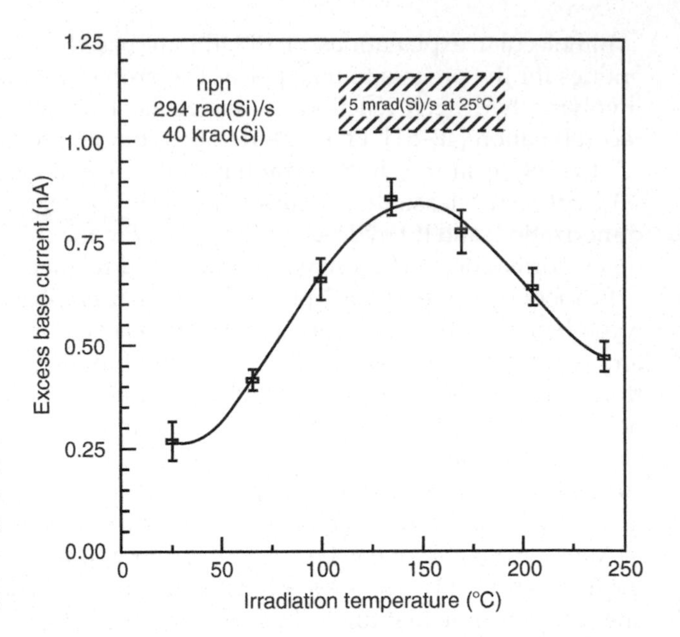

FIGURE 19.10
Radiation-induced excess base current versus irradiation temperature for irradiated LPNP transistors. Also shown is the degradation measured for room temperature irradiation at a LDR of 5 mrad(SiO_2)/s. (After Witczak, S.C. et al., *IEEE Trans. Nucl. Sci.*, 44, 1989, 1997. With permission.)

degradation arises from the variation of the probability of radiation-generated electrons in the oxide recombining with the hole that was generated simultaneously as part of the pair [50,61,62]. In the case of binary reaction rate theory, the increased degradation is the result of accelerated transport of the second species (reactant B, likely hydrogen) at elevated temperature [52].

19.6.2.3 Electric Field Effects on ELDRS

The electric field in the oxide affects the occurrence of ELDRS through its effect on the motion of the charge species in the oxide over the base region. The effects of electric field can be studied by including an independent gate above the oxide that covers the surface of the base region. Standard BJTs do not have a gate over the base region, but inclusion of the gate allows independent control of the surface potential. Figure 19.11 is a plot of interface-trap density versus total dose for gated LPNP transistors irradiated at two different dose rates (LDR and HDR) and with two different biases (0 and −100 V) during irradiation. These results do not include additional annealing time for the high dose-rate irradiations, because there were no directly comparable data at both dose rates. However, these data are sufficient to suggest that the effect of dose rate is much larger than the effect of irradiation bias for these samples and experimental conditions. The samples irradiated at −100 V had a negative field on the oxide side of the interface during irradiation (because the radiation bias was more negative than the flatband voltage), while the samples irradiated at 0 V probably had a positive field in the same region because of the high density of preirradiation positive-trapped charge.

These results suggest that there may be rate-limiting processes that affect ELDRS, other than transport of protons in the bulk of the oxide to the interface, where they react to form interface traps. If the dose-rate dependence is caused only by processes that affect the transport of protons to the interface from the bulk of the oxide, we should see significant suppression of interface-trap formation at negative biases, both at HDR and LDR; this suppression is not observed in the data of Figure 19.11. As discussed in Section 19.7, hydrogen released from the substrate or near-interfacial region may play a significant role in ELDRS.

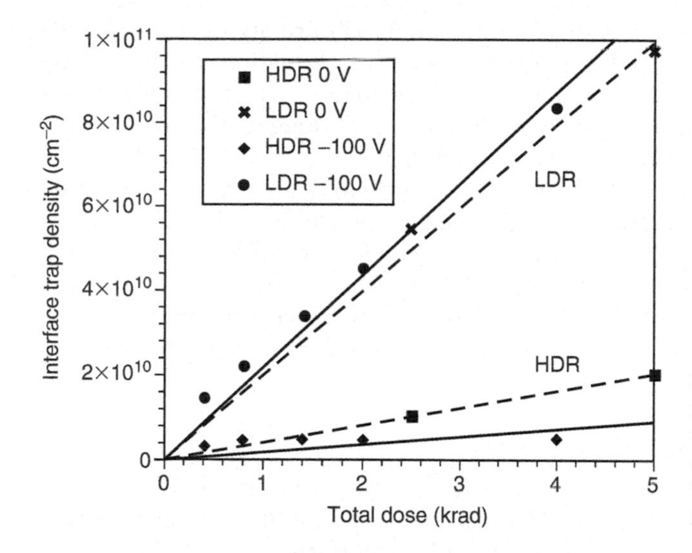

FIGURE 19.11
Interface-trap density versus total dose for gated LPNP transistors irradiated at two different dose rates and with two different irradiation biases (0 and −100 V). The circles represent LDR data and the squares correspond to HDR.

19.7 Impact of Hydrogen on Defects in BJTs

Recent work has suggested that the presence of hydrogen in packages containing bipolar ICs can have a significant impact on the total dose radiation response and the existence of ELDRS [63]. Also, calculations have shown that the physical mechanisms responsible for ELDRS may be related to the processes that produce bias-temperature instability (BTI) in MOSFETs [64], and that both of these degradation phenomena are related to hydrogen. BTI occurs when an electric field in the range of 2–6 MV/cm is applied at temperatures of 100°C–200°C [65]. Instabilities that occur at negative BTI (NBTI) in p-channel MOSFETs have been studied extensively. The n-type substrate in p-channel MOSFETs is analogous to the n-type base region of pnp BJTs for the purpose of considering degradation mechanisms. The kinetics of the degradation process have been analyzed with a reaction–diffusion (RD) model [66] that includes a reaction that generates interface traps and releases mobile reaction products in SiO_2. There are two possible reactions involving H that can create interface traps, i.e., Si dangling bonds, at the Si/SiO_2 interface:

$$\equiv \text{Si–H} + \text{H} \rightarrow \equiv \text{Si}^* + \text{H}_2 \tag{19.17}$$

and

$$\equiv \text{Si–H} \rightarrow \equiv \text{Si}^* + \text{H} \tag{19.18}$$

Here ≡Si–H (≡Si*) is a passivated Si bond at the Si/SiO_2 interface. Equations 19.17 and 19.18 are described as depassivation and dissociation reactions, respectively. Although the dissociation process (Equation 19.18) has been proposed as the BTI reaction [65], a detailed study of the kinetics concluded that reaction (Equation 19.18) is unlikely to play a dominant role for BTI degradation [67]. The reaction energy of reaction (Equation 19.18) is 1.9 eV. When holes are present at the interface, as occurs during NBTI stressing, the reaction energy decreases to 1.6 eV. Even with this reduced value, the reaction energy of reaction (Equation 19.18) is too high for this reaction to be important at typical BTI temperatures.

Unlike dissociation, the depassivation reaction (Equation 19.17) can reach quasiequilibrium during BTI stress, which is one of the requirements for the RD model to be applicable. The reaction energy and barrier of (Equation 19.17) are about 0.5 and 1 eV, respectively [68]. The combination of the reaction energy of (Equation 19.17) and the diffusion barrier of 0.45 eV for the product H_2 in SiO_2 gives an apparent activation energy of 0.35 eV, in agreement with experiments [65,69]. These results suggest that BTI degradation proceeds through depassivation reactions [67,70,71].

The reaction described by (Equation 19.17) requires the arrival of an extra H atom in the vicinity of the passivated Si–H bond. The Si–Si bonds next to interfacial Si–H complexes are trapping sites for an extra H atom, with a binding energy of 0.2 eV [68,70]. The extra H atom may get trapped at this site before or during BTI stress. For H to arrive at the interface during stressing, there must be a source of hydrogen that can reach the interface under negative bias. One possibility is that the H is released from dopants in the substrate. This dopant-release mechanism is plausible because:

1. H-dopant complexes are known to exist in silicon.
2. Under negative bias there are holes at the interface, and the electric field attracts positively charged species from the substrate to the interface. In the presence of holes, H is positively charged, so it drifts toward the interface.
3. Holes facilitate the release of H from dopants in the depletion region of an n-type substrate. For negative bias, the ionized donor atoms are positive and H is negative because the n-type substrate is inverted. The dopant-release process described here does not exclude other possible release mechanisms, for example, H_2O cracking at the Si/SiO_2 interface in the presence of holes [72].

The role of hydrogen in ELDRS may be similar to that described above for NBTI [64]. ELDRS under positive or zero bias has been attributed to electrochemical reactions of H^+ in the SiO_2 gate oxide and ensuing creation of traps at the Si/SiO_2 interface [45,49]. In the case of negative bias, however, H^+ species in the oxide drift away from the interface, so they cannot create interface traps. On the other hand, when H is released in the substrate, it becomes positively charged and drifts toward the interface where it can create interface traps. The difference between ELDRS under negative bias and NBTI is that the role of temperature in the release mechanism is replaced by radiation.

19.8 Electron Capture, Hydrogen Release, and ELDRS

The competition between electron capture by transporting or metastably trapped holes in the base oxide of a linear bipolar device with the release of hydrogen is one factor that may determine in large part whether or not a device exhibits significant ELDRS [73]. When a typical linear bipolar transistor is irradiated under worst-case conditions for ELDRS, the electric field in the oxide is low; effective charge yields are small; and electron and hole transport occur primarily via diffusion. As discussed in Section 19.6, charge transport and trapping are more strongly affected by radiation-induced charge trapping (local space charge) than applied fields. The cross section for electron capture by a trapped hole is approximately the same under low field conditions as the probability for hole trapping, $\sim 10^{-13}$ cm^2 [74–76]. In contrast, the cross section for electron capture by a proton in SiO_2 is $\sim$100 times smaller, as illustrated in Figure 19.12 [77]. Approximately the same charge neutralization that occurs for trapped holes in 60 s of UV irradiation in Figure 19.12 occurs in $\sim$6000 s for protons under the same exposure conditions. Hence, the cross section for radiation-induced charge neutralization for protons in SiO_2 is $\sim 10^{-15}$ cm^2. This lower cross

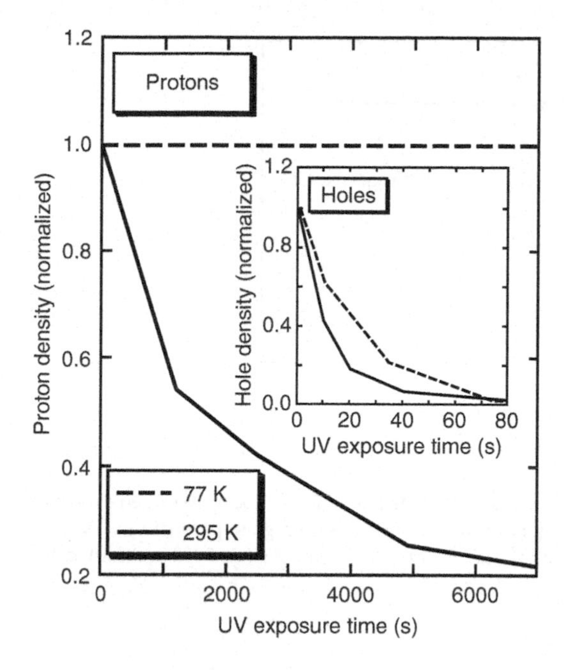

FIGURE 19.12
Relative charge neutralization rates for trapped holes or protons in ~1 μm oxides grown on zone-melt recrystallized materials during 5 eV ultraviolet illumination for the same exposure conditions at room temperature and 77 K. (After Vanheusden, K. et al., *Appl. Phys. Lett.*, 72, 28, 1998. With permission.)

section is consistent with density functional theory (DFT) calculations that show that a proton is not easily neutralized by electrons from the Si as it approaches the Si/SiO_2 interface [11,12]. The proton experiences a low barrier (~0.3 eV) for lateral motion of the H^+ along the Si/SiO_2 interface, increasing the probability that it can react with a passivated dangling bond (Si–H) to form an interface trap via the reaction: $Si\text{–}H + H^+ = Si^+ + H_2$ [11,12]. In contrast, the cross section for an electron to be captured by a slowly transporting or metastably trapped hole in an E'_δ center (most likely an unrelaxed O vacancy in SiO_2 [78,79]) is much higher still, $\sim 10^{-12}$ cm^2 [80,81]. Hence, the probability of a metastably trapped or slowly transporting hole in an E'_δ center being annihilated or compensated by electron capture is $\sim 10^3$ times greater than the probability of a proton being neutralized by an electron in SiO_2. Hence, the positive space charge [41–45] that most strongly determines the direction of electron transport at low fields has a much higher probability of leading to charge neutralization when it is in the form of a transporting or trapped hole than after the charge has been transferred to H to release a proton.

If a diffusing hole in SiO_2 encounters a site at which proton release is energetically favored before being captured in an E'_δ center or a deep hole trap, the overall probability for interface-trap creation to occur is enhanced significantly because a trapped or neutralized hole cannot transfer its charge to hydrogen. At HDRs, there is an abundance of electrons available in a given unit of time to be captured by transporting or metastably trapped holes in E'_δ centers. At lower rates, there is more time for the hole to be released before it is neutralized, because the density of electrons in the oxide is reduced accordingly [48–51]. DFT calculations have demonstrated a broad range of bonding and reaction energies for hydrogen-related defects in SiO_2. For example, Bunson et al. [82] have described the reaction energies for hydrogen release in SiO_2 for a variety of defect complexes incorporating hydrogen. For example, an O vacancy bridged by a single H atom (Si–H–Si) strongly favors proton release upon hole capture [82]. Thus, the simple substitution of a Si–H–Si defect for a Si–Si defect in SiO_2 can make an oxide more prone to exhibit ELDRS. This hydrogen incorporation can occur during device processing; or during device irradiation, preirradiation elevated-temperature stress, and storage; depending on fabrication details, passivation layers, and storage environment [47,53,55,56,63,83,84].

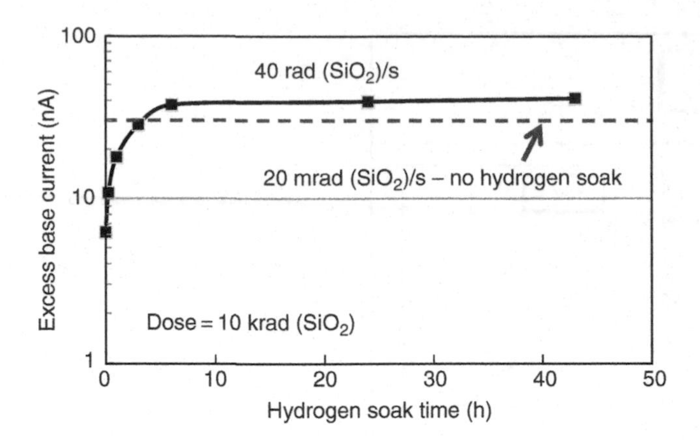

FIGURE 19.13
Excess base current at zero gate voltage and base-emitter voltage = 0.5 V as a function of 100% hydrogen exposure at room temperature for gated LPNP devices built by National Semiconductor, with p-glass passivation. The low rate results are from Ref. [47]. (From Fleetwood, D.M. et al., *IEEE Trans. Nucl. Sci.*, 55(6), 2008, submitted for publication. With permission.)

As an example of the modification of the radiation response of a linear bipolar transistor base oxide that can occur with exposure to hydrogen, Figure 19.13 shows a comparison of hydrogen exposure data (room temperature, 100% H_2, using an apparatus described in Ref. [63]) with low-dose rate test results from Ref. [47] for gated LPNP transistors from National Semiconductor with p-glass passivation [73]. For these devices, irradiation and exposure conditions, ~3 h of exposure leads to about the same degradation as a reduction of dose rate from 40 rad(SiO_2)/s to 20 mrad(SiO_2)/s. Additional hydrogen exposure time only leads to a small amount of additional degradation. Hence, the effect has nearly saturated in a few hours. These results suggest that H_2 has diffused into the oxide and reacted efficiently at O vacancy sites, thereby enhancing the probability that H^+ is released during the high dose-rate irradiations. In the absence of this excess, weakly bonded hydrogen, transporting or metastably trapped holes are trapped or neutralized instead, neither of which leads to excess base current in these devices. Similar hydrogen motion and reactions have been inferred to contribute to time-dependent irradiation and aging effects [9–12,46,47,52–57,63,64,83,84], as well as preirradiation elevated-temperature stress [55,56].

It is not just the probability for the release of hydrogen that influences whether a device shows ELDRS. If the H^+ becomes trapped in SiO_2, e.g., at an O vacancy or suboxide bond, the resulting increase in trapped positive charge moderates the gain degradation for a pnp transistor, as compared to interface-trap creation, which leads to more gain degradation for a pnp transistor [85]. If H^+ is released from its trapping site in the SiO_2 and later forms an interface trap, one can see a significant increase in gain degradation at extreme LDRs or during postirradiation annealing [52,54,55,86]. This introduces additional variability into pnp device response. Depending on the nature of the charge trapping and defect creation, an oxide with a large O vacancy density and significant hydrogen concentration can appear hard at all dose rates, soft at all dose rates, or exhibit ELDRS. This depends on the relative fractions of holes and H^+ that are trapped and uncompensated versus those that contribute to interface-trap creation [73]. The removal of hydrogen in device processing likely would reduce ELDRS, although almost certainly at the expense of lower pre-irradiation gain, for typical lateral or substrate pnp transistors. Hence, it is likely that ELDRS will continue as a concern for linear bipolar devices used in space systems.

19.9 Summary

Degradation of bipolar transistors due to irradiation is a serious problem, particularly for devices in space systems where the total dose is accumulated at a low rate. Degradation caused by TID results from the combined effects of defects at the Si/SiO_2 interface and from the electrostatic effects of charge trapped in the oxide that covers the base region. Both npn and pnp transistors degrade due to interface traps, but positive charge trapping tends to moderate the degradation in pnp BJTs. However, pnp transistors are usually more sensitive to TID because the current flows laterally under the sensitive oxide interface.

The formation of interface traps is a complex process in which hydrogen and hydrogen-related species play key roles. In particular, ELDRS can be produced by several phenomena that originate with hydrogen release in different regions of the device. The key physical processes related to ELDRS include

1. Space-charge effects in the oxide due to slowly transporting or trapped holes
2. Differences in the mobilities, and therefore the transport times, of holes and protons
3. Increased electron capture and reduced charge yield at HDRs
4. Differences in electron capture cross sections for holes and protons in SiO_2
5. Proton release from dopant atoms in the substrate

Bipolar transistors also degrade due to the effects of radiation-induced displacement damage. Defects in the bulk of the Si contribute to the displacement-related degradation, as well as defects at the Si/SiO_2 interface. Cluster defects produced by high-NIEL particles produce the greatest amount of displacement-related degradation.

Acknowledgments

Technical contributions from Hugh Barnaby, Ken Galloway, Steve Kosier, Lloyd Massengill, Matt Beck, and Sergey Rashkeev are gratefully acknowledged.

References

1. Schrimpf, R.D., Recent advances in understanding total-dose effects in bipolar transistors, *IEEE Trans. Nucl. Sci.*, 43, 787, 1996.
2. Kosier, S.L. et al., Physically based comparison of hot-carrier-induced and ionizing-radiation-induced degradation in BJTs, *IEEE Trans. Electron Devices*, 42, 436, 1995.
3. Cressler, J.D., SiGe HBT technology: a new contender for Si-based RF and microwave applications, *IEEE Trans. Microwave Theory Tech.*, 46, 572, 1998.
4. Pellish, J.A. et al., A generalized SiGe HBT single-event effects model for on-orbit event rate calculations, *IEEE Trans. Nucl. Sci.*, 54, 2322, 2007.
5. Schrimpf, R.D., Gain degradation and enhanced low-dose-rate sensitivity in bipolar junction transistors, *Int. J. High Speed Electron. Syst.*, 14, 503, 2004.
6. Turowski, M., Raman, A., and Schrimpf, R.D., Nonuniform total-dose-induced charge distribution in shallow-trench isolation oxides, *IEEE Trans. Nucl. Sci.*, 51, 3166, 2004.
7. Fleetwood, D.M., 'Border Traps' in MOS devices, *IEEE Trans. Nucl. Sci.*, 39, 269, 1992.

8. Derbenwick, G.F. and Gregory, B.L., Process optimization of radiation hardened CMOS circuits, *IEEE Trans. Nucl. Sci.*, 22, 2151, 1975.
9. Saks, N.S. and Brown, D.B., Interface trap formation via the two-stage H^+ process, *IEEE Trans. Nucl. Sci.*, 36, 1848, 1989.
10. Pantelides, S.T. et al., Reactions of hydrogen with Si-SiO_2 interfaces, *IEEE Trans. Nucl. Sci.*, 47, 2262, 2000.
11. Rashkeev, S.N. et al., Proton-induced defect generation at the Si-SiO_2 interface, *IEEE Trans. Nucl. Sci.*, 48, 2086, 2001.
12. Rashkeev, S.N. et al., Defect generation by hydrogen at the Si-SiO_2 interface, *Phys. Rev. Lett.*, 87, 165506.1, 2001.
13. Srour, J.R., Marshall, C.J., and Marshall, P.W., Review of displacement damage effects in silicon devices, *IEEE Trans. Nucl. Sci.*, 50, 653, 2003.
14. Srour, J.R., A framework for understanding displacement damage mechanisms in irradiated silicon devices, *IEEE Trans. Nucl. Sci.*, 53, 3610, 2006.
15. Lindhard, J. et al., Integral equations governing radiation effects (notes on atomic collisions, III), *Mat. Fys. Medd. Dan. Vid. Selsk.*, 33, 1, 1963.
16. Morin, J. et al., Measuring 1 MeV (Si) equivalent neutron fluences with PIN silicon diodes, in *RADECS Proceedings*, Saint-Malo, France, 1993, pp. 20–26.
17. Beck, M.J. et al., Atomic-scale mechanisms for low-NIEL dopant-type dependent damage in Si, *IEEE Trans. Nucl. Sci.*, 53, 3621, 2006.
18. Pease, R.L. et al., Total dose effects in recessed oxide digital bipolar microcircuits, *IEEE Trans. Nucl. Sci.*, 30, 4216, 1983.
19. Titus, J.L. and Platteter, D.G., Wafer mapping of total dose failure thresholds in a bipolar recessed field oxide technology, *IEEE Trans. Nucl. Sci.*, 34, 1751, 1987.
20. Enlow, E.W. et al., Total dose induced hole trapping in trench oxides, *IEEE Trans. Nucl. Sci.*, 36, 2415, 1989.
21. Raymond, J.P., Gardner, R.A., and LaMar, G.E., Characterization of radiation effects on trench-isolated bipolar analog microcircuit technology, *IEEE Trans. Nucl. Sci.*, 39, 405, 1992.
22. Jenkins, W.C., Dose-rate-independent total dose failure in 54F10 bipolar logic circuits, *IEEE Trans. Nucl. Sci.*, 39, 1899, 1992.
23. Shockley, W. and Read, W.T. Jr., Statistics of the recombination of holes and electrons, *Phys. Rev. B*, 87, 835, 1952.
24. Hall, R.N., Electron-hole recombination in germanium, *Phys. Rev. B*, 87, 387, 1952.
25. Messenger, G.C. and Spratt, J.P., The effects of neutron irradiation on germanium and silicon, *Proc. IRE*, 46, 1038, 1958.
26. Barnaby, H.J. et al., Test structures for analyzing proton radiation effects in bipolar technologies, *IEEE Trans. Semicond. Manuf.*, 16, 253, 2003.
27. Barnaby, H.J. et al., Proton radiation response mechanisms in bipolar analog circuits, *IEEE Trans. Nucl. Sci.*, 48, 2074, 2001.
28. Barnaby, H.J. et al., Analytical model for proton radiation effects in bipolar devices, *IEEE Trans. Nucl. Sci.*, 49, 2643, 2002.
29. Wei, A. et al., Excess collector current due to an oxide-trapped-charge-induced emitter in irradiated NPN BJTs, *IEEE Trans. Electron Devices*, 42, 923, 1995.
30. Barnaby, H.J. et al., The effects of emitter-tied field plates on lateral PNP ionizing radiation response, in *IEEE BCTM Proceedings*, Minneapolis, MN, 1998, pp. 35–38.
31. Messenger, G.C., A summary review of displacement damage from high energy radiation in silicon semiconductors and semiconductor devices, *IEEE Trans. Nucl. Sci.*, 39, 468, 1992.
32. Enlow, E.W. et al., Response of advanced bipolar processes to ionizing radiation, *IEEE Trans. Nucl. Sci.*, 38, 1342, 1991.
33. Kosier, S.L. et al., Bounding the total-dose response of modern bipolar transistors, *IEEE Trans. Nucl. Sci.*, 41, 1864, 1994.
34. Nowlin, R.N. et al., Trends in the total-dose response of modern bipolar transistors, *IEEE Trans. Nucl. Sci.*, 39, 2026, 1992.
35. Nowlin, R.N. et al., Hardness-assurance and testing issues for bipolar/BiCMOS devices, *IEEE Trans. Nucl. Sci.*, 40(6), 1686, 1993.

36. Nowlin, R.N., Fleetwood, D.M., and Schrimpf, R.D., Saturation of the dose-rate response of BJTs below 10 rad(SiO_2)/s: Implications for hardness assurance, *IEEE Trans. Nucl. Sci.*, 41, 2637, 1994.
37. Schrimpf, R.D. et al., Hardness assurance issues for lateral PNP bipolar junction transistors, *IEEE Trans. Nucl. Sci.*, 42, 1641, 1995.
38. Wei, A. et al., Dose-rate effects on radiation-induced bipolar junction transistor gain degradation, *Appl. Phys. Lett.*, 65, 1918, 1994.
39. Witczak, S.C. et al., Accelerated tests for simulating low dose rate gain degradation of lateral and substrate PNP bipolar junction transistors, *IEEE Trans. Nucl. Sci.*, 43, 3151, 1996.
40. Witczak, S.C. et al., Hardness assurance testing of bipolar junction transistors at elevated irradiation temperatures, *IEEE Trans. Nucl. Sci.*, 44, 1989, 1997.
41. Witczak, S.C. et al., Space charge limited degradation of bipolar oxides at low electric fields, *IEEE Trans. Nucl. Sci.*, 45, 2339, 1998.
42. Fleetwood, D.M. et al., Radiation effects at low electric fields in thermal, SIMOX, and bipolar-base oxides, *IEEE Trans. Nucl. Sci.*, 43, 2537, 1996.
43. Fleetwood, D.M. et al., Physical mechanisms contributing to enhanced bipolar gain degradation at low dose rates, *IEEE Trans. Nucl. Sci.*, 41, 1871, 1994.
44. Graves, R.J. et al., Modeling low-dose-rate effects in irradiated bipolar-base oxides, *IEEE Trans. Nucl. Sci.*, 45, 2352, 1998.
45. Rashkeev, S.N. et al., Physical model for enhanced interface-trap formation at low dose rates, *IEEE Trans. Nucl. Sci.*, 49, 2650, 2002.
46. Rashkeev, S.N. et al., Effects of hydrogen motion on interface trap formation and annealing, *IEEE Trans. Nucl. Sci.*, 51, 3158, 2004.
47. Pease, R.L. et al., Characterization of enhanced low dose rate sensitivity (ELDRS) effects using gated lateral PNP transistor structures, *IEEE Trans. Nucl. Sci.*, 51, 3773, 2004.
48. Belyakov, V.V. et al., Use of MOS structure for investigation of low-dose-rate effect in bipolar transistors, *IEEE Trans. Nucl. Sci.*, 42, 1660, 1995.
49. Hjalmarson, H.P. et al., Mechanisms for radiation dose-rate sensitivity of bipolar transistors, *IEEE Trans. Nucl. Sci.*, 50, 1901, 2003.
50. Boch, J. et al., Physical model for the low-dose-rate effect in bipolar devices, *IEEE Trans. Nucl. Sci.*, 53, 3655, 2006.
51. Boch, J. et al., Dose rate effects in bipolar oxides: Competition between trap filling and recombination, *Appl. Phys. Lett.*, 88, 232113, 2006.
52. Freitag, R.K. and Brown, D.B., Study of low-dose-rate radiation effects on commercial linear bipolar ICs, *IEEE Trans. Nucl. Sci.*, 45, 2649, 1998.
53. Fleetwood, D.M. et al., $1/f$ noise, hydrogen transport, and latent interface-trap buildup in irradiated MOS devices, *IEEE Trans. Nucl. Sci.*, 44, 1810, 1997.
54. Freitag, R.K. and Brown, D.B., Low dose rate effects on linear bipolar IC's: Experiments on the time dependence, *IEEE Trans. Nucl. Sci.*, 44, 1906, 1997.
55. Shaneyfelt, M.R. et al., Thermal-stress effects and enhanced low dose rate sensitivity in linear bipolar ICs, *IEEE Trans. Nucl. Sci.*, 47, 2539, 2000.
56. Shaneyfelt, M.R. et al., Impact of passivation layers on enhanced low-dose-rate sensitivity and pre-irradiation elevated-temperature stress effects in bipolar linear ICs, *IEEE Trans. Nucl. Sci.*, 49, 3171, 2002.
57. Pease, R.L. et al., A proposed hardness assurance test methodology for bipolar linear circuits and devices in a space ionizing radiation environment, *IEEE Trans. Nucl. Sci.*, 44, 1981, 1997.
58. Pease, R.L. and Gehlhausen, M., Elevated temperature irradiation of bipolar linear microcircuits, *IEEE Trans. Nucl. Sci.*, 43, 3161, 1996.
59. Pease, R.L., Johnston, A.H., and Azarewicz, J.L., Radiation testing of semiconductor devices for space electronics, *Proc. IEEE*, 76, 1510, 1988.
60. Pease, R.L. et al., Evaluation of proposed hardness method for bipolar linear circuits with enhanced low dose rate sensitivity (ELDRS), *IEEE Trans. Nucl. Sci.*, 45, 2665, 1998.
61. Onsager, L., Initial recombination of ions, *Phys. Rev.*, 54, 554, 1938.
62. Boch, J. et al., Temperature effect on geminate recombination, *Appl. Phys. Lett.*, 89, 042108, 2006.
63. Pease, R.L. et al., The effects of hydrogen in hermetically sealed packages on the total dose and dose rate response of bipolar linear circuits, *IEEE Trans. Nucl. Sci.*, 54, 2168, 2007.

64. Tsetseris, L. et al., Common origin for enhanced low-dose-rate sensitivity and bias temperature instability under negative bias, *IEEE Trans. Nucl. Sci.*, 52, 2265, 2005.
65. Stathis, J.H. and Zafar, S., The negative bias temperature instability in MOS devices: a review, *Microelectron. Reliab.*, 46, 270, 2006.
66. Jeppson, K.O. and Svensson, C.M., Negative bias stress of MOS devices at high electric fields and degradation of MNOS devices, *J. Appl. Phys.*, 48, 2004, 1977.
67. Tsetseris, L. et al., Physical mechanisms of negative-bias temperature instability, *Appl. Phys. Lett.*, 86, 142103, 2005.
68. Tsetseris, L. and Pantelides, S.T., Migration, incorporation and passivation reactions of molecular hydrogen at the Si-SiO_2 interface, *Phys. Rev. B*, 70, 245320, 2004.
69. Zhou, X.J. et al., Negative bias-temperature instabilities in metal-oxide-silicon devices with SiO_2 and SiO_xN_y/HfO_2 gate dielectrics, *Appl. Phys. Lett.*, 84, 4394, 2004.
70. Tsetseris, L. and Pantelides, S.T., Hydrogenation/deuteration of the Si-SiO_2 interface: Atomic-scale mechanisms and limitations, *Appl. Phys. Lett.*, 86, 112107, 2005.
71. Houssa, M. et al., Insights on the physical mechanism behind negative bias temperature instabilities, *Appl. Phys. Lett.*, 90, 043505, 2007.
72. Tsetseris, L. et al., Hole-enhanced reactions of water at the Si–SiO_2 interface, *Mater. Res. Soc. Symp. Proc.*, 786, 171, 2004.
73. Fleetwood, D.M. et al., Electron capture, hydrogen release, and ELDRS, *IEEE Trans. Nucl. Sci.*, 55(6), submitted for publication, 2008.
74. McLean, F.B., Boesch, Jr. H.E., and Oldham, T.R., Electron-hole generation, transport, and trapping in SiO_2, in *Ionizing Radiation Effects in MOS Devices & Circuits*, Ma, T.P. and Dressendorfer, P.V. (Eds.), John Wiley & Sons, New York, 1989, pp. 87–192.
75. Fleetwood, D.M. and Scofield, J.H., Evidence that similar point defects cause $1/f$ noise and radiation-induced-hole trapping in MOS devices, *Phys. Rev. Lett.*, 64, 579, 1990.
76. Fleetwood, D.M., Radiation-induced charge neutralization and interface-trap buildup in MOS devices, *J. Appl. Phys.*, 67, 580, 1990.
77. Vanheusden, K. et al., Thermally activated electron capture by mobile protons in SiO_2 thin films, *Appl. Phys. Lett.*, 72, 28, 1998.
78. Chavez, J.R. et al., Microscopic structure of the E'_δ center in amorphous SiO_2: A first principles quantum mechanical investigation, *IEEE Trans. Nucl. Sci.*, 44, 1799, 1997.
79. Lu, Z.Y. et al., Structure, properties, and dynamics of oxygen vacancies in amorphous SiO_2, *Phys. Rev. Lett.*, 89, 285505, 2002.
80. Warren, W.L. et al., Paramagnetic defect centers in irradiated BESOI and SIMOX buried oxides, *IEEE Trans. Nucl. Sci.*, 40, 1755, 1993.
81. Conley, Jr. J.F. et al., Observation and electronic characterization of new E′ center defects in technologically relevant thermal SiO_2 on Si—an additional complexity in oxide charge trapping, *J. Appl. Phys.*, 76, 2872, 1994.
82. Bunson, P.E. et al., Hydrogen-related defects in irradiated SiO_2, *IEEE Trans. Nucl. Sci.*, 47, 2289, 2000.
83. Fleetwood, D.M. et al., Effects of device aging on microelectronics radiation response and reliability, *Microelectron. Reliab.*, 47, 1075, 2007.
84. Chen, X.J. et al., Mechanisms of enhanced radiation-induced degradation due to excess molecular hydrogen in bipolar oxides, *IEEE Trans. Nucl. Sci.*, 54, 1913, 2007.
85. Schmidt, D.M. et al., Modeling ionizing radiation induced gain degradation of the lateral pnp bipolar junction transistor, *IEEE Trans. Nucl. Sci.*, 43, 3032, 1996.
86. Johnston, A.H., Rax, B.G., and Lee, C.I., Enhanced damage in linear bipolar integrated circuits at low dose rate, *IEEE Trans. Nucl. Sci.*, 42, 1650, 1995.

20

Silicon Dioxide–Silicon Carbide Interfaces: Current Status and Recent Advances

S. Dhar, Sokrates T. Pantelides, J.R. Williams, and L.C. Feldman

CONTENTS

20.1 Introduction

Research and development throughout the last decade has led to the emergence of silicon carbide (SiC) electronics. SiC's potential for high-temperature, high-power, and high-frequency electronics arises out of attractive material properties such as wide band gap, high critical breakdown field, high thermal conductivity, and high electron saturation velocity. These properties, coupled with extreme chemical inertness and mechanical hardness, make SiC extremely attractive for electronics operating under extreme conditions.

In general, it has been widely recognized that replacement of conventional Si by a wide band gap material would lead to substantial socioeconomic gains in niche application such as hybrid electric vehicles, power distribution on the grid, power distribution in military vehicles, and sensing technology. A 1995 report by the National Research Council [1] addressed many of the issues related to the development of "next-generation" semiconductors that could replace Si-based devices and operate reliably under extreme conditions. Since then, significant advances have taken place with regards to wide band semiconductor materials such as SiC, GaN, and diamond (see Appendix). Among all these competing materials, SiC technology is currently the most mature with respect to device processing. Currently, SiC Schottky diodes and metal-semiconductor field-effect transistors (MESFETs) are available commercially. Power metal-oxide-semiconductor field-effect transistors (MOSFETs) are expected to be available within the next couple of years. The market for SiC diodes and transistors is expected to exceed $50 million by the year 2009 [2].

SiC has numerous polytypes determined by the stacking sequence of the Si–C bilayers in the crystal structure, with each polytype possessing significantly different electronic properties [3,4]. The 4H and 6H polytypes are most common from a commercial standpoint. Some of the properties of 4H-SiC and 6H-SiC compared to Si are summarized in Table 20.1. 4H-SiC has a wider band gap and a higher and more isotropic bulk electron mobility compared to 6H-SiC. These characteristics make 4H the preferred polytype for high-power and high-temperature applications. From a technological standpoint, the success of any novel metal-oxide-semiconductor (MOS) technology primarily depends on the following factors—the fourth is a particular advantage for SiC over many competing technologies:

1. Availability of large area single crystal wafers of the and high-quality epitaxial layers
2. Effective doping processes for producing n- and p-type materials
3. Suitable metallization schemes for ohmic contact formation
4. Formation of a defect-free and reliable gate dielectric

SiC meets each of these demands as a result of significant breakthroughs and continual development in each of these areas. Although SiC device processing has not yet attained the degree of perfection of Si processing, rapid developments are underway for each of the above issues. Wafer size and quality has been steadily growing while the cost per unit area has been decreasing. At present, 4″ diameter wafers with low defect densities (<15 micropipes cm^{-2}) are commercially available. Ion implantation of N and P (for *n*-type) and B and Al (for *p*-type) are used for producing extrinsic SiC, and various metallization schemes are available for reliably producing ohmic and rectifying contacts. Many issues remain regarding wafer quality, implantation, activation annealing processes and ohmic contact formation, particularly on *p*-SiC. Each of these topics defines separate and extensive research problems, well beyond the scope of discussion in this chapter

TABLE 20.1

Selected Properties of Si, 4H-SiC, and 6H-SiC

Property	Si	4H-SiC	6H-SiC
Band gap (eV)	1.1	3.26	3.0
Critical field (MV cm^{-1})	0.3	2.0	2.4
Saturation carrier velocity ($\times 10^7$ cm s^{-1})	1.0	2.0	2.0
Electron mobility (cm^2 V^{-1} s^{-1})	1350	850	370
Thermal conductivity (W cm^{-1} s^{-1})	1.5	4.5	4.5

(see also Chapter 21). We confine ourselves wholly to the final factor listed above, i.e., the formation of high-quality gate dielectrics for SiC MOS devices. In Section 20.2, we will introduce the classic mobility problem in 4H-MOSFETs. We will discuss in detail the impact of the dielectric–semiconductor interface in relation to SiC MOSFET operation. Next we will take a closer look at the oxidation process in SiC, followed by a discussion of bulk and interfacial composition of oxides grown on SiC. This will be followed by a basic discussion of interface characterization in SiC MOS devices using electrical measurements. In Section 20.6, we will review state-of-the-art interface-trap passivation processes. This will be followed by a discussion on the current understanding of the origin of interface defects and atomic-scale passivation mechanisms. Finally, we will comment on the most relevant problems for state-of-the-art oxides grown on SiC with regards to future device development.

We also note that studies of the SiO_2/SiC interface on different polytypes with different band gaps and employing different crystal faces provide a unique "laboratory" for interface physics. There is no other semiconductor/dielectric system that allows these broad variations, particularly involving SiO_2, the most widely used gate dielectric. Historically this ability to vary the properties of the semiconductor has played a large role in formulating models of the interface and the nature of the interfacial defects.

20.2 SiC MOSFET and the Channel Mobility Problem

A schematic diagram of a typical *n*-channel SiC power MOSFET is shown in Figure 20.1. The highly doped ($>10^{19}$ cm^{-3}) *n*-type source and the moderately doped ($\sim 10^{16}-10^{17}$ cm^{-3}) *p*-type base regions are formed by implantation of N and Al ions, respectively. The absence of a positive gate voltage characterizes the "off-state" where no current flows. In this state of operation, the depletion region formed between the *p*-base and *n*-drift region blocks the large voltage V_B (blocking voltage) applied between the source and drain. When a positive gate voltage is applied, the *p*-surface region at the oxide–semiconductor interface inverts to form a conducting electron channel. As a result, a current I_{SD} flows between the source and drain characterizing the "on-state" of the device. In the on-state, the total resistance R_{ON} to the current flow arises from the sum of the various internal

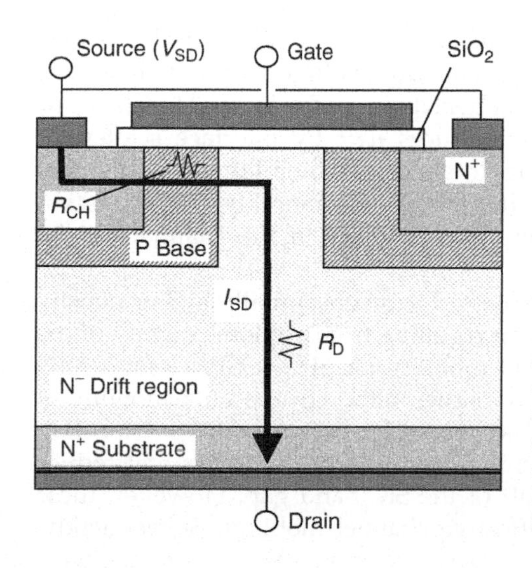

FIGURE 20.1
Schematic diagram of a vertical power MOSFET. The inversion channel is formed by the application of a positive gate voltage. The arrow indicates the flow of electrons from source to drain. R_{CH} and R_D represent the resistance of the inversion channel and the resistance of the drift (blocking) layer respectively. (Adapted from Dhar, S. et al., *MRS Bull.*, 30, 288, 2005.)

resistances of various portions of the device. This quantity is of critical importance in any power device as it dictates the internal power dissipation, which is required to be minimal. Two of the most important internal resistances for SiC MOSFETs are the channel resistance R_{CH} and the drift layer resistance R_D as labeled in Figure 20.1. A figure of merit for evaluating the potential of a vertical MOSFET is the "specific on-resistance" or "resistance-area product" of the device given by [5]:

$$R_{SP} \sim \frac{V_B^2}{\mu_N \varepsilon_S E_C^3} \tag{20.1}$$

where
V_B is the voltage blocked by the MOSFET
μ_N is the bulk electron mobility perpendicular to the semiconductor/oxide interface
ε_S is the permittivity of SiC
E_C is the critical electric field

This parameter represents the ideal on-resistance of the device, which is the resistance of the drift layer in the absence of any other resistances in the device. Owing to an approximately seven times higher E_C of SiC compared to Si, the specific on-resistance of SiC MOSFETs is expected to be ~300 times lower than that of their Si counterparts for the same blocking voltage. This is the primary advantage of SiC over Si for power MOSFET applications [6]. In practice, however, on-resistance may be dominated by other resistance components in the device. For SiC devices (especially for applications at blocking voltages ≤5 kV), one of the most important components is the resistance of the inversion channel. In general, the channel resistance R_{CH} at low electric fields is given by

$$R_{CH} = \frac{L}{W Q_N \mu_N^{CH}} \tag{20.2}$$

where
L and W are the length and width of the channel, respectively
Q_N is the inversion layer charge available for current conduction
μ_N^{CH} is the mobility of carriers in the inversion channel

The later two quantities μ_N^{CH} and Q_N depend critically on the quality of the oxide–semiconductor interface. It is desirable to have Q_N and μ_N^{CH} as high as possible so that $R_{CH} \ll R_D$. In the case of SiC, a large interface defect density in oxides formed by standard oxidation methods reduces Q_N by trapping and reduces μ_N^{CH} by interface scattering. Historically, 4H-MOSFETs exhibited surprisingly low-channel mobility, typically less than 10 $cm^2\ V^{-1}\ s^{-1}$, which is approximately 1/100 of the bulk mobility. This situation may be compared to Si MOSFETs where the electron mobility in the channel can be 40%–50% of the bulk value.

The severe mobility degradation in SiC was attributed to a substantially higher density of interface traps (D_{it}) across the wide SiC band gap resulting from a poorer quality of the SiO_2/SiC interface [7–9]. Interestingly, the channel mobility for 6H-MOSFETs fabricated using similar oxidation methods was considerably higher, although the bulk mobility of the 6H polytype is lower than 4H (300 $cm^2\ V^{-1}\ s^{-1}$ for 6H; ~850 $cm^2\ V^{-1}\ s^{-1}$ for 4H). Initial approaches to solve this problem were focused on processing schemes that reduce D_{it} near the mid gap region and in the lower half of the SiC band gap. However, these methods were not effective in improving the effective channel mobility. Subsequently,

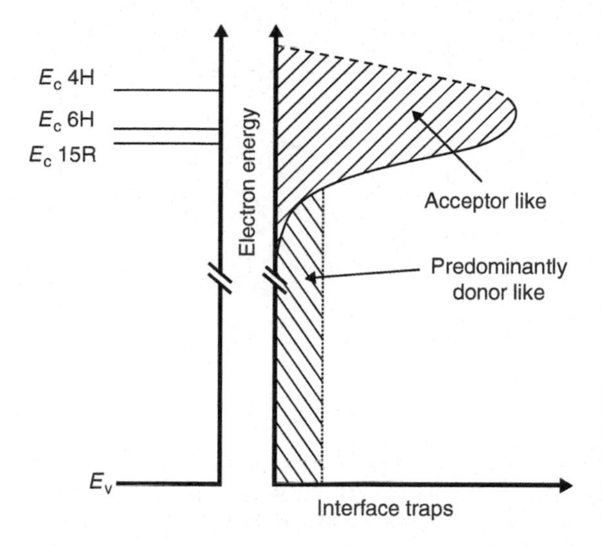

FIGURE 20.2
Schorner's model for the nature of interface traps at the SiO_2/SiC interface for the 4H, 6H, and 15 R polytypes. (From Schomer, R., et al., *IEEE Electron Dev. Lett.*, 20, 241, 1999. With permission.)

Schorner et al. [10] compared the D_{it} distribution and effective channel mobility (μ_{eff}) of *n*-channel MOSFETs fabricated on different SiC polytypes with different band gaps (namely 15R, 6H, and 4H with band gaps of 2.7, 3.0, and 3.23 eV, respectively) and found that μ_{eff} decreased with increasing band gap among the different polytypes. With this experimental data, they suggested that the mobility degradation is caused by an exponentially increasing D_{it} distribution near the conduction band edge of SiC. They pointed out that such acceptor-like traps were energetically located $\leq$0.2 eV from the conduction band edge of 4H-SiC (Figure 20.2). (Note that in MOS nomenclature, acceptor-like traps are negative when filled with an electron and neutral when empty, while donor-like traps are neutral when filled with an electron and positive when empty or filled with a hole). With the valence band edges (VBE) of the different polytypes more or less aligned, the majority of acceptor-like D_{it} distribution lies above the conduction band of 6H-SiC ($E_g \sim 3$ eV), as well as 15R-SiC, and hence do not affect carrier mobility in the inversion layer. However, a substantial fraction of these states lie within the band gap of 4H-SiC ($E_g \sim 3.3$ eV), so that in inversion, a major fraction of the carrier electrons get trapped. This large interface trap density for 4H-SiC was subsequently observed in several laboratories [11–14]. At 0.1–0.2 eV below the 4H conduction band edge, trap densities for as-oxidized interfaces are greater than $\sim 10^{13}$ cm^{-2} eV^{-1} compared to $\sim 10^{12}$ cm^{-2} eV^{-1} near mid gap for unpassivated SiO_2 on Si and 10^{10} cm^{-2} eV^{-1} following passivation with H_2.

When an inversion channel is formed in an *n*-channel MOSFET (*p*-type SiC), a large fraction of carrier electrons are trapped in the acceptor-like interface or near-interface traps energetically located below the position of the surface Fermi level. As higher positive gate voltage is applied for strong inversion conditions, the surface Fermi level moves closer and closer to the conduction band. This results in enhanced carrier trapping due to the exponentially increasing D_{it} close to the band edge, an effect that is most severe for the 4H polytype. The effective mobility is reduced as a major fraction of the mobile inversion charge is rendered immobile due to trapping. In addition, the negatively charged acceptor-like traps undergo Coulomb scattering with the remaining free carriers, which further degrades the effective mobility. It is important to note that, using standard MOSFET current–voltage characterization, the inversion charge carrier density Q_N and the "true" carrier mobility μ_N^{CH} (or Hall mobility) cannot be separated. Rather, an effective mobility that is a function of the product of the two quantities is usually measured. These quantities can be independently determined in SiC inversion layers by Hall measurements on MOS

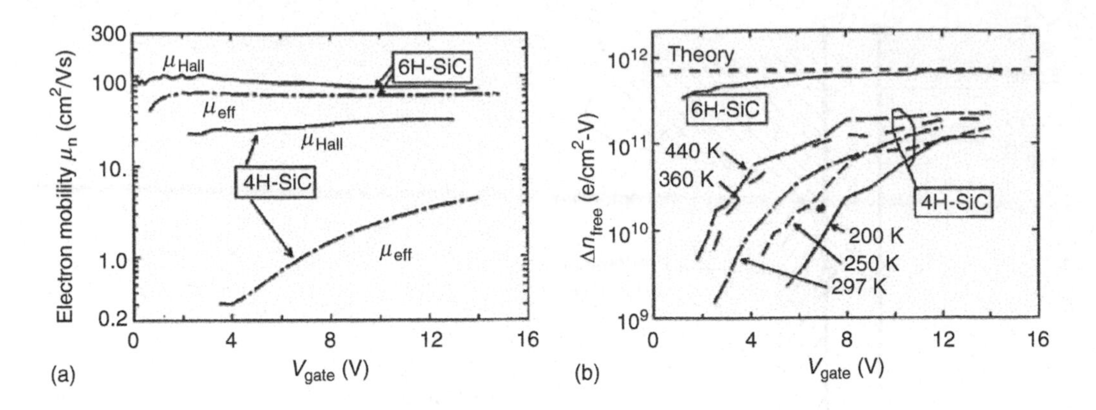

FIGURE 20.3
(a) Comparison of Hall and field-effect mobility in 4H and 6H-SiC MOSFETs. The large discrepancy between μ_{hall} and μ_{eff} for 4H arises because of the much higher electron trapping in 4H. (b) Free electron concentration measured using Hall effect as a function of gate voltage and temperature for 4H-SiC and 6H-SiC. (From Saks. N.S. and Agarwal, A.K., *Appl. Phys. Lett.*, 77, 3281, 2000. With permission.)

gated Hall bars [15,16]. Key results obtained from the work by Saks et al. are shown in Figure 20.3. The Hall measurements reveal that the Hall mobility of inversion electrons in 6H-SiC is approximately equal to the MOSFET effective mobility as shown in Figure 20.3a. On the other hand, in the case of 4H-SiC, the effective mobility was found to be significantly lower than the Hall mobility. Independent measurement of free carrier density in the inversion channel (Figure 20.3b) was shown to be significantly lower in 4H compared to 6H due to enhanced electron trapping at the SiO_2/4H-SiC interface. The temperature dependence of the free carrier concentration, also shown in Figure 20.3b, indicates an increase in free carrier density at higher temperature that is consistent with enhanced interface trapping near the conduction band edge of 4H-SiC. This experiment highlights the importance of interface trap passivation processes for oxides grown on SiC.

20.3 Oxidation of SiC

As in the case of Si, SiC can be converted to silicon dioxide under the influence of oxidizing agents. This is a unique advantage in SiC compared to other compound semiconductors, as it opens up the possibility of MOS fabrication along the lines of Si technology. At first glance, the oxidation of Si and SiC are very similar. However, the oxidation of SiC is significantly more complicated. This is primarily because oxidation of SiC involves not only the formation of bonds between Si and O but also requires the removal of C from the system in the form of CO. In the macroscopic scale, the overall oxidation reaction can be expressed as

$$SiC + 1.5O_2 \rightarrow SiO_2 + CO.$$

Typically the wafers are loaded onto a quartz tube and oxidized at a high temperature (~1100°C–1200°C) in a resistively heated furnace. As in the case of Si, dry oxygen or a wet ambient ($O_2 + H_2O$) can be used for oxidation. From the point of view of oxidation kinetics, the most notable difference between oxidation of Si and SiC is the substantially lower oxidation rate of SiC. Another striking difference is the significantly different oxidation

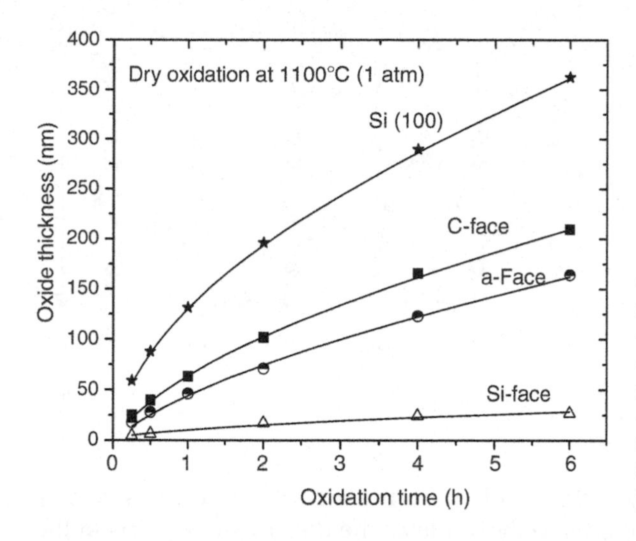

FIGURE 20.4
Oxidation kinetics of different crystal faces of SiC compared to (100) Si for 1100°C dry oxidation. C-face, and Si-face denote the (000–1) C-terminated face and the (0001) Si-terminated face, respectively. The "a-face" denotes the (11–20) face which ideally consists of 50% Si and 50% C atoms on the surface. (From Song, Y. et al., *J. Appl. Phys.*, 95, 4953, 2004. With permission.)

kinetics between the different crystal faces of SiC. Oxide growth curves for 1 atm (flowing gas), dry oxidation at 1100°C are shown in Figure 20.4. Growth kinetics of three 4H-SiC crystal faces—the (0001) Si-terminated face, (000–1) C-terminated face, and the (11–20) face (50% Si and 50% C surface termination) are shown in comparison with (100) Si. The general trend follows the well-documented growth of Si, namely a liner-parabolic type of growth kinetics. As can be seen from the figure, under these oxidation conditions the C-face exhibits ~5 times higher oxidation rate compared to the Si-face, which in turn is a factor of ~2 slower than that of (100) Si. The large oxidation anisotropy is not completely understood and is not explainable using a simple Deal–Grove oxidation model [17,18]. Qualitatively, the oxidation reaction rate scales with the carbon surface areal densities of the different crystal faces. Song et al. have argued that, on the C-face, there is a large driving force for the first layer of C atoms to easily break Si–C bond and forming CO. Following the C removal, the Si atoms in the second layer may react relatively easily with O since there is only one Si–C bond connected to the third layer of C atoms. On the other hand, on the Si-face, it is more difficult to break Si–C bonds and form Si–O since there are three Si–C bonds connected to the second C layer. An alternative explanation has been suggested by Hornetz et al. [19] where the slower oxidation rate of the Si-face was attributed to a thicker oxycarbide transition layer which acts a diffusion barrier to oxygen.

A number of theoretical and experimental studies [20–25] have dealt with atomic-scale mechanisms of Si and SiC oxidation. Herein, we compare some of the fundamental differences between Si and SiC oxidation. Let us consider the Si case first. In Figure 20.5, we show a schematic diagram of an "ideal" abrupt SiO_2/(001) Si interface, where the top layer of Si atoms has two Si back bonds and two "dangling bonds" are connected to the oxide by Si–O–Si bridges. The bond lengths and bond angles are quite close to the preferred values, so that such an abrupt structure can, in principle, be realized. First-principles density functional calculations (DFT) show that "suboxide bonds," i.e., Si–Si bonds on the oxide side of the interface, are not energetically favored [26]. As a consequence, during oxidation, oxygen atoms prefer to insert suboxide bonds on the oxide side of the interface instead of the nearest Si–Si bond on the Si side. This is a key fundamental process that minimizes the chances of oxygen "protrusions" into the semiconductor and a disruption of the oxide ring structure. Once the oxidation process is initiated, the interface advances as O_2 molecules arrive at the interface, break up, and O atoms are inserted into available Si–Si bonds. As a result, (001) Si oxidation proceeds smoothly from the buried interface without any disruptions of the bonding structure, resulting in a fairly abrupt

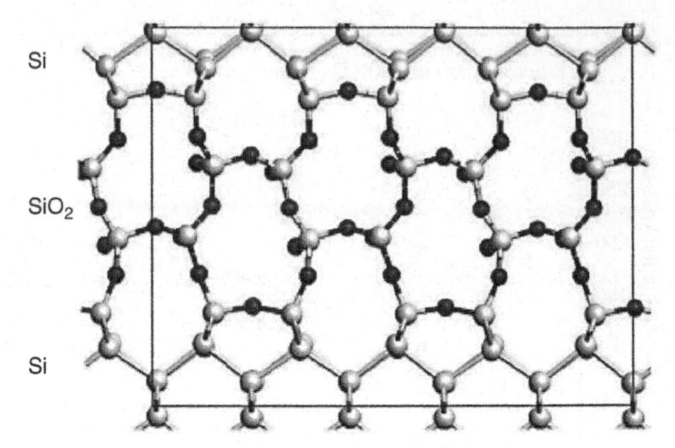

FIGURE 20.5
Schematic of an abrupt SiO_2/(100) Si interface. The oxide is amorphous. An ordered SiO_2 structure is shown for purposes of understanding only. (From Pantelides, S.T. et al., *Mater. Sci. Forum*, 527–529, 935, 2006. With permission.)

SiO_2/Si interface as shown in Figure 20.5. In addition, from density requirements, a very small number of Si atoms must be emitted into the oxide as the density of Si atoms in the oxide is roughly half of the density of Si atoms in crystalline Si.

In contrast, fundamental energy considerations for interfaces formed on the (0001) Si terminated surface of SiC (the preferred crystal face for device fabrication) do not favor an abrupt interface. This is because the oxidation of SiC is fundamentally different from the oxidation of Si as the process now entails emission of C atoms. The basic atomic-scale oxidation step for SiC is shown in Figure 20.6. A locally abrupt SiO_2/SiC interface is shown in Figure 20.6a. The next two O atoms that arrive can remove a C atom and form two Si–O–Si bridges as shown in Figure 20.6c. The process is enabled by the fact that the Si–Si distance in SiC is 3.1 Å, essentially the same as the preferred Si–Si distance in a Si–O–Si bridge. The ejection of the C atom can be facilitated by a third O, whereby a CO is emitted. The energy barriers for these processes have not been calculated, but they are expected to be large, resulting in the slow rate-limiting step that is responsible for

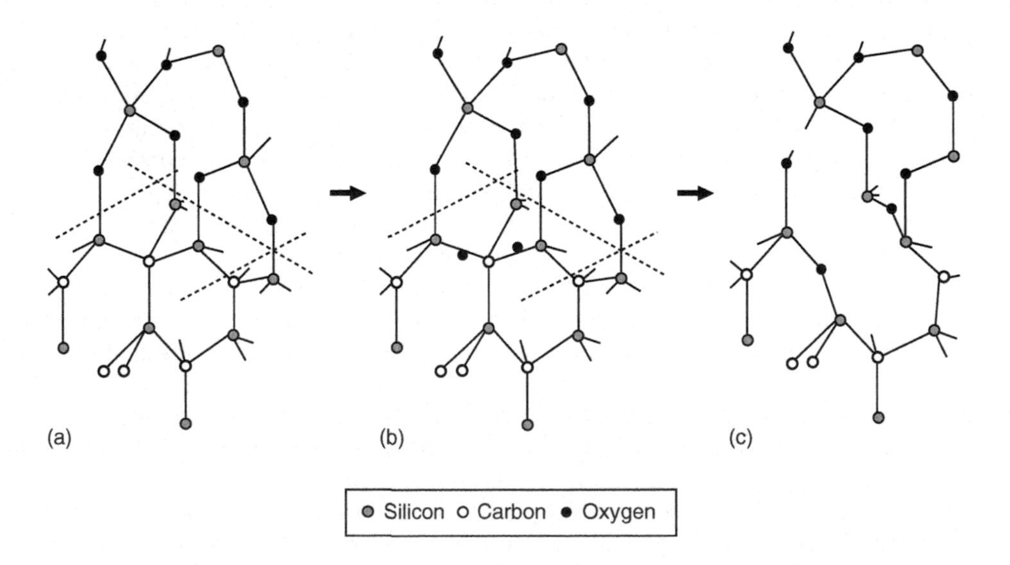

FIGURE 20.6
Basic atomic-scale steps for SiC oxidation: (a) an initially abrupt interface region, (b) two O atoms arrive at the interface as shown, and (c) a C atom is ejected and two Si–O–Si bridges are formed. (From Pantelides, S.T. et al., *Mater. Sci. Forum*, 527–529, 935, 2006. With permission.)

the slower oxidation rates of SiC, compared to Si. The process of ejecting C atoms continues as the oxidation proceeds but as in the case of Si, every once in a while, a Si atom must also be emitted because of density considerations. Overall, it is energetically favorable to oxidize suboxide bonds first (as in the case of Si), so emitted C atoms do not enter the SiC side, and arriving O atoms do not enter the SiC. These are positive aspects of SiC oxidation that favor an abrupt SiO_2/SiC interface. However, there are two main factors that negate this: (1) disruption of the SiO_2 ring structure by the emission of C atoms and (2) incorporation of a minute fraction of the emitted C atoms in the interfacial region. As a result, oxidation of SiC leads to a more complex interface structure.

20.4 Oxide and Interface Composition

The key point is that, since the removal of carbonaceous species through the oxide occurs during oxidation, the bulk of the oxide formed is essentially carbon free. Characterization of the oxide by various optical, x-ray, electron, and ion beam spectroscopy techniques show that the bulk of the oxide is stoichiometric SiO_2 and "carbon free" within the sensitivity of the analytical techniques. A typical secondary ion mass spectroscopy (SIMS) profile of C in the oxide is shown in Figure 20.7. A small trace of C is detected with an average concentration of about $\sim 5.0 \times 10^{19}$ cm^{-3} (corresponding to 5×10^{13} cm^{-2} for a 50 nm oxide with uniform C distribution), which is close to the detection limit of the experiment (10^{19} cm^{-3}). Note that such a small concentration can arise from various SIMS experimental factors such as background carbon in the SIMS chamber, ion-induced carbon deposition, and pin holes in the oxide. Such small levels of C contamination have also been observed in oxides grown on Si. In any case, SIMS sets an upper limit to the concentration of C in the bulk of the oxide to $\leq 1.0 \times 10^{20}$ cm^{-3}. For oxides grown on SiC, bulk properties such as dielectric constant, refractive index, density, and breakdown field are very similar to thermal oxides

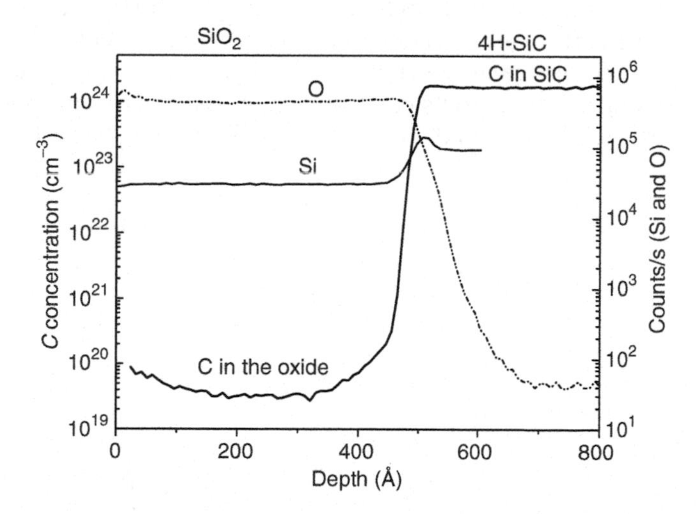

FIGURE 20.7
Typical elemental depth profiles of C, O, and Si obtained from SIMS on SiO_2 grown on 4H-SiC. Data shown are for a ~50 nm oxide grown at 1150°C on the C-face. The measurement shows that the bulk of the oxide is essentially C free. Measured C concentration in the bulk of the oxide is $\sim 5 \times 10^{19}$ cm^{-3} which is close to the sensitivity of the experiment (10^{19} cm^{-2}). Experimental factors such as background carbon in the SIMS chamber, ion-induced carbon deposition and pin holes in the oxide, etc., may also contribute to the C SIMS signal.

grown on Si [27]. However, the situation is not the same at or near the oxide–carbide interface. Numerous studies utilizing a variety of techniques such as ion scattering spectrometry (RBS) [28–31], x-ray photoelectron spectroscopy (XPS) [19,23,32–34], transmission electron microscopy (TEM), and electron energy loss spectroscopy (EELS) [26,35–37], surface-enhanced Raman spectroscopy (SERS) [38], and spectroscopic ellipsometry [39] indicate the presence of a nonstoichiometric carbon containing $Si_xC_yO_z$ layer very close to the interface. Two of these experiments will be discussed here.

20.4.1 Ion Scattering Spectrometry

A typical high-resolution medium energy ion channeling spectrum of a thin oxide (~1 nm) film on 4H-SiC performed with 100 keV H^+ ions is shown in Figure 20.8. Under optimal ion channeling conditions, the intensity peaks associated with the Si, C, and O can be attributed to the following: (1) O, the intensity associated with O atoms in the oxide; (2) Si, the sum of the contributions from atoms in the oxide, "excess" Si in the interfacial region, and the intrinsic Si surface intensity associated with the ideal SiC crystal; and (3) C, the sum of the contributions from excess atoms in the oxide/interfacial region and the intrinsic C surface intensity associated with the ideal SiC crystal. In ion channeling experiments, the intrinsic surface intensity arises from the interaction of the probing ion beam with the first few monolayers of substrate atoms [40]. "Excess" atoms denote the number of Si or C atoms that are in excess as compared to an ideal SiO_2/SiC structure. Taking the intrinsic surface peaks into account, an upper limit for nonstoichiometric oxide (or oxycarbide) transition layer between stoichiometric SiO_2 and ideal SiC can be determined to be <10 Å. In another related experiment, the stoichiometry of oxide films of different thicknesses on 6H-SiC measured by Rutherford back scattering (RBS) and channeling using 1.8 MeV He^+ ions is shown in Figure 20.9 (thickness is represented as atomic areal density of O in the figure, 10^{16} cm^{-2} of O ≈2.2 nm of SiO_2) [41]. The slope of the straight line clearly shows that the ratio of Si:O in the films is close to 1:2. The excess C concentration is independent of the film thickness, indicating that the source of C is at or near the SiO_2/SiC interface. The interfacial C or Si concentration in these measurements (represented by the intercept of the straight line on the vertical axis) is close to that expected for an ideal SiC surface. These experiments set a limit to the maximum amount of excess carbon at the interface to about a monolayer (considering 1×10^{15} atoms cm^{-2} as a monolayer). It is important to note that such deviations (~one monolayer of disorder) from an ideal SiO_2/SiC interface is almost at

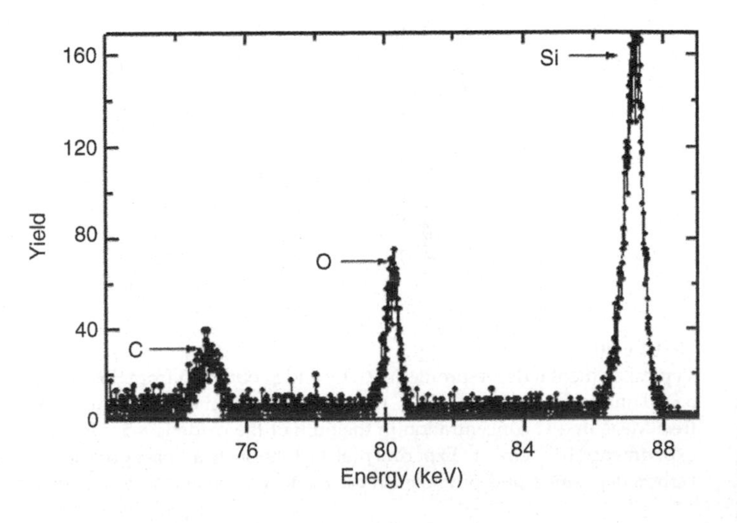

FIGURE 20.8
Typical medium energy ion scattering spectrum of SiO_2/SiC in ion channeling mode using 100 keV protons. In this case the oxide was ~1 nm thick after etch back of a ~30 nm oxide. The peak positions for C, O, and Si are indicated. Area under the peak indicates the areal density of the elements at the surface. (From Dhar, S. et al., *Appl. Phys. Lett.*, 84, 1498, 2004. With permission.)

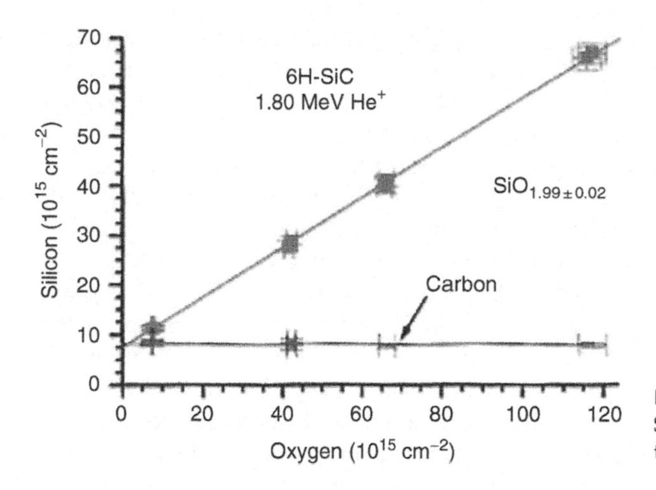

FIGURE 20.9
Stoichiometry measurements on oxide layers thermally grown on 6H-SiC.

the detection limits of most physical analysis tools. However, if electrically active, such defects can be important from the standpoint of device physics.

20.4.2 TEM and EELS

TEM and EELS measurements have also detected nonstoichiometric carbon-containing transition layer at the SiO_2/SiC interface. Figure 20.10 shows Z-contrast images of Si/SiO_2

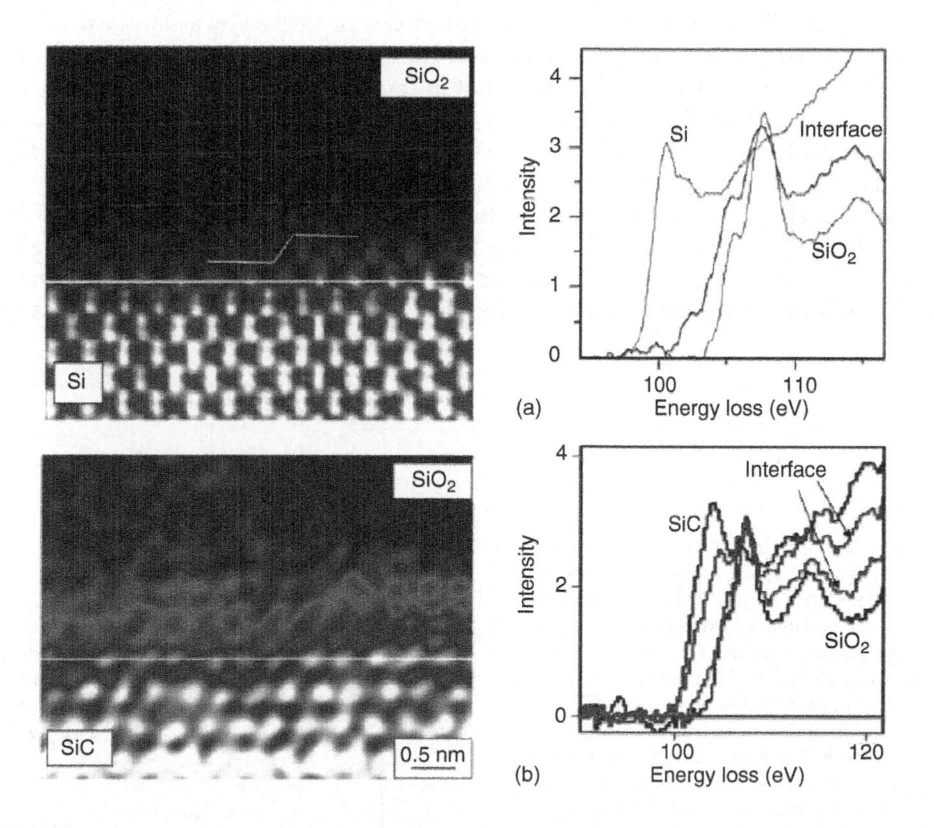

FIGURE 20.10
Z-contrast images of and EELS spectra from (a) a SiO_2/Si interface and (b) a SiO_2/SiC interface. (From Pantelides, S.T. et al., *Mater. Sci. Forum*, 527–529, 935, 2006. With permission.)

and SiC/SiO_2 interfaces using ultra-high resolution scanning transmission electron microscopy (STEM) and EELS [26]. In both the Si/SiO_2 and SiC/SiO_2 interfaces, a step is clearly visible. Such steps could be responsible for macroscopic roughness that contributes to carrier scattering and mobility degradation in the channel relative to the bulk material. Away from the step, the Si/SiO_2 interface appears very abrupt. In contrast, in the SiC/SiO_2 image, a 1–2 atomic layers thick transition layer can be clearly discerned, where the intensity undergoes a transition from values typical in the SiC side to values typical in the SiO_2 side, consistent with the ion scattering measurements mentioned above.

The presence of the interlayer is confirmed by EELS collected from the various regions. In Figure 20.10, the Si L_{23} EELS spectra (transitions from the spin–orbit split Si 2p level to available empty states) are also shown for SiO_2/Si and SiO_2/SiC. In both cases, differences between the spectra from the semiconductor side and the oxide side are clear. For the Si case, in the interface region, distinct spectra that correspond to different oxidation states of Si are obtained. In the SiO_2/SiC interfacial region, however, a very wide variation of signal intensities is observed, which indicates very diverse bonding environments for Si atoms in the interfacial region.

20.5 SiO_2/SiC Interface Trap Density Measurements

Typical interface trap distributions across the entire band gap for as-processed wet oxides grown on 4H-SiC are shown in Figure 20.11. The data have been adapted from the work by Chung et al. [42]. The two sets of data represent oxides processed by a standard 1100°C wet oxidation and oxides processed similarly but terminated by a low-temperature (950°C) wet reoxidation anneal. The figures show that in the case of 4H-SiC the distribution in the upper and lower half of the band gap is asymmetric, with $D_{it} \sim 10^{13}$ cm^{-2} eV^{-1} near the conduction band edge and $\sim 10^{12}$ cm^{-2} eV^{-1} close to the valence band. The D_{it} near the band edges is comparable for dry oxides and wet oxides. But in the case of dry oxides, D_{it} of energetically deep donor states in the lower half of the gap (energetically located between 0.6 eV above the valence band, up to mid gap) is usually higher compared to the deep acceptor-like state density in the upper half of the gap. Similar results are reported by different studies that employ a variety of electrical analysis techniques. There are variations

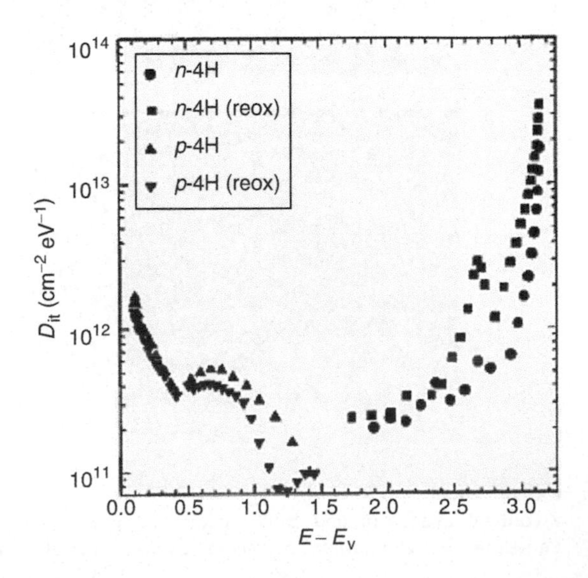

FIGURE 20.11
SiO_2/4H-SiC interface trap density as a function of energy level with respect to the valence band of 4H-SiC for interfaces formed by wet oxidation at 1100°C and wet oxidation followed by a low temperature wet reoxidation anneal at 950°C (denoted "reox"). The D_{it} profile across the band gap was extracted using hi–lo *C–V* technique on *n*-type and *p*-type MOS capacitors for the top and bottom half of the band gap, respectively. D_{it} below and above 0.5 eV from the majority carrier band edge was extracted using measurements at room temperature and at 350°C, respectively. Thicknesses of the oxides were about ~40 nm. (From Chung, G.Y. et al., *Appl. Phys. Lett.*, 76, 1713, 2000. With permission.)

in the absolute value of D_{it} depending on the measurement technique employed (as in the case of Si) and the details of the of oxidation employed (dry, wet, deposited with or without inert ambient postannealing), but all point toward an extremely high D_{it} ($\sim 10^{13}$ cm^{-2}) near the conduction band edge. In addition to different variations of capacitance–voltage (C–V) and conductance–admittance techniques (G–ω) [43–45], a variety of other techniques used widely for Si MOS studies have been employed such as deep level transient spectroscopy (DLTS) [7], thermally stimulated current spectroscopy (TSC) [46,47], and Hall effect [14] for interface characterization in SiC. Additionally, techniques such as internal photoemission spectroscopy [48], positron annihilation spectroscopy [49,50], photon stimulated tunneling [51], and XPS [52] have also been used to characterize the electronic properties of SiO_2/SiC interface traps.

Electrical characterization remains the most sensitive probe for interface imperfections. Typically MOS capacitor structures, MOS gated diodes, or lateral test MOSFETs are used for these measurements. However, the applicability and limitations of each method must be evaluated, considering the larger band gap of SiC, before applying these techniques [9,53]. The underlying physics for the different methods are based on an exchange of charges between the bulk of the semiconductor and its interface in equilibrium. The equilibrium is maintained only if the interface traps follow the dc bias of the probing voltage sweep quasistatically. In other words, an interface trap is detected only if it charges by capturing carriers from the semiconductor and subsequently discharges within the timescale of measurements. If the interface trap captures charge but does not discharge within the measurement cycle, it would appear as a fixed charge. The capture rate is given by

$$c_n = \sigma_n v_T n \tag{20.3}$$

where

c_n is the capture rate (s^{-1})
σ_n represents the electron capture cross section (cm^2)
v_T is the thermal velocity of electrons (cm s^{-1})
n is the density of electrons at the semiconductor surface

The time scale of the capture process in wide band gap MOS systems is not expected to be significantly different than Si as the measured cross sections are of same order of magnitude [9]. However, the time scale of the discharge process is significantly different at any temperature. During measurement, there are essentially two ways by which interface traps can discharge trapped majority carriers: (1) emission to the majority carrier band and (2) capturing a minority carrier from the inversion layer. The former process leads to extremely long response times of interface traps energetically located deep in the semiconductor band gap. This can be understood as follows: the time constant for electron emission from an interface trap at energy E to the conduction band is given by the following expression:

$$\tau_n(E) = \frac{1}{\sigma_n N_C v_T} \exp\left(\frac{E_C - E}{k_B T}\right) \tag{20.4}$$

where

σ_n represents the electron capture cross section of the trap (cm^2)
N_C is the density of states in the semiconductor conduction band
v_T is the thermal velocity of electrons
E_C and E are the conduction band minimum and energy of the trap with respect to the VBE of SiC, respectively

A similar expression holds for trapped hole emission with respect to the VBE. As a consequence of this exponential energy dependence, energetically deep interface traps in wide band gap semiconductors such as SiC have extremely long response times (years) at room temperature. For 4H-SiC at room temperature, only interface traps which are at most 0.6–0.7 eV away from the majority carrier band edge can emit their trapped charge during the measurement, and thus be detected using conventional techniques. Interface traps deeper in the gap, in the broad mid gap region show a fixed-charge-like behavior. At elevated measurement temperatures (~350°C and higher), the emission rate of interface traps increases sufficiently, and almost all interface traps located in the upper or lower half of the gap are detectable, depending on whether *n*-type or *p*-type MOS capacitors are used for the measurement. This is illustrated in Figure 20.12 for interface traps in the upper half of the 4H-SiC band gap, using energy dependent capture cross sections values reported in recent DLTS studies [54]. The measurement frequency (1 MHz) was selected as the lower cut-off for the time constant (high-frequency limit), while the upper cut-off was set to be k_BT times the sweep rate of 10 s V^{-1} (low-frequency limit). It can be seen from the figure that, using electrical measurements at 350°C, interface traps energetically located down to ~1.3 eV below the conduction band edge may be probed accurately.

As mentioned above the alternative way by which interface traps can exchange charge is by capturing opposite charges from the semiconductor. But this process is not favorable in SiC even at high temperatures. An extremely low intrinsic carrier concentration in SiC at room temperature ($n_i \sim 10^{-9}$ cm^{-3} in 4H-SiC, compared to $\sim 10^{10}$ cm^{-3} in Si) leads to orders of magnitude lower minority carrier generation rates than in Si [9,55,56]. As a consequence, inversion layers do not form in SiC MOS capacitors (within the finite time scale of measurements) and *C–V* curves show deep depletion behavior. Therefore, in *n*-type (*p*-type) SiC MOS capacitors, interface traps in the lower (upper) half of the band gap, which would normally exchange their charge with free holes (electrons), are not

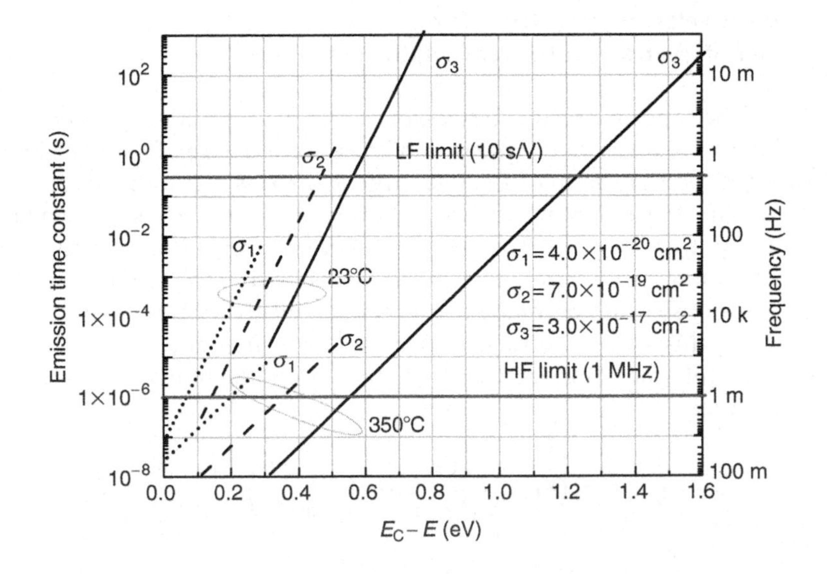

FIGURE 20.12
Emission time constants from interface traps in *n*-type 4H-SiC at 23°C (dashed lines) and 350°C (solid lines). Energy-dependent capture cross sections are $\sigma_1 \sim 4.0 \times 10^{-20}$ cm^2, $\sigma_1 \sim 7.0 \times 10^{-19}$ cm^2, and $\sigma_1 \sim 3.0 \times 10^{-17}$ cm^2. Horizontal lines indicate the time and energy limits of several common MOS analysis techniques. HF and LF stand for high and low frequency, respectively. (Plot generated using capture cross-section values from Chen, X.D. et al., *J. Appl. Phys.*, 103, 033701, 2008. With permission.)

detectable. These interface traps can be analyzed only if an external minority carrier source such as UV light or the *p–n* junction in a gated diode are used.

For studies using MOS capacitors, *n*- and *p*-type samples are used to characterize D_{it} in the upper and lower half of the band gap, respectively. For the remainder of this chapter, we will mostly confine ourselves to D_{it} measurements near the conduction band edge of 4H-SiC, as this is most pertinent to the inversion mobility problem. There are large errors associated with the measured absolute values of D_{it} very close to the band edges ($\leq$0.1 eV) using conventional *C–V* techniques, as in the case of Si [44]. Therefore, comparisons of D_{it} between different samples measured under identical conditions are a more definitive measure of relative D_{it} as opposed to the absolute values.

20.6 Interface Trap Passivation in SiO_2/4H-SiC

Attempts to passivate traps at the SiO_2/4H-SiC interface have focused primarily on oxidation procedures (e.g., dry, wet, pyrogenic, reoxidation, deposited), and postoxidation anneals in various ambients such as NO, N_2O, NH_3, H_2, as well as postmetallization annealing in hydrogen containing ambients. However, the most efficient passivation schemes for SiO_2/4H-SiC involve interfacial nitridation in some form [57]. Herein, we will focus on NO postoxidation annealing process, which is the most established among interface passivation schemes. We will also discuss the effects of hydrogen on the SiO_2/SiC interface as, traditionally, it is the important passivation agent in Si MOS technology. This will be followed by a review on interface passivation trends on 4H-SiC as a function of wafer orientation. We will conclude this section by discussing intriguing results with regards to interface trap density and inversion layer mobility obtained recently by employing process that involves oxidation in the presence of sintered alumina.

20.6.1 Interface Trap Passivation by NO

In 1997, Dimitrijev and Li et al. [58,59] demonstrated beneficial aspects of rapid thermal annealing in NO and N_2O of oxides grown on 6H-SiC. A breakthrough in interface trap passivation for 4H-SiC was achieved around 1999–2000 by Chung et al. [42] Using an optimized postoxidation annealing process in NO, they reported an order of magnitude reduction of interface trap density near the conduction band edge of 4H-SiC. Subsequently, Chung et al. [60] demonstrated an order of magnitude higher effective mobility in lateral test MOSFETs fabricated by this process. Typical processing using NO involves standard oxidations followed by postoxidation annealing in flowing NO at high temperatures ($\geq$1150°C). D_{it} profiles near the conduction band edge obtained using high–low *C–V* at room temperature on *n*-type (0001) 4H-SiC capacitors before and after NO annealing are shown in Figure 20.13a. In this particular case, dry oxidation was performed at 1150°C, followed by NO annealing at 1175°C for 2 h. The figure shows that the treatment results in an order of magnitude reduction of D_{it} close to the band edge, from $\sim 10^{13}$ to $\sim 10^{12}$ cm^{-2} eV^{-1}. The reduction is most dramatic near the conduction band edge, but is consistently observed throughout the entire band gap. D_{it} reduction associated with NO close to the valence band (measured in *p*-type samples processed similarly) is shown in Figure 20.13b. The reduction of trap states close to the valence band is visibly smaller compared to the conduction band edge. However, it is important to note that, for dry oxides, NO annealing results in an order of magnitude reduction of donor-like traps deeper in the gap. This shows up as a substantial reduction of the apparent fixed positive charge in *p*-type 4H-MOS capacitors. D_{it} obtained with NO postoxidation annealing is independent

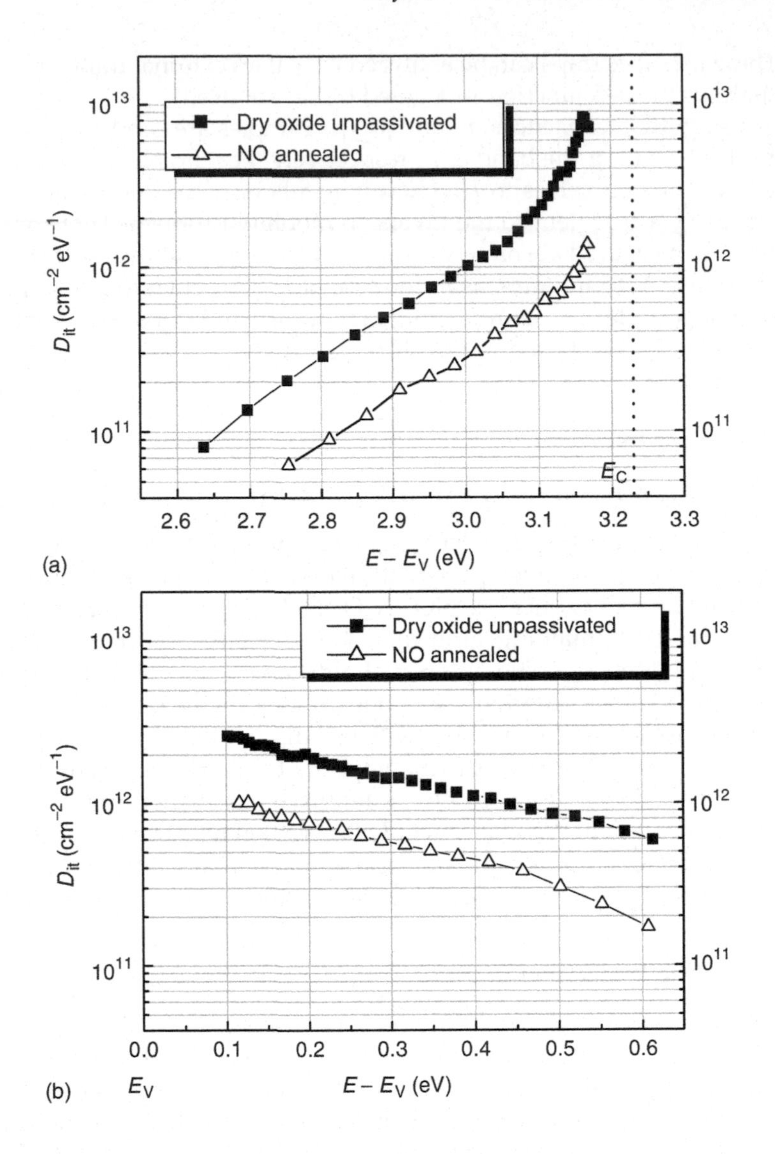

FIGURE 20.13
Effect of NO postoxidation annealing on D_{it} of ~40 nm thick oxides grown on 4H-SiC (a) close to the conduction band edge and (b) close to the VBE of 4H-SiC. The energy axis represents energy of the trap level with respect to the valence band. Unpassivated dry oxide samples were made by dry oxidation at 1150°C. The NO passivated samples underwent postoxidation annealing in pure NO at 1175°C for 2 h. D_{it} was extracted by hi–lo C–V measurements at room temperature on n- and p-type 4H-SiC MOS capacitors for measuring D_{it} near the conduction and valence bands, respectively.

of the oxide thickness in the 20–100 nm range and relatively insensitive to the type of oxidation performed prior to the passivation anneal (dry, wet, or deposited) [61]. The reduction in acceptor-like states correlates with the effective channel mobility in n-channel MOSFETs, consistent with Schorner's mobility degradation model [10] discussed earlier in this chapter.

Figure 20.14 shows room temperature field-effect mobility (μ_{fe}) [43,62] as a function of gate voltage for lateral test MOSFETs fabricated by employing NO postoxidation annealing, compared to mobility obtained from MOSFETs fabricated utilizing a standard dry oxidation process. Also shown are the results for interfaces processed by sequential

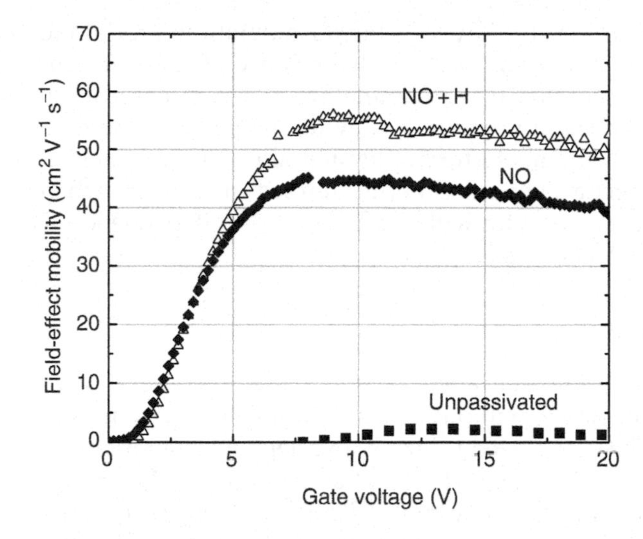

FIGURE 20.14
Field-effect mobility before and after passivation for lateral (0001) MOSFETs. Passivation with NO increases the mobility by an order of magnitude. A postmetallization (Pt) anneal in H_2 following the NO anneal further increases the mobility by 25%–30%. (From Dhar, S. et al., *MRS Bull.*, 30, 288, 2005. With permission.)

annealing in NO and atomic hydrogen. These results will be discussed shortly. Note that the NO annealed device has a peak mobility of around ~45 cm^2 V^{-1} s^{-1}, an order of magnitude higher than the unpassivated device (~5 cm^2 V^{-1} s^{-1}). Typical maximum μ_{fe} values for NO annealed lateral MOSFETs fabricated on epitaxial *p*-type 4H-SiC material vary between 35 and 50 cm^2 V^{-1} s^{-1}; this variation may be related to other aspects of MOSFET processing, such as source–drain implant annealing conditions and ohmic contact anneals. The temperature dependence of the maximum μ_{fe} for unpassivated and NO annealed lateral MOSFETs as reported by Chung et al. in Ref. [59] is shown in Figure 20.15. For the unpassivated device, μ_{fe} increases with temperature, a behavior that is characteristic of MOSFETs processed using unpassivated oxides on 4H-SiC [15,63–65]. This effect is a consequence of extreme inversion charge trapping at the interface. At higher temperatures, the emission rates of interface traps increases in accordance with Equation 20.4, which results in the thermal emission of a fraction of electrons from traps close to the surface Fermi level (close to the conduction band for strong inversion in *n*-channel MOSFET) into the conduction band. The net effect is an increase in effective mobility due to an increase in number of conduction electrons in the channel as well as a reduction in the number of scattering centers. Additionally, the Coulomb scattering interaction is also reduced at higher temperatures because the faster moving carriers interact less effectively

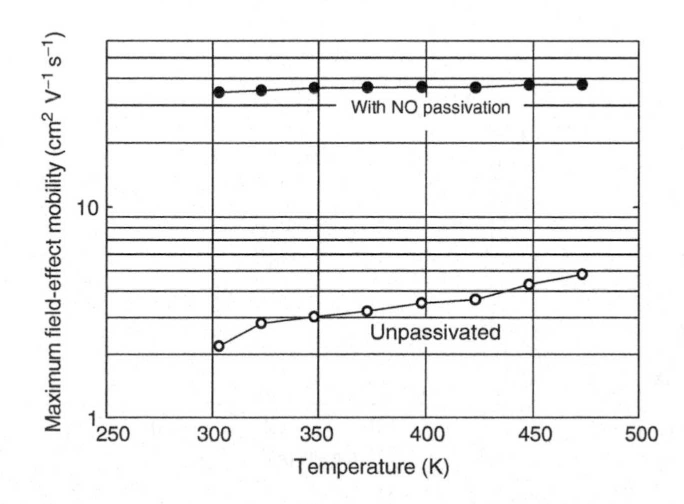

FIGURE 20.15
Temperature dependence of the maximum field-effect mobility with and without NO passivation. (From Chung, G.Y. et al., *J. Phys. Condens. Matter*, 16, 1857, 2004. With permission.)

with the stationary trapped charge, which in turn favors a higher mobility in a Coulomb scattering limited mobility. For the NO annealed case, the dependence of μ_{fe} with temperature is significantly weaker, consistent with the substantial interface trap passivation. It is important to note, however, that the slight increase of μ_{fe} with temperature indicates that the mobility is still interface trap/Coulomb scattering limited after NO treatment. If this were not the case, one would expect a phonon scattering limited mechanism where the mobility decreases with increasing temperature with a T^{-x} $(1 \leq x \leq 2)$ dependence, as typically observed in state-of-the-art Si MOSFETs [43].

20.6.2 Interfacial Nitridation

The main chemical aspect of postoxidation annealing in NO is the incorporation of N atoms exclusively at the SiO_2/SiC interface, without any measurable changes in bulk-oxide stoichiometry. Interfacial N has been analyzed utilizing a variety of techniques such as SIMS [66,67], XPS [66,68,69], MEIS [28,29], NRA [29,70,71] and EELS [37]. For typical anneal conditions (say 1175°C, 2 h), about a monolayer $\sim 10^{15}$ cm^{-2} of N is incorporated at the interface, which is distributed in an extremely sharp profile (~1 nm) located within ~1 nm from the interface. Interestingly, under similar annealing conditions, N incorporation at SiO_2/Si interfaces is an order of magnitude higher [70].

Depth profiles obtained by SIMS and EELS for NO annealing of oxides grown on 4H-SiC at 1175°C are shown in Figures 20.16 and 20.17. The measured widths of the N profile, i.e., ~3.6 nm for SIMS and ~1.5 nm for EELS, which are essentially the resolutions of the techniques, indicating that the N profile is even sharper.

The mechanism of trap passivation using N is not completely understood, but the incorporation of nitrogen atoms at or near the interface plays a key role in the trap passivation process. To first order, the reduction of traps is a function of interfacial nitrogen incorporation only and does not depend on the nitridation kinetics, i.e., the anneal temperature and time. Trap density reduction continues for increasing interfacial nitrogen content until the reduction saturates at a critical nitrogen areal density of about 2.5×10^{14} atoms cm^{-2}. McDonald et al. [72] quantitatively modeled the D_{it} near the conduction band as a function of interfacial N content. In this model, interfacial defects were considered as clusters of atoms (such as carbon clusters or Si–Si suboxide bonds) with a variety of sizes that produce trap levels in the band gap at energies depending on the size of the clusters. Larger clusters occupied levels closest to the conduction band edge, while smaller clusters constituted trap levels deeper in the gap. When an atom in the cluster is passivated

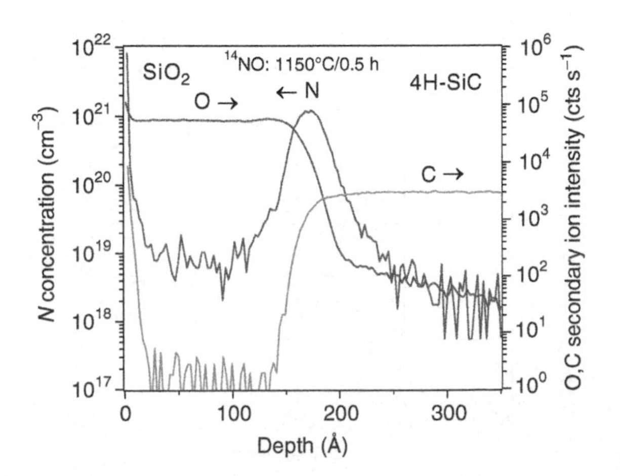

FIGURE 20.16
Interfacial nitrogen profile measured at the SiO_2/(0001) 4H-SiC interface using SIMS: (0001) 4H-SiC, NO anneal: 1175°C, 0.5 h. The FWHM of the N peak is ~3.6 nm which is limited by the resolution of SIMS. (From McDonald, K. et al., *J. Appl. Phys.*, 93, 2257, 2003. With permission.)

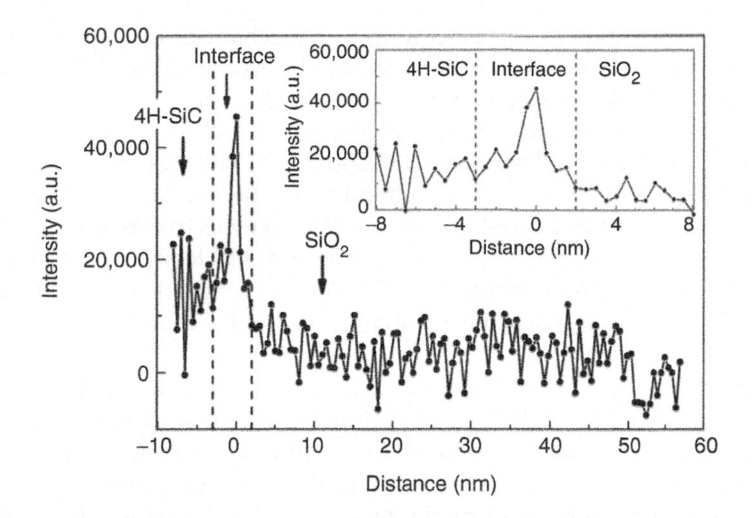

FIGURE 20.17
Interfacial nitrogen profile obtained from EELS by monitoring probing electron's energy loss associated with N K core shell ionization. The FWHM of the peak is <2 nm, limited by the depth resolution of the technique. The sample is an oxide grown on the (000$\bar{1}$) C terminated face of 4H-SiC, annealed in NO at 1175°C for 2 h. (From Chang, K.-C. et al., *J. Appl. Phys.*, 97, 1, 2005. With permission.)

by a N atom, the atom is removed from the cluster and results in a reduction of size of the original defect cluster. As a result, the energy of the defect level shifts lower in the semiconductor band gap. The dissolution of the clusters continues with the addition of N, until the energy level of the trap moves out of the band gap and becomes electrically inactive. Figure 20.18 shows the results from this model taken from Ref. [68]. D_{it} at energies 0.2–0.6 eV from the conduction band edge has been plotted as a function of interfacial N areal density and fit to the model. At any energy level, the trap density initially remains constant until a critical amount of N uptake (about $\sim 2.5 \times 10^{14}$ cm^{-2}) occurs, when the trap density reduces significantly. Within the framework of McDonald's model, this can be interpreted as a competition between trap passivation at any energy level and the generation of traps at this level associated with the dissolution of larger defect clusters at

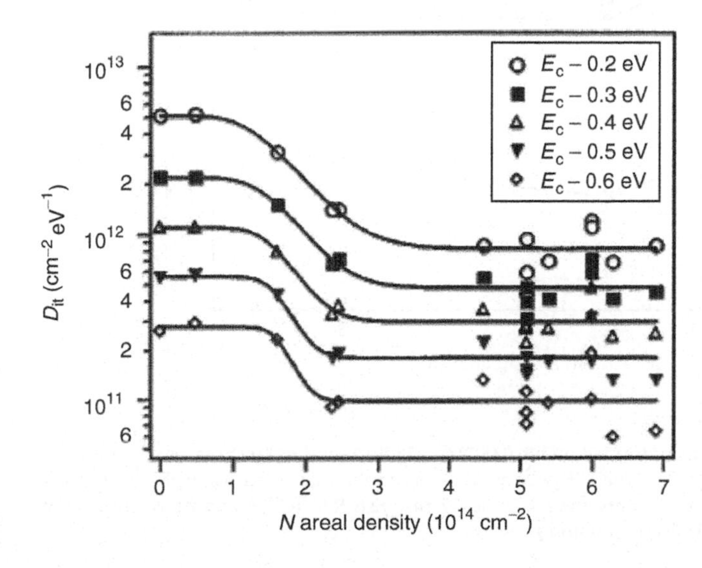

FIGURE 20.18
Interface trap density D_{it} at 0.2–0.6 eV from the conduction band as a function of nitrogen areal density. The solid line indicates fits to the model proposed in Ref. [72]. (From McDonald, K. et al., *J. Appl. Phys.*, 93, 2719, 2003. With permission.)

higher energy levels. Although the exact atomistic mechanism of trap passivation associated with N remains elusive, the model successfully explains the "nitrogen passivation threshold," and provides a possible quantitative explanation why $\sim 2.5 \times 10^{14}$ cm^{-2} of N atoms are necessary to passivate a trap density of $\sim 10^{12}$ cm^{-2}.

The interfacial chemistry of the nitrided $SiO_2(N)/SiC$ interface is complex and not yet completely understood. Si–N and Si–N–O bonds have been observed by XPS [66,68], but the coexistence of C–N bonds cannot be ruled out. In fact, electron paramagnetic resonance (EPR) studies on porous 4H-SiC show a decrease in the EPR line associated with carbon dangling bonds following NO nitridation [73]. Also, recent SERS experiments on oxides formed on the $(000\bar{1})$ C-terminated face of 4H-SiC suggest a reduction of graphitic carbon clusters upon NO treatment [74].

Nitridation kinetics of SiO_2/SiC by NO are dictated by a competition between two processes: (1) reaction of NO with Si, C, or O atoms at the interface to incorporate nitrogen; and (2) additional oxidation associated with O_2 produced by the thermal decomposition of NO ($\sim$1 nm SiO_2 for 1175°C, 2 h NO anneal) [67,71]. The latter process is a result of NO decomposition at high temperatures by the reaction $NO \rightarrow N_2 + O_2$ [75,76]. N_2 is expected to be inert at these temperatures ($\sim$1175°C), but the additional oxidation associated with the partial pressure of O_2 removes N already incorporated at the interface. Therefore, as the oxide grows, nitrogen is incorporated at a new interface while it is removed from its previous location. Interfacial N incorporation as a function of annealing time for NO annealing at 1175°C for oxides grown on different crystal faces is shown in Figure 20.19. It is interesting to note that the N incorporation is crystal face dependent and appears to follow the same trend as the oxidation rate. It can be seen from the figure that, for all crystal faces, the nitridation kinetics are characterized by an initial rapid nitridation phase followed by a decrease in N content. This behavior can be modeled by a competition

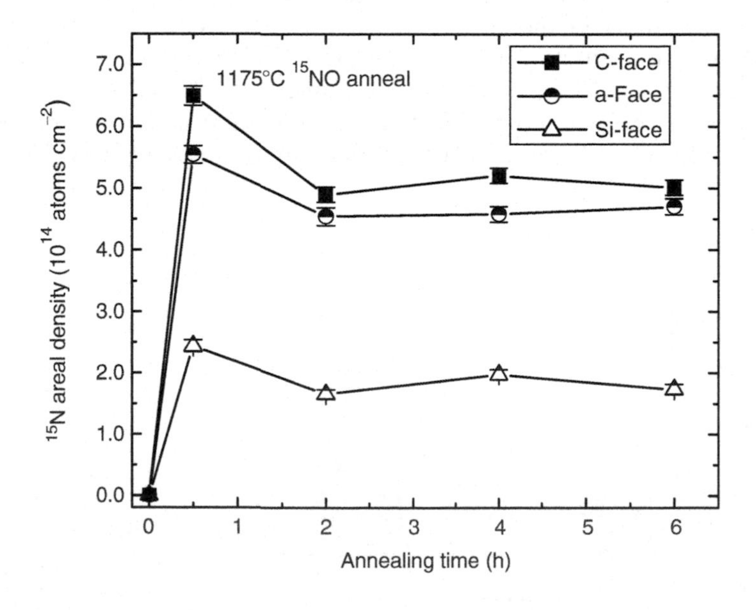

FIGURE 20.19
Nitridation kinetics for SiO_2/4H-SiC interfaces formed on the $(000\bar{1})$ C-, $(11\bar{2}0)$ a- and (0001) Si-faces of 4H-SiC. All samples were annealed at 1175°C, in 100 Torr ^{15}NO. The N areal density was quantified using the $^{15}N(p,\alpha)^{12}C$ nuclear reaction. Thickness of oxides prior to annealing was 30–37 nm. See Ref. [69] for more details. (From Dhar, S. et al., *J. Appl. Phys.*, 97, 1, 2005. With permission.)

between nitridation and additional oxidation during NO anneals as discussed above [71]. When the two reactions reach equilibrium, the interfacial nitrogen content saturates.

The additional oxide growth that occurs during the NO anneal may be the limiting factor for further interface trap passivation using NO, as it promotes a competition between defect passivation by nitrogen and new defect generation by oxidation. Lower trap densities and higher channel mobilities have been reported by Cree, Inc., using a proprietary process that introduces nitrogen with minimal additional oxidation [77]. In this regard, it is relevant to briefly mention processes involving N_2O and NH_3. Postoxidation annealing in N_2O leads to an order of magnitude lower interfacial N in SiO_2/SiC compared to NO [70]. At elevated temperatures, N_2O decomposes quickly to N_2 (~60%), O_2 (~30%), and NO (~10%), where the exact compositions depend on annealing parameters such as temperature and flow rate. It has been suggested that the species responsible for nitridation of SiO_2/Si during N_2O treatment is really the NO that forms by the decomposition of N_2O [78]. The relatively small amount of NO (~10%) produced during N_2O annealing, compared to a 100% NO anneal, accounts for the significantly lower nitridation rate in the case of SiO_2/SiC. Another factor is the higher additional oxidation associated with N_2O. This process is very likely to aid the removal of N from the interface during N_2O annealing. These factors are believed to be the main reasons for the mixed results obtained via N_2O processes with respect to interface trap passivation. Although a few articles report interfacial improvement associated with N_2O annealing or direct oxide growth in N_2O [79–81], it is widely accepted that N_2O passivation is significantly less effective than NO [58,82,83]. Ammonia treatment results in the reduction of SiO_2 to $SiO_xN_yH_z$ oxide which places nitrogen throughout the oxide layer as well as at the interface [84]. It has been demonstrated that NH_3 annealing results in D_{it} comparable to the NO process, but the breakdown strength of NH_3 nitrided oxides are significantly lower, possibly due to the excess H in the dielectric.

20.6.3 SiO_2/SiC Interface Trap Passivation Using Hydrogen

It is well known that hydrogen can have significant impact, both positive and negative, on various device processing steps in Si technology (Chapters 7 and 13). One of fundamental positive aspects of H is the ability to passivate Si dangling bonds or P_b centers on Si surfaces at the SiO_2/Si interface. One of the last steps of MOS fabrication on Si comprises a forming gas (10% H_2, 90% N_2) postmetallization anneal at around 400°C, which results in an order of magnitude reduction of D_{it} across the band gap (from $\sim 10^{11}$ to 10^{10} cm^{-2} eV^{-1} at mid gap). Hydrogen annealing under similar conditions is ineffective for interface trap passivation in SiO_2/4H SiC interfaces [7,85]. High-temperature (~1000°C) postoxidation annealing in hydrogen has been reported to be effective for interface passivation for oxides grown on the C-terminated face of SiC, but not on the Si-face [86,87]. Theoretical calculations suggest that there could be a class of defects at the SiO_2/SiC interface that should susceptible to atomic hydrogen and not to molecular hydrogen [88]. This will be discussed in more detail in Section 20.6.4.

Recently, atomic hydrogenation was pursued by performing a postmetallization anneal of 4H-SiC devices using Pt as the gate metal [88,89]. Pt is well known for its catalytic ability for cracking H_2 molecules into monoatomic H. The atomic H is expected to diffuse through the oxide, arrive at the SiO_2/SiC interface, and bind to possible interface defects during the anneal. Annealing at 500°C for 1 h using this process leads to a substantial incorporation of H in the oxide and at the interface, as shown in the SIMS profile in Figure 20.20. The corresponding interface trap density reduction is about factor of ~2 at $E_C - E \sim 0.1$ eV, compared to the unpassivated dry oxides, a relatively small effect compared to nitridation (Figure 20.20). Also shown in Figure 20.21 is the typical D_{it} profile (denoted NO + H) of an

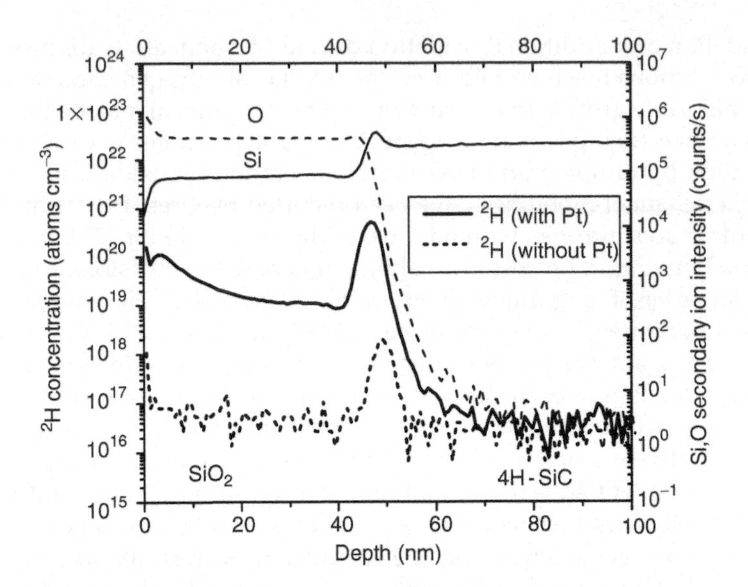

FIGURE 20.20
SIMS profiles for $SiO_2(N)/4H\text{-}SiC$ samples with comparable oxide thickness annealed in ^2H (deuterium) at 500°C for 1 h with and without a Pt overlayer. (The O and Si profiles are not shown for the "with Pt" sample) About 80% (2.5×10^{14} atoms cm^{-2}) of the ^2H is at or near the SiO_2/SiC interface with a distribution having FWHM of ~3.5 nm within the depth resolution of SIMS. Deuterium was used in place of hydrogen as it offers high sensitivity for detection. (From Dhar, S. et al., *Mater. Sci. Forum*, 527–529, 949, 2006. With permission.)

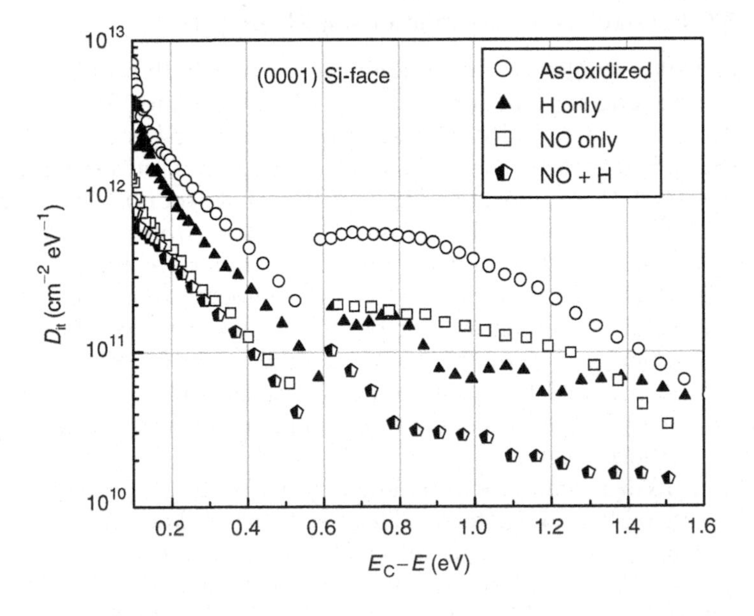

FIGURE 20.21
Interface trap densities in the upper half of the 4H-SiC band gap obtained for (0001) Si-face 4H-SiC MOS capacitors processed using the different interface passivation treatments as described in the text. Oxide thicknesses are in the range 45–55 nm. Discontinuities near $E_C - E$ ~0.6 eV are the result of *C–V* measurements at 23°C and 300°C, respectively, below and above 0.6 eV. (From Wang, S. et al., *Phys. Rev. Lett.*, 98, 026101, 2007. With permission.)

interface processed with atomic hydrogen after undergoing standard NO POA. Comparison with the "NO only" data reveals a ~30% lower D_{it} at $E_C - E \approx 0.1$ eV and deeper in the band gap, i.e., from $E_C - E = 0.6$ –1.6 eV. In the lower half of the band gap (data not shown), no significant difference is observed between "NO only" and "NO + H." The improvement in D_{it} associated with H is reflected in the effective mobility of MOSFETs fabricated using the same process.

Maximum field-effect mobilities for MOSFETs fabricated using the NO + H process is 25%–30% higher compared to mobilities for "NO only" devices, as shown in Figure 20.14. The overall conclusion from these results is that hydrogenation using atomic hydrogen can lead to additional improvement of channel mobility in nitrided SiO_2/4H-SiC interfaces. However, the process is yet to be optimized with respect to device reliability for commercial processing [89]. Apparently, the amount of hydrogen incorporated in the oxide (~10^{14} cm^{-2}) via the Pt-induced process is substantially higher than the amount required for the trap passivation (most likely around ~10^{12} cm^{-2}), which could have a detrimental influence on the long-term device operation.

20.6.4 Alternate Wafer Orientations

Electrical properties of semiconductor–dielectric interfaces are a manifestation of atomic-level interface composition and structure that result from complex solid-state chemical reactions. The differences in structure and properties of interfaces formed by oxidation of different crystal faces of Si have been widely studied. Among the major crystal orientations, the SiO_2/(001) Si interface exhibits the best electrical properties. This is usually attributed to the lower density of electrically active Si dangling bonds centers on this surface [43,44]. In the case of SiC, different crystal faces having different surface stoichiometries (Si:C ratios), are particularly interesting for interface studies where it is well known that the oxidation rates differ greatly on the crystal faces. The anisotropy in oxidation kinetics (Figure 20.4) suggests different rate-limiting reactions on these crystal faces [18], which could possibly result in different interfacial or near-interfacial defects.

Some of the major crystal orientations that are being investigated for MOS electronics are shown schematically in Figure 20.22. All of the discussion so far has been on the (0001) Si-terminated face, which is the most widely studied and commercially relevant crystal face for SiC. Ideally, the (0001) Si-face surface consists of 100% Si atoms, while the (000$\bar{1}$) C-face consists of 100% C atoms. Wafers are usually processed with the surface normals for both surfaces approximately 8° off the <0001> and <000$\bar{1}$> directions in order to improve the quality of epitaxial layers. Therefore, in practice, a small fraction of Si or C atoms will exist on the C-face and Si-face, respectively, due to the miscut. The (11$\bar{2}$0) face has equal numbers of carbon and silicon atoms in a plane, and epitaxial growth on the a-face is carried out on-axis. Interface studies on the (11$\bar{2}$0) face are also important from the standpoint of UMOSFETs. The UMOSFET is a specialized vertical power MOSEFT design fabricated on {0001} oriented wafers, but the inversion channel lies on trench sidewalls defined by the {11$\bar{2}$0} planes [5]. Epilayer quality has historically been better for the Si-face compared to other crystal faces. In recent years, high-quality epilayers have been reported on unconventional crystal faces [90]. As a consequence, interface studies and MOSFET fabrication using alternate crystal faces of 4H-SiC are also being pursued. We will discuss some aspects of interfaces formed on the (000$\bar{1}$) and (11$\bar{2}$0) orientations and compare with results on the (0001) face, as discussed in the previous sections. For simplicity, we will refer to the (0001), (000$\bar{1}$), and (11$\bar{2}$0) as the Si-face, C-face, and a-face.

Trap densities near the conduction band edge for dry oxides grown on the three faces are shown in Figure 20.23. Close to the conduction band edge, D_{it} is somewhat comparable. But deeper in the gap, D_{it} is significantly higher for the C-face and the a-face. Also, the

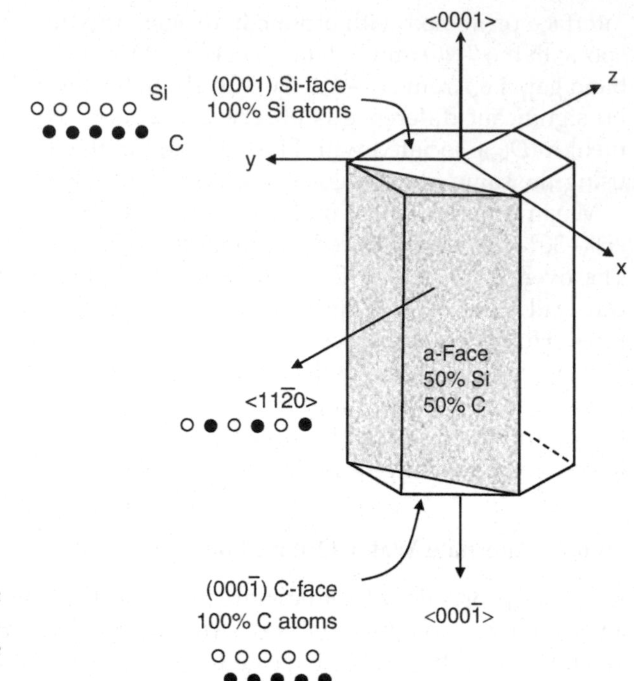

FIGURE 20.22
Three of the most common crystallographic planes for MOS interface studies on 4H-SiC. Si-face is conventional face for device fabrication.

density of energetically deep traps in lower half of the gap for the C-face is typically very high (~10^{14} cm^{-2}) [7], which even results in the pinning of the surface Fermi level on *p*-type C-face MOS capacitors [91]. Interestingly, the total interface trap density across the band gap qualitatively scales with the surface C density, i.e., C-face > a-face > Si-face. This has

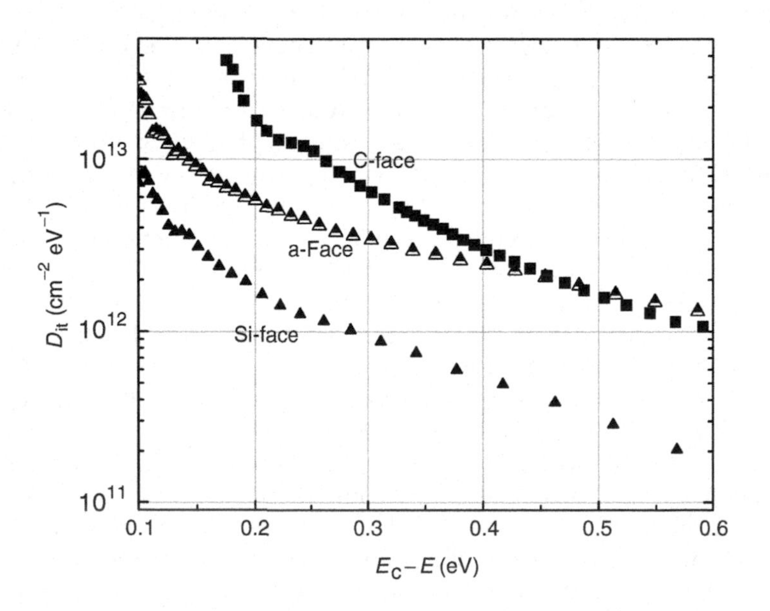

FIGURE 20.23
Interface trap density near the conduction band edge as a function of crystal face for 4H-SiC for dry oxidation at 1150°C. D_{it} profile was extracted using hi–lo *C–V* at room temperature on *n*-type MOS capacitors. (From Dhar, S. et al., *J. Appl. Phys.*, 98, 14902, 2005. With permission.)

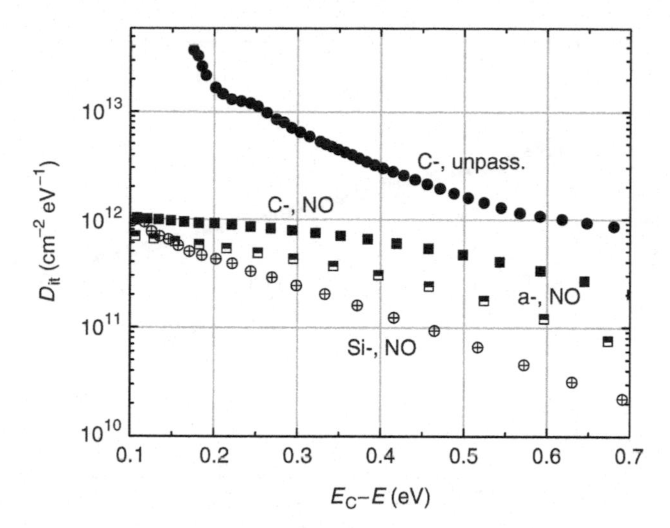

FIGURE 20.24
D_{it} near the conduction band edge of 4H-SiC for oxides grown on the Si-, a-, and C-faces using dry oxidation at 1150°C followed by annealing in NO post-oxidation at 1175°C for 2 h. The data are an average of 10 capacitors. D_{it} for C-face dry oxide without NO POA is also shown for comparison. D_{it} profile was extracted using hi–lo C–V at room temperature on n-type MOS capacitors. (From Dhar, S. et al., *J. Appl. Phys.*, 98, 14902, 2005. With permission.)

been attributed to higher C-related defects such as C clusters on C-rich SiC surfaces, although the exact reason remains unclear [7].

Recent surface enhanced raman spectroscopy (SERS) studies have detected higher density of graphitic carbon (possibly clusters) on the C-face compared to the Si-face [38]. While this result is consistent with the carbon clusters hypothesis, at this stage there is no concrete evidence to suggest a direct correlation between the measured D_{it} and the graphitic carbon detected by SERS. It should be noted that the compositions of the bulk oxides grown on the different faces are identical. The excess interfacial carbon density measured by ion scattering, as described in Section 20.4, for the C-face and a-face is similar to that of the Si-face, i.e., $\leq 10^{15}$ cm^{-2} [28]. Interestingly, XPS studies have detected different classes of suboxides (possibly C containing) in the near-interfacial regions of thin oxides (~5 nm) grown on different faces [23]. Such differences in interfacial chemistry may potentially have a role in the crystal face dependency of D_{it}, but at this stage, the detailed relationship between such suboxides and electrical response of the interface remain unknown.

The effect of interfacial nitridation has been recently studied for the a-face and C-face [28,92]. Interface trap density profiles after NO postoxidation annealing (1175°C, 2 h) are shown in Figure 20.24. Typical D_{it} for unpassivated dry oxide on the C-face and nitrided Si-face are also shown for comparison. The passivation effect associated with nitridation for the a-face and C-face is comparable to the Si-face, resulting in almost an order of magnitude reduction of total trap density across the band gap. A substantial reduction of states is also prevalent deeper in the band gap and in the lower half of the band gap. A comparison between the C-face and Si-face for energetically deeper ($E_C - E \sim 0.4$ eV to mid gap) interface traps in the upper half of the gap after NO nitridation is shown in Figure 20.25. It is important to note that even after nitridation the crystal face dependency of D_{it} follows same trend as that of as-oxidized interfaces, with the C-face oxide having the highest D_{it}. Not only do these results highlight the beneficial nature of nitridation for different SiO_2/SiC interfaces in general, but they also suggest the presence of at least one kind of electrically active defect with similar chemical behavior on these different crystal faces. The field-effect mobilities in lateral n-channel MOSFETs fabricated on the different faces with similar NO passivation procedures are shown in Figure 20.26. The peak mobility values are ~45, ~57, and ~35 cm^{-2} V^{-1} s^{-1} for the Si-, a-, and C-face MOSFETs, respectively. The higher peak mobility for the a-face compared to the Si-face may be attributed to the anisotropy in the bulk mobility of 4H-SiC ($\mu_{(0001)}/\mu_{(11-20)} \sim 0.8$) [93]. Therefore, it is

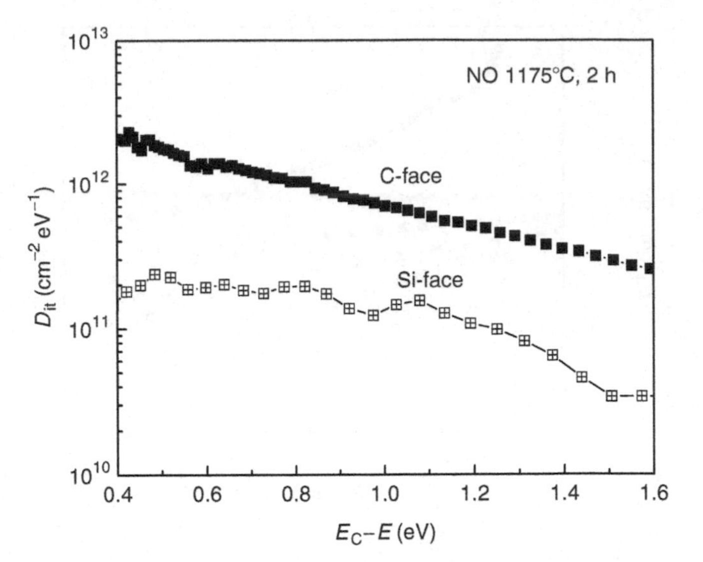

FIGURE 20.25
Comparison of D_{it} deeper into the band gap (from $E_C - E = 0.4$ eV up to mid gap) for nitrided Si-face and C-face interfaces shows almost an order of magnitude higher integrated density of traps for the C-face, even after nitridation. D_{it} profile was extracted using hi–lo *C–V* at 300°C on *n*-type MOS capacitors.

evident that the field-effect mobility (measured for *p*-type material) scales qualitatively with the results of the trap density measurements near the conduction band (measured in *n*-type material), once again highlighting the importance of eliminating electron trapping by acceptor-like states near the conduction band of 4H-SiC. It is surprising, however, that the significantly higher density of energetically deep interface traps for the C-face does not have a big impact on the effective mobility of C-face MOSFETs.

As in the case of the Si-face, NO annealing results in nitrogen incorporation exclusively at or near the SiO_2/SiC interface for the other crystal faces. Interestingly, the interfacial N uptake by NO postoxidation annealing is crystal face dependent and has the same trend as oxidation, i.e., C-face $\geq$ a-face > Si-face [71]. For the same anneal temperature and time, the C-face and a-face interfaces incorporate about a factor of ~3 higher N than the Si-face. Furthermore, the additional oxidation during NO anneals is higher on these faces due to their 3–5 times higher oxidation rates compared to the Si-face. This is advantageous for performing oxide growth directly in NO and bypassing the initial dry oxidation step [94]. As the oxidation rate in NO is about an order of magnitude lower than in pure O_2, such a process is impractical for the Si-face, but is potentially promising for the a- and C- faces.

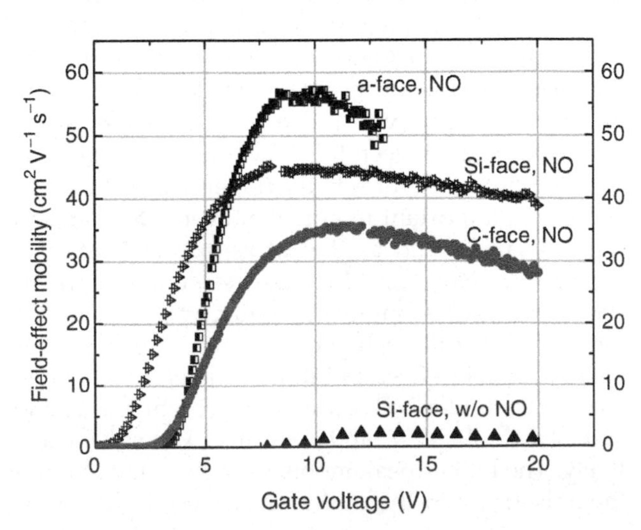

FIGURE 20.26
Field-effect mobilities in *n*-channel lateral test MOSFETs for the corresponding SiO_2/4H-SiC interfaces as in Figure 20.20. (From Dhar, S. et al., *Mater. Sci. Forum*, 527–529, 949, 2006. With permission.)

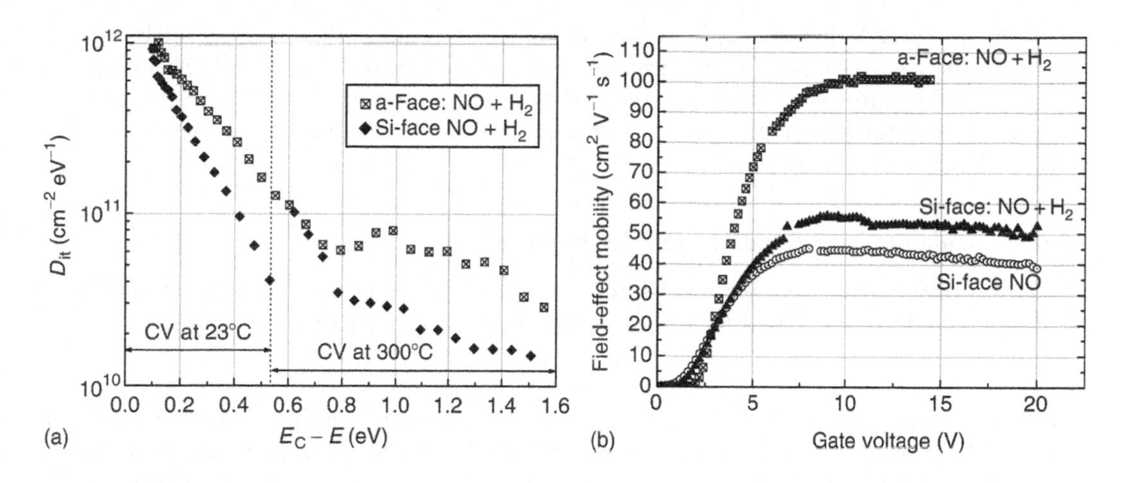

FIGURE 20.27
(a) D_{it} profiles obtained from MOS capacitors fabricated on the Si-face and a-face using sequential annealing in NO and H_2 as described in the text. (b) Field-effect mobilities in *n*-channel MOSFETs for the corresponding SiO_2/4H-SiC interfaces, compared to the "NO only" Si-face. (From Dhar, S. et al., *Mater. Sci. Forum*, 527–529, 949, 2006. With permission.)

Postmetallization annealing of nitrided oxides in H_2 in the presence of Pt has been studied for the a-face and C-face. This process has been discussed previously in this chapter. Additional passivation associated with atomic hydrogen is also observed for these faces [92]. D_{it} values measured on a-face and Si-face MOS capacitors processed with the (NO + H) process are shown in Figure 20.27a. Corresponding field-effect mobility values obtained from lateral test MOSFETs fabricated are ~55 and ~100 cm^{-2} V^{-1} s^{-1} for the Si-face and a-face, respectively (Figure 20.27b). Note that the ratio of peak mobility values for the two faces is 0.65, which deviates from the bulk mobility ratio (0.85). This indicates a relatively greater improvement for the a-face as a result of the H treatment. This result is intriguing, because the measured D_{it} of the a-face is higher than the Si-face throughout the band gap. The higher mobility may be due to lower roughness on the a-face surface following implantation and/or activation, or to a lower trap density very near the conduction band edge ($E_C - E < 0.1$ eV) [95,96]. Recall that accurate measurements of trap distributions so close to the band edge are difficult to achieve experimentally [44]. There have been other instances where higher channel mobilities have been measured for MOSFETs fabricated on alternate crystal faces compared to the standard Si-face. Senzaki et al. [97] used wet pyrogenic oxidation without any additional passivation treatment and determined a field-effect mobility of ~110 cm^2 V^{-1} s^{-1} for a-face MOSFETs, which was significantly higher than Si-face devices fabricated similarly. Fukuda et al., [98] have reported field-effect mobilities as high as 127 cm^2 V^{-1} s^{-1} for the C-face following post-oxidation passivation anneals in H_2. These results strongly suggest that more studies are required to conclusively determine the detailed nature of mobility degradation mechanisms for these alternate crystal faces.

At this stage, the limiting factor for using nonstandard crystal orientations for SiC MOS devices is the oxide reliability. In general, nitridation with NO improves the breakdown characteristics of oxide layers on the Si-face. Typically, excellent breakdown fields greater than 10 MVcm^{-1} are obtained [61]. However, for nitrided oxides on the a-face and C-face, the breakdown strength is considerably lower, averaging approximately 6 MV cm^{-1} [94]. The reason for the lower dielectric strength is not clearly understood. One possibility could be related to the higher density of extrinsic defects on substrates prior to oxidation. This is not unlikely considering that epitaxy on nonstandard crystal orientation is fairly new. If this were the case, one may expect improvement of the breakdown characteristics as the material quality improves. The other possibility could be the softness of oxides grown on these exotic crystal faces with respect to charge injection and trapping—issues that definitely require further investigation.

20.6.5 Oxidation in Sintered Alumina Environment

Oxidation of SiC in the presence of sintered alumina is a process that is causing a lot of excitement within the SiC MOS community lately. This process was discovered by researchers at Phillips [99] and subsequently investigated by groups at Chalmers University [100,101] and Cree, Inc. [102] The process involves standard dry oxidation of 4H-SiC in a tube or boat made with a specific variety of sintered alumina, without any postoxidation passivation treatment. The alumina contains a high density of alkali metals such as Na, K, Ca, and Mg and transition metals such as Fe, Cr, Ni, and Ti, mostly in the form of oxides. Excellent field-effect mobilities of about 150 [101] and 69 $cm^2\ V^{-1}\ s^{-1}$ [102] have been reported on lateral Si-face MOSFETs with epitaxial channels. These values are considerably higher than typical mobilities in MOSFETS with state-of-the-art NO nitrided SiO_2/4H-SiC interfaces. Sveinbjörnsson et al. [101] have attributed the mobility enhancement to reduction of near interface traps or border traps associated with the alumina process. They used a thermally stimulated current (TSC) technique to show that these oxides have an order of magnitude lower density of energetically shallow near-interface traps compared to standard wet oxides and about a factor 4–5 higher near-interface trap density than N_2O oxidized samples. To the authors' knowledge, no comparison with TSC has been reported to date between the NO nitrided interfaces and interfaces formed by the alumina process. Using ac conductance measurements, Das et al. [102] have shown that the D_{it} values near the conduction band (at least up to $E_C - E \sim 0.2$ eV) in these oxides are very similar to NO nitrided oxides. They have suggested that a possible reason for the enhanced mobility could be a significantly lower D_{it} at shallower energies with respect to the conduction band, where standard electrical C–V techniques are unreliable. The exact reason for the mobility improvement remains an open question.

Some typical features associated with the alumina process have been identified. First, the oxidation is characterized by an order of magnitude higher oxidation rate and nonuniform oxide growth. This is associated with the catalytic properties of metallic impurities in the presence of the alumina. In fact, it is well known that alkali metals (Na, K, Cs, etc.) enhance the oxidation rate of Si [103]. It is possible that an analogous process is happening in the case of SiC. The other main characteristic of these oxides is a high density ($\sim 10^{12}$–10^{13} cm^{-2}) of mobile ion contamination, most likely in the form of Na^+ distributed in the oxide. It is well known from Si technology that mobile ion contamination leads to severe voltage and temperature related device stability problems. This feature severely limits the use of the alumina for practical applications with SiC. However, from the standpoint of science, the role of Na and other metal impurities on the mobility enhancement needs to be understood in greater detail. Sveinbjörnsson et al.'s measurements indicate that the location of the mobile Na+ ions in the oxide has an effect on the peak field-effect mobility, but does not influence the mobility ($\sim$100 $cm^2\ V^{-1}\ s^{-1}$) at high electric fields. However, the issue of Na is far from resolved. One cannot rule out the possibility that the negative interface-trapped charges are effectively shielded by positively charged Na, which is rendered immobile at the SiO_2/4H-SiC interface. Such a process could potentially reduce Coulomb scattering and enhance the effective mobility.

The higher oxidation rate associated with the alumina process also has been suspected as a reason for the improvement in interface properties. Recent experiments on high-pressure oxidation, however, indicate otherwise [104]. High pressure (>1 atm) O_2 was used to enhance the oxidation rate of 4H-SiC (at any temperature) in the absence of metallic impurities. Samples oxidized at 1150°C under 4 atm O_2 had a factor of 3–4 times higher oxidation rate than samples oxidized under standard conditions, i.e., 1 atm O_2 at the same temperature. However, no difference in D_{it} was observed between these samples, which strongly suggests that the oxidation rate enhancement is not the primary mechanism for the interfacial improvement associated with the alumina process.

In summary, the alumina process is unacceptable from the standpoint of semiconductor fabrication; however, it certainly poses important questions for understanding the mobility degradation mechanisms at the SiO_2/SiC interface. Most importantly, this process sets a minimum limit for the maximum effective mobility achievable in 4H-SiC MOSFETs.

20.7 Atomic-Level Defects at the SiO_2/SiC Interface and Passivation Mechanisms

To discuss possible atomic-scale defects at SiO_2/SiC interfaces it is important to review the well-known defects at SiO_2/Si interface. Although still not completely understood, the primary defect structures associated with this interface have been identified and widely studied. The most important defects at (100) SiO_2/Si interfaces are threefold coordinated Si dangling bonds (P_{b0} centers) and partially oxidized Si–Si dimer defects (P_{b1} centers), which can be detected by EPR [105,106]. These defects are discussed extensively in Chapter 6. Recall that, from fundamental atomic-level considerations, the oxidation of Si naturally can lead to a nearly perfectly abrupt interface. The amorphous nature of the oxide, however, inevitably leads to an occasional bonding mismatch and leads to the formation of such dangling bonds. In addition, the primary intrinsic deviations from an ideally abrupt interface are "Si–Si suboxide bonds" on the oxide side and "oxygen protrusions" on the Si side, which are a consequence of the amorphous nature of the oxide and finite temperature-effects (entropy). Suboxide bonds that lie deeper in the oxide may be viewed as oxygen vacancies, also known as E' centers. These defects are primarily important for charge trapping in the oxide for long-term device operation and/or exposure to ionizing radiation. Other defects in the SiO_2/Si materials system are overcoordinated Si and O, but these are most likely of secondary importance. Finally, the presence of the oxide, especially the presence of oxygen protrusions, causes strain in the last few atomic layers of crystalline Si [107].

Typically, the D_{it} in the mid gap region is attributed to the Si dangling bonds. These can be essentially eliminated completely by appropriate annealing in hydrogen gas. The other main contributions to D_{it} are from suboxide bonds of varying lengths. First-principles DFT calculations show that those that are longer than the normal 2.35 Å result in smaller bonding–antibonding splitting and could end up as localized states near the band edges of Si [26]. The high quality of the (001)Si/SiO_2 interface in commercial MOSFETs is due to passivation of dangling bonds by hydrogen. Hydrogen can also passivate the longer-than-normal suboxide bonds by inserting two H atoms into a long Si–Si bond and forming two electrically inactive Si–H bonds. Normal-length suboxide bonds are not passivated by hydrogen, and in fact can potentially trap a single H and act like fixed positive charge [108].

In comparison to SiO_2/Si, the SiO_2/SiC materials system is in its infancy stage, with regards to understanding the physical nature of interface defects. EPR, which is the main experimental arsenal for understanding the physical identity and nature of interfacial defects in SiO_2/Si, has only achieved limited success recently in the case of SiC. In 2004,

Cantin et al. [109] reported the first successful observation of a paramagnetic C dangling bond center (P_{bC}) in oxidized porous 4H-SiC using EPR. Prior to this, only Macfarlane et al. [110] had observed a probable C dangling bond defect in nonporous 4H and 3C-SiC induced by postoxidation annealing of wet oxides in dry ambients. To date, Si dangling bonds have not been observed in SiO_2/SiC. SiC oxidation is fundamentally more complex than Si oxidation and does not naturally favor an abrupt SiO_2/SiC interface. As a consequence, a variety of defect structures are potentially possible, which makes theoretical studies of this interface extremely difficult. The main theoretical findings on SiC oxidation, possible interface defect structures, and defect passivation mechanisms can be found in the work led by Pantelides et al. [20,21,88,111] and Knaup et al. [22,112]. Based on the current state of knowledge, defects at the SiO_2/SiC interface can be classified into two broad categories: (1) carbon related and (2) defects in SiO_2. A summary of the most important theoretical findings with regards to defect structures and passivation will be presented here along with correlative experimental observations based on this broad classification.

20.7.1 Carbon-Related Defects at or Near the SiO_2/SiC Interface

Historically, the biggest concern with SiC oxidation has been inefficient removal of C from the interfacial region. Interfacial C, if inadequately bonded can lead to states in the band gap of SiC and act as interface traps. As discussed in Section 20.4, various experimental techniques have detected a $Si_xC_yO_z$ transition layer (~1 nm) at the interface, where such defects could indeed exist, although a complete atomic-scale structure of this transition layer is yet to be deciphered. Si dangling bonds are not believed to be the primary contribution to the D_{it} because (1) no Si dangling bonds have been detected by EPR to date and (2) annealing in H_2 at moderate temperatures (~400°C) does not lead to a substantial reduction in D_{it} [7]. On the other hand, paramagnetic C dangling bond centers (P_{bC}) have been recently detected by EPR on dry oxides grown on porous 3C and 4H-SiC [113]. However, subsequently it was demonstrated that these dangling bonds are also successfully passivated by H_2 at 400°C. In general, it has been suspected that graphitic carbon clusters can form at or near the interface as a by-product of oxidation [114,115]. Using internal electron photoemission, Afanas'ev et al. [7] observed a similarity between the spectrum of states within the band gap of SiC at SiO_2/SiC interfaces and at interfaces between amorphous hydrogenated carbon films (consisting of a mixture of *sp2* and *sp3* carbon) and SiO_2. It was argued that the π-bonds of the *sp2* bonded carbon atoms or C clusters would have states in the band gap of SiC, and the size of such a cluster would determine the position of the defect state in the band gap. It was also noted that D_{it} was higher in crystal faces with higher surface C density (e.g., the C-face), where the formation of C clusters could be more probable. From these observations, it was proposed that C clusters of different sizes resulting from the oxidation of SiC could account for continuously rising D_{it} profile at SiO_2/SiC interfaces.

Subsequently, it was found that low-temperature (~950°C) annealing of oxides in oxygen, i.e., reoxidation, leads to a reduction of D_{it} in the lower half of the gap [116]. (Refer to Figure 20.11 for a comparison between wet oxidation and reoxidation.) The effect was attributed to the dissolution of C clusters under the influence of oxygen at temperatures where SiC does not oxidize, i.e., no new C clusters form; but the existing C clusters oxidize to CO or CO_2. On the theoretical front, mechanisms involving reactions between out-diffusing CO molecules were identified as the primary source of O deficient C clusters and the reoxidation process was also validated using first-principles DFT calculations [111]. With the discovery of the NO anneal process, the passivation mechanism associated with nitridation was also explained in the light of the C cluster model [42]. C clusters were modeled as a group of C interstitials at the interface. It was found that isolated

C interstitials have an energy level in the upper part of the band gap, and the level goes higher in energy with each additional C atom added to a cluster. It was proposed that nitrogen added to the interface can remove C from a cluster by binding to it and decreasing the cluster size. As a result, the level associated with the original cluster drops deeper into the band gap (possibly moving out of the band gap), which could be a possible explanation for the significant reduction of D_{it} associated with nitridation [72].

While the definitive detection of small atomic-scale clusters is probably impossible with current analytical techniques, larger clusters (~1 nm) may be expected to be detected within sensitivity limits. Various experiments report the presence of excess carbon in the interfacial region (Section 20.4), but direct experimental observation of C clusters has not been reported at interfaces formed on the Si-face of 4H-SiC. SERS measurements performed after etching the oxide with HF have detected peaks associated with graphitic nanoclusters ($\leq$2 nm) on the C-face of 4H-SiC [38]. No significant signal was detected on the Si-face, consistent with the earlier hypothesis that C clusters form more favorably in C-rich surfaces. Recently, it has been shown for the C-face intensities that Raman peaks associated with the C clusters are reduced after NO annealing [74]. More experiments, however, are required to understand the detailed nature of these Raman peaks. While C-clusters remain elusive, for the Si-face, they remain as one of the most important candidates for electrically active defects at SiO_2/SiC interfaces.

In a first-principles DFT treatment by Knaup et al., [22] the mostly likely atomic-level reaction routes for the oxidation of 4H-SiC were suggested and defects that form along the reaction routes were identified. Among the defects investigated, single C interstitials sharing a C lattice site at the interface and pairs of such C interstitials were found to be electrically active. The former defect results in filled states (donor-like traps) and empty states (acceptor-like traps) near the mid gap region of SiC. The later defect, described as a pair of C interstitials double bonded together as a dimer, lead to a band of donor like traps close to the VBE, but no unoccupied state in the band gap of SiC. They pointed out that such C pairs were extremely stable and could possibly aggregate to form larger clusters. Such defects could be the source of D_{it} near the VBE and account for the very large fixed positive charge in dry oxidized *p*-type 4H-SiC MOS capacitors, which is usually attributed to trapping in deep donor-like interface traps. Interestingly, Knaup et al. did not find any C related defect at the interface that could be assigned to the acceptor-like states near the conduction band of 4H-SiC. However, they suggested in a later report [112] that C–C dimers located in the oxide could be possible candidates for such defects.

Very recently, the work by Wang et al. [88] gave new insights into the nature of defects at the SiO_2/SiC interface with regards to their susceptibility to hydrogen and nitrogen passivation. Using first-principles DFT calculations, it was determined that threefold coordinated single C interstitials or isolated C dangling bonds can be completely passivated by molecular H_2. On the other hand, correlated C dangling bonds or neighboring pairs threefold coordinated C with different combinations of back-bonds, can only be passivated by monatomic H. It was suggested that such defects were dominant over isolated carbon dangling bond type of defects. A stable configuration of a single C interstitial (isolated C dangling bond) on the oxide side of the interface is shown in Figure 20.28. The C atom is bonded to two Si atoms and a "dangling" O atom was found to have a doubly occupied level in the gap (0.5 eV above the VBE), and thus act as a hole trap. Upon exposure to molecular hydrogen (H_2), this defect would be fully passivated without any states in the gap. This result was found to be a general trend. It was found that isolated excess C atoms in a variety of configurations lead to states in the mid gap region, but can capture a H from a H_2 molecule (with the other H migrating away) and be fully passivated. Since molecular H_2 is not effective at passivating traps at the

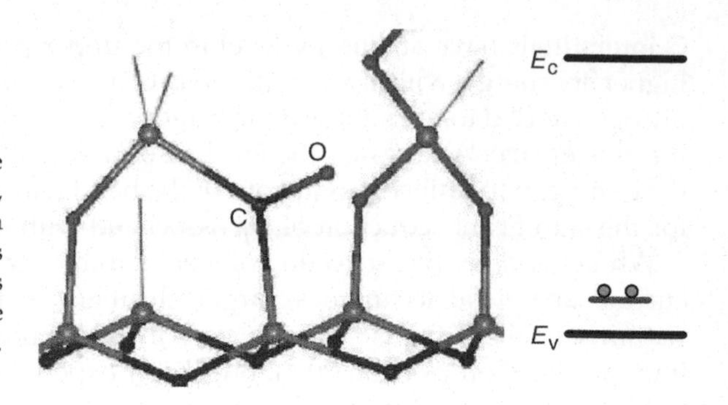

FIGURE 20.28
Example of a single C interstitial defect on the oxide side of the interface. In this particular case, a C atom is sharing a C site with another C atom on the oxide side of the interface. This defect has a filled state near the valence band of 4H-SiC, as a result it would act like a donor like interface trap or hole trap. (From Wang, S. et al., *Phys. Rev. Lett.*, 98, 026101, 2007. With permission.)

as-oxidized SiO_2/SiC interface, it was concluded that C atoms are not incorporated in the interface as single interstitials.

An example of "correlated" C dangling bond defect is shown in Figure 20.29a. The two C atoms share a nominal Si site and the structure was found to be stable. The two central C atoms are threefold coordinated and each produces a level in the gap (about 1 eV below the conduction band edge). The key point here is that the three neighbors of each of the two C atoms define different planes. As a consequence, if each of the C atoms were passivated with H atoms, the two C–H bonds would point in opposite directions (Figure 20.29b). The calculations revealed that a large energy barrier would be involved if a H_2 molecule is introduced to passivate the defect. Further passivating only one of the C atoms by a H_2 molecule, letting the other H migrate (hopping from O to O in the oxide) was found to be energetically unfavorable. Therefore, H_2 molecules are not effective in passivating correlated C dangling bonds. On the other hand, monatomic H was found to favorably bind to both single C dangling bonds as well as correlated C dangling bonds and result in passivation. A variety of local geometries for correlated C dangling bonds are possible, and in all cases this result is expected to hold. This result explains the effectiveness of atomic hydrogen annealing (annealing in H_2 in the presence of catalytic Pt; see Section 20.6) in the reduction of D_{it} observed in 4H-SiC MOS structures. As mentioned earlier, EPR has detected the passivation of C dangling bonds in oxides grown on porous SiC by annealing in forming gas (H_2), although typically such anneals do not reduce the D_{it} efficiently. Wang et al. argued that this seemingly contrasting result between electrical measurements and EPR may be accounted for by considering that only isolated C dangling bonds are passivated by molecular H_2, whereas correlated dangling bonds are not.

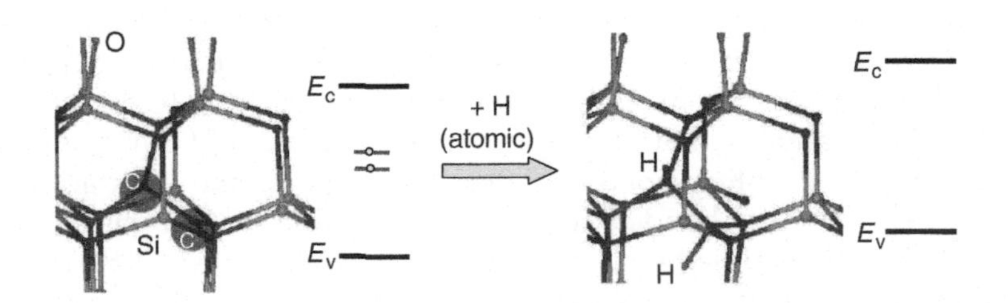

FIGURE 20.29
Example of a correlated C dangling bond defect on the SiC side of the interface. This defect has two closely spaced level in the SiC band gap. It is completely passivated using two H atoms as shown on the right. (From Wang, S. et al., *Phys. Rev. Lett.*, 98, 026101, 2007. With permission.)

The correlated C dangling bonds, with a much higher density but with much more variation in geometry and back-bond structure, are likely to give a very broad EPR signal and thus be undetected. That large surface areas for required detection by EPR [109,113] (realized only by using porous SiC, which has a lot of surface area associated with a porous structure) strongly hints to the relatively lower density of EPR active isolated C dangling bonds. It was suggested that the SiO_2/SiC interface be viewed as a bonded Si–C–O interlayer (~1–2 nm) with a mix of threefold and fourfold coordinated C atoms. Energy levels of the correlated threefold coordinated C atoms span the mid gap region and give rise to the observed D_{it}.

In this work, it was also shown that N can also passivate defects in this interlayer, but by a different mechanism. Calculations show that, instead of simply binding to a dangling bond, N can replace a threefold C (or Si) atom (isolated or correlated) and remove the level in the gap. As a result, N is an effective passivator of both C and Si dangling bonds. However, the threefold N that forms induces an energy level (due to its lone pair state) in the gap (about ~0.5 eV above the VBE) when it has Si–N or C–N back-bonds. When N has O–N back-bonds, on the other hand, it does not have any states in the gap. This result may explain why, for oxides processed by wet reoxidation followed by NO annealing, N is effective at passivating interface traps in the upper half of the band gap and less effective in the lower part [42]. Such behavior is not observed for as-processed NO annealed dry oxides, where a significant reduction in D_{it} is consistently observed for both halves of the gap (see Figure. 20.13). However, enhanced hole trapping in NO annealed dry oxides subjected to ionizing radiation has recently been observed [117]. This has indeed been attributed to threefold coordinated N in the near-interfacial region, which can potentially act as a hole trap due to the N lone pair state discussed above.

20.7.2 Defects in the Oxide

The picture that has emerged from the available experimental and theoretical work is that, while C-related defects can be used to explain the D_{it} distribution for the major portion of the SiC band gap, they are less likely to cause the extremely high D_{it} ($\geq 10^{13}$ cm^{-2} eV^{-1}) very close to the conduction band edge. Intrinsic defects in the oxide are the most likely candidates in this regards. Such defects could be common to both the SiO_2/Si and SiO_2/SiC system. But due to the much wider band gap of SiC, interface traps associated with such defects would lie within the band gap of SiC, but not in Si. Defects in the oxide with energy levels lying close to the conduction band of 4H-SiC have been reported experimentally for 4H-SiC MOS structures. Afanas'ev et al. used photon stimulated tunneling current measurements to determine electron trap levels at ~2.8 eV below the conduction band edge of SiO_2 (which would be close to conduction band of 4H-SiC) in comparable densities on SiO_2/Si and SiO_2/SiC structures [51]. It was suggested that these traps were intrinsic to SiO_2 and related to excess Si or oxygen deficiency in the near-interfacial region. Evidence of near interface oxide traps or border traps has also been reported in SiC MOS structures using the TSC technique [46,47]. It was determined that the density of energetically shallow border traps was significantly higher in 4H-SiC than 6H-SiC, consistent with the higher band gap of the 4H polytype. The electron capture process of these traps was found to be strongly dependent on oxide electric field and temperature, which indicates a spatial distribution of these traps away from the metallurgical interface and into the oxide.

It was mentioned before that oxide traps intrinsic to SiO_2 that do not play a significant role in Si devices can be critically important for 4H-SiC. We will now elaborate on this point here using Si–Si suboxide bonds as an example. Suboxide bonds are among the most probable defects in the near-interfacial region of SiO_2/Si and SiO_2/SiC. The bonding–antibonding splitting of these defects is very large compared with the Si band gap if the

Si–Si bond is normal length (2.35 Å) [26]. As a result, the normal length suboxide bonds cannot contribute to the D_{it} in the case of Si. Only unusually long Si–Si bonds have small enough bonding–antibonding splittings to result in "tail states" at the Si band edges. These very long bonds are effectively passivated by two H atoms. In contrast, at the SiC/SiO_2 interface, all Si–Si bonds are defects. Even those with normal bond length have bonding–antibonding splittings that are comparable to the SiC energy gap. Calculations show that the Si–Si bonding level is near the top of the valence bands and moves into the band gap for longer Si–Si bonds. The antibonding level remains in the conduction band near the edge, but the uncertainties in the calculations are large and this level may well be in the band gap of SiC.

It is generally believed that nitridation results in passivation of C-related defects that appear more or less across the band gap, as well as shallow near-interfacial oxide defect states [82,88]. The later may be related to the passivation of Si–Si suboxide bonds by N. Theoretical calculations have predicted that N is more effective for passivation of Si–Si suboxide bonds compared to H. While H can successfully passivate only normal length suboxide bonds, N can passivate all suboxide bonds. Two mechanisms have been identified. In the first, a single N atom can be inserted in Si–Si bonds to form a Si–N–Si bridge, with a localized state at the VBE. This localized level contains only electron, which would suggest that every such defect would become negatively charged in *n*-type SiC. As this is not observed in practice, such a process probably can be ruled out. The other mechanism involves the replacement of both Si atoms by N, which results in no states in the gap. This result, namely that N is more effective than H at passivating the Si–Si interface traps near the band edges, is in accord with the experimental data of Figure 20.21.

20.8 Remaining Issues at the SiO_2/SiC Interfaces and Concluding Remarks

With the discovery of the efficient interface trap passivation involving nitridation, channel mobilities in 4H-SiC MOSFETs have increased by an order of magnitude. As a consequence, the on-resistance of high power devices with blocking voltages of ~2 kV or greater are no longer dominated by the channel resistance (on-resistance is dominated by thick drift layers). The best 4H-SiC power MOSFETs reported in the literature have been demonstrated to operate near the theoretical limit of 4H-SiC, which is far superior than the theoretical limit for Si. From the standpoint of technology and fundamental materials science, however, many issues remain. Herein, we will briefly discuss some of these issues.

One the most challenging problems for state-of-the-art SiC MOSFETs is threshold voltage control. From a commercial standpoint, normally off power MOSFETs with positive threshold voltages are desirable. Unfortunately, the interface passivation processes that are critical for channel mobility improvement have an undesirable side-effect. The loss of trapped interfacial negative charge as a result of passivation exposes the positive fixed charge in the oxide. This causes the MOSFET threshold voltage to shift toward zero, so the operating characteristics tend toward normally-on. The effect can be understood by considering the following expression for threshold voltage in the absence of mobile charges.

$$V_T = V_T^{ideal} - \frac{1}{C_{ox}}(Q_f^+ - Q_{it}^-) \quad (20.5)$$

where

V_T^{ideal} is the ideal positive threshold voltage in the absence of any charges
C_{ox} is the oxide capacitance
Q_f^+ is the fixed positive charge assuming its location to be the oxide–semiconductor interface
Q_{it}^- is the negative charge trapped in interface traps at threshold

Passivation processes such as nitridation with NO reduce Q_{it}^-. As Q_f^+ is not reduced as well, V_T shifts toward zero. At this stage, the origin of the positive fixed charge and its correlation with processing parameters remain largely unknown. Processes that reduce both fixed charge and interface charge will lead not only to a desirable threshold voltage, but also to higher channel mobility and transconductance [118].

In addition, there are problems related to oxide growth on implanted material. Note that, in a typical power *n*-channel MOSFET, the gate oxide is grown over a *p*-well (Figure 20.1). Morphological disorder commonly known as "step bunching" occurs during the high-temperature implant activation anneal (~1550°C–1650°C) and translates to roughness at the SiO_2/SiC interface. This raises a lot of concerns with respect to the long-term stability and reliability of the gate oxide [119,120]. Also, for the same interface passivation scheme, effective mobility in implanted channels is typically ~30% lower compared to epitaxial channels. This is related to the interface roughness described above, or more likely, inadequate healing of the crystal damage associated with ion implantation. For MOSFETs with blocking voltages less than ~2 kV, further interface trap passivation is necessary to improve channel mobility and on-resistance. Although channel mobility improvement is not a big concern for high power MOSFETs, further interface trap passivation is still important from the point of view of oxide reliability and stable device operation.

Under high electric fields, the most important mechanism that causes dielectric degradation is Fowler–Nordheim (F–N) tunneling. The smaller band offsets for SiO_2/4H-SiC compared to SiO_2/Si makes F–N tunneling a severe concern for SiC due to the exponential dependence of tunneling current on the barrier height. A high interface trap density contributes to a worst case scenario for trap-assisted F–N tunneling of carriers from the channel into the oxide, with subsequent degradation of the oxide breakdown characteristics [121]. It has been reported that nitridation not only reduces the interface trap density, but also significantly increases oxide mean time to failure for temperatures up to 300°C [120]. This indeed highlights the importance of interface trap passivation with regards to reliability.

From the standpoint of fundamental materials science, several issues require further investigation. One of the most pertinent questions is "what is the ultimate inversion channel mobility in 4H-SiC?" In order to answer this question, mobility limiting mechanisms in addition to interface trapping and scattering need to be identified and quantified. Recent simulation results suggest surface roughness as a critical factor for mobility degradation at higher electric fields [118]. In this regards, the role of surface roughness needs to be investigated in greater detail. Additionally, details of the interfacial chemistry at the nanoscale, the identity of the interfacial defects, and role of N and H with respect to passivation need to be determined conclusively.

Acknowledgments

This work has been supported over the years by DARPA, ONR, and ARO. We are pleased to acknowledge useful discussions with J.A. Cooper (Purdue University), M. K. Das (Cree Inc.), and J. Rozen (Vanderbilt University).

References

1. *Materials for High Temperature Semiconductor Devices, National Research Council Report NMAB-747,* National Academy Press, Washington, DC, 1995.
2. *Silicon Carbide Electronics Markets 2004–2009: New Horizons for Power Electronics, Market study by Wicht Technologie Consulting,* Munich, Germany, 2005.
3. Iwami, M., Silicon carbide: Fundamentals, *Nucl. Instrum. Meth. Phys. Res. A,* 466, 406, 2001.
4. Harris, G.L., *Properties of Silicon Carbide,* IEE, London, 1995.
5. Cooper, J.A. Jr. et al., Status and prospects for SiC power MOSFETs, *IEEE Trans. Electron Dev.,* 49, 658, 2002.
6. Baliga, B.J., Power semiconductor device figure of merit for high-frequency applications, *IEEE Electron Dev. Lett.,* 10, 455, 1989.
7. Afanasev, V.V. et al., Intrinsic SiC/SiO_2 interface traps, *Phys. Status Solidi A,* 162, 321, 1997.
8. Ouisse, T., Electron transport at the SiC/SiO_2 interface, *Phys. Status Solidi A,* 162, 339, 1997.
9. Cooper, J.A. Jr., Advances in SiC MOS technology, *Phys. Status Solidi A,* 162, 305, 1997.
10. Schomer, R. et al., Significantly improved performance of MOSFETs on silicon carbide using the 15R-SiC polytype, *IEEE Electron Dev. Lett.,* 20, 241, 1999.
11. Das, M.K., Um, B.S., and Cooper, A.J. Jr., Anomalously high density of interface traps near the conduction band in SiO_2/4H-SiC MOS devices, in *ICSCRM '99: The International Conference on Silicon Carbide and Related Materials,* Oct 10–Oct 15, 1999, *Materials Science Forum,* Trans Tech Publications, Stafa-Zurich, Switzerland, 338 (II, 1069, 2000).
12. Chung, G.Y. et al., Effect of Si:C source ratio on SiO_2/SiC interface trap density for nitrogen doped 4H and 6H-SiC, in *ICSCRM '99: The International Conference on Silicon Carbide and Related Materials,* Oct 10–Oct 15, 1999, *Materials Science Forum,* Trans Tech Publications, Stafa-Zurich, Switzerland, 338 (II, 1097, 2000).
13. Saks, N.S., Mani, S.S., and Agarwal, A.K., Interface trap profile near the band edges at the 4H-SiC/SiO_2 interface, *Appl. Phys. Lett.,* 76, 2250, 2000.
14. Saks, N.S., Ancona, M.G., and Rendell, R.W., Using the Hall effect to measure interface trap densities in silicon carbide and silicon metal-oxide-semiconductor devices, *Appl. Phys. Lett.,* 80, 3219, 2002.
15. Saks, N.S. and Agarwal, A.K., Hall mobility and free electron density at the SiC/SiO_2 interface in 4H-SiC, *Appl. Phys. Lett.,* 77, 3281, 2000.
16. Chatty, K. et al., Comparative Hall measurements on wet- and dry-oxidized 4H-SiC MOSFETs, *J. Electron. Mater.,* 31, 356, 2002.
17. Ueno, K., Orientation dependence of the oxidation of SiC surfaces, *Phys. Status Solidi A,* 162, 299, 1997.
18. Song, Y. et al., Modified Deal Grove model for the thermal oxidation of silicon carbide, *J. Appl. Phys.,* 95, 4953, 2004.
19. Hornetz, B., Michel, H.J., and Halbritter, J., ARXPS studies of SiO_2–SiC interfaces and oxidation of 6H SiC single crystal Si-(0001) and C-(000–1) surfaces, *J. Mater. Res.,* 9, 3088, 1994.
20. Di Ventra, M. and Pantelides, S.T., Atomic-scale mechanisms of oxygen precipitation and thin-film oxidation of SiC, *Phys. Rev. Lett.,* 83, 1624, 1999.
21. Buczko, R., Pennycook, S.J., and Pantelides, S.T., Bonding arrangements at the Si–SiO_2 and SiC–SiO_2 interfaces and a possible origin of their contrasting properties, *Phys. Rev. Lett.,* 84, 943, 2000.
22. Knaup, J.M. et al., Theoretical study of the mechanism of dry oxidation of 4H-SiC, *Phys. Rev. B Condens. Matter,* 71, 235321, 2005.
23. Virojanadara, C. and Johansson, L.I., Studies of oxidized hexagonal SiC surfaces and the SiC/SiO_2 interface using photoemission and synchrotron radiation, *J. Phys. Condens. Matter,* 16, 1783, 2004.
24. Vickridge, I.C. et al., Limiting step involved in the thermal growth of silicon oxide films on silicon carbide, *Phys. Rev. Lett.,* 89, 256102, 2002.
25. Amy, F. et al., Si-rich 6H- and 4H-SiC(0001) 3×3 surface oxidation and initial SiO_2/SiC interface formation from 25° to 650°C, *Phys. Rev. B Condens. Matter,* 65, 165323, 2002.

26. Pantelides, S.T. et al., Si/SiO_2 and SiC/SiO_2 interfaces for MOSFETs-challenges and advances, *Mater. Sci. Forum*, 527–529, 935, 2006.
27. Raynaud, C., Silica films on silicon carbide: A review of electrical properties and device applications, *J. Non-Cryst. Solids*, 280, 1, 2001.
28. Dhar, S. et al., Effect of nitric oxide annealing on the interface trap density near the conduction bandedge of 4H-SiC at the oxide/(11–20) 4H-SiC interface, *Appl. Phys. Lett.*, 84, 1498, 2004.
29. Okawa, T. et al., Kinetics of oxynitridation of 6H-SiC(11–20) and the interface structure analyzed by ion scattering and photoelectron spectroscopy, *Surf. Sci.*, 601, 706, 2007.
30. Hayton, D.J. et al., Optical and ion-scattering study of SiO_2 layers thermally grown on 4H-SiC, *Semicond. Sci. Technol.*, 17, 29, 2002.
31. Radtke, C., Baumvol, I.J.R., and Stedile, F.C., Effects of ion irradiation in the thermal oxidation of SiC, *Phys. Rev. B Condens. Matter*, 66, 155437, 2002.
32. Jernigan, G.G., Stahlbush, R.E., and Saks, N.S., Effect of oxidation and reoxidation on the oxide–substrate interface of 4H- and 6H-SiC, *Appl. Phys. Lett.*, 77, 1437, 2000.
33. Virojanadara, C. and Johansson, L.I., Interfacial investigation of in situ oxidation of 4H-SiC, *Surf. Sci.*, 472, 145, 2001.
34. Johansson, L.I. and Virojanadara, C., Synchrotron radiation studies of the SiO_2/SiC(0001) interface, *J. Phys. Condens. Matter*, 16, 3423, 2004.
35. Chang, K.C. et al., High-carbon concentrations at the silicon dioxide-silicon carbide interface identified by electron energy loss spectroscopy, *Appl. Phys. Lett.*, 77, 2186, 2000.
36. Chang, K.C. et al., Electrical, structural, and chemical analysis of silicon carbide-based metal-oxide-semiconductor field-effect-transistors, *J. Appl. Phys.*, 95, 8252, 2004.
37. Chang, K.-C. et al., High-resolution elemental profiles of the silicon dioxide-4H-silicon carbide interface, *J. Appl. Phys.*, 97, 1, 2005.
38. Lu, W. et al., Graphitic features on SiC surface following oxidation and etching using surface enhanced Raman spectroscopy, *Appl. Phys. Lett.*, 85, 3495, 2004.
39. Tomioka, Y. et al., Characterization of the interfaces between SiC and oxide films by spectroscopic ellipsometry, in *Silicon Carbide and Related Materials 2001. ICSCRM2001. International Conference*, Oct. 28–Nov. 2, 2001, *Materials Science Forum*, Trans Tech Publications, 2002, p. 1029.
40. Feldman, L.C., Mayer, J.W., and Picraux, S.T., *Materials Analysis by Ion Channeling: Submicron Crystallography*, Academic Press, New York, 1982.
41. McDonald, K., Nitrogen incorporation and trap reduction in SiO_2/4H-SiC, PhD thesis, Vanderbilt University, 2001.
42. Chung, G.Y. et al., Effect of nitric oxide annealing on the interface trap densities near the band edges in the 4H polytype of silicon carbide, *Appl. Phys. Lett.*, 76, 1713, 2000.
43. Sze, S.M., *Physics of Semiconductor Devices*, Wiley, New York, 1981.
44. Nicollian, E.H. and Brews, J.R., *MOS (Metal Oxide Semiconductor) Physics and Technology*, Wiley, New York, 1982.
45. See reference 9 and the references therein for a review of electrical characterization of SiC MOS structure.
46. Olafsson, H.O. et al., Border traps in 6H-SiC metal-oxide-semiconductor capacitors investigated by the thermally-stimulated current technique, *Appl. Phys. Lett.*, 79, 4034, 2001.
47. Rudenko, T.E. et al., Interface trap properties of thermally oxidized *n*-type 4H-SiC and 6H-SiC, *Solid State Electron.*, 49, 545, 2005.
48. Afanas'ev, V.V. et al., Band alignment and defect states at SiC/oxide interfaces, *J. Phys. Condens. Matter*, 16, 1839, 2004, and references therein.
49. Dekker, J. et al., Observation of interface defects in thermally oxidized SiC using positron annihilation, *Appl. Phys. Lett.*, 82, 2020, 2003.
50. Afanas'ev, V.V. et al., Shallow electron traps at the 4H-SiC/SiO_2 interface, *Appl. Phys. Lett.*, 76, 336, 2000.
51. Afanas'ev, V.V. and Stesmans, A., Interfacial defects in SiO_2 revealed by photon stimulated tunneling of electrons, *Phys. Rev. Lett.*, 78, 2437, 1997.

52. Kobayashi, H. et al., Interface traps at SiO_2/6H-SiC(0001) interfaces observed by x-ray photoelectron spectroscopy measurements under bias: Comparison between dry and wet oxidation, *Phys. Rev. B Condens. Matter*, 67, 115305, 2003.
53. Friedrichs, P., Burte, E.P., and Schorner, R., Interface properties of metal-oxide-semiconductor structures on *n*-type 6H and 4H-SiC, *J. Appl. Phys.*, 79, 7814, 1996.
54. Chen, X.D. et al., Electron capture and emission properties of interface traps in thermally oxidized and NO-annealed SiO_2/4H-SiC, *J. Appl. Phys.*, 103, 033701, 2008.
55. Neudeck, P. et al., Measurement of *n*-type dry thermally oxidized 6H-SiC metal-oxide-semiconductor diodes by quasistatic and high-frequency capacitance versus voltage and capacitance transient techniques, *J. Appl. Phys.*, 75, 7949, 1994.
56. Pan, J.N., Cooper, J.A., Jr., and Melloch, M.R., Extremely long capacitance transients in 6H-SiC metal-oxide-semiconductor capacitors, *J. Appl. Phys.*, 78, 572, 1995.
57. Dhar, S. et al., Interface passivation for silicon dioxide layers on silicon carbide, *MRS Bull.*, 30, 288, 2005.
58. Li, H.-F. et al., Interfacial characteristics of N_2O and NO nitrided SiO_2 grown on SiC by rapid thermal processing, *Appl. Phys. Lett.*, 70, 2028, 1997.
59. Dimitrijev, S. et al., Nitridation of silicon-dioxide films grown on 6H silicon carbide, *IEEE Electron Dev. Lett.*, 18, 175, 1997.
60. Chung, G.Y. et al., Improved inversion channel mobility for 4H-SiC MOSFETs following high temperature anneals in nitric oxide, *IEEE Electron Dev. Lett.*, 22, 176, 2001.
61. Chung, G.Y. et al., 4H-SiC oxynitridation for generation of insulating layers, *J. Phys. Condens. Matter*, 16, 1857, 2004.
62. Pierret, R.F., Field effect devices, in *Modular Series on Solid State Devices*, Neudeck, G.W. and Pierret, R.F., Eds., Addison-Wesley, Reading, MA, 1990.
63. Sridevan, S. and Baliga, B.J., Lateral N-channel inversion mode 4H-SiC MOSFET's, *IEEE Electron Dev. Lett.*, 19, 228, 1998.
64. Singh, R., Ryu, S.-H., and Palmour, J.W., High temperature, high current, 4H-SiC Accu-DMOSFET, in *International Conference on Silicon Carbide and Related Materials, ICSRM'99*, Oct. 10–15, 1999, *Materials Science Forum*, Trans. Tech Publications, 2000, p. 1271.
65. Harada, S. et al., Temperature dependences of channel mobility and threshold voltage in 4H- and 6H-SiC MOSFETs, in *Silicon Carbide—Materials, Processing and Devices Symposium*, Nov. 27–29, 2000, Mater. Res. Soc, Warrendale, PA, 2001, p. 5.
66. Jamet, P. and Dimitrijev, S., Physical properties of N_2O and NO-nitrided gate oxides grown on 4H SiC, *Appl. Phys. Lett.*, 79, 323, 2001.
67. McDonald, K. et al., Kinetics of NO nitridation in SiO_2/4H-SiC, *J. Appl. Phys.*, 93, 2257, 2003.
68. Virojanadara, C. and Johansson, L.I., Studies of NO on 4H-SiC(0001) using synchrotron radiation, *J. Phys. Condens. Matter*, 16, 3435, 2004.
69. Jamet, P., Dimitrijev, S., and Tanner, P., Effects of nitridation in gate oxides grown on 4H-SiC, *J. Appl. Phys.*, 90, 5058, 2001.
70. McDonald, K. et al., Comparison of nitrogen incorporation of SiO_2/SiC and SiO_2/Si structures, *Appl. Phys. Lett.*, 76, 568, 2000.
71. Dhar, S. et al., Nitridation anisotropy in SiO_2/4H-SiC, *J. Appl. Phys.*, 97, 1, 2005.
72. McDonald, K. et al., Characterization and modeling of the nitrogen passivation of interface traps in SiO_2/4H-SiC, *J. Appl. Phys.*, 93, 2719, 2003.
73. Von Bardeleben, H.J. et al., Modification of the oxide/semiconductor interface by high temperature NO treatments: A combined EPR, NRA and XPS study on oxidized porous and bulk *n*-type 4H-SiC, in *5th European Conference on Silicon Carbide and Related Materials*, Aug 31.–Sept. 4, 2004, *Materials Science Forum*, Trans. Tech Publications, Stafa-Zurich, Switzerland, 2005, p. 277.
74. Choi, S.H. et al., Nitridation of the SiO_2/4H-SiC interface studied by surface-enhanced Raman spectroscopy, *Appl. Surf. Sci.*, 253, 5411, 2007.
75. Laurendeau, N.M., Fast nitrogen dioxide reactions: Significance during NO Decomposition and NO_2 formation, *Combust. Sci. Technol.*, 11, 89, 1975.
76. Wu, R.J. and Yeh, C.T., Activation energy for thermal decomposition of nitric oxide, *Int. J. Chem. Kinet.*, 28, 89, 1996.

77. Das, M.K., Recent advances in (0001) 4H-SiC MOS device technology, in *Proceedings of the 10th International Conference on Silicon Carbide and Related Materials, ICSCRM 2003, Oct 5–10, 2003,* Trans. Tech Publications Ltd, Zurich-Ueticon, CH-8707, Switzerland, 2004, p. 1275.
78. Gupta, A. et al., Nitrous oxide gas phase chemistry during silicon oxynitride film growth, *Prog. Surf. Sci*, 59, 103, 1998.
79. Singh, R. et al., Development of high-current 4H-SiC ACCUFET, *IEEE Trans. Electron Dev.*, 50, 471, 2003.
80. Lipkin, L.A., Das, M.K., and Palmour, J.W., N_2O processing improves the 4H-SiC:SiO_2 interface, in *Silicon Carbide and Related Materials 2001, ICSCRM2001, International Conference*, Oct. 28–Nov. 2, 2001, *Materials Science Forum*, Trans. Tech Publications, Stafa-Zurich, Switzerland, 2002, p. 985.
81. Lai, P.T., Xu, J.P., and Chan, C.L., Effects of wet N_2O oxidation on interface properties of 6H-SiC MOS capacitors, *IEEE Electron Dev. Lett.*, 23, 410, 2002.
82. Afanas'ev, V.V. et al., Mechanisms responsible for improvement of 4H-SiC/SiO_2 interface properties by nitridation, *Appl. Phys. Lett.*, 82, 568, 2003.
83. Dimitrijev, S., Tanner, P., and Harrison, H.B., Slow-trap profiling of NO and N_2O nitrided oxides grown on Si and SiC substrates, *Microelectron. Reliab.*, 39, 441, 1999.
84. Chung, G. et al., Effects of anneals in ammonia on the interface trap density near the band edges in 4H-silicon carbide metal-oxide-semiconductor capacitors, *Appl. Phys. Lett.*, 77, 3601, 2000.
85. Williams, J.R. et al., Nitrogen passivation of the interface traps near the conduction band edge in 4H-silicon carbide, in *Silicon Carbide-Materials, Processing and Devices, Nov. 27–29, 2000*, Materials Research Society, Warrendale, PA, 2001, p. 3.
86. Fukuda, K. et al., Effect of oxidation method and post-oxidation annealing on interface properties of metal-oxide-semiconductor structures formed on *n*-type 4H-SiC C(000–1) face, *Appl. Phys. Lett.*, 77, 866, 2000.
87. Fukuda, K. et al., Reduction of interface-trap density in 4H-SiC *n*-type metal-oxide-semiconductor structures using high-temperature hydrogen annealing, *Appl. Phys. Lett.*, 76, 1585, 2000.
88. Wang, S. et al., Bonding at the SiC–SiO_2 interface and the effects of nitrogen and hydrogen, *Phys. Rev. Lett.*, 98, 026101, 2007.
89. Dhar, S. et al., Nitrogen and hydrogen induced trap passivation at the SiO_2/4H-SiC interface, *Mater. Sci. Forum*, 527–529, 949, 2006.
90. Nakamura, D. et al., Ultrahigh-quality silicon carbide single crystals, *Nature*, 430, 1009, 2004.
91. Ciobanu, F. et al., Nitrogen implantation—an alternative technique to reduce traps at SiC/SiO_2 interfaces, *Mater. Sci. Forum*, 527–529, 991, 2006.
92. Dhar, S. et al., Interface trap passivation for SiO_2/(000–1) C-terminated 4H-SiC, *J. Appl. Phys.*, 98, 14902, 2005.
93. Casady, J.B. and Johnson, R.W., Status of silicon carbide (SiC) as a wide-bandgap semiconductor for high-temperature applications: A review, *Solid State Electron.*, 39, 1409, 1996.
94. Dhar, S. et al., unpublished.
95. Yano, H., Kimoto, T., and Matsunami, H., Interface traps of SiO_2/SiC on (11–20) and (0001) Si faces, in *3rd European Conference on Silicon Carbide and Related Materials*, Sep 3–Sep 7, 2000, *Materials Science Forum*, 353–356, 627, 2001.
96. Yano, H. et al., A cause for highly improved channel mobility of 4H-SiC metal-oxide-semiconductor field-effect transistors on the ($\bar{1}$120) face, *Appl. Phys. Lett.*, 78, 374, 2001.
97. Senzaki, J. et al., Excellent effects of hydrogen postoxidation annealing on inversion channel mobility of 4H-SiC MOSFET fabricated on (1 $\bar{1}$20) face, *IEEE Electron Dev. Lett.*, 23, 13, 2002.
98. Fukuda, K. et al., 4H-SiC MOSFETs on C(000–1) face with inversion channel mobility of 127cm^2/Vs, in *Proceedings of the 10th International Conference on Silicon Carbide and Related Materials, ICSCRM 2003,* Oct 5–10, 2003, Trans. Tech Publications Ltd., Zurich-Ueticon, 2004, p. 1417.
99. Alok, D. et al., US Patent and Trademark Office, 6,559,068, 2003.
100. Gudjonsson, G. et al., High field-effect mobility in *n*-channel Si Face 4H-SiC MOSFETs with gate oxide grown on aluminum ion-implanted material, *IEEE Electron Dev. Lett.*, 26, 96, 2005.
101. Sveinbjornsson, E.O. et al., High channel mobility 4H-SiC MOSFETs, *Mater. Sci. Forum*, 527–529, 961, 2006.

102. Das, M.K. et al., Improved 4H-SiC MOS interfaces produced via two independent processes: Metal enhanced oxidation and 1300°C NO anneal, *Mater. Sci. Forum*, 527–529, 967, 2006.
103. Soukiassian, P. et al., SiO_2–Si interface formation by catalytic oxidation using alkali metals and removal of the catalyst species, *J. Appl. Phys.*, 60, 4339, 1986.
104. Ray, E. et al., Pressure dependence of SiO_2 growth kinetics and electrical properties of SiC, *J. Appl. Phys.*, 103, 023522, 2008.
105. Poindexter, E.H. et al., Interface traps and electron spin resonance centers in thermally oxidized (111) and (100) silicon wafers, *J. Appl. Phys.*, 52, 879, 1981.
106. Stirling, A. et al., Dangling bond defects at Si-SiO_2 interfaces: Atomic structure of the P_b center, *Phys. Rev. Lett.*, 85, 2773, 25.
107. Bongiorno, A. et al., Transition structure at the Si(100)–SiO_2 interface, *Phys. Rev. Lett.*, 90, 186101, 2003.
108. Rashkeev, S.N. et al., Defect generation by hydrogen at the Si–SiO_2 interface, *Phys. Rev. Lett.*, 87, 165506, 2001.
109. Cantin, J.L. et al., Identification of the carbon dangling bond center at the 4H-SiC/SiO_2 interface by an EPR study in oxidized porous SiC, *Phys. Rev. Lett.*, 92, 15502, 2004.
110. Macfarlane, P.J. and Zvanut, M.E., Characterization of paramagnetic defect centers in three polytypes of dry heat treated, oxidized SiC, *J. Appl. Phys.*, 88, 4122, 2000.
111. Wang, S. et al., Atomic-scale dynamics of the formation and dissolution of carbon clusters in SiO_2, *Phys. Rev. Lett.*, 86, 5946, 2001.
112. Knaup, J.M. et al., Defects in SiO_2 as the possible origin of near interface traps in the SiC/SiO_2 system: A systematic theoretical study, *Phys. Rev. B*, 72, 115323, 2005.
113. Cantin, J.L. et al., Hydrogen passivation of carbon P_b like centers at the 3C- and 4H-SiC/SiO_2 interfaces in oxidized porous SiC, *Appl. Phys. Lett.*, 88, 92108, 2006.
114. Afanas'ev, V.V. et al., Elimination of SiC/SiO_2 interface traps by preoxidation ultraviolet-ozone cleaning, *Appl. Phys. Lett.*, 68, 2141, 1996.
115. Bassler, M., Pensl, G., and Afanas'ev, V., Carbon cluster model for electronic states at SiC/SiO_2 interfaces, *Diamond Relat. Mater.*, 6, 1472, 1997.
116. Lipkin, L.A. and Palmour, J.W., Improved oxidation procedures for reduced SiO_2/SiC defects, in *37th Electronic Materials Conference. III–V Nitrides and Silicon Carbide, 1995, Journal of Electronic Materials*, TMS, 1996, p. 909.
117. Dixit, S.K. et al., Total dose radiation response of nitrided and non-nitrided SiO_2/4H-SiC MOS capacitors, *IEEE Trans. Nucl. Sci.*, 53, 3687, 2006.
118. Potbhare, S. et al., Numerical and experimental characterization of 4H-silicon carbide lateral metal-oxide-semiconductor field-effect transistor, *J. Appl. Phys.*, 100, 44515, 2006.
119. Senzaki, J. et al., A long-term reliability of thermal oxides grown on *n*-type 4H-SiC wafer, *Mater. Sci. Forum*, 457–460, 1269, 2004.
120. Krishnaswami, S. et al., A study on the reliability and stability of high voltage 4H-SiC MOSFET devices, *Mater. Sci. Forum*, 1313–1316, 2006.
121. Singh, R., Reliability and performance limitations in SiC power devices, *Microelectron. Reliab.*, 46, 713, 2006.

21

Defects in SiC

E. Janzén, A. Gali, A. Henry, I.G. Ivanov, B. Magnusson, and N.T. Son

CONTENTS

21.1 Introduction

More than 200 polytypes of SiC are known but only a few have been explored so far for industrial applications. These are the hexagonal 4H and 6H polytypes, as well as the cubic 3C polytype, which has the zinc blende structure. Since these are also the most studied polytypes, our attention will mainly be restricted to them.

Since both Si and C are of valence four, substitutional atoms of the third and fifth group of the periodic table produce shallow acceptor and donor levels, respectively. Thus, nitrogen is the common donor introduced unintentionally during growth of SiC. Except for the cubic 3C-SiC and the hexagonal 2H polytype (wurtzite structure), all the other polytypes possess inequivalent sites in their unit cells. Therefore, the ionization energy of a substitutional donor (e.g., N) or acceptor (e.g., Al) depends on the site of the host atom, which has been substituted. For example, 4H-SiC has two inequivalent lattice sites both for the silicon and carbon atoms in a unit cell, commonly termed as cubic (*k*) and hexagonal (*h*) sites. The nitrogen donor is shown to substitute carbon; hence, doping 4H-SiC with N produces two donor levels (one for each inequivalent site) with substantially different ionization energies, as will be seen later. Similarly, 6H-SiC has three inequivalent sites (denoted usually h, k_1, and k_2) and doping with a single donor (or acceptor) species produces three donor (or acceptor) levels.

Band-structure calculations have been carried out for several polytypes of SiC [1–6] and, since experimental data are available mainly for 3C-, 4H-, and 6H-SiC, we briefly review the main electronic band features in these polytypes.

The Brillouin zones (BZs) of the cubic 3C polytype (space group T_d^2) and the hexagonal 4H and 6H polytypes (space group C_{6v}^4) are shown in Figure 21.1. A common feature of all SiC polytypes is the indirect band gap, i.e., the valence band maximum (VBM) is always at the center Γ of the BZ, whereas the conduction band minimum (CBM) is at different other points within the BZ, X for 3C-SiC, M for 4H-SiC, and U (along the M–L line) for 6H-SiC (cf. Figure 21.1). The electron and hole effective masses and other electronic band parameters have been measured in several polytypes [7–16]. A summary of the band parameters of the three considered polytypes is given in Table 21.1. The top of the valence band is double degenerate in 3C-SiC due to the cubic symmetry, and the next valence band below it is shifted from the top by the spin–orbit interaction by approximately 10 meV. The crystal field, which exists in all hexagonal polytypes, splits the aforementioned degeneracy. The value of this crystal-field splitting is significantly larger than the spin–orbit splitting in 4H- and 6H-SiC (cf. Table 21.1), but it is still significantly below the acceptor ionization energies in these polytypes, as will be seen later. The following remarks refer to Table 21.1. The theoretical calculations conducted using the local-density approximation (LDA) always underestimate the value of the energy band gap. The data given in the table reflect the scatter of the theoretical results, and it was shown in Ref. [6] that a major reason for this

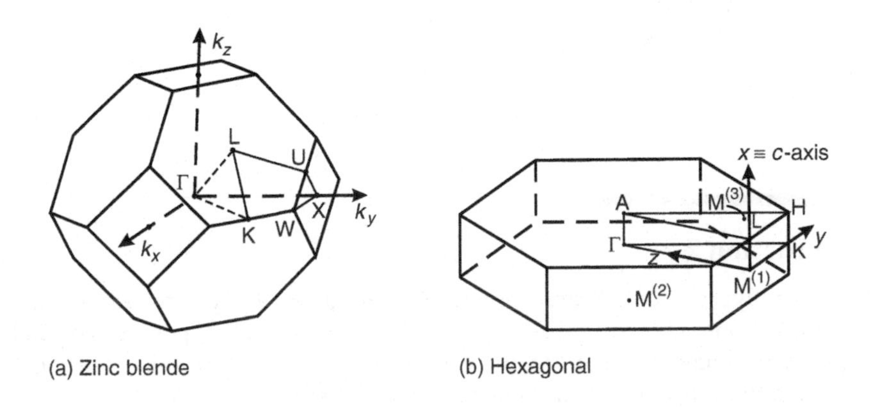

FIGURE 21.1
The BZ of (a) 3C-SiC, and (b) 4H-SiC (this is the zone also for other hexagonal polytypes; however, the height of the prism will be different). The high-symmetry points are denoted, and in the case of 4H-SiC, the three equivalent M-points and the coordinate system used in Equation 21.2 are drawn.

TABLE 21.1

Band-Structure Parameters of 3C-, 4H-, and 6H-SiC

	3C-SiC		4H-SiC		6H-SiC	
Parameter	**Theory**	**Exp. [Ref.]**	**Theory**	**Exp. [Ref.]**	**Theory**	**Exp. [Ref.]**
Energy band gap (eV)						
At room temperature		2.39		3.26		3.06
At 2 K	1.33–2.07	2.416 [18]	2.16–2.90	3.287 [17,19,20]	1.96–2.69	3.083 [18,21]
Free exciton-binding energy (meV)	—	~27 [18]	—	~21 [17,19,20]	—	~60 [18,21]
Electron effective masses	$m_{\perp}=0.24$	0.247 [7,8]	$m_{\Gamma M}=0.61$	0.58 [9]	$m_{\parallel \Gamma M}=0.81$	[a]
	$m_{\parallel}=0.73$	0.667 [7,8]	$m_{MK}=0.29$	0.31 [9]	$m_{\parallel MK}=0.25$	[a]
			$m_{ML}=0.33$	0.33 [9]	$m_{ML}=2.07$	1.8–2.2 [10]
Hole effective mass	$m \sim 1.15$	—	$m_{\perp}=0.59$	0.66 [12]	$m_{\perp}=0.58$	0.66 [13]
			$m_{\parallel}=1.60$	1.75 [12]	$m_{\parallel}=1.59$	1.85 [13]
Spin–orbit splitting (meV)	~14	~10 [14]	~10	6.8 [16]	~10	7.2 [15]
Crystal-field splitting (meV)	n.a.	n.a.	72	~60 [16]	53	42–44 [15]

Note: The mass unit is the electron rest mass, m_0. The energy band gaps at room temperature assume the same temperature dependence as for the excitonic band gap presented in Ref. [17]. The theoretical results for the energy band gap and the crystal-field splitting are from Refs. [3–6]; the energy intervals are given. The theoretical electron and hole effective masses are the polaron masses of Ref. [5]; the rest of the theoretical data are from Refs. [5,6]. n.a., not applicable.

[a] The geometrical average $\sqrt{m_{\parallel \Gamma M} m_{\parallel MK}}=0.42$ of Ref. [10] agrees with the theoretical value 0.45.

scatter is the choice of different exchange-correlation potentials. The scatter of the values for the crystal-field splitting in the hexagonal polytypes (also shown in the table) has a different background, since the calculated values of this parameter strongly depend on the exact atomic position used in the calculation [6]. The remaining theoretical values are quite consistent among the different calculations and only the set of Refs. [5,6] is given.

In Section 21.2, we will first review the experimental and theoretical data on shallow levels in SiC. Intrinsic defects including vacancies, interstitials, and combination of them will be discussed in detail in Section 21.3. Section 21.4 will be devoted to unknown defects found in semi-insulating (SI) materials. Finally, Section 21.5 will review metallic impurities, irradiation-induced defects, and other unknown defects. An earlier review on optical properties of defects in SiC can be found in Ref. [22].

21.2 Shallow Levels

21.2.1 Electronic Structure of the Shallow Donors and Acceptors

In a rough approximation, a shallow single donor in an insulating crystal can be seen as a hydrogen atom (the positively charged donor core attracts the extra donor electron via the Coulomb interaction). Thus, roughly, the bound electronic states of the donor are anticipated to be similar to the bound states of the hydrogen atom. However, two scaling factors arise for the donor. First, the Coulomb interaction is screened by the dielectric constant of the media ε, and, second, the electron rest mass m_0 must be replaced by the effective mass

m in the crystal. Thus, within this simple model the hydrogen-like energy levels of the donor electron are described by the formula:

$$E_n = \frac{1}{\varepsilon^2} \frac{m}{m_0} \frac{Ry}{n^2}, \tag{21.1}$$

where

n is the main quantum number

$Ry = m_0 e^4/2\hbar^2 = 13.605$ eV is the usual Rydberg, the ionization energy of the hydrogen atom

The energy levels of a single acceptor can be treated in a similar way, with the electron effective mass replaced by that of the hole.

However, such a simple picture usually does not provide an adequate description. According to the effective-mass theory (EMT), the electronic structure of a shallow donor (acceptor) is stipulated by the structure of the conduction (valence) band of the semiconductor near the extremum. In addition, the electron and hole effective masses are tensors, not scalars, and in all uniaxial polytypes the dielectric constant is also a tensor. Due to the complicated valence band structure, the one-band EMT cannot be applied for calculating the acceptor levels in 3C-, 4H-, and 6H-SiC, and no attempts have been made so far in this direction, to the best of our knowledge. One remark is due on the CBM of 6H-SiC. It shows strong nonparabolicity near the minimum at the U-point (the so-called camelback structure, see Ref. [5]); therefore, the effective masses cannot be defined in the usual way as the inversed curvature of the parabola describing the minimum. Thus, while the EMT works well for shallow donors in 3C- and 4H-SiC (see below), it is considered to be inapplicable to 6H-SiC in its traditional form and, to the best of our knowledge, no attempts have been made to calculate the donor energy levels in this polytype. Far-infrared (IR) absorption measurements have been performed on 6H-SiC [23]; however, due to the lack of adequate theory, the interpretation of the excited states observed is unreliable.

In general, two methods have been used in the literature for calculation of the donor electron states: the EMT and first-principles calculations based on minimization of the energy in a large enough supercell of the crystal. These two methods are complementary to each other. While the EMT is known to provide accurately the energies of the excited states of the donor, it fails to describe correctly the ground state. This is because the ground-state envelope function of the donor electron has nonvanishing amplitude in the vicinity and at the donor site. In this region (the central-cell region) the potential deviates considerably from the screened Coulomb potential used all over the space, the electron experiences the influence of this (usually unknown) central-cell potential, and the energy of the ground state calculated using the common screened Coulomb potential becomes inaccurate. Since higher s-like states (e.g., 2s, 3s, etc.) also have nonvanishing amplitudes at the donor core, their energies are also affected by the central-cell potential to a much smaller extent, because the weight of the wave function at the central cell rapidly decreases with increasing the main quantum number of the state. On the other hand, the first-principles calculation relies on the atomic potentials and is expected to provide accurate results, especially for the ground state since this is also the most localized one. Unfortunately, the main factor limiting the accuracy of this calculation is the size of the supercell. In these calculations, the real crystal with a single donor is modeled using a fragment of the crystal (supercell) containing several unit cells with a single donor in it. This supercell is then periodically repeated all over the space to represent the infinite crystal. Larger supercells will provide better accuracy, because most of the wave function of the donor electron will be within the supercell and interaction with the donors in neighboring supercells will become negligible. However, the calculation efforts grow enormously with the size of the supercell. Currently, the largest supercells used have about 580 atoms, which means that their extensions are of order of a few lattice constants, clearly not enough for a shallow impurity, the ground-state function of which extends over tens of lattice constants in any direction. Nevertheless, these calculations provide important guiding information about the character of the wave function and the energy of the ground state.

In Section 21.2.2, we shall give a brief account of the EMT applied to shallow donors in 3C- and 4H-SiC, and compare the theory with the N donor excitation spectra measured in 3C- and 4H-SiC. Afterward, we shall review the results of the ionization energies of the N and P donors and the Al acceptors obtained by analyzing the donor–acceptor pair (DAP) luminescence in 4H-SiC and compare the data with the results of first-principle calculations. We also briefly review the experimental data on Raman scattering from donor states, the recent results of electron paramagnetic resonance (EPR) techniques applied for studying the N and P donors, and the photoluminescence (PL) related to excitons bound to the N and P donors, as well as the Al, Ga, and B-acceptors.

21.2.2 EMT

The presence of several equivalent minima (valleys) of the conduction band within the first BZ is the reason for the so-called valley–orbit splitting of each donor level. Effective-mass equations can be written for the envelope function of the donor electron for each valley. If the attractive potential of the donor core is approximated by a Coulomb potential in the whole space, the equation corresponding to one of the three valleys in 4H-SiC (M_1, cf. Figure 21.1) is

$$H_1\varphi_1 = \left[-\frac{\hbar^2}{2}\left(\frac{1}{m_{\mathrm{ML}}}\frac{\partial^2}{\partial x^2}+\frac{1}{m_{\mathrm{MK}}}\frac{\partial^2}{\partial y^2}+\frac{1}{m_{\Gamma\mathrm{M}}}\frac{\partial^2}{\partial z^2}\right)-\frac{e^2}{\sqrt{\varepsilon_\perp\varepsilon_\parallel}\sqrt{(\varepsilon_\perp/\varepsilon_\parallel)x^2+y^2+z^2}}\right]\varphi_1 = E\varphi_1 \tag{21.2}$$

Here $m_{\Gamma\mathrm{M}} = (0.58 \pm 0.01)m_0$, $m_{\mathrm{MK}} = (0.31 \pm 0.01)m_0$, and $m_{\mathrm{ML}} = (0.33 \pm 0.01)m_0$ [9] are the components of the electron effective-mass tensor in the Γ–M, M–K, and M–L directions of the BZ, respectively, and the values of the dielectric constant along and perpendicular to the crystal **c**-axis are $\varepsilon_\parallel = 10.36 \pm 0.1$ and $\varepsilon_\perp = 9.55 \pm 0.1$ [24]. The equations corresponding to the other minima are identical, except for the choice of the coordinate axes. Thus, any linear combination of the three envelope functions will have the same energy and the valley–orbit splitting is neglected. In fact, the three wave functions are coupled together by the central-cell potential; therefore, states with nonvanishing amplitude of the wave function in the central cell will exhibit sensible valley–orbit splitting. This is the case of the ground state and, to a smaller extent, higher s-like states. The rest of the excited donor states do have negligible amplitudes in the central-cell region and, therefore, Equation 21.2 is a very good approximation.

A similar equation can be written in the case of 3C-SiC; however, there are only two components of the effective-mass tensor ($m_\parallel = m_{\Gamma\mathrm{X}} = 0.667m_0$ and $m_\perp = 0.247m_0$ [7] in directions parallel and perpendicular to the Γ–X line; cf. Figure 21.1), and the dielectric constant $\varepsilon = 9.82$ [25] is a scalar due to the cubic symmetry. The resulting Hamiltonian has cylindrical symmetry ($D_{\infty h}$), and the theory worked out by Faulkner [26] is applicable. Transitions from the ground state to various p-like excited states for the nitrogen and another unidentified donor have been observed in 3C-SiC [25], and the energies of the excited states obtained using Faulkner's theory are in excellent agreement with the experiment.

Similarly, the EMT has been generalized for the case of three different effective masses and anisotropic dielectric constant [24] (the case of Equation 21.2 describing 4H-SiC with Hamiltonian of D_{2h} symmetry). Transitions from the ground to excited states of the N donor have also been measured for high-quality samples of 4H-SiC [27–30], and the energies of the excited states are also in excellent agreement with the EMT of Ref. [24]. Knowing from the theory the binding energies of the excited states and the energies of the experimentally observed transitions, it is easy to deduce the energy of the ground state, as well as of its valley–orbit split-off counterpart (populated at elevated temperatures). We note that the spectra of Refs. [27–30] have been obtained using a modification of the absorption measurement, a photoconductivity measurement in which the sample itself serves as a detector. In these measurements, electrons are freed from the donor in a two-step process. First an infrared photon is absorbed transferring the electron from the ground state to an excited state, and afterward the electron is thermally excited into the conduction band. The results of the measurements for 3C- and 4H-SiC are compared to theory in Table 21.2.

We shall consider now briefly the symmetry of the total wave functions of the donor electron and the resulting selection rules, considered in detail in Ref. [30] for the case of 4H-SiC. The site symmetry of the nitrogen donor in the environment of 4H crystal lattice is C_{3v} (this is also the symmetry of the Hamiltonian, which includes the valley–orbit interaction, see e.g., Ref. [31] for the case of 4H-SiC). Therefore, the total envelope function must transform as one of the irreducible representations of C_{3v}, $A_1(\Gamma_1)$, $A_2(\Gamma_2)$, or $E(\Gamma_3)$. (The notations in parentheses are according to Ref. [32], but the other set is most commonly used in the literature.) On the other hand, the one-valley envelope functions of Equation 21.2 transform as the irreducible representations of D_{2h}. The one-valley wave functions Ψ_k are approximated in Kohn–Luttinger's theory [33] by the product

$$\Psi_k = \varphi_k' \varphi_k \tag{21.3}$$

where

φ_k denotes the envelope function corresponding to the kth valley as in Equation 21.2
φ_k' is the Bloch function at the kth minimum

The latter transforms according to C_{2v}, the group of the wave vector at the CBM for 4H-SiC. The band-structure calculation [5] yields that φ_k' transforms as the irreducible representation M_4 in C_{2v} (cf. Ref. [30]), which means that the product Ψ_k (Equation 21.3) transforms in C_{2v} as shown in Table 21.3. Appropriate linear combinations transforming according to the C_{3v} symmetry have to be formed from the functions Ψ_k, not the envelope functions φ_k alone, which leads to the symmetry classification of the donor states in 4H-SiC also shown in Table 21.3 (cf. Ref. [30]). Note that the ground state has two components (split by the valley–orbit interaction), $1s(A_1)$ and $1s(E)$, and no component of A_2 symmetry in C_{3v}. Ref. [30] was in error in the interpretation of the symmetry of the Bloch-part of the wave function; as a consequence A_1 and A_2 are wrongly interchanged throughout this reference. The correct results are obtained by interchanging A_1 with A_2 and vice versa everywhere in Ref. [30], but the conclusions on the selection rules remain unchanged.

The resulting selection rules for direct dipole transitions from the ground states $1s(A_1)$ and $1s(E)$ to various excited states in 4H-SiC are also given in Table 21.3. Surprisingly, $1s - ns$ $(n > 1)$ transitions are not forbidden by symmetry for polarization of light **E** parallel to the crystal **c**-axis (**E** $\parallel$ **c**). However, as discussed in Ref. [30] they still may have vanishing probability, depending on the properties of the Bloch function φ_k'. Unfortunately, the

TABLE 21.2

Comparison between the Experimental Excitation Spectra of the N Donor in 3C (Ref. [25]) and 4H-SiC (hexagonal site) (Refs. [27–29]) and the Theory of Ref. [24], which Is Used to Calculate the Theoretical Binding Energies of the States

	4H-SiC			
	Experimental Energy (meV)		Theoretical Energy	
Transition	**Refs. [27,28]**	**Ref. [29]**	**Binding**	**Transition**
Polarization **E** ⊥ **c**				
1s(E)-$2p_0$(A_1,E)	38.1	38.3	15.67	38.23
1s(E)-$2p_-$(A_2,E)	41.8	41.8	12.25	41.65
1s(A_1)-$2p_0$(E)	45.6	45.6	15.67	45.70
1s(E)-$3p_0$(A_1,E)		46.8	7.05	46.85
1s(E)-$3p_-$(A_2,E)	48.4	≈ 48.5	5.50	48.40
1s(A_1)-$2p_-$(E)	49.2	49.2	12.25	49.12
1s(A_1)-$3p_0$(E)	54.2	Shoulder ≈ 54.7	7.05	54.32
1s(A_1)-$3p_-$(E)	55.9–56.3	55.9	5.50	55.87
Polarization E ∥ c				
1s(A_1)-2s(A_1) + 1s(A_1)-$2p_+$(A_1)	48.1 (broad)		13.72	47.65
			12.79	48.58
1s(E)-2s(E) + 1s(E)-$2p_+$(E)	40.7 (broad)		13.72	40.18
			12.79	41.11

	3C-SiC		
		Theoretical Energy	
Transition	**Experimental Energy (meV) [25]**	**Binding**	**Transition**
1s(E)-$2p_0$	30.63	15.21	30.67
1s(E)-$2p_\pm$	35.5	10.39	35.49
1s(A_1)-$2p_0$	38.99	15.21	38.99
1s(E)-$3p_\pm$	41.1	4.77	41.11
1s(E)-$4p_\pm$	42.61	3.26	42.62
1s(E)-$4f_\pm$	43.17	2.75	43.13
1s(A_1)-$2p_\pm$	43.84	10.39	43.81
1s(A_1)-$3p_0$	47.19	6.98	47.22
1s(A_1)-$3p_\pm$	49.43	4.77	49.43
1s(A_1)-$4p_0$	50.09	4.08	50.12
1s(A_1)-$4p_\pm$	50.95	3.26	50.94
1s(A_1)-$4f_\pm$	51.45	2.75	51.45
1s(A_1)-$5p_\pm$	52.1	2.12	52.09
1s(A_1)-$6p_\pm$	52.69	1.52	52.68

Note: The values for the static dielectric constant used in the calculation are $\varepsilon = \sqrt{\varepsilon_\perp \varepsilon_\parallel} = 9.93$ for 4H-SiC, and $\varepsilon = 9.80$ for 3C-SiC. The theoretical transition energies are calculated by subtracting the calculated binding energy from the energy of the initial state, which places the initial states for the nitrogen donor in 4H-SiC at 61.37 meV for 1s(A_1) and 53.9 meV for 1s (E). The corresponding values for 3C-SiC are 54.20 meV for 1s(A_1) and 45.88 meV for 1s(E). Thus the valley–orbit splittings for the nitrogen donor (N_h in 4H-SiC) are 8.32 ± 0.05 meV and 7.47 ± 0.1 meV for 3C and 4H-SiC, respectively.

absorption spectra with this polarization [28] are too broad to allow any conclusion concerning experimental evidence on this point.

The application of the EMT to the far-IR absorption spectra has established that the binding energies of the two states of the ground-state manifold of the N donor in 3C-SiC

TABLE 21.3

Symmetries of the One-Valley Wave Functions Ψ_k in C_{2v} and the Total Wave Function Ψ in C_{3v} (the Latter Has Two Degenerate States within the EMT Components, which Are Associated with the Valley–Orbit Splitting, Neglected in the EMT)

Type of Ψ_k	Symmetry of Ψ_k in C_{2v}	Symmetry of Ψ in C_{3v}
ns state	M_4	A_1 or E
np_0 state	M_1	A_1 or E
np_- state	M_3	A_2 or E
np_+ state	M_4	A_1 or E

Selection rules for transition between the donor states

Initial state	Allowed final states ($n>1$)
Photon polarization $\mathbf{E} \perp \mathbf{c}$	
$1s(A_1)$	$np_0(E)$, $np_-(E)$
1s(E)	$np_0(E)$, $np_-(A_2)$, $np_0(A_1)$, $np_-(E)$
Photon polarization $\mathbf{E} \parallel \mathbf{c}$	
$1s(A_1)$	$ns(A_1)$, $np_+(A_1)$
1s(E)	ns(E), $np_+(E)$

Note: The selection rules for light-induced transitions from the ground state and its valley–orbit split-off counterpart are also shown.

are 54.2 and 45.8 meV, respectively [25]; see also Table 21.2. The corresponding values for N at hexagonal site (N_h) in 4H-SiC are 61.4 and 53.9 meV [24,29,30]. N at cubic site (N_k) has not been identified in absorption, probably because the transition energies fall within the Restrahlen band of the crystal. An alternative donor, phosphorus (P), has been studied extensively in recent years, but no absorption measurements are available at present. Nevertheless, accurate data about the ionization energies of both N and P donors at cubic site, as well as Al acceptors at cubic and hexagonal sites (Al_k and Al_h) have been extracted recently from analysis of the DAP spectra.

21.2.3 DAP Luminescence from N–Al and P–Al Pairs

DAP spectra from N–Al pairs have been measured in 3C-SiC long ago [34,35]. The spectra consist of a large number of sharp lines corresponding to recombination of carriers bound at donors and acceptors at various distances and are identified as type II (donors and acceptors substitute different host atoms in the lattice). The spectra arising from N–Al pairs have been analyzed, emission lines from close pairs have been identified, and the quantity $E_D+E_A=324$ meV has been accurately obtained (using the value 2416 meV for the energy band gap) [36]. Here E_D and E_A denote the nitrogen donor and aluminum acceptor ionization energies, respectively. At that time the ionization energy of the N donor was determined to be 53.6 ± 0.5 meV using the two-electron satellites (see Section 21.2.6) observed in the PL spectrum of 3C-SiC [37]. If the more accurate value 54.2 meV is used for E_D [25] instead of 53 meV as in Ref. [36], we obtain that the ionization energy of the Al acceptor is about 270 meV (271 meV in Ref. [36]).

A later study [38] examines the DAP emission in several polytypes (4H, 6H, and 15R) and with various acceptors (Al, Ga, and B), the donor species being nitrogen. The spectra obtained in this study show mainly broad emission bands, typical for samples of relatively high doping levels, as explained in Ref. [39]. The authors, however, demonstrate the anticipated site dependence of the ionization energies and show that it is much more pronounced for the donors than for the acceptors in all studied polytypes. Several other studies [40–46] all deal with samples showing broad bands. Attempts to identify the observed lines are usually restricted to close pairs [34].

Recently an attempt to fit the DAP PL in samples of 4H-SiC doped with N and Al [39,47] and P and Al [47] was made, and it was shown that a part of the spectrum corresponding to the so-called intermediate pairs (pairs at intermediate distances in the range 25–37 Å approximately) can be fit rather well with a model spectrum using the well-known equation for the energy of recombination $\hbar\omega_{PL}$ (R) of a DAP at a distance R:

$$\hbar\omega_{PL}(R) = E_g - (E_D + E_A) + E_C(R) + J(R) \tag{21.4}$$

Here E_g, E_D, and E_A are the energy band gap, the donor, and the acceptor ionization energies, and $E_C(R)$ is the Coulomb interaction between the ionized donor and acceptor. $J(R)$ denotes all corrections to the initial and final states of the pair and is negligible for intermediate pairs, as is justified by the good fit of the synthesized spectrum with the experimental one. The situation in a uniaxial crystal with two inequivalent lattice sites (h and k) is complicated by the presence of two actual donors and acceptors per species (N, resp. Al) with different ionization energies; thus, four combinations are possible for $E_D + E_A$ in Equation 21.4. Arguments were given in Ref. [39] that a contribution from the most shallow donor–acceptor species is expected to dominate the spectrum. In addition, the dielectric constant is a tensor, hence, the Coulomb interaction E_C is anisotropic and depends on the orientation of the pair axis with respect to the crystal axis, but this circumstance simply makes the calculation during synthesizing the spectrum somewhat more elaborate. As a result, a good fit was illustrated with a synthetic spectrum of type II using only the N donor and the Al acceptor at hexagonal sites (N_h–Al_h). Since the ionization energy of N_h is known (61.4 meV, see above) and the fit provides the quantity $\hbar\omega_\infty = E_g - (E_D + E_A)$, the ionization energy of Al_h could be estimated (about 199 meV, using the value $E_g = 3287$ meV [19]). In a subsequent publication [47], P and Al codoped samples were examined in the same way, and it was shown that the emission now correlates with two sets of type I: hexagonal–cubic and cubic–cubic. It was quite straightforward to identify the donor and the acceptor in these sets, namely, P_k–Al_h and P_k–Al_k. Using both sets to synthesize the spectrum results in a very convincing fit, the separation between the two corresponding quantities $\hbar\omega_\infty$ was found to be 3.4 meV. This is simply the difference in the ionization energies between Al_h and Al_k, the former being shallower. In addition, the fit proves that Al and P reside on the same sublattice (the silicon one, since the spectra are of type I and N resides on the C sublattice, as will be illustrated later). Using the justification concerning the difference in the ionization energies of the two Al acceptors Al_h and Al_k, a new fit was attempted on the N–Al pair spectra. The fit actually improved when the, properly shifted by 3.4 meV, N_h–Al_K set was added to the N_h–Al_h set. A successful fit was obtained also with both sets involving N_k, which enabled the determination of its ionization energy, 125.5 meV. The ionization energy of P_h remained undetermined, since no sharp features were observable in the corresponding part of the spectrum. The broad bands visible in the P–Al pair spectrum due to recombination of remote pairs suggest that the binding energy of P_h is significantly larger than that of P_k, similar to the relation between the ionization energies of N_h and N_k. In contrast to that, the first-principle calculations yield similar ionization energies for P_h and P_k [48]. Therefore, this disagreement remains to be resolved. The experimental data on the ionization energies of the species considered so far is summarized in Table 21.4.

TABLE 21.4

Ionization Energies of P, N Donors, and Al Acceptors in 4H-SiC

Species	Identification	Ionization Energy (meV)
N_h	On C *h*-site	61.4
N_k	On C *k*-site	125.5
Al_h	On Si *h*-site	197.9
Al_k	On Si *k*-site	201.3
P_k	On Si *k*-site	60.7
P_h	On Si *h*-site	120 ± 20

Note: The errors are ± 0.1 meV for the donor ionization energies (except for P_h). The accuracy of the acceptor ionization energies depends on the accuracy of the electronic band gap, the value $E_g = 3287$ meV is used [19]. The value for P_h is estimated by analogy with N–Al remote DAP emission [47].

21.2.4 Raman Scattering from Donor States

Raman scattering from the donor electron can be observed in a process in which the donor electron in the ground state interacts with an incident photon. As a result the electron is left in an excited state (usually, the valley–orbit split-off counterpart of the ground state), and a photon at a Raman shift equal to the energy difference between the ground and excited states is emitted. The first observation of this process in 6H-SiC is due to Colwell and Klein [49]. Three Raman lines have been observed (at 13.0, 60.3, and 62.6 meV) and associated with the three inequivalent sites for the N donor in this polytype. A comprehensive review containing data on the Raman scattering from the N donor is available [50], where the polytypes 3C, 4H, 6H, and 15R are considered. Obviously, the Raman spectrum yields information on the valley–orbit splitting of the ground state, with several Raman lines in the case when several inequivalent lattice sites (consequently, different donors) are present. Thus, one line is expected in 3C-SiC, two in 4H, three in 6H, and five in 15R. However, for unknown reasons not all expected lines are always observed [50]. Thus, only one line at 7.1 meV is observed in 4H-SiC, most probably due to scattering from the shallower N_h donor. This energy is in good agreement with the value 7.4 meV for the valley–orbit splitting in this donor; cf. Table 21.2. The line from the second donor N_k is either overlapping an intrinsic Raman line, or too weak to be observed. The valley–orbit splitting from the Raman spectrum of 3C-SiC (8.2 meV) is also in very good agreement with the absorption data (8.4 meV; cf. Table 21.2). All observed Raman lines in the hexagonal polytypes are of E symmetry, consistent with transitions between the $1s(A_1)$ and 1s(E) states.

Recently, Raman scattering due to the P donor in 4H-, 6H-, and 15R-SiC was reported [51]. The authors observe a peak at 4.3 meV related to the valley–orbit splitting of the shallowest P donor in 4H-SiC, and two peaks at 3.5 and 5.3 meV in 6H-SiC, all of E symmetry. The proximity of the two peaks observed in 6H-SiC is, in our opinion, an indication that they originate from the P donors at cubic sites. This is consistent with the assignment of P_k in 4H, and P_{k1}, P_{k2} in 6H being shallower than P_h, in contrast to the situation for the N donor. For both shallow donors considered of the 4H polytype, the valley–orbit split-off counterpart of the ground state has a binding energy that is very close to the effective-mass theoretical 1s-energy of 54 meV; namely, it is 54 meV for N_h and 56.4 meV for P_k. Thus, both donors can be considered as effective-mass-like shallow donors in the sense that their excited states are described well by the EMT.

21.2.5 EPR Measurements and Related Theoretical Studies

A comprehensive account of most of the EPR and electron nuclear double resonance (ENDOR) investigations of shallow impurities in the SiC polytypes up to 1996 can be found in Ref. [52]. Concerning the shallow acceptors, several works on boron have appeared in the literature since this review. High-frequency (95 GHz) EPR and ENDOR studies using ^{13}C enriched 6H- and 4H-SiC crystals [53,54] provide rather detailed information on the electronic structure of the shallow B-acceptor. It was found that around 40% of the spin density is localized at the C–B dangling bond in the *c*-direction and there is no direct spin density on B. The rest of the spin density is distributed in the crystal with a Bohr radius of ~2.2 Å. The spatial distribution of the wave function of the shallow B was found to be highly anisotropic and is different for the different sites in 4H- and 6H-SiC [54].

The material quality has improved significantly in the recent years, leading to spectra with better resolution showing greater detail. We summarize new findings concerning the shallow donors here. In the high-frequency (95 GHz) EPR and ENDOR studies of the shallow N donor in ^{13}C—enriched 4H- and 6H-SiC [55], the hyperfine (hf) interactions with many C and Si neighboring shells have been observed. Due to the line broadening, however, the hf interactions with the nearest Si neighbors were not observed in EPR, and for all ENDOR measurements the magnetic field was set on the main line. Therefore, the hf interactions with the nearest Si neighbors were not detected. This leads to an incorrect picture of the spin localization in Si and C sublattices. In a subsequent work on 4H-SiC [56], the hf interactions with the nearest Si neighbors were observed in both EPR and pulsed-ENDOR, providing a more complete picture of the spin localization of the shallow N donors in 4H-SiC. The N donors were also unambiguously shown to occupy the carbon sublattice, a result which was widely accepted since the earliest studies of silicon carbide.

Earlier EPR and ENDOR studies of the P donor have been performed only on neutron transmutation P-doped samples [57–60] or sublimation-grown material [61] containing high concentrations of residual N donors. The spectra showed overlapping contributions from both N and P, but it was clear that five different P-related centers were observed in 6H-SiC instead of the expected three corresponding to the three inequivalent lattice sites. The initial interpretations were quite controversial until a recent work conducted on samples, doped with P in situ during chemical vapor deposition (CVD) and containing residual N-doping below the EPR detection limit, provided a consistent picture of the P donor in 3C-, 4H-, and 6H-SiC [48]. In this work, two-dimensional EPR and electron spin-echo envelope modulation experiments confirm that the *d*Pc1 and *d*Pc2 doublets in 6H-SiC [60] are related to different allowed and forbidden transitions of the same P center with $S=1/2$ and $I=1/2$. Based on the observed ^{31}P hf interaction for three polytypes and the ^{13}C hf interaction with nearest neighbors for 4H- and 6H-SiC, the P-related spectra with small ^{31}P hf splitting are assigned to the ground states of the isolated shallow P at Si site. It seems that only P centers at the quasicubic sites (k in 4H-SiC and k_1 and k_2 in 6H-SiC) have been observed. In 3C-SiC, the shallow P center has D_{2d} symmetry ($g_{\|}=2.0051$, $g_{\perp}=2.0046$) and very small ^{31}P hf constants ($A_{\|}=0.53$ MHz and $A_{\perp}=-0.13$ MHz). Thus, in as-grown 3C-, 4H-, and 6H-SiC doped with P during the crystal growth, P is confirmed to be at the Si site [48]. This experimental observation is in agreement with the results of theoretical work [62] considering the incorporation of P in the two sublattices during physical vapor transport (PVT) deposition. It is shown that the incorporation into the Si sublattice dominates over that into the carbon one by at least one order of magnitude (often two orders of magnitude or more, depending on the growth conditions). Another theoretical work [63] shows that this result can be extended also to CVD-grown material. It should be mentioned, however, that the appearance of some new weak lines in the low-temperature PL of P-doped material was reported [64], which might be the signature of P at the C sublattice.

A different situation arises when P is introduced in SiC by neutron transmutation or implantation and subsequent annealing. Theory [65] predicts that in this case a significant amount of P can reside on the carbon sublattice. Recent experimental data [66] obtained on high-purity SI (HPSI) 6H-SiC substrate implanted with P and annealed seem to validate the theory [65]. Three new P-related centers were observed (named P_a, P_b, and P_c in this work), two of which (P_a and P_b) were associated with P on the C sublattice, occupying the hexagonal and the two cubic (unresolved) sites, respectively. The microscopic structure of P_c is not as yet identified. However, whether phosphorus can be incorporated by a certain doping procedure in the C sublattice in significant concentrations remains an open question.

It is interesting to note that the valley–orbit splitting of the shallow P donors in 3C-, 4H-, and 6H-SiC were estimated from the temperature dependence of the spin–lattice relaxation time and of the EPR signal intensities: 5.4 meV for P in 3C-SiC, 4.2–4.8 meV for P_k in 4H-SiC, and 4.2–4.3 and 6–7 meV, respectively, for the two quasicubic sites in 6H-SiC [48]. These values are in good agreement with the Raman data of Ref. [51] presented above.

We have to address also the results of ab initio supercell calculations concerning P and N donors in 4H-SiC presented in Ref. [48]. The theoretical estimates of ionization energies are 76 meV/56 meV for P_k/P_h, and 142 meV/44 meV for N_k/N_h, respectively. Thus, while the theory provides qualitative agreement with experiments for N_k/N_h, it predicts similar energies for P_k and P_h, with the latter donor shallower, which seems to contradict the data from the P–Al DAP emission. An interesting feature is the possibility of inversed order of the states in P_h pointed out in Ref. [48]; the calculated 1s(E) state lies below 1s(A_1).

21.2.6 PL from Excitons Bound to Shallow Donors and Acceptors

The PL from N donor-bound excitons (BEs) has been studied long ago in several polytypes of SiC: 3C [67], 4H [68], 6H [69], and many other polytypes. A detailed review of the optical properties of SiC polytypes is available [70], which reflects most of the important results obtained up to 1997. Since all these polytypes are indirect band gap semiconductors, the direct recombination of the free exciton is forbidden and only phonon-assisted transitions are allowed. The phonons involved in the recombination must have the same momentum as the exciton; therefore, they are from this point of the BZ where the CBM occurs. When the excitons are bound to defects, the crystal *k*-vector is not a good quantum number anymore, the exciton wave function is spread out in the reciprocal space, and direct recombination becomes allowed, giving rise to the so-called no-phonon (NP) (zero-phonon) lines. Typical spectra of excitons bound to the N donors in 3C-, 4H-, and 6H-SiC are shown in Figure 21.2. The spectrum of 3C-SiC exhibits a single NP line (NPL) and four phonon replicas shifted from the former by the energies of the phonons assisting the recombination. This is consistent with a single N donor and the nondegenerate LO and LA and doubly degenerate TO and TA phonons at the X-point of the BZ of 3C-SiC (the approximate phonon energies in meV are given as subscripts to the labels of the lines). The situation is more complicated in 4H- and 6H-SiC. The number of NPLs corresponds to the number of inequivalent lattice sites, two (three) in 4H (6H) SiC, respectively. The number of phonons is also larger, 24 (36) in 4H (6H), respectively. However, not all phonon replicas are observed with the common experimental geometry, because of the polarization selection rules present in uniaxial crystals; i.e., some phonon replicas are polarized perpendicular to the crystal *c*-axis, and some parallel to it [71]. Their numbers are equal in 4H-SiC (12 replicas with each polarization). The NPLs are also polarized, and a simple group-theoretical analysis [71] shows that the observed perpendicular to **c** polarization is

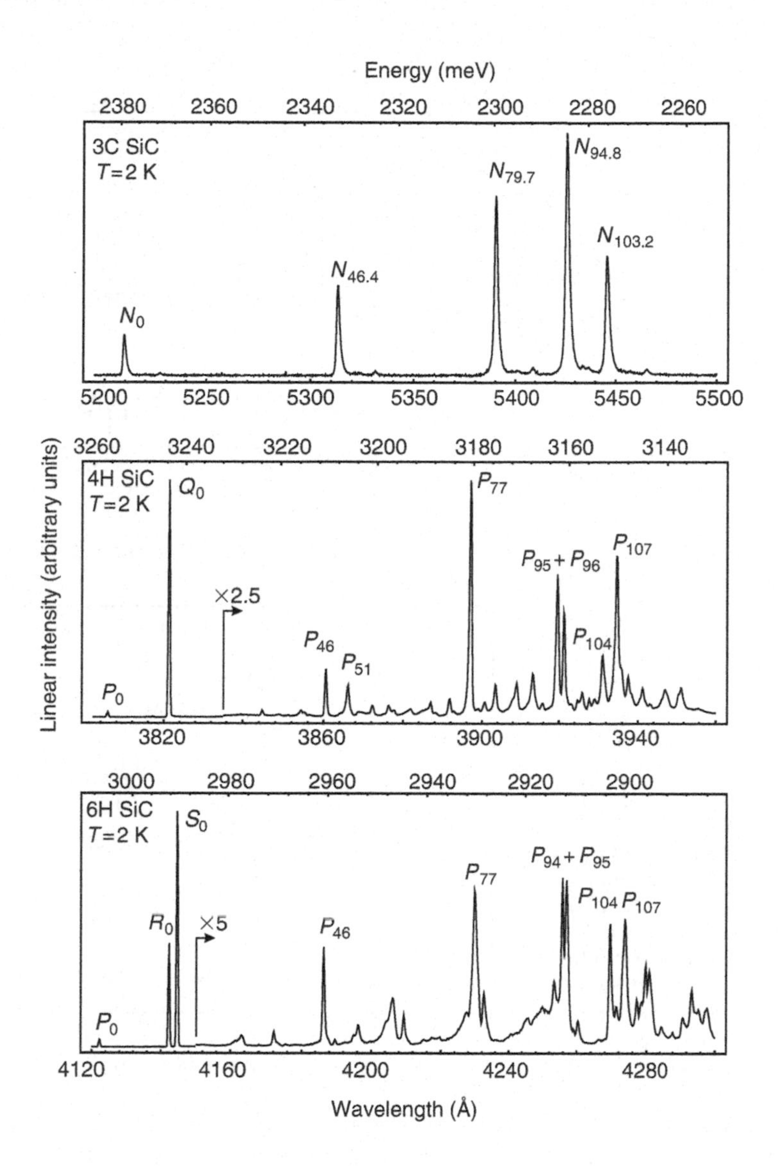

FIGURE 21.2
The PL spectra due to N-bound-exciton recombination in 3C-, 4H-, and 6H-SiC. The spectra are recorded with $\mathbf{E} \perp \mathbf{c}$ polarization in the case of 4H- and 6H-SiC. The lines with index 0 denote the NPLs; their number is equal to the number of inequivalent lattice sites in each polytype. Some of the prominent phonon replicas of P_0 (N_0 in 3C-SiC) are marked with P (N for 3C-SiC) and an index showing the approximate energy of the momentum-conserving phonon in meV.

compatible with donor ground state of $1s(A_1)$ symmetry. Note that the replicas of the shallowest line (denoted P_0 in both 4H- and 6H-SiC) dominate the spectra. This is consistent with the weak binding of the exciton; the spreading out of the wave function in reciprocal space is much less than for the strongly BEs, hence results a weak NPL and strong replicas. It should be noted that only lines with perpendicular to **c** polarization are observed in the usual geometry of exciting and registering the PL through the surface of the sample, which is usually perpendicular to the **c**-axis. The remaining lines have been observed and classified in Ref. [71] using the appropriate experimental geometry. The relative intensity of the N-BEs depends on the N concentration [72].

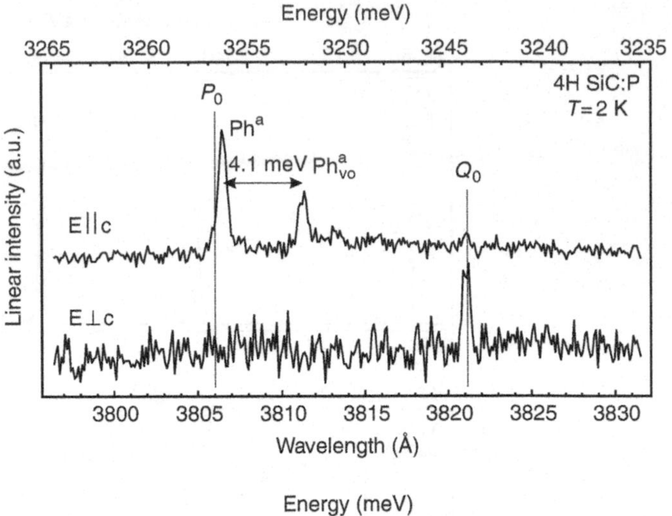

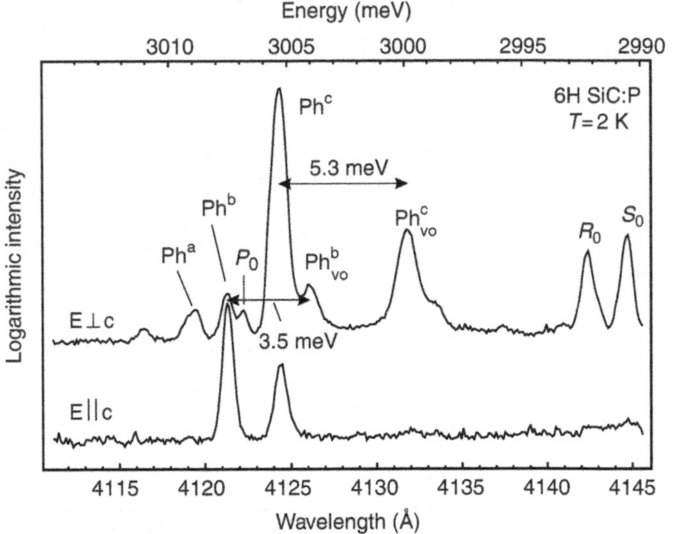

FIGURE 21.3
The NP part of the low-temperature PL spectra related to P-BEs in 4H- and 6H-SiC, recorded with $\mathbf{E} \perp \mathbf{c}$ and $\mathbf{E} \parallel \mathbf{c}$ polarizations of the detected light, as labeled for each spectrum. The positions of the NPLs (or the lines themselves when observable) of the N-BE are marked in the usual way (cf. Figure 21.2). The markers with subscript VO denote the satellite lines due to two-electron transitions as discussed in text.

Phosphorus BEs have also been observed, first in 6H-SiC [73] and recently in 3C-, 4H-, and 6H-SiC samples doped in situ during the CVD growth [74]. Apart from the intensity ratio between the different phonon replicas, the set of replicas is similar to that of the N-BE, especially in the case of 3C- and 4H-SiC. The structure of the NPLs, however, is significantly different from that of N. The NP parts of the spectra with $\mathbf{E} \perp \mathbf{c}$ and $\mathbf{E} \parallel \mathbf{c}$ polarizations are shown in Figure 21.3 for 4H- and 6H-SiC. The main difference from the spectra of N-BE NPLs is the presence of satellites due to the so-called two-electron transitions. These satellites come up when a part of the energy of the recombination of the exciton is transferred to the donor electron, leaving the neutral donor in excited final state. Usually this is the counterpart of the ground state split-off by the valley–orbit splitting, therefore the index VO for the two-electron satellites. The valley–orbit splittings deduced from the spectra and denoted in Figure 21.3 are in excellent agreement with the Raman data [51]. Another difference from the N-BE is the presence of NPLs with $\mathbf{E} \parallel \mathbf{c}$ polarization, which might be the confirmation of the existence of a P donor with ground state of 1s(E) symmetry. The structure of this part of the spectrum, however, needs further understanding.

Let us turn now to the BE spectra related to the Al-, Ga-, and B-acceptors in 3C-, 4H-, and 6H-SiC. Group-theoretical consideration [75–77] predicts 12, 3, and 6 NPLs per inequivalent site in 3C, 4H, and 6H, respectively. Typical spectra of the Al- and Ga-BE NPLs in

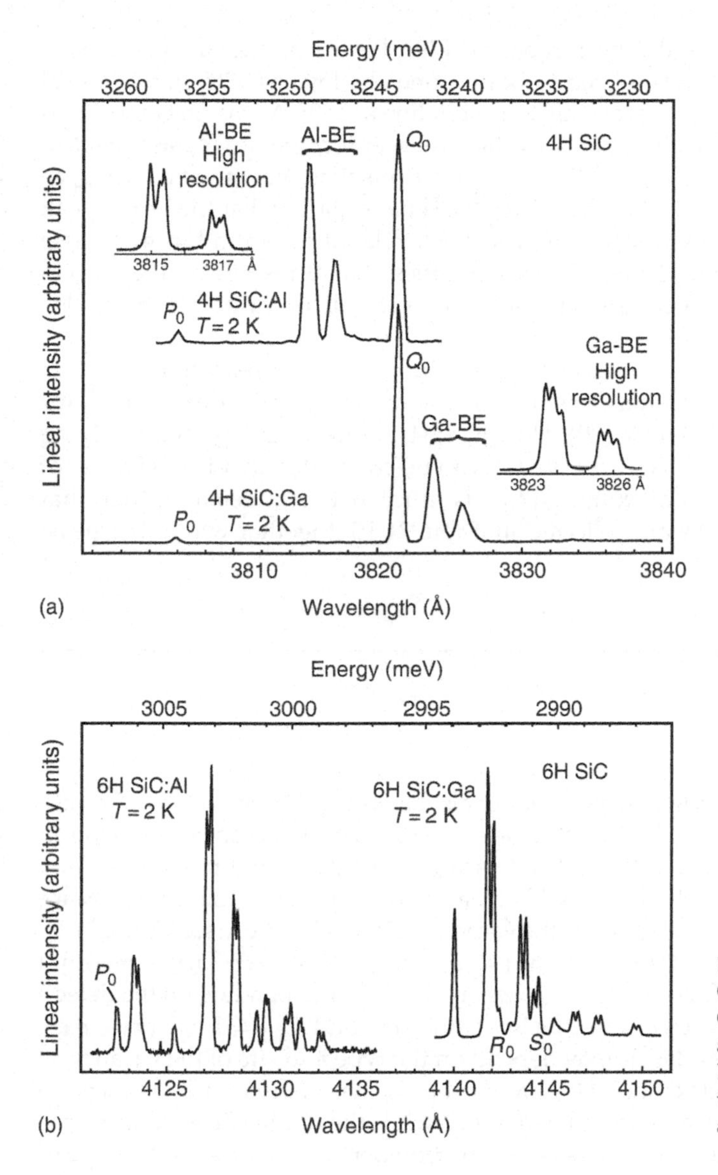

FIGURE 21.4
Comparison of the NPLs of the Al- and Ga-BEs in 4H- and 6H-SiC. P_0, Q_0, R_0, and S_0 denote the NPLs of the nitrogen BEs; the rest of the lines are related to NP emission of the corresponding acceptor BEs.

4H- and 6H-SiC are shown in Figure 21.4. Already the first observation of Al-BEs [75] revealed the complicated structure of the NP part of the spectrum. All 12 lines predicted for 3C could not be resolved in this reference; however, the structure of the peak in the region 5240–5250 Å was demonstrated. The simplest structure seems to be observed in 4H-SiC for both Al- and Ga-BEs. The Al-BE spectrum consists of two relatively narrow NPLs, corresponding to the two inequivalent sites, each of which actually splits into three lines, as revealed by the high-resolution spectra (see the insert of Figure 21.4). Eighteen lines are expected for Al- and Ga-BEs in 6H (six per each of the three inequivalent sites), and up to 16 have actually been observed. The resemblance in the structure of the NP part of the spectra from Al- and Ga-BEs has been illustrated in Ref. [77]. It has been shown that these spectra in 6H-SiC are spot-dependent (different spots exhibit different splitting), meaning that the spectrum is quite sensitive to the local stress built in the sample (although the effect was found to be less pronounced for the Ga-BEs than for the Al ones). More recently the Al-BE structure in 6H-SiC has been shown to differ for the Si- and C-faces [78].

Boron-related PL in 4H-SiC has also been reported [79]. Electrical measurements have revealed two electrically active levels related to boron, one shallower (~285 meV) [80,81] and one deep (~500–550 meV) [79]. A sharp NP emission line at 3838 Å was first associated with the exciton recombination at the shallowest boron acceptor, tentatively assigned to substitutional boron at silicon site (B_{Si}) [79]; however, later this was shown to be wrong, as discussed further in Section 21.5.2.2. Other studies [82,83] have shown that the broad peak around 3838 Å observed in highly doped samples is not related to boron but may arise from N- or Al-BEs at high doping levels. The possibilities that B resides on the carbon sublattice, and that it easily forms complexes with C interstitials or vacancies, have also been proposed in the literature as candidates for the deep boron-associated centra [84–87]. A structured DAP PL also has been observed from N donor—deep B-acceptor pairs in 3C-SiC [88] and 4H-SiC [79]. In the latter case the ionization energy of the deep B-acceptor involved has been estimated ~650 meV [79]. However, structured pair spectra involving the shallow B-acceptor have never been observed. It can be concluded that the behavior of Al and Ga as single substitutional acceptors in SiC is much better understood than that of B. Results from the shallow BEs are collected in Table 21.13, together with other near-band gap emissions.

21.3 Intrinsic Defects

21.3.1 Introduction

The theory of vacancies is well-known in cubic materials. However, it might not be the case for hexagonal crystals like 4H- or 6H-SiC. Therefore, we analyze the vacancies in detail in 4H-SiC. The simplest vacancy defect is the isolated vacancy. The point group of the isolated vacancy is C_{3v} in 4H-SiC. Taking the first neighbor atoms (I, II, III, IV) around the vacant site, four dangling bonds ($\sigma_1, \sigma_2, \sigma_3, \sigma_4$) point to the vacancy before relaxation (see Figure 21.5). These four dangling bonds form two a_1 states ($\psi_{a_1(1)} = \alpha\sigma_1 + \sqrt{1-\alpha^2/3}\,(\sigma_2 + \sigma_3 + \sigma_4)$ and $\psi_{a_1(2)} = -\sqrt{1-\alpha^2}\,\sigma_1 + \alpha/\sqrt{3}\,(\sigma_2 + \sigma_3 + \sigma_4)$, where $0 \leq \alpha \leq 1$ depending on the defect) and one double degenerate e state ($\psi_{e(3)} = 1/\sqrt{6}\,(2\sigma_2 - \sigma_3 - \sigma_4)$ and $\psi_{e(4)} = 1/\sqrt{2}\,(\sigma_3 - \sigma_4)$). The position and the order of these levels may depend on the type and site of the vacancy as well as its charge state. During relaxation of the atoms, the C_{3v} symmetry may be reduced to C_{1h} symmetry. In that case the a_1 states appear as a' states while the double degenerate e state splits to a' and a'' states. The occurrence of this reconstruction depends not just on the type of the vacancy and its charge state, but also on the site of the vacancy in the hexagonal 4H-SiC. Obviously, the group theory itself cannot predict the order and the position of the defect levels. Usually, density functional theory calculations within the local-density functional theory

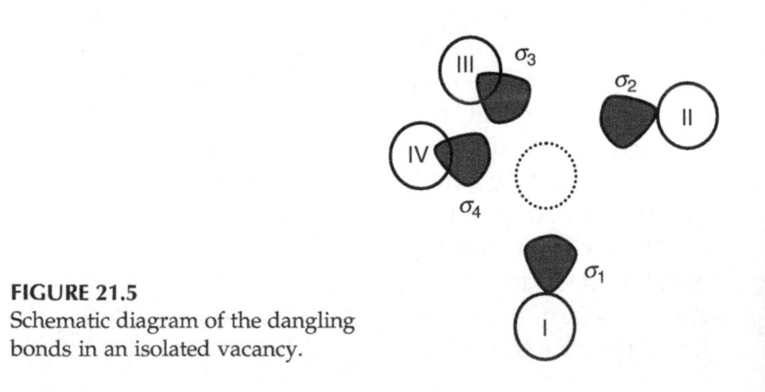

FIGURE 21.5
Schematic diagram of the dangling bonds in an isolated vacancy.

TABLE 21.5

Possible Multiplets of $(ea_1)^{(2)}$, $e^{(2)}$, and $e^{(1)}e^{(1)}$ Electron Configurations in C_{3v} Symmetry

Configurations	Multiplets
$(ea_1)^{(2)}$	$^3A_2$, ^1E, ^1E, $^1A_1$, $^1A_1$, ^3E
$e^{(2)}$	$^3A_2$, ^1E, $^1A_1$
$e^{(1)}e^{(1)}$	$^1A_1$, $^1A_2$, ^1E, $^3A_1$, $^3A_2$, ^3E

(DFT-LDA) are used to predict the positions of the levels. Due to the approximations in DFT–LDA the ionization energies obtained by DFT–LDA cannot be compared directly with the experiment, but the order of the levels can usually be predicted well. For instance, the calculated adiabatic transition energies of the carbon vacancy, i.e., the occupation levels, are widely scattered in the literature [89–92]. The reason is due to the different treatment of the LDA gap correction and the applied charge correction of the supercell method. At the moment, there is not any accurate methodology to handle these problems with a sufficient number of atoms needed to describe the defect states in solids. Therefore, the calculated occupation levels within these approximations should be taken with great caution. Especially, convergent transition energy between high charge states can be found only in very large supercells, and scaling methodology is needed to treat the energy of the charged supercell correctly [93]. Therefore, we provide the calculated ionization energies, if obtained with large supercells (~300 atoms or above) and self-consistent gap corrections. The schematic diagram of the calculated LDA levels are depicted for each defect in the next subsections.

The counterpart of the silicon vacancy, the carbon antisite–carbon vacancy pair, can be treated as simple vacancies. The four dangling bonds give rise to two a_1 and one e states which, depending on the symmetry, may split to a' and a'' states. The divacancy has six dangling bonds, which give rise to two a_1 and two e states. These may again split due to lower symmetry. In the vacancies the nonoverlapping dangling bonds can be strongly correlated with each other; in the Hartree–Fock language the configurational interaction (CI) between the electrons cannot be neglected any more, and the resulting states cannot be described by simple independent single-particle schemes. These states, called multiplets, cannot be calculated precisely by simple DFT–LDA. For instance, the CI can be significant in the neutral silicon vacancy and divacancy. The allowed multiplets in C_{3v} symmetry are shown for $(ea_1)^{(2)}$, $e^{(2)}$ and $(e^{(1)}e^{(1)})$ states in Table 21.5. These are characteristic of the ground states of the neutral silicon vacancy and divacancy, and the excited states of the neutral divacancy, respectively.

21.3.2 Carbon Vacancy V_C

Theory predicts that one of the most abundant defects is the carbon vacancy (V_C) in SiC [89–92]. V_C can be double positively and negatively ionized; i.e., it is amphoteric. In the unrelaxed neutral state the first a_1 level falls into the valence band. The second a_1 level is in the gap occupied by two electrons, while the e level is empty in 4H-SiC. After relaxation the symmetry is reduced to C_{1h} and the silicon dangling bonds form long bonds pair-wise (see Figure 21.6a and b).

Interestingly, in the (+) state V_C conserves its C_{3v} symmetry at the h-site, while it reconstructs to C_{1h} symmetry at the k-site, resulting in a significantly different spin density distribution (see Figure 21.6b and c). Indeed, the symmetry lowering from C_{3v} to C_{1h} was observed for V_C^+ (k) [94,95], while V_C^+ (h) kept its C_{3v} symmetry even at low temperature

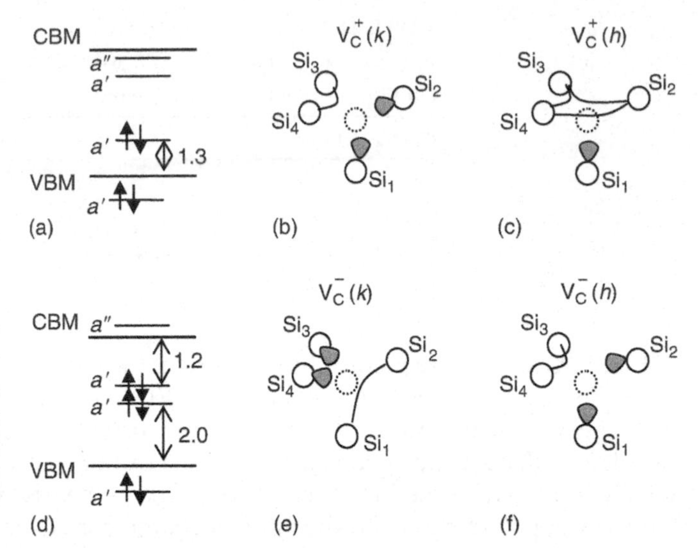

FIGURE 21.6
(a) and (d) are the calculated LDA + GW one-electron levels of V_C^0 and V_C^{2-} in eV (published first here); (b) and (c) show the spin distribution of V_C^+ at different sites; (e) and (f) show the spin distribution of V_C^- at different sites. CBM, conduction band minimum; VBM, valence band maximum.

[94]. The spin density is mainly localized on one Si atom in V_C^+ (*h*), while it is more evenly spread out on the first neighboring Si atoms in V_C^+ (*k*) (see Table 21.6) [94–96]. In both cases, the total spin density localized at the first four neighboring Si atoms is about 68% (see Table 21.6), where it is also shown that V_C^+ in 4H- and 6H-SiC are very similar. The V_C^+ signals are often observed in irradiated p-type 4H- and 6H-SiC [97,98], and are also common in HPSI 4H- and 6H-SiC substrates grown by PVT or high-temperature CVD (HTCVD) [99–101]. V_C^+ is shown to be the dominating defect in HPSI 4H-SiC substrates with thermal activation energies, as determined from the temperature dependence of the resistivity, $E_a \sim 1.5$ eV [102,103]. An EPR spectrum of V_C^+ with detailed ^{29}Si hf structures observed at 77 K in a HPSI 4H-SiC substrate grown by HTCVD for $\mathbf{B} \perp \mathbf{c}$ is shown in Figure 21.7a. Photoexcitation EPR (photo-EPR) suggests the (+/0) donor level of V_C in 4H-SiC to be at about 1.47 eV above the valence band [104].

In the negatively charged states the atoms relax very strongly with bending to each other, and the two *a*′ defect levels appear close to each other in the gap. The empty *a*″ level

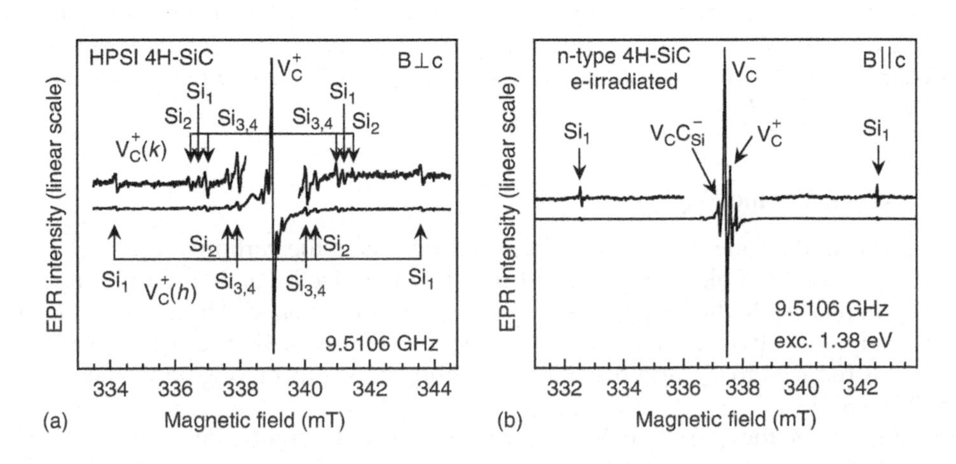

FIGURE 21.7
EPR spectra of (a) V_C^+ in HPSI 4H-SiC substrates, and (b) V_C^- in electron-irradiated n-type 4H-SiC. The spectra were measured at 77 K (a) for $\mathbf{B} \perp \mathbf{c}$ and (b) for $\mathbf{B} \parallel \mathbf{c}$ under illumination with light of photon energy of 1.38 eV.

TABLE 21.6

Spin-Hamiltonian Parameters of V_C^+ and V_C^- Centers in 4H- and 6H-SiC

Center	Parameters	X (⊥)	Y (⊥)	Z (∥)	η^2%	θ	Ref.
V_C^+ (k) in 4H-SiC	**g** (EI5) 5 K	2.0056	2.0048	2.0030		22.7°	[94]
	g (EI5) 138 K	2.00484	2.00484	2.00322		0°	[97]
	A (Si_1) 5 K	4.44	4.36	6.46	19.9	7.7°	[94]
	A (Si_2) 5 K	3.25	3.19	4.74	14.7	121.5°	—
	A ($Si_{3,4}$) 5 K	3.85	3.81	5.52	16.5	103.2°	—
	$\Sigma\eta^2$ (Si_{1-4})				67.6		
V_C^+ (h) in 4H-SiC	**g** (EI6) 10 K	2.0052	2.0052	2.0026		0°	[94]
	g (EI6) 293 K	2.0046	2.0046	2.0032		0°	—
	A (Si_1) 5 K	10.61	10.61	15.48	47.3	0°	—
	A (Si_{2-4}) 5 K	1.40	1.40	2.11	6.8	98°	—
	$\Sigma\eta^2$ (Si_{1-4})				67.6		
V_C^+ (k) in 6H-SiC	**g** (EI5) 102 K	2.00461	2.00461	2.00316		0°	[97]
	g (EI5) 23 K	2.0053	2.0042	2.0026		25°	[105]
	g (Ky1) 4.2 K	2.0060	2.0026	2.0025		34°	[95]
	A (Si_1) 4.2 K	3.94	3.94	5.74	17.5	8°	—
	A (Si_2) 4.2 K	3.79	3.79	5.49	16.5	119°	—
	A ($Si_{3,4}$) 4.2 K	3.82	3.82	5.55	16.8	104°	—
	$\Sigma\eta^2$ ($Si_{1,2,3,4}$)				67.7		
	g (Ky2) 4.2 K	2.0050	2.0040	2.0023		27°	—
	A (Si_1) 4.2 K	4.33	4.33	6.44	20.3	8°	—
	A (Si_2) 4.2 K	3.09	3.09	4.60	14.5	121°	—
	A ($Si_{3,4}$) 4.2 K	3.85	3.85	5.52	16.3	104°	—
	$\Sigma\eta^2$ ($Si_{1,2,3,4}$)				67.5		
V_C^+ (h) in 6H-SiC	**g** (EI6) 23 K	2.00487	2.00487	2.00236		0°	[106]
	g (EI6) 102 K	2.00478	2.00478	2.00286		0°	—
	g (Ky3) 15 K	2.0046	2.0046	2.0020		0°	[95]
	g (Ky3) 77 K	2.0045	2.0045	2.0025		0°	—
	A (Si_1) 77 K	10.09	10.09	14.65	44.3	0°	—
	A (Si_{2-4}) 77 K	1.45	1.45	2.23	7.4	101°	—
	$\Sigma\eta^2$ (Si_{1-4})				66.6		
V_C^- (h) in 4H-SiC	**g** (C_{3v}) 150 K	2.00381	2.00381	2.00401		0°	[107]
	g (C_{1h}) 60 K	2.00287	2.00407	2.00459		38°	—
	A (Si_1) 60 K	7.76	7.76	10.07	24.1	7°	—
	A (Si_2) 60 K	11.78	11.67	15.19	36.2	101°	—
	$\Sigma\eta^2$($Si_{1,2}$)				60.2		

Note: X, Y, and Z are the directions of the principal axes of **g** and **A** tensors. θ is the angle between the principal Z-axis and the c-axis. The principal A values are given in mT. η^2 is the fraction of the total spin density that is localized at one of the neighboring Si atoms.

goes above the CBM (see Figure 21.6d). Theoretical calculations found that the geometry differs for V_C^- at different sites resulting in different spin density distribution [107] (Figure 21.6e and f). V_C^- (h) has already been identified by a combined EPR and theory study in electron-irradiated n-type 4H-SiC [107]. Figure 21.7b shows the spectrum observed at 77 K for **B** ∥ **c** under illumination with light of photon energy of 1.38 eV. The V_C^- (h) signal becomes dominating under illumination with photon energies larger than ~1.1 eV. Photo-EPR experiments [107] suggest that, in irradiated n-type samples, $V_C(h)$ may be in the double negative charge state lying at ~1.1 eV below the conduction band E_C. Illumination with photon energies larger than ~1.1 eV excites electrons from the (−/2−) level to the conduction band and leaves the center in the single-negative charge state being detected by EPR. This scenario is supported by theory (see Figure 21.6d). Photo-EPR of the V_C^- signal in HPSI 4H-SiC samples [101] also suggests that the (0/−) level of V_C is at ~$E_C - 1.1$ eV, in agreement with the result obtained in n-type samples. At low temperatures, the center has C_{1h} symmetry. Above ~70 K, due to the thermal reorientation the symmetry becomes C_{3v}.

An activation energy of ~20 meV was estimated for this thermal reorientation process [107]. Parameters for the V_C^- center in both C_{3v} and C_{1h} configurations are also given in Table 21.6. So far V_C has not been identified in 3C-SiC.

In deep-level transient spectroscopy (DLTS) studies of 4H-SiC CVD layers irradiated with low energy electrons (80–210 keV) [108], the EH7 component of the EH6/EH7 level [109] at $\sim E_C - 1.54$ eV and the $Z_{1/2}$ level at $\sim E_C - 0.68$ eV were observed simultaneously and suggested to be related to V_C. This was supported by a recent DLTS study [110]. Taking into account the band gap narrowing at the measured temperature in DLTS (~3.15 eV for 4H-SiC at ~600 K [17]), the EH7 level is close to the (+/0) level of V_C.

21.3.3 Silicon Vacancy V_{Si}

In the silicon vacancy (V_{Si}) carbon dangling bonds point to the vacant site. These carbon dangling bonds are too localized to overlap with each other; therefore, no long bonds are created. Instead, the carbon atoms relax outward and basically keep the C_{3v} symmetry independently of its charge state or its site in 4H-SiC [89,90,111,112]. The order of the defect levels is a_1, a_1, e in this case. The first a_1 level is resonant with the valence band not far from the top of it, while the remaining a_1 and e levels are close to each other at around 0.5–0.7 eV above the valence band edge in the neutral charge state (Figure 21.8a and c).

In the neutral charge state the a_1 and e levels in the gap are occupied by two electrons. In this case the CI between these levels (and even with the first a_1 level in the valence band) may not be neglected. In cubic SiC it was shown that the CI can result in a spin-singlet ^{1}E ground state that can be described by two Slater-determinants, so the simple independent single-particle scheme cannot be used [113]. Later, it has been confirmed by an independent LDA calculation that the triplet $^3T_1$ state does not possess the lowest energy in 3C-SiC [114].

Taking only the defect levels in the gap the electronic configuration of V_{Si}^0 in 4H-SiC is $(a_1e)^{(2)}$, where the possible multiplets can be found in Table 21.5. Two triplet states and several singlet states are possible. The $M_S = \pm 1$ spin states of the ^{3}E and $^3A_2$ states can be described by a single Slater-determinant. In the case of the ^{3}E state the a_1 and e levels are each occupied by a single electron with the same spin state, while for the $^3A_2$ state the e level is occupied by two electrons with the same spin state and the upper a_1 level is empty. Naturally, the single-particle LDA calculations resulted in an $S = 1$ spin state as the ground state of V_{Si}^0 [89,111]; nevertheless, it was not reported which $S = 1$ state was found. We found in our very recent LDA calculations in a 576-atom supercell that the ^{3}E state is slightly more stable than the $^3A_2$ state at both sites, but the energy difference between the

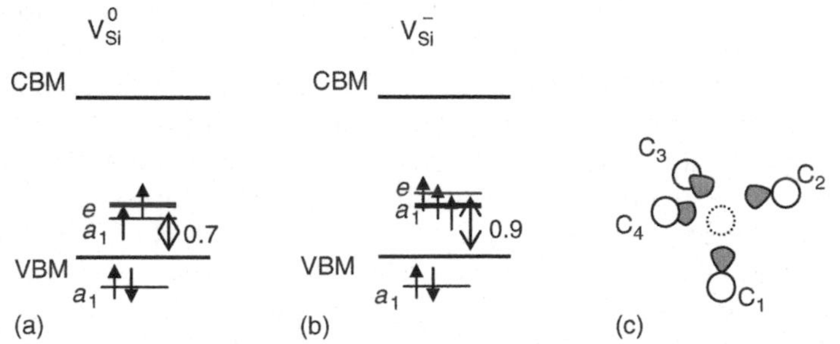

FIGURE 21.8
(a) and (b) are the calculated LDA + scissor one-electron levels of V_{Si}^0 and V_{Si}^- in eV; (c) schematic picture of the near neighbor C atoms of V_{Si} with their dangling bonds.

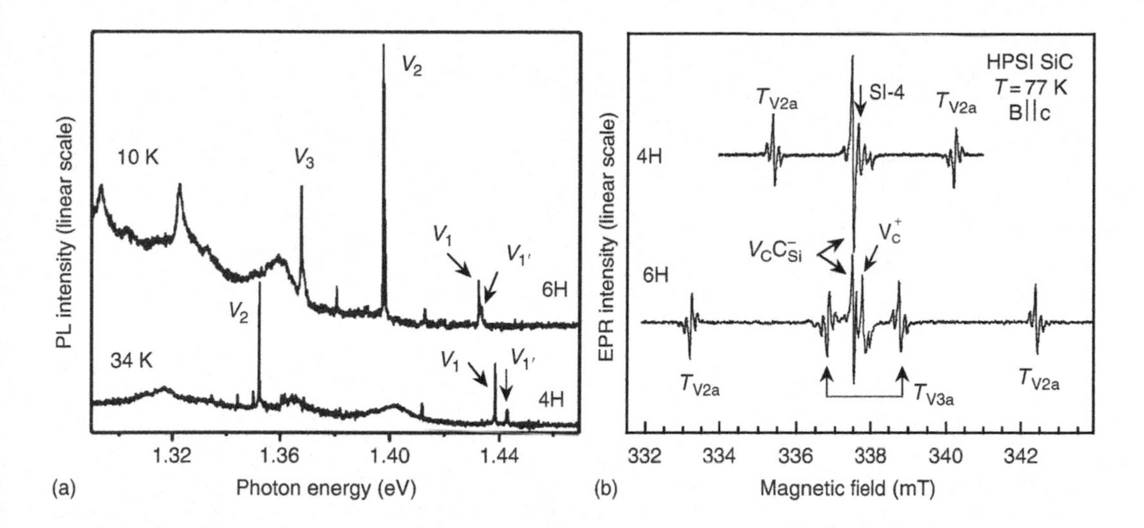

FIGURE 21.9
(a) PL spectra of V_{Si}-related centers in electron-irradiated 4H- and 6H-SiC measured at moderate temperatures (34 K for 4H-SiC and 10 K for 6H-SiC) to make the V1′ line visible. (From Wagner, Mt. et al., *Phys. Rev. B*, 62, 16555, 2000. With permission.) (b) EPR spectra of the T_{V2a} and T_{V3a} centers in as-grown HPSI 4H- and 6H-SiC. The microwave frequency is 9.4715 GHz.

^3E and $^3A_2$ states is smaller at the *h*-site than at the *k*-site. We note, however, that one of the singlet states as a possible ground state cannot be disregarded and a precise CI calculation is needed to investigate the true ground state of V_{Si}^0 in 4H-SiC.

The excitation scheme of V_{Si} is very complicated due to the CI. PL centers in the near-infrared spectral region (Figure 21.9a) with two NPLs V_1 (1.438 eV) and V_2 (1.352 eV) in 4H-SiC and three NPLs V_1 (1.438 eV), V_2 (1.366 eV), and V_3 (1.398 eV) in the 6H polytype were observed [116]. Four and six different triplet states detected by optical detection of magnetic resonance (ODMR) by monitoring this PL band in 4H- and 6H-SiC, respectively, were earlier attributed to V_{Si}^0 [116], but are now believed to be due to V_{Si}^-. The T_{V2a} ODMR/EPR center with its observed spin triplet state ($S=1$) was a possible candidate for the isolated V_{Si}^0 [116–119]. Figure 21.9b shows the EPR spectra of the T_{V2a} center in HPSI 4H-SiC, and T_{V2a} and T_{V3a} centers in the 6H polytype. The hf interactions with four nearest C neighbors observed by ODMR [117] and EPR [120] for T_{V2a} suggest that the center is the isolated V_{Si}. However, pulsed-EPR and ENDOR studies [120,121] suggested the quartet spin state $S=3/2$ for T_{V2a} and the center is now believed to be V_{Si}^-. The high-spin state $S=3/2$ has been theoretically predicted for V_{Si}^- [89,90,96,111,122]. In 3C-SiC the t_2 state in the gap is occupied by three electrons with parallel spins yielding T_d symmetry. The T_1 EPR center showed indeed these properties with four carbon hf lines identifying V_{Si}^- in 3C-SiC [123]. In hexagonal SiC the *e* and a_1 states are occupied by three electrons with parallel spins yielding a $^4A_2$ state with C_{3v} symmetry. In this case the gap levels are shifted up by about 0.2 eV with respect to those of the neutral defect (Figure 21.8b). (Theoretically, multideterminant doublet states are also possible, but it has neither been calculated nor has such a state been detected.) The quartet state was measured by EPR with four equivalent carbon dangling bonds possessing C_{3v} symmetry by Wimbauer et al. in 4H-SiC [122]. This center is isotropic and has a zero or negligible fine structure parameter [118] (see Table 21.7). In the case of the spin quartet state of the T_{V2a} center, the crystal-field splitting is nonzero. It was speculated that this was caused by a perturbation of the crystal field, possibly due to the presence of an accompanying impurity or defect [120]. However,

TABLE 21.7

Spin-Hamiltonian Parameters of Silicon Vacancy Centers in 3C-, 4H-, and 6H-SiC

Center	Parameters	X (⊥)	Y (⊥)	Z (‖)	η^2%	θ	Ref.
V_{Si}^- in 3C-SiC	**g**	2.0029	2.0029	2.0029		0°	[123]
	A (C_{1-4})	1.18	1.18	2.86	15.9	0°	—
	A (Si) ×12	0.292	0.292	0.292	0.18	0°	—
	$\Sigma\eta^2$ (C_{1-4})				63.6		
	$\Sigma\eta^2$ (Si_{1-12})				2.1		
V_{Si}^- in 4H-SiC	**g**	2.0032	2.0032	2.0032		0°	[118]
	g	2.0028	2.0028	2.0028		0°	[120]
	D		<0.05			0°	[118]
	A (C_{1-4})	1.21	1.21	2.86	15.7	0°	[122]
	A (Si) ×12	0.298	0.298	0.298	0.18	0°	—
	$\Sigma\eta^2$ (C_{1-4})				62.6		
	$\Sigma\eta^2$ (Si_{1-12})				2.2		
V_{Si}^- in 6H-SiC	**g**	2.0015	2.0015	2.0015		0°	—
	g	2.0032	2.0032	2.0032		0°	[124]
	g	2.0024	2.0024	2.0024		0°	[a]
	A (C_{1-4})	1.15	1.15	2.87	16.2	0°	[122]
	A (Si) ×12	0.297	0.297	0.297	0.18	0°	—
	$\Sigma\eta^2$ (C_{1-4})				64.9		
	$\Sigma\eta^2$ (Si_{1-12})				2.2		
	g	2.0029	2.0029	2.0029		0°	[120]
T_{V2a} in 4H-SiC	D ($S=3/2$)		1.252			0°	—
	D ($S=1$)		2.498			0°	[116]
	A (C_1)	1.24	1.24	2.87	15.5	0°	[120]
	A (C_{2-4})	0.97	1.12	2.70	15.6	107.5°	—
	A (Si) ×12	0.31	0.31	0.31	0.19	0°	—
	$\Sigma\eta^2$ (C_{1-4})				62.2		
	$\Sigma\eta^2$ (Si_{1-12})				2.3		
	g	2.0038	2.0038	2.0035		0°	[116]
T_{V2a} in 6H-SiC	D ($S=1$)		4.58			0°	—
	A (C_1)	1.34	1.34	2.86	14.6	0°	[117]
	A (C_{2-4})	1.1	1.1	2.86	16.7		—
	A (Si) ×12	0.28	0.28	0.28	0.17	0°	—
	$\Sigma\eta^2$ (C_{1-4})				64.6		
	$\Sigma\eta^2$ (Si_{1-12})				2.0		

Note: The A and D values are given in mT. η^2 is the fraction of the total spin density that is localized at one of the neighboring C or Si atoms.

[a] We have checked carefully and found that the g-value of V_{Si}^- in 6H-SiC is 0.0004 less than that in the 4H polytype.

the increase of the EPR signal with a dose of irradiation (to an electron fluence of 2×10^{18} cm^{-2}) in HPSI SiC substrates does not support the model involving a perturbation of an impurity.

The lines shown in Figure 21.9a are visible both in PL and absorptions, which indicates that the transitions giving rise to the lines are allowed. If the ground state of V_{Si}^- is $^4A_2$ (that is, two electrons with opposite spins in the lower a_1 level degenerate with the valence band, two electrons in the e level, and one electron in the upper a_1 level; the three electrons in the band gap having parallel spins), the only allowed transitions are promoting an electron from the lower a_1 level to either the e level (the excited state will be ^4E) or the upper a_1 level (the excited state will be $^4A_2$). Both the initial and final states of the transitions should be ODMR active. Taking into account also the inequivalent sites, it is not surprising to find several different ODMR signals associated with this PL band. The similar g-values obtained for the different ODMR signals could be the reason why no Zeeman splitting of any line in magnetic field (up to 5 T) has been observed [115], since both initial and final state split in a similar way.

Theoretical calculations predict that V_{Si} can be easily negatively ionized even to a (4−) state in 4H-SiC (where the a_1 and e levels are fully occupied) [89,90]. Since no bonds are created in the ionized defect, the carbon dangling bonds will repulse each other, which results in about 0.2 eV upward shift of the gap levels for each ionization step. The paramagnetic (3−) and (+) states, though, have not yet been identified by EPR. By negatively ionizing the defect, its formation energy decreases rapidly in n-type hexagonal SiC. From EPR and PL, it is known that the T_{V2a} center is the dominant defect in irradiated n-type SiC. It is therefore expected that in n-type material irradiated with low doses of electrons (10^{14}–10^{15} cm^{-2}), when the Fermi level still lies at the N donor level, the levels corresponding to high charge states of V_{Si} in the upper half of the band gap should be filled. These levels work as electron traps and should be detectable by DLTS. Among the reported DLTS levels in irradiated n-type 4H-SiC, the EH2 level [109] at $E_C - 0.68$ eV has an annealing behavior similar to that of V_{Si} [116]; i.e., it is annealed out at ~750°C, and is a possible candidate for the (3−/4−) level of V_{Si}, although the measured capture coefficient seems to be too high.

21.3.4 Carbon Antisite–Carbon Vacancy Complex $C_{Si}V_C$

In a compound semiconductor, the antisite–vacancy complex (AV) is a fundamental defect [125]. Theory predicts that the carbon AV pair ($C_{Si}V_C$), the counterpart of the silicon vacancy, is stable in p-type SiC and metastable in n-type SiC [92,126]. In the $C_{Si}V_C$ defect three silicon dangling bonds come from the V_C part, while one carbon dangling bond comes from C_{Si}. The unrelaxed defect possesses C_{3v} symmetry if C_{Si} is along the *c*-axis (on-axis configurations, $C_{Si}(k)V_C(k) = C_{Si}V_C(kk)$ and $C_{Si}V_C(hh)$), while it has C_{1h} symmetry if C_{Si} is out-of *c*-axis (off-axis configurations, $C_{Si}V_C(kh)$ and $C_{Si}V_C(hk)$) in 4H-SiC.

In the neutral charge state of the on-axis configurations the first a_1 level falls into the valence band, while the second a_1 level, occupied by two electrons with an empty e level, lies in the band gap. The second a_1 state is mostly localized on C_{Si} while the e level is localized on the silicon dangling bonds (Figure 21.10a). In the case of off-axis configurations the e state splits into a' and a''. The former can be mixed with the lower a' state in the gap.

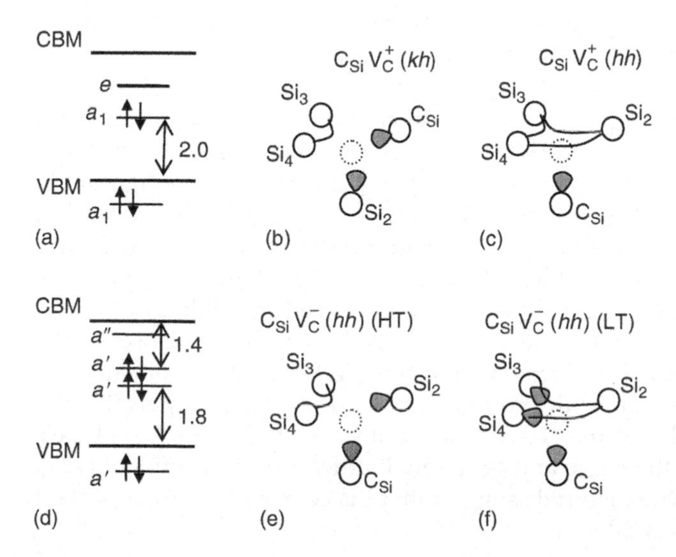

FIGURE 21.10
(a) and (d) are the calculated LDA + scissor one-electron levels of C_{Si} V_C^0 and C_{Si} V_C^{2-} in eV (published first here); (b) and (c) show the spin distribution of C_{Si} V_C^+ in off-axis and on-axis configurations; (e) and (f) show the spin distribution of C_{Si} V_C^- in high-temperature (HT) and low-temperature (LT) configurations.

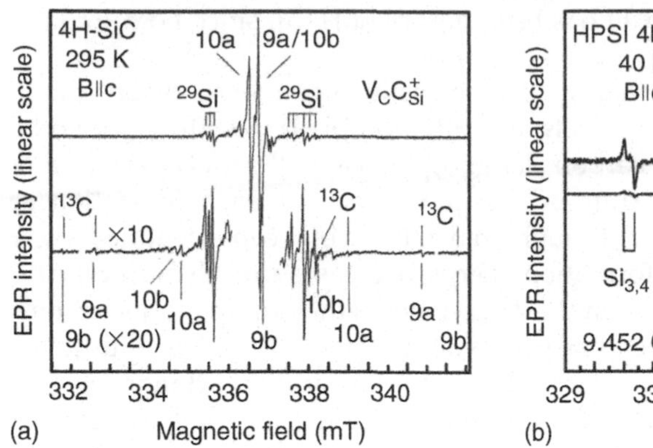

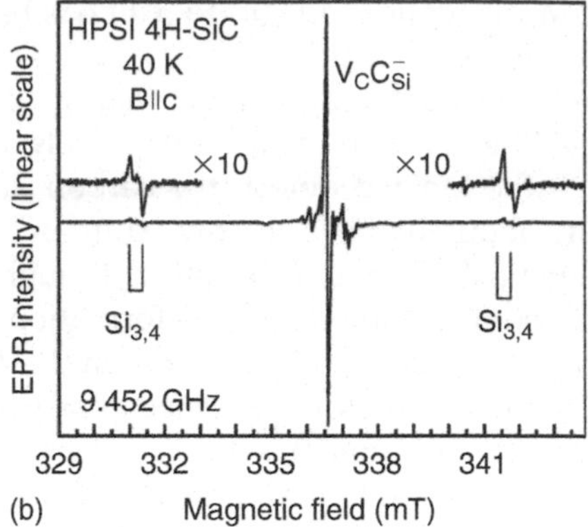

FIGURE 21.11
EPR spectra of (a) $C_{Si}V_C{}^+$ in electron-irradiated p-type 4H-SiC with the HEI9a,b and HEI10a,b centers corresponding to the C_{3v} and C_{1h} defect configurations, respectively. (From Umeda, T. et al, *Phys. Rev. B*, 75, 245202, 2007. With permission.) (b) $C_{Si}V_C{}^-$ in HPSI 4H-SiC substrates measured at 40 K for **B** || **c**.

The (+) charge state is paramagnetic, and recently has recently been identified by EPR in 4H-SiC [127,128]. Figure 21.11a [129] shows the $C_{Si}V_C^+$ spectrum called *HEI9/10ab* observed in p-type 4H-SiC irradiated by electrons. As expected, the spin density is mostly localized on C_{Si} (see Table 21.8), and the carbon dangling bond is slightly mixed with the silicon dangling bonds in off-axis configurations (Figure 21.10b and c).

The SI-5 EPR center, which often is detected in HPSI 4H- and 6H-SiC substrates [101], has recently been identified by EPR and supercell calculations [125] as the carbon antisite–carbon vacancy pair in the negative charge state ($C_{Si}V_C^-$). The EPR spectrum of the center in HPSI 4H-SiC observed at 40 K for the magnetic field parallel to the *c*-axis is shown in Figure 21.11b. The center can be detected after irradiation, and its concentration reaches a maximum after annealing at 800°C–850°C. In electron-irradiated n-type samples or in as-grown HPSI SiC substrates with the activation energies $E_a \sim 1.1$ eV or less, the $C_{Si}V_C$ center is in the double negative charge state and is not detected in the dark. The $C_{Si}V_C^-$ signal appears after illuminating the samples with light of photon energies larger than ~1.1 eV. Once illuminated, the signal keeps its intensity even in the dark. In this case the spin density is localized in the V_C part of the defect. Since the *e* level is occupied by one electron in $C_{Si}V_C^-$, this system is Jahn–Teller unstable and it reconstructs to C_{1h} symmetry (Figure 21.10d). Two possible C_{1h} configurations exist with close energies at each site, resulting in a low- and high-temperature configuration (Figure 21.10e and f). The high-temperature configuration shows an effective C_{3v} symmetry due to a dynamical reorientation effect. It was indeed shown that the symmetry of the SI-5 center lowers from C_{3v} to C_{1h} at temperatures below 60 K [125]. In the C_{1h} configuration, additional large hf splitting due to the interaction with two ^{29}Si (>10 mT) were detected. These can be associated with the low-temperature configuration. Such hf structures can also be detected for the SI-5 center in HPSI 4H-SiC substrates (Figure 21.10b). The spin-Hamiltonian parameters for $C_{Si}V_C^+$ and $C_{Si}V_C^-$ are summarized in Table 21.8.

Comparing to the reported DLTS data, the (1−/2−) level of $C_{Si}V_C$ is close to the EH5 level at $\sim E_C - 1.13$ eV observed in irradiated n-type 4H-SiC after annealing at ~750°C [109]. This is also in agreement with the calculations using large supercells and reliable gap correction, which predicted the (1−/2−) level of $C_{Si}V_C$ to be at $\sim E_C - 1.1$ eV and $\sim E_C - 0.9$ eV for the defect at the cubic and hexagonal lattice site, respectively [125]. Annealing studies of irradiated n-type materials suggest that the center plays an important role in compensating the N donors [125].

TABLE 21.8

Spin-Hamiltonian Parameters of $C_{Si}V_C^-$ and $C_{Si}V_C^+$ Centers in 4H- and 6H-SiC

Center	Parameters	X (⊥)	Y (⊥)	Z (‖)	η^2%	θ	Ref.
$C_{Si}V_C^-$, 4H-SiC C_{3v}, 100 K	**g** (C_{3v})	2.00339	2.00339	2.00484		0°	[125]
	A (Si_{2-4})$_{average}$	6.38	6.38	8.02	17.6	109°	—
	A (C_{II} ×3)$_{average}$	1.77	1.77	2.22	3.9	0°	—
	$\Sigma\eta^2$ ($C_{Si}+Si_{2-4}$)				52.9		
	$\Sigma\eta^2$ (C_{II} ×3)				11.7		
$C_{Si}V_C^-$, 4H-SiC C_{1h}, 30 K	**g** (C_{1h})	2.00372	2.00259	2.00534		10°	—
	A ($Si_{3,4}$)	10.16	10.16	12.99	29.9	111°	—
	A (C_I) ×2	1.3	1.3	1.8	5.4		—
	$\Sigma\eta^2$ ($C_{Si}+Si_{2-4}$)				58.8		
	$\Sigma\eta^2$ (C_I ×2)				10.9		
$C_{Si}V_C^-$, 6H	**g** (C_{3v}) 77 K	2.0030	2.0030	2.0048		0°	[101]
$C_{Si}V_C^+$ in 4H-SiC C_{3v}, 295 K	**g** (*hh*) (HEI9a)	2.00408	2.00408	2.00227		0°	[128]
	A (C_{Si})	2.27	2.27	8.25	55.2	0°	—
	$\Sigma\eta^2$ $(C_{Si}+Si_{2-4})_{hh}$				55.2		
	g (*kk*) (HEI9b)	2.00379	2.00379	2.00195		0°	—
	A (C_{Si})	3.71	3.71	9.95	58.6	0°	—
	$\Sigma\eta^2$ $(C_{Si}+Si_{2-4})_{kk}$				58.6		
$C_{Si}V_C^+$ in 4H-SiC C_{1h}, 295 K	**g** (*kh*) (HEI10a)	2.00348	2.00258	2.00226		145°	—
	A (C_{Si})	2.65	2.60	8.75	56.7	109°	—
	$\Sigma\eta^2$ $(C_{Si}+Si_{2-4})_{kh}$				56.7		
	g (*hk*)(HEI10b)	2.00399	2.00345	2.00263		116°	—
	A (C_{Si})	2.45	2.31	8.44	53.9	109°	—
	$\Sigma\eta^2$ $(C_{Si}+Si_{2-4})_{hk}$				53.9		

Note: The principal A values are given in mT. η^2 is the fraction of the total spin density that is localized at one of the neighboring C or Si atoms.

21.3.5 Divacancy $V_{Si}V_C$

The closest pair of V_{Si} and V_C, the divacancy, can form during the migration of the isolated vacancies [130,131]. As has been explained for $C_{Si}V_C$, four possible configurations exist for the divacancy due to the inequivalent sites in 4H-SiC. We analyze the on-axis configurations in detail; it has C_{3v} symmetry with three Si dangling bonds (V_C part) and three C dangling bonds (V_{Si} part). These six dangling bonds form two a_1 and two e levels. According to the calculations the two a_1 levels are resonant with the valence band, while the e levels appear in the gap (Figure 21.12a and b). The lower e level is localized on V_{Si}, while the upper e level is localized on V_C. In the ground state of $V_{Si}V_C^0$ the a_1 levels are fully

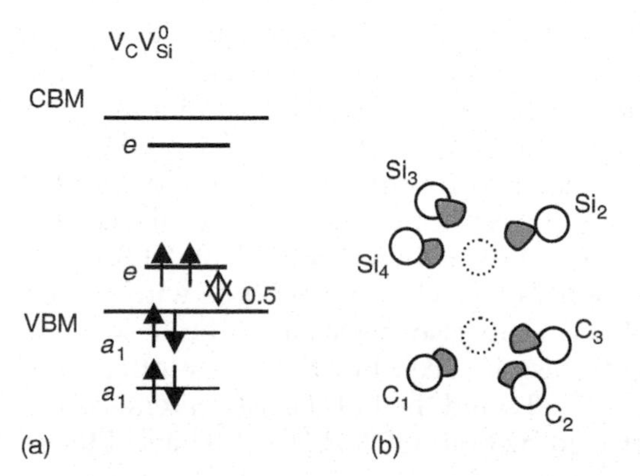

FIGURE 21.12
(a) The calculated LDA + scissor one-electron levels of $V_CV_{Si}^0$ in eV (published first here). (b) Schematic picture of the near neighbor atoms of the divacancy and their dangling bonds.

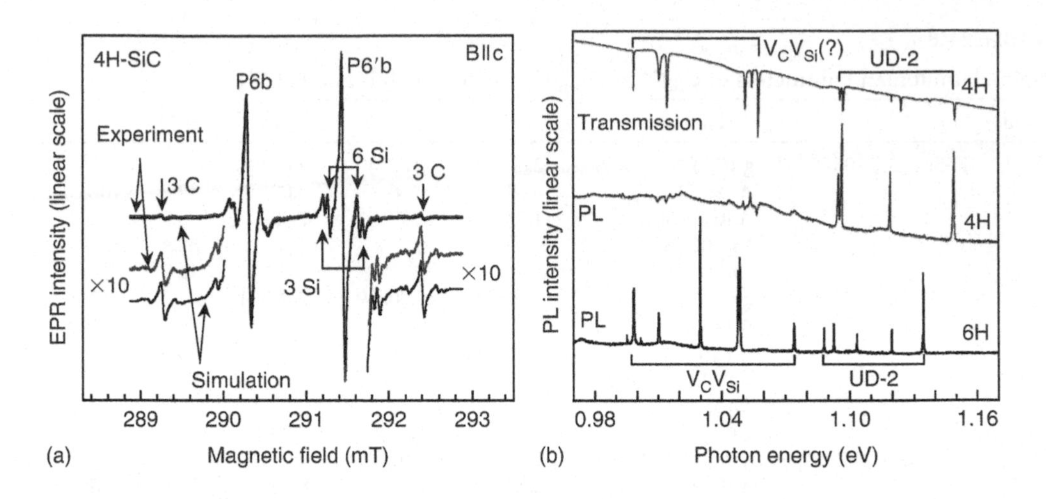

FIGURE 21.13
(a) EPR spectra of P6/P7 centers ($V_CV_{Si}^0$) in a HPSI 4H-SiC substrate with $E_a = 1.53$ eV measured at 77 K under illumination with light of photon energies ~2.0–2.8 eV for **B** || **c**. (From Son, N.T. et al., *Phys. Rev. Lett.*, 96, 055501, 2006. With permission.) (b) The PL spectrum related to the divacancy in 6H-SiC. The relation to V_CV_{Si} of the PL in 4H-SiC has not been confirmed.

occupied, while the lower *e* level is occupied by two electrons. In principle, the C_{3v} symmetry can be conserved by $^3A_2$, ^1E, and $^1A_1$ multiplets by assuming an $e^{(2)}$ electron configuration, as can be inferred from Table 21.5. The latter two $S = 0$ states cannot be described with an independent single-particle scheme, while the $^3A_2$ state can be described by putting two electrons with parallel spins on the *e* level. Alternatively, one can put two electrons with antiparallel spins on the *e* level, which is Jahn–Teller unstable and reconstructs to C_{1h} symmetry. The latter two situations can be described by independent single-particle schemes (like the usual LDA calculation). The LDA calculations found [131] that the high-spin state ($S = 1$) is favorable and should be the ground state. We note here that ^1E and $^1A_1$ states should be also calculated to prove this unambiguously. Figure 21.13a shows the P6/P7 EPR spectra, first reported in Ref. [132], obtained from a HPSI 4H-SiC substrate with $E_a = 1.53$ eV. By comparing the calculated and measured hf tensors, the P6/P7 EPR centers (see Table 21.9) have been identified as the neutral divacancy, $V_{Si}V_C^0$ [133]. Since P6/P7 EPR centers were also detected in the dark at very low temperature (1.2–8 K), the $^3A_2$ is indeed the ground state of $V_{Si}V_C^0$ [133,134]. The excitation scheme of $V_{Si}V_C^0$ is very complicated due to the possible configuration interaction between the states. The a_1 levels are not far from the lower *e* level. Taking this electron configuration, $a_1^{(2)}e^{(2)}$, the possible multiplets are shown in the first row of Table 21.5. Resonant transition from one of the a_1 level is possible, forming the ^3E excited state ($^3A_2 \rightarrow {}^3E$); however, the defect level within the VBM should be filled by electrons, which may hinder reemission. Further excitation is possible from the lower *e* level to the upper *e* level ($e^{(1)}e^{(1)}$ configuration of the excited state in Table 21.5: $^3A_2 \rightarrow {}^3E$). Only careful CI calculation can predict this excitation energy. Magnetic circular dichroism of the absorption (MCDA) and MCDA-detected EPR studies in irradiated n-type 6H-SiC [135] confirmed that P6/P7 triplet centers are related to the PL band with the NPLs at 0.9887, 1.0119, 1.0300, 1.0487, and 1.0746 eV (Figure 21.13b). In 4H-SiC with a strong EPR signal of $V_CV_{Si}^0$, several NPLs with similar energies at 0.9975, 1.0136, 1.0507, and 1.0539 eV were also detected in the absorption spectra [136]. These strong absorption lines can also be seen in PL, but are often very weak [136]. In 4H-SiC, the relation between these NPLs and the EPR P6/P7 centers has not been confirmed. The P6/P7 centers can be detected by both ODMR [135,137] and EPR in

TABLE 21.9

Spin-Hamiltonian Parameters of the Neutral Divacancy Centers in 4H-SiC

Center	Parameters	X (⊥)	Y (⊥)	Z (∥)	η^2%	θ	Ref.
$[V_{Si}(h)-V_C(h)]^0$ 6b	D		$D=47.80$				[133]
in 4H-SiC C_{3v}	A ($C_{1\text{-}3}$)	1.89	1.78	3.93	20.1	73°	—
	A ($Si_{2\text{-}4}$)	~0.1	~0.1	~0.1	0.06		—
	$\Sigma\eta^2$ ($C_{1\text{-}3}+Si_{2\text{-}4}$)				60.5		
$[V_{Si}(k)-V_C(k)]^0$ 6′b	D		46.62			—	
in 4H-SiC C_{3v}	A ($C_{1\text{-}3}$)	1.68	1.61	3.71	19.7	73°	—
	A ($Si_{2\text{-}4}$)	~0.1	~0.1	~0.1	0.06		—
	$\Sigma\eta^2$ ($C_{1\text{-}3}+Si_{2\text{-}4}$)				59.3		
$[V_{Si}(h)-V_C(k)]^0$ 7b	D and E		$D=47.80$, $E=9.62$			70.5°	—
in 4H-SiC C_{1h}	A (C_1)		Not observed				—
	A ($C_{3,4}$)		Not observed				—
$[V_{Si}(k)-V_C(h)]^0$ 7′b	D and E		$D=43.63$, $E=1.07$			71°	—
in 4H-SiC C_{1h}	A (C_1)	1.86	1.86	3.93	19.9	2°	—
	A ($C_{2,3}$)	1.71	1.61	3.89	21.2	70°	—
	$\Sigma\eta^2$ ($C_{1\text{-}3}+Si_{2\text{-}4}$)				62.3		

Note: The g-values for the divacancy in all the C_{3v} and C_{1h} configurations are isotropic $g=2.003$. The A, D, and E values are given in mT. Here $D=(3/2)D_{zz}$ and $E=(D_{xx}-D_{yy})/2$. η^2 is the fraction of the total spin density that is localized at one of the neighboring C or Si atoms.

6H-SiC. In 4H-SiC, the centers have not been detected by ODMR; the reason may be due to their too weak PL.

The divacancy, like V_{Si}, can be ionized at least up to (3−) in 4H-SiC. In the (3−) state the upper e defect level is occupied by an electron that is mostly localized on the carbon vacancy. The atoms reconstruct in this case, reducing the symmetry to C_{1h} even for on-axis configurations. While this defect can be singly positively ionized as well, it is much more probable that it can capture a hole (h^+), possibly forming a pseudoacceptor state, like ($V_CV_{Si}^- + h^+$), which might also result in transitions in the PL band. Theoretical calculations predicted that the neutral divacancy can recombine with an injected hole with an excitation energy of ~1.0 eV, which is close to the measured transition energies [131]. Recent EPR studies [138] in irradiated HPSI 4H-SiC substrates show that in the dark at 77 K only the V_C^+ signal was detected. Increasing the temperature to 293 K, the $V_CV_{Si}^0$ signal could be weakly detected in the dark. This suggests that the (0/−) level of V_CV_{Si} may lie slightly below the (+/0) level of V_C at $\sim E_V+1.47$ eV [104], so that at room temperature it could be partly thermally ionized and changed to the neutral charge state. Supercell calculations [133] predicted the (0/−) level to be at $\sim E_V+1.4$ eV, close to the estimation by EPR. The EH6/EH7 level observed by DLTS in n-type irradiated 4H-SiC [109] seems to fit very well the (0/−) level of V_CV_{Si}. In DLTS studies by Storasta et al. [108], only the EH7 signal was observed in samples irradiated with 210 keV electrons, which can create V_C, and the EH6 level could only be detected in samples irradiated by higher energy electrons. This may indicate that the EH6 and EH7 levels may be not related to the same defect. The annealing behavior of the EH6 level shown in this work is similar to that of the P6/P7-related PL band in irradiated 6H-SiC [139]. The EH6 level is a possible candidate for the (0/−) level of V_CV_{Si}.

21.3.6 Interactions between Defects and Their Role in SI Substrates

HPSI SiC substrates for high-frequency power devices have been developed in recent years by PVT [140,141] and by HTCVD [142,143]. In HPSI materials grown by HTCVD and PVT, intrinsic defects are believed to be responsible for compensating the residual shallow

N donors and B-acceptors, creating the SI properties. Vacancy-related defects and several other defects have been observed in HTCVD and PVT materials [99–101,142–145]. It was found that crystals grown by HTCVD under conditions favoring the incorporation of V_{Si} are characterized by the thermal activation energy of the resistivity $E_a \sim 0.8$–0.85 eV or smaller, as determined from the temperature dependence of the resistivity [142,143]. After annealing at 1600°C, the resistivity was reduced by several orders of magnitude, and E_a decreases to ~0.6 eV [142–144]. Under growth conditions favoring the formation of V_C, crystals exhibit a higher resistivity, which is stable upon annealing at 1600°C, and has an activation energy $E_a \sim 1.4$ eV [142–144]. HPSI 4H-SiC substrates grown by PVT have several thermal activation energies of ~1.1, ~1.3, and ~1.5 eV [140,141]. For some of them the resistivity reduces by several orders of magnitude after annealing at 1600°C [100,101]. Recent studies of a large amount of samples [102,103] revealed that HPSI SiC substrates can be classified into three different groups characterized by their thermal activation energies: (1) $E_a \sim 0.8$–0.9 eV; (2) $E_a \sim 1.1$–1.3 eV; and (3) $E_a \sim 1.5$ eV. In samples with $E_a \sim 0.8$–0.9 eV, the V_{Si}^- center is dominating, while the $C_{Si}V_C^-$ signal is moderate (Figure 21.14a). The signal of V_{Si}^- is weaker in the samples with $E_a \sim 1.1$–1.3 eV,

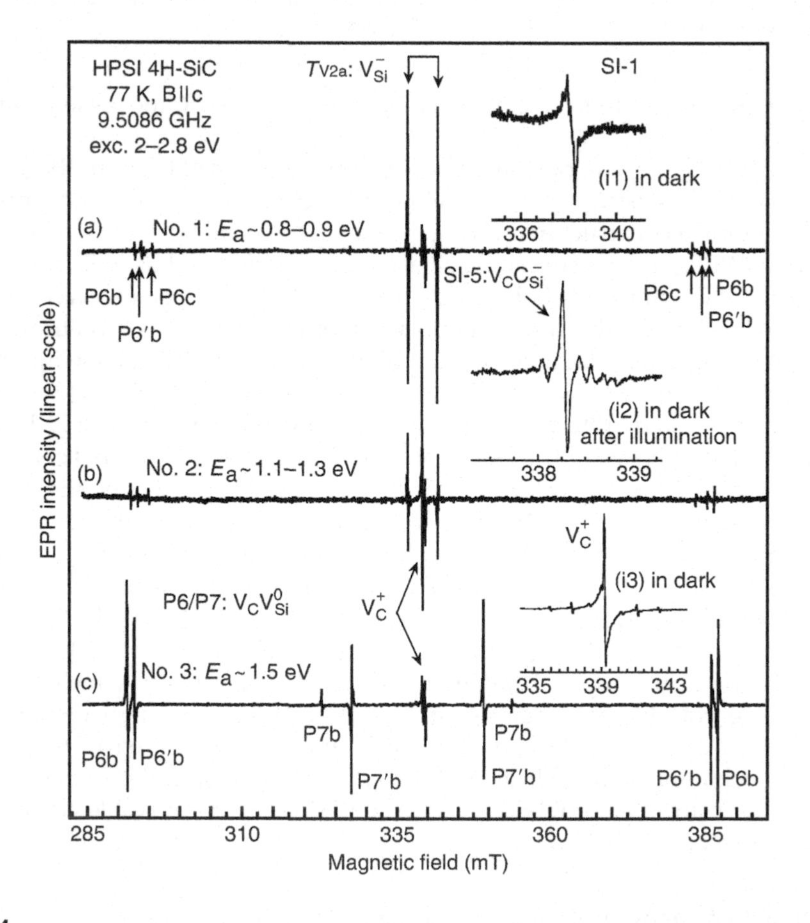

FIGURE 21.14
Typical EPR spectra measured under illumination with light of photon energies of ~2–2.8 eV in HPSI 4H-SiC samples sets (a) No. 1: $E_a \sim 0.8$–0.9 eV, (b) No. 2: $E_a \sim 1.1$–1.3 eV, and (c) No. 3: $E_a \sim 1.5$ eV. The inserts show the spectra measured (i1) in the dark and (i2) in the dark after illumination in sample sets No. 1 and No. 2, and (i3) in the dark in the sample set No. 3. (From Son, N.T. et al., *Phys. Rev. B*, 75, 155204, 2007. With permission.)

TABLE 21.10

Concentrations of Intrinsic Defects in the Sample Sets Nos. 1–3 Estimated by Comparing the Intensities of Their EPR Signals with That of the N in a Calibrated Sample with n[N] $\sim 1 \times 10^{16}$ cm^{-3}

E_a (eV)	V_{Si} (cm^{-3})	$C_{Si}V_C$ (cm^{-3})	V_C (cm^{-3})	V_CV_{Si} (cm^{-3})
~0.8–0.9	~2–3 × 10^{15}	~1 × 10^{15}	Mid 10^{14}	Mid 10^{14}
~1.1–1.3	~1 × 10^{15}	~3–4 × 10^{15}	~2–3 × 10^{15}	Mid 10^{14}
~1.5	Not detectable	Not detectable	~3–4 × 10^{15}	~5–6 × 10^{15}

Note: The concentrations of N and B measured by SIMS in these samples are in the range: n[N] $\sim$ 4–5 × 10^{15} cm^{-3} and n[B] $\sim$ 1–3 × 10^{15} cm^{-3} (~3 × 10^{15} cm^{-3} for PVT samples). Defects that are not detectable by our EPR experiments may have concentrations in the low 10^{14} cm^{-3} range.

which are shown to have strong signals of V_C^+ and $C_{Si}V_C^-$ (Figure 21.14b). In the samples with $E_a \sim 1.5$ eV, neither the V_{Si}^- nor the $C_{Si}V_C^-$ signals are detected, whereas the signals of V_C^+ and $V_CV_{Si}^0$ are dominating (Figure 21.14c). The concentrations of these intrinsic defects were estimated by comparing the intensity of their EPR signals with that of the EPR signal of the shallow N donor in an n-type sample with the N concentration of n[N] $\sim 1 \times 10^{16}$ cm^{-3}, as determined from secondary ion mass spectrometry (SIMS). The estimated concentrations of these vacancy-related defects are given in Table 21.10.

The annealing behaviors of vacancy-related defects in three different sample sets studied in the temperature range 600°C–1600°C [138] are shown in Figure 21.15. In sample set No. 1, V_{Si} and $C_{Si}V_C$ are the only two dominating defects. It is possible that $E_a \sim 0.85$ eV is related to either the (1−/2−) levels of $C_{Si}V_C$ or the (2−/3−) level of V_{Si}. After annealing at 1600°C, E_a was reduced to ~0.6 eV, and the resistivity reduces to ~10^5 Ω cm. These energy levels may be related to the (3−/4−) levels of different configurations of V_{Si}, which were filled when the total concentrations of V_{Si} and $C_{Si}V_C$ were reduced (Figure 21.15a). The reduction of the concentration was observed for all defects (V_{Si}, V_C, $C_{Si}V_C$, and V_CV_{Si}). In addition, the transformation between V_{Si} and $C_{Si}V_C$ during high-temperature annealing

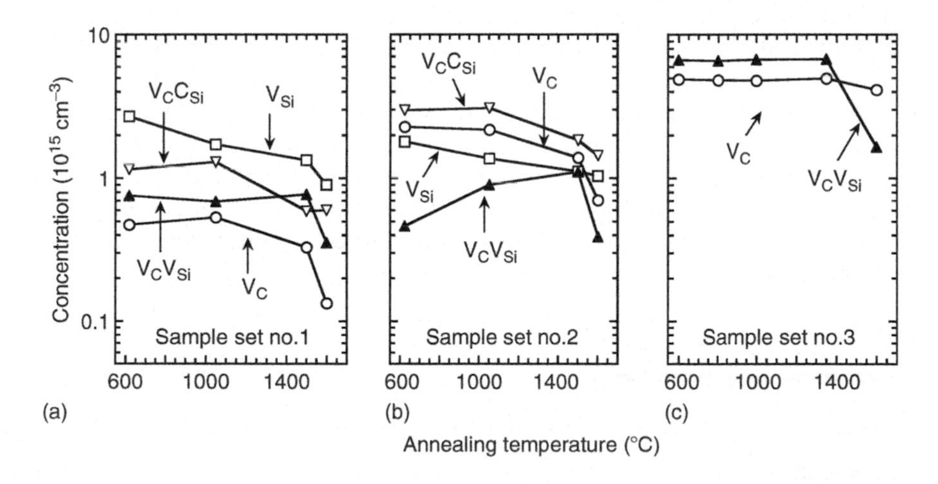

FIGURE 21.15
Annealing behavior of intrinsic defects in different types of HPSI 4H-SiC substrates: (a) sample set No. 1 with $E_a \sim$ 0.8–0.9 eV, (b) sample set No. 2 with $E_a \sim$ 1.1–1.3 eV, and (c) sample set No. 3 with $E_a \sim 1.5$ eV. (From Son, N.T. et al., *Phys. Rev. B*, 75, 155204, 2007. With permission.)

could also be seen (Figure 21.15a). With low concentrations of V_C, the formation of the divacancy via the interaction $V_C + V_{Si} \rightarrow V_C V_{Si}$ seems to be an inefficient process.

In sample set No. 2, the concentrations of V_C and $C_{Si}V_C$ are high (Figure 21.15b). In the case of relatively low concentrations of N, it is possible that electrons may fill up to the (0/−) level of $C_{Si}V_C$ at the energy range $\sim E_C - (1.24\text{–}1.35)$ eV, resulting in activation energies $E_a \sim 1.25\text{–}1.28$ eV as measured before annealing for some samples in set No. 2. For higher concentrations of N, the (0/−) level of V_C and (1−/2−) level of $C_{Si}V_C$ may also be filled, and E_a will be reduced (to ~1.1 eV in this case). It is difficult to determine if the activation energy E_a ~1.1 eV is related to the acceptor levels of V_C or $C_{Si}V_C$ since both defects have high concentrations. After annealing at 1600°C, E_a was reduced to ~0.6 eV and the resistivity from $\sim 10^8\text{–}10^9$ to $\sim 10^5$ Ω cm. In this sample set, $E_a \sim 0.6$ eV may be related to higher negative charge states of V_C or $C_{Si}V_C$. With reasonable concentrations of both V_C and V_{Si}, interaction between the vacancies to form the divacancy was observed (Figure 21.15b).

In sample set No. 3, both V_C and $V_C V_{Si}$ have high concentrations (Figure 21.15c). The B-acceptors and the (0/−) level of $V_C V_{Si}$ seem to be enough to compensate the N donors in these samples. Depending on the N concentration, the Fermi level can be pinned at different (0/−) levels corresponding to four configurations of $V_C V_{Si}$ in 4H-SiC. If the concentration of N is low, only the deeper (0/−) level of the divacancy may be filled and, consequently, a larger activation energy of $E_a \sim 1.6$ eV is obtained. For higher N concentrations, other higher lying (0/−) levels corresponding to shallower configurations of the divacancy may also be filled, resulting in smaller activation energies of $E_a \sim 1.4\text{–}1.5$ eV. This may explain the map of the activation energy in PVT substrates [140,141] with the values $E_a \sim 1.55\text{–}1.6$ eV observed only near the edge of the wafer, where the N concentration is often lower, and $E_a \sim 1.4\text{–}1.5$ eV in the typically higher doping center region of the wafer. After annealing at 1600°C, the concentration of V_C is only slightly decreased, whereas the reduction of the divacancy concentration is more pronounced (Figure 21.15c). After annealing at 1600°C, E_a is reduced from ~1.53 to ~1.06 eV, but the resistivity is still $\geq 10^9$ Ω cm. The divacancy concentration is reduced to $\sim 1.5 \times 10^{15}$ cm^{-3}. The reduction of the divacancy concentration leads to the electron occupancy of the higher lying (0/−) level of V_C, which may be at $\sim E_C - 1.1$ eV. This explains the change of E_a from ~1.53 to ~1.06 eV. If the total concentration of the divacancy and B-acceptors is still larger than that of the N donors, the Fermi level and hence E_a remains unchanged as seen in some HPSI SiC samples grown by HTCVD [142–144]. In absence of V_{Si}, the process $V_C + V_{Si} \rightarrow V_C V_{Si}$ does not occur. The migration of V_C via the nearest neighbors to form vacancy-antisite pairs is unlikely since $V_{Si}Si_C$ is predicted to be unstable [146]. Therefore, in sample set No. 3, V_C is more thermally stable.

Annealing studies suggest that V_C and the divacancy are suitable defects for controlling the SI properties in SiC. Since V_C is stable only when V_{Si} is absent or present in negligible concentration, optimizing the crystal growth conditions to minimize the formation of V_{Si} is therefore important.

21.4 Unknown Defects Found in SI Substrates

21.4.1 PL Results

The interest of defects in SI SiC has been growing since 1999 when vanadium-free SI SiC was introduced [147,148]. A number of defect centers of unknown origin is reported in the literature. To bring order into the unidentified emission and absorption lines, the most

TABLE 21.11

Energy Positions of Optical Transition of Defects in SiC in the Near-IR Region [136], UD-4 in 6H-SiC from Ref. [150]

Defect	Emission [meV], Label in Bracket	
	4H-SiC	**6H-SiC**
Erbium	808.2	808.2
Vanadium	928.4 ($\beta_{1,2}$)	893.1 (γ)
	967.3 (α_1)	917.0 (β)
	968.1 (α_2)	945.3 (α_1)
	969.5 (α_3)	946.0 (α_2)
	970.3 (α_4)	947.5 (α_3)
		948.2 (α_4)
$V_C - V_{Si}$	997.5	998.7
	1013.6	1010.9
	1050.7	1030.0
	1053.9	1048.7
		1074.6
UD-1	1058.6	995.1
	1059.5	1001.4
		1001.9
UD-2	1095.3	1088.2
	1096.8	1092.8
	1119.4	1103.3
	1149.6	1119.6
		1134.2
		1134.5
Cr	1158.3 (Cr_A)	1155.7 (Cr_A)
	1189.8 (Cr_C)	1179.7 (Cr_B)
		1188.6 (Cr_C)
Mo	1152.1	1105.7
UD-3	1355.7	1343.3
V_{Si}	1352.8	1368.2
	1439.0	1398.0
		1434.0
UD-4	1394.4	1371
	1395.2	1426
	1464.6	

common defect centers were labeled UD-1, UD-2, UD-3, and UD-4 [136,149]. All these defects are present in both 4H- and 6H-SiC. Table 21.11 summarizes the energy positions of the emissions of the UD defects, together with the identified intrinsic defects discussed in Section 21.3 and transition-metal related V, Cr, Mo, and Er, which are discussed in Section 21.5.

The UD-1 defect has its optical transition at 1.0 eV [151] as shown in Figure 21.16; the defect was first found in HTCVD-grown material. The intensity of the PL can be increased by annealing at 1600°C [151], most likely due to the disappearance of other defects [152], and not due to an increase of the UD-1 defect. The defect is also found in both HPSI and low-vanadium-doped SiC grown with the sublimation method. Low-resolution Zeeman results for the UD-1 defect in 6H-SiC have earlier been presented [153]. Figure 21.17 shows the Zeeman splitting in high resolution of the two NPLs in 4H-SiC. Each line splits into four lines in a magnetic field. The splitting of the two lines is schematically described in Figure 21.18. UD-1 might be an impurity, since the defect cannot be annealed out at high temperatures [154] or created by irradiation in epitaxial SiC using different irradiation

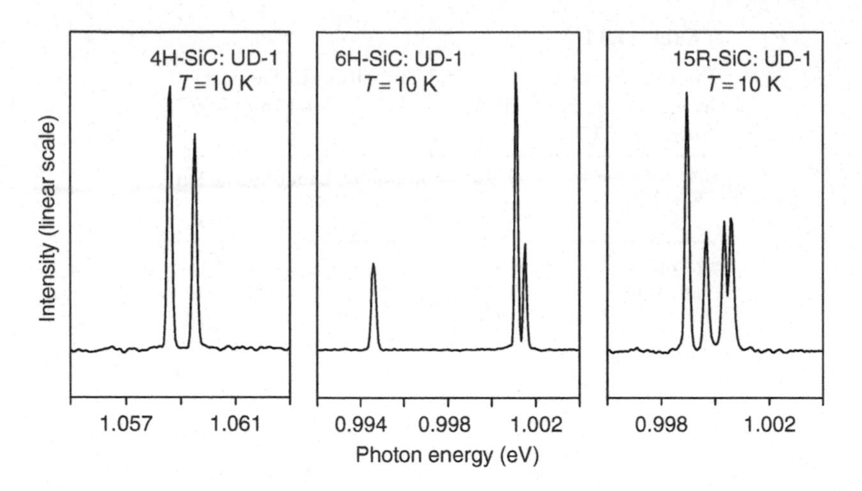

FIGURE 21.16
PL at low temperature (10 K) showing the UD-1 defect in 4H, 6H, and 15R-SiC.

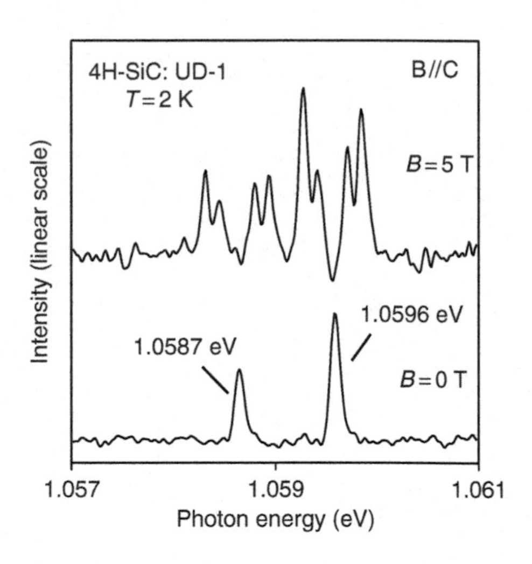

FIGURE 21.17
Zeeman splitting of the UD-1 defect in 4H-SiC at 5 T. Each NPL is split into four lines.

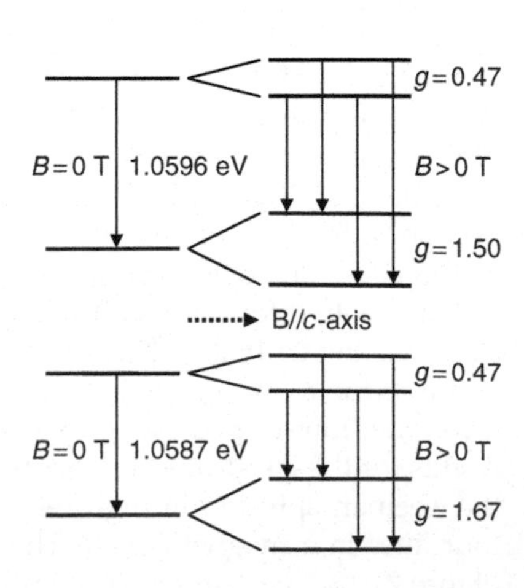

FIGURE 21.18
Schematics of the levels in the internal transitions of UD-1 defect. g-Values (assuming $\Delta m = 1$) and splitting from Zeeman measurements.

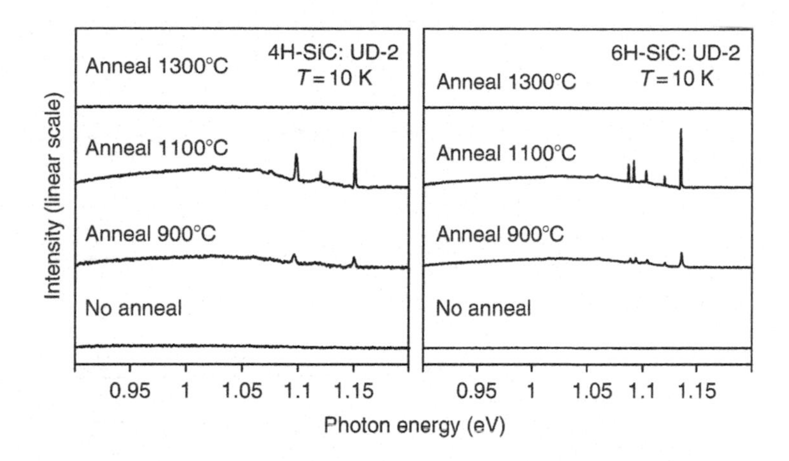

FIGURE 21.19
PL at low temperature (10 K) showing the UD-2 defect after different annealing temperatures in 4H- and 6H-SiC.

sources like electrons, positrons, ions, and neutrons. Instead a sample showing UD-1 PL initially often loses all such PL after high dose irradiation but gets it back after subsequent annealing when the irradiation-induced defects are reduced. ODMR has not been successful, and the origin of the defect may only be revealed if successful intentional doping of SiC is made.

The UD-2 defect with its emission around 1.1 eV can be found in as-grown bulk material, but has so far not been reported in CVD-grown material at 1600°C. Zeeman measurements up to 5 T were made on all the four lines in 4H-SiC, but even with a line width of 0.5 meV or better, no splitting, broadening, or shift can be observed. The UD-2 defect is also easily created by irradiation of epitaxial, low-doped bulk or high resistivity bulk material. If the irradiated samples are annealed at 800°C to 1100°C, the PL intensity of the defect increases, but if higher temperatures are used it disappears completely [155]; see Figure 21.19. This applies also to bulk material with the defect present without irradiation. No sample has had the UD-2 defect still present after a 1600°C annealing in hydrogen. From the annealing cycle of the defect in irradiated 4H-SiC, it has been suggested that the defect is a silicon vacancy-related complex such as the divacancy [155]. Based on photo-EPR measurements in 4H-SiC, Carlos et al. [145] suggested a relationship with the P6/P7 EPR center, which is now identified as the divacancy in the neutral charge state [133,156]; see Section 21.3.5. Lingner had earlier used MCDA measurements, which connect the energy position of an optical transition with an EPR signal, to connect another set of PL lines (not UD-2) between 0.99 and 1.08 eV (see Table 21.11), with the P6/P7 EPR center in 6H-SiC [135]. The UD-2 defect could still be related to the divacancy, but in another charge state than the neutral one. The two sets of PL lines can be observed at the same time in a few samples, but the divacancy PL only dominates if the sample is slightly n-type, while a sample with dominating UD-2 PL typically is highly resistive.

The UD-3 defect was first reported in 1973 by Gorban and Slobodyanyuk [157]. The defect center was studied in more detail in Refs. [158,159]. It was concluded that the number of emission lines is the same as the number of hexagonal sites in the 6H, 15R, and 33R samples. The excited state of the UD-3 defect at cubic sites is believed to be in the conduction band, and the optical transition is therefore not possible. Wagner et al. made an in-depth study of the defect center and suggested both a level scheme and candidates for the origin of the defect [160]. The NPLs are followed by a phonon-assisted emission with both some sharp lines and broad features; see Figure 21.20. The sharp lines are local phonon replicas at an energy distance of 85–90 meV from the NPL. Similar local phonon

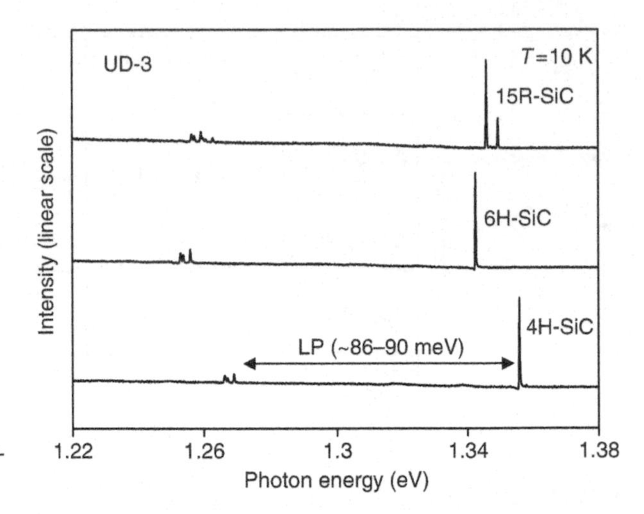

FIGURE 21.20
PL at low temperature (10 K) showing the UD-3 defect in 4H-, 6H-, and 15R-SiC.

energies can be found in PL spectra from Cr and V impurities in SiC. At elevated temperatures emissions from additional excited states close to the main UD-3 line are visible. Zeeman splitting gives a *g*-value very close to 2, with a magnetic field parallel to the *c*-axis of the crystal. The temperature dependence and PLE measurement at 5 T magnetic field reveal that the excited state is split and that the ground state is a singlet as summarized in the level scheme of the defect in Figure 21.21. The defect behaves very similarly in all the studied polytypes (4H, 6H, and 15R), and no large differences in excited states or Zeeman splitting can be observed. This supports the idea that the UD-3 defect in the different polytypes has the same origin. The defect has not to the authors' knowledge been possible to neither create by irradiation nor to anneal out. This, together with the local vibrational modes, supports the model of UD-3 as an impurity on a lattice site. Based on the level scheme and Zeeman splitting, it is suggested that the chemical origin of the defect would be Fe^{3+}, Co^{3+}, or Ni^{4+}. The UD-3 defect would thus have intra-3d-shell transitions similar to those for V and Cr.

The UD-4 PL has been identified in 4H-SiC [136] and recently also observed in the 6H polytype [150]. The PL in 4H-SiC consists of three lines in the range 1.39–1.46 eV at low temperatures. The high-energy line at 1.46 eV has been extensively studied with different optical techniques [161], and the temperature dependence of the intensity gives a thermal activation energy of 55 meV. The origin of the UD-4 defect is still unknown and no good candidates can be suggested. A number of other absorption and emission lines of unknown origin are also present in the near-IR region [162].

21.4.2 EPR Results

Beside the vacancies, divacancy, and vacancy-antisite pairs, there are several other unidentified EPR centers, SI-1 to SI-4, commonly detected in HPSI SiC substrates [100,101]. Among these, the most common defect is the SI-1 center, which gives rise to an EPR spectrum consisting of a broad line without any structure (Figure 21.22a). The SI-1 spectra

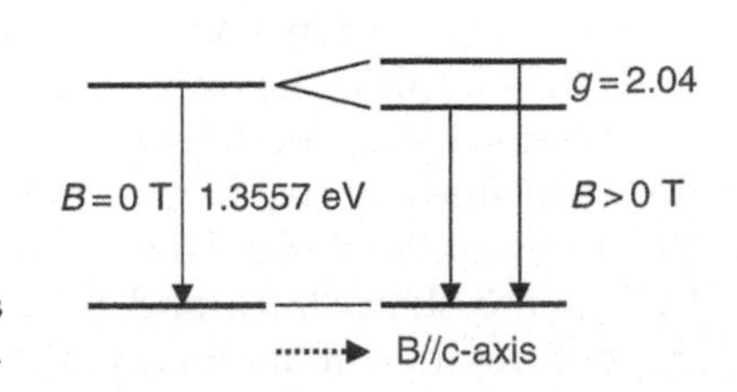

FIGURE 21.21
Schematic diagrams of the levels in the internal transition of UD-3.

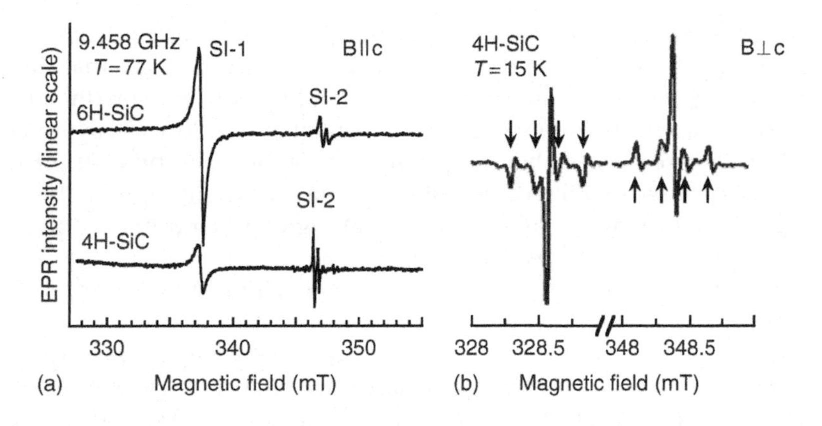

FIGURE 21.22
(a) EPR spectra of the SI-1 and SI-2 centers in HPSI 4H- and 6H-SiC substrates measured in the dark for **B** || **c** at 77 K, and (b) EPR spectrum observed in HPSI 4H-SiC after annealing at 1800°C at 15 K and for **B** ⊥ **c**. The arrows indicate the ^{29}Si hf lines due to two sets of three Si atoms. (Reprinted from Carlos, W.E. et al., *Phys. Rev. B*, 74, 235201, 2006. With permission.)

in 4H- and 6H-SiC are the same with an electron spin $S=1/2$ and an isotropic g-value of 2.0026. The SI-1 spectrum is often observed in all types of HPSI SiC substrates with different thermal activation energies E_a. The SI-1 signal is stable at very high temperatures and even increases its intensity after annealing at 1800°C, indicating that the center may be related to a more extended defect. Such high-temperature annealing also promotes clustering of vacancy defects as reported in Ref. [163], although there are no direct proofs that the SI-1 signal and the vacancy clusters are related. That the SI-1 signal is stable or even increases after high-temperature annealing when the substrates are already changed to n-type suggests that the center does not play an important role in carrier compensation.

Another common defect in HPSI SiC substrates is the SI-2 center [100,101], which is often observed in samples with E_a below 1.1–1.3 eV. The spectra in the 4H and 6H polytypes are also shown in Figure 21.22a. This center has a spin $S=1/2$ and C_{1h} symmetry with very anisotropic g-values (see Table 21.12). The hf structure of the SI-2 spectrum consists of

TABLE 21.12
Spin-Hamiltonian Parameters of EPR Centers in HPSI 4H-SiC Grown by HTCVD and PVT

Center	Symmetry	Spin	g_{xx}	g_{yy}	g_{zz}	α	D	Ref.
SI-1	Isotropic	1/2	2.0026	2.0026	2.0026	0°		[101]
SI-2	C_{1h}	1/2				3°		—
SI-3	C_{1h}	1/2	1.7302	1.9882	1.9502	68°		—
SI-4	C_{3v}	1/2				0°		—
SI-9	C_{3v}	1/2	2.0012	1.9995	2.0188	0°		—
SI-11	Isotropic	1/2				0°		—
P_{u1}	C_{3v}	1	2.0024	2.0024	2.0040	0°	35.14	[165]
$V_{gi}V_C$ (?)	C_{3v} (?)	1				0°	11.12	[166]
			2.0021	2.0021	2.0017			
			2.0037	2.0037	2.0037			
			2.0052	2.0052	2.0052			
			2.0029	2.0029	2.0038			

Note: α is the angle between the principal z-axis of the g tensor and the c-axis. The fine structure parameter D is given in mT.

several lines with unequal intensities and splitting with rather high intensity ratios corresponding to natural abundances of about 15% or lower. This indicates that the SI-2 center may be related to impurities. The hf structure was detected only in main directions, and the identification of the impurity involved is still not possible. The low C_{1h} symmetry and the complicated hf structure suggest that the center may contain even two impurities. Annealing studies showed that the HPSI 4H-SiC substrates having strong signals of the SI-2 center did not change to slightly n-type after annealing at high temperatures, as often happened to HPSI SiC samples with E_a below 1.1–1.3 eV.

The SI-3 and SI-4 centers were observed in some HPSI 4H-SiC samples grown by HTCVD [100,101]. The SI-3 center has C_{1h} symmetry and anisotropic g-values: $g_{xx} = 2.0012$, $g_{yy} = 1.9995$, and $g_{zz} = 2.0188$. The angle between the principal z of the **g** tensor and the c-axis is ~68°. The signal is rather weak and no clear hf structures have been detected. The SI-4 signal has C_{3v} symmetry with $g_{\parallel} = 2.0024$ and $g_{\perp} = 2.0040$. The signal appears in the same magnetic field region as many other centers (SI-1, SI-2, V_C^+ and $C_{Si}V_C^-$); it is not possible to observe any clear hf structures due to overlapping with other signals. Both centers were detected in samples with somewhat higher concentrations of titanium (Ti) than normal substrates. However, there is no clear evidence indicating the involvement of Ti. After annealing at 1600°C, these spectra were not detected. Their identification and role in carrier compensation in HPSI SiC is still not clear.

Recently, it became clear that the SI-5 center, which is commonly detected in HPSI SiC substrates and was previously attributed to the divacancy [101], is $C_{Si}V_C^-$ [125], and that the SI-6 center [101] is the V_C^- center. Another two centers, SI-9 and SI-11, with C_{3v} symmetry and spin $S = 1/2$ are also commonly observed in HPSI 4H-SiC substrates grown by HTCVD. Both signals are usually weak compared to the known vacancy-related centers. The SI-9 has g-values $g_{\parallel} = 2.0021$ and $g_{\perp} = 2.0017$ [101]. The SI-11 center is often observed in HPSI 4H-SiC substrates together with the $C_{Si}V_C^-$ signal. The SI-11 center is isotropic with $g = 2.0037$. Its signal nearly coincides with the $C_{Si}V_C^-$ line for $\mathbf{B} \perp \mathbf{c}$ and almost overlaps with the high-field ^{29}Si hf line of $C_{Si}V_C^-$ for $\mathbf{B} \parallel \mathbf{c}$ [101]. The concentration of the SI-11 center in HPSI SiC substrates is low. The intensity of the SI-11 line is just slightly higher than that of the ^{29}Si hf line of $C_{Si}V_C^-$. EPR measurements under illumination (above and below band gap excitation) show no change in the intensity of the SI-11 signal. With the concentrations estimated to be in the low 10^{14} cm^{-3} range, the SI-11 center will not play an important role in the SI behavior of HPSI SiC substrates.

In a recent study of the thermal evolution of defects in HPSI 4H-SiC substrates, Carlos and coworkers [166] observed an EPR spectrum related to an excited $S = 1$ state with the principal axis in the basal plane. The spectrum was detected after annealing the sample at ~1400°C; its intensity increases with annealing temperature up to ~1800°C. Figure 21.22b [164] shows the spectrum measured at 15 K under illumination with light of wavelength $\lambda > 450$ nm for $\mathbf{B} \perp \mathbf{c}$. As indicated in the figure, two sets of hf structures, each corresponding to the interaction with three Si atoms with the splitting of ~17.2 and 4.6 MHz, could be observed. Although no hf structures due to interaction with one ^{13}C nucleus were detected, the center was tentatively assigned to the V_C–V_{Si}–V_C defect, an annealing product of the divacancy, which was proposed as a metastable configuration of the divacancy [167].

In early developed HPSI 4H-SiC substrates grown by PVT (the concentrations of B and N were in the low and mid 10^{16} cm^{-3}, respectively, and V from 10^{13} cm^{-3} up to mid 10^{16} cm^{-3}), Zvanut et al. [165] observed an axial center with $S = 1$ labeled P_{u1}. The parameters of the center are given in Table 21.12. The signal is stable up to 1200°C and decreases after annealing at higher temperatures (up to 1600°C). The center was suggested to be related to intrinsic defects. Due to the lack of information on the hf interactions, the center is still not conclusively identified.

21.5 Metallic Impurities, Irradiation-Induced Defects, and Unknown Defects

21.5.1 Metallic Impurities

Transition metals strongly affect electrical and optical properties of SiC. They are commonly present in SiC, especially in bulk-grown material where they also are used as dopants to obtain high resistivity material. The titanium-related emission was reported in SiC already in the beginning of 1960s [168]. Vanadium, chromium, molybdenum, and erbium have been studied since the early 1990s, all creating deep levels with optical absorption and PL due to intrashell transitions.

Metallic impurities introduce deep levels in SiC. For many impurities, their levels have been conclusively identified. Detailed information on the energy levels of the elements Be, Zn, Cd, In, Ti, V, Cr, Ta, W, Er, and Sm in common SiC polytypes can be found in a recent survey of radiotracer DLTS studies [169]. Transition metal impurities are often paramagnetic centers and sometimes are also optically active and have been extensively studied by EPR and ODMR. The electronic structures of Ti, V, Mn, Cr, Sc, and Mo have been investigated by EPR and ODMR since the 1990s [170–178].

The Ti-related PL lines in the literature called the ABC lines in 6H-SiC were among the first reported in SiC [168]; they were first called X and Y type spectra. Primarily due to a lack of understanding about the mixing of polytypes, a wrong interpretation was proposed. Many years later, this spectrum was proven to be Ti-related [179] with isotopic line splitting. The 0.1 ms observed lifetimes [180] are consistent with an exciton residing at an isoelectronic center. EPR in n-type 6H-SiC [181] indicated that the $(TiN)^+$ pairs were the centers of recombination of BEs. However, in a detailed study using DLTS [182], it was concluded that the PL is coming from an exciton bound to the neutral Ti acceptor. Finally, models of isoelectronic centers composed of either individual Ti atoms or Ti–N pair representing a complex of the electronic trap and recombination center have also been proposed [183]. The previous studies, however, have mainly been done with the 6H polytype, and the origin of the observed PL is still unclear. Figure 21.23 shows typical

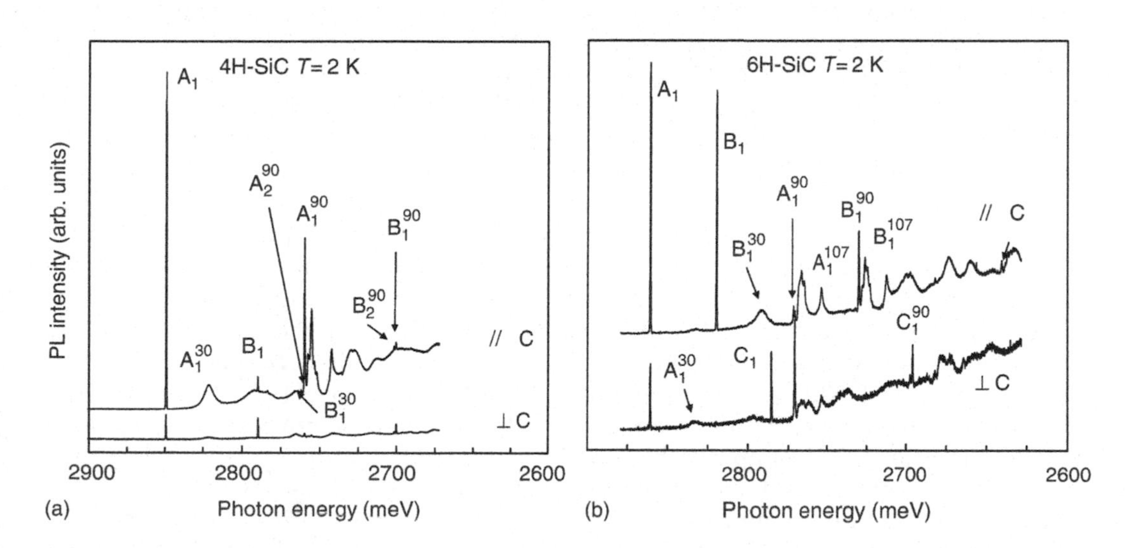

FIGURE 21.23
PL of (a) 4H and (b) 6H samples, respectively, recorded at 2 K and showing the Ti-related spectra for polarization parallel and perpendicular to the *c*-axis. The NP-BE lines are A_1, B_1, and C_1 and the other lines are phonon-related lines.

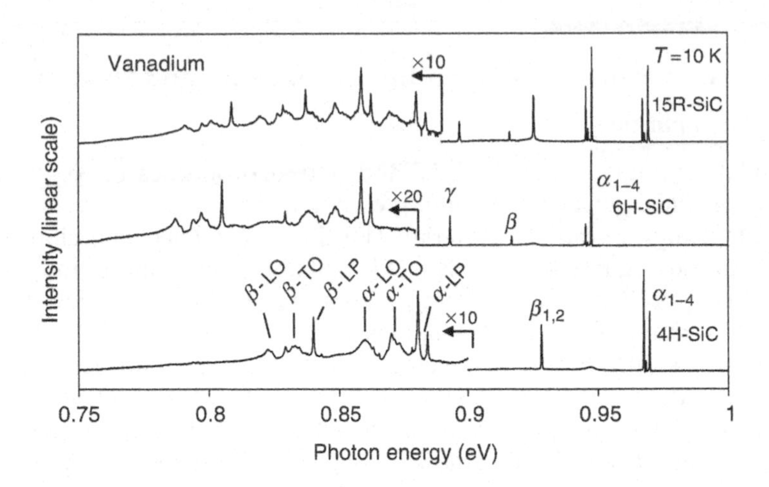

FIGURE 21.24
PL from vanadium at low temperature (10 K) in 4H-, 6H-, and 15R-SiC.

PL spectra recorded at low temperature (2 K) for both 4H and 6H polytypes. The spectra, which depend strongly on the polarization, consist of the NPLs, two in 4H and three in 6H, with phonon structure containing both broad and sharp replicas. At higher temperature, higher energy NPLs appear, some very close to those observed at 2 K.

PL from the neutral vanadium defect V^{4+} ($3d^1$) was identified in both 4H and 6H-SiC by Schneider et al. [184] using correlation with EPR measurements. PL at 0.9–1.0 eV at low temperatures consists of NPLs belonging to the different equivalent sites in SiC and a phonon-assisted spectrum with both local and lattice phonons [185]; see Figure 21.24. In 4H-SiC the cubic site gives rise to a doublet $\beta_{1,2}$ (928.4 meV) and the hexagonal site to four lines α_1 (967.3 meV), α_2 (968.1 meV), α_3 (969.5 meV), and α_4 (970.3 meV). The two cubic sites in 6H-SiC give rise to a line γ (893.1 meV) and a doublet β (917.0 meV); the hexagonal site gives rise to four lines α_1 (945.3 meV), α_2 (946.0 meV), α_3 (947.5 meV), and α_4 (948.2 meV). The splitting and origin of the vanadium PL from the hexagonal site have been investigated in detail and can be explained by crystal-field splitting of the internal transition of the $3d^1$ electron [171,186]. The V^{3+} ($3d^2$) charge state has also been identified with MCDA measurement [187] and the optical transitions at 602 and 620 meV [188,189].

The internal transitions in Cr^{4+} ($3d^2$) have been investigated in both absorption [190] and PL (Figure 21.25) [191,192]. The PL lines have been identified in both 4H-SiC with NPLs Cr_A (1158.3 meV) and Cr_C (1189.8 meV), and in 6H-SiC with NPLs Cr_A (1155.7 meV), Cr_B (1179.7 meV), and Cr_C (1188.6 meV) belonging to the different equivalent sites and a phonon-assisted spectrum with both local and lattice phonons all very similar to the phonon-assisted spectrum of vanadium.

Mo^{4+} ($4d^2$) has been investigated with MCDA in 6H-SiC [177]; the internal transitions are between 1.04 and 1.24 eV. The presence of the PL center in this region labeled I-1 [136] with one NPL at 1152.1 meV in 4H-SiC and at 1105.7 meV in 6H-SiC correlates with intentional molybdenum doping; see Figure 21.26. The molybdenum doping was measured by SIMS in doped sample to be 5×10^{14} cm^{-3}, while the undoped sample grown under the same conditions showed a concentration below 1×10^{14} cm^{-3}, which is the SIMS detection limit for molybdenum.

PL from erbium-implanted samples has been investigated for a number of different polytypes [193]. The emission is dominated by one line at 808 meV in both 4H- and 6H-SiC. The intense and temperature stable Er^{3+} intra-f emission due to Er is unlikely to be due to an isolated Er defect [194].

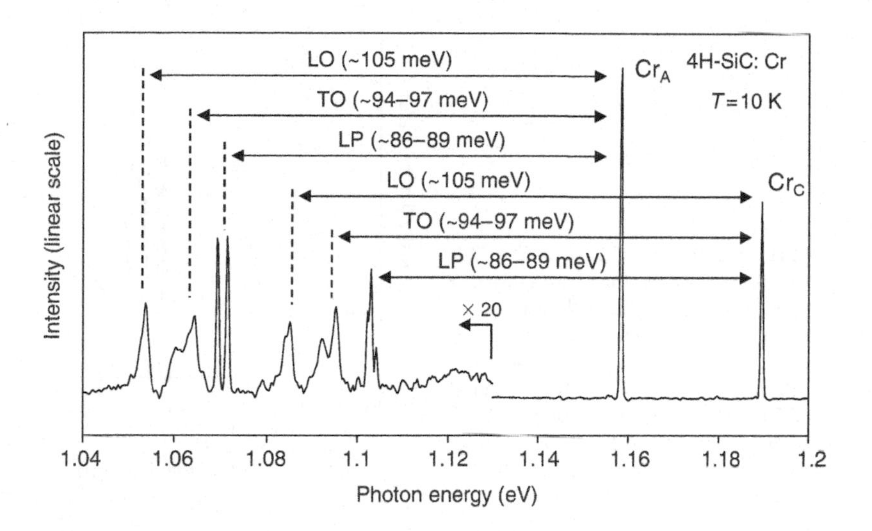

FIGURE 21.25
PL from chromium at low temperature (10 K) in 4H-SiC.

21.5.2 Irradiation-Induced Defects and Unknown Defects

21.5.2.1 EPR and DLTS

Except a few vacancy-related centers as described in Section 21.3, most EPR defects introduced by irradiation have not been conclusively identified. EPR centers in irradiated 3C-SiC were mostly studied during the last decade [123,195]. In recent years, defect studies were mostly focused on the 4H and 6H polytypes. The EI1/EI1′ and EI3/EI3′ EPR centers are the dominating defects observed in irradiated p-type 4H- and 6H-SiC [97]. Being annealed out at rather low temperatures (~200°C), these centers were previously suggested to be related to vacancy-hydrogen complexes [97]. The later calculations of the hf constants of H-related and C interstitial-related defects, suggested the EI1/EI1′ and EI3/EI3′ EPR centers to be related to different configurations and charge states of the C split interstitials $(C_2)_C$ [112,196]. The hf constants of the nearest Si neighbors obtained from calculations are indeed close to the experimental values. The predicted principal hf values of the ^{13}C interstitial are in the range of 3–6 mT. However, this C hf has not been detected by EPR so far. Our recent study seems to indicate that, within a clear spectral region of ~16 mT, there are no signs of a hf structure that could be related to the interaction with one ^{13}C atom (with 1.16% of natural abundance, its intensity should be ~1/10 the intensity of the Si hf line). As can be seen in Figure 21.27, such a hf line, if present, should be clearly detected.

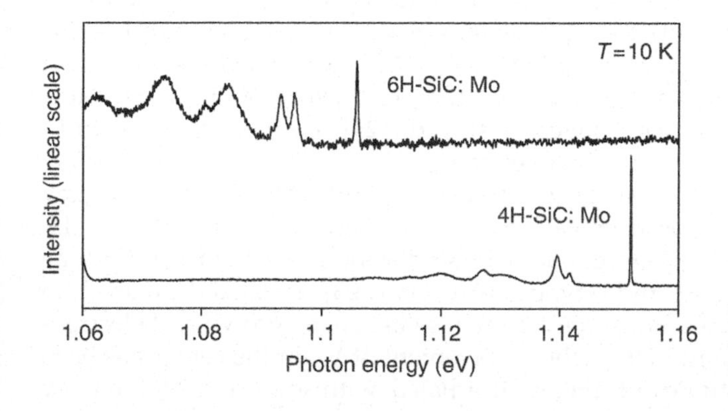

FIGURE 21.26
PL from molybdenum at low temperature (10 K) in 4H and 6H-SiC.

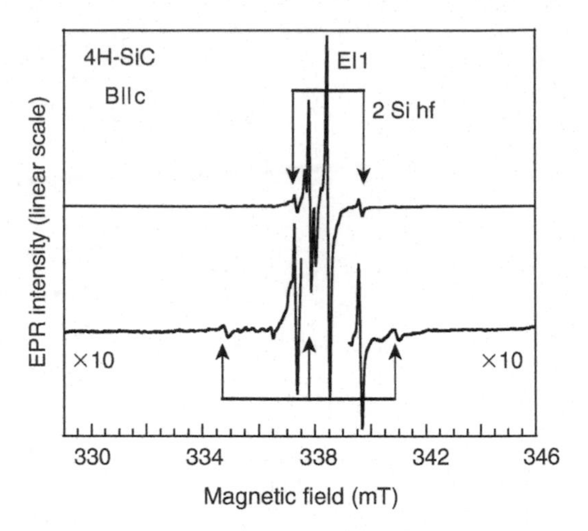

FIGURE 21.27
EPR spectrum of EI1 center in 4H-SiC measured at 77 K for **B** || **c**. The × 10 scale parts show the hf structure due to the interaction with two Si atoms of the EI1 center and the weaker hf structure of another signal at ~338 mT. There are no signs of a hf structure due to the interaction with one ^{13}C atom within the spectral region of 9 mT below and 7.5 mT above the main line. If present, such an hf line with an expected intensity of ~1/10 of the Si hf line should be clearly observed.

The C hf structure expected from the theoretical model of the C split interstitial has so far not been detected for the EI1 center.

Recently, Pinheiro et al. [197] observed a spin $S = 1$ center in n-type irradiated 6H-SiC. Its parameters ($g_{\|} = 2.0107$, $g_{\perp} = 2.0090$, and $D = 481 \times 10^{-4}$ cm^{-1}) are different from that of the EI3 and EI3′ centers [97]. In some directions, a weak hf structure was detected and attributed to the interaction with one ^{13}C. The center was attributed to the C split interstitial in the double positive charge state $(C_2)_C^{2+}$.

Also in electron-irradiated n-type 6H-SiC, other EPR signals, P6′ and P6″ in the family of P6 centers, were observed [139]. These signals were observed together with a series of P6 centers, which today is attributed to the divacancy in the neutral charge state [133]. For one of the EPR line (P6′c), a weak hf structure was detected and attributed to the interaction with two ^{13}C atoms. This led to the suggestion that the P6′ and P6″ are different configurations of the $V_CC_{Si}(Si_CC_{Si})$ defects, which are the second products of the annealing of V_{Si} [139]. It was further suggested that the dissociation of V_C from this defect leads to the formation of the $C_{Si}Si_CC_{Si}$ defect. Based on the calculations of the local vibrational modes, the $C_{Si}Si_CC_{Si}$, defect was assigned to the D_I PL center. However, later calculations [198] show that the close-by antisite pair adjacent to a carbon antisite defect is not stable enough to account for the D_I center. Further simulations on the stability of the antisite pair are reported in Refs. [199,200]. As shown recently in the annealing studies of HPSI SiC substrates (see Section 21.3.6 and Ref. [138]), no signs of the P6′ and P6″ centers were detected in connection with the annealing out of V_C, V_{Si}, divacancies (P6/P7 centers), and vacancy-antisite pairs ($V_CC_{Si}^-$). It seems that more data are needed for the conclusive identification of the P6′ and P6″ centers.

In p-type 6H-SiC irradiated with low-energy electrons (300–350 keV and close to the threshold for the Si atom displacement), two spin quartet ($S = 3/2$) V_{Si}-related spectra, one with C_{3v} symmetry ($g_{\|} = 2.0032$, $g_{\perp} = 2.0028$, and $D = 68.7 \times 10^{-4}$ cm^{-1}) and the other with C_{1h} symmetry ($g_{xx} = 2.0015$, $g_{yy} = 2.0039$, $g_{zz} = 2.0035$ and zero-field splitting parameters $D = 76 \times 10^{-4}$ cm^{-1} and $E = 19 \times 10^{-4}$ cm^{-1}) were observed [124]. The defects were attributed to the on-axis and 24° off-axis $V_{Si}^- + Si_i$ Frenkel pairs.

Radiation-induced deep levels in SiC have been intensively studied by DLTS, and many levels with ionization energies ranging from ~0.4 to ~1.65 eV have been observed in 4H- and 6H-SiC (see reviews [105,182]). More recent DLTS studies of irradiated n- and p-type 4H-SiC [108,201,202] and as-grown n- and p-type 4H-SiC [140,203] revealed deep levels in a similar range of ionization energies. For some common centers in as-grown and electron-irradiated materials such as $Z_{1/2}$ and EH6/7, the involvement of V_C in the center has been demonstrated [108,110]. DLTS studies of 4H-SiC irradiated with low- and high-energy electrons [108] indicate that the EH6 and EH7 centers are not from the same defects.

21.5.2.2 PL Results

After particle irradiation a PL spectrum known as the alphabet lines appears in the 2.80–2.91 eV range for the 4H-SiC [162,204] and 2.52–2.62 eV range for 6H-SiC [162,205] (Figure 21.28). In 4H-SiC it comports 12 families of NPLs together with their associated phonon structures, and the nine first NPLs show a dependency on which face the irradiation was done. Carbon Frenkel pairs with different orientations and distances have thus been proposed to be the origin of these lines [206]. An important feature of the observed spectrum is the presence of localized modes (LM) or gap modes between the acoustic and longitudinal lattices modes. Recent theoretical calculations suggested the four first alphabet lines to be the four forms of the close-by antisite pair [198]. The alphabet lines are observed when using above band gap excitation. However, when exciting below the SiC band gap, new PL lines appear in 6H-SiC within the 2.36–2.47 eV range [207]. Six different optical centers are observed that have two high-energy LM (higher than the longitudinal lattices modes) at about 133 and 179 meV from the NPLs. They were associated with six closely related centers that have properties consistent with C–C dumbbell, and five of them also with a C–Si dumbbell. These new spectra as well as the alphabet lines appear when irradiating the sample with electron energy well below the Si displacement threshold. Whereas the alphabet lines start to anneal out at about 650°C and disappear after annealing at 1000°C [204,206], the lines observed in the 2.36–2.47 eV range seem to have different behaviors depending on the material conductivity, and some of them are stable up to 1300°C, the highest annealed temperature used in the study [207]. Further investigations on carbon-related defects can be found in Refs. [208–211].

Progressively when the alphabet lines anneal out, the so-called D_I center is observed. This PL has been observed in different polytypes and can even be seen in as-grown material; after irradiation and subsequent annealing, the defect is stable at as high temperature as 1700°C. It is one of the most studied centers in SiC probably due to its high PL efficiency. The binding energies of the associated exciton range from 350 to 450 meV in the different polytypes. In some polytypes several recombinations of excitons with different binding energies are observed. This is a commonly observed feature in SiC polytypes other than 3C and is due to the multiplicity of inequivalent lattices sites in the crystal; in 6H-SiC there are three excitons observed and one in 3C. However, there is only

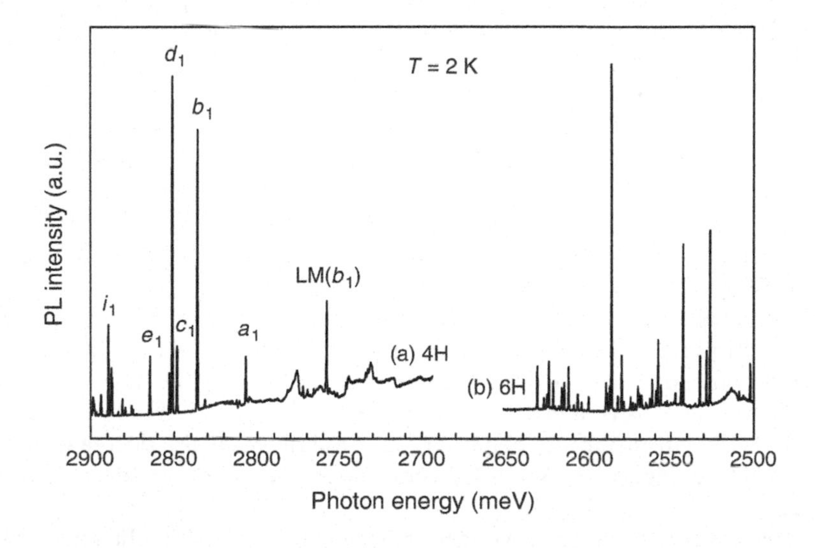

FIGURE 21.28
PL spectra of the alphabet lines recorded at 2 K of (a) 4H- and (b) 6H-SiC samples. The samples were low-doped epilayers irradiated with electrons with a dose of 10^{17} cm^{-3}.

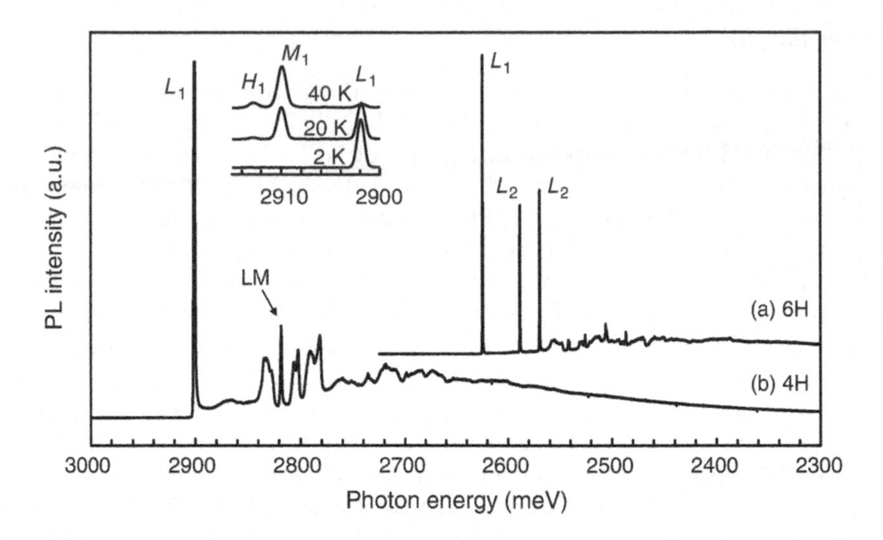

FIGURE 21.29
PL spectra recorded at 2 K showing the D_I center in (a) 4H- and (b) 6H-SiC. The NPLs are labeled as L_1 in 4H and L_1, L_2, and L_3 in 6H. The inset shows an enlargement view of the NP part in 4H for three indicated temperatures.

one in 4H-SiC, when two would be expected and this with an NPL at about 2.901 eV [212] (see Figure 21.29). This isoelectronic defect has been shown to have a pseudodonor character. The model predicts a ground-state level acting as a hole trap at about 0.33 eV above the valence band [212]. This hole trap has been identified using minority carrier transient spectroscopy [213]. In addition, trapped carrier electroluminescence has been used to correlate this minority carrier trap with the luminescence measurement [214]. The phonon structure comports also two sharp gap modes at about 83 meV from the NPL. At elevated temperatures lines at higher energy than the NPLs are observed (see the inset of Figure 21.29). In Ref. [215], based on calculated occupation levels, it was proposed that the D_I center was V_{Si} related, however, since V_{Si} can be highly negatively ionized, this model does not support the pseudodonor character of the D_I center. Other models associate the D_I center either to antisite pairs, Si_CC_{Si} [216], or to antisite complexes, $Si_C(C_{Si})_2$ [139], although the thermal stability was questioned, see previous section, and more recently to an isolated Si_C [198].

The second defect stable at high temperature is the so-called D_{II} center. Whereas the D_I defect can be observed after any kind of irradiation, the D_{II} center is only observed after implantation with ions, suggesting that it requires an excess of atoms in the lattices. This spectrum contains gap modes at about 84 and 89 meV and a large number (up to 21 in 4H) of high-energy LM above the lattice phonons from 121 to 167 meV [206]. Larger carbon antisite clusters have been proposed to be a possible origin of this PL [217]. Typical spectra are shown in Figure 21.30.

The coexistence of lattice damages and hydrogen (or deuterium) in SiC can give rise to efficient PL spectra [218]. The observation of a 370 (274) meV C–H (C–D) bond stretching vibration mode has suggested a model for the center with an H (D) atom bonded to a C atom at a Si vacancy. The isoelectronic center has also been shown to have a pseudodonor character [219] with a metastability behavior [220], mainly seen as a decrease of the PL with excitation time (see Figure 21.31). Different models of this center are discussed in Refs. [221,222].

The last center reported having LM at 83–87 meV from the NPLs has been observed in a boron-doped epilayer after SIMS [223] or after hydrogenation [224]. During SIMS

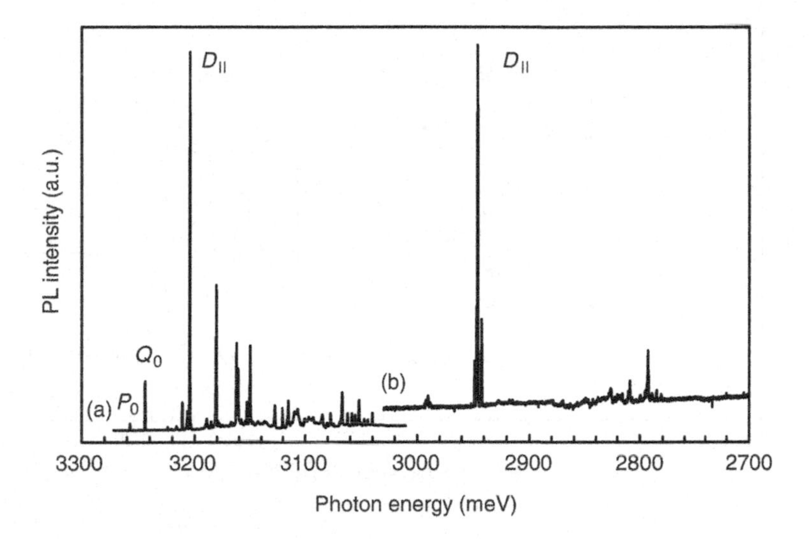

FIGURE 21.30
PL spectra recorded at 2 K showing the D_{II} spectrum in (a) 4H- and (b) 6H-SiC.

damage creation can occur outside the analysis crater itself since the alphabet lines have been observed there [225]. The center was attributed to a complex of boron and hydrogen. In 4H-SiC the related NPL is near 3838 Å, whereas it is located close to 4182 Å for the 6H polytype (Figure 21.32). The NPL has two components for the 4H polytype, but three for the 6H case with an energy separation of about 0.6 meV (see inset of Figure 21.32). Our PL results can be explained by an excitonic recombination at an isoelectronic center. In 4H ab initio supercell calculation [226] indicated that the complex involving B at the Si site with hydrogen at the bond center position between boron and one of its carbon neighbors (B_{Si}–H_{BC}) was the most stable in 4H. Calculation of the C–H stretch mode vibration frequencies for this complex yielded 386 meV. We should thus

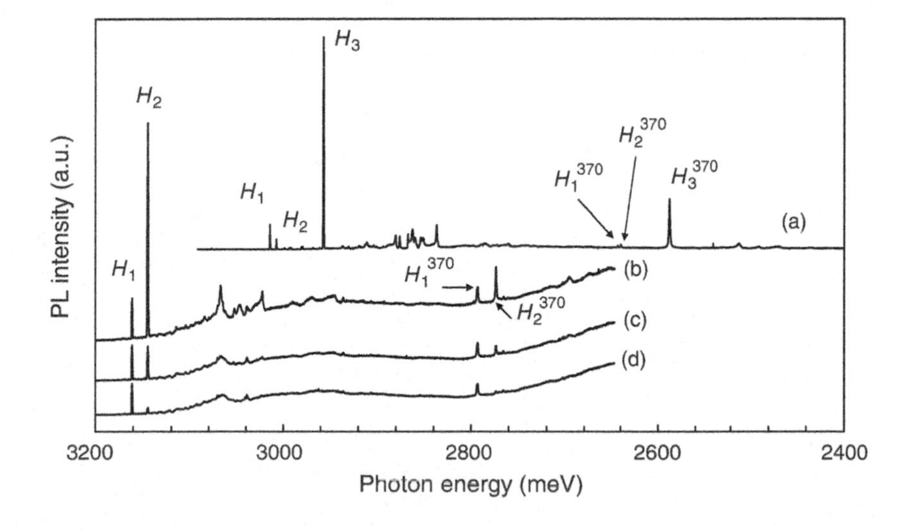

FIGURE 21.31
PL spectra showing hydrogen-related lines recorded at 2 K from (a) 6H and (b–d) 4H-SiC. The (c) and (d) spectra have been recorded 4 and 30 min after the (b) spectrum.

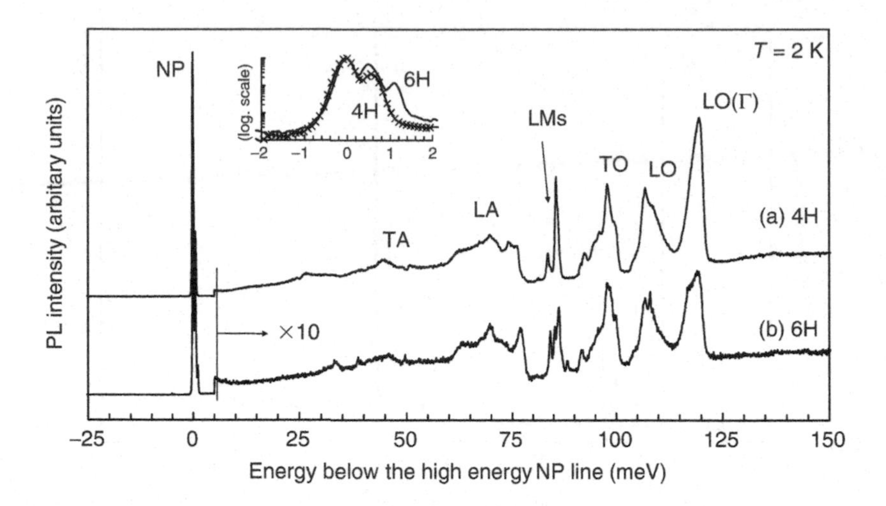

FIGURE 21.32
Boron-related PL spectra recorded at 2 K of (a) a 4H-SiC and (b) 6H-SiC sample. The samples were boron-doped epilayer and have been investigated by SIMS. The NPL in 4H is at 3838 Å (3229.5 meV), whereas in 6H it is at 4182 Å (2963.9 meV). The inset shows more in detail the NPLs.

TABLE 21.13

Energies of the Main NP PL Emissions in Some SiC Polytypes Observed at Low Temperatures (2 K) and in the Visible Range

	4H		6H		3C	
	Label	meV	Label	meV	meV	Ref.
N-BE	P_0	3256	P_0	3006.9	2379	[68,69]
	Q_0	3243	R_0	2992.3		
			S_0	2990.5		
P-BE	Ph[a]	3256.3	Ph[a]	3008.9	2378	[73,74]
			Ph[b]	3007.6		
			Ph[c]	3005.4		
Al-BE		3246.5–3249		2998.9–3004.5	2360–2365	[75,78]
Ga-BE		3240–3242		2988–2994	2346	[77]
		3229.5		2963.9		[82,223]
D_{II}		3204.5		2945.9	2305	[206]
D_I	L_1	2901	L_1	2624.5	1973.8	[212]
			L_2	4788.7		
			L_3	4823.7		
H related	H_1	3161.1	H_1	3013.4		[219]
	H_2	3144.6	H_2	3006.5		
			H_4	2999.1		
			H_3	2955.9		
			H_5	2979.5		
Ti related	A_1	2849.2	A_1	2860.8		[168,179,180]
	B_1	2789.8	B_1	2819.7		
			C_1	2785.4		
	a_1–l_1	2810–2910		2524–2620		[162,204,205]
			P_0–T_0	2360–2467		[207]

Note: In some cases a range of energies is given when many unresolved lines are observed.

expect to observe this mode at about 386 meV from the NPL toward lower energies. A similar mode is expected for the 6H polytype with about the same energy difference (within a few meV). However, this has not been observed until now. The center starts to anneal out at about 400°C and is very weakly observed after annealing at 600°C; the deactivation energy is found to be rather low and close to 1.2 eV. Some of the results reported here are collected in Table 21.13.

21.5.3 Others

Oxygen is one of the most common impurities in many semiconductors and can form electrically active defects. Its incorporation into Si and the formation of O-related defects have been studied extensively for more than 50 years. Very limited information on optical properties about O in SiC is available. However, electrically active defects related to O have been observed in O^+ implanted or O-doped 4H and 6H epilayers [227,228]. Shallow double donors with energy ionization in the range 129–360 meV below the conduction band have been associated with O on the C side (O_C), which is also the most stable O-related defect expected for n-type material [229]. Three very deep acceptor-like defects have also been reported at E_c−480, 560, and 610 meV, which could not be assigned to O_{Si}. The theoretical calculations predicted O_{Si} to be a hyperdeep double donor in 3C, but less stable than O_C. Interstitial oxygen (O_i) has also been investigated theoretically; however, no gap level is associated with these defects [230].

Recently sulfur ions have been implanted into n-type SiC, and electrical investigations revealed that S atoms form double donors in SiC. The number of observed double donors corresponded to the number of inequivalent sites in 6H [231]. For both impurities, oxygen and sulfur, no optical characterization has been reported to the best of our knowledge.

There are some optical features that have not been understood in SiC. Figure 21.33 shows a typical example observed very often in high-purity low-doped 4H epilayer. Note the high resolution used in this case. Although the PL is weak, it appears almost at the same energy position as Al-BE (see above), which might lead to mistakes when analyzing the PL spectrum.

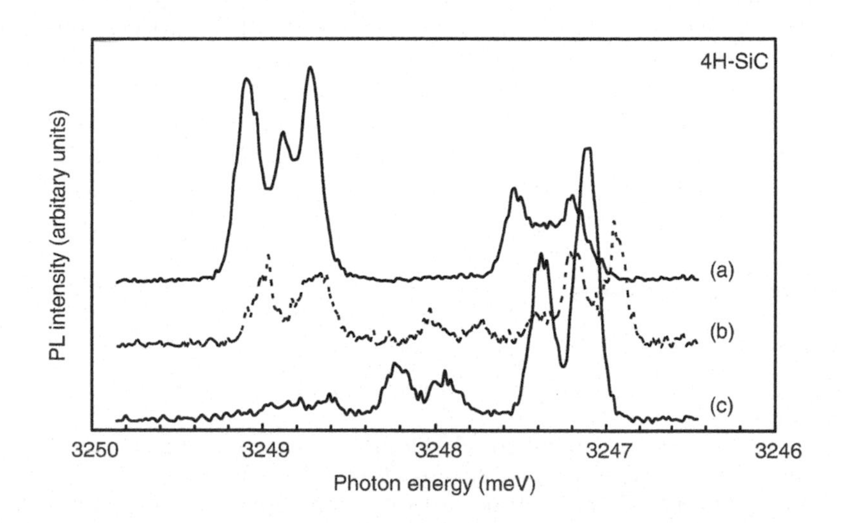

FIGURE 21.33
(a) Typical PL spectrum of Al-BE in 4H recorded with high resolution. (b) Possible superposition of the two BE PL spectra. (c) PL spectrum of an unknown BE with energy position very close to the Al-BE.

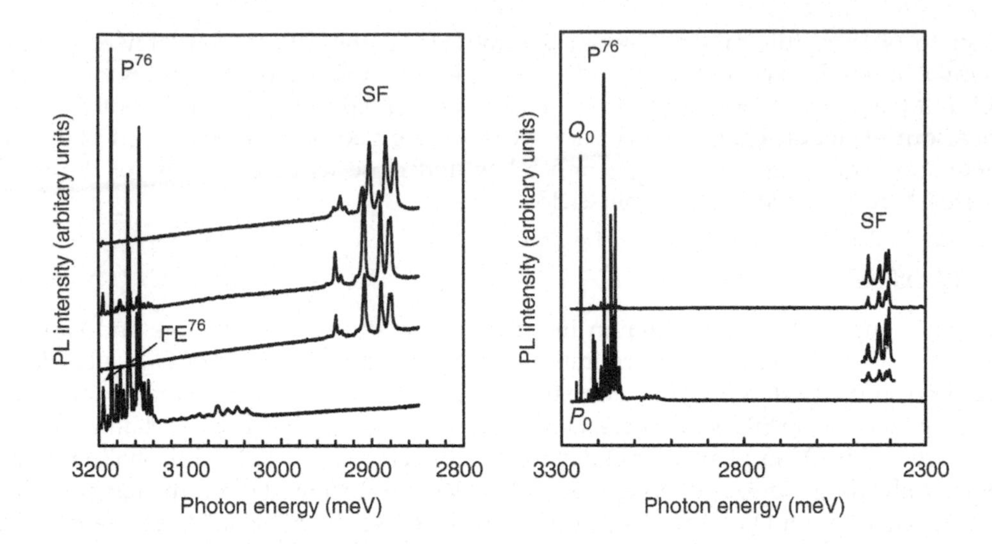

FIGURE 21.34
PL spectra recorded at 2 K for two different 4H-SiC epilayers (left and right). For each epilayer PL spectra were recorded at various localizations: the bottom spectrum in both cases is typically for low-doped ($n \sim 2.10^{15}$ cm^{-3}) layers with near-band gap emission as the dominant feature. The other spectra (or part of spectra) show signature of the stacking faults.

Structural defects also play an important role in SiC, since they affect doping, carrier lifetime, as well as morphology (including killing structural defects). Using the hot-wall CVD technique, growth of high-purity layers with a residual carrier concentration below 10^{14} cm^{-3} has been demonstrated. The carrier lifetime varies between 0.3 and 1 μs and can occasionally be several microseconds [232]. It is not known which defects determine the lifetime or why the lifetime varies on a wafer [233,234]. Lifetime-killing defects may be related to intrinsic defects, but impurity-related (metals, etc.) cannot be ruled out. Structural imperfections have been shown to affect the carrier lifetime. A well-established example is the stacking faults in SiC. They can be created during operation of diodes [235], but are also generated during growth, as for example when the nitrogen doping concentration exceeds 2×10^{19} cm^{-3} [236]. They have been the subjects of many studies during the last years and have been shown to give rise to PL [237–241], as early proposed [242] and deeply studied [243–245]. The observed PL spectrum is complicated and can be at various energy positions in the PL spectra (Figure 21.34). When this type of spectrum is observed, a decrease or absence of the near-band gap emission is always observed.

References

1. Park, C.H. et al., Structural and electronic properties of cubic, 2H, 4H, and 6H SiC, *Phys. Rev. B*, 49, 4485, 1994.
2. Käckel, P., Wenzien, B., and Bechstedt, F., Electronic properties of cubic and hexagonal SiC polytypes from ab initio calculations, *Phys. Rev. B*, 50, 10761, 1994.
3. Lambrecht, W.R.L. et al., Electronic band structure of SiC polytypes: a discussion of theory and experiment, *Phys. Status Solidi B*, 202, 5, 1997.
4. Wellenhofer, G. and Rössler, U., Global band structure and near-band-edge states, *Phys. Status Solidi B*, 202, 107, 1997.

5. Persson, C. and Lindefelt, U., Relativistic band structure calculation of cubic and hexagonal SiC polytypes, *J. Appl. Phys.*, 82, 5496, 1997.
6. Persson, C. and Lindefelt, U., Dependence of energy gaps and effective masses on atomic positions in hexagonal SiC, *J. Appl. Phys.*, 86, 5036, 1999.
7. Kaplan, R. et al., Electron cyclotron resonance in cubic SiC, *Solid State Commun.*, 55, 67, 1985.
8. Kono, J. et al., High-field cyclotron resonance and impurity transition in n-type and p-type 3C-SiC at magnetic fields up to 175 T, *Phys. Rev. B*, 48, 10909, 1993.
9. Volm, D. et al., Determination of the electron effective-mass tensor in 4H SiC, *Phys. Rev. B*, 53, 15409, 1996.
10. Son, N.T. et al., Electron effective masses and mobilities in high-purity 6H-SiC chemical vapor deposition layers, *Appl. Phys. Lett.*, 65, 3209, 1994.
11. Meyer, B.K. et al., Optically detected cyclotron resonance investigations on 4H and 6H SiC: band-structure and transport properties, *Phys. Rev. B*, 61, 4844, 2000.
12. Son, N.T. et al., Hole effective masses in 4H SiC, *Phys. Rev. B*, 61, R10544, 2000.
13. Son, N.T., Hallin, C., and Janzén, E., Hole effective masses in 6H-SiC from optically detected cyclotron resonance, *Phys. Rev. B*, 66, 045304, 2002.
14. Humphreys, R.G., Bimberg, D., and Choyke, W.J., Wavelength modulated absorption in SiC, *Solid State Commun.*, 39, 163, 1981.
15. Choyke, W.J., Devaty, R.P., and Sridhara, S.G., A survey of conduction and valence band edges in SiC, *Phys. Scr.*, T79, 9, 1999.
16. Sridhara, S.G. et al., Differential absorption measurement of valence band splittings in 4H SiC, *Mater. Sci. Forum*, 338–342, 567, 2000.
17. Choyke, W.J., Optical properties of polytypes of SiC: interband absorption, and luminescence of nitrogen-exciton complexes. *Mater. Res. Bull.*, 4, S141, 1969.
18. Bimberg, D., Altarelli, M., and Lipari, N.O., A calculation of valence band masses, exciton and acceptor energies and the ground state properties of the electron-hole liquid in cubic SiC, *Solid State Commun.*, 40, 437, 1981.
19. Ivanov, I.G. et al., Photoconductivity of lightly-doped and semi-insulating 4H-SiC and the free exciton binding energy, *Mater. Sci. Forum*, 389–393, 613, 2002.
20. Dubrovskii, G.B. and Sankin, V.I., Structure of exciton electroabsorption in 4H SiC, *Sov. Phys. Solid State*, 17, 1847, 1976.
21. Ivanov, I.G. et al., Intrinsic photoconductivity of 6H SiC and the free exciton binding energy, *Mater. Sci. Forum*, 353–356, 405, 2001.
22. Egilsson, T. et al., Exciton and defect photoluminescence from SiC, in *Silicon Carbide: Materials, Processing, and Devices*, Feng, Z.C. and Zhao, J.H. (eds.), Taylor & Francis, NewYork and London, 2003, Chap. 3, pp. 81–120.
23. Suttrop, W. et al., Hall effect and infrared absorption measurements on nitrogen donors in 6H-silicon carbide, *J. Appl. Phys.*, 72, 3708, 1992.
24. Ivanov, I.G. et al., Effective-mass approximation for shallow donors in uniaxial indirect band-gap crystals and application to 4H-SiC, *Phys. Rev. B*, 73, 045205, 2006.
25. Moore, W.J. et al., Nitrogen donor excitation spectra in 3C-SiC, *Phys. Rev. B*, 48, 12289, 1993.
26. Faulkner, R.A., Higher donor excited states for prolate-spheroid conduction bands: a reevaluation of silicon and germanium, *Phys. Rev.*, 184, 713, 1969.
27. Chen, C.Q. et al., Photothermal ionization spectroscopy of shallow nitrogen donor states in 4H-SiC, *J. Appl. Phys.*, 87, 3800, 2000.
28. Chen, C.Q. et al., Infrared absorption of 4H silicon carbide, *Appl. Phys. A*, 72, 717, 2001.
29. Ivanov, I.G. et al., Effective-mass theory of shallow donors in 4H-SiC, *Mater. Sci. Forum*, 483–485, 511, 2005.
30. Ivanov, I.G., Magnusson, B., and Janzén, E., Optical selection rules for shallow donors in 4H-SiC and ionization energy of the nitrogen donor at the hexagonal site, *Phys. Rev. B*, 67, 165212, 2003.
31. Ivanov, I.G. and Janzén, E., Theory of the Stark effect on the donor levels in 4H silicon carbide, *Mater. Sci. Forum*, 556–57, 435, 2007.
32. Koster, G.F. et al., *Properties of the Thirty-Two Point Groups*, MIT Press, Cambridge, MA, 1963.
33. Luttinger, J.M. and Kohn, W., Motion of electrons and holes in perturbed periodic fields, *Phys. Rev.*, 97, 869, 1955.

34. Choyke, W.J. and Patrick, L., Luminescence of donor-acceptor pairs in cubic SiC, *Phys. Rev. B*, 2, 4959, 1970.
35. Long, N.N. et al., Line spectrum of donor-acceptor pairs in β–SiC doped with aluminum, *Opt. Spectrosc.*, 29, 388, 1970.
36. Choyke, W.J., Optical and electronic properties of SiC, in *The Physics and Chemistry of Carbides, Nitrides and Borides* (NATO ASI Series, Series E: Applied Sciences, Vol. 185), Robert Freer (ed.), Kluwer Academic Publishers, Dordrecht, 1990, 563.
37. Dean, P.J., Choyke, W.J., and Patrick, L., The location and shape of the conduction band minima in cubic silicon carbide, *J. Lumin.*, 15, 299, 1977.
38. Ikeda, M., Matsunami, H., and Tanaka, T., Site effect on the impurity levels in 4H, 6H, and 15R SiC, *Phys. Rev. B*, 22, 2842, 1980.
39. Ivanov, I.G., Magnusson, B., and Janzén, E., Analysis of the sharp donor-acceptor pair luminescence in 4H-SiC doped with nitrogen and aluminum, *Phys. Rev. B*, 67, 165211, 2003.
40. Ikeda, M., Matsunami, H., and Tanaka, T., Site-dependent donor and acceptor levels in 6H-SiC, *J. Lumin.*, 20, 111, 1979.
41. Suzuki, A., Matsunami, H., and Tanaka, T., Photoluminescence due to Al, Ga, and B acceptors in 4H-, 6H-, and 3C-SiC grown from a Si melt, *J. Electrochem. Soc.*, 124, 241, 1977.
42. Long, N.N. et al., Interimpurity recombination and band-impurity transitions in β–SiC crystals doped with aluminum, *Opt. Spectrosc.*, 30, 165, 1971.
43. Hagen, S.H., van Kemenade, A.W.C., and van der Does de Bye, J.A.W., Donor-acceptor pair spectra in 6H and 4H SiC doped with nitrogen and aluminum, *J. Lumin.*, 8, 18, 1973.
44. Lisitsa, M.P. et al., Low-temperature photoluminescence of α-SiC (6H) single crystals, *Opt. Spectrosc.*, 28, 264, 1970.
45. Kuwabara, H. and Yamada, S., Free-to-bound transitions in β–SiC doped with boron, *Phys. Status Solidi A*, 30, 739, 1975.
46. Kuwabara, H., Yamada, S., and Tsunekawa, S., Radiative recombination in β–SiC doped with boron, *J. Lumin.*, 12/13, 531, 1976.
47. Ivanov, I.G., Henry, A., and Janzén, E., Ionization energies of phosphorus and nitrogen donors and aluminum acceptors in 4H silicon carbide from the donor-acceptor pair emission, *Phys. Rev. B*, 71, 241201(R), 2005.
48. Son, N.T. et al., Electron paramagnetic resonance and theoretical studies of shallow phosphorous centers in 3C-, 4H-, and 6H-SiC, *Phys. Rev. B*, 73, 075201, 2006.
49. Colwell, P.J. and Klein, M.V., Raman scattering from electronic excitations in n-type silicon carbide, *Phys. Rev. B*, 6, 498, 1972.
50. Nakashima, S. and Harima, H., Raman investigation of SiC polytypes, *Phys. Status Solidi A*, 162, 39, 1997.
51. Püsche, R. et al., Electronic Raman studies of shallow donors in silicon carbide, *Mater. Sci. Forum*, 527–529, 579, 2006.
52. Greulich-Weber, S., EPR and ENDOR investigations of shallow impurities in SiC polytypes, *Phys. Status Solidi A*, 162, 95, 1997.
53. Matsumoto, T. et al., Electronic structure of the shallow boron acceptor in 6H-SiC: a pulsed EPR/ENDOR study at 95 GHz, *Phys. Rev. B*, 55, 2219, 1997.
54. van Duijn-Arnold, A. et al., Spatial distribution of the electronic wave function of the shallow boron acceptor in 4H- and 6H-SiC, *Phys. Rev. B*, 60, 15829, 1999.
55. van Duijin-Arnold, A. et al., Electronic structure of the N donor center in 4H-SiC and 6H-SiC, *Phys. Rev. B*, 64, 085206, 2001.
56. Son, N.T. et al., Hyperfine interaction of the nitrogen donor in 4H-SiC, *Phys. Rev. B*, 70, 193207, 2004.
57. Veinger, A.I. et al., Paramagnetic and electrical properties of transmutation-generated phosphorus impurity in 6H-SiC, *Sov. Phys. Solid State*, 28, 917, 1986.
58. Kalabukhova, E.N., Lukin, S.N., and Mokhov, E.N., Electron spin resonance in 2 mm range of transmutation phosphorus impurity in 6H-SiC, *Phys. Solid State*, 35, 361, 1993.
59. Greulich-Weber, S. et al., On the microscopic structures of shallow donors in 6H SiC: Studies with EPR and ENDOR, *Solid State Commun.*, 93, 393, 1995.

60. Baranov, P.G. et al., EPR study of shallow and deep phosphorous centers in 6H-SiC, *Phys. Rev. B*, 66, 165206, 2002.
61. Pinheiro, M.V.B., Greulich-Weber, S., and Spaeth, J.M., Magnetic resonance investigation on P-doped 6H-SiC, *Physica B*, 340–342, 146, 2003.
62. Bockstedte, M., Mattausch, A., and Pankratov, O., Solubility of nitrogen and phosphorus in 4H-SiC: a theoretical study, *Appl. Phys. Lett.*, 85, 58, 2004.
63. Hornos, T. et al., Doping of phosphorus in chemical-vapor-deposited silicon carbide layers: a theoretical study, *Appl. Phys. Lett.*, 87, 212114, 2005.
64. Yan, F. et al., Evidence for phosphorus on carbon and silicon sites in 6H and 4H SiC, *Mater. Sci. Forum*, 527–529, 585, 2006.
65. Rauls, E. et al., Reassignment of phosphorus-related donors in SiC, *Phys. Rev. B*, 70, 085202, 2004.
66. Isoya, J. et al., Shallow P donors in 3C-, 4H-, and 6H-SiC, *Mater. Sci. Forum*, 527–529, 593, 2006.
67. Choyke, W.J., Hamilton, D.R., and Patrick, L., Optical properties of cubic SiC: luminescence of nitrogen-exciton complexes, and interband absorption, *Phys. Rev.*, 133, A1163, 1964.
68. Patrick, L., Choyke, W.J., and Hamilton, D.R., Luminescence of 4H SiC, and location of conduction-band minima in SiC polytypes, *Phys. Rev.*, 137, A1515, 1965.
69. Choyke, W.J. and Patrick, L., Exciton recombination radiation and phonon spectrum of 6H SiC, *Phys. Rev.*, 127, 1868, 1962.
70. Devaty, R.P. and Choyke, W.J., Optical characterization of silicon carbide polytypes, *Phys. Status Solidi A*, 162, 5, 1997.
71. Ivanov, I.G. et al., Phonon replicas at the M point in 4H-SiC: a theoretical and experimental study, *Phys. Rev. B*, 58, 13634, 1998.
72. Henry, A. et al., Determination of nitrogen doping concentration in doped 4H-SiC epilayers by low temperature photoluminescence, *Phys. Scr.*, 72, 254, 2005.
73. Sridhara, S.G. et al., Phosphorus four particle donor bound exciton complex in 6H SiC, *Mater. Sci. Forum*, 264–268, 465, 1998.
74. Henry, A. and Janzén, E., Photoluminescence of phosphorus doped SiC, *Mater. Sci. Forum*, 527–529, 589, 2006.
75. Clemen, L.L. et al., Aluminum acceptor four particle bound exciton complex in 4H, 6H, and 3C SiC, *Appl. Phys. Lett.*, 62, 2953, 1993.
76. Devaty, R.P. et al., Neutral aluminum and gallium four particle complexes in silicon carbide polytypes, *Mater. Sci. Eng., B*, 61–62, 187, 1999.
77. Henry, A. et al., Ga-bound excitons in 3C-, 4H-, and 6H-SiC, *Phys. Rev. B*, 53, 13503, 1996.
78. Pedersen, H. et al., Growth and photoluminescence study of aluminium doped SiC epitaxial layers, *Mater. Sci. Forum*, 556–557, 97, 2007.
79. Sridhara, S.G. et al., Photoluminescence and transport studies of boron in 4H SiC, *J. Appl. Phys.*, 83, 7909 1998.
80. Lomakina, G.A., Electrical properties of hexagonal SiC with N and B impurities, *Sov. Phys. Solid State*, 7, 475, 1965.
81. Troffer, T. et al., Boron-related defect centers in 4H silicon carbide, *Inst. Phys. Conf. Ser.*, 142, 281, 1996.
82. Henry, A. et al., The 3838 Å photoluminescence line in 4H-SiC, *J. Appl. Phys.*, 94, 2901, 2003.
83. Henry, A. et al., Characterization of bulk and epitaxial SiC material using photoluminescence spectroscopy, *Mater. Sci. Forum*, 389–393, 593, 2002.
84. Gali, A. et al., Boron-vacancy complex in SiC, *Phys. Rev. B*, 60, 10620, 1999.
85. Bockstedte, M., Mattausch, A., and Pankratov, O., Different roles of carbon and silicon interstitials in the interstitial-mediated boron diffusion in SiC, *Phys. Rev. B*, 70, 115203, 2004.
86. Aradi, B. et al., Boron centers in 4H-SiC, *Mater. Sci. Forum*, 353–356, 455, 2001.
87. Gali, A. et al., Activation of shallow boron acceptor in C/B coimplanted silicon carbide: a theoretical study, *Appl. Phys. Lett.*, 86, 102108, 2005.
88. Yamada, S. and Kuwabara, S., Photoluminescence of β-SiC doped with boron and nitrogen, in *Silicon Carbide 1973*, Marshall, R.C., Faust, J.W., and Ryan, C.E., (eds.), University of South Carolina Press, Columbia, 1974, 305.
89. Zywietz, A., Furthmüller, J., and Bechstedt, F., Vacancies in SiC: influence of Jahn-Teller distortions, spin effects, and crystal structure, *Phys. Rev. B*, 59, 15166, 1999.

90. Torpo, L. et al., Comprehensive ab initio study of properties of monovacancies and antisites in 4H-SiC, *J. Phys.: Condens. Matter*, 13, 6203, 2001.
91. Aradi, B. et al., Ab initio density-functional supercell calculations of hydrogen defects in cubic SiC, *Phys. Rev. B*, 63, 245202, 2001.
92. Bockstedte, M., Mattausch, A., and Pankratov, O., Ab initio study of the migration of intrinsic defects in 3C-SiC, *Phys. Rev. B*, 68, 205201, 2003.
93. Castleton, C.W.M., Höglund, A., and Mirbt, S., Managing the supercell approximation for charged defects in semiconductors: finite-size scaling, charge correction factors, the band-gap problem, and the ab initio dielectric constant, *Phys. Rev. B*, 73, 035215, 2006.
94. Umeda, T. et al., EPR and theoretical studies of positively charged carbon vacancy in 4H-SiC, *Phys. Rev. B*, 70, 235212, 2004.
95. Bratus', V.Ya. et al., Positively charged carbon vacancy in three inequivalent lattice sites of 6H-SiC: combined EPR and density functional theory study, *Phys. Rev. B*, 71, 125202, 2005.
96. Bockstedte, M., Heid, M., and Pankratov, O., Signature of intrinsic defects in SiC: ab initio calculations of hyperfine tensors, *Phys. Rev. B*, 67, 193102, 2003.
97. Son, N.T., Hai, P.N., and Janzén, E., Carbon vacancy-related defect in 4H and 6H SiC, *Phys. Rev. B*, 63, 201201(R), 2001.
98. Bratus', V.Ya. et al., Positively charged carbon vacancy in 6H-SiC: EPR study, *Physica B*, 308–310, 621, 2001.
99. Zvanut, M.E. and Konovalov, V.V., The level position of a deep intrinsic defect in 4H-SiC studied by photoinduced electron paramagnetic resonance, *Appl. Phys. Lett.*, 80, 410, 2002.
100. Son, N.T. et al., Defects in semi-insulating SiC substrates, *Mater. Sci. Forum*, 433–436, 45–50, 2003.
101. Son, N.T. et al., Defects in high-purity semi-insulating SiC, *Mater. Sci. Forum*, 457–460, 437, 2004.
102. Son, N.T. et al., Characterization of semi-insulating SiC, *Mater. Res. Soc. Symp. Proc.*, 911, 201, 2006.
103. Son, N.T. et al., Intrinsic defects in semi-insulating SiC: deep levels and their roles in carrier compensation, *Mater. Sci. Forum*, 556–57, 465, 2007.
104. Son, N.T., Magnusson, B., and Janzén, E., Photoexcitation-electron-paramagnetic resonance studies of the carbon vacancy in 4H-SiC, *Appl. Phys. Lett.*, 81, 3945, 2002.
105. Son, N.T. et al., Electronic structure of deep defects in silicon carbide, in *Recent Major Advances in SiC*, Choyke, J.W., Matsunami, H., and Pensl, G. (eds.), Springer-Verlag, Berlin Heidelberg, 2004, 461.
106. Son, N.T., Hai, P.N., and Janzén, E., Intrinsic defects in silicon carbide polytypes, *Mater. Sci. Forum*, 353–356, 499, 2001.
107. Umeda, T. et al., EPR and theoretical studies of negatively charged carbon vacancy in 4H-SiC, *Phys. Rev. B*, 71, 193202, 2005.
108. Storasta, L. et al., Deep levels created by low energy electron irradiation in 4H-SiC, *J. Appl. Phys.*, 96, 4909, 2004.
109. Hemmingsson, C. et al., Deep-level defects in electron-irradiated 4H SiC epitaxial layers, *J. Appl. Phys.*, 81, 6155, 1997.
110. Kimoto, T. et al., Growth and electrical characterization of SiC epilayers, *Mater. Sci. Forum*, 556–57, 35, 2007.
111. Torpo, L. et al., Silicon vacancy in SiC: a high-spin state defect, *Appl. Phys. Lett.*, 74, 221, 1999.
112. Bockstedte, M. et al., Identification and annealing of common intrinsic defect centers, *Mater. Sci. Forum*, 433–436, 471, 2003.
113. Deák, P. et al., The spin state of the neutral silicon vacancy in 3C-SiC, *Appl. Phys. Lett.*, 75, 2103, 1999.
114. Zywietz, A., Furthmüller, J., and Bechstedt, F., Spin state of vacancies: from magnetic Jahn-Teller distortions to multiplets, *Phys. Rev. B*, 62, 6854, 2000.
115. Wagner, Mt. et al., Electronic structure of the neutral silicon vacancy in 4H and 6H SiC, *Phys. Rev. B*, 62, 16555, 2000.
116. Sörman, E. et al., Silicon vacancy related defect in 4H and 6H SiC, *Phys. Rev. B*, 61, 2613, 2000.
117. Wagner, Mt. et al., Ligand hyperfine interaction at the neutral silicon vacancy in 4H- and 6H-SiC, *Phys. Rev. B*, 66, 155214, 2002.
118. Orlinski, S.B. et al., Silicon and carbon vacancies in neutron-irradiated SiC: a high-field electron paramagnetic resonance study, *Phys. Rev. B*, 67, 125207, 2003.

119. Son, N.T., Zolnai, Z., and Janzén, E., Silicon vacancy related TV2a center in 4H-SiC, *Phys. Rev. B*, 68, 205211, 2003.
120. Mizuochi, N. et al., Continuous-wave and pulsed EPR study of the negatively charged silicon vacancy with $S = 3/2$ and C3v symmetry in n-type 4H-SiC, *Phys. Rev. B*, 66, 235202, 2002.
121. Mizuochi, N. et al., Spin multiplicity and charge state of a silicon vacancy TV2a in 4H-SiC determined by pulsed ENDOR, *Phys. Rev. B*, 72, 235208, 2005.
122. Wimbauer, T. et al., Negatively charged Si vacancy in 4H SiC: a comparison between theory and experiment, *Phys. Rev. B*, 56, 7384, 1997.
123. Itoh, H. et al., Intrinsic defects in cubic silicon carbide, *Phys. Status Solidi A*, 162, 173, 1997.
124. von Bardeleben, H.J. et al., Vacancy defects in p-type 6H-SiC created by low-energy electron irradiation, *Phys. Rev. B*, 62, 10841, 2000.
125. Umeda, T. et al., Identification of the carbon antisite-vacancy pair in 4H-SiC, *Phys. Rev. Lett.*, 96, 145501, 2006.
126. Rauls, E. et al., Theoretical study of vacancy diffusion and vacancy-assisted clustering of antisites in SiC, *Phys. Rev. B*, 68, 155208, 2003.
127. Gali, A. et al., Point defects and their aggregation in silicon carbide, *Mater. Sci. Forum*, 556–557, 439, 2007.
128. Umeda, T. et al., Electron paramagnetic resonance study of carbon antisite-vacancy pair in p-type 4H-SiC, *Mater. Sci. Forum*, 556–557, 453, 2007.
129. Umeda, T. et al, Identification of positively charged carbon antisite-vacancy pairs in 4H-SiC, *Phys. Rev. B*, 75, 245202, 2007.
130. Torpo, L., Staab, T.E.M., and Nieminen, R.M., Divacancy in 3C- and 4H-SiC: an extremely stable defect, *Phys. Rev. B*, 65, 085202, 2002.
131. Gali, A. et al., Divacancy and its identification: theory, *Mater. Sci. Forum*, 527–529, 523, 2006.
132. Vainer, V.S. and Il'in, V.A., Electron spin resonance of exchange-coupled vacancy pairs in hexagonal silicon carbide, *Fiz. Tverd. Tela*, 23, 3659, 1981; *Sov. Phys. Solid State*, 23, 2126, 1981.
133. Son, N.T. et al., Divacancy in 4H-SiC, *Phys. Rev. Lett.*, 96, 055501, 2006.
134. Ilyin, I.V. et al., Evidence of the ground triplet state of silicon-carbon divacancies, P6, P7 centers in 6H-SiC: an EPR study, *Mater. Sci. Forum*, 527–529, 535, 2006.
135. Lingner, Th. et al., Structure of the silicon vacancy in 6H-SiC after annealing identified as the carbon vacancy–carbon antisite pair, *Phys. Rev. B*, 64, 245212, 2001.
136. Magnusson, B. and Janzén, E., Optical characterization of deep level defects in SiC, *Mater. Sci. Forum*, 483–485, 341, 2005.
137. Son, N.T. et al., Optically detected magnetic resonance studies of intrinsic defects in 6H SiC, *Semicond. Sci. Technol.*, 14, 1141, 1999.
138. Son, N.T. et al., Defects and carrier compensation in semi-insulating 4H-SiC substrates, *Phys. Rev. B*, 75, 155204, 2007.
139. Pinheiro, M.V.B. et al., Silicon vacancy annealing and D_I luminescence in 6H-SiC, *Phys. Rev. B*, 70, 245204, 2004.
140. Müller, St.G. et al., Sublimation-grown semi-insulating SiC for high frequency devices, *Mater. Sci. Forum*, 433–436, 39, 2003.
141. Jenny, J.R. et al., Development of large diameter high-purity semi-insulating 4H-SiC wafers for microwave devices, *Mater. Sci. Forum*, 457–460, 35, 2004.
142. Ellison, A. et al., HTCVD growth of semi-insulating 4H-SiC crystals with low defect density, *Mater. Res. Soc. Symp.*, 640, H1.2, 2001.
143. Ellison, A. et al., HTCVD grown semi-insulating SiC substrates, *Mater. Sci. Forum*, 433–436, 33, 2003.
144. Ellison, A. et al., SiC crystal growth by HTCVD, *Mater. Sci. Forum*, 457–460, 9, 2004.
145. Carlos, W.E., Glaser, E.R., and Shanabrook, B.V., Optical and magnetic resonance signatures of deep levels in semi-insulating 4H SiC, *Physica B*, 340–342, 151, 2003.
146. Mattausch, A., Bockstedte, M., and Pankratov, O., Self diffusion in SiC: the role of intrinsic point defects, *Mater. Sci. Forum*, 353–356, 323, 2001.
147. Ellison, A. et al., Fast SiC epitaxial growth in a chimney CVD reactor and HTCVD crystal growth developments, *Mater. Sci. Forum*, 338–342, 131, 2000.

148. Mitchel, W.C. et al., Vanadium-free semi-insulating 4H-SiC substrates, *Mater. Sci. Forum*, 338–342, 21, 2000.
149. Janzén, E. et al., Defects in SiC, *Physica B*, 340–342, 15, 2003.
150. Gällström, A. et al.; The electronic structure of the UD-4 defect in 4H, 6H and 15R SiC, in *Proceedings of the International Conference of Silicon Carbide and Related Material 2007*, to be published.
151. Magnusson, B., Ellison, A., and Janzén, E., Properties of the UD-1 deep-level center in 4H-SiC, *Mater. Sci. Forum*, 389–393, 505, 2002.
152. Magnusson, B. et al., Optical studies of deep centers in semi-insulating SiC, *Mater. Sci. Forum*, 527–529, 455, 2006.
153. Magnusson, B. et al., Deep-level luminescence at 1.0 eV in 6H SiC, *Mater. Res. Soc. Symp.*, 640, H7.11.1, 2001.
154. Magnusson, B. et al., Infrared absorption and annealing behavior of semi-insulating 4H SiC HTCVD substrates, *Mater. Res. Soc. Symp. Proc.*, 680, 172, 2001.
155. Magnusson, B. et al., As-grown and process-induced intrinsic deep-level luminescence in 4H SiC, *Mater. Sci. Forum*, 353–356, 365, 2001.
156. Son, N.T. et al., Identification of divacancies in 4H-SiC, *Physica B*, 376–377, 334, 2006.
157. Gorban, I.S. and Slobodyanyuk, A.V., Luminescence of excitons localized at donor–acceptor dipoles in α SiC (6H), *Sov. Phys. Solid State*, 15, 548, 1973.
158. Gorban, I.S. and Slobodyanyuk, A.V., Infrared luminescence and energy levels of deep centers in silicon carbide, *Sov. Phys. Semicond.*, 10, 668, 1976.
159. Hagen, S.H. and van Kemenade, A.W.C, Infrared luminescence in silicon carbide, *J. Lumin.*, 9, 9, 1974.
160. Wagner, Mt. et al., UD-3 defect in 4H, 6H, and 15R SiC: electronic structure and phonon coupling, *Phys. Rev. B*, 66, 115204, 2002.
161. Thuaire, A. et al., Investigation of the electronic structure of the UD-4 defect in 4H-SiC by optical techniques, *Mater. Sci. Forum*, 527–529, 461, 2006.
162. Steeds, J.W. et al., Transmission electron microscope radiation damage of 4H and 6H SiC studied by photoluminescence spectroscopy, *Diamond Relat. Mater.*, 11, 1923, 2002.
163. Aavikko, R. et al., Clustering of vacancy defects in high-purity semi-insulating SiC, *Phys. Rev. B*, 75, 085208, 2007.
164. Carlos, W.E. et al, Annealing of multivacancy defects in 4H-SiC, *Phys. Rev. B*, 74, 235201, 2006.
165. Zvanut, M.E. et al., Observation of a spin one native defect in as-grown high-purity semi-insulating 4H SiC, *J. Appl. Phys.*, 97, 123509, 2005.
166. Carlos, W.E. et al., Thermal evolution of defects in semi-insulating 4H SiC, *Mater. Sci. Forum*, 527–529, 531, 2006.
167. Gerstmann, U., Rauls, E., and Overhof, H., Annealing of vacancy-related defects in semi-insulating SiC, *Phys. Rev. B*, 70, 201204(R), 2004.
168. Choyke, W.J., Hamilton, D.R., and Patrick, L., Polarized edge emission of SiC, *Phys. Rev.*, 117, 1430, 1960.
169. Achtziger, N. and Witthuhn, W., Radiotracer deep level transient spectroscopy, in *Silicon Carbide: Recent Major Advances*, Choyke, W.J., Matsunami, H., and Pensl, G. (eds.), Springer-Verlag, Berlin Heidelberg, 2004, 537.
170. Maier, K., Müller, H.D., and Schneider, J., Transition metals in silicon carbide, SiC: vanadium and titanium, *Mater. Sci. Forum*, 83–87, 1183, 1992.
171. Kunzer, M., Müller, H.D., and Kaufmann, U., Magnetic circular dichroism and site-selective optically detected magnetic resonance of the deep amphoteric vanadium impurity in 6H-SiC, *Phys. Rev. B*, 48, 10846, 1993.
172. Dombrowski, K.F. et al., Deep donor state of vanadium in cubic silicon carbide (3C-SiC), *Appl. Phys. Lett.*, 65, 1811, 1994.
173. Dombrowski, K.F. et al., Identification of the neutral V4 + impurity in cubic 3C-SiC by electron-spin resonance and optically detected magnetic resonance, *Phys. Rev. B*, 50, 18034, 1994.
174. Feege, M., Greulich-Weber, S., and Spaeth, J.M., Observation of an interstitial manganese impurity in 6H-SiC, *Semicond. Sci. Technol.*, 8, 1620, 1993.

175. Baranov, P.G., Khramtov, V.A., and Mokhov, E.N., Chromium in silicon carbide: electron paramagnetic resonance studies, *Semicond. Sci. Technol.*, 9, 1340, 1994.
176. Baranov, P.G. and Romanov, N.G., Acceptors in silicon carbide: ODMR data, *Mater. Sci. Forum*, 83–87, 1207, 1992.
177. Kunzer, M. et al., Identification of optical and electrical active molybdenum trace impurities in 6H-SiC substrates, *Inst. Phys. Conf. Ser.*, 142, 385, 1996.
178. Baur, J., Kunzer, M., and Schneider, J., Transition metals in SiC polytypes as studied by magnetic resonance techniques, *Phys. Status Solidi A*, 162, 153, 1997.
179. van Kemenade, A.W.C. and Hagen, S.H., Proof of the involvement of Ti in the low-temperature ABC luminescence spectrum of 6H SiC, *Solid State Commun.*, 14, 1331, 1974.
180. Henry, A. and Janzén, E., Titanium related luminescence in SiC, *Superlattices Microstruct.*, 40, 328, 2006.
181. Vainer, V.S. et al., Electron spin resonance of (TiN)0 impurity pairs in the 6H polytype of silicon carbide, *Sov. Phys. Solid State*, 28, 201, 1986.
182. Dalibor, T. et al., Deep defect centers in silicon carbide monitored with deep level transient spectroscopy, *Phys. Status Solidi A*, 162, 199, 1997.
183. Suleimanov, Yu.M., Zaharchenko, I., and Ostapenko, S., Luminescence characterization of titanium related defects in 6H-SiC, *Physica B*, 308, 714, 2001.
184. Schneider, J. et al., Infrared spectra and electron spin resonance of vanadium deep level impurities in silicon carbide, *Appl. Phys. Lett.*, 56, 1184, 1990.
185. Magnusson, B. et al., Vanadium-related center in 4H silicon carbide, *Mater. Sci. Forum*, 338–342, 631, 2000.
186. Kaufmann, B., Dörnen, A., and Ham, F.S., Crystal-field model of vanadium in 6H silicon carbide, *Phys. Rev. B*, 55, 13009, 1997.
187. Kunzer, M. et al., Magnetic circular dichroism and electron spin resonance of the A^- acceptor state of vanadium, V^{3+}, in 6H-SiC, *Mater. Sci. Eng.*, B29, 118, 1995.
188. Lauer, V. et al., Electrical and optical characterisation of vanadium in 4H and 6H–SiC, *Mater. Sci. Eng.*, B61–62, 248, 1999.
189. Bickermann, M. et al., Preparation of semi-insulating silicon carbide by vanadium doping during PVT bulk crystal growth, *Mater. Sci. Forum*, 433–436, 51, 2003.
190. Dörnen, A. et al., Optical absorption and Zeeman study of 6H-SiC:Cr, *Mater. Sci. Forum*, 258–263, 697, 1997.
191. Son, N.T. et al., Chromium in 4H and 6H SiC: Photoluminescence and Zeeman studies, *Mater. Sci. Forum*, 264–268, 603, 1998.
192. Son, N.T. et al., Photoluminescence and Zeeman effect in chromium-doped 4H and 6H SiC, *J. Appl. Phys.*, 86, 4348, 1999.
193. Choyke, W.J. et al., Intense erbium-1.54-μm photoluminescence from 2 to 525 K in ion-implanted 4H, 6H, 15R, and 3C SiC, *Appl. Phys. Lett.*, 65, 1668, 1994.
194. Prezzi, D. et al., Optical and electrical properties of vanadium and erbium in 4H-SiC, *Phys. Rev. B*, 69, 193202, 2004.
195. Son, N.T. et al., Optically detected magnetic resonance studies of defects in electron-irradiated 3C SiC layers, *Phys. Rev. B*, 55, 2863, 1997.
196. Gali, A. et al., Calculation of hyperfine constants of defects in 4H-SiC, *Mater. Sci. Forum*, 433–436, 511, 2003.
197. Pinheiro, M.V.B. et al., Carbon related split-interstitials in electron-irradiated n-type 6H-SiC, *Mater. Sci. Forum*, 527–529, 551, 2006.
198. Eberlein, T.A.G. et al., Density functional theory calculation of the D_I optical center in SiC, *Phys. Rev. B*, 74, 144106, 2006.
199. Posselt, M., Gao, F., and Weber, W.J., Atomistic simulations on the thermal stability of the antisite pair in 3C- and 4H-SiC, *Phys Rev. B*, 73, 125206, 2006.
200. Gao, F. et al., Ab initio atomic simulations of antisite pair recovery in cubic silicon carbide, *Appl. Phys. Lett.*, 90, 221915, 2007.
201. Cavallini, A. et al., Deep levels in 4H silicon carbide epilayers induced by neutron irradiation up to 1016 n/cm, *Mater. Res. Soc. Symp. Proc.*, 911, 237, 2006.

202. Danno, K. and Kimoto, T., Deep levels in as-grown and electron-irradiated p-type 4H-SiC, *Mater. Res. Soc. Symp. Proc.*, 911, 247, 2006.
203. Danno, K., Kimoto, T., and Matsunami, H., Midgap levels in both n- and p-type 4H–SiC epilayers investigated by deep level transient spectroscopy, *Appl. Phys. Lett.*, 86, 122104, 2005.
204. Egilsson, T. et al., Photoluminescence of electron-irradiated 4H-SiC, *Phys. Rev. B*, 59, 8008, 1999.
205. Ling, C.C. et al., Electron energy dependence on inducing the photoluminescence lines of 6H-SiC by electron irradiation, *Physica B*, 376–377, 374, 2006.
206. Carlsson, F.H.C., Spectroscopic Studies of Irradiation Induced Defects in SiC, Dissertation No 803, Linköping University, Linköping, 2003.
207. Evans, G.A. et al., Identification of carbon interstitials in electron-irradiated 6H-SiC by use of a ^{13}C enriched specimen, *Phys. Rev. B*, 66, 035204, 2002.
208. Gali, A. et al., Aggregation of carbon interstitials in silicon carbide: A theoretical study, *Phys. Rev. B*, 68, 125201, 2003.
209. Mattausch, A., Bockstedte, M., and Pankratov, O., Structure and vibrational spectra of carbon clusters in SiC, *Phys. Rev. B*, 70, 235211, 2004.
210. Gali, A., Son, N.T., and Janzén, E., Electrical characterization of metastable carbon clusters in SiC: a theoretical study, *Phys. Rev. B*, 73, 033204, 2006.
211. Mattausch, A., et al., Thermally stable carbon-related centers in 6H-SiC: Photoluminescence spectra and microscopic models, *Phys. Rev. B*, 73, 161201(R), 2006.
212. Egilsson, T. et al., Properties of the D_I bound exciton in 4H-SiC, *Phys. Rev. B*, 59, 1956, 1999.
213. Storasta, L. et al., Pseudodonor nature of the D_I defect in 4H-SiC, *Appl. Phys. Lett.*, 78, 46, 2001.
214. Carlsson, F.H.C. et al., Trapped carrier electroluminescence (TraCE)—a novel method for correlating electrical and optical measurements, *Physica B*, 308, 1165, 2001.
215. Fissel, A. et al., On the nature of the D_I center in SiC: A photoluminescence study of layers grown by solid-source molecular-beam epitaxy, *Appl. Phys. Lett.*, 78, 2512, 2001.
216. Gali, A. et al., Correlation between the antisite pair and the D_I center in SiC, *Phys. Rev. B*, 67,155203, 2003.
217. Mattausch, A., Bockstedte, M., and Pankratov, O., Carbon antisite clusters in SiC: A possible pathway to the D_{II} center, *Phys. Rev. B*, 69, 045322, 2004.
218. Choyke, W.J., A review of radiation damage in SiC, *Inst. Phys. Conf. Ser.*, 31, 58, 1977.
219. Egilsson, T. et al., Excitation properties of hydrogen-related photoluminescence in 6H-SiC, *Phys. Rev B*, 62, 7162, 2000.
220. Henry, A. et al., Metastability of a hydrogen-related defect in 6H-SiC, *Mater. Sci. Forum*, 338–342, 651, 2000.
221. Gali, A. et al., Anharmonicity of the C-H stretch mode in SiC: Unambiguous identification of hydrogen-silicon vacancy defect, *Appl. Phys. Lett.*, 80, 237, 2002.
222. Prezzi, D. et al., Hydrogen-related photoluminescent centers in SiC, *Phys. Rev. B*, 70, 205207, 2004.
223. Henry, A., Jansson, M.S., and Janzén, E., Properties of the bound excitons associated to the 3838A line in 4H-SIC and the 4182A line in 6H-SiC, *Mater. Sci. Forum*, 457–460, 549, 2004.
224. Koshka, Y., Optically induced formation of the hydrogen complex responsible for the $4B_0$ luminescence in 4H-SiC, *Appl. Phys. Lett.*, 82, 3260, 2003.
225. Henry, A. et al., Presence of hydrogen in SiC, *Mater. Sci. Forum*, 353–356, 373, 2001.
226. Aradi, B. et al., Impurity-controlled dopant activation—the role of hydrogen in p-type doping of SiC, *Mater. Sci. Forum*, 389–393, 561, 2002.
227. Dalibor, T. et al., Oxygen in silicon carbide: shallow donors and deep acceptors, *Mater. Sci. Eng.*, B61–62, 454, 1999.
228. Klettke, O. et al., Oxygen-related defects centers observed in 4H/6H-SiC epitaxial layers grown under CO_2 ambient, *Mater. Sci. Forum*, 353–356, 459, 2001.
229. Gali, A. et al., Isolated oxygen in 3C- and 4H-SiC: A theoretical study, *Phys. Rev. B*, 66, 125208, 2002.
230. Deák, P. et al., Theoretical studies on defects in SiC, *Mater. Sci. Forum*, 264–268, 279, 1998.
231. Pensl, G. et al., Defect-engineering in SiC by ion implantation and electron irradiation, *Microelectron. Eng.*, 83, 146, 2006.

232. Henry, A. et al., Thick silicon carbide homoepitaxial layers grown by CVD techniques, *Chem. Vap. Deposition*, 12, 475, 2006.
233. Bergman, J.P., Kordina, O., and Janzén, E., Time resolved spectroscopy of defects in SiC, *Phys. Status Solidi A*, 162, 65, 1997.
234. Bergman, J.P. et al., The role of defects on optical and electrical properties of SiC, in *2000 International Semiconducting and Insulating Materials Conference*. SIMC-XI (Cat. No.00CH37046), 283, 2000.
235. Lendenmann, H. et al., Degradation in SiC bipolar devices: sources and consequences of electrically active dislocations in SiC, *Mater. Sci. Forum*, 433–436, 901, 2003.
236. Irmscher, K. et al., Formation and properties of stacking faults in nitrogen-doped 4H-SiC, *Physica B*, 376–377, 338, 2006.
237. Sridhara, S.G. et al., Luminescence from stacking faults in 4H SiC, *Appl. Phys. Lett.*, 79, 3944, 2001.
238. Galeckas, A., Linnros, J., and Pirouz, P., Recombination-induced stacking faults: evidence for a general mechanism in hexagonal SiC, *Phys. Rev. Lett.*, 96, 025502, 2006.
239. Bai, S. et al., Spectra associated with stacking faults in 4H-SiC grown in a hot-wall CVD reactor, *Mater. Sci. Forum*, 389–393, 589, 2002.
240. Izumi, S. et al., Structure on in-grown stacking faults in the 4H-SiC epitaxial layers, *Mater. Sci. Forum*, 483–485, 323, 2005.
241. Camassel, J. and Juillaguet, S., Intensity ratio of the doublet signature of excitons bound to 3C-SiC stacking faults in a 4H-SiC matrix, *Mater. Sci. Forum*, 483–485, 331, 2005.
242. Gorban, I.S. and Mishinova, G.N., Basics of luminescent diagnostics of the dislocation structure of SiC crystals, *Proc. SPIE—Int. Soc. Opt. Eng.*, 3359, 187, 1998.
243. Iwata, H., Stacking Faults in Silicon Carbide, Dissertation No. 817, Linköping University, Linköping, 2003.
244. Lindefelt, U. et al., Stacking faults in 3C-, 4H-, and 6H-SiC polytypes investigated by an ab initio supercell method, *Phys. Rev. B*, 67, 155204, 2003.
245. Iwata, H.P. et al., Ab initio study of 3C inclusions and stacking fault-stacking fault interactions in 6H-SiC, *J. Appl. Phys.*, 94, 4972, 2003.

22

Defects in Gallium Arsenide

J.C. Bourgoin and H.J. von Bardeleben

CONTENTS

22.1 Introduction

GaAs is an important electronic material for high-frequency and optoelectronic applications that are not achievable with Si-based microelectronics. The two main advantages of GaAs over Si are: (a) GaAs has a considerably higher electron mobility, and (b) since GaAs is a direct bandgap material, it allows efficient light emission, a prerequisite for applications in optoelectronics. Both semi-insulating (SI) bulk crystals and highly conductive n-type bulk crystals and n- and p-type epitaxial layers are needed for different applications and can be obtained with specific growth techniques. It is possible to obtain an intrinsic material of high resistivity (larger than 10^6 Ωcm) at room temperature due to the presence of a dominating defect associated with a midgap level that compensates the residual donors and acceptors. In bulk GaAs this defect can be introduced through the control of the growth conditions; it is the so-called EL2 defect, a deep double donor with a first 0/+ ionization state at $E_C - 0.86$ eV. In contrast, high conductivities are required in other applications and can be obtained by doping with Si for n-type or Zn for p-type properties. More recently, it has been shown that Mn doping in the atomic% range adds an additional functionality to this material: it gives rise to the formation of ferromagnetic thin films with interesting fundamental properties.

In this brief review, we describe the present state of knowledge of the electrically active point defects in GaAs. Since their natures and concentrations vary with the growth process, we treat separately the following three categories of crystals: bulk crystals grown from the melt, high-temperature epitaxially grown layers, and low-temperature molecular beam epitaxial (MBE) grown thin films. An extensive list of references on defects in GaAs, GaN, and ZnO materials is provided in Appendix A, as well.

In addition to the specific native defects induced by the growth process, defects generated by particle bombardment and dopant-related defects also have been investigated. In the present state of growth technology, native defects are generally well controlled and purposely used to obtain particular material properties. After extensive research in the 1970s and 1980s, the microscopic and electronic properties of the native and the main intrinsic defects have been reviewed on different occasions [1–4]. Later, further theoretical investigations have shown that some original defect models had to be revised [5]. The electronic properties of the 3d transition metals in GaAs also have been reviewed [6]. New developments in crystal growth following this period were concerned with low-temperature MBE growth, which allowed the development of highly resistive thin epitaxial layers [7] and doping with magnetic ions to obtain ferromagnetic properties [8]. Studies on highly transition-metal doped GaAs are still underway and aimed to the achievement of room temperature ferromagnetism. At the end of 2006, the maximum critical temperature obtained was in the 200 K range.

The principal experimental techniques used for defect detection, characterization, and microscopic identification are deep level transient spectroscopy (DLTS), electron paramagnetic resonance (EPR), optical absorption, and positron annihilation. In some cases the defects seen by these techniques can be correlated, but this has unfortunately not always been possible. One reason is the difference in sensitivity between DLTS and EPR techniques, the latter being limited to the $>10^{15}$ cm^{-3} concentration range for bulk samples and to even higher ones for thin epitaxial layers. On the contrary, DLTS can detect defect concentrations at a level of 1% of the doping level, which corresponds to 10^{13} cm^{-3} in practice. EPR spectroscopy in GaAs is limited by the larger linewidth encountered, as compared to Si or SiC. Thus, superhyperfine interactions, which provide important information for defect identification, are often not resolved in an EPR spectrum. However, these interactions can be assessed by electron nuclear double resonance (ENDOR) spectroscopy, which together with hyperfine structure calculations, has largely contributed to the establishment of defect models [9]. The formation energies and electronic structures of the main intrinsic and extrinsic defects have also been modeled in detail [10]. The theoretical predictions have revealed inconsistencies with previously adopted models for some of the defects, and further investigations are required to solve these questions.

The chapter is organized as follows: the relevant defects in bulk crystals are discussed in Section 22.2. Sections 22.3 and 22.4 deal with the case of standard epitaxial layers and low-temperature MBE grown layers, respectively. In Section 22.5, we discuss the special case of ferromagnetic GaAsMn thin films and multistructures.

22.2 Defects in Bulk Crystals

GaAs bulk crystals are used for high-frequency microelectronics and optoelectronics applications. The respective requirements are different: a highly resistive material in the first case, and heavily n-type doped material in the second case. To obtain SI material, the two dominating native defects in high purity material, the carbon acceptor, and the EL2 donor defect, have to be controlled by the growth conditions [11]. The Czochralski growth technique (LEC) generally uses a boron nitride encapsulation of the melt; it is the main technique for growing SI material. Less frequently, the vertical gradient freeze (VGF) technique is also used. For recent reviews on the bulk growth of GaAs, see refs. [3,4,12,13]. In addition to point defects, extended defects also must be controlled. Due to its lower thermal conductivity as compared to Si and because its bond strength is also weaker, the production of large (>150 mm) diameter ingots with low dislocation densities is more difficult to obtain. The result is that, contrary to the case of Si, large diameter substrates contain still a large concentration of nonuniformly distributed dislocations [14].

The control of the stoichiometry of the ingot is a major issue. Nonstoichiometry leads to the formation of point defects, i.e., vacancy, interstitial, and antisite defects which all introduce electrically and optically active levels. It has been shown that, in particular, both antisite defects, Ga_{As} and As_{Ga}, are important native defects [15–18]. Surprisingly, the isolated arsenic antisite defect is a rather rare defect [19]; instead, As antisite-related complexes $As_{Ga}As_4Ga_{12}$-X of various constituents are generally formed. For this reason the term of the As_{Ga} defect family has been introduced [20]. The isolated As antisite has only been observed in low-temperature electron-irradiated crystals; it transforms in this case when the sample is warmed to room temperature [19]. The As_{Ga} complexes can be distinguished from each other by their different central hyperfine interaction constants [21,22] (Figure 22.1); their different superhyperfine parameters can be determined by ENDOR spectroscopy [23] or their optical response can be assessed via low temperature photoexcitation [16,24]. These complexes also possess different thermal activation energies for electron emission [21,25]. The assignment of these defects to particular models has changed considerably due to improved model calculations. For a recent update, see Ref. [20]. Among the different complexes investigated, only the native As antisite complex EL2 displays optically induced metastability phenomena. Bulk materials may contain a rather large concentration ($\sim 10^{16}$ cm^{-3}) of EL2 defects voluntarily introduced for electrical compensation. In crystals with high dislocation densities, the EL2 defects are not randomly distributed but "decorate" the dislocations, leading to an inhomogeneous distribution and consequently to inhomogeneous electronic properties.

The atomic configuration of the most important native arsenic antisite defect, EL2, has been studied in great detail, but its atomic configuration has not yet been completely established despite considerable efforts over more than 20 years [26]. The most recent high-frequency ENDOR measurements indicate that it is not an isolated As antisite, as had been suggested by theory some years ago [27,28]. This finding is not in contradiction with theory, which has shown that the isolated As_{Ga} defect can already undergo an optically

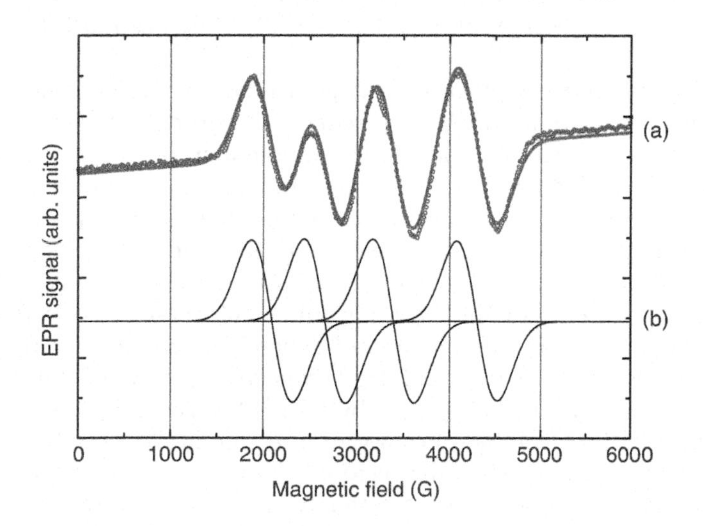

FIGURE 22.1
EPR spectrum of the arsenic antisite-related defect observed in n-type electron irradiated GaAs: (a) points—experimental spectrum, line—simulation by four Gaussian lines; (b) with parameters $\Delta B = 400$ G, $g = 1.97$, $A = 680 \times 10^{-4}$ cm^{-1}. (After von Bardeleben, H.J., Bourgoin, J.C., and Miret, A., *Phys. Rev. B* 34, 1360, 1986. With permission.)

induced metastability transition. The point is that the As antisite-related defect complexes are separated in most cases by distances larger than the first nearest neighbor shell and second, isolated As antisite defects are difficult to generate.

The nonuniform distribution of these defects results in a material containing SI regions and conductive ones [29]. This situation made the production of microelectronic devices (using ion implantation or diffusion) directly on wafers, i.e., on a large area, impossible in practice because of parasitic effects, such as side and back gating effects [30]. The way to overcome this difficulty has been to grow an epitaxial layer on top of a SI wafer. Indeed, epitaxy allows one to obtain layers with uniform electronic properties, nearly free of defects and dislocations, as we shall see in Section 22.3. This has been investigated by comparative photoluminescence mapping on bulk Czochralski grown SI material and on epitaxial layers. Photoluminescence reveals the presence of EL2 defects [31], which are efficient nonradiative recombination centers, distributed along the dislocations. Since much lower dislocation densities can be obtained in epitaxial layers, the SI wafers are only used as a support for the layer that is the electronically active material.

The growth of n-type bulk crystals is generally performed by the Bridgman technique, which proceeds without encapsulation and under controlled As pressure. Once again the As antisite turns out to be the dominant native defect [17]. Its concentration varies with the As pressure under which the crystal is grown and further with the shallow donor dopant concentration (Si, S, Te). Increasing the donor concentration rapidly suppresses the formation of As antisite defects that are formed during the cooling of the melt from high temperature.

22.3 Defects in Standard Epitaxial Layers

There are a number of growth techniques for the growth of thin epitaxial layers: MBE, metallorganic chemical vapor deposition (MOCVD), vapor phase epitaxy (VPE), atomic layer epitaxy (ALE), and, for the growth of thick layers, close space vapor transport (CSVT). All of these have been applied to GaAs, with MOCVD and MBE being the most used ones.

For standard microelectronic purposes, the electronic properties have to be optimized, which implies that the concentrations of the defects, in particular the ones associated with deep levels that act as efficient nonradiative recombination centers, has to be minimized. Indeed, large mobilities and negligible compensation are obtained only in the case of low defect concentration; radiative recombination necessitates the absence of deep levels. Consequently, the growth conditions are adjusted for this purpose. Using MOCVD growth, very low defect concentrations, of the order of 10^{13} cm^{-3}, can be obtained [32]. For a review on MOCVD growth of GaAs, see Ref. [33]. It is only in alloys, such as GaAlAs [34], that the defect concentration is significantly higher and depends on the alloy composition [35].

Given their low concentration (residual defect concentrations are in the 10^{14} cm^{-3} range), these defects can generally only be assessed by DLTS. The As antisite-related EL2 defect is still present [32], but at a level that does not reduce significantly the radiative recombination properties. As a consequence, the residual shallow impurities are no longer sufficiently compensated and highly resistive material cannot be obtained.

The concentration of the intrinsic defects, such as EL2, in epitaxial layers depends on the growth conditions—in particular, the deposition temperature and the growth rate. Indeed, the mobility of the atoms deposited on the surface, as well as the cracking of molecules near the surface, varies strongly with temperature [36]. A detailed study shows that the incorporation of EL2 defects is directly related to the concentration of As_2 molecules on the surface, suggesting that this defect is associated with two As atoms occupying a Ga site [37]. The concentration of intrinsic defects should increase with the growth rate since the time necessary for an atom to find a regular site depends on the time to grow one additional atomic layer. The concentrations of intrinsic defects as well as of impurities increase with increasing growth rate [38]. An attempt to obtain SI layers by increasing the concentration of EL2 defects using the growth rate has been made with some success [39].

22.4 Defects in Low-Temperature MBE Grown Layers

Although high-quality epitaxial layers with residual defect concentrations below 10^{14} cm^{-3} can be grown by MBE at 600°C [40], a lowering of the growth temperature in the 200°C–300°C range leads to a quite different material with a high concentration of intrinsic defects. Defect densities higher than 10^{20} cm^{-3} have been reported [41,42]. The interest to grow such layers lies in their modified electrical properties (SI) and the very short (<ps) photocarrier lifetimes. The initial state of such layers grown in the 250°C range is quasi metastable and can be modified by annealing at temperatures as low as 300°C. Both undoped and heavily Si and Be low-temperature MBE (LTMBE) doped layers have been studied. For given growth conditions, the doping influences the concentration of the native intrinsic defects. The two main intrinsic defects detected were the As antisite [41,43] and the Ga vacancy defects [44]. These defects are related to the As-rich growth conditions generally applied. Their concentrations can be determined by EPR (As_{Ga}^{+}) and optical absorption (As_{Ga}^{+}, As_{Ga}^{0}); concentrations of about 10^{19} cm^{-3} are typical [43,45]. For recent reviews on native defects in LTMBE, see Bliss et al. [44], Pritchard et al. [46], and Stellmacher et al. [47]. Hydrogen passivation of dopants (Si, Be) and native defects in LTMBE films also has equally investigated [46].

The layers are often highly nonstoichiometric and may contain excess As up to 1% (18) in the form of As antisites, As interstitials, and As precipitates (Figure 22.2) [48].

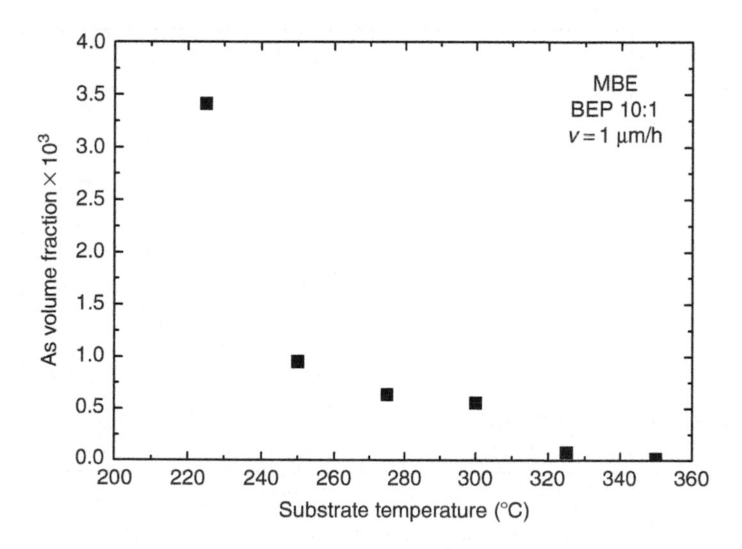

FIGURE 22.2
Arsenic precipitate volume fraction in LTMBE GaAs as a function of the growth temperature; growth conditions BEP 10:1 and growth rate 1 µm/h. (From Melloch, M.R. et al., *MRS Symposium Proceedings*, Materials Research Society, Pittsburgh, 1992. With permission.)

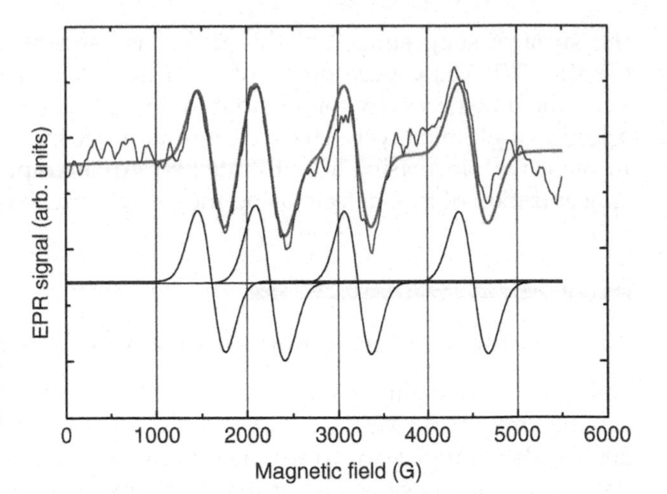

FIGURE 22.3
EPR spectrum of the arsenic antisite-related defect As_{Ga}^{+} in LTMBE GaAs : points—experimental spectrum, line—simulation by a sum of four individual Gaussian lines; line width $\Delta B = 297$ G, $g = 2.04$, $A = 866 \times 10^{-4}$ cm^{-1}.

The deviation from stoichiometry has been evaluated by Auger spectroscopy and energy dispersive x-ray emission (EDX). It has been shown that Auger measurements may overestimate the As excess by a factor of 10 due to selective sputtering and different surface and bulk compositions. The neutral As antisite concentration has been measured by infrared (IR) absorption, which had been previously calibrated for the $EL2^0$ defect in SI bulk material [15,16]. However, the native antisite (Figure 22.3) in LTMBE GaAs is different from the EL2 defect. Interestingly, the high defect concentration leads to a change of the lattice constants (Figure 22.4), which increase up to $\Delta a/a = 14 \times 10^{-4}$. Both the presence of As antisites and As interstitials will lead to lattice expansion [46]. Correlation between observations made by x-ray diffraction (XRD), IR absorption, Hall effect, electron diffraction, and Auger spectroscopy allow one to conclude reasonably well that the As-related defects contain two As atoms [47]. In materials with very high defect concentrations, these antisite defects are not stable; arsenic precipitates are formed following annealing

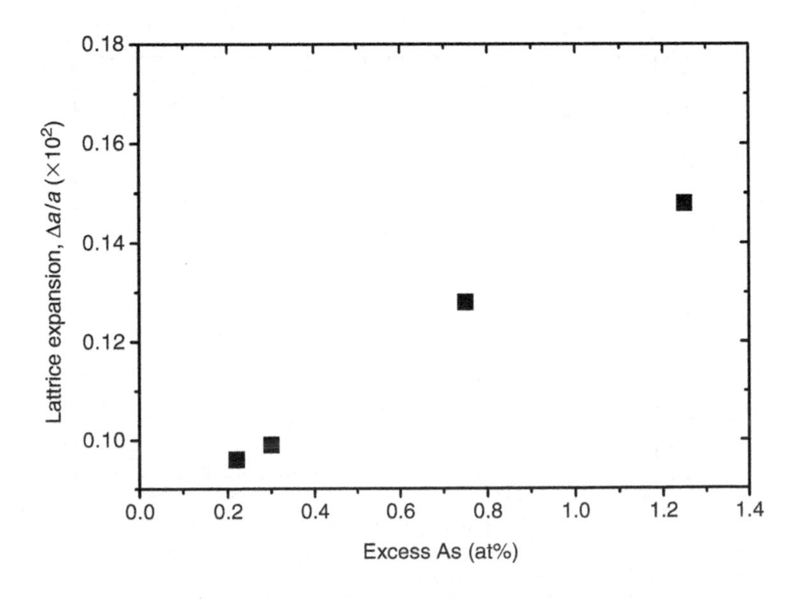

FIGURE 22.4
Lattice expansion $\Delta a/a$ as a function of excess arsenic in LTMBE GaAs. (After Liliental-Weber, Z., *MRS Symposium Proceedings*, Materials Research Society, Pittsburgh, 1992. With permission.)

temperatures above 200°C. The concentration and size of these precipitates depend strongly on the temperature [48–50]. A detailed analysis of the kinetics associated with the growth of the precipitates show that their growth is governed by the mobility of interstitial As [51].

Conduction in LTMBE layers occurs via hopping in the impurity band associated with EL2 [42,52]. It is characterized by a percolation regime around insulating regions. The measurement of the associated relaxation time allows an evaluation of the average distance between these regions and, by comparison with the concentration of As precipitates, demonstrates that depleted zones are formed around these precipitates [48,53].

Because As antisites are efficient recombination centers, with concentrations that can be adjusted through the growth conditions, layers containing a given concentration can be used to produce photodetectors exhibiting a specific response time. Fast photodetectors require large As antisite concentrations, which can be obtained by decreasing drastically the temperature at which the growth is performed. Detailed studies have been made to master the introduction of As antisite defects, to investigate the thermal stability, to correlate its concentration with the minority carrier lifetime, to measure the breakdown voltage, and to understand the mechanism of conduction; these have been discussed in different conferences devoted to this material [54–56].

Thermal annealing allows one to modify strongly the properties of LTMBE GaAs layers. As the high defect concentration observed in the as-grown samples are close to their solubility limit, they can be strongly reduced by post-growth thermal annealing in the 300°C–600°C range. The annealing is accompanied by a sharp rise of the electrical resistivity, by orders of magnitude. It is only the 600°C annealed material that displays high resistivities in the 10^6 Ωcm range [43,44]. As mentioned above, the annealing modifies also the size and distribution of the native arsenic precipitates [48].

22.5 Defects in Ferromagnetic Mn-Doped Low-Temperature MBE Grown Layers

Since the reports by Ohno et al. [59] and Dietl et al. [60] of ferromagnetism in Mn-doped GaAs, numerous experimental studies and theoretical studies, demonstrating the carrier-induced origin of the ferromagnetism, have been performed. Detailed reviews of this subject are already available [61–63]. The key to ferromagnetic properties is a Mn doping in the $\geq 5\%$ range which, under normal growth conditions, is above the solubility limit in GaAs. It has been shown that Mn doping at $x = 0.05$ can be achieved by LTMBE growth, and that it does indeed lead to a long-range ferromagnetic phase (Figure 22.5). Contrary to the conventional metallic ferromagnets, the magnetic properties in the diluted magnetic semiconductor GaMnAs are dominated by the magnetocrystalline anisotropies and not by the demagnetization fields. This allows an engineering of different material properties as the anisotropies can be modified by heteroepitaxy-induced strain.

Ferromagnetic GaMnAs layers with a thickness between 5 and 300 nm can now reproducibly be grown. Their magnetic properties can be studied easily by ferromagnetic resonance spectroscopy (FMR); see Figure 22.6. The critical temperatures of the as-grown layers are still low, in the 50 K range, but they can be significantly increased by post-deposition annealing to the >100 K range, and a highest temperature of 170 K has been reported for optimally annealed samples [64]. A typical sample structure consists of a bulk GaAs substrate, an undoped LTMBE GaAs buffer layer, followed by a 50 nm thick GaMnAs layer. Due to the lattice mismatch between the GaMnAs layer and the undoped LTMBE GaAs layer, the GaMnAs layer is biaxially strained. Even though this strain is

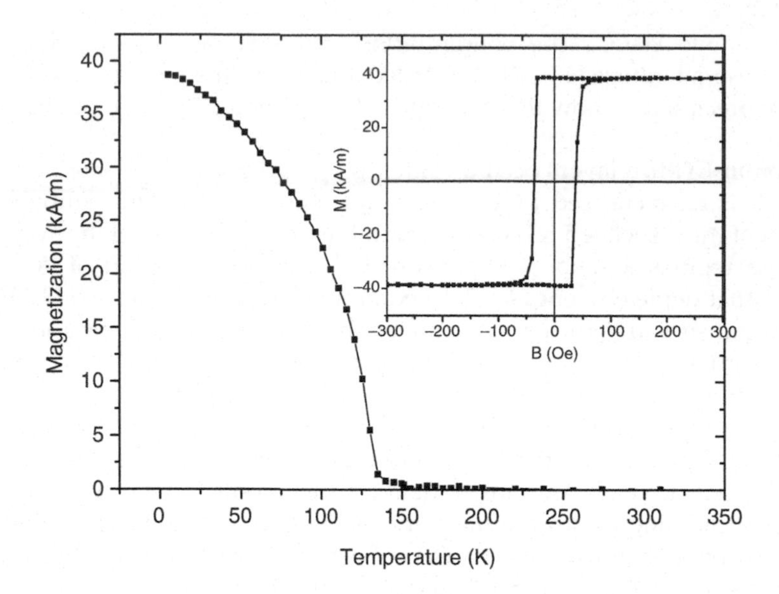

FIGURE 22.5
Temperature dependence of the static magnetization of an annealed 50 nm thick $Ga_{0.93}$ $Mn_{0.07}$ As epitaxial layer grown by LTMBE on GaAs. The insert shows the hysteresis cycle observed at $T = 4$ K.

small (10^{-4}), it is sufficient to determine and modify the orientation of the easy and hard axes of magnetization [65]. If the buffer layer is changed, from GaAs to GaInAs, for example, the strain is reversed from compressive to tensile and the magnetocrystalline anisotropy constants are drastically changed [65]. In addition to this basic GaMnAs/GaAs structure, more complex structures such as GaMnAs/GaInAs/GaMnAs magnetic multilayers with anisotropic tunneling magnetoresistance [66,67] properties, as well as patterned films, have equally been investigated (Figure 22.5) [68].

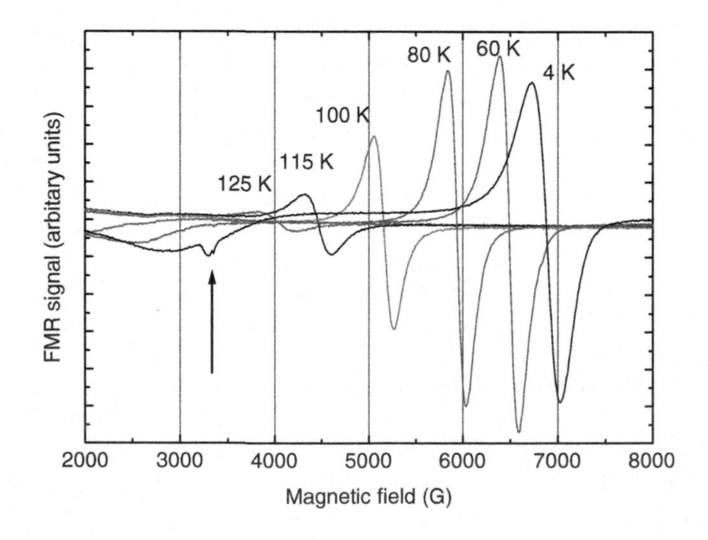

FIGURE 22.6
X-band FMR spectra of a 5 nm thick epitaxial ferromagnetic $Ga_{0.95}Mn_{0.05}As$ layer on (100) GaAs for B//[001] as a function of the measurement temperature. The Curie temperature for this layer is estimated to $T_C = 130$ K. The black arrow indicates the paramagnetic resonance position of the Mn^{2+} center.

The ferromagnetic properties of GaMnAs layers have been modeled in the mean field Zener model [60]. The ferromagnetism is carrier (hole) mediated with an antiferromagnetic interaction between the spin $S = 5/2$ Mn ions and the heavy and light holes, which gives rise to a positive exchange interaction between the Mn ions. The Mn dopant in GaAs shows simultaneous magnetic and electrical activity related to its spin $S = 5/2$ magnetic ground state [69] and the Mn^{2+}, Mn^{2+}-h^+ $-/+$ shallow acceptor states with an acceptor level of $E_v + 0.1$ meV. This is different from the case of the other III–V compounds, such as GaN, where the Mn acceptor is deep and changes its electron configuration with its charge state. In GaN, for example, it will present a $3d^5$ Mn^{2+} $S = 5/2$ ground state, n-type material, and a $3d^2$ $S = 2$ ground state in SI and p-type material. Since the first reports on ferromagnetism in GaMnAs, many details of this system have been studied. The results obtained show the importance of the magnetically active Mn concentration, the strain and the related valence band splitting, the hole concentration, and the role of compensating defects for the observed magnetic properties.

As has been shown in the preceding paragraph, LTMBE layers are characterized by native defect concentrations in the %range, which is not negligible as compared to the dopant concentration. Thus, their influence cannot be neglected in the modeling of the electrical and magnetic properties of the GaMnAs layers. Direct evidence for the effects of defects other than substitutional Mn_{Ga} in the magnetic properties is the low value of the saturation magnetization that might be expected to be just the sums of M_{Mn} and M_{holes}. However, experimental values obtained by SQUID measurements indicate much lower values of about 60% of the Mn doping concentration [70]. Further, this value is significantly modified by postdeposition annealing, as shown above in Figure 22.5.

As expected from previous knowledge on undoped LTMBE layers, the relevant defects in LTMBE GaMnAs layers are As antisites, As interstitials, and to a lesser extent Ga vacancies. Hole traps with activation energies of $E_V + 0.38$ eV, $E_V + 0.43$ eV, and $E_V + 0.65$ eV have been observed by DLTS and attributed to a Ga vacancy-related defect and the As antisite, respectively [71]. The As antisite has also been observed via its optical absorption band at $E = 1.2$ eV [72,73]. These authors [72,73] give an As antisite concentration of 1.3×10^{20} cm^{-3} for $[Mn] = 0.05$, which varies with the Mn doping level. Arsenic antisite defects have also been observed by scanning tunneling spectroscopy [74]. The Ga vacancy-related defects have been observed by slow positron beam spectroscopy. The corresponding defect is believed not to be the isolated Ga vacancy, but a complex with a donor like defect [75]. Its concentration decreases with the Mn doping level, and is only in the 10^{16} cm^{-3} range for $[Mn] > 0.01$.

In addition, it has been shown that only a fraction of the Mn dopant is incorporated as a substitutional electrically and magnetically active defect, and that a fraction of up to 30% is incorporated in the form of Mn interstitials [76,77]. The Mn interstitials are expected to reduce the hole concentration due to their double donor character and to passivate Mn_{Ga} ions by the formation of antiferromagnetic coupled Mn_{Ga}–Mn_i pair defects.

In principle, Mn doping levels higher than $x = 0.05$ would be desirable. In fact, room temperature ferromagnetism has been predicted for $\geq$10% Mn doping. However, up to now, 8% seems to be the limiting value achievable for the current LTMBE growth conditions. Higher doping levels lead to second phase formation and material with degraded properties. The maximum Curie temperature of 170 K is still insufficient for spintronic applications, and defect optimization is evidently required in order to achieve higher critical temperatures. Postdeposition low-temperature annealing in the 200°C range under neutral or As atmospheres of either unprotected layers or capped layers allow an important improvement in all magnetic properties, but the bottleneck of a $T_C < 300$ K has not been overcome. The beneficial effect of the postdeposition thermal annealing is generally ascribed to the out-diffusion of interstitial Mn ions [78]. However, more complex processes involving intrinsic defect modifications may occur in addition.

22.6 Conclusions

The intrinsic, dopant-related, and native point defects in GaAs bulk crystals and epitaxial layers have been studied both experimentally and theoretically in quite detail. Their electronic and microscopic structures have been identified in most cases. The formation of these defects as a function of growth conditions seems also well controlled. Whereas this is true for high purity material with defect concentrations in the range below 10^{16} cm^{-3}, the situation is less clear for the more recently developed thin films with very high defect concentrations ($>10^{19}$ cm^{-3}), generally obtained by LTMBE. The new functionality of ferromagnetism in Mn-doped GaAs can be expected to induce more studies in this field.

References

1. Martin, G.M. and Makram-Ebeid S., in *Deep Centers in Semiconductors*, Pantelides, S.T., Ed., Gordon and Breach, New York, 1986, Chap. 6.
2. Bourgoin, J.C., von Bardeleben, H.J., and Stievenard, D., Native defects in gallium arsenide, *J. Appl. Phys.* 64, R65, 1988.
3. Rudolph, P. and Jurisch, M., Bulk growth of GaAs: An overview, *J. Cryst. Growth* 198/199, 325, 1999.
4. Hurle, D.T.J., A comprehensive thermodynamic analysis of native point defects and dopant solubilities in GaAs, *J. Appl. Phys.* 85, 6957, 1999.
5. Delerue, C., Electronic structure and electron paramagnetic resonance properties of intrinsic defects in GaAs, *Phys. Rev. B* 44, 10535, 1991.
6. Delerue, C., Lannoo, M., and Allan, G., New theoretical approach of transition metal impurities in semiconductors, *Phys. Rev. B* 39, 1669, 1989.
7. Smith, F.W. et al., New MBE buffer used to eliminate backgating in GaAs MESFETS, *IEEE Electron Dev. Lett.* 9, 77, 1988.
8. Ohno, H., Making nonmagnetic semiconductors ferromagnetic, *Science* 281, 4991, 1998.
9. Spaeth, J.M. and Overhof, H., Analysis of ENDOR spectra, in *Point Defects in Semiconductors and Insulators*, Spaeth, J.M. and Overhof, H., Eds., Springer, Berlin, 2003, Chap. 6, p. 197.
10. Baraff, G.A. and Schlüter, M., Binding and formation energies of native defect pairs in GaAs, *Phys. Rev. B* 33, 7346, 1986.
11. Blakemore, J.S., Semiconducting and other major properties of GaAs, *J. Appl. Phys.* 53, R123, 1982.
12. Kirkpatrick, C.G. et al., Compensation mechanism in LEC GaAs, in *Gallium Arsenide: Materials, Devices and Circuits*, Howes, M.J. and Morgan, D.V., Eds., Wiley, New York, 1985, chap. 2.
13. Rudolph, P., Nonstoichiometry related defects at the melt growth of semiconductor compound crystals: A review, *Cryst. Res. Technol.* 38, 542, 2003.
14. Rudolph, P.et al., Studies on dislocation patterning and bunching in semiconducting compound crystals of GaAs, *J. Cryst. Growth* 265, 331, 2004.
15. Martin, G.M. et al., Compensation mechanisms in GaAs, *J. Appl. Phys.* 51, 2840, 1980.
16. Martin, G.M., Optical assessment of the main electron trap in bulk semi-insulating GaAs, *Appl. Phys. Lett.* 39, 747, 1981.
17. Lagowski, J. et al., Origin of the 0.82 eV electron trap and its annihilation by shallow donors, *Appl. Phys. Lett.* 40, 342, 1983.
18. Wagner, R.J. et al., Submillimeter EPR evidence for the As antisite defect in GaAs, *Solid State Commun.* 36, 15, 1980.

19. Krambrock, K. et al., Identification of the isolated arsenic antisite defect in electron irradiated GaAs, *Phys. Rev. B* 45, 1481, 1992.
20. Overhof, H. and Spaeth, J.M., Defect identification in the [As]Ga family in GaAs, *Phys. Rev. B* 72, 11205, 2005.
21. von Bardeleben, H.J., Bourgoin, J.C., and Miret, A., Identification of the arsenic antisite arsenic vacancy defect in electron irradiated GaAs, *Phys. Rev. B* 34, 1360, 1986.
22. von Bardeleben, H.J., Lim, H., and Bourgoin, J.C., Arsenic antisite formation by electron irradiation of n-type GaAs, *J. Phys. C Solid State Phys.* 20, 1353, 1987.
23. Tkach, I. et al., High field ODMR investigation of the EL2 defect in semi-insulating GaAs, *Physica B* 340, 353, 2003.
24. Skowronski, M., Lagowski, J., and Gatos, H.G., Metastability of the midgap level EL2 in GaAs, relationship with the arsenic antisite defect, *Phys. Rev. B* 32, 4264, 1985.
25. Look, D.C. et al., New {As}Ga related center in GaAs, *Phys. Rev. Lett.* 70, 465, 1993.
26. Spaeth, J.M. et al., High field optically detected EPR and ENDOR of semiconductor defects using W band microwave Fabry-Perot resonators, *Mag. Reson. Chem.* 43, S153, 2005.
27. Dabrowski, J. and Scheffler, M., Theoretical evidence for an optically induced structural transition of the isolated antisite in GaAs, *Phys. Rev. Lett.* 60, 2183, 1988.
28. Chadi, D.J. and Chang, K.J., Metastability of the isolated antisite defect in GaAs, *Phys. Rev. Lett.* 60, 2187, 1988.
29. Khirouni, K. and Bourgoin, J.C., Free electron transport in semi-insulating GaAs, *J. Appl. Phys.* 82, 1656, 1997.
30. Koyama, R.Y. et al., Parasitic effects and their impact on gallium arsenide circuits, in *Proc. 5th Conf. on Semi-insulating III–V Materials*, Grossman, G. and Lebedo, M. Eds., 1988, p. 203.
31. Tajima, M., Mapping of EL2 related luminescence on semi-insulating GaAs wafers at room temperature, *Jpn. J. Appl. Phys.* 27, L1323, 1988.
32. Feng, S.L., Bourgoin, J.C., and Razeghi, M., Defects in high purity GaAs grown by low pressure metallorganic chemical vapor deposition, *Semicond. Sci. Technol.* 6, 229, 1991.
33. Ludowise, M.J., Metallorganic chemical vapor deposition of III–V semiconductors, *J. Appl. Phys.* 58, R31, 1985.
34. Feng, S.L., Zazoui, M., and Bourgoin, J.C., Defects in molecular beam epitaxy grown GaAlAs layers, *Appl. Phys. Lett.* 55, 68, 1989.
35. Mejri, H. et al., Photoluminescence study of planar doped AlGaAs, *Semicond. Sci. Technol.* 5, 900, 1990.
36. Gandouzi, M. et al., Antisite incorporation during epitaxial growth of GaAs, *J. Cryst. Growth* 234, 279, 2002.
37. Bourgoin, J.C. et al., As antisite incorporation in epitaxial GaAs, *Physica B* 273, 725, 1999.
38. Samic, H. and Bourgoin, J.C., Defects in thick epitaxial layers, *Mater. Sci. Forum* 258–263, 997, 1997.
39. Castenedo, R., Mimila-Arroyo, J., and Bourgoin, J.C., Semi-insulating epitaxial GaAs, *J. Appl. Phys.* 68, 6274, 1990.
40. Lang, D.V. et al., Study of electron traps in n-GaAs grown by molecular beam epitaxy, *J. Appl. Phys.* 47, 2558, 1976.
41. Kaminska, M. et al., Structural properties of As rich GaAs grown by molecular beam epitaxy at low temperatures, *Appl. Phys. Lett.* 54, 1881, 1989.
42. Look, D.C. et al., Anormalous Hall effect results in low temperature molecular beam epitaxy GaAs: hopping in a dense EL2 like band, *Phys. Rev. B* 42, 3578, 1990.
43. von Bardeleben, H.J. et al., Electron paramagnetic resonance study of GaAs grown by low temperature molecular beam epitaxy, *Phys. Rev. B* 45, 3372, 1992.
44. Bliss, D.E. et al., Annealing studies of low temperature grown GaAs:Be, *J. Appl. Phys.* 71, 1699, 1992.
45. Manasreh, M.O. et al., Infrared absorption of deep defects in molecular beam epitaxial GaAs layers grown at 200°C: Observation of an EL2 like defect, *Phys. Rev. B* 41, 10272, 1990.
46. Pritchard, R.E. et al., Native defects in low temperature grown GaAs and the effect of hydrogenation, *J. Appl. Phys.* 78, 2411, 1995.

47. Stellmacher, M. et al., Defects and defect behavior in GaAs grown at low temperature, *Semicond. Sci. Technol.* 16, 440, 2001.
48. Warren, A.C. et al., Arsenic precipitates and the semi-insulating properties of GaAs buffer layers grown by low temperature molecular beam epitaxy, *Appl. Phys. Lett.* 57, 1333, 1990.
49. Mahalingam, K. et al., Substrate temperature dependence of arsenic precipitate formation in AlGaAs and GaAs, *J. Vac. Sci. Technol. B* 9, 2328, 1991.
50. Melloch, M.R. et al., Arsenic cluster engineering for excitonic electro-optics, *J. Vac. Sci. Technol. B* 11, 795, 1993.
51. Bourgoin, J.C., Khirouni, K., and Stellmacher, M., The behavior of As precipitates in low temperature grown GaAs, *Appl. Phys. Lett.* 72, 442, 1998.
52. Khirouni, K., Nagle, J., and Bourgoin, J.C., Electron transport in low temperature grown GaAs, *Mater. Sci. Eng. B* 44, 334, 1997.
53 Khirouni, K. et al., Electron conduction in low temperature grown GaAs, *Solid State Electron.* 43, 589, 1999.
54. LTMBE materials, in *EMRS Symposium Proceedings*, Vol. 40, Hirtz, J.P. et al., Eds., North Holland, Amsterdam, 1993, pp 1–97.
55. Low temperature GaAs and related materials, in *MRS Symposium Proceedings*, Vol. 241, Witt, G.L. et al., Eds., Materials Research Society, Pittsburgh, 1992.
56. Physics and Applications of Advanced Semiconductors, in *MRS Symposium Proceedings* Vol. 329, Manasreh, M.O. et al., Eds., Materials Research Society, Pittsburgh, 1998.
57. Melloch, M.R. et al., Incorporation of excess arsenic in GaAs and AlGaA epilayers grown at low substrate temperatures by molecular beam epitaxy, in *MRS Symposium Proceedings*, Vol. 241, Witt, G.L. et al., Eds., Materials Research Society, Pittsburgh, 1992, p. 113.
58. Liliental-Weber, Z., Crystal structure of LT GaAs layers before and after annealing, ibid., p. 101.
59. Ohno, H. et al., GaMnAs: A new diluted magnetic semiconductor based on GaAs, *Appl. Phys. Lett.* 69, 363, 1996.
60. Dietl, T. et al., Zener model description of ferromagnetism in zincblende magnetic semiconductors, *Science* 287, 1019, 2000.
61. Jungwith, T. et al., Theory of ferromagnetic IIIMnV semiconductors, *Rev. Mod. Phys.* 78, 809, 2006.
62. Dietl, T., Ferromagnetic semiconductors, *Semicond. Sci. Technol.* 17, 377, 2002.
63. Liu, X. and Furdyna, J.K., Ferromagnetic resonance in GaMnAs diluted magnetic semiconductors, *J. Phys. Condens. Matter* 18, R245, 2006.
64. Edmonds, K.W. et al., High Curie temperature GaMnAs obtained by resistance monitored annealing, *Appl. Phys. Lett.* 81, 4991, 2002.
65. Liu, X. and Furdyna, J.K., Ferromagnetic resonance in GaMnAs: The effect of magnetic anisotropy, *Phys. Rev. B* 67, 205204, 2003.
66. Elsen, M. et al., Spin transfer experiments in GaMnAs/GaAs/GaMnAs magnetic tunnel junctions, *Phys. Rev. B* 73, 035303, 2006.
67. Rüster, C. et al., Very large tunneling anisotropic magnetoresistance of a GaMnAs/GaAs/GaMnAs stack, *Phys. Rev. Lett.* 94, 027203, 2005.
68. Rüster, C. et al., Very large magnetoresistance in lateral ferromagnetic GaMnAs wires with nanoconstrictions, *Phys. Rev. Lett.* 91, 216602, 2003.
69. Schneider, J. et al., Electronic structure of the manganese acceptor in gallium arsenide, *Phys. Rev. Lett.* 59, 240, 1987.
70. Kirby, B.J. et al., Magnetic and chemical nonuniformity in GaMnAs films as probed by polarized neutron and X-ray reflectometry, *Phys. Rev. B* 74, 245304, 2006.
71. Yoon, I.T. et al., Native hole traps of ferromagnetic GaMnAs layers on (100) GaAs substrates, *Appl. Phys. Lett.* 83, 4354, 2003.
72. Tuomisto, F. et al., Gallium sublattice defects in GaMnAs: Thermodynamical and kinetic trends, *Phys. Rev. Lett.* 93, 055505, 2004.
73. Wolos, A. et al., Properties of Arsenic antisite defects in GaMnAs, *J. Appl. Phys.* 96, 530, 2004.
74. Mahieu, G. et al., Compensation mechanisms in low temperature grown GaMnAs monitored by scanning tunneling microscopy, *Appl. Phys. Lett.* 82, 712, 2003.
75. Yu, K.M. et al., Effects of the location of Mn sites in ferromagnetic GaMnAs on the Curie temperature, *Phys. Rev. B* 65, 201303(R), 2002.

76. Yu, K.M. et al., Lattice location of Mn and fundamental Curie temperature limit in GaMnAs, *Nucl. Instrum. Meth. Phys. B* 219, 636, 2004.
77. Blinowski, J. and Kacman. P., Spin interactions of interstitial Mn ions in ferromagnetic GaMnAs, *Phys. Rev.B* 67, 121204(R), 2003.
78. Sadowski, J. et al., High ferromagnetic phase transition temperatures in GaMnAs layers annealed under As capping, ArXiv cond-mat/0601623, 2006.

Appendix A: Selected High-Impact Journal Articles on Defects in Microelectronic Materials and Devices

Daniel M. Fleetwood

CONTENTS

A.1 Introduction

In this appendix several tables and brief descriptions are provided for selected journal articles that have had significant impact on the field of defects in microelectronic materials and devices. Nearly all of the papers listed and discussed have received more than 100 citations. Citation counts are taken from the Institute for Scientific Information (ISI) Web of Science in a series of keyword or targeted searches performed in May and June, 2007. These searches were guided by a broad range of sources, including several previous review articles. There inevitably will be a large number of significant papers with high impact on the field that are not listed, owing to the search and selection processes employed. Contributions from some of the most significant and prolific contributors to the field have been somewhat more lightly sampled than contributions from others. To help make this task more manageable, the results focus primarily on topics of most relevance to this book. Studies of many interesting and significant topics, such as the optical properties of bulk silica, amorphous and porous silicon devices and materials, and fundamental properties of bare Si or SiO_2 surfaces mostly were not included. Materials and device systems are limited to Si, SiGe, SiC, GaAs, GaN, and ZnO. This choice includes the vast majority of commercial and defense microelectronics technologies, but also omits many other systems that have been investigated intensively. Papers on device fabrication or performance that do not include a significant discussion of defect properties are mostly not included. Also, recent work on emerging nanostructures such as nanotubes and quantum dots has been excluded. Fundamental materials studies have been given less emphasis than work directed toward supporting microelectronics technologies. Studies of the growth and mechanical properties of defects (e.g., threading dislocations, stacking faults, strain) also are mostly not included, with the primary exception of strained SiGe layers that are increasingly used in commercial complementary MOS (CMOS) technologies.

For this appendix, complete references are in alphabetical order by the last name of the first author. Throughout the text, publications are listed in tables and discussed in the text by first author and year of publication. Many topics are discussed in detail elsewhere in this book; others supplement the material in this volume. In performing surveys in this way, some interesting limitations of keyword searches are quickly found. For example, an inclusive search on the four keywords associated with this volume (defects, microelectronic, materials, devices) returns the amazingly low number of 20 publications that meet the full set of search criteria! The most highly cited article from that particular search on the ISI database is Mooney (1996, 41 citations), which is a review of work on strain relaxation and dislocations in SiGe/Si structures. Hence, more highly targeted and focused searches were required to generate the series of tables that are provided here. These surveys are intended primarily to provide a starting point for deeper investigation into several rich and vibrant areas of research, and are not intended to provide the kind of in-depth analysis one will find elsewhere in this volume.

Although numbers of citations of journal articles are used as a guiding metric, this is only one measure of impact of contributions to a field. To a significant extent, impact in the microelectronics industry is measured by progress in technology, and market share. Citations are often a poor measure of the factors that are most directly useful to the microelectronics industry. For example, the original report of the transistor to the *Physical Review* by Bardeen and Brattain was only cited 202 times for the period included in the ISI Web of Science search (1965–2007), yet a general search on the keyword transistor returned nearly 25,000 hits in the same publication database. Persuasive arguments could be made for patents as a measure of impact of new inventions, but these tend not to focus on defect properties. Instead, with the exceptions of devices that depend on defects that are introduced in a controlled fashion, modes of operation under conditions of manageable (or optimal) defect densities typically are presumed.

On the more academic front, there are some well-known limitations of the citation metric. Review articles are often highly cited in lieu of multiple citations of studies by other workers. Recent work with high impact will not have had the time to reach the citation count level it ultimately will achieve. Very early work is not as easily accessible or as thoroughly indexed as is more recent work. Journals and papers that already are known to have high-impact factors or citation counts tend to be read more than journals with lower impact factors and papers with lower citation counts, owing to the increasing uses of databases to build reference lists (like this one). This is an example of the Matthew Effect, developed by Merton (1968, 531 citations), which points out the tendency for credit to accrue most easily to those who already are well known.

Popular theoretical methods are applied across a broad range of scientific and engineering applications, not just calculations of defect microstructure and energy levels. This leads to some of the exceedingly large citation counts of many papers describing calculation methods, but one cannot overstate the significance of these and other techniques by which one tests ideas against first-principles theory. Journal articles that appear at the beginning of work in a particular area of interest tend to dominate citation statistics, even though later, follow-on publications may present more refined treatments. Papers that end an active area of investigation by resolving a problem can be victimized by their own success—if a problem is considered to be solved, after some checking, researchers properly go on to other topics. And of course some areas of research are just more popular than others; it is quite possible to do first-rate work with high impact in a fairly narrow field without resulting in a high citation count. Finally, we neglect books and nonindexed conference publications entirely; many of these have been extremely influential in the field. Nevertheless, I personally have found this to be an extremely useful exercise.

A.2 SiO_2 and the Si/SiO_2 Interface

One of the most significant reasons for the huge commercial success of the silicon-based microelectronics industry is that Si surfaces can be passivated effectively with thermally grown SiO_2 to form metal–oxide–semiconductor (MOS) transistors with low, as-processed defect densities. The most popular model of the oxidation of Si was developed by Deal and Grove (1965, 1557 citations). The long-term reliability and radiation response of MOS devices and integrated circuits (ICs) that survive the complex design, layout, and manufacturing processes are limited most by defects in the gate insulator, at the Si/SiO_2 interface, or in the near-surface Si bulk. The structure and properties of many of these defects are examined in detail elsewhere in this volume.

A.2.1 Measurement, Calculation, and Analysis Techniques

Tables A.1 and A.2 list papers that describe methods to estimate or calculate defect densities or energies in microelectronic devices and materials. Table A.1 includes mostly theoretical methods (Chapter 4); Table A.2 includes mostly experimental methods (Chapter 5). Early theoretical work that underlies much of this theory but often is used without citation now includes foundational work of the theory of semiconductor band structure by Dirac (1930, 1390 citations), Bloch (1928, 1077 citations), Thomas (1927, 729 citations), Fock (1930, 673 citations), Fermi (1928, 653 citations), Hartree (1928, 627 citations), Schrodinger (1926, 471 citations), Slater (1937, 376 citations), Tamm (1932, 355 citations), Sommerfeld (1928, 340 citations), and other pioneers of the quantum mechanical theory of semiconductors.

Of particular note for calculations of defect configurations and energy levels are the extraordinarily influential theoretical works of Lee et al. (1988a, 18,220 citations), who restated the Colle–Salvetti correlation-energy formula as a functional of the electron density and local kinetic energy density; Kohn and Sham (1965, 12,336 citations), who formulated an electron density-functional theory that includes exchange and correlation effects; Perdew and coworkers, who developed a generalized gradient approximation for calculations of exchange and correlation (Perdew and Zunger, 1981, 6139 citations; Perdew et al., 1996, 5874 citations; Perdew, 1986, 5706 citations; Perdew et al., 1992, 4707 citations; Perdew and Wang, 1992, 4667 citations); Ceperly and Alder (1980, 5009 citations), who provided an exact simulation of the Schroedinger equation to calculate electron correlation energies; Chadi and Cohen (1973, 1497 citations) and later Monkhorst and Pack (1976, 4623 citations), who developed methods to efficiently integrate periodic functions of wave

TABLE A.1

Citation Summary (# Cit ≥130) for Journal Articles that Describe Theoretical Methods to Estimate or Calculate Defect Densities and Energies in Microelectronic Materials and Devices

1st Author	Year	Journal	Abridged Title	# Cit.
Lee, C.	1988	*Phys. Rev. B*	Development of the Colle–Salvetti correlation-energy formula	18,220
Perdew, J.P.	1981	*Phys. Rev. B*	Self-interaction correction to density-functional approximations	6,139
Perdew, J.P.	1996	*Phys. Rev. Lett.*	Generalized gradient approximation made simple	5,874
Perdew, J.P.	1986	*Phys. Rev. B*	Density-functional approximation for the correlation energy	5,706
Ceperley, D.M.	1980	*Phys. Rev. Lett.*	Ground state of the electron gas by a stochastic method	5,009
Perdew, J.P.	1992	*Phys. Rev. B*	Generalized gradient approximation for exchange correlation	4,707
Perdew, J.P.	1992	*Phys. Rev. B*	Accurate and simple analytic representation of electron gas correlation	4,667
Monkhorst, H.J.	1976	*Phys. Rev. B*	Special points for Brillouin zone integrations	4,623
Vanderbilt, D.	1990	*Phys. Rev. B*	Soft self-consistent pseudopotentials in a generalized eigenvalue formalism	4,388
Car, R.	1985	*Phys. Rev. Lett.*	Unified approach for molecular dynamics and density-functional theory	4,199
Kresse, G.	1994	*Phys. Rev. B*	Efficient iterative schemes for ab initio total energy calculations	3,474
Payne, M.C.	1992	*Rev. Mod. Phys.*	Iterative minimization techniques for ab initio total energy calculations	2,838
Bachelet, G.B.	1982	*Phys. Rev. B*	Pseudopotentials that work—from H to Pu	2,305
Blöchl, P.	1994	*Phys. Rev. B*	Projector augmented wave method	1,625
Chadi, D.J.	1973	*Phys. Rev. B*	Special points in Brillouin zone	1,497
Hamann, D.R.	1979	*Phys. Rev. Lett.*	Norm-conserving pseudopotentials	1,409
Dirac, P.A.M.	1930	*Proc. Camb. Philos. Soc.*	Note on exchange phenomena in the Thomas atom	1,390
Chelikowsky, J.R.	1976	*Phys. Rev. B*	Nonlocal pseudopotential calculations for electronic structure	1,265
Cohen, M.L.	1966	*Phys. Rev.*	Band structures and pseudopotential for factors for 14 semiconductors	1,250
Kresse, G.	1994	*J. Phys.: Condens. Matt.*	Norm- and ultrasoft-conserving pseudopotentials	1,229
Bloch, F.	1928	*Z. Phys. A*	Uber die quantenmechanik der elektronen in kristallgittern	1,077
Bardeen, J.	1947	*Phys. Rev.*	Surface states and rectification at a metal–semiconductor contact	1,030
Lehmann, G.	1972	*Phys. Status Solidi B*	Numerical calculation of density of states and related properties	946
Shockley, W.	1949	*Bell Syst. Tech. J.*	Theory of p–n junctions in semiconductors and p–n junction transistors	930
Thomas, L.H.	1927	*Proc. Camb. Philos. Soc.*	The calculation of atomic fields	729
Fock, V.	1930	*Z. Phys.*	Naherungmethode zur losung des quantenmechanischen mehrkorperproblem	673
Fermi, E.	1928	*Z. Phys.*	Eine statische method zur bestimmung einiger eigenschaten des atoms	653
Hartree, D.R.	1928	*Proc. Camb. Philos. Soc.*	The wave mechanics of an atom with a non-Coulomb central field	627
Laasonen, K.	1993	*Phys. Rev. B*	Car–Parrinello molecular dynamics with Vanderbilt ultrasoft pseudopotentials	569
Schottky, W.	1942	*Z. Phys.*	Vereinfachte und erweiterte theorie der randschichtgleichrichter	515

TABLE A.1 (continued)

Citation Summary (# Cit ≥130) for Journal Articles that Describe Theoretical Methods to Estimate or Calculate Defect Densities and Energies in Microelectronic Materials and Devices

1st Author	Year	Journal	Abridged Title	# Cit.
Schrodinger, E.	1926	*Ann. Phys.*	Quantization as an eigenvalue problem	471
Schottky, W.	1939	*Z. Phys.*	Halbleitertheorie der sperrschicht und spitzengleichrichter	377
Slater, J.C.	1937	*Phys. Rev.*	Wave functions in a periodic potential	376
Tamm, I.E.	1932	*Phys. Z. Sowietunion*	On the possible bound states on a crystal surface	355
Sommerfeld, A.	1928	*Z. Phys.*	Zur elektronen theorie der metalle auf grund der Fermischen statistik	340
Bernholc, J.	1980	*Phys. Rev. B*	Scattering theoretic method for defects in semiconductors; applied to vacancy in Si	225
Bernholc, J.	1978	*Phys. Rev. B*	Scattering theoretic method for defects in semiconductors (Si, Ge, GaAs)	219
Bernholc, J.	1978	*Phys. Rev. Lett.*	Self-consistent method for point defects in semiconductors	214
Simmons, J.G.	1971	*Phys. Rev. B*	Nonequilibrium steady-state statistics and associated effects for insulators	194
Simmons, J.G.	1971	*J. Phys. D—Appl. Phys.*	Conduction in thin dielectric films	189

TABLE A.2

Citation Summary (# Cit ≥130) for Journal Articles that Describe Experimental Methods to Estimate or Calculate Defect Densities and Energies in Microelectronic Materials and Devices

1st Author	Year	Journal	Abridged Title	# Cit.
Lang, D.V.	1974	*J. Appl. Phys.*	DLTS—new method to characterize traps in semiconductors	2289
Fowler, R.H.	1928	*Proc. R. Soc. London A*	Electron emission in intense electric fields	1585
Deal, B.E.	1965	*J. Appl. Phys.*	General relationship for thermal oxidation of Si	1557
Terman, L.M.	1962	*Solid-State Electron.*	An investigation of surface states at a Si/SiO_2 interface using MOS diodes	792
Lenzlinger, M.	1969	*J. Appl. Phys.*	Fowler–Nordheim tunneling into thermally grown SiO_2	728
Dutta, P.	1981	*Rev. Mod. Phys.*	Low-frequency fluctuations in solids—$1/f$ noise	713
Lang, D.V.	1974	*J. Appl. Phys.*	Fast capacitance apparatus—application to ZnO and O centers in GaP	551
Groeseneken, G.	1984	*IEEE Trans. Electron Devices*	A reliable approach to charge-pumping measurements in MOS transistors	478
Kuhn, M.	1970	*Solid-State Electron.*	A quasistatic technique for MOS C–V and surface state measurements	451
Sah, C.T.	1970	*Solid-State Electron.*	Thermal and optical emission and capture rates and capture cross sections	431
Weeks, R.A.	1956	*J. Appl. Phys.*	Paramagnetic resonance of lattice defects in irradiated quartz	424
Miller, G.L.	1977	*Annu. Rev. Mater. Sci.*	Capacitance transient spectroscopy	423
Snow, E.H.	1965	*J. Appl. Phys.*	Ion transport phenomena in insulating films	421

(*continued*)

TABLE A.2 (continued)

Citation Summary (# Cit ≥130) for Journal Articles that Describe Experimental Methods to Estimate or Calculate Defect Densities and Energies in Microelectronic Materials and Devices

1st Author	Year	Journal	Abridged Title	# Cit.
Grove, A.S.	1965	*Solid-State Electron.*	Investigation of thermally oxidized structures using MOS structures	418
Kirton, M.J.	1989	*Adv. Phys.*	Noise in solid state microstructures—a new perspective on individual defects	381
Berglund, C.N.	1966	*IEEE Trans. Electron Devices*	Surface states at steam-grown Si–SiO_2 interfaces	354
Gray, P.V.	1966	*Appl. Phys. Lett.*	Density of SiO_2–Si interface states	315
Jaccodine, R.J.	1966	*J. Appl. Phys.*	Measurement of strains at Si–SiO_2 interface	301
Chantre, A.	1981	*Phys. Rev. B*	Deep-level optical spectroscopy in GaAs	298
Grunthaner, F.J.	1979	*Phys. Rev. Lett.*	High-resolution XPS as a probe of SiO_2 and the Si/SiO_2 interface	290
McWhorter, P.J.	1986	*Appl. Phys. Lett.*	Technique for separating effects of interface traps and trapped oxide charge	256
Grunthaner, F.J.	1979	*J. Vac. Sci. Technol.*	Local atomic and electronic structure of oxide-GaAs and SiO_2–Si using XPS	255
Lefevre, H.	1977	*Appl. Phys.*	Double correlation technique (DDLTS) for analysis of deep-level profiles	205
Bardeen, J.	1948	*Phys. Rev.*	The transistor, a semiconductor triode	202
Winokur, P.S.	1984	*IEEE Trans. Nucl. Sci.*	Correlating the radiation response of MOS capacitors and transistors	198
Heremans, P.	1989	*IEEE Trans. Electron Devices*	Analysis of the charge-pumping technique and application to MOSFETs	191
Watkins, G.D.	1959	*J. Appl. Phys.*	Spin resonance in electron-irradiated Si	183
Castagne, R.	1971	*Surf. Sci.*	Description of SiO_2–Si properties by means of very low-frequency *C–V*	170
DiMaria, D.J.	1976	*J. Appl. Phys.*	Determination of trapped charge densities and centroids via photo-IV	169
Hurtes, C.	1978	*Appl. Phys. Lett.*	Deep-level spectroscopy in high resistivity materials	169
Brugler, J.S.	1969	*IEEE Trans. Electron Devices*	Charge pumping in MOS devices	165
Schroder, D.K.	1971	*Solid-State Electron.*	Interpretation of surface and bulk defects using pulsed MIS capacitor	157
Nicollian, E.H.	1969	*Appl. Phys. Lett.*	Avalanche injection currents and charging phenomena in thermal SiO_2	150
Heiman, F.P.	1967	*IEEE Trans. Electron Devices*	On determination of minority carrier lifetime from transistor response	146
Deuling, H.	1972	*Solid-State Electron.*	Interface states in Si–SiO_2 structures	143
Fleetwood, D.M.	1993	*J. Appl. Phys.*	Effects of oxide, interface, and border traps on MOS devices	138

vectors using special points in the Brillouin zone, to facilitate density-functional theory calculations; Vanderbilt (1990, 4388 citations), who developed soft pseudopotentials to support density-functional theory calculations; and Car and Parrinello (1985, 4199 citations), who developed a unified approach for molecular dynamics and density-functional theory. The implementation of Car–Parrinello molecular dynamics with Vanderbilt ultrasoft pseudopotentials was described by Laasonen et al. (1993, 569 citations). Kresse and coworkers described a plane wave basis set and norm-conserving pseudopotentials that can be effectively applied with as few as 75–100 atoms (Kresse and Furthmuller, 1994, 3474 citations; Kresse and Hafner, 1994, 1229 citations). Payne et al. (1992, 2838 citations) reviewed the total energy pseudopotential method and described a wide range of applications. Bachelet et al. (1982, 2305 citations) and Hamann et al. (1979, 1409 citations) also have presented a complete set of pseudopotentials to support calculations of electronic and structural properties of microelectronic materials. A complementary projector augmented wave method was developed by Blöchl (1994, 1625 citations). Pseudopotentials for Si, GaAs, and 12 other diamond and zinc blende structure semiconductors were developed by Chelikowsky and Cohen (1976, 1265 citations) and Cohen and Bergstresser (1966, 1250 citations). An alternative method of integration over the Brillouin zone in density of states calculations was developed by Lehmann and Taut (1972, 946 citations).

Bardeen (1947, 1030 citations) developed a theory of interface traps at a metal–semiconductor contact, which along with Shockley's theory of p–n junctions (1949, 930 citations), was influential in developing the transistor (Bardeen and Brattain, 1948, 202 citations). A theory of metal–semiconductor contacts was developed by Schottky (1942, 515 citations; 1939, 377 citations). A self-consistent technique involving tight binding calculations was developed and applied to calculate the energy levels of vacancies in Si and other semiconducting materials by Bernholc et al. (1980, 225 citations; 1978, 214 citations) and Bernholc and Pantelides (1978, 219 citations). These and other popular theoretical techniques make it possible to calculate the static and dynamic properties of a wide variety of materials, and have often been applied to calculate the properties of defects in semiconductors and insulators.

On the experimental front, Lang (1974a, 2289 citations; 1974b, 551 citations) developed deep-level transient spectroscopy, which has been applied broadly to characterize defects in a wide range of semiconductor devices and materials. This is a high-frequency capacitance technique that exploits the time dependence of typical capture and emission processes for defects in microelectronic devices and materials to enable one to extract a broad range of trap parameters as a function of temperature. A general review of capacitance transient spectroscopy was performed by Miller et al. (1977, 423 citations). Chantre et al. (1981, 298 citations) and Hurtes et al. (1978, 169 citations) combined optical injection with deep-level capacitance transient spectroscopy (DLTS) to develop deep-level optical spectroscopy techniques. A double deep-level technique based on DLTS (DDLTS) was developed by Lefevre and Schulz (1977, 205 citations).

In the Si/SiO_2 system, Terman (1962, 792 citations) demonstrated a simple method to vary the surface potential of MOS structures to extract interface-trap energies and densities. The use of Fowler and Nordheim (1928, 1585 citations) tunnel injection to generate hot carriers in SiO_2 was pioneered by Lenzlinger and Snow (1969, 728 citations). Dutta and Horn (1981, 713 citations) developed a self-consistent theoretical approach that links low-frequency noise in materials to defects; in particular, the temperature and frequency dependence of the noise can be used to characterize defect energy distributions. The utility of electron spin (paramagnetic) resonance (ESR/EPR) in evaluating radiation damage in oxides and in silicon was illustrated by Weeks (1956, 424 citations—this paper reports on the E_1 defect in quartz, which is similar to the E' moiety in thin-film amorphous SiO_2) and by Watkins et al. (1959, 183 citations). They and others applied ESR to obtain an extraordinary amount of structural information about defects in microelectronic materials, as discussed further in the next section. Photocurrent, dark current, and capacitance transient techniques were developed by Sah et al. (1970, 431 citations) to characterize defect electrical properties in bulk semiconductors. Changes in capacitance–voltage characteristics of MOS devices as a function of time, temperature, and applied electric field were used by Snow et al. (1965, 421 citations) to profile mobile ion effects in SiO_2. This technique has been adopted widely within the microelectronics industry, and was used to help reduce contamination levels by more than three to five orders of magnitude in finished MOS devices and ICs over the last ~40 years. This technique builds on basic capacitance–voltage techniques described by Grove et al. (1965, 418 citations).

The charge-pumping technique to estimate interface-trap densities in MOS transistors was described by Brugler and Jespers (1969, 165 citations) and refined and popularized by Groeseneken et al. (1984, 478 citations). This technique allows one to estimate interface-trap charge densities and energy distributions by first capturing electrons (holes) and then holes (electrons) at recombination centers that are located at or near the Si/SiO_2 interface. The charge-pumping technique was further analyzed and extended by Heremans et al. (1989, 191 citations). A quasistatic technique to estimate interface-trap charge densities and energies was developed by Kuhn (1970, 451 citations). Kirton and Uren (1989, 381 citations) illustrated how random telegraph noise can be used to analyze defect capture and emission properties at or near the Si/SiO_2 interface. Berglund (1966, 354 citations) showed that one could estimate interface-trap densities and energies from a comparison of experimental and theoretical *C–V* curves. The Gray–Brown technique (Gray and Brown, 1966, 315 citations) exploits changes in interface-trap capture and emission characteristics with temperature, typically room temperature and 77 K, to estimate interface-trap densities. For example, when large stretchout is observed in capacitance–voltage (*C–V*) curves at room temperature, but not at low temperatures, that is a typical signature of interface-trap buildup, as opposed to other causes of *C–V* stretchout (e.g., charge lateral nonuniformities). Jaccodine and Schlegel (1966, 301 citations) developed methods to induce and measure strain at the Si/SiO_2 interface, and evaluated defect formation as a function of tensile or compressive stress. Grunthaner et al. (1979a, 290 citations; 1979b, 255 citations) applied x-ray photoelectron spectroscopy (XPS) to investigate the local atomic structure and defects at or near the Si/SiO_2 interface.

A simple method that exploits the approximate charge neutrality of interface traps at midgap surface potential to separate the effects of interface traps and oxide-trap charge on the characteristics of MOS capacitors and transistors via capacitance–voltage or current–voltage measurements was developed by Winokur et al. (1984, 198 citations) and McWhorter and Winokur (1986, 256 citations). In a series of papers, Simmons and colleagues developed a theoretical basis and experimental methodology for investigating defects in insulators via the thermally stimulated current technique (Simmons and Taylor, 1971, 194 citations; Simmons 1971, 189 citations; Simmons and Taylor, 1972, 123 citations).

A low-frequency *C–V* technique to estimate Si/SiO_2 interface properties was developed by Castagne and Vapaille (1971, 170 citations). The use of photocurrent–voltage techniques to estimate bulk trapped charge densities and centroids in SiO_2 was described by DiMaria (1976, 169 citations). Schroder and coworkers developed a pulsed metal–insulator–capacitor technique to separate the bulk and surface contributions to carrier lifetimes and surface generation velocities (Schroder and Guldberg, 1971, 157 citations; Schroder and Nathanson, 1970, 108 citations). Nicollian and coworkers pioneered the use of avalanche injection to investigate hole and electron injection into SiO_2 to create oxide- and interface-trap charge (Nicollian et al., 1969, 150 citations; Nicollian and Berglund, 1970, 114 citations). Heiman (1967, 146 citations) illustrated how one could extract minority carrier lifetime from MOS capacitor transient response to a voltage change. Deuling et al. (1972, 143 citations) describe a method to characterize interface-trap properties, including density, capture cross section, surface potential, and dispersion parameters that is based on conductance and capacitance measurements as a function of temperature. Fleetwood et al. (1993, 138 citations) developed and applied low-frequency noise and thermally stimulated current techniques to provide separate estimates of MOS interface and near-interface oxide (border) trap charge densities (Chapter 7). This work follows a guest editorial by Fleetwood (1992, 96 citations) who introduced the term border trap to describe near-interface oxide traps that exchange charge with the Si during the timescale of the measurements, and affirmed the practical utility of this concept.

A.2.2 Microstructure and Electrical Properties

Tables A.3 and A.4 summarize a large number of papers that describe the microstructure and properties of defects in the Si/SiO_2 system. Table A.3 focuses primarily on microstructural properties of defects, and Table A.4 focuses mostly on electrical properties, but of course there is significant overlap of these two areas. An extensive review of the fundamental conduction mechanisms in two-dimensional systems like the near-interfacial silicon surface of a (100) oxidized Si wafer, including the effects of defects at low and high temperatures, was provided by Ando et al. (1982, 3811 citations). Shockley–Read–Hall recombination, where carriers of opposite signs are captured in turn at a defect site, with highest probability at the middle of the band gap, was reported in an extremely influential series of papers by Shockley and Read (1952, 2290 citations), Read and Shockley (1950, 681 citations), and Hall (1952, 839 citations). Electron–hole recombination in p–n junctions was characterized by Sah et al. (1957, 1266 citations) and later by Grove and Fitzgerald (1966, 310 citations). The fundamental properties of defect-modulated conduction in disordered systems, including the introduction of localized states associated with defects in the material, was reviewed by Mott (1967, 1233 citations). The microstructural properties of the Si/SiO_2 interface were characterized in detail via high-resolution core-level spectroscopy by Himpsel et al. (1988, 883 citations).

Hot carrier damage to MOS transistors was studied extensively by Hu et al. (1985, 571 citations), and attributed to the breaking of Si–H bonds during electrical stress. A first-principles treatment of electron transport using Monte Carlo techniques was performed by Fischetti and Laux (1988, 403 citations). Design constraints on CMOS devices due to hot carrier effects were discussed by Ning (1979, 160 citations). A transistor design that included lightly doped drain extenders was described by Ogura et al. (1980, 196 citations), to help offset hot carrier-induced charge trapping in MOS devices. Early evidence of the link between interface traps at the Si/SiO_2 interface and Si dangling bonds was provided by Deal et al. (1967, 454 citations). The effects of defects at or near the Si/SiO_2 interface on MOS carrier mobility was investigated by Sun and Plummer (1980, 397 citations). A theory of defect-modulated conduction in SiO_2 that attributed the anomalous Poole–Frenkel effect to conduction via the Schottky mechanism was developed by Simmons (1967, 357 citations). The oxygen vacancy model for the E' defect in SiO_2, a dominant hole trapping center, was presented by Feigl et al. (1974, 347 citations), and further refined by Rudra and Fowler (1987, 185 citations) to include the effects of SiO_2 network relaxation. The dominant hole trap in the SiO_2 gate oxides of irradiated MOS devices was identified as an E' defect by Lenahan and Dressendorfer (1984, 365 citations), who also identified a trivalent Si defect (the P_b, first observed by Nishi, 1971, 191 citations) as the dominant interface trap in irradiated MOS devices (Lenahan and Dressendorfer, 1983, 99 citations). Properties of interface traps and fixed oxide charge were investigated using ESR and electrical characterization techniques by Poindexter et al. (1981, 299 citations) and Caplan et al. (1979, 278 citations). They identified the P_b defect on (111) Si as a trivalent Si defect at the Si/SiO_2 interface; the P_b (often called P_{b0} at the 100 Si/SiO_2 interface) and a second defect (usually called Pb_{b1}) was found by Poindexter et al. (1981, 299 citations). Gerardi et al. (1986, 156 citations) found that the P_{b0} correlated more closely with electrically active interface traps, and exhibited different energy levels from the P_{b1} defect, the nature of which remains under investigation (Chapter 6). The Si-29 hyperfine structure associated with the P_b defect was measured by Brower (1983, 174 citations).

TABLE A.3

Citation Summary (# Cit ≥100) for Journal Articles that Describe Defects in the Si/SiO_2 System, Mostly from a Microstructural Perspective

1st Author	Year	Journal	Abridged Title	# Cit.
Himpsel, F.J.	1988	*Phys. Rev. B*	Microscopic structure of the SiO_2/Si interface	883
Lenahan, P.M.	1984	*J. Appl. Phys.*	Hole traps and trivalent Si centers in MOS devices	365
Griscom, D.L.	1980	*J. Non-Cryst. Sol.*	ESR in glasses—magnetic properties	352
Feigl, F.J.	1974	*Solid State Commun.*	Oxygen vacancy model for $E'/1$ center in SiO_2	347
Poindexter, E.H.	1981	*J. Appl. Phys.*	Interface states and ESR centers in (111) and (100) Si wafers	299
Caplan, P.J.	1979	*J. Appl. Phys.*	ESR centers, interface states, and oxide fixed charge in thermally oxidized Si	278
O'Reilly, E.P.	1983	*Phys. Rev. B*	Theory of defects in vitreous SiO_2	253
EerNisse, E.P.	1977	*Appl. Phys. Lett.*	Viscous flow of thermal SiO_2	197
Goodnick, S.M.	1985	*Phys. Rev. B*	Surface roughness at the Si/SiO_2 interface	197
Nishi, Y.	1971	*Japan. J. Appl. Phys.*	Study of Si–SiO_2 structure by ESR	191
Asokakumar, P.	1994	*J. Appl. Phys.*	Characterization of defects in Si and SiO_2–Si using positrons	189
DiStefano, T.H.	1971	*Phys. Rev. Lett.*	Photoemission measurements of valence levels of amorphous SiO_2	189
Tromp, R.	1985	*Phys. Rev. Lett.*	High-temperature SiO_2 decomposition at the SiO_2/Si interface	188
Rudra, J.K.	1987	*Phys. Rev. B*	Oxygen vacancy and the $E'1$ center in crystalline SiO_2	185
Brower, K.L.	1983	*Appl. Phys. Lett.*	Si-29 hyperfine structure of unpaired spins at the Si/SiO_2 interface	183
Daum, W.	1993	*Phys. Rev. Lett.*	Identification of strained Si layers at Si/SiO_2 interfaces	181
Griscom, D.L.	1974	*Solid State Commun.*	Observation and analysis of primary Si-29 hyperfine structure of E' center	175
Grunthaner, F.J.	1982	*IEEE Trans. Nucl. Sci.*	Radiation-induced defects in SiO_2 as determined by XPS	166
Griscom, D.L.	1979	*Phys. Rev. B*	E′ center in glassy SiO_2—microwave saturation properties and Si-29 hyperfine	158
Poindexter, E.H.	1984	*J. Appl. Phys.*	Electronic traps and P_b centers: band gap energy distribution	157
Gerardi, G.J.	1986	*Appl. Phys. Lett.*	Interface traps and P_b centers in oxidized (100) Si wafers	156
Edwards, A.H.	1982	*Phys. Rev. B*	Theory of the peroxy-radical defect in a-SiO_2	153
Iwata, S.	1996	*J. Appl. Phys.*	Electron spectroscopic analysis of the SiO_2/Si system	153
Lucovsky, G.	1979	*Philos. Mag. B—Phys. Condens. Matter*	Spectroscopic evidence for valence-alternation-pair defect states in SiO_2	151
Stathis, J.H.	1987	*Phys. Rev. B*	Time-resolved photoluminescence in a-SiO_2	146
Griscom, D.L.	1980	*Phys. Rev. B*	E′ center in glassy SiO_2—O-17, H-1, and Si-29 hyperfine	134
Helms, C.R.	1994	*Rep. Prog. Phys.*	The Si–SiO_2 system—its microstructure and imperfections	130
Poindexter, E.H.	1983	*Prog. Surf. Sci.*	Characterization of Si/SiO_2 interface defects by ESR	116
Boero, M.	1997	*Phys. Rev. Lett.*	Structure and hyperfine parameters of $E'(1)$ centers in alpha quartz and a-SiO_2	105
Griscom, D.L.	1981	*Phys. Rev. B*	Si-29 hyperfine structure of nonbridging oxygen hole center and peroxy radical	105
Pasquarello, A.	1998	*Nature*	Interface structure between Si and SiO_2 by first-principles molecular dynamics	101

TABLE A.4

Citation Summary (# Cit ≥100) for Journal Articles that Describe Defects in the Si/SiO_2 System, Mostly from an Electrical Perspective

1st Author	Year	Journal	Abridged Title	# Cit.
Ando, T.	1982	*Rev. Mod. Phys.*	Electronic properties of two-dimensional systems	3811
Shockley, W.	1952	*Phys. Rev.*	Statistics of the recombination of holes and electrons	2290
Mott, N.F.	1967	*Adv. Phys.*	Electrons in disordered systems	1233
Hall, R.N.	1952	*Phys. Rev.*	Electron–hole recombination in Ge	839
Read, W.T.	1950	*Phys. Rev.*	Dislocation models of crystal grain boundaries	681
Weissman, M.B.	1988	*Rev. Mod. Phys.*	$1/f$ noise and other slow, nonexponential kinetics in condensed matter	619
Hu, C.M.	1985	*IEEE Trans. Electron Devices*	Hot electron-induced MOSFET degradation	571
Deal, B.E.	1967	*J. Electrochem. Soc.*	Characteristics of surface-state charge of thermally oxidized Si	454
Hooge, F.N.	1969	*Phys. Lett. A*	$1/f$ noise is no surface effect	439
Hooge, F.N.	1981	*Rep. Prog. Phys.*	Experimental studies on $1/f$ noise	429
Fischetti, M.V.	1988	*Phys. Rev. B*	Monte Carlo analysis of electron transport in small semiconductor devices	403
Sun, S.C.	1980	*IEEE Trans. Electron Devices*	Electron mobility in inversion and accumulation layers on oxidized Si	397
Simmons, J.G.	1967	*Phys. Rev.*	Poole–Frenkel effect and Schottky effect in metal–insulator–metal systems	357
Nicollian, E.H.	1971	*J. Appl. Phys.*	Electrochemical charging of thermal SiO_2 films by injected electrons	332
Ralls, K.S.	1984	*Phys. Rev. Lett.*	Discrete resistance switching in submicrometer Si inversion layers	322
Grove, A.S.	1966	*Solid-State Electron.*	Surface effects on p–n junctions under nonequilibrium conditions	310
Lai, S.K.	1983	*J. Appl. Phys.*	Interface-trap generation in SiO_2 when electrons are captured by trapped holes	303
Skuja, L.	1998	*J. Non-Cryst. Solids*	Optically active O-deficiency-related centers in a-SiO_2	301
Snow, E.H.	1967	*Proc. IEEE*	Effects of ionizing radiation on oxidized Si surfaces and planar devices	279
Takagi, S.	1994	*IEEE Trans. Electron Devices*	On the universality of inversion layer mobility in Si MOSFETs	275
Hooge, F.N	1976	*Physica B & C*	$1/f$ noise	267
Takeda, E.	1983	*IEEE Electron Device Lett.*	An empirical model for device degradation due to hot carrier injection	241
Deal, B.E.	1974	*J. Electrochem. Soc.*	Current understanding of charges in thermally oxidized Si structure	218
Schwank, J.R.	1984	*IEEE Trans. Nucl. Sci.*	Physical mechanisms contributing to device rebound	205
Young, D.R.	1979	*J. Appl. Phys.*	Electron trapping in SiO_2 at 295 and 77 K	205
Ogura, S.	1980	*IEEE Trans. Electron Devices*	Design and characteristics of the LDD insulated gate FET	196
Maserjian, J.	1982	*J. Appl. Phys.*	Behavior of the Si/SiO_2 interface observed by Fowler–Nordheim tunneling	192
Heiman, F.P.	1965	*IEEE Trans. Electron Devices*	Effects of oxide traps on MOS capacitance	189
Hofmann, K.R.	1985	*IEEE Trans. Electron Devices*	Hot electron and hole emission effects in short *n*-channel MOSFETs	189
Hooge, F.N.	1994	*IEEE Trans. Electron Devices*	$1/f$ noise sources	188
Ning, T.H.	1974	*J. Appl. Phys.*	Optically induced injection of hot electrons into SiO_2	188

(*continued*)

TABLE A.4 (continued)

Citation Summary (# Cit ≥100) for Journal Articles that Describe Defects in the Si/SiO_2 System, Mostly from an Electrical Perspective

1st Author	Year	Journal	Abridged Title	# Cit.
Feigl, F.J.	1981	*J. Appl. Phys.*	The effects of water on oxide and interface-trap charge in SiO_2	180
Sah, C.T.	1972	*Surf. Sci.*	Scattering of electrons by surface oxide charges and lattice vibrations	176
Hooge, F.N.	1972	*Physica*	Discussion of recent experiments on $1/f$ noise	172
Hughes, R.C.	1977	*Phys. Rev. B*	Time-resolved hole transport in a-SiO_2	171
Fang, F.F.	1970	*J. Appl. Phys.*	Hot electron effects and saturation velocities in Si inversion layers	170
Derbenwick, G.F.	1975	*IEEE Trans. Nucl. Sci.*	Process optimization of rad-hard CMOS ICs	164
Powell, R.J.	1970	*J. Appl. Phys.*	Interface barrier energy determination from photoinjected currents	161
Hung, K.K.	1990	*IEEE Trans. Electron Devices*	A unified model for the flicker noise in MOSFETs	160
Ning, T.H.	1979	*IEEE Trans. Electron Devices*	1 μm MOSFET VLSI technology: Hot electron design constraints	160
Hughes, R.C.	1973	*Phys. Rev. Lett.*	Charge carrier transport phenomena in a-SiO_2—drift mobility and lifetime	159
Berglund, C.N.	1971	*J. Appl. Phys.*	Photoinjection into SiO_2—electron scattering in image force potential well	154
Kar, S.	1972	*Solid-State Electron.*	Interface states in MOS structures with 2–4 nm SiO_2 films on Si	154
Ning, T.H.	1977	*J. Appl. Phys.*	Emission probability of hot electrons from Si into SiO_2	154
Fair, R.B.	1981	*IEEE Trans. Electron. Devices*	Threshold voltage instability in MOSFETs due to channel hot holes	147
Tam, S.	1984	*IEEE Trans. Electron Devices*	Lucky electron model of channel hot electron injection in MOSFETs	146
Razouk, R.R.	1979	*J. Electrochem. Soc.*	Dependence of interface state density on Si thermal oxidation process variables	142
Zaininger, K.H.	1967	*RCA Rev.*	A survey of radiation effects in metal–insulator–semiconductor devices	142
Sah, C.T.	1976	*IEEE Trans. Nucl. Sci.*	Origin of interface states and oxide charges generated by ionizing radiation	141
Deal, B.E.	1969	*J. Electrochem. Soc.*	Characteristics of fast surface states [at] SiO_2–Si and Si_3N_4–SiO_2–Si [interfaces]	138
Powell, R.J.	1971	*IEEE Trans. Nucl. Sci.*	Vacuum ultraviolet radiation effects in SiO_2	138
DiMaria, D.J.	1977	*J. Appl. Phys.*	Location of positive charges in SiO_2 films on Si	134
Liao, L.S.	1996	*Appl. Phys. Lett.*	Blue luminescence from Si^+ implanted SiO_2 films thermally grown on Si	133
Lai, S.K.	1981	*Appl. Phys. Lett.*	Two-carrier nature of interface state generation in hole trapping	131
Wright, P.J.	1989	*IEEE Trans. Electron Devices*	The effect of fluorine in SiO_2 gate dielectrics	131
DiMaria, D.J.	1985	*J. Appl. Phys.*	Electron heating in SiO_2 and off-stoichiometric SiO_2 films	129
Doyle, B.	1990	*IEEE Trans. Electron Devices*	Interface state creation and charge trapping during hot carrier stress	129
Vandamme, L.K.J.	1994	*IEEE Trans. Electron Devices*	Noise as a diagnostic tool for quality and reliability of electronic devices	128
Aitken, J.M.	1978	*J. Appl. Phys.*	Electron trapping in electron-beam irradiated SiO_2	127

TABLE A.4 (continued)

Citation Summary (# Cit ≥100) for Journal Articles that Describe Defects in the Si/SiO_2 System, Mostly from an Electrical Perspective

1st Author	Year	Journal	Abridged Title	# Cit.
Ning, T.H.	1976	*J. Appl. Phys.*	High-field capture of electrons by Coulomb-attractive centers in SiO_2	123
Powell, R.J.	1971	*J. Appl. Phys.*	Photoinjection studies of charge distributions of MOS structures	123
Ghibaudo, G.	1991	*Phys. Status Solidi A*	Improved analysis of low-frequency noise in MOSFETs	122
Handel, P.H.	1980	*Phys. Rev. A*	Quantum approach to $1/f$ noise	122
Oldham, T.R.	1986	*IEEE Trans. Nucl. Sci.*	Spatial dependence of trapped holes determined from tunneling analysis	122
Cheng, Y.C.	1973	*Surf. Sci.*	Role of scattering by surface roughness in Si inversion layers	120
Reimbold, G.	1984	*IEEE Trans. Electron Devices*	$1/f$ trapping noise theory from weak to strong inversion	119
Johnson, N.M.	1983	*Appl. Phys. Lett.*	Characteristic electronic defects at the $Si–SiO_2$ interface	115
Hsu, F.C.	1984	*IEEE Electron Device Lett.*	Hot-electron-induced interface state generation	115
Hooge, F.N.	1978	*Phys. Lett. A*	Lattice scattering causes $1/f$ noise	114
Arnold, G.W.	1973	*IEEE Trans. Nucl. Sci.*	Ion implantation effects in noncrystalline SiO_2	114
Fischetti, M.V.	1985	*J. Appl. Phys.*	Generation of positive charge in SiO_2 during avalanche and tunnel injection	113
Mott, N.F.	1977	*Adv. Phys.*	SiO_2 and the chalcogenide semiconductors	111
Sah, C.T.	1966	*Phys. Rev. Lett.*	Evidence of surface origin of $1/f$ noise	111
Aitken, J.M.	1976	*J. Appl. Phys.*	Electron trapping by radiation-induced charge in MOS devices	110
Deal, B.E.	1968	*J. Electrochem. Soc.*	Electrical properties of vapor-deposited Si_3N_4 and SiO_2 films on Si	110
Ning, T.H.	1978	*Solid-State Electron.*	Hot electron emission from Si into SiO_2	110
Terry, F.L.	1983	*IEEE Electron Device Lett.*	Radiation effects in nitrided oxides	110
Hsu, F.C.	1984	*IEEE Electron Device Lett.*	Structure-enhanced MOSFET degradation due to hot electron injection	109
Lelis, A.J.	1989	*IEEE Trans. Nucl. Sci.*	The nature of the trapped-hole annealing process	109
Gwyn, C.W.	1969	*J. Appl. Phys.*	Model for radiation-induced charge trapping in oxide layer of MOS devices	105
Woods, M.H.	1976	*J. Appl. Phys.*	Hole traps in SiO_2	105
Montillo, F.	1971	*J. Electrochem. Soc.*	High-temperature annealing of oxidized Si surfaces	104
Nissan-Cohen, Y.	1986	*J. Appl. Phys.*	Dynamic model of trapping–detrapping in SiO_2	104
Buchanan, D.	1990	*J. Appl. Phys.*	Interface and bulk trap generation in MOS capacitors	104
Aberle, A.G.	1992	*J. Appl. Phys.*	Impact of illumination level and oxide parameters on SRH recombination	103
Cottrell, P.E.	1979	*IEEE Trans. Electron Devices*	Hot electron emission in *n*-channel IGFETs	103
Boesch, H.E.	1976	*IEEE Trans. Nucl. Sci.*	Charge yield and dose effects in MOS capacitors at 80 K	102
Deal, B.E.	1965	*J. Electrochem. Soc.*	Observation of impurity redistribution during thermal oxidation	102
Hu, C.	1980	*Appl. Phys. Lett.*	Relationship between trapped holes and interface states in MOS capacitors	102

(*continued*)

TABLE A.4

Citation Summary (# Cit ≥100) for Journal Articles that Describe Defects in the Si/SiO_2 System, Mostly from an Electrical Perspective

1st Author	Year	Journal	Abridged Title	# Cit.
Hartstein, A.	1981	*Appl. Phys. Lett.*	Identification of electron traps in thermal SiO_2 films	101
Uren, M.J.	1985	*Appl. Phys. Lett.*	$1/f$ and random telegraph noise in Si MOSFETs	101
Castro, P.L.	1971	*J. Electrochem. Soc.*	Low-temperature reduction of fast surface states associated with SiO_2/Si	100

Early work on the effects of oxide traps on MOS capacitance was performed by Heiman and Warfield (1965, 189 citations). Methods to reduce the densities of oxide- and interface-trap charge in irradiated MOS devices were described by Derbenwick and Gregory (1975, 164 citations). Surveys of very early work in this area were performed by Snow et al. (1967, 279 citations) and Zaininger and Holmes-Seidle (1967, 142 citations). A strong correlation between oxide-trap charge and interface-trap generation was noted by Lai (1983, 303 citations), in a detailed study of charge injection into SiO_2. A similar link was noted by Hu and Johnson (1980, 102 citations). Lai (1981, 131 citations) also reported the formation of an interface dipole by first trapping and then the subsequent compensation of the trapped hole by electron injection. It is likely that these defects are similar to the border traps studied in detail by Fleetwood et al. (1993, 138 citations). Characteristics of E' defects in SiO_2 were reviewed by Skuja (1998, 301 citations). The Si-29 hyperfine lines associated with the E' defect were characterized in detail in a series of studies by Griscom et al. (1974, 175 citations) and Griscom (1979, 158 citations; 1980b, 352 citations; 1980a, 134 citations). Spectroscopic evidence supporting the concept of valence-alternation pair defects in SiO_2 was presented by Lucovsky (1979, 151 citations). Characteristic emission bands in SiO_2 were discussed in relation to intrinsic and new defect centers created during irradiation by Stathis and Kastner (1987, 146 citations). A theoretical description of the peroxy-radical defect in SiO_2 was developed by Edwards and Fowler (1982, 153 citations). The Si-29 hyperfine structures of the peroxy radical and nonbridging oxygen hole center in SiO_2 were measured by Griscom and Friebele (1981, 105 citations). The structure and hyperfine properties of E' centers in alpha quartz and amorphous SiO_2 were calculated by Boero et al. (1997, 105 citations). The postirradiation, reversible charging and discharging of hole traps was discovered by Schwank et al. (1984, 205 citations) in a study of the effects of trapped-hole annealing and interface-trap buildup on the long-term response of irradiated MOS devices. This phenomenon was later studied extensively by Lelis et al. (1989, 109 citations; 1988, 98 citations), who attributed the charge exchange to a stable dipole state associated with the E' defect (Chapters 6 and 7). The increased stability of trapped holes at low temperature, relative to room temperature, was investigated in detail by Boesch and McGarrity (1976, 102 citations).

Nicollian et al. (1971, 332 citations) investigated a water-related electron trap in SiO_2, with a relatively low capture cross section ($\sim 10^{-17}$ cm^2). The temperature and oxide thickness dependence of electron trapping in SiO_2 was investigated by Young et al. (1979, 205 citations), Aitken et al. (1978, 127 citations), and Aitken and Young (1976, 110 citations); bulk traps were found to dominate at ~300 K, with interface traps dominating at 77 K. The enhancement of Fowler–Nordheim tunnel current due to the buildup of oxide and interface traps was evaluated by Maserjian and Zamani (1982, 192 citations). Hot electron emission from Si into SiO_2, the subsequent charge trapping, and trapped charge neutralization rates were studied in detail in a series of publications by Ning (1974, 188 citations), Ning et al. (1977, 154 citations), Ning (1976, 123 citations; 1978, 110 citations). Electron transport in SiO_2 after a radiation pulse was studied by Hughes (1973, 159 citations), and found to exhibit an effective mobility of ~20 $cm^2 V^{-1} s^{-1}$ at room temperature. Later he measured the mobility of transporting holes after similar radiation pulses and found it to be dispersive, a much stronger function of temperature, and 5–12 orders of magnitude smaller than the electron drift mobility in SiO_2 (Hughes, 1977, 171 citations). The threshold for electron injection into the SiO_2 from Si was measured by Powell (1970, 161 citations) to be ~3 eV, consistent with theoretical predictions. Powell and coworkers (Berglund and Powell, 1971, 154 citations; Powell and Derbenwick, 1971, 138 citations; Powell and Berglund, 1971, 123 citations) performed early studies of trapped oxide charge in SiO_2 using photoinjection and vacuum ultraviolet irradiation techniques. Fluorine treatment was observed to improve Si/SiO_2 interface properties by Wright and Saraswat (1989, 131 citations). Electron traps with capture cross sections of ~10^{-18} to 10^{-17} cm^2 were characterized by Hartstein and Young (1981, 101 citations), which were especially prominent after H_2O diffusion.

O'Reilly and Robertson (1983, 245 citations) performed detailed theoretical calculations of defect structures and energy levels in amorphous SiO_2. Deal (1974, 218 citations) provided a review of MOS oxide and interface-trap charge properties in early 1970s vintage technology. EerNisse (1977, 197 citations) suggested that the viscous flow of SiO_2 above ~925°C–960°C led to defect (e.g., O vacancy) creation. Goodnick et al. (1985, 197 citations) studied the Si surface roughness at the Si/SiO_2 interface using a high-resolution transmission electron microscope, and related the statistical properties of the roughness to mobility degradation. The effects of interface roughness on Si inversion layer mobility also were investigated by Takagi et al. (1994, 275 citations) and Cheng and Sullivan (1973, 120 citations).

The effects of water on MOS oxide- and interface-trap charge were investigated by Feigl et al. (1981, 180 citations). Of particular interest to MOS oxide- and interface-trap charge, the silicon dangling bond, the neutral oxygen vacancy, and the positively charged oxygen vacancy (E' center) all were found to have deep states near midgap. Asokakumar et al. (1994, 189 citations) review the use of positron annihilation spectroscopy to identify and characterize the properties of defect complexes in SiO_2 and at the Si/SiO_2 interface. DiStefano and Eastman (1971, 189 citations) examined the SiO_2 valence band using photoelectron spectroscopy, and observed a narrow, nonbonding level at the band edge that facilitates lattice trapping of valence band holes. Tromp et al. (1985, 188 citations) illustrate the out-diffusion of oxygen during high-vacuum annealing of SiO_2, with the concomitant formation of voids in the SiO_2 network. Daum et al. (1993, 181 citations) apply nonlinear optical techniques (second harmonic generation; sum-frequency spectra) to identify strained Si layers at (100) and (111) Si/SiO_2 interfaces. Grunthaner et al. (1982, 166 citations) studied the properties of defects in the strained, near-interfacial layer of SiO_2 with XPS, as a function of different oxide processing. Enhanced strain was found for oxides that were not radiation hardened, as compared to hardened oxides. Poindexter et al. (1984, 157 citations) used EPR and capacitance–voltage techniques that identify two prominent peaks in the energy distribution of interface traps in the Si band gaps, which are attributed to two distinct charge states of Si dangling bonds at the interface. This work reinforced the conclusions of Johnson et al. (1983, 115 citations), who used deep-level transient spectroscopy to find similar interface-trap energy distributions. A reduction in radiation-induced interface-trap density via interface nitridation was demonstrated by Terry et al. (1983, 110 citations).

Although low-frequency noise in MOS devices is often attributed to the interactions of carriers with defects in the near-interfacial SiO_2 (e.g., Sah and Hielscher, 1966, 111 citations; Uren et al., 1985, 101 citations; Fleetwood et al., 1993, 138 citations), a series of papers by

Hooge and coworkers (Hooge, 1969, 439 citations; Hooge et al., 1981, 429 citations; Hooge, 1976, 267 citations; Hooge, 1994, 188 citations; Hooge, 1972, 172 citations; Hooge and Vandamme, 1978, 114 citations) challenged this viewpoint and led to a significant number of publications that investigated the relative roles of defects and lattice scattering on MOS $1/f$ noise. Handel (1980, 122 citations) argued for a quantum origin for $1/f$ noise. The cases against a dominant role for lattice scattering in MOS $1/f$ noise and against a quantum origin of $1/f$ noise were presented by Weissman (1988, 619 citations). These issues were also considered by van der Ziel (1988, 117 citations). Ralls et al. (1984, 322 citations) characterized random telegraph signals (RTS) associated with single carrier-defect interactions at the Si/SiO_2 interface at cryogenic temperatures. In more highly scaled devices, these RTS can be observed at room temperature quite easily. Sah et al. (1972, 176 citations) evaluated the effects of interface-trap charge on carrier scattering rates in MOS devices; Sah (1976, 141 citations) also presented an early treatment of the origin of oxide- and interface-trap charge in irradiated MOS devices; this followed a previous report by Gwyn (1969, 105 citations). The effects of carrier number and mobility fluctuations due to defects near the Si/SiO_2 interface on MOS transistor low-frequency noise was modeled by Hung et al. (1990, 160 citations); another treatment of this topic was performed by Ghibaudo et al. (1991, 122 citations). A model of MOS field effect transistor (MOSFET) noise in weak and strong inversion that includes the effects of defects was developed by Reimbold (1984, 119 citations). The use of low-frequency noise as a diagnostic tool to identify defective electronic devices was reviewed by Vandamme (1994, 128 citations).

Kar and Dahlke (1972, 154 citations) identify interface-trap peaks associated with metallic impurities (Mg, Cu, Cr, Au) that evidently diffuse from the device contacts to the Si/SiO_2 interface. Iwata and Ishizaka (1996, 153 citations) apply electron spectroscopy for chemical analysis to characterize Si/SiO_2 interface-trap properties, and correlate this technique with conventional electrical measurements. Threshold voltage instabilities associated with hot hole emission were characterized by Fair and Sun (1981, 147 citations). The effects of oxidation parameters on interface-trap densities and impurity redistribution were also investigated by Razouk and Deal (1979, 142 citations), Deal et al. (1965, 102 citations), and Castro and Deal (1971, 100 citations); Deal et al. (1969, 138 citations; 1968, 110 citations) also evaluated the characteristics of interface traps at interfaces between Si and SiO_2 and SiO_2/Si_3N_4. DiMaria et al. (1977, 134 citations) compared the properties of trapped charge in SiO_2 that were created by irradiation and high-field stress; charge trapping was found to occur primarily near the interfaces in each case. A detailed study of the spatial dependence of radiation-induced oxide-trap charge in hard and soft oxides was performed by Oldham et al. (1986, 122 citations). The effects of electron heating on trap creation in SiO_2 were studied by DiMaria et al. (1985, 129 citations). This topic was also examined theoretically by Fischetti (1985, 113 citations). Hot carrier effects on MOS devices were studied by Takeda and Suzuki (1983, 241 citations); Hofmann et al. (1985, 187 citations); Fang and Fowler (1970, 178 citations); Tam et al. (1984, 146 citations), Hsu and Tam (1984, 115 citations), and Hsu and Grinolds (1984, 109 citations); and Doyle et al. (1990, 129 citations). Helms and Poindexter (1994, 130 citations) performed a review and analysis of data on the microstructure and chemical physics of defects in the Si/SiO_2 system that are obtained using a wide variety of materials and electrical characterization techniques; an earlier review of ESR characterization of defects at the Si/SiO_2 interface was provided by Poindexter and Caplan (1983, 116 citations). Similarities and differences among defects in SiO_2 and chalcogenide semiconductors were reviewed by Mott (1977, 111 citations). Defect generation was studied in detail in MOS capacitors by Buchanan and DiMaria (1990, 104 citations); in this article, the authors evaluate oxide, interface, and near-interface oxide-trap charge under a variety of stress conditions.

The effects of ion implantation on defects in SiO_2 were evaluated by Arnold (1973, 114 citations). Blue luminescence was observed in Si^+ implanted SiO_2 films on Si by Liao et al. (1996, 133 citations). Trapped-hole energies in SiO_2 were estimated after high-field stress by Woods and Williams (1976, 105 citations). Montillo and Balk (1971, 104 citations) demonstrated that fixed charge densities in the near-interfacial SiO_2 could be decreased with high temperature annealing. Nissan-Cohen et al. (1986, 104 citations) studied trap generation in SiO_2 under high-field stress conditions. Substrate current and charge trapping associated with hot carrier injection were measured by Cottrell et al. (1979, 103 citations). Aberle et al. (1992, 103 citations) evaluate the effects of defects and experimental conditions on the measured surface recombination velocity associated with Shockley–Read–Hall recombination at the Si/SiO_2 interface. Pasquarello et al. (1998, 101 citations) simulate the oxidation of Si via molecular dynamics to investigate how a Si-rich interface can evolve during the oxidation process without the creation of dangling bond defects.

A.2.3 Effects of Hydrogen

Table A.5 lists papers that discuss the effects of hydrogen on interface-trap formation, which has been a particularly active area of investigation for more than 30 years (Chapters 6, 7, and 13), as well as the interactions of hydrogen with defects in bulk silicon (Chapter 2). Pearton et al. (1987, 690 citations) have reviewed the properties of hydrogen in Si and other bulk semiconductors. McLean (1980, 322 citations) provided the first complete model of the effects of protons on interface-trap formation, primarily in the context of the

TABLE A.5

Citation Summary (# Cit $\geq$ 100) for Journal Articles that Discuss the Effects of Hydrogen on Interface-Trap Formation in MOS Devices, as Well as Hydrogen Interactions with Bulk Si Defects

1st Author	Year	Journal	Abridged Title	# Cit.
Pearton, S.J.	1987	*Appl. Phys. A—Mater. Sci. Proc.*	Hydrogen in crystalline semiconductors	690
Van de Walle, C.	1989	*Phys. Rev. B*	Theory of hydrogen diffusion and reactions in crystalline Si	369
Bruel, M.	1995	*Electron. Lett.*	Silicon on insulator material technology	353
DiMaria, D.J.	1989	*J. Appl. Phys.*	Trap creation in SiO_2 produced by hot electrons	346
McLean, F.B.	1980	*IEEE Trans. Nucl. Sci.*	Framework for understanding radiation-induced interface states	322
Johnson, N.M.	1987	*Phys. Rev. B*	Defects in single crystal Si induced by hydrogenation	303
Pankove, J.I.	1983	*Phys. Rev. Lett.*	Neutralization of shallow acceptor levels in Si by atomic hydrogen	290
Griscom, D.L.	1985	*J. Appl. Phys.*	Diffusion of radiolytic hydrogen as a mechanism for [interface-trap buildup]	276
Johnson, N.M.	1986	*Phys. Rev. Lett.*	Interstitial hydrogen and neutralization of shallow donor impurities in Si	265
Sah, C.T.	1983	*Appl. Phys. Lett.*	Deactivation of the boron acceptor in Si by hydrogen	222
Pankove, J.I.	1985	*Appl. Phys. Lett.*	Hydrogen localization near boron in Si	219
Van de Walle, C.	1994	*Phys. Rev. B*	Energies of various configurations of hydrogen in Si	212
Lyding, J.W.	1996	*Appl. Phys. Lett.*	Reduction of hot electron degradation in MOS transistors by deuterium	196
Jeppson, K.O.	1977	*J. Appl. Phys.*	Negative bias stress of MOS devices at high electric fields and degradation	189

(continued)

TABLE A.5 (continued)

Citation Summary (# Cit $\geq$ 100) for Journal Articles that Discuss the Effects of Hydrogen on Interface-Trap Formation in MOS Devices, as Well as Hydrogen Interactions with Bulk Si Defects

1st Author	Year	Journal	Abridged Title	# Cit.
Van de Walle, C.	1988	*Phys. Rev. Lett.*	Theory of hydrogen diffusion and reactions in crystalline Si	180
Johnson, N.M.	1985	*Phys. Rev. B*	Mechanism for hydrogen compensation of shallow acceptor impurities	175
Myers, S.M.	1992	*Rev. Mod. Phys.*	Hydrogen interactions with defects in crystalline solids	173
Weldon, M.K.	1997	*J. Vac. Sci. Technol. B*	On the mechanism of the H-induced exfoliation of Si	161
Bergman, K.	1988	*Phys. Rev. B*	Donor–hydrogen complexes in passivated Si	157
Chang, K.J.	1989	*Phys. Rev. Lett.*	Diatomic hydrogen complex diffusion and self-trapping in crystalline Si	152
Stein, H.J.	1975	*J. Electron. Mater.*	Bonding and thermal stability of implanted hydrogen in Si	151
Chang, K.J.	1988	*Phys. Rev. Lett.*	Theory of hydrogen passivation of shallow level dopants in crystalline Si	149
Estreicher, S.K.	1995	*Mater. Sci. Engrg. R—Rep.*	Hydrogen-related defects in crystalline semiconductors—a theorist's perspective	141
Pankove, J.I.	1984	*Appl. Phys. Lett.*	Neutralization of acceptors in Si by atomic hydrogen	140
Deleo, G.G.	1985	*Phys. Rev. B*	Hydrogen–acceptor pairs in Si—effect on hydrogen vibrational frequency	139
Deák, P.	1988	*Phys. Rev. B*	State and motion of hydrogen in crystalline Si	138
Holbech, J.D.	1993	*Phys. Rev. Lett.*	H_2* defect in crystalline Si	133
Zundel, T.	1989	*Phys. Rev. B*	Dissociation energies of shallow acceptor–hydrogen pairs in Si	132
Bruel, M.	1996	*Nucl. Instrum. Methods Phys. Res., Sect. B*	Application of hydrogen ion beams to SOI material technology	126
Cartier, E.	1993	*Appl. Phys. Lett.*	Passivation and depassivation of Si dangling bonds by atomic hydrogen	125
Singh, V.A.	1977	*Phys. Status Solidi B*	Vibrational and electronic structure of hydrogen-related defects in Si	124
Revesz, A.G.	1979	*J. Electrochem. Soc.*	Role of hydrogen in SiO_2 films on Si	123
Winokur, P.S.	1979	*J. Appl. Phys.*	Two-stage process for buildup of radiation-induced interface states	122
Gale, R.	1983	*J. Appl. Phys.*	Hydrogen migration under avalanche injection of electrons in MOS capacitors	120
Benton, J.L.	1980	*Appl. Phys. Lett.*	Hydrogen passivation of point defects in Si	117
Corbett, J.W.	1983	*Phys. Lett. A*	Atomic and molecular hydrogen in the Si lattice	116
Denteneer, P.J.H.	1989	*Phys. Rev. B*	Microscopic structure of the hydrogen–boron complex in crystalline Si	115
Saks, N.S.	1989	*IEEE. Trans. Nucl. Sci.*	Interface-trap formation via 2-stage H^+ process	115
Brower, K.L.	1988	*Phys. Rev. B*	Kinetics of H_2 passivation of P_b centers at the (111) Si/SiO_2 interface	112
Brower, K.L.	1990	*Phys. Rev. B*	Dissociation kinetics of H-passivated (111) Si–SiO_2 interface defects	112
Reed, M.L.	1988	*J. Appl. Phys.*	Chemistry of Si–SiO_2 interface-trap annealing	105
Winokur, P.S.	1977	*IEEE. Trans. Nucl. Sci.*	Field- and time-dependent radiation effects at the SiO_2 interface	103

results of experiments on MOS performed by Winokur et al. (1979, 122 citations; 1977, 103 citations). DiMaria and Stasiak (1989, 346 citations) showed that hydrogen release and reactions during hot carrier stress played a key role in trap creation in the SiO_2 bulk and near its interfaces. Griscom (1985, 276 citations) developed a diffusion model for interface-trap buildup in irradiated MOS devices; later he collaborated with Saks and Brown to refine this model to include the effects of proton motion on radiation-induced interface-trap generation (Griscom et al., 1988, 86 citations). Lyding et al. (1996, 196 citations) showed one could reduce hot carrier buildup significantly by replacing hydrogen annealing in an MOS process with deuterium annealing. Jeppson and Svensson (1977, 189 citations) showed the critical role of hydrogen reactions and diffusion on the phenomenon of negative bias-temperature stress (Chapters 12 through 14). The topic of hydrogen transport and reactions in SiO_2 was the subject of intensive investigation by Saks and Brown (1989, 115 citations) as well. Myers et al. (1992, 173 citations) have reviewed extensively the interactions of hydrogen with defects at the Si/SiO_2 interface and in bulk semiconductors and metals. A theoretical review on this topic was also provided by Estreicher (1995, 141 citations).

A theory of hydrogen diffusion and reactions in Si was developed by Van de Walle et al. (1989, 369 citations; 1988, 180 citations), based on the pseudopotential-density-functional method in a supercell geometry. Defects introduced into single crystal Si that are induced by hydrogenation were characterized by Johnson et al. (1987, 303 citations). These defects can form electrically active gap states that can function as acceptors in Si. The deactivation of boron acceptors by hydrogen was demonstrated in capacitors by Sah et al. (1983, 222 citations), and shown by Pankove et al. (1983, 290 citations; 1984, 140 citations) to lead to up to a sixfold increase in Si resistivity. Hydrogen localization near boron was studied in Si by Pankove et al. (1985, 219 citations) and by Johnson (1985, 175 citations). Energies of hydrogen in several different configurations in Si were calculated by Van de Walle (1994, 212 citations). The deactivation of donors in Si was demonstrated by Johnson et al. (1986a, 265 citations). Donor–hydrogen complexes in Si were studied by Bergman et al. (1988, 157 citations). Chadi and Chang (1989, 152 citations) studied the diffusion and self-trapping of a metastable diatomic hydrogen complex in Si, and also the passivation of acceptors and donors by hydrogen in Si (Chadi and Chang, 1988a,b, 149 citations). Hydrogen–boron and hydrogen–aluminum pairs in Si were studied by Deleo and Fowler (1985, 139 citations). Relatively low barriers for the diffusion of atomic hydrogen or H^+ in Si by found by Deák et al. (1988, 138 citations); Corbett et al. (1983, 116 citations) also predicted that H_2 would form in Si as a result of atomic hydrogen motion. The H_2^* defect in Si was studied in detail by Holbech et al. (1993, 133 citations). The dissociation of acceptor–hydrogen complexes was studied by Zundel and Weber (1989, 132 citations). Cartier et al. (1993, 125 citations) evaluated in detail the passivation and depassivation of Si dangling bonds at the Si/SiO_2 interface by atomic hydrogen. The vibrational properties and electronic structure of hydrogen-related defects in Si were calculated by Singh et al. (1977, 124 citations). An extensive review of the various roles hydrogen can play in defect generation in SiO_2 was provided by Revesz (1979, 123 citations). Hydrogen redistribution in SiO_2 as a result of electrical stress, leading to interface-trap formation, was demonstrated by Gale et al. (1983, 120 citations). Hydrogen passivation of point defects in Si was studied by Benton et al. (1980, 117 citations). The microstructure of the hydrogen–boron complex in Si was studied theoretically by Denteneer et al. (1989, 115 citations). The dissociation kinetics of hydrogen-passivated Si dangling bonds and the H_2 passivation of P_b centers at the Si/SiO_2 interface were both studied with EPR techniques by Brower (1990, 112 citations; 1988, 112 citations). The chemistry of interface-trap annealing was investigated by Reed and Plummer (1988, 105 citations), with a strong role envisioned for hydrogen in interface-trap annealing, as well as interface-trap buildup.

In extremely high densities, e.g., after hydrogen implantation, defects and the subsequent buildup of H_2 facilitate the exfoliation and cracking of Si wafers, which has been exploited in layer transfer techniques such as bonded silicon-on-insulator technologies, as demonstrated by Bruel (1995, 353 citations; 1996, 126 citations). The mechanisms of the defect buildup and exfoliation have been studied by Weldon et al. (1997, 161 citations). The bonding and stability of implanted hydrogen in Si was studied by Stein (1975, 151 citations).

A.2.4 Thin Oxides, Leakage, and Breakdown

Defects in ultrathin SiO_2 affect the leakage through the dielectric layer, and also can limit device reliability because of enhanced leakage current or dielectric breakdown. In addition, defects in MOS dielectric layers can reduce the lifetime of charge storage devices, such as flash memories. In contrast, charge trapping at defects is also exploited to create nonvolatile memory elements. Table A.6 lists papers that describe defect properties in ultrathin SiO_2 layers, as well as in the critical gate dielectrics of MOS-based charge storage devices, in which leakage current is a primary failure mechanism (Chapters 15 through 18). Tiwari, Hanafi, and coworkers describe a nonvolatile memory technology based on $\sim$5 nm Si nanocrystals in SiO_2 that serve as charge traps when devices are programmed by a voltage pulse (Tiwari et al., 1996, 526 citations; Hanafi et al., 1996, 166 citations). DiMaria et al. (1993, 388 citations) found in a detailed study using a large number of complementary characterization techniques that the degradation and eventual breakdown of SiO_2 during high-field electrical stress is triggered by trap creation near the Si and gate interfaces of SiO_2 films on Si. These results were obtained primarily on oxides thicker than $\sim$4.5 nm. Green et al. (2001, 269 citations) have extensively reviewed the processing, structural, physical, and electrical limits of ultrathin SiO_2 as a dielectric layer. An impact ionization model of oxide breakdown was developed by DiStefano and Shatzkes (1974, 103 citations); impact ionization models of breakdown in SiO_2 were questioned by Harari (1978, 274 citations), who found more evidence supporting the electron trap model of breakdown. In turn, Chen, Lee, Hu, and coworkers showed evidence supporting impact ionization and hole trapping models of SiO_2 breakdown (Chen et al., 1985, 266 citations), and characterized the resulting breakdown distribution with an effective oxide thinning model (Lee et al., 1988b, 143 citations).

Defect densities and energy levels play a crucial role in determining the ultimate scaling limits of thermal SiO_2 as a manufacturable gate dielectric layer. In thinner oxides, DiMaria and Cartier (1995, 255 citations) attribute stress-induced leakage current in thin SiO_2 gate dielectric layers to the generation of neutral electron traps in the SiO_2. These defects in the SiO_2 bulk are attributed to hydrogen release, transport, and reactions within the films that are not unlike the reactions known to cause interface-trap formation in MOS devices. These results are generally consistent with an earlier study by Olivo et al. (1988, 188 citations). A percolation model that accounts for oxide breakdown due to electrical stress-induced electron trap generation in thin oxides was developed by Degraeve et al. (1998, 186 citations); the potential limitations on oxide scaling due to the generation of defects during high-field stress were discussed in the context of this percolation model by Stathis (1999, 103 citations); a hole injection model of oxide breakdown was developed by Schuegraf and Hu (1994, 180 citations). The dependence of the quality and resistance to breakdown of SiO_2 films on the surface roughness was characterized by Ohmi et al. (1992, 164 citations). A hydrogen release and electron-trapping model of oxide breakdown was presented by Arnold et al. (1994, 131 citations). Depas et al. (1996, 134 citations) characterize the soft breakdown characteristics of ultrathin oxides, and attribute the increase in leakage current with electrical stress to electron traps in the oxide. Stress-induced trap

TABLE A.6

Citation Summary (# Cit ≥100) for Journal Articles that Discuss Defects in Ultrathin SiO_2, and in MOS Memory Structures that are Sensitive to Leakage Current-Related Failures

1st Author	Year	Journal	Abridged Title	# Cit.
Tiwari, S.	1996	*Appl. Phys. Lett.*	A Si nanocrystals-based memory	526
DiMaria, D.J.	1993	*J. Appl. Phys.*	Impact ionization, trap creation, degradation and breakdown in SiO_2 on Si	388
Harari, E.	1978	*J. Appl. Phys.*	Dielectric breakdown in electrically stressed thin films of thermal SiO_2	274
Green, M.L.	2001	*J. Appl. Phys.*	Ultrathin SiO_2 and Si–O–N gate dielectric layers for Si microelectronics	269
Chen, I.C.	1985	*IEEE Trans. Electron Devices*	Electrical breakdown in thin gate and tunneling oxides	266
DiMaria, D.J.	1995	*J. Appl. Phys.*	Mechanism for stress-induced leakage currents in thin SiO_2 films	255
DiMaria, D.J.	1984	*J. Appl. Phys.*	Electroluminescence studies in SiO_2 films containing tiny Si islands	207
Olivo, P.	1988	*IEEE Trans. Electron Devices*	High-field induced breakdown in ultrathin SiO_2 films	188
Degraeve, R.	1998	*IEEE Trans. Electron Devices*	Relation between electron trap generation and oxide breakdown	186
Schuegraf, K.F.	1994	*IEEE Trans. Electron Devices*	Hole injection SiO_2 breakdown model for low voltage lifetime extrapolation	180
Hanafi, H.I.	1996	*IEEE Trans. Electron Devices*	Fast and long retention time nanocrystal memory	166
Frohman-Bentchkowsky, D.	1969	*J. Appl. Phys.*	Charge transport and storage in MNOS structures	165
Ohmi, T.	1992	*IEEE Trans. Electron Devices*	Dependence of thin oxide films on surface microroughness	164
Buchanan, D.A.	1999	*IBM J. Res. Dev.*	Scaling the gate dielectric: materials, integration, and reliability	159
Lee, J.C.	1988	*IEEE Trans. Electron Devices*	Modeling and characterization of gate oxide reliability	143
Depas, M.	1996	*IEEE Trans. Electron Devices*	Soft breakdown of ultrathin gate oxide layers	134
Arnold, D.	1994	*Phys. Rev. B*	Theory of high-field electron transport and impact ionization in SiO_2	131
Eitan, B.	2000	*IEEE Electron Device Lett.*	NROM: A localized trapping, 2-bit nonvolatile memory cell	130
Dumin, D.	1993	*IEEE Trans. Electron Devices*	Correlation of stress-induced leakage current with trap generation	127
Shi, Y.	1998	*J. Appl. Phys.*	Effects of traps on charge storage characteristics of MOS memory	117
Nakashima, S.	1993	*J. Mater. Res.*	Analysis of buried oxide layer formation and mechanism of dislocation generation	116
Stein, H.J.	1979	*J. Electrochem. Soc.*	Properties of plasma-deposited silicon nitride	111
Suñé, J.	1990	*Thin Solid Films*	On the breakdown statistics of very thin SiO_2 films	106
DiStefano, T.H.	1974	*Appl. Phys. Lett.*	Impact ionization model for dielectric instability and breakdown	103
Stathis, J.H.	1999	*J. Appl. Phys.*	Percolation models for gate oxide breakdown	103

distributions in thin oxides were evaluated in detail by Dumin and Maddux (1993, 127 citations). Nakashima and Izumi (1993, 116 citations) evaluated the properties of separation by implantation of oxygen (SIMOX) buried oxides on threading dislocations that form during buried oxide growth and annealing. The statistics of oxide breakdown in very thin oxides were characterized by Suñé et al. (1990, 106 citations). Buchanan (1999, 159 citations) discusses impurity penetration, leakage, and other processing and reliability concerns that ultimately limit the ability of SiO_2 and oxynitride-based dielectrics to continue to scale down in thickness.

The properties of SiO_2 layers with nanocrystalline Si inclusions, which have been studied as potential memory storage devices, were characterized by DiMaria et al. (1984, 207 citations). Charge storage after injection was demonstrated as a memory device in metal–nitride–oxide–Si structures by Frohman-Bentchkowsky and Lenzlinger (1969, 165 citations). Shi et al. (1998, 117 citations) investigate charge buildup and decay in Si nanocrystal-based memories, and find that interface traps reduce the retention time of the devices. A new type of flash nonvolatile memory cell based on channel hot electron injection was described by Eitan et al. (2000, 130 citations). Defects in silicon nitride films intended for use in nonvolatile memory structures were characterized as a function of processing conditions by Stein et al. (1979, 111 citations).

A.2.5 Bulk Silicon

Here we consider the extensive literature on defects in bulk Si and related materials (Chapters 1 through 3) (Table A.7). These defects can affect the yield, minority carrier lifetime, dopant distribution, and many other properties of Si-based microelectronic circuits and devices as they are processed, or when used in a high-radiation environment. Early calculations on radiation-induced displacement damage in Si and Ge were performed by Kinchin and Pease (1955, 870 citations) and by Gossick (1959, 260 citations). The stopping of ions in compounds was surveyed and categorized by Ziegler and Manoyan (1988, 185). The results of these experiments and calculations have been incorporated into the computer codes transport of ions in matter (TRIM)/stopping and range of ions in matter (SRIM; see www.srim.org), which are commonly applied to make first-order calculations of displacement damage in semiconductor materials and devices. Summers (1993, 158 citations) has shown the utility of the nonionizing energy loss calculations (e.g., in TRIM/SRIM) in the analysis of the effects of displacement damage on semiconductor device and materials properties degradation.

Fahey et al. (1989, 652 citations) review point defects and diffusion in Si; an earlier review of this topic was performed by Seeger and Chik (1968, 363 citations). The Si E center (a vacancy-phosphorus defect) and the Si A center (a vacancy-oxygen center) were investigated using ESR by Watkins (1964, 578 citations; 1961, 569 citations). The activation energy for the diffusion of the divacancy defect in crystalline Si was found to be $\sim$1.3 eV by Watkins and Corbett (1965, 541 citations). The divacancy and vacancy defects in Si were further studied via EPR by Cheng et al. (1966, 260 citations) and Watkins (1963, 255 citations). The transient-enhanced diffusion of boron and phosphorus in Si was reviewed by Stolk et al. (1997, 311 citations), who discuss quantitatively the roles of interstitials in this process. The interactions between interstitials and boron during postimplant annealing processes was also characterized by Eaglesham (1994, 298 citations). The transient-enhanced diffusion of phosphorus during processing and postimplant annealing was studied as well by Giles (1991, 176 citations). Point defects, diffusion, and swirl defects in Si were studied by Tan and Gosele (1985, 302 citations). The effects of growth conditions on swirl defect formation was investigated in detail by Voronkov (1982, 254 citations). The effects of vacancies and interstitials on Si self-diffusion were evaluated by Car et al. (1984, 207 citations). A review of the electronic structure and point defects in semiconductors was performed by Pantelides (1978, 396 citations). The Si vacancy was demonstrated to be an Anderson negative-U center by Baraff et al. (1980, 343 citations), Baraff and Schluter

TABLE A.7

Citation Summary (# Cit ≥ 130) of Journal Articles that Discuss Defects Primarily in Bulk Si

1st Author	Year	Journal	Abridged Title	# Cit.
Kinchin, G.H.	1955	*Rep. Prog. Phys.*	The displacement of atoms in solids by radiation	870
Fahey, P.M.	1989	*Rev. Mod. Phys.*	Point defects and dopant diffusion in Si	652
Weber, E.R.	1983	*Appl. Phys. A—Mater. Sci. Proc.*	Transition metals in Si	642
Watkins, G.D.	1964	*Phys. Rev. A*	Defects in irradiated Si—EPR and ENDOR of Si E center	578
Watkins, G.D.	1961	*Phys. Rev.*	Defects in irradiated Si—ESR of Si A center	569
Watkins, G.D.	1965	*Phys. Rev.*	Defects in irradiated Si—EPR of divacancy	541
Pantelides, S.T.	1978	*Rev. Mod. Phys.*	Electronic structure of impurities and other point defects in semiconductors	396
Seeger, A.	1968	*Phys. Status Solidi*	Diffusion mechanisms and point defects in Si and Ge	363
Baraff, G.A.	1980	*Phys. Rev. B*	Theory of the Si vacancy—an Anderson negative U center	343
Stolk, P.A.	1997	*J. Appl. Phys.*	Physical mechanisms of transient-enhanced dopant diffusion	311
Tan, T.Y.	1985	*Appl. Phys. A—Mater. Sci. Process.*	Point defects, diffusion processes, and swirl defect formation in Si	302
Eaglesham, D.J.	1994	*Appl. Phys. Lett.*	Implantation and transient B-diffusion in Si—the source of the interstitials	298
Jones, K.S.	1988	*Appl. Phys. A—Mater. Sci. Process.*	A systematic analysis of defects in ion-implanted Si	269
Cheng, L.J.	1966	*Phys. Rev.*	1.8, 3.3, and 3.9 μm bands in irradiated Si—correlations with divacancy	260
Gossick, B.R.	1959	*J. Appl. Phys.*	Disordered regions in semiconductors bombarded by fast neutrons	260
Watkins, G.D.	1963	*J. Phys. Soc. Jpn*	An EPR study of the lattice vacancy in Si	255
Voronkov, V.V.	1982	*J. Cryst. Growth*	The mechanism of swirl defects formation in Si	254
Borghesi, A.	1995	*J. Appl. Phys.*	Oxygen precipitation in Si	227
Bourgoin, J.C.	1972	*Phys. Lett. A*	New mechanism for interstitial migration	216
Baraff, G.A.	1978	*Phys. Rev. Lett.*	Self-consistent Green's function calculation of ideal Si vacancy	210
Car, R.	1984	*Phys. Rev. Lett.*	Microscopic theory of atomic diffusion mechanisms in Si	207
Sauer, R.	1985	*Appl. Phys. A—Mater. Sci. Proc.*	Dislocation-related photoluminescence in Si	204
Watkins, G.D.	1980	*Phys. Rev. Lett.*	Negative-U properties for point defects in Si	202
Lee, Y.H.	1976	*Phys. Rev. B*	EPR studies of defects in electron-irradiated Si—V–O complexes	199
Watkins, G.D.	1975	*Phys. Rev. B*	EPR and ENDOR study of interstitial boron in Si	195
Watkins, G.D.	1976	*Phys. Rev. Lett.*	EPR observation of isolated C in Si	192
Ziegler, J.F.	1988	*Nucl. Instrum. Methods Phys. Res., Sect. B*	The stopping of ions in compounds	185
Mooney, P.M.	1977	*Phys. Rev. B*	Defect energy levels in B-doped Si irradiated with 1-MeV electrons	182
Giles, M.D.	1991	*J. Electrochem. Soc.*	Transient P diffusion below the amorphization threshold	176
Omling, P.	1985	*Phys. Rev. B*	Electrical properties of dislocations and point defects in plastically deformed Si	170
Hu, S.M.	1991	*J. Appl. Phys.*	Stress-related problems in Si technology	168

(*continued*)

TABLE A.7 (continued)

Citation Summary (# Cit $\geq$ 130) of Journal Articles that Discuss Defects Primarily in Bulk Si

1st Author	Year	Journal	Abridged Title	# Cit.
Blöchl, P.E.	1993	*Phys. Rev. Lett.*	First principles calculations of self-diffusion constants in Si	164
Elkin, E.L.	1968	*Phys. Rev.*	Defects in irradiated Si—EPR and ENDOR study of As and Sb vacancies	163
Lang, D.V.	1980	*Phys. Rev. B*	Complex nature of gold-related deep levels in Si	163
Baraff, G.A.	1979	*Phys. Rev. Lett.*	Si vacancy—possible Anderson negative U center	161
Antoniadis, D.A.	1982	*J. Appl. Phys.*	Diffusion of substitutional impurities in Si at short oxidation times	159
Lee, Y.H.	1973	*Phys. Rev. B*	EPR studies in neutron-irradiated Si—negative charge state of vacancy cluster	159
Summers, G.P.	1993	*IEEE Trans. Nucl. Sci.*	Damage correlations in semiconductors exposed to radiation	158
Corbett, J.W.	1981	*Nucl. Instrum. Methods Phys. Res., B*	Ion-induced defects in semiconductors	152
Stavola, M.	1983	*Appl. Phys. Lett.*	Diffusivity of oxygen in Si at the donor formation temperature	152
Justo, J.F.	1998	*Phys. Rev. B*	Interatomic potential for Si defects and disordered phases	149
Kwon, I.	1994	*Phys. Rev. B*	Transferrable tight binding models for Si	147
Chen, J.W.	1980	*Annu. Rev. Mater. Sci.*	Energy levels in Si	142
Tasch, A.F.	1970	*Phys. Rev. B*	Recombination-generation and optical properties of gold acceptor in Si	139
Kimerling, J.C.	1983	*Physica B and C*	Electronically controlled reactions of interstitial Fe in Si	137
Sugino, O.	1992	*Phys. Rev. Lett.*	Vacancy in Si—successful description within the local density approximation	135
Lee, Y.H.	1976	*Phys. Rev. B*	EPR studies in neutron-irradiated Si—positive charge state of di-interstitial	133
Coutinho, J.	2000	*Phys. Rev. B*	Oxygen and dioxygen centers in Si and Ge	132
Feldman, L.C.	1970	*J. Appl. Phys.*	Depth profiles of lattice disorder resulting from ion bombardment of Si	132
Kimerling, L.C.	1979	*Appl. Phys. Lett.*	Defect states associated with dislocations in Si	132
Pantelides, S.T.	1974	*Phys. Rev. B.*	Theory of localized states in semiconductors	132
Baraff, G.A.	1984	*Phys. Rev. B*	Migration of interstitials in Si	131
Sah, C.T.	1969	*Appl. Phys. Lett.*	Thermal emission rates of carriers at gold centers in Si	131

(1978, 210 citations), and Baraff et al. (1979, 161 citations). A review and analysis of defects in ion-implanted Si was performed by Jones et al. (1988, 269 citations); a review and analysis of oxygen precipitation and properties of oxide precipitates in Si was performed by Borghesi et al. (1995, 227 citations).

Mechanisms of interstitial migration in semiconductors was studied by Burgoin and Corbett (1972, 216 citations). Photoluminescence associated with dislocations in Si was studied by Sauer et al. (1985, 204 citations). The negative-U properties of the isolated vacancy and interstitial boron in Si were emphasized by Watkins and Troxell (1980, 202 citations). The triplet state of the vacancy-oxygen center was investigated using EPR by Lee and Corbett (1976, 199 citations), who also studied the negative charge state of a nonplanar vacancy cluster (Lee and Corbett, 1973, 159 citations) and a di-interstitial (Lee et al., 1976, 133 citations) in neutron-irradiated Si. Watkins and Brower (1976, 192 citations) used

EPR to observe isolated C in Si; EPR and electron nuclear double resonance (ENDOR) methods were used by Watkins (1975, 195 citations) to study interstitial boron in Si, as well as As and Sb vacancy pairs in irradiated Si (Elkin and Watkins, 1968, 163 citations). Mooney et al. (1977, 182 citations) investigated the defect energy levels of 1-MeV electron-irradiated Si. The energy levels of point defects and dislocations introduced by the plastic deformation of Si were studied via capacitance transient spectroscopy by Omling et al. (1985, 170 citations). Stress-related issues in Si-based CMOS technologies were reviewed by Hu (1991, 168 citations). Defect-mediated self-diffusion in Si was studied by Blöchl et al. (1993, 164 citations); the self-interstitial mechanism was found to be dominant over other mechanisms. Point defect kinetics were studied in oxidized Si by Antoniadis and Moskowitz (1982, 159 citations), with a focus on the diffusion of substitutional impurities. The diffusion of oxygen in Si at the donor formation temperature was studied by Stavola et al. (1983, 152 citations). Defect-induced amorphization and other ion-induced effects were described by Corbett et al. (1981, 152 citations). An empirical potential was developed for Si by Justo (1998, 149 citations) to facilitate the calculation of properties such as the relaxation of point defects, the core properties of partial dislocations, and the structure of disordered phases. A tight binding model was developed and applied to defect formation energies for vacancies and interstitials in Si by Kwon et al. (1994, 147 citations). Properties of the vacancy in Si were calculated using the local density approximation by Sugino and Oshiyama (1992, 135 citations).

Extensive reviews of impurity energy levels in Si were provided by Weber (1983, 642 citations) and by Chen and Milnes (1980, 142 citations). The gold center in Si was studied by Lang et al. (1980, 163 citations) and by Tasch and Sah (1970, 139 citations) and Sah et al. (1969, 131 citations). Interstitial iron in Si was studied by Kimerling and Benton (1983, 137 citations). Calculations of the properties of localized states in semiconductors were performed by Pantelides and Sah (1974, 132 citations). The formation energies, local vibrational modes, and diffusion or reorientational energies of a variety of defect centers in Si and Ge were calculated by Coutinho et al. (2000, 132 citations) using density-functional theory; defect states associated with dislocations in Si have been studied by Kimerling and Patel (1979, 132 citations). Ion-induced lattice disorder was profiled in depth in Si by Feldman and Rodgers (1970, 132 citations). The migration of interstitial Si and Al in Si was studied theoretically by Baraff and Schluter (1984, 131 citations).

A.3 High-*K* Dielectrics on Silicon

In the last 10 years, it has become increasingly clear that future MOS devices will need to incorporate dielectric layers with dielectric constants higher than that of SiO_2. This will enable gate oxides to be physically thick enough to minimize the high tunneling currents that would flow through SiO_2 layers with equivalent capacitance (Chapters 9 through 12). High-*K* dielectrics are also of great interest as materials for capacitors in dynamic random access memories, as well as nonvolatile memory devices (e.g., ferroelectrics).

Table A.8 lists papers that discuss defect densities and properties of MOS devices with high-*K* gate dielectrics. Wilk et al. (2001, 1550 citations) provide an extensive review of early progress in developing high-*K* replacements for SiO_2 gate dielectrics; achieving manageable defect densities remains a significant challenge for the manufacturability of high-*K* dielectrics. Polaron formation and binding to impurities and dopant atoms in transition metal oxides was addressed by Austin and Mott (1969, 1511 citations). The thermodynamic stability of binary oxides on Si was evaluated by Hubbard and Schlom (1996, 548 citations). Hafnium and zirconium silicates were identified by Wilk

TABLE A.8

Citation Summary (# Cit ≥ 100) for Journal Articles that Describe Defects in High-*K* Dielectrics on Si

1st Author	Year	Journal	Abridged Title	# Cit.
Wilk, G.D.	2001	*J. Appl. Phys.*	High-*K* gate dielectrics: Current status and materials properties	1550
Austin, I.G.	1969	*Adv. Phys.*	Polarons in crystalline and noncrystalline materials	1511
Hubbard, K.J.	1996	*J. Mater. Res.*	Thermodynamic stability of binary oxides in contact with Si	548
Wilk, G.D.	2000	*J. Appl. Phys.*	Hf and Zr silicates for advanced gate dielectrics	477
Robertson, J.	2000	*J. Vac. Sci. Technol*	Band offsets of wide band gap oxides and implications for devices	465
Copel, M.	2000	*Appl. Phys. Lett.*	Structure and stability of ultrathin ZrO_2 on Si(100)	373
Sze, S.M.	1967	*J. Appl. Phys.*	Current transport and maximum dielectric strength of silicon nitride films	305
Wilk, G.D.	1999	*Appl. Phys. Lett.*	Electrical properties of hafnium silicate gate dielectrics on Si	281
Lee, B.H.	2000	*Appl. Phys. Lett.*	Thermal stability and electrical characterization of ultrathin HfO_2 on Si	277
Chaneliere, C.	1998	*Mater. Sci. Eng. R-Rep.*	Ta_2O_5 thin films for advanced dielectric applications	182
Alers, G.B.	1998	*Appl. Phys. Lett.*	Intermixing at the Ta_2O_5/Si interface in gate dielectric structures	176
Campbell, S.A.	1997	*IEEE Trans. Electron Devices*	MOSFETs fabricated with high permittivity TiO_2 dielectrics	175
Robertson, J.	1991	*Philos. Mag. B*	Electronic structure of silicon nitride	166
Eisenbeiser, K.	2000	*Appl. Phys. Lett.*	FETs with $SrTiO_3$ gate dielectric on Si	162
Ma, T.P.	1998	*IEEE Trans. Electron Devices*	Making silicon nitride a viable gate dielectric	134
Fischetti, M.V.	2001	*J. Appl. Phys.*	Effective electron mobility in Si inversion layers in MOS with high-*K*	118
Gusev, E.P.	2001	*Microelectron. Eng.*	Ultrathin high-*K* metal oxides on Si	114
Houssa, M.	2000	*J. Appl. Phys.*	Trap-assisted tunneling in high permittivity gate dielectric stacks	113
Callegari, A.	2001	*J. Appl. Phys.*	Physical and electrical characterization of HfO_2 and Hf silicate sputtered films	100

et al. (2000, 477 citations) as excellent candidates for high-*K* dielectrics that are compatible with Si MOS processing. Band offsets and metal-induced gap states for wide band gap oxides on Si were calculated by Robertson (2000, 465 citations). The structure and stability of thin ZrO_2 layers on Si was demonstrated by Copel et al. (2000, 373 citations). Sze (1967, 305 citations) characterized the charge transport and dielectric strength of silicon nitride films on Si. The electrical properties of Hf silicate dielectrics deposited directly on Si were evaluated by Wilk and Wallace (1999, 281 citations). The thermal stability and leakage current of HfO_2 films on Si were characterized by Lee et al. (2000, 277 citations). The properties of Ta_2O_5 for microelectronics application were reviewed by Chaneliere et al. (1998, 182 citations). Intermixing between the Ta_2O_5 and the interfacial SiO_2 layer was demonstrated by Alers et al. (1998, 176 citations). High interface-trap densities were observed by Campbell et al. (1997, 175 citations) to limit the mobilities of MOSFETs with TiO_2 dielectrics. Robertson (1991, 166 citations) has calculated the bonding and electronic structure of silicon nitride to estimate its band gap and characteristic defect energies. Eisenbeiser et al. (2000, 162 citations) describe MOS transistors with low interface-trap densities fabricated with $SrTiO_3$ gate dielectrics. Ma (1998, 134 citations) compares the insulating properties and defect densities of MOS structures fabricated with jet vapor deposited (JVD) silicon nitride as the gate dielectric to devices with thermal SiO_2 gate oxides, and concludes that JVD nitrides are a viable option for advanced integrated circuit technologies with highly scaled dimensions. Scattering rate effects on carrier mobilities in MOS devices with high-K dielectrics were calculated (Chapter 4) by Fischetti et al. (2001, 118 citations). Processing, characterization, and integration issues for high-K dielectrics on Si were reviewed by Gusev et al. (2001, 114 citations). Trap-assisted tunneling through SiO_x/ZrO_2 and SiO_x/Ta_2O_5 gate dielectrics was investigated by Houssa et al. (2000, 113 citations). Callegari et al. (2001, 100 citations) performed detailed materials and electrical characterization of HfO_2 and Hf silicate films, and identified process conditions that produced negligible flatband voltage shifts and low interface defect densities.

A.4 Nonsilicon Devices and Materials

Although silicon dominates the commercial microelectronics industry, there remains intense interest in semiconductor technologies that can provide higher speed, dissipate higher power, operate at higher temperatures, etc. Materials used for MOS devices other than Si include SiC, GaN, diamond, Ge, etc. Dielectrics employed in these technologies include SiO_2 and high-K materials.

A.4.1 MOS Devices Built on Non-Si Substrates

Table A.9 lists papers that describe defects in MOS devices that are built on substrates or materials layers other than Si. Dislocation-free growth of Ge on the (100) surface of Si was demonstrated by Eaglesham and Cerullo (1990, 1139 citations). Bean et al. (1984, 504 citations) demonstrated dislocation-free growth of Ge_xSi_{1-x} films on Si with $x \leq 0.5$ by means of strained-layer epitaxy; People and Bean (1985, 856 citations) calculated the critical thickness for dislocation-free growth for Ge_xSi_{1-x} films on Si for $0 \leq x \leq 1$. The properties and applications of these structures were reviewed by People (1986, 479 citations). Dodson and Tsao (1987, 376 citations) developed a model of Ge_xSi_{1-x} strained-layer relaxation via plastic flow. Legoues et al. (1991, 254 citations; 1992, 185 citations) showed that defect-free SiGe layers containing up to 60% Ge could be grown on Si, with the aid of a relaxation process assisted by dislocations in the Si substrate. The band structure and carrier mobilities for strained Si, Ge, and SiGe alloy structures were calculated by Fischetti and Laux (1996, 231 citations). A kinetic model for strain relaxation in SiGe/Si heterostructures was developed by Houghton (1991, 204 citations); the band gap and doping properties of these heterostructures were measured by Lang et al. (1985, 203 citations) and People et al. (1984, 203 citations). The effects of vacancies and interstitials on Ge and B diffusing in SiGe was studied by Cowern et al. (1994, 118 citations). The growth of GaAs and other compound semiconductors on Si for device applications was reviewed by Fang et al. (1990, 231 citations).

The fabrication, characterization, and defect properties of electronic devices in SiC (Chapter 21) were reviewed by Davis et al. (1991, 297 citations). Donor and acceptor levels of defects in SiC were investigated by Pensl and Choyke (1993, 199 citations). Deep-level transient spectroscopy has been applied to investigate deep defect centers in 3C, 4H, and 6H SiC polytypes (Dalibor et al., 1997, 179 citations). Lipkin and Palmour (1996, 161 citations) show that a reduction in oxidation temperature and a subsequent postoxidation annealing treatment can reduce significantly the oxide and interface-trap densities of oxides on SiC (Chapter 22). Afanase'ev et al. (1997, 143 citations) have characterized the

TABLE A.9

Citation Summary (# Cit ≥100) for Journal Articles that Discuss Defects in MOS Structures Built on Semiconductors Other than Si

1st Author	Year	Journal	Abridged Title	# Cit.
Eaglesham, D.J.	1990	*Phys. Rev. Lett.*	Dislocation-free Stranski–Krastanow growth of Ge on Si (100)	1139
People, R.	1985	*Appl. Phys. Lett.*	Calculation of critical layer thickness vs. lattice mismatch for $Ge_{1-x}Si_x$	856
Bean, J.C.	1984	*J. Vac. Sci. Technol. A*	$Ge_{1-x}Si_x$ strained-layer superlattice grown by molecular beam epitaxy	504
People, R.	1986	*IEEE J. Quantum Electron.*	Physics and applications of Ge_xSi_{1-x}/Si strained-layer heterostructures	479
Dodson, B.W.	1987	*Phys. Rev. Lett.*	Relaxation of strained-layer semiconductor structures via plastic flow	376
Legoues, F.K.	1991	*Phys. Rev. Lett.*	Anomalous strain relaxation in SiGe thin films and superlattices	254
Fang, S.F.	1990	*J. Appl. Phys.*	GaAs and other compound semiconductors on Si	231
Fischetti, M.V.	1996	*J. Appl. Phys.*	Band structure and carrier mobility in strained Si, Ge, and SiGe	231
Landstrass, M.I.	1989	*Appl. Phys. Lett.*	Resistivity of CVD deposited diamond films	229
Houghton, D.C.	1991	*J. Appl. Phys.*	Strain relaxation kinetics in $Si_{1-x}Ge_x$/Si heterostructures	204
Lang, D.V.	1985	*Appl. Phys. Lett.*	Measurement of band gap of Ge_xSi_{1-x}/Si strained-layer heterostructures	203
People, R.	1984	*Appl. Phys. Lett.*	Modulation doping in Ge_xSi_{1-x}/Si strained-layer heterostructures	203
Pensl, G.	1993	*Physica B*	Electrical and optical characterization of SiC	199
Legoues, F.K.	1992	*J. Appl. Phys.*	Mechanism and conditions for anomalous strain relaxation in SiGe	185
Landstrass, M.I.	1989	*Appl. Phys. Lett.*	Hydrogen passivation of electrically active defects in diamond	181
Dalibor, T.	1997	*Phys. Status Solidi A*	Deep defect centers in SiC monitored with DLTS	179
Lipkin, L.A.	1996	*J. Electron. Mater.*	Improved oxidation procedures for reduced SiO_2/SiC defects	161
Afanase'ev, V.V.	2000	*Appl. Phys. Lett.*	Intrinsic SiC/SiO_2 interface states	143
Neudeck, P.G.	1994	*IEEE Electron Device Lett.*	Performance limiting micropipe defects in SiC wafers	138
Cowern, N.E.B.	1994	*Phys. Rev. Lett.*	Diffusion in strained Si(Ge)	118
Shenoy, J.N.	1995	*J. Electron. Mater.*	Characterization and optimization of the SiO_2/SiC MOS interface	114
Schneider, J.	1993	*Physica B*	Point defects in SiC	111
Geis, M.	1991	*IEEE Trans. Electron Devices*	*C–V* measurements on metal–SiO_2–diamond structures	100

energy distribution of defects at the SiC/SiO_2 interface via standard electrical techniques, as well as internal photoemission spectroscopy. Neudeck and Powell (1994, 138 citations) report the characteristics of micropipe defects in SiC, and their impact on junction failures in SiC devices. Shenoy et al. (1995, 114 citations) describe the special care that is required on wide band gap semiconductors such as SiC to obtain reliable estimates of interface-trap densities, owing to the very long time constants associated with these defects even at room temperature. Hence, conventional techniques that often are applied to MOS devices on Si must be modified in application. A variety of point defects in SiC are reviewed by Schneider and Maier (1993, 111 citations).

Landstrass and Ravi (1989a, 229 citations; 1989b, 181 citations) show that hydrogen can passivate defects in thin-film diamond, increasing its resistivity by up to six orders of magnitude. Geis et al. (1991, 100 citations) evaluate the quality of MOS capacitors built on diamond substrates, and find a low electron generation rate in the diamond and a small barrier for electrons to enter the SiO_2 from the diamond conduction band.

A.4.2 Defects in GaAs

Table A.10 lists papers that discuss defects in GaAs and closely related materials and devices (Chapter 22). Matthews and Blakeslee (1974, 2531 citations) characterized misfit dislocations in GaAs-based epitaxial multilayer structures. Henry and Lang (1977, 702 citations) and Lang and Henry (1975, 132 citations) characterize the capture and recombination properties of defects in GaAs and GaP. Capture cross section and energy-level data are analyzed in detail; the importance of lattice relaxation is noted for ZnO and O centers in GaP. A unified defect model for Schottky barrier formation on GaAs was developed by Spicer et al. (1979, 696 citations; 1980a, 565 citations; 1980b, 561 citations), which identifies defects responsible for interface traps and Fermi-level pinning in GaAs and other III–V materials and devices; follow-on work by Spicer et al. (1988, 302 citations) updates this model and identifies the As antisite as the defect that pins the Fermi level at 0.75 and 0.5 eV above the valence band maximum. Photocapacitance measurements were performed by Lang et al. (1979, 701 citations) to determine the capture cross sections for the DX center in $Al_xGa_{1-x}As$. This followed earlier work by Lang and coworkers on lattice-defect coupling in the $Al_xGa_{1-x}As$ system (Lang and Logan, 1977, 447 citations), deep levels in GaAs (Lang and Logan, 1975, 318 citations), and recombination-enhanced defect reactions in GaAs (Lang and Kimerling, 1974, 249 citations). Surface states and barrier heights for GaAs and other semiconductor–metal systems were calculated by Cowley and Sze (1965, 572 citations). Impurity levels in GaAs were analyzed by Sze and Irvin (1968, 425 citations). Recombination-enhanced defect reactions in GaAs were investigated further by Kimerling (1978, 256 citations) and Weeks et al. (1975, 212 citations).

Fifteen different electron traps in bulk GaAs were extensively characterized and categorized by Martin et al. (1977, 542 citations), including the most prominent electron trap, EL2. Hole traps in GaAs were also studied by Mitonneau et al. (1977, 200 citations). Near-infrared optical absorption techniques were employed by Martin (1981, 417 citations) to demonstrate the lattice relaxation associated with the EL2 defect. The atomic and electronic structure of the DX center in GaAs and $Al_xGa_{1-x}As$ was calculated by Chadi and Chang (1988a,b, 519 citations; 1989, 387 citations), who suggest this defect is a negatively charged defect center associated with donor atoms. This followed work by Mizuta et al. (1985, 231 citations) who provided evidence the DX is substitutional donor in AlGaAs, and by Morgan (1986, 139 citations), who argued that the DX center is a substitutional donor that is displaced from its centered lattice position. A review of DX centers in III–V semiconductors was provided by Mooney (1990, 476 citations). Baraff and Schluter (1985, 377 citations) calculate the electronic structures and total energies of eight isolated point defects in GaAs with a self-consistent Green's function technique. Arsenic antisite defects were identified in plastically deformed GaAs by Weber et al. (1982, 326 citations). The importance of defects on the As sublattice to GaAs radiation damage was demonstrated by Pons and Bourgoin (1985, 319 citations). The observation of a photoelectric memory effect led Vincent et al. (1982, 291 citations) to conclude that the EL2 defect has at least two states. The metastability of the isolated As antisite defect in GaAs was compared with properties of the EL2 defect by Chadi and Chang (1988a,b, 276 citations). Compensation of dopants in GaAs by the EL2 defect was reported by Martin et al. (1980, 276 citations). The As pressure dependence and the annihilation of the EL2 defect level at 0.82 eV by shallow donors led

TABLE A.10

Citation Summary (# Cit ≥130) for Journal Articles that Describe Defects in GaAs and Closely Related Devices and Materials

1st Author	Year	Journal	Abridged Title	# Cit.
Matthews, J.W.	1974	*J. Cryst. Growth*	Defects in epitaxial multilayers—misfit dislocations	2531
Henry, C.H.	1977	*Phys. Rev. B*	Nonradiative capture and recombination in GaAs and GaP	702
Lang, D.V.	1979	*Phys. Rev. B*	Trapping characteristics and donor-complex (DX) model for AlGaAs	701
Spicer, W.E.	1979	*J. Vac. Sci. Technol.*	Uniform defect model for Schottky barrier and III–V insulator interface traps	696
Spicer, W.E.	565	*J. Vac. Sci. Technol.*	Unified defect model and beyond	565
Spicer, W.E.	561	*Phys. Rev. Lett.*	Unified mechanism for Schottky barrier formation and III–V oxide interface traps	561
Cowley, A.M.	557	*J. Appl. Phys.*	Surface states and barrier height of metal–semiconductor systems	557
Martin, G.M.	1977	*Electron. Lett.*	Electron traps in bulk and epitaxial GaAs crystals	542
Chadi, D.J.	1988	*Phys. Rev. Lett.*	Theory of the atomic and electronic structure of DX centers	519
Mooney, P.M.	1990	*J. Appl. Phys.*	Deep donor levels (DX) in III–V semiconductors	476
Lang, D.V.	1977	*Phys. Rev. Lett.*	Large lattice relaxation model for persistent photoconductivity	447
Sze, S.M.	1968	*Solid-State Electron.*	Resistivity, mobility, and impurity levels in GaAs, Ge, and Si at 300 K	425
Martin, G.M.	1981	*Appl. Phys. Lett.*	Optical assessment of the main electron trap in bulk semi-insulating GaAs	417
Chadi, D.J.	1989	*Phys. Rev. B*	Energetics of DX center formation in GaAs and AlGaAs alloys	387
Baraff, G.A.	1985	*Phys. Rev. Lett.*	Electronic structure, total energies, and abundances of point defects in GaAs	377
Weber, E.R.	1982	*J. Appl. Phys.*	Identification of As_{Ga} antisites in plastically deformed GaAs	326
Pons, D.	1985	*J. Phys. C—Solid State Phys.*	Irradiation-induced defects in GaAs	319
Lang, D.V.	1975	*J. Electron. Mater.*	Study of deep levels in GaAs by capacitance spectroscopy	318
Spicer, W.E.	1988	*J. Vac. Sci. Technol. B*	The advanced unified defect model for Schottky barrier formation	302
Zhang, S.B.	1991	*Phys. Rev. Lett.*	Chemical potential dependence of defect formation energies in GaAs	296
Vincent, G.	1982	*J. Appl. Phys.*	Photoelectric memory effect in GaAs	291
Williams, E.W.	1968	*Phys. Rev.*	Evidence for self-activated luminescence in GaAs—Ga vacancy donor center	280
Chadi, D.J.	1988	*Phys. Rev. Lett.*	Metastability of the isolated As antisite defect in GaAs	276
Martin, G.M.	1980	*J. Appl. Phys.*	Compensation mechanisms in GaAs	276
Kimerling, L.C.	1978	*Solid-State Electron.*	Recombination-enhanced defect reactions	256
Pavesi, L.	1994	*J. Appl. Phys.*	Photoluminescence of AlGaAs alloys	252
Lagowski, J.	1982	*Appl. Phys. Lett.*	Origin of the 0.82-eV electron trap in GaAs and annihilation by shallow donors	250
Lang, D.V.	1974	*Phys. Rev. Lett.*	Observation of recombination-enhanced defect reactions in semiconductors	249
Vincent, G.	1979	*J. Appl. Phys.*	Electric field effect on thermal emission of traps in semiconductor junctions	245
Mizuta, M.	1985	*Jpn. J. Appl. Phys.*	Direct evidence for the DX center being a substitutional donor in AlGaAs	231

TABLE A.10 (continued)

Citation Summary (# Cit ≥130) for Journal Articles that Describe Defects in GaAs and Closely Related Devices and Materials

1st Author	Year	Journal	Abridged Title	# Cit.
Dabrowski, J.	1988	*Phys. Rev. Lett.*	Theoretical evidence [that] the isolated As antisite in GaAs [is the] EL2	217
Weeks, J.D.	1975	*Phys. Rev. B*	Theory of recombination-enhanced defect reactions in semiconductors	212
Bourgoin, J.C.	1988	*J. Appl. Phys.*	Native defects in GaAs	205
Mitonneau, A.	1977	*Electron. Lett.*	Hole traps in bulk and epitaxial GaAs crystals	200
Wagner, R.J.	1980	*Solid-State Commun.*	Submillimeter EPR for the As antisite defect in GaAs	191
von Bardeleben, H.J.	1986	*Phys. Rev. B*	Identification of a defect in a semiconductor—EL2 in GaAs	190
Lang, D.V.	1977	*Phys. Rev. B*	Identification of a defect state associated with a Ga vacancy in GaAs and AlGaAs	179
Chevallier, J.	1985	*Appl. Phys. Lett.*	Donor neutralization in GaAs(Si) by atomic hydrogen	175
Kaminska, M.	1983	*Appl. Phys. Lett.*	Intracenter transitions in the dominant deep level (EL2) in GaAs	175
Hurle, D.T.J.	1979	*J. Phys. Chem. Solids*	Revised calculation of point defect equilibria and nonstoichiometry in GaAs	155
Omling, P.	1983	*J. Appl. Phys.*	DLTS evaluation of nonexponential transients in semiconductor alloys	154
Kaminska, M.	1985	*Phys. Rev. Lett.*	Identification of the EL2 in GaAs as an isolated antisite As defect	153
Harrison, J.W.	1976	*Phys. Rev. B*	Alloy scattering in ternary III–V compounds	151
Pearton, S.J.	1986	*J. Appl. Phys.*	Hydrogenation of shallow donor levels in GaAs	146
Pons, D.	1980	*J. Appl. Phys.*	Energy dependence of deep-level introduction in irradiated GaAs	145
Northrup, J.E.	1993	*Phys. Rev. B*	Dopant and defect energetic—Si in GaAs	141
Littlejohn, M.A.	1978	*Solid-State Electron.*	Alloy scattering and high-field transport in ternary and quaternary III–V	140
Lang, D.V.	1976	*J. Appl. Phys.*	Study of electron traps in n-GaAs grown by molecular beam epitaxy (MBE)	139
Morgan, T.N.	1986	*Phys. Rev. B*	Theory of the DX center in $Al_xGa_{1-x}As$ and GaAs crystals	139
Dabrowski, J.	1989	*Phys. Rev. B*	Isolated As antisite defect in GaAs and the properties of EL2	138
Johnson, N.M.	1986	*Phys. Rev. B*	Hydrogen passivation of shallow acceptor impurities in p-type GaAs	138
Theis, T.N.	1988	*Phys. Rev. Lett.*	Electron localization by a metastable donor level in n-GaAs	138
Tan, T.Y.	1991	*Crit. Rev. Solid State Mater. Sci.*	Point defects, diffusion mechanisms, and superlattice disordering in GaAs	135
Silverberg, P.	1988	*Appl. Phys. Lett.*	Hole photoionization cross sections of EL2 in GaAs	133
Lang, D.V.	1975	*Phys. Rev. Lett.*	Nonradiative recombination at deep levels in GaAs and GaP	132

Lagowski (1982, 250 citations) to propose that EL2 is an As_{Ga} antisite defect. The isolated As antisite defect was studied via EPR by Wagner et al. (1980, 191 citations), and was found to be similar in structure to the P antisite defect in GaP. Kaminska et al. (1983, 175 citations) studied intracenter transitions, and presented uniaxial stress and magnetic field experiments that also suggested the EL2 defect is an isolated As antisite defect (Kaminska et al., 1985, 153 citations). EPR and DLTS results were presented by Bourgoin and von Bardeleben et al. (Chapter 22), which suggested instead that the EL2 defect in GaAs is a complex formed by an As antisite and an intrinsic As or Ga interstitial (von Bardeleben et al., 1986, 190 citations). Further evidence in support of this interpretation was presented by Burgoin et al. (1988, 205 citations). In turn, Dabrowski and Scheffler (1988, 217 citations; 1989, 138 citations) performed detailed calculations of the As antisite, an As-interstitial/Ga-vacancy defect pair, and various other configurations, and concluded that the isolated As antisite in GaAs could account for all experimentally observed properties of the EL2. The hole photoionization cross sections of the EL2 defect were studied by Silverberg et al. (1988, 133 citations). A 10 eV threshold energy for creation of the E2–E5 traps in GaAs by electron irradiation was found by Pons et al. (1980, 145 citations).

The dependence of defect formation energies in GaAs on chemical potential and the formation energies of Si donors, acceptors, and defect complexes in GaAs were studied by Zhang and Northrup (1991, 296 citations) and Northrup and Zhang (1993, 141 citations). Luminescence from a Ga donor in GaAs was studied by Williams (1968, 280 citations). The photoluminescence of $Al_xGa_{1-x}As$ alloys was reviewed by Pavesi and Guzzi (1994, 252 citations), who catalog a wide variety of defect centers associated with defects and impurities in the material. The effects of electric field on emission from traps in GaAs diodes was examined by Vincent et al. (1979, 245 citations). The E3 defect in GaAs and $Al_xGa_{1-x}As$ was studied by Lang et al. (1977, 179 citations), and attributed to a Ga vacancy. Shallow donor neutralization in Si-doped GaAs was studied by Chevallier et al. (1985, 175 citations) and Pearton et al. (1986, 146 citations). Hurle (1979, 155 citations) showed the significance of Frenkel defect formation on the As sublattice in GaAs, with the As vacancy functioning as a donor at high temperatures. Omling et al. (1983, 154 citations) calculated DLTS spectra for broadened energy levels associated with electron traps in $GaAs_{1-x}P_x$. Harrison, Littlejohn, Hauser, and coworkers calculated the effects of alloy scattering on high-field transport in GaAs and related materials (Harrison and Hauser, 1976, 151 citations; Littlejohn et al., 1978, 140 citations). Lang et al. (1976, 139 citations) applied DLTS to characterize at least nine electron traps in GaAs. The passivation of shallow acceptors in p-type GaAs was demonstrated by Johnson et al. (1986b, 138 citations). Electron localization associated with the donor-related DX center was demonstrated by Theis et al. (1988, 138 citations). Tan et al. (1991, 135 citations) review the properties of a variety of point defects in GaAs.

A.4.3 Defects in GaN

Table A.11 lists papers that discuss defects in GaN materials and devices. An extensive review of the properties of SiC, III–nitride, and II–VI ZnSe semiconductor devices was presented by Morkoc et al. (1994, 1301 citations). The limiting role of defects in these materials is discussed. This followed an earlier review of III–nitrides by Davis (1991, 273 citations). Much of the focus of defect studies in GaN are focused on the use of III–nitride devices as blue light emitting diodes or laser diodes, as demonstrated by Nakamura et al. (1994, 1291 citations; 1996, 686 citations), but transistor applications are also of increasing significance. Pearton et al. (1999, 607 citations) have reviewed the roles of extended and point defects in GaN, including key impurities such as C, O, and H. The formation of acceptor–hydrogen complexes in GaN was characterized by Nakamura et al. (1992, 467 citations). These complexes are found to cause hole compensation and modify the photoluminescence spectrum of GaN. Neugebauer and Van de Walle studied the electronic structure and formation energies in GaN using total energy calculations and found vacancies to be the dominant defects in GaN (Neugebauer and Van de Walle, 1994, 415 citations). They also find that Ga vacancies are responsible for the yellow luminescence in GaN (Neugebauer and Van de Walle, 1996a, 414 citations), which was studied experimentally by a number of groups, including Lester et al. (1995, 510 citations) and Ponce et al. (1996,

TABLE A.11

Citation Summary (# Cit ≥130) for Journal Articles that Describe Defects in GaN Devices and Materials

1st Author	Year	Journal	Abridged Title	# Cit.
Morkoc, H.	1994	*J. Appl. Phys.*	SiC, III–V, III–nitride, and II–VI ZnSe based semiconductor device technologies	1301
Nakamura, S.	1994	*Appl. Phys. Lett.*	Candela class high-brightness InGaN/AlGaN blue LEDs	1291
Nakamura, S.	1996	*Jpn. J. Appl. Phys.*	InGaN based multiquantum well-structured laser diodes	686
Pearton, S.J.	1999	*J. Appl. Phys.*	GaN: Processing, defects, and devices	607
Lester, S.D.	1995	*Appl. Phys. Lett.*	High dislocation densities in high efficiency GaN-based LEDs	510
Nakamura, S.	1992	*Jpn. J. Appl. Phys.*	Hole compensation mechanism of p-type GaN films	467
Neugebauer, J.	1994	*Phys. Rev. B*	Atomic geometry and electronic structure of native defects in GaN	415
Neugebauer, J.	1996	*Appl. Phys. Lett.*	Ga vacancies and the yellow luminescence in GaN	414
Ponce, F.A.	1996	*Appl. Phys. Lett.*	Spatial distribution of the luminescence in GaN think films	238
Pankove, J.I.	1976	*J. Appl. Phys.*	Photoluminescence of ion-implanted GaN	225
Neugebauer, J.	1995	*Phys. Rev. Lett.*	Hydrogen in GaN—novel aspects of a common impurity	215
Wu, X.H.	1996	*J. Appl. Phys.*	Defect structure of MOCVD grown GaN on sapphire	207
Weimann, N.G.	1998	*J. Appl. Phys.*	Scattering of electrons at threading dislocations in GaN	201
Sakai, A.	1997	*Appl. Phys. Lett.*	Defect structure in selectively grown GaN films with low dislocation density	189
Gotz, W.	1996	*Appl. Phys. Lett.*	Activation energies of Si donors in GaN	185
Look, D.C.	1999	*Phys. Rev. Lett.*	Dislocation scattering in GaN	170
Look, D.C.	1997	*Phys. Rev. Lett.*	Defect donor and acceptor in GaN	155
Binari, S.C.	2001	*IEEE Trans. Electron Devices*	Trapping effects and microwave power performance in AlGaN/GaN HEMTs	149
Mattila, T.	1997	*Phys. Rev. B*	Point defect complexes and broadband luminescence in GaN and AlN	146
Reynolds, D.C.	1997	*Solid State Commun.*	Similarities in the photoluminescence of ZnO and GaN	146
Vetury, R.	2001	*IEEE Trans. Electron Devices*	The impact of surface states on the DC and RF characteristics of AlGaN/GaN	146
Hansen, P.J.	1998	*Appl. Phys. Lett.*	Scanning capacitance microscopy imaging of threading dislocations in GaN	132
Neugebauer, J.	1996	*Appl. Phys. Lett.*	Role of hydrogen in doping of GaN	132

238 citations). Mattila and Nieminen (1997, 146 citations) suggest that donors in GaN that are bound to cation vacancies can account for much of the broadband luminescence observed. Pankove and Hutchby (1976, 225 citations) performed a wide-ranging study of the photoluminescence spectra of 35 different elements implanted into GaN. Hydrogen in GaN was studied in detail by Neugebauer and Van de Walle, who find significant difference of behavior from that of hydrogen in Si or GaAs, owing to the strongly ionic GaN bond (Neugebauer and Van de Walle, 1995, 215 citations); they also explore the role of hydrogen in doping of GaN (Neugebauer and Van de Walle, 1996b, 132 citations). Defects in GaN films on sapphire were studied by Wu et al. (1996, 207 citations) and by Sakai et al. (1997, 189 citations). The activation energies of Si donors in GaN were investigated by Gotz et al. (1996, 185 citations). Weimann et al. (1998, 201 citations), Look et al. (1999a, 170 citations), and Ng et al. (1998, 157 citations) studied the scattering of electrons by dislocations in GaN. Look et al. (1997, 155 citations) also presented evidence that an N vacancy is a shallow donor in irradiated GaN, while the N interstitial serves as an acceptor. Similarities

between the yellow luminescence band in GaN and the green luminescence band in ZnO were described by Reynolds et al. (1997, 146 citations), and attributed to common defect-related mechanisms. Threading dislocations in GaN were studied by scanning capacitance microscopy by Hansen et al. (1998, 132 citations).

In transistor applications, the properties of the two-dimensional electron gas that forms spontaneously at the AlGaN/GaN interface was characterized by Ambacher et al. (1999, 406 citations; 2000, 231 citations). Binari et al. (2001, 149 citations) characterized trapping effects in AlGaN/GaN high electron mobility transistors (HEMTs). The effects of surface traps on the DC and radio frequency current–voltage response of AlGaN/GaN field effect transistors was studied by Vetury et al. (2001, 146 citations).

A.4.4 Defects in ZnO

Table A.12 lists papers that discuss defects in ZnO materials and devices; recent advances in this area were discussed by Look (2001, 509 citations). Defects play a prominent role in determining the properties of ZnO. Vanheusden et al. (1996b, 695 citations; 1996a, 329 citations) associate the prominent green photoluminescence of ZnO with an O vacancy defect. Van de Walle (2000, 312 citations) details the unexpected effects of hydrogen on doping in ZnO, and concludes that, instead of passivating dopants as it does in Si and GaAs, it can be incorporated in high concentrations in ZnO and function as a shallow

TABLE A.12

Citation Summary (# Cit $\geq$100) for Journal Articles that Describe Defects in ZnO Devices and Materials

1st Author	Year	Journal	Abridged Title	# Cit.
Vanheusden, K.	1996	*J. Appl. Phys.*	Mechanisms behind green photoluminescence in ZnO	695
Vanheusden, K.	1996	*Appl. Phys. Lett.*	Correlation between photoluminescence and O vacancies in ZnO phosphors	329
Look, D.C.	2002	*Appl. Phys. Lett.*	Characterization of homoepitaxial p-type ZnO grown by MBE	321
Van de Walle, C.G.	2000	*Phys. Rev. Lett.*	Hydrogen as a cause of doping in ZnO	312
Lin, B.X.	2001	*Appl. Phys. Lett.*	Green photoluminescent center in undoped ZnO films deposited on Si	246
Kohan, A.F.	2000	*Phys. Rev. B*	First principles study of native point defects in ZnO	237
Look, D.C.	1999	*Phys. Rev. Lett.*	Residual shallow donor in ZnO	218
Zhang, S.B.	2001	*Phys. Rev. B*	Intrinsic n-type vs. p-type doping asymmetry and defects physics of ZnO	185
Ozgur, U.	2005	*J. Appl. Phys.*	A comprehensive review of ZnO materials and devices	161
Pearton, S.J.	2005	*Prog. Mater. Sci.*	Recent progress in processing and properties of ZnO	153
Laks, D.B.	1992	*Phys. Rev. B*	Native defects and self-compensation in ZnSe	152
Reynolds, D.C.	1998	*Phys. Rev.*	Neutral donor-bound exciton complexes in ZnO crystals	147
Meyer, B.K.	2004	*Phys. Status Solidi B*	Bound exciton and donor–acceptor pair recombinations in ZnO	143
Jin, B.J.	2000	*Thin Solid Films*	Violet and UV luminescence emitted from ZnO thin films on sapphire	106
Yan, Y.F.	2001	*Phys. Rev. Lett.*	Control of doping by impurity chemical potentials: Predictions for p-type ZnO	103
Pearton, S.J.	2004	*J. Vac. Sci. Technol. B*	Recent advances in processing of ZnO	100

donor. Lin et al. (2001, 246 citations) attribute the green emission observed during photoluminescence to an oxide antisite defect O–Zn, as opposed to an O vacancy. The electronic structure, atomic geometry, and formation energy of native point defects in ZnO was calculated by Kohan et al. (2000, 237 citations). They find Zn and O vacancies to be the dominant native point defects. Look et al. (1999b, 218 citations) characterize a shallow donor in ZnO that the authors attribute to interstitial Zn–I or a Zn–I related complex. Zhang et al. (2001, 185 citations) evaluate the intrinsic defect physics of ZnO, and explore the reasons that ZnO cannot be doped p-type via native defects such as vacancies and interstitials. This has been reviewed extensively by Ozgur et al. (2005, 161 citations) and Pearton et al. (2005, 153 citations; 2004, 100 citations). Previously, Laks et al. (1992, 152 citations) showed that deviations from stoichiometry cannot explain why ZnSe can only be doped one way. On the basis of theoretical calculations, it was predicted by Yan et al. (2001, 103 citations) that the incorporation of NO or N_2O into the processing of ZnO might enable p-type doping; this was confirmed by Look (2002, 321 citations). Isolated neutral donors are found by Reynolds et al. (1998, 147 citations) to be made up of defect pair complexes. Bound exciton and donor–acceptor pair recombination was studied in ZnO by Meyer et al. (2004, 143 citations), who find a key role for hydrogen and *Y*-line defects. Jin et al. (2000, 106 citations) attribute yellow–green luminescence in ZnO to a defect level at the grain boundary of ZnO crystals.

Acknowledgments

I would like to thank my coeditors, Ron Schrimpf and Sokrates Pantelides, for helpful discussions, as well as the other contributors to this volume for their insights into these areas. This appendix is inspired by a similar list of high-impact papers in the area of radiation effects on electronics that was published in the June 2003 special issue of the *IEEE Transactions on Nuclear Science* by Ken Galloway.

References

Aberle, A.G., Glunz, S., and Warta, W., Impact of illumination level and oxide parameters on Shockley-Read-Hall recombination at the Si/SiO_2 interface, *J. Appl. Phys.*, 71, 4422, 1992.

Afanase'ev, V.V. et al., Intrinsic SiC/SiO_2 interface states, *Phys. Status Solidi A*, 162, 321, 1997.

Aitken, J.M. and Young, D.R., Electron trapping by radiation induced charge in MOS devices, *J. Appl. Phys.*, 47, 1196, 1976.

Aitken, J.M., Young, D.R., and Pan, K., Electron trapping in electron-beam irradiated SiO_2, *J. Appl. Phys.*, 49, 3386, 1978.

Alers, G.B. et al., Intermixing at the Ta_2O_5/Si interface in gate dielectric structures, *Appl. Phys. Lett.*, 73, 1517, 1998.

Ambacher, O. et al., Two-dimensional electron gases induced by spontaneous and piezoelectric polarization charges in N- and Ga-face AlGaN/GaN heterostructures, *J. Appl. Phys.*, 85, 3222, 1999.

Ambacher, O. et al., Two-dimensional electron gases induced by spontaneous and piezoelectric polarization in undoped and doped AlGaN/GaN heterostructures, *J. Appl. Phys.*, 87, 334, 2000.

Ando, T., Fowler, A.B., and Stern, F., Electronic properties of two-dimensional systems, *Rev. Mod. Phys.*, 437, 1982.

Antoniadis, D.A. and Moskowitz, I., Diffusion of substitutional impurities in Si at short oxidation times—an insight into point-defect kinetics, *J. Appl. Phys.*, 53, 6788, 1982.

Arnold, G.W., Ion implantation effects in noncrystalline SiO_2, *IEEE Trans. Nucl. Sci.*, 20(6), 220, 1973.

Arnold, D., Cartier, E., and DiMaria, D.J., Theory of high-field electron transport and impact ionization in SiO_2, *Phys. Rev. B*, 49, 10278, 1994.

Asokakumar, P., Lynn, K.G., and Welch, D.O., Characterization of defects in Si and SiO_2–Si using positrons, *J. Appl. Phys.*, 76, 4935, 1994.

Austin, I.G. and Mott, N.F., Polarons in crystalline and non-crystalline materials, *Adv. Phys.*, 18, 41, 1969.

Bachelet, G.B., Hamann, D.R., and Schluter, M., Pseudopotentials that work—from H to Pu, *Phys. Rev. B*, 26, 4199, 1982.

Baraff, G.A. and Schluter, M., Self consistent Green's function calculation of ideal Si vacancy, *Phys. Rev. Lett.*, 41, 892, 1978.

Baraff, G.A. and Schluter, M., Migration of interstitials in Si, *Phys. Rev. B*, 30, 3460, 1984.

Baraff, G.A. and Schluter, M., Electronic structure, total energies, and abundances of the elementary point defects in GaAs, *Phys. Rev. Lett.*, 55, 1327, 1985.

Baraff, G.A., Kane, E.O., and Schluter, M., Si vacancy—possible Anderson negative-U center, *Phys. Rev. Lett.*, 43, 956, 1979.

Baraff, G.A., Kane, E.O., and Schluter, M., Theory of the Si vacancy—an Anderson negative U center, *Phys. Rev. B*, 21, 5662, 1980.

Bardeen, J., Surface states and rectification at a metal semiconductor contact, *Phys. Rev.*, 71, 717, 1947.

Bardeen, J. and Brattain, W.H., The transistor, a semiconductor triode, *Phys. Rev.*, 74, 230, 1948.

Bean, L.C. et al., $Ge_{1-x}Si_x$ strained layer superlattice grown by molecular beam epitaxy, *J. Vac. Sci. Technol., A*, 2, 436, 1984.

Benton, J.L. et al., Hydrogen passivation of point defects in Si, *Appl. Phys. Lett.*, 36, 670, 1980.

Berglund, C.N., Surface states at steam grown Si-SiO_2 interfaces, *IEEE Trans. Electron Devices*, 13, 701, 1966.

Berglund, C.N. and Powell, R.J., Photoinjection into SiO_2—electron scattering in image force potential well, *J. Appl. Phys.*, 42, 573, 1971.

Bergman, K. et al., Donor-hydrogen complexes in passivated Si, *Phys. Rev. B*, 37, 2770, 1988.

Bernholc, J. and Pantelides, S.T., Scattering theoretic method for defects in semiconductors—tight-binding description of vacancies in Si, Ge, and GaAs, *Phys. Rev. B*, 18, 1780, 1978.

Bernholc, J., Lipari, N.O., and Pantelides, S.T., Self consistent method for point-defects in semiconductors: application to vacancy in Si, *Phys. Rev. Lett.*, 41, 895, 1978.

Bernholc, J., Lipari, N.O., and Pantelides, S.T., Scattering theoretic method for defects in semiconductors: self consistent formulation and application to the vacancy in Si, *Phys. Rev. B*, 21, 3545, 1980.

Binari, S.C. et al., Trapping effects and microwave power performance in AlGaN/GaN HEMTs, *IEEE Trans. Electron Devices*, 48, 465, 2001.

Bloch, F., Uber die quantenmechanik der elektronen in kristallgittern, *Z. Phys. A*, 52, 555, 1928.

Blöchl, P.E. et al., 1st-principles calculations of self-diffusion constants in Si, *Phys. Rev. Lett.*, 70, 2435, 1993.

Blöchl, P.E., Projector augmented-wave method, *Phys. Rev. B*, 50, 17953, 1994.

Boero, M. et al., Structure and hyperfine parameters of E′(1) centers in a-quartz and in vitreous SiO_2, *Phys. Rev. Lett.*, 78, 887, 1997.

Boesch, H.E. Jr. and McGarrity, J.M., Charge yield and dose effects in MOS capacitors at 80 K, *IEEE Trans. Nucl. Sci.*, 23, 1520, 1976.

Borghesi, A. et al., Oxygen precipitation in Si, *J. Appl. Phys.*, 77, 4169, 1995.

Brower, K.L., Si-29 hyperfine structure of unpaired spins at the Si/SiO_2 interface, *Appl. Phys. Lett.*, 43, 1111, 1983.

Brower, K.L., Kinetics of H_2 passivation of P_b centers at the (111) Si/SiO_2 interface, *Phys. Rev. B*, 38, 9657, 1988.

Brower, K.L., Dissociation kinetics of hydrogen-passivated (111) Si-SiO_2 interface defects, *Phys. Rev. B.*, 42, 3444, 1990.

Bruel, M., Silicon on insulator material technology, *Electron. Lett.*, 31, 1201, 1995.

Bruel, M., Application of hydrogen ion beams to SOI materials technology, *Nucl. Instrum. Methods Phys. Res., Sect. B*, 108, 313, 1996.

Brugler, J.S. and Jespers, P.G.A., Charge pumping in MOS devices, *IEEE Trans. Electron Devices*, 16, 297, 1969.

Buchanan, D.A., Scaling the gate dielectric: materials, integration, and reliability, *IBM J. Res. Dev.*, 43, 245, 1999.

Buchanan, D.A. and DiMaria, D.J., Interface and bulk trap generation in MOS capacitors, *J. Appl. Phys.*, 67, 7439, 1990.

Burgoin, J.C. and Corbett, J.W., New mechanism for interstitial migration, *Phys. Lett. A*, 38, 135, 1972.

Burgoin, J.C., von Bardeleben, H.J., and Stievenard, D., Native defects in GaAs, *J. Appl. Phys.*, 64, 65, 1988.

Callegari, A. et al., Physical and electrical characterization of HfO_2 and Hf silicate sputtered films, *J. Appl. Phys.*, 90, 6466, 2001.

Campbell, S.A. et al., MOSFETs fabricated with high permittivity TiO_2 dielectrics, *IEEE Trans. Electron Devices*, 44, 104, 1997.

Caplan, P.J. et al., ESR centers, interface states, and oxide fixed charge in thermally oxidized Si wafers, *J. Appl. Phys.*, 50, 5847, 1979.

Car, R. and Parrinello, M., Unified approach for molecular dynamics and density functional theory, *Phys. Rev. Lett.*, 55, 2471, 1985.

Car, R. et al., Microscopic theory of atomic diffusion mechanism in Si, *Phys. Rev. Lett.*, 52, 1814, 1984.

Cartier, E., Stathis, J.H., and Buchanan, D.A., Passivation and depassivation of Si dangling bonds at the Si/SiO_2 interface by atomic hydrogen, *Appl. Phys. Lett.*, 63, 1510, 1993.

Castagne, R. and Vapaille, A., Description of SiO_2-Si interface properties by means of very low frequency MOS capacitance measurements, *Surf. Sci.*, 28, 157, 1971.

Castro, P.L. and Deal, B.E., Low-temperature reduction of fast surface states associated with thermally oxidized Si, *J. Electrochem. Soc.*, 118, 280, 1971.

Ceperly, D.M. and Alder, B.J., Ground state of the electron gas by a stochastic method, *Phys. Rev. Lett.*, 45, 566, 1980.

Chadi, D.J. and Chang, K.J., Metastability of the isolated As antisite defect in GaAs, *Phys. Rev. Lett.*, 60, 2187, 1988a.

Chadi, D.J. and Chang, K.J., Theory of the atomic and electronic structure of DX centers in GaAs and $Al_xGa_{1-x}As$ alloys, *Phys. Rev. Lett.*, 61, 873, 1988b.

Chadi, D.J. and Chang, K.J., Energetics of DX center formation in GaAs and $Al_xGa_{1-x}As$ alloys, *Phys. Rev. B*, 39, 10063, 1989.

Chadi, D.J. and Cohen, M.L., Special points in Brillouin zone, *Phys. Rev. B*, 8, 5747, 1973.

Chaneliere, C. et al., Ta_2O_5 thin films for advanced dielectric applications, *Mater. Sci. Eng. R-Rep.*, 22, 269, 1998.

Chang, K.J. and Chadi, D.J., Theory of hydrogen passivation of shallow level dopants in crystalline Si, *Phys. Rev. Lett.*, 60, 1422, 1988.

Chang, K.J. and Chadi, K.J., Diatomic hydrogen complex diffusion and self trapping in crystalline Si, *Phys. Rev. Lett.*, 62, 937, 1989.

Chantre, A., Vincent, G., and Bois, D., Deep level optical spectroscopy in GaAs, *Phys. Rev. B*, 23, 5335, 1981.

Chelikowsky, J.R. and Cohen, M.L., Nonlocal pseudopotential calculations for electronic structure of 11 diamond and zincblende semiconductors, *Phys. Rev. B*, 14, 556, 1976.

Chen, J.W. and Milnes, A.G., Energy levels in Si, *Annu. Rev. Mater. Sci.*, 10, 157, 1980.

Chen, I.C., Holland, S.E., and Hu, C.M., Electrical breakdown in thin gate and tunneling oxides, *IEEE Trans. Electron Devices*, 32, 413, 1985.

Cheng, Y.C. and Sullivan, E.A., Role of scattering by surface roughness in Si inversion layers, *Surf. Sci.*, 34, 717, 1973.

Cheng, L.J. et al., 1.8, 3.3 and 3.9 μm bands in irradiated Si—correlations with divacancy, *Phys. Rev.*, 152, 761, 1966.

Chevallier, J. et al., Donor neutralization in GaAs(Si) by atomic hydrogen, *Appl. Phys. Lett.*, 47, 108, 1985.

Cohen, M.L. and Bergstresser, T.K., Band structures and pseudopotential form factors for 14 semiconductors of the diamond and zincblende structures, *Phys. Rev.*, 141, 789, 1966.

Copel, M., Gribelyuk, M., and Gusev, E., Structure and stability of ultrathin ZrO_2 on Si(001), *Appl. Phys. Lett.*, 76, 436, 2000.
Corbett, J.W., Karins, J.P., and Tan, T.Y., Ion induced defects in semiconductors, *Nucl. Instrum. Methods Phys. Res., B*, 182, 457, 1981.
Corbett, J.W. et al., Atomic and molecular hydrogen in the Si lattice, *Phys. Lett. A*, 93, 303, 1983.
Cottrell, P.E., Troutman, R.R., and Ning, T.H., Hot electron emission in *n*-channel IGFETs, *IEEE Trans. Electron Devices*, 26, 520, 1979.
Coutinho, J. et al., Oxygen and dioxygen centers in Si and Ge: Density functional calculations, *Phys. Rev. B*, 62, 10824, 2000.
Cowern, N.E.B. et al., Diffusion in strained Si(Ge), *Phys. Rev. Lett.*, 72, 2585, 1994.
Cowley, A.M. and Sze, S.M., Surface states and barrier height of metal-semiconductor systems, *J. Appl. Phys.*, 36, 3212, 1965.
Dabrowski, J. and Scheffler, M., Theoretical evidence for an optically inducible structural transition of the isolated as antisite in GaAs—identification and explanation of EL2, *Phys. Rev. Lett.*, 60, 2183, 1988.
Dabrowski, J. and Scheffler, M., Isolated As antisite defect in GaAs and the properties of EL2, *Phys. Rev. B*, 40, 10391, 1989.
Dalibor, T. et al., Deep defect centers in SiC monitored with deep level transient spectroscopy, *Phys. Status Solidi A*, 162, 199, 1997.
Daum, W. et al., Identification of strained Si layers at Si/SiO_2 interfaces and clean Si surfaces by nonlinear-optical spectroscopy, *Phys. Rev. Lett.*, 71, 1234, 1993.
Davis, R.F., III–V nitrides for electronic and optoelectronic applications, *Proc. IEEE*, 79, 702, 1991.
Davis, R.F. et al., Thin film deposition and microelectronic and optoelectronic device fabrication and characterization in monocrystalline α and β SiC, *Proc. IEEE*, 79, 677, 1991.
Deák, P., Snyder, L.C., and Corbett, J.W., State and motion of hydrogen in crystalline Si, *Phys. Rev. B*, 37, 6887, 1988.
Deal, B.E., Current understanding of charges in thermally oxidized Si structure, *J. Electrochem. Soc.*, 121, 198, 1974.
Deal, B.E. and Grove, A.S., General relationship for thermal oxidation of Si, *J. Appl. Phys.*, 36, 3770, 1965.
Deal, B.E. et al., Observation of impurity redistribution during thermal oxidation of Si using MOS structure, *J. Electrochem. Soc.*, 112, 308, 1965.
Deal, B.E. et al., Characteristics of surface-state charge (Q_{ss}) of thermally oxidized Si, *J. Electrochem. Soc.*, 114, 266, 1967.
Deal, B.E., Fleming, P.J., and Castro, P.L., Electrical properties of vapor-deposited SiO_2 and Si_3N_4 films on Si, *J. Electrochem. Soc.*, 115, 300, 1968.
Deal, B.E., MacKenna, E.L., and Castro, P.L., Characteristics of fast surface states associated with SiO_2-Si and Si_3N_4-SiO_2-Si structures, *J. Electrochem. Soc.*, 116, 997, 1969.
Degraeve, R. et al., New insights in the relation between electron trap generation and the statistical properties of oxide breakdown, *IEEE Trans. Electron Devices*, 45, 904, 1998.
Deleo, G.G. and Fowler, W.B., Hydrogen-acceptor pairs in Si—pairing effect on the hydrogen vibrational frequency, *Phys. Rev. B*, 31, 6861, 1985.
Denteneer, P.J.H., Van de Walle, C.G., and Pantelides, S.T., Microscopic structure of the hydrogen-boron complex in crystalline Si, *Phys. Rev. B*, 39, 10809, 1989.
Depas, M., Nigam, T., and Heyns, M.M., Soft breakdown of ultra-thin gate oxide layers, *IEEE Trans. Electron Devices*, 43, 1499, 1996.
Derbenwick, G.F. and Gregory, B.L., Process optimization of radiation-hardened CMOS integrated circuits, *IEEE Trans. Nucl. Sci.*, 22, 2151, 1975.
Deuling, H. et al., Interface states in Si-SiO_2 structures, *Solid-State Electron.*, 15, 559, 1972.
DiMaria, D.J., Determination of insulator bulk trapped charge densities and centroids from photo-current-voltage characteristics of MOS structures, *J. Appl. Phys.*, 47, 4073, 1976.
DiMaria, D.J. and Cartier, E., Mechanism for stress-induced leakage currents in thin SiO_2 films, *J. Appl. Phys.*, 78, 3883, 1995.
DiMaria, D.J. and Stasiak, J.W., Trap creation in SiO_2 produced by hot electrons, *J. Appl. Phys.*, 65, 2342, 1989.

DiMaria, D.J., Weinberg, Z.A., and Aitken, J.M., Location of positive charges in SiO_2 films on Si generated by VUV photons, X-rays, and high-field stress, *J. Appl. Phys.*, 48, 898, 1977.
DiMaria, D.J. et al., Electroluminescence studies in SiO_2 films containing tiny Si islands, *J. Appl. Phys.*, 56, 401, 1984.
DiMaria, D.J. et al., Electron heating in SiO_2 and off-stoichiometric SiO_2 films, *J. Appl. Phys.*, 57, 1214, 1985.
DiMaria, D.J., Cartier, E., and Arnold, D., Impact ionization, trap creation, degradation, and breakdown in SiO_2 films on Si, *J. Appl. Phys.*, 73, 3367, 1993.
Dirac, P.A.M., Note on exchange phenomena in the Thomas atom, *Proc. Camb. Philos. Soc.*, 26, 376, 1930.
DiStefano, T.H. and Eastman, D.E., Photoemission measurements of valence levels of a-SiO_2, *Phys. Rev. Lett.*, 27, 1560, 1971.
DiStefano, T.H. and Shatzkes, M., Impact ionization model for dielectric instability and breakdown, *Appl. Phys. Lett.*, 25, 685, 1974.
Dodson, B.W. and Tsao, J.Y., Relaxation of strained-layer semiconductor structures via plastic flow, *Appl. Phys. Lett.*, 51, 1325, 1987.
Doyle, B. et al., Interface state creation and charge trapping in the medium to high voltage range during hot carrier stressing of *n*MOS transistors, *IEEE Trans. Electron Devices*, 37, 744, 1990.
Dumin, D. and Maddux, J.R., Correlation of stress-induced leakage current in thin oxides with trap generation inside the oxides, *IEEE Trans. Electron Devices*, 40, 986, 1993.
Dutta, P. and Horn, P.M., Low-frequency fluctuations in solids: $1/f$ noise, *Rev. Mod. Phys.*, 53, 497, 1981.
Eaglesham, D.J., Implantation and transient B-diffusion in Si—the source of the interstitials, *Appl. Phys. Lett.*, 65, 2305, 1994.
Eaglesham, D.J. and Cerullo, M., Dislocation-free Stranski-Krastanow growth of Ge on Si (100), *Phys. Rev. Lett.*, 64, 1943, 1990.
Edwards, A.H. and Fowler, W.B., Theory of the peroxy-radical defect in a-SiO_2, *Phys. Rev. B*, 26, 6649, 1982.
Eisenbeiser, K. et al., FETs with $SrTiO_3$ gate dielectric on Si, *Appl. Phys. Lett.*, 76, 1324, 2000.
Eitan, B. et al., NROM: A novel localized trapping, 2-bit nonvolatile memory, *IEEE Electron Device Lett.*, 21, 543, 2000.
EerNisse, E.P., Viscous flow of thermal SiO_2, *Appl. Phys. Lett.*, 30, 290, 1977.
Elkin, E.L. and Watkins, G.D., Defects in irradiated Si—EPR and ENDOR of As and Sb vacancy pairs, *Phys. Rev.*, 174, 881, 1968.
Estreicher, S.K., Hydrogen-related defects in crystalline semiconductors—a theorist's perspective, *Mater. Sci. Eng. R-Rep.*, 14, 319, 1995.
Fahey, P.M., Griffin, P.B., and Plummer, J.D., Point-defects and dopant diffusion in Si, *Rev. Mod. Phys.*, 61, 289, 1989.
Fair, R.B. and Sun, R.C., Threshold voltage instability in MOSFETs due to channel hot-hole emission, *IEEE Trans. Electron Devices*, 28, 83, 1981.
Fang, F.F. and Fowler, A.B., Hot electron effects and saturation velocities in Si inversion layers, *J. Appl. Phys.*, 41, 1825, 1970.
Fang, S.F. et al., GaAs and other compound semiconductors on Si, *J. Appl. Phys.*, 68, R31, 1990.
Feigl, F.J., Fowler, W.B., and Yip, K.L., Oxygen vacancy model for E_1' center in SiO_2, *Solid State Commun.*, 14, 225, 1974.
Feigl, F.J. et al., The effects of water on oxide and interface trapped charge generation in thermal SiO_2 films, *J. Appl. Phys.*, 52, 5665, 1981.
Feldman, L.C. and Rodgers, J.W., Depth profiles of lattice disorder resulting from ion bombardment of Si single crystals, *J. Appl. Phys.*, 41, 3776, 1970.
Fermi, E., Eine statische method zur bestimmung einiger eigenschaten des atoms und ihre anwendung auf die theorie des periodischen systems der elemente, *Z. Phys.*, 48, 73, 1928.
Fischetti, M.V., Generation of positive charge in SiO_2 during avalanche and tunnel electron injection, *J. Appl. Phys.*, 57, 2860, 1985.
Fischetti, M.V. and Laux, S.E., Monte Carlo analysis of electron transport in small semiconductor devices including band structure and space charge effects, *Phys. Rev. B*, 38, 9721, 1988.

Fischetti, M.V. and Laux, S.E., Band structure, deformation potentials, and carrier mobility in strained Si, Ge, and SiGe alloys, *J. Appl. Phys.*, 80, 2234, 1996.

Fischetti, M.V., Neumayer, D.A., and Cartier, E.A., Effect of electron mobility in Si inversion layers in MOS systems with a high-κ insulator: The role of remote phonon scattering, *J. Appl. Phys.*, 90, 4587, 2001.

Fleetwood, D.M., Border traps in MOS devices, *IEEE Trans. Nucl. Sci.*, 39, 269, 1992.

Fleetwood, D.M. et al., Effects of oxide traps, interface traps, and "border traps" on MOS devices, *J. Appl. Phys.*, 73, 5058, 1993.

Fock, V., Naherungmethode zur losung des quantenmechanischen mehrkorperproblems, *Z. Phys.*, 61, 126, 1930.

Fowler, R.H. and Nordheim, L.W., Electron emission in intense electric fields, *Proc. R. Soc. London A*, 119, 173, 1928.

Frohman-Bentchkowsky, D. and Lenzlinger, M., Charge transport and storage in MNOS structures, *J. Appl. Phys.*, 40, 3307, 1969.

Gale, R. et al., Hydrogen migration under avalanche injection of electrons in Si MOS capacitors, *J. Appl. Phys.*, 54, 6938, 1983.

Geis, M.W., Gregory, J.A., and Pate, B.B., Capacitance-voltage measurements on metal-SiO_2-diamond structures fabricated with (100)-oriented and (111)-oriented substrates, *IEEE Trans. Electron Devices*, 38, 619, 1991.

Gerardi, G.J. et al., Interface traps and P_b centers in oxidized (100) Si wafers, *Appl. Phys. Lett.*, 49, 348, 1986.

Ghibaudo, G. et al., Improved analysis of low-frequency noise in MOSFETs, *Phys. Status Solidi A*, 124, 571, 1991.

Giles, M.D., Transient phosphorus diffusion below the amorphization threshold, *J. Electrochem. Soc.*, 138, 1160, 1991.

Goodnick, S.M. et al., Surface roughness at the Si(100)-SiO_2 interface, *Phys. Rev. B*, 32, 8171, 1985.

Gossick, B.R., Disordered regions in semiconductors bombarded by fast neutrons, *J. Appl. Phys.*, 30, 1214, 1959.

Gotz, W. et al., Activation energies of Si donors in GaN, *Appl. Phys. Lett.*, 68, 3144, 1996.

Gray, P.V. and Brown, D.M., Density of SiO_2 interface states, *Appl. Phys. Lett.*, 8, 31, 1966.

Green, M.L. et al., Ultrathin ($<$4 nm) SiO_2 and Si-O-N gate dielectric layers for Si microelectronics: understanding the processing, structure, and physical and electrical limits, *J. Appl. Phys.*, 90, 2057, 2001.

Griscom, D.L., E′ center in glassy SiO_2—microwave saturation properties and confirmation of the primary Si-29 hyperfine structure, *Phys. Rev. B*, 20, 1823, 1979.

Griscom, D.L., E′ center in glassy SiO_2—O-17, H-1, and very weak Si-29 hyperfine structure, *Phys. Rev. B*, 22, 4192, 1980a.

Griscom, D.L., Electron spin resonance in glasses—magnetic properties, *J. Non-Cryst. Solids*, 40, 211, 1980b.

Griscom, D.L., Diffusion of radiolytic molecular hydrogen as a mechanism for the post-irradiation buildup of interface states in SiO_2-on-Si structures, *J. Appl. Phys.*, 58, 2524, 1985.

Griscom, D.L. and Friebele, E.J., Fundamental defect centers in glass—Si-29 hyperfine structure of the non-bridging oxygen hole center and the peroxy-radical in a-SiO_2, *Phys. Rev. B*, 24, 4896, 1981.

Griscom, D.L., Friebele, E.J., and Sigel, G.H., Observation and analysis of primary Si-29 hyperfine structure of E′ center in non-crystalline SiO_2, *Solid State Commun.*, 15, 479, 1974.

Griscom, D.L., Brown, D.B., and Saks, N.S., Nature of radiation-induced point defects in amorphous SiO_2 and their role in SiO_2-on-Si structures, in *The Physics and Chemistry of SiO_2 and the Si-SiO_2 Interface*, Helms, C.R. and Deal, B.E. (Eds.), Plenum Press, New York, 1988, pp. 287–297.

Groeseneken, G. et al., A reliable approach to charge pumping measurements in MOS transistors, *IEEE Trans. Electron Devices*, 31, 42, 1984.

Grove, A.S. et al., Investigation of thermally oxidized Si surfaces using MOS structures, *Solid-State Electron.*, 8, 145, 1965.

Grove, A.S. and Fitzgerald, D.J., Surface effects on *p-n* junctions—characteristics of surface space-charge regions under nonequilibrium conditions, *Solid-State Electron.*, 9, 783, 1966.

Grunthaner, F.J. et al., High-resolution X-ray photoelectron spectroscopy as a probe of local atomic structure—application to amorphous SiO_2 and the Si/SiO_2 interface, *Phys. Rev. Lett.*, 1683, 1979a.

Grunthaner, F.J. et al., Local atomic structure and electronic structure of oxide-GaAs and SiO_2-Si interfaces using high resolution XPS, *J. Vac. Sci. Technol.*, 16, 1443, 1979b.

Grunthaner, F.J., Grunthaner, P.J., and Maserjian, J., Radiation-induced defects in SiO_2 as determined with XPS, *IEEE Trans. Nucl. Sci.*, 29, 1462, 1982.

Gusev, E.P. et al., Ultrathin high-K metal oxides on Si: Processing, characterization, and integration issues, *Microelectron. Eng.*, 59, 341, 2001.

Gwyn, C.W., Model for radiation-induced charge trapping and annealing in oxide layer of MOS devices, *J. Appl. Phys.*, 40, 4886, 1969.

Hall, R.N., Electron-hole recombination in Ge, *Phys. Rev. B*, 87, 387, 1952.

Hamann, D.R., Schluter, M., and Chiang, C., Norm conserving pseudopotentials, *Phys. Rev. Lett.*, 43, 1494, 1979.

Hanafi, H.I., Tiwari, S., and Khan, I., Fast and long retention-time nano-crystal memory, *IEEE Trans. Electron Devices*, 43, 1553, 1996.

Handel, P.H., Quantum approach to $1/f$ noise, *Phys. Rev. A*, 22, 745, 1980.

Hansen, P.J. et al., Scanning capacitance microscopy imaging of threading dislocations in GaN films grown on (0001) sapphire by MOCVD, *Appl. Phys. Lett.*, 72, 2247, 1998.

Harari, E., Dielectric breakdown in electrically stressed thin films of thermal SiO_2, *J. Appl. Phys.*, 49, 2478, 1978.

Harrison, J.W. and Hauser, J.R., Alloy scattering in ternary III-V compounds, *Phys. Rev. B*, 13, 5347, 1976.

Hartree, D.R., The wave mechanics of an atom with a non-Coulomb central field, Part I—theory and methods, *Proc. Camb. Philos. Soc.*, 24, 89, 1928.

Hartstein, A. and Young, D.R., Identification of electron traps in thermal SiO_2 films, *Appl. Phys. Lett.*, 38, 631, 1981.

Heiman, F.P., On determination of minority carrier lifetime from transient response of an MOS capacitor, *IEEE Trans. Electron Devices*, 14, 781, 1967.

Heiman, F.P. and Warfield, G., Effects of oxide traps on MOS capacitance, *IEEE Trans. Electron Devices*, 12, 167, 1965.

Helms, C.R. and Poindexter, E.H., The Si/SiO_2 system—its microstructure and imperfections, *Rep. Prog. Phys.*, 57, 791, 1994.

Henry, C.H. and Lang, D.V., Nonradiative capture and recombination by multiphonon emission in GaAs and GaP, *Phys. Rev. B*, 15, 989, 1977.

Heremans, P. et al., Analysis of the charge pumping technique and its application for the evaluation of MOSFET degradation, *IEEE Trans. Electron Devices*, 36, 1318, 1989.

Himpsel, F.J. et al., Microscopic structure of the SiO_2/Si interface, *Phys. Rev. B*, 38, 6084, 1988.

Hofmann, K.R. et al., Hot electron and hole emission effects in short n-channel MOSFETs, *IEEE Trans. Electron Devices*, 32, 691, 1985.

Holbech, J.D. et al., H_2^* defect in crystalline Si, *Phys. Rev. Lett.*, 71, 875, 1993.

Hooge, F.N., $1/f$ noise is no surface effect, *Phys. Lett. A*, 29, 139, 1969.

Hooge, F.N., Discussion of recent experiments on $1/f$ noise, *Physica*, 60, 130, 1972.

Hooge, F.N., $1/f$ noise, *Physica B & C*, 83, 14, 1976.

Hooge, F.N., $1/f$ noise sources, *IEEE Trans. Electron Devices*, 41, 1926, 1994.

Hooge, F.N. and Vandamme, L.K.J., Lattice scattering causes $1/f$ noise, *Phys. Lett. A*, 66, 315, 1978.

Hooge, F.N., Kleinpenning, T.G.M., and Vandamme, L.K.J., Experimental studies on $1/f$ noise, *Rep. Prog. Phys.*, 44, 479, 1981.

Houghton, D.C., Strain relaxation kinetics in $Si_{1-x}Ge_x/Si$ heterostructures, *J. Appl. Phys.*, 70, 2136, 1991.

Houssa, M. et al., Trap-assisted tunneling in high permittivity gate dielectric stacks, *J. Appl. Phys.*, 87, 8615, 2000.

Hsu, F.C. and Grinolds, H.R., Structure-enhanced MOSFET degradation due to hot-electron injection, *IEEE Electron Device Lett.*, 5, 71, 1984.

Hsu, F.C. and Tam, S., Relationship between MOSFET degradation and hot-electron-induced interface-state degradation, *IEEE Electron Device Lett.*, 5, 50, 1984.

Hu, C.M. et al., Hot electron induced MOSFET degradation—model, monitor, and improvement, *IEEE Trans. Electron Devices*, 32, 375, 1985.
Hu, S.M., Stress-related problems in Si technology, *J. Appl. Phys.*, 70, R53, 1991.
Hu, G. and Johnson, W.C., Relationship between trapped holes and interface states in MOS capacitors, *Appl. Phys. Lett.*, 36, 590, 1980.
Hubbard, K.J. and Schlom, D.G., Thermodynamic stability of binary oxides in contact with Si, *J. Mater. Res.*, 11, 2757, 1996.
Hughes, R.C., Charge carrier transport phenomena in a-SiO_2—direct measurement of drift mobility and lifetime, *Phys. Rev. Lett.*, 30, 1333, 1973.
Hughes, R.C., Time-resolved hole transport in a-SiO_2, *Phys. Rev. B*, 15, 2012, 1977.
Hung, K.K. et al., A unified model for the flicker noise in MOSFETs, *IEEE Trans. Electron Devices*, 37, 654, 1990.
Hurle, D.T.J., Revised calculation of point defect equilibria and non-stoichiometry in GaAs, *J. Phys. Chem. Solids*, 40, 613, 1979.
Hurtes, C. et al., Deep level spectroscopy in high-resistivity materials, *Appl. Phys. Lett.*, 32, 821, 1978.
Iwata, S. and Ishizaka, A., Electron spectroscopic analysis of the SiO_2/Si system and correlation with MOS device characteristics, *J. Appl. Phys.*, 79, 6653, 1996.
Jaccodine, R.J. and Schlegel, W.A., Measurement of strains at Si/SiO_2 interface, *J. Appl. Phys.*, 37, 2429, 1966.
Jeppson, K.O. and Svensson, C.M., Negative bias stress of MOS devices at high electric-fields and degradation of MNOS devices, *J. Appl. Phys.*, 48, 2004, 1977.
Jin, B.J., Im, S., and Lee, S.Y., Violet and UV luminescence emitted from ZnO thin films grown on sapphire by pulsed laser deposition, *Thin Solid Films*, 366, 107, 2000.
Johnson, N.M., Mechanism for hydrogen compensation of shallow acceptor impurities in single crystal Si, *Phys. Rev. B*, 31, 5525, 1985.
Johnson, N.M. et al., Characteristic electronic defects at the Si-SiO_2 interface, *Appl. Phys. Lett.*, 43, 563, 1983.
Johnson, N.M., Herring, C., and Chadi, D.J., Interstitial hydrogen and neutralization of shallow donor impurities in single crystal Si, *Phys. Rev. Lett.*, 56, 769, 1986a.
Johnson, N.M. et al., Hydrogen passivation of shallow acceptor impurities in *p*-type GaAs, *Phys. Rev. B*, 33, 1102, 1986b.
Johnson, N.M. et al., Defects in single-crystal Si induced by hydrogenation, *Phys. Rev. B*, 35, 4166, 1987.
Jones, K.S., Prussin, S., and Weber, E.R., A systematic analysis of defects in ion-implanted Si, *Appl. Phys. A – Mater. Sci. Process.*, 45, 1, 1988.
Justo, J.F., Interatomic potential for Si defects and disordered phases, *Phys. Rev. B*, 58, 2539, 1998.
Kaminska, M. et al., Intracenter transitions in the dominant deep level (EL2) in GaAs, *Appl. Phys. Lett.*, 43, 302, 1983.
Kaminska, M., Skowronski, M., and Kuszko, W., Identification of the 0.82 eV electron trap, EL2 in GaAs, as an isolated antisite As defect, *Phys. Rev. Lett.*, 55, 2204, 1985.
Kar, S. and Dahlke, W.E., Interface states in MOS structures with 20–40 Å thick SiO_2 films on nondegenerate Si, *Solid-State Electron.*, 15, 154, 1972.
Kimerling, L.C., Recombination enhanced defect reactions, *Solid-State Electron.*, 21, 1391, 1978.
Kimerling, L.C. and Benton, J.L., Electronically controlled reactions of interstitial iron in Si, *Physica B & C*, 116, 297, 1983.
Kimerling, L.C. and Patel, J.R., Defect states associated with dislocations in Si, *Appl. Phys. Lett.*, 34, 73, 1979.
Kinchin, G.H. and Pease, R.S., The displacement of atoms in solids by radiation, *Rep. Prog. Phys.*, 18, 1, 1955.
Kirton, M.J. and Uren, M.J., Noise in solid-state microstructures—a new perspective on individual defects, interface states and low-frequency (1/*f*) noise, *Adv. Phys.*, 38, 367, 1989.
Kohan, A.F. et al., First principles study of native point defects in ZnO, *Phys. Rev. B*, 61, 15019, 2000.
Kohn, W. and Sham, L.J., Self-consistent equations including exchange and correlation effects, *Phys. Rev. B*, 140, A1133, 1965.

Kresse, G. and Furthmuller, J., Efficient iterative schemes for ab initio total energy calculations using a plane wave basis set, *Phys. Rev. B*, 54, 11169, 1994.
Kresse, G. and Hafner, J., Norm conserving and ultrasoft pseudopotentials for first row and transition elements, *J. Phys.: Condens. Matter*, 6, 8245, 1994.
Kuhn, M., A quasi-static technique for MOS C-V and surface state measurements, *Solid-State Electron.*, 13, 873, 1970.
Kwon, I. et al., Transferrable tight binding models for Si, *Phys. Rev. B*, 49, 7242, 1994.
Laasonen, K. et al., Car-Parrinello molecular dynamics with Vanderbilt ultrasoft pseudopotentials, *Phys. Rev. B*, 47, 10142, 1993.
Lagowski, J., Origin of the 0.82 eV electron trap in GaAs and its annihilation by shallow donors, *Appl. Phys. Lett.*, 40, 342, 1982.
Lai, S.K., 2-Carrier nature of interface-state generation in hole trapping and radiation damage, *Appl. Phys. Lett.*, 39, 58, 1981.
Lai, S.K., Interface trap generation in SiO_2 when electrons are captured by trapped holes, *J. Appl. Phys.*, 54, 2540, 1983.
Laks, D.B. et al., Native defects and self compensation in ZnSe, *Phys. Rev. B*, 45, 10965, 1992.
Landstrass, M.I. and Ravi, K.V., Resistivity of chemical vapor deposited diamond films, *Appl. Phys. Lett.*, 55, 975, 1989a.
Landstrass, M.I. and Ravi, K.V., Hydrogen passivation of electrically active defects in diamond, *Appl. Phys. Lett.*, 55, 1391, 1989b.
Lang, D.V., Deep-level transient spectroscopy—new method to characterize traps in semiconductors, *J. Appl. Phys.*, 45, 3023, 1974a.
Lang, D.V., Fast capacitance transient apparatus—application to ZnO and O centers in GaP para-normal junctions, *J. Appl. Phys.*, 45, 3014, 1974b.
Lang, D.V. and Henry, C.H., Nonradiative recombination at deep levels in GaAs and GaP by lattice relaxation multiphonon emission, *Phys. Rev. Lett.*, 35, 1525, 1975.
Lang, D.V. and Kimerling, L.C., Observation of recombination-enhanced defect reactions in semiconductors, *Phys. Rev. Lett.*, 33, 489, 1974.
Lang, D.V. and Logan, R.A., Study of deep levels in GaAs by capacitance spectroscopy, *J. Electron. Mater.*, 4, 1053, 1975.
Lang, D.V. and Logan, R.A., Large lattice relaxation model for persistent photoconductivity in compound semiconductors, *Phys. Rev. Lett.*, 39, 635, 1977.
Lang, D.V. et al., Study of electron traps in n-GaAs grown by molecular beam epitaxy, *J. Appl. Phys.*, 47, 2558, 1976.
Lang, D.V., Logan, R.A., and Kimerling, L.C., Identification of defect state associated with a Ga vacancy in GaAs and $Al_xGa_{1-x}As$, *Phys. Rev. B*, 15, 4874, 1977.
Lang, D.V., Logan, R.A., and Jaros, M., Trapping characteristics and a donor complex (DX) model for the persistent photoconductivity trapping center in Te-doped $Al_xGa_{1-x}As$, *Phys. Rev. B*, 19, 1015, 1979.
Lang, D.V. et al., Complex nature of gold-related deep levels in Si, *Phys. Rev. B*, 22, 3917, 1980.
Lang, D.V. et al., Measurement of the band gap of Ge_xSi_{1-x}/Si strained layer heterostructures, *Appl. Phys. Lett.*, 47, 1333, 1985.
Lee, Y.H. and Corbett, J.W., EPR studies in neutron irradiated Si—negative charge state of a nonplanar 5-vacancy cluster (V5-), *Phys. Rev. B*, 8, 2810, 1973.
Lee, Y.H. and Corbett, J.W., EPR studies of defects in electron-irradiated Si—triplet state of vacancy-oxygen complexes, *Phys. Rev. B*, 13, 2653, 1976.
Lee, Y.H., Gerasimenko, N.N., and Corbett, J.W., EPR studies of defects in neutron-irradiated Si—positive charge state of (100) split di-interstitial, *Phys. Rev. B*, 14, 4506, 1976.
Lee, C., Yang, W., and Parr, R.G., Development of the Colle-Salvetti correlation-energy formula into a functional of the electron density, *Phys. Rev. B*, 37, 785, 1988a.
Lee, J.C., Chen, I.C., and Hu, C.M., Modeling and characterization of gate oxide reliability, *IEEE Trans. Electron Devices*, 35, 2268, 1988b.
Lee, B.H. et al., Thermal stability and electrical characteristics of ultrathin HfO_2 gate dielectric reoxidized with rapid thermal annealing, *Appl. Phys. Lett.*, 76, 1926, 2000.

Lefevre, H. and Schulz, M., Double correlation technique (DDLTS) for analysis of deep level profiles in semiconductors, *Appl. Phys.*, 12, 45, 1977.
Legoues, F.K., Meyerson, B.S., and Morar, J.F., Anomalous strain relaxation in SiGe thin films and superlattices, *Phys. Rev. Lett.*, 66, 2903, 1991.
Legoues, F.K. et al., Mechanism and conditions for anomalous strain relaxation in graded thin films and superlattices, *J. Appl. Phys.*, 71, 4230, 1992.
Lehmann, G. and Taut, M., Numerical calculation of density of states and related properties, *Phys. Status Solidi B*, 54, 469, 1972.
Lelis, A.J. et al., Reversibility of trapped hole annealing, *IEEE Trans. Nucl. Sci.*, 1186, 1988.
Lelis, A.J. et al., The nature of the trapped hole annealing process, *IEEE Trans. Nucl. Sci.*, 36, 1808, 1989.
Lenahan, P.M. and Dressendorfer, P.V., An ESR study of radiation-induced electrically active paramagnetic centers at the Si/SiO_2 interface, *J. Appl. Phys.*, 54, 1457, 1983.
Lenahan, P.M. and Dressendorfer, P.V., Hole traps and trivalent Si centers in MOS devices, *J. Appl. Phys.*, 55, 3495, 1984.
Lester, S.D. et al., High dislocation densities in high efficiency GaN-based LEDs, *Appl. Phys. Lett.*, 66, 1249, 1995.
Lenzlinger, M. and Snow, E.H., Fowler-Nordheim tunneling into thermally grown SiO_2, *J. Appl. Phys.*, 40, 278, 1969.
Liao, L.S. et al., Blue luminescence from Si^+ implanted SiO_2 films thermally grown on crystalline Si, *Appl. Phys. Lett.*, 68, 850, 1996.
Lin, B.X., Fu, Z.X., and Jia, Y.B., Green luminescent center in undoped ZnO films on deposited on Si substrates, *Appl. Phys. Lett.*, 79, 943, 2001.
Lipkin, L.A. and Palmour, J.W., Improved oxidation procedures for reduced SiO_2/SiC defects, *J. Electron. Mater.*, 25, 909, 1996.
Littlejohn, M.A. et al., Alloy scattering and high field transport in ternary and quaternary III-V semiconductors, *Solid-State Electron.*, 21, 107, 1978.
Look, D.C., Recent advances in ZnO materials and devices, *Mater. Sci. Eng. B*, 80, 383, 2001.
Look, D.C., Characterization of homoepitaxial p-type ZnO grown by molecular beam epitaxy, *Appl. Phys. Lett.*, 81, 1830, 2002.
Look, D.C. et al., Defect donor and acceptor in GaN, *Phys. Rev. Lett.*, 79, 2273, 1997.
Look, D.C. et al., Dislocation scattering in GaN, *Phys. Rev. Lett.*, 82, 1237, 1999a.
Look, D.C., Hemsky, J.W., and Sizelove, J.R., Residual shallow donor in ZnO, *Phys. Rev. Lett.*, 82, 2552, 1999b.
Lucovsky, G., Spectroscopic evidence for valence-alternation-pair defect states in vitreous SiO_2, *Philos. Mag. B Phys. Condens. Matter*, 39, 513, 1979.
Lyding, J.W., Hess, K., and Kizilyalli, I.C., Reduction of hot electron degradation in MOS transistors by deuterium processing, *Appl. Phys. Lett.*, 68, 2526, 1996.
Ma, T.P., Making silicon nitride film a viable gate dielectric, *IEEE Trans. Electron Devices*, 45, 680, 1998.
Martin, G.M., Optical assessment of the main electron trap in bulk semi-insulating GaAs, *Appl. Phys. Lett.*, 39, 747, 1981.
Martin, G.M., Mitonneau, A., and Mircea, A., Electron traps in bulk and epitaxial GaAs, *Electron. Lett.*, 13, 191, 1977.
Martin, G.M. et al., Compensation mechanisms in GaAs, *J. Appl. Phys.*, 51, 2840, 1980.
Maserjian, J. and Zamani, N., Behavior of the Si/SiO_2 interface observed by Fowler-Nordheim tunneling, *J. Appl. Phys.*, 53, 559, 1982.
Matthews, J.W. and Blakeslee, A.E., Defects in epitaxial multilayers—misfit dislocations, *J. Cryst. Growth*, 27, 118, 1974.
Mattila, T. and Nieminen, R.M., Point defect complexes and broadband luminescence in GaN and AlN, *Phys. Rev. B*, 55, 9571, 1997.
McLean, F.B., A framework for understanding radiation-induced interface states in SiO_2 MOS structures, *IEEE Trans. Nucl. Sci.*, 27, 1651, 1980.
McWhorter, P.J. and Winokur, P.S., Simple technique for separating the effects of interface traps and trapped-oxide charge in MOS transistors, *Appl. Phys. Lett.*, 48, 133, 1986.
Merton, R.K., Matthew effect in science, *Science*, 159, 56, 1968.

Meyer, B.K. et al., Bound exciton and donor-acceptor pair recombinations in ZnO, *Phys. Status Solidi B*, 241, 231, 2004.
Miller, G.L., Lang, D.V., and Kimerling, L.C., Capacitance transient spectroscopy, *Annu. Rev. Mater. Sci.*, 7, 377, 1977.
Mitonneau, A., Martin, G.M., and Mircea, A., Hole traps in bulk and epitaxial GaAs crystals, *Electron. Lett.*, 13, 666, 1977.
Mizuta, M. et al., Direct evidence for the DX center being a substitutional donor in AlGaAs alloy systems, *Jpn. J. Appl. Phys.*, 24, L143, 1985.
Monkhorst, H.J. and Pack, J.D., Special points for Brillouin zone integrations, *Phys. Rev. B*, 13, 5188, 1976.
Montillo, F. and Balk, P., High-temperature annealing of oxidized Si surfaces, *J. Electrochem. Soc.*, 118, 1463, 1971.
Mooney, P.M., Deep donor levels (DX centers) in III-V semiconductors, *J. Appl. Phys.*, 67, R1, 1990.
Mooney, P.M., Strain relaxation and dislocations in SiGe/Si structures, *Mater. Sci. Eng. R-Rep.*, 17, 105, 1996.
Mooney, P.M. et al., Defect energy-levels in boron-doped Si irradiated with 1-MeV electrons, *Phys. Rev. B*, 15, 3836, 1977.
Morgan, T.N., Theory of the DX center in $Al_xGa_{1-x}As$ and GaAs crystals, *Phys. Rev. B*, 34, 2664, 1986.
Morkoc, H. et al., Large band gap SiC, III-V nitride, and II-VI ZnSe based semiconductor device technologies, *J. Appl. Phys.*, 76, 1363, 1994.
Mott, N.F., Electrons in disordered structures, *Adv. Phys.*, 16, 49, 1967.
Mott, N.F., SiO_2 and the chalcogenide semiconductors: similarities and differences, *Adv. Phys.*, 26, 363, 1977.
Myers, S.M. et al., Hydrogen interactions with defects in crystalline solids, *Rev. Mod. Phys.*, 64, 559, 1992.
Nakamura, S. et al., Hole compensation mechanism of *p*-type GaN films, *Jpn. J. Appl. Phys.*, 31, 1258, 1992.
Nakamura, S., Mukai, T., and Senoh, M., Candela class high-brightness InGaN/AlGaN double heterostructure blue light emitting diodes, *Appl. Phys. Lett.*, 64, 1687, 1994.
Nakamura, S. et al., InGaN-based multi-quantum-well structure laser diodes, *Jpn. J. Appl. Phys.*, 35, L74, 1996.
Nakashima, S. and Izumi, K., Analysis of buried oxide layer formation and mechanism of threading dislocation generation in the substoichiometric oxygen dose region, *J. Mater. Res.*, 8, 523, 1993.
Neudeck, P.G. and Powell, J.A., Performance limiting micropipe defects in SiC wafers, *IEEE Electron Device Lett.*, 15, 63, 1994.
Neugebauer, J. and Van de Walle, C.G., Atomic geometry and electronic structure of native defects in GaN, *Phys. Rev. B*, 50, 8067, 1994.
Neugebauer, J. and Van de Walle, C.G., Hydrogen in GaN—novel aspects of a common impurity, *Phys. Rev. Lett.*, 75, 4452, 1995.
Neugebauer, J. and Van de Walle, C.G., Ga vacancies and the yellow luminescence in GaN, *Appl. Phys. Lett.*, 69, 503, 1996a.
Neugebauer, J. and Van de Walle, C.G., Role of hydrogen in doping of GaN, *Appl. Phys. Lett.*, 68, 1829, 1996b.
Ng, H.M. et al., The role of dislocation scattering in *n*-type GaN films, *Appl. Phys. Lett.*, 73, 821, 1998.
Nicollian, E.H. and Berglund, C.N., Avalanche injection of electrons into insulating SiO_2 using MOS structures, *J. Appl. Phys.*, 41, 3052, 1970.
Nicollian, E.H., Goetzberger, A., and Berglund, C.N., Avalanche injection currents and charging phenomena in thermal SiO_2, *Appl. Phys. Lett.*, 15, 174, 1969.
Nicollian, E.H. et al., Electrochemical charging of thermal SiO_2 films by injected electron currents, *J. Appl. Phys.*, 42, 5654, 1971.
Ning, T.H., Optically induced injection of hot electrons into SiO_2, *J. Appl. Phys.*, 45, 5373, 1974.
Ning, T.H., High-field capture of electrons by Coulomb-attractive centers in SiO_2, *J. Appl. Phys.*, 47, 3203, 1976.
Ning, T.H., Hot electron emission from Si into SiO_2, *Solid-State Electron.*, 21, 273, 1978.

Ning, T.H., 1 μm MOSFET VLSI technology: Hot-electron design constraints, *IEEE Trans. Electron Devices.*, 26, 346, 1979.
Ning, T.H., Osburn, C.M., and Yu, H.N., Emission probability of hot electrons from Si into SiO_2, *J. Appl. Phys.*, 48, 286, 1977.
Nishi, Y., Study of Si-SiO_2 structure by electron spin resonance, *Jpn. J. Appl. Phys.*, 10, 52, 1971.
Nissan-Cohen, Y., Shappir, J., and Frohman-Bentchkowsky, D., Dynamic model of trapping-detrapping in SiO_2, *J. Appl. Phys.*, 60, 2024, 1986.
Northrup, J.E. and Zhang, S.B., Dopant and defect energetics—Si in GaAs, *Phys. Rev. B*, 47, 6791, 1993.
Ogura, S. et al., Design and characteristics of the lightly doped drain-source insulated gate FET, *IEEE Trans. Electron Devices*, 27, 1359, 1980.
Ohmi, T. et al., Dependence of thin oxide films quality on surface microroughness, *IEEE Trans. Electron Devices*, 39, 537, 1992.
Oldham, T.R., Lelis, A.J., and McLean, F.B., Spatial dependence of trapped holes determined from tunneling analysis and measured annealing, *IEEE Trans. Nucl. Sci.*, 33, 1203, 1986.
Olivo, P., Nguyen, T.N., and Ricco, B., High field induced degradation in ultrathin SiO_2 films, *IEEE Trans. Electron Devices*, 35, 2259, 1988.
Omling, P., Samuelson, L., and Grimmeiss, H.G., DLTS spectroscopy evaluation of non-exponential transients in semiconductor alloys, *J. Appl. Phys.*, 54, 5117, 1983.
Omling, P. et al., Electrical properties of dislocations and point defects in plastically deformed Si, *Phys. Rev. B*, 32, 6571, 1985.
O'Reilly, E.P. and Robertson, J., Theory of defects in vitreous SiO_2, *Phys. Rev. B*, 27, 3780, 1983.
Ozgur, U. et al., A comprehensive review of ZnO materials and devices, *J. Appl. Phys.*, 98, article no. 041301, 2005.
Pankove, J.I. and Hutchby, J.A., Photoluminescence of ion-implanted GaN, *J. Appl. Phys.*, 47, 5387, 1976.
Pankove, J.I. et al., Neutralization of shallow acceptor levels in Si by atomic hydrogen, *Phys. Rev. Lett.*, 51, 2224, 1983.
Pankove, J.I., Wance, R.O., and Berkeyheiser, J.E., Neutralization of acceptor in Si by atomic hydrogen, *Appl. Phys. Lett.*, 45, 1100, 1984.
Pankove, J.I. et al., Hydrogen localization near boron in Si, *Appl. Phys. Lett.*, 421, 1985.
Pantelides, S.T., Electronic structure of impurities and other point defects in semiconductors, *Rev. Mod. Phys.*, 50, 797, 1978.
Pantelides, S.T. and Sah, C.T., Theory of localized states in semiconductors—new results using an old method, *Phys. Rev. B*, 10, 621, 1974.
Pasquarello, A., Hybertsen, M.S., and Car, R., Interface structure between Si and its oxide by first-principles molecular dynamics, *Nature*, 396, 58, 1998.
Pavesi, L. and Guzzi, M., Photoluminescence of $Al_xGa_{1-x}As$ alloys, *J. Appl. Phys.*, 75, 4779, 1994.
Payne, M.C. et al., Iterative minimization techniques for ab initio total energy calculations—molecular dynamics and conjugate gradients, *Rev. Mod. Phys.*, 64, 1045, 1992.
Pearton, S.J. et al., Hydrogenation of shallow donor levels in GaAs, *J. Appl. Phys.*, 59, 2821, 1986.
Pearton, S.J., Corbett, J.W., and Shi, T.S., Hydrogen in crystalline semiconductors, *Appl. Phys. A—Mater. Sci. Process.*, 43, 153, 1987.
Pearton, S.J. et al., GaN: processing, defects, and devices, *J. Appl. Phys.*, 86, 1, 1999.
Pearton, S.J. et al., Recent advances in processing of ZnO, *J. Vac. Sci. Technol. B*, 22, 932, 2004.
Pearton, S.J. et al., Recent progress in processing and properties of ZnO, *Prog. Mater. Sci.*, 50, 293, 2005.
Pensl, G. and Choyke, W.J., Electrical and optical characterization of SiC, *Physica B*, 185, 264, 1993.
People, R., Physics and applications of Ge_xSi_{1-x}/Si strained layer heterostructures, *IEEE J. Quantum Electron.*, 22, 1696, 1986.
People, R. and Bean, J.C., Calculation of critical layer thickness vs. lattice mismatch for Ge_xSi_{1-x}/Si strained layer superlattices, *Appl. Phys. Lett.*, 47, 322, 1985.
People, R. et al., Modulation doping in Ge_xSi_{1-x}/Si strained layer heterostructures, *Appl. Phys. Lett.*, 45, 1231, 1984.
Perdew, J.P., Density functional approximation for the correlation energy of the inhomogeneous electron gas, *Phys. Rev. B*, 33, 8822, 1986.

Perdew, J.P. and Wang, Y., Accurate and simple analytic representation of the electron gas correlation energy, *Phys. Rev. B*, 45, 13244, 1992.

Perdew, J.P. and Zunger, A., Self interaction correction to density-functional approximations for many electron systems, *Phys. Rev. B*, 23, 5048, 1981.

Perdew, J.P. et al., Atoms, molecules, solids, and surfaces—applications of the generalized gradient approximation for exchange and correlation, *Phys. Rev. B*, 46, 6671, 1992.

Perdew, J.P., Burke, K., and Ernzerhof, M., Generalized gradient approximation made simple, *Phys. Rev. Lett.*, 77, 3865, 1996.

Poindexter, E.H. and Caplan, P.J., Characterization of Si/SiO_2 interface defects by electron-spin resonance, *Prog. Surf. Sci.*, 14, 201, 1983.

Poindexter, E.H. et al., Interface states and electron-spin resonance centers in thermally oxidized (111) and (100) Si wafers, *J. Appl. Phys.*, 52, 879, 1981.

Poindexter, E.H. et al., Electronic traps and P_b centers at the Si/SiO_2 interface—band-gap energy distribution, *J. Appl. Phys.*, 56, 2844, 1984.

Ponce, F.A. et al., Spatial distribution of the luminescence in GaN thin films, *Appl. Phys. Lett.*, 68, 57, 1996.

Pons, D. and Bourgoin, J.C., Irradiation induced defects in GaAs, *J. Phys. C—Solid State Phys.*, 18, 3839, 1985.

Pons, D., Mooney, P.M., and Bourgoin, J.C., Energy dependence of deep level introduction in electron irradiated GaAs, *J. Appl. Phys.*, 51, 2038, 1980.

Powell, R.J., Interface barrier energy determination from voltage dependence of photoinjected currents, *J. Appl. Phys.*, 41, 2424, 1970.

Powell, R.J. and Berglund, C.N., Photoinjection studies of charge distributions in oxides of MOS structures, *J. Appl. Phys.*, 42, 4390, 1971.

Powell, R.J. and Derbenwick, G.F., Vacuum ultraviolet radiation effects in SiO_2, *IEEE Trans. Nucl. Sci.*, 18(6), 99, 1971.

Ralls, K.S. et al., Discrete resistance switching in submicrometer Si inversion layers: individual interface traps and low-frequency (1/*f*?) noise, *Phys. Rev. Lett.*, 52, 228, 1984.

Razouk, R.R. and Deal, B.E., Dependence of interface state density on Si thermal oxidation process variables, *J. Electrochem. Soc.*, 126, 1573, 1979.

Read, W.T. and Shockley, W., Dislocation models of crystal grain boundaries, *Phys. Rev.*, 78, 275, 1950.

Reed, M.L. and Plummer, J.D., Chemistry of Si-SiO_2 interface trap annealing, *J. Appl. Phys.*, 63, 5776, 1988.

Reimbold, G., Modified 1/*f* trapping noise theory and experiments in MOS transistors biased from weak to strong inversion: Influence of interface states, *IEEE Trans. Electron Devices*, 31, 1190, 1984.

Revesz, A.G., The role of hydrogen in SiO_2 films on Si, *J. Electrochem. Soc.*, 126, 122, 1979.

Reynolds, D.C. et al., Similarities in the band edge and deep centre photoluminescence mechanisms of ZnO and GaN, *Solid State Commun.*, 101, 643, 1997.

Reynolds, D.C. et al., Neutral donor bound exciton complexes in ZnO crystals, *Phys. Rev.*, 57, 12151, 1998.

Robertson, J., Electronic structure of silicon nitride, *Philos. Mag. B*, 63, 47, 1991.

Robertson, J., Band offsets of wide band gap oxides and implications for future electronic devices, *J. Vac. Sci. Technol. B*, 18, 1785, 2000.

Rudra, J.K. and Fowler, W.B., Oxygen vacancy and the E'_1 center in crystalline SiO_2, *Phys. Rev. B*, 35, 8223, 1987.

Sah, C.T., Origin of interface states and oxide charges generated by ionizing radiation, *IEEE Trans. Nucl. Sci.*, 23, 1563, 1976.

Sah, C.T. and Hielscher, F., Evidence of surface origin of 1/*f* noise, *Phys. Rev. Lett.*, 17, 956, 1966.

Sah, C.T., Noyce, R.N., and Shockley, W., Carrier generation and recombination in *p-n* junctions and *p-n* junction characteristics, *Proc. IRE*, 45, 1228, 1957.

Sah, C.T. et al., Thermal emission rates of carriers at gold centers in Si, *Appl. Phys. Lett.*, 15, 145, 1969.

Sah, C.T. et al., Thermal and optical emission and capture rates and cross sections of electrons and holes at imperfection centers in semiconductors from photo and dark junction current and capacitance experiments, *Solid-State Electron.*, 13, 759, 1970.

Sah, C.T., Tschopp, L.L., and Ning, T.H., Scattering of electrons by surface oxide charges and by lattice vibrations at Si/SiO_2 interface, *Surf. Sci.*, 32, 561, 1972.
Sah, C.T., Sun, J.Y., and Tzou, J.J., Deactivation of the boron acceptor in Si by hydrogen, *Appl. Phys. Lett.*, 43, 204, 1983.
Sakai, A., Sunakawa, H., and Usui, A., Defect structure in selectively grown GaN films with low threading dislocation density, *Appl. Phys. Lett.*, 71, 2259, 1997.
Saks, N.S. and Brown, D.B., Interface trap formation via the two-stage H^+ process, *IEEE Trans. Nucl. Sci.*, 36, 1848, 1989.
Sauer, R. et al., Dislocation related photoluminescence in Si, *Appl. Phys. A Mater. Sci. Process.*, 36, 1, 1985.
Schneider, J. and Maier, K., Point defects in SiC, *Physica B*, 185, 199, 1993.
Schottky, W., Halbleitertheorie der sperrschicht und spitzengleichrichter, *Z. Phys.*, 113, 367, 1939.
Schottky, W., Vereinfachte und erweiterte theorie der randschichtgleichrichter, *Z. Phys.*, 118, 539, 1942.
Schroder, D.K. and Guldberg, J., Interpretation of surface and bulk effects using pulsed MIS capacitor, *Solid-State Electron.*, 14, 1285, 1971.
Schroder, D.K. and Nathanson, H.C., On the separation of bulk and surface components of lifetime using the pulsed MOS capacitor, *Solid-State Electron.*, 13, 577, 1970.
Schrodinger, E., Quantisation as an eigenvalue problem, *Ann. Phys.*, 79, 361, 1926.
Schuegraf, K.F. and Hu, C.M., Hole injection SiO_2 breakdown model for very-low voltage lifetime extrapolation, *IEEE Trans. Electron Devices*, 41, 761, 1994.
Schwank, J.R. et al., Physical mechanisms contributing to device "rebound," *IEEE Trans. Nucl. Sci.*, 31, 1434, 1984.
Seeger, A. and Chik, K.P., Diffusion mechanisms and point defects in Si and Ge, *Phys. Status Solidi*, 29, 455, 1968.
Shenoy, J.N. et al., Characterization and optimization of the SiO_2/SiC MOS interface, *J. Electron. Mater.*, 24, 303, 1995.
Shockley, W., The theory of p-n junctions in semiconductors and p-n junction transistors, *Bell Syst. Tech.* J., 27, 435, 1949.
Shockley, W. and Read, W.T. Jr., Statistics of the recombination of holes and electrons, *Phys. Rev. B*, 87, 835, 1952.
Shi, Y. et al., Effects of traps on charge storage characteristics in MOS memory structures based on Si nanocrystals, *J. Appl. Phys.*, 84, 2358, 1998.
Silverberg, P., Omling, P., and Samuelson, L., Hole photoionization cross sections of EL2 in GaAs, *Appl. Phys. Lett.*, 52, 1689, 1988.
Simmons, J.G., Poole-Frenkel effect and Schottky effect in metal-insulator-metal systems, *Phys. Rev.*, 155, 657, 1967.
Simmons, J.G., Conduction in thin dielectric films, *J. Phys. D Appl. Phys.*, 4, 613, 1971.
Simmons, J.G. and Taylor, G.W., Nonequilibrium steady-state statistics and associated effects for insulators and semiconductors containing an arbitrary distribution of traps, *Phys. Rev. B*, 4, 502, 1971.
Simmons, J.G. and Taylor, G.W., High-field isothermal currents and thermally stimulated currents in insulators having discrete trapping levels, *Phys. Rev. B*, 5, 1619, 1972.
Singh, V.A. et al., Vibrational and electronic structure of hydrogen-related defects in Si calculated by extended Huckel theory, *Phys. Status Solidi B*, 81, 637, 1977.
Skuja, L., Optically active oxygen-deficiency-related centers in amorphous SiO_2, *J. Non-Cryst. Solids*, 239, 16, 1998.
Slater, J.C., Wave functions in a periodic potential, *Phys. Rev.*, 51, 846, 1937.
Snow, E.H. et al., Ion transport phenomena in insulating films, *J. Appl. Phys.*, 36, 1664, 1965.
Snow, E.H., Grove, A.S., and Fitzgerald, D.J., Effects of ionizing radiation on oxidized Si surfaces and planar devices, *Proc. IEEE*, 55, 1168, 1967.
Sommerfeld, A., Zur elektronen theorie der metalle auf grund der Fermischen statistic, *Z. Phys.*, 47, 1, 1928.
Spicer, W.E. et al., New and unified defect model for Schottky barrier and III-V insulator interface states formation, *J. Vac. Sci. Technol.*, 16, 1422, 1979.
Spicer, W.E. et al., Unified defect model and beyond, *J. Vac. Sci. Technol.*, 17, 1019, 1980a.

Spicer, W.E. et al., Unified mechanism for Schottky barrier formation and III-V-oxide interface states, *Phys. Rev. Lett.*, 44, 420, 1980b.

Spicer, W.E. et al., The advance unified defect model for Schottky barrier formation, *J. Vac. Sci. Technol. B*, 6, 1245, 1988.

Stathis, J.H., Percolation models for gate oxide breakdown, *J. Appl. Phys.*, 86, 5757, 1999.

Stathis, J.H. and Kastner, M.A., Time-resolved photoluminescence in amorphous SiO_2, *Phys. Rev. B*, 35, 2972, 1987.

Stavola, M. et al., Diffusivity of oxygen in Si at the donor formation temperature, *Appl. Phys. Lett.*, 42, 73, 1983.

Stein, H.J., Bonding and thermal stability of implanted hydrogen in Si, *J. Electron. Mater.*, 4, 159, 1975.

Stein, H.J., Wells, V.A., and Hampy, R.E., Properties of plasma-deposited silicon nitride, *J. Electrochem. Soc.*, 126, 1750, 1979.

Stolk, P.A. et al., Physical mechanisms of transient enhanced dopant diffusion in ion-implanted Si, *J. Appl. Phys.*, 81, 6031, 1997.

Sugino, O. and Oshiyama, A., Vacancy in Si—successful description within the local density approximation, *Phys. Rev. Lett.*, 68, 1858, 1992.

Summers, G.P., Damage correlations in semiconductors exposed to gamma radiation, electron radiation, and proton radiation, *IEEE Trans. Nucl. Sci.*, 40(6), 1372, 1993.

Sun, S.C. and Plummer, J.D., Electron mobility in inversion and accumulation layers on thermally oxidized Si surfaces, *IEEE Trans. Electron Devices*, 27, 1497, 1980.

Suñé, J. et al., On the breakdown statistics of very thin SiO_2 films, *Thin Solid Films*, 185, 347, 1990.

Sze, S.M., Current transport and maximum dielectric strength of silicon nitride films, *J. Appl. Phys.*, 38, 2951, 1967.

Sze, S.M. and Irvin, J.C., Resistivity, mobility, and impurity levels in GaAs, Ge, and Si at 300 K, *Solid-State Electron.*, 11, 599, 1968.

Takagi, S. et al., On the universality of inversion layer mobility in Si MOSFETs: effects of substrate impurity concentration, *IEEE Trans. Electron Devices*, 41, 2357, 1994.

Takeda, E. and Suzuki, N., An empirical model for device degradation due to hot-carrier injection, *IEEE Electron Device Lett.*, 4, 111, 1983.

Tam, S., Ko, P.K., and Hu, C.M., Lucky electron model of channel hot-electron injection in MOSFETs, *IEEE Trans. Electron Devices*, 31, 1116, 1984.

Tamm, I.E., On the possible bound states of electrons on a crystal surface, *Phys. Z. Sowietunion*, 1, 733, 1932.

Tan, T.Y. and Gosele, U., Point-defects, diffusion processes, and swirl defect formation in Si, *Appl. Phys. A—Mater. Sci. Process.*, 37, 1, 1985.

Tan, T.Y., Gosele, U., and Yu, S., Point defects diffusion mechanisms, and superlattice disordering in GaAs-based materials, *Crit. Rev. Solid State Mater. Sci.*, 17, 47, 1991.

Tasch, A.F. and Sah, C.T., Recombination-generation and optical properties of gold acceptor in Si, *Phys. Rev. B*, 1, 800, 1970.

Terman, L.M., An investigation of surface states at a Si/SiO_2 interface employing MOS diodes, *Solid-State Electron.*, 5, 285, 1962.

Terry, F.L., Aucoin, R.J., and Naiman, M.L., Radiation effects in nitrided oxides, *IEEE Electron Device Lett.*, 4, 191, 1983.

Theis, T.N., Mooney, P.M., and Wright, S.L., Electron localization by a metastable donor level in n-GaAs—a new mechanism limiting the free carrier density, *Phys. Rev. Lett.*, 60, 361, 1988.

Thomas, L.H., The calculation of atomic fields, *Proc. Camb. Philos. Soc.*, 23, 542, 1927.

Tiwari, S. et al., A Si nanocrystals based memory, *Appl. Phys. Lett.*, 68, 1377, 1996.

Tromp, R., et al., High-temperature SiO_2 decomposition at the SiO_2/Si interface, *Phys. Rev. Lett.*, 55, 2332, 1985.

Uren, M.J., Day, D.J., and Kirton, M.J., $1/f$ and random telegraph noise in Si MOSFETs, *Appl. Phys. Lett.*, 47, 1195, 1985.

Vandamme, L.K.J., Noise as a diagnostic tool for quality and reliability of electronic devices, *IEEE Trans. Electron Devices*, 41, 2176, 1994.

Vanderbilt, D., Soft self-consistent pseudopotentials in a generalized eigenvalue formalism, *Phys. Rev. B*, 41, 7892, 1990.

van der Ziel, A., Unified presentation of 1/*f* noise in electronic devices: Fundamental 1/*f* noise sources, *Proc. IEEE*, 76, 233, 1988.
Van de Walle, C.G., Energies of various configurations of hydrogen in Si, *Phys. Rev. B*, 49, 4579, 1994.
Van de Walle, C.G., Hydrogen as a cause of doping in ZnO, *Phys. Rev. Lett.*, 85, 1012, 2000.
Van de Walle, C.G., Baryam, Y., and Pantelides, S.T., Theory of hydrogen diffusion and reactions in crystalline Si, *Phys. Rev. Lett.*, 60, 2761, 1988.
Van de Walle, C.G. et al., Theory of hydrogen diffusion and reactions in crystalline Si, *Phys. Rev. B*, 39, 10791, 1989.
Vanheusden, K. et al., Correlation between photoluminescence and O vacancies in ZnO phosphors, *Appl. Phys. Lett.*, 68, 403, 1996a.
Vanheusden, K. et al., Mechanisms behind green photoluminescence in ZnO phosphor powders, *J. Appl. Phys.*, 79, 7983, 1996b.
Vetury, R. et al., The impact of surface states on the DC and RF characteristics of AlGaN/GaN HFETs, *IEEE Trans. Electron Devices*, 48, 560, 2001.
Vincent, G., Chantre, A., and Bois, D., Electric field effect on the thermal emission of traps in semiconductor junctions, *J. Appl. Phys.*, 50, 5484, 1979.
Vincent, G., Bois, D., and Chantre, A., Photoelectric memory effect in GaAs, *J. Appl. Phys.*, 53, 3643, 1982.
von Bardeleben, H.J. et al., Identification of a point defect in a semiconductor—EL2 in GaAs, *Phys. Rev. B*, 34, 7192, 1986.
Voronkov, V.V., The mechanism of swirl defect formation in Si, *J. Cryst. Growth*, 59, 625, 1982.
Wagner, R.J. et al., Submillimeter EPR for the As antisite defect in GaAs, *Solid-State Commun.*, 36, 15, 1980.
Watkins, G.D., Defects in irradiated Si—ESR of Si-A center, *Phys. Rev. B*, 121, 1001, 1961.
Watkins, G.D., An EPR study of the lattice vacancy in Si, *J. Phys. Soc. Jpn.*, 18(Supp. II), 22, 1963.
Watkins, G.D., Defects in irradiated Si—ESR and ENDOR of Si E center, *Phys. Rev. A*, 12, 5824, 1964.
Watkins, G.D., Defects in irradiated Si—EPR and electron-nuclear double-resonance of interstitial boron, *Phys. Rev. B*, 12, 5824, 1975.
Watkins, G.D. and Brower, K.L., EPR observation of isolated interstitial carbon atom in Si, *Phys. Rev. Lett.*, 36, 1329, 1976.
Watkins, G.D. and Corbett, J.W., Defects in irradiated Si: electron paramagnetic resonance of the divacancy, *Phys. Rev. A*, 138, 543, 1965.
Watkins, G.D. and Troxell, J.R., Negative-U properties for point-defects in Si, *Phys. Rev. Lett.*, 44, 593, 1980.
Watkins, G.D., Corbett, J.W., and Walker, R.M., Spin resonance in electron irradiated Si, *J. Appl. Phys.*, 30, 1198, 1959.
Weber, E.R., Transition metals in Si, *Appl. Phys. A—Mater. Sci. Process.*, 30, 1, 1983.
Weber, E.R. et al., Identification of As_{Ga} antisites in plastically deformed GaAs, *J. Appl. Phys.*, 53, 6140, 1982.
Weeks, R.A., Paramagnetic resonance of lattice defects in irradiated quartz, *J. Appl. Phys.*, 27, 1381, 1956.
Weeks, J.D., Tully, J.C., and Kimerling, L.C., Theory of recombination enhanced defect reactions in semiconductors, *Phys. Rev. B*, 12, 3286, 1975.
Weimann, N.G. et al., Scattering of electrons at threading dislocations in GaN, *J. Appl. Phys.*, 83, 3656, 1998.
Weissman, M.B., 1/*f* noise and other slow, nonexponential kinetics in condensed matter, *Rev. Mod. Phys.*, 60, 537, 1988.
Weldon, M.K. et al., On the mechanism of the hydrogen-induced exfoliation of Si, *J. Vac. Sci. Technol. B*, 15, 1065, 1997.
Wilk, G.D. and Wallace, R.M., Electrical properties of Hf silicate gate dielectrics deposited directly on Si, *Appl. Phys. Lett.*, 74, 2854, 1999.
Wilk, G.D., Wallace, R.M., and Anthony, J.M., Hf and Zr silicates for advanced gate dielectrics, *J. Appl. Phys.*, 87, 484, 2000.
Wilk, G.D., Wallace, R.M., and Anthony, J.M., High-K gate dielectrics: current status and materials properties considerations, *J. Appl. Phys.*, 89, 5243, 2001.

Williams, E.W., Evidence for self-activated luminescence in GaAs—Ga vacancy donor center, *Phys. Rev.*, 168, 922, 1968.
Winokur, P.S. et al., Field- and time-dependent radiation effects at the SiO_2/Si interface of hardened MOS capacitors, *IEEE Trans. Nucl. Sci.*, 24, 2113, 1977.
Winokur, P.S. et al., Two-stage process for buildup of radiation-induced interface states, *J. Appl. Phys.*, 50, 3492, 1979.
Winokur, P.S. et al., Correlating the radiation response of MOS capacitors and transistors, *IEEE Trans. Nucl. Sci.*, 31, 1453–1460, 1984.
Woods, M.H. and Williams, R., Hole traps in SiO_2, *J. Appl. Phys.*, 47, 1082, 1976.
Wright, P.J. and Saraswat, K.C., The effect of fluorine in SiO_2 gate dielectrics, *IEEE Trans. Electron Devices*, 36, 879, 1989.
Wu, X.H. et al., Defect structure of MOCVD grown epitaxial (0001) GaN/Al_2O_3, *J. Appl. Phys.*, 80, 3228, 1996.
Yan, Y.F., Zhang, S.B., and Pantelides, S.T., Control of doping by impurity chemical potentials: Predictions for p-type ZnO, *Phys. Rev. Lett.*, 86, 5723, 2001.
Young, D.R. et al., Electron trapping in SiO_2 at 295 and 77 K, *J. Appl. Phys.*, 50, 6366, 1979.
Zaininger, K.H. and Holmes-Seidle, A.G., A survey of radiation effects in metal-insulator-semiconductor devices, *RCA Rev.*, 28, 208, 1967.
Zhang, S.B. and Northrup, J.E., Chemical potential dependence of defect formation energies in GaAs—application to Ga self diffusion, *Phys. Rev. Lett.*, 67, 2339, 1991.
Zhang, S.B., Wei, S.H., and Zunger, A., Intrinsic n-type vs. p-type doping asymmetry and the defect physics of ZnO, *Phys. Rev. B*, article no. 075205, 2001.
Ziegler, J.F. and Manoyan, J.F., The stopping of ions in compounds, *Nucl. Instrum. Meth. B*, 35, 215, 1988.
Zundel, T. and Weber, J., Dissociation energies of shallow acceptor hydrogen pairs in Si, *Phys. Rev. B*, 39, 13549, 1989.

Index

C

D

E

F

I

J

K

L

M

P

Q

R

S

U

V

W

X

Y

Z